Advances in Intelligent Systems and Computing

Volume 1383

The series “Advances in Intelligent Systems and Computing” contains publications on theory, applications, and design methods of Intelligent Systems and Intelligent Computing. Virtually all disciplines such as engineering, natural sciences, computer and information science, ICT, economics, business, e-commerce, environment, healthcare, life science are covered. The list of topics spans all the areas of modern intelligent systems and computing such as: computational intelligence, soft computing including neural networks, fuzzy systems, evolutionary computing and the fusion of these paradigms, social intelligence, ambient intelligence, computational neuroscience, artificial life, virtual worlds and society, cognitive science and systems, Perception and Vision, DNA and immune based systems, self-organizing and adaptive systems, e-Learning and teaching, human-centered and human-centric computing, recommender systems, intelligent control, robotics and mechatronics including human-machine teaming, knowledge-based paradigms, learning paradigms, machine ethics, intelligent data analysis, knowledge management, intelligent agents, intelligent decision making and support, intelligent network security, trust management, interactive entertainment, Web intelligence and multimedia.

The publications within “Advances in Intelligent Systems and Computing” are primarily proceedings of important conferences, symposia and congresses. They cover significant recent developments in the field, both of a foundational and applicable character. An important characteristic feature of the series is the short publication time and world-wide distribution. This permits a rapid and broad dissemination of research results.

Indexed by DBLP, EI Compendex, INSPEC, WTI Frankfurt eG, zbMATH, Japanese Science and Technology Agency (JST), SCImago.

All books published in the series are submitted for consideration in Web of Science.

More information about this series at http://www.springer.com/series/11156

Ajith Abraham · Yukio Ohsawa ·
Niketa Gandhi · M. A. Jabbar ·
Abdelkrim Haqiq · Seán McLoone ·
Biju Issac
Editors

Proceedings of the 12th International Conference on Soft Computing and Pattern Recognition (SoCPaR 2020)

Springer

Editors
Ajith Abraham
Scientific Network for Innovation and Research Excellence
Machine Intelligence Research Labs (MIR Labs)
Auburn, WA, USA

Niketa Gandhi
Scientific Network for Innovation and Research Excellence
Machine Intelligence Research Labs (MIR Labs)
Auburn, WA, USA

Abdelkrim Haqiq
Faculty of Sciences and Techniques
Hassan 1st University
Settat, Morocco

Biju Issac
Northumbria University
Newcastle, UK

Yukio Ohsawa
Systems Innovation, School of Engineering
The University of Tokyo
Bunkyo-ku, Tokyo, Japan

M. A. Jabbar
Vardhaman College of Engineering
Hyderabad, Telangana, India

Seán McLoone
Queen's University
Belfast, UK

ISSN 2194-5357 ISSN 2194-5365 (electronic)
Advances in Intelligent Systems and Computing
ISBN 978-3-030-73688-0 ISBN 978-3-030-73689-7 (eBook)
https://doi.org/10.1007/978-3-030-73689-7

This Springer imprint is published by the registered company Springer Nature Switzerland AG
The registered company address is: Gewerbestrasse 11, 6330 Cham, Switzerland

Preface

Welcome to the 12th International Conference on Soft Computing and Pattern Recognition (SoCPaR 2020) and the 16th International Conference on Information Assurance and Security (IAS 2020) held online during December 15–18, 2020. SoCPaR 2019 and IAS 2019 were held in Hyderabad, India, and Bhopal, India, respectively.

SoCPaR 2020 is organized to bring together worldwide leading researchers and practitioners interested in advancing the state of the art in soft computing and pattern recognition, for exchanging knowledge that encompasses a broad range of disciplines among various distinct communities. The themes for this conference are thus focused on "Innovating and Inspiring Soft Computing and Intelligent Pattern Recognition". SoCPaR 2020 received submissions from 26 countries, and each paper was reviewed by at least five reviewers in a standard peer review process. Based on the recommendation by five independent referees, finally sixty-two papers were presented during the conference (acceptance rate of 31%).

Information assurance and security has become an important research issue in the networked and distributed information sharing environments. Finding effective ways to protect information systems, networks, and sensitive data within the critical information infrastructure is challenging even with the most advanced technology and trained professionals. The 16th International Conference on Information Assurance and Security (IAS) aims to bring together researchers, practitioners, developers, and policy makers involved in multiple disciplines of information security and assurance to exchange ideas and to learn the latest development in this important field. IAS 2020 received submissions from 21 countries, and each paper was reviewed by at least five reviewers in a standard peer-review process. Based on the recommendation by five independent referees, finally thirty-five papers were presented during the conference (acceptance rate of 33%).

Many people have collaborated and worked hard to produce this year successful SoCPaR–IAS joint conferences. First and foremost, we would like to thank all the authors for submitting their papers to the conference, for their presentations and discussions during the conference. Our thanks to program committee members and reviewers, who carried out the most difficult work by carefully evaluating the submitted papers. Our special thanks to the following plenary speakers, for their exciting plenary talks:

- Prof. Dr. Sushmita Mitra, Indian Statistical Institute, Kolkata, India
- Prof. Dr. Bassem Jarboui, Sfax University, Tunisia
- Prof. Dr. Tzung-Pei Hong, National University of Kaohsiung, Taiwan
- Prof. Dr. André Rossi, Université Paris-Dauphine, France
- Prof. Dr. Amine Nait-Ali, University Paris-Est Créteil, France
- Dr. Rudy A. Oude Vrielink, University of Twente & TwiQel, Netherlands
- Prof. Dr. Frederico Gadelha Guimaraes, Federal U. of Minas Gerais, Brazil
- Prof. Dr. Fakhri Karray, University of Waterloo, Canada
- Prof. Dr. Saman K. Halgamuge, University of Melbourne, Australia
- Prof. Dr. Salvador García, University of Granada, Granada, Spain
- Prof. Dr. Vincenzo Piuri, Università degli Studi di Milano, Italy
- Prof. Dr. Dries F. Benoit, Ghent University, Belgium
- Prof. Dr. Dijiang Huang, Arizona State University, USA
- Prof. Dr. Ketan Kotecha, Symbiosis Institute of Technology, India
- Prof. Dr. Sheela Ramanna, University of Winnipeg, Canada

Our special thanks to the Springer Publication team for the wonderful support for the publication of these proceedings. We express our sincere thanks to the session chairs and organizing committee chairs for helping us to formulate a rich technical program. We express our sincere thanks to the organizing committee chairs for helping us to formulate a rich technical program. Enjoy reading the articles!

Ajith Abraham
Yukio Ohsawa
Seán McLoone
Abdelkrim Haqiq
Biju Issac
Isabel Jesus
M. A. Jabbar
Kanwalinderjit Gagneja
Álvaro Herrero

SoCPaR-IAS Organization

General Chairs

Ajith Abraham	Machine Intelligence Research Labs, USA
Yukio Ohsawa	The University of Tokyo, Japan
Seán McLoone	Queen's University Belfast, UK
Abdelkrim Haqiq	Hassan 1st University, Morocco

Program Chairs

Biju Issac	Northumbria University, UK
Isabel Jesus	ISEP, Portugal
M. A. Jabbar	Vardhaman College of Engineering, Hyderabad, India
Kanwalinderjit Gagneja	California State University, Bakersfield, USA
Álvaro Herrero	Universidad de Burgos, Spain

Publication Chairs

Niketa Gandhi	Machine Intelligence Research Labs, USA
Kun Ma	University of Jinan, China

Publicity Chairs

Jyotshna Dongardive	University of Mumbai, India
Pooja Manghirmalani Mishra	University of Mumbai, India

International Program Committee

Abdelkrim Haqiq	Hassan 1st University, Morocco
Adelaide Cerveira	INESC–TEC and UTAD, Portugal
Akila Muthuramalingam	KPR Institute of Engineering and Technology, India
Angel Arroyo	UBU, Spain
António Pinto	ESTG, Portugal
Azza Ali	Sinai University, Egypt
C. Alberto Ochoa-Zezatti	Universidad Autónoma de Ciudad Juárez, Mexico
Carlos Pereira	ISEC, Portugal
Chian C. Ho	National Yunlin University of Science and Technology, Taiwan
Constantino Malagón	Nebrija University, Spain
Dabiah Alboaneen	Imam Abdulrahman Bin Faisal University (IAU), Saudi Arabia
David Herrera-Sánchez	University of Veracruz, Mexico
Devottam Gaurav	Indian Institute of Technology Delhi, India
Dilip Pratihar	Indian Institute of Technology Kharakpur, India
Eduardo Pires	University of Tras-os-Montes and Alto Douro, Portugal
Efrén Mezura-Montes	University of Veracruz, Mexico
Fernando Tricas	Universidad de Zaragoza, Spain
Francisco Valera	Universidad Carlos III de Madrid, Spain
Frantisek Zboril	Brno University of Technology, Czech Republic
Gabriella Casalino	Università degli Studi di Bari, Italy
Gai-Ge Wang	Jiangsu Normal University, China
Ghizlane Orhanou	Mohammed V University in Rabat, Morocco
Gloria Cerasela Crisan	"Vasile Alecsandri" University of Bacau, Romania
Gouse Baig Mohammad	Vardhaman College of Engineering, India
Guadalupe Carmona Arroyo	Universidad Veracruzana, Mexico
Hector-Gabriel Acosta-Mesa	Universidad Veracruzana, Mexico
Hicham Triba	Abdelmalek Essaadi University, Morocco
Iman El Mir	Université Hassan, Morocco
Isabel S. Jesus	Instituto Superior de Engenharia do Porto, Portugal
Jabbar Meerja	Vardhaman College of Engineering, India
Jerry Chun-Wei Lin	Western Norway University of Applied Sciences, Norway
Jesús-Adolfo Mejía-de-Dios	University of Veracruz, Mexico
João Paulo Magalhaes	ESTGF, Porto Polytechnic Institute, Portugal
José Everardo Bessa Maia	State University of Ceará, Brazil

Jose M. Molina	Universidad Carlos III de Madrid, Spain
Jose Vicent	Universidad de Alicante, Spain
José-Clemente Hernández-Hernández	University of Veracruz, Mexico
Jose-Luis Ferrer-Gomila	University of the Balearic Islands, Spain
Katarzyna Kaczmarek	Polish Academy of Sciences, Poland
Keun Ho Ryu	Chungbuk National University, South Korea
Konstantinos Parsopoulos	University of Ioannina, Greece
Lin Wang	University of Jinan, China
Mahendra Kanojia	Sheth L.U.J and Sir M.V. College, India
Maleh Yassine	Hassan 1st University, Morocco
Mario Graff	Consejo Nacional de Ciencia y Tecnología, Mexico
Mauro Gaggero	National Research Council of Italy, Italy
Meera Ramadas	University College of Bahrain, Bahrain
Mohammed Ridouani	Hassan II University, Morocco
Nageswararao Sirisala	Vardhaman College of Engineering, India
Najid Abdellah	Institut National des Postes et Télécommunications, Morocco
Nancy Pérez Castro	University of Papaloapan, Mexico
Neera Jeyamohan	Northumbria University, England
Niketa Gandhi	Machine Intelligence Research Labs (MIR Labs), USA
Oscar Castillo	Tijuana Institute Technology, Mexico
Ozgur Koray Sahingoz	Istanbul Kultur University, Turkey
Pedro Coelho	UERJ, Brazil
Pedro Juan Roig	Miguel Hernández University, Spain
Pooja Manghirmalani Mishra	University of Mumbai, India
Prasadkantipudi M. V. V.	Sreyas Institute of Engineering and Technology, India
Radu-Emil Precup	Politehnica University of Timisoara, Romania
Raghunandan Reddy Alugubelli	H. Lee Moffitt Cancer Center & Research Institute, USA
Rolf Oppliger	eSECURITY Technologies, Switzerland
Ronak Panchal	Vidyabharti Trust College of BCA, India
Said El Kafhali	Hassan 1st University, Morocco
Salma Benazzouza	Hassan II University, Morocco
Salvador Alcaraz	Miguel Hernandez University, Spain
Salvatore Venticinque	Seconda Università di Napoli, Italy
Sanju Tiwari	National Institute of Technology, India
Shankarshan Prasad Tiwari	National Institute of Technology, India
Simone Ludwig	North Dakota State University, USA
Sorin Stratulat	Université de Lorraine, Metz, France
Subodh Deolekar	REDx Innovation Lab WeSchool, India

Sye Loong Keoh	University of Glasgow, Singapore
Teruaki Hayashi	The University of Tokyo, Japan
Tzung-Pei Hong	National University of Kaohsiung, Taiwan
Varun Ojha	University of Reading, UK
Wenjian Luo	University of Science and Technology of China, China
Zahi Jarir	Cadi Ayyad University, Morocco

Contents

Mask RCNN for Human Motion and Actions Recognition

Soumaya Zaghbani[1(✉)] and Med Salim Bouhlel[2]

[1] ENIG, SETIT, Gabes, Tunisia
[2] ISBS, SETIT, Sfax, Tunisia

Abstract. Human actions recognition (HAR) and understanding become very popular topics in the field of computer vision and signal processing. The purpose of human activities recognition is to automatically examine and characterize actions from a video sequence. The main goal of a HAR system is to identify simple actions of everyday life (like walking, running, jumping ...) from videos. Each of these actions, performed by one person or many persons within a specific period of time, must be represented by a simple movement model. In recent years, a large number of applications of HAR have been proposed in literature such as video surveillance, human-computer interaction and video indexing. In this line, we present our method for HAR using Mask Region Based CNN, MRCNN. This technique will help us to make the accent on the body of individual and this step facilitates the recognition of the current action. With the mask RCNN we used key frame extraction and background extraction. This framework was tested on two datasets KTH and WEIZMANN datasets and experimental results showed the performance of the proposed technique.

Keywords: Human motion detection · HAR · Mask RCNN · CNN

1 Introduction

The particularity of human physics encourages many researchers to exploit its features in different domains such as user identification, behavior analysis, computer technology, security, and psychology. Human body analysis is completed in three main stages: firstly, detection, normalization, feature extraction and classification. This depends in particular on the nature of the application, the size of the learning base, and the shooting conditions (noise, luminosity, etc.). The first step in human factor analysis is capturing: a person's body is digitized by sensors such as digital cameras for the 2D representation, 3D scanners to have depth information or also infrared or near-infrared sensors. The captured images or videos constitute a dataset or information to be used in different applications for analyzing human actions. The purpose of these applications is to observe people and identify automatically what they do as actions in the video sequence. HAR and understanding become very popular topics in the field of computer vision

A. Abraham et al. (Eds.): SoCPaR 2020, AISC 1383, pp. 1–9, 2021.
https://doi.org/10.1007/978-3-030-73689-7_1

and signal processing [1]. The goal of a human activity/action recognition system is to identify simple actions of everyday life (like walking, running, jumping ...) from videos. Each of these actions, performed by one person or many persons within a specific period of time, must be represented by a simple movement model. In recent years, many methods have been proposed for the recognition and understanding of human actions. A large number of applications of HAR have been proposed in literature such as video surveillance, human-computer interaction and video indexing [2]. The main advantage of these methods is their importance in critical application such as surveillance in public places, abnormal and normal activities detection fall detection for elderly people assistance. We distinguish several types of actions: gestures, actions, interactions and group of activities [3]. In this work we present a new framework for HAR to recognize some daily life actions such as walking and running.

2 Previous Work

HAR is an area of research very interesting and different researches have been proposed in literature. Hoang et al. [4] presents a framework for HAR based on Selected Active Frames from videos to construct datasets. In this work, authors proposed a Vote fusion method for resembling multi-stream networks. Results on KTH dataset showed an accuracy of 93.8%. Kuang-Pen et al. [5], presented another framework for action recognition. In their work they compared three interesting and practical architectures for multi-view video. These systems are the nearest neighbor classifier (NNC), Gaussian mixture model classifier (GMMC) and the nearest mean classifier (NMC). The work was tested on two datasets KTH and WEIZMANN and MuHAVi. Found accuracies for the NNC are 89.31%, 93.21% and 97.50% for the three cited datasets respectively. For the GMMC architecture founded results are 90.21%, 91.11% and 93.21% respectively. Finally for NMC accuracies respectively are 90.58%, 95.56% and 97.50%. In 2019, Salah et al. [6], present a method for HAR using temporal salience. Researchers Local Dense Descriptor (LDD) to capilatize features. Method performances were tested on three datasets KTH, DHA and WEIZMANN. Result showed the performance of this method achieved 99.06%, 99.39 and 99.65 for the three datasets respectively. Quingquan et al. [7] presented in 2019 approach for HAR based on multi-layer visual features. In this method, authors combined the motion information and apparent information using coding layer. The motion information was encoded with dynamic pose flow fiels based on human trajectory and classification was made by Support Vector Machine (SVM) model. Accuracy achieved 94.5% on KTH, 89.3% on Penn and 66.4% on HMDB datasets. Nguyen et al. [8], present a novel method for HAR using hierarchical Bayesian non-parametric model (HBNPM). In this method, researchers suggested a schema for complex actions using generative topic model (GTP) and Multi-label Hierarchical Dirichlet Process (ML-HDP). The model the co-occurrence relationship of actions and motion units by processing three level representation by connecting the low level representation to high-level action using mid-level atomic actions.

The proposed architecture was tested on four datasets: KTH, MSR-II, Hollywood2 and UCF101. Results found on these datasets respectively were: 96.50%, 68.8% and 89.3%. Denver et al. [9], presented in 2018 a novel method for HAR generating Motion History Images (MHI) and extracting features using Bag of features. Finally, they used SVM to train obtained vectors. Their work was evaluated on Weizmann and the KTH datasets. The work achieved the 90% and 82% respectively.

3 Proposed Method

In this article we present our method for actions recognition using MRCNN and CNN architectures. The work is divided into three important phases: the first one is the application of the MRCNN on the input video; the second one is a step of treatment where we are looking to extract only the human action and subtract the background. The final phase is the classification of actions using CNN architecture. Figure 1 described the pipeline of the presented work.

3.1 Mask RCNN

RCNN presented in [10] shows his performance in machine learning tasks. RCNN or Region based CNN is a method of objects classification based on deep networks was presented by Ross Girshick et al. in 2013. The method uses a model of region of interest independent. RCNN is a method deep networks; it is a bounding-box approach, used for object detection. The main principle of this approach is to propose and manage a number of regions of interest in the image independently. After introducing an image in the network, an algorithm of selective search is used to extract region proposal (region proposal it is a regions of an image). After, for each proposal, feature extraction is calculated using a pre-trained CNN. Finally, SVM model is employed to classify each proposal through its extracted features. Although the effectiveness of RCNN in object classification, it presented big problems of time, due to the number of regions in one image. To overcome those problems Shaoqing Ren et al. [11] proposed the Fast Region CNN (FRCNN) detection process, another version of RCNN which is able to solve the issue of speed in the training by sharing the computation of all convolutional layers between different proposals and swich the priority of generating region proposals and run the CNN. After that Kaiming He et al. [12] presented a new version called Mask RCNN extended from Faster R-CNN by adding a section to predect an object mask besides bounding box recognition. The main purpose of this work is to exploit the efficiency of MRCNN to firstly extract human motion in the video sequence and secondly, make the accent on the human body. The MRCNN will help us to detect humans in the sequence and to apply a mask on it which is very important to limit the action area. Firstly, for every frame of the video we applied the MRCNN, the MRCNN is divided in two main steps. The first step is the extraction of features map in the goal to define proposals. This step is made using Region Proposal Network

(RPN) and help us to detect proposals and proposes candidate object bounding boxes. As the MRCNN is pre-trained model using transfer learning, it is simple to extract human area. The second step consist of the extraction of features using RoIPool from each nominee box and performs classification and bounding-box predection. As conclusion, the MRCNN is used in our work for two goals: the first one is the human motion extraction and the second one to exploit mask area to highlight the current action to improve the classification process.

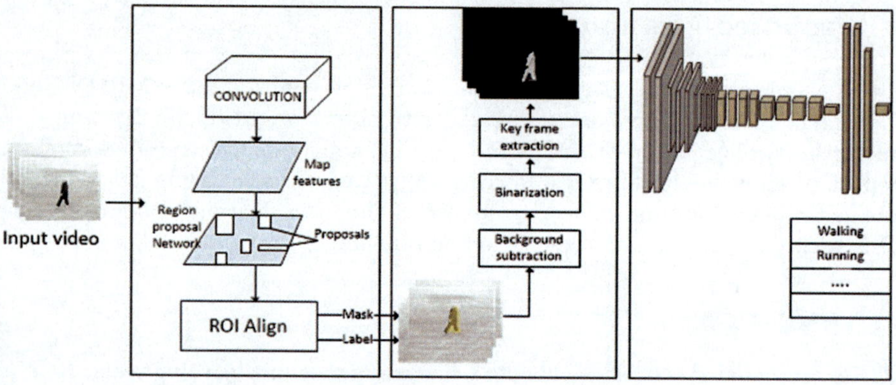

Fig. 1. Pipeline of the proposed method

3.2 Background Extraction

The result of the first stage MRCNN is a mask on the human silhouette will help us to facilitate the second stage. In the next step we moved to background subtraction, this step is very important in video processing and especially in HAR. This classic approach is employed to subtract an image that contains the target object, with the previous background image that has no foreground objects of interest [13]. The area of the founded masks on the human body is significant and indicates the pixel location of the moving human. Actions are separated using differencing technique. After the mask RCNN image M(l, m) is obtained, we subtract the background image B(l, m) from the current frame F(l, m). If the pixel appears in the mask as background pixel, we subtract the mask from the image as following:

$$Bk(I, m) = 1 when M(l, m), 0 otherwise \quad (1)$$

3.3 Key-Frame Extraction

Key-frame extraction also called a summary of the video because it presents only important content inside video. Hence it is very important for video processing and video data applications. This step is very important since it help to remove redundancy. Different approaches have been presented in literature in the purpose to detect Key-frames only. They could be determined by considering motion as important features which can be calculated by using inter frame difference [14]. Our goal from this step is not to summarize the video sequence but to highlight it in the goal to extract apex frames and concentrate on the action. This step will help us to reduce redundancy and number of unnecessary samples inside the class.

3.4 Classification

A CNN model is composed of a set of important and successive layers. The first one called convolutional layer, it is the key component of CNN and is still at least their first layer. It detects notable characteristics in the input data. The first layers, the deep layers that are closest to the input raw data, learn to detect simple features, while the upper layers combine simple features to produce complex features. Between the convolutional layers, max-pooling layers are introduced which reduce the size of the representation by under-sampling while accentuating the strong signals, which allows essential gains in computing power [15]. This process participate to loose some important characteristics detected in the first layers such as orientation, scale, position. Various loss functions suitable for different tasks can be used. The "Softmax" function is used to calculate the probability distribution on the output classes [16]. A common use of the Softmax function (SF) appears in the field of machine learning, especially in logistic regression: we associate with each output possibility a score, which is converted into probability with the SF. Specifically, we have a vector input, which is a column matrix, denoted x, N rows. We will multiply it by a matrix called "weight" W of T lines and N columns, to transform x into a vector of T elements. The SF is used to convert the logits into a probability vector, indicating the probability that x belongs to each of the output classes T. Here we compute probability value and assimilate this result with threshold θ. This threshold can be converted into a simple binary classification problem. The expression of the logical function is shown as follows:

$$h_\theta(x) = g(\theta^T x)\frac{1}{1+e^{-\theta T x}} = p(y=1|x;\theta) \tag{2}$$

Where $h_\theta(x)$ is the probability of 1, and θ is the model parameter. The optimization of the parameters is presented through successive tuning to minimize the loss function. The loss function is expressed in the following form:

$$j(\theta) = -\frac{1}{m}\left[\sum_{i=1}^{m}\sum_{j=1}^{k} 1\left\{y^{(i)}=j\right\}\log\frac{e^{\theta_j^T x^{(i)}}}{\sum_{l=1}^{k} e^{\theta_j^T x^{(i)}}}\right] \tag{3}$$

For multiple Softmax classifiers, the expansion reads in the form:

$$\begin{aligned} h_\Theta(x^i) &= \left[p(y^{(i)}=1|x^{(i)};\Theta), p(y^{(i)}=2|x^{(i)};\Theta)...p(y^{(i)}=k|x^{(i)};\Theta)\right] \\ &= \frac{1}{\sum_{j=1}^{k} e^{\Theta_j^T}x(i)}\left[e^{\Theta_1^T}x(i), e^{\Theta_2^T}x(i)...e^{\Theta_k^T}x(i)\right] \end{aligned} \tag{4}$$

Where the output value is a k dimensional vector of classes, and the model parameters are $\theta_1, \theta_2 ... \theta_k$.

4 Experimental Results

This section is dedicated to evaluate the proposed method and compared the found results with various methods have been presented in literature. The work was tested on KTH and WEIZMANN datasets. KTH is a dataset used to recognize five actions (Running, Walking, Jogging, hand-clapping and hand-waving). KTH database contains 2391 video sequences. WEIZMAN is database of video sequences showing 9 different people, each one presenting 10 natural actions. In our work we used seven actions: "run," "walk," "skip", "jump", "side", "waveone- hand", or "bend." Results showed the effectiveness of the proposed method in comparison with other architectures proposed in literature as demonstrated in Table 1. The accuracy for 80 epohs presents 98.41%, is an interesting results and exceeded many recent researches working in action recognition especially on KTH dataset. Experimental results shown an accuracy of 98.15% for WEIZMANN dataset.

Table 1. Comparison of founded result (accuracy%) with state of the art methods for KTH and WEIZMANN datasets

Authors	Year	KTH	WEIZMANN
Hoang et al. [4]	2018	93.80%	–
Kuang-Pen et al. [5]	2018	90.58%	95.56%
Quingquan et al. [15]	2019	94.50%	–
Nguyen et al. [12]	2018	96.5%	–
Denver et al. [16]	2018	82%	90%
Proposed method	–	**98.41%**	**98.15%**

The MRCNN plays an important role to enhance the visibility of the action and concentrate on the silhouette of the human. The contribution of the MRCNN against the previous version of this architecture such as the RCNN and FRCNN is that the output of this model is a mask on the object instead of a bonding box and this mask help us to limit the area of the object (in our experiment is a human). The founded mask limit as possible and accentuated the executed action helped us to prepare useful inputs to the classification process. Figure 2 shows the result of applying the two first stages of our proposed method. In the above work we used the key frame extraction to limit the number of the unnecessary frames and keep only the apex frames. This step helps us to improve the recognition process and eliminate the redundant frames in the same class or between classes. After applying the key frame extraction the accuracy moved from 97.22% to 98.41% The confusion matrix showns the effectiveness

Fig. 2. Human motion detection using MRCNN: initial frames, MRCNN application, background subtraction and binarisation for KTH dataset for running action

of this method; the best rate presented concern the hand-waving class. After that, we found the hand clapping class with 99.11%, hand waving with 99.17% walking class with 98.15%, running class with 97.81% and finally the jogging class with97.82%. In this research we used the CNN architecture to enhance the classification process and optimize action recognition. After some iteration we found the best accuracy with 80 epochs. The model was trained with Adam optimizer and we used model with four blocks. Block composed by convolutional layer with kernel $3 \times 3 \times 32$, followed by activation layer ReLU and by batch normalization layer. The second block contains two convolutional layers with $3 \times 3 \times 64$ followed by ReLU and batch-normalization. The next block contains two convolutional layers with $3 \times 3 \times 128$. After every block, we applied the max pooling and drop out a technique to randomly disconnect nodes from the current layer to the next layer in the goal to decrease the over-fitting. Finally we have two FCL and we used Softmax function to vote for every class.

For the WEIZEMANN dataset, founded results achieved the 91.15% of accuracy after training the network with 80 epochs. As demonstrated in the table, the presented method exceeded many models have been presented in literature recently. Figure 3 shows the result of applying the two first stages of our proposed method. The founded result is explained by the ability of the proposed method to resolve different problems related with HAR in video such as human motion detection and background subtraction. The combination of these metrics with the classification using an efficient model such as CNN ameliorates the recognition process. The presented method is simple, easy to build and rapid and present interesting result in the opposite of different methods presented recently which are very costly and complex (Fig. 4).

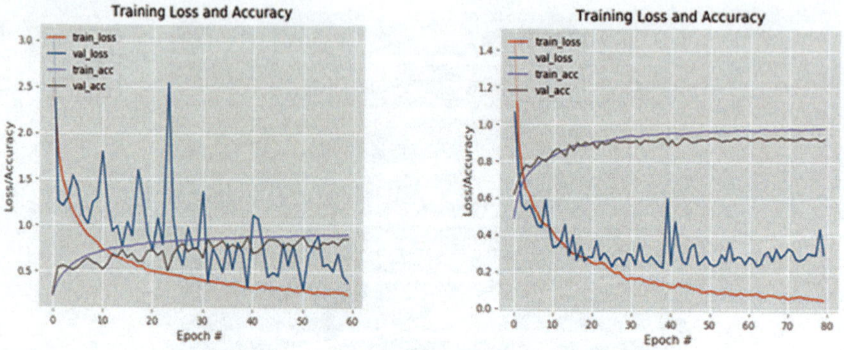

Fig. 3. Accuracy of HAR after 60 and 80 epochs of KTH dataset

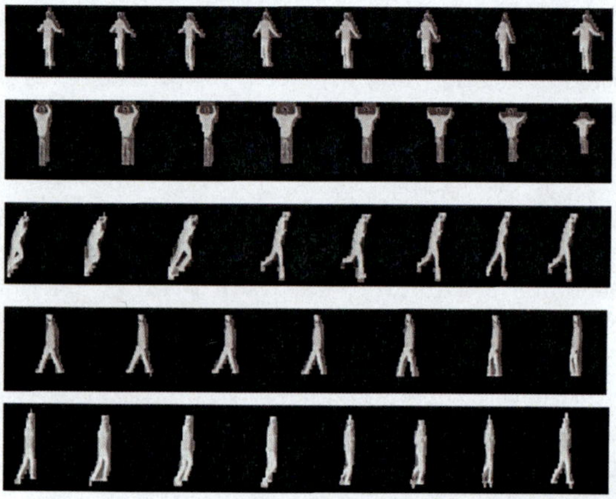

Fig. 4. The result of applying the two first stages of our proposed method for hand-waving, run, walk and skip actions for WEIZMANN dataset

5 Conclusion

In this paper, we have concretely described an approach for HAR. In this method we used the MRCNN not only to extract human motion but also to make the accent on his body using the mask. This mask helped us to contour the body area in the goal to highlight the action. For the classification we used the CNN with four blocks to enhance the final result.

References

1. Willems, G., Tuytelaars, T., Van, L.: An efficient dense and scale-invariant spatio-temporal interest point detector. In: Computer Vision ECCV 2008 10th European Conference on Computer Vision Marseille France (2008)

2. Wang, H., Wang, L., Member, S.: Beyond joints: learning representations from primitive geometries for skeleton-based action recognition and detection. J. Latex Class Files (2017)
3. ter Haar, F., Veltkamp, R.: A 3d face matching framework, vol. 2008 (2008)
4. Binh, H.T., Chau, M.T., Sugimoto, A., Duyp, B.T.: Selecting active frames for action recognition with vote fusion method. In: International Conference on Computer and Communication engineering ICCCE (2018)
5. Chou, K.P., Prasad, M., Wu, D., Sharma, N., Li, D.L., Lin, Y.F., Michael Blumenstein, W.C.: Feature-based automated multi-view human action recognition system. IEEE Access **6**, 15283–15296 (2018)
6. Obaidi, S.A., Abhayaratne, C.: Action recognition based on multi layer visual features. In: 10th International Conference on Intelligent Human Machine Systems and Cybernetics (2018)
7. Feng, Q., Ni, G., Dai, B., Dai, J.: Action recognition based on multi-layer visual features. In: 10th International Conference on Intelligent Human Machine Systems and Cybernetics (2018)
8. Tu, N.A., HuynhThe, T., Lee, K.U.K.Y.K.: Ml-HDP: a hierarchical bayesian nonparametric model for recognizing human actions in video. IEEE Transactions Circuits Systems Video Technology **27**(7), 1478–1490 (2018)
9. Naidoo, D., Tapamo, J.R., Walingo, T.: Human action recognition using spatial-temporal analysis and bag of visual words In: 14th International Conference on Signal-Image Technology Internet-Based Systems (SITIS) (2018)
10. Girshick, R., Donahue, J., Darrell, T., Malik, J.: Rich feature hierarchies for accurate object detection and semantic segmentation. In: CVPR 2014 (2013)
11. Girshick, R.: Fast r-cnn. In: ICCV (2015)
12. He, K., Gkioxari, G., Dollár, P., Girshick, R.: Mask r-cnn. In: Computer Vision and Pattern Recognition (2017)
13. Lu, D., Hou, Z.: Mimo model-free adaptive control color background image extraction to video. In: IEEE 8th Data Driven Control and Learning Systems Conference (DDCLS) (2019)
14. Momin, B., Rupnar, G.: Keyframe extraction in surveillance video using correlation (2016)
15. Zhang, R., Ni, B.: Learning behavior recognition and analysis by using 3d convolutional neural network. In: nternational Conference on Engineering, Applied Sciences and Technology (ICEAST) (2019)
16. Lian, M.: Captured multi-label relations via joint deep supervised autoencoder. Appl. Soft Comput. J. **74**, 709–728 (2018)

Intelligent and Secure Autofill System in Web Browsers

Joy Bose(✉)

Ericsson, Bangalore, India
joy.bose@ieee.org

Abstract. This paper describes an application of machine learning in a web browser feature called autofill. Automatic form fill or autofill is a time saving feature present in web browsers. This feature automatically fills text values in a new web form based on the text values filled by the user for the same or similar fields in other forms on the web, which are stored locally. This is dependent on the correct prediction of the labels of the form fields. In this paper, we describe an intelligent and secure solution for autofill. This includes fixing some security vulnerabilities related to hidden fields in a web form, resulting in a secure auto fill. Our solution also intelligently predicting the field labels in a form using machine learning. We describe the system architecture and some results for the accuracy of the system.

Keywords: Web forms · Intelligent autofill · Security · Hidden fields

1 Introduction

Autofill [1–3] is a common feature in most web browsers such as Google Chrome, Microsoft Edge and Mozilla Firefox. Using this feature, the fields in a web form such as name, address, credit card info etc. are filled with stored user data automatically upon loading of the form.

The autofill feature results in productivity enhancements for the user, since the user does not have to fill the same form field repeatedly for multiple forms. This saves user effort and makes it convenient to avoid repeated typing, which is especially useful for small form factor devices such as mobile phones.

In a web form, every field has a corresponding field label. In order to fill the information correctly in the form, the field label of each field must be predicted and the corresponding data retrieved from memory. The prediction is commonly made by storing the field labels from the web forms and the associated data values for those field labels in the local memory of the web browser, that stores the mapping of each field label and the corresponding info. Sometimes, this information is hashed and securely stored in a remote server. However, this method of storing each unique field separately has the disadvantage of not being scalable for similar looking web form field labels. Using a machine learning based solution would make the system more robust and accurate for multiple web forms.

A. Abraham et al. (Eds.): SoCPaR 2020, AISC 1383, pp. 10–19, 2021.
https://doi.org/10.1007/978-3-030-73689-7_2

In this paper, we describe a system for making the autofill feature in web browsers more secure and intelligent. The security is enhanced by plugging a common loophole, by detecting nested forms and disabling autofill for the nested forms. The intelligent part of the system uses machine learning to predict the field labels of previously unseen forms.

Fig. 1. An illustration of autofill suggestions for username and password in a yahoo login form in a web browser.

The screenshot in Fig. 1 illustrates the user interface for autofill in a web form.

The rest of the paper is described as follows: in Sect. 2 we describe the problem of accurately predicting the field labels in a form, along with different candidate approaches. Section 3 describes related work in this area. Section 4 gives the system architecture for the proposed ML based intelligent autofill solution. Section 5 gives the problem statement for the secure autofill, and Sect. 6 gives a rules based solution for secure autofill. Section 7 gives the implementation details and Sect. 8 the model for classification. Section 9 gives experimental results. Section 10 concludes the paper.

2 Intelligent Autofill: Problem Statement

As mentioned, the problem in devising an intelligent autofill solution is to accurately predict the labels of different fields in a new web form. These predicted labels are used to fill the stored values from the internal database of the web browser.

In the following subsection, we consider some candidate approaches for intelligent autofill:

2.1 A Naïve Approach for Autofill Prediction

A naïve approach for predicting the field labels in web forms in autofill is to keep a file with the mappings of {*URL, field name, predicted label*} for each field in each form. When the user browses a previously unseen web form, there is no prior storage for the field name and so the system cannot fill the form. However, the next time the same user visits the same form on the web, the previously stored values are filled automatically.

This naïve approach is simple and easy to implement. However, such an approach would not be scalable, since it would not work for new field names. Also, since the web has billions of forms and many forms may be similar to each other, making storing details of each form redundant. Besides, if the HTML code for a form changes or is dynamic, resulting in a changed field name the next time the user visits the page, it would have to be stored again.

2.2 A Rules Based Approach with Regular Expressions

A slightly more sophisticated approach to predict the field labels is to use a combination of heuristic rules for the purpose of predicting a field label. An example of such a rule may be as follows:

If < Id > or < name > or < label > follows < regular expression >, then set < predicted label > to < given value >

In order to implement this, one can store a table in the internal database of the web browser with fixed values for each of these fields. The use of regular expressions provides a bit more flexibility for similar looking fields. This stored table keeps expanding as the number of forms browsed by the user goes up. This is the approach taken by many common web browsers.

However, such an approach may not give desired results for some kinds of forms, such as the following:

- Forms where the id of the form field may be one of multiple categories. For example, where any of the following are valid: username or mobile number or email
- Forms that are dynamic, such as forms behind a login or paywall. In these cases, the HTML keeps changing and so does the field id.

2.3 A Machine Learning (ML) Based Approach

In a machine learning (ML) based approach for the prediction of the autofill values, the fields of a web form are predicted based on training a machine learning model using a labeled dataset. As earlier, the model is trained to learn the same mapping of domain, field name, predicted label, and value, but the difference is that this time an ML model is used instead of fixed mapping or regular expression for the mapping.

The ML approach may fix some of the problems with the previous approaches. It can learn the patterns behind the forms and form fields and hence generalize better. Therefore, it can give a better performance with dynamic webpages, than with the naïve or rule-based approaches. Hence, in this paper we use the ML based approach.

3 Related Work for Intelligent Autofill

Automatic form-fill in web browsers such as Edge [1], Chrome and Firefox has been present as a feature since the early versions of the browsers. A few patents such as by Goodman et. al [2] are also available covering the basic idea of autofill, as well as enhanced features.

Liddle [3] proposed an early solution for automatic form fill, focusing on e-commerce related queries on web forms. Their focus was on getting a better aggregate accuracy for the queries, rather than accuracy of autofill on individual web forms.

Myers [4] did an early study of heuristics for web interfaces in general.

Kantorski [5] did a survey on approaches of autofill for accessing information via search forms for the dark web. They surveyed both heuristic based and machine learning based approaches, focusing on different kinds for complex combinations for the web forms. However, the focus of this survey was to retrieve information efficiently using web forms for the dark web, not to store and autofill common web forms.

Winckler [6] focused on getting an accurate autofill solution by leveraging patterns of contextual information in different web forms, such as names and addresses occurring together. The solution by Hartmann [7] focused on the similar issue of context awareness to increase the accuracy of predicted labels. Wang [8] focused on contexts as well as user usage patterns across different web forms. Troiano [9] modelled the problem of predicting the field labels as a Bayesian network, treating the whole form as a sequence of related entries instead of treating each field as a single entry.

While these are interesting approaches, they do not focus on our problem of guessing the field values from similar looking names or other features in different web forms.

There have also been a few studies on the privacy issues surrounding the autofill feature, such as by Lin [10], Zhao [11], Gutmann [12] and Oesch [13]. These focus on frameworks to deal with privacy and other security issues as a whole.

In the following sections, we detail the steps of preparing the dataset and training the machine learning model.

4 Architecture of the Machine Learning (ML) Based Intelligent Autofill

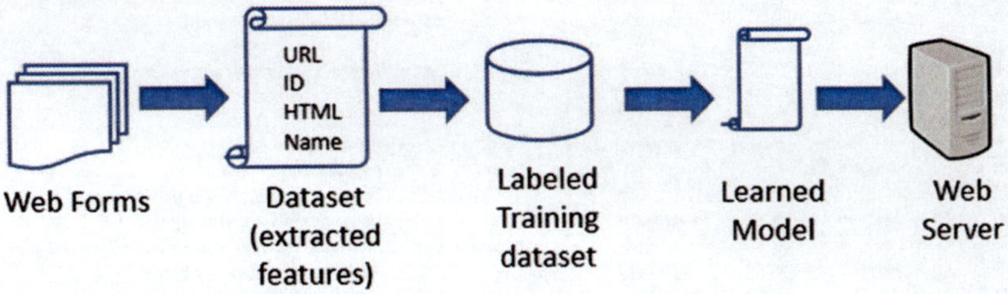

Fig. 2. Steps for training the Machine Learning (ML) model to predict the form fields

The machine learning (ML) solution for intelligent autofill has the following architecture: it comprises a web service to predict the field labels for the fields in each web form, and a trained machine learning model hosted in a backend server.

The ML model is trained on the server with the help of a dataset created out of extracted features from common web forms. The trained model predicts the label of the form field in real time, given a set of features extracted from the HTML of the web form.

Alternatively, the trained model file can also be included as part of the web browser executable or by using a web browser extension. However, in this paper we focus on the web service approach, since it is easier to update the model when new data is acquired. It must be noted that the actual form data remains in the web browser client since that data is private, only the features related to the form and field details extracted from the HTML are sent to the server and used to predict the field labels.

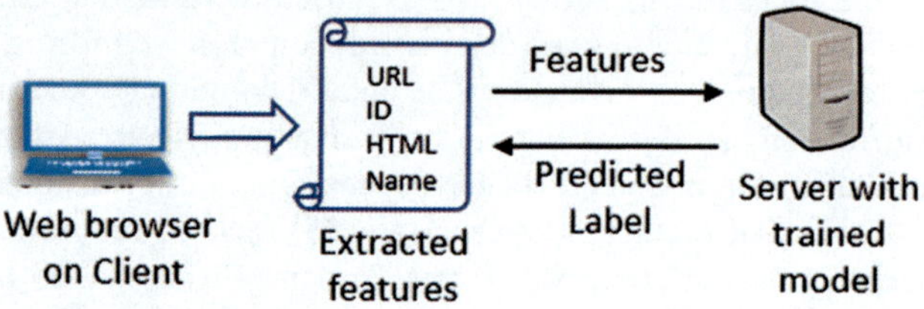

Fig. 3. Architecture of the web service to predict the labels for the form fiedls

Figure 2 shows the steps of training the machine learning model, and Fig. 3 shows the architecture of the web service to predict the field labels.

5 Dataset Preparation

We generate a training dataset for our ML model by extracting features a few web forms and manually providing a label for each of the fields. Common labels used in most web forms are name, address, username, password, age etc. (Fig. 4)

label	name	id	type	url	
Email:	email	email2stress	email	https://ie-snap/testsnew/redstone/edge/autoformfill/autoformfilltests.html	1
Email or mobile phone number	email	ap_email	email	https://www.amazon.in/ap/signin	1
Enter your email address	username	login-username	text	https://login.yahoo.com/	1
Email or phone	session_key		text	https://www.linkedin.com/	1
Email or phone	session_key	username	text	https://www.linkedin.com/login?trk=guest_homepage-basic_nav-header-signin	1
Password	session_password	password	password	https://www.linkedin.com/uas/login?fromSignIn=true&trk=uno-reg-join-sign-in	0
Email or Phone	email	email	email	https://www.facebook.com/	1
First name	firstname	u_0_n	text	https://www.facebook.com/	0
Surname	lastname	u_0_p	text	https://www.facebook.com/	0
Mobile number or email address	reg_email__	u_0_s	text	https://www.facebook.com/	1
New password	reg_passwd__	u_0_z	password	https://www.facebook.com/	0
Password (6 or			passwor	https://www.linkedin.com/start/join?trk=gue	

Fig. 4. A dataset for autofill prediction. The first few features are extracted from the web form such as label, name. The last column gives a value of 1 if the predicted label is email, 0 if not.

For labeling the fields, we use crowdsourcing, with a custom service similar to Amazon mechanical turk, to generate a dataset of around 4000 values from commonly used web forms. Our input feature are label, name, id and URL.

Our preprocessing includes stop word removal from the HTML of the form fields. We train a separate model for each form field. For example, to predict if a form field is email, we have a separate trained model than for predicting if it is username or address. We also experimented with a multi-class classifier.

We used a one hot encoding approach to encode the dataset as a feature vector. We use a dictionary of possible values and each of the features was a value in that dictionary. For example, we got all the possible field names from our dataset (such as username, username_01 etc.) and built a feature vector. The final feature vector is built by concatenating the individual encodings of the different features.

6 Prediction of Labels Using Azure ML Studio and WebML

Microsoft's Azure Machine Learning (ML) studio [14] provides a flexible interface for training and testing machine learning models. Using this interface, one can add blocks representing steps such as preprocessing, different choice of ML algorithms, evaluation of model accuracy etc. It can also convert the ML model to a web service hosted on Azure, making it very easy to deploy as an API. Figure 5 gives a sample interface.

Another option for using ML on Windows platforms is WinML with ONNX [15].

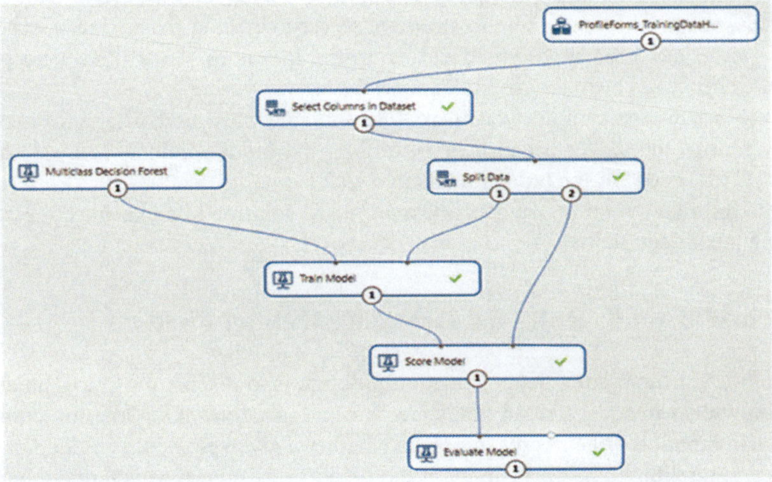

Fig. 5. Interface for training a machine learning model for predicting field labels using Azure ML studio

We trained different ML models on our dataset for each feature (email, id, label etc.). We experimented with multiple machine learning models to optimize the accuracy. Once we had obtained a desired level of accuracy for each of the labels (email classifier, state classifier, or multi class classifier etc.), we exposed the model as a web service to call from the web browser client using a browser extension.

7 Experimental Results for Intelligent Autofill Solution

For our dataset, we configured a train:test ratio of 70:30 for our solution.

After experimenting with different models including linear regression, support vector machines and decision trees, we found that a multi class decision forest (one of the off-the-shelf ML algorithm libraries available in the Azure ML studio) gave the best results for our dataset.

The parameters we used for the decision forest are as follows: resampling method = bagging, number of decision trees = 16, maximum depth = 100, random splits per node = 128, maximum samples per leaf node = 1.

After tuning the model parameters to improve the accuracy, we obtained an overall precision of around 95% for the email classifier. We obtained comparable results for the multi-class classifier as well.

8 Secure Autofill: Hidden Fields Problem

An existing problem with current autofill systems in web browsers is that there may be hidden fields in a web form, i.e. fields that are part of the form but which are not visible to the user.

In case the user goes to one of the forms having hidden fields, the browser might fill those fields using autofill while the user may be unaware of it since the fields are not visible to them. Sensitive personal details of the user, such as credit card info or passwords, might be secretly sent to a remote server and being phished or misused.

This issue has been highlighted earlier in the literature [16–19]. But no general purpose solution has been devised to combat this privacy issue.

Figure 6 shows a sample web form with hidden fields, and what information is actually sent to the server because of autofilling the hidden fields. Figure 7 shows a sample HTML code for the hidden fields in a web form.

In the following section, we describe a proposed solution to make autofill secure by detecting the hidden fields.

9 Secure Autofill: Rules for Detecting Hidden Fields

Our solution for the hidden fields detection problem is to disable the autofill feature in a web browser, whenever hidden fields are detected in a form. Our solution comprises using a combination of heuristics to decide whether a given element is hidden or not.

A hidden field may be an attribute of the field itself (input type is hidden, height or width is 0, style is display: none, etc.). Or else, it can be an attribute of the container DIV element.

The heuristic rules will have to consider each of the possible ways in which an element can be hidden. One such rule might be: Height or width of the div element is 0px. Another such case is where the div element is outside the user's viewable area so he or she cannot view it on opening the form, in this sense it is hidden.

To detect and disable autofill for individual fields with hidden attributes in a form, we use a combination of the following heuristics to detect and disable the autofill.

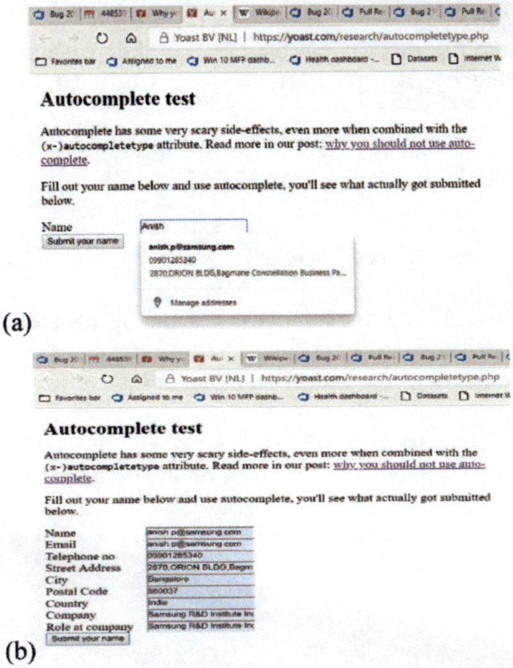

Fig. 6. (a) A sample web form with autofill for hidden fields (b) Information that actually got submitted

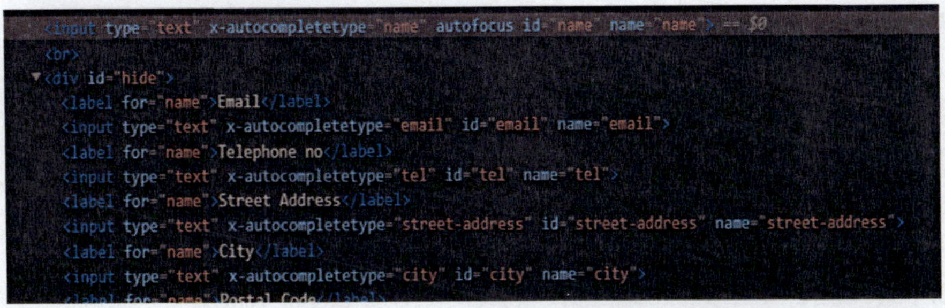
```
<input type="text" x-autocompletetype="name" autofocus id="name" name="name"> == $0
<br>
▼<div id="hide">
  <label for="name">Email</label>
  <input type="text" x-autocompletetype="email" id="email" name="email">
  <label for="name">Telephone no</label>
  <input type="text" x-autocompletetype="tel" id="tel" name="tel">
  <label for="name">Street Address</label>
  <input type="text" x-autocompletetype="street-address" id="street-address" name="street-address">
  <label for="name">City</label>
  <input type="text" x-autocompletetype="city" id="city" name="city">
```

Fig. 7. Sample HTML code for hidden fields in a web form

- CSS property says visibility = "hidden".hidden {display: none;}
- Input type = hidden < input id = "prodId" name = "prodId" type = "hidden" value = "xm234jq" > < input type = "hidden" class = "form-control" id = "hiddenfirstname1" placeholder = "First Name" name = "firstName" >
- Input type is display none < input id = "second" type = "text" style = "display:none;"/ > < input type = "tel" class = "form-control" id = "hiddenphone" placeholder = "Enter Mobile Number" name = "mobile" style = "display:none;" >
- Style says width or height is 0 < input type = "text" id = "hiddenpin1" placeholder = "Enter PinCode" name = "pincode" style = "width:0px;height:0px" >

- JQuery.visible selector is false
- The field is outside the user's viewable area, so they cannot see it element.style {margin-left: -500px;}

In our solution, we check for each of the above conditions whenever a web form is loaded. We disable autofill if any of the conditions are met.

10 Conclusion and Future Work

In this paper, we discussed a solution for secure and intelligent autofill for web browsers. A rules based approach is used to detect hidden fields and disable autofill, thus enhancing security, and ML is used for predicting the form fields. The system is deployed as a web service.

An early preprint of the paper is uploaded on arxiv [20].

Our planned future work includes: training on a variety of form fields, building a more robust and bigger dataset, incorporate contextual information to better predict the form fields and finally to incorporate a more secure framework by fixing more kinds of vulnerabilties.

References

1. Wahle, A., Prause, G.P.M., von Birgelen, C., Erbel, R., Sonka, M.: Fusion of angiography and intravascular ultrasound in vivo: establishing the absolute 3-D frame orientation. IEEE Trans. Biomed. Eng. **46**, 1176–1180 (1999)
2. Microsoft Edge Support. Automatically fill info in Microsoft Edge. https://support.microsoft.com/en-us/microsoft-edge/automatically-fill-info-in-microsoft-edge-81da697c-9910-d9b8-d50a-1712d96f3db8
3. Goodman, J.T., Kadie, C.M., Chickering, D.M., Bradford, D.E., Glasgow, D.A.: Microsoft Corp, assignee. Intelligent autofill. United States patent US 7,660,779. 2010 Feb 9 (2010)
4. Liddle, S.W., Embley, D.W., Scott, D.T., Yau, S.H.: Extracting Data behind Web Forms. In: Olivé, A., Yoshikawa, M., Yu, E.S.K. (eds.) Advanced Conceptual Modeling Techniques. ER 2002. LNCS, vol. 2784. Springer, Heidelberg (2003). https://doi.org/10.1007/978-3-540-45275-1_35
5. Myers, B.A., Wolf, R., Potosnak, K., Graham, C.: Heuristics in real user interfaces. In: Proceedings of the INTERACT '93 and CHI '93 Conference on Human Factors in Computing Systems (CHI '93), pp. 304–307. Association for Computing Machinery, New York (1993). https://doi.org/10.1145/169059.169240
6. Kantorski, G.Z., Moreira, V.P., Heuser, C.A.: Automatic filling of hidden web forms: a survey. SIGMOD Rec. **44**(1) (2015), 24–35 (2015). https://doi.org/10.1145/2783888.2783898
7. Winckler, M., Gaits, V., Vo, D.B., Sergio, F., Rossi, G.: An approach and tool support for assisting users to fill-in web forms with personal information. In: 29th ACM international conference on Design of communication (SIGDOC '11), pp. 195–202. ACM, New York (2011)
8. Hartmann, M., Muhlhauser, M.: Context-aware form filling for web applications. In: 2009 IEEE International Conference on Semantic Computing (ICSC '09). IEEE Computer Society, Washington, DC (2009)

9. Wang, S., Zou, Y., Upadhyaya, B., Ng, J.: An intelligent framework for auto-filling web forms from different web applications. In: IEEE Ninth World Congress on Services June 2013, pp. 175–179. IEEE Computer Society, Washington, DC (2013)
10. Troiano, L., Birtolo, C., Armenise, R.: (2017) Modeling and predicting the user next input by Bayesian reasoning. Soft. Comput. **21**(6), 1583–1600 (March 2017)
11. Lin, X., Ilia, P., Polakis, J.: Fill in the blanks: empirical analysis of the privacy threats of browser form autofill. In: Proceedings of the 2020 ACM SIGSAC Conference on Computer and Communications Security (CCS '20), pp. 507–519. Association for Computing Machinery, New York (2020). https://doi.org/10.1145/3372297.3417271
12. Zhao, R., Yue, C.: All your browser-saved passwords could belong to us: a security analysis and a cloud-based new design. In: Proceedings of the third ACM conference on Data and application security and privacy (CODASPY '13), pp. 333–340. Association for Computing Machinery, New York (2013). https://doi.org/10.1145/2435349.2435397
13. Gutmann, A., Murdoch, S.J.: Taken out of context: security risks with security code AutoFill in iOS & macOS. In: USENIX Security Symposium 2019 (2019)
14. Oesch, S., Ruoti, S.: That was then, this is now: a security evaluation of password generation, storage, and autofill in browser-based password managers. In: Proceedings of USENIX Security Symposium 2020 (2020)
15. Azure Machine Learning: Microsoft Azure. https://azure.microsoft.com/en-in/services/machine-learning/
16. Microsoft Docs. Windows Machine Learning, 29 May 2020. https://docs.microsoft.com/en-us/windows/ai/windows-ml/
17. Allan, P.: Your Browser's Autofill Data Can Be Phished, Here's How to Stay Safe. Lifehacker (2017). https://lifehacker.com/your-browsers-autofill-data-can-be-phished-heres-how-t-1791084371
18. Coppock, M.: Your browser might be filling in hidden fields and giving away your secrets. Digital Trends (2017) https://www.digitaltrends.com/computing/browser-bug-can-fill-in-personal-information-in-hidden-fields/
19. Browser autofill phishing. https://anttiviljami.github.io/browser-autofill-phishing/
20. Bose, J.: A field label prediction for autofill in web browsers. Arxiv preprint (2019). https://arxiv.org/abs/1912.08809

Synthesis Biometric Materials Based on Cooperative Among (DSA, WOA and gSpan-FBR) to Water Treatment

Samaher Al-Janabi(✉) and Ghydaa Kadhum

Department of Computer Science, Faculty of Science for Women (SCIW), University of Babylon, Babylon, Iraq
samaher@itnet.uobabylon.edu.iq

Abstract. Starting from a training point in the chemical laboratory, whose events take place in the chemical laboratory in addition to wasting time and money, in addition to the revolution, the dream in the world of technology and electronic computing is intelligence and feasibility, so that they can be creative and innovative, in addition to that, reduce the material costs and efforts in this field. Therefore, our research aims to analyze the complex chemical materials to generate new materials from them and to verify them from a computerized validity before performing this experiment in a laboratory. The proposed model consists of three basic steps: First: Dismantling the chemical compound, which represents a complex network, and converting it into a group of subgraphs through the use of one of the graph mining algorithms. Second: After that, each subgraph is taken and the stimulating substance is added to it to speed up the work of the chemical reaction until it reaches a state of stability, by adding new nodes and connections to that subgraph which leads to the production of New subgraph. Third: One of the optimization techniques is applied to each new set of subgraphs to produce a set of patterns that represent the best materials that can be produced from the original spoilers. Thus, we have determined in advance to the researcher what materials can be added to accelerate the chemical reaction and what percentages are accurately mixed using electronic computing before this reaction is carried out in his laboratory.

Keywords: Synthesis · Whale Optimization Algorithm (WOA) · Deterministic selection algorithm (DSA) · gSpan · FBRule

1 Introduction

Biochmestrical analysis also called biological chemistry, Biochemistry processes complicate life and represent chemical processes within living organisms. It enables scientists to analyze chemical reactions and substances found in living organisms where it refers to a set of methods assays and procedures [5].

Synthesis is a process of creating a new substance by reacting two or more feedstocks to form new compounds. This is caused by physical and chemical manipulation and usually has one or more reactions. The compounds used in chemical synthesis are

A. Abraham et al. (Eds.): SoCPaR 2020, AISC 1383, pp. 20–33, 2021.
https://doi.org/10.1007/978-3-030-73689-7_3

called reagents or reagents and when exposed to certain conditions these compounds will transform. As a result, we can define Synthesis as the procedure opposite analysis procedure, where Synthesis refers to merging components and separate elements to form a consistent component [7, 9, 11].

Intelligent Computation represents the ability of the device to learn from data or observations of experiments to perform a specific task. Although it's synonymous with soft computing, there is no agreed definition of computational intelligence. The feature "intelligence" is usually attributed to humans. Recently, many devices and items are referred to as "smart", an inherent characteristic with interpretation and decision-making [1, 9].

Graph mining has taken great interest and is an important area of research. Large amounts of graph data can be used to process, analyze, and extract useful information. The graph can be modeled in the form of a set of sub-graphs. Large networks can then generate by use sets of models of frequent subgraphs. Frequent subgraph can be used to compress graphs by calculating a frequency-based pressure gauge and replacing sub-graphics with the highest values in one node. Although useful, the problem of finding mathematically repetitive sub-graphics depends primarily on NP-Hard space. This is because finding the complete computational solution to extract the recurring graph of large graphs requires solving the maximum problem of independent counting of groups, NP-Hard. It is traceable for static graphs, but it usually becomes impossible for dynamic graphs. As most real-world charts enlarge over time, we supposed that our chart data contains continuous updates of node and edge extensions. Because most real-world graphs contain data related to nodes and edges, we allow naming and edges to be named [3, 20, 21].

Optimization represents that under specific conditions the best results can be obtained. Engineers have several stages to make many technological and administrative decisions to design any engineering system. The objective of these decisions is to either increase interest or reduce the effort required. Since we have a function of a specific decision, optimization can be characterized as the way toward discovering conditions that give the maximum or minimum value of a function [22, 23]. All in all, we will have more than one acceptable design, and the motivation behind optimization is to pick the best one of the numerous acceptable designs. Thus, a specific standard must be chosen to look at different designs and choose the best. The criterion used to improve the design, when expressed as a function of design factors, it is known as the model or legitimacy or objective function. There are two kinds of objective functions optimization, single-objective function, and multiple objective functions. In single-objective optimization, according to the value of the objective function will accept or reject the solutions, and there is just one search space. While multiobjective optimization may have many possible conflicting objectives. Therefore, there exists a trade-off among objectives, i.e., making concessions to the objective to an improvement another objective. There is no optimum solution simultaneously for all m objective functions. Therefore multiple-objective functions will be under a specified set of constraints [9, 17, 18].

2 Materials and Tools

In this section will description the main tools and materials used to solve the problem.

2.1 Comlex Network (CN)

CN consists of a set of elements known as nodes or vertices and the edges connected between them. A system that takes the structure of networks is called graphs. Complex networks have distinct topology and patterns that are not present in a simple network. A complex network is very broad and consists of a large number of elements and link patterns that are irregular or simply random. Complex network analyses are a better methodology expectant of representing and extracting information about the structure of frameworks characterized by interacting components, therefore, giving another purpose of seeing to data mining as the undertaking of classification and regression [3, 4].

2.2 Synthesis Analysis

In general, analysis is characterized as the procedure by which we separate an intellectual or substantial whole into parts or segments. While synthesis is characterized as the opposite technique that is a procedure of merging split parts to make a coherent composition. [9].

2.3 Optimal Subgraph

For any given subgraphs, the one-of-a-kind optimizer portrays the degrees of the nodes that together span the subgraph. We locate that each subgraph happens typically between vertices with explicit degree ranges. Subgraphs uncover information about the geometry and functionalities of complex networks [11, 13, 16].

2.4 Optimal Patterns

Characterized as the way toward getting the best patterns dependent on a given client preference or specific conditions [9].

2.5 Whale Optimization Algorthm (WOA)

WOA is a new population-based algorithm that is inspired by the bubble-net hunting strategy which mimics the social behavior of humpback whales. Similar to other population-based algorithms, WOA uses a set of random candidate solutions (population) and uses three rules to update and improve the position of candidate solutions in each step which are encircling prey, spiral updating position and search for prey. WOA has the ability to avoid local optima and get a globally optimal solution and its appropriate for solving different constrained or unconstrained optimization problems for practical applications. This algorithm has primary parameters and secondary parameters. The main parameters are X, X*, C, A, D, while secondary are t, n, p, a, r,l, b. X is input dataset, X* = Y are the best solution (goal), Xrand is randomly selected solution vector, F is fitness function, C & A are coefficient vectors, D is direction of motion, t is current iteration, r is random vector, a is vector decreasing from 2 to 0 throughout the iteration, p is probability of motion, spiral or square, b is logarithmic spiral constant and l is random number in the range [−1, 1] [7].

$$D = \left|C * X_{k_{rand}} - X(t)\right| X(t+1) = X_{k_{rand}} - A * D$$

$$X(t+1) = \mathrm{D} * \exp(b * l) * \cos(2 * pi * l) + x(t)$$

```
Algorithm #1: WOA
Input data, Number of maxiter and Population etc
Initialize the whales population Xi (i = 1, 2, ..., n)
Initialize a, A, C, l and p
Output:optimal solution
1: Calculate the fitness of each search agent
2: X*= the best search agent
3: while (it < Maxiter)
4:     For each search agent
5:       IF (p < 0.5)
6:           IF (|A| < 1)
7:                Update the position of the current searchagent
                  by the equation (1)
8:           Else if (|A| ≥ 1)
9:                  Select a random search agent (X_rand)
10:                 Update the position of the current search agent
                    by the equation (3)
11:          End if
12:      Else if (p ≥ 0.5)
13:          Update the position of the current search by the by
             the equation (2)
14:      End if
15:    End for
16:    Calculate the fitness of each search agent
17:    Update X* if there is a better solution
18:    it=it+1
19:    Update a, A, C, l and p
20: End while
```

2.6 Deterministic Selection Algorithm

The median-of-medians calculation is a deterministic direct time choice calculation. The calculation works by partitioning a rundown into sub lists and afterward decides the surmised median in every one of the sub lists. At that point, it takes those medians and places them into a rundown and finds the median of that rundown. It utilizes that median incentive as a rotate and looks at different components of the rundown against the turn [24, 25]. In the event that a component is not exactly the rotate esteem, the component is put to one side of the turn, and if the component has a worth more prominent than the rotate, it is put to one side. As explain in Algorithm 2.

```
Algorithm #2 Dselect
Input: Graph
Output: Set of optimal seed
// calculate activation function for each compound
1: For each raw in data set
2: |    compute activation function
3: |    save results in end of the same raw
4: End for
// split data set into 5 groups and sort each group
5: Break data set into group of 5
6: For each one of this 5 groups
7: |  sort component of group i based on value of
   |   activation function
8: |  create new array and push median of each group
   |   in this new array // array of optimal seed
9: End for
End Dselect
```

2.7 gSpan

The gSpan algorithm is designed to reduce the generation of duplicate graphs. It need not search previously discovered frequent graphs for duplicate detection. It does not extend any duplicate graph, yet still guarantees the discovery of the complete set of frequent graphs as explain in Algorithm 3 [27]. The rules generation from gSpan can verification from it by forward backward rules as shown in Algorithm 4 [26].

Algorithm #3: gSpan 27

```
Input:S, a DFS code; D, a graph data set;
      min sup, the minimum support threshold.
Output:The frequent graph set, S.
1: S =
2: Call gSpan(s, D, min sup, S)
Procedure Pattern Growth Graph (s, D, min sup, S)
1: IF s ≠dfs(s), then
2:      return;
3: insert s into S;
4: set C to ∅;
5: scan D once; find all the edges e such that s can be
right-most extended to s _r e;
      Insert s _r e into C and count its frequency;
6: sort C in DFS lexicographic order;
7: For each frequent s _r e in C do
8:      gSpan(s _r e, D, min sup, S);
9: return
```

Algorithm #4: gSpan-FBR **26**

Input: Optimal Cluster

Output: S set of patterns (recommenders)

```
S =
For each optimal cluster
    s = Call DFS
    Insert s into S
    Call BFS
    Find all the edges e such that s can be right-most extended
    insert in c
    If the trees have max number of edges
        Remove duplication tree
    Else
        kept the tree has minimal number of edges
    End if
End for
//forward & backward Rules
For each value of set patterns (S)
    If v1 of e1 < v2 of e1
        If v1 of e2 < v2 of e2
            If v1 of e2 <= v2 of e1 & v2 of e2 == v2 of e1+1
                this is an acceptable e2
            Else
                e2 being considered isn't valid
        Else
            e2 is a backward edge
            If v1 of e2 == v2 of e1 & v2 of e2 <v1 of e1
                this is an acceptable e2
            Else
                e2 being considered isn't valid
    Else
        e1 is a backward edge
        If v1 of e2 < v2 of e2 (forward edge)
            If v1 of e2 <= v1 of e1& v2 of e2 == v1 of e1+1
                this is an acceptable e2
            Else
                e2 being considered isn't valid
        Else
            e2 is a backward edge
            If v1 of e2 == v1 of e1& v2 of e1<v2 of e2
                this is an acceptable e2
            Else
                e2 being considered isn't valid
End forward/backward edge
End gSpan-FBR
```

3 Experiments

The main steps of propose method called cooperative among deep graph mining techniques explain in Fig. 1. While, this method will apply on dataset comes from the daily measures of sensors in a urban waste water treatment plant. The objective is to classify the operational state of the plant in order to predict faults through the state variables of the plant at each of the stages of the treatment process. This domain has been stated as an ill-structured domain in Table 4. In general, that dataset.

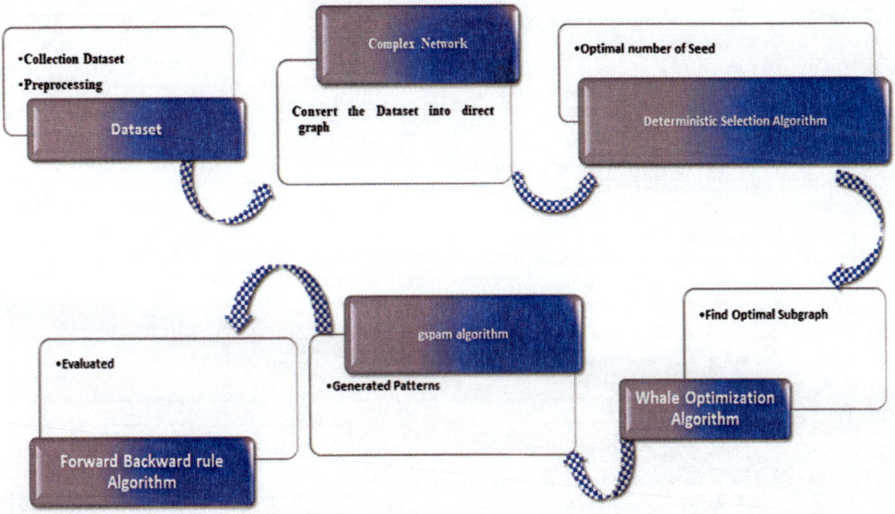

Fig. 1. Main steps of proposed system

Rules Generation Through gSpan-FBR

In this section we will explain the total number of rules generation are 509 rules without apply WOA while through apply it result 118 rules only by gSpan then pass that rules on FBR to verification from it with determined the true rules only equal 7 rules as explain here:

IF (PH-E IS 4.35) AND (Q-E IS 30040.5) AND (ZN-E IS 16.75) AND (DBO-E IS 219) AND (DQO-E IS 470.5) AND (SED-E IS 18) AND (SS-E IS 1004) AND (SSV-E IS 42.5) AND (COND-E IS 1615) AND (PH-P IS 4.25) AND (DBO-P IS 258.5) AND (SED-P IS 23) AND (SS-P IS 846) AND (SSV-P IS 46.75) AND (COND-P IS 1585) AND (PH-D IS 4.2) AND (DBO-D IS 142.5) AND (DQO-D IS 255.5) AND (SED-D IS 1.75) AND (SS-D IS 122) AND (SSV-D IS 50) AND (COND-D IS 1845) AND (DBO_S1 IS 160) AND (DQO-S IS 175) AND (SED-S IS 1.75) AND (SS-S IS 119) AND (SSV-S IS 50) AND (COND-S IS 1975) THEN PH-S is 4.85.

IF (PH-E is 4.35) AND (Q-E is 30040.5) AND (ZN-E is 16.75) AND (DBO-E is 219) AND (DQO-E is 470.5) AND (SED-E is 18) AND (SS-E is 1004) AND (SSV-E is 42.5) AND (COND-E is 1615) AND (PH-P is 4.25) AND (DBO-P is 258.5) AND (SED-P is 23) AND (SS-P is 846) AND (SSV-P is 46.75) AND (COND-P is 1585) AND

Table 1. Main parameters of dataset.

1: Q-E	(flow to plant)
2: ZN-E	(Zinc to plant)
3: PH-E	(pH to plant)
4: DBO-E	(Biological demand of oxygen to plant)
5: DQO-E	(chemical demand of oxygen to plant)
6: SS-E	(suspended solids to plant)
7: SSV-E	(volatile supended solids to plant)
8: SED-E	(sediments to plant)
9: COND-E	(conductivity to plant)
10: PH-P	(pH to primary settler)
11: DBO-P	(Biological demand of oxygen to primary settler)
12: SS-P	(suspended solids to primary settler)
13: SSV-P	(volatile supended solids to primary settler)
14: SED-P	(sediments to primary settler)
15: COND-P	(conductivity to primary settler)
16: PH-D	(pH to secondary settler)
17: DBO-D	(Biological demand of oxygen to secondary settler)
18: DQO-D	(chemical demand of oxygen to secondary settler)
19: SS-D	(suspended solids to secondary settler)
20: SSV-D	(volatile supended solids to secondary settler)
21: SED-D	(sediments to secondary settler)
22: COND-D	(conductivity to secondary settler)
23: PH-S	(pH)
24: DBO-S	(Biological demand of oxygen)
25: DQO-S	(chemical demand of oxygen)
26: SS-S	(suspended solids)
27: SSV-S	(volatile supended solids)
28: SED-S	(sediments)
29: COND-S	(conductivity)
30: RD-DBO-P	(input Biological demand of oxygen in primary settler)
31: RD-SS-P	(input suspended solids to primary settler)
32: RD-SED-P	(input sediments to primary settler)
33: RD-DBO-S	(input Biological demand of oxygen to secondary settler)
34: RD-DQO-S	(input chemical demand of oxygen to secondary settler)
35: RD-DBO-G	(performance input Biological demand of oxygen)
36: RD-DQO-G	(performance input chemical demand of oxygen)
37: RD-SS-G	(performance input suspended solids)
38: RD-SED-G	(performance input sediments)

(PH-D is 4.2) AND (DBO-D is 142.5) AND (DQO-D is 255.5) AND (SED-D is 1.75) AND (SS-D is 122) AND (SSV-D is 50) AND (COND-D is 1845) AND (PH-S is 4.85) AND (DQO-S is 175) AND (SED-S is 1.75) AND (SS-S is 119) AND (SSV-S is 50) AND (COND-S is 1975) THEN DBO_S1 is 160.

IF (PH-E IS 4.35) AND (Q-E IS 30040.5) AND (ZN-E IS 16.75) AND (DBO-E IS 219) AND (DQO-E IS 470.5) AND (SED-E IS 18) AND (SS-E IS 1004) AND (SSV-E IS 42.5) AND (COND-E IS 1615) AND (PH-P IS 4.25) AND (DBO-P IS 258.5) AND (SED-P IS 23) AND (SS-P IS 846) AND (SSV-P IS 46.75) AND (COND-P IS 1585) AND (PH-D IS 4.2) AND (DBO-D IS 142.5) AND (DQO-D IS 255.5) AND (SED-D IS 1.75) AND (SS-D IS 122) AND (SSV-D IS 50) AND (COND-D IS 1845) AND (PH-S IS

Table 2. Initial Root of each subgraph generate by DSA.

Sub-G	1	2	3	4	5
Y1	37844.121	37405.544	37763.012	36273.412	36848.36
Y2	2.352	2.327	2.621	2.085	2.237
Y3	7.808	7.787	7.770	7.783	7.828
Y4	189.150	182.565	181.662	196.880	192.137
Y5	399.125	412.801	388.773	421.247	412.796
Y6	212.644	246.083	212.431	245.092	222.286
Y7	62.199	59.969	61.027	60.810	62.871
Y8	4.429	4.790	4.427	5.251	4.499
Y9	1529.898	1478.240	1442.931	1471.894	1462.153
Y10	7.808	7.798	7.794	7.802	7.848
Y11	210.399	194.201	201.726	216.014	206.278
Y12	240.390	269.313	246.000	274.912	239.347
Y13	60.884	59.540	59.982	60.277	61.077
Y14	4.936	5.320	5.128	5.906	5.056
Y15	1547.415	1509.448	1458.686	1485.142	1472.459
Y16	7.791	7.790	7.796	7.788	7.813
Y17	126.462	118.949	119.099	119.629	127.245
Y18	282.458	276.501	264.834	267.710	278.408
Y19	94.636	94.031	93.941	92.331	96.398
Y20	73.099	72.778	72.583	73.039	73.318
Y21	0.417	0.357	0.410	0.377	0.400
Y22	1550.381	1494.188	1462.971	1455.726	1483.908
Y23	7.686	7.715	7.693	7.703	7.692
Y24	21.885	21.273	20.337	19.977	20.134
Y25	88.028	88.718	85.773	87.741	86.916
Y26	23.157	23.786	21.873	21.429	22.357
Y27	81.198	79.041	80.435	80.135	79.705
Y28	0.039	0.076	0.040	0.044	0.059
Y29	1544.102	1529.323	1464.245	1441.432	1495.051
Y30	38.461	38.247	39.306	42.027	37.040
Y31	58.593	59.478	58.175	61.071	54.903
Y32	90.510	91.393	90.328	90.862	89.666
Y33	84.345	82.337	82.576	83.656	84.124
Y34	69.221	67.285	67.103	67.383	67.894
Y35	89.596	88.387	88.370	89.244	89.329
Y36	78.089	77.799	77.335	78.308	77.654
Y37	88.775	88.935	88.947	89.470	88.650
Y38	98.959	98.748	99.310	99.381	98.999

4.85) AND (DBO_S1 IS 160) AND (SED-S IS 1.75) AND (SS-S IS 119) AND (SSV-S IS 50) AND (COND-S IS 1975) THEN DQO-S is 175.

IF (PH-E IS 4.35) AND (Q-E IS 30040.5) AND (ZN-E IS 16.75) AND (DBO-E IS 219) AND (DQO-E IS 470.5) AND (SED-E IS 18) AND (SS-E IS 1004) AND (SSV-E IS 42.5) AND (COND-E IS 1615) AND (PH-P IS 4.25) AND (DBO-P IS 258.5) AND (SED-P IS 23) AND (SS-P IS 846) AND (SSV-P IS 46.75) AND (COND-P IS 1585) AND (PH-D IS 4.2) AND (DBO-D IS 142.5) AND (DQO-D IS 255.5) AND (SED-D IS 1.75) AND (SS-D IS 122) AND (SSV-D IS 50) AND (COND-D IS 1845) AND (PH-S IS 4.85) AND (DBO_S1 IS 160) AND (DQO-S IS 175) AND (SED-S IS 1.75) AND (SSV-S IS 50) AND (COND-S IS 1975) THEN SS-S is 119.

IF (PH-E IS 4.35) AND (Q-E IS 30040.5) AND (ZN-E IS 16.75) AND (DBO-E IS 219) AND (DQO-E IS 470.5) AND (SED-E IS 18) AND (SS-E IS 1004) AND (SSV-E IS 42.5) AND (COND-E IS 1615) AND (PH-P IS 4.25) AND (DBO-P IS 258.5) AND (SED-P IS 23) AND (SS-P IS 846) AND (SSV-P IS 46.75) AND (COND-P IS 1585) AND (PH-D IS 4.2) AND (DBO-D IS 142.5) AND (DQO-D IS 255.5) AND (SED-D IS 1.75) AND (SS-D IS 122) AND (SSV-D IS 50) AND (COND-D IS 1845) AND (PH-S IS 4.85) AND (DBO_S1 IS 160) AND (DQO-S IS 175) AND (SED-S IS 1.75) AND (SS-S IS 119) AND (COND-S IS 1975) THEN SSV-S is 50.

IF (PH-E IS 4.35) AND (Q-E IS 30040.5) AND (ZN-E IS 16.75) AND (DBO-E IS 219) AND (DQO-E IS 470.5) AND (SED-E IS 18) AND (SS-E IS 1004) AND (SSV-E IS 42..5) AND (COND-E IS 1615) AND (PH-P IS 4.25) AND (DBO-P IS 258.5) AND

Table 3. Optimal Graphs generated by WOA.

Optimal-G	1	2	3	4	5
Y1	42934.704	37362.248	33394.544	27913.925	50224.510
Y2	2.479	2.484	2.420	1.709	2.019
Y3	7.846	7.775	7.774	7.758	7.865
Y4	170.627	192.047	200.654	202.433	159.528
Y5	358.081	430.036	450.360	408.373	299.306
Y6	215.426	244.015	231.464	211.403	216.694
Y7	58.221	61.622	64.270	65.934	51.629
Y8	4.162	5.229	4.938	4.447	3.787
Y9	1483.537	1500.627	1495.136	1470.343	1361.959
Y10	7.858	7.791	7.797	7.765	7.863
Y11	186.027	215.007	217.752	216.812	172.307
Y12	237.685	273.746	256.959	243.537	239.551
Y13	58.010	61.288	62.510	64.465	50.084
Y14	4.888	5.907	5.409	5.166	4.071
Y15	1505.750	1518.515	1508.669	1490.478	1377.163
Y16	7.819	7.778	7.803	7.752	7.822
Y17	114.763	125.501	129.522	122.692	105.240
Y18	256.899	289.822	287.675	274.210	221.469
Y19	91.558	95.493	95.089	93.746	91.204
Y20	69.993	74.224	74.995	76.797	63.869
Y21	0.372	0.389	0.409	0.429	0.346
Y22	1499.815	1509.799	1507.024	1494.299	1355.755
Y23	7.695	7.675	7.708	7.710	7.712
Y24	23.704	22.238	19.765	18.570	16.427
Y25	89.878	97.252	85.230	83.974	67.692
Y26	24.903	24.728	20.820	21.873	17.806
Y27	79.308	80.714	80.567	81.358	77.391
Y28	0.068	0.044	0.044	0.064	0.036
Y29	1515.537	1505.903	1502.077	1503.370	1382.102
Y30	36.921	39.586	39.238	40.886	39.504
Y31	57.135	59.269	59.912	56.908	56.914
Y32	89.633	91.752	91.231	89.509	88.403
Y33	82.942	81.822	84.441	84.198	84.562
Y34	65.424	66.918	69.786	68.150	68.307
Y35	87.684	88.013	89.975	90.238	89.689
Y36	74.804	76.900	80.566	78.663	76.753
Y37	86.961	88.319	90.403	88.718	90.509
Y38	98.315	99.476	99.275	98.703	99.593

Table 4. Relationship between the optimal Root of Subgraphs

	1	2	3	4	5
1	0	5573.405	9540.832	15020.970	7295.002
2	5573.405	0	3967.870	9448.588	12866.405
3	9540.832	3967.870	0	5480.959	16833.091
4	15020.970	9448.588	5480.959	0	22312.338
5	7295.002	12866.405	16833.091	22312.338	0

(SED-P IS 23) AND (SS-P IS 846) AND (SSV-P IS 46.75) AND (COND-P IS 1585) AND (PH-D IS 4.2) AND (DBO-D IS 142.5) AND (DQO-D IS 255.5) AND (SED-D IS 1.75) AND (SS-D IS 122) AND (SSV-D IS 50) AND (COND-D IS 1845) AND (PH-S IS 4.85) AND (DBO_S1 IS 160) AND (DQO-S IS 175) AND (SS-S IS 119) AND (SSV-S IS 50) AND (COND-S IS 1975) THEN SED-S is 1.75.

IF (PH-E is 4.35) AND (Q-E is 30040.5) AND (ZN-E is 16.75) AND (DBO-E is 219) AND (DQO-E is 470.5) AND (SED-E is 18) AND (SS-E is 1004) AND (SSV-E is 42.5) AND (COND-E is 1615) AND (PH-P is 4.25) AND (DBO-P is 258.5) AND (SED-P is 23) AND (SS-P is 846) AND (SSV-P is 46.75) AND (COND-P is 1585) AND (PH-D is 4.2) AND (DBO-D is 142.5) AND (DQO-D is 255.5) AND (SED-D is 1.75) AND (SS-D is 122) AND (SSV-D is 50) AND (COND-D is 1845) AND (PH-S is 4.85) AND (DBO_S1 is 160) AND (DQO-S is 175) AND (SED-S is 1.75) AND (SS-S is 119) AND (SSV-S is 50) THEN COND-S is 1975.

4 Discussion

One of the main problems is the living water pollution problem that the world suffers from today, so this research attempted to find a smart way to generate optimal patterns to reduce that phenomenon was represented by using an intelligent data analysis algorithm where it was first to perform a preliminary manipulation of those data obtained in real-time, it was suffering from missing data and was processed by the using mean, where it was found that the percentage of loss is 15% of the total, Then an algorithm was proposed to find the best seed or center for each cluster by using Deterministic selection algorithm.

Where seven seeds were found and their characteristics. Then the whale optimization algorithm was used. Was it basic parameter are X, X*, C, A, D, F, and secondary parameter are t, n, p, a, r, l, b. The accuracy rate Within-class 10.32% and Between-classes 89.68%.

Then the pattern was generated using its algorithm D gSpan. Where we found 200 rules was generated and validated using FBR, we now have 50 rules, this method is characterized by being highly complex in terms of time and calculations as being better in terms of accuracy.

5 Conclusions

Whale optimization algorithm which is inspired by bubble-net the haunting strategy of humpback whales is a new, simple and robust optimization approach. In this paper, the whale optimization algorithm developed to solve the Water pollution problem. Clustering is gathering data into clusters such that the data in the same cluster have a high degree of similarity and data from different clusters being as possible as dissimilar. The whale optimization algorithm can successfully be applied to solve clustering problems. Moreover, these results proposed algorithm was effective, easy to implement and robust as compared with other approaches. Some directions can improve the performance of the proposed algorithm in the future. The combination of the WOA clustering algorithm with other clustering approaches and using other fitness functions in the clustering approach can be further researched.

References

1. Al-Janabi, S., Alkaim, A.F.: A nifty collaborative analysis to predicting a novel tool (DRFLLS) for missing values estimation. Springer, Soft Comput. **24**(1), 555–569 (2020). https://doi.org/10.1007/s00500-019-03972-x
2. Raya, A., Holdera, L.B., Bifetb, A.: Efficient frequent subgraph mining on large streaming graphs. Intell. Data Anal. **23**(1), 103–132 (2019). https://doi.org/10.3233/IDA-173705
3. Bechrouri, S., Monir, A., Mraoui, H., Sebbar, E.-H., Saalaoui, E., Choukri, M.: Big Data and Smart Digital Environment, Springer International Publishing, vol. 53, pp. 215–224 (2019). https://doi.org/10.1007/978-3-030-12048-1
4. Al-Janabi, S., Al-Shourbaji, I., Shojafar, M., Abdelhag, M.: Mobile cloud computing: challenges and future research directions. In: IEEE, 2017 10th International Conference on Developments in eSystems Engineering (DeSE), Paris, pp. 62–67 (2017). https://doi.org/10.1109/DeSE.2017.21
5. Ghosh, S.K., Ehnbom, A., Lewis, K.G., Gladysz, J.A.: Hydrogen bonding motifs in structurally characterized salts of the tris(ethylenediamine) cobalt trication, [Co(en)3]3+; An interpretive review, including implications for catalysis. Chem. Soc. Rev. **350**, 30–48 (2017). https://doi.org/10.1016/j.ccr.2017.04.002
6. Ehnbom, A., Ghosh, S.K., Lewis, K.G., Gladysz, J.A.: Octahedral Werner complexes with substituted ethylenediamine ligands: a stereochemical primer for a historic series of compounds now emerging as a modern family of catalysts. Roy. Soc. Chem. **45**(24), 6799–6811 (2016). https://doi.org/10.1039/c6cs00604c
7. Mirjalili, S., Lewis, A.: The whale optimization algorithm. Adv. Eng. Softw. **95**, 51–67 (2016). https://doi.org/10.1016/j.advengsoft.2016.01.008
8. Alkaim, A.F., Al_Janabi, S.: Multi objectives optimization to gas flaring reduction from oil production. In: Farhaoui, Y. (eds.) Big Data and Networks Technologies. BDNT 2019. Lecture Notes in Networks and Systems, vol 81. Springer, Cham (2020). https://doi.org/10.1007/978-3-030-23672-4_10
9. Mahnke, L.K., Wendt, M., Näther, C., Bensch, W.: Rational syntheses of three new {V 14 Sb 8 } clusters applying a water-soluble high-nuclearity cluster as synthon. Crystal Growth Des. **18,** 6100–6106. (2018). https://doi.org/10.1021/acs.cgd.8b00987
10. Lisa, K., Ulrike, W., Mengxi, L., Christian, N., Christoph, A., Wolfgang, B.: New Water-Soluble Cluster Compound {Zn(en)3}3[V15Sb6O42(H2O)]· (Ethylenediamine)3·10 H2O as a Synthon for the Generation of Two New Antimonato Polyoxovanadates. Chem. Eur. J. **24**(21), 5522–5528 (2018). https://doi.org/10.1002/chem.201705732
11. Al-Janabi, S., Alkaim, A.F., Adel, Z.: An Innovative synthesis of deep learning techniques (DCapsNet & DCOM) for generation electrical renewable energy from wind energy. Soft Comput. (2020). https://doi.org/10.1007/s00500-020-04905-9
12. Kalajdzic, K.: Samaher Hussein Ali, Ahmed Patel, Rapid lossless compression of short text messages. Comput. Stand. Interf. **37**, 53–59 (2015). https://doi.org/10.1016/j.csi.2014.05.005
13. Ali, S.H.: Miner for OACCR: Case of medical data analysis in knowledge discovery. In: IEEE, 2012 6th International Conference on Sciences of Electronics, Technologies of Information and Telecommunications (SETIT), Sousse, 2012, pp. 962–975 (2012). https://doi.org/10.1109/SETIT.2012.6482043
14. Srinivasan, B.R., Naik, A.R., Na, C.: Ther, Herbert Pausch And Wolfgang Bensch, (2009), Synthesis and structural characterization of tris(ethylenediamine)nickel(II) dichromate. J. Coordination Chem. **62**(22), 3583–3591. https://doi.org/10.1080/00958970903114433
15. Chung, F., Tsiatas, A.: Finding and visualizing graph clusters using pagerank optimization. Springer, pp. 86–97 (2010). https://doi.org/10.1007/978-3-642-18009-5_9

16. Al-Janabi, S., Mohammad, M., Al-Sultan, A.: A new method for prediction of air pollution based on intelligent computation. Soft Comput. **24**, 661–680 (2020). https://doi.org/10.1007/s00500-019-04495-1
17. Meenakshi, S., Renukadevi, R.: A review on spectral clustering and its applications. Int. J. Innov. Res. Comput. Commun. Eng. **4**(8), August 2016. https://doi.org/10.15680/IJIRCCE.2016. 0408050
18. Breur, T.: Statistical Power Analysis and the contemporary "crisis" in social sciences. J. Mark. Anal., 61–65 (2016). https://doi.org/10.1057/s41270-016-0001-3.
19. Sabeur Aridhia and Engelbert Mephu Nguifob: Big Graph Mining: Frameworks and Techniques. Elsevier (2016). https://doi.org/10.1016/j.bdr.2016.07.002
20. Mahdi, M.A., Al_Janabi, S.: A novel software to improve healthcare base on predictive analytics and mobile services for cloud data centers. In: Farhaoui, Y. (eds.) Big Data and Networks Technologies. BDNT 2019. Lecture Notes in Networks and Systems, vol 81. Springer, Cham (2020). https://doi.org/10.1007/978-3-030-23672-4_23
21. Chehouri, A., Younes, R., Ilinca, A., Perron, J.: Review of performance optimization techniques applied to wind turbines. Appl. Energy **142**, 361–388 (2015). https://doi.org/10.1016/j.apenergy.2014.12.043
22. Deng, W., Zhao, H., Zou, L., Li, G., Yang, X., Wu, D.: A novel collaborative optimization algorithm in solving complex optimization problems. Soft. Comput. **21**(15), 4387–4398 (2016). https://doi.org/10.1007/s00500-016-2071-8
23. Jahed, H., Farshi, B., Eshraghi, M.A., Nasr, A.: A numerical optimization technique for design of wheel profiles. Wear **264**(1–2), 1–0 (2008). https://doi.org/10.1016/j.wear.2006.06.006
24. Deb, K., Pratap, A., Agarwal, S., Meyarivan, T.: A fast and elitist multiobjective genetic algorithm: NSGA-II. IEEE Trans. Evol. Comput. **6**(2), 182–197 (2002). https://doi.org/10.1109/4235.996017
25. Al-Janabi S., Al-Shourbaji, I.: A Hybrid Image steganography method based on genetic algorithm. In: 2016 7th International Conference on Sciences of Electronics, Technologies of Information and Telecommunications (SETIT), Hammamet, 2016, pp. 398–404. https://doi.org/10.1109/SETIT.2016.793990
26. Al-Janabi, S., Noor, K.: Intelligent computation to build a novel recommender of products through (PageRank-Clustering and DgSpan- FBR). In: 2020 International Conference on Decision Aid Sciences and Application (DASA), Sakheer, Bahrain, pp. 547–554 (2020). https://doi.org/10.1109/DASA51403.2020.9317269
27. Al_Janabi, S., Kadiam, N.: Recommendation system of big data based on PageRank clustering algorithm. In: Farhaoui, Y. (ed.) BDNT 2019. LNCS, vol. 81. Springer, Cham (2020). https://doi.org/10.1007/978-3-030-23672-4_12

An Initial Parameter Search for Rapid Concept Drift Adaptation in Deep Neural Networks

Muhammad Zafran Bin Muhammad Zaly Shah(✉) and Anazida Binti Zainal

School of Computing, Faculty of Engineering, Universiti Teknologi Malaysia, 81310 Johor Bahru, Malaysia

muhammad.zafran@graduate.utm.my, anazida@utm.my

Abstract. Concept drift is a common issue in data stream mining algorithms that causes prediction models to lose its original performance gradually or abruptly due to the non-stationarity of the data distribution and decision boundaries. To combat concept drifts, prediction models need to be updated periodically or when concept drifts occur to adapt to the current concept. Unfortunately, training deep neural networks often require a large amount of data samples and high computational resource consumption, making adaptation slow when concept drifts occur. This paper proposes an approach by searching for an optimum initial parameter that could be adapted quickly to all possible concept drift situations. The initial parameter search is based on the Reptile [1] algorithm, which had been successfully applied in image classification, which allows a neural network model to learn from a few samples and minimal gradient steps. We argue that using an optimum initial parameter allows prior information to be embedded and makes the prediction model less reliant on training exclusively from new data when concept drift occurs. Experimental results show that this approach performs at least as well as current data streaming algorithms but with the lowest computational overhead.

Keywords: Data stream mining · Concept drift · Global optimization · Deep learning

1 Introduction

Data stream mining is a subfield of machine learning that deals with an environment where data arrives at a fast rate. Moreover, data in this environment are often non-stationary, which means that their data distribution and relationship between features might change from time to time. This phenomenon is known as concept drift. The occurrences of concept drifts may cause the model's prediction performance to deteriorate, which makes the prediction unreliable. Thus, in concept drifts, the prediction model needs to be updated to adapt to the current concept to achieve its maximum prediction performance. It is also crucial that the prediction model adapts to new concepts quickly to minimize the prediction model's downtime.

A. Abraham et al. (Eds.): SoCPaR 2020, AISC 1383, pp. 34–42, 2021.
https://doi.org/10.1007/978-3-030-73689-7_4

The current approach to combat concept drift can be divided into either using a structure update or an incremental learning approach. The structure update approach usually updates the structure of an ensemble of classifiers by replacing the classifier with the worst performance and replacing a new classifier trained with the most recent data. This way, the classifier can adapt to sudden or abrupt drifts since the classifier unable to cope with the new concept can be removed and replaced quickly.

The incremental learning approach is preferred to cope with slow but constantly changing concepts (gradual drifts). The incremental learning approach constantly trains the classifier on every arriving example, allowing it to learn and adapt to an ever-changing environment continually. At the same time, the incremental learning approach constantly forgets old concepts to ensure the classifier makes its prediction based on the current concept.

However, abrupt and gradual drifts are not the only type of concept drifts that are encountered. Another concept drifts, such as a reoccurring concept drift where a previous concept is reappearing, requires data from a previous time-step to be reused to reupdate the model. However, reusing data from the previous time step is an undesirable characteristic of the data stream mining algorithm because it can dramatically increase memory consumption [2].

Data stream mining algorithms have an architectural constraint that prevents it from having a too high dimension in the hypothesis space because of the data storage restriction. This had been proven by Vapnik [3] that the larger the dimension of the classifier, the more data it requires to be more accurate in its prediction. Therefore, using a very high dimensional algorithm such as deep neural networks is impractical for data stream mining.

To solve these issues, this paper proposes a concept drift adaptation approach by finding an optimum initial parameter for the neural network to adapt to all possible concept drift situations. The initial parameter is searched based on the Reptile algorithm [1], which had been successfully applied in meta-learning for image classification. The intuition behind using an optimum initial parameter is that instead of learning exclusively from new data, prior information can be embedded in the initial parameter to minimize the data required when the classifier needs to be updated in the event of concept drift.

This proposed approach's only drawback is that there will be a delay in deployment because it requires enough data to be collected initially to search for an optimum initial parameter. However, this trade-off is acceptable in data stream mining because this procedure is conducted only once and does not affect the deployment [4]. Based on the experimental results, our approach performs at least as well as current approaches for data streaming applications but with the lowest data and computational resource consumption.

2 Literature Review

This section reviews the important works conducted to provide approaches to model adaptation in concept drifts. Overall, concept drift adaptation approaches can be categorized as either a structure update approach or an incremental learning approach.

2.1 Structure Update Approach

The structure update approach is usually adopted by ensemble-based approaches such as decision trees. The ensemble approach's motivation is to improve the diversity of the models by training a different subset of the training set on members of the ensemble to better adapt to concept drifts [5]. The structure update approach is usually achieved by removing the weakest classifier in the ensemble.

For example, the Adaptive Random Forest (ARF) [6] attaches a concept drift detection mechanism on each classifier in the ensemble to determine which classifier performs poorly. The classifier that triggers the drift alarm will be replaced by a classifier trained on the most recent data. The Fast Learn ++.NSE algorithm [7] proposes adapting the voting weights based on the classifiers' moving average performance. Another approach is to select only the best classifiers to bring forward to the next time step, which is adopted by Diversity and Transfer based Ensemble Learning (DTEL) [8].

2.2 Incremental Learning

Incremental learning approaches use a single learner trained on every arriving example that allows it to adapt to gradual concept drifts. Using a single learner minimizes the system's complexity since it does not require a robust model and data management approaches since the model can update itself when new data is available [9].

Incremental learning approaches usually employ a forgetting mechanism to forget old concepts and prevents the parameters from becoming saturated. This can be observed in approaches by Very Fast Decision Trees (VFDT) [10] and OzaBoost [11], which uses a parameter $0 < \lambda < 1$, that gradually reduces the influence of old concepts. To minimize computational resource consumption, Extremely Fast Decision Trees (EFDT) [12] applies the Hoeffding's Anytime Tree (HATT) [13] that only splits attributes that provide significant information to the classification.

This work differs from the current approaches for concept drift adaptation by proposing an approach to find optimum initial parameters for initialization that is highly adaptable to all concept drift situations. It requires only a few data samples and minimal gradient descent steps. This approach also does not impose any restriction on model architecture, such as decision trees or decision rules, which means that any neural network model architecture could be used using this approach.

3 Concept Drift Adaptation from an Optimum Initial Parameter

For a neural network to adapt to concept drifts with minimal gradient descent steps, we propose to search for an optimum neural network initial parameter θ^*_{Meta} that minimizes the least squared distance to all possible concept specific parameter ϕ. A naïve approach is to sample an interval $\mathcal{D}_w$ of size w from the dataset $\mathcal{D}$ and train the neural network to obtain the concept specific parameter ϕ using a gradient-based optimization method, e.g., Stochastic Gradient Descent (SGD) but with a small number of gradient descent steps. The concept specific parameter will be used to update the initial parameter using $\theta_{Meta} \leftarrow \theta_{Meta} + \beta(\phi - \theta_{Meta})$ where β is the meta-learning rate. This step is repeated many times until the initial parameter θ_{Meta} converge to the point that minimizes the least squared distance to all concept specific parameter ϕ.

However, using a pure windowing method to partition the dataset is risky because choosing an inappropriate window size might cause the initial parameter θ_{Meta} to converge to a suboptimal value. For example, if the dataset contains many abrupt drifts, choosing a too large window w will cause each window to contain too many concepts and prevents the concept specific parameter ϕ from specializing to a single concept. This will, in turn, cause the θ_{Meta} to converge to a suboptimal value.

To mitigate this issue, we propose to use the Pruned Exact Linear Time (PELT) [14] changepoint detection algorithm to partition the dataset according to the output of the loss function $\mathcal{L}(.)$ at each time step. To use the offline changepoint detection algorithm, the first part of the window $\mathcal{D}_{train} \subset \mathcal{D}_w$ is used to train the neural network with many gradient descent steps[1] to induce overfitting. The rest of the window $\mathcal{D}_w$ is then predicted, and the performance is recorded using some loss function $\mathcal{L}(.)$. By inducing overfitting, even a slight change in the feature relationship will cause an abrupt change in the output of the loss function $\mathcal{L}(.)$ and making it easier for the changepoint detection algorithm to detect a changing of concept. The changepoint detection algorithm is then used to obtain the set of optimum changepoints $\mathcal{T} = \{\tau_i, .., \tau_k\}$.

A single interval from the data window, $D_{\tau_{i-1}+1:\tau_i} \subset \mathcal{D}_w$ is obtained by sampling a changepoint index, τ_i to obtain the concept specific parameter $\phi \leftarrow SGD_j(\theta_{Meta}, D_{\tau_{i-1}+1:\tau_i})$ using the current initialization parameter θ_{Meta} with j gradient descent steps. Sampling only a single changepoint index τ_i at each iteration is to prevent the initial parameter update $\theta_{Meta} \leftarrow \theta_{Meta} + \beta(\phi - \theta_{Meta})$ from being biased to a segment of the training set and prevents the initial parameter, θ_{Meta} to converge to a value that minimizes the squared distance to all possible concept specific parameters, ϕ. The whole algorithm is summarized in Algorithm 1.

Updating the neural network during deployment is straightforward. When a concept drift is detected, the parameter of the neural network is restored using θ_{Meta} and retrained using the current data in the buffer with j gradient descent steps.

[1] This optimization procedure can be regarded as a dummy optimization not used to search for the optimum initial parameter θ_{Meta}.

Algorithm 1 Optimum Initial Parameter Search. **Input:** dataset $\mathcal{D} = \{(x_1, y_1), \ldots, (x_n, y_n)\}$, window size w, optimization learning rate α, meta-learning rate β, number of gradient descent steps j **Output:** Optimum initial parameter θ^*_{Meta}.

function InitialParameterSearch (α, β)
Randomly initialize the initial parameter vector, θ.
  while not converge:
    $D_w \leftarrow$ Sample a data interval of size, w, from the data stream.
    Save current initial parameter $model\_state \leftarrow \theta$
    Train for many gradient descent steps $\phi \longleftarrow SGD(\theta_{Meta}, D_{train} \subset D_w)$
    Restore initial parameter $\theta \leftarrow model\_state$
    Predict samples in the window $\hat{y} \leftarrow model(\phi, D_w)$
    Evaluate running performance $performance\_array \leftarrow \mathcal{L}(\hat{y}, y)$
    Obtain optimum changepoint indices $\mathcal{T} \leftarrow PELT(performance\_array)$
    Sample changepoint index $\tau_i \sim uniform(\mathcal{T})$
    Obtain the concept specific parameter $\phi \leftarrow SGD_j(\theta_{Meta}, D_{\tau_{i-1}+1:\tau_i})$
    Update $\theta_{Meta} + \beta(\phi - \theta_{Meta})$
  output θ^*_{Meta}
end function

4 Experiment and Results

This section will discuss the experimental setup and discuss the results obtained to evaluate the proposed approach's performance.

4.1 Experimental Setup

The proposed approach will be evaluated on benchmark datasets in data stream mining listed in Table 1, along with their characteristics. The artificial datasets are infinite datasets generated using some function, and the real-world datasets are finite datasets collected from real-world case studies.

Table 1. Benchmark datasets

Dataset type	Dataset	Drift type	# of attributes	# of classes	Characteristics	# of instances
Artificial	SEA	Abrupt/Recurring	3	2	Noisy	–
	Stagger	Abrupt/Recurring	3	2	Class imbalance	–
	Rotating Hyperplane	Gradual/Recurring	5	2	Noisy	–
Real-World	Airline	Mixed	7	2	High-Dimensional	539384
	Electricity	Mixed	8	2	–	45312
	Cover type	Mixed	54	10	Multi-class	581012

The experimental setup is conducted on neural network architectures with different number of layers depending on the number of attributes in the dataset. However, due to space constraints, Table 2 only lists the number of parameters for each of the neural networks, which is the more important variable since we need to evaluate how Algorithm 1 performs with different numbers of parameters. Other parameters for Algorithm 1 are also listed in Table 2.

Table 2. Experimental setup and parameters for algorithm 1

Dataset	# of Parameters	α	β	j	w	Batch size
SEA, 5% Noise	8	0.001	0.1	5	300	50
SEA, 15% Noise	56	0.001	0.1	5	300	50
STAGGER	20	0.01	0.01	4	300	50
Rotating Hyperplane (d = 0.1)	6017	0.01	0.01	6	10000	50
Rotating Hyperplane (d = 0.2)	6,017	0.01	0.01	6	10000	50
Airline	5,063,192	0.01	0.02	6	1000	10
Electricity	220,202	0.001	0.2	6	1000	10
Cover type	46,237	0.01	0.2	6	1000	50

For classification performance evaluation, one instance is presented to the neural network classifier to predict one at a time. The concept drift method used is the Hoeffding's Drift Detection Method (HDDM), which has a theoretical bound on the false positive/negatives [15]. The performance of this approach is evaluated against other state of the art (SOTA) approaches such as the Extremely Fast Decision Trees (EFDT) [12], Adaptive Random Forest (ARF) [6], and Very Fast Decision Rules (VFDR) [16]. Finally, we will use the GMean measure for quantitative evaluation, which is a well-known quantitative measure for imbalanced datasets.

4.2 Results and Discussion

Based on Fig. 1(a)–(c), the classification performance generally achieves good performance compared to other SOTA approaches. Despite this good performance, this approach can be inconsistent depending on the concept it encounters. For example, in the SEA dataset, there is a drop in performance between the 500,000th and the 600,000th sample. A possible explanation for this issue is that this approach assumes that the best parameters for the concept specific parameters, ϕ could be reached within the same number of gradient descent steps for all concepts. However, some concepts might require more gradient steps to reach the best concept specific parameters, ϕ. Thus, resulting in inconsistent performance using this approach.

While analyzing Fig. 1(b), this approach performs very well on imbalanced datasets (STAGGER) despite using a deep neural network known to overfit. Note that ARF and VFDR performances are identical, which causes their plots to overlap. The good performance on an imbalanced dataset is due to the few gradient descent steps, which acts as a form of regularization and is a well-known approach to training models on imbalanced datasets [17–19].

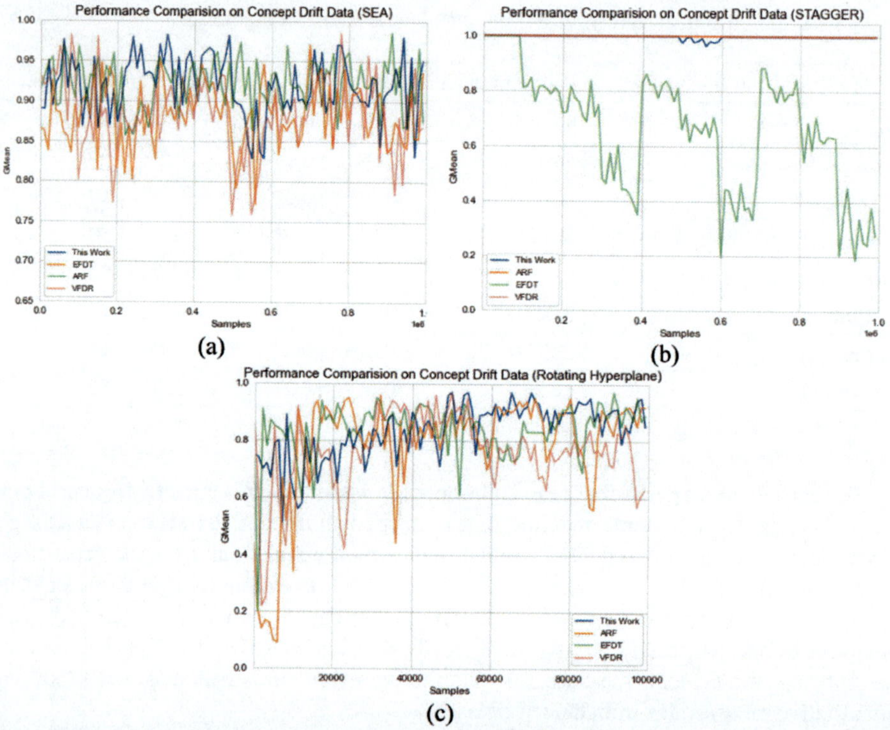

Fig. 1. Running classification performance on artificial datasets against current benchmark algorithms. (a) SEA, (b) STAGGER, and (c) Rotating Hyperplane.

Figure 2(a)–c) shows the running performance comparison on the real-world dataset. Comparing Fig. 1 (a)–(c) and Fig. 2 (a)–(c), this approach consistently outperforms the current SOTA on the real-world dataset. For real-world dataset which is high-dimensional and has more features compared to artificial datasets, this approach has a clear advantage by applying deep neural network which has a larger hypothesis space that allows it to capture more complex relationships between features compared to decision trees and decision rules that is popular with the current approach for data stream mining.

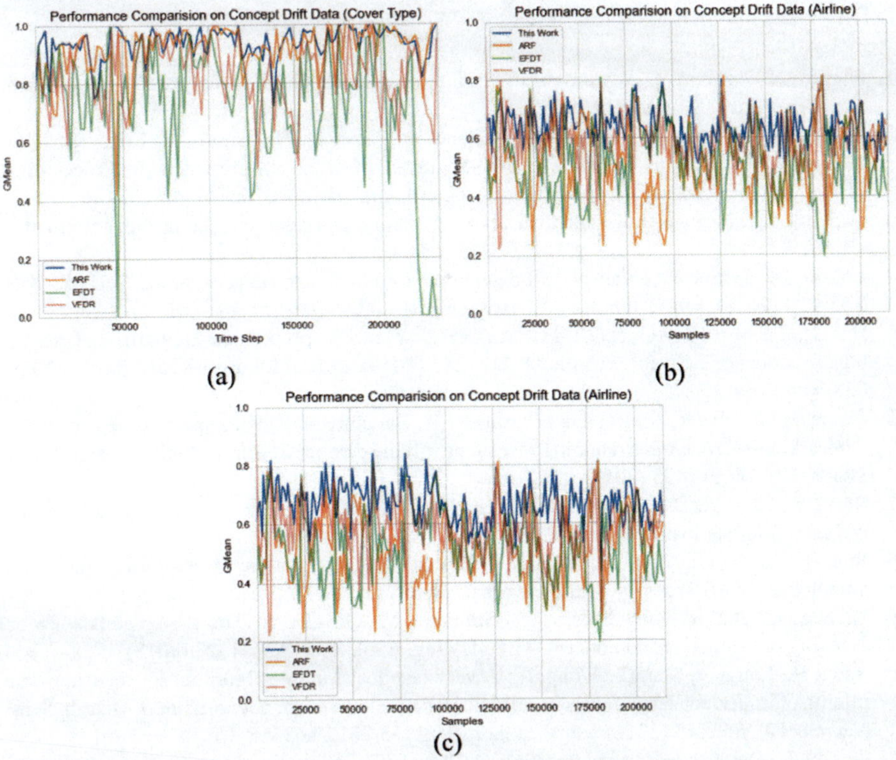

Fig. 2. Running classification performance on real-world dataset. (a) Cover type, (b) Electricity, and (c) Airlines dataset.

5 Conclusion

This study aims to study how large and complex neural network architectures could be adapted in the event of concept drift with minimal data and few gradient descent steps to minimize the downtime of the prediction model. This paper proposed an approach for concept drift adaptation via optimum parameter initialization. The optimum parameter is built upon the Reptile algorithm that had been successful in the meta-learning field for supervised classification by learning new examples with few class samples and few gradient descent steps.

Based on the experiments conducted, the optimum initialization parameter search is successfully obtained on real-world and artificial datasets by observing the convergence during the training. For classification results, this approach performs the best on real-world datasets due to its larger and more complex architectures, which allows it to model high-dimensional datasets more effectively.

For future work, we desire that this approach require fewer initial data and obtain the initial parameter with fewer iterations. Also, we would like this approach to be incrementally trained to better adapt to gradual concept drifts.

References

1. Nichol, A., Achiam, J., Schulman, J.: On first-order meta-learning algorithms. arXiv 2018. arXiv preprint arXiv:1803.02999 (2018)
2. Bifet, A., Holmes, G., Pfahringer, B., Kranen, P., Kremer, H., Jansen, T., Seidl, T.: Moa: Massive online analysis, a framework for stream classification and clustering. In: Proceedings of the First Workshop on Applications of Pattern Analysis 2010, pp. 44–50. PMLR
3. Vapnik, V.: The nature of statistical learning theory. Springer science & business media, (2013)
4. Cabral, D., Barros, R.: Concept drift detection based on fisher's exact test. Inf. Sci. **442–443**, 220–234 (2018). https://doi.org/10.1016/j.ins.2018.02.054
5. Barros, R.S.M.d., Santos, S.G.T.d.C.: An overview and comprehensive comparison of ensembles for concept drift. Inf. Fusion **52**, 213–244 (2019). https://doi.org/10.1016/j.inffus.2019.03.006
6. Gomes, H.M., Bifet, A., Read, J., Barddal, J.P., Enembreck, F., Pfharinger, B., Holmes, G., Abdessalem, T.: Adaptive random forests for evolving data stream classification. Mach. Learn. **106**(9–10), 1469–1495 (2017)
7. Shen, Y., Zhu, Y., Du, J., Chen, Y.: A fast learn++. NSE classification algorithm based on weighted moving average. Filomat **32**(5), 1737–1745 (2018).
8. Sun, Y., Tang, K., Zhu, Z., Yao, X.: Concept drift adaptation by exploiting historical knowledge. IEEE Trans. Neural Networks Learn. Syst. **29**(10), 4822–4832 (2018)
9. Khamassi, I., Sayed-Mouchaweh, M., Hammami, M., Ghédira, K.: Discussion and review on evolving data streams and concept drift adapting. Evol. Syst. **9**(1), 1–23 (2018)
10. Yang, H., Fong, S., Sun, G., Wong, R.: A very fast decision tree algorithm for real-time data mining of imperfect data streams in a distributed wireless sensor network. Int. J. Distrib. Sens. Netw. **8**(12), 863545 (2012). https://doi.org/10.1155/2012/863545
11. Oza, N.C.: Online bagging and boosting. In: 2005 IEEE International Conference on Systems, Man and Cybernetics 2005, pp. 2340–2345. IEEE (2005)
12. Bifet, A., Zhang, J., Fan, W., He, C., Zhang, J., Qian, J., Holmes, G., Pfahringer, B.: Extremely fast decision tree mining for evolving data streams. In: Proceedings of the 23rd ACM SIGKDD International Conference on Knowledge Discovery and Data Mining 2017, pp. 1733–1742 (2017)
13. Manapragada, C., Webb, G.I., Salehi, M.: Extremely fast decision tree. In: Proceedings of the 24th ACM SIGKDD International Conference on Knowledge Discovery & Data Mining, pp. 1953–1962 (2018)
14. Killick, R., Fearnhead, P., Eckley, I.A.: Optimal detection of changepoints with a linear computational cost. J. Am. Stat. Assoc. **107**(500), 1590–1598 (2012)
15. Frías-Blanco, I., del Campo-Ávila, J., Ramos-Jimenez, G., Morales-Bueno, R., Ortiz-Díaz, A., Caballero-Mota, Y.: Online and non-parametric drift detection methods based on Hoeffding's bounds. IEEE Trans. Knowl. Data Eng. **27**(3), 810–823 (2014)
16. Kosina, P., Gama, J.: Very fast decision rules for classification in data streams. Data Min. Knowl. Disc. **29**(1), 168–202 (2015)
17. Cao, P., Liu, X., Zhang, J., Zhao, D., Huang, M., Zaiane, O.: $\ell 2$, 1 norm regularized multi-kernel based joint nonlinear feature selection and over-sampling for imbalanced data classification. Neurocomputing **234**, 38–57 (2017)
18. Hong, X., Chen, S., Harris, C.J.: A kernel-based two-class classifier for imbalanced data sets. IEEE Trans. Neural Networks **18**(1), 28–41 (2007)
19. Yuan, X., Xie, L., Abouelenien, M.: A regularized ensemble framework of deep learning for cancer detection from multi-class, imbalanced training data. Pattern Recogn. **77**, 160–172 (2018)

A Comparison of Swarm Intelligence Algorithms Exploiting a Novel Neighbour Generation Technique

Gözde Alp(✉) and Ali Fuat Alkaya

Engineering Faculty, Computer Engineering Istanbul,
Marmara University, Istanbul, Turkey
{gozde.alp,falkaya}@marmara.edu.tr

Abstract. This study aims at introducing a new perspective on a widely known Capacitated Vehicle Routing Problem (CVRP). Migrating birds optimization (MBO) algorithm is applied to solve the CVRP for the first time, to the best of our knowledge. Besides, a novel multi directional neighbourhood search heuristic is presented. MBO is compared with the state-of-the-art algorithms that have already shown a good performance on CVRP on a number of benchmark instances and a real data set. Experimental results indicate that the MBO is 3.89% better than the other algorithms on obtaining shorter routes.

1 Introduction

Regional or local transportation systems are the fundamental constituent of people's daily life. Increase in the world population indicates that the need for transportation and solutions to the transportation problems will never end. Scientific research about transportation problems have been a significant topic of interest in combinatorial optimization research society, inherently.

One of the transportation problems is the Vehicle Routing Problem (VRP) [4]. VRP aims to discover the optimal group of routes for a set of vehicles, in order to deliver a good or item to a given set of customers. Generally, all routes start at a central depot and goods are delivered to the customers starting from the depot location. Many VRP variants are introduced in the literature due to the distinctive requirements in real life.

The problem we tackle in this study is CVRP and it is a VRP variation formulated as a graph theoretic problem by Laporte [9]. The main goal of CVRP is minimizing the total route cost by minimizing total distance to deliver goods. It assigns customers (delivery points) to vehicles so that vehicle's capacities are not exceeded. Each customer belongs to exactly one vehicle. Each vehicle has a route starting from the depot and ends at the depot.

VRP is NP-hard, so the size of problems that can be optimally solved using mathematical programming or combinatorial optimization algorithms may be limited. In this study, a new problem specific neighbour generation technique

A. Abraham et al. (Eds.): SoCPaR 2020, AISC 1383, pp. 43–52, 2021.
https://doi.org/10.1007/978-3-030-73689-7_5

and implemented the migrating birds optimization (MBO) algorithm for solving CVRP. To our best, MBO has not been applied on any VRP variant before. MBO is a nature inspired, swarm based and promising solution technique that has become popular in recent years. Four additional algorithms are implemented which are already shown to present good performance in the literature. Those are genetic algorithm (GA), tabu search (TS), differential evolution (DE) and simulated annealing (SA). Computational experiments are conducted on instances taken from the classical benchmark sets of CVRP [1].

The remainder of the manuscript is organized as follows: the studies investigated in the literature are summarized in Sect. 2. Implemented solution methods are explained detaily in Sect. 3. Experimental setup is demonstrated and the results are discussed in Sect. 4. Section 5 concludes and comprises ideas for future research.

2 Literature Review

A brief summary of remarkable or current studies on VRP [4] is given in this section. A survey of some of the recently employed solution methodologies in the field of VRP and its variants are presented in [7]. There are many VRP variations that had been identified over the time, due to diversity of problem constraints and requirements. One of them, VRP with time windows, is studied in [11], where each customer has a time limit to be visited. VRP with uncertain demands is another variation, stochastic in customer demand [3]. Open VRP is another open route VRP variant [13].

Particle swarm optimization algorithm was implemented for VRP in [10]. GA containing TS movements is applied to OVRP in [20]. A meta-heuristic approach is proposed for VRP in [15]. The proposed technique actually works similar with SA in point of giving chance to worse results. A local search like heuristic improvement procedure, that utilizes integer linear programming improvements is proposed in [12]. New CVRP benchmark instances were proposed in 2017 in [17]. DE that has local search movements inside is applied on CVRP in [16] in 2015. GA based approach is proposed to solve two-echelon CVRP in 2017 [18]. A novel SA framework is improved to solve CVRP in 2018 [19]. A meta-heuristic algorithm based on a probabilistic granular TS applied on CVRP in [2] and a real data set taken from a collaborator company.

Many different methods have been applied for solving the VRP including local search algorithms, evolutionary algorithms, and swarm based techniques. Even though MBO has been recently proposed but presents promising performance on the problems it has been applied to, MBO has not been implemented on any variant of VRP. Therefore, this paper is expected to fill an important gap in this research area.

3 Solution Methods

This section comprises the essential information about implemented algorithms. Firstly, the problem specific versatile neighbour generation technique we have introduced is presented, and the recombination operator we had used for the problem is explained. The solution techniques that we have implemented in this study are summarized subsequently.

3.1 Versatile Neighbour Generation Technique

In this subsection we describe our Versatile neighbour generation technique (VNG) in detail. We have applied VNG in all exploited algorithms. Sample solutions (individuals) are illustrated in Figs. 1, 2 and 3; the illustration in left hand side of the figures is the individual before applying the VNG technique, while the one in the right hand side is the neighbour solution. The total customer number is assumed to be n, so customers vary from one to n and are indicated as $C_1, ..., C_n$. An individual comprises a number of vehicles, each has its own route starting from the depot (D_0) and ends at the depot. Customer locations and requirements are known a priori. The distances between customers are computed using Euclidean distance as specified in the benchmark dataset. Customers are assigned to vehicles by not exceeding their capacities.

VNG, changes customer assignments in a solution and generates neighbour individuals. VNG primarily selects two random numbers between one and the total number of vehicles. The numbers may turn out to be equal or not. If those random numbers are equal to each other, it means that only one vehicle is selected as shown in Fig. 1 (e.g. vehicle two is selected). In such a case another two random numbers are picked between one and the total number of customers in that particular vehicle. Then, the positions of these two numbers are swapped. For the illustration in Fig. 1 C_{14} and C_{20} are selected and swapped.

Vehicle1 Route →	D_0	C_1	C_{16}	C_{17}	C_{16}	D_0
Vehicle2 Route →	D_0	C_2	C_{14}	C_{19}	C_{20}	D_0
Vehicle3 Route →	D_0	C_{10}	C_3	C_{18}	C_{15}	D_0
Vehicle4 Route →	D_0	C_9	C_8	C_4	C_{21}	D_0
Vehicle5 Route →	D_0	C_{11}	C_{12}	C_{17}	C_5	D_0
Vehicle6 Route →	D_0	C_{13}	D_0			

⇨

Vehicle1 Route →	D_0	C_1	C_{16}	C_{17}	C_{16}	D_0
Vehicle2 Route →	D_0	C_2	C_{20}	C_{19}	C_{14}	D_0
Vehicle3 Route →	D_0	C_{10}	C_3	C_{18}	C_{15}	D_0
Vehicle4 Route →	D_0	C_9	C_8	C_4	C_{21}	D_0
Vehicle5 Route →	D_0	C_{11}	C_{12}	C_{17}	C_5	D_0
Vehicle6 Route →	D_0	C_{13}	D_0			

Fig. 1. Illustration of the operation performed when randomly selected vehicles are the same

If primarily selected random numbers are not equal to each other (different vehicles are selected), VNG checks the loads of picked vehicles. If one of those vehicle's load is less than the specified minimum, then a random customer is transferred from the full one to the one being less loaded. We set minimum load parameter as 60% of the vehicle capacity. A sample case is given in Fig. 2. Vehicle two and vehicle six are selected randomly, and vehicle six is loaded less than the minimum. A random customer (C_{19}) is picked with in vehicle two. C_{19} extracted from vehicle two and added to the route of vehicle six.

Apart from the above conditions; if different vehicles are selected primarily and there is no load issue then, one customer is selected from each of two vehicles

Vehicle1 Route →	D_0	C_1	C_{16}	C_{17}	C_{16}	D_0
Vehicle2 Route →	D_0	C_2	C_{14}	C_{19}	C_{20}	D_0
Vehicle3 Route →	D_0	C_{10}	C_3	C_{18}	C_{15}	D_0
Vehicle4 Route →	D_0	C_9	C_8	C_4	C_{21}	D_0
Vehicle5 Route →	D_0	C_{11}	C_{12}	C_{17}	C_5	D_0
Vehicle6 Route →	D_0	C_{13}	D_0			

⇨

Vehicle1 Route →	D_0	C_1	C_{16}	C_{17}	C_{16}	D_0
Vehicle2 Route →	D_0	C_2	C_{14}	C_{20}	D_0	
Vehicle3 Route →	D_0	C_{10}	C_3	C_{18}	C_{15}	D_0
Vehicle4 Route →	D_0	C_9	C_8	C_4	C_{21}	D_0
Vehicle5 Route →	D_0	C_{11}	C_{12}	C_{17}	C_5	D_0
Vehicle6 Route →	D_0	C_{13}	C_{19}	D_0		

Fig. 2. llustration of the operation performed when randomly selected vehicles are different with load issue

Vehicle1 Route →	D_0	C_1	C_{16}	C_{17}	C_{16}	D_0
Vehicle2 Route →	D_0	C_2	C_{14}	C_{19}	C_{20}	D_0
Vehicle3 Route →	D_0	C_{10}	C_3	C_{18}	C_{15}	D_0
Vehicle4 Route →	D_0	C_9	C_8	C_4	C_{21}	D_0
Vehicle5 Route →	D_0	C_{11}	C_{12}	C_{17}	C_5	D_0
Vehicle6 Route →	D_0	C_{13}	D_0			

⇨

Vehicle1 Route →	D_0	C_1	C_{16}	C_{17}	C_{16}	D_0
Vehicle2 Route →	D_0	C_2	C_{14}	C_9	C_{20}	D_0
Vehicle3 Route →	D_0	C_{10}	C_3	C_{18}	C_{15}	D_0
Vehicle4 Route →	D_0	C_{19}	C_8	C_4	C_{21}	D_0
Vehicle5 Route →	D_0	C_{11}	C_{12}	C_{17}	C_5	D_0
Vehicle6 Route →	D_0	C_{13}	D_0			

Fig. 3. Illustration of the operation performed when selected vehicles are different without load issue

randomly. Selected customer assignments are swapped. An illustrative example is given in Fig. 3. Vehicle two and vehicle four are selected randomly, C_9 and C_{19} are picked inside those vehicles and swapped.

All procedures in VNG operator are implemented by caring the constraints in the CVRP formulation. Versatile neighbour generation method contains; in-vehicle swap, inter-vehicle swap, capacity balancing abilities. In-vehicle swap enables route healing, inter-vehicle swap expands the solution space and capacity control balances the load on vehicles.

3.2 One Point Crossover

GA and DE are evolutionary algorithms which require a recombination operator. We have applied one point crossover operator for these algorithms. One point crossover selects a random cut point and copies the part before cut point from the first parent to the offspring and the rest is completed by the second parent starting from the cut point.

Parent one and parent two are CVRP instances (individuals) those illustrated in Fig. 4. Customers are named from C_1 to C_{10}, and there are totally three vehicles for each solution. Depot node is "D_0". For parent one, vehicle one starts with zero and C_5, C_7 C_8 are visited respectively, route ends in depot D_0. Following route for second vehicle starts at index five again with depot D_0 and so on. Random cut point is six for this illustration. When we analyze offspring one and offspring two, nodes from index zero to six are taken from parent one and parent two, respectively. The rest is taken from the remaining parent starting from the cut point by caring the vehicle capacities and avoiding the repetitions. If a node exceeds vehicle capacity; it is added to a new vehicle.

Index	0	1	2	3	4	5	6	7	8	9	10	11	12	13	14	15
Parent 1	D_0	C_5	C_7	C_8	D_0	D_0	C_9	C_{10}	C_2	C_3	D_0	D_0	C_1	C_4	C_6	D_0
Parent 2	D_0	C_2	C_5	C_8	D_0	D_0	C_9	C_4	C_{10}	D_0	D_0	C_3	C_6	C_1	C_7	D_0

Index	0	1	2	3	4	5	6	7	8	9	10	11	12	13	14	15
Offspring 1	D_0	C_5	C_7	C_8	D_0	D_0	C_9	C_4	C_{10}	D_0	D_0	C_3	C_6	C_1	C_2	D_0
Offspring 2	D_0	C_2	C_5	C_8	D_0	D_0	C_9	C_{10}	C_3	C_1	D_0	D_0	C_4	C_6	C_7	D_0

Fig. 4. One point crossover illustration

3.3 Exploited Algorithms

The algorithms those are we have exploited while solving CVRP is mentioned briefly in this subsection. Those algorithms are MBO, TS, SA, DE and GA.

MBO is a swarm based search algorithm inspired from V formation of wild geese. Each member in the flock is actually an individual of the problem. The flock is initialized similar to the birds of a V shape migration. The best individual in the flock is assigned as the leader. A number of neighbour solutions for each flock member starting from the leader are produced. This algorithm allows ensuring rewarding effect of a promising individual. Except the main or leading solution, all individuals are influenced by the solutions in their front. When this process is done a number of times, a new flock is constructed by moving leader solution to the end of the flock and a new leader is selected. Same transaction is repeated for the new flock until the termination condition satisfied and the best solution is presented as the result solution. More detailed information about MBO may be found in [5]. Neighbour solutions are generated by applying the versatile neighbour generation technique defined in Sect. 3.1.

TS [6] is a meta-heuristic search technique uses memory structures commonly applied and achieved very successful results for VRP. Initial solution is generated and initializations are done firstly. Neighbour solutions which generated by versatile heuristic are accepted if they improve the solution and they are not in tabu list. While a solution is accepted then it is forbidden to accept it again during *tabu tenure*. Tabu list is updated at each iteration, new solution is added to the list and the one that is situated during after number of iterations as many as *tabu tenure* removed from the list.

SA is a well-known local search algorithm [8], inspired from the annealing process of metal work. A starting temperature T is determined. Neighbour generations are generated by same temperature until reaching to equilibrium condition R. When equilibrium condition is satisfied, cooling process is done by decreasing T with a specific cooling rate (a) until the temperature converges to a steady 'frozen' state. When temperature became less than the stopping condition SA stops. In our implementation R value changes and increases by an initially specified b value. When a random neighbour is produced, it is compared with the previous solution. If it is better than the previous one, it is accepted otherwise it may still be accepted with a small probability, computed based on the difference of old and new solution and the temperature. R is the number of iteration time in each temperature and it is updated by the time by a number b.

DE is a meta-heuristic algorithm [14], optimizes a problem by maintaining a population of candidate solutions and creating new candidate solutions by mutation and recombination procedures. Versatile heuristic is adapted as mutation operator for DE. One point crossover used as recombination operator.

GA is a widely applied evolutionary algorithm inspired by the nature. GA is widely used to generate high quality solutions to optimization and search problems by relying on nature-inspired mutation and recombination operators and selection mechanisms. Versatile heuristic is used as mutation operator for GA. One point crossover used as recombination operator. Roulette wheel selection is preferred while selecting parents.

4 Experimental Setup and Results

In this section, empirical setup, essential implementation details and experimental results in terms of cost is provided and discussed. All algorithm implementations are done using Java programming language on Eclipse IDE running on Intel i7-5500U CPU, and 16 GB memory computer operated by Windows 10 operating system. We conducted our experiments on a real data set and 24 instances taken from Augerat et al. (set P) where many CVRP benchmark instances are provided. The input data are available online in [1]. The best parameters for all exploited algorithms except TS are determined as in [5].

MBO: The number of birds (nob) is set as 51, number of flakes (nof) is 10, the number of neighbours (non) is 3 and the overlap factor (olf) is set to 1. VNG method used for generating neighbour solutions.

SA: SA parameters are determined admitting terminating SA execution by the same instance generation number. Temperature (T) is set to 20.000, the cooling rate (a) is set to 1.5, R initially set to 2000 and it increases by the time by a predetermined ratio (b) 1.1. VNG technique used for generating neighbour solutions.

TS: The *tabu tenure* is set to 10, and the VNG previously described is used to find out neighbouring solutions for MBO, SA and TS. DE: The number of

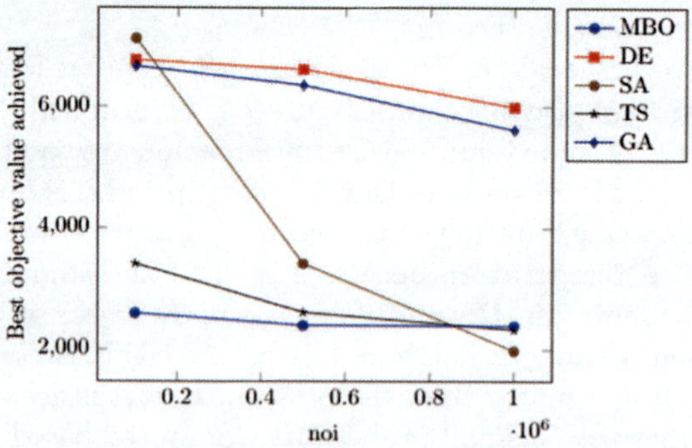

Fig. 5. Performances of the algorithms on real data.

generations (*nog*) is set to 1,000, number of solutions (*nos*) is set to 200, crossover probability is set to 0.5 and mutation probability is set to 0.5. One point crossover is used as recombination operator. VNG used as mutation operator.

GA: Number of generations is set to 100, population size is 100, crossover and mutation probabilities both are set to 0.5. One point crossover is used as recombination operator and VNG applied as mutation operator.

The real data set is provided by a collaborative company that needs a special help for their employee service planning problem. The data set contains 200 nodes and the company has at most 30 vehicles with capacity 20. Ten independent runs are conducted for each algorithm when (*noi*) is set to 100,000, 500,000 and 1,000,000. Results are presented in Fig. 5. MBO achieves the best solutions when *noi* is 100,000 and 500,000 whereas SA outperforms MBO when (*noi*) is 1,000,000. MBO is capable on discovering promising solutions within the restricted boundaries, whereas SA may find better results when it has an excessive exploration possibility.

Each benchmark instance is solved using stated solution techniques. Ten independent runs are conducted for each case and the average performances on these ten runs are tabulated. Number of instances (*noi*) is defined as the stopping criterion for the algorithms and the algorithms are designed to stop when they create *noi* instances. The results obtained when (*noi*) is set to 100,000 and set to 500,000 are presented in Table 1. Cost values in the table correspond to the total

Table 1. The objective values when *noi* is equal to 100,000 and 500,000.

Instance	*noi*=100,000					*noi*=500,000				
	MBO	DE	SA	TS	GA	MBO	DE	SA	TS	GA
P-n101-k4	1091.9	1110.5	2552.7	**968.9**	1159.7	**908**	1029.2	945.9	1051.2	1023.3
P-n16-k8	445	449	454.9	458.9	**444.7**	**444.5**	450.9	454	450.3	444.7
P-n19-k2	**215.8**	234.2	301.3	248.5	217.2	**212.7**	236.9	216.2	257.4	214.6
P-n20-k2	**213.9**	241.3	310.9	259.6	221.2	**217.1**	246.3	225.2	263.1	220.4
P-n21-k2	**213.5**	240.5	314.4	248.5	218	212	228.6	214.1	255.2	**211.9**
P-n22-k2	**216.6**	246.3	334.6	254.3	218.1	**216.3**	248.5	219.8	266.9	219
P-n22-k8	594.5	595.6	640.6	660.6	**582.3**	594.7	596.1	609.4	638.1	**584.4**
P-n23-k8	533.3	540	570.1	553.8	**530.2**	532.9	533.5	558.4	557.4	**531.2**
P-n40-k5	**500.1**	554	925.7	584.4	522.4	**485.4**	574.6	580.7	587.8	517.3
P-n45-k5	**559.8**	668.1	1093.1	662.7	599.8	**561.7**	652.5	638.6	674.2	590.7
P-n50-k10	**740.8**	811.3	931.2	826	771.3	**726.4**	836.2	780.9	806.8	758.1
P-n50-k7	**612.2**	684.1	1181.1	673.1	641.8	**612.6**	689	684.5	722.7	615.6
P-n50-k8	**684**	742	1203.3	743.6	687.7	**661.4**	744.3	727.7	718	683.5
P-n51-k10	**818.6**	878.2	1319.3	876.5	822.2	**802.5**	873.6	881.6	898.6	820.3
P-n55-k10	**734.7**	819.3	1343.9	818.7	762.9	**730.2**	808	815.4	809.3	748.4
P-n55-k15	**975.7**	1039.6	1065.8	1056.3	997.9	**964.8**	1031.7	965	1044	991.4
P-n55-k7	**633.9**	709.5	1302.6	704	673.7	**614.7**	719.8	685.3	677.2	637.9
P-n55-k8	**640.2**	766.2	1285.3	729.3	669.3	**626.1**	721.4	675.3	735.3	662.9
P-n60-k10	**801.4**	938.6	1504.6	919.5	860.8	**804.8**	887.7	877.2	906.4	824.5
P-n60-k15	**1015.5**	1077.3	1126.8	1096.9	1053.1	**1010.7**	1098.6	1103.6	1094.3	1036.4
P-n65-k10	**885.8**	986.8	1696	968.6	905.9	**854.2**	979.4	944.4	980.2	908
P-n70-k10	**948.2**	1042.8	1030.1	1059.4	986.5	**917.7**	1079.6	980.5	1061.7	939.5
P-n76-k4	**797.8**	911	1157.5	866.8	856.6	**727.8**	856	775.8	857.8	792.1
P-n76-k5	**837**	904.3	1070.8	906.1	874.1	**775.1**	899	793.4	935.8	854.1

route lengths those found by each algorithm. For *noi*=100,000 MBO is able to beat its rivals for 20 instances out of 24. When the *noi* is increased to 500,000 MBO produces the best result for 21 of 24 instances. GA produces the second best result for nearly all instances.

Average cost values for *noi*=100,000 and *noi*=500,000 for all applied techniques are illustrated in Fig. 6. MBO is able to produce lower cost in average.

We have also applied t-test to check if the differences are statistically significant. MBO and its closest follower GA are tested and the average t-test result for *noi*=100,000 is 0.0152%, the average t-test result for *noi*=500,000 is 0.00573% implying that those two result sets are statistically different from each other.

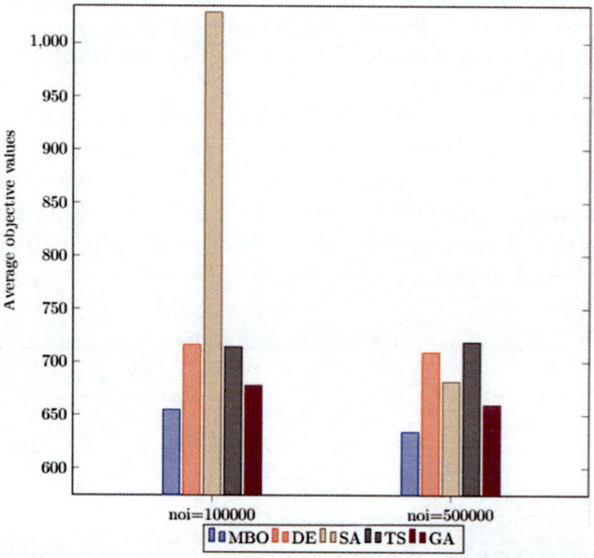

Fig. 6. Average costs illustration.

5 Conclusion and Further Research

We have introduced a versatile neighbour generation method and applied a solution technique that has not been implemented before for solving CVRP. The versatile neighbour generation technique enables multi-directional solution alteration containing in-vehicle swap, inter-vehicle swap, capacity balancing abilities. Besides, it is the first time that MBO is applied to VRP and is undeniably successful in terms of cost according to the experimental results. MBO gained 3.89% better results than its closest follower algorithm (GA) on the average in terms of cost.

For the future research, we plan to apply experiences gained in this study to for solving transportation issues while constructing employee schedules in a real company. Employee transportation is generally maintained by service buses. Due to the frequent changes in employee schedules, service schedules are also updated frequently caring the transportation costs. Another research direction may be constructing weekly schedules by considering employees' home locations; where a second goal will be decreasing transportation costs.

References

1. Augerat.: Cvrplib (1995). http://vrp.atd-lab.inf.puc-rio.br/index.php/en/
2. Bernal, J., Escobar, J.W., Paz, J.C., Linfati, R., Gatica, G.: A probabilistic granular tabu search for the distance constrained capacitated vehicle routing problem. Int. J. Ind. Syst. Eng. **29**(4), 453–477 (2018)
3. Cao, E., Lai, M., Yang, H.: Open vehicle routing problem with demand uncertainty and its robust strategies. Expert Syst. Appl. **41**(7), 3569–3575 (2014)
4. Dantzig, G.B., Ramser, J.H.: The truck dispatching problem. Manage. Sci. **6**(1), 80–91 (1959)
5. Duman, E., Uysal, M., Alkaya, A.F.: Migrating birds optimization: a new metaheuristic approach and its performance on quadratic assignment problem. Inf. Sci. **217**, 65–77 (2012)
6. Glover, F.: Future paths for integer programming and links to artificial intelligence. Comput. Oper. Res. **13**(5), 533–549 (1986)
7. Goel, R., Maini, R.: Vehicle routing problem and its solution methodologies: a survey. Int. J. Logistics Syst. Manage. **28**(4), 419–435 (2017)
8. Kirkpatrick, S., Gelatt, C.D., Vecchi, M.P.: Optimization by simulated annealing. Science **220**(4598), 671–680 (1983)
9. Laporte, G., Nobert, Y., Desrochers, M.: Optimal routing under capacity and distance restrictions. Oper. Res. **33**(5), 1050–1073 (1985)
10. MirHassani, S., Abolghasemi, N.: A particle swarm optimization algorithm for open vehicle routing problem. Expert Syst. Appl. **38**(9), 11547–11551 (2011)
11. Redi, A.P., Maghfiroh, M.F., Vincent, F.Y.: An improved variable neighborhood search for the open vehicle routing problem with time windows. In: 2013 IEEE International Conference on Industrial Engineering and Engineering Management, pp. 1641–1645. IEEE (2013)
12. Salari, M., Toth, P., Tramontani, A.: An Ilp improvement procedure for the open vehicle routing problem. Comput. Oper. Res. **37**(12), 2106–2120 (2010)
13. Sariklis, D., Powell, S.: A heuristic method for the open vehicle routing problem. J. Oper. Res. Soc. **51**(5), 564–573 (2000)
14. Storn, R., Price, K.: Differential evolution-a simple efficient adaptive scheme for global optimization. Technical report, Technical Report over continuous spaces. International Computer Science Institute (1995)
15. Tarantilis, C.D., Ioannou, G., Kiranoudis, C.T., Prastacos, G.P.: Solving the open vehicle routeing problem via a single parameter metaheuristic algorithm. J. Oper. Res. Soc. **56**(5), 588–596 (2005)
16. Teoh, B.E., Ponnambalam, S.G., Kanagaraj, G.: Differential evolution algorithm with local search for capacitated vehicle routing problem. Int. J. Bio-Inspired Comput. **7**(5), 321–342 (2015)

17. Uchoa, E., Pecin, D., Pessoa, A., Poggi, M., Vidal, T., Subramanian, A.: New benchmark instances for the capacitated vehicle routing problem. Eur. J. Oper. Res. **257**(3), 845–858 (2017)
18. Wang, K., Lan, S., Zhao, Y.: A genetic-algorithm-based approach to the two-echelon capacitated vehicle routing problem with stochastic demands in logistics service. J. Oper. Res. Soc. **68**(11), 1409–1421 (2017)
19. Wei, L., Zhang, Z., Zhang, D., Leung, S.C.: A simulated annealing algorithm for the capacitated vehicle routing problem with two-dimensional loading constraints. Eur. J. Oper. Res. **265**(3), 843–859 (2018)
20. Yu, S., Ding, C., Zhu, K.: A hybrid GA-TS algorithm for open vehicle routing optimization of coal mines material. Expert Syst. Appl. **38**(8), 10568–10573 (2011)

Skin Cancer Automatic Detection Based on Image Characteristics of Shape, Colour, and Texture

Hugo W. Maia[1] and Marcos E. Cintra[2(✉)]

[1] Federal Institute of Rio Grande do Norte, Mossoró, Brazil
[2] The Brazilian University of the Semiarid Region, Mossoró, Brazil
https://www.researchgate.net/profile/Hugo_Wendell_Maia,
https://www.researchgate.net/profile/Marcos_Cintra

Abstract. Skin cancer is the most common type of cancer. Melanoma is a malign type of skin cancer responsible for several deaths in the world. An early diagnosis increases the probability of cure. Dermatoscopy is one of the procedures for such diagnosis. Using dermatoscopy, it is possible to recognize skin structures that are not visible to the human sight. The dermatoscope zooms in images for a more precise diagnostic. Although less efficient than the analysis of experts, it is possible to find several proposals for the automatic classification of skin lesions in the literature, some of them using image characteristics of shape, colour, and texture. Our proposal aims at supporting doctors using dermatoscopy for the automatic classification of skin lesions. In our experiments, we used 21 classification algorithms with different inducing paradigms. The best results were obtained using the FuzzyDT algorithm, based on a set of image characteristics related to shape, colour, and texture. One of the main contributions of this proposal, is the characteristic selection procedure using FuzzyDT, which obtained high accuracy, as well as a dataset of preprocessed cancer skin images. The proposal is detailed and the results are presented and discussed.

Keywords: Artificial intelligence · Machine learning · Skin cancer · Fuzzy systems · Fuzzy decision trees

1 Introduction

Skin cancer cases have increased significantly in the last years. According to the World Health Organization (WHO) [1], around 132,000 new cases of skin cancer occur every year in the world. In Brazil, according to INCA (Brazilian National Cancer Institute) [2], skin cancer is the number one type of cancer, responsible for about 25% of all diagnosed cancer cases. Statistics are similar for other countries and, thus, this disease is quite relevant and worrisome.

We thank CAPES for the financial support.

A. Abraham et al. (Eds.): SoCPaR 2020, AISC 1383, pp. 53–62, 2021.
https://doi.org/10.1007/978-3-030-73689-7_6

It is important to notice that among the types of skin cancer, melanoma is the one with lowest incidence, however, with the highest death rates. According to [1], melanoma is responsible for just 4% of all skin cancer cases in Brazil, but with a death rate of 70%. In this sense, an early diagnosis is essential for the reduction of the death rates, as the success percentage of this type of cancer is high when it is detected and treated during its early stages. In the literature, it is reported that the survival rates for melanoma when early diagnosed reaches 90 to 97%. On the other hand, the survival rates drop to 10 to 15% when the cancer is diagnosed in more advanced stages.

There are, basically, two ways to perform such diagnostic, using invasive or non-invasive methods. The invasive method uses a biopsy, i.e., the laboratory analysis of a piece of skin that is extracted from the patient. This method is costly in financial terms, as well as the inconvenience for the patient. The non-invasive method consists in the visual analysis of the lesion, including some common characteristics of this type of cancer. This procedure can be carried out in different forms. One of them is the dermatoscopy, introduced by [3], which consists in the recognition of the morphological structures of the skin that are not visible by the human sight. Dermatoscopy uses a dermatoscope to analyse the lesions based on certain predefined characteristics. Although the analysis is carried out using enlarged images and well defined extracted characteristics, it generates a subjective diagnosis, as it suffers the interference of the experience and tiredness of the dermatologist, irregular illumination, early stage lesions, among others.

Classification algorithms can be used to support the process of evaluating skin images. The task of classification has been widely researched in both, the Machine Learning and Soft Computing communities. Classification is a relevant task, widely used for pattern recognition, data mining, and decision making [4]. Classification can be performed by Neural Networks [5], fuzzy systems [6], genetic algorithms [7], decision trees [8], among others.

Decision trees (DT) algorithms usually produce simple models with competitive classification rates and simple but powerful graphical representations. DTs are highly interpretable and can handle continuous and discrete attributes. C4.5 [8] is a popular DT based on entropy and information gain. Fuzzy decision trees have also been proposed in the literature [6,9,10]. Fuzzy DTs combine the high interpretability of classic DTs with the fuzzy capability of dealing with imprecision and uncertainty in data. The FuzzyDT [9] algorithm is based on the classic C4.5 algorithm.

In this proposal, we used the FuzzyDT algorithm to automatize the classification of skin lesions based on the image characteristics of shape, colour, and texture. The FuzzyDT algorithm was compared to other classification algorithms and obtained the highest classification rates. The proposal and the experiments are presented and discussed. This paper is organized as follows: Sect. 2 presents the FuzzyDT algorithm. Section 3 presents related proposals. Section 4 details the proposed method. Section 5 presents and discusses the obtained results, followed by the final considerations in Sect. 6.

2 FuzzyDT

The fuzzy logic [11] allows the natural treatment of uncertainty and imprecision present in data. Fuzzy systems use a reasoning mechanism which is able to represent the imprecision of the human thought. Any system that includes at least one variable defined in linguistic terms which, in turn, are represented by fuzzy sets, according to the fuzzy set and fuzzy logic theories, is considered a fuzzy system. Rule-based fuzzy systems are formed by a knowledge base, which contains a fuzzy data base, a fuzzy rule base, and an inference mechanism. The fuzzy data base contains the definitions of the variables in terms of fuzzy sets. The fuzzy rule base contains a set of fuzzy rules representing the knowledge of the domain. The inference mechanism uses both, the fuzzy rule base and the fuzzy data base to classify an input example.

The FuzzyDT algorithm, proposed in [9] and available in [12], uses the same measures of the classic C4.5 algorithm [8], entropy and information gain, to decide on the importance of the dataset characteristics, also named as attributes or features. However, for FuzzyDT, continuous attributes are defined in terms of fuzzy sets before the induction of the tree. This way, the induction process uses a set of discretized attributes. In other words, instead of continuous values, FuzzyDT uses the fuzzy set with highest membership degree to induce the final tree. This way, FuzzyDT uses each attribute only once, avoiding the repeated use of attributes with different, in many cases several, splitting points.

3 Related Proposals

In [13], the authors present an approach to automatically quantify the colours that are clinically meaningful in dermatoscopic images. The first step of the proposal is to reduce/optimize the number of colours of the image using a K-Means clustering algorithm [4]. In the sequel, a regression algorithm is trained on a set of 617 images. Finally, a mathematical equation given by the algorithm is used for two classes (benign and malignant cancer). The proposal obtained a 62% sensibility and a 76% specificity rates. In [14], the authors propose the complete analysis of the images using a real time system to support melanoma detection. The first stage of this framework treats noise in the images and segments the Region of Interest (ROI) of the skin. Then they extract the image features like 2-D Fast Fourier Transform features set (FFT2), 2-D Discrete Cosine Transform features set (DCT2), size and complexity features sets. Then such features are used by a KNN classifier. In [15], the authors use SVM to extract information regarding the format of the border of the images. They are classified in symmetric or not, smooth or not. Such information is used by an expert to classify the lesions. The proposal obtained a 73% accuracy. In [16], the authors use a border detector method proposed in [17] to pre-process the images. They use a modified decision tree algorithm to classify the images with a 90% accuracy.

The ABCD characteristics are widely used and focus on the format and the colours of the lesions. [18] use this approach to extract 24 characteristics.

Then two combinations of these characteristics are formed. The first is defined using the *Sequential Floating Forward Selection* (SFFS). This technique selects the best characteristic, then the second best, and so forth. The second combination uses the *Sequential Floating Backward Selection* (SFBS), which starts with all characteristics and removes the least important, and so forth. SFFS obtained the best results with the following characteristics: shape factor, symmetry, tone extension and means, saturation and lightness of the colour model. The authors used three classifiers, KNN, a Bayes-based method and a MLP, and the combination all three algorithms. The best accuracy rate was obtained with the combination of the classifiers reaching 87,76%. Other approaches use only texture characteristics, such as [19]. Using a MLP, the approach obtained a 92.0% accuracy with 102 images, 51 from Dermquest [20] and 51 from Dermnet [21].

In [22], the authors were able to achieve 91.5% accuracy for the same PH2 database we used in this work, based on shape and colour features. In [23], the authors achieved 92.5% accuracy combining VGG19 and Logistic Regression for the same PH2 dataset.

4 Proposal for the Classification of Skin Lesion Images

The main steps of this proposal are:

1. Selection of an image database;
2. Extraction of the images in shades of gray;
3. Segmentation;
4. Extraction of characteristics;
5. Classification of the images using the FuzzyDT algorithm.

These steps are detailed next.

4.1 Selection of an Image Database

We adopted the PH [24] image database. The selection was based on the facts that it is widely used in the literature, it was created by health professionals, it contains 200 images with appropriate resolution, and the images contain 80 examples of common nevus, 80 of atypical nevus, and 40 of melanoma. We used all images in the database, which are in the RGB format (*Red*, *Green*, and *Blue*), with a 767 $\times$ 576 pixel resolution. The database contains the images and their respective masks. The masks show the lesion in white and the remaining part of the image in black. Figure 1(a) shows one of the original images with its respective mask in Fig. 1(b).

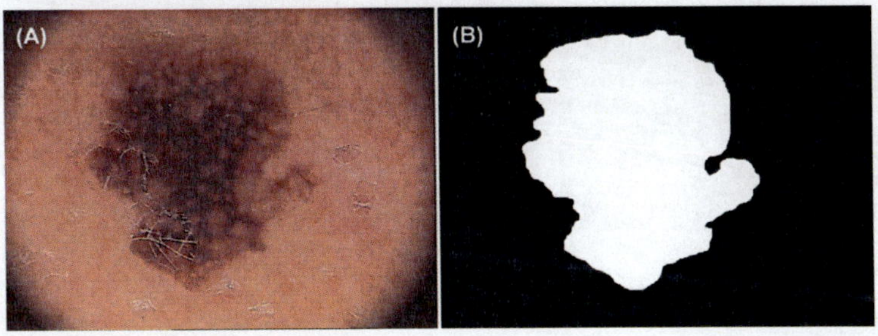

Fig. 1. An example of an (A) Original image and its respective (B) mask.

4.2 Conversion of the Images into Shades of Grey

In order to extract the image in shades of gray, the original image is separated into its primitive colours (*Red*, *Green*, and *Blue*). Each value is multiplied by 0.30, 0.59, and 0.11, respectively (these are default values). The result of the multiplication is summed and defines the value of the corresponding pixel in shades of grey. This process was performed using MATLAB [25]. Figure 2 illustrates this extraction process.

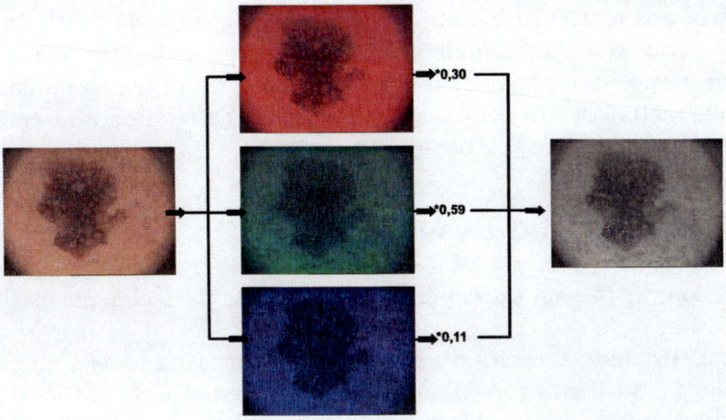

Fig. 2. An example of conversion of images from RGB to shades of grey.

4.3 Segmentation

In order to perform the segmentation of the images, each pixel was analysed. In fact, the pixels of the converted image was compared to its mask. From this comparison, a third image was generated. Figure 3 illustrates this procedure.

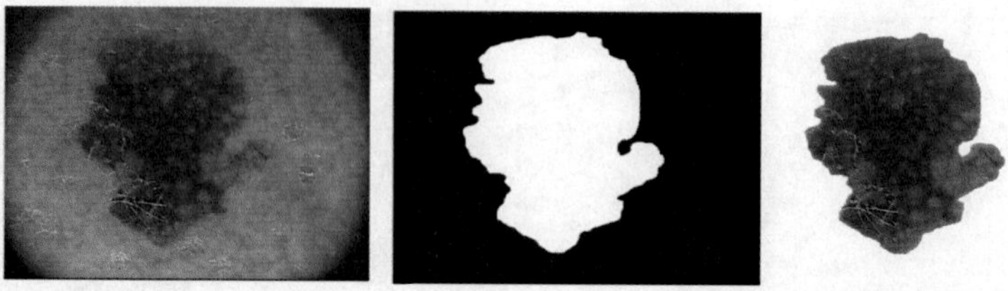

Fig. 3. Segmentation of the images in shades of grey.

In Fig. 3, the left figure is the original one; the middle figure represents the mask defined in the previous step, when the primitive colours were multiplied by three different values, generating 9 masks (three masks for each of the red, green, and blue primitive colours), allowing the removal of the relevant segment of the original figure; the right image is the extracted region of interest from the original figure, using the shades of green technique.

4.4 Extraction of Characteristics

The following step of this proposal consists of the extraction of characteristics from the segmented images. This process was performed based on co-occurrence GLCM matrices [19]. In fact, 8 GLCM matrices were extracted for each image. Each matrix was extracted based on a different angle: 0°, 45°, 90°, 135°, 180°, 225°, 270°, and 315° (all default values). After extracting the matrices, they were summed, position by position, for each image. From the remaining matrix, the four characteristics of contrast, homogeneity, correlation and energy were extracted from the lesions, all based on texture.

5 Experiments and Results

In this Section, we present the experiments as well as the technique used to select characteristics.

In total, 16 characteristics were extracted from each image, all related to shape, colour, and texture. A Wrapper-based [26] approach was used to select the best subset of characteristics in order to classify the images. The 16 characteristics generates 65,535 combinations. The wrapper was used with a best-first search strategy. In order to exemplify the process, each algorithm was executed using all 16 characteristics; next, the same algorithm was executed removing one of the attributes at a time; if the removal of one attribute produced a model with smaller error rate, the wrapper continued generating models removing one single attribute of the remaining ones, until no improvement was found.

We performed experiments using 20 different classifiers, besides FuzzyDT [9], all available in the WEKA [27] framework. The tested algorithms were BayesNet, NaiveBayes, NaiveBayesUpdateable, MultiLayerPeceptron (MLP), SMO, Logistic, SimpleLogistic, IBK, KStar, LWL, DecisionStump,

(C4.5) J48, LMT, RandomForest, RandomTree, REPTree, DecisionTable, JRip, OneR, and PART. These algorithms were executed using their default parameters. For FuzzyDT, we used three fuzzy sets defining each attribute with a 25% confidence rate for the post pruning. All algorithms were performed using a 10-fold cross validation strategy. The carried out experiments are listed next.

It is important to justify that other state-of-the-art Fuzzy classification algorithms, such asFURIA [28] and IVTURS [29] were not selected for our experiments presented in this paper due to the fact that they produce rules with specific definitions for the fuzzy sets of each rule. In other words, the fuzzy database used for FuzzyDT cannot be compared to the fuzzy database of such algorithms because these algorithms do not use one single definition of fuzzy partition for all rules. As an example of this specificity of such algorithms, let us take the linguistic value defining ***high speed*** for cars in highways: for one rule of the dataset automatically generated by these algorithms, such linguistic value can be defined by a triangular membership limited from 80 to 100 kilometres per hour, while for another rule of the same dataset, the same linguistic value might be defined (automatically) by a triangular membership in the interval of 90 to 120 kilometres per hour. The comparison of the results of such algorithms cannot be considered directly using the same linguistic values. Thus, we intend to use such methods in our comparisons in future work.

The images were classified into three classes, common nevus, atypical nevus, and melanoma. Nevus is a non-specific medical term for a visible, circumscribed, chronic lesion of the skin. The list of experiments carried out are next:

- Experiments with all attributes of shape and colour;
- Experiments with all attributes of texture;
- Experiments with all attributes of shape, colour, and texture;
- Experiments with selection of attributes of shape and colour;
- Experiments with all attributes of texture;
- Experiments with all attributes of shape, colour, and texture.

We used 12 characteristics of shape and colour (asymmetry, pigment network, regression areas, white blue, points, stretch marks, white, red, light brown, dark brown, blue, and black) and four characteristics of texture (contrast, homogeneity, correlation, and energy).

Due to space limitations, we present only the results obtained by all algorithms using the characteristics of shape, colour, and texture (the last combination previously presented. Table 1 shows the obtained error rates. The best (smallest) error rates are highlighted.

The best results, from all combinations of experiments, were obtained by FuzzyDT using the following characteristics: homogeneity, correlation, energy, pigment network, stretch marks, white, red, dark brown, blue, and black. In fact, FuzzyDT obtained 100% accuracy with such characteristics.

Table 1. Error rates obtained using the characteristics of shape, colour, and texture.

Group	Algorithm	Error rate
Bayes	BayesNet	9.5%
Bayes	NaiveBayes	9.5%
Bayes	NaiveBayesUpdateable	9.5%
Function	MLP	10.0%
Function	SMO	7.0%
Function	Logistic	11.0%
Function	SimpleLogistic	10.5%
Lazy	IBK	11.0%
Lazy	KStar	17.5%
Lazy	LWL	19.5%
Trees	DecisionStump	21.5%
Trees	J48	11.5%
Trees	LMT	10.0%
Trees	RandomForest	8.0%
Trees	RandomTree	13.5%
Trees	REPTree	11.5%
Trees	**FuzzyDT**	**5.0%**
Rules	DecisionTable	9.0%
Rules	JRip	7.5%
Rules	OneR	21.5%
Rules	PART	9.0%

6 Final Considerations

In face of the great incidence of skin cancer, several researchers have been exploring different techniques to improve and accelerate its diagnostics. An early diagnostic increases the chances of cure of the disease considerably. Dermatoscopy is a possible resource for skin cancer detection. However, although this diagnostic analyses lesions based on extracted characteristics, such method is still subjective as it suffers interference from a number of factors. This way, the computational analysis of skin images have been researched as a means of supporting experts.

In this work, we focus on improving the precision of dermatoscopy using a methodology to classify skin lesion images with classification algorithms.

Several classification methods were tested with different combinations of characteristics. The best results were obtained by the FuzzyDT algorithm based on selected characteristics of shape, colour, and texture.

As a contribution of this work, we have created a new dataset [30] with the following characteristics: contrast, correlation, homogeneity, energy, asymmetry, pigment network, regression areas, white blue, points, stretch marks, white, red, light brown, dark brown, blue, and black.

As future work, we intend to compare the statistical relevance of the results using significance statistical tests. We also intend to expand the number of algorithms tested, including, the aforementioned FURIA and IVTURS, for instance. We also intend to adopt more skin lesion datasets. We also intend to include other search strategies for the Wrapper approach.

References

1. WHO: World Health Organization (2020). https://www.who.int/news-room/q-a-detail/protecting-oneself-from-skin-cancer. Accessed Oct 2020
2. INCA: Brazilian National Cancer Institute (2020). https://www.inca.gov.br/. Accessed Oct 2020
3. Soyer, H., Smolle, J., Kerl, H., Stettnre, H.: Early diagnosis of malignant melanoma by surface microscopy. Lancet **330**(8562), 803 (1987)
4. Alpaydin, E.: Introduction to Machine Learning: Adaptive Computation and Machine Learning. 3 edn. MIT Press, Cambridge (2014)
5. Bishop, C.M.: Neural Networks for Pattern Recognition. Oxford University Press - Claredon Press (2004)
6. Senge, R., Hüllermeier, E.: Fast fuzzy pattern tree learning for classification. IEEE Trans. Fuzzy Syst. **23**(6), 2024–2033 (2015)
7. Cintra, M.E., Monard, M.C., Camargo, H.A.: FCA-Based Rule Generator, a framework for the genetic generation of fuzzy classification systems using formal concept analysis. In: Proceedings of the FuzzIEEE 2015 - IEEE International Conference on Fuzzy Systems, vol. 1, pp. 1–8. IEEE (2015)
8. Quinlan, J.R.: C4.5: programs for machine learning. Morgan Kaufmann Publishers Inc., San Francisco (1993). http://portal.acm.org/citation.cfm?id=152181
9. Cintra, M.E., Camargo, H.A.: Feature subset selection for fuzzy classification methods. In: International Conference on Information Processing and Management of Uncertainty in Knowledge-Based Systems - IPMU, vol. 80, pp. 318–327 (2010)
10. Pedrycz, W., Sosnowski, Z.A.: Genetically optimized fuzzy decision trees. IEEE Trans. Syst. Man Cybern. Part B (Cybernetics) **35**(3), 633–641 (2005)
11. Klir, G.J., Yuan, B.: Fuzzy Sets and Fuzzy Logic - Theory and Applications. Prentice-Hall, Upper Saddle River (1995)
12. Cintra, M.E., Camargo, H.A., Monard, M.C.: Implementation of the fuzzyDT algorithm. https://www.researchgate.net/publication/341069396_FuzzyDT (2020)
13. Celebi, M.E., Zornberg, A.: Automated quantification of clinically significant colors in dermoscopy images and its application to skin lesion classification. IEEE Syst. J. **8**(3), 980–984 (2014)
14. Abuzaghleh, O., Barkana, B.D., Faezipour, M.: SKINcure: a real time image analysis system to aid in the malignant melanoma prevention and early detection. In: Southwest Symposium on Image Analysis and Interpretation (SSIAI), IEEE (2014)
15. Oliveira, R.B., Junior, C.R.D.C., Guido, R.C., Marranghello, N., Pereira, A.S.: Classificação de assimetria em lesoes de pele por meio de imagens usando maquina de vetor de suporte. VIII Workshop de Visão Computacional (2012)
16. Ganzeli, H.S., Bottesini, J.G., Paz, L.O., Ribeiro, M.F.S.: SKAN: skin scanner - system for skin cancer detection using adaptive techniques. IEEE Latin Am. Trans. **9**(2), 206–212 (2011)

17. Canny, J.: A computation approach to edge detection. IEEE Trans. Pattern Anal. Mach. Intell. **8**, 679–698 (1986)
18. Ruiz, D., Berenguer, V., Soriano, A., Sánchez, B.: A decision support system for the diagnosis of melanoma: a comparative approach. Exp. Syst. Appl. **38**(12), 15217–15223 (2011). http://www.sciencedirect.com/science/article/pii/S0957417411008633
19. Maurya, R., Singh, S.K., Maurya, A.K., Kumar, A.: Glcm and multi class support vector machine based automated skin cancer classification. In: 2014 International Conference on Computing for Sustainable Global Development (INDIACom), pp. 444–447 (2014)
20. Dermquest: Dermquest. https://dermaquestinc.com/ (Mar 2019)
21. Dermnet: Dermnet Skin Disease Atlas. http://www.dermnet.com/images/Melanoma-Skin-Cancer-Nevi-and-Moles/ (2020)
22. Albay, E., Kamaşak, M.E.: Improved classification of skin lesions using shape and color features. In: 2016 24th Signal Processing and Communication Application Conference (SIU), pp. 733–736 (2016)
23. Maia, L.B., et al.: Evaluation of melanoma diagnosis using deep features. In: 2018 25th International Conference on Systems, Signals and Image Processing (IWSSIP), pp. 1–4 (2018)
24. Mendonça, T., Ferreira, P.M., Marques, J.S., Marcal, A.R.S., Rozeira, J.: Ph2 - a dermoscopic image database for research and benchmarking. In: 2013 35th Annual International Conference of the IEEE Engineering in Medicine and Biology Society (EMBC), pp. 5437–5440 (2013). http://www.fc.up.pt/addi/ph2database.html
25. Vieira, C., Moraes, W.: MATLAB - Curso Completo. FCA, 1 edn. (2013)
26. Facelli, K., Lorena, A.C., Gama, J., Carvalho, A.C.P.L.F.: Inteligência Artificial: uma abordagem de aprendizado de máquina. LTC, 1 edn. (2011)
27. Witten, I.H., Frank, E., Hall, M.A.: Data Mining: Practical Machine Learning Tools and Techniques, 3rd edn. Morgan Kaufmann, Amsterdam (2011)
28. Hühn, J., Hüllermeier, E.: Furia: an algorithm for unordered fuzzy rule induction. Data Min. Knowl. Disc. **19**(3), 293–319 (2009)
29. Sanz, J., Fernández, A., Bustince, H., Herrera, F.: IVTURS: a linguistic fuzzy rule-based classification system based on a new interval-valued fuzzy reasoning method with tuning and rule selection. IEEE Trans. Fuzzy Syst. **21**(3), 399–411 (2013)
30. Maia, H.W.: Dataset with extracted characteristics of skin cancer images. http://dx.doi.org/10.13140/RG.2.2.12759.09120 (2020)

Breast Cancer Detection in Infrared Thermal Images Using Convolutional Neural Networks: A Transfer Learning Approach

Swapna Davies(✉) and S Anand

Department of Electronics and Communication Engineering,
Rajagiri School of Engineering and Technology, Kochi, India
anands@rajagiritech.edu.in

Abstract. Cancer is the most deadly disease by nature. Breast cancer is the top cancer in women globally and led to crucial increase in cancer fatality. Initial stage cancers do not cause death in human beings. Excessive work is carried out for the early diagnosis so that patients can get successful treatment. Even though health authorities recommend regular screening with mammography in women, infrared thermography can be a potential adjunct mechanism for identifying breast cancer. Deep learning techniques have opened up a new age in the image classification especially in the computer aided medical image analysis. A novel attempt combining deep learning and thermography for early stage breast cancer diagnosis is proposed in this paper. A deep learning scheme is adopted for the classification of cancerous and normal thermal rat images based on the transfer learning technique. For extracting features multiple deep pre-trained networks are utilized and classification is done with a support vector machine. The proposed method is also compared with a standard softmax classifier.

Keywords: Deep learning · Transfer learning · Pre-trained network · SVM · Softmax classifier

1 Inroduction

Cancer is a collective term for the group of diseases due to unrestrained expansion and bisection of cells. Breast cancer is very common in women ubiquitously in the world. Its rate of occurrence has regularly increased and led to a significant rise of cancer fatality in women [1]. Generally, breast cancer diagnosis is done at the later stages due to the inadequacy of organized breast cancer screening programs, lack of diagnostic aids and awareness, in developing and third world countries.

Cancer is healable if it is detected at the early stages. But if the cancer is spread out or metastasized, it is harder to treat successfully and creates heavy

A. Abraham et al. (Eds.): SoCPaR 2020, AISC 1383, pp. 63–74, 2021.
https://doi.org/10.1007/978-3-030-73689-7_7

burdens on the existing health systems. According to the survey of World Health Organization, around 14 million people develop cancer every year and it may reach above 21 million by 2030. The cancer survival rate can be increased by ensuring the health services focusing on diagnosis and treatment of the disease at the early stages. Early detection reduces the financial hardship of cancer. Treatment cost is very less in the initial stages and also people can pursue their job and make a living.

Breast cancer need not have any symptoms in early stages and in many cases a tumor is too small to sense. So a consistent screening is very crucial for early detection. Multiple methods are feasible for identifying breast cancer, such as mammograms, positron emission mammography, breast ultrasound scanning, magnetic resonance imaging etc. Among these, mammography being treated as the gold standard, is the most generally used breast cancer detection technique. But it exhibits low sensitivity in dense breast tissues or in young women. It needs breast compression during screening which is very painful and also needs exposure to harmful radiation. In mammography, 2 views are taken with an average radiation dose of 0.4 millisievert. Frequent exposure to this amount of radiation can even cause cancer. The setup and installation of such equipment is complex and costly. As a result, it is not viable and cannot cover huge population like in India.

Infrared thermography is a potential adjunct tool for detecting the breast cancer. Thermographic screening is especially beneficial for the people below 50 years, because mammography is less sensitive in younger ages. The cancer cells divide relentlessly, and hence biochemical reaction rate and blood flow are increased in tumorous area leading to a rise in temperature. The heat produced is transported to the skin surface. Infrared cameras with high precision are used to sense the skin temperature and produce high resolution images in thermography. The temperature irregularities can be the earliest demonstrations of malignant growth in breast.

Thermography results can be assorted into five, known as thermo-biological (TH) grading system as: TH-1: no unusual features, normal, TH-2: rise in heat, sensitive to cold challenge, normal, TH-3: typical rise in heat, insensitive to cold challenge, TH-4: abnormal rise in heat, insensitive to cold challenge and TH-5: severely abnormal rise in heat, insensitive to cold challenge [2]. If abnormalities are detected then further investigations are needed which may include other medical examinations. Thermography can be carried out in conjunction with mammography and a clinical exam which can detect around 95% of early stage cancers.

Since our body exhibits contralateral symmetry, one common method for breast abnormality detection is the asymmetry analysis between the right and left segments. Thermography identifies clear breast anomalies on the basis of symmetry. It is very simple, extremely easy as taking a photograph, ensures 100% safety because no harmful radiation is involved, so can be repeated any number of times. It is painless since bosom compression is not needed during thermographic test. Thermography can detect the physiological changes related with the breast cancer a few years in advance compared to other screening modalities [3]. For women belonging to any age with breast implants and or with thick bosom tissues, thermography can be used. It is insensitive to hormonal changes. Thermography has got some flaws too. Being the thermal picture of the skin, it cannot localize a tumor area for surgical biopsy. The false negative cases are relatively high since it cannot detect microcalcifications [4] which avoids further screening leading to fatality.

In this paper we explore the possibilities of deep learning techniques for the investigation of breast cancer using in vivo thermal images. The feature extraction is done with deep pre-trained networks AlexNet, VGG 16, GoogLeNet (Inception V3), Residual Network (ResNet 50) and Inception ResNet V2 and classification is done with a support vector machine or SVM. The SVM classifier is compared with the standard fully connected layer succeeded by a softmax classifier and the performance evaluation is done in each case.

Deep learning techniques particularly exceptional progress have been accomplished by convolutional neural networks or CNN in medical imaging domain. Some of the selected previous works in this area are listed below.

Sahiner et al. in 1996 classified the region of interest (ROIs) obtained from mammograms as malignant or non-malignant using CNN [5]. The extracted texture features from multiple sub regions of a single ROI were given as the CNN inputs. The Grey Level Difference Statistics features and Spatial Grey Level Dependence features contains the texture changes in an ROI. The classification accuracy was compared for various network architectures and texture features. Since the study was done in the pre-GPU time, the training times are not given.

Meinel et al. in 2007 designed a Computer Assisted Diagnosis (CAD) system to assist the medical practitioners in the investigation of breast MRI lesions [6]. The system has a semiautomated segmentation approach, comprising of feature computation and selection mechanism and neural network for classifying the biopsy proven lesions. A comparative study using Receiver Operating Characteristic or ROC curves was made on the classification done by the radiologists with and without the support of the CAD system. Remarkable performance improvement can be attained if human readers are assisted by the developed system irrespective of their experiences. A minor drawback is that more CPU time is spent during the training stage of the network.

Bhandari H. in 2015 used bag of features approach to detect breast cancer in histopathology images [7]. Scale Invariant Feature Transform was used to draw out the features and clusters of local features of training data were formed using K-means algorithm. The centroids of clusters represent visual codewords. Classification is performed by comparing the codeword frequency histogram with that of the training images based on Euclidean distance. In the experiment 74.28% normal samples and 90.39% cancerous samples are classified correctly.

Sun et al. in 2017 designed a deep CNN to identify breast cancer using a Semi-Supervised Learning (SSL) technique based on graph theory [8]. One major issue in medical imaging domain is the non-availability of labelled training data. The labelled data used was small in number compared to un-labelled data.

The developed SSL model comprises of 3 modules to label the unknown data: data weighing to mitigate the effects of outliers, feature selection and a dividing co-training which straggles and partitions the available labelled data into subgroups to train a set of classifiers. The final labelled data set is used for training as well as testing the CNN. The accuracy reported is 0.8243 and area under the curve is 0.8818 from the developed network established on LeCun's model [9].

Khan et al. in 2019 [10] developed a deep learning structure for classifying the cytology images of breast using transfer learning approach to achieve higher accuracy and low cost using few images. The most common CNN architectures ResNet, VGGNet and GoogLeNet are used for extracting the features. The cells are classified with a fully connected or FC layer from the combined features. This method achieves a better accuracy of detection compared to the individual architectures.

Javaria Amin et al. in 2019 used stacked Autoencoders in brain cancer detection in MR images [11]. Segmentation is done with seed growing mechanism and slices are classified using a model with two hidden layers and a softmax layer. The system achieves very high accuracy on different datasets with a minimum value of 0.9.

2 Methodology

A deep learning scheme is proposed in the study for classifying the malignant and normal thermal rat images as shown in Fig. 1. It is based on the transfer learning technique because of the unavailability of large data sets. As an initial step, data augmentation is used to add more number of images since a large data set is crucial for a deep learning architecture to perform well. The most powerful CNN architectures AlexNet, Inception V3, VGG 16, ResNet 50 and Inception ResNet V2 are employed for feature extraction. Finally classification is done from the extracted features with a linear SVM.

2.1 Transfer Learning

In transfer learning [10] a pretrained model designed for a task is used to develop another network for a related computer vision task. i.e. first a base structure is trained with a base task and data and then transfer the learned features to a target network with a target task and data.

Transfer learning is practiced in predictive modelling problems in two ways: develop model approach and pre-trained model approach. In the first approach a related source task with large amount of available data sets is selected and a proficient model is developed. This is the base of the model to be designed for the target task. Voluntarily it can be fine tuned with the target data sets. In the pre-trained model approach a suitable pretrained model is selected from the existing models and used as base network for the target task. This can be also refined with the target data sets. The second type is more prevalent in the domain of deep learning.

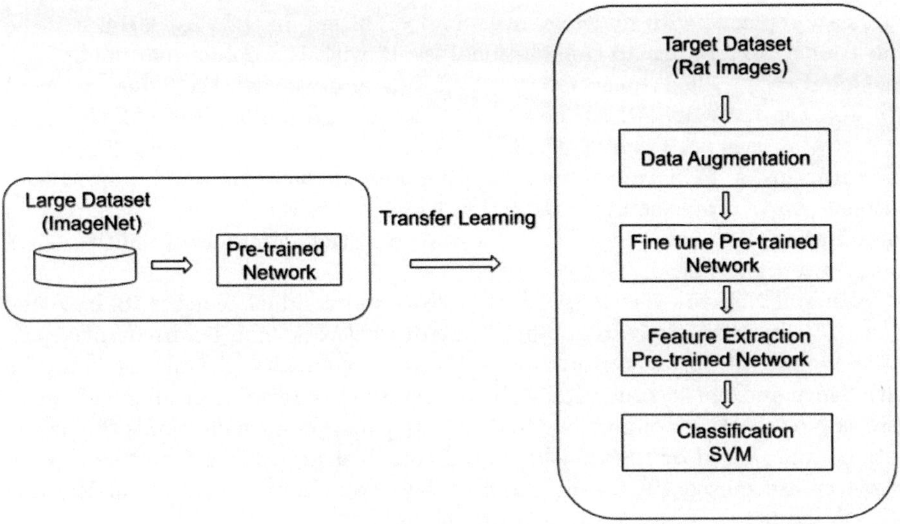

Fig. 1. Block diagram

2.2 Data Augmentation

Image augmentation is required to develop a dominant classifier when images in training data set is small [12]. With a small data set, convolutional neural networks learn irrelevant features in training and performance declines due to overfitting. Overfitting means the generated model operates well on training data but fails on test data. More data can be generated by making minor alteration to the existing data such as flips, rotation, shear and translation or a sequence of these multiple processing methods. A CNN is invariant to the geometric transformations, so it can robustly classify the objects with augmented data.

2.3 Pretrained Architectures for Feature Extraction

We exploit five different deep CNN architectures having different width and depth built up using the ImageNet for the feature extraction. ImageNet is a huge image dataset used for ocular object identification research organized as per the WordNet hierarchy. AlexNet, VGG 16, Inception V3, ResNet 50 and Inception ResNet V2 are the pre-trained networks used for extraction of features.

AlexNet is among the first deep CNN architectures to win ImageNet Large Scale Visual Recognition Challenge (ILSVRC) with an accuracy 84.7% in 2012. It is 8 layers deep, formed by stacking 5 Convolutional layers (11×11, 5×5, 3×3) and 3 FC layers with 60 million parameters. Rectified Linear Unit is the activation function used as it converges faster.

VGGNet is built to curtail the parameter count in CNN and all the convolutional kernels are of size 3×3. It is similar to AlexNet but 11×11, 5×5, 3×3

filters are replaced with multiple layers of 3×3 filters. In ILSVRC 2014, it is the first runner up and has 16 convolutional layers with 138 million parameters.

GoogLeNet called Inception V1 (V2, V3 and V4 are the later versions) has won the ILSVRC 2014. The architecture is quite unlike from AlexNet and VGGNet, it uses 1×1 convolutions at the middle, global average pooling at the end rather using FC layers and an inception module technique. The 1×1 convolutions help in dimensionality reduction which again reduces the complexity in computation. The same input is given to convolutional kernels of multiple sizes in the inception module.

Residual Network has won ILSVRC 2015 and residual block is its base element [13]. In principle as the depth increases networks can learn complex patterns leading to better performance. But the complexity of training increases with depth and the accuracy starts degrading due to vanishing gradients. ResNet uses skip or shortcut connections to do identity mapping which avoids the degradation issue caused by the vanishing gradients. The activations from the preceding layer are reused till the neighboring layer weights are learnt and learning takes place faster compared to a plain network.

The performance of ResNet stimulated the introduction of a hybrid Inception-ResNet architecture. Residual connections are added to inception architecture to speed up training.

2.4 Classification

There are different approaches for building the classifier. 1) The standard practice is to use a heap of FC layers and a softmax layer [14]. The highest probability class is given by softmax layer from the probability distribution computed over the possible classes. 2) Global average pooling and softmax layers. 3) SVM which returns a score for every input data as the basis of evaluation [15]. The proposed method uses SVM classifier and a performance comparison is made between SVM and softmax layer in succession with FC layer.

Softmax Layer. Softmax classification layer transforms all activations from the last layer into probabilities. The softmax function has got several advantages compared to the hard max function as: 1) It is differentiable and hence used in gradient descent networks. 2) The output values are from 0 to 1 and thresholding can be done easily. 3) The outputs are bounded and can be considered as relative probability distributions.

For a K class problem, the softmax layer has K output nodes p_i where $i = 1, 2, .., K$ and $\sum_{i=1}^{K} p_i = 1$, p_i being the probability.

Let $h_1, h_2, ..., h_N$ be the activations from the output layer and W is the weight vector connecting the output layer and the softmax layer. Let the softmax layer inputs are given by

$$a_i = \sum_{n=1}^{N} h_n w_{ni} \tag{1}$$

Then the outputs are given by

$$p_i = \frac{\exp(a_i)}{\sum_{j=1}^{K} \exp(a_j)} \tag{2}$$

The predicted class would be

$$i = arg\, max_i\, p_i = arg\, max_i\, a_i \tag{3}$$

Support Vector Machines. The SVM is a supervised training model utilized in regression as well as classification analysis. In classification problems every input data is mapped to a point in the D dimensional feature space. The algorithm creates a maximum margin hyper-plane which separate the input vectors to the respective classes.

Let (x_n, t_n) represents training sample set where $x_n \in R^D$ is an input pattern and $t_n \in (-1, +1)$ represents the corresponding label. The hyperplane is constructed by minimizing a cost function,

$$\phi(W, \zeta) = \frac{1}{2} W^T W + C \sum_{n=1}^{N} \zeta_n \tag{4}$$

subjected to the constraints $W^T x_n t_n \geq 1 - \zeta_n, \forall n$ and $\zeta_n \geq 0, \forall n$, ζ_n are the slack variables to account for the data points that are not within the correct margin boundary, W is the weight vector and C is a regularizing parameter.

3 Results and Discussion

3.1 Dataset Description

The thermal images used are obtained from Clinical Animal Laboratory of the Faculty of Bioscience and Medical Engineering (FBME) at the Universiti Teknologi Malaysia in Johor, Malaysia. Sprague-Dawley female rats of two classes comprising of fifteen experimental (mammary cancer induced) rats and fifteen control (normal) rats are availed in the experiment. A thermal camera having 640 × 480 pixel resolution, is used for taking snapshots (7640 P-series). The rats were screened for 12 weeks and histology test was used to confirm that the hot spots found in pathological group are due to cancer in mammary gland. Thermal image models used in the investigation are shown in Fig. 2. The dataset and source code are available at https://drive.google.com/drive/folders/1Y_RcH1kHzFaoidAefx4LzdE8fs6lnFwj.

3.2 Result Analysis

In a binary classification problem, the outcomes can be labelled as positive and negative i.e. pathological and normal in this study. There are four possible outcomes, true positive TP, true negative TN, false positive FP (False alarm or

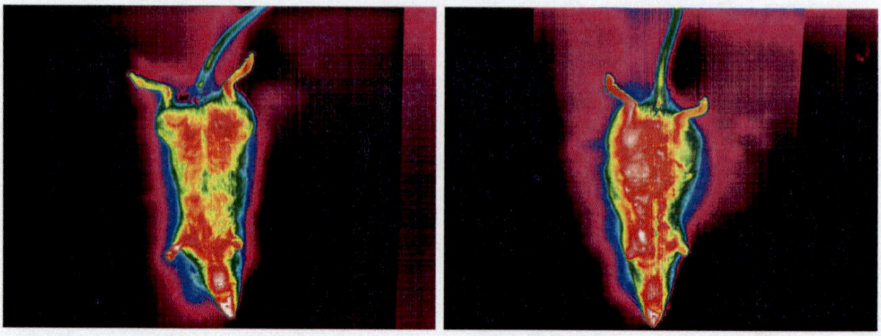

(a) Normal Rat (b) Pathological Rat

Fig. 2. Thermal images

type I error) and false negative FN (Missed detection or type II error). Here the true negatives are those actual healthy subjects which are also being predicted as healthy and true positives are those sick subjects which are being predicted as sick itself. False negatives are those sick subjects which are being predicted as healthy and false positives are those healthy subjects which are being predicted as sick.

The performance comparison of various pre-trained networks of the suggested model with SVM classifier and standard softmax classifier are listed in Table 1 and Table 2. The performance metrics used are Accuracy, Sensitivity, Specificity, Precision, F1 score and Area Under Curve or AUC. Accuracy is percentage of subjects which are rightly classified to the total of all subjects, and in deciding a classifier performance it is the most perceptive one. Count of positive labelled subjects divided with the total actual positive subjects gives sensitivity. Negative labelled subject count to total count of actual healthy subjects gives specificity. Precision is the ratio of subjects rightly positive labelled to all positively labelled subjects by the algorithm. The harmonic mean of sensitivity and precision is the F1 Score, which is a balance between sensitivity and precision. F1 Score is a better performance measure than accuracy, particularly for uneven class distribution. The numerical values for each of these parameters as defined below are obtained by plotting the Confusion Matrix which is a table describing the performance parameters of a classifier.

$$Accuracy = \frac{TP+TN}{TP+FN+FP+TN}, \quad Sensitivity = \frac{TP}{TP+FN}, \quad Specificity = \frac{TN}{FP+TN}$$

$$Precision = \frac{TP}{TP+FN}, \quad F1Score = \frac{2*Sensitivity*Precision}{Sensitivity+Precision}$$

ROCs are plotted in Fig. 3 and AUC is computed for each case. ROC curve or AUC measures the discrimination potential of a predictive classification model. AUC computes the two dimensional area beneath ROC curve from (0,0) to (1,1), calculated by integrating true positive rate with respect to false positive

Table 1. SVM classifier

Serial number	Performance metrics	AlexNet 2012	VGG 16 2014	Inception V3 2015	ResNet 50 2015	Inception ResNet V2 2016
1	Accuracy	0.89	0.83	0.82	0.93	0.81
2	Sensitivity	0.938	0.891	0.797	0.938	0.844
3	Specificity	0.844	0.781	0.859	0.922	0.781
4	Precision	0.857	0.803	0.85	0.923	0.794
5	AUC	0.947	0.908	0.849	0.978	0.889
6	F1 Score	0.895	0.873	0.822	0.93	0.818
7	Time(s)	16.93	93.81	82.55	48.72	119.26

rate using the trapezoidal method. Its value ranges from 0 to 1. The results are obtained using processor Intel i7 -7500u @2.7 GHz, 8 GB RAM, 1TB Hard disk in MATLAB R2018b. A fixed seed for generating random number is used to get reproducible results.

Table 2. Softmax classifier

Serial number	Performance metrics	AlexNet 2012	VGG 16 2014	Inception V3 2015	ResNet 50 2015	Inception ResNet V2 2016
1	Accuracy	0.86	0.83	0.888	0.888	0.833
2	Sensitivity	1	0.944	0.833	0.944	0.889
3	Specificity	0.722	0.722	0.944	0.833	0.778
4	Precision	0.783	0.773	0.938	0.855	0.80
5	AUC	0.988	0.978	0.935	0.99	0.959
6	F1 Score	0.878	0.849	0.882	0.897	0.842
7	Time(m)	1.53	53.27	22.35	17.39	492.37

With the proposed SVM classifier, ResNet 50 is giving the best result with regard to Accuracy, Sensitivity, Specificity, Precision, F1 Score and AUC by setting a threshold of 90% for each metric, compared to the other pretrained networks Alexnet, VGG 16, Inception V3 and Inception ResNet V2. ResNet 50 provides an Accuracy of 93%, Sensitivity 93.8%, Specificity 92.2%, Precision 92.3%, F1 score 93% AUC 0.978 and needs processing time of 48.72s with SVM classifier. In cancer diagnosis minimizing the risk is not bounded to simply minimizing the number of misclassifications. If a patient with no disease is incorrectly diagnosed as having disease, the consequences are less serious than if a patient with disease is diagnosed as healthy. The latter error can be fatal. 93.8% of Sensitivity ensures remarkable accomplishment as the occurrence of false negatives is highly intolerable. The performance of the pretrained architectures with standard softmax classifier is also given for comparison. It is noted that the pre-trained networks ensures better performance with SVM classifier.

A balanced dataset i.e. equal number of normal and cancerous images, is used for conducting the experiments. With an imbalanced dataset i.e. 20% normal and 80% cancerous images, the performance metrics are given for SVM classifier with ResNet 50 in Table 3. The specificity or true negative rate is highly reduced even

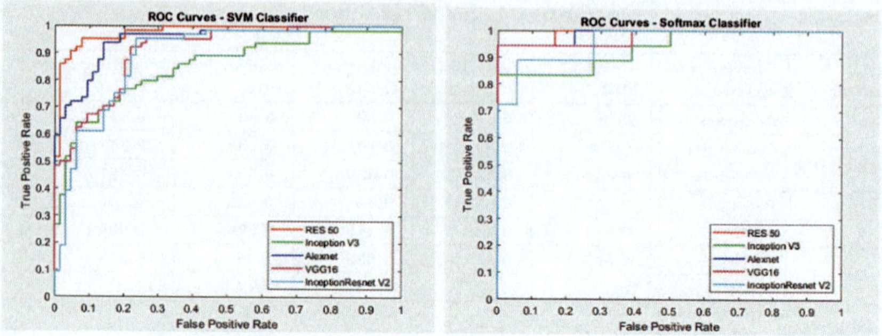

(a) SVM (b) Softmax

Fig. 3. ROC curves

Table 3. ResNet 50 and SVM classifier performance with imbalanced dataset

Accuracy	Sensitivity	Specificity	Precision	AUC	F1 Score	Time(s)
0.884	0.96	0.579	0.9	0.938	0.929	52.58

though other metrics remain good. Since the class distribution is skewed with imbalanced data, a balanced dataset is always preferred.

The classification using Bag of Features and Stacked Autoencoders are also given in Table 4, which shows that the projected technique in this paper achieves impressive performance.

Table 4. Comparison with other existing methods

	Accuracy	Sensitivity	Specificity	Precision	F1 Score	Time(m)
Bag of Features 2015	0.82	0.71	0.93	0.91	0.797	5.32
Stacked Auto-encoders 2019	0.758	0.75	0.768	0.763	0.756	22.56

4 Conclusion

Deep learning techniques have started a major breakthrough in the medical field. In this paper we have explored transfer learning technique using five different deep pre-trained networks for removing features and classification is done with SVM. A performance comparison is given with standard softmax classifier. ResNet 50 performs the best with SVM classifier. Even though comparable performance is achieved with softmax classifier with most of the pretrained networks, SVM classifier proves to achieve remarkable performance. The results are also compared with classification using bag of features and stacked autoencoders methods.

Analyzing the performance of various pretrained networks, in future a deep learning structure exclusively for investigating breast cancer from scratch is to be developed from a large data set. The difference between healthy and malignant tissue is very subtle in medical images. To bring out the hidden minor details, images to be enhanced to provide better input to the designed network.

Acknowledgement. Authors thank Clinical Animal Laboratory of the Faculty of Bioscience and Medical Engineering (FBME) at the Universiti Teknologi Malaysia in Johor, Malaysia for providing the data set.

References

1. Wahab, A.A., Salim, M.I.M., Aziz, M.N.C.: In vivo thermography-based image for early detection of breast cancer using two-tier segmentation algorithm and artificial neural network. In: Ng, E., Etehadtavakol, M. (eds.) Application of Infrared to Biomedical Sciences. Series in Bio-Engineering, vol. 1, pp. 109–131. Springer, Singapore (2017)
2. Ramesh O., Ali K., Sadaf A., Masoume N., Mansour A., Mitra N., Afsaneh A., Nasrin A., Khojasteh B., Shahrzad I.: Comparison of the accuracy of thermography and mammography in the detection of breast cancer. Breast Care Multi. J. Res. Diagn. Ther. Breast Dis. **11**(4), 260-264 (2016)
3. Tan, T.Z., Quek, C., Ng, G.S., Ng, E.Y.K.: A novel cognitive interpretation of breast cancer thermography with complementary learning fuzzy neural memory structure. Expert Syst. Appl. **33**(3), 652–666 (2007)
4. Ng, E.Y.K., Fok, S.C., Peh, Y.C., Ng, F.C., Sim, L.S.J.: Computerized detection of breast cancer with artificial intelligence and thermograms. J. Med. Eng. Technol. **26**(4), 152–157 (2002)
5. Sahiner, B., et al.: Classification of mass and normal breast tissue: a convolution neural network classifier with spatial domain and texture images. IEEE Trans. Med. Imaging **15**(5), 598–610 (1996)
6. Meinel et al.: Breast MRI lesion classification: improved performance of human readers with a backpropagation neural network computer-aided diagnosis (CAD) system. J. Magn. Magn. Mater. **25**(1), 89–95 (2007)
7. Bhandari H: A bag of features approach for malignancy detection in breast histopathology images. In: IEEE International Conference on Image Processing, IEEE, Quebec City, QC, Canada (2015). https://doi.org/10.1109/ICIP.2015.7351745
8. Sun, W., Tseng, T.L.B., Zhang, J. and Qian, W.: Enhancing deep convolutional neural network scheme for breast cancer diagnosis with unlabeled data. Comput. Med. Imaging Graph. **57**, 4–9 (2017)
9. LeCun, Y., Bottou, L., Bengio, Y., Haffner, P.: Gradient-based learning applied to document recognition. Proc. IEEE **86**(11), 2278–2323 (1998)
10. Khan, S., Islam, N., Jan, Z., Din, I.U., Rodrigues, J.J.C.: A novel deep learning based framework for the detection and classification of breast cancer using transfer learning. Pattern Recogn. Lett. **125**, 1-6 (2019)
11. Amin, J., et al.: Brain tumor detection by using stacked autoencoders in deep learning. Image Signal Proc. J. Med. Syst. **44**, 32 (2019)
12. Shorten, C., Khoshgoftaar, T.M.: A survey on image data augmentation for deep learning. J. Big Data **6**(1), 1–48 (2019)

13. He, K., Zhang, X., Ren, S., Sun, J.: Deep Residual Learning for Image Recognition. Computer Science, Computer Vision and Pattern Recognition, arXiv:1512.03385. (2015)
14. Qi, X., Wang, T., Liu, J.: Comparison of support vector machine and softmax classifiers in computer vision. In: International Conference on Mechanical, Control and Computer Engineering. IEEE (2017)
15. Yichuan Tang: Deep Learning using Linear Support Vector Machines. Computer Science, Machine Learning, arXiv:1306.0239. (2013)

Implementation of a System for Collecting and Analyzing Public and Open Access Information from Brazilian Federal Legislative Houses

Iván José Mestre Fernández(✉) and Leonardo de Souza Mendes

Universidade Estadual de Campinas UNICAMP, Campinas, SP 13083-970, Brazil
i209446@dac.unicamp.br, lmendes@unicamp.br

Abstract. The overspread of corruption in today's world is considered one of the major global problems, one that has affected almost every country in the world, developed or underdeveloped, throughout history. The specific case of institutionalized corruption, normally associated with the political structure of a country, is particularly heinous because it can be very difficult to expose and fight. In the case of Brazil, in recent years, major corruption scandals have arisen and completely shaken the national political environment. For centuries society was left defenseless in the face of such problems; however, supported by recent legislation on government compliance and transparency, the use of Data Mining and Data Analytics can offer important tools for society in the combat of these old and condemned practices of corruption. This article presents the implementation of an information retrieval system using Web Data Mining techniques, to collect information of public interest from politicians and political processes in Brazil, using public and open access information from the Brazilian federal legislative houses.

Keywords: Web mining · Data mining · Politics · Web crawling

1 Introduction

Political corruption is a global problem that has affected almost every country in the world, developed or underdeveloped, throughout history. In recent times, due to the media globalization and the emergence of social networks, political corruption scandals in various parts of the world have become events of people's daily life. In the specific case of Brazil, in recent years, major corruption scandals have arisen that and completely shaken the national political environment. These scandals gave rise to gigantic investigations that covered all levels of Brazilian politics. Examples of these scandals are the "Mensalão Scandal" and Operation "Lava Jato". Corruption has been mentioned as one of the main factors that triggered the 2013 protests across the country [1].

A. Abraham et al. (Eds.): SoCPaR 2020, AISC 1383, pp. 75–84, 2021.
https://doi.org/10.1007/978-3-030-73689-7_8

The "Mensalão" Scandal was the name given to the judicial process that investigated the illegal purchase of votes in the Brazilian Federal Congress during the first term of former President Luiz Inácio Lula da Silva. The scandal happened in 2005 when a congressman publicly accused the Partido dos Trabalhadores (PT – Workers Party) of having paid R$ 30,000 (about USD $11,000) a month to political allies since 2003 to ensure that they voted in favor of laws and amendments proposed by the PT. It was considered, at the time, the biggest case of corruption in the recent history of Brazil and was about to collapse the government of Lula [2].

Operation "Lava Jato" is an ongoing criminal investigation carried out by the Brazilian Federal Police and presided over by Judge Sérgio Moro, since March 17, 2014 and by Judge Luiz Antônio Bonat, since 2019. Its goal is to investigate a money laundering and bribing scheme suspected of moving more than R$ 10 billion (about USD $1.8 billion). It is considered by the Federal Police as the largest corruption investigation in the history of Brazil [3, 4].

According to the Corruption Perception Index, in 2019, Brazil occupied position 106 in the rank of corruption, tied with several other countries, like Albania, Ivory Coast, and Egipt. This index is prepared annually by the German Non-Governmental Organization *Transparency International*, whose objective is to combat global corruption and prevent criminal activities that are the product of corruption [5].

The democratization of Internet access has allowed the sharing of information at unprecedented levels in history. This has been used by different groups, motivated by economic or other interests, for the massive generation of fake news. Fake News has become a major global problem that has caused a considerable decline in public trust in traditional media. Many people today believe that mainstream media is more concerned with attracting a larger audience than with suitably informing the population, sacrificing accuracy to be the first to report a story or to distort the news to support a specific ideology or narrative.

For all these reasons, it is necessary to develop technological tools that allow citizens to obtain factual knowledge based on real data about the main members and the functioning of the government. For example, it is necessary to have ways to track donations made to political election campaigns and the relationship these donations have to subsequent actions taken by the politicians. In addition, they must help people exercise their right and duty as citizens to vote for their representatives, but also to track these representatives acts afterwards in order to ensure the fulfillment of their campaign promises. For the creation of such a system, it would be necessary to use various information retrieval and analysis techniques, mainly regards to Web Data Mining and Web Data Analytics.

2 Web Data Mining

There is no consensus on the exact definition of Web Data Mining. According to Etzioni [6], one of the first concepts we can find, this can be defined as:

- Integration of information obtained through traditional data mining methods with information collected on the web.
- Discovering interesting patterns in the structure, content and use of websites.

Therefore, a more detailed definition of Web Data Mining can be stated as: Web Data Mining is a complex process that includes analyzing a wide variety of information, such as the content and structure of web documents (*HTML*, *XML*), text files, databases, user access logs, reference logs from other servers, user profiles, among many others, for the purpose of finding useful and relevant information, as well as finding new knowledge that will serve a user's needs and decision making process.

Web Data Mining mainly deals with the following problems [7]:

- **Discover relevant information:** People use search engines to find specific information. Current search tools have low accuracy problems due to the irrelevance of a large number of search results, as well as the inability to index all the information available on the web. This makes it difficult to find relevant and useful information for users.
- **Creating new knowledge based on available information:** It appears as a subproblem of the previous one. It is assumed that there is a large collection of data available on the web and there is a need to extract potentially useful knowledge from it.
- **Information personalization:** There is a difference in the preferences and content presentation of the different users who interact with the web.
- **Learn about users or consumers:** This problem is about understanding the actions and desires of users and consumers. According to the way in which they use the web, it is possible to discover information about users that allows the performance of various actions such as the classification of users, personalization of services, prediction and suggestion of actions, among many others.

The Web Data Mining process is similar to the traditional Data Mining process, which is composed of the following steps [8]:

- **Resource collection:** In this phase, we gather from the web the resources to which Data Mining will be carried out. These resources can be collected from several web pages or complete websites. Within these resources there may be different types of data of interest, such as text documents, images, audio, or videos. At this stage, it is customary to use various information retrieval techniques, this task is known as *crawling*.
- **Selection of information:** This step performs the process of extracting the specific information found in each resource and the information contained on the page that is not relevant to the Web Data Mining process is discarded. This step is composed of selection, filtering, and pre-processing of information tasks, it is known as *scraping*.
- **Generalization:** In this phase, we extract the patterns of interest, trends and new knowledge from the information repositories identified and collected during the first two stages. These patterns and trends can be extracted using supervised and unsupervised Data Mining techniques.
- **Analyze:** This step performs the validation and interpretation of the patterns found in the previous phase. It is at this stage that it is possible to distinguish relevant patterns, trends and new knowledge and add value to the understanding of the information repositories used during the process.

Depending on the source of information used and the knowledge you want to obtain, Web Data Mining can be classified into three main areas of interest. These areas are [7, 8]:

- **Content Mining:** Web Content Mining is the process of extracting useful information from web documents' content, such as pages or websites. In this area, Web Data Mining techniques are applied directly to the content of pages or websites, be it text, images, videos, etc., to identify patterns that allow the creation of new knowledge or complement existing information. It is particularly useful in the tasks of classification, grouping, retrieving specific information, and analyzing content created by users. This allows the understanding of the perception of different elements, such as products or people.
- **Structure Mining:** Web Structure Mining is the process of discovering information about the web structure using the hyperlinks that connect the different pages and websites. When performing an analysis of links between different pages and sites, it is possible to extract relevant information that allows the understanding on how various sites relate to each other, as well as discovering attempts to deceive search engine positioning algorithms.
- **Usage Mining:** Web Usage Mining is the process of extracting useful information about users' browsing and access patterns on pages and websites using access logs. This information is useful for understanding the user's actions within a web system, which can be used to improve a system's structure, prediction, and recommendation of actions, among others.

2.1 Web Content Mining

Web Content Mining can be defined as [8]: "The process of extracting useful information and new knowledge from content on pages and sites available on the *World Wide Web*". These pages usually contain objects with content in different formats such as text, images, audio, or video. Sometimes, each page can correspond to a single document, where the content is represented mainly in a single type of data. But in other cases, the pages contain multiple content objects, each of which can be found in different formats.

Generally, Web Content Mining focuses on exploring, transforming, and extracting information from text objects. This is because the exploration, processing, and extraction of information from other formats, such as images, audio and video are non-trivial tasks. That is why the content mining process often uses text-based techniques, even when content objects are in multimedia formats. Typically, multimedia objects are labeled with text describing the contents of these objects, this can be done manually, semi-automatically or fully automatically. So, when doing the process of exploring, processing, and extracting knowledge from these multimedia objects, Content Mining techniques based on the texts of these labels are used [8].

The Web Content Mining process generally addresses four major tasks [8]:

- **Extraction of specific information from pages and websites**
- **Grouping web documents into similar document sets**

- **Categorizing of web documents into predefined categories**
- **Performing Sentiment Analysis or Opinion Mining on blogs and forums**

These tasks require a first step of collecting the resources needed for the process, where the Web is scanned to collect the resources that are of interest to the Data Mining process. When resources have been collected, a pre-processing of the resources is done to be able to build a representation of the content that allows the application of Data Mining techniques.

3 Implementation

In order to solve the problems presented above, it is first necessary to implement an information retrieval system that satisfies the requirements to feed the mining and analysis processes. This system, currently under development, is called L5P and uses several of the techniques for retrieving and analyzing information described above.

Before starting the implementation of tools that allow to collect information from Web sources or other remote repositories, it is necessary to choose the sources of information that will feed the system. After extensive research, the following were found:

- **The "Câmara dos Deputados" website** (https://www.camara.gov.br): This website is an official source of government information and has a wide variety of information about the congressmen. Data on political trajectory as well as legislative performance can be obtained, including voting records, speeches, etc.
- **The "Divulgação de Candidaturas e Contas do Tribunal Superior Eleitoral" website** (https://divulgacandcontas.tse.jus.br): This website is an official source of government information and has a wide variety of information about the elections and their candidates, as well as the accounts and assets that the candidates declared.
- **The "Prestação de Contas do Tribunal Superior Eleitoral" website** (https://inter01.tse.jus.br): This website is an official source of government information and contains information on donations for political election campaigns.
- **"Atlas Político"** (https://www.atlaspolitico.com.br): This website is a private site and contains information that complements the information collected from official sources.
- **The "Portal Brasileiro de Dados Abertos"** (https://dados.gov.br/): This website is an official source of government information and contains a wide variety of information about the functioning of the government.
- **The "Meu Congresso Nacional" website** (https://meucongressonacional.com/): This website is a private site and contains information that complements the information collected from official sources.

Once the sources of information to be used have been identified, they can be passed on to the information collection and consolidation phase.

During the information collection process, different sources were analyzed, and various strategies were developed to collect the necessary data. Several of the above sources have web services or data query APIs. This is the case of the **Câmara de Deputados** website, **O Portal Brasileiro de Dados Abertos** website and the **Meu Congresso Nacional** website. These services are found in different formats and structures, so it was necessary to implement customized functions for collecting and structuring data. The **Meu Congresso Nacional** website and the **Câmara de Deputados** website use web services in XML format, while the **Portal Brasileiro de Dados Abertos** website uses web services in JSON format.

Other sources, such as the **Prestação de Contas do Tribunal Superior Eleitoral** website, provide the public with spreadsheets in CSV format. Specific methods were created to download and subsequently process each of these worksheets. However, the rest of the information sources do not have structured data query methods, but they do have a large amount of information within the content of their pages. It was necessary to implement custom *crawlers* to extract information from the **Atlas Político** and the **Divulgação de Candidaturas e Contas do Tribunal Superior Eleitoral** websites. These *crawlers* have the function of identifying the relevant information within the HTML content of the analyzed pages and extracting the information that may be useful for further analysis. To do this, it is necessary to first perform an analysis of the page structure to identify the HTML *tags* where the information to be collected is found. It is sometimes necessary for *crawlers* to navigate through a hierarchy of these HTML *tags* until they reach the desired content. Each page of these sites needs a *crawler* specifically implemented for them; this is because of the difference in the presentation of data on each page.

After the end of this process, the information is passed to the data structuring phase. The data were structured in the following entities, which will represent tables in the Database:

- **Pessoa Física (Natural person):** This entity represents a natural person; this entity will store the data of congressmen and natural person donors.
- **Pessoa Jurídica (Legal entity):** This entity represents a legal entity; it will store the data of donors that are companies.
- **Fato Social (Social Fact):** This entity represents an event or fact that occurs, and it will store the data from Election Campaigns, Legal Proceedings and Donations.
- **Ato (Act):** This entity represents a document, and it will store the data of documents, as well as websites that have information about other entities.
- **Lugar (Place):** This entity represents a place; it will store the data of the States or Places mentioned in the data.
- **Medida (Measure):** This entity represents a measure; it will store the donations data and other values present in the data.
- **Coisa (Thing):** This entity represents a thing; it will store the data that cannot be categorized in other entities.

With these entities it is possible to represent the whole known world, so it is a general solution that allows to adapt to any type of data that needs to be saved. Each of these entities can have many-to-many relationships with any of the other entities, as well as with itself, as seen in Fig. 1.

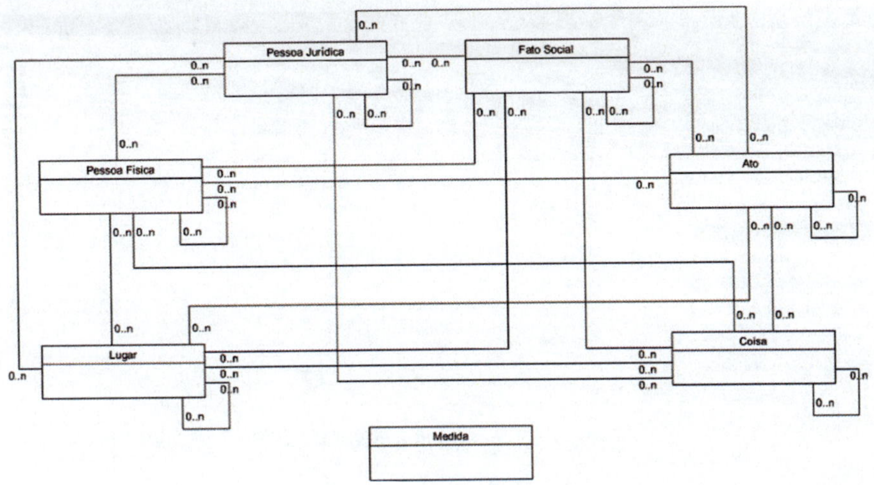

Fig. 1. System entities.

4 Results

Once the data is structured, it is possible to proceed to the process of visualizing and analyzing the data. Several screen shots of the application where the collected data is displayed will be shown. Personal data was censored to protect people's privacy.

Figure 2 shows the general data collected by L5P, such as: Name, Nickname, CPF, Party, Education, State, among others.

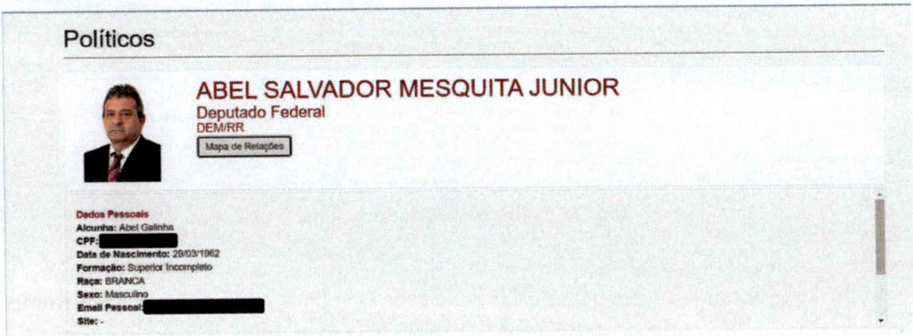

Fig. 2. Congressman's general data.

Figure 3 shows the data of donations received by politicians during election campaigns. In addition to collecting the donations' data, general data of donors is also collected.

Figure 4 shows the expenses incurred by the deputy or senator during his term. Data such as the name of the supplier, the cost of the expense, the description, and the date on which it occurred are presented.

DOAÇÕES

Campanha 2014

Doador	Valor	Data
PARTIDO DEMOCRATICO TRABALHISTA	R$ 2.050,00	15/07/2014
PARTIDO DEMOCRATICO TRABALHISTA	R$ 2.214,00	15/07/2014
PARTIDO DEMOCRATICO TRABALHISTA	R$ 2.548,97	15/07/2014
PARTIDO DEMOCRATICO TRABALHISTA	R$ 2.582,64	15/07/2014
PARTIDO DEMOCRATICO TRABALHISTA	R$ 4.646,67	15/07/2014
[redacted]	R$ 4.174,62	15/07/2014
PARTIDO DEMOCRATICO TRABALHISTA	R$ 500.000,00	15/07/2014
PARTIDO DEMOCRATICO TRABALHISTA	R$ 3.006,67	15/07/2014
PARTIDO DEMOCRATICO TRABALHISTA	R$ 1.776,67	15/07/2014
PARTIDO DEMOCRATICO TRABALHISTA	R$ 1.940,12	15/07/2014

Mostrando de 1 até 10 de 351 registros

Anterior 1 2 3 ... 36 Próximo

Fig. 3. Campaign donations' data.

DESPESAS

Fornecedor	Valor	Descrição	Data
JRM COMERCIO DE DERIVADO DE PETROLEO	R$ 100,00	COMBUSTÍVEIS E LUBRIFICANTES.	04/01/2015
BORGES CAHER COM. DE DERIVADOS DE PETROLEO LTDA	R$ 177,00	Id de Documento: 5663205 \n Número de Documento: 091284 \n Id de Lote: 1192297 \n Número de Ressarcimento: 5036 \n Tipo de Despesa: COMBUSTÍVEIS E LUBRIFICANTES.	04/01/2015
BORGES CAHER COM. DE DERIVADOS DE PETROLEO LTDA	R$ 177,00	COMBUSTÍVEIS E LUBRIFICANTES.	04/01/2015
JRM COMERCIO DE DERIVADO DE PETROLEO	R$ 100,00	Id de Documento: 5658626 \n Número de Documento: 014774 \n Id de Lote: 1184294 \n Número de Ressarcimento: 5005 \n Tipo de Despesa: COMBUSTÍVEIS E LUBRIFICANTES.	04/01/2015
Cia Aérea - GOL	-R$ 640,93	Id de Documento: Não definido \n Número de Documento: Bilhete: RLD9PX \n Id de Lote: Não definido \n Número de Ressarcimento: 0 \n Tipo de Despesa: Emissão Bilhete Aéreo	05/01/2015
Cia Aérea - GOL	-R$ 640,93	Emissão Bilhete Aéreo	05/01/2015
CASCOL COMBUSTIVEIS	R$ 40,00	COMBUSTÍVEIS E LUBRIFICANTES.	09/01/2015
CASCOL COMBUSTIVEIS	R$ 40,00	Id de Documento: 5804453 \n Número de Documento: 26377 \n Id de Lote: 1229908 \n Número de Ressarcimento: 5173 \n Tipo de Despesa: COMBUSTÍVEIS E LUBRIFICANTES.	09/01/2015
CASCOL COMBUSTÍVEIS PARA VEÍCULOS LTDA	R$ 80,00	COMBUSTÍVEIS E LUBRIFICANTES.	10/01/2015
CASCOL COMBUSTÍVEIS PARA VEÍCULOS LTDA	R$ 80,00	Id de Documento: 5832165 \n Número de Documento: 026458 \n Id de Lote: 1238293 \n Número de Ressarcimento: 5212 \n Tipo de Despesa: COMBUSTÍVEIS E LUBRIFICANTES.	10/01/2015

Mostrando de 1 até 10 de 2.611 registros

Anterior 1 2 3 ... 262 Próximo

Fig. 4. Politician expenditure data

Some traditional analysis techniques based on Database queries have been implemented. One of the main objectives of the tool is to dynamically visualize the data, especially the financial data and the contributions to the political campaigns. Figure 5 shows a map of the relationship between the congressmen and the 10 highest donors, as well as the amount donated.

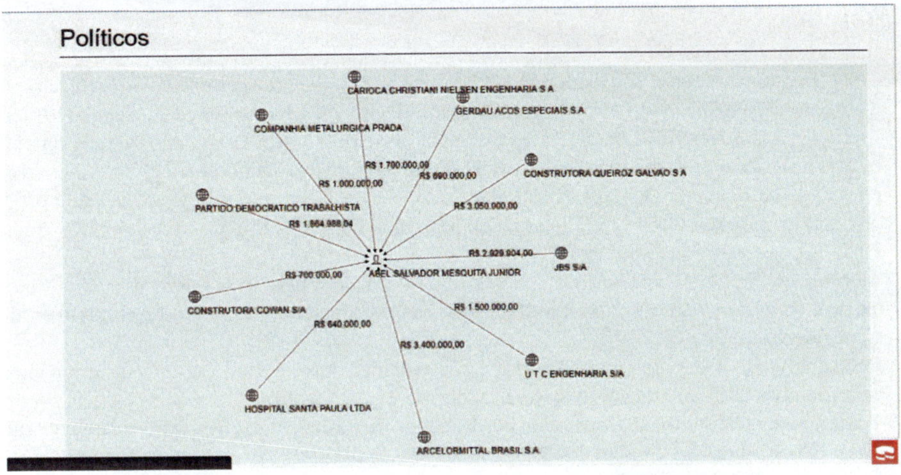

Fig. 5. Relationship map of the congressman and his donors.

5 Conclusions

After carrying out this work, several conclusions or considerations can be made, as well as recommendations for future work:

- Technological tools of political transparency are important for the people to remain informed about the functioning of their governments and the actions of their representatives.
- Web Data Mining techniques can be used for the collection, processing, and analysis of the information necessary for the implementation of a political transparency system.
- The development of L5P, implementing Data Mining techniques that allow extracting patterns, trends, relationships, and new knowledge from the collected data are important to allow deeper insight into the work of people's political representants.

Some future work that can be developed are:

- Implementation of Text Mining techniques that allow analyzing the speeches and public statements of politician.
- Implementation of Sentiment Analysis techniques for the analysis of statements and publications on social networks and contrast these statements with the actions of the politicians.
- Creation of several visualization methods (graphs, relationship maps, etc.) that capture the users' attention and that facilitate the understanding of the data.
- Implementation of Web Usage Mining techniques to analyze user interaction with L5P and customize the tool to the users' needs.

References

1. CNN (Internet Archive). What's REALLY behind the Brazilian riots? https://web.archive.org/web/20170723002811/http://ireport.cnn.com/docs/DOC-988431. Accessed 21 Sept 2020
2. Journal of Latin American Studies, Cambridge University. A Great Leap Forward for Democracy and the Rule of Law? Brazil's Mensalão Trial. https://www.cambridge.org/core/journals/journal-of-latin-american-studies/article/great-leap-forward-for-democracy-and-the-rule-of-law-brazils-mensalao-trial/DE95F43232D5531BDA22F2010551B4AD. Accessed 21 Sept 2020
3. Bloomberg. No One Has Ever Made a Corruption Machine Like This One. https://www.bloomberg.com/news/features/2017-06-08/no-one-has-ever-made-a-corruption-machine-like-this-one. Accessed 21 Sept 2020
4. Deutsche Welle. Odebrecht bribed across Latin America. https://www.dw.com/en/odebrecht-bribed-across-latin-america/a-36887600. Accessed 21 Sept 2020
5. Transparency International. Corruption Perceptions Index 2019. https://www.transparency.org/en/cpi/2019. Accessed 21 Sept 2020
6. Etzioni, O.: The world-wide web: quagmire or gold mine? Commun. ACM **39**(11), 65–68 (1996)
7. Singh, B., Singh, H.K.: Web data mining research: a survey. In: 2010 IEEE International Conference on Computational Intelligence and Computing Research, p. 10. IEEE, December 2010
8. Cacheda, F., Fernández, J., Huete, J.: Recuperación de información: Un enfoque práctico y multidisciplinar, capítulo 19. Minería de datos en la web, Ra-Ma (2011)

Data Origination: Human-Centered Approach for Design, Acquisition, and Utilization of Data

Teruaki Hayashi(✉) and Yukio Ohsawa

The University of Tokyo, 7-3-1, Hongo, Bunkyo-ku, Tokyo 113-8656, Japan

Abstract. The development of artificial intelligence and the global emergence of big data have provided access to data from different fields. However, while the reuse and sharing of data resources is vital for cost cutting, the data potentially reflects the design intent of those who design and obtain the data. It is necessary to establish a mechanism to quantify the data quality by sharing information regarding who, for what purpose, and how the target data was acquired. In this study, we discuss the methodology to observe and digitize unobserved events and propose the concept of data origination. Further, we introduce two tools to realize and support data origination: variable quest and TEEDA. Moreover, we explain the limitations of the current approach in achieving the data origination and discuss the approaches to overcome them.

Keywords: Data origination · Data design · Metadata · Variable

1 Introduction

Recently, the development of artificial intelligence and the global emergence of big data have been providing access to data from different fields. The expectation to use secondary data sources has been rising [1, 2]. At the government level, open data have been released for secondary use, linked to various services, and have generated value [3]. In the academic field, language corpora and learning data for image analysis, such as Reuters-21578 text categorization collection data set and ImageNet, are open to the public. Certain companies publish large-scale data for public use, e.g., Open Images Dataset by Google. Moreover, there is a requirement and expectation for interdisciplinary data collaboration, and market and services to exchange and combine different types of data have emerged [4–7].

However, while the reuse and sharing of data resources is vital for cost cutting, the data potentially reflects the design intent of those who design and obtain the data. The exchange and analysis of data without considering the design intent involves a risk of recognizing false facts or verifying wrong hypothesis. It is necessary to establish a mechanism to quantify the data quality by sharing information regarding who, for what purpose, and how the target data were acquired. Furthermore, the advancement of big data and sensors and the development of infrastructure for knowledge sharing have enabled acquiring and sharing data with high accuracy and granularity. However, the

A. Abraham et al. (Eds.): SoCPaR 2020, AISC 1383, pp. 85–93, 2021.
https://doi.org/10.1007/978-3-030-73689-7_9

methodology and technology to observe those events that have not yet been converted into data are undeveloped.

Moreover, the COVID-19 phenomenon, which is an unprecedented crisis, has been an opportunity to reconsider the importance of data understanding. Indeed, the importance of interdisciplinary data exchange and sharing has been recognized, which is essential for understanding and controlling unknown events. To overcome the corona-related confusion, publicly available large data have been highlighted and gained social attention, but the intention and background of data acquisition are often unclear. Such data may be insufficient to grasp facts and decide accurately. Silver said, "the number of reported COVID-19 cases is not a very useful indicator of anything—unless you also know something about how tests are being conducted" [8]. He warned against observing statistical data without understanding the testing method and the background of data acquisition. The corona-related confusion is ongoing, and although several international organizations and companies have published the data, the data that we really require have not been fully shared [9]. Particularly, it is limited to unilateral information provision from data providers, and there are few discussions on how to design, acquire, and use data of unobserved events and methods to support them.

In this study, we discuss the methodology to observe and digitize unobserved events and propose the concept of data origination. Further, we introduce two tools to realize and support data origination and present the utilization method and the knowledge obtained by considering corona-related confusion as an example.

The remainder of this paper is organized as follows. In Sect. 2, we discuss the issues in data origination and the goal of this approach with relevant studies. In Sect. 3, we present the two support tools for data origination from unobserved events. Further, we discuss the limitations of the current approach and present the challenging areas for data origination. Finally, we provide the conclusion in Sect. 4.

2 Data Origination

In the research fields of natural sciences and social sciences, we attempt to apply human interpretations to the data obtained from observations and experiments of phenomena and understand the mechanism behind the data. For example, there are attempts to understand the orbital period by observing the movement of stars and to learn the mechanism of earthquakes from the occurrence of earthquakes and the observation of plates. Currently, two stages of information loss are known to occur.

It is easier to understand using the agent-in-the-world model by Boisot and Canals [10] (Fig. 1). This is a model of the process used by an observing agent, such as a human, to acquire knowledge from the world. First, the agent receives the real-world stimuli through perceptual filters and converts them into data. This corresponds to a sensory organ for humans and a sensor for machines. Second, the data are converted into information through conceptual filters. Conceptual filters correspond to a mechanism for interpreting data by providing meaning to electrical signals and symbolized data and making them knowledgeable. The process involves attempting to understand the mechanism of the events by interpreting the data obtained by observing and experimenting the phenomena. Notably, although knowledge can further strengthen the perceptual and conceptual filters by expectations, and there is a mechanism that works on the real-world data similar to actions in the model, we do not discuss these functions in detail in this paper.

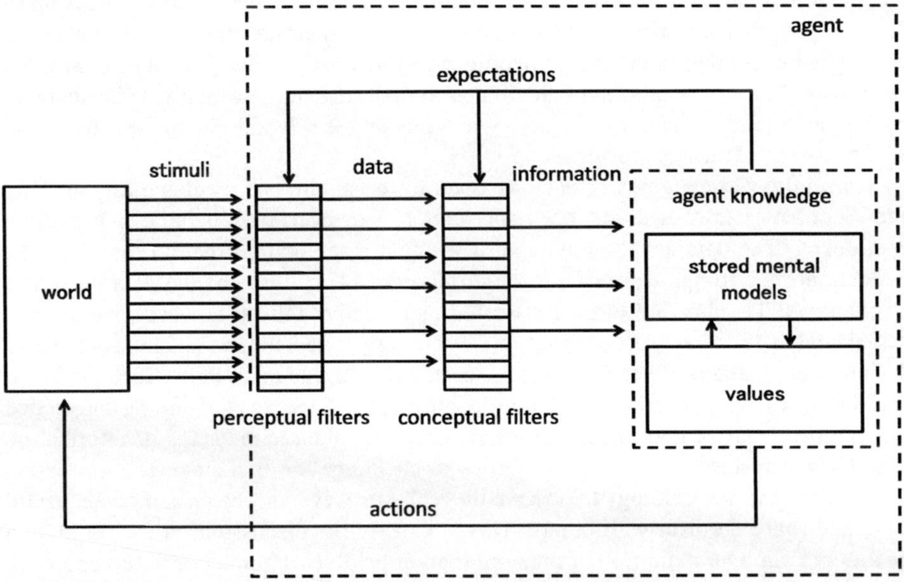

Fig. 1. Agent-in-the-world model [10] (partially modified by authors)

The first information loss occurs in the process of digitizing the target event through observation and experiment. This corresponds to the part where stimuli is converted into data by perceptual filters, as depicted in Fig. 1. Data corresponds to an abstraction of the real-world events by reconstructing them with strings or symbols. In this first stage, significant information is lost. Because the human eye cannot perceive ultraviolet light, the stimulus cannot be received as a signal. In terms of sensors, it is difficult to convert all stimuli into data owing to the limitation of optical resolution, which is the ability to measure or identify an object with a device. However, this information loss is not necessarily pessimistic. For example, if we use telematics to comprehend the congestion situation of the road, we can abstract the moving car as a point, enabling us to calculate the density of roads and estimate the extent of traffic jams. Subsequently, we can obtain the number of people in the cars or the types of vehicle; however, in several cases these information are not converted into data because they are not vital for hypothesis testing. In addition, in several cases, the risk of addition of noise to the excessively collected data makes it difficult to extract the essence of the event we want to understand. Therefore, moderate information loss makes the data analysis easier and increases the likelihood of obtaining useful insights.

The second information loss occurs in the process of analyzing data, estimating, predicting, and testing hypotheses. This corresponds to the informatization of data through the conceptual filters, as illustrated in Fig. 1. This process is necessary because even digitized events are complex and require further information extraction, summarization, and organization to obtain useful knowledge. If the essence of the thing or the event to be observed is accurately converted into data, the hypothesis can be verified using a data analysis method in this process. Once a complex event is quantified and digitized, statistical methods can be applied despite certain missing values or noise inclusion in the data. Several powerful tools have been proposed that are effective against errors caused by human measurement or phenomena with variations that cannot be controlled by humans. Indeed, the process of obtaining information from data involves considerable human interpretation, and it is necessary to focus on the selection of analysis tools and the discussion of analysis results.

Currently, various types of data are open to the public and available on the Web, and secondary use of data has been accelerated. However, the potential risk is often overlooked. The data are the outcome of the first stage of information loss. That is, the data are the things converted through the perceptual filters of the data designers or obtainers. The data are often considered as objective things; however, we have to understand that data are highly biased, as they strongly reflect the design intent of the data designers or obtainers. Therefore, data are more subjective than we think. Techniques for data analysis such as statistical modeling are aimed at the process of obtaining knowledge through the acquired data, which primarily challenge the second stage of information loss. However, there are a few approaches to challenge the first stage of information loss, such as the methodology to observe the real-world events, design and obtain useful data, and share the data with others. This is exactly the discussion on the origin and quality of data. Our definition of data origination involves a human-centered approach to design, obtain, and use the data by emphasizing the diversity of subjective knowledge to externalize and overcome the interdisciplinary issues.

According to the relevant studies, chance discovery involves the same motivational approach as data origination. To discover a chance implies to become aware of and understand the significance of a chance, which is an event with significant impact on a human's decision-making [11]. As an application of chance discovery, the human-computer interactive approach has been proposed to discover the invisible dark events [12]. Dark events are the invisible events that play a vital role in the dynamics of visible events, such as the detection of terrorist attacks [13] or the earthquake risks [14]. Data mining is an approach to gain knowledge from acquired data that implicitly assumes that the data being analyzed include meaningful knowledge, which may not be the case always. Both chance discovery and data origination are different from the data mining approach. Moreover, chance discovery focuses on understanding the significance of a chance, which is an event with significant impact on a human's decision-making. However, data origination is different in aiming to support and control data generation by observing unobserved events.

Although data origination aims to guarantee the data quality and overcome challenges by utilizing data, it is difficult to establish a methodology to support and realize it. This is because the world or the events to be observed has substantial and diverse information. Therefore, it is nearly impossible for human beings to observe and record all the stimuli. Though advanced sensing technologies enable acquiring a wide variety of data beyond the human cognition, it would be difficult to obtain all the data from all the events in the world. Because the sensors themselves are designed by humans, it is not possible to acquire data beyond the range recognized and designed by humans. Moreover, there are various methods for observing events depending on the domains. Therefore, observing unknown and unobservable events is an extremely challenging task.

One possible approach to observe the unobserved events is to use collective intelligence related to observations. From the perspective of a collective intelligence approach that focuses on the variables in the data, in the next section, we introduce the methods that implement data origination.

3 Support Tools of Data Origination

The data of various formats and structures are acquired in various fields, and it is difficult to discuss them on the same format. Therefore, we use data of data, that is, metadata, to overcome this challenge. Here, the metadata are composed primarily of a set of variables. Variables are the logical set of attributes of a real-world event [15, 16], e.g., in the case of healthcare data, "name," "weight," "height," "sex," or "blood type" are the variables. The variables define the elements that make up the events, and each variable has a value represented by numbers and strings.

By collecting metadata that describes data with a common notation from a wide variety of domains, we can obtain perceptual and conceptual filters, as depicted in Fig. 1, for the events observed in the domains as a set of variables. By treating the collected metadata as collective intelligence of observation, several analysis tools and methods have been proposed. In this section, we explain the overview of each tool, discuss where to support the data origination process, and clarify the type of support tools to be developed in the future.

In the analysis tools, data jacket (DJ) is used to describe metadata. The DJ is a form of metadata that enables to understand the contents of data while keeping the data secret, and has been proposed as a technology to support the data marketplace [17]. The advantage for using DJ is that it supports both human comprehension and computer readability and includes a description of variables.

3.1 Variable Quest

From the outlook of data design, it is essential to determine the data to be collected to meet our goals. The risk of missing data exists subsequent to the collection of data. For example, failing to obtain the nationality of the participants after obtaining the attributes of the attendees of an event in a questionnaire. Data design is furthermore vital for events that can be observed only once, such as earthquakes and the Olympic games.

A variable quest (VQ) is a support tool for decision makers to obtain new data comprising a set of variables [18, 19]. VQ provides an interface to search with keywords or sentences used in the search engines such as Google and Yahoo!. The natural language search adopted by VQ enables acquisition of variable sets extracted from various data, without requiring a user to possess special knowledge. For example, assume that one inputs the following sentence: "I want to know annual population trends in Japan." In this case, one can learn the type of variables that can and should be acquired, such as "number of births," "number of deaths," "number of transferees," "population," and "number of newborn children."

VQ models the similarity of data outlines and the co-occurrence of variables. Figure 2 depicts a visualization of the estimation result based on co-occurrence by inputting "risk of contact infection with new coronavirus." The size of the node represents the similarity with the query, and the thickness of the edge represents the co-occurrence frequency between variables. Notably, not only the variables such as "the number of travelers," "length of stay," "the number of flights," and "the number of passengers," but also the variables such as "annual income," "family structure," "values for work/health," "satisfaction of life," and "self-help consciousness" are estimated. The data with these variables are currently discussed worldwide to understand the corona-related confusion.

Fig. 2. Visualization of the estimation result by inputting the query "risk of contact infection with new coronavirus."

3.2 TEEDA

In existing methods, only the data providers or someone who knows the location of the data describe metadata and provide them on the data marketplace. To overcome this problem, treasuring every encounter of data affairs (TEEDA) facilitates communication and matching between data holders and users by capturing and presenting requests for data (call for data) [20]. The data holders register the metadata of their own datasets, and the data users provide information regarding the purpose and structure of the data in the form of data requests in TEEDA. The data requests work as the call for data with variables of unobserved events.

It is known that the providable data and the data requests share only a few variables, and the data requests are composed of variables that have not yet been observed and digitized [9, 20]. Figure 3 illustrates a co-occurrence graph of the variables in data items

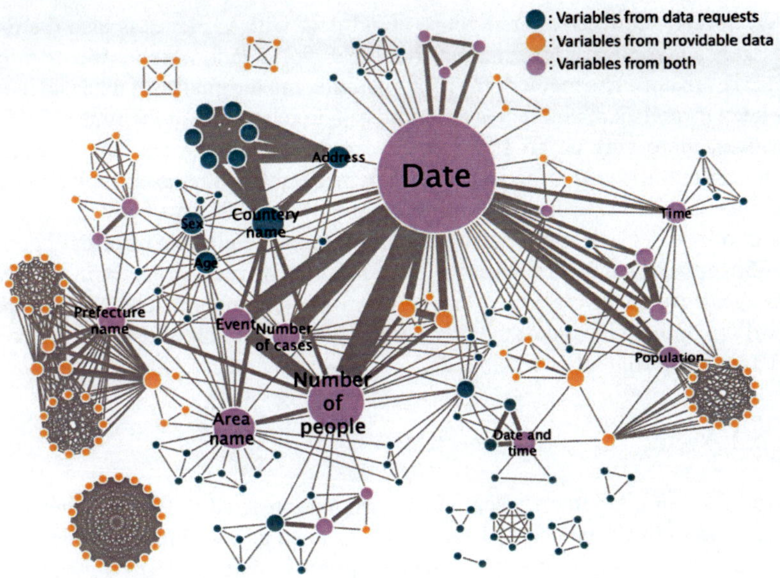

Fig. 3. Co-occurrence graph of the variables in data items (data requests and providable data) obtained by the TEEDA workshop.

(data requests and providable data) obtained by the TEEDA workshop whose theme is corona-related confusion. The size of the node represented the frequency of occurrence, and the thickness of the edge represented the frequency of co-occurrence of the variables. Labels were attached to the top 13 frequently appearing variables. The variables that appeared in both data requests and providable data, frequently appeared across data items and played a central role in the graph. However, the graph was composed of numerous infrequent variables that created several local clusters. It can be said that the providable data did not satisfy the data requests. Moreover, Hayashi et al. proposed three data acquisition scenarios from the outlook of "phenomenon understanding," "individual decision making," and "organizational decision making" to overcome the corona-related confusion, based on the analysis of data requests collected by TEEDA. That is, it was important to externalize the outlines of unobserved events by sharing the outlook of others as collective intelligence, and link them to actions such as data acquisition and analysis.

3.3 Limitations and Future Work

This section focuses on VQ and TEEDA and discusses approaches to data origination. VQ is a tool that supports design process when acquiring data, and TEEDA is the one that supports the externalization and matching of data requirements. The technologies that support the current data origination are significantly based on metadata and collective intelligence. Humans are the subjects who design and observe the stimuli of the real-world events. There are certain purposes for acquiring data. An issue is what type of data design and acquisition should be supported according to each purpose of data use.

However, notably, collective intelligence about data with variables is also the dataset, and is merely a mapping of information about real-world data, which is biased when the metadata are created. The methodology for guaranteeing the quality of data and metadata has not been established, and this is a major issue to overcome in the future.

However, there may be an approach to acquire all observable data substantially. However, such excessive data acquisition is not preferable from the outlook of data minimization. Data minimization refers to minimizing the possibility of collecting personal data about others [21]. Because personal data contains identifiable information, obtaining significant personal information involves a risk of privacy violation in the event of data leakage. If we incorporate this perspective in data origination, it would be important to support the optimal design of the appropriate set and amount of data for the purpose, rather than considering all obtainable data.

4 Conclusion

In this study, we proposed the concept of data origination and discussed the methodology to observe and digitize unobserved events. It has been a long time since the advent of the age of big data, however, it is not well understood, in that, we know only a part of the events and convert them into data. The world is full of unknown events that have not yet been recognized, verbalized, and observed. We hope that research on data origination will raise awareness on the existence of such unknown events and contribute to valuable discoveries and data utilization.

Acknowledgement. This study was supported by JSPS KAKENHI (JP20H02384), the "Startup Research Program for Post-Corona Society" of Academic Strategy Office, School of Engineering, the University of Tokyo, and the Artificial Intelligence Research Promotion Foundation. We wish to thank Editage for providing English language editing.

References

1. Rabinovich, E., Cheon, S.: Expanding horizons and deepening understanding via the use of secondary data sources. J. Bus. Logist. **32**(4), 303–316 (2011)
2. Ellram, M.L., Tate, L.W.: The use of secondary data in purchasing and supply management research. J. Purch. Supply Manag. **22**(4), 250–254 (2016)
3. Manyika, J., Chui, M., Groves, P., Farrell, D., Kuiken, S.V., Doshi, E.A.: Open data: Unlocking innovation and performance with liquid information, McKinsey Global Institute (2013)
4. Balazinska, M., Howe, B., Suciu, D.: Data markets in the cloud: an opportunity for the database community. VLDB Endowment **4**(12), 1482–1485 (2011)
5. Stahl, F., Schomm, F., Vossen, G.: Data marketplaces: an emerging species. Frontiers in Artificial Intelligence and Applications, pp. 145–158 (2014)
6. Liang, F., Yu, W., An, D., Yang, Q., Fu, X., Zhao, W.: A survey on big data market: pricing, trading and protection. IEEE Access **6**, 15132–15154 IEEE (2018)
7. Spiekermann, M.: Data marketplaces: trends and monetisation of data goods. Intereconomics **54**(4), 208–216 (2019)
8. Silver, N.: Coronavirus Case Counts Are Meaningless, FiveThirtyEight. https://fivethirtyeight.com/features/coronavirus-case-counts-are-meaningless/. Accessed 12 Oct 2020

9. Hayashi, T., Uehara, N., Hase, D., and Ohsawa, Y.: Data Requests and Scenarios for Data Design of Unobserved Events in Corona-related Confusion Using TEEDA, arXiv:2009.04035 (2020)
10. Boisot, M., Canals, A.: Data, Information and knowledge: have we got it right? J. Evol. Econ. **14**, 43–67 (2004)
11. Ohsawa, Y., McBurney, P.: Chance Discovery. Springer, Heidelberg (2003)
12. Maeno, Y., Ohsawa, Y.: Human-Computer Interactive Annealing for Discovering Invisible Dark Events. IEEE Trans. Industr. Electron. **54**(2), 1184–1192 (2007)
13. Maeno, Y., Ohsawa, Y.: Intuitive visualization of the intelligence for the run-down of terrorist wire-pullers, arXiv:0805.3972 (2008)
14. Ohsawa, Y.: Detection of Earthquake Risks with KeyGraph, Chance Discovery, Springer-Verlag Berlin Heidelberg, 339–350, (2003).
15. Babbie, E.R.: The basics of social research (7th Edition), Cengage Learning, (2016).
16. Hayashi, T., Ohsawa, Y.: Understanding the structural characteristics of data platforms using metadata and a network approach. IEEE Access **8**, 35469–35481 (2020)
17. Ohsawa, Y., Kido, H., Hayashi, T., Liu, C.: Data jackets for synthesizing values in the market of data. Procedia Comput. Sci. **22**, 709–716 (2013)
18. Hayashi, T., Ohsawa, Y.: VARIABLE QUEST: network visualization of variable labels unifying co-occurrence graphs. In: ICDM Workshops, pp. 577–583 (2017)
19. Hayashi, T., Ohsawa, Y.: Inferring variable labels using outlines of data in data jackets by considering similarity and co-occurrence. Int. J. Data Sci. Anal. **6**(4), 351–361 (2018)
20. Hayashi, T., Ohsawa, Y.: TEEDA: an interactive platform for matching data providers and users in the data marketplace. Information **11**(4), 218 (2020)
21. Pfitzmann, A, Hansen, M.: A terminology for talking about privacy by data minimization: anonymity, unlinkability, undetectability, unobservability, pseudonymity, and identity management (2010). https://dud.inf.tu-dresden.de/Anon_Terminology.shtml. Accessed 12 Oct 2020

Coarse-Grained Genetic Algorithm for Flexible Job Scheduling Problem

Eliot Bytyçi, Artan Berisha(✉), and Fis Geci

University of Prishtina, Prishtina, Kosova
{eliot.bytyci,artan.berisha,fis.geci}@uni-pr.edu
http://www.uni-pr.edu

Abstract. Flexible Job Scheduling Problem presents a relaxation of the Job Scheduling Problem. It falls into the category of NP-hard problems. Our approach is to tackle the problem through usage of a coarse-grained genetic algorithm. Coarse-grained extends the genetic algorithm and thus increasing chances of more fitness landscape to be examined. Solutions are benchmarked with other approaches, through two of the most used datasets. Even though the results are less favourable than other approaches, in few cases our approach reaches the upper bound of the benchmarked solutions, thus impelling us to further investigate on the problem solution.

Keywords: Coarse grained · Flexible job scheduling problem · Genetic algorithm

1 Introduction

Today machinery is an integral part of every manufacturing process. In order for the process to be more efficient and to use the machinery as efficiently as possible, one needs to manage resources. The management of the resources should be as competent as possible, ensuring that the products are created without excessive time spent. This issue and other ones related to it, are object of the study in the field of scheduling problems [1].

Scheduling problems aim to find the optimal solution, taking into the consideration different characteristics of the job. Job Scheduling Problem (JSP) is an optimization problem, in which jobs are scheduled in a job shop [2]. Nowadays, JSP has wider application and therefore it is related to its assignment to certain resources at specific time. For example, there are k jobs of different operational times, which have to be assigned to j machines of different operational power. One of the variations of the problem is the Flexible Job Scheduling Problem (FJSP), in which a relaxation is presented from the original problem of JSP [3].

In FJSP, each job can be operated on any machine or set of machines, where machines are identical. It represents one of the well-known combinatorial optimization problems, with complexity presented as NP-hard, since one of its special version, Travelling Salesman Problem or JSP with only one job, have NP-hard

A. Abraham et al. (Eds.): SoCPaR 2020, AISC 1383, pp. 94–102, 2021.
https://doi.org/10.1007/978-3-030-73689-7_10

complexity [3]. A number of researchers have tackled the problem through different algorithms.

Our approach tends to find comparable solutions by using coarse-grained genetic algorithm. It should be emphasized that genetic algorithms have been used also before, as presented in section of related work. Furthermore, there exist an approach of using coarse-grained algorithm [4] but differs from our approach as explained in related work section.

This paper is structured as follows: Sect. 2 presents related work, while Sect. 3 presents problem formulation and description of algorithms used. Section 4 presents results of the experiments and Sect. 5 concludes the paper and presents future steps.

2 Related Work

Authors in [5] were the first to address FJSP as one of the most important multi objective optimization problems. Even though the polynomial algorithm for two job problem was developed by them, it couldn't lead to useful solution when dealing with three or more jobs. Their conclusion was that the deterministic algorithms are not effective for solving FJSP with small number of instances.

Another, more effective solution, was proposed by [6], where authors have used Tabu search. Their approach is described as a hierarchical algorithm for flexible job schedule. Other heuristic procedures for more effective solutions, were genetic algorithm [3,4,7,8] and simulated annealing [9].

Authors in [10–12] used fuzzy logic approach for multi-objective optimization problem, thus to handle the uncertainty around the choice of evolutionary operators [10], for adaptation of gap generation in genetic algorithm [11] and optimizing advertising text [12].

Authors in [3] introduced a number of strategies for generating the initial population and selection of the individuals for duplication and reproduction. That, usage of a mix of different selection criteria and process constraints, led to better results. Another approach using genetic algorithms, was presented by [7], where the initial assignment for generating initial population with different selection criteria, improved the speed and quality of the solution. An hybridized genetic algorithm is proposed in [8] with innovative local search procedure, allowing numerical experiments on large number of representative problems. Authors have proposed and developed an effective decoding method for interpreting each chromosome. The representation of the chromosome was done through two-vector permutation for the heritability of crossover operation. Yet the most similar approach to ours is the paper presented by authors in [4]. They have used multi population genetic algorithm, in which each population has its independent evolution process. By doing that, they have enabled the use of parallel genetic algorithm and thus achieving better effectiveness of the proposed model. The difference of our approach is in the representation of the solution and in the exploration of the solution set. The representation of solution in [4] is done sequentially while in our approach the representation is handled by a separate algorithm, which checks the order of jobs to be performed and places them appropriately. Furthermore, our algorithm sets its individuals' probability of selection by using gamma distribution and thus improving the performance of the system.

3 Problem Formulation and Algorithm Description

Flexible Job Scheduling Problem consists of k independent jobs, each with its own processing time, which have to be processed in m machines. In order to complete a job j, a number of ordered operations have to be completed. An operation of the job j can be processed on any machine [18]. Mathematical formulation for the FJSP can be seen as an objective to minimize the function describing makespan – maximum completion time of all jobs. There are several constraints presented in the problem formulation:

- A specific operation is assigned to only one machine.
- The time to finish an operation is described as the difference between time of completion and the time of start. That time has to equal the least of the processing time.
- Two operations cannot be processed at the same time in the same machine.
- An operation cannot start before the preceding operation, related to it, has not finished.
- The sum of completion time of operations is the completion time of the job.
- The makespan is greater or equal to the job time completion.

Figure 1 shows a very simple instance of the FJSP. The problem has three machines (M) and two jobs (J). Each job has two operations (O) that have their own processing time (numeric value) for each machine.

		M_1	M_2	M_3
J_1	$O_{1,1}$	5	2	1
	$O_{1,2}$	3	1	2
J_2	$O_{2,1}$	1	5	2
	$O_{2,2}$	3	4	1

Fig. 1. A 2 × 3 flexible job shop problem.

A. Genetic Algorithm. In order to have a bigger search space and avoid local maxima, our approach used genetic algorithm (GA) [13]. Genetic algorithm has several features such as: individual solution representation, population, fitness, crossover, mutation and selection, each presented in following sub-sections.

a) Individual's Solution Representation: The FJSP is composed of two sub-problems: the assignment of machines to the operations and the sequencing of operations in each machine. Our approach uses two data structures to tackle these problems. The first data structure is a vector of length n, where n represents the total number of operations. This vector is the machine vector (MV). The operations of each job are assigned to the entries in the MV in the format as presented in Fig. 2. The value of an entry presents the machine in which the operation is being processed at. For example, in the MV shown in Fig. 2, $O_{1,1}$ (operation 1 of job 1) is to be processed at machine 1.

1	3	1	2
$O_{1,1}$	$O_{1,2}$	$O_{2,1}$	$O_{2,2}$

Fig. 2. A Machine vector for a 2 × 3 FJSP.

The second data structure is also a vector of length n, where n now represents the total number of operations. This vector is the operations vector (OV). In the OV, each entry is one of the operations available for processing. The order in which the operations appear in the OV is the order that the operations will be processed. Solutions of the problem are built by using the combination of MV and OV. In Fig. 3, a solution is described, where each row represents schedule of a machine, which is contains the operations. The last row represents the time the operations finished processing, where makespan represents the time last operation finished processing.

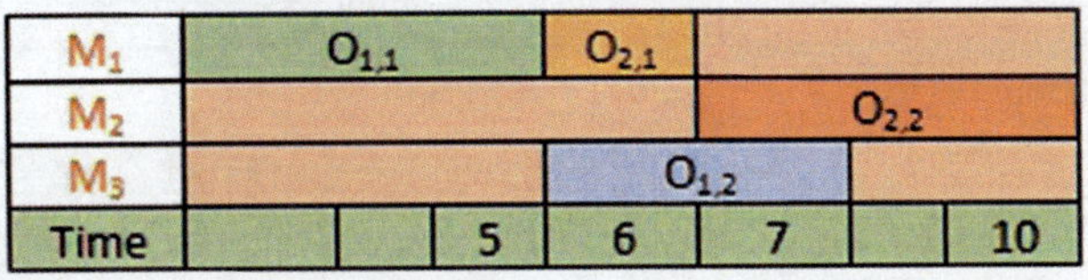

Fig. 3. A solution of the 2 × 3 FJSP with a makespan of 10.

b) Initial population: representing the subset of the solutions, initial population is the first step in the GA. Usually it is created by a random function and consists of set of chromosomes. In our approach, as described earlier every individual of the population consists of two vectors. The approach creates a random MV and a random OV and then applies perturbation to the OV. The reason to apply perturbation is that is introduces variety in the population. This function does one of two things at random:

- It either selects a random operation from the OV and it changes the machine that the operation will be processed on,
  or

– It selects a random operation from a machine and switches place with its neighbour to its right.

Through experiment trials, it was concluded that the optimal size of the initial population should range from 600 to 1000.

c) Fitness: since the main objective is to minimize the makespan of a solution, our approach uses makespan as the value of the fitness. Every individual is assigned this value which is used to order them. After that a gamma's distribution probability density function given in (1) with parameters $\alpha = 1.2$ and $\lambda = 0.78$ is used to set each individuals' probability of selection for the selection phase.

$$f_T(t) = \gamma(t; \alpha, \lambda) = \begin{cases} \frac{\lambda^\alpha t^{\alpha-1} e^{-\lambda t}}{\Gamma(\alpha)} & \text{for } t > 0, \\ 0 & \text{for } t < 0. \end{cases} \tag{1}$$

After a number experiments with different parameters for α and λ the one we chose worked best for our case.

d) Crossover: represents a two-step process. After the selection of two parents, a new offspring is created, as an MV, from their parents MV's. Then an OV of the offspring is created from the parents' OV's. The chance for creation of the MV is 50% for a gene from a parent to be selected and inserted in the same place in the offspring MV, as presented in Fig. 4. The offspring OV, is then generated in a bit more complex manner. Initially from the first Parent a random job is selected and a sub-set of available operations from that job is created. The position that they were inserted has to be remembered and its content is inserted in offspring OV, in the same position as on the parent.

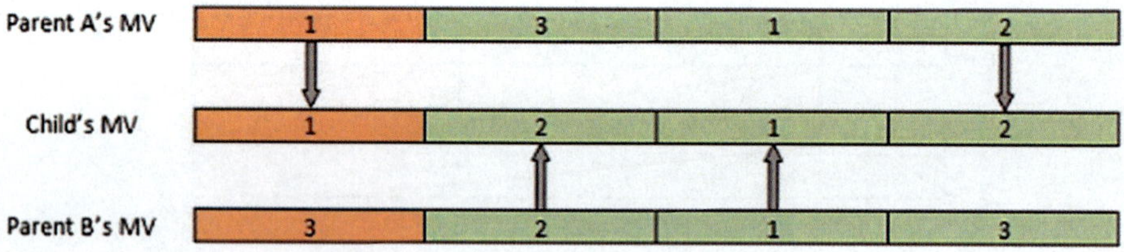

Fig. 4. Crossover of machine vector.

After the process is completed, elements added in the offspring are deleted-from the second parent's OV. Then empty spaces in the child's OV are filled with the remainingelementsof the second parent's OVas shown in Fig. 5.

e) Mutation: Just like crossover, mutation consists of two steps. The first step mutates the offspring's MV while the other step mutates OV. To mutate the MV we select a random operation from the offspring's MV and select another random machine that the operation can be processed on. The offspring's OV is mutated by first selecting two random operations from the offspring OV and then exchanging those elements places with each other, as presented in Fig. 6.

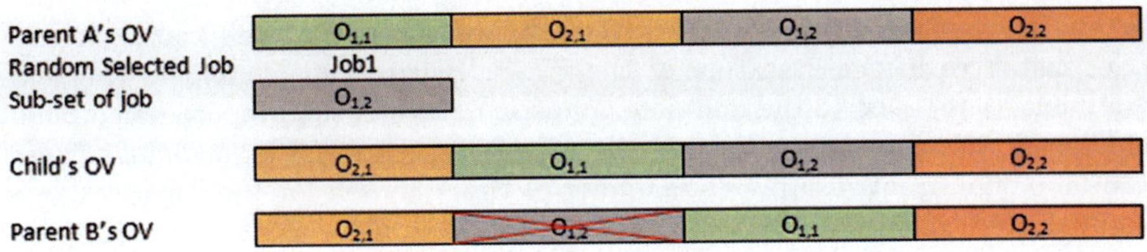

Fig. 5. Crossover of operation vector.

f) Selection: is performed through tournament selection, with two members. These randomly selected members of the population are pinned against each other. The member with the lower fitness is selected and removed from the population.

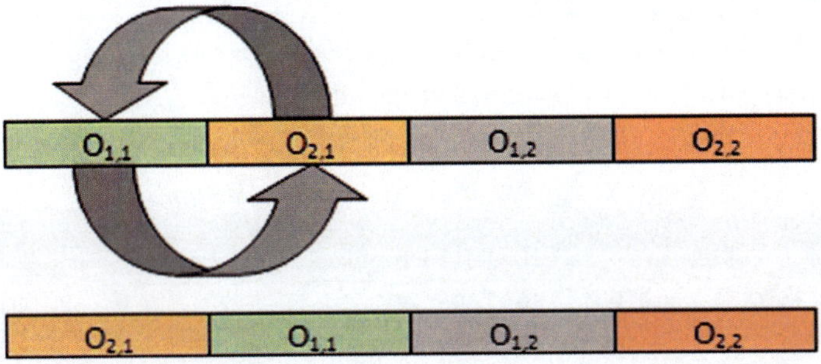

Fig. 6. Mutation of operation vector.

B. Coarse Grained Algorithm. Our approach uses coarse-grained algorithm to solve the FJSP. Coarse-grained algorithm or island algorithm, creates several instances of tournament based genetic algorithm. These are called islands because their population can communicate only through the managing algorithm. In every island, on every n-th generation, best instance migrates in all other islands. This increases the chance that the population in general, doesn't get stuck on local maxima. On the other hand, it allows the best solution to become part of other islands but also allows islands to diverge considerably from each other. Through experiments, it was observed that a great diversity and better results were achieved by setting the amount of islands from 20–25 and migration mark every 5th generation.

4 Results of the Experiments

The proposed algorithm is implemented in Java programming language on a 2.6 GHz Intel Core I7-9750H processor and 16 Gb of RAM memory. To evaluate the efficiency of the algorithm, tests are based on two sets of well-known benchmark instances in the literature of the FJSP: Barnes and Chambers [14] and Brandimarte [6]. Due to the non-deterministic nature of the proposed algorithm and to allow it as much variation, the algorithm is run 50 independent times to obtain significant results. The number of times chosen for the runs was as a result of pre experimental trials. The computation results show the known upper and lower bounds for a given instance and show the proposed algorithm's best reached solutions. The parameters chosen for the algorithm, after an excessive number of trials, are:

- Mutation Probability is 0.8
- Maximum number of iterations is 1000
- The population size is 600–1000
- Crossover Probability is 0.8

Results of our approach using 10 Bradimarte instances are shown in Table 1, compared to state of the art low bound [15] and upper bounds[3,15–17]. Our approach, has reached to be concurrent with three of the results from state of art and underperforming in seven other results.

Table 1. Results from brandimarte instances.

	LB	UP	Proposed algorithm
Mk01	36	39	40
Mk02	24	26	27
Mk03	204	204	204
Mk04	48	60	62
Mk05	168	172	173
Mk06	33	58	65
Mk07	133	139	143
Mk08	523	523	523
Mk09	299	307	307
Mk10	165	197	233

In Table 2, results of 14 Barnes and Chambers instances are presented. Proposed algorithm in this case underperforms, finding only one result, which is comparable to low bound and upper bound [6,15,17] state of the art results.

Table 2. Results from Barnesand Chambres instances.

	LB	UP	Proposed algorithm
Mt10c1	655	927	927
Mt10cc	655	910	919
Mt10x	655	918	934
Mt10xx	655	918	934
Mt10xxx	655	918	934
Mt10xy	655	905	929
Mt10xyz	655	847	858
Setb4c9	857	914	959
Setb4cc	857	909	932
Setb4x	846	925	941
Setb4xx	846	925	959
Setb4xxx	846	925	959
Setb4xy	845	916	941
Setb4xyz	838	905	930

5 Conclusion and Future Work

In this paper we have presented a different approach on solving flexible job scheduling problem through coarse- grained genetic algorithm. Our results, even though promising, underperform in most of the cases, when compared with state of the art. However, in at least four cases, of both instances our approach reaches the results of the state of the art. In the future, our approach will be refined by improving the selection and the initial generation of the population. Parallelization will also be incorporated in order to achieve better results and efficiency.

References

1. Taillard, E.: Benchmarks for basic scheduling problems. Eur. J. Oper. Res. **64**(2), 278-285 (1993)
2. Burstall, R.M.: A heuristic method for a job-scheduling problem. J. Oper. Res. Soc. **17**(3), 291–304 (1966)
3. Pezzella, F., Morganti, G., Ciaschetti, G.: A genetic algorithm for the flexible job-shop scheduling problem. Comput. Oper. Res. **35**(10), 3202–3212 (2008)
4. Gaham, M., Bouzouia, B., Achour, N.: An agent-based multi-population genetic algorithm for the flexible job shop scheduling problem. In: 8ème conférence internationale conception & production intégrées, CPI (2013)
5. Brucker, P., Schlie, R.: Job-shop scheduling with multi-purpose machines. Computing **45**(4), 369–375 (1990)
6. Brandimarte, P.: Routing and scheduling in a flexible job shop by tabu search. Ann. Oper. Res. **41**(3), 157–183 (1993)

7. Zhang, G., Gao, L., Shi, Y.: An effective genetic algorithm for the flexible job-shop scheduling problem. Expert Syst. Appl. **38**(4), 3563–3573 (2011)
8. Gao, J., Gen, M., Sun, L., Zhao, X.: A hybrid of genetic algorithm and bottleneck shifting for multiobjective flexible job shop scheduling problems. Comput. Ind. Eng. **53**(1), 149–162 (2007)
9. Kolonko, M.: Some new results on simulated annealing applied to the job shop scheduling problem. Eur. J. Oper. Res. **113**(1), 123–136 (1999)
10. A novel multi-objective evolutionary algorithm with fuzzy logic based adaptive selection of operators: FAME. Inf. Sci. **471**, 233-251 (2019)
11. Cervantes, L., Castillo, O., Hidalgo, D., Martinez-Soto, R.: Fuzzy dynamic adaptation of gap generation and mutation in genetic optimization of type 2 fuzzy controllers. Adv. Oper. Res. **2018**, 9570410:1-9570410:13 (2018)
12. Madera, Q., Castillo, O., García-Valdez, M., Mancilla, A.: A method based on Interactive Evolutionary Computation and fuzzy logic for increasing the effectiveness of advertising campaigns. Inf. Sci. **414**, 175–186 (2017)
13. Whitley, D.: A genetic algorithm tutorial. Stat. Comput. **4**(2), 65–85 (1994)
14. Barnes, J., Chambers, J.: Flexible job shop scheduling by tabu search, Graduate Program in Operations and Industrial Engineering, The University of Texas at Austin, Technical Report Series, ORP96-09 (1996)
15. Mastrolilli, M., Gambardella, L.M.: Effective neighbourhood functions for the flexible job shop problem. J. Sched. **3**(1), 3–20 (2000)
16. Xing, L.-N., Chen, Y.-W., Wang, P., Zhao, Q.-S., Xiong, J.: A knowledge-based ant colony optimization for flexible job shop scheduling problems. Appl. Soft Comput. **10**(3), 888–896 (2010)
17. Gao, J., Sun, L., Gen, M.: A hybrid genetic and variable neighborhood descent algorithm for flexible job shop scheduling problems. Comput. Oper. Res. **35**(9), 2892–2907 (2008)
18. Fattahi, P., Saidi Mehrabad, M., Jolai, F. Mathematical modeling and heuristic approaches to flexible job shop scheduling problems. J. Intell. Manuf. **18**, 331–342 (2007)

AI Improving the Lives of Physically Disabled

Hemshree Madaan and Shubham Gupta(✉)

Bharati Vidyapeeth's College of Engineering, New Delhi 110063, India

Abstract. Physical disability is one of the significant concerns that hamper individuals to access the web easily. Most of the physically disabled people cannot use technology because of the limitation of accessibility tools and techniques. It is required that the websites should be made compliant with the requirements of every citizen in a country; that's why they should cater to the needs of the differently-abled citizens as well. Features have to be introduced in the websites so that they are easy to use, readily accessible, understandable, and convenient to everyone including best practices/standards and global innovation techniques. At times, accessibility is confused with providing solutions to disabled people, but the fact is accessibility is not only for differently-abled people, but it's also there for everyone. The matter is every person needs accessibility and uses it when in need.

Keywords: Deep learning · Computer vision · Image processing · Natural language processing · Artificial intelligence · Machine learning · Perceptual user interfaces · Web accessibility · AI for social good

1 Introduction

According to the World Health Organisation statistics, over 15% of the human population suffers from disabilities. A majority of these disabilities confine a person's effectiveness in interacting with the web in general. In the recent scenario, most of the residents of developed countries are well-equipped electronics gadgets such as computers which they use in their day to day lives to achieve their tasks and goals conveniently. Hence, there is a need for devising proper solutions to counter the obligatory requirement of differently-abled citizens which can help them in their personal as well as professional life. Technology has been providing support to enhance the quality of life of people.

Therefore, a human needs to access the web and technology freely without any hassle. There is a need for incorporating several changes in the web technologies through the means of Artificial Intelligence and Computer Vision which can help in providing the considerable potential to support the needs of this group (Fig. 1).

Physical disabilities consist of disabilities like Motor impairment which is a kind of disability that includes people who suffer from muscular dystrophy,

A. Abraham et al. (Eds.): SoCPaR 2020, AISC 1383, pp. 103–112, 2021.
https://doi.org/10.1007/978-3-030-73689-7_11

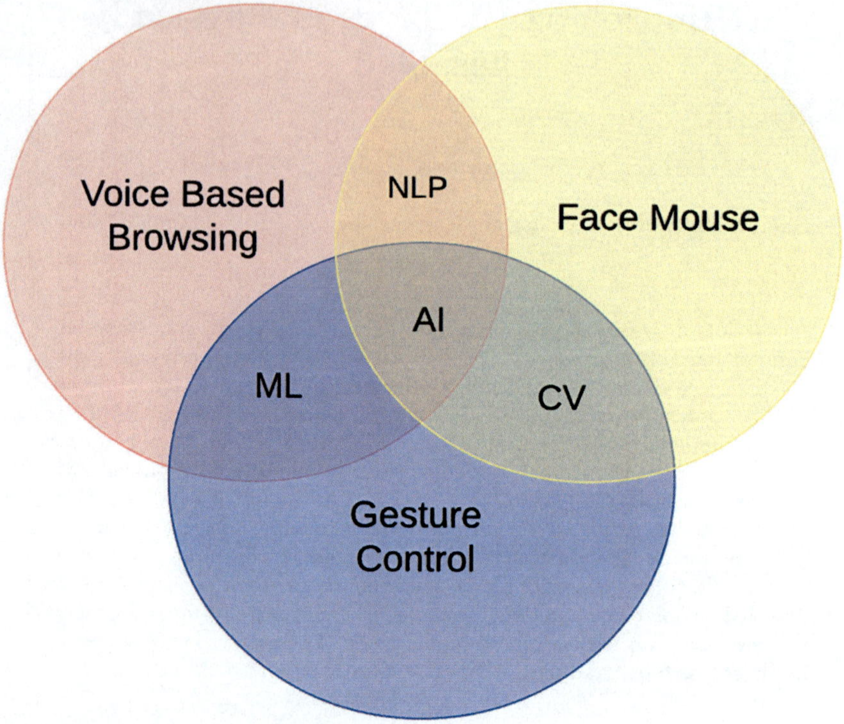

Fig. 1. Venn diagram depicting the areas of research work

broken bones, or other conditions that cause tremors or loss of fine muscle control. These people face difficulty in using their hands on a website. Given these limitations, this group of people can use a variety of assistive Computer Vision technologies to gain access to the web.

Elimination of hand-controlled input devices can cater to the needs of these groups of individuals to a much larger extent. For which, Perceptual user interfaces come into play which can help in interacting with the computers naturally. Human gestures are the motion of the special part of the human body such as the face and hand. It is most likely for a hand gesture to replace the spoken language and communicate with each other. This is the reason that deaf people usually use hand gestures to express their means, for example, American sign language is one of the hand gesture communication methods. Therefore, hand gestures could be a communication method between the Human body and the Machine.

With this fast-moving world and developing science, people cannot live well without a computer. Thus, it is becoming more and more important to be able to interact with the computer naturally in the area of Human-Computer Interaction [1].

The human-computer interaction which makes use of input devices such as keyboard or mouse for communication lacks natural communication between humans and machines. Traditional input devices such as a mouse, a keyboard can no longer meet user interactive requirements, there is always a strong barrier between the computer and user through the traditional technology. For this purpose, it is important to develop applications or devices that support intuitive communication.

Voice commands can also help to instantiate the perceptual behavior of computers. Speech provides an accessibility feature to such groups and can be used as an interacting entity with the computer. To achieve this, Natural Language Processing is used to process the paraphrases spoken by an individual, and results are generated based on that.

The overview of this research lies somewhere between the Venn Diagram of Natural Language Processing, Machine Learning, and Computer Vision which simply indicates that it revolves around the domain of Artificial Intelligence. The research is segmented into three major categories to provide ease to the motor impaired citizens.

- Voice-based browsing
- Face mouse
- Gesture recognition

2 Related Work

A substantial amount of research has been done in this area using image processing. Different hand and fingers angle2s were used and set as different gestures. A simple webcam was used in the study for the interaction between human and computer (Athiya Marium et al. 2017) [2]. Hand gesture control has been achieved before by using different hardware techniques like stereo cameras, depth cameras, Microsoft Kinect that uses IR for hand tracking, and many more (AL. Liu et al. 2013) [3].

(Naguri et al. 2017) [4], they present fast and highly accurate gesture recognition systems based on long short-term memory (LSTM) and convolutional neural networks (CNN). The model was trained to process input sequences of 3D hand positions and velocities acquired from infrared sensors. In real time recognition, the proposed architectures obtain 97% F-measure, demonstrating a significant potential for practical applications in novel human-computer interfaces.

(L. Pigou et al. 2015) [5], they showed that adding bidirectional recurrence and temporal convolutions improves frame-wise gesture recognition in video significantly. It was seen that those Recurrent Neural Networks which responded to high-level spatial features have greater performance than that of single-frame and temporal pooling architectures.

Facial tracking is a very common application of Perceptual User Interfaces. Six-Segmented Rectangular Filter (SSR) along with a support vector machine (SVM) for face verification is the basic strategy that many researchers use for detection of faces. (M. Turk et al. 2002) [6] The patterns of between-the-eyes are tracked with update template matching in face tracking. Facial recognition can help to a large extent in providing an alternative mouse control option to the users. (Ravikrishna R et al. 2003) [7].

The first process for the gesture recognition is known as Segmentation. It divides the hand gesture image into regions that are separated by fine boundaries.(Khan et al. 2012) [8] If the segmentation process is followed in a proper way, the results of the segmentation process can enhance.(Gao Long et al. 2017) [9]. Some methods and algorithms are applied for the process of feature extraction. (Nutipalli et al. 2017) [10], (Liu Shi-lei et al. 2017) [11].

There are certain guidelines by the World Wide Web Consortium (W3C) to make the web accessible to all. These guidelines should be followed to design any project related to accessibility. (Ballantyne et al. 2018) [12]. The styling should be in accordance with those protocols. (Venkateswarlu et al. 2011) [13] Speech Recognition can be used to provide accessibility to those who can't access the web easily. (Ismail et al. 2010) [14], (Ballantyne et al. 2018) [15] The speech recognition features can be incorporated to mobile devices as well. (Corbett et al. 2016) [16], (Kim Patch et al. 2015) [17] If applied in real-life, this technology can help the people having cognitive disabilities to a great extent. (Young et al. 2018) [18], (Deli et al. 2014) [19]

3 Research Methodology

3.1 Voice-Based Browsing

The field of Computer Science that deals with the design of systems that recognize words expressed orally is called Voice Recognition. It is important to mention that the recognition implies the literal identification of words present in the analyzed sound and the understanding of the meaning of them. Right now, the idea of maintaining a conversation with an apparent computer can be a theme of fictional science. But this circumstance is changing very quickly. A large number of people have spoken, at this moment, with their mobile devices. For example, send an email, text messages, create notifications on its calendar, search for people on your contact list, hold an alarm, or search for information.

Due to the predominance of manual interaction required at the time to interact with both browsers as with websites, it is of interest to formulate an alternative interaction without resign the supported functionality. A model is presented whose objective is to offer an alternative way of web browsing completely dispensing with manual interaction through the use of exclusive voice commands.

Much like Visually analyzing the data in front of us and then performing certain actions to reach a goal for an individual. Challenges are faced by Visually impaired and physically disabled to work with websites to invoke certain actions and hinder their ability to browse through the network on websites. In addition to Visual and Touch, it is necessary to add more means to work with a website that brings voice into the question.

Voice-based browsing, as it sounds is about achieving all you can from the web just by using your voice. One can look up URLs, skim through the content, ask for specific content, fill in details for forms perhaps, and a lot more. It is said that voice-based browsing has the capabilities to reduce the number of keyboards in near future.

The current advent of ML on the web using Javascript and Natural language Processing tools like Google Actions, Dialogflow, Amazon Alexa opens up many boundaries by allowing the use of classifiers to understand what is expected from it. The possibilities of merging Dialogflow's kernel with browser and using customized reinforcement learning ideology to improve and add more humanity into the responses/requests. The idea holds a step by step process for associating actions to works. The steps are as follows:

- Receive the user's sentence between pauses. (voice)
- Run it through a reinforced Dialogflow model. (voice to text
- Return the identifier expected which points to associated action (text)
- Perform the associated action (text)
- Return feedback from associated actions and respond to the user (text to voice)

3.2 Hand Gestures

The hand gesture can be defined as the communication method which could be used for the human and machine interface. Hand gesture provides a non-verbal communication that can plausibly replace verbal communication such as speech or writing as it is universal i.e. same for the whole world. Even in new generation games, AR and VR, gestures are used primely. In this study, we will be using a simple webcam as an input camera. We designed a 3D CNN model that can determine different hand gestures. The dataset used is Jester Dataset which contains 27 different static as well as dynamic hand gestures in total. When evaluated on real-time recognition of six types of hand gestures, the proposed architectures obtain 94% accuracy in demonstrating the significant potential for human-computer interfaces.

A substantial amount of research has been done in this area using machine learning on traditional video and depth camera input, as in [10]. Only a simple webcam is used in this study for the interaction between human and computer, no external hardware is needed as in the State-of-the-Art techniques. Hand gesture control has been achieved before by using different hardware techniques like stereo cameras, depth cameras, Microsoft Kinect that uses IR for hand tracking, and many more. In this study, we provide Hand Gesture control by just the use of a webcam.

This study provides a deep learning method to recognize hand gestures by just using a simple webcam. We designed a 3D convolution neural network model for hand gesture recognition using the Jester V1.0 dataset. According to the result of the training experiment, it got an accuracy of 94%.

Model Architecture. Figure 2 shows the architecture of the model we use for gesture recognition, which includes, in total, 6 layers including the last layer. The First 3 layers are the 3D-COV layers, the next layer is an LSTM layer, then the last 2 layers are the Dense Layers which provide us with an output. In all the layers activation is Relu but Softmax is used in the output layer. There are 2 dropouts used after the 2nd and 3rd layers and batch normalization is also used.

The input which is fed to the model is in the form of an image. The frames are fetched from and then preprocessing of the data frames is performed. The frames are then inserted into the hand gesture model. When the model detects which gesture it is, the navigation is performed. PyAutoGUI, which is a cross-platform GUI Python module enables to automate and simulate some more actions and is reiterated till the hand gestures are recognized.

3.3 Facemouse

A facial mouse is used to replicate the features of a normal mouse. By tracing the movement of the face, one can perform actions like up-scroll, down-scroll, click, and so on. This technique is very beneficial for that group of people who have certain kinds of physical impairment due to which they cannot control the mouse through their hands (Fig. 3).

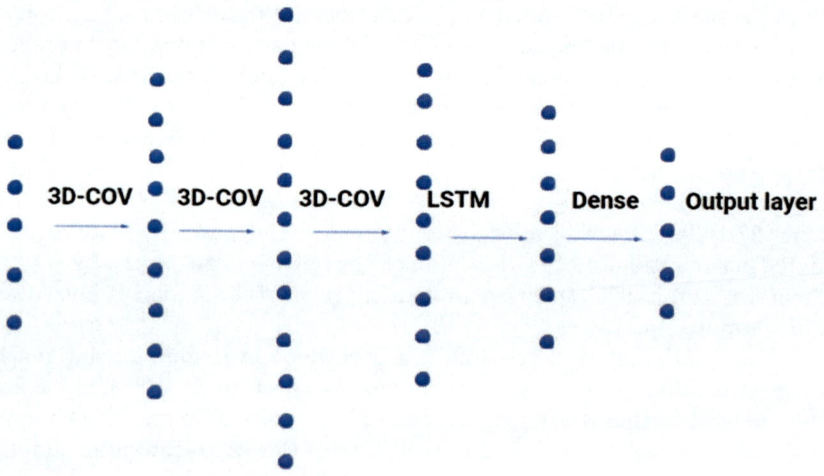

Fig. 2. Model architecture

Face mouse controls the mouse cursor with your facial movements and works just with a regular webcam. It is hands-free, with no wearable hardware or sensors needed. It is made by using Computer vision techniques and just image processing, no deep learning is used to achieve it. It runs smoothly at 30 frames per second.

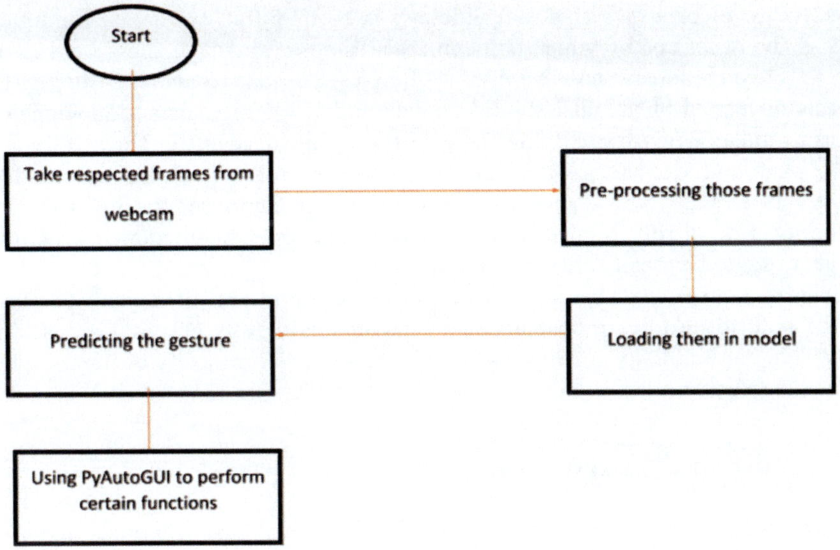

Fig. 3. System flowchart

- Get respected frames
- Pre-process them
- Find facial landmarks, for example, nose, lips, etc.
- Track the movement of nose for the mouse cursor
- Decide the ratio of open mouth for switching into modes
- Put wait time for the left click (Fig. 4).

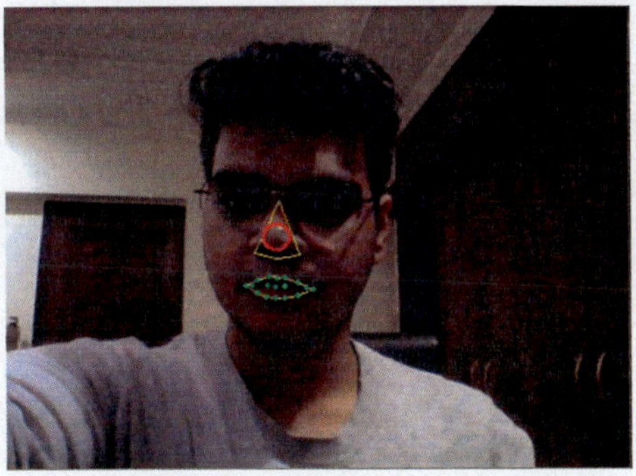

Fig. 4. Facial landmarks for face mouse

Facial recognition follows the standard approach of fetching the respective frames from the video input. After that, these frames are preprocessed. The key objective to incorporate this methodology is to first find the facial landmarks such as the nose, lips by which one can track the movement of these landmarks. Say, we have to detect that the user wants to scroll. So, to achieve this, first of all, the movement of the nose should be detected to identify the action which one wants to make. The upward movement of the nose will result in the scrolling-up of the page and vice-versa. In addition to this, an open mouth feature is added to the model which will decide the mode of action. Suppose, one has to switch the action from scroll to click, the user has to first open their mouth which will act as a toggle between the two actions.

In this manner, just by the tracking of facial landmark movement, the users will be able to perform mouse actions without having any external mouse support. This methodology is a novel approach to solve the majority of the problems of the disabled group.

4 Result and Discussions

This research is based on the integration of Computer Vision models and Artificial Intelligence on the Web. The technique gave a successful result when the model ran simultaneously with a website. The accuracy of the model is also pretty good but we are working to enhance it more. In addition to this, the voice-based browsing engine can run and detect local languages as well which makes it highly compatible with the group of differently-abled citizens that don't know how to communicate in English. The results not only boost the cognitive abilities of the target audience but also empower them. Without the usage of much computational power, the models based on Perceptual User Interfaces run smoothly on the system, giving excellent results (Fig. 5) (Table 1).

Table 1. Result comparison

Method	Modality	Accuracy%
Liu and Shao [3]	Color + depth	88.7
Pigou et al. [5]	Color + depth + skeleton	97.2
Wenjin and Wang [20]	3D-CNN	86
Chinmaya and Razvan [4]	3D-CNN + LSTM + depth	91
Ours	**3D-CNN + LSTM**	**94**

For the experiment purpose, we created an application that can control video through gestures like right swipe can fast forward the video, and many more like it. The model we created gave us an accuracy of about 94%, which is considerable for only webcam videos as all the work done in this field uses either IR images or depth images but, in this study, only RGB image frames were used taken from a simple webcam. Comparison between the state-of-the-art techniques is displayed below with their modality:

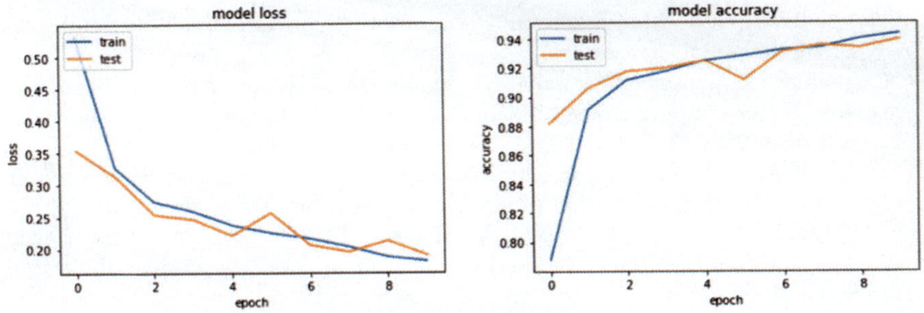

Fig. 5. Model accuracy and Loss

5 Conclusion

Looking at the challenges faced by differently-abled citizens, the proper implementation of this research will help them to use the web in an enhanced manner. The aim of creating a holistic web experience for differently-abled citizens can be accomplished very conveniently. Based on the usage of AI for social good, this solution is a versatile combination of all required things in one single place. Moreover, this methodology can prove to be an emerging technology and hence can simplify the tasks of the group we are targeting. To conclude, this research is the future of AI-based Websites.

References

1. Turk, M., Robertson, G.: Perceptual user interfaces. In: Communications of the ACM, vol. 43 (2000)
2. Marium, A., Rao, D., Crasta, D.R., Acharya, K., D'Souza, R.: Hand gesture recognition using webcam. Am. J. Intell. Syst. **7**(3), 90–94 (2017). https://doi.org/10.5923/j.ajis.20170703.11
3. AL, L., Shao, L.: Learning discriminative representations from RGB-D video data. In: IJCAI (2013)
4. Naguri, C.R., Bunescu, R.C.: Recognition of dynamic hand gestures from 3D motion data using LSTM and CNN architectures. In: 2017 16th IEEE International Conference on Machine Learning and Applications (ICMLA). IEEE (2017)
5. Pigou, L., Oord, A.V.D., Dieleman, S., Herreweghe, M.V., Dambre, J.: Beyond temporal pooling: recurrence and temporal convolutions for gesture recognition in video. arXiv (2015)
6. Turk, M., Hu, C., Feris, R., Lashkari, F., Beall, A.: TLA based face tracking. In: Intern Conference on Vision Interfacec Calgary, pp. 229–235 (2002)
7. Ruddarraju, R., et al.: Perceptual user interfaces using vision-based eye tracking. In: The 5th International Conference on Multimodal Interfaces, British Columbia, Canada, Vancouver, pp. 227–233 (2003)
8. Khan, R.Z., Ibraheem, N.: Hand gesture recognition: a literature review. Int. J. Artif. Intell. Appl. (IJAIA). **3**, 161–174 (2012). https://doi.org/10.5121/ijaia.2012.3412

9. Gao, L. Research on Static Gesture Recognition Algorithm Based on Neural Network. Ning Xia University (2017)
10. Nutipalli, P.: A comparative analysis on hand gesture recognition using deep learning. Int. J. Res. Appl. Sci. Eng. Technol. **8**, 1197–1203 (2020). https://doi.org/10.22214/ijraset.2020.6195
11. Liu, S.: The study concerning human-computer interaction used in the manual segmentation and identification of key techniques. Shan Dong University (2017)
12. Ballantyne, M., Jha, A., Jacobsen, A., Hawker, J., El-Glaly, Y.: Study of accessibility guidelines of mobile applications, pp. 305–315 (2018). https://doi.org/10.1145/3282894.3282921
13. Venkateswarlu, L.K., Rambatla, V., Ande, K.: A novel approach to speech recognition by using generalized regression neural networks. Int. J. Comput. Sci. Issues **8**, 484–489 (2011)
14. Ismail, N., Zaman, H.B.: Search Engine Module in Voice Recognition Browser to Facilitate the Visually Impaired in Virtual Learning (MGSYS VISI-VL) (2010)
15. Ballantyne, M., Jha, A., Jacobsen, A., Hawker, J., El-Glaly, Y.: Study of accessibility guidelines of mobile applications, pp. 305-315 (2018). https://doi.org/10.1145/3282894.3282921
16. Corbett, E., Weber, A.: What can I say?: addressing user experience challenges of a mobile voice user interface for accessibility, pp. 72–82 (2016). https://doi.org/10.1145/2935334.2935386
17. Kim, P., Spellman, J., Wahlbin, K.: Mobile Accessibility: How WCAG 2.0 and Other W3C/WAI Guidelines Apply to Mobile (2015)
18. Young, M., Courtad, C.A., Douglas, K., Chung, Y.-C.: The effects of text-to-speech on reading outcomes for secondary students with learning disabilities. J. Spec. Educ. Technol. **34**, 016264341878604 (2018). https://doi.org/10.1177/0162643418786047
19. Delić, V., Sečujski, M., Sedlar, N.V., Miskovic, D., Mak, R., Bojanic, M.: How Speech Technologies Can Help People with Disabilities (2014). https://doi.org/10.1007/978-3-319-11581-8_30
20. Zhang, W., Wang, J.: Dynamic hand gesture recognition based on 3D convolutional neural network Models. In: IEEE 16th International Conference on Networking, p. 2019. Sensing and Control (ICNSC). IEEE (2019)

Subspace Guided Collaborative Clustering Based on Optimal Transport

Fatima-Ezzahraa Ben-Bouazza[1,2,3](✉), Younès Bennani[1,2], Abdelfettah Touzani[3], and Guénaël Cabanes[1,2]

[1] LIPN UMR 7030 CNRS, Université Sorbonne Paris Nord, Villetaneuse, France
{fatima-ezzahraa.ben-bouazza,younes.bennani,guenael.cabanes}@sorbonne-paris-nord.fr

[2] LaMSN, La Maison des Sciences Numériques, Saint Denis, France

[3] LAMA-FSDM, Université Sidi Mohamed Ben Abdellah, Fes, Morocco
abdelfettah.touzani@usmba.ac.ma

Abstract. Collaborative clustering is a promising approach in the learning from other learners research area. Although extensive research have been done to improve the collaborative approaches, they still suffer from several issues, including the mechanism of exchanging the information and how to measure the quality of this information. In this paper we introduce a new model of collaboration guided by feature selection, where the main idea is to choose the features that give the best representation for each collaborator and guarantee the communication between them, while preserving the privacy of each collaborator. Collaborative clustering will be developed within the framework of the theory of optimal transport. Indeed, this theory offers a formalism that is highly adapted to collaboration between members of a set of collaborators. Extensive experiments were conducted on multiple data-sets to evaluate the proposed approach and demonstrate its utility.

Keywords: Collaborative clustering · Subspace clustering · Features selection · Optimal transport

1 Introduction

Clustering is one of the main exploratory framework in machine learning. The idea is to cluster the data on different clusters according to a similarity criteria, where the objects that belong to the same cluster are more similar than the ones that belong to other clusters. In the literature, several approaches have been proposed based on the idea of maximizing the similarity intra-cluster and dissimilarity inter-cluster. With the development of hardware technology, a huge amount of data represented in different sites and different structures have been generated in real word applications. This kind of data is considered as a new challenge to develop the existing clustering algorithms, designed for single view data, to be more adaptable to multi-view data. Several approaches have been

A. Abraham et al. (Eds.): SoCPaR 2020, AISC 1383, pp. 113–124, 2021.
https://doi.org/10.1007/978-3-030-73689-7_12

proposed in the context of Multi-View Clustering, where the main idea is to train different models originating from different views to form a global model called the consensus clustering [2]. The consensus leads to a global minimization, or to aggregate different information between views in order to find a local minimum for each learner, i.e. learn from other learners.

However, one of the most challenging problem in Multi-view clustering is how to make models provided from each view comparable. In other word, the diversity between views is very high, which can lead to lose the sense of learning between sites or views, especially when the characteristics of each view are very different. Furthermore, some of the views may be of a high dimensionality, which leads to a high computational complexity and could decreased the quality of the models of each view. Feature selection is one of the most popular approach to address this type of problem, and can both simplify the calculation and make the models more accurate to comparison.

Collaborative learning is one of the new area of research in Multi-models analysis that aims to exchange different information between models in order to enhance the performances of learning models [3]. The collaborative learning categorized up to now in unsupervised learning. The main interest of this discipline is to help local clustering algorithms to make decision about how to improve there local cluster by using different information provided from other algorithms trained on different sites that could either have the same instances represented by different characteristics, or different samples that sharing the same characteristics. This kind of algorithms have proved a very useful to learn from distant representation of data spread in multiples sites while preserving the privacy of each sites. The collaborative clustering was introduced as by [8] as a fuzzy clustering framework, but lately extended to prototype based clustering using Self-Organized-Maps(Co-SOMs) and generalized using Generative-Topographic-Maps (Co-GTM) [7]. Another studies was done on satellites images using a collaborative objective function based on entropy and prove the effectiveness of the collaborative clustering on this type of data [11]. This studies have given a better results comparing to the first one but the main idea behind changing the model is to find better mechanism that resolves the paradigm of "how to choose the right collaborator?" in other word how to odder the collaboration so it we can avoid the over-fitting of the models. Recently new approach have proposed using optimal transport (Co-OT) [1] which gives the better results to comparing to the recent prototype based approach, and the main idea about choosing the collaborators was based on the comparison of the local distribution using Wasserstein distance. On the other hand, unsupervised feature selection aims to reduce the time complexity and improve the generalization ability of learning machines by removing the redundant, irrelevant and noisy features. The selection of the best features to be used in learning systems is a key issue in obtaining a satisfactory performance. Feature selection methods have proved a significant results in Multi-views consensus clustering [14]. For this reason we had the intuition to adapt this approach on collaborative clustering to guide the diversity between local models and to get more accurate local models. We propose in this paper a new algorithm of collaborative learning guided by feature selection, where the main idea is to improve the learning from the collaborative by selecting the most informative feature according to evolution of the considered collaborator.

The rest of the paper will be organized as follow: the next section will explain the motivation behind the proposed approach, and will be followed by section where we details the proposed approach. In Sect. 3 we will investigate and discuss experimentally the quality of our approach, and finally we will conclude by a summary about the proposed approach and some perspective work.

2 Proposed Approach

2.1 Motivation

Recent work have been done on the impact of diversity between the collaborators and the quality gain brought from each collaboration [9]. The main idea of this of this work was to analyse the evolution of the quality during the collaboration, this work proved that the collaborators should not be very diverse otherwise will decrease the quality or will completely lose there own structure. From this point, we had the intuition to guide this diversity by using feature selection technique. The idea behind is to make the collaboration strengthened by selecting not only the most informative features for each site but also the features that will increase the quality of the distant collaborator. This selection will be done iteratively for each selected collaborator. This way we will control diversity during the collaboration. Consequently the the choice of the collaborator will be more accurate and the collaboration will be more controlled. Collaborative clustering will be developed within the framework of the theory of optimal transport. This theory offers a formalism that is highly adapted to collaboration between members of a set of collaborators. In what follows we will explain the proposed algorithm and detail the mechanism of the method.

2.2 Background and Notations

We remind that this work aims to improve the collaborative algorithm based on optimal transport which had already proved higher performance comparing the collaborative prototype based method. To do so,we will briefly introduce some needed optimal transport background to follow algorithm.

Optimal Transport
Let consider $\Omega \subseteq \mathbb{R}^d$ Given two families of data sets in Ω, $X = \{x_i\}_{i=1}^n$ and $Y = \{y_i\}_{i=1}^m$, let μ and ν be their respective distributions over Ω.

The Monge-Kantorovich problem is a convex relaxation which aims to find optimal transport maps as coupling γ defined as a joint probability measure over $\Omega \times \Omega$ with marginals μ and ν that minimizes the cost of transport w.r.t $c : \Omega \times \Omega \rightarrow \mathbb{R}_+$:

$$\gamma^* = \underset{\gamma \in \Pi}{\text{argmin}} \int_{\Omega \times \Omega} c(x, y) d\gamma(x, y) \tag{1}$$

We consider the discrete setting of Optimal Transport problem. This case arises where μ and ν are only accessible through discrete samples. The empirical measures can be defined as:

$$\mu = \sum_{i=1}^{n} p_i^1 \delta_{x_i} \qquad \text{and} \qquad \nu = \sum_{i=1}^{m} p_i^2 \delta_{y_i} \tag{2}$$

With δ_{x_i} the Dirac function at $x_i \in \mathbb{R}^n$ p_i^1 and p_i^2 the probability masses associated to the *ith* sample, s.t $\sum_{i=1}^{n} p_i^1 = \sum_{i=1}^{m} p_i^2 = 1$.

The Optimal Transport plan consist to find γ^* a joint probability that minimizes 3, the problem admits a solution fue its convexity, and called *Wasserstein distance.*

$$W(\mu, \nu) = \underset{\gamma \in \Pi(\mu_s, \mu_t)}{\text{argmin}} \; <\gamma, C>_F \tag{3}$$

with:

$<.,.>_F$ the Frobenius dot product

$C \in \mathbb{R}_+^{n \times m}$ the transport cost with C_{ij} given by: $C : X \times Y \rightarrow \mathbb{R}_+$

$\Pi(\mu, \nu) = \left\{\gamma \in \mathbb{R}_+^{n \times m} \mid \gamma \mathbf{1}_n = \mu, \gamma^T \mathbf{1}_m = \nu\right\}$ the transportation polytope where $\mathbf{1}_n$ is a n-dimensional vectors of ones.

The Wasserstein distance has known a successful application in machine learning. However this distance still suffer in high dimensional in term of complexity, low convergence and stability. For this reason in our proposed approach we choose to use the regularized Wassesretein distance proposed by Cuturi [5], where they add an entropic penality not only to make the function more smoothed with high stability, but also to make the problem resolvable by using Sinkhorn-Knopp matrix Scaling algorithm [4].

This is represented formally by the following minimization problem:

$$\gamma_\lambda^* = \underset{\gamma \in \Pi(\mu_s, \mu_t)}{\text{argmin}} \; <\gamma, C> - \frac{1}{\lambda} E(\gamma) \tag{4}$$

where $E(\gamma) = -\sum_{i,j}^{n_s, n_t} \gamma_{ij} \log(\gamma_{ij})$ and $\lambda > 0$ the entropy regularization parameter and C is the cost matrix.

2.3 Proposed Algorithm

Let consider r collaborators, where the data of each collaborator v, $X^v = \{x_i^v\}_{i=1}^{n^v}$ with $x_i^v \in \mathbb{R}^{d^v}$ and d^v the dimension of the space representation of v, and corresponds to a distribution $\mu^v = \frac{1}{n^v} \sum \delta_{x_i^v}$. Starting with the first step of the proposed approach which is selecting the features. The main idea of this step is to choose the best feature for the collaboration so we can reduce the complexity and the diversity between collaborator. Thus, in this step we will first rank each feature according to its discrimination, to do so, we will use the improved F-score 5 which measures the discrimination among several samples [13].

Considering the dth vector, $d = 1, ..., d^v$ $f_{i,d}, i = 1, \ldots, n$, and l samples of the data-set, the improved $F-$score is defined as follow :

$$F_d = \frac{\sum_{j=1}^{l}(\overline{f_d}^{(j)} - \overline{f_d})^2}{\sum_{j=1}^{l}\frac{1}{n_j-1}\sum_{k=1}^{n_j}(f_{i,d}^{(j)} - \overline{f_d}^{(j)})^2} \tag{5}$$

Where n_j is the size of the jth sample, $\overline{f_d}, \overline{f_d}^{(j)}$ are the average of the dth feature of the whole data-set, and the jth sample respectively. Then numerator indicates the discrimination between each sample, and the denominator indicates the one within each of sample.The larger the improved $F-$score is, the feature is more likely to be discriminative.

The idea behind ranking each feature before the threshold step is to avoid the biased selection because some feature my have some overlaps. Therefor, considering only threshold selection my affect the performance of the proposed algorithm. given for example of the following Scenario where there are two features f_1 and f_2, using both of them will increase the quality by 5%, while using just f_1 would increase the quality by 4%, and 5% for f_2. The percentage threshold is 3% imagine that the algorithm selects first f_1, in this case it will select f_2 because the quality will be increased only by 1%. Consequently, the quality will be less than its actual potential. For this purpose we chose the rank each feature sort then in a descending order according to their discrimnation rank.

Algorithm 1 details the mechanism of this hybrid technique of feature selection that combines between ranking the importance of the feature using F-score 5, and forward selection basing on the variation of the clustering quality index $Q_{S_d^v}$ by measuring it using the Threshold p^v. It should be mentioned that the feature selection algorithm depends also on the quality of the collaborator that needs information. More precisely when the algorithm chose a collaborator that he will exchange the information with him, the selection of the feature will take in consideration if the exchange is beneficial or not for the distant collaborator, this will be detailed in Algorithm 2. Considering the same notation, in the collaboration step, we seek to minimize the following objective function:

$$\underset{L^v, M^v}{\text{argmin}}\{< L^v, C(X^v, M^v) >_F - \frac{1}{\lambda}E(L^v) + \sum_{v'=1.v'\neq v}^{r} \alpha_{v',v}(< L^{v,v'}, C(M^v, M^{v'}) >_F - \frac{1}{\lambda}E(L^{v,v'}))\} \tag{6}$$

Where, $M^v = \{m_1^v, .., m_{k^v}^v\}$, corresponding to a distribution ν^v, that represents the local clusters of each collaborator v, $L^v = \{l_{ij}^v\}_{i,j=1}^{i,j=n^v,k^v}$ is OT plan between the the local data X^v and the centroids M^v, and and $L^{v,v'}$ is the OT plan between the centroids of the vth and v'^{th} collaborators.

The first term refers to the first step on the collaboration where it aims to find the distribution of the local centroids. The second term of the objective function refers to second step which consists to exchange to the information between the collaborators, in order to increase the quality, this exchange based on influencing the local distribution of the centoirds by the distant distribution. $\alpha_{v',v'}$ are non-negative coefficients proportional to a mutual information which measure the diversity between the collaborators, called "confidence coefficients". The last term of the of the objective function is a combination of a penalty term based on entropy regularization locally between instances and local centroids, and between centroids distant.

Algorithm 2 details the computation of proposed approach. The idea behind is to make the collaboration more blinded en the exchange of the information is more punctual and targeted by introducing the forward feature selection for each collaborators based on its local quality of clustering as a threshold, and then chose the median collaborator based on the comparison of similarity between the distribution of local centroids to the distant collaborators. Following [9] we choose at each iteration the median one to exchange with, in this case if the quality is increased, the centroids are updated, otherwise, we give another shot to the chosen collaborator to execute another shot of forward selection basing on the quality index of the collaborator. In this case either the quality will be increased or this collaborator will be removed from the list and the process will be repeated until the quality of the collaborators stops increasing. It must highlighted that proposed algorithm is adaptable with all the prototype based algorithms, more precisely instead of using the Sinkhorn-Means as a local algorithm to get the centroids, we can use other prototype models like k-means, SOM, EM, etc. The proposed algorithm has therefore the capability to work with hybrid models. However, in our case we chose to use Sinkhorn-means as local model to keep all the mechanism based on OT.

Algorithm 1: Threshold-Forward Feature selection

Input : $F^v = \{f_d^v\}_{d=1}^{d^v}$: Set of the features of collaborator v
p^v : the parameter of selected percentage
Output: The the set of the selected feature S_{best}^v

Initialize the set of the best selected feature$S_{best}^v = S_0^v = \emptyset$
The quality index of clustering $Q_{best}^v = Q_0^v$
repeat
    Calculate the rank value for each feature f_d^v, $1 < d < d_v$
    Stored features in descending order according to their rank.
    $S_d^v = S_{d-1}^v \cup f_d^v$
    Perform the clustering algorithm.
    if $| Q_{S_{d+1}^v} - Q_{S_d^v} | > p^v$ **then**
        $Q_{best}^v = Q_{S_{d+1}^v}$
        $S_{best}^v = S_{d+1}^v$
until *convergence*;
return S_{best}^v

Algorithm 2: Collaborative clustering (Co-FS-OT)

Input : $\{X^v\}_{v=1}^r$: the r collaborators' data with distributions $\{\mu^v\}_{v=1}^r$
$\{k^v\}_{v=1}^r$: the numbers of clusters
λ : the entropic constant
$\{\alpha_{v,v'}\}_{v,v'=1}^{v,v'=r}$: the confidence coefficient matrix

Output: The partition matrix $\{(L^v)^*\}_{v=1}^r$ and the centroids $\{M^v\}_{v=1}^r$

Initialize the centroids $M^v = \left\{m_j^v\right\}_{j=1}^{k^v}$ randomly
Compute the associated distribution $\nu^v = \frac{1}{k^v}\sum_{j=1}^{k^v}\delta_{m_j^v}$, $\forall v \in \{1, ..., r\}$
repeat
  for $v = 1, ..., r$ **do**
    Update the centroids M^v and the partition matrix $(L^v)^*$ using a **local algorithm** (e.g. Sinkhorn-Means, SOM, GTM, k-Means...)
    Update the centroids distribution $\nu^v = \frac{1}{k^v}\sum_{j=1}^{k^v}\delta_{m_j^v}$
    for $v' = 1, ..., r$ *and* $v \neq v'$ **do**
      Compute **Feature Selection Algorithm** 1 on $X^{v'}$
      Compute the OT matrix $(L^{v,v'})^* = \{l_{jj'}\}_{j,j'=1}^{j,j'=k^v,k^{v'}}$ between the centroids of collaborators v and v':

$$(L^{v,v'})^* = \underset{L^{v,v'}\in\Pi(\nu^v,\nu^{v'})}{\operatorname{argmin}} < L^{v,v'}, C(M^v, M^{v'}) >_F - \frac{1}{\lambda}E(L^{v,v'})$$

    Chose the median collaborator :

$$v^* = median_{v'}\left\{(L^{v,v'})^*\right\}_{v'=1}^r, v' \neq v$$

    if *Quality increased* **then**
      Update the local centroids based on the collaborator X^{v*}:

$$m_j^v = \alpha_{v,v*}\sum_{j'} l_{jj'}^{v,v*} m_{j'}^{v*} \quad 1 \leq j \leq k^v$$

    else
      Compute **Feature Selection Algorithm 1** on the median collaborator X^{v*}
      Update the local centroids if the quality is increased.

$$m_j^v = \alpha_{v,v*}\sum_{j'} l_{jj'}^{v,v*} m_{j'}^{v*} \quad 1 \leq j \leq k^v$$

until *convergence*;
return $\{(L^v)^*\}_{v=1}^r$ *and the centroids* $\{M^v\}_{v=1}^r$

3 Experiments Results

3.1 Protocol Experiments

In order to evaluate the proposed approach experimentally, we start with a pre-processing of the data, where we create the collaborators in the form of the local subsets that share the same instances but represented with different characteristics. We split the data horizontally into 10 collaborators (subets), each collaborators X^v that will be considered as the inputs of Algorithm 2. The first step consists to applied forward selection followed by Sinkhorn-means [4] to get the centroids. The, the coefficient is computed based on diversity between the distributions of each collaborator. Algorithm 2 in trained 20 times to estimate the mean quality of the collaboration and a 95% confidence interval for the 20 experiments. The comparison will be done between a collaboration with OT only, and by using the feature selection during the training. Both approaches will be tested on the same subsets and the same local model(Sinkhorn-Means). due to the limit of pages we chose to use only two index of comparison, an internal and external one that will be defined in the next section.

3.2 Experiments Results

In this section we evaluate the proposed approach using several data sets, described in Table 1. We chose for this evaluation two type of quality index, an internal one which is *Silhouette* index [10], and an external one since original data is labeled which the Adjusted Rand Index(ARI) [12]. It must be mentioned that during the training of the algorithm we used another unsupervised quality index to control the collaboration and feature selection which is Davis-Bouldin [6]. The reason why we chose to add another unsupervised index instead of using only DB index is that the collaboration depends on the variation of this index, which means it iterates while DB is decreasing and converges when the index stops changing.

Table 1. Some characteristics of the experimental real-world data-sets

Datasets	#instances	#Attributes	#Classes
Glass	214	9	7
Spambase	4601	57	6
Waveform-noise	5000	40	3
WDBC	569	30	2

The *Silhouette* index is based on measurement of the difference between the average of distances between the instances belonging to the same cluster and the average distance between to other clusters. The closer the *silhouette* value is to 1, means that the instances belongs the right cluster. Table 2 chose

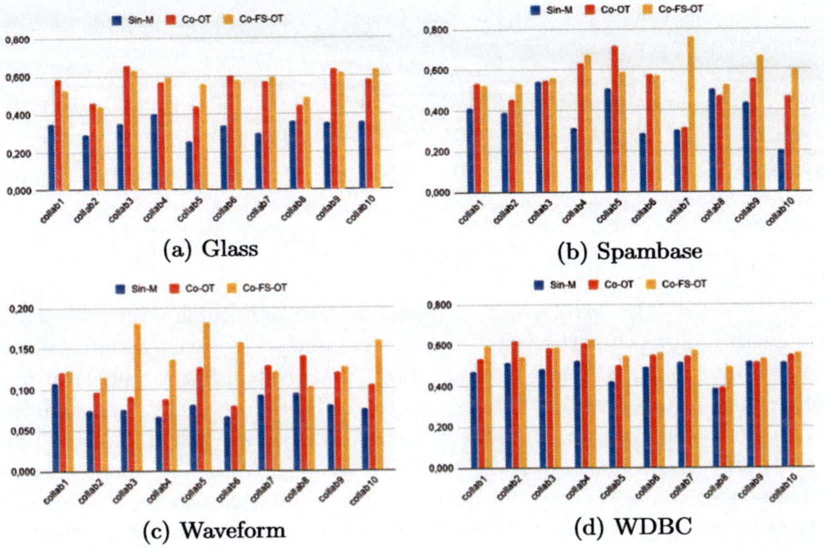

(a) Glass (b) Spambase

(c) Waveform (d) WDBC

Fig. 1. Histogram that compares *Silhouette* index values for 10 collaborators for each data sets.

the result of this index comparing the quality of the clusters using Sinkhorn-Means locally, the collaboration based on OT interactions (Co-OT) and the collaboration guided by the feature selection (Co-FS-OT). As one can see, over 20 executions the average value does prove that that forward feature selection improve the quality of the cluster on different data sets. Moreover, Fig. 1 shows the details of each collaborator and proves that each collaborator succeeds in improving its local quality by giving the possibility to feature selection shots during the collaboration. Further more, since the data that we used provide labels we use *ARI* index which measures the agreement between two partitions, one provided from the origin labels and the second one provided from the proposed approach. The value of the *ARI* is between 0 and 1, the more closer the *ARI* value to 1, the better agreement is, consequently better cluster. Table 3 shows result of the average of this index over 20 executions of the 10 collaborators, the results in Fig. 2 confirm that guidance of the collaboration by the feature

Table 2. Average values ($\pm CI_{95\%}$) of Silhouette index for each data set over 20 executions.

Models	Sin-Means	Co-OT	Gain	Co-FS-OT	Gain
Glass	0.336 ± 0.025	0.553 ± 0.04	0.217 ± 0.04	0.563 ± 0.03	0.227 ± 0.03
Spambase	0.390 ± 0.06	0.526 ± 0.06	0.135 ± 0.07	0.599 ± 0.04	0.208 ± 0.01
Waveform	0.082 ± 0.01	0.110 ± 0.01	0.028 ± 0.01	0.140 ± 0.01	0.059 ± 0.02
WDBC	0.484 ± 0.02	0.540 ± 0.04	0.056 ± 0.02	0.562 ± 0.01	0.078 ± 0.02

Table 3. Average values ($\pm CI_{95\%}$) of ARI index for each data set over 20 executions.

Models	Sin-Means	Co-OT	Gain	Co-FS-OT	Gain
Glass	0.155 ± 0.04	0.237 ± 0.02	0.120 ± 0.05	0.257 ± 0.01	0.103 ± 0.03
Spambase	0.072 ± 0.02	0.136 ± 0.01	0.063 ± 0.02	0.148 ± 0.01	0.076 ± 0.02
Waveform	0.179 ± 0.01	0.218 ± 0.02	0.039 ± 0.01	0.252 ± 0.02	0.073 ± 0.02
WDBC	0.219 ± 0.02	0.439 ± 0.02	0.198 ± 0.02	0.494 ± 0.03	0.274 ± 0.02

selection increases the agreement between the two partitions which means that we are getting close the true label of each instance. This proves that even that we are working on unsupervised clustering, the proposed approach its quality to cluster the data correctly comparing to its labels. It must mentioned that although up to our comparisons, the collaboration based on OT (Co-OT) is the best prototype based method, the Co-FS-OT has improved that feature selection algorithm chose the best representation of each collaborator to learn from. In this way we gave the opportunity to the collaborator to control the distant clustering, while preserving privacy and anonymity. Sensitivity Box-Whiskers plots Fig. 3 are drawn for the 20 experiments scores for each data-set for the collaboration Co-OT and Co-FS-OT, They enable us to study the distributional characteristics of scores as well as the level of the scores. To begin with, scores are sorted over the 20 tests. Then four equal sized groups are made from the ordered scores.

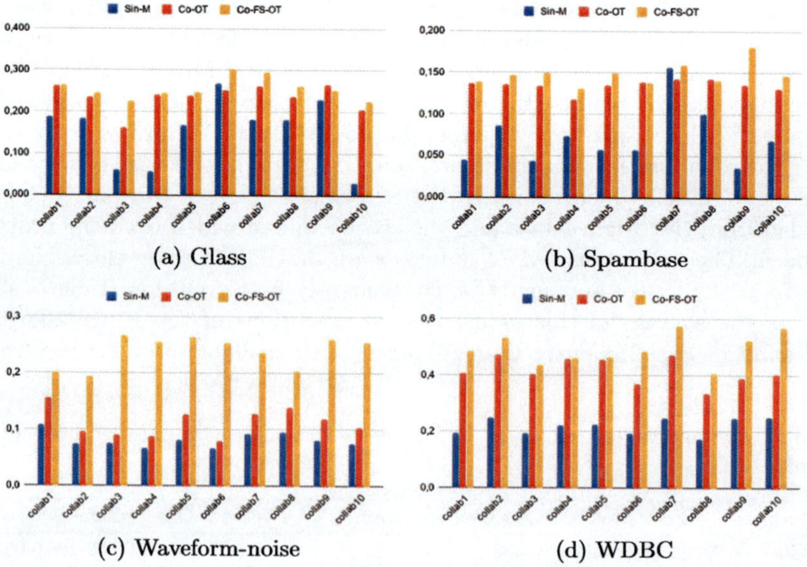

Fig. 2. Histogram that compares Adjusted Rand Index (ARI) values for 10 collaborators for each data sets.

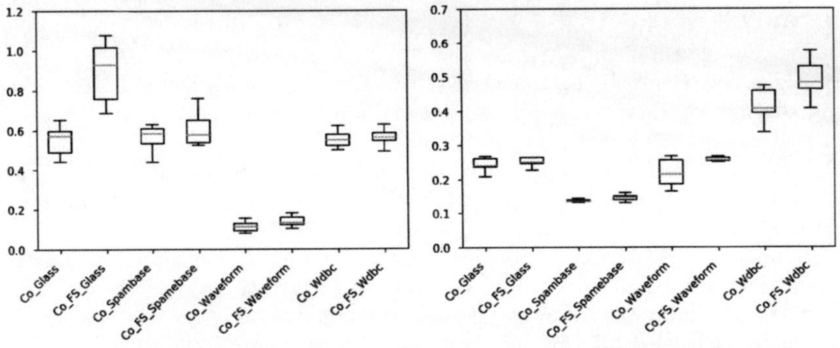

(a) Silhouette (b) ARI

Fig. 3. Sensitivity box plost CO-OT and CO-FS-OT comparaison

That is, 25% of all scores are placed in each group. The lines dividing the groups are called quartiles, and the groups are referred to as quartile groups. Usually we label these groups 1 to 4 starting at the bottom. The median (middle quartile) marks the mid-point of the scores and is shown by the line that divides the box into two parts. Half the scores are greater than or equal to this value and half are less. The middle "box" represents the middle 50% of scores for the group. The range of scores from lower to upper quartile is referred to as the inter-quartile range. The middle 50% of scores fall within the inter-quartile range. As can be seen from these graphs, the overall performance behaviour shows a clear improvement using the feature selection. Moreover, we remark that for *ARI* box the diversity between the collaborators is reduced, which mean that the Co-FS-OT improve the distribution to be nearer to the real distribution of the data-set.

4 Conclusion

To summarise, in this paper we improved a recent prototype-based collaborative approach by introducing a forward feature selection who contribute in the control of the collaboration and enforce interactions between the collaborators in each iteration while preserving the private data of each learners. Compared to the previous collaborative approach based on OT, the proposed algorithm had given positive results and improved the quality of the exchange of information. Its strength is highlighted by good experimental results both for artificial and real data-sets. For the perspective work, we are still working on the improvement of the notion of confidence between the collaborators. We believe that this notion has a strong relation with the diversity between collaborators and their local quality. Thus, we want to build a system that is capable to measure this diversity and quality at the same time, which would lead to create better interaction between learners, reduce the training time of the algorithm and guarantee better convergence.

References

1. Bouazza, B., Bennani, F.E., Cabanes, Y., Touzani, G.A.: Collaborative clustering through optimal transport. In: International Conference on Artificial Neural Networks, pp. 873–885. Springer (2020)
2. Ben Bouazza, F.E., Bennani, Y., El Hamri, M., Cabanes, G., Matei, B., Touzani, A.: Multi-view clustering through optimal transport. Aust. J. Intell. Inf. Process. Syst. **15**(3), 1–9 (2019)
3. Cornuéjols, A., Wemmert, C., Gançarski, P., Bennani, Y.: Collaborative clustering: why, when, what and how. Inf. Fusion **39**, 81–95 (2018)
4. Cuturi, M.: Sinkhorn distances: lightspeed computation of optimal transport. In: Advances in Neural Information Processing Systems, pp. 2292–2300 (2013)
5. Cuturi, M., Doucet, A.: Fast computation of Wasserstein barycenters. In: ICML, pp. 685–693 (2014)
6. Davies, D.L., Bouldin, D.W.: A cluster separation measure. IEEE Trans. Pattern Anal. Mach. Intell. **2**, 224–227 (1979)
7. Ghassany, M., Grozavu, N., Bennani, Y.: Collaborative clustering using prototype-based techniques. Int. J. Comput. Intell. Appl. **11**(03), 1250017 (2012)
8. Pedrycz, W., Rai, P.: Collaborative clustering with the use of fuzzy c-means and its quantification. Fuzzy Sets Syst. **159**(18), 2399–2427 (2008)
9. Rastin, P., Cabanes, G., Grozavu, N., Bennani, Y.: Collaborative clustering: how to select the optimal collaborators? In: 2015 IEEE Symposium Series on Computational Intelligence, pp. 787–794. IEEE (2015)
10. Rousseeuw, P.J.: Silhouettes: a graphical aid to the interpretation and validation of cluster analysis. J. Comput. Appl. Math. **20**, 53–65 (1987)
11. Sublime, J., Cabanes, G., Matei, B.: Study on the influence of diversity and quality in entropy based collaborative clustering. Entropy **21**(10), 951 (2019)
12. Wu, J., Xiong, H., Chen, J.: Adapting the right measures for k-means clustering. In: SIGKDD, pp. 877–886. ACM (2009)
13. Xie, J., Wang, C.: Using support vector machines with a novel hybrid feature selection method for diagnosis of erythemato-squamous diseases. Expert Syst. Appl. **38**(5), 5809–5815 (2011)
14. Zhu, P., Zhu, W., Hu, Q., Zhang, C., Zuo, W.: Subspace clustering guided unsupervised feature selection. Pattern Recogn. **66**, 364–374 (2017)

Exploring Automatic Assessment-Based Features for Clustering of Students' Academic Performance

Anita Qoiriah(✉), Yuni Yamasari, Asmunin, Andi Iwan Nurhidayat, and Rina Harimurti

Informatics Department, Universitas Negeri Surabaya, Jln. Ketintang, Surabaya, Indonesia
{anitaqoiriah,yuniyamasari,asmunin,andy134ks, 5rinaharimurti}@unesa.ac.id

Abstract. Many students have difficulty learning computer programming languages. Programming is a skill that requires a lot of practice, not just theory. Students are required to have the ability to all processes; analyze problems, design algorithms, translate algorithms into program code, and write program code with the correct syntax. The number of time students spend coding is the factor that has the highest impact on their programming skills. A basic programming lab work is implemented to increase students' programming skills with a lot of practice writing a code directly. Providing feedback on programming assignments is an integral part of a class on basic programming and requires substantial effort with personal teaching. It is needed an automatic assessment tool that can help the task of lecturers in evaluating assignments. Performance evaluation is one of the basics to monitor the progress of student performance. Grouping students according to their level of performance makes it easy for lecturers to monitor student performance levels and can provide learning according to the abilities of students in these groups. The clustering is used to group students based on the point of each lab work. The classification method used is K-Means Clustering. The cluster describes groups of students according to their performance. Based on the results of a case study with 31 students, it clusters students into 3 groups: 39% of people are in moderate ability, 45% of people have high abilities, and 16% of people are students whose programming abilities are still lacking system.

Keywords: K-Means · Clustering · Automatic assessment · Students' performance

1 Introduction

Programming is a basic principle of computer science. Most importantly, practical knowledge of computer programming is a pre-requisite for higher-level computing courses [1], but programming is a course that for most students is considered rather hard. Many students have difficulty learning computer programming languages. Especially because of a lack of knowledge in the field of mathematics and informatics and

A. Abraham et al. (Eds.): SoCPaR 2020, AISC 1383, pp. 125–134, 2021.
https://doi.org/10.1007/978-3-030-73689-7_13

does not develop the ability to think abstractly and logically. Besides, there is also a lack of motivation and fear of programming because there are already difficult assumptions.

Most students learn programming languages by reading books or listening to lecturers' explanations. So it does not provide maximum results because programming is a skill that requires a lot of practice, not just theory.

The programming method begins by converting the issue into an algorithm and translating the algorithm into a program code. The hardest part is the application of algorithm specifications. The right algorithm will produce a program that is as expected. Consequently, students are required to have the ability to all processes; analyze problems, design algorithms, translate algorithms into a program code, and write a program code with the correct syntax.

Research [2] findings indicate that the number of time students spend coding is the aspect that has the largest effect on the results of the examination. Students' performance is associated with their programming experience and confidence in programming [3]. There is an interaction between practical learning and theory learning when students do programming tasks in the laboratory.

To help students and lecturers in the educational environment, the existence of educational tools is very important [4]. So are students understand the basic concepts of programming and develop problem-solving skills, programming basic courses are given learning programming through practical activities. A basic programming lab work is implemented to increase students' programming skills with a lot of practice writing a code directly. In this activity, students are given lab work modules which are consisted of lab work assignments and 1 last project. Each lab module has lab work assignments, both personal and group. Individual assignments worked live in class, and the results are directly demonstrated to the lecturer/assistant. Group assignments are worked on together. Finally, the students have to make lab work reports and collected them. The students do the last project at the last meeting with larger and more complex problem cases.

Lab work activities in the class are not enough time for students to practice some problem-solving skills in more complex programming. Because of that, many lab work assignments finally become homework for students. Furthermore, students will show it to the lecturer. The lecturer will give some tests and corrections on assignments given by students. Then students will improve and the results will be given back to the lecturer. This process requires a lot of time and energy from both the lecturer's side and the students' side. Feedback on programming assignments is an important part of an introductory programming class and involves significant effort on the part of the teacher [3]. Managing and assessing the results of the student's lab work assignments is very complex and time-consuming.

With the growth of internet technology and program analysis techniques, web-based tutoring systems can help the role of lecturers in teaching and training programming. Making automatic programming assessments can ease the task of instructors in evaluating students' lab work assignments so that lecturers have more time to concentrate on the material that will be given to students to improve their skills and follow the progress of student progress.

Based on the aforementioned problems, an application/tool is needed that can help ease the tasks of lecturers/assistants in evaluating assignments given to students as training materials to improve skills in programming and their ability to solve problems. This tool is expected to replace most of the tasks of lecturers/assistants without eliminating characterize of the lab work itself. Tools must be able to display assignments, live coding, upload program code, evaluate uploaded program code, compile, display compilation results, and grades. This tool has to display scores and rankings from all lab work participants who have uploaded the program code.

Performance evaluation is one of the basics to monitor the progress of student performance. Grouping students according to their level of performance makes it easy for lecturers to monitor student performance levels and can provide learning according to the abilities of students in these groups. Each group can have different treatment according to what is needed. Grouping can be done by clustering based on the similarity of the performance of each student. With this grouping, the assignment of practical work and monitoring student performance can be more targeted. In the traditional grouping, students are grouped based on their average scores. In this way, it is difficult to get a comprehensive view of the state of student performance.

In the educational domain, data mining techniques are very useful for improving current education standards and management. These techniques provide routes to various levels of ranking, determinations that provide new perceptions of how the masses can become experts in this education sector. Educational data mining (EDM), based on tasks, is divided into many categories, namely: classification [5–8], clustering [9–11], association analysis [12–14] etc. Classification usually is applied for the prediction [15–18], for example, students' performance has long been an interesting area of research and it aids to distinguish the weak students or students at risk [2, 19].

Data clustering is a method by which large data sets extract previously unknown, actual, positionally useful, and invisible patterns. The amount of data found in the education database continues to increase. The clustering technique is the most widely used technique for mapping [20–22]. The primary purpose of clustering is to classify students according to their capabilities and skill into homogeneous classes. This application can help instructors and students to improve the quality of education [23]. This study uses cluster analysis to segment students into groups according to their characteristics.

2 Material and Method

This application was built using Open-Source based technology including Ubuntu Linux, Laravel, Nginx, and MySQL. The process of compiling and evaluating a program code uses GCC and G ++ that is preinstalled on the operating system. The stag-es of application developers use the waterfall model approach.

The tools are arranged as shown in the flowchart in Fig. 1. Starting from the program code that has been sent by the student, the Memory Limit, Size Limit, Time Limit, and TotTimeLimit are determined. The next step is to check the Limit Exceeded and Restricted Function, if it is correct then delete the temporary files and if it is wrong then set Outfile_File and stringCmd_Compile.

The main modules in this application are evaluation, assessment, and clustering. The evaluation module will conduct an in depth evaluation of the program code sent, and be

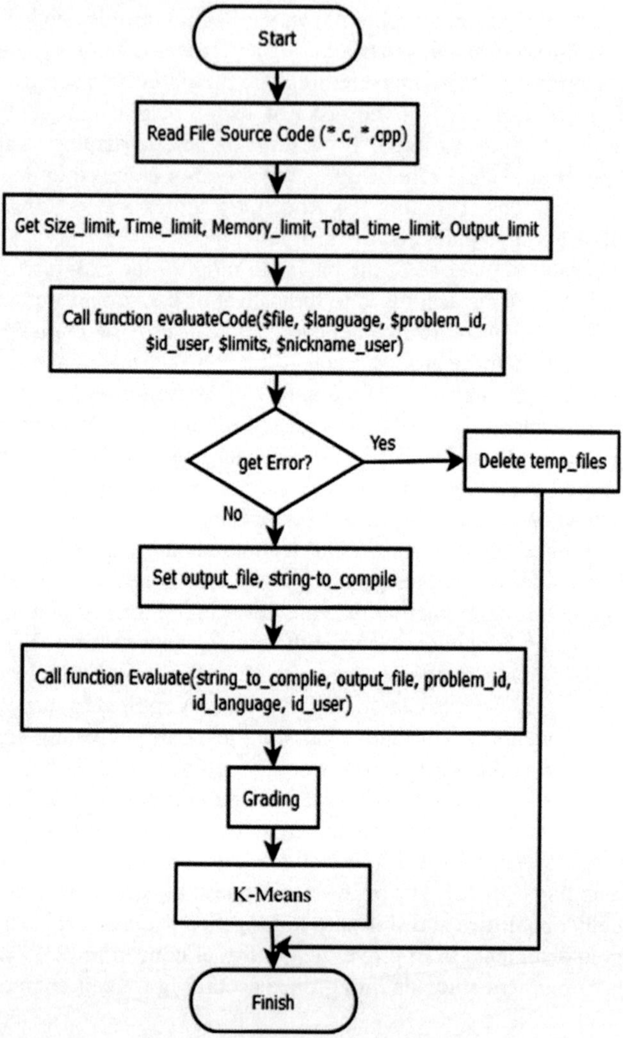

Fig. 1. Flowchart automatic assessment

a determinant for the next process. The results of the process in the evaluation module are then processed in the assessment module. In the assessment module, a comparison process with the test case is carried out and a point calculation is performed. The clustering module is used to group students based on the point of each lab work. The clustering method used is K-Means Clustering.

2.1 Scoring

The assessment process is carried out by reading the input of test cases and comparing them with the output of test cases. The score lab works assignment is calculated as in

Eqs. (1) and (2).

$$\text{Score} = \text{weight of assignment x accuracy} \tag{1}$$

$$\text{Accuracy} = \text{total accepted / number of submit} \tag{2}$$

2.2 Data Set Clustering

The data in this study were collected from 110 first semester students of the Department of Informatics. Students are participants in the Basic Programming course. The data collected is the result of an Automatic Programming Assessment Tools assessment from a basic programming lab work that uses C++ Language. Data collection was carried out from the value of 3 lab work assignments that must be completed by each student.

2.3 K-Means Clustering

K-Means aims to minimize the Sum of Squared Error (SSE) between data objects with many k-centroids. K centroids in this study amounted to 3, they are a group into low, medium, and high programming abilities. The steps for the K-Means algorithm [8]:

1. From the data set to be clustered many k objects are chosen randomly as the initial centroid, cj.
2. Every object that is not centroid is inserted into the nearest cluster based on the Euclidean distance measure as in Eq. (3).

$$\text{Euclidean distance measure} = ||\text{ xi (j)} - \text{cj }||2 \tag{3}$$

3. Each centroid is updated based on the average of the objects in each cluster
4. Iterate for the second and third steps until all centroids are converging and stable, where all the centroids produced in the current iteration are the same as all the centroids produced previously.

Several distance measurements are used as a measure of similarity of data, one of which is the Euclid distance. The Sum of Squared Error (SSE) as in an Eq. (4).

$$j = \sum\nolimits_{j=1}^{k} \sum\nolimits_{i=1}^{n} \|x_i^{(j)} - c_j\|^2 \tag{4}$$

Where J is the objective function, k is the number of clusters that is 3, n is the number of data objects that is 41, *xi* is the value of the data object to *i*, *cj* is the centroid cluster *j*, $||x_i^{(j)} - c_j||^2$ is a distance function.

3 Result and Discussion

This study was carried out in the Department of Informatics Engineering, FT Unesa, for the lab programming of basic courses with 110 students. The programming language used is C++. While doing a lab work assignment, every student must register for the application and doing the assignment provided. The score obtained by all of the students will be processed with the K-Means algorithm to classified students by their cluster.

3.1 Implementation

Before students start doing lab work assignments with automatic assessment applications, the lecture must create assignments first. Several things that must be included are the description of the assignments, the level of difficulty, the group of material assignments, points, or weight of the assignments, the output of the assignments. In addition to each assignment made, there are some time settings and program sizes.

To use this program, a student has to log in, it will proceed with working on the problems. The assignments that students work on can be chosen based on the problem group and the level of difficulty. Furthermore, students can work on the selected assignments like Fig. 2, directly write the program code on the application or use another editor.

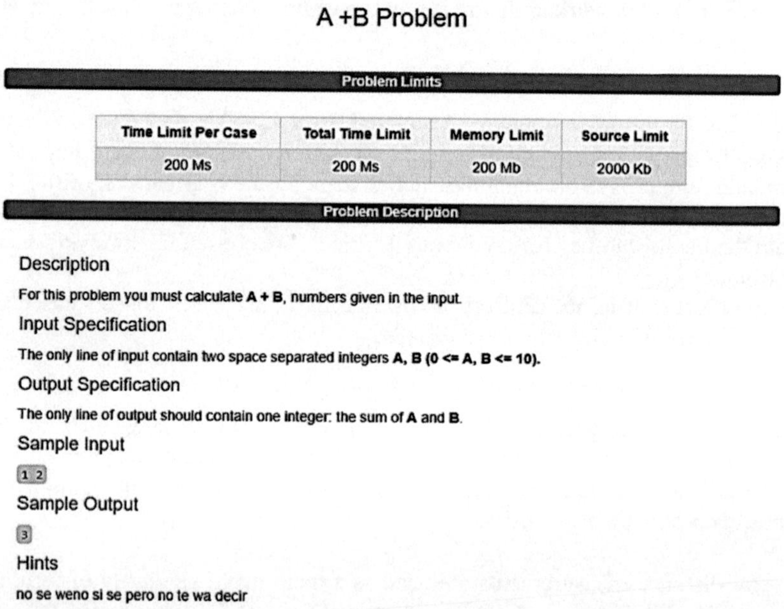

Fig. 2. Problems description

The completed program is then uploaded to the application and submitted. After the assignment has finished submitting, the execution results will come out, whether the program was received or there was still an error like in Fig. 3. Code Testing is done by unit testing conducted during the development and acceptance testing, that is one by students.

An assessment of student work scores is calculated based on the percentage of accuracy multiplied by the weight of each question. Based on the results of a trial of 330 submitted answers, 124 answers have 100% accuracy, meaning that with 1 time the answer is correct. And 41 answers have accuracy below 50%.

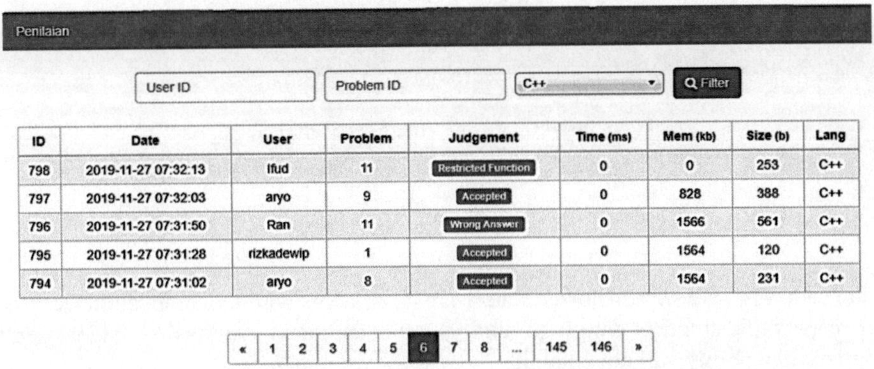

ID	Date	User	Problem	Judgement	Time (ms)	Mem (kb)	Size (b)	Lang
798	2019-11-27 07:32:13	Ifud	11	Restricted Function	0	0	253	C++
797	2019-11-27 07:32:03	aryo	9	Accepted	0	828	388	C++
796	2019-11-27 07:31:50	Ran	11	Wrong Answer	0	1566	561	C++
795	2019-11-27 07:31:28	rizkadewip	1	Accepted	0	1564	120	C++
794	2019-11-27 07:31:02	aryo	8	Accepted	0	1564	231	C++

Fig. 3. List of programs subscribed.

3.2 Clustering

The clustering feature of this tool is very important, especially for grouping students according to their abilities. So the lecturer can give learning according to the ability group of each student. In this way, it is hoped that maximum learning will be produced.

Raw data clustering is the scores of 3 lab work assignments generated by the Automatic programming assessment tool. So that the total data is 330 values. Furthermore, the data will be processed so that it becomes 3 clusters. The cluster describes groups of students according to their performance, groups that have weak, medium, and high abilities. Clustering with K-Means begins by determining the initial random centroid. By the number of groupings, each work lab assignment will be taken 3 initial centroids. The initial centroids are produced as in Table 1.

Table 1. Initial centroid

Centroid 1			Centroid 2			Centroid 3		
81	65	65	65	81	65	65	65	81

Furthermore, from non-centroid data, the minimum distance from the initial centroid will be sought. The next process is to update the centroid value based on the average value of the object minimum values. The same process applies to all subsequent data objects. The process is repeated until a stable centroid is produced, i.e. the centroid value now and then remains the same. A stable midpoint or centroid value will be generated as can be seen in Table 2.

By using this centroid, the data will be grouped into 3 clusters. Clustering is determined based on the minimum distance between objects and centroids. The process will be repeated by determining the new centroid based on clustering results. The loop will stop until you get the same clustering result as before. Clustering results were obtained

Table 2. Centroid data

Cluster	P1	P2	P3
Cluster 0	42	68.000	61.400
Cluster 1	50.357	51.071	80.500
Cluster 2	35.833	68.333	79.417

as shown in Fig. 4, wherein cluster 3 there are 43 students with moderate ability, cluster 1 as many as 50 students with high ability, and cluster 0 as many as 17 students with programming ability still not enough.

The experimental result indicates that poor students of this subject are 16%. Therefore, the teacher needs to improve the learning and teaching process to reduce the number of low students by a suitable instructional method. Furthermore, this system clustering of students is very useful because the system can support the teacher as a decision-maker to enhance better students' performance as soon as possible.

K means

Proses Iterasi Selanjutnya

ID	Nama	P1	P2	P3	Centroid 1	Centroid 2	Centroid 3	C1	C2	C3
					70.00 57.50 71.75	35.56 74.44 72.67	40.00 53.53 80.29			
32	M. FAHRUR ROZI	30	50	80	41.524841962372	26.114289191935	10.608722826052	0	0	1
33	MAULIDA NURSANTI	20	80	60	56.074169632728	20.822009989432	38.888880158729	0	1	0
34	INTAN PRASTIKA	30	70	60	43.523700440105	14.531211236507	27.981154372184	0	1	0
35	KURNITA SARI	40	40	80	35.697513919039	35.490225414894	13.533107551483	0	0	1
36	LILING APRISTIANA	20	60	80	50.737683234456	22.457873897589	21.022487959326	0	0	1
37	IIS TIANAH	40	50	73	30.94854600785	24.842224135532	8.0996913521442	0	0	1
38	ARIFATUS SHOLEHA	50	40	73	26.604745817241	37.346165800521	18.3358937606	0	0	1
39	IKRIMATUS SHOLEHA	30	60	73	40.097537330864	15.476953834654	13.96441907134	0	0	1
40	DEFFA ERLIAN SARI	40	50	77	31.36578549949	25.214600928827	4.8254533465779	0	0	1
41	NAZILA QORIATUN NISA	40	50	80	32.004882440028	25.898959438557	3.541892149685	0	0	1
42	DEWI LUTFIANA	30	60	80	40.918363848033	17.121801891156	11.914067315573	0	0	1
43	TIWI ROBIANINGSIH	40	50	83	32.90611645272	26.902343764066	4.4502808900113	0	0	1
44	AMALIYAH	30	60	83	41.627064513367	18.604733268714	12.214949856631	0	0	1
45	RULLI BAGUS PRATAMA	40	50	83	32.90611645272	26.902343764066	4.4502808900113	0	0	1

Fig. 4. Clustering results

4 Conclusions

The automatic assessment tool makes it easy for the user evaluation of lab work assignments. From the student side, the ease is inside testing and collection of program code, while from the lecturer side also makes it easy to manage questions and answers and get student grades automatically. Questions are made with several settings, namely the time, file size, and the weight of each question. Assessment questions are the product of the process of the results received multiplied by the weight of the questions.

To make it easier for lecturers to provide appropriate learning, the grouping is carried out in class according to students' abilities. Grouping using the K-Means method. Based on the results of trials with 110 students, The results of clustering can classify students into 3 groups: 39% of people are in moderate ability groups, 45% of people are students with high abilities, and 16% of people are students whose programming abilities are still lacking.

The system can give the early warning for the teacher to solve relating to low students' performance in a specific subject compared to the conventional system.

References

1. Ayalew, Y., Tshukudu, E., and Lefoane, M.: Factors Affecting Programming Performance of First-Year Students at a University in Botswana. African J. Res. Math. Sci. Technol. Educ. **22**(3), 363–373 (2018)
2. Satratzemi, M., Xinogalos, S., Tsompanoudi, D., and Karamitopoulos, L.: Examining Student Performance and Attitudes on Distributed Pair Programming. Sci. Program. **2018**, 1–8 (2018)
3. Thuné, M., Eckerdal, A.: Analysis of Students' learning of computer programming in a computer laboratory context. Eur. J. Eng. Educ. **44**(5), 769–786 (2019)
4. Kautsar, I.A., Sarno, R.: A supportive tool for project-based learning and laboratory-based education. Int. J. Adv. Sci. Eng. Inf. Technol. **9**(2), 630–639 (2019)
5. Yamasari, Y., Mardi, S., Nugroho, S., Yoshimoto, K., Takahashi, H., Purnomo, M.H.: Information and control ICIC international. Int. J. Innov. Comput. **15**(6), 2085–2107 (2019)
6. Shanmugarajeshwari, V., Lawrance, R.: Analysis of students' performance evaluation using classification techniques. In: 2016 International Conference on Computing Technologies and Intelligent Data Engineering (ICCTIDE 2016), pp. 1–7. IEEE, Kovilpatti, India (2016)
7. Yahya, A.A.: Swarm intelligence-based approach for educational data classification. J. King Saud Univ. – Comput. Inf. Sci. **31**(1), 35–51 (2019)
8. Mhetre, V., Nagar, M.: Classification based data mining algorithms to predict slow, average, and fast learners in an educational system using WEKA. In: 2017 International Conference on Computing Methodologies and Communication (ICCMC), pp. 475–479. IEEE, Erode, India (2017)
9. Wang, S., Hu, Y., Liu, Y., Guo, Z., Shou, G.: Student clustering and learning atmosphere analysis based on trajectory data. In: 2017 12th International Conference on Computer Science and Education (ICCSE), pp. 208–213. IEEE, Houston, TX, USA (2017)
10. Park, Y., Yu, J.H., Jo, I.-H.: Clustering blended learning courses by online behavior data: a case study in a Korean higher education institute. Internet High. Educ. **29**, 1–1 (2016)
11. Yamasari, Y., Nugroho, S.M.S., Harimurti, R., Purnomo, M.H.: Improving the cluster validity on student's psychomotor domain using feature selection. In: 2018 International Conference on Information and Communications Technology (ICOIACT), pp. 460–465. IEEE, Yogyakarta, Indonesia (2018)
12. Yang, G., Mabu, S., Shimada, K., Hirasawa, K.: A novel evolutionary method to search interesting association rules by keywords. Expert Syst. Appl. **38**(10), 13378–13385 (2011)
13. Abdullah, Z., Herawan, T., Ahmad, N., Mat Deris, M.: Extracting highly positive association rules from students' enrollment data. Procedia Soc. Behav. Sci. **28**, 107–111 (2011)

14. García, E., Romero, C., Ventura, S., De Castro, C.: A collaborative educational association rule mining tool. Internet High. Educ. **14**(2), 77–88 (2011)
15. Pradeep, A., Das, S., Kizhekkethottam, J.J.: Students dropout factor prediction using EDM techniques. In: 2015 Proceedings of the IEEE International Conference on Soft-Computing and Network Security(ICSNS). IEEE, Coimbatore, India (2015)
16. Altujjar, Y., Altamimi, W., Al-Turaiki, I., Al-Razgan, M.: Predicting critical courses affecting students performance: a case study. Procedia Comput. Sci. **82**, 65–71 (2016)
17. Harimurti, R., Yamasari, Y., Ekohariadi, Munoto, Asto, B.I.G.P.: Predicting student's psychomotor domain on the vocational senior high school using linear regression. In: 2018 International Conference on Information and Communications Technology (ICOIACT), pp. 448–453. IEEE, Yogyakarta, Indonesia (2018)
18. Shahiri, A.M., Husain, W., Rashid, N.A.: A review on predicting student's performance using data mining techniques. Procedia Comput. Sci. **72**, 414–422 (2015)
19. Gulwani, S., Radiček, I., Zuleger, F.: Automated clustering and program repair for introductory programming assignments. In: Proceedings of the ACM SIGPLAN Conference on Programming Language Design and Implementation (PLDI), pp. 465–480. ACM, Philadelphia, USA (2018)
20. Harwati, H., Alfiani, A.P., Ayu Wulandari, F.: Mapping Student's Performance Based on Data Mining Approach (A Case Study). Procedia - Social and Behavioral Sci. **38**(1), 173–177 (2015).
21. Pang, Y., Xiao, F., Wang, H., and Xue, X.: A clustering-based grouping model for enhancing collaborative learning. In: 13th International Conference on Machine Learning and Applications, pp. 562–567. IEEE, Detroit, USA (2014)
22. Huang, L.: Teaching management data clustering analysis and implementation on ideological and political education of college students. In: 2016 International Conference on Smart Grid and Electrical Automation (ICSGEA), pp. 308–311. IEEE, Zhangjiajie, China (2016)
23. Shashikant, M., Borgavakar, P.: Evaluating student's performance using K-Means clustering. Int. J. Eng. Res. Technol. (IJERT) **6**(5), 114–116 (2017)

Quantum Semi Non-negative Matrix Factorization

Kaoutar Benlamine[1,2(✉)], Younès Bennani[1,2], Basarab Matei[1,2], and Nistor Grozavu[1,2]

[1] LIPN UMR 7030 CNRS, Université Sorbonne Paris Nord, Villetaneuse, France
benlamine@lipn.univ-paris13.fr
[2] LaMSN, La Maison des Sciences Numériques, USPN, Villetaneuse, France

Abstract. Semi Non-negative Matrix Factorization (SNMF) is a machine learning algorithm that is used to decompose large data matrices where the data matrix is unconstrained (i.e., it may have mixed signs). We develop the quantum version of SNMF using quantum gradient descent, and we show that the quantum version of SNMF provides an exponential speedup compared to its classical version. Our approach is well adapted for high-dimensional problems since in the context of using a quantum approach only a few number of iterations is enough to obtain a close approximation of the factorization. From the algorithmic analysis point of view, the proposed algorithm for Quantum Semi Non-negative Matrix Factorization (QSNMF) yields a performance $T\mathcal{O}(polylogK(M+N))$, while the classical version of Semi Non-negative Matrix Factorization (SNMF) yields a performance $T\mathcal{O}(poly(KMN))$ where M and N are the dimensions of the matrix, K is the number of clusters and T is the maximum number of iterations to prepare the quantum states.

Keywords: Quantum machine learning · Semi non-negative matrix factorization · Quantum gradient descent

1 Introduction

Machine learning techniques are applied for solving a wide variety of problems such as: classification, regression, pattern recognition, etc. There are two main types of machine learning: supervised and unsupervised machine learning. In supervised machine learning, the learner is provided a set of training examples with features presented in the form of high-dimensional vectors and with corresponding labels to mark its category. The aim is to classify new examples based on these training sets. In unsupervised machine learning, the system aims to classify the data into different groups without prior information.

Clustering has received a lot of attention as a significant issue with many applications, and a wide range of different algorithms and methods have emerged

A. Abraham et al. (Eds.): SoCPaR 2020, AISC 1383, pp. 135–144, 2021.
https://doi.org/10.1007/978-3-030-73689-7_14

over the years. Much interest has been shown using NMF for clustering of non-negative data [2,15], it has been applied in different areas such as pattern recognition, segmentation, clustering and text mining. SNMF is an unsupervised algorithm for clustering where a data matrix is factorized into (usually) two matrices with the property that the matrix G is constrained to be non negative whereas the data matrix X and the basis vectors of F are unconstrained, so as to make the interpretation of the resulting matrices much more easier. This factorization leads to two matrices one of them contains the prototypes of a data set while the other one is a data partition matrix.

NMF [8,12] is equivalent to K-means clustering, more details in [3]. Kim and Park [7] proposed sparse NMF algorithm for data clustering and they compare it with K-means. They showed that their algorithm outperforms the K-means and the ordinary NMF in terms of the consistency of the results.

Clustering could pose a challenge for classical computers especially with the current speed growing data size. If our vectors are in N dimensional vector spaces, this would take time $\mathcal{O}(polyN)$ on classical computers, while with quantum computers this would take time $\mathcal{O}(\log N)$. Thus, quantum clustering can provide exponential speed-ups for problems involving large vectors.

In this paper, we are interested in the quantum version of SNMF that we can implement using quantum gradient descent. Quantum Semi Non-negative Matrix Factorization (QSNMF) gives an exponential speedup compared to its classical version. The main idea of this quantum gradient descent is to improve the objective function in order to have an algebraic form. Indeed, both homogeneous and inhomogeneous function over large dimensional spaces can be optimized by using quantum gradient descent [14]. Then, we use multiple copies of the current quantum state $|current\rangle$ to produce multiple copies of another new quantum state $|new\rangle$ by using the objective function of the gradient that has been optimized.

The paper is organized as follows: Sect. 2 presents the classical SNMF algorithm. Section 3 and 4 describe our proposed approach QSNMF where we give the details of transforming a vector to a quantum state and we show also the quantum gradient descent of SNMF. Finally, the conclusion summarizes the advantages of our contribution and directions for future works.

2 Classical Semi Non-negative Matrix Factorization

Let $X = (\mathbf{x}_1, \mathbf{x}_2, \ldots, \mathbf{x}_N) \in \mathbb{R}_+^{M \times N}$, be a non-negative data matrix with M rows and N columns, here $\mathbf{x}_n \in \mathbb{R}_+^{M \times 1}$ represents the nth column of X. In what follows, $\|\cdot\|$ stays for the Euclidean norm and $\|\cdot\|_F$ for the Frobenius norm. Let K be a fixed input parameter.

2.1 Non-negative Matrix Factorization

The Non-negative Matrix Factorization (NMF)gives a low rank approximation of X by the product of two non-negative matrices FG where the factors are $F = (\mathbf{f}_1, \mathbf{f}_2, \ldots, \mathbf{f}_K) \in \mathbb{R}_+^{M \times K}$, $G = (\mathbf{g}_1, \mathbf{g}_2, \ldots, \mathbf{g}_K)^T \in \mathbb{R}_+^{K \times N}$ and T denotes the transpose operator. The NMF decomposition could be formulated as a constrained optimization problem by minimizing $\|X - FG\|_F^2$ under positivity constraints $F, G \geq 0$.

The NMF decomposition has been studied early by Golub and Paatero [12,16]. Several families of algorithms are proposed to solve this matrix approximation problem. The approach proposed by Lee and Seung [9] is based on a gradient descent strategy. In the Lee and Seung's approach, this algorithm could converge to a stationary point which is not necessarily local minima. There is no guarantee that we can exactly recover the original matrix from F and G, so we will approximate it in terms of the approximation error measured in Frobenius norm as best as possible. The work of Lee and Seung [9] revealing that NMF has an inherent clustering property, i.e., it automatically clusters the columns of input matrix X since the matrix columns vector factor $(\mathbf{f}_1, \mathbf{f}_2, \ldots, \mathbf{f}_K)$ could be considered as cluster centroids while the matrix rows vector factor $(\mathbf{g}_1, \mathbf{g}_2, \ldots, g_K)^T$ could be considered as cluster indicators.

2.2 Semi Non-negative Matrix Factorization

The Semi Non-negative Matrix Factorization (SNMF) [4] relaxes the non-negativity constrains of NMF and allows the data matrix X and the loadings matrix F to have mixed signs, while restricting only the features matrix G to comprise of strictly non-negative components. More precisely we define the following optimization:

$$(F, G) = \operatorname*{argmin}_{G \geq 0} \mathcal{L}(X, F, G), \quad \text{where} \quad \mathcal{L}(X, F, G) = \|X - FG\|_F^2. \tag{1}$$

Note that the domain of the real valued functional $\mathcal{L}$ is $\mathbb{R}^{M \times N} \times \mathbb{R}^{M \times K} \times \mathbb{R}^{K \times N}$. The SNMF also allows to define a clustering algorithm that aims to factorize a given matrix into (usually) two matrices, where we consider one of the matrices as a matrix containing the prototypes of a data set and the other one as a data partition matrix.

The aim of clustering is to cluster the columns of X, so as to optimize the difference between X and the clustered matrix revealing significant block structure. More formally, we seek to partition the set of columns $\{1, \ldots, N\}$ into K clusters $C = \{C_1, \ldots, C_K\}$. The partitioning naturally induces clustering prototypes matrix $F \in \mathbb{R}^{M \times K}$, and a reduced partition matrix $G \in \mathbb{R}_+^{K \times N}$ specifying the cluster representation. Basically, we can interpret X to be a linear combination of some columns of F, where each column in F is a component, and each row in G contains the corresponding coefficient for each component.

To solve the optimization problem (1), with the only matrix constraint $G \geq 0$, we use gradient descent, which starts with a random initialization of F and after usually unfolds in two phases: at first the functional is minimized with respect to G while in the second phase, the functional is minimized with respect to F.

Let us mention several particular properties of the binary indicator matrix G. If the clusters $C_1, \ldots, C_K$ have distinct cluster centroids $\mathbf{f}_1, \ldots, \mathbf{f}_K$, each of the

N columns of G will contain a single 1 and $K-1$ elements that are 0. Therefore, the columns of G will sum to one

$$\sum_{k=1}^{K} g_{kn} = 1, \text{ for all column } n. \tag{2}$$

And its row sums will indicate the number of elements per cluster

$$\sum_{n=1}^{N} g_{kn} = n_k, \text{ for all row } k. \tag{3}$$

It follows that the rows of G are pairwise perpendicular because $g_{kn}g_{k'n} = \delta_{kk'}$, δ being the Kronecker symbol, which is then to say that the matrix GG^T is a diagonal matrix where: $(GG^T)_{kk'} = n_k\delta_{kk'}$ for all $k, k' = 1, \ldots, K$. Our strategy to solve the problem (1) is to write it in terms of polynomial optimization. Then we use several results of [14] to derive our quantum algorithm.

3 Problem Statement

In this section, we formulate the gradient descent procedure which will be thereafter simulated in the quantum context. Here, we use some general linear algebra results and also some well known results about the NMF [9]. The gradient descent method unfolds in the two following phases:

$$F = F - \eta_F \circ [\nabla_F(\|X - FG\|_F^2)] \tag{4}$$

and

$$G = G - \eta_G \circ [\nabla_G(\|X - FG\|_F^2)]. \tag{5}$$

Due to the symmetric role of F and G we will detail only the phase corresponding to the optimization with respect to the matrix F. Also, in order to include $G \geq 0$ at each step of the optimization process, all the negative entries of G are redefined as 0.

Lemma 1. *The gradient of* $[\nabla_F(\|X - FG\|_F^2)]$ *is composed of two terms: the gradient of the homogenous quadratic form with a diagonal matrix and the gradient of an inhomogenous affine form.*

Proof. Let us remark that as squared Frobenius norm of a matrix difference, the real valued functional $\mathcal{L}$ can be written as:

$$\begin{aligned} \|X - FG\|_F^2 &= tr[(FG)^T(FG)] - 2tr[X^TFG] + tr[X^TX] \\ &= [vec(F)]^TB^TBvec(F) - 2vec(X)^TBvec(F) + tr[X^TX], \end{aligned}$$

where $B = (G^T \otimes I_M)$ is the $MN \times MK$ matrix the Kronecker product of the matrix G^T with I_M the M dimensional identity matrix. Also for any $M \times N$

matrix A, $vec(A)$ is the $MN \times 1$ column vector; the vectorization of A which obtained by stacking the columns of the matrix A on top of one another.

Note that $B^T B \in \mathbb{R}_+^{MK \times MK}$ symmetric matrix by construction. Moreover we have:

$$B^T B = (G^T \otimes I_M)^T (G^T \otimes I_M) = (G \otimes I_M)(G^T \otimes I_M) = (GG^T \otimes I_M I_M),$$

where we have used the mixed product property of the Kronecker product. Since GG^T is diagonal, see Eqs. (3) and (2), it follows that $B^T B$ is diagonal. We also have:

$$\nabla_F[[vec(F)]^T B^T Bvec(F)] = 2vec(F)B^T B = 2B^T Bvec(F). \tag{7}$$

So the gradient of the quadratic function acts as an operator on F. By construction $vec(X)^T Bvec(F) + tr[X^T X]$ is an affine form with respect to F, we have:

$$\nabla_F[-2vec(X)^T Bvec(F) + tr[X^T X]] = -2vec(X)^T B. \tag{8}$$

Note that $[vec(F)]^T B^T Bvec(F)$ is a quadratic form with $B^T B$ symmetric matrix and that $vec(X)^T Bvec(F) + tr[X^T X]$ is an affine form which prove our claim.

The same conclusion stays for the gradient of $[\nabla_G(\|X - FG\|_F^2)]$ which will be composed of two terms: the gradient of the homogenous quadratic form with a diagonal matrix and the gradient of an inhomogenous affine form. There exists a difference between the two computations since FF^T is no more diagonal, nevertheless FF^T is diagonalizable as real symmetric matrix. Then to estimate G we will work with the diagonal form of FF^T.

4 Quantum Gradient Descent Algorithm for SNMF

In this section, we present the gradient descend strategy to estimate F, the same strategy could be used to estimate G. Let us denote $y = vec(F) \in \mathbb{R}^{MK \times 1}$, $A = 2B^T B \in \mathbb{R}_+^{MK \times MK}$, $C^T = -2vec(X)^T B \in \mathbb{R}^{MK \times 1}$. So the functional $\mathcal{L}(X, F, G)$ for fixed G could be written as $f : \mathbb{R}^{MK \times 1} \to \mathbb{R}$ where:

$$f(y) = \frac{1}{2} y^T Ay + C^T y + tr[X^T X], \tag{9}$$

where A is diagonal. To estimate G the functional $\mathcal{L}(X, F, G)$ for fixed F has the same form as the functional in Eq. (9) with the matrix A is no more diagonal but sparse.

4.1 Data Preparation: From a Vector to a Quantum State

We assume that our data sets consist of N data vectors stored as M dimensional column vectors, i.e.: $X = (\mathbf{x}_1, \mathbf{x}_2, \ldots, \mathbf{x}_N) \in \mathbb{R}_+^{M \times N}$. We also assume without loss of generality that $MK = 2^{n_0}$ where n_0 is an integer. A vector that not satisfy this condition can be always padded to 2^{n_0} as long as the corresponding entries of A in Eq. (9) are set to 0.

Therefore the data matrix X elements are presented as the amplitudes of a n_O qubit quantum state (see [10,13]) as follows:

$$|x\rangle = \frac{1}{|x|}\sum_{i=1}^{n_O} x_i |i\rangle,$$

where $|i\rangle$ is the ith computational basis state. In what follows, all the produced vectors y^0 by the quantum algorithm are normalized by $\sum_{i=1}^{n_O} |y_i^0|^2 = 1$, this constraint is known as spherical constraint. We make the following main assumptions: (The goal of these assumptions is to ensure an efficient simulation of e^{-iAt}, e.g.: [10,13]).

- A_1 We assume that there exists a quantum algorithm such that the operation $|\mathbf{0}\rangle \rightarrow |y^0\rangle$ performs n_O qubits. Recall that the initial quantum state $|y^0\rangle$ is defined via amplitude encoding (see [14] or [6].
- A_2 Let $j, k = 1, \dots, MK$. We assume that there exists a quantum algorithm that performs the operation $|j,k\rangle|\mathbf{0}\rangle \rightarrow |j,k\rangle|A_{j,k}\rangle$ on $2n_O + n_\alpha$ qubits where $A_{j,k}$ is encoded to accuracy $\alpha = 2^{-n0}$. We assume that the accuracy error is smaller than other errors involved in the optimization. To take advantage of sparsity, we also suppose the following oracle, which allows to choose the non-zero matrix elements.
- A_3 The input matrix is sparse, in order to use the sparsity of A, let $j = 1, \dots, MK$ and $l = 1, \dots, K$, and we assume that there exists a function $g_A(j,l)$ which compute the column index of the l-th non-zero element of row j of matrix A. Then we assume that the operation $|j,l\rangle \rightarrow |j, g_A(j,l)\rangle$ performs on 2^{n_O} qubits.

4.2 Computing the Gradient

By using Lemma (1) we have:

$$f = f_{hom} + f_{inhom}, \quad \text{where } f_{hom} = \frac{1}{2} y^T A y \quad \text{and} \quad f_{inhom} = C^T y + tr[X^T X]. \tag{10}$$

We want to simulate the time evolution under the respective gradient of the function f. More precisely we assume that we have multiple copies of the state $y^{(t)}$ at step t we want to define the state:

$$\left|y^{(t+1)}\right\rangle = \left|y^{(t)}\right\rangle - \eta\,|\nabla_y f(y)\rangle \tag{11}$$

with some given accuracy. Again by using Lemma (1) we have:

$$\nabla_y f = \nabla_y f_{hom} + \nabla_y f_{inhom}, \quad \text{where } \nabla_y f_{hom} = Ay \quad \text{and} \quad \nabla_y f_{inhom} = C^T. \tag{12}$$

Therefore the simulation of the time evolution is based on quantum state exponentiation and the use of additional subroutines to simulate terms that contain inner products and outer products.

Computing $\nabla_y f_{hom}$: Note that $f_{hom} = \frac{1}{2}y^T Ay$ can be written as:

$$\langle y^T | A | y \rangle = tr(A\rho)$$

where $\rho = |y\rangle \langle y|$ is the corresponding density matrix, and tr is the trace operator. By using Lemma (1) the gradient $D = \nabla_y f_{hom}$ is an operator applied to y. Thus, the gradient operator D can be implemented as a Hamiltonian and it can be represented as:

$$D = tr\{(\rho \otimes I)A\}$$

where ρ is the joint quantum state of one copy of ρ. The matrix exponential $e^{-iA\tau}$ can be simulated efficiently on a quantum computer.

The idea is to exponentiate the matrix $e^{-iD\Delta t}$ [11]. So, we can exponentiate D by the mapping from D to A and the exponentiation of A. For a short time Δt, multiple copies of ρ are used and matrix exponentiation of M is performed. In the reduced space of the last copy of ρ, we have:

$$tr\{e^{-iA\Delta t}\rho^{\otimes 2}e^{iA\Delta t}\} = e^{-iD\Delta t}\rho^{\otimes 2}e^{iD\Delta t} + \epsilon$$

where ϵ is the error term that contains the error simulation of $e^{-iA\Delta t}$ and the error of the Hamiltonian simulation.

The multiplication $D|y\rangle = |\nabla f(y)\rangle$ can be implemented via phase estimation. We implement series of powers of the exponential in the equation above controlled on a time register in order to extract the eigenvalues of D. By conditional rotating and measuring an extra ancilla qubit [5], the matrix multiplication with D is performed:

$$\sum_j \lambda_j(D)\beta_j |u_j(D)\rangle$$

where $|y\rangle = \sum_j \beta_j |u_j(D)\rangle$ is the original state $|y\rangle$ written in the eigenbasis $\{|u_j(D)\rangle\}$ of D, and $\lambda_j(D)$ is the additional register encoding the corresponding eigenvalue in binary representation.

Computing $\nabla_y f_{inhom}$: Additional simulation terms and vector additions could be done in order to optimize the functions containing terms of the form f_{inhom}. The following elements are needed so as to perform the non-homogeneous updates.

1. We make additional vector additions related to the states C . For instance in the gradient of the inhomogenous part, we need one additional vector addition by C. Before adding the vectors we require to conditionally apply the gradient operator to the y and C^Trespectively. This requires at each iteration a certain number of copies of the c_{ij} and y_j. As we need in each state another copy of the states, this adds another tree of states to the resource requirements, because we need to build them in parallel with the main algorithm.
2. We perform operations similar to homogeneous matrix multiplications and matrix inversions through conditional rotations on the registers of the eigenvalue and post selection.
3. In the end, we carry out a measurement in the yes/no-basis as before and carry out the vector addition of the current solution and the update of the step.

4.3 Gradient Descent Procedure

Assume we have multiple copies of the state $y^{(t)}$ at step t with an accuracy $\epsilon^{(t)}$. Our purpose is to prepare the state $\left|y^{(t+1)}\right\rangle = \left|y^{(t)}\right\rangle - \eta \left|\nabla f(y)\right\rangle$ with an accuracy $\epsilon^{(t)} \geq \epsilon^{(t+1)}$. Let's begin by preparing the state:

$$\left|y^{(t)}\right\rangle (\cos\theta \left|\mathbf{0}\right\rangle - i \sin\theta \left|\mathbf{1}\right\rangle)$$

where θ is an external parameter determining the size of the gradient step and an ancilla qubit is used for the gradient step in the second register.

We multiply the operator D by $\left|y^{(t)}\right\rangle$, provided that the first ancilla is in the state $\left|\mathbf{1}\right\rangle$. After the conditional phase estimation, we get:

$$\left|\psi\right\rangle = \cos\theta \left|y\right\rangle \left|\mathbf{0}\right\rangle - i \sin\theta \sum_j \lambda_j(D) \left|\mathbf{1}\right\rangle \beta_j \left|u_j(D)\right\rangle$$

where $\left|u_j(D)\right\rangle$ are the eigenstates of D and $\lambda_j(D)$ are the eigenvalues.

We use another ancilla initialized to $\left|\mathbf{1}\right\rangle$, and perform a joint measurement of the final two ancillas. The first ancilla is measured in the basis $\left|yes\right\rangle = \frac{1}{\sqrt{2}}(\left|\mathbf{0}\right\rangle + i\left|\mathbf{1}\right\rangle)$ and $\left|no\right\rangle = \frac{1}{\sqrt{2}}(i\left|\mathbf{0}\right\rangle + \left|\mathbf{1}\right\rangle)$ and the measurement of the second ancilla on the state $\left|\mathbf{1}\right\rangle$:

$$\cos\theta \left|y^{(t)}\right\rangle - \xi_D \sin\theta D \left|y^{(t)}\right\rangle$$

where $\xi_D \geq \eta$ and θ can be chosen like:

$$\cos\theta = \frac{1}{\sqrt{1+\frac{\eta^2}{\xi_D^2}}}, \quad \sin\theta = \frac{\eta}{\xi_D\sqrt{1+\frac{\eta^2}{\xi_D^2}}}$$

To obtain the state: $\left|y^{(t+1)}\right\rangle = \frac{1}{C^{(t+1)}}\left|y^{(t)}\right\rangle - \eta\left|\nabla f(y)\right\rangle$ where η is the learning rate and: $(C_D^{(t+1)})^2 = 1 - 2\eta \left\langle y^{(t)}\right| D \left|y^{(t)}\right\rangle + \eta^2 \left\langle y^{(t)}\right| D^2 \left|y^{(t)}\right\rangle$. The probability success of the measurement of $\left|yes\right\rangle \left|\mathbf{1}\right\rangle$ is given by:

$$P = \frac{1}{2(1+\frac{\eta^2}{\xi_D^2})}(C^{(t+1)})^2$$

Note that by construction there exists $0 < \epsilon_P < 1$ such that the success probability satisfy: $P > \epsilon_P$, therefore the probability is never close to zero. More precisely as in Result 1 in [14] the success probability P is lower bounded by 1/16. Therefore the proposed algorithm performs the gradient computation with the post selection procedure.

The quantum state $\left|y^{(t+1)}\right\rangle$ is an approximation to the classical vector $y^{(t+1)}$. The approximation error in the updated step is $\mathcal{O}(\epsilon^{(t)} + \eta\epsilon_D^{(t)})$.

4.4 Complexity

We presented the quantum version of the gradient descent for polynomial optimization for SNMF. So as to analyze the complexity of the procedure, we first note that the sparsity factor which is the number of non-zero matrix elements in each row and column of order K of A is $\mathcal{O}(K)$ by definition of A. Furthermore, the L^2 norm of A is bounded by $\|A\| \leq 2KMN$, again by definition of A.

Let's assume the setting of optimization of polynomials. For that task we use the quantum gradient descent method for $T \geq 0$ steps to prepare a solution $|y^{(t+1)}\rangle$ to final accuracy $0 < \delta < 1$, with step-size schedule $\eta_>0$. Suppose that a quantum algorithm exists at every step $0 < t < T$ and the error $\epsilon \geq 0$. We suppose as well that the space explored with the algorithms is bounded by a constant, i.e. $T\eta_t = \mathcal{O}(1)$. Using the Result 3 (Multiple gradient steps) from [14], we have: that there is a quantum algorithm for multiple gradient steps that requires $\mathcal{O}(poly(K(M+N)^T)$ copies of the initial state $|y^{(0)}\rangle$. The gate requirement is $\mathcal{O}(poly(K(M+N)^T n_0)$ quantum gates.

Recall that the computational complexity for the classic Semi-NMF is of order $T\mathcal{O}(K(MN))$ see [4], T recall is the maximum number of iterations. It follows that: The result on the complexity of the proposed algorithm follow from the properties of the gradient descend procedure obtained in [14]. In [14] the authors show that the number of computational steps grows exponentially with the number of iterations T while it grows logarithmically in the dimension N of the solution vector. Due to construction of our solutions vectors as $vec(F)$ and $vec(G)$ we obtain that the complexity of our algorithm is of $T\mathcal{O}(polylogK(M+N))$ (Table 1).

Table 1. QSNMF versus SNMF

	SNMF	QSNMF
Complexity	$T\mathcal{O}(KMN)$	$T\mathcal{O}(polylogK(M+N))$

5 Conclusion

In the present work, we implement quantum SNMF using quantum gradient descent. We showed how an improvement in terms of complexity can be gained through quantum computation. We used an alternate optimization method for quantum gradient descent for both homogeneous and inhomogeneous part of the objective function. Our algorithm Quantum Semi Non-negative Matrix Factorization (QSNMF) leads to an exponential speed-up compared to its classical version: the complexity of the problem becomes $T\mathcal{O}(polylogK(M+N))$ instead of $T\mathcal{O}(KMN)$ where M and N are the dimension of the matrix and K is the number of clusters.

In addition, the sparsity of matrices has the potential to have better optimization. In future work, we are intending to use Collaborative Quantum NMF that consists on the collaboration of different clustering methods, so as to reach an agreement on the partitioning of a common dataset [1]. This latter doesn't have only the advantage of speed up but also by collaborating several clustering solutions, each one with its own bias and imperfections, one will get a better overall solution in terms of classification and also stability.

References

1. Benlamine, K., Grozavu, N., Bennani, Y., Matei, B.: Collaborative non-negative matrix factorization. In: International Conference on Artificial Neural Networks, pp. 655–666. Springer (2019)
2. Cichocki, A., Phan, A.-H.: Fast local algorithms for large scale nonnegative matrix and tensor factorizations. IEICE Trans. Fundam. Electron. Commun. Comput. Sci. **92**(3), 708–721 (2009)
3. Ding, C., He, X., Simon, H.D.: On the equivalence of nonnegative matrix factorization and spectral clustering. In: Proceedings of the 2005 SIAM International Conference on Data Mining, pp. 606–610. SIAM (2005)
4. Ding, C.H.Q., Li, T., Jordan, M.I.: Convex and semi-nonnegative matrix factorizations. IEEE Trans. Pattern Anal. Mach. Intell. **32**(1), 45–55 (2008)
5. Harrow, A.W., Hassidim, A., Lloyd, S.: Quantum algorithm for linear systems of equations. Phys. Rev. Lett. **103**(15), 150502 (2009)
6. Iten, R., Colbeck, R., Kukuljan, I., Home, J., Christandl, M.: Quantum circuits for isometries. Phys. Rev. A, **93**, 032318 (2016)
7. Kim, J., Park, H.: Sparse nonnegative matrix factorization for clustering. Technical report, Georgia Institute of Technology (2008)
8. Lee, D.D., Seung, H.S.: Learning the parts of objects by non-negative matrix factorization. Nature **401**(6755), 788 (1999)
9. Lee, D.D., Seung, H.S.: Algorithms for non-negative matrix factorization. In: Advances in Neural Information Processing Systems, pp. 556–562 (2001)
10. Lloyd, S., Mohseni, M., Rebentrost, P.: Quantum algorithms for supervised and unsupervised machine learning. arXiv preprint arXiv:1307.0411 (2013)
11. Lloyd, S., Mohseni, M., Rebentrost, P.: Quantum principal component analysis. Nat. Phys. **10**(9), 631 (2014)
12. Paatero, P., Tapper, U.: Positive matrix factorization: a non-negative factor model with optimal utilization of error estimates of data values. Environmetrics **5**(2), 111–126 (1994)
13. Rebentrost, P., Mohseni, M., Lloyd, S.: Quantum support vector machine for big feature and big data classification. arXiv preprint arXiv:1307.0471 (2013)
14. Rebentrost, P., Schuld, M., Wossnig, L., Petruccione, F., Lloyd, S.: Quantum gradient descent and newton's method for constrained polynomial optimization. New J. Phys. **21**(7) (2019)
15. Shahnaz, F., Berry, M.W., Pauca, V.P., Plemmons, R.J.: Document clustering using nonnegative matrix factorization. Inf. Process. Manage. **42**(2), 373–386 (2006)
16. Xie, Y.-L., Hopke, P.K., Paatero, P.: Positive matrix factorization applied to a curve resolution problem. J. Chemom. **12**(6), 357–364 (1999)

A New Genetic-Based Hyper-Heuristic Algorithm for Clustering Problem

Mohammad Babrdel Bonab[1,2](✉), Goi Bok-Min[2], Madhavan a/l Balan Nair[1,2], Chua Kein Huat[2], and Wong Chim Chwee[1]

[1] Center for Artificial Intelligence and Computing Applications, Universiti Tunku Abdul Rahman, Petaling Jaya, Selangor, Malaysia
{babrdel,madhavam,wongcc}@utar.edu.my

[2] Lee Kong Chian Faculty of Engineering and Science, Universiti Tunku Abdul Rahman, Petaling Jaya, Malaysia
{goib,chuakh}@utar.edu.my

Abstract. Observations over recent studies indicate that most of the methods and algorithms used to deal with clustering problems are based on hybrid metaheuristic and metaheuristic algorithms to rectify the solutions. However, these approaches are restricted by the number of heuristics. Hyperheuristic algorithms are new generation of metaheuristic algorithm that use a collection of low-level search strategies and high-level heuristics which works in heuristic space and solution spaces, while metaheuristic algorithms just work in solution space to find better solutions. The main goal of this research is to propose a hyperheuristic framework for clustering problem which is able to optimize the difference of all data objects of one cluster from their respective cluster centres. The proposed hyperheuristic clustering algorithm has used from pool of meta-heuristic and heuristic approaches and mapping between solution space and heuristic spaces is one of the prevalent and powerful methods in the optimization domains. By mapping the solution spaces into heuristic space, it would be possible to make easy decision to settle data clustering problems. Our suggested hyperheuristic clustering works into three major spaces including high-level space, low-level space and problem space. The experiments of this study have proven that the suggested method has successfully generated efficient and robust clustering results.

Keywords: Hyper-heuristic algorithms · Genetic clustering · Clustering algorithms

1 Introduction

Data clustering is commonly used in machine learning, pattern recognition, and data mining. Clustering refers to completing operations over datasets to categorize them into discrete groups with least similarity among the intergroup members and the most similarity among the intragroup members. Since the samples didn't carry any information about their actual classes, hence the analysis of cluster is one of the most important unsupervised learning approaches. On the other hand, the clustering algorithms are breakdown

A. Abraham et al. (Eds.): SoCPaR 2020, AISC 1383, pp. 145–155, 2021.
https://doi.org/10.1007/978-3-030-73689-7_15

into different categories such as density-based, network-based, partitioning, and hierarchal methods [1, 2]. K-means is a one of the well-known methods in data clustering because of its preference, ease of implementation and high operational speed. The main idea behind the k-means approach is to optimize and minimize the dissimilarity of all data pints into a cluster from their cluster centres [3]. Several metaheuristic and hybrid of metaheuristic algorithms are proposed for initialization problem of k-means approach but, problems such as the risk in getting trapped at local optima and sensitivity to initial solutions hinder its performance. Therefore, in order to obtain better cluster method, one should focus on finding a solution to first overcome the problems from getting trapped into local optima [4]. There are multiple researches to deal with k- means problems. In recent years, there are many reported studies on data clustering using k-means. For instance, Babrdel and his colleagues have suggested new swarm based hyperheuristic framework, which employs swam based high-level heuristic to generate new sequence of low- level heuristic to perform on solution space. In another research, Babrdel and his colleagues introduced another hyperheuristic framework which is efficient and robust for clustering problem. The framework use hybrid metaheuristic algorithm as high-level heuristic and employs fifteen low-level heuristics to work with solution space [5, 6]. Another study of mentioned research team has done on hybrid algorithm which employs bee and differential evolution algorithms to solve clustering problem [7]. Niknam and his colleagues have proposed a hybrid approach by combining the ant colony optimization and partial swarm optimization with k-means approach for data clustering but it is sensitivity to the initialization and get stuck into local optimality [8]. Cao and his team have presented a combination method based on the hybrid of genetic algorithm, k-means algorithm, and maximization of logarithmic regression expectation but get stuck into local optimality and sensitive to data behavior [9]. Kao and his colleagues have suggested a hybrid algorithm which combine the genetic algorithm, partial swarm optimization, and Nelder-Mead simplex search altogether. However, the proposed algorithm has premature convergence problem [10]. Fathian and his colleagues in another research, have introduced an algorithm to settle this problem based on the Honey-Bees Mating Optimization (HBMO) [11]. Zalik has proposed a technique for clustering without pre-assigning cluster numbers. The suggested algorithm includes two separate stage. The first step is a preprocessing procedure which applies initial clustering and assigns for each cluster at least one seed-point [12]. Krishna his colleague have introduced an algorithm for cluster analysis called genetic k-means algorithm. They combine the genetic algorithm with a gradient descent algorithm which is used in K-means clustering [13]. Mualik and his colleague Bandyopadhyay introduced a genetic- based approach to deal with clustering problems and evaluate the performance of their algorithm using a real data. In the proposed algorithm, genetic algorithms is used in the feature space to look for proper cluster centres. In the algorithm, chromosomes are strings of real numbers which is used to encode the centres of clusters. They defined the spatial distance-based mutation according to mutation operator for clustering [14]. Laszlo his colleagues suggested another genetic-based algorithm for clustering problem which exchanges neighboring cluster centres. The proposed algorithm uses a new crossover operator which exchanges the neighboring centers [15]. Shelokar et al. proposed the Ant-Colony-Optimization algorithm to deal with k-means clustering problem. The suggested algorithm applies distributed factors which mimic the real ants behavior to find the shortest path between ants nest and food source [16].

In the rest of this study, in Sect. 2 the data clustering and in Sect. 2.1 the k-means problem are reviewed. In Sect. 3, the proposed hyper-heuristic framework is introduced for clustering problem. In the Sect. 4, results of suggested hyper-heuristic framework are illustrated and evaluated with other algorithms on benchmark data sets finally Sect. 5 is contained the conclusions.

2 Clustering

Clustering is classifying of data objects inside clusters. Cluster is a set of objects similar with together and different from the other clusters and other classes. In cluster analysis or unsupervised learning, objects are divided into different groups with maximum similarity between the objects in a cluster and minimum similarity among clusters. In traditional definition of hard type clustering, $D = \{d_1, d_2 ... d_n\}$ is a set of data with n members which are to be clustered or grouped into k clusters as $CP = \{cp_1, cp_2 \ldots cp_k\}$ so that each group satisfies the following:

1. $\cup_{i=1}^{k} \quad cp_i = D$
2. $cp_i \neq \phi \quad \mathrm{i} = 1 \ldots \mathrm{k}$
3. $cp_i \cap cp_i \neq \phi, \ \mathrm{j} \neq \mathrm{i\,j}, \mathrm{i} = 1 \ldots \mathrm{k},$

Accordingly various states for clustering of n data point into k group will be:

$$NP(k, cp) = \frac{1}{k!} \sum_{i=0}^{k} (-1)^i . \binom{k}{i} .(k - i)^n \tag{I}$$

Where k indicate number of clusters which is manually determined in Eq. (I), and n is total number of objects. The relation shows that even with a given k, finding the optimum state for clustering of data is complicated and not too easy. Also clustering states from n objects to k clusters growth s by $k^n/k!$. Therefore, finding the optimum scenario for clustering n samples into k cluster s is a $NP - Complete$ problem, which needs to be settled by metaheuristic algorithms.

2.1 K-means Algorithm

In different studies, there have been reported different methods to illustrate the clustering problem, however among presented methods the k- means clustering is one of the most famous one. The algorithm is one of the most known techniques to clustering problem due to its easy implementation, simple and fast technique, which generally required only smaller number of iterations. The idea behind k-means is to find cluster centres $(cp_1, cp_2 \ldots cp_k)$ by minimizing the squared sum of the distance of each sample d_i fromthe nearest cluster centre cp_j. In k-means, initial selection of cluster centres influences the performance of the k-means approach. The primary trend of k- means algorithm to cluster all objects come as follows:

i. Select k samples randomly from $D = \{d_1, d_2 \ldots d_n\}$ as cluster centers $(cp_1, cp_2 \ldots cp_k)$
ii. According to the relation (II) assign each data pint d_i from $D = \{d_1, d_2 \ldots d_n\}$ to mp_j as below:

$$\left\| d_i - cp_j \right\| < \left\| xi - cp_q \right\|, \; q \neq j, \quad 1 \leq q \leq k \tag{II}$$

iii. According to the step (ii), the new cluster centre s are updated and evaluated based on relation (III) in the way that n_i is the number of sample s in i^{th} cluster:

$$s_i^* = \frac{1}{n_i} \sum_{d_i \in cp_j}^{k} d_i \quad k \geq j \geq 1 \tag{III}$$

iv. If the centres of cluster updated, start the algorithm from step (ii), else do the clustering base d on the reported centres.

K-mean s algorithm performance relies on the initial cluster center s and determining the cluster center is the major challenge in k-means approach. Random initialization at first stage, makes the approach yield out different result s for several runs on the same data and this random initialization is considered as one of potential risks of k-means approach, due to its convergence tendency towards local optimum [17].

2.2 Cost Function

In this research, the cost function is calculated based on the distance of each data pint and cluster centres. Accordingly, in the first stage, a set of centres are elected randomly, in the second step, clustering are estimated according to the Eq. (II). Then according to the assigned cluster centers in the interaction step, new center of clusters and cost of solutions are computed according to the Eq. (III).

$$costs = \sum_{i=1}^{n} \sum_{j=1}^{k} \left\| d_i - cp_j \right\|^2 \tag{4}$$

Where d_i is input vector or observation, cp_j is jth cluster centre, K is number of clusters, $\|d_i - cp_j\|$ is the similarity index to calculate similarity between all samples from the corresponding cluster centre, and n is number of observations.

3 Proposed Algorithm

Hyper-heuristics is new generation of metaheuristics algorithms which attracted an increasing amount of research attention. These kinds of algorithms instead of designing heuristic algorithm with pre-defined mechanism or parameters, hyperheuristics look for specific low-level heuristic and learn the selections mechanism to choose suitable low-level heuristic to settle the problem. Hyperheuristics algorithms search heuristics space instead of solutions space. They therefore are self-adaptive algorithms, and are able to deal with a class of problem instead of just one problem without extensive development effort. Hyper-heuristics due to their self-adaptive nature have been successfully used within intelligent systems to concern various applications including knapsack, timetabling, assembly line, routing, 2D/3D strip packing, job shop scheduling, personnel scheduling, and many other complex problems. The goal of a hyper heuristic framework is to decrease the amount of domains knowledge by using the capabilities of high-level heuristic and the abilities of low-level search strategies parallelly in the search space. In hyper heuristics, the high-level technique, depending upon the search conditions and problem state, elects which low-level heuristic should be considered at any given time. Figure 1 illustrates the several parts of one hyper-heuristic framework.

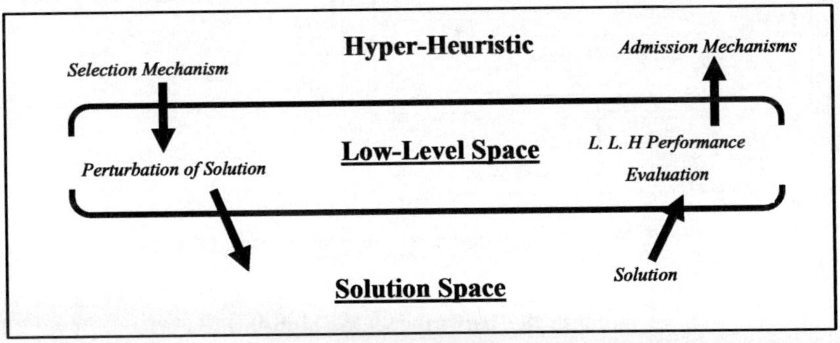

Fig. 1 Hyper-heuristic framework

3.1 Proposed Genetic-Based Hyper-Heuristic

A hyper-heuristic algorithm can generate and produce new heuristics based on the available heuristics. Two main feature of each hyper-heuristic framework are heuristics to choose heuristics and also heuristic to generate heuristics. In the first task, hyper-heuristic algorithm is able to generate new heuristic order based on existing heuristics and as a second task hyper-heuristic algorithm is able to select proper heuristics form heuristic pool. Therefore, the idea is to design a high-level mechanism which is intelligent enough to specify the weaknesses and strengths of a set of low-level heuristics. Each hyper-heuristic algorithm include two main spaces including high-level and low-level spaces. The proposed framework uses genetic-based high-level algorithm to select proper low-level heuristics (L.L.H). The proposed high-level algorithm works based on the genetic algorithm which investigates and evaluates the low-level heuristics according to their performances and generate and specify the order of low-level heuristics. In the first step, it select a low-level heuristic based on its merit. If number of selected low-level heuristics are less then specified number, then it selected another low-level heuristic. Heuristic selection part investigates the merit of each low-level heuristic based on low-level performance evaluation section which has interaction with low-level space to select heuristics from heuristic pool. Figure 2 shows the different steps of genetic-based high-level heuristic. After investigation step, admission mechanism is applied to make a decision about low-level heuristic to select better heuristics.

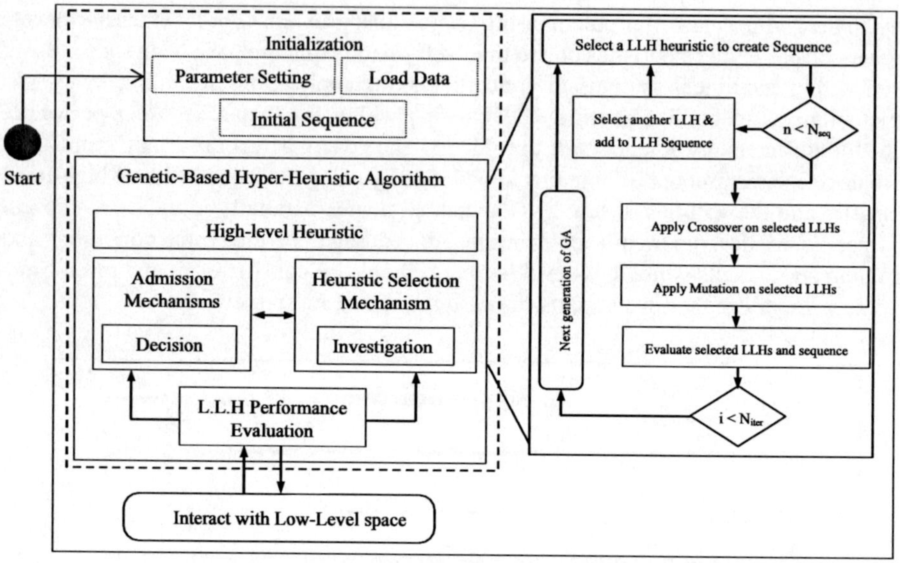

Fig. 2 Proposed genetic-based hyper-heuristic algorithm

In this research, to increase the performance of high-level heuristic, in each step we use a competition on the selected heuristic to increase the diversity of algorithm. After specifying the sequence of selected heuristics, we need to apply this heuristics on the solution space to find beret solution s in clustering domain. After finding the new solutions for clustering problem, the selected low-level search strategies are evaluated according to the number of best solutions ever found which specify the low-level search strategies performances to prepare information for performances evaluation section in high-level space. We iterate this algorithm until reaching certain number of iterations.

3.2 Low-Level Heuristic (L.L.H)

One of the biggest advantage of hyper-heuristic algorithms is taking the privileges of different low-level heuristics (L.L.H). The low-level heuristics space uses the pool of heuristics to keep different heuristic into pool of heuristics and there is no limitation to use low-level heuristics that is another advantage of hyper-heuristic algorithms. The proposed method takes the advantages of different low-level heuristics to search new solution and would lead to superior result s compare d to when applying each one individually. One of the disadvantages of metaheuristic and hybrid-metaheuristic methods is confined to apply number of different heuristic approaches however in hyper heuristic there is a low-level heuristics pool which can employ different number of low-level heuristic algorithms as mentioned earlier. This advantage helps the algorithm to enhance the quality of result s and solution s. In this framework, 19 low-level heuristics are employed to work in solution space.

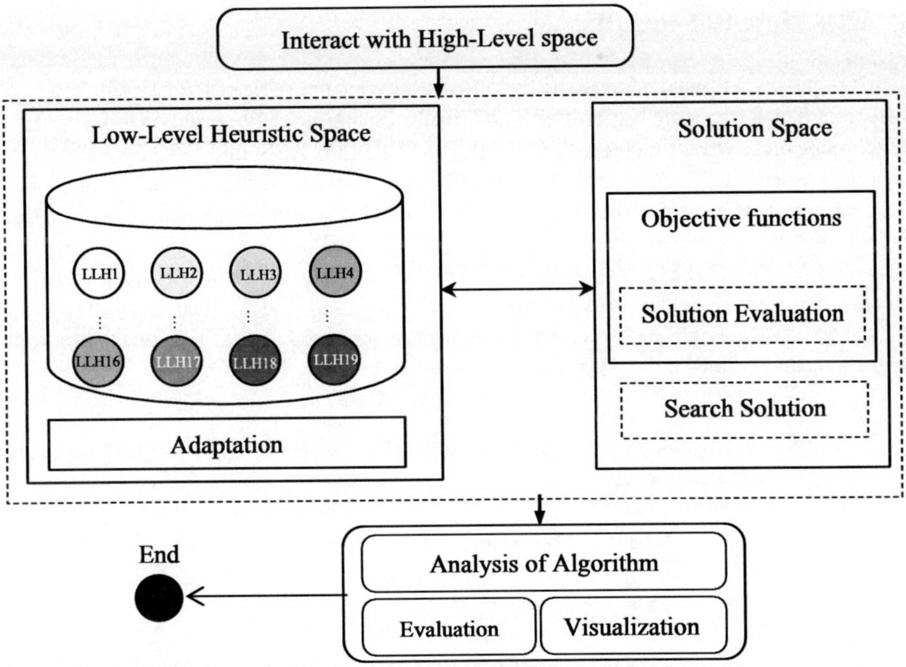

Fig. 3 Low-level heuristic space

This research use two kind of low-level search strategies to apply on continues domain which includes 5 mutation and 14 crossover heuristics. Following Table 1 illustrates the low-level search strategies which are employed in proposed framework. In this framework we have proposed three new low-level heuristics also.

Table 1. L.LH search strategies lists

Seq.	L.L. Heuristics	Seq.	L.L. Heuristics	Seq.	L.L. Heuristics
1	Bit string mutation	8	Uniform crossover	14	Multi-point crossover
2	Boundary mutation	9	Double point crossover	15	Heuristic uniform crossover
3	Flip bit mutation	10	Shuffle crossover	16	Discrete crossover
4	Non-uniform mutation	11	Blend crossover	17	Reduced surrogate crossover
5	Modified Gaussian mutation	12	Flat Crossover	18	Single point crossover
6	Hybrid crossover	13	Modified arithmetical crossover	19	Heuristic crossover/intermediate crossover
7	Average crossover				

4 Experimental Results

In this step, we compare and evaluate the proposed hyper-heuristic efficiency with regards to specified cost function. Experiments and results have been deliberated on four real-life benchmark data, which includes CANCER, VOWEL, GLASS, and IRIS datasets from UCI repository [18]. The suggested approach were run using MATLAB on Windows 8.1 OS running on 4GB of RAM and core i5 CPU. To measure the suggested algorithm, the execution results of the suggested hyper-heuristic framework is evaluated with TS, GA, HBMO, PSO–ACO, PSO–SA, ACO–SA, SA, ACO, PSO, and k-means result which is presented in reference [7] are tabulated in Tables (2, 3, 4 and 5). The values reported are the standard deviations, execution time, best, worst and mean of the fitness function over mentioned datasets for 100 runs.

Table 2. Comparison of proposed algorithm for 100runs over IRIS data set

Methods	Results			CPU-times	STD
	Worsts	Averages	Bests		
GA	139.778	125.197	113.986	~140	4.62
TS	98.569	97.868	97.365	~135	13.84
SA	102.01	99.957	97.457	~32	0.98
ACO	97.808	97.171	97.100	~75	0.29
PSO	97.897	97.232	96.8942	~30	0.57
HBMO	97.757	96.953	96.752	~82	0.18
ACOA	96.863	96.731	96.660	~25	0.19
PSO–SA	96.678	96.671	96.660	~17	0.65
PSO–ACO	96.674	96.654	96.654	~17	0.17
K–Means	120.45	106.05	97.333	0.4	85.5
Proposed Alg.	**96.5403**	**96.5403**	**96.5403**	**~14**	**0**

Table 3. Comparison of proposed algorithm for 100runs over GLASS data set

Methods	Results			CPU-times	STD
	Worsts	Averages	Bests		
GA	286.77	282.32	278.37	~410	4.138712
TS	286.47	283.79	279.87	~410	4.192734
SA	287.18	282.19	275.16	~410	4.238458
ACO	280.08	273.46	269.72	~395	3.584829
PSO	283.52	275.71	270.57	~400	4.557134
HBMO	249.54	247.71	245.73	~390	2.438120
ACO–SA	202.76	201.89	200.71	~49	0.887234
PSO–SA	202.45	201.45	200.14	~38	0.892430
PSO–ACO	200.01	199.61	199.57	~35	0.13914
K–Means	255.38	235.5	215.74	~1	12.47107
Proposed Alg.	**210.46**	**210.43**	**210.42**	**~27**	**0.0078305**

Table 4. Comparison of proposed algorithm for 100runs over VOWEL data set

Methods	Results			CPU-times	STD
	Worsts	Averages	Bests		
GA	165,991.6520	159,153.498	149,513.735	~39	3,105.5445
TS	165,996.4280	162,108.5381	149,468.268	~33	2,846.23516
SA	165,986.4200	161,566.2810	149,370.4700	~32	2,847.08594
ACO	165,939.8260	159,458.1438	149,395.602	~27	3,485.3816
PSO	149,121.1834	148,999.8251	148,976.0152	~30	28.8134692
HBMO	165,804.671	161,431.0431	149,201.632	~32	2,746.0416
ACO–SA	149,364.30	149,141.41	149,005.0	~35	120.38
PSO–SA	154,751.26	150,343.45	149,001.85	~33	351.45
PSO–ACO	149,101.6800	148,995.2032	148,976.0100	~17	24.5420931
K–Means	161,236.81	159,242.89	149,422.26	~0.45	916
Proposed Alg.	**149,087.3864**	**149,005.3904**	**148,967.2408**	**~13**	**38.7983**

Table 5. Comparison of proposed algorithm for 100runs over CANCER dataset

Methods	Result			CPU-times	STD
	Worsts	Averages	Bests		
GA	3,427.43	3,249.46	2,999.32	~35	229.734
TS	3,434.16	3,251.37	2,982.84	~130	232.217
SA	3,421.95	3,239.17	2,993.45	~126	230.192
ACO	3,242.01	3,046.06	2,970.49	~123	90.50028
PSO	3,318.88	3,050.04	2,973.50	~123	110.8013
HBMO	3,210.79	3,112.43	2,989.95	~136	103.470
ACO–SA	2,968.29	2,966.63	2,967.83	~40	1.7732
PSO–SA	2,967.41	2,966.32	2,965.17	~28	1.7201
PSO–ACO	2,964.50	2,964.39	2,964.38	~17	0.037947
K–Means	3,521.59	3,251.21	2,999.19	~0.5	251.14
Proposed Alg.	**2,964.381**	**2,964.381**	**2,964.381**	**~22**	**0**

5 Conclusions

The major goal of current research was to introduce a novel genetic-based hyper-heuristic framework which generate high quality performance and robust results across different low-level search strategies by minimizing the dissimilarity of all data items in a cluster. The proposed heuristics algorithm is able to elect proper low-level heuristics order based on their merits. For increasing the exploitation of hyper-heuristic method, several low-level heuristics are employed. The proposed algorithm works on clustering problem efficiently.

Acknowledgment. This study was funded through the university grant with the number of (IPSR/RMC/UTARRF/2019-C2/M01. The authors would like to express their deepest gratitude to Universiti Tunku Abdul Rahman (UTAR) and Centre for Artificial Intelligence and Computing Applications (CAICA) for their supports and comments to make this research a meaningful one.

Reference s

1. Yang, F., Sun, T., Zhang, C.: An efficient hybrid data clustering method based on K-harmonic means and particle swarm optimization. Exp. Syst. Appl. **36**(6), 9847–9852 (2009)
2. Güngör, Z., Ünler, A.: K-harmonic means data clustering with tabu-search method. Appl. Math. Model. **32**(6), 1115–1125 (2008)
3. Bonab, M.B., Hashim, S.Z.M., Alsaedi, A.K.Z., Hashim, U.R.: Modified k-means combined with artificial bee colony algorithm and differential evolution for color image segmentation. In: Phon-Amnuaisuk, S., Thien Wan, Au. (eds.) Computational Intelligence in Information Systems, pp. 221–231. Springer, Cham (2015)

4. Hamerly, G., Elkan, C.: Alternatives to the k-means algorithm that find better clusterings. In: Proceedings of the 11th International Conference on Information and Knowledge Management, McLean, Virginia, USA, pp. 600–607. ACM (2002)
5. Bonab, M.B., et al.: An Efficient Robust Hyper-Heuristic Algorithm to Clustering Problem. Springer, Cham (2019)
6. Bonab, M.B., et al.: A new swarm-based simulated annealing hyper-heuristic algorithm for clustering problem. Procedia Comput. Sci. **163**, 228–236 (2019)
7. Babrdel Bonab, M., et al.: An effective hybrid of bees algorithm and differential evolution algorithm in data clustering. Math. Probl. Eng. **2015**, 17 (2015)
8. Niknam, T., Amiri, B.: An efficient hybrid approach based on PSO, ACO and k-means for cluster analysis. Appl. Soft Comput. **10**(1), 183–197 (2010)
9. Nguyen, C.D., Cios, K.J.: GAKREM: a novel hybrid clustering algorithm. Inf. Sci. **178**(22), 4205–4227 (2008)
10. Kao, Y.-T., Zahara, E., Kao, I.W.: A hybridized approach to data clustering. Exp. Syst. Appl. **34**(3), 1754–1762 (2008)
11. Afshar, A., et al.: Honey-bee mating optimization (HBMO) algorithm for optimal reservoir operation. J. Franklin Inst. **344**(5), 452–462 (2007)
12. Žalik, K.R.: An efficient k'-means clustering algorithm. Pattern Recogn. Lett. **29**(9), 1385–1391 (2008)
13. Krishna, K., Murty, M.N.: Genetic K-means algorithm. IEEE Trans. Syst. Man Cybern. Part B Cybern. **29**(3), 433–439 (1999)
14. Maulik, U., Bandyopadhyay, S.: Genetic algorithm-based clustering technique. Patt. Recogn. **33**(9), 1455–1465 (2000)
15. Laszlo, M., Mukherjee, S.: A genetic algorithm that exchanges neighboring centers for k-means clustering. Patt. Recogn. Lett. **28**(16), 2359–2366 (2007)
16. Shelokar, P.S., Jayaraman, V.K., Kulkarni, B.D.: An ant colony approach for clustering. Anal. Chim. Acta **509**(2), 187–195 (2004)
17. Bonab, M.B.: Modified k modified k-means algorithm for genetic clustering means algorithm for genetic clustering means algorithm for genetic clustering. IJCSNS **11**(9), 24 (2011)
18. Bonab, M.B., Hashim, S.Z.M.L Image segmentation with genetic clustering using weighted combination of particle swarm optimization. In: 14th International Conference on Applied Computer and Applied Computational Science, ACACOS 2015 (2015)

LSTM vs. GRU for Arabic Machine Translation

Nouhaila Bensalah[1(✉)], Habib Ayad[1], Abdellah Adib[1], and Abdelhamid Ibn El Farouk[2]

[1] Team Networks, Telecoms and Multimedia, University of Hassan II Casablanca, Casablanca 20000, Morocco
nouhaila.bensalah@etu.fstm.ac.ma, abdellah.adib@fstm.ac.ma
[2] Teaching, Languages and Cultures Laboratory Mohammedia, Casablanca, Morocco

Abstract. The same Machine Translation (MT) approach may not work for European languages as for Arabic, because of its structure. MT based on Neural Networks methods has recently become an alternative approach to the statistical MT. In this paper, a case study is presented on how different sequence to sequence Deep Learning (DL) models perform in the task of Arabic MT. A comprehensive comparison between these models based mainly on: Long Short Term Memory (LSTM), Gated Recurrent Unit (GRU), Bidirectional LSTM (BiLSTM) and Bidirectional GRU (BiGRU) is presented. Specifically, each input sequence will be translated into English one using an Encoder-Decoder model based on the four architectures with an attention mechanism. Furthermore, we study the impact of different preprocessing techniques on Arabic MT.

Keywords: Arabic Machine Translation · LSTM · GRU · Bidirectional LSTM · Bidirectional GRU · Arabic preprocessing · Farasa

1 Introduction

MT is an intricate process that uses a computer application to translate texts or speech or even captures from one natural language to another [2]. Many approaches from traditional rule-based approaches to the recent statistical methods have been applied since the introduction of MT.

Since the 1950 s, research in MT has been an active topic [16,23]. In 1982, the Nagao paper [18] introduced a rules-based MT approach for transferring grammatical concepts between English and Japanese languages. In [9], it has also been suggested to use a phrase-based statistical MT systems between Arabic and English with remarkable improvements compared to the rules-based MT systems.

Recently, Neural Networks were very strong because of their excellent performance on difficult problems such as speech recognition [21], Biomedical Engineering [6], Question Answering [4] as well as MT [3].

A. Abraham et al. (Eds.): SoCPaR 2020, AISC 1383, pp. 156–165, 2021.
https://doi.org/10.1007/978-3-030-73689-7_16

To enhance translation quality, many MT systems based on Recurrent Neural Networks (RNNs) schemes have been employed and often use an Encoder-Decoder model. The Encoder-Decoder model consists of three components:

1. An Encoder module to encode the input vector sequence.
2. An attention module to extract related information from the input sequence and transfer it to the decoder.
3. A Decoder to generate the output sequence.

However, RNN schemes suffer from vanishing and exploding gradient problems [14]. A popular alternative is either to use LSTM [15] or GRU [7] architectures in order to fix these issues and to perform well in capturing long-term dependencies.

In this paper, four types of RNNs are used; namely, LSTM, GRU, BiLSTM and BiGRU. Moreover, we will give a comprehensive comparison of the performance of Arabic MT under these widely used architectures using an Encoder-Decoder model with the attention mechanism [3]. Furthermore, we study the impact of different preprocessing techniques on Arabic MT, such as the Penn Arabic Treebank (ATB) tokenization which has shown to have a significant impact on Statistical Machine Translation in Arabic [13].

The rest of the paper will be organized as follows. In Sect. 2, we will detail the different components that embody the DL models studied in this work. Sect. 3 presents the experiments and the results achieved. Finally, a conclusion in Sect. 4 ends the paper.

2 The Encoder-Decoder Model with Attention Mechanism

In the Encoder-Decoder model, an encoder maps the input vector sequence $\mathbf{x} = (x_1, x_2, .., x_n)$ to the hidden vector sequence $\mathbf{h} = (h_1, h_2, .., h_n)$. To generate the output sequence $\mathbf{y} = (y_1, y_2, .., y_n)$, a mechanism of attention then a decoder are applied on the generated vector sequence $\mathbf{h}$. Specifically, each time the model generates a word of the output sentence, the model focuses on the words in the source sentence where the most relevant information is concentrated. Then, the model predicts the next word of the output sentence based on the context as well as the previous hidden vector generated respectively by the attention mechanism and the decoder. Finally, this process is repeated until the end of the source sentence.

To study the performance of the Arabic MT under different RNNs using the Encoder-Decoder model with attention mechanism, 16 different combinations of RNNs will be used:

1. LSTM as an encoder and LSTM as a decoder.
2. LSTM as an encoder and BiLSTM as a decoder.
3. LSTM as an encoder and GRU as a decoder.
4. LSTM as an encoder and BiGRU as a decoder.
5. BiLSTM as an encoder and LSTM as a decoder.

6. BiLSTM as an encoder and BiLSTM as a decoder.
7. BiLSTM as an encoder and GRU as a decoder.
8. BiLSTM as an encoder and BiGRU as a decoder.
9. GRU as an encoder and LSTM as a decoder.
10. GRU as an encoder and BiLSTM as a decoder.
11. GRU as an encoder and GRU as a decoder.
12. GRU as an encoder and BiGRU as a decoder.
13. BiGRU as an encoder and LSTM as a decoder.
14. BiGRU as an encoder and BiLSTM as a decoder.
15. BiGRU as an encoder and GRU as a decoder.
16. BiGRU as an encoder and BiGRU as a decoder.

Hereafter, the RNN variants as well as the attention mechanism used in this work are described.

2.1 The Architecture of LSTM

To fix the vanishing gradient problem, LSTM [15] was created. It has internal mechanism called gates, that can regulate the flow of information into and out of the cell. Figure 1 shows LSTM's overall architecture.

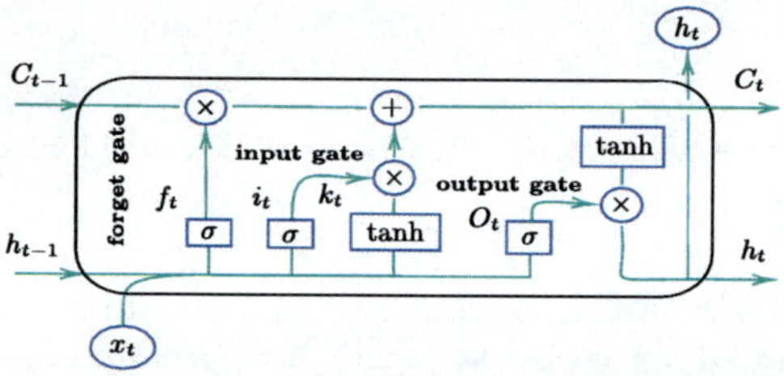

Fig. 1. The architecture of LSTM

The forget gate is used to remember the significant information from the past steps. Then, the input gate decodes the data to be added from the present step. Finally, the output gate determines the next hidden state. To compute the states of the LSTM, we act as the following:
First, we calculate at the forget gate:

$$f_t = \sigma(W_f h_{t-1} + U_f x_t + b_f) \quad (1)$$

Then, we evaluate at the input gate:

$$i_t = \sigma(W_i h_{t-1} + U_i x_t + b_i) \quad (2)$$

$$k_t = tanh(W_k h_{t-1} + U_k x_t + b_k) \quad (3)$$

We calculate C_t using the above three vectors, which represents the LSTM's memory state:

$$C_t = C_{t-1} \bigodot f_t + i_t \bigodot k_t \tag{4}$$

At the output gate, we compute:

$$O_t = \sigma(W_O h_{t-1} + U_O x_t + b_O) \tag{5}$$

Finally, the hidden state at step t is calculated as follows:

$$h_t = \begin{cases} O_t \odot tanh(C_t) & \text{if} \quad t > 0 \\ 0 & \text{if} \quad t = 0 \end{cases} \tag{6}$$

where: $\odot$ is the element-wise Hadamard product, x_t is the word representation at time t, $W_O, U_O, W_k, U_k, W_i, U_i, W_f, U_f$ are weight matrices, b_O, b_k, b_i, b_f are the bias, σ and $tanh$ represent respectively the element-wise sigmoid activation and the tangent hyperbolic functions.

2.2 The Architecture of GRU

GRU [8] utilizes an update and a reset gates to avoid the issue of the vanishing gradient that penalizes the standard RNN. These are basically two vectors that agree which data will be transferred to the output.

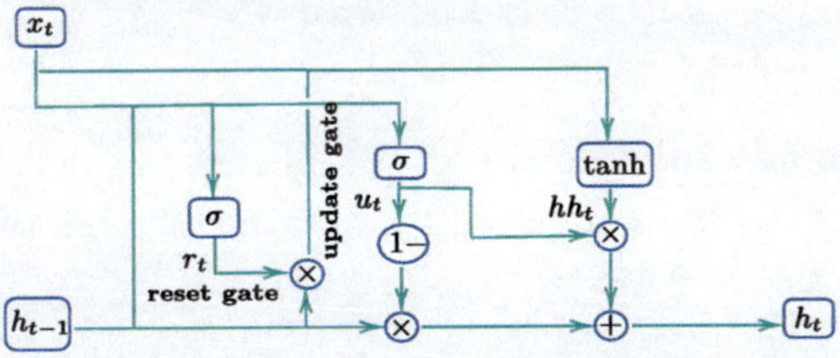

Fig. 2. The architecture of GRU

As shown in Fig. 2, the update gate enables the system to determine the amount of the past data that needs to be passed, and which is computed as follows:

$$u_t = \sigma(U_u h_{t-1} + W_u x_t) \tag{7}$$

Essentially, the reset gate is used to determine the amount of data that needs to be forgotten, and it is evaluated hence:

$$r_t = \sigma(U_r h_{t-1} + W_r x_t) \tag{8}$$

Finally, the hidden state h_t at step t is computed using hh_t:

$$hh_t = tanh(U_{hh}(r_t \bigodot h_{t-1}) + W_{hh} x_t) \tag{9}$$

$$h_t = \begin{cases} (1 - u_t) \odot tanh(h_{t-1}) + u_t \odot hh_t & \text{if} \quad t > 0 \\ 0 & \text{if} \quad t = 0 \end{cases} \tag{10}$$

where: W_r, U_r,W_u, U_u, W_{hh}, U_{hh} are weight matrices.

2.3 Bidirectional LSTM/ GRU

In Arabic MT, the GRU and LSTM architectures are motivated by their ability to capture both semantic and syntactic structures of the input sentences and also to represent the sequential data by taking into account the previous data. However, they are only able to make use of the amount of information seen in the previous steps. To exploit the future information as well, Bidirectional RNNs [20] do this by processing the data in two opposite directions. The Bidirectional RNNs architecture is illustrated in Fig. 3.

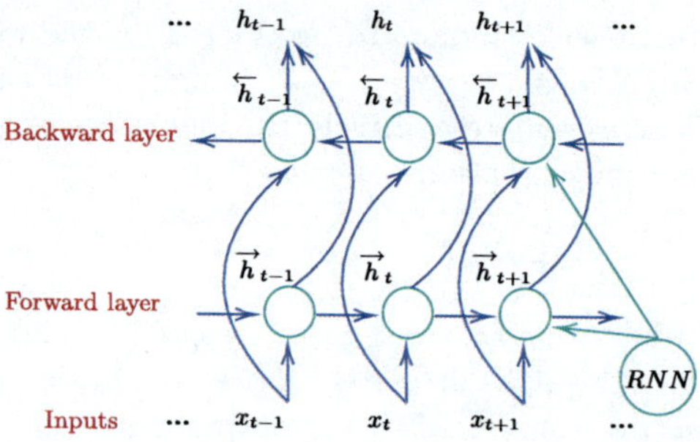

Fig. 3. Bidirectional RNN

In this case, the forward hidden vector sequence $\overrightarrow{h}_t$ and the backward hidden vector sequence $\overleftarrow{h}_t$ are computed by iterating the backward layer and the forward layer from $t = 1$ to $t = n$.

$$\overrightarrow{h}_t = \tanh(W_{x\overrightarrow{h}}x_t + U_{\overrightarrow{h}\overrightarrow{h}}h_{t-1} + b_{\overrightarrow{h}}) \tag{11}$$

$$\overleftarrow{h_t} = \tanh(W_{x\overleftarrow{h}}x_t + U_{\overleftarrow{h}\overleftarrow{h}}h_{t-1} + b_{\overleftarrow{h}}) \tag{12}$$

In this way, a sentence could be represented as $\mathbf{h} = (h_1, h_2, .., h_n)$ where $h_t = [\overrightarrow{h_t}, \overleftarrow{h}_t]$. Combining Bidirectional RNN with GRU/LSTM gives Bi(GRU/LSTM) [12] that could exploit long-range context in both input directions.

2.4 Attention Mechanism

The attention mechanism [3] was developed intuitively to attend to different parts of the source sentence at each step of the output generation.

As illustrated in Fig. 4, the context vector z_t $t \in (1, 2, .., n)$ depends on a sequence of annotations $(h_1, h_2, .., h_n)$ and the hidden state e_{t-1} (generated by the decoder) and is calculated as follows; see Fig. 5:

$$z_t = \sum_{j=1}^{n} s_{tj}h_j \tag{13}$$

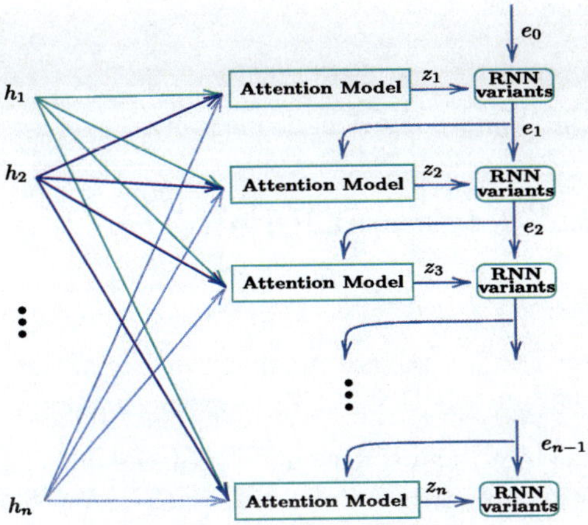

Fig. 4. Attention mechanism

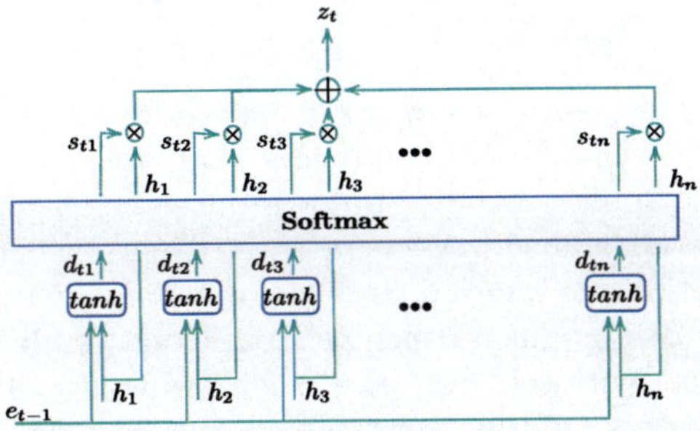

Fig. 5. Attention model

where: the weight s_{tj} of each annotation h_j is computed according to:

$$s_{tj} = \frac{exp(d_{tj})}{\sum_{k=1}^{n} exp(d_{tk}))} \tag{14}$$

and with: $d_{tj} = a(e_{t-1}, h_j)$ is an alignment model that shows how the input at position j and the output at position $t-1$ match. Three different methods can be used to evaluate it:

– **Additive attention:**

$$d_{tj} = W_a^{\intercal} tanh(W_h h_j + U_h e_{t-1}) \tag{15}$$

– **Multiplicative attention:**

$$d_{tj} = h_j^{\intercal} W_m e_{t-1} \tag{16}$$

– **Dot product:**

$$d_{tj} = h_j^\intercal e_{t-1} \tag{17}$$

where: W_a, W_h, U_h, W_m are the weight matrices.

3 Experimental Setup and Results

3.1 Dataset and Preprocessing

In this paper, we aimed to investigate the performance of the Arabic MT system with an OPUS dataset, namely UN [22]. In our experiments, we normalized some arabic letters such as (آ, أ, إ) and ة, which are replaced with ا and ه, respectively. Moreover, we removed any special characters such as (#$%) and the tashkeel symbols. Furthermore, we used Farasa [1] as tokenization scheme which has shown to have a significant effect on phrase-based translation system in [10].

3.2 Experimental Results

We train our model by using sentences of length 20 for both the source and the output sentences, and for the sentences that are short (length < 20), a padding technique is applied in order to have the same length for all the source sentences. Arabic word embeddings were trained via FastText model [5] that takes into consideration the internal structure of words, which makes it suitable for morphologically rich languages such as Arabic one. Arabic embeddings were trained on UN dataset with settings: (size = 256, window = 10, min count = 5).

The hyperparameters are the cornerstone of the learning of the training process. In this study, the proposed model parameters were optimized by Adam [17] with a learning rate of 10^{-2}. We train our model for 20 epochs with a batch size of 64. Finally, the early stopping technique based on the validation loss is added to stop the training process after 4 epochs, if this loss starts to increase [11]. As an evaluation metric, the BLEU score [19] was used.

As introduced earlier, the main idea of this study is to give a comprehensive comparison of the performance of all the RNN variants in the context of Arabic-English translation. Hence 16 DL models (discussed in Sect. 2) are used, which are based on LSTM, GRU, BiLSTM and BiGRU. Table 1 provides the performance measurement of the 16 DL models with and without preprocessing.

The reported results in Table 1 showed a greater gain performance of the Encoder-Decoder with attention mechanism based on LSTM architecture in terms of BLEU score (BLEU score = 41.42%). However, in terms of training time or computational speed, the model composed of BiGRU as an Encoder and LSTM as a Decoder was the faster (training time = 1187 s). While training the different models, it was also observed that the model which achieves the best performance in terms of BLEU score and computational speed is composed of BiGRU as an Encoder and BiLSTM as a Decoder (BLEU score = 41.11%, training time = 1474 s). These results can be explained by the fact that the

Table 1. The overall performance for each DL model with and without preprocessing

	Before preprocessing		After preprocessing	
	BLEU score (%)	Time (s)	BLEU score (%)	Training time (s)
1	41.42	2180	41.87	2093
2	40.98	2493	41.31	2303
3	40.71	1567	40.83	1509
4	40.18	2184	40.23	2113
5	39.50	2579	40.05	2480
6	40.78	1714	41.02	1643
7	40.50	2336	40.70	2250
8	40.84	1501	41.10	1454
9	39.16	1763	39.80	1701
10	41.00	2648	41.52	2540
11	40.27	1320	40.52	1240
12	40.75	2175	40.95	2002
13	39.95	1187	40.12	1036
14	41.11	1474	41.62	1343
15	40.88	2116	41.16	1980
16	40.43	1483	40.67	1396

model (the 14th DL model) in this case benefits from BiGRU which is known by its ability to train faster than BiLSTM and also from BiLSTM which generally outperforms BiGRU in terms of BLEU score.

As mentioned in Sect. 3.1, we used Arabic preprocessing which includes many steps such as: normalization and tokenization (ATB scheme) using Farasa. Examining the results, it seems that Arabic preprocessing was effective in being applied to Arabic sentences. The 14th DL model in this case achieved a BLEU score of 41.62% and was faster (training time = 1343 s) compared to 41.11% and1474 s obtained without preprocessing. These findings may be explained by the ability of Arabic preprocessing to deal with data sparsity and also handle the tokens that are not present in the training corpus.

As a result of this analysis, the best appropriate combination to be used in this study is BiGRU as an Encoder, BiLSTM as a Decoder, the attention mechanism and the Arabic preprocessing. Table 2 illustrates three Arabic sentences and their translations in English in this case. In example 1, the system translates the sentence fluently. Example 2, which represents much longer sentence, the system does not translate the Arabic sentence as the truth, but instead it preserves the original meaning of the source sentence. However, in example 3, the system fails to select the better prediction.

Table 2. A few examples of translations generated by our model

Source	الاستراتيجية الدولية للحد من الكوارث
Our model	international strategy for disaster reduction
Truth	international strategy for disaster reduction
Source	١٥ تؤكد أن عمل الوكالة لا يزال ضروريا في جميع ميادين العمليات
Our model	15 notes that the agency's working remains important in all areas of service
Truth	15 affirms that the functioning of the agency remains essential in all the fields of operation
Source	و إذ ترحب بقرار المجلس التنفيذي أن يشجع بصورة نشطة استراتيجية جمع الأموال للمعهد
Our model	welcoming the board in promoting and development of the institute
Truth	welcoming the decision of the executive board to actively promote a fund-raising strategy for the institute

4 Conclusion

This paper achieves the goal of translating Arabic sentences using 16 different DL models. More precisely, the models were based on LSTM, BiLSTM, GRU and BiGRU. Experimental results showed that BiGRU as an Encoder, BiLSTM as a Decoder and the attention mechanism achieved the best results in terms of BLEU score and computational speed. We also notice that the morphology-based tokenization and the normalization improve the performance of Arabic MT system. One possibility for future research would be to find a better architecture to improve the translation quality.

References

1. Abdelali, A., Darwish, K., Durrani, N., Mubarak, H. Farasa: a fast and furious segmenter for Arabic. In: Proceedings of the 2016 Conference of the North American Chapter of the Association for Computational Linguistics: Demonstrations, pp. 11–16 (2016)
2. Alqudsi, A., Omar, N., Shaker, K.: Arabic machine translation: a survey. Artif. Intell. Rev. **42**(4), 549–572 (2014)
3. Bahdanau, D., Cho, K., Bengio, Y.: Neural machine translation by jointly learning to align and translate (2014). Accepted at ICLR 2015 as oral presentation
4. Bensalah, N., Ayad, H., Adib, A., El farouk, A.I.: Combining word and character embeddings in Arabic chatbots. In: Advanced Intelligent Systems for Sustainable Development, AI2SD 2020, Tangier, Morocco (2020)
5. Bojanowski, P., Grave, E., Joulin, A., Mikolov, T.: Enriching word vectors with subword information. Trans. Assoc. Comput. Linguist. **5**, 135–146 (2017)

6. Bouny, L.E., Khalil, M., Adib, A.: ECG heartbeat classification based on multiscale wavelet convolutional neural networks. In: IEEE International Conference on Acoustics, Speech and Signal Processing (ICASSP), Barcelona, Spain, Mai 4–8 (2020)
7. Cho, K., van Merriënboer, B., Bahdanau, D., Bengio, Y.: On the properties of neural machine translation: encoder–decoder approaches. In: Proceedings of SSST-8, Eighth Workshop on Syntax, Semantics and Structure in Statistical Translation (2014)
8. Cho, K., van Merriënboer, B., Gülçehre, a., Bahdanau, D., Bougares, F., Schwenk, H., Bengio, Y.: Learning phrase representations using RNN encoder-decoder for statistical machine translation. In: EMNLP ACL (2014)
9. Devlin, J., Zbib, R., Huang, Z., Lamar, T., Schwartz, R., Makhoul, J.: Fast and robust neural network joint models for statistical machine translation. In: Proceedings of the 52nd Annual Meeting of the Association for Computational Linguistics (2014)
10. El Kholy, A., Habash, N. Orthographic and morphological processing for English-Arabic statistical machine translation. Mach. Transl. **26**, 25–45 (2012)
11. Feurer, M., Hutter, F.: Hyperparameter optimization. In: Automated Machine Learning. Springer, pp. 3–33 (2019)
12. Graves, A., Schmidhuber, J.: Framewise phoneme classification with bidirectional lstm and other neural network architectures. Neural Netw. **18**(5–6), 602–610 (2005)
13. Habash, N., Sadat, F.: Arabic preprocessing schemes for statistical machine translation. In: Proceedings of Human Language Technology Conference of the North American Chapter of the Association of Computational Linguistics (2006)
14. Hanin, B.: Which neural net architectures give rise to exploding and vanishing gradients? In: NeurIPS (2018)
15. Hochreiter, S., Schmidhuber, J.: Long short-term memory. Neural Comput. (1997)
16. Hutchins, W.J., Somers, H.L.: An Introduction to Machine Translation. Academic Press, Cambridge (1992)
17. Kingma, D.P., Ba, J.: Adam: a method for stochastic optimization. In: 3rd International Conference on Learning Representations, ICLR (2015)
18. Nagao, M.: A framework of a mechanical translation between Japanese and English by analogy principle. In: Artificial and Human Intelligence, North-Holland (1984)
19. Papineni, K., Roukos, S., Ward, T., Zhu, W.-J.: Bleu: a method for automatic evaluation of machine translation. In: Proceedings of the 40th Annual Meeting on Association for Computational Linguistics, pp. 311–318. Association for Computational Linguistics (2002)
20. Schuster, M., Paliwal, K.K.: Bidirectional recurrent neural networks. IEEE Trans. Sign. Process. **45**(11), 2673–2681 (1997)
21. Sekkate, S., Khalil, M., Adib, A., Jebara, S.B.: An investigation of a feature-level fusion for noisy speech emotion recognition. Comput. **8**, 91 (2019)
22. Tiedemann, J.: Parallel data, tools and interfaces in OPUS. In: Proceedings of the Eighth International Conference on Language Resources and Evaluation (LREC 2012) Istanbul, Turkey, pp. 2214–2218 (2012)
23. Turing, A.M.: Computing machinery and intelligence. Mind (1950)

Towards Lip Motion Based Speaking Mode Detection Using Residual Neural Networks

Sadia Afroze and Mohammed Moshiul Hoque(✉)

Department of Computer Science and Engineering, Chittagong University of Engineering and Technology, Chittagong 4349, Bangladesh
moshiul_240@cuet.ac.bd

Abstract. Speaking mode (i.e., talking or non-talking) detection is a significant research problem in the area of HCI and computer vision. Detecting the speaking mode of a speaker is quite challenging owing to the coarse-resolution images, interaction styles, and various noises. This paper proposes a vision-based technique to identify a human's speaking mode in terms of talking and non-talking state by using the residual neural networks. Visual lip motions is a prominent cue and play a pivotal role in detecting the speaking mode of the human. Thus, we consider the vision-based technique rather voice-based, which is noisy or interrupted. The evaluation result in two datasets shows better performance (of 99.56% accuracy) in mouth state detection than previous approaches. Moreover, analysis with 36 min long video data of 15 participants reveals that the proposed technique acquired an accuracy of 98.88% in detecting speaking mode.

Keywords: Human computer interaction · Computer vision · Speaking mode detection · Lip motion detection · Mouth state classification

1 Introduction

Detecting of speaking mode provides salient knowledge in HCI such as automatic visual speech recognition and speaker verification [1]. It has various applications in human-robot interaction and fatigue detection. Recently, several AI-based systems employed speaking mode detection and voice recognition in unmanned aerial vehicles, robot, interactive games and some clinical tools. The voice-driven techniques are malfunctioning due to the encompassing sound. Thus, the vision-based approach can improve the performance of speaking mode detection. Moreover, visual cues for speaking mode detection play a significant role when sound is interrupted. The visual cue, such as lip movements played a vital role and considered an essential parameter in various interactive systems. The visual analysis of lip gesture can be used to identify the speaking mode of speakers in video conferencing, and hypo vigilance inspection. Several surveys reported the majority

A. Abraham et al. (Eds.): SoCPaR 2020, AISC 1383, pp. 166–175, 2021.
https://doi.org/10.1007/978-3-030-73689-7_17

of accidents happens due to the talking between the passenger and the driver. Therefore, the driver's speaking mode identification can be integrated into driving assistance system, which may design alarm system to create awareness and helps in reducing the street accident rate.

Most of the previous survey considered audio-based speaking mode detection, which is much complicated due to surrounding noise. Moreover, previous vision-based speaking mode detection considers a statistical approach which does not work well in different facial orientation and consumes much execution time. To address this issue, we proposed a vision based framework that can detect speaking mode into two classes: talking and non-talking. A residual neural network (ResNet) technique with ResNet layer architecture is used to develop the model [2]. ResNet can remove the vanishing gradient problem and obtain better performance than machine learning based techniques. A self-developed data repository and YawDD dataset [3] are utilized to train the classifier model.

2 Related Work

Identification of speaking mode (i.e., talking or non-talking) in real-time is one of the challenging tasks in computer vision that has numerous HCI implications. Several strategies used to classify talking or non-talking mode using lip motions analysis [4]. This system used Viola and Jones face detection algorithm [5] to detect the face and STASM technique [6] for localizing 68 facial landmarks. The developed technique is not reliable in classifying the rare face, and expressive face while the mouth is widely open or shrink. A lip motion-based system is developed which used mean squared differences to classify the talking face in TV contents [7]. Ji et al. [8] proposed a technique to estimate the left and right corner of the mouth region with a bounding box. Their system used a convolutional neural network [9] to classify mouth state and mouth gesture detection.

Lip segmentation method used contour features to analyze the mouth state. Nevertheless, this work cannot recognize the mouth state when the mouth opens widely, and the mouth internal contour extraction is harrowing when the mouth shrinks. A progressive learning-based method used to extract mouth region in [10] but this technique entirely depends on the grey distribution of the facial region. Few studies used neural networks or SVM to detect the mouth region [11]–[12], which achieved higher accuracy. Nevertheless, these techniques require much training data and consume high time in training. Huang et al. [13] proposed a mouth detector using a face recognition technique. They considered skin colour segmentation to filter out the non-skin colour region to improve mouth localization. A recent method used a supervised machine learning approach to detect speaking mode [14]. Another scheme used pattern recognition to classify talking mode [15]. However, these systems are not performed well for large datasets and different lighting conditions.

Most of the existing statistical techniques suffer in low lighting environments and rare faces. Other existing methods require an elevated level of lips representation and considered only mouth motion to detect the speaking mode. In our work, a deep leaning technique is used to detect talking and non-talking modes which alleviate the shortcomings of the existing methods.

3 Speaking Mode Detection Framework

The suggested framework comprises of four major modules: mouth region extraction, training, testing, and speaking mode prediction. Figure 1 represents the abstract view of the proposed speaking mode detection.

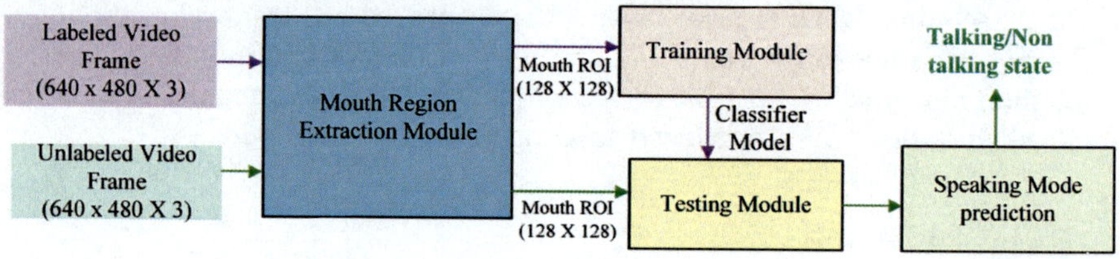

Fig. 1. Speaking mode detection framework.

3.1 Mouth Region Extraction Module

The system detects the face region using a multi-task cascaded convolutional neural network (MTCNN) [16] for extracting the mouth region. This network provides five facial key points, namely left mouth corner, right mouth corner, the nose tip, and two middle points (one for each eye). The left and right mouth points used as the points to determine the mouth region. The MTCNN network comprises of three sub-networks: proposal network (P-Net), refine network (R-Net), and output network (O-Net). The P-Net calculates the regression vectors of the candidate windows and tunes the candidate windows to select the bounding box of the face region. The overlapping bounding box is removed using non-maximum suppression (NMS). The R-Net structure performs the same task as P-Net, but it adjusted the output size of the P-Net into 24×24. Finally, O-Net structure filters the false-positive bounding box and adjusts output size of R-Net into 48×48, which provides five facial points. Four border points in the mouth frame produced based on the left and right corner points of the mouth. In case of talking and non-talking mode, the left-hand and right-hand border points vary significantly. Thus, these points used to find the mouth region of interest (ROI). Figure 2 depicted the selection of the mouth region.

To find the mouth rectangle, four facial key points are considered such as nose tip point ($n(x_1, y_1)$), mouth left corner point ($m_1(x_2, y_2)$), mouth right corner point ($m_2(x_3, y_3)$), and bottom right corner point ($B(x_4, y_4)$). $\alpha(x_2, y_1)$ indicates the mouth rectangle top left corner point and $\beta(x_3, (y_4 - y_3)/2)$ indicates the bottom right corner point. The height (H) and width (W) of the mouth rectangle will be $y_4 - y_1$ and $x_4 - x_2$ respectively. The detected mouth region passes to the classification module.

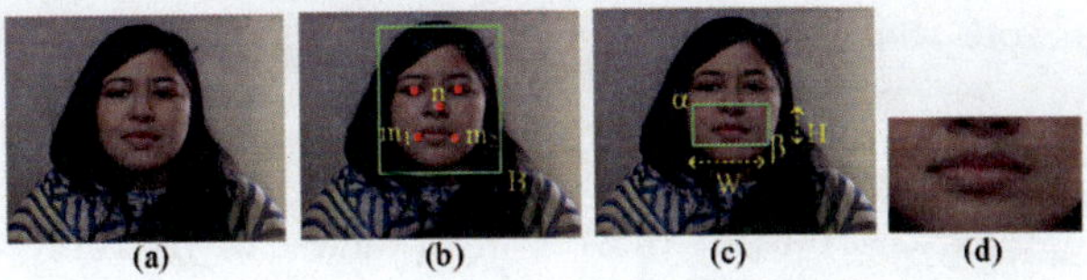

Fig. 2. Mouth region selection: (a) Input image (b) Detected face rectangle with five facial key points (c) Mouth rectangle selection (d) Extracted mouth region.

3.2 Training Module

The training module takes $128 \times 128 \times 3$ mouth region as the input and produces a classification model. ResNet18 layer architecture is used to develop the classifier model. Figure 3 illustrates the training module for mouth state classification.

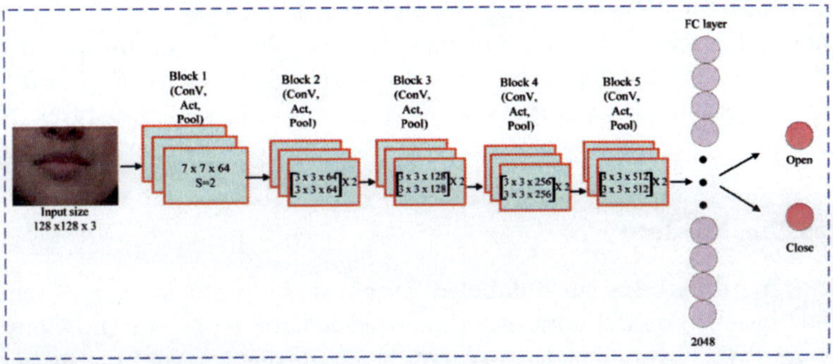

Fig. 3. Architecture of the training module for mouth state classification.

ResNet18 network contains five cascade blocks, and each block containes batch normalization (B_n), convolution layer (Conv), activation layer (Act), and pooling layer (Pool). The size of the initial mouth region is $128 \times 128 \times 3$ which pass to the Conv layer. Conv layer consists of 64 number of 3×3 kernels used for convolution operation. Conv extracts local features and generates a feature map matrix (FM) using the Eq. (1).

$$M(i,j) = \sum_{k=i,p=0}^{R=i+WF_h} \sum_{l=j,q=0}^{C=j+WF_w} M(k,l) \times WF(p,q) \tag{1}$$

here, $M(i,j)$ denotes the present input position, R represents the row of the kernel, C indicates columns of the image. The parameters F, F_h and F_w indicates the number of kernels, height of the kernel, and width of the kernel respectively.

The FM passes to BN layer for normalization. BN layer performs mean and variance operation to speed up the training time [17]. Act layer reduces the high volume of data from the FM by using the Eq. (2).

$$P(i,j) = max(P(i,j), 0) \tag{2}$$

here, P(i,j) indicates the output of Act layer and can be represented as the rectified feature map matrix (RFMM).

The RFMM passed to the Pooling layer to perform average pooling operation (Eq. 3) and max pooling operation (Eq. 4). These operations are useful to minimize the high volume of data to produce pooled feature map matrix (PFMM).

$$P(i,j) = \frac{1}{K_h \times K_w} \sum_{k=i}^{i+K_h} \sum_{l=j}^{j+K_w} P(k,l) \tag{3}$$

$$P(i,j) = \max{}_{k=1,l=1}^{k=h,l=w} P(k,l) \tag{4}$$

The similar operation continues in the rest of the blocks of ResNet18. Two strides are used in blocks 1 and 2. Each block kernel size was 3 × 3 whereas the number of kernel increases from block 1 to block 5 for improved performance. Finally, PFMM flattened in a fully connected layer (FC layer). The number of feature in FC layer is 2048, which passes to the softmax layer (SM) for classification.

3.3 Testing Module

The testing module takes an unlabeled frame as the input for the classification task. The classifier model contains the saved weight kernel in the Conv layer, FC layer and SM layer, which used to initialize weight kernel of the classifier. The unlabeled input frame passes to the classifier model and finally produce a score for each frame by using the Eq. (5).

$$M_{s_i} = \frac{e_i^x}{\sum_{k=0}^{n} e^{x_k}} \tag{5}$$

Here, M_{s_i} denotes the mouth state, and n represents the number of mouth state (for closed mouth $n = 0$, and open mouth $n = 1$).

3.4 Speaking Mode Prediction Module

The prediction modules predict the label of a frame, whether as talking or non-talking modes. In talking mode, there are consecutive mouth open or close frames with short intervals. Thus, we observe the talking frames around five seconds. Based on the empirical observations on mouth open/close frames, we design rules (Rules 1–3) to determine the talking mode: (i) Rule 1: If two successive opening mouth frames follow by a closing mouth frame or vice versa (ii) Rule 2: If the ratio between an open mouth frame (60%), and a close mouth frame (40%) or vice versa, and (iii) Rule 3: If absolute difference between M_o and M_c

varies from 22 to 27 for last 125 frames. M_c indicates the number of frames for the close mouth and M_o denotes the number of frames for the open mouth.

In non-talking mode, there are consecutive frames of close mouth only. Thus, the system detects non-talking mode when it gets continuous close mouth frame.

4 Experiments

The proposed vision-based speaking mode detection system implemented on Intel I7-7700K (LGA-1151) 7TH GEN, 4.20GHZ processor. We used deep learning library Keras [18] and Tensorflow as back end framework. The framework was running on an NVIDIA GTX 1070 GPU with 32 GB RAM on Ubuntu16.04 platform.

4.1 Datasets

To identify speaking mode, we consider ResNet18 for mouth classification. Two datasets used to evaluate the proposed system: (i) YawDD dataset [3] and (ii) Self-built dataset. YawDD is a video dataset consisting of 351 video sequences with various facial features of male/female. We convert this video datasets into $164,389$ images and classified the images according to open or close mouth state manually. The self-built dataset contained $230,456$ images. Table 1 illustrates the distribution of two datasets. It is indicated that the total number of training images are 260000, validation images are 75000 and testing images are 59845. Speaking mode prediction module evaluated by 36 minutes live video from 15 undergraduate students.

Table 1. Distribution of mouth state classification datasets.

Dataset	#Training images	#Validation images	#Testing images
YawDD	120 000	25 000	19 389
Self-built	140 000	50 000	40 456
Total	260 000	75 000	59 845

5 Results

The results of the proposed framework considers mouth state classification performance measure and speaking mode prediction performance measure. Mouth state classification training module computes loss and accuracy as performance measure using the Eq. 6 which is represented in Fig. 4.

$$Loss(I) = -\sum_{i=0}^{C} T_i \times log(pr_i) \tag{6}$$

where, $Loss(I)$ denotes the loss value, T_i represents the true labeled for i^{th}, pr_i denotes probability of that sample, and C indicates the number of classes. The training accuracy is calculated from the actual and predicted labeled using the ratio, $\frac{\#actula\_predict}{\#sample}$.

The accuracy of the training/validation phase shown in Fig. 4a. The training accuracy starts from 0.61, whereas the validation accuracy begins from 0.66. After two epochs, both accuracy values increase sharply and stable at the epoch 42. At this epoch, the training phase achieved almost 100.00% accuracy, and the validation phase gained about 99.55% accuracy. Thus, it observed that the classifier model is not overfitted or under-fitted.

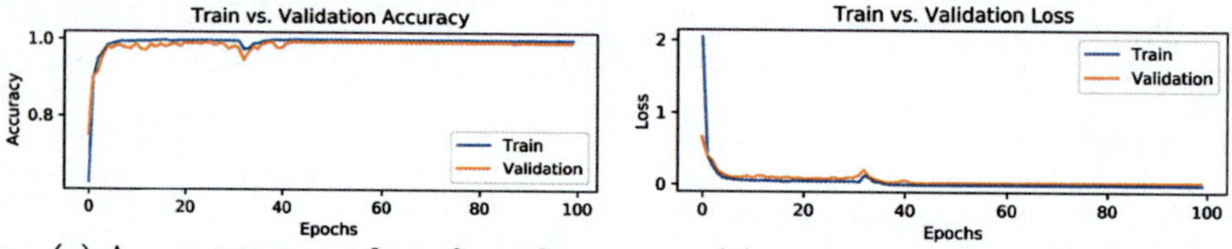

(a) Accuracy vs. number of epochs. (b) Loss vs. number of epochs.

Fig. 4. Accuracy/losses during the training and validation phases.

Figure 4b illustrates the loss in the training and validation phases. The training loss begins from 2.11, whereas the loss of validation starts from 0.73. After a few epochs, both losses are mostly the same. The loss curves gradually retain a steady distant from the epochs 40, which reveals that the technique is free from the problem of overfitting/underfitting.

To evaluate the testing phase of mouth state classification, we used several statistical measures such as sensitivity (S_t), specificity (S_f), true positive rate (TPR), false-positive rate (FPR), accuracy (A) and precision (P). Figure 5a represents a graph for precision versus recall. The maximum precision and recall values for the close mouth class are 99.89 and 99.87. The maximum precision and recall values for the open mouth class are 99.77 and 97.99. Figure 5b shows the receiver operating characteristics (ROC) of the suggested technique. This curve illustrates the performance of the classifier model graphically concerning accuracy (99%). The curve indicates that the best threshold for the open and close mouth classes achieved at 0.99.

Figure 6a depicts the various kinds of statistical outcomes of the mouth state detection. In this research, the sensitivity (S_t) is 99.87%, specificity (S_f) is 99.66%, accuracy (A) is 99.56%, precision (P) is 99.47% and Recall is 99.55%. Figure 6b shows the performance comparison of ResNet18 with other algorithms in testing time. Here, the ResNet18 achieves maximum accuracy (99.56%) for mouth state classification. When we select VGG16 and VGG19 the model is underfitted because this model gives a poor performance on training samples. The model becomes overfitted when we select ResNet34 because of less amount of training data. Thus, we selected ResNet18 for the proposed system.

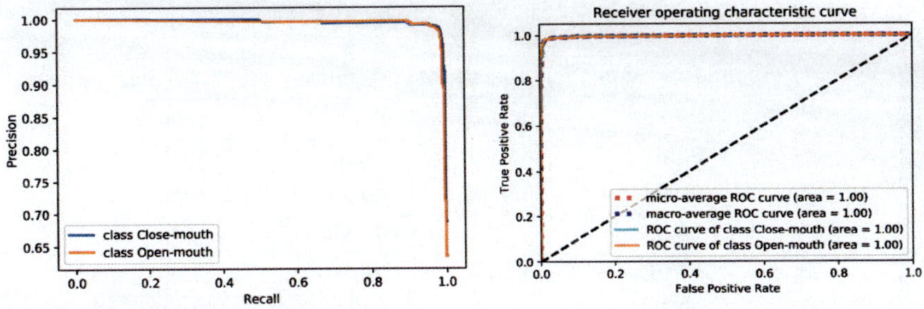

(a) Precision vs recall curve (b) ROC curve

Fig. 5. Testing phase evaluation for the mouth state detection.

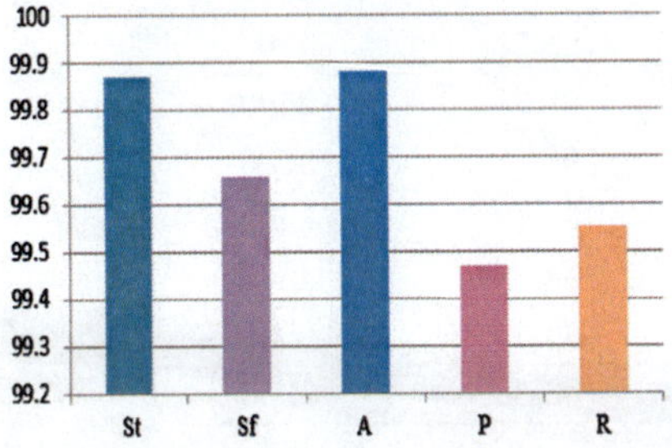

(a)

Model	Accuracy (%)
VGG16	**97.66**
VGG19	**98.23**
ResNet18	**99.56**
ResNet34	**98.65**

(b)

Fig. 6. (a) Statistical measurements (b) Performance comparison for mouth state detection.

Speaking mode performance is measured from 36 minutes video where 15 under grade students has participated. Table 2 shows the class-wise average precision, recall and accuracy. We consider consecutive 125 frames for evaluating speaking mode performance. That means the system runs on 25 fps and after the 5 sec period, we verify the talking or non-talking state. Table 2 shows that the proposed system achieved maximum 98.88% accuracy.

Figure 7a shows the sample output of the proposed framework for talking state detection. A video clip is collected for the talking mode detection[1]. Figure 7b represents the non-talking state frame sequence which are selected from self-built dataset. To assess the performance, the suggested technique compared with existing approaches [14,19]. Table 3 represents the result of comparison. The proposed technique achieved the highest of 98.88% than the previous approaches.

[1] https://www.youtube.com/watch?v=u0zP9TPMfNg&t=19s.

Table 2. Summary of the analysis

#Person	Precision (%)	Recall (%)	Accuracy (%)	#Frame
P_1	99.23	98.23	99.33	2000
P_2	98.67	99.32	98.56	2500
P_3	99.23	99.45	99.43	3000
P_4	99.55	98.56	97.89	3500
P_5	97.22	99.88	98.89	4000
P_6	98.11	97.92	99.08	4500
P_7	99.55	98.68	98.67	5000
P_8	98.22	99.18	99.78	5500
P_9	99.68	98.85	97.68	6000
P_{10}	97.88	97.82	98.76	4500
P_{11}	98.18	99.65	99.26	2500
P_{12}	98.45	98.32	98.78	3500
P_{13}	99.67	99.11	99.78	1500
P_{14}	98.22	98.23	98.88	4500
P_{15}	99.33	99.01	99.46	2500
Avg.	98.75	98.81	98.88	3666

(a) Talking state detection sequence (b) Non-talking state detection sequence

Fig. 7. Speaking mode detection.

Table 3. Performance comparison

Method	Accuracy(%)
Log-polar + SVM [19]	97.4
HOG + SVM [14]	98.00
MTCNN + ResNet-18 (Proposed)	98.88

6 Conclusion

The primary concern of our work is to develop a real-time computational framework that can detect the human's speaking mode concerning talking and non-talking. For this purpose, this work developed a deep learning-based technique using ResNet with MTCNN algorithm. Speakers lip motions are analyzed to extract patterns of the mouth open or close states in recognizing talking mode. Evaluation results revealed the effectiveness of the proposed system with 98.88%

detection accuracy, which is higher concerning the previous methods. The proposed system can be extended by considering the speaking modes in multi-person scenarios.

References

1. Zhou, Z., Zhao, G., Hong, X., Pietikainen, N.: A review of recent advances in visual speech decoding. Image Vis. Comput. **32**(9), 590–605 (2014)
2. He, K., Zhang, X., Ren, S., Sun, J.: Deep residual learning for image recognition. In: IEEE Conference on Computer Vision and Pattern Recognition (CVPR), pp. 770-778. Las Vegas (2016)
3. Abtahi, S., Omidyeganeh, M., Shirmohammadi, S., Hariri, B.: YawDD: a yawning detection dataset. In: Proceedings of the 5th ACM Multimedia Systems Conference, pp. 24-28, New York, USA (2014)
4. Bendris, M., Charlet, D., Chollet, G.: Lip activity detection for talking faces classification in TV-content. In: The 3rd International Conference on Machine Vision (2010)
5. Viola, P., Jones, M.: Rapid object detection using a boosted cascade of simple features. In: Computer Vision and Pattern Recognition, April 2001
6. Milborrow, S., Nicolls, F.: Locating facial features with an extended active shape model, pp. 504–513 (2008)
7. Everingham, M., Sivic, J., Zisserman, A.: Hello! my name is... buffy – automatic naming of characters in TV video. In: Proceedings of the British Machine Vision Conference (2006)
8. Ji, Y., et al.: Fatigue state detection based on multi-index fusion and state recognition network. IEEE Access **7**, 64136–64147 (2019)
9. Hossain, M.R., Afroze, S., Hoque, M.M., Siddique, N.: Automatic detection of eye cataract using deep convolution neural networks (DCNNs). In: 2020 IEEE Region 10 Symposium (TENSYMP), pp. 1333-1338, June 2020
10. Jia, L.N., et al.: Smartphone-based fatigue detection system using progressive locating method. Inst. Eng. Technol. **10**(3), 148–156 (2016)
11. Punitha, A., Geetha, M.K., Sivaprakash, A.: Driver fatigue monitoring system based on eye state analysis. In: International Conference on Circuits, Power and Computing Technologies, pp. 1405–1408. IEEE, India (2014)
12. Du, C., Gao, S.: Image segmentation-based multi-focus image fusion through multi-scale convolutional neural network. IEEE Access **5**(99), 15750–15761 (2017)
13. Huang, H.Y., Lin, Y.C.: An efficient mouth detection based on face localization and edge projection. Int. J. Comput. Theor. Eng. **5**(3), 514 (2013)
14. Afroze, S., Hoque, M.M.: Talking vs Non-Talking: a vision based approach to detect human speaking mode. In: 2019 International Conference on Electrical, Computer and Communication Engineering (ECCE). Cox'sBazar, Bangladesh (2019)
15. Chang, C.C., Lin, C.J.: LIBSVM : a library for support vector machines. ACM Transactions on Intelligent Systems and Technology (2011)
16. Zhang, K.P., Zhang, Z.P., Li, Z.F., Qiao, Y.: Joint face detection and alignment using multitask cascaded convolutional networks. IEEE Sig. Process. Lett. **23**(10), 1499–1503 (2016)
17. Ioffe, S., Szegedy, C.: Batch normalization: accelerating deep network training by reducing internal covariate shift. Int. Conf. Mach. Learn. **37**, 448–456 (2015)
18. Chollet, F.: Keras (2015). https://github.com/keras-team/keras
19. Mandal, B., et al.: Towards detection of bus driver fatigue based on robust visual analysis of eye state. IEEE Trans. Intell. Transp. Syst. **18**(3), 545–557 (2017)

Reducing the Error Mapping of the Students' Performance Using Feature Selection

Yuni Yamasari[1](✉), Naim Rochmawati[1], Anita Qoiriah[1], Dwi F. Suyatno[1], and Tohari Ahmad[2]

[1] Department of Informatics, Universitas Negeri Surabaya, Surabaya, Indonesia
{yuniyamasari,naimrochmawati,anitaqoiriah, dwifatrianto}@unesa.ac.id

[2] Department of Informatics, Institut Teknologi Sepuluh Nopember, Surabaya, Indonesia
tohari@if.its.ac.id

Abstract. In an educational environment, classifying the cognitive aspect of students is critical. It is because an accurate classification is needed by a lecturer to take the right decision for enhancing a better educational environment. To the best of our knowledge, there is no previous research that focuses on this classification process. In this paper, we propose discretization and feature selection methods before the classification. For this purpose, we adopt the equal frequency for the discretization whose result is evaluated by using logistic regression with two regularizations: lasso and ridge. The experimental result shows that four-intervals on the ridge achieve the highest accuracy. It is to be the base to determine the level of the student's performance: excellent, good, fair, and poor. Next, we remove unnecessary features, by using the Gain Ratio and Gini Index. Also, we build classifiers to evaluate our proposed methods by using k-Nearest Neighbors (k-NN), Neural Network (NN), and CN2 Rule Induction. The experimental result indicates that both discretization and feature selection can enhance the performance of the classification process. Concerning the accuracy level, there is an increase of about 35%, 2.14%, and 3.8% on average of k-NN, NN, and CN2 Rule Induction respectively, from those with original features.

Keywords: Data mining · Features selection · Classification · Performance · Student

1 Introduction

In an educational environment, learning and evaluation systems based on the web are a consequence of the fast growth of Information and Communication Technology (ICT). In this digital era, a web-based learning system is developed to support the learning process. So, students can learn about the subjects they take anytime and anywhere. Also, a web-based evaluation system is built to evaluate the understanding and capability of students in a certain subject. All systems are developed to simplify the learning and evaluation processes, regarding time, cost, and easiness [1].

A. Abraham et al. (Eds.): SoCPaR 2020, AISC 1383, pp. 176–185, 2021.
https://doi.org/10.1007/978-3-030-73689-7_18

The stored data in those systems consist of many features that represent the characteristics of students. They are useful for users to do further processing which is done by data mining. There are many tasks in data mining, including classification, clustering and association analysis, etc. [2]. In the data mining environment, tasks applied to educational data are called educational data mining (EDM). Here, EDM uses statistical, data mining methods, and machine learning [3] for obtaining the required information. Therefore, it can be inferred that EDM focuses more on methodologies and techniques [4].

There is much research relating to the student data processing using machine learning methods as a subset of Artificial Intelligence [5, 6]. In the beginning, the previous research exploits student data to classify sex, age, family size to some groups by applying machine learning methods for the classifiers like Nearest Neighbors, Support Vector Machines, and Naive Bayes algorithms [7]. In other research, classifying student cognitive is done by using Bayesian Network, Naïve Bayes, and J48 [8]. Decision Tree as a classification method is implemented to assess students' skills relating to the designing and conducting experiments within a complex systems micro world [9]. Another research also uses a Decision Tree to mine student data [10]. The goal of this research is to explore Students' Self-Regulation in Learning Contexts. In further research, a binary classifier is built to predict performance on student activity patterns in Virtual Learning Environment (VLE) [11].

All the previous research depicts that classification has been an important factor in EDM. As far as we know, there has been no research focusing on improving the accuracy of classification tasks, whereas, accurate classification of student data gives advantages for the education field. For example, the classification results make it easier for the users to support the best decision because the main goal of EDM is to extract educational data to produce information for supporting the decision making in the educational system [4]. For example: identifying students based on their learning achievement, modeling their achievement, and predicting their learning results. Moreover, this accurate information can be used to design a better academic environment.

Because of those factors, to achieve higher accuracy, much research has worked on the enhancement of the classification performance. Ramaswami et al. study techniques to select certain features to find the best predictive performance of the student achievement classification [12]. Here, they use Correlation, Chi-Square, Gain-Ratio, Information-Gain, Relief, and Symmetrical Uncertainty which are executed with some classifiers, such as Voted Perceptron, Naïve Bayes, one R, and PART. In this case, the best results are generated by information gain and correlation with ROC value = 0,729 and F1-Measure = 0.592. A comparison of feature selection is also made in [13] which explores the student performance classification. Various methods are applied in this research: Genetic Algorithm, Support Vector Machine, Information Gain, and Minimum Redundancy and Maximum Relevance (MRMR). Then, four classifiers are operated to evaluate these techniques: D-tree, Naïve Bayes, k-NN, and ANN. The result shows that MRMR and k-NN are the best methods whose accuracy is 91.12. Regha et al. study about selecting features in educational data mining by using a new technique called Artificial Fish Swarm-Cuckoo Search Optimization. It is done to remove redundancy and irrelevant feature, so the classification performance is more effective [14]. Here, all research does not explain the effect of the accuracy level on student modeling, especially on the classification.

By considering the scope of previous research, here we intend to work on those problems by proposing methods to optimize the classification process and to evaluate its effects on the classification of student's domain cognitive. Furthermore, different from existing research, we investigate not only increasing the accuracy by implementing the Gini Index and Gain Ratio but also evaluate their effect on the performance of the classification.

2 Method

In this section, we explain our proposed method whose overall process is shown in Fig. 1. It comprises three stages: pre-processing, classification, and post-processing stages. The details of those stages are as follows.

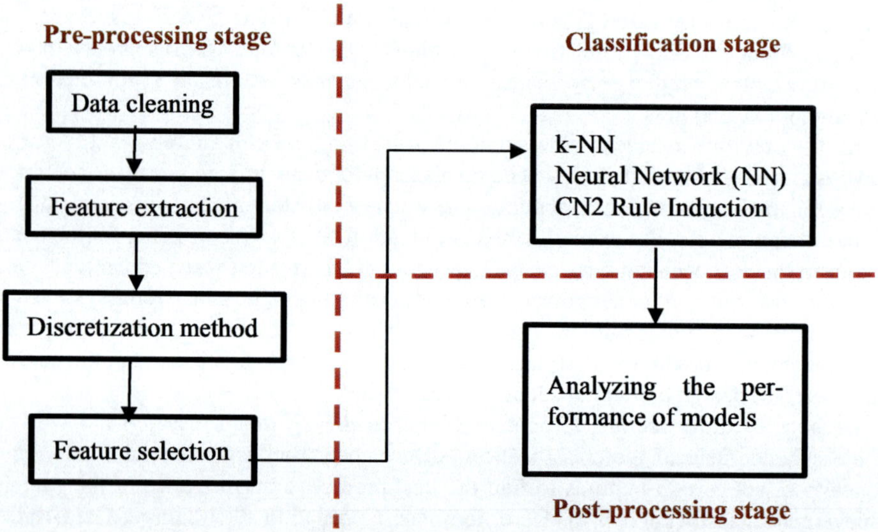

Fig. 1. The proposed method

In the pre-processing stage, stage comprises many steps, namely: data cleaning, category-based features extraction, exploring of discretization method, and feature selection. The data cleaning step is done to avoid redundancy, null variable or feature, noise, etc. for our student data as suggested by [15]. Then, Feature extraction is extracted based on the category, as in our previous research [16] to improve the performance. Discretization Method is explored, namely: equal frequency. We do discretization of student data using three, four, and five-intervals. Then, we build a model by using logistic regression to evaluate this discretization method. Here, there are two regularizations implemented. They are lasso and ridge. In this step, firstly we compute the score of all features, which is then used for evaluating their impact on the dependent (class) variable. For this purpose, we propose to make use of both the Gini Index and Gain Ratio. In the case of the Gini index, two randomly chosen instances may have different classes. This condition can be depicted in (2), where denote the fraction of records belonging to class i at node t.

$$Gini(t) = 1 - \sum_{i=0}^{c=i} [p(i|t)]^2 \quad (1)$$

$$Entropy(t) = - \sum_{i=0}^{c=i} p(i|t)\log_2 p(i|t) \quad (2)$$

Next, Gain Ratio is defined as the information gain divided by the entropy of the feature's value, and it is introduced by [17]. This research uses the Gain Ratio which is described in (3), (4). In those formulas, and represent the training set, the number of partitions, and the attribute of the data set, respectively.

$$Gain\ Ratio\ (B) = \frac{Gain\ (B)}{Split\ Info\ (B)} \quad (3)$$

$$Split\ Info_B(D) = - \sum_{j=1}^{x} \frac{|D_j|}{|D|} \log_2 \left(\frac{|D_j|}{D} \right) \quad (4)$$

Once the scores of all features have been computed, they are ranked. By considering that some have represented all features, we take of them, with is the number of available features. Therefore, we have 4 out of 10 features.

In the next stage, we construct a classification model by using machine learning: k-NN, Neural Network (NN), and CN2 Rule Induction [18]. Furthermore, we apply Euclidean for the distance parameter, and the number of neighbors is set to 5. Additionally, we design to use a supervised Neural Network with sklearn's Multi-Layers Perceptron (MLP) algorithm which utilizes backpropagation for the training [19, 20]. According to [21, 22], MLP has been able to optimize the log-loss function by using stochastic gradient descent. Lastly, we propose a classification model by using CN2, which is introduced in [23, 24].

Lastly, in the post-Processing Stage, we do analyze the experimental result for investigating whether the previously selected features can produce better performance than that of original features (all features). The performance is measured by utilizing classifier metrics: accuracy, precision, recall, and F1.

3 Result and Discussion

In this section, we analyze the experimental results which are obtained from the proposed method. After data cleaning and feature extraction, student data is generated as depicted in Table 1.

Then, we focus to observe the number of classes. Especially on the classification of student's performance, the number of the class target is very important. Because it indicates the mapping of student's abilities, so we propose the discretization method on the labeling process. Here, we only depict the experimental result on the training set = 60% because almost all intervals achieve the highest accuracy level on this training set.

This is represented in Fig. 2. Additionally, the four-intervals have achieved the Fig. 2. In detail, respectively, the highest accuracy level on the lasso and ridge is reached by three-intervals about 85.9% and by four-intervals about 100%. For that, we decide on four classes on the labeling process for student's performance, namely: poor, fair, good, and excellent.

Table 1. Data type and category-based extraction features

No.	Feature	Data type	Description
1	MID	Numeric	Number of the answered main items
2	MID%_True	Numeric	The proportion of the correct answered to all answered main items
3	Time_MID	Numeric	Time is taken for answering the main item
4	Score_MID	Numeric	A score of the student from the main items
5	GID	Numeric	Number of the answered scaffolding/guide
6	GID%_True	Numeric	The proportion of the correct answered to guide items
7	Time_GID	Numeric	Time is taken for answering the scaffolding/guide items
8	Hint	Numeric	Number of the used hints
9	Score_GID	Numeric	A score of the student from the guide items
10	TotalScore	Numeric	A total score of MID and GID

In the next process, we calculate the score of each feature by implementing the Gini Index and Gain Ratio, as in (2)–(4). In the result using Gini Index, the highest four features are Score_GID, GID, Time_GID, and MID%_True whose value is respectively 0.489, 0.448, 0.41, 0.407. Meanwhile, with Gain Ratio, the highest four features are Hint_GID, Score_GID, GID, and MID%_True having values 0.741, 0.689, 0.613, and 0.613, respectively. For the next process, we ignore the rest features, by assuming that those are irrelevant.

In the process of classification, as described in the previous section, we experiment by using cross-validation. In this research, we apply to 2, 3, 5, 10, and 20 folds. These results are then evaluated by using accuracy, precision, recall, and F1. To investigate the effect of this proposed method, we also implement it to all features (without selecting the feature). We compute the average of all metrics as shown in Table 2. The feature selection can increase significantly the accuracy level average on k-NN by about 24%–26.1%. In contrast, feature selection enhances slightly the accuracy level average on the NN, namely: about 0.4%–2.4%. In the remaining, the feature selection increases moderately the performance on the CN2 Rule Induction method by about 3.9%.

From those three classifiers which have been applied to different sets of features, we only visualize the lowest and the highest of performance as illustrated in Fig. 3(a) and (c). Overall, this figure describes that students' performance is classified into four classes: excellent, good, fair, and poor. It is found that many students are incorrectly classified. This misclassification happens mostly in k-NN with Original_features depicted in Fig. 3(a). Contrary, when this NN classifier is applied to Gini_Features, there is a significant improvement. That is, the number of misclassified students can be reduced depicted in Fig. 3(c).

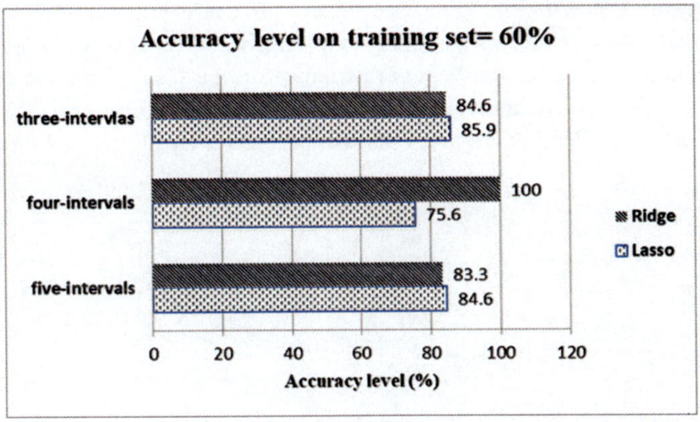

Fig. 2. Comparison of accuracy level on the combination of the discretization method and the logistic regression for all intervals on size of training set = 60%

Table 2. The comparison of performance average on all methods

Method	Accuracy	Precision	F1	Recall
k-NN_original	56.3	0.76	0.73	0.79
k-NN_Gain	92.4	0.97	1	0.94
k-NN_Gini	90.3	0.99	1	0.98
NN_original	92.2	1	1	1
NN_Gain	92.6	0.97	1	0.95
NN_Gini	94.6	1	1	1
CN2 Rule Induction_original	87.9	0.94	0.94	0.93
CN2 Rule Induction_Gain	91.8	0.98	1	0.96
CN2 Rule Induction_Gain	91.8	0.98	1	0.96

For visualizing that classifier performance, we use the ROC (Receiver Operating Characteristics) whose evaluation is performed by calculating its AUC (Area Under Curve). Indirectly, this also expresses the value of the respective confusion matrix. This illustration of the narrowest of AUC occurs on the combination of original features and all classifiers provided in Fig. 3(b). In contrast, the widest of AUC is achieved by the Gini_features on all classifiers illustrated in Fig. 3(d). In detail, a combination of Gini_Features with every classifier is larger than Gain_Features and Original_Features, where CN2 Rule Induction, k-NN, and NN have 0.996, 1, and 1, respectively. On contrary, the smallest AUC is generated by Original_Features. That is, the AUC of k-NN,

NN and CN2 are 0.946, 1, and 0.986. Moreover, we can infer that the use of feature selection (Gain_Features and Gini_Features) can optimally improve the classification performance, especially on k-NN. It is indicated by an increase of AUC by about 0.047–0.054.

For analyzing the error mapping, we observe the best performance of classifiers for analyzing its impact on the students' mapping based on the performance. On k-NN, Gain_Features achieves the best performance on the fold of 10. We compare the number of misclassified students obtaining from Original_Features and Gain_Features. It is shown that Gain_Features reduces the mapping error by 37, from 43 to 6 students.

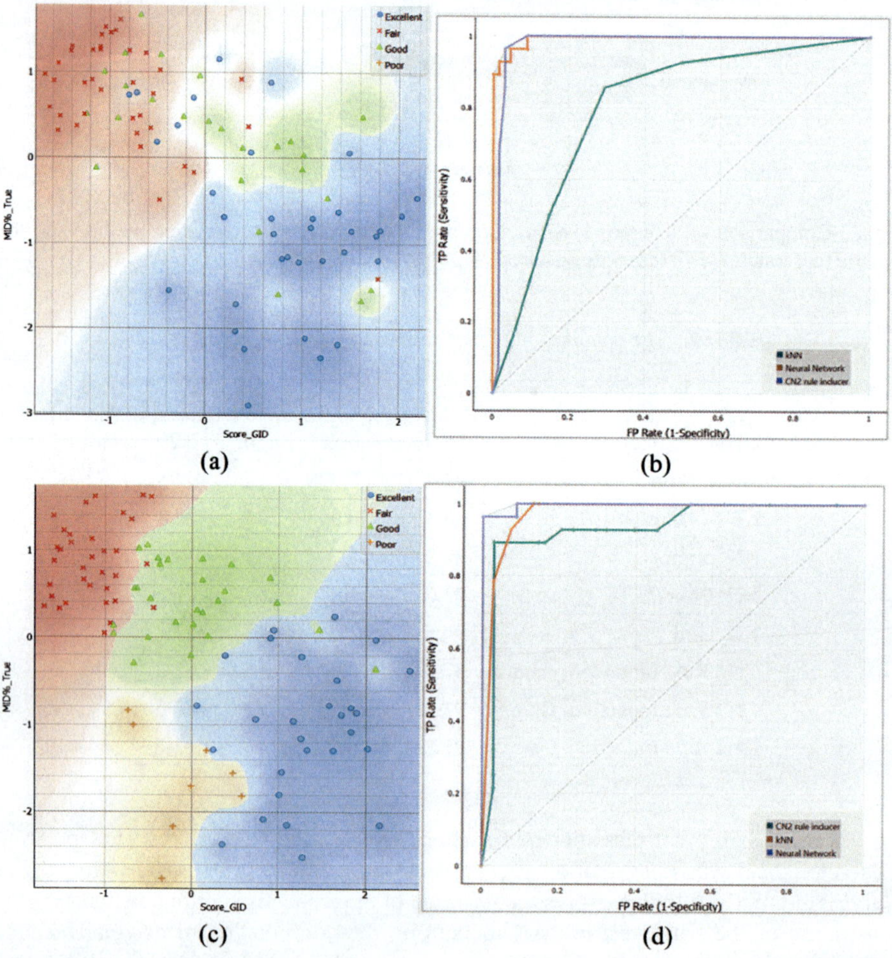

Fig. 3. The visualization of the students' performance mapping on (a) k-NN and Original_features (b) ROC of Original_features on all classifiers (c) NN and Gini features (d) ROC of Gini_features on all classifiers

Gini_Features obtain the best performance on folds of 3 and 5. Gini_Features decreases the error by 34, from 43 to 9, while Gain_Features is by 37, from 43 to 6.

For NN Method, we measure the performance by using NN. It is found that both Gain_Features and Gini_Features have the best performance on the fold of 3. Gain_Features and Gini_Features can reduce the number of misclassified students to 6 and 3, respectively, from 7 students who are generated by Original_Features. It is shown that different from that with k-NN, the combination of NN and Gain_Features holds better results. Lastly, on the CN2 Rule Induction Method, the performance evaluation by using CN2 Rule Induction is done with the fold of 3 since this is better than others. It can be seen that by using CN2 Rule Induction, the number of misclassified students is also lower for both Gain_Features and Gini_Feature than that for Original_Features.

Table 3. Average of accuracy numbers and its corresponding number of students for all sets of features and classification methods

Classification method	Features	Average of accuracy (%)	Number of misclassified students
k-NN	Original_Features	56.3	43
	Gain_Features	92.38	6
	Gini_Features	90.3	9
NN	Original_Features	92.2	7
	Gain_Features	92.58	6
	Gini_Features	94.6	3
CN2 rule induction	Original_Features	87.98	15
	Gain_Features	91.78	7
	Gini_Features	91.78	7

Overall, Gain_Features and Gini_Features can reduce the student mapping errors of all classifiers as provided in Table 3. We can infer from Table 3 that features selection can reduce the mapping error on the students 'performance classification. Overall, the best performance is achieved by NN which is applied to Gini_Features. This combination reaches 94.6% accuracy and produces 3 misclassified students. The second-lowest misclassification, which is 6, is obtained by NN with Gain_Features whose accuracy is 92.58%. On the other hand, k-NN that is applied to Gain_Features also generates 6 misclassified students with slightly lower accuracy, which is 92.38%. It is also shown that CN2 Rule Induction generates lower accuracy for both Gain_Features and Gini_Features. Also, the use of Original_Features leads to the lowest performance, which is represented by lower accuracy and higher misclassification than both Gain_Features and Gini_Features.

4 Conclusion

Our research employs two methods of feature selection: Gain Ratio and Gini Index to reduce the unsuitable mapping on students' performance domain. Those methods have been proven to increase the classifier's performance regarding the Area Under Curve (AUC), they can extend AUC very well. Moreover, feature selection can improve the accuracy level significantly.

References

1. Wanarti, P., Ismayanti, E., Peni, H., Yamasari, Y.: The enhancement of teaching-learning process effectiveness through the development of instructional media based on e-learning of Surabaya's vocational student. In: Proceedings of the 6th International Conference on Educational, Management, Administration and Leadership, pp. 342–346 (2016). https://doi.org/10.2991/icemal-16.2016.71
2. Tan, P.-N., Steinbach, M., Kumar, V.: Introduction to Data Mining. Pearson Addison Wesley, Boston (2005)
3. Dillenbourg, P.: The evolution of research on digital education. Int. J. Artif. Intell. Educ. **26**(2), 544–560 (2016). https://doi.org/10.1007/s40593-016-0106-z
4. Liñán, L.C., Pérez, Á.A.J.: Educational data mining and learning analytics: differences, similarities, and time evolution. RUSC. Univ. Knowl. Soc. J. **12**(3), 98 (2015). https://doi.org/10.7238/rusc.v12i3.2515
5. Koza, J.R., Bennett, F.H., Andre, D., Keane, M.A.: Automated design of both the topology and sizing of analog electrical circuits using genetic programming. In: Gero, J.S., Sudweeks, F. (eds.) Artificial Intelligence in Design 1996, pp. 151–170. Springer, Dordrecht (1996). https://doi.org/10.1007/978-94-009-0279-4_9
6. Samuel, A.L.: Some studies in machine learning using the game of checkers. IBM J. Res. Dev. **3**(3), 210–229 (1959). https://doi.org/10.1147/rd.33.0210
7. Troussas, C., Virvou, M., Mesaretzidis, S.: Comparative analysis of algorithms for student characteristics classification using a methodological framework (2015)
8. Sukajaya, N., Purnama, K.E., Purnomo, M.H.: Intelligent classification of learner's cognitive domain using Bayes Net, Naïve Bayes, and J48 utilizing bloom's taxonomy-based serious game. Int. J. Emerg. Technol. Learn. **10**(2), 46–52 (2015). https://doi.org/10.3991/ijet.v10i1.4451
9. Gobert, J.D., Kim, Y.J., Sao, M.A., Pedro, M.K., Betts, C.G.: Using educational data mining to assess students' skills at designing and conducting experiments within a complex systems microworld. Think. Skills Creativity **18**, 81–90 (2015). https://doi.org/10.1016/j.tsc.2015.04.008
10. Ko, C.-Y., Leu, F.-Y.: Applying data mining to explore students' self-regulation in learning contexts. In: 2016 IEEE 30th International Conference on Advanced Information Networking and Applications (AINA), March 2016, pp. 74–78 (2016). https://doi.org/10.1109/AINA.2016.123.
11. Casey, K., Azcona, D.: Utilizing student activity patterns to predict performance. Int. J. Educ. Technol. High. Educ. **14**(1), 4 (2017). https://doi.org/10.1186/s41239-017-0044-3
12. Ramaswami, M., Bhaskaran, R.: A study on feature selection techniques in educational data mining. J. Comput. **1**(1), 2151–9617 (2009). Accessed 16 Aug 2017. https://pdfs.semanticscholar.org/d11c/46515632f3e462d1a952e67fd4657a5f009e.pdf

13. Punlumjeak, W., Rachburee, N.: A comparative study of feature selection techniques for classify student performance. In: 2015 7th International Conference on Information Technology and Electrical Engineering (ICITEE), October 2015, pp. 425–429 (2015). https://doi.org/10.1109/ICITEED.2015.7408984
14. Sasi Regha, R., Uma Rani, R.: Optimization feature selection for classifying student in educational data mining. Int. J. Innov. Eng. Technol. **490**(4) (2016). https://ijiet.com/wp-content/uploads/2017/01/65.pdf. Accessed 16 Aug 2017
15. Han, J., Kamber, M., Pei, J.: Data Mining Concepts and Techniques. Elsevier, USA (2012)
16. Yamasari, Y., Nugroho, S.M.S., Sukajaya, I.N., Purnomo, M.H.: Features extraction to improve performance of clustering process on student achievement. In: 2016 International Computer Science and Engineering Conference (ICSEC), December 2016, pp. 1–5 (2016). https://doi.org/10.1109/ICSEC.2016.7859946
17. Quinlan, J.R.: Induction of decision trees. Mach. Learn. **1**(1), 81–106 (1986). https://doi.org/10.1023/A:1022643204877
18. Altman, N.S.: An introduction to Kernel and nearest-neighbor nonparametric regression. Am. Stat. **46**(3), 175–185 (1992). https://doi.org/10.1080/00031305.1992.10475879
19. Rumelhart, D.E., Hinton, G.E., Williams, R.J.: Learning representations by back-propagating errors. Nature **323** (1986). https://www.iro.umontreal.ca/~pift6266/A06/refs/backprop_old.pdf
20. LeCun, Y., Bottou, L., Orr, G.B., Muller, K.-R.: Efficient BackProp. Springer, New York (1998)
21. Xu, W.: Towards optimal one pass large scale learning with averaged stochastic gradient descent, July 2011. https://arxiv.org/abs/1107.2490. Accessed 4 April 2018
22. Kingma, D.P., Lei Ba, J.: ADAM: a method for stochastic optimization (2015)
23. Clark, P., Niblett, T.: The CN2 induction algorithm. Mach. Learn. **3**(4), 261–283 (1989). https://doi.org/10.1023/A:1022641700528
24. Clark, P., Boswell, R.: 'Rule induction with CN2: some recent improvements. In: Clark, P., Boswell, R. (eds.) Machine Learning - Proceedings of the 5th European Conference, EWSL-91, pp. 151–163. Springer, Heidelberg (1991)https://www.cs.utexas.edu/users/pclark/papers/newcn.pdf

Spatial Graph Regularized Correlation Filters for Visual Object Tracking

Sajid Javed, Xiao Zhang, Jorge Dias, Lakmal Seneviratne, and Naoufel Werghi(✉)

KUCARS, Khalifa University, Abu Dhabi, UAE
naoufel.werghi@ku.ac.ae

Abstract. In Visual object tracking (VOT), Discriminative correlation filters trackers have achieved promising results for VOT in many complex scenarios. However, because of the unwanted boundary effects and lack of structural constraints, these methods suffer from performance degradation. In the current work, we propose a spatial graph-regularized correlation filter for robust VOT. In this method, we transform the circulant shifted target samples to a particular subspace such that the target and the background become linearly separable. For this purpose, we encode pairwise similarities among the circulant shifted target samples as a spatial graph via a learnt correlation filter constrained to act as an eigenvector of the Laplacian of this spatial graph. We propose an objective function which incorporates this spatial constraint into the DCFs learning framework, which we solve using ADMM with a closed-form solution. Evaluated on a set of four datasets, our framework showed superior performance when compared to competitive state-of-the-art (SOTA) methods.

Keywords: Visual tracking · Deep features · Correlation filters

1 Introduction

VOT is an essential task across many applications, including video surveillance [12], human activity recognition [22], and vehicle navigation [6]. The main aim of VOT is to estimate the target position between the consecutive frames given its position in the first frame [31]. VOT becomes challenging when a target object suffers from appearance variations challenges, especially in the presence of illumination variation, occlusion, and scale variation, to name a few [30]. Over the past few years, many methods have been proposed to address the VOT challenges [3,10,11]. Several survey studies [7,27,31] and challenging datasets [20,23,29,30] have also been proposed for VOT. Despite these efforts, VOT remains a hard problem [27].

Among the most representative trackers [27,31], the Discriminative correlation filters (DCFs) methods have been widely employed in VOT for their efficiency and remarkable performance [1]. In DCFs methods, the filter is first

A. Abraham et al. (Eds.): SoCPaR 2020, AISC 1383, pp. 186–195, 2021.
https://doi.org/10.1007/978-3-030-73689-7_19

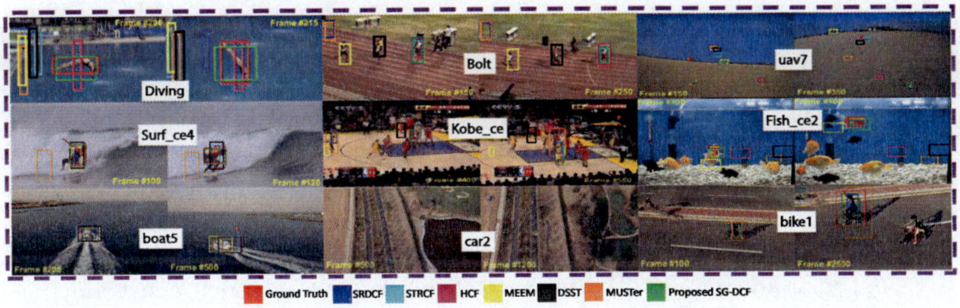

Fig. 1. Qualitative comparison of our SG-DCF tracker with SOTA methods on nine challenging videos from the datasets OTB50 [29], OTB100 [30], TC128 [20], and UAV123 [23]. The proposed SG-DCF tracker has shown excellent localization and tracking accuracy.

trained on circularly shifted samples of the target by a least-squares minimization. Afterwards, convolution of trained filter is applied between frames is applied to detect the target. As the computations are performed in the Fourier domain, the DCFs benefits of the high computational efficiency [10]. The success of the DCFs mainly lies in the circular shifted sampling operation of the target; however, this underlying assumption at both training and testing process produces unwanted boundary effects [3]. As a matter of fact , SOAT DCFs trackers degrade in challenfing scenarios. Figure 1 shows examples of VOT b using SOAT DCF trackers in scale variations and target drifting.

Several attempts have been made to improve the VOT performance of DCFs [3,19]. Methods proposed spatial regularization [3], CNN features [21], background-aware DCFs [8], and spatio-temporal regularizations [19]. While these methods showcased a significant progress they could not cope well with target drifting challenge (Fig. 1), and have been criticized for their computational complexity. Other works proposed proposed methods based on robust principal component analysis methods [13–15] and deep features methods [16].

To solve the above shortcomings, we propose a novel algorithm, dubbed as Spatial Graph-regularized DCF (SG-DCF) for efficient VOT. In the proposed algorithm, we start by extracting deep hierarchical features from VGG-19 model, afterwards we compute, from each layer, the circularly shifted versions of the target. The spatial consistency of the target object is maintained by integrating, in a graph structure, the pairwise similarities between circularly shifted versions of the target in a graph structure. Since the shifted versions are highly correlated, they span a low-dimensional space. This correlation is reflected in the eigenvectors of the Laplacian matrix containing the graph clustering information. Here we advocate that a proper DFC should meet the consistency requirement. We propose implementing this constraint by enforcing the resulting correlation filter to act as an eigenvector of the spatial graph-Laplacian matrix. By incorporating this spectral clustering-based spatial regularization into the DCFs framework, we ensure that the resulting correlation filter maintains the spatial structural consistency of the target object.

By incorporating a structural constraint into the objective function, our proposed SG-DCF algorithm allows to track target object in a more robust manner, in many complex tracking scenarios, even when the appearance of the target changes suddenly, a scenario which is difficult to handle by current SOTA DCFs tackers (see Fig. 1). We solve our proposed objective function using ADMM method, where the spatial graph-regularization is incorporated into the joint DCFs optimization model. To the best of our knowledge, such constraint has not been proposed before in the DCF framework.

Compared to these methods, our SG-DCF tracker incorporates spatial-graph Laplacian component in the DCFs framework. The graph-Laplacian matrix penalizes the CF weights such that they discrimination between target and background is highlighted and structural information is to the best of our knowledge, such constraint has not been proposed before in the CF framework. Our objective function is novel and it is solved efficiently using ADMM method. With the introduction of the spatial-graph regularization, SG-DCF tracker can learn a more robust appearance model and hence leading to superior VOT performance.

We conducte a series of experiments on four publicly available datasets including OTB100 [30], TC128 [20], VOT2018 [17], and UAV123 [23]. Our results demonstrate the excellent performance of the proposed SG-DCF tracker as compared to many existing SOTA methods in terms of accuracy and low computational complexity.

2 Proposed SG-DCF Algorithm

Our proposed algorithm consists of three main steps: deep features extraction, the construction of spatial-graph, and DCF objective function minimization.

2.1 Deep Features Extraction:

Our method starts by cropping the target object in the first frame in a region $\mathbf{A} \in \mathbb{R}^{m \times n}$, where m and n are the height and he width of the cropped region, respectively. The we employ the VGG19 model to extract feature from the layers $conv3_4$, $conv4_4$, and $conv5_5$. Using VGG-19 as features extractor [21], we derive from deep features from the layers: *conv3-4*, *conv4-4*, and *conv5-5*. Afterwards, features matrices $\mathbf{X}_l \in \mathbb{R}^{m \times n \times d}$ are constructed from $\mathbf{A}$, where l denotes the l-th layer of VGG19 while d os the number of channels in each layer.

2.2 Discriminative Correlation Filters (DCFs):

DCFs allow for dense sampling around the target at meager computational cost which is achieved by using all possible translations of the target in a search window as circulant shifts form the data matrix $\mathbf{X}_0 \in \mathbb{R}^{p \times p \times d}$, where $p = m \times n$. The circulant structure of this matrix facilitates a very efficient solution to the ridge regression problem in the Fourier domain [10]. The multi-channel correlation filter can then be formulated as [9]:

$$\arg\min_{\mathbf{w}} ||\sum_{i=1}^{d} \mathbf{X}_{0i}\mathbf{w}_i - \mathbf{y}||^2 + \lambda_1 \sum_{i=1}^{d} ||\mathbf{w}_i||^2, \tag{1}$$

where, the learned correlation filter is denoted by $\mathbf{w}_i \in \mathbb{R}^p$, $\mathbf{X}_{0i}$ contains all circulant shifts of the i-th channel of the feature map, $\mathbf{y} \in \mathbb{R}^p$ is the vectorized Gaussian response, and λ_1 is the regularization parameter.

2.3 Saptial-Graph Regularized DCF:

Classical DCFs formulation (Eq. 1) observes undesirable overwhelming boundary effects because of the periodic assumption resulting in performance degradation [3]. To tackle this issue, we propose to incorporate a new constraint to maintian the spatial appearance structure of the target in the DCFs model (Eq. 1).

The motivation of the above constraint comes from our hypothesis that DCFs learning should be consistent with the Riemann manifold structure. That is, if two circulant shifts $\mathbf{x}_i$ and $\mathbf{x}_j$ are close on the data manifold, then their corresponding coefficients in $\mathbf{w}$ will also be close [32]. Here, we consider spatial closeness among the circulant shifts of the feature map, encoded, in the graph $\mathbf{G} = (\mathbf{X}_o, \mathbf{A}_o)$ using the k-nearest neighbor strategy [24]. In this graph, the vertices $\mathbf{x} \in \mathbf{X}_o$ are the columns of matrix $\mathbf{X}_o$ and $\mathbf{A}_o$ is the adjacency matrix, which encodes the weights and connectivity of the graph.

Incorporating the structural constraint using the graph $\mathbf{G}$ in the computation of the correlation filter $\mathbf{w}$ can be performed by first estimating a graph-Laplacian matrix $\mathbf{L}$ and then enforcing $\mathbf{w}$ to act as eigenvector of $\mathbf{L}$ along with DCFs objective function. The first step involves searching for the closest neighbors for all the columns in the circulant shifted features based on the Euclidean distances. Here, each node is connected to its k-nearest neighbors, so that if $\mathbf{x}_i$ and $\mathbf{x}_j$ are in the k-nearest neighbours of each other, we set:

$$A_o(i,j) = exp\left(-\frac{||\mathbf{x}_i - \mathbf{x}_j||_2^2}{2\sigma^2}\right), \tag{2}$$

where σ is a normalizing parameter set as the average distance among the nodes. If $\mathbf{x}_i$ and $\mathbf{x}_j$ are connected then $A_o(i,j) > 0$, otherwise, $A_o(i,j) = 0$. Based on the weighted adjacency matrix $\mathbf{A}_o$, we compute the normalized Laplacian matrix $\mathbf{L}$ of the graph $\mathbf{G}$ by

$$\mathbf{L} = \mathbf{I} - \mathbf{D}^{-\frac{1}{2}}\mathbf{A}_o\mathbf{D}^{-\frac{1}{2}}, \tag{3}$$

where $\mathbf{I}$ is an $p \times p$ identity matrix, and $\mathbf{D}$ is an $p \times p$ diagonal matrix with its i-th diagonal element being equal to the sum of the i-th row of $\mathbf{A}_o$ (i.e. $\sum_j A_o(i,j)$). The correlation filter encoding graph-Laplacian matrix is then formulated as follows:

$$\arg\min_{\mathbf{w}} ||\sum_{i=1}^{d} \mathbf{X}_{0i}\mathbf{w}_i - \mathbf{y}||^2 + \frac{\lambda_1}{2} \sum_{i=1}^{d} \mathbf{w}_i^\top \mathbf{L}_i \mathbf{w}_i \tag{4}$$

The second term is the proposed structural constraint, which means that a good correlation filter must encode the structure of the graph $\mathbf{G}$. Such a structure plays a very important role in discriminating between the target and the background resulting in robustness to target appearance variations. The normalized Laplacian matrix $\mathbf{L}$ can be symmetrically decomposed as:

$$\mathbf{L} = \mathbf{V}\Sigma\mathbf{V}^{\top} = (\Sigma^{\frac{1}{2}}\mathbf{V}^{\top})^{\top}\Sigma^{\frac{1}{2}}\mathbf{V}^{\top} = \mathbf{B}^{\top}\mathbf{B}, \tag{5}$$

where $\mathbf{V}$ is an $p \times p$ orthonormal matrix having each column as an eigenvector of $\mathbf{L}$, and Σ is an $p \times p$ diagonal matrix with its diagonal element Σ_{ii} being an singular value of $\mathbf{L}$ (sorted as $0 \leq \Sigma_{ii} \leq ...\Sigma_{pp}$). In addition, $\mathbf{B} = \Sigma^{\frac{1}{2}}\mathbf{V}^{\top}$ which is computed with all the eigenvectors of $\mathbf{L}$, we may consider $\mathbf{B}$ defined based upon the manifold structure of the circulant shifted target samples. With this regularization, the proposed SG-DCF tracker using l_2 norm is formulated as follows:

$$\arg\min_{\mathbf{w}} ||\sum_{i=1}^{d} \mathbf{X}_{0i}\mathbf{w}_i - \mathbf{y}||^2 + \frac{\lambda_1}{2}\sum_{i=1}^{d}||\mathbf{B}_i\mathbf{w}_i||^2 \tag{6}$$

Our proposed structural constraint can be interpreted as enforcing $\mathbf{w}$ to be orthogonal to the eigenvectors of $\mathbf{L}$.

SG-DCF Optimization: We solve the Eq. (6) using the iterative ADMM method [2] where we introduce two auxiliary variable $\mathbf{g}$ and $\mathbf{h}$, convert (6) into a Lagrangian formulation, and solving one variable while fixing the others, through the following iterative equations:

$$\mathbf{w}^{(k+1)} = \min_{\widehat{\mathbf{w}}} \frac{1}{2}||\sum_{i=1}^{d} \widehat{\mathbf{X}}_{0i}\widehat{\mathbf{w}}_i - \widehat{\mathbf{y}}||^2 + \frac{\gamma}{2}\sum_{i=1}^{d}||\widehat{\mathbf{w}}_i - \widehat{\mathbf{g}}_i + \widehat{\mathbf{h}}_i||^2,$$

$$\mathbf{g}^{(k+1)} = \min_{\mathbf{g}} \frac{\lambda_1}{2}\sum_{i=1}^{d}||\mathbf{B}\mathbf{g}_i||^2 + \frac{\gamma}{2}\sum_{i=1}^{d}||\mathbf{w}_i - \mathbf{g}_i + \mathbf{h}_i||^2,$$

$$\mathbf{h}^{(k+1)} = \mathbf{w}^{(k+1)} - \mathbf{g}^{(k+1)} + \mathbf{h}^{(k)},$$

$$\gamma^{(k+1)} = \min(\gamma^{\max}, \rho\gamma^{k})$$

where γ is the penalty factor, $\widehat{\mathbf{w}}$ denotes the DFT of the filter $\mathbf{w}$, and ρ is the scale factor.

Solving Sub-problem w: From the above equations, we can see that the j-th element of the label $\widehat{\mathbf{y}}$ only depends on the j-th element of the filter $\widehat{\mathbf{w}}$ and the sample $\widehat{\mathbf{x}}$ across all the d channels. Let by $\mathcal{V}_j(\mathbf{w}) \in \mathbb{R}^d$ the vector consisting of the j-th elements of $\mathbf{w}$ along all d channels. It can be further decomposed into mn subproblems, each of them is defined as:

$$\min_{\nu V_j(\widehat{\mathbf{w}})} ||\mathcal{V}_j(\widehat{\mathbf{X}}_0)^{\top}\mathcal{V}_j(\widehat{\mathbf{w}}) - \widehat{\mathbf{y}}||^2 + \gamma||\mathcal{V}_j(\widehat{\mathbf{w}}_i) - \mathcal{V}_j(\widehat{\mathbf{g}}_i) + \mathcal{V}_j(\widehat{\mathbf{h}}_i)||^2$$

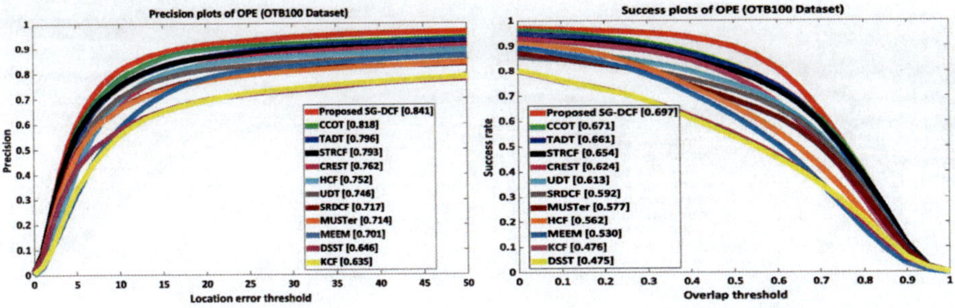

(a) Precision Plot (b) Success Plot

Fig. 2. OTB100 dataset: comparison of performance for the precision and success.

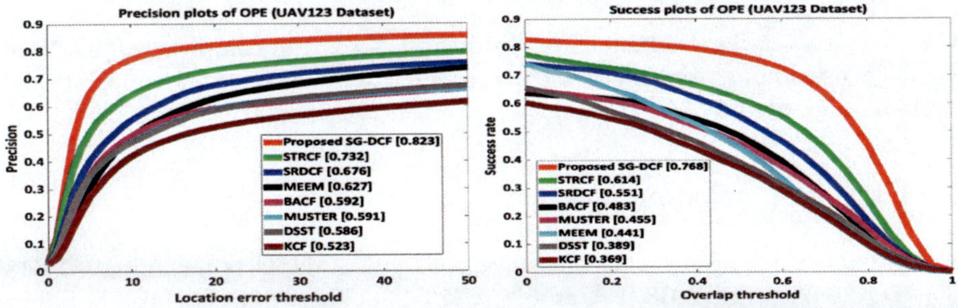

(a) Precision Plot (b) Success Plot

Fig. 3. UAV123 dataset: comparison of performance for the precision and success.

By taking the derivative and setting it zero, we can get a closed-form solution for $\mathcal{V}_j(\widehat{\mathbf{w}})$, using Sherman Morrsion formula [25], as:

$$\begin{aligned} \mathcal{V}_j(\widehat{\mathbf{w}}) &= \frac{1}{\gamma}(\mathbf{I} - \frac{\mathcal{V}_j(\widehat{\mathbf{X}}_0)\mathcal{V}_j(\widehat{\mathbf{X}}_0)^\top}{\gamma + \mathcal{V}_j(\widehat{\mathbf{X}}_0)^\top \mathcal{V}_j(\widehat{\mathbf{X}}_0)})\mathbf{q} \\ \mathbf{q} &= \mathcal{V}_j(\widehat{\mathbf{X}}_0)\widehat{\mathbf{y}}_j + \gamma\mathcal{V}_j(\widehat{\mathbf{g}}) - \gamma\mathcal{V}_j(\widehat{\mathbf{h}}). \end{aligned} \tag{7}$$

Solving Subproblem g: From the above equation, each element of $\mathbf{g}$ can be computed independently, and thus the closed-form solution of $\mathbf{g}$ can be computed by

$$\mathbf{g} = (\mathbf{B}\mathbf{B}^\top + \gamma\mathbf{I})^{-1}(\gamma\mathbf{w} + \gamma\mathbf{h}) \tag{8}$$

where $\mathbf{B}$ represents the $dmn \times dmn$ diagonal matrix.

Detection Step: Given an image patch in the next frame, the l-th correlation response R_l is then computed by:

$$R_l = \mathscr{F}^{-1}(\sum_{i=1}^{d} \mathbf{W}_d \odot \overline{\mathbf{X}}_d) \tag{9}$$

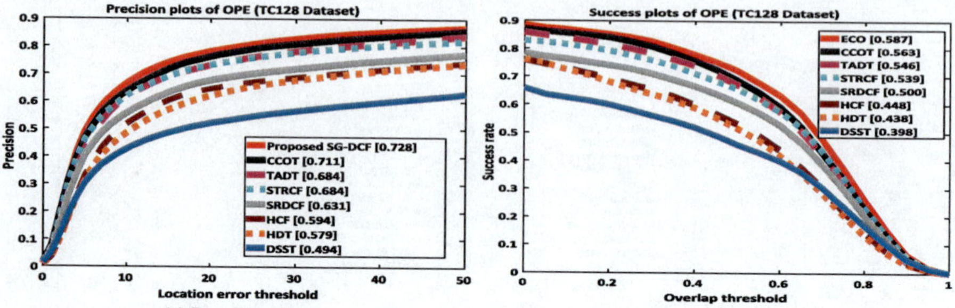

(a) Precision Plot (b) Success Plot

Fig. 4. TC128 dataset:comparison of performance for the precision and success.

where $\odot$ denotes the Hadamard operator and $\overline{\mathbf{X}}_d$ is the complex conjugate of $\mathbf{X}_d$. The maximum response is then computed among all the l convolutional layers as suggested by Ma *et al.* [21].

3 Experimentation

We conducted a series of experiments on four public datasets including OTB15 [30], TC128 [20], VOT2018 [17], and UAV123 [23]. These datasets pose 11 different tracking challenges such as Occlusion (Occ), Fast Motion (FM), and Illumination Variation (IV). The comparison involved the following 12 SOTA trackers: SRDCF [3], KCF [10], MUSTer [11],DSST [4], STRCF [19], MEEM [33], BACF [8], TADT [18], UDT [28], CREST [26], HCF [21], and CCOT [5].

Following the procedure suggested in [3], we crop the target object from the centre about $\sqrt{5mn}$, where (m, n) is the size of the target patch size. We also weight the deep features extracted from the target region with a a cosine window. In our optimizing model (6), we used $\gamma_o = 10$, $\gamma_{\max} = 100$, $\lambda_1 = 0.8$, and $\rho = 1.2$.

Our tracker is able to track a target at 26.8 fps on OTB15 dataset. The optimization method converges within 3 iterations for each frame. For quantitative evaluations, we use the one-pass evaluation (OPE) with precision and success plots [30].

3.1 Qualitative Evaluations

In Fig. 1 we report the visual results comparing our tracker with six SOTA trackers on nine challenging sequences. Majority of the compared trackers do not perform well in all cases except for the *boat5* and *car7* sequences. In comparison to SRDCF and STRCF, our SG-DCF tracker handled these sequences well under fast motion, motion blur, and OCC scenarios. The HCF tracker incorporates hierarchical deep features therefore, it performs comparatively on some challenges. The deep features based spatial graph facilitates the proposed tracker

Table 1. Performance evaluation on the VOT2018 dataset using EAO, A, and R measures. The best two results are shown in red and blue colors.

Trackers	CCOT	UDT	TADT	STRCF	SRDCF	HCF	KCF	DSST	Proposed SG-DCF
EAO↑	0.33	0.22	0.29	**0.34**	0.24	0.22	0.19	0.18	**0.35**
A↑	0.53	0.54	**0.55**	0.52	0.53	0.45	0.48	0.53	**0.57**
R↓	0.23	0.34	0.31	**0.21**	0.41	0.39	0.56	0.70	**0.20**

to handle complex scenes. As a result, the response map predicted by SG-DCF is more accurate for target localization, especially in *diving*, *Bolt*, and *Fish*. Overall, the visual evaluation indicates that proposed SG-DCF tracker performs excellently on these sequences.

3.2 Quantitative Evaluations

OTB100 Dataset: This dataset contain 100 challenging video sequences. In Figs. 2(a)-(b) we report a comparison with SOTA trackers. We can notice that the performance of our tracker exceeds its best competitor by 2.30% and 2.60% margin in the precision and success, respectively.

Our proposed tracker behaves quite better than the SRDCF and STRCF trackers. Note that by putting $\lambda_1 = 0$ in (6), our r SG-DCF tracker becomes the classical KCF tracker.

TC128 Dataset: This dataset contain 128 challenging video sequences. Figures 4(a)-(b) shows that the proposed SG-DCF tracker significantly outperformed all compared trackers.

UAV123 Dataset: This dataset encmpasses 123 video sequence taken with UAV flying at a low attitude. The results reported in Figs. 3(a)-(b) indicate the superiority of the SG-DCF tracker with a precision and success scores of 0.823 and 0.768, preceding its competitors with a large margin.

VOT2018 Dataset: We compared our proposed SG-DCF on VOT2018 dataset containing 60 challenging sequences using the following metrics: Accuracy (A), Robustness (R), and Expected Average Overlap (EAO) as suggested in [17] (Table 1). Overall, the proposed SC-DCF has achieved the best performance in all three measures while STRCF and CCOT trackers also obtained the comparative performance.

4 Conclusion

We proposed a DCF-based tracking algorithm variant enhanced with graph-based spatial structure consistency constraint. We presented a framework, which uses circulant shifted version of the target to construct a graph and enforce the

DCF to act as the eigen- vectors of its Laplacian matrix. We evidenced that encoding the graph structure into the correlation filter resulted in a significant improvement over the baseline as well as other previously proposed constraints in the current SOTA trackers. The proposed SG-DCF is evaluated on four benchmark datasets and compared with 12 SOTA trackers. Our results demonstrate the excellent performance of the proposed tracker. In future, we will consider the inclusion of re-detection module to improve the long term tracking performance.

References

1. Bolme, D.S.: Visual object tracking using adaptive correlation filters. In: IEEE CVPR (2010)
2. Boyd, S., Parikh, N., Chu, E., Peleato, B., Eckstein, J., et al.: Distributed optimization and statistical learning via the alternating direction method of multipliers. Found. Trends ML **3**(1), 1–122 (2011)
3. Danelljan, M., Hager, G., Khan, F.S., Felsberg, M.: Learning spatially regularized correlation filters for visual tracking. In: IEEE ICCV (2015)
4. Danelljan, M., Häger, G., Khan, F.S., Felsberg, M.: Discriminative scale space tracking. IEEE T-PAMI **39**(8), 1561–1575 (2016)
5. Danelljan, M., Robinson, A., Khan, F.K., Felsberg, M.: Beyond correlation filters: learning continuous convolution operators for visual tracking. In: ECCV (2016)
6. Ess, A., Schindler, K., Leibe, B., Gool, L.V.: Object detection and tracking for autonomous navigation in dynamic environments. IJRR **29**(14), 1707–1725 (2010)
7. Fiaz, M., Mahmood, A., Javed, S., Jung, S.K.: Handcrafted and deep trackers: a review of recent object tracking approaches. ACM Surv. **52**(2), 43 (2018)
8. Galoogahi, H.K., Fagg, A., Lucey, S.: Learning background-aware correlation filters for visual tracking. In: IEEE CVPR (2017)
9. Hamed, G., Sim, T., Lucey, S.: Multi-channel correlation filters. In: IEEE CVPR (2013)
10. Henriques, J.F., Caseiro, R., Martins, P., Batista, J.: High-speed tracking with kernelized correlation filters. IEEE T-PAMI **37**(3), 583–596 (2015)
11. Hong, Z., Chen, Z., Wang, C., Mei, X., Prokhorov, D., Tao, D.: Multi-store tracker (muster): a cognitive psychology inspired approach to object tracking. In: IEEE CVPR (2015)
12. Javed, O., Shah, M.: Tracking and object classification for automated surveillance. In: ECCV (2002)
13. Javed, S., Mahmood, A., Dias, J., Werghi, N.: Structural low-rank tracking. In: 2019 16th IEEE International Conference on Advanced Video and Signal Based Surveillance (AVSS), pp. 1–8 (2019)
14. Javed, S., Mahmood, A., Dias, J., Werghi, N.: CS-RPCA: clustered sparse RPCA for moving object detection. In: 2020 IEEE International Conference on Image Processing (ICIP), pp. 3209–3213 (2020)
15. Javed, S., Mahmood, A., Dias, J., Werghi, N.: Robust structural low-rank tracking. IEEE Trans. Image Process. **29**, 4390–4405 (2020)
16. Javed, S., Zhang, X., Seneviratne, L., Dias, J., Werghi, N.: Deep bidirectional correlation filters for visual object tracking. In: 2020 IEEE 23rd International Conference on Information Fusion (FUSION), pp. 1–8 (2020)

17. Kristan, M., Leonardis, A., Matas, J., Felsberg, M., Pflugfelder, R., Zajc, L.C., Vojir, T., Bhat, G., Lukezic, A., Eldesokey, A., et al.: The sixth visual object tracking vot2018 challenge results. In: ECCV-W (2018)
18. Li, X., Ma, C., Wu, B., He, Z., Yang, M.H.: Target-aware deep tracking. In: IEEE CVPR (2019)
19. Liand, F., Tian, C., Zuo, W., Zhang, L., Yang, M.H.: Learning spatial-temporal regularized correlation filters for visual tracking. In: IEEE CVPR (2018)
20. Liang, P., Blasch, E., Ling, H.: Encoding color information for visual tracking: algorithms and benchmark. IEEE T-IP **24**(12), 5630–5644 (2015)
21. Ma, C., Huang, J.B., Yang, X., Yang, M.H.: Hierarchical convolutional features for visual tracking. In: IEEE ICCV (2015)
22. Matikainen, P., Hebert, M., Sukthankar, R.: Trajectons: action recognition through the motion analysis of tracked features. In: IEEE ICCV-W (2009)
23. Mueller, M., Smith, N., Ghanem, B.: A benchmark and simulator for UAV tracking. In: ECCV (2016)
24. Muja, M., Lowe, G.D.: Scalable nearest neighbor algorithms for high dimensional data. IEEE T-PAMI **36**(11), 2227–2240 (2014)
25. Petersen, K.B., Pedersen, M.S., et al.: The matrix cookbook, vol. 7. TUD, vol. 15 (2008)
26. Song, Y., Ma, C., Gong, L., Zhang, J., Lau, R.W., Yang, M.H.: Crest: convolutional residual learning for visual tracking. In: IEEE CVPR (2017)
27. Wang, J., Zheng, L., Tang, M., Feng, J.: A comparison of correlation filter based trackers and struck trackers. IEEE T-CSVT **30**(9), 3106–3118 (2019)
28. Wang, N., Song, Y., Ma, C., Zhou, W., Liu, W., Li, H.: Unsupervised deep tracking. In: IEEE CVPR (2019)
29. Wu, Y., Lim, J., Yang, M.H.: Online object tracking: a benchmark. In: IEEE CVPR (2013)
30. Wu, Y., Lim, J., Yang, M.H.: Object tracking benchmark. IEEE T-PAMI **37**(9), 1834–1848 (2015)
31. Yilmaz, A., Javed, O., Shah, M.: Object tracking: a survey. ACM-CS **38**(4), 13 (2006)
32. Yin, M., Gao, J., Lin, Z.: Laplacian regularized low-rank representation and its applications. IEEE T-PAMI **38**(3), 504–517 (2015)
33. Zhang, J., Ma, S., Sclaroff, S.: Meem: robust tracking via multiple experts using entropy minimization. In: ECCV (2014)

Three Dimensional Wireframe Model of Medical and Complex Images Using Cellular Logic Array Processing Techniques

Shilpa Rani[1](✉), Kamlesh Lakhwani[1], and Sandeep Kumar[2]

[1] School of Computer Science and Engineering, Lovely Professional University, Phagwara, India
Shilpa.41900568@lpu.in, kamlesh.20980@lpu.co.in
[2] Sreyas Institute of Engineering and Technology, Hyderabad, India
sandeepkumar@sreyas.ac.in

Abstract. The main aim of Computer vision is to provide the representation of images which is helpful for a better understanding of Two Dimensional and Three-dimensional world. Considering this, in this paper we proposed a method called cellular logic array processing for Three Dimensional wireframe representation of 3D images. This approach detects the junctions, straight lines, curves, depth of the 3D image and constructs the 3D wireframe model for the given dataset. The proposed algorithm has been evaluated on example-based Synthesis of 3D Object Arrangements dataset, PASCAL 3D+ dataset, and our own medical images dataset. The time complexity of the proposed method is $O(n^3r^3) + O(n^3)$ and the Average processing time is 1.12 s. The number of passes required for converting an input image into a wireframe model is one that is less than the other methods. The proposed method is simpler and has better accuracy compared to other wireframe based methods.

Keywords: Wireframe model · CLAP · 3D images

1 Introduction

The surface of the 3D object contains a lot of information about the object and it's always an important and prominent research area in the field of computer vision with a variety of applications. Several researchers have made efforts in this field [1]. A 3D image that is displayed on the computer screen contains different light intensities and it's difficult to visualize the hidden parts of the image to get some meaningful information. Therefore, the processing of the 3D image is all about the extraction of the physical, structural, and morphological information from the hidden or invisible part of the image. The precise and useful information can be obtained only when the technique which we are applying on the 3D image is more reliable in terms of accuracy. The point, lines, curves, and closed surfaces are called volumetric features of 2D and 3D images, and extracting these features

A. Abraham et al. (Eds.): SoCPaR 2020, AISC 1383, pp. 196–207, 2021.
https://doi.org/10.1007/978-3-030-73689-7_20

from the image is always remains a difficult task. These volumetric features are useful to construct the CAD model of different physical objects which is also called a wireframe model of 3D images. Most of the feature extraction algorithm has been developed for the processing of 2D images and according to the literature survey, there is no algorithm which can extract all the visible and hidden features of 3D image appropriately. The Volumetric intersection approach was first introduced by Martin et al. [2] and further investigated and modified by other researchers [3–7].

In this area, Rajan et al. [8] introduced a mathematical framework with the name "Cellular Logic Array Processing" method. This method can process 2D and 3D images as well. CLAP is a logical and mathematical framework which is developed with the help of the fundamental concept of "Markov's Normal Algorithms" and "von Neumann's Cellular Automata". Markov's Normal Algorithms Works as a Turing machine and cellular automata is a parallel processor. It is represented in Fig. 1 Cellular Logic Array Processing (CLAP) can detect the different features of 3D images such as 3D points, 3D lines, 3D skeleton, and 3D surface. In the proposed method we are adding a new approach into the CLAP which can find the wireframe of the 3D images.

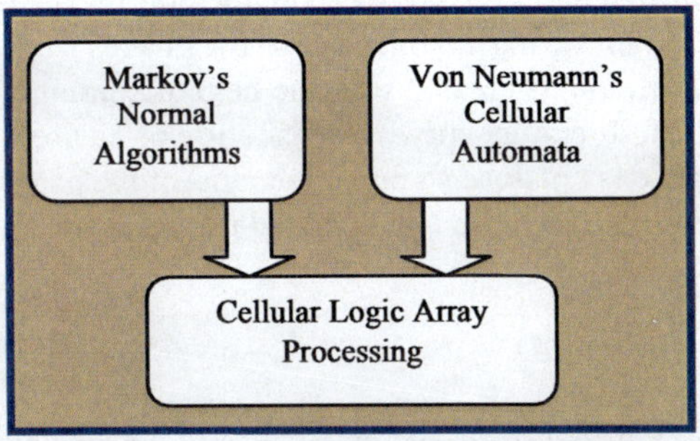

Fig. 1. Cellular logical array processing

The objective of the proposed methodology is to find a feasible solution to extract the volumetric features of a 3D image in the form of a wireframe model. To retrieve the wireframe model of a 3D image CLAP based approach is used. Further, the paper is divided into the following section. Section 2 consists of a literature survey; Sect. 3 explains the proposed methodology. Section 4 describes the experimental setup, evaluation parameters, and Results. Section 5 combines the conclusion and future scope of the proposed work.

2 Literature Survey

Superficial and volumetric features are always important for the processing of any 3D image. The surface, points, and lines of an image are called superficial features. Edges, Contour, a skeletal representation of an image are called volumetric features. In a 3D image, Edges are also two types. The edges which intersect the two planes or surfaces are called volumetric edges and the edges which separate the two regions are called superficial edges. The extended version of 3D edges from the 2D edges is introduced by H. Lui [9]. The author had given the approach which generalized the detection of 3D edges using a gradient-based method. Further, Zucker et al. [10] have given the extension of the Hueckel et al. [11], and rather than considering edge as a line they considered edge as a plane that passes through the center of the particular voxel. This approach has given an optimal view of representing the image. Rosenfeld et al. [12] proposed a new method for finding the 3D edge using the magnitude of the surface. O. Monga et al. [13] proposed the canny detector algorithm for 3D images. Most of the 3D edge detection algorithms are an extension of the 2D edge detection algorithm. The algorithm for extracting the wireframe contour of a 3D object was introduced by Mukherjee et al. [14]. The author has considered a 3D input image as a binary array. The first thinning of the image had been done and then the segmentation method was applied. Further, each segment is polygonized using the Graham Polygonized method and Boundary traversal method and if segments are adjacent then they would be connected through vertex merging methods. Finally, the obtained results are processed to remove unwanted values. Xiaoyuan Ren et al. [15] have proposed a 3D viewpoint & shape estimation-based method and conducted the wireframe modeling using the maximum probability distribution method.CNN model is used to detect the viewpoint of the 3D objects. Simant Prakoonwit and Ralph Benjamin [16] proposed a method for the reconstruction of 3D surface points and wireframe using 2D images. All the images were taken from different viewpoints and the wireframe is created with the help of combining the contour of 2D images. According to our literature survey, the creation of a wireframe model from the volumetric-based method is not done so far and in most of the papers, authors have used different viewpoints of 3D images and constructed the wireframe model of 3D images.

3 Proposed Methodology

The proposed method is designed for 3D images and an image could be a simple image that consists of several objects or it could be medical images that are more complex and contains a lot of hidden information. To understand and analyze a 3D image in a proper way, we can have a different view of the 3D image which is represented in Fig. 2.

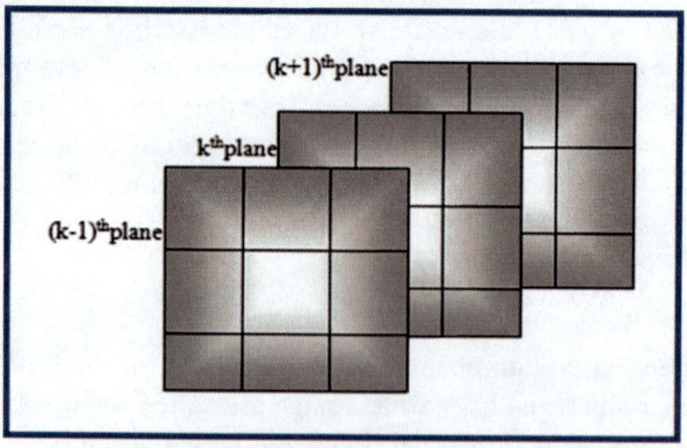

Fig. 2. Cellular logical array processing

In the proposed method, first, the data is acquired from different sources. In this approach, three different datasets are used. Example-based Synthesis of 3D Object Arrangements is Synthesis dataset which consists of different scenes of 3D objects, models, and texture related image. This dataset is provided by Fisher et al. [17]. PASCAL 3D+ is benchmark dataset for 3D object reconstruction. This dataset consists of more than 3000 instances of each category. Our own dataset consists of medical images and this dataset is gathered from the Pentagram research center private limited, India. First, the image enhancement has been done for the entire image, and then images are converted from RGB to Grayscale. A further detailed description of the workflow of the proposed method is represented in Fig. 3.

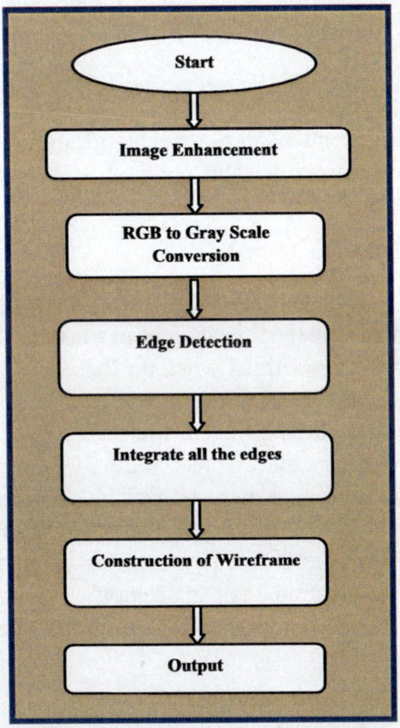

Fig. 3. Basic steps of proposed work

3.1 Image Enhancement

Image enhancement is a preprocessing technique that helps to improve the quality of the image. Generally, the Image enhancement technique will remove unwanted information to reduce the noise of the image. This technique enhanced the quality of the image and will sharpen the edges of the image. The proposed methodology is worked on the contrast enhancement method which is used to enhance the image. First Gaussian convolution is applied using Gaussian function with the input value channel image in the HSV space. This process can be expressed as:

$$P_{CON}(x, y) = P_I(x, y) \otimes G(x, y) \tag{1}$$

Here P_{CON} is represented in the convolution result which consists of brightness information of neighborhood pixels. Now, the center pixel value is compared with the Gaussian convolution to find the contrast enhancement of the center pixel. This is described as follows:

$$P_{CE}(x, y) = 255\, P_{LE}(x, y)^{E(x,y)} \tag{2}$$

Here $PCE(x, y)$ is the result of contrast enhancement, and $E(x, y)$ can be expressed as:

$$E(x, y) = \left(P_{CON}(x, y)/P_I(x, y)\right)^g \tag{3}$$

Here g is the image independent parameter. If illumination of the center pixel is greater than the neighborhood pixels then contrast will be pulled up otherwise pull down.

3.2 Edge Detection

Many edge detection algorithms used the concept of the neighborhood principle for the detection of the edges. These algorithms work on the abrupt change in the gray level value of the surrounding pixels. Keeping the same concept in mind, our algorithm will try to locate those points where boundaries of the region over which gray level are not changing. This process is done for all 3 planes of the 3D image which is represented in Fig. 2. And it is done in the following manner: For the scanning of the input image, the size of the sliding window is 3 × 3 × 3.In each movement of the sliding window 3 × 3 × 3 size of the input image is covered. In each movement, the algorithm will check that neighborhood pixels are forming a convex region. It means at least there should be at least one pixel in each row and column. The possibility of such type of region is 16 and it is shown in Fig. 4.

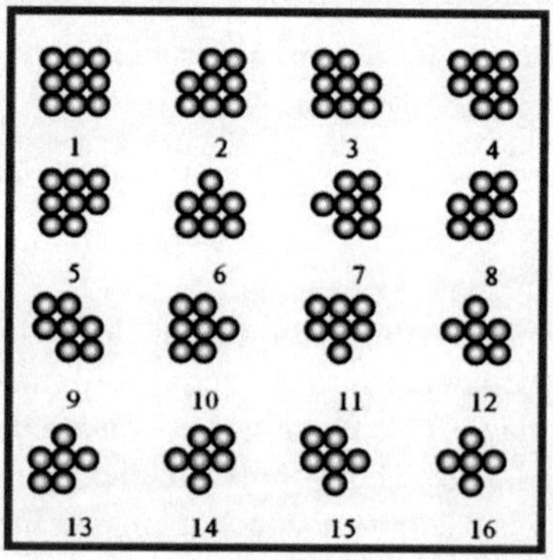

Fig. 4. Sixteen convex patterns in a 3 × 3 × 3 window

Pattern number sixteen is available in the rest of the other patterns so any one of the patterns has occurred or not, we need to check the sixteenth pattern alone. If the sub-image is containing the 16th pattern it means the algorithm will assign 0 to the central cell. So, the boundaries of gray level regions in the sub-image will be retained and it will erase all the interior parts of the sub-image. This procedure is continued until it covers all the neighborhood pixels of the front plane and then the same procedure will be repeated for the central plane and rear/backward plane. The effect of this method is that it gives the boundary of the given 3D gray level image. A detailed description of the procedure is mentioned in the Following Algorithm.

3.3 Construction of Wireframe

The edges and corners which are discovered from the 3 Dimensional structure using edge detection methods are now integrated. The tracked edge features are integrated to make a wireframe model of static objects.

4 Experimental Setup and Evaluation Parameters

The above-mentioned algorithm is implemented in python using the Anaconda tool. The experiments are performed on Acer Aspire 7, core i5, 9th generation laptop with NVIDIA Geforce GTX 1650/60 Hz 4 GB dedicated graphics card, 8 GB memory, and windows operating system, and only one core is used for the computation purpose. In the proposed method, we are looking for the accuracy of edge detection and the wireframe model.

4.1 Datasets

The proposed method is evaluated on Example-based Synthesis of 3D Object Arrangements dataset, PASCAL 3D+ dataset, and our own created medical dataset. Example-based Synthesis of 3D Object Arrangements dataset is a synthesis dataset that consists of different scenes, models, and textures. PASCAL 3D+ is benchmark dataset for 3D object reconstruction. This dataset consists of more than 3000 instances of each category. Our own dataset is a collection of medical images that consists of MRI of the Lungs, brain, and skull.

4.2 Evaluation Parameters

The proposed method is evaluated on 2 Different parameters. The first evaluation parameter is Accuracy. Accuracy means that how accurately our algorithm can find the wireframe model of the given image and the results of the wireframe model are represented in Figs. 5, 6, and 7. The other parameter is the processing time. Here, the processing time is nothing but the time which is required for converting the input image into the wireframe model. To calculate the Processing time following formula is used:

$$\text{Processing time} = 1/(\text{Performance of the Processor}) \tag{4}$$

The other evaluation parameter is the number of passes i.e. No. of passes required for detecting/ constructing the wireframe model of the 3D image.

Algorithm for the Proposed Methodology

Input: 3D image Dataset.
Output: Wireframe model.

Step 1. Read image from the dataset from i=0 to i<=n
Step 2: Use a sliding window of size 3X3X3 to read the pixels of 3D image.
Step 3: Repeat the sliding window until it's not finished reading the pixels of all plane.
{
Step 3a: Start Reading the pixel of front plane and repeat until it's not finished reading the pixels of Front plane and then go to step 3b.
{
If all the pixel value which is covered by 3X3X3 sliding window are same then Central pixel value would be replaced by 0 else move the sliding window to read the next pixels.
}
Step 3b: Start Reading the pixel of central plane and repeat until it's not finished reading the pixels of central plane and then go to step 3b.
{
If all the pixel value which is covered by 3X3X3 sliding window are same then Central pixel value would be replaced by 0 else move the sliding window to read the next pixels.
}
Step 3c: Start Reading the pixel of rear plane and repeat until it's not finished reading the pixels of rear plane and then go to step 3b.
{
If all the pixel value which is covered by 3X3X3 sliding window are same then Central pixel value would be replaced by 0 else move the sliding window to read the next pixels.
}
}
Step 4: Integrate all the tracked edges and create a wireframe model.
Step 5: Display the result.
Step 6: End.

4.3 Results

Comparative analysis has been done with an edge detection algorithm because most of the work is related to the wireframe model, authors have worked on the viewpoints and the reconstruction of the 3D image using the wireframe model. So the comparison is difficult with those papers. Considering this, we have chosen some edge detection algorithms which are as follows: Dilation based Three Dimensional edge detection (DBED), Erosion based three-dimensional edge detection (EBED), Morphological filtering of three

Dimensional images for edge detection (MFED), Opening based three Dimensional edge detection (OBED).

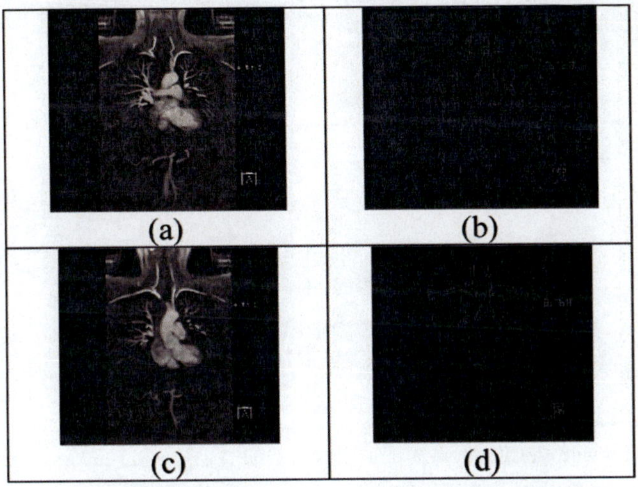

Fig. 5. Original image (left) and wireframe model (right) of the 3D image of our own dataset.

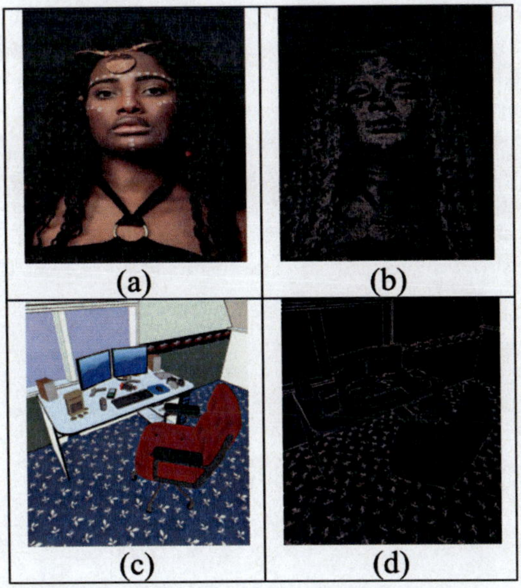

Fig. 6. Original image (left) and wireframe model (right) of the 3D image of example-based Synthesis of 3D object arrangements dataset.

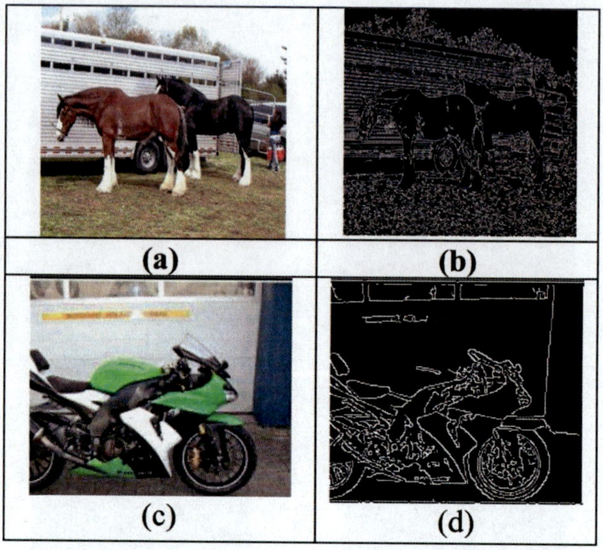

Fig. 7. Original image (left) and wireframe model (right) of the 3D image of the Pascal 3D+ dataset.

The dataset which is used here for the representation of comparative analysis consists of MRI images and the width, height, and depth of the images are 256,256, and 43. The maximum grayscale level is 256. The Proposed work outperformed the standard database as well as its own database. The results are shown in Fig. 5, 6, and 7.

The result of these algorithms is compared with the proposed method in terms of processing time and the size of the voxel after processing the image. The time complexity of the proposed algorithm is O (n^3r^3) + O (n^3) where n is input size in units and r is the number of passes to get the wireframe model. One pass is required to find the wireframe model of the input image. Table 1 shows the input image and their corresponding processed image using different algorithms. The results are represented in Table 2 as well as Fig. 8.

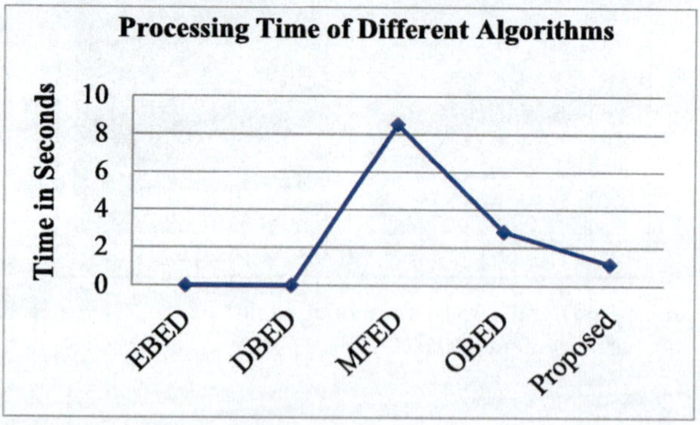

Fig. 8. Processing time of different algorithms on own dataset.

Table 1. The input image and its processed results

Algorithm	Input Image	Processed Image
EBED	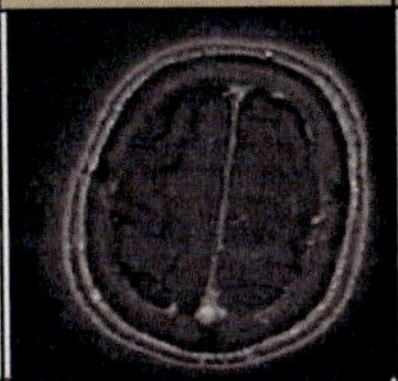	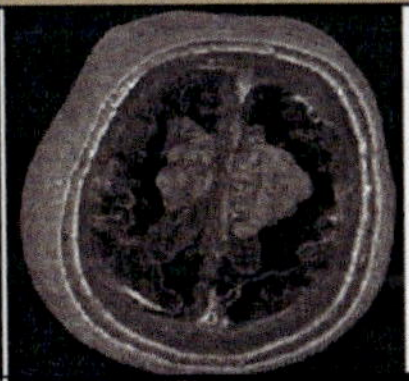
DBED	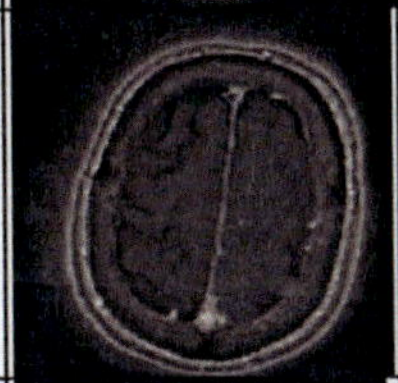	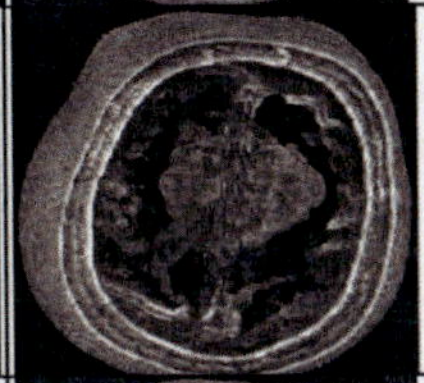
MFED	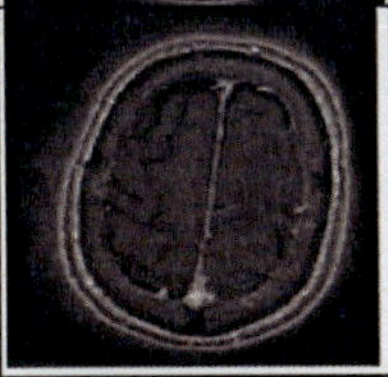	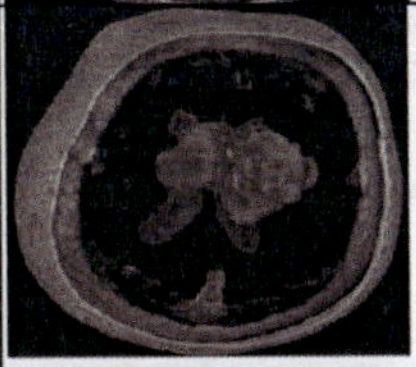
+OBED	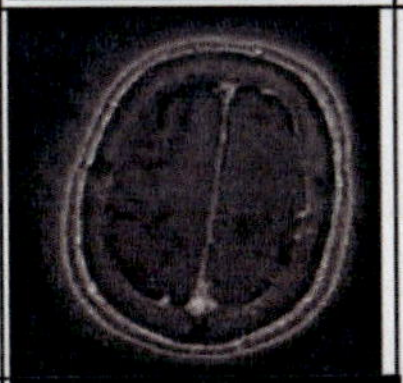	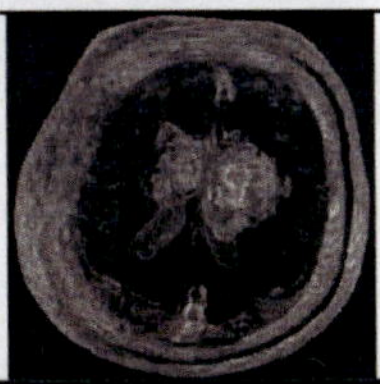
Proposed Method	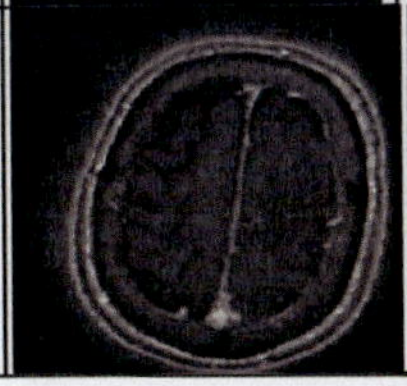	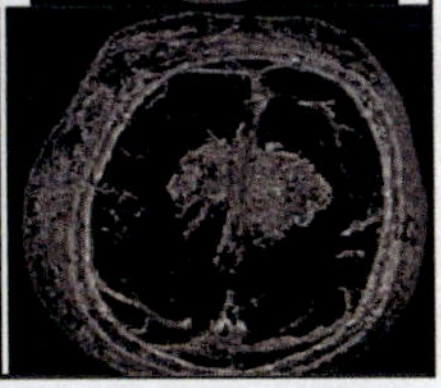

Table 2. Comparative analysis of the proposed algorithm on own dataset

Comparative analysis			
Algorithm	Number of voxels (before processing)	Number of voxels (after processing)	Processing time (in seconds)
EBED	1205998	327815	<1
DBED	1205998	360752	<1
MFED	1205998	107257	8.54
OBED	1205998	124252	2.85
Proposed method	1205998	153059	1.12

5 Conclusion and Future Work

According to the performance of our proposed work a cellular logic array processing for the 3D wireframe model performed well. The performance of the algorithm is compared with EBED, DBED, OBED, and MFED. The proposed method can find the wireframe model in one pass and the average processing time is 1.12 s that is less than the other existing algorithms. For a better analysis of the algorithm, the proposed method is evaluated on 3 different types of Datasets and the results are outperformed. The algorithm can find the wireframe model for simple, complex image and the image which consists of more than one object. The proposed methodology could be useful for the application where 3D image processing is used. This algorithm can be used for Electronic surgery and laparoscopy surgery. The speed of the algorithm could be improved more with the help of a parallel processing system. In the future, this algorithm can be modified for the processing of color 3D images and 3D video as well.

Acknowledgments. The constant support of Prof. Dr. E. G. Rajan, President, Pentagram Research, and Director, Avatar MedVision US LLC NC USA has resulted to bring the results of the research paper. The authors express their sincere thanks to Mr. Pawan Rao, a Data Scientist in the Pentagram Research Centre, and Hyderabad for his valuable suggestion during the preparation of this research paper.

References

1. Eisert, P., Steinbach, E., Girod, B.: Automatic reconstruction of stationary 3-D objects from multiple uncalibrated camera views. IEEE Trans. Circuits Syst. Video Technol. **10**(2), 261–277 (2000)
2. Martin, W.N., Aggarwal, J.K.: Volumetric descriptions of objects from multiple views. IEEE Trans. Patt. Anal. Mach. Intell. **PAMI-5**(2), 150–158 (1983). https://doi.org/10.1109/TPAMI.1983.4767367
3. Boyer, E.: Object models from contour sequences. In: Buxton, B., Cipolla, R. (eds.) Computer Vision — ECCV '96: 4th European Conference on Computer Vision Cambridge, UK, April 15–18, 1996 Proceedings Volume II, pp. 109–118. Springer, Heidelberg (1996)
4. Sullivan, S., Ponce, J.: Automatic model construction and pose estimation from photographs using triangular splines. IEEE Trans. Pattern Anal. Mach. Intell. **20**(10), 1091–1097 (1998)
5. Niem, W., Wingbermuhle, J.: Automatic reconstruction of 3D objects using a mobile monoscopic camera. In: Proceedings of the International Conference on Recent Advances in 3-D Digital Imaging and Modeling (Cat. No. 97TB100134), pp. 173–180. IEEE (1997)
6. Szeliski, R.: Rapid octree construction from image sequences. CVGIP: Image Underst. **58**(1), 23–32 (1993). https://doi.org/10.1006/ciun.1993.1029
7. Fitzgibbon, A.W., Cross, G., Zisserman, A.: Automatic 3D model construction for turn-table sequences. In: Koch, R., Van Gool, L. (eds.) 3D Structure from Multiple Images of Large-Scale Environments, pp. 155–170. Springer, Heidelberg (1998)
8. Rajan, E.G.: Cellular logic array processing techniques for high-throughput image processing systems. Sadhana **18**(2), 279–300 (1993)
9. Liu, H.K.: Two-and three-dimensional boundary detection. Comput. Graph. Image Process. **6**(2), 123–134 (1977)

10. Zucker, S.W., Hummel, R.A.: A three-dimensional edge operator. IEEE Trans. Pattern Anal. Mach. Intell. **3**, 324–331 (1981)
11. Heuckel, M.H.: An operator which locates edges in digital pictures. J. Assoc. Comput. Mach. **18**, 113–125 (1971)
12. Morgenthaler, D.G., Rosenfeld, A.: Multidimensional edge detection by hypersurface fitting. IEEE Trans. Pattern Anal. Mach. Intell. **4**, 482–486 (1981)
13. Monga, O., Deriche, R.: A new three-dimensional boundary detection. In: Proceedings of the International Conference on Pattern Recognition, Paris (1986)
14. Jayanta Mukherjee, P.P., Das, B.N.C.: An algorithm for the extraction of the wire frame structure of a three-dimensional object. Patt. Recogn. **23**(9), 999–1010 (1990)
15. Ren, X., Jiang, L., Tang, X., Liu, W.: 3D Wireframe modeling and viewpoint estimation for multi-class objects combining deep neural network and deformable model matching. Appl. Sci. **9**(10), 1975 (2019)
16. Prakoonwit, S., Benjamin, R.: 3D surface point and wireframe reconstruction from multiview photographic images. Image Vis. Comput. **25**(9), 1509–1518 (2007)
17. Fisher, M., Ritchie, D., Savva, M., Funkhouser, T., Hanrahan, P.: Example-based synthesis of 3D object arrangements. ACM Trans. Graph. (TOG) **31**(6), 1–1 (2012)
18. Eason, G., Noble, B., Sneddon, I.N.: On certain integrals of Lipschitz-Hankel type involving products of Bessel functions. Philos. Trans. R. Soc. Lond. Ser. A Math. Phys. Sci. **247**(935), 529–551 (1955)

Suspicious Activity Detection Using Transfer Learning Based ResNet Tracking from Surveillance Videos

Shubhangi Kale[1,2](✉) and Raghunathan Shriram[1](✉)

[1] SCSE, VIT, Bhopal, Madhya Pradesh, India
spkale@comp.maepune.ac.in, {spkale.2019, shriram.r}@vitbhopal.ac.in
[2] SCET, MIT AOE, Pune, India

Abstract. Tracking objects in video surveillance is a challenging task for public security. The advent of the semantic approach has greatly raised the growth of anomaly detection. However, the current anomaly detection methods typically experience problems like inadequate use of movement patterns and inconsistency on various datasets. This research proposed a system to enhance the efficiency of anomaly detection in video surveillance. The proposed system consists of two parts that involves object tracking and suspicious activity detection. The overall framework detects and tracks the abnormal objects in video surveillance. The transfer learning-based ResNet tracking has been used for object tracking. Distance Metric Learning (DML) method has been used for detecting suspicious activities in video surveillance. The results are estimated to analyze the efficiency of the proposed method. The proposed network classifier is compared with the existing ResNet and VGG-16 network. The proposed method provides 99% accuracy that had high performance compared to other existing methods.

Keywords: ResNet network · Transfer learning · Distance metric learning · Video surveillance

1 Introduction

Nowadays, Artificial Intelligence (AI) makes a person works more efficiently with high accuracy and precision results. The Security surveillance of the public is important to individual and organization security. The abnormal event detection and localization is the most important challenge in computer vision. The surveillance video becomes the largest source that is an increased number of surveillance cameras in many public places such as hospitals, airports, railway stations, ATM, elevators, and shopping malls. The visual surveillance defines the dynamic scenes using video analysis [1]. In the core of automatic video surveillance, the anomaly detection methods are most essential to determine unusual or abnormal events without prior knowledge about these events. Some of the abnormal events are street fights, robberies, breaks-in, acts of vandalism, and unusual crowd movements. The security cannot be import by surveillance cameras. The

A. Abraham et al. (Eds.): SoCPaR 2020, AISC 1383, pp. 208–220, 2021.
https://doi.org/10.1007/978-3-030-73689-7_21

intelligent video analytics provides more security measures and impacts in preventing anomalies [2, 3]. The anomaly detection methods can be divided into two types such as machine learning methods and pattern recognition methods. The machine learning methods classify the "normal" behavior from video sequences and the moving objects' behavior derivate from normal behavior. One of the most important requirements is tracking an object to ensure security and safety management [4]. In anomaly detection, object detection is the most essential part.

There are various researches and studies proposed various automatic video surveillance methods and object tracking methods with a lack of accuracy and limited realistic surveillance videos [5, 6]. Used probabilistic graphical model for detecting irregular behavior of video sequences. In [7], the covariance feature descriptor was estimated in the video frame and the SVM classifier was applied to detect the abnormal event. In this research [8], the feature descriptor extraction method was employed to capture the abnormal samples from the testing process. These methods' performances are analyzed by computational complexity, training and processing time, memory requirements, ROC, EER, AUC, and energy requirements. Most of the traditional algorithms are presented based on hand-crafted features.

In this research, we proposed a new framework with object-oriented anomaly detection. The proposed method involves the stages of frame conversion, mean filter, background suppression, foreground extraction, transfer learning based ResNet tracking, pattern prediction, and distance metric learning. Initially, the frames are converted from the obtained video that is called as images. The duplicate images and unnecessary images are removed which means the images contain without moving activity. In the pre-processing step, the mean filter is used for the smoothening image. Smoothen image involves the background suppression technique to remove the background. After the background is removed, the foreground extraction technique is applied to extract the moving object from the image and obtained the binary mask of the moving region. Then we used transfer learning-based ResNet tracking to track the moving objects in each frame. It detects the exact object in the network. Then the obtained results are applied to the Distance Metric Learning (DML) method. In this stage, the theft object is selected by user feedback. The DML technique calculates the shortest distance for all data patterns. The proposed system determines the suspicious activities based on this calculated shortest distance between objects. We have used 10 youtube videos as a dataset to evaluate the experimental analysis. The contribution of this research is described below,

- To enhance frames smoothening through the mean filter.
- To obtain the binary mask of the moving region using background suppression and foreground extraction.
- To improve the object tracking by means of transfer learning based ResNet tracking.
- To determine the suspicious activities in video surveillance by the employment of Distance Metric Learning (DML) method.

The remaining organization of the paper is as follows. Section 2 deals with the review of existing work related to suspicious activity detection in video surveillance. Section 3 discusses the implementation process of the proposed method Python. Section 4 discusses the performance analysis and Sect. 5 concludes this research work.

2 Review of Existing Works

The abnormal events detection in the video is an active research area and so many works had done in this area. This section contains various existing methods and techniques for anomaly detection in surveillance videos. [9] utilized the statistical measures and highly motion techniques for abnormal detection of online video. In existing researches, accuracy is one of the important challenges. To rectify this challenge, this study presented a new framework to maintain a high accuracy rate. By using two motion techniques, the proposed method achieves the efficiency between video processing time and accuracy. Furthermore, they used various statistical analysis measures of motion features to detect the abnormal behavior action from normal incidents. [10] utilized the deep learning classifier for abnormal event detection in video. To achieve high efficiency, this research proposed the anomaly detection method based on deep learning in video processing. The proposed method used various modules to achieve the complete detection of abnormal action. The utilized modules are Object Segmentation (OS) Module, Background Estimation (BE) Module, Activity Recognition (AR) Module, and Feature Extraction (FE) Module. [11] used trajectory features for the detection of abnormal maritime activity detection from satellite image sequences. In results, the recall rate provides 78% for object detection, 91% for tracking, and 78% for anomaly detection. The proposed method provides the most benefits for national defense and labor-saving.

Furthermore, [12] used CNN network and a 1-class SVM classifier for abnormal event detection. They presented the deep learning method for localization of spatial and temporal abnormal action by using training samples of normal actions. The proposed method of evaluation results was conducted on the UCSD Ped2 dataset, and it contains complex because of its presence of many occlusions and low resolution. The proposed method outperforms compared to other methods. [13] used a spectral clustering technique for the detection of trajectory-based abnormal action in vide traffic surveillance. The disadvantage of the impact factor was recovered by using anisotropic impact parameter Bmat. For the classification of normal or abnormal events, the Bayesian decision model was used. The proposed method was implemented using GPU in real-time and the results provide 12% accuracy in abnormalities detection. [14] this research designed Intelligent Video Analytics Model (IVAM) for identifying and detecting abnormal human activities and video-based objects. The proposed model was implemented using MATLAB and the obtained results are proved that the efficiency with high accuracy.

[15] used Interface Prostheses with classifier feedback. In myoelectric prosthetic device control, user training affects the efficiency of pattern recognition. The dimensionality reduction was achieved by reducing the number of features. The clustering feedback guides users to modify muscle contraction forces and motion gestures. The efficient results are evaluated with hand motion recognition accuracy. [16] used two approaches named as a frame-by-frame method as SVM and sequence-based RNN network to detect eh similar and dissimilar video sequences. The pre-trained network was employed to extract high-level abstract features from overall video frames. Both proposed approaches used for transfer learning. The experimental results show that accuracy under the right circumstances. [17] presented the feedback based metric learning for abnormal determination in video sequences. This study proposed the Mixture Expert

Model based on crowdsourcing and feedback samples. This method continuously determined the variation among crowdsourcing samples and user activity. Then updated the activity recognition models with determined variations. [18] employed Semantic Substance Extraction (SSE) in video surveillance. The object extraction is the most important from video data. The proposed method results at 78% faster than other exiting neural and fuzzy methods [19]. Discussed various intelligent surveillance techniques for detecting suspicious activities. It concludes the various researches with techniques, research problems, advantages, and disadvantages. [20] used the deep learning approach for detecting suspicious activities from surveillance video [21]. In the prediction of suspicious activities, the proposed method sends the alert message to the corresponding authority. Initially, the features are extracted from video frames and these features are used for efficient neural network classification. [22] proposed the CNN-DBNN algorithm to detect suspicious human activity in video surveillance applications. In the end, the proposed method provides 90% of accuracy for classification. [23] used correlation filters for suspicious tracking across multiple cameras. The errors are prevented by using the YOLO object detection framework that reduces the information loss and overlapping. The experimental results proved that the proposed method of efficiencies.

Problem Identification – There are various existing researches and studies are discussed for detecting abnormal events and activities in video surveillance. However, it had disadvantages like,

- Limited labeled data
- Insensitive to static changes
- Time-consuming
- Difficulties in noisy data

Due to the above demerits, the solving of the problem of abnormal events detection is still a challenging issue. Most of the existing methods solve a specific formulation of the problem motivated by various factors such as the nature of data, type of anomalies and availability of labeled data, etc. [7]. Researches are utilized in the application domain based on these problem characteristics.

3 Proposed Technique

In various applications, object tracking is the most important part which includes security, surveillance, and augmented reality. Generally, the object detection use extracted objects and learning algorithms to classify instances of an object category. Object detection determines the location of the object over a sequence of frames. One way of determining suspicious activities is to identify the objects inserted in it and how they move. By observing the suspect object's location and their behavior, it is possible to know the abnormal activities in video surveillance. This research related to the automatic detection of moving objects and tracking of useful information from a video sequence.

To determine the human suspicious activities in video surveillance, the proposed system involves various stages like,

- Pre-processing
- Binary mask of moving region
- Pattern prediction and tracking
- Suspect object prediction

The overall flow of proposed method is given (see Fig. 1). The various stages of proposed system is detail described as below.

Pre-processing – Surveillance Video is a collection of images that are called frames. Initially, the video is converted into frames to initiate the process. The pre-processing step removes the duplicate frames from overall images. To attain high efficiency, we require to remove these frames from total frames. Some images vary in size and it should establish a base size for all frames. The pre-processing step achieved by using a mean filter, smoothen image, background suppression, and foreground extraction.

Mean Filter – It filter is a smoothing filter that enhances the brightness by using the mean of the brightness of the pixel and its neighbor. It is simple and easy to implement for smooth images. This filter reduces the intensity variation among pixels. It replaces each pixel value by neighbor pixel of mean value in an image. It eliminates the unrepresentative pixel values of their surroundings that reduce the intensity variation among pixels. Smoothening of the image reduces the noises that produce a less pixelated image. Also, it enhances the quality of an image to use further processing. For further improving smoothening, a 3×3 square kernel is utilized.

Background Suppression – Background suppression determines the moving objects from extracted frames. The background suppression is detected by calculating the difference between the reference frame and the current frame. This technique is implemented based on the variation of complexity. When image consists of low complexity, the frame difference method is used otherwise the Gaussian method is used for background suppression.

$$|t - Rt| > T \tag{1}$$

Where, "T" represents the threshold value, "t" current frame, "Rt" represents the reference frame.

The moving objects are identified and stored as foreground pixels. Then background pixels are identified that are occupied as the foreground object moves.

After the complete background is recognised, it is subtracted from other frames to determine the foreground and moving objects. In the image, the foreground part is the essential part and it extracts to identify the location of objects in an image. Once the background suppression is achieved, the foreground is extracted. The automatic foreground extraction method is used to accurately determine the artificial areas in the image.

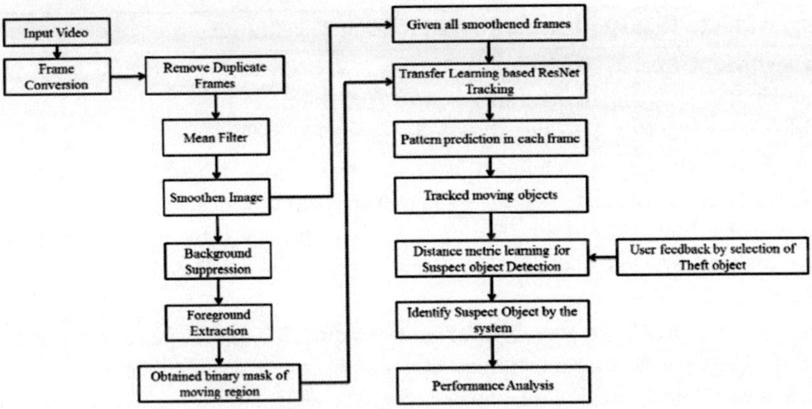

Fig. 1. Overall flow of proposed method

Binary Mask of Moving Region – The binary mask also called an ROI (Region of Interest) of the original image. The pixel value 1 expressed as the image pixel to ROI and 0 expressed as the image pixel is part of the background. After the pre-processing stage, the binary mask of the moving region is obtained. Some morphological operations on obtained binary mask image.

Pattern Prediction and Tracking – The smoothened images and obtained binary mask of a moving region are used for pattern prediction in each frame. The ResNet model is used for pattern prediction and object tracking. The ResNet tracking performs the object tracking in each frame. This method avoids saturation performance and improves tracking performance when training deeper layers. The ResNet model implemented by identifying the shortcut connection that avoids more than one layers and learns the residual mapping of the layer before the original mapping. To detect the exact object, we employed transfer learning in the network. The transfer learning can group into the feature-based transfer, sample-based transfer, relationship-based transfer, and model-based transfer. It transfers the learned model parameters into new model training. In the ResNet network, each layer performs local predictions and deviations from subsequent network layers. The network generates the convolutional feature map that identifies the region of the proposal and then reshapes into fixed-size by using the ROI pooling layer. It further fed into a fully connected layer and the bounding box is determined by using the softmax layer. The ResNet tracking and bounding box track the objects in subsequent frames.

Then determine the patterns from the obtained location of objects for making the predictions.

Algorithm I: Transfer learning based ResNet Tracking

Procedure:

The input layer is initialized with specified dimension,

$Input_{sz} = [96\ 128\ 1]$

Step 1: Include the Convolutional layer which is a number and it decides the dimensionality of the yield space with 7*7 kernel size and 2 strides.

Step 2: Add the Batch normalization layer is a strategy for preparing extremely profound neural systems that normalizes the contributions to a layer for every small scale group.

Step 3: Then ReLU activation layer is incorporated, a piecewise direct capacity that will yield the information legitimately if is sure, else, it will yield zero.

Step 4: Next incorporate, a maximum pooling layer accomplishes down-inspecting by partitioning the contribution to rectangular pooling locales, and figuring the limit of every district. $\left\{\begin{matrix} Conv1 \\ BatchNorm1 \\ relu1 \\ MaxPool1 \\ . \\ Conv2 \\ IdBlock1 \\ IdBlock2 \\ . \\ UpsampleConv1 \\ IdBlock3 \\ IdBlock4 \\ BatchNorm2 \end{matrix}\right\}$

Step 5: Generate the convolutional block $Conv\ block\ \{\begin{matrix} Conv1 \\ BatchNorm1 \\ relu1 \end{matrix}\}$

Step 6: Then structured the identity block where the convolutional layer, group standardization, and enactment layer is utilized with portion size is 3.

$IdBlock\quad \{\begin{matrix} Conv1 \\ BatchNorm1 \\ relu1 \end{matrix}\}$

Step 7: Design upsampling convolutional block for the forthcoming block portrayal,

$IdBlock\quad \{\begin{matrix} UpsampleConv1 \\ Conv1 \\ BatchNorm1 \\ relu1 \end{matrix}\}$

Step 8: Train the structured model with ground truth images and train images.

Step 9: Testing the model to obtain the segmented test images.

Suspect Object Prediction – Once the pattern is predicted, the suspected object is determined by using distance calculation. The object detection and calculation of its distance are the most important to determine the anomaly events. Distance metric learning (DML) is used to learn the distance from data. The DML learns the new distance that satisfies the pairwise limitations by using class label information. Distance between two data points are calculated by,

$$dist_{pattern} = \sqrt{\left(I_{pattern} - I_{object}\right)^2} \quad (2)$$

The new distance is calculated by Euclidean distance of data into subspace. The new distance will satisfy symmetry, non-negativity, and the triangle inequality. It provides the smallest distance for all data patterns. The suspected activities will be detected based on these smallest distance with the target object pattern.

Algorithm II: distance metric learning for Suspect object Detection
Input: Frame I_{frame} Output: Suspect object $I_{suspect}$ Step 1: Collect pattern using from previous location$(a, b + \Delta b)$. Step 2: Perform pattern extraction from the moving object which was obtained during the tracking process. $I_{pattern} = I_{frame}(a{:}\, b, a{:}\, b + \Delta b)$ Step 3: Estimate distance to all pattern elements from target object I_{object} which is selected by a user and return the small distance which is achieved for all patterns. Step 4: Update the new bounding box which is equal to the central region and calculate the distance with the target object pattern. Step 5: update the distance metric matrix $G_{pattern} = dist_{pattern}$ for all available moving patterns with the targeted object. Step 6: Predict the least distance index among the distance matrix. Step 7: Obtained suspect region from the resultant of the index. $index_{suspect} = \underset{p_t \to 1\ to\ P}{\text{argmax}}\, G_{pattern}(p_t)$ and $I_{suspect} = I_{pattern}(index_{suspect})$

ResNet Classifier – The average pooling layer followed from the ResNet network comprised of three convolutional layers. To each successive layer, each layer outputs generate an input. One fully connected flowed residual block are presented. From all previous convolutional layers, the inputs are retrieved from each convolutional layer in the residual model. The empirical number of convolutional layers are determined and for our model three optimal convolutional layer are required. To the input data, the convolution operations are applied from convolutional layers. Before feeding to classifier, average pooling operation applied to final pooling layer to data.

4 Result and Discussion

This section describes the proposed method of results with various metrics like accuracy, sensitivity, specificity, precision, and recall. Table 1 shows Output for various stages involved in the proposed method for 1–6 videos. Same ways we got the output for remaining videos.

Dataset Description – We have obtained 10 youtube videos for the dataset and the proposed method is applied to these videos. From each video, 80% of samples taken for training, and 20% of samples taken for testing. Table 2 shows the utilized 10 youtube video links.

Performance Measures – When we applying the proposed method on the dataset, there are various performance measures are evaluated such as accuracy, precision, recall etc.

Table 1. Output for various stages involved in the proposed method for 1–6 videos.

Steps	Video 1	Video 2	Video 3	Video 4	Video 5	Video 6
Input Image	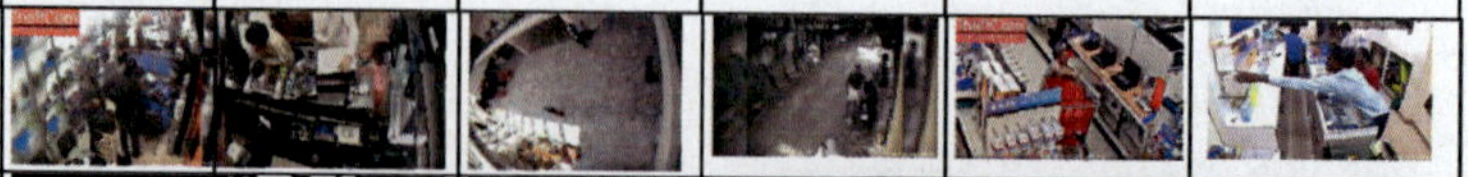					
Background Suppression	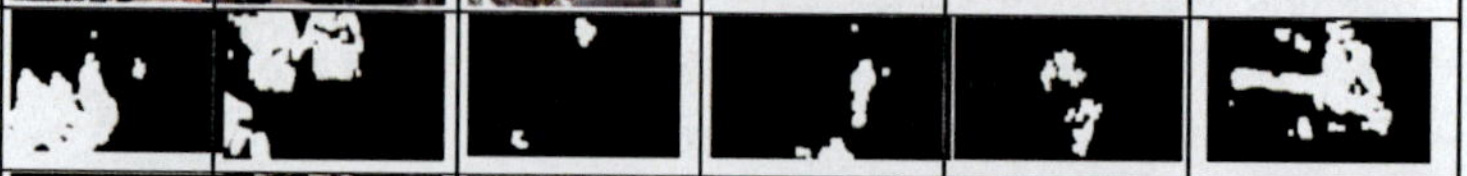					
Forground Extraction						
Tracking Result	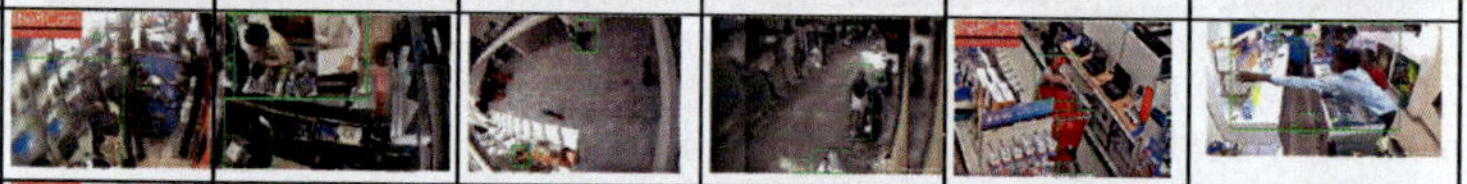					
Selection of object	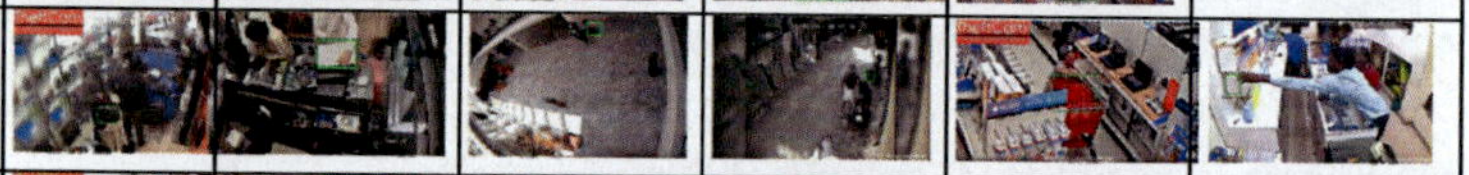					
Suspect Object						

Table 2. 10 Youtube video links

Sr. N.	Url
1	https://www.youtube.com/watch?v=flULrR0NJo8
2	https://www.youtube.com/watch?v=x9uiqBjYrlY
3	https://groups.inf.ed.ac.uk/vision/CAVIAR/CAVIARDATA1
4	https://www.youtube.com/watch?v=Sci0K-90wtA
5	https://www.youtube.com/watch?v=mNwmx3uUq8A
6	https://www.youtube.com/watch?v=1w6Wj7ld3Hs
7	https://www.youtube.com/watch?v=F2WszYqo0bs
8	https://www.youtube.com/watch?v=7st_jf-rGdw
9	https://www.youtube.com/watch?v=HmAmr2iRNCs
10	https://www.youtube.com/watch?v=7st_jf-rGdw

Table 3 represents the various performance measures for tracking that is estimated for 10 videos. For video 1, the recognition accuracy contains 95%, video 2 provides 98% of accuracy, video 3 contains 95% of accuracy, video 4 gives 93% of accuracy, video 5 provides 98% of accuracy, video 6 contains 90% of accuracy, video 7 gives 96% of accuracy, video 8 provides 95% of accuracy, video 9 contains 95% of accuracy, video 10 gives 93% of accuracy. Likely, other measures provide more efficiencies with minimum error rate.

Table 3. Performance measures of tracking for 10 videos

Video tracking										
Metric	1	2	3	4	5	6	7	8	9	10
Accuracy	94.730	97.653	94.7586	92.531	97.114	90.440	96.119	94.585	94.605	95.147
Sensitivity	90.466	95.517	90.5118	87.0031	94.543	83.949	92.798	90.229	90.261	91.153
Specificity	100	100	100	100	100	100	100	100	100	100
Recall	90.466	95.517	90.5118	87.003	94.543	83.949	92.798	90.229	90.261	91.153
F-Score	0.9499	0.977	0.9502	0.931	0.972	0.9127	0.964	0.949	0.949	0.954
Jaccard	0.905	0.955	0.9051	0.87	0.945	0.8395	0.928	0.902	0.903	0.912
PPV	100	100	100	100	100	100	100	100	100	100
NPV	89.462	95.307	89.5171	85.062	94.229	80.880	92.239	89.170	89.21	90.294
Kappa	0.930	0.969	0.9299	0.900	0.961	0.8722	0.948	0.928	0.928	0.935
Dice	0.950	0.977	0.9502	0.931	0.972	0.9127	0.963	0.949	0.949	0.954

Table 4. Performance measures of of the proposed method based on Epochs

Epochs	10	20	30	40	50	60	70	80	90
Accuracy	85.135	87.901	89.5035	91.298	92.911	94.542	95.171	96.120	98.224
Sensitivity	85.022	87.665	89.1995	90.927	92.495	94.126	94.788	95.78	98.048
Specificity	85.023	87.665	89.1995	90.927	92.495	94.126	94.788	95.78	98.048
Precision	83.870	87.086	88.947	91.028	92.883	94.673	95.290	96.244	98.281
Jaccard	0.7307	0.7763	0.8036	0.835	0.8638	0.8936	0.9052	0.923	0.964
F-score	0.843	0.8735	0.8907	0.910	0.9267	0.9437	0.9501	0.9599	0.982
Missed classification	0.149	0.121	0.105	0.087	0.071	0.055	0.048	0.039	0.018
Epochs	10	20	30	40	50	60	70	80	90
Accuracy	85.135	87.901	89.5035	91.298	92.911	94.542	95.171	96.120	98.224
Sensitivity	85.022	87.665	89.1995	90.9265	92.495	94.126	94.788	95.78	98.048

Similarly, Table 4 represents the various performance measures of the proposed network model based on Epochs like 10, 20, 30, 40, 50, 60, 70, 80, and 90. The accuracy is improved when the number of epochs increased. This analysis proved that the proposed method of enhanced results based on epochs.

Table 5 describes the comparison of various tracking classifier. The proposed Transfer learning based ResNet tracking is compared with the existing VGG-16 and basic ResNet tracking network. The proposed network classifier provides 99% of accuracy, 99% of sensitivity, 99% of specificity, 99% of precision, and minimum error rate. The VGG provides 97% of accuracy and basic ResNet provides 96% of accuracy. These performance measures proved that the proposed method had high efficiencies compared to another network classifier.

Table 5. Comparison of various tracking classifier

	ResNet tracking	VGG-16	Basic ResNet
Accuracy	98.819	97.5647	96.3364
Sensitivity	98.711	97.3376	96.0084
Specificity	98.711	97.3376	96.0084
Precision	98.8425	97.6364	96.4593
Jaccard	0.9758	0.9509	0.9271
F-score	0.9878	0.9748	0.9622
Missed classification	0.0118	0.0244	0.0366

It represents the comparison of proposed DML in the Transfer learning ResNet network (see Fig. 2(a)). The proposed distance metric learning gives 0.9% of precision, 0.8% of recall, and 0.8% of f-measure. It represents the comparison of the proposed DML method in the VGG-16 Network (see Fig. 2(b)). The proposed distance metric learning gives 0.9% of precision, 0.8% of recall, and 0.8% of f-measure in the existing VGG-16 network classifier [24].

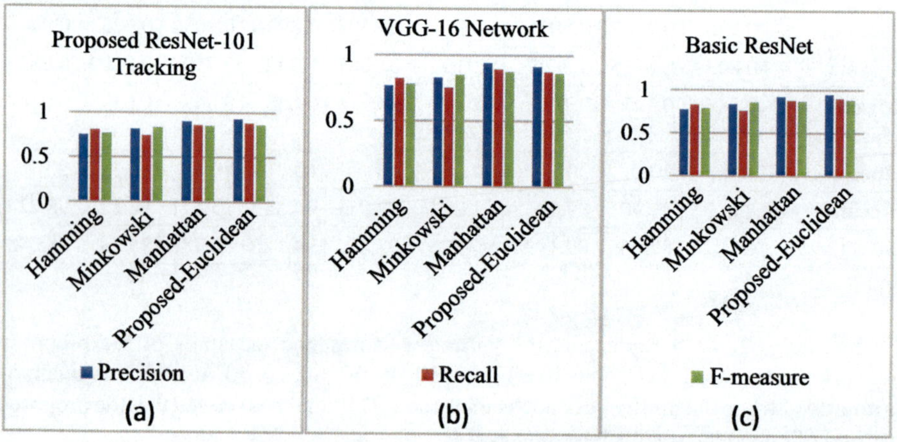

Fig. 2. Comparison of proposed DML method

The comparison of the proposed DML method in the VGG-16 Network (see Fig. 2(c)). The proposed distance metric learning gives 0.9% of precision, 0.9% of recall, and 0.8% of f-measure in the existing basic ResNet network classifier [25]. Various existing distance metrics provide minimum performance measures in the proposed network classifier. This statistical analysis proved that the proposed DML method had high performance compared to other methods.

5 Conclusions

In this research, we detect and tracking abnormal events in video surveillance using transfer learning based ResNet tracking. Initially, the proposed method tracking the object and determine the suspicious activities. The combination of ResNet network and distance metric learning technique provides more efficiencies to determine the abnormal action in video surveillance. The experimental analysis conducted based on the various measures such as accuracy, precision, recall, sensitivity, specificity and f-measure. Based on these measures, the proposed method proved that the proposed method had high performances compared than other existing VGG-16 and basic ResNet network.

References

1. Zhou, J.T., Du, J., Zhu, H., Peng, X., Liu, Y., Goh, R.S.M.: Anomalynet: an anomaly detection network for video surveillance. IEEE Trans. Inf. Forensics Secur. **14**, 2537–2550 (2019)
2. Wang, T., Qiao, M., Deng, Y., Zhou, Y., Wang, H., Lyu, Q.: Abnormal event detection based on analysis of movement information of video sequence. Optik **152**, 50–60 (2018)
3. Wang, T., Qiao, M., Zhu, A., Niu, Y., Li, C., Snoussi, H.: Abnormal event detection via covariance matrix for optical flow based feature. Multimedia Tools Appl. **77**, 17375–17395 (2018)
4. Feng, Y., Yuan, Y., Lu, X.: Learning deep event models for crowd anomaly detection. Neurocomputing **219**, 548–556 (2017)
5. Li, A., Miao, Z., Cen, Y., Zhang, X.-P., Zhang, L., Chen, S.: Abnormal event detection in surveillance videos based on low-rank and compact coefficient dictionary learning. Pattern Recogn. **108**, 107355 (2020)
6. Slavic, G., Campo, D., Baydoun, M., Marin, P., Martin, D., Marcenaro, L.: Anomaly detection in video data based on probabilistic latent space models. arXiv preprint arXiv:2003, 07623 (2020)
7. Yu, J., Yow, K.C., Jeon, M.: Joint representation learning of appearance and motion for abnormal event detection. Mach. Vis. Appl. **29**, 1157–1170 (2018)
8. Wang, T., Qiao, M., Zhu, A., Shan, G., Snoussi, H.: Abnormal event detection via the analysis of multi-frame optical flow information. Front. Comp. Sci. **14**, 304–313 (2020)
9. Al-Dhamari, A., Sudirman, R., Mahmood, N.H., Khamis, N.H., Yahya, A.: Online video-based abnormal detection using highly motion techniques and statistical measures. Telkomnika **17**, 2039–2047 (2019)
10. Revathi, A., Kumar, D.: An efficient system for anomaly detection using deep learning classifier. SIViP **11**, 291–299 (2017)
11. Fahn, C., Ling, J., Yeh, M., Huang, P., Wu, M.: Abnormal maritime activity detection in satellite image sequences using trajectory features. Int. J. Future Comput. Commun. **8**(1), 29–33 (2019)
12. Bouindour, S., Hittawe, M.M., Mahfouz, S., Snoussi, H.: Abnormal event detection using convolutional neural networks and 1-class SVM classifier. In: 8th International Conference on Imaging for Crime Detection and Prevention (ICDP 2017) (2017)

13. Athanesious, J.J., Chakkaravarthy, S.S., Vasuhi, S., Vaidehi, V.: Trajectory based abnormal event detection in video traffic surveillance using general potential data field with spectral clustering. Multimedia Tools Appl. **78**, 19877–19903 (2019)
14. Balasundaram, A., Chellappan, C.: An intelligent video analytics model for abnormal event detection in online surveillance video. J. Real Time Image Proc. **17**, 915–930 (2018)
15. Fang, Y., Zhou, D., Li, K., Liu, H.: Interface prostheses with classifier-feedback-based user training. IEEE Trans. Biomed. Eng. **64**, 2575–2583 (2016)
16. Jönsson, H.: Motion event recognition using user feedback. Master's theses in Mathematical Sciences (2018)
17. Wang, X., Xu, Y., Hu, H., Liu, M., Li, G.: Feedback-based metric learning for activity recognition. Expert Syst. Appl. **162**, 112209 (2018)
18. Manju, A., Valarmathie, P.: Video analytics for semantic substance extraction using OpenCV in python. J. Ambient Intell. Humaniz. Comput. 1–10 (2020)
19. Mathur, G., Bundele M.: Intelligent video surveillance techniques for suspicious activity detection critical review. In: International Conference on Recent Advances and Innovations in Engineering (ICRAIE), pp. 1–8 (2016)
20. Amrutha, C., Jyotsna, C., Amudha, J.: Deep learning approach for suspicious activity detection from surveillance video. In: 2nd International Conference on Innovative Mechanisms for Industry Applications (ICIMIA), pp. 335–339 (2020)
21. Agrawal, M., Agrawal, S.: Suspicious event detection in real-time video surveillance system. In: Shukla R., Agrawal J., Sharma S., Chaudhari N., Shukla K. (eds) Social Networking and Computational Intelligence. LNNS, vol. 100, pp. 509–516. Springer, Singapore (2020)
22. Basha, A., Parthasarathy, P., Vivekanandan, S.: Detection of suspicious human activity based on CNN-DBNN algorithm for video surveillance applications. In: Innovations in Power and Advanced Computing Technologies (i-PACT), pp. 1–7 (2019)
23. Sheu, R.-K., Pardeshi, M., Chen, L.-C., Yuan, S.-M.: STAM-CCF: suspicious tracking across multiple camera based on correlation filters. Sensors **19**, 3016 (2019)
24. Sahoo, S.R., Dash R., Mahapatra, R.K., Sahu, B.: Unusual event detection in surveillance video using transfer learning. In: International Conference on Information Technology, pp. 319–324 (2019)
25. Dubey, S., Boragule, A., Jeon, M.: 3D ResNet with ranking loss function for abnormal activity detection in videos. In: International Conference on Control, Automation and Information Sciences (ICCAIS), pp. 1–6 (2019)

Vehicle Detection in UAV Videos Using CNN-SVM

Najiya Koderi Valappil and Qurban A. Memon(✉)

EE Department, College of Engineering, UAE University, 15551 Al Ain, United Arab Emirates
{201990093,qurban.memon}@uaeu.ac.ae

Abstract. Conventional monitoring devices are usually kept at fixed locations which yields a fixed surveillance coverage. Unmanned aerial vehicles (UAVs) are receiving much attention from researchers in traffic monitoring due to their low cost, high flexibility, and wide view. Unlike stationary surveillance, the camera platform of UAVs is in constant motion and makes it difficult to process for data extraction. The inaccuracy in detection rates of vehicles from UAV videos becomes the motivation for combining optical flow methods with supervised learning algorithms. The proposed method incorporates steps that make use of the Kanade–Lucas optical flow method for moving object detection, connected graphs theory and CNN-SVM for further classification. Optical flow generated contains some background objects detected as vehicle when the camera platforms are moving. The classifier rules out the presence of any other moving objects to be detected as vehicles. The proposed method is tested on few stationary and moving aerial videos. The system is found to be 100% accurate in case of stationary aerial videos and 98% accurate in moving videos.

Keywords: Aerial video · Traffic surveillance · Traffic video analysis · Unmanned aerial vehicle · Kanade-Lucas optical flow · CNN-SVM

1 Introduction

Th use of videos from unmanned aerial vehicle (UAV) for traffic monitoring applications is gaining more and more popularity in research as well as industry. Compared to conventional monitoring devices such as inductive loop detectors, microwave radar detectors, infrared detectors, ultrasonic detectors etc., UAV is considered more cost-effective and convenient [1–3]. They can capture traffic conditions on fixed and discrete locations only, hence a single road segment needs to be monitored by many such units [4–7]. However, a few UAV(s) can cover a large area of roadway or a traffic network. It is also difficult to use these sensors to provide other useful information such as vehicle trajectories, routing information, and paths through the network. Additionally, the maintenance of any of the fixed detectors leads to additional fees and would inevitably interrupt the normal traffic. Aerial view gives a better top-view perspective, by covering a larger area and by avoiding occlusions between road users. For surveillance video, the tilt in fixed cameras creates an angle in frames which makes the pixels at different locations different from the real-world pixels. However, the pixels of an UAV video frame

A. Abraham et al. (Eds.): SoCPaR 2020, AISC 1383, pp. 221–232, 2021.
https://doi.org/10.1007/978-3-030-73689-7_22

are very close to each other, which makes the camera calibration and estimations from an UAV video much easier. Compared to manned aircrafts, UAVs are more flexible due to their high mobility and the ability to provide immediate response to emergencies [1, 2]. UAVs are more robust to adverse weather conditions as they can fly at low altitudes. When coupled with ground-based sensors, more absolute data can be obtained.

The complication involved in the aerial surveillance from an UAV is due to camera motions such as panning, tilting and rotation. Also, the different camera heights largely affect the detection results. In addition to a couple of practical entanglements like short battery life and privacy trouble, the most important technical issue in programmed UAV-based traffic monitoring is the ego-motion issue, which can be generated either deliberately (e.g., control at base station) or accidently (e.g., by wind). The video background movement, caused by UAV ego-motion, makes the customary vehicle detection and traffic stream estimation techniques intended for fixed reconnaissance methods not to function admirably. Thus, background subtraction [36, 37] and blob detection [38] cannot be applied for UAV-based videos. To simplify the examination of UAV applications in traffic monitoring, UAV videos with little ego-motion were used in few initial studies. This gives a scope of introducing a better classification system by reducing the detection of background objects as vehicles.

Even though the purpose of monitoring task varies, the studies related to detection of vehicles from UAV videos can be categorized into two: image registration [8–13] methods and optical flow [12–17] based estimation methods. Image registration goals to change the moving background into fixed background, as a result it is possible to apply traditional methods such as background subtraction. However, it becomes invalid in dense traffic situations, as majority of the feature points belong to the video background. Hence the background motion will be inaccurately anticipated. Optical flow is an effective method for traffic video analysis because of its ability to extract the foreground motion pattern. This makes optical flow one of the best methods in UAV video processing. However, in current research, it is usually combined with image registration or unsupervised learning such as K-means to estimate the foreground motion. When optical flow is collaborated with clustering methods, separation of vehicles and environment takes places. But the methods that employ optical flow are still not efficient for dense traffic conditions. One of the drawbacks of optical flow-based clustering is that it would classify some low velocity vehicles as background, which may cause errors as well. This becomes the motivation for integrating optical flow methods with supervised learning.

In this paper, the proposed method exploits the capability of Convolutional Neural Network (CNN) based deep learning methods to study more descriptive features. CNN picks up the interest points from the optical flow and rule out the detection of background object as vehicle. An additional support vector machine (SVM) classifier is employed to carry out the classification process. Thus, the CNN acts as feature extractor and SVM as classifier. This gives more accurate classification results when compared to Softmax classifier. The target is to propose a new framework to achieve more accurate detection of vehicles in UAV video. The framework starts with video from an UAV as input, uses Kanade-Lucas optical flow for motion estimation [34], connected graph method to find the centroids of Region of interest and morphological operations to reduce the number of false positives. Finally, the images undergo feature extraction with CNN and classification by SVM.

2 Related Work

UAV videos are the new type of data source forming the base of recent research. All works related to this area point to the estimation of traffic parameters efficiently. In terms of research objectives, literature is roughly classified into three categories: road detection, vehicle detection and parameter estimation [17, 18]. This paper focuses on the second category of vehicle detection and tracking in UAV videos. The existing vehicle detection methods can be broadly classified into two categories: image processing techniques and machine learning based techniques.

Methods in the first category include image registration [8, 9, 19], background subtraction [10], optical flow [13], and stereo vision [17]. For example, Shastry and Schowengerdt [9] explained video registration technique using KLT tracker [20] to automatically determine the frame to frame correspondence of control points. The errors by the motion of UAV is reduced and corrected by mapping video frames into common coordinate system. In [21], frame differencing was performed for detection of moving objects. The object detection and tracking were improved by using spatial and motion relationship [22]. Yalcin et al. [13] proposed an approach based on optical flow to detect moving vehicles by segmenting the optical flow fields. However, the complexity of environment caused the optical flow fields to be dense and lead to large computational load. Angel and Hickman [10] used the location, angle and scale statistics of automobiles expected in a reference frame to align to the adjoining frames. But the assumption about the velocity of the vehicles might not be legitimate in urban surroundings and the computational load in the method is intensive.

Machine learning based detection methods [23–25] form the second category which have gained the most of attention in recent research. These methods learn the patterns of the objects to be detected by extracting various features. Generally, these systems are noise insensitive. Rosenbaum et al. [26] used mild AdaBoost with Haar-like features to obtain a confidence image and generate regions of interest by means of clustering pixels. SVM was used to refine the detection results. Zhao and Nevatia [24] combined features like the shadows of vehicles to detect both moving and stationary vehicles. A Bayesian network was trained with vehicle and background samples to classify the objects. The vehicle boundaries, windshield, and shadows were used as features. However, in aerial videos, these boundaries are blurred and is unreliable due to the camera platform motion. Teney and Hebert [27] proposed a method for extraction of dense optical flow with convolutional neural networks. The paper showed a network of fully convolutional layers with each location in the output linked to a spatially- limited receptive field. The method did emphasis on detection of motion and not on classification of vehicles. Wang et al. [28] proposed a four-stage vehicle detection and tracking, which combined image registration, image feature extraction, vehicle shape detecting and vehicle tracking. The study proposed a method to combine image registration methods with machine learning methods. Ruimin et al. [18] used ensemble classifier with optical flow to estimate real-time traffic flow parameters. The first two stages of the system employed Haar cascade classifier with CNN for vehicle detection. The next stage used KLT based motion estimation for parameter estimation. Hansen et al. [29] proposed an affine model for motion detection. The method used background subtraction followed by binary SVM. HOG and size are considered as the prominent features. In [30], a combination of SIFT, SVM and Affinity Propagation (AP) algorithm was used. SIFT technique is invariant to scaling, rotation and illumination changes. AP was used as the key feature for classification using SVM. Xianbin et al. [31] used SVM classifier for vehicle detection in low altitude UAV videos. The paper could obtain a global optimal solution using BLPS-HOG feature extraction method. However, the method could only reach a detection rate of 89% and a high false positive rate. In paper [32], a framework for traffic flow parameter estimation was proposed of which the vehicle detection consisted of two stages. The first stage was KLT tracker to follow the movement of the vehicles and the second stage was a binary SVM classifier to separate vehicles from other moving objects. The method used seven handcrafted features including Gray Level Co-occurrence Matrix (GLCM) and Contourlet Transform coefficients. Handcrafted features make the system less accurate compared to deep learning techniques, and that leads to insufficient learning.

3 Methodology

3.1 Overview

The proposed method contains three stages: Motion Vector Estimation, Connected Graph development and Classification using CNN-SVM. In the first step, Kanade-Lucas optical flow is exploited to analyze the motion vector. Connected graph theory, second stage, is used to cluster the moving points on each vehicle. Pixels around the centroids, after certain morphological operations, are cropped and fed into the third stage. The motion vector contains false positives as the moving camera platform creates an illusion of moving foreground. Classification stage is carried out with CNN-SVM, where the features are extracted with CNN layers and the binary classification is done with SVM. Figure 1 shows workflow of the proposed method.

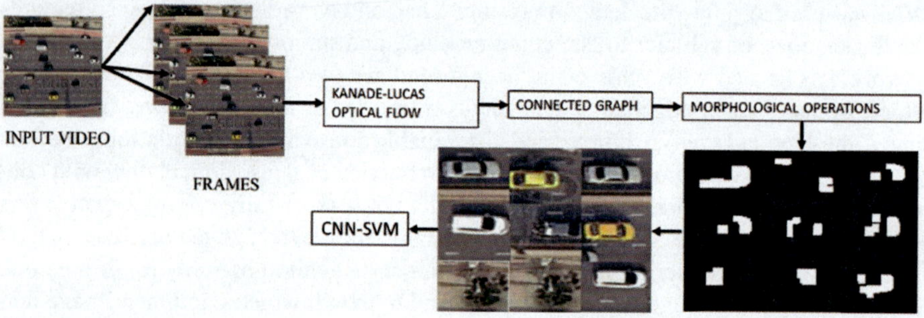

Fig. 1. Workflow of the proposed system.

3.2 Motion Vector Estimation

The primary goal of optical flow estimation is to split the moving foreground objects from the background and generate optical flow field vector for the moving object. Background motion in UAV videos, makes many conventional methods like Kalman filter, particle filter etc., hard to gain the desired accuracy and efficiency for real-time use. This points to the importance of Kanade-Lucas optical flow, which is a two-frame differential method. The method assumes that the displacement of the image contents between two consecutive frames is small and approximately constant in proximity of the point.

Consider a voxel at the location (x, y, t),with intensity I(x, y, t) and Δx, Δy and Δt are the movement between two consecutive frames, under the assumption that there is no drastic change in brightness. Since the movement is small, this can be expressed in Taylor Series [33] by:

$$I(x+\Delta x, y+\Delta y, t+\Delta t) = I(x,y,t) + \frac{\partial I}{\partial x}\Delta x + \frac{\partial I}{\partial y}\Delta y + \frac{\partial I}{\partial t}\Delta t + O\left(\Delta x^2, \Delta y^2, \Delta t^2\right) \quad (1)$$

Rewriting Eq. (1),

$$\frac{\partial I}{\partial x}V_x + \frac{\partial I}{\partial y}V_y + \frac{\partial I}{\partial t} \quad (2)$$

where Vx and Vy are the velocities in x and y directions, respectively. Equation (2) is known as the optical flow equation with two unknown variables. For solving equation, Kanade- Lucas operator [34] is applied assuming that the flow is constant in immediate neighborhood of the pixel under consideration. The flow vector (Vx, Vy) can be calculated by obtaining the minimum of the following [34]:

$$\sum\nolimits_{x,y} W^2(x)\left(I_x V_x + I_y V_y + I_t\right)^2 \quad (3)$$

where W is a n × n diagonal matrix of weights ω_i. To create a compromise in weightage between neighboring pixels, Kanade-Lucas uses least square criterion [34]:

$$A^T W^2 A V = A^T W^2 b \quad (4)$$

where A = $(\Delta I(q_i),\ldots,\Delta I(q_n))^T$, W = diag$(W(q_1),\ldots,W(q_n))$, b = $-(\frac{\partial I(q_1)}{\partial t},\ldots,\frac{\partial I(q_n)}{\partial t})^T$ and $q_1,\ldots,q_2$ are neighboring pixels. V can be found as following:

$$V = (A^T W^2 A)^{-1} A^T W^2 b \quad (5)$$

$$\begin{bmatrix} V_x \\ V_y \end{bmatrix} = \begin{bmatrix} \sum_i I_x(q_i)^2 & \sum_i I_x(q_i) I_y(q_i) \\ \sum_i I_y(q_i) I_x(q_i) & \sum_i I_y(q_i)^2 \end{bmatrix}^{-1} \begin{bmatrix} -\sum_i \frac{I_x(q_i)}{I_t(q_i)} \\ \sum_i I_y(q_i) I_t(q_i) \end{bmatrix} \quad (6)$$

The Vx and Vy obtained from (6) can be combined with their orientations to determine the velocity of a vehicle as:

$$V = V_x cos\theta + V_y sin\theta \quad (7)$$

The regions with a velocity value are given to Connected Graph to obtain the centroids of the moving object.

3.3 Connected Graph Theory

Pixels with same Vx and Vy in proximity are considered as interest points on the same vehicle. This is under the assumption that vehicles are rigid objects and hence the pixels move in proximity in 2D image space [17]. These pixels undergo morphological operations like erosion and dilation to eliminate the presence of other small moving objects like pedestrians and motorcycles. Next is the grouping process, where the interest points with same x and y velocities are grouped. Ideally each group represents a vehicle, and its motion can be estimated by average motion of all interest points and its location can be determined by locating the centroid of the interest points.

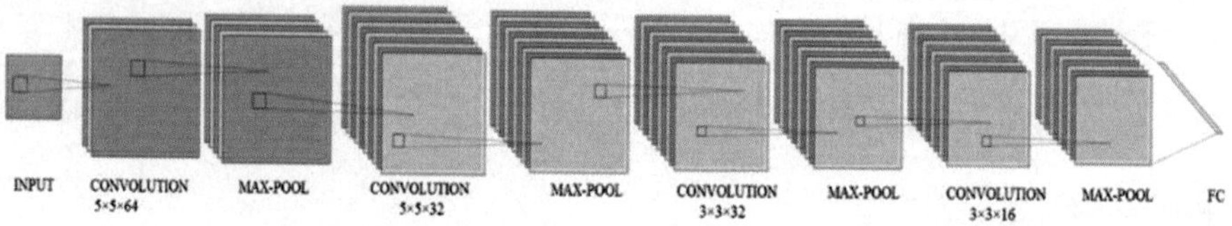

Fig. 2. The proposed CNN architecture for feature extraction.

3.4 Classification Using CNN-SVM

The proposed regions of interest undergo the final step of classification by CNN-SVM. This step proceeds with utilizing the capability of Convolutional Neural Networks to successfully extract features and the efficiency of Support Vector Machine in binary classification. One of the advantage of deep learning techniques is its capacity to execute feature engineering by its own, i.e., it can scan the data to search for features that correlate and combine them to enable faster learning with restricting itself to user's knowledge domain. CNNs constitutes of several layers of linear and non-linear operators that are joined end-to-end to solve specific tasks. Specifically, they are stacks of convolutional, normalization, pooling, and fully connected layers. The convolutional layer is the prominent block made up of a set of learnable filters. Spatially small filters convolve through the full depth of the input image, which gives feature maps that are then fed into a non-linear gating function such as the rectified linear unit (ReLU) [35]. Normalised output of ReLU is fed into the pooling layer. The pooling layer subsamples a small rectangular block of convolutional layer to single output to avoid over-fitting. After several convolutional and pooling layers, a fully connected layer which gives the final probabilities for each label. The weights of the whole layers in the networks are fine-tuned to produce specific classes. This completes the feature extraction part. Later, these extracted features are used to train an external SVM classifier. A SVM is a discriminative classifier for supervised learning, formally defined by a separating hyperplane. It classifies data by finding the best hyperplane that can separate all data points in one class from those in the other class. The best hyperplane for an SVM means the one with the largest margin between the classes.

The dataset consisted of 100 car images and 100 background images, which was divided in 8:2 ratio for training and testing, respectively. The dataset is made general as possible by including all possible coloured cars. By trial and error method, the architecture of CNN was found to be of four convolutional layers, four max-pool layer and a single fully connected layer that gave 100% validation accuracy for the chosen dataset. Figure 2 shows the proposed CNN architecture. Compared to pre-trained CNN architectures, the proposed structure is lighter and has fewer parameters, thus making the system faster to train.

4 Results and Discussions

The testing dataset consists of a few videos of stationary and moving aerial cameras, which was resized to 336 × 592 as this made the calculation of optical flow easier. The stationary videos are of cameras that are fixed on poles, traffic signals etc. whereas moving videos are taken by a UAV travelling above a freeway segment. The videos vary from low to medium high traffic flow and vary in altitudes of UAV at which the videos are collected. In our study, the program was developed in Matlab R 2018a on a 64-bit i5 Intel core CPU. The method exploited optical flow and deep learning libraries of Matlab. Figure 3 shows the detection results of Kanade-Lucas optical flow. Figure 3(a) shows the detection results from a video fixed on a pole. Figure 3(b) shows the detection results from a video captured by a UAV moving over a freeway segment. These images were then processed by CNN-SVM to classify them into vehicle and objects in background. The CNN was trained with a dataset of total 200 images. The dataset for cars included images of cars in different angles. This made the system suitable to process all types of aerial videos. Also, the background dataset was generalized as possible. To increase the efficiency of the system, image augmentation (random rotation) was done between 0 to 360°. This makes the system insensitive to vehicle orientation. The input image size was set to 227 × 227 × 3, in which 3 implies that it is a RGB image. A cent per cent validation accuracy for the dataset was obtained with four convolution layers, of sizes 5 × 5 × 64, 5 × 5 × 32, 3 × 3 × 32 and 3 × 3 × 16 in order. Each convolution layer was followed by a max pooling layer of size 2 × 2 with stride 2.

Fig. 3. (Left) Detection of moving objects in video with fixed camera (Right) video from UAV platform.

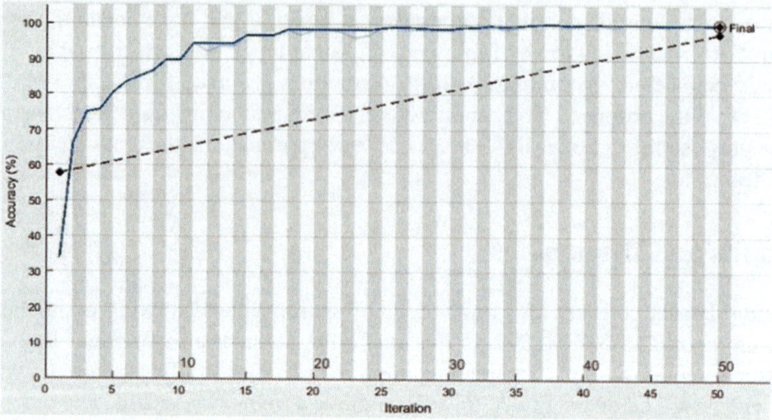

Fig. 4. Training progress plot for 50 epochs showing 100% accuracy in training

The fully connected layer has an input size 16 and an output size of 2 implying two classes of classification. The activation from the fully connected layer is transferred in SVM. The training was done for 50 epochs as the system attained maximum efficiency after 45th epoch. The four-layer CNN architecture is much efficient than the pre-trained architectures such as VGG, AlexNet etc. This decreases the computational load due to reduced number of parameters. Also, max-pool layer after each convolutional layer significantly reduced the number of parameters and convolutional layers required.

The stationary video was processed with a noise threshold of 0.025 for noise reduction in Kanade-Lucas optical flow. This was found by trial and error for many stationary aerial videos. An increase or decrease in the threshold resulted in false positives and false negatives. The morphological actions decided the point from which the car is to be detected. Beyond a point, when the view of the vehicle is very small, the detection is done as background. An accuracy of 100% was obtained in detection and classification of vehicles in videos from fixed cameras.

In videos from cameras with moving platform, the instability of UAV flying in air creates a lot of false positives. The vehicle motion in the video is the sum of motion of vehicle combined with that of the UAV. This results in detection of background objects as moving vehicles. To reduce the effect of motion of camera platform, the noise threshold was set to be 0.010 for noise reduction in optical flow. This can vary for different videos according to the density of urban environment. Also, morphological actions played a great role in reducing false positives. A few true negatives happened when the color of the car matched with that of the road segment. This gives a cumulative accuracy of 98% in vehicle detection and classification.

Figure 4 shows the accuracy graph of training progress of CNN for the given dataset. It contains the plots for training accuracy, smoothed training accuracy and validation accuracy. First one is the classification accuracy on each individual mini batch. Smoothed training accuracy is obtained by applying a smoothing algorithm to the training accuracy. It is less noisy than the unsmoothed accuracy which makes it easier to spot trends. Classification accuracy on the entire validation set gives the validation accuracy. Figure 5

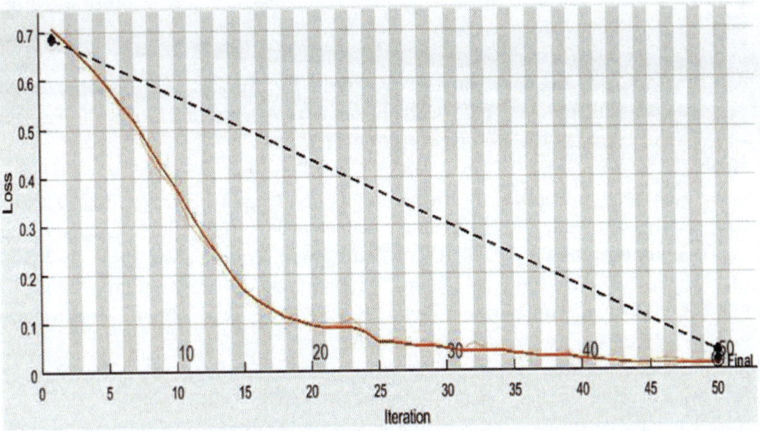

Fig. 5. Plot representing backward loss in training CNN

shows the loss graph of the training progress which includes the loss on each mini batch, its smoothed version, and the loss on the validation set. The activation from the fully connected layer is used to train SVM. Later, the moving objects detected by the optical flow are given for prediction by SVM. This eliminates the classification of any other moving or background objects as vehicles.

Comparing with existing methods, the system shows much more accuracy. The framework for vehicle counting with the combination of KLT tracker and k-means clustering [17] shows an accuracy of 87%. Previous work that uses KLT+SVM [32] with a dataset size of 100 images shows an accuracy of 90%. A work that incorporates ensemble classifier with KLT tracker [18] for motion estimation with a large dataset of 20,000 images shows a vehicle count accuracy of 95.8%. The limitation of this comparison is that the accuracies are defined on datasets of different sizes. For fairer comparison, the simulation scenario can be extended to include larger dataset and/or different datasets can be used to estimate accuracy averaged over such calculations.

5 Conclusion

The study proposed an efficient method for detection of vehicles from UAV videos by combining Kanade-Lucas optical flow and CNN-SVM. In the first stage, optical flow was employed to detect the moving objects in the aerial videos. The second stage progressed by using four layered CNN for feature extraction and SVM as classifier. The four-layer CNN is much efficient than pre-trained CNNs. The system works well for low to medium congested traffic flow. The future of the work aims in creating a framework for traffic parameter estimation along with a multi class classification using SVM for detecting cars based on colors. Vehicles can also be classified based on direction of motion. Also, the work would be extended to apply the system in real time by including methods for decreasing the processing time. The dataset will be expanded by increasing the number to include more generalized images.

References

1. Coifman, B., McCord, M., Mishalani, R., Redmill, K. : Surface transportation surveillance from unmanned aerial vehicles. In: Proceedings of 83rd Annual Meeting of the Transportation Research Board, Washington, DC, USA, pp. 1–9 (2004)
2. Kyoungah, C., Lee, I., Hong, J., Oh, T., Shin, S.W.: Developing a UAV-based rapid mapping system for emergency response. In: Proceedings of SPIE, vol. 7332, p. 733209, April 2009
3. Zhou, H., Kong, H., Wei, L., Creighton, D., Nahavandi, S.: Efficient road detection and tracking for unmanned aerial vehicle. IEEE Trans. Intell. Transp. Syst. **16**(1), 297–309 (2015)
4. Tang, J., Liu, F., Zou, Y., Zhang, W., Wang, Y.: An improved fuzzy neural network for traffic speed prediction considering periodic characteristic. IEEE Trans. Intell. Transp. Syst. **18**(9), 2340–2350 (2017)
5. Zhang, S., Tang, J., Wang, H., Wang, Y., An, S.: Revealing intra-urban travel patterns and service ranges from taxi trajectories. J. Transp. Geogr. **61**, 72–86 (2017)
6. Wang, Y., Ke, R., Zhang, W., Cui, Z., Henrickson, K.: Digital roadway interactive visualization and evaluation network applications to WSDOT operational data usage. University of Washington, Seattle, WA, USA, Technical report. WA-RD 854.1 (2016)
7. Ke, R., Pan, Z., Pu, Z., Wang, Y.: Roadway surveillance video camera calibration using standard shipping container. In: Proceedings of IEEE International Smart Cities Conference, Wuxi, China, September 2017, pp. 1–6 (2017)
8. Yamazaki, F., Wen, L., Vu, T.: Vehicle extraction and speed detection from digital aerial images. In: Proceedings of the IEEE International Conference on Geoscience and Remote Sensing Symposium, Boston, MA, USA, July 2008, pp. III-1334–III-1337 (2008)
9. Shastry, A.C., Schowengerdt, R.A.: Airborne video registration and traffic-flow parameter estimation. IEEE Trans. Intell. Transp. Syst. **6**(4), 391–405 (2005)
10. Angel, A., Hickman, M., Mirchandani, P., Chandnani, D.: Methods of analyzing traffic imagery collected from aerial platforms. IEEE Trans. Intell. Transp. Syst. **4**(2), 99–107 (2003)
11. Du, Y., Zhao, C., Li, F., Yang, X.: An open data platform for traffic parameters measurement via multirotor unmanned aerial vehicles video. J. Adv. Transp. **2017**, 8324301 (2017)
12. Cao, X., Lan, J., Yan, P., Li, X.: Vehicle detection and tracking in airborne videos by multi-motion layer analysis. Mach. Vis. Appl. **23**(5), 921–935 (2012)
13. Yalcin, H., Hebert, M., Collins, R., Black, M.J.: A flow-based approach to vehicle detection and background mosaicking in airborne video. In: Proceedings of the IEEE Computer Society Conference on Computer Vision and Pattern Recognition, June 2005, p. 1202 (2005)
14. Cao, X., Gao, C., Lan, J., Yuan, Y., Van, P.: Ego motion guided particle filter for vehicle tracking in airborne videos. Neurocomputing **124**, 168–177 (2014)
15. Yu, Q., Medioni, G.: Motion pattern interpretation and detection for tracking moving vehicles in airborne video. In: Proceedings of the IEEE Conference on Computer Vision and Pattern Recognition, Miami, FL, USA, June 2009, pp. 2671–2678 (2009)
16. Ke, R., Kim, S., Li, Z., Wang, Y.: Motion-vector clustering for traffic speed detection from UAV video. In: Proceedings of the. IEEE Conference on Smart Cities, Guadalajara, Mexico, October 2015, pp. 1–5 (2015)
17. Ke, R., Li, Z., Kim, S., Ash, J., Cui, Z., Wang, Y.: Real-time bidirectional traffic flow parameter estimation from aerial videos. IEEE Trans. Intell. Transp. Syst. **18**(4), 890–901 (2017)
18. Ke, R., Li, Z., Tang, J., Pan, Z., Wang, Y.: Real-time traffic flow parameter estimation from uav video based on ensemble classifier and optical flow. IEEE Trans. Intell. Transp. Syst. **20**(1), 54–64 (2019). https://doi.org/10.1109/TITS.2018.2797697

19. Cohen, I., Medioni, G.: Detecting and tracking moving objects in video from an airborne observer. In: Proceedings of the IEEE Image Understand. Workshop, pp. 217–222, November 1998
20. Lucas, B., Kanade, T.: Detection and tracking of point features 1981. Department of Computer Science, Carnegie Mellon University, Pittsburgh, PA, Tech report. CUM-CS-91–132, April 1991
21. Ali, S., Reilly, V., Shah, M.: Motion and appearance contexts for tracking and reacquiring targets in aerial video. In: Proceedings of the European Conference on Computer Vision (CVPR), pp. 1–6 (2007)
22. Memon, Q.: On assisted living of paralyzed persons through real-time eye features tracking and classification using support vector machines. Med. Technol. J. **3**(1), 316–333 (2019)
23. Cucchiara, R., Piccardi, M., Mello, P.: Image analysis and rule-based reasoning for a traffic monitoring system. IEEE Trans. Intell. Transport. Syst. **1**(2), 119–130 (2000)
24. Zhao, T., Nevatia, R.: Car detection in low resolution aerial image. In: Proc. IEEE International Conference on Computer Vision, vol. 1, pp. 710–717, July 2001
25. Al-Kassim, Z., Memon, Q.: Designing a low-cost eyeball tracking keyboard for paralyzed people. Comput. Electr. Eng. **58**, 20–29 (2017)
26. Rosenbaum, D., Leitloff, J., Kurz, F., Meynberg, O., Reize, T.: Realtime image processing for road traffic data extraction from aerial images. In: Proceedings of the ISPRS Commission VII Symposium, June 2010, pp. 372–388 (2010)
27. Teney, D., Hebert, M.: Learning to extract motion from videos in convolutional neural networks. In: Asian Conference on Computer Vision (2016). https://doi.org/10.1007/978-3-319-54193-8_26
28. Wang, L., Chen, F., Yin, H.: Detecting and tracking vehicles in traffic by unmanned aerial vehicles. Autom. Constr. **72**(3), 294–308 (2016)
29. Hansen, M., Anadan, P., Dana, K., van de Wal, G., Burt, P.: Realtime scene stabilization and Mosaic Construction. In: Proceedings of IEEE CVPR, 54–62 (1994)
30. Sahli, S., Ouyang, Y., Sheng, Y., Lavigne, D.A.: Robust vehicle detection in low-resolution aerial imagery. a Image Science group
31. Cao, X., Wu, C., Lan, J., Yan, P., Li, X.: Vehicle detection and motion analysis in low-altitude airborne video under urban environment. IEEE Trans. Circuits Syst. Video Technol. **21**(10), 1522–1533 (2011). https://doi.org/10.1109/TCSVT.2011.2162274
32. Najiya, K.V., Archana, M.: UAV video processing for traffic surveillance with enhanced vehicle detection. In: 2018 Second International Conference on Inventive Communication and Computational Technologies (ICICCT), Coimbatore, pp. 662–668 (2018). https://doi.org/10.1109/ICICCT.2018.8473204
33. Rodriguez-Canosa, G.R., Thomas, S., del Cerro, J., Barrientos, A., MacDonald, B.: A realtime method to detect and track moving objects (DATMO) from unmanned aerial vehicles (UAVs) using a single camera. Remote Sens. **4**(4), 1090–1111 (2012). https://doi.org/10.3390/rs4041090(ISSN:2072-4292)
34. Lucas, B.D., Kanade, T.: An iterative image registration technique with an application to stereo vision. In: Proceedings of the 7th International Joint Conference on Artificial Intelligence, vol. 2, pp. 674–679. Morgan Kaufmann Publishers Inc., San Francisco (1981)
35. Krizhevsky, A.; Sutskever, I.; Hinton, G.E.: ImageNet classification with deep convolutional neural networks. In: Advances in Neural Information Processing Systems. MIT Press, Cambridge (2012)
36. Yang, H., Qu, S.: Real-time vehicle detection and counting in complex traffic scenes using background subtraction model with low-rank decomposition. IET Intell. Transp. Syst. **12**(1), 75–85 (2018). https://doi.org/10.1049/iet-its.2017.0047

37. Mandellos, N.A., Keramitsoglou, I., KiranoudisA, C.T.: Background subtraction algorithm for detecting and tracking vehicles. Expert Syst. Appl. **38**(3), 1619–1631 (2011)
38. Li, X., Yao, X., Murphey, Y.L., Karlsen, R., Gerhart, G.: A real-time vehicle detection and tracking system in outdoor traffic scenes. In: Proceedings of the 17th International Conference on Pattern Recognition, ICPR 2004, Cambridge, vol. 2, pp. 761–764 (2004). https://doi.org/10.1109/ICPR.2004.1334370

Underwater Moving Object Detection Using GMG

Vrushali Pagire and Anuradha Phadke(✉)

Department of ECE, MIT World Peace University, Pune, Maharashtra, India
{vrushali.pagire,anuradha.phadke}@mitwpu.edu.in

Abstract. The development in ocean exploration and observation make the demand for moving object detection in underwater increasingly urgent. The moving object detection in underwater with dynamic or moving background is the challenging and difficult task for the researchers. In dynamic environment, foreground object as well as background of the scene both are in motion condition, thus it is very difficult to detect foreground moving object in underwater medium. In proposed work, background is categorized into three types i.e. static, moderate dynamic and high dynamic backgrounds. The GMG algorithm invented by Andrew B. Godbehere, Akihiro Matsukawa, Ken Goldberg is implemented to detect moving object and compare the results for the three background categories. The sensitivity and accuracy parameter values are considered for the comparison of the results.

Keywords: Underwater object detection · Static background · Dynamic background · GMG

1 Introduction

The moving object detection is one of the significant and leading task in video and image processing applications. Underwater object detection is very useful in ecologist, aqua culturist, marine biologist etc. The automatic object detection analysis opens a new direction in research and surveillance applications. The automation of moving object detection is used in security, identification of threads, unauthorized intruders in defense and navy. The images and videos captured in underwater are more deteriorated as compared to air medium [1]. The physical properties of water medium affect quality of the images and videos captured in underwater. When light enters into the water medium, light is strongly attenuated due to water medium. As the depth increases, the obscurity and poor contrast is experienced in videos [3]. Depending on wavelength of the colours, the colours are degraded and luminosity reduces as the depth increases.

Underwater object detection has a wide range of applications. The detection of foreground object in moderate dynamic and static environment is one of the challenging tasks in the analysis of the videos. The object detection in static background does not carriage major problem as compared to the object detection in dynamic background condition [8]. The background is stationary throughout the video and the foreground goes

A. Abraham et al. (Eds.): SoCPaR 2020, AISC 1383, pp. 233–244, 2021.
https://doi.org/10.1007/978-3-030-73689-7_23

on changing its position in the static environment. The foreground is considered as a moving object maybe it single or multiple objects. The underwater videos are composed of several dynamic entities. The moving objects are moving along with backgrounds that are themselves moving in dynamic environment. Both foreground and background have motion in dynamic environment. The differentiation between foreground and background is very difficult task. To detect moving object in dynamic environment lot of many issues have to be considered in the analysis of the video frames.

Numerous algorithms are available to detect moving object in static, dynamic and complex background environments. Hence, the aim of the proposed work is to detect moving object in dynamic background using image processing techniques for underwater environment. The proposed work is organized as follows: In Sect. 1, Introduction of the proposed work is included. Literature survey is included in Sect. 2 and understanding of underwater moving object detection research work at the national and international scenario, Sect. 3 gives the methodology of the proposed work, analysis results shows in Sect. 4, conclusion is explained in Sect. 5.

2 Related Work

The oceanic environment is non-uniform and dense due to which light scattering, attenuation is occurring in underwater. Images and videos captured in underwater suffer from deuteriation and low visibility of light. It has been observed that degradation of color and contrast takes place for underwater videos and images. In underwater environment, as depth increases the illuminance is lowers and the visibility is dependent on the density of suspended matters [12]. It is very difficult to detect underwater moving object to the researchers because of the absorption of the light, density, scattering and restrictive access to human beings [4].

In complex scenario, most of the time background modelling is widely used for moving object detection by utilizing different features. Single Gaussian model was used on each pixel of the frame for background modelling [2]. The **Single Gaussian model** cannot handle the variations in dynamic background but it is very useful in static background condition. The variations in dynamic background can be addressed by Gaussian mixture model (GMM).

Lu Zhang, Xu Yang [2], have developed algorithm based on neural network. They have introduced multi-box feature pyramid module. The module shows semantic cues for more accurate object detection. Zhu Y., et al. [5], they proposed structure for video segmentation of object extraction in underwater medium which is a combination of DeepLab, LDOF and CLAHE. Yoon K., Kim W. [6], proposed enhancement method for sonar images. Cheng M., et al. [7], presented a method by which to compare feature detectors on sonar imagery. Prabowo M., et al. [10], used background modeling technique for moving object detection. The performance results show satisfactory values because a recall value and precision value are above 50%. Tueller P., et al. [9], presented work on underwater sonar imagery. They have presented a comparison of feature detectors for sonar imagery. Alexandri T., Diamant R. [11], presented work on target motion Analysis. Md. Moniruzzaman [13], worked on deep learning on underwater marine object detection. N. Muhammad [14], proposed a technique based on map-based localization and loop-closure detection. They have used a continuously moving platform in work. For semantic object segmentation in video, Zhang Y. [15], proposed an approach based on segmentation-by-detection. Pont-Tuset J., Marques F. [16], presented estimation of segmentation of image and object proposal techniques. Yang Y., et al. [17], proposed a spatiotemporal model for object detection. Akshaya B., et al. [18] proposed a sparse representation to localize objects images. Leil F., et al. [19], proposed a codebook model for underwater object detection. Barbosa W., et al. [20], presented enhancement in images based on quality learning. Pagire V., Phadke A. [21], presented FPGA based system which detects a foreground object.

In the Sect. 3, GMG algorithm is explained for moving object detection in underwater environment.

3 Methodology

3.1 GMG Algorithm

GMG (invented by Andrew B. Godbehere, Akihiro Matsukawa, Ken Goldberg) is based on Gaussian Mixture model. In GMG algorithm first few frames are used for background modelling. In GMG algorithm, the combination of Bayesian Inference and Kalman Filters is used for background modelling. The weighted values of each pixels are used to model the background and the modified value is updated into model. For determined amount of time, color values which are static are considered for background.

As shown in Fig. 1, P(K) is image and $\widehat{B}$(K) is statistical background model. P(K) is quantized in color-space. The quantized output is compared against $\widehat{B}$(K) model and generated a posterior probability image. Then the morphological filtering operations are applied on the image. Connected components algorithm is used to segmentation and it form a bounding boxes set, Q(K). Kalman filter bank is used to keep pursued visitors $\widehat{R}$(K). It has forecast R(K), bounding boxes for each time k. To pair the elements of Q(K) and R(K), Gale-Shapley matching algorithm is used. The pairs updated the Kalman Filter bank. Finally, the foreground pixels are identified and the result is $\widehat{R}$(K). The collected pixels $\widehat{R}$(K) and P(K) are used to modify the background model to $\widehat{R}$(K + 1).

In the initialization process, the frame is always black. To stabilize the initialization model, large number frames are used in this process. To classify the pixels as foreground and background the threshold parameter is used into the model. For the initial stage, color static for the determined time with lower weighted pixel value. The threshold parameter considered it as a part of the background. It is good to select low value for threshold parameter because high value of threshold may result in loss of object detections.

Median filtering is applied on images to remove noise. Filter effectively removes the 'salt and pepper' type noise from images. In this filter, median value is calculated and pixel value is replaced by calculated value. The morphological dilation, erosion operations are applied on images to make the object more visible.

3.2 Types of Background

For the analysis of results of implemented algorithms, the background is divided into three categories:

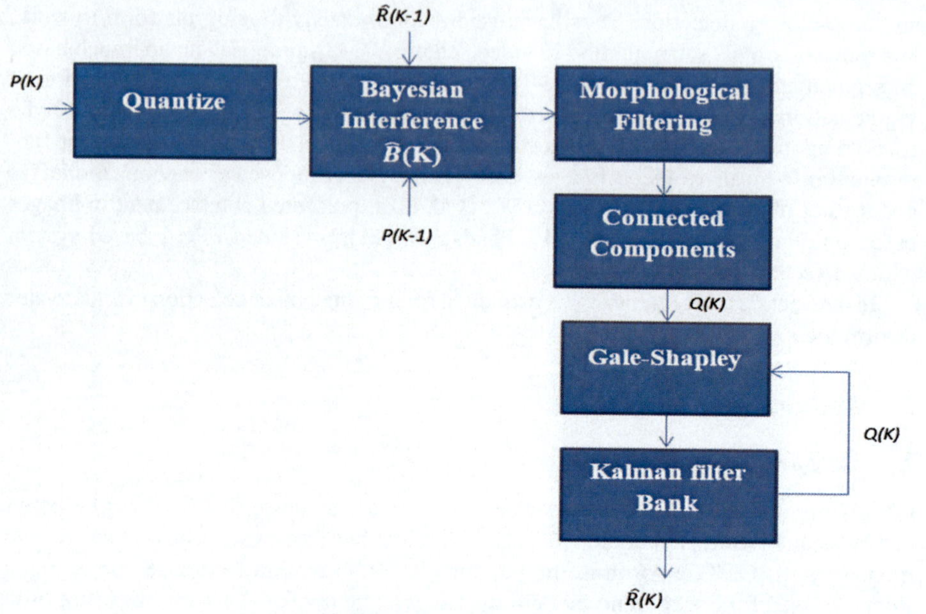

Fig. 1. GMG approach

Static Background
In the static background category, the foreground object is moving and the background is static. An example of static background is shown in Fig. 2.

Fig. 2. Static background

Moderate Dynamic Background

In the moderate dynamic background, the foreground object is moving and some part of background is dynamic. An example of moderate dynamic background is shown in Fig. 3.

Fig. 3. Moderate dynamic background

High Dynamic Background

The foreground object is moving but at the same time maximum part of background is dynamic in video. An example of high dynamic background is shown in Fig. 4.

Fig. 4. High dynamic background

3.3 Performance Metric

For testing and analysis of implemented algorithm various videos from Fish4Knowlege dataset are used. These videos from the data base are divided into three categories First is static or stationary background, Moderated dynamic background and High dynamic background. For each category 10 videos are used. Thus total 30 videos are analyzed. From every video total 20 frames are selected, out of which in 10 frames object is present and in 10 frames object is not present. Thus, to compare the analysis results of the implemented methodology of background modelling, in all 600 frames are used. Above category videos are analyzed for GMG approach. Here fish is considered as an object.

A confusion matrix is used to describe the performance of the proposed model. The true values are known for the test data. The Fig. 5 shows the confusion matrix.

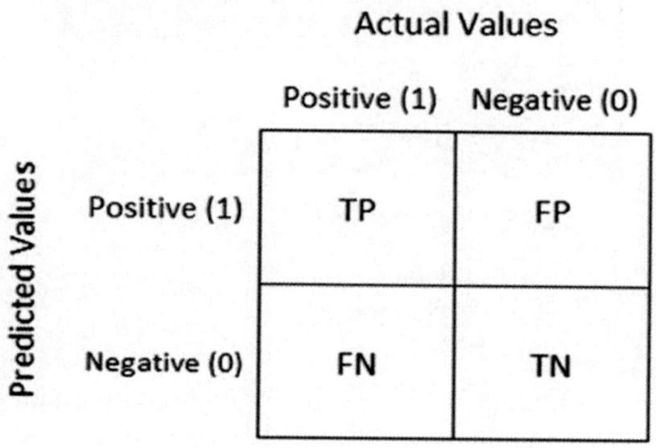

Fig. 5. Confusion matrix

The parameters considered for the matrix are shown in the Table 1.

Table 1. Parameters for confusion matrix

Parameters	Form	Description
TP	True Positive	Fish is existing and detected
TN	True Negative	Fish is not existing and not detected
FP	False Positive	Fish is existing and not detected
FN	False Negative	Fish is not existing and detected

The sensitivity, accuracy, precision, Specificity, recall, F-measure parameters are calculated using confusion matrix parameters.

4 Results

The GMG algorithm is implemented for videos with three types of backgrounds namely, static, moderate dynamic and high dynamic background. The results of the implemented algorithm are shown below for different background categories.

4.1 Qualitative Analysis

In the static background, only fish is moving and the background is static. The algorithm is tested for static background video database. The results for the implemented algorithm are shown below. The Fig. 6, shows an example of image from static background category video. The Fig. 7, gives the results of applying GMG algorithm on the sample video. Expected results are indicated in the original image by red circles. From Fig. 7, it is clear that, all the moving objects in the frame i.e. five fishes at different locations indicated by red circles, are detected correctly.

Fig. 6. Results for static background: Original Image with ground truth marked by red colored circle

For the moderate dynamic background, object is moving and some part of background is dynamic like corals are moving in the video. The algorithm is tested for moderate dynamic background video database. An example of input frame of type moderate dynamic background and the result of the implemented GMG algorithm for the sample input frame are shown in Fig. 8 and Fig. 9 respectively. Figure 9 indicates that the moving fish is detected along with some false object detection has also occurred.

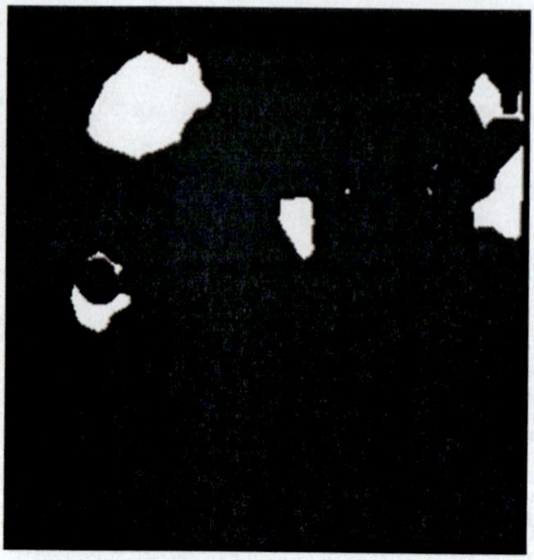

Fig. 7. Results for static background: Detected object

Fig. 8. Results for moderate dynamic background: Original Image with ground truth marked by red colored circle

In the high dynamic background, object is moving and the maximum background is dynamic like corals are moving in the video. The algorithm is tested for high dynamic background video database. The results for the implemented algorithm are shown below. The Fig. 10, shows the original image. The results of GMG algorithm are shown in Fig. 11. The moving fish is detected but maximum false object detection is occurred.

4.2 Quantitative Analysis

The previous study [1] concluded that, the precision for dynamic background is 73% and F-score measure is 64%. The performance parameters values are calculated for GMG algorithm and are stated in Table 2. The analysis of the results for the implemented algorithm is done for static background, moderate dynamic background and high dynamic

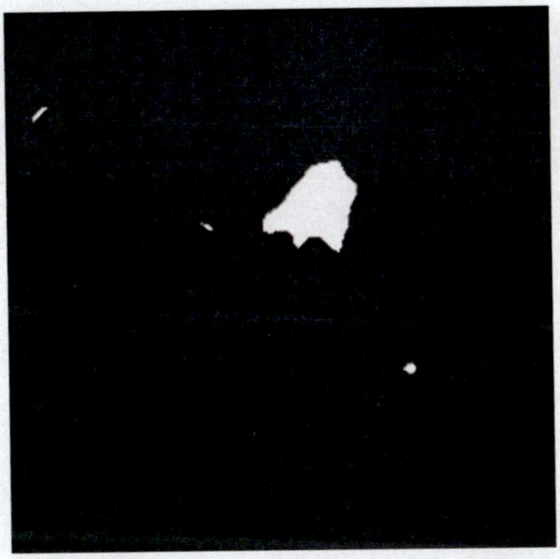

Fig. 9. Results for moderate dynamic background: Detected object

Fig. 10. Results for high dynamic background: Original Image with ground truth marked by red colored circle

background. In Table 2, the sensitivity, specificity, precision, F-measure, accuracy and recall are measured and shown.

For comparison, the sensitivity and accuracy parameters are considered. The sensitivity and accuracy are 79.36% and 87.73% respectively for static background. For moderate dynamic background sensitivity and accuracy is 68.27% and 79.01% respectively and for high dynamic background the sensitivity is 50.76% and accuracy is 55.29% respectively. The sensitivity and accuracy are plotted in Fig. 12 for GMG algorithm. The sensitivity and accuracy for all the three types of backgrounds are plotted in Fig. 12 for the implemented GMG algorithm.

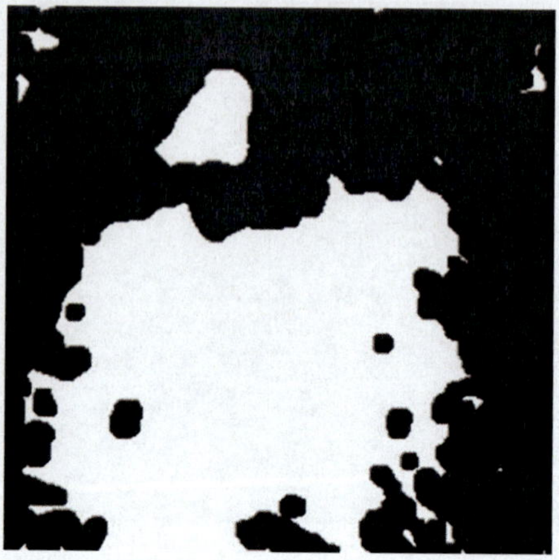

Fig. 11. Results for high dynamic background: Detected object

Table 2. Performance parameters of GMG algorithm

Background	Sensitivity (%)	Specificity (%)	Accuracy (%)	Precision (%)	Recall (%)	F -Measure (%)
Static	79.36	99.8	87.73	99.8	79.36	88.49
Moderate dynamic	68.27	99.8	79.01	99.8	68.27	80.97
High dynamic	57.6	99.8	55.29	99.8	50.76	67.33

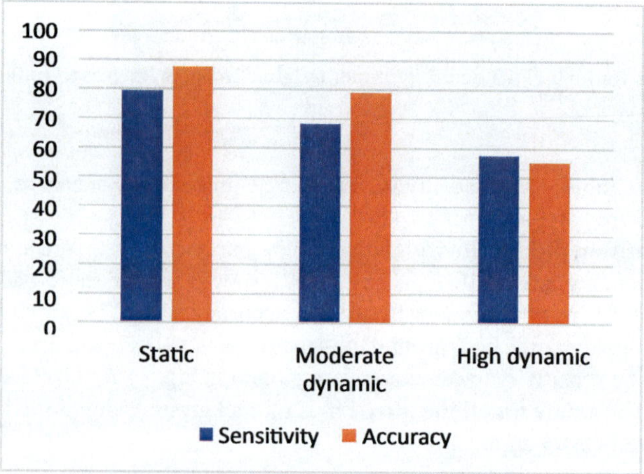

Fig. 12. Graphical representation for GMG algorithm.

5 Conclusion

In this paper, we proposed GMG algorithm for underwater moving object detection. The underwater background is divided into static background, moderate dynamic background and high dynamic background categories. The sensitivity and accuracy are good for the static and moderate dynamic backgrounds. For high dynamic background the sensitivity and accuracy are poor as compared to that for static and moderate dynamic backgrounds. Thus, we can accomplish that proposed GMG algorithm is appropriate for moving objects detection in presence of static or moderate dynamic backgrounds. In direction to notice objects in the existence of high dynamic background it is necessary to modify the GMG algorithm.

References

1. Vasamsetti, S., Setia, S., Mittal, N., Sardana, H., Babbar, G.: Automatic underwater moving object detection using multi- feature integration framework in complex backgrounds. IET Comput. Vision **12**(6), 770–778 (2018)
2. Zhang, L., Yang, X., Liu, Z., Qi, L.U, Zhou, H., Chiu, C.: Single shot feature aggregation network for underwater object detection. In: 24th International Conference on Pattern Recognition (ICPR), Beijing, China, 20–24 August 2018
3. Mi, Z., Liang, Z., Wang, Y., Fu, X.: Multi-scale gradient domain underwater image Enhancement. IEEE, 978-1-5386-1654-3/18 (2018)
4. Mengnan, S., Zhaorui, G., Zheng, H., Zheng, B., Watson, J.: Underwater Wide-Area Layered Light Field for Underwater Detection, Digital Object Identifier. 2877591, vol. 6 (2018)
5. Zhu, Y., Song, Y., Zhang, X., Pengfei, L., Guangliang, L., Bo, H., Tianhong, Y.: Segmentation of underwater object in videos. IEEE. 978-1-5386-1654-3/2018
6. Yoon, K., Kim, W.: Efficient edge-preserved sonar image enhancement method based on CVT for object recognition. IET Image Process. **13**(1), 15–23 (2019)
7. Cheng, M., Mitra, N., Huang, X., Torr, P., Hu, S.: Global contrast based salient region detection. IEEE Trans. Pattern Anal. Mach. Intell. **37**(3), March 2015
8. Hemavathy R., Shobha G.: Object detection and tracking under static and dynamic environment: a review. Int. J. Adv. Res. Comput. Commun. Eng. **2**(10) (2013)
9. Tueller, P., Kastner, R., Diamant, R.: A Comparison of Feature Detectors for Underwater Sonar Imagery. IEEE, 978-1-5386-4814-8 (2018)
10. Prabowo, M., Hudayani, N., Purwiyanti, S., Sulistiyanti, S., Setyawan, F.: A moving objects detection in underwater video using subtraction of the background model. In: Proceedings of the EECSI 2017, Yogyakarta, Indonesia, 19–21 September (2017)
11. Alexandri, T., Diamant, R.: A reverse bearings only target motion analysis for autonomous underwater vehicle navigation. IEEE Trans. Mob. Comput. **18**, 494–506 (2018)
12. Shahrizan, G., Fakhri, A., Nasir, A., Faizura, W., Tarmizi, W.: Integration of enhanced background filtering and wavelet fusion for high visibility and detection rate of deep sea underwater image of underwater vehicle. In: Fifth International Conference on Information and Communication Technology (2017)
13. Moniruzzaman, M., Mohammed, S., Bennamoun, M., Lavery, P.: Deep Learning on Underwater Marine Object Detection: A Survey. Springer International Publishing ACIVS 2017, LNCS, vol. 10617, pp. 150–160 (2017)
14. Muhammad, N., Francisco, J., Tuhtan, J., Toming, G., Musall, M., Kruusmaa, M.: Map-based localization and loop-closure detection from a moving underwater platform using flow features. LLC, part of Springer Nature, Springer Science (2018)

15. Zhang, Y., Chen, X., Jia, L., Wang, C., Xia, C., Li, J.: Semantic object segmentation in tagged videos via detection. IEEE Trans. Pattern Anal. Mach. Intell. **40**, 1741–1754 (2017)
16. Pont-Tuset J., Marques F.: Supervised evaluation of image segmentation and object proposal techniques. IEEE Trans. Pattern Anal. Mach. Intell. **38**, 1465–1478 (2018)
17. Yang, Y., Zhang, Q., Wang, P., Hu, X., Wu, N.: Moving object detection for dynamic background scenes based on spatiotemporal model. Hindawi Advances in Multimedia, Volume 2017, Article ID 5179013
18. Akshaya, B., Narmadha, Y., Sree, T., Rajendran, Y.: Sparse Representation to Localize Objects in Underwater Acoustic Images. IEEE, 978-1-4799-6085-9 (2015)
19. Lei, F., Huang, W., Zhang, Z.: Underwater Moving Object Detection Based on Codebook Model. International Conference on Computational and Information Sciences, 978-0-7695-5004-6 (2013)
20. Barbosa, W., Amaral, H., Rocha, T., Nascimento, E.: Visual-Quality-Driven Learning for Underwater Vision Enhancement. IEEE, 978-1-4799-7061-2/2018
21. Pagire, V., Phadke, A.: Detection of Moving Object Based on FPGA. IEEE Pune Section International Conference, 978-1-7281-1924-3, Dec 18–20 (2019)

Linear Dilation-Erosion Perceptron Trained Using a Convex-Concave Procedure

Angelica Lourenço Oliveira and Marcos Eduardo Valle(✉)

Universidade Estadual de Campinas,
Campinas 13083-970, Brazil
{ra211686,valle}@ime.unicamp.br

Abstract. Mathematical morphology (MM) is a theory of non-linear operators used for the processing and analysis of images. Morphological neural networks (MNNs) are neural networks whose neurons compute morphological operators. Dilations and erosions are the elementary operators of MM. From an algebraic point of view, a dilation and an erosion are operators that commute respectively with the supremum and infimum operations. In this paper, we present the *linear dilation-erosion perceptron* (ℓ-DEP), which is given by applying linear transformations before computing a dilation and an erosion. The decision function of the ℓ-DEP model is defined by adding a dilation and an erosion. Furthermore, training an ℓ-DEP can be formulated as a convex-concave optimization problem. We compare the performance of the ℓ-DEP model with other machine learning techniques using several classification problems. The computational experiments support the potential application of the proposed ℓ-DEP model for binary classification tasks.

Keywords: Mathematical morphology · Concave-convex optimization · Binary classification · Continuous piece-wise linear function

1 Introduction

Machine learning techniques play an important role in both pattern recognition and soft computing. Neural networks, which are inspired by the biological nervous system, are among the most efficient machine learning techniques used to solve pattern recognition tasks including, for example, computer vision and natural language processing [3]. Due to page constraints, we will not provide an extensive review of the many interesting and effective neural network models

This work was supported in part by the São Paulo Research Foundation (FAPESP) under grant no 2019/02278-2, the National Council for Scientific and Technological Development (CNPq) under grant no 310118/2017-4, and the Coordenação de Aperfeiçoamento de Pessoal de Nível Superior - Brasil (CAPES) - Finance Code 001.

A. Abraham et al. (Eds.): SoCPaR 2020, AISC 1383, pp. 245–255, 2021.
https://doi.org/10.1007/978-3-030-73689-7_24

which have been proposed in the last years but focus on the few models relevant to the development of the linear dilation-erosion perceptron addressed in this paper.

First of all, it is widely known that the Perceptron, introduced by Rosenblatt in the late 1950s, can be used for binary classification tasks [18]. Precisely, the perceptron yields one class if the weighted sum of the inputs is greater than or equal to a threshold and returns the other class otherwise. Furthermore, Rosemblatt proposed a learning rule that converges whenever the samples of the two classes are linearly separable [10]. The multi-layer perceptron (MLP) network, with at least one hidden layer and trained using backpropagation algorithms, overcomes the limitations of the Rosenblatt's perceptron for non-linearly separable classification tasks [10]. The support vector classifiers (SVCs) developed by Vapnik and collaborators also overcomes the limitations of the perceptrons for binary classification tasks. Like the perceptron, an SVC also predicts the class of a sample by computing a weighted sum. However, training an SVC is performed by maximizing the margin of separation between the two classes [25]. Furthermore, the classification performance of an SVC increases significantly by using the kernel trick. The kernel trick allows an SVC model to be effectively applied to non-linearly classification problems. For comparison purposes, in this paper, we consider both linear and radial-basis function SVC models as well as an MLP network.

In the late 1980s, Ritter and collaborators developed the so-called image algebra as an attempt to provide a unified framework for image processing and analysis techniques [16]. Using the image algebra framework, both linear and morphological operators are defined analogously. By replacing the usual linear dot-product in Rosemblatt's perceptron model with the corresponding lattice-based operation used to define morphological operators, Ritter and Sussner derived the morphological perceptron in the middle 1990s [14,15]. The elementary operations from mathematical morphology are dilations and erosions [20]. The maximum of additions yields a dilation while the minimum of sums yields an erosion. Hence, a dilation-based perceptron classifies a pattern by computing the maximum of its components plus the corresponding weights. Dually, the decision function of an erosion-based morphological perceptron is given by the minimum of the pattern's components plus the weights.

Such as the MLP, a multi-layer morphological perceptron (MLMP) with at least one hidden layer with both dilation-based and erosion-based neurons can theoretically solve any binary classification tasks [17]. From a geometric point of view, an MLMP discriminates two classes by enclosing patterns in hyper-boxes [17,22]. The non-differentiability of the lattice-based operations, however, makes training morphological neural networks difficult for gradient-based algorithms. Thus, some researchers developed training algorithms in which the network structure grows by adding hyper-boxes to fit the training data [17,22,23]. On the downside, growing algorithms usually overfit the training data.

Recently, Charisopoulos and Maragos formulated the training of a morphological perceptron as a concave-convex optimization problem which, in some sense, resembles the training of a linear SVC [4]. Despite the encouraging performance in simple classification tasks, both dilation-based and erosion-based morphological perceptrons implicitly assume a relationship between the input features and their classes [24]. Using concepts from vector-valued mathematical morphology, one of us proposed the reduced dilation-erosion perceptron (r-DEP), which overcomes the limitations of the morphological perceptron [24]. In a few words, the r-DEP first transforms the input into a feature vector. The feature vector is then classified using a linear combination of a dilation-based and an erosion-based perceptron.

In this paper, a linear transformation is used to map the input to the feature vector. The resulting model is referred to as a *linear dilation-erosion perceptron* (ℓ-DEP). We point out that an ℓ-DEP model can also theoretically solve any binary classification task. Furthermore, inspired by the works of Charisopoulos and Maragos, we formulate the training of an ℓ-DEP classifier as a convex-concave optimization problem.

The paper is organized as follows. The next section provides the mathematical background necessary to understand the models presented in this paper. Section 3 briefly reviews the morphological perceptron as well as the r-DEP model. The ℓ-DEP model and its training algorithm are introduced in Sect. 4. Computational experiments comparing the performance of the ℓ-DEP model with other classical models from the literature are given in Sect. 5. The paper finishes with some concluding remarks in Sect. 6.

2 Basic Concepts on Mathematical Morphology

Mathematical morphology is a non-linear theory widely used for image processing and analysis [12,20]. From the theoretical point of view, mathematical morphology is very well-defined on complete lattices. A partially ordered set $(\mathbb{L}, \preceq)$ is a complete lattice if any subset of $\mathbb{L}$ has both a supremum and an infimum, which are denoted respectively by $\sup X$ and $\inf X$.

The elementary operators from mathematical morphology commute with the lattice-based operations. Formally, given complete lattices $\mathbb{L}$ and $\mathbb{M}$, a dilation $\delta : \mathbb{L} \to \mathbb{M}$ and an erosion $\varepsilon : \mathbb{L} \to \mathbb{M}$ satisfy respectively the following identities for all $X \in \mathbb{L}$ [11]:

$$\delta(\sup X) = \sup_{\mathbf{x} \in X}\{\delta(\mathbf{x})\} \quad \text{and} \quad \varepsilon(\inf X) = \inf_{\mathbf{x} \in X}\{\varepsilon(\mathbf{x})\}. \tag{1}$$

Example 1. Let $\bar{\mathbb{R}} = \mathbb{R} \cup \{-\infty, +\infty\}$ denote the extended real-numbers. The cartesian product $\bar{\mathbb{R}}^n$ is a complete lattice with the partial ordering given by $\mathbf{x} \preceq \mathbf{y} \Leftrightarrow x_i \leq y_i, \forall i = 1 : n$. Given $\mathbf{a}, \mathbf{b} \in \mathbb{R}^n$, the operators $\delta_{\mathbf{a}}, \varepsilon_{\mathbf{b}} : \bar{\mathbb{R}}^n \to \bar{\mathbb{R}}$ given by

$$\delta_{\mathbf{a}}(\mathbf{x}) = \max_{j=1:n}\{a_j + x_j\} \quad \text{and} \quad \varepsilon_{\mathbf{b}}(\mathbf{x}) = \min_{j=1:n}\{b_j + x_j\}, \tag{2}$$

for all $\mathbf{x} \in \bar{\mathbb{R}}^n$, are respectively a dilation and an erosion [23].

The lattice-based elementary operations of mathematical morphology can be extended to more abstract sets using the concept of reduced orderings [8]. Let us briefly review the concepts of reduced dilation and reduced erosion as defined recently in [24]. Let $\mathbb{V}$ and $\mathbb{W}$ be arbitrary sets, not necessarily complete lattices. Also, let $\rho : \mathbb{V} \to \mathbb{L}$ and $\sigma : \mathbb{W} \to \mathbb{M}$ be surjective mappings, where $\mathbb{L}$ and $\mathbb{M}$ are both complete lattices. An operator $\delta^r : \mathbb{V} \to \mathbb{W}$ is a reduced dilation and an operator $\varepsilon^r : \mathbb{V} \to \mathbb{W}$ is a reduced erosion if there exist a dilation $\delta : \mathbb{L} \to \mathbb{M}$ and an erosion $\varepsilon : \mathbb{L} \to \mathbb{M}$ such that

$$\sigma\left(\delta^r(\mathbf{x})\right) = \delta\left(\rho(\mathbf{x})\right) \quad \text{and} \quad \sigma\left(\varepsilon^r(\mathbf{x})\right) = \varepsilon\left(\rho(\mathbf{x})\right), \quad \forall \mathbf{x} \in \mathbb{V}. \tag{3}$$

Reduced morphological operators, such as the reduced dilation and the reduced erosion defined above, have been effectively used for processing vector-valued images such as color and hyperspectral images [26,27].

3 Reduced Dilation-Erosion Perceptron

Morphological neural networks (MNNs) are neural networks whose induced local field of the neurons are given by an operation from mathematical morphology [23]. The morphological perceptrons, introduced by Ritter and Sussner in the middle 1990s for binary classification tasks, are one of the earliest MNNs [14]. In a few words, the morphological perceptrons are obtained by replacing the usual affine transformation $A(\mathbf{x}) = \langle \mathbf{w}, \mathbf{x} \rangle + b$, for all $\mathbf{x} \in \mathbb{R}^n$, by an elementary morphological operator in Rosenblatt's perceptron given by $y = f\big(A(\mathbf{x})\big) \equiv fA(\mathbf{x})$, where f denotes a hard limiter activation function [9]. Formally, the two morphological perceptrons are given by the equations $y = f\big(\delta_{\mathbf{a}}(\mathbf{x})\big) \equiv f\delta_{\mathbf{a}}(\mathbf{x})$ and $y = f\big(\varepsilon_{\mathbf{b}}(\mathbf{x})\big) \equiv f\varepsilon_{\mathbf{b}}(\mathbf{x})$, for all $\mathbf{x} \in \bar{\mathbb{R}}^n$. Specifically, the dilation-based perceptron and the erosion-based perceptron are given respectively by

$$y = f\left(\max_{j=1:n}\{a_j + x_j\}\right) \quad \text{or} \quad y = f\left(\min_{j=1:n}\{b_j + x_j\}\right), \quad \mathbf{x} \in \bar{\mathbb{R}}^n. \tag{4}$$

For simplicity, in this paper we shall consider the signal function as the hard limiter activation function of the morphological perceptrons, i.e., $f(x) = +1$ if $x \geq 0$ and $f(x) = -1$, otherwise. Note that a morphological perceptron model is given by either the composition $f\delta_{\mathbf{a}}$ or the composition $f\varepsilon_{\mathbf{b}}$, where $\delta_{\mathbf{a}} : \bar{\mathbb{R}}^n \to \bar{\mathbb{R}}$ and $\varepsilon_{\mathbf{b}} : \bar{\mathbb{R}}^n \to \bar{\mathbb{R}}$ denote respectively the dilation and the erosion given (1), for $\mathbf{a}, \mathbf{b} \in \mathbb{R}^n$. Because of the maximum operation, a dilation-based perceptron given by $y = f\delta_{\mathbf{a}}$ favors the positive class whose label is $+1$ [24]. Dually, an erosion-based perceptron defined by $y = f\varepsilon_{\mathbf{b}}$ favors the negative class, i.e., the class label -1, because of the minimum operation.

In [1], Araujo proposed a hybrid MNN called dilation-erosion perceptron (DEP). For binary classification tasks, a DEP computes the convex combination of a dilation and an erosion followed by a hard-limiter function. Intuitively, the DEP model allows for a balance between the two classes. In mathematical terms, given $\beta \in [0, 1]$, a DEP model is defined by $y = f(\tau(\mathbf{x})) \equiv f\tau(\mathbf{x})$, where

$$\tau(\mathbf{x}) = \beta\delta_{\mathbf{a}} + (1 - \beta)\varepsilon_{\mathbf{b}}, \quad \forall \mathbf{x} \in \mathbb{R}^n, \tag{5}$$

is the decision function of the DEP model.

Despite the successful application of the DEP model for times series prediction [1], the DEP classifier has a serious drawback: as a lattice-based model, the DEP classifier presupposes a partial ordering on the input space as well as on the set of classes. To overcome this problem, Valle recently proposed the so-called *reduced dilation-erosion perceptron* (r-DEP) using concepts from vector-valued mathematical morphology [24]. An r-DEP classifier is defined as follows: Let $\mathbb{V}$ be the input space and let $\mathbb{C} = \{c_1, c_2\}$ be the set of class labels of a binary classification task. The input space $\mathbb{V}$ is usually a subset of $\mathbb{R}^n$, but we may consider more abstract input spaces. Also, consider the complete lattice $\mathbb{L} = \bar{\mathbb{R}}^r$, equipped with the usual component-wise ordering, and let us denote $\mathbb{M} = \{-1, +1\}$. Given a one-to-one correspondence $\sigma : \mathbb{C} \to \mathbb{M}$ and a surjective mapping $\rho : \mathbb{V} \to \bar{\mathbb{R}}^r$, an r-DEP model is defined by the equation $y = \sigma^{-1}\big(f\big(\tau^r(\mathbf{x})\big)\big) \equiv \sigma^{-1} f \tau^r(\mathbf{x})$, where $\tau^r : \mathbb{V} \to \mathbb{R}$ is the decision function given by the following convex combination for $\beta \in [0, 1]$:

$$\tau^r(\mathbf{x}) = \beta\delta_{\mathbf{a}}\big(\rho(\mathbf{x})\big) + (1-\beta)\varepsilon_{\mathbf{b}}\big(\rho(\mathbf{x})\big), \quad \forall \mathbf{x} \in \mathbb{V}. \tag{6}$$

Finally, let us briefly comment on the training of an r-DEP classifier. Consider a training set $\mathcal{T} = \{(\mathbf{x}_i, d_i) : i = 1, \ldots, m\} \subseteq \mathbb{V} \times \mathbb{C}$. From (6), training an r-DEP is done simply training a DEP classifier using the transformed training data $\mathcal{T}^r = \{(\rho(\mathbf{x}_i), \sigma(d_i)) : i = 1, \ldots, m\} \subseteq \mathbb{L} \times \mathbb{M}$. At this point, we would like to recall that Charisopoulos and Maragos formulated the training of the morphological perceptrons as well as the hybrid DEP classifier as the solution of a convex-concave programming (CCP) problem [4]. Training r-DEP classifier using convex-concave procedures yielded encouraging results on classification tasks [24].

4 Linear Dilation-Erosion Perceptron

The main challenge in defining a successful r-DEP classifier is how to determine the surjective mapping $\rho : \mathbb{V} \to \mathbb{R}^r$. In this paper, we propose linear transformations as the mapping ρ, resulting in the so-called *linear dilation-erosion perceptron* (ℓ-DEP) classifier. Precisely, an ℓ-DEP model is defined as follows. Let the input space be the n-dimensional vector space $\mathbb{V} = \mathbb{R}^n$ and let $\mathbb{C} = \{c_1, c_2\}$ be the set of class labels of a binary classification task. Also, let $\mathbb{L}_1 = \bar{\mathbb{R}}^{r_1}$ and $\mathbb{L}_2 = \bar{\mathbb{R}}^{r_2}$ be complete lattices with the usual component-wise partial ordering. Given linear mappings $\rho_1 : \mathbb{V} \to \mathbb{L}_1$ and $\rho_2 : \mathbb{V} \to \mathbb{L}_2$, the application of linear mappings before the evaluation of an elementary morphological mapping yields the decision function $\tau^\ell : \mathbb{V} \to \mathbb{R}$ given by

$$\tau^\ell(\mathbf{x}) = \beta\delta_{\mathbf{a}}\big(\rho_1(\mathbf{x})\big) + (1-\beta)\varepsilon_{\mathbf{b}}\big(\rho_2(\mathbf{x})\big), \quad \mathbf{x} \in \mathbb{V}.$$

Finally, such as the previous model, the ℓ-DEP model is defined by the equation

$$y = \sigma^{-1}\big(f\big(\tau^\ell(\mathbf{x})\big)\big) \equiv \sigma^{-1} f \tau^\ell(\mathbf{x}), \tag{7}$$

where f is the sign function and $\sigma : \mathbb{C} \to \{-1, +1\}$ is a one-to-one mapping.

In contrast to the decision function τ^r of the r-DEP model given by (6), the decision function τ^ℓ of the ℓ-DEP is defined using two linear mappings ρ_1 and ρ_2. As a consequence, τ^ℓ is a continuous piece-wise linear function which is able to approximate any continuous-valued function from a compact on $\mathbb{R}^n$ to $\mathbb{R}$ [7,21]. Let us formalize this remark. The linear mappings ρ_1 and ρ_2 satisfies $\rho_1(\mathbf{x}) = R_1\mathbf{x}$ and $\rho_2(\mathbf{x}) = R_2\mathbf{x}$ for real-valued matrices $R_1 \in \mathbb{R}^{r_1 \times n}$ and $R_2 \in \mathbb{R}^{r_2 \times n}$. Defining $W = \beta R_1 \in \mathbb{R}^{r_1 \times n}$, $M = (\beta - 1)R_2 \in \mathbb{R}^{r_2 \times n}$, $\mathbf{c} = \beta\mathbf{a}$, and $\mathbf{d} = (\beta - 1)\mathbf{b}$, the decision function of the ℓ-DEP classifier can be alternatively written as

$$\begin{aligned}
\tau^\ell(\mathbf{x}) &= \beta\delta_{\mathbf{a}}(R_1\mathbf{x}) + (1-\beta)\varepsilon_{\mathbf{b}}(R_2\mathbf{x}) \\
&= \beta \max\{R_1\mathbf{x} + \mathbf{a}\} - (\beta - 1)\min\{R_2\mathbf{x} + \mathbf{b}\} \\
&= \max\{\beta R_1\mathbf{x} + \beta\mathbf{a}\} - \max\{(\beta-1)R_2\mathbf{x} + (\beta-1)\mathbf{b}\} \\
&= \max\{W\mathbf{x} + \mathbf{c}\} - \max\{M\mathbf{x} + \mathbf{d}\} \\
&= \max_{i=1:r_1}\{\mathbf{w}_i^T\mathbf{x} + c_i\} - \max_{j=1:r_2}\{\mathbf{m}_j^T\mathbf{x} + d_j\},
\end{aligned} \tag{8}$$

where $\mathbf{w}_i^T$ and $\mathbf{m}_i^T$ are rows of W and M, respectively. From the last identity, we can identify τ^ℓ with a piece-wise linear function [28]. Moreover, from Theorem 4.3 in [7], τ^ℓ is a universal approximator. Thus, an ℓ-DEP classifier can theoretically solve any binary classification problem. Let us now address the training of an ℓ-DEP classifier.

Given a training dataset $T = \{(\mathbf{x}_k, d_k) : k = 1, \ldots, m\} \subseteq \mathbb{V} \times \mathbb{C}$, define the sets $C^+ = \{\mathbf{x}_k : \sigma(d_k) = +1\}$ and $C^- = \{\mathbf{x}_k : \sigma(d_k) = -1\}$ of input samples. Inspired by the linear SVC and the concave-convex procedure developed by Charisopoulos and Maragos [4,10], the parameters W, $\mathbf{c}$, M and $\mathbf{d}$ of an ℓ-DEP classifier can be determined by minimizing the hinge loss function subject to the constraints

$$\tau^\ell(\mathbf{x}_k) \leq -1 + \xi_k, \text{ if } \mathbf{x}_k \in C^-, \quad \text{and} \quad \tau^\ell(\mathbf{x}_k) \geq 1 - \xi_k, \text{ if } \mathbf{x}_k \in C^+. \tag{9}$$

The terms -1 and $+1$ in the left-hand side of the inequalities imposes a margin of separation between the two classes. The slack variables ξ_k, for $k = 1, \ldots, m$, allows some classification errors in the training set. Note that, regardless of its class, the input sample $\mathbf{x}_k$ is well classified if $\xi_k \leq 0$. The hinge loss minimizes the positive values of ξ_k, for $\xi = 1, \ldots, m$.

Concluding, from (8), an ℓ-DEP classifier is trained by solving the following disciplined convex-concave programming (DCCP) problem [29]:

$$\begin{cases}
\underset{W,\mathbf{c},M,\mathbf{d},\xi}{\text{minimize}} & \sum_{k=1}^{m} \max(\xi_k, 0), \\
\text{s.t.} & \max_{i=1:r_1}(\mathbf{w}_i^T\mathbf{x}_k + c_i) + 1 \leq \max_{j=1:r_2}(\mathbf{m}_j^T\mathbf{x}_k + d_j) + \xi_k, \forall \mathbf{x}_k \in C^- \\
& \max_{i=1:r_1}(\mathbf{w}_i^T\mathbf{x}_k + c_i) + \xi_k \geq \max_{j=1:r_2}(\mathbf{m}_j^T\mathbf{x}_k + d_j) + 1, \ \forall \mathbf{x}_k \in C^+
\end{cases} \tag{10}$$

We would like to point out that we solved the disciplined convex-concave programming problem (10) using the algorithm proposed by Shen et al. [19]. Precisely, in our computational implementations, we solved (10) using the `CVXPY`

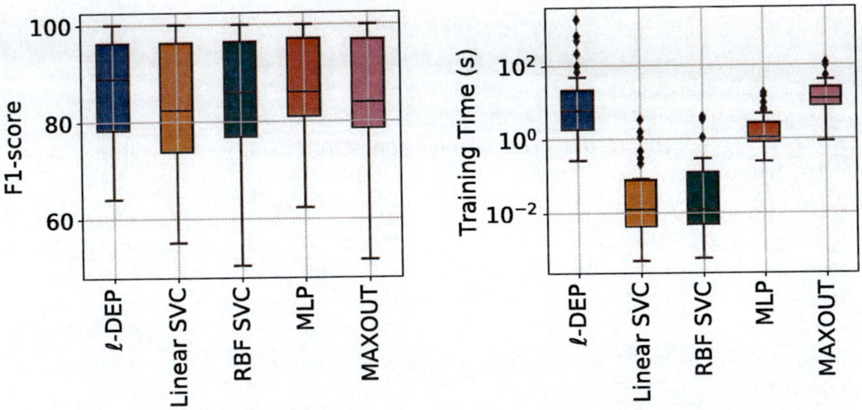

Fig. 1. Boxplot of the F1-score and time required to train the binary classifiers

package [5], which has an extension for solving DCCP problems, combined with the `MOSEK` solver [2].

5 Computational Experiments

Let us briefly evaluate the performance of the proposed ℓ-DEP classifier on several datasets from the OpenML repository available at https://www.openml.org/. For simplicity, we fixed the parameters $r_1 = r_2 = 10$ of the ℓ-DEP classifier for all datasets. Let us also compare the proposed classifier with a multi-layer perceptron (MLP), the maxout network, and the linear and the RBF SVCs [6,7,9]. We used `python's scikit-learn (sklearn)` implementations of the MLP and SVC classifiers with their default parameters [13]. The maxout network has been implemented using `tensorflow` with extra functionalities, that is, the maxout layer of `tensorflow-addons`. In our experiments, we used 10 locally affine regions in the maxout network for all datasets [7].

We would like to point out that we handled missing data using sklearn's `SimpleImputer()` command. Furthermore, we partitioned the data set into training and test sets using the sklearn's `StratifiedKFold()` command with $k = 5$. Finally, since some datasets are not balanced, we used the F1-score to measure the performance of the classifiers.

Table 1 contains the mean and the standard deviation of the F1-score obtained from the five classifiers using stratified 5-fold cross-validation. The boxplot shown in Fig. 1 summarizes the scores depicted on this table as well as the execution time for training the classifiers. Although the MPL yielded the largest average performance, the ℓ-DEP model produced the largest median performance. Furthermore, it is clear from the boxplot on the left of Fig. 1 that the ℓ-DEP is comparable to the other classifiers. In particular, ℓ-DEP achieved slightly higher performance than the maxout network. As to the training

Table 1. Average and standard deviation of the F1-Score.

	ℓ-DEP	Linear SVC	RBF SVC	MLP	MAXOUT
Accute inflammations	**100 ± 0.0**	**100 ± 0.0**	**100 ± 0.0**	**100 ± 0.0**	**100 ± 0.0**
Australian	83.3 ± 1.9	85.2 ± 3.6	**86.1 ± 2.9**	**86.1 ± 1.9**	85.5 ± 3.1
Banana	89.2 ± 3.1	55.2 ± 0.0	**90.4 ± 1.1**	90.1 ± 1.4	84.0 ± 10.4
Banknote	**100 ± 0.0**	98.5 ± 0.5	**100 ± 0.0**	**100 ± 0.0**	99.5 ± 1.0
Blood transfusion	77.7 ± 2.4	76.1 ± 0.5	76.6 ± 1.1	**79.8 ± 1.5**	79.5 ± 2.0
Breast cancer Wisconsin	96.7 ± 1.1	97.2 ± 0.7	**97.7 ± 0.5**	97.5 ± 1.4	97.2 ± 0.7
Chess	**99.3 ± 0.4**	96.8 ± 0.6	98.3 ± 0.7	99.2 ± 0.4	99.1 ± 0.1
Colic	81.5 ± 4.0	81.3 ± 6.6	**85.1 ± 3.7**	84.8 ± 3.2	81.6 ± 5.9
Credit approval	80.4 ± 3.3	85.5 ± 2.8	**85.7 ± 3.0**	84.9 ± 0.9	83.9 ± 2.8
Credit-g	68.4 ± 5.0	75.8 ± 3.0	**76.5 ± 4.4**	73.8 ± 3.7	68.5 ± 4.6
Cylinder bands	76.7 ± 5.9	69.4 ± 2.4	77.2 ± 2.8	**77.4 ± 5.8**	72.8 ± 4.7
Diabetes	70.8 ± 3.6	76.4 ± 3.4	75.2 ± 4.7	**76.7 ± 3.7**	75.1 ± 3.3
EEG-Eye-State	**89.3 ± 1.5**	61.7 ± 1.9	68.1 ± 5.0	85.6 ± 3.1	74.1 ± 7.5
Haberman	70.3 ± 3.8	73.2 ± 0.8	73.5 ± 2.6	**74.5 ± 2.4**	74.2 ± 1.8
Hill-Valley	**93.6 ± 1.3**	60.6 ± 1.9	50.5 ± 1.2	62.4 ± 2.7	51.7 ± 2.0
Ilpd	64.1 ± 4.8	71.4 ± 0.4	71.2 ± 0.3	**73.1 ± 3.0**	70.2 ± 3.1
Internet advertisements	93.7 ± 1.8	96.0 ± 0.7	96.6 ± 0.3	**97.2 ± 0.4**	97.0 ± 0.6
Ionosphere	88.9 ± 4.5	88.6 ± 2.7	**94.3 ± 2.3**	92.6 ± 3.1	91.7 ± 4.5
MOFN-3-7-10	**100 ± 0.0**	**100 ± 0.0**	**100 ± 0.0**	**100 ± 0.0**	**100 ± 0.0**
Monks-2	**88.0 ± 3.7**	65.7 ± 0.2	72.7 ± 2.3	82.4 ± 3.1	83.5 ± 3.2
Mushroom	**100 ± 0.0**	97.8 ± 0.9	**100 ± 0.0**	**100 ± 0.0**	**100 ± 0.0**
Phoneme	84.9 ± 1.2	77.3 ± 1.6	84.5 ± 1.5	**85.6 ± 1.4**	81.5 ± 3.5
Pishing websites	95.7 ± 0.6	90.5 ± 0.7	95.0 ± 0.9	**96.9 ± 0.5**	95.6 ± 0.6
Sick	97.0 ± 0.7	96.6 ± 1.0	96.3 ± 0.4	**97.1 ± 0.7**	96.8 ± 1.0
Sonar	77.4 ± 3.5	74.5 ± 3.9	**84.2 ± 4.2**	**84.2 ± 3.5**	80.8 ± 4.4
Spambase	92.8 ± 1.5	92.9 ± 0.5	93.3 ± 0.5	**94.5 ± 0.9**	93.5 ± 0.6
Steel plates fault	**100 ± 0.0**	**100 ± 0.0**	99.6 ± 0.5	99.9 ± 0.1	99.9 ± 0.1
Thoracic surgery	76.8 ± 3.1	**85.1 ± 0.0**	**85.1 ± 0.0**	83.6 ± 1.8	81.9 ± 2.7
Tic-Tac-Toe	92.6 ± 5.0	65.3 ± 0.2	88.8 ± 2.1	86.3 ± 3.2	**95.3 ± 4.5**
Titanic	**79.0 ± 2.3**	77.6 ± 2.0	77.8 ± 1.9	77.7 ± 1.9	77.6 ± 1.7
MEAN ± STD	86.9 ± 10.6	82.4 ± 13.3	86.0 ± 12.0	**87.5 ± 10.0**	85.7 ± 12.0
MEDIAN ± MAD	**88.9 ± 9.0**	82.4 ± 11.3	86.1 ± 9.4	86.1 ± 8.3	84.0 ± 9.8

execution time, the linear and the RBF SVCs are the fasted models. Despite its longer, the ℓ-DEP is not quite different from the MLP and maxout models in training time.

6 Concluding Remarks

In this paper, we introduced the linear dilation-erosion perceptron (ℓ-DEP) classifier which is given by a convex combination of the composition of linear transformations and the two elementary morphological operations. Specifically, given a one-to-one mapping σ from the set of class labels $\mathbb{C}$ to $\{+1,-1\}$, an ℓ-DEP classifier is defined by means of the equation $y = \sigma^{-1} f\tau^{\ell}(\mathbf{x})$, where the decision function $\tau^{\ell} : \mathbb{R}^n \to \mathbb{R}$ is given by (7). Alternatively, τ^{ℓ} can be expressed by means of (8) for matrices $W \in \mathbb{R}^{r_1 \times n}$ and $M \in \mathbb{R}^{r_2 \times n}$ and vectors $\mathbf{c} \in \mathbb{R}^{r_1}$ and $\mathbf{d} \in \mathbb{R}^{r_2}$. Except for the one-to-one mapping $\sigma : \mathbb{C} \to \{-1,+1\}$, which can be defined arbitrarily, the other parameters W, M, $\mathbf{c}$, and $\mathbf{d}$ are self-adjusted in the training of the ℓ-DEP model. Moreover, the classifier is trained by solving the disciplined convex-concave programming problem given by (10), which has been inspired by the works of Charisopoulos and Maragos [4]. From a theoretical point of view, the decision function τ^{ℓ} of the ℓ-DEP is a continuous piece-wise linear function [28]. As a consequence, an ℓ-DEP model can in principle solve any binary classification task [7]. Computational experiments with 30 binary classification problems revealed comparable performance of the proposed ℓ-DEP model with other classifiers from the literature, namely, the linear and RBF SVCs, MLP, and the maxout network.

In the future, we plan to investigate further in detail the concave-convex programming problem used to train an ℓ-DEP classifier. In particular, we intend to include regularization terms in the objective function to improve the generalization capability of the ℓ-DEP classifier.

References

1. de A. Araújo, R.: A class of hybrid morphological perceptrons with application in time series forecasting. Knowl.-Based Syst. **24**(4), 513–529 (2011). https://doi.org/10.1016/j.knosys.2011.01.001
2. ApS, M.: MOSEK Optimizer API for Python (2020). https://docs.mosek.com/9.2/pythonapi.pdf
3. Géron, A.: Hands–On Machine Learning with Scikit–Learn and TensorFlow: Concepts, Tools, and Techniques to Build Intelligent Systems. O Reilly, Sebastopol, California, USA., 2nd edn (2019)
4. Charisopoulos, V., Maragos, P.: Morphological perceptrons: geometry and training algorithms. In: International Symposium on Mathematical Morphology and Its Applications to Signal and Image Processing, pp. 3–15. Springer (2017). https://doi.org/0.1007/978-3-319-57240-61
5. Diamond, S., Boyd, S.: CVXPY: A Python-Embedded Modeling Language for Convex Optimization. - (2016). https://dl.acm.org/doi/10.5555/2946645.3007036

6. Goodfellow, I., Bengio, Y., Courville, A.: Deep Learning. MIT Press, Cambridge (2016)
7. Goodfellow, I., Warde-Farley, D., Mirza, M., Courville, A., Bengio, Y.: Maxout etworks. In: International Conference on Machine Learning, pp. 1319–1327 (2013)
8. Goutsias, J., Heijmans, H.J.A.M., Sivakumar, K.: Morphological operators for image sequences. Comput. Vis. Image Underst. **62**, 326–346 (1995)
9. Haykin, S.: Neural Networks: A Comprehensive Foundation. Prentice Hall, Upper Saddle River, NJ (1999)
10. Haykin, S.: Neural Networks and Learning Machines. Prentice-Hall, Upper Saddle River, NJ, 3rd edition edn. (2009)
11. Heijmans, H.J.A.M., Ronse, C.: The algebraic basis of mathematical morphology I. Dilations and erosions. Comput. Vis. Graph. Image Process. **50**(3), 245–295 (1990). https://doi.org/10.1016/0734-189X(90)90148-O
12. Heijmans, H.J.A.M.: Mathematical morphology: a modern approach in image processing based on algebra and geometry. SIAM Rev. **37**(1), 1–36 (1995)
13. Pedregosa, F., Varoquaux, G., Gramfort, A., Michel, V., Thirion, B., Grisel, O., Blondel, M., Prettenhofer, P., Weiss, R., Dubourg, V., Vanderplas, J., Passos, A., Cournapeau, D., Brucher, M., Perrot, M., Duchesnay, E.: Scikit-learn: machine learning in python. J. Mach. Learn. Res. **12**(null), 2825–2830 (2011)
14. Ritter, G.X., Sussner, P.: An introduction to morphological neural networks. In: Proceedings of the 13th International Conference on Pattern Recognition, Vienna, Austria, pp. 709–717 (1996)
15. Ritter, G.X., Sussner, P.: Morphological erceptrons. In: ISAS 1997, Intelligent Systems and Semiotics. Gaithersburg, Maryland (1997)
16. Ritter, G.X., Wilson, J.N., Davidson, J.L.: Image algebra: an overview. Comput. Vis. Graph. Image Process. **49**(3), 297–331 (1990)
17. Ritter, G.X., Urcid, G.: Lattice algebra approach to single-neuron computation. IEEE Trans. Neural Netw. **14**(2), 282–295 (2003)
18. Rosenblatt, F.: The perceptron: a probabilistic model for information storage and organization in the brain. Psychol. Rev. **65**, 386–408 (1958)
19. Shen, X., Diamond, S., Gu, Y., Boyd, S.: Disciplined convex-concave programming. In: 2016 IEEE 55th Conference on Decision and Control (CDC), pp. 1009–1014 (2016). https://doi.org/10.1109/CDC.2016.7798400
20. Soille, P.: Morphological Image Analysis. Springer Verlag, Berlin (1999)
21. Stone, M.H.: The generalized Weierstrass approximation theorem. Math.. Mag. **21**(4), 167–184 (1948). https://doi.org/10.2307/3029750, https://www.jstor.org/stable/3029750, publisher: Mathematical Association of America
22. Sussner, P.: Morphological perceptron learning. In: IEEE International Symposium on Intelligent Control - Proceedings, pp. 477–482. IEEE (1998). https://doi.org/10.1109/isic.1998.713708
23. Sussner, P., Esmi, E.L.: Morphological perceptrons with competitive learning: lattice-theoretical framework and constructive learning algorithm. Inform. Sci. **181**(10), 1929–1950 (2011). https://doi.org/10.1016/j.ins.2010.03.016
24. Valle, M.E.: Reduced dilation-erosion perceptron for binary classification. Mathematics **8**(4), 512 (2020). https://doi.org/10.3390/math8040512
25. Vapnik, V.N.: Statistical Learning Theory. John Wiley and Sons, New York, NY, USA (1998)
26. Velasco-Forero, S., Angulo, J.: Supervised ordering in R^p: application to morphological processing of hyperspectral images. IEEE Trans. Image Process. **20**(11), 3301–3308 (2011). https://doi.org/10.1109/TIP.2011.2144611

27. Velasco-Forero, S., Angulo, J.: Vector ordering and multispectral morphological image processing. In: Celebi, M.E., Smolka, B. (eds.) Advances in Low-Level Color Image Processing, pp. 223–239. Springer Netherlands, Dordrecht (2014). https://doi.org/10.1007/978-94-007-7584-8_7
28. Wang, S.: General constructive representations for continuous piecewise-linear functions. IEEE Trans. Circ. Syst. Regul. Pap. **51**(9), 1889–1896 (2004). https://doi.org/10.1109/TCSI.2004.834521
29. Yuille, A.L., Rangarajan, A.: The concave-convex procedure. Neural Computat. **15**(4), 915–936 (2003). https://doi.org/10.1162/08997660360581958

A Performance Evaluation of Machine Learning Models on Human Activity Identification (HAI)

Taki Hasan Rafi(✉) and Faisal Farhan

Ahsanullah University of Science and Technology, 141-142 Love Road, Tejgaon, Dhaka 1208, Bangladesh

Abstract. Human activity identification (HAI) is presently a promising field for artificial intelligence researchers. HAI can be applied in various areas in our daily life such as surveillance, health care, etc. There are several machine learning and deep learning techniques that are applied to recognize the human activity. Various multi-dimensional sensors from smartphones, smart tablets, smartwatches can record different types of human activity. Some extensive sensors in smartphones such as accelerometer, gyroscope, microphone, GPS, and camera which can respond to record human gestures. These gestures can be sitting, lying, standing, walking, etc. In this study, the authors analyze the performance of traditional machine learning techniques as well as 1D-CNN architecture to recognize human activity using the UCI human activity identification dataset. PCA has been enforced to reduce the dimension and feature selection to get better results. 1D-CNN performs better in this study with an accuracy of 97.30%. Where Logistic regression achieves 96.00%, Linear Support Vector Machine achieves 93.71%, Kernel-Support Vector Machine achieves 94.85%, Random Forest achieves 90.59% and Decision Tree achieves 84.46% accuracy. All the accuracies are considered by taking the average of individual class accuracy. However, the 1D-CNN model is proposed by the authors to implement in less computational powered devices such as smartphones, smartwatches, etc.

Keywords: Human activity identification · HCI · Machine learning · 1D-CNN

1 Introduction

Human activity identification (HAI) is an emerging area in past few years of advancement of technology. Due to getting enormous applications in various fields of health care, surveillance, automated smart homes, activity recognition is now an attractive area for artificial intelligence researchers. There are some ongoing applications such as human security, administrative purpose, human behavior monitoring, elder care of human activity recognition [2]. It is also used in hospitals, shopping mails, subway stations to tackle any unwanted occurrence [3].

A. Abraham et al. (Eds.): SoCPaR 2020, AISC 1383, pp. 256–265, 2021.
https://doi.org/10.1007/978-3-030-73689-7_25

Another most widely used application of HAI is Ambient Assisted Living (AAL), which enables older people to live securely [4].

Machine learning models are now very useful to recognize human activity in a more efficient approach. Some daily life sensors such as smart phone sensors, environmental sensors, CCTV video sensors, wearable sensors are utilized to record human actions [5]. These activity data are collected from wearable sensors, images, video frames. Nowadays smart phone and smart watches are extensively used in our daily life. These smart phones and watches have several sensors such as accelerometer, gyroscope, microphone, GPS and camera. These sensors can quantify our movements, gestures. But sometimes these sensors can record some undesirable signals which causes noise in the data generation [12].

In this paper, authors have compared several machine learning techniques such as Logistic Regression, Linear and Kernel Support Vector Machine, Random Forest, Decision Trees and 1D-CNN in the human activity identification dataset. The classification of human activity such as sitting, standing, walking, lying, walking downstairs and walking upstairs. Where authors have found that 1D-CNN architecture has slightly better accuracy than other machine learning techniques. The models are evaluated in terms of the average of prediction accuracy, precision, f1-score and recall.

Key contributions of this study are,

- **Authors have enforced various machine learning models to determine an efficient algorithm for human activity identification.**
- **Data pre-processing and var-PCA based selection of features are exploited to validate a satisfactory outcome.**

The following paper is oriented as, Sect. 2 represents the related works of this research, Sect. 3 represents the methods and brief introduction of machine learning techniques. Section 4 represents the experimental results and our recommendations.

2 Related Works

In recent years, several works have been done on human activity identification. To train the machine learning models holistically, we need a large-scale dataset. Which is a real challenge in ambient assisted living field.

In the work of [1], they used publicly available dataset of human activity identification and implemented CNN model VGG-16 and a baseline model random forest. They achieved 92.71% accuracy. In [2], they utilized a feed-forward multi-layer perceptron to identify the human activity in video surveillance data. They achieved 94% identification rate overall. In [3], they used KTH dataset in their human activity identification task. They extracted the features by using Kalman Filter. Naive Bayes and CNN have been tested in their experiment, CNN achieved 92%, 96% and 100% accuracy, which is higher than Naive Bayes. In the work of [4], they used LSTM model to identification the human activity. However, they compared their model with traditional machine learning models such as

Logistic Regression, SVM, RF, KNN, ANN, where their proposed LSTM model has 92% accuracy. In [5], authors have utilized the exercise activity identification dataset using surface electromyography sensor. They have implemented Decision Tree and multi-layer perceptron models. Where multi-layer perceptron has achieved 99.97% accuracy. In the work of [6], they applied data parallelism technique to develop an user friendly Ambient Assisted Living (AAL) by using LSTM model. In [7], authors have utilized bidirectional LSTM (BiLSTM) model on top of PCA technique. Their proposed model has achieved an accuracy of 97.64%. They also utilized SVM model. In the work of [8], they have worked with handling missing data. However, they have tested various machine learning techniques to predict HAR. They achieved 89.4752% accuracy in K-NN model. In [9], authors have worked with Weizmann vision-based activity identification dataset. However, they utilized transfer learning CNN model VGG-16 and achieved 96.95% accuracy. In [10], they have utilized UCI dataset of human activity. They implemented RNN and achieved the highest accuracy of 97.55%. In [11], authors have worked with missing data. They used HASC dataset and Single Chest-Mounted accelerometer dataset in their experiment. They have found that 8% of missing data can reduce the prediction accuracy drastically.

Considering all the related works, we have enforced several machine learning and 1D-CNN techniques to estimate the performance outcome of each individual model. However, to enhance the performance of machine learning models, feature selection is employed.

3 Methods

In this study, authors have utilized various state-of-art machine learning techniques in their experiment. Some extensively used techniques such as Logistic Regression, Support Vector Machine, Random Forest, Decision Tree and deep convolutional neural network to compare the prediction accuracy. Prior data pre-processing and feature selection using PCA have been done to conduct this experiment. Work flow diagram of this experiment is shwon in Fig. 1.

3.1 Dataset Description

In this experiment, a publicly available UCI dataset [13,23] has been utilized. This dataset has been generated by two smart phone sensors such as accelerometer and gyroscope. To generated the dataset, it has been carried out 30 volunteers with an age range of 19–48. Six physical activities such as sitting, standing, laying, walking, walking downstairs and walking upstairs. To validate the dataset, 70% volunteers was selected and 30% volunteers was selected to generate test set. The whole dataset generation was conducted 50 Hz constant speed in 3-axial angular velocity. Table 1 has shown the no. of Samples in the dataset for each class.

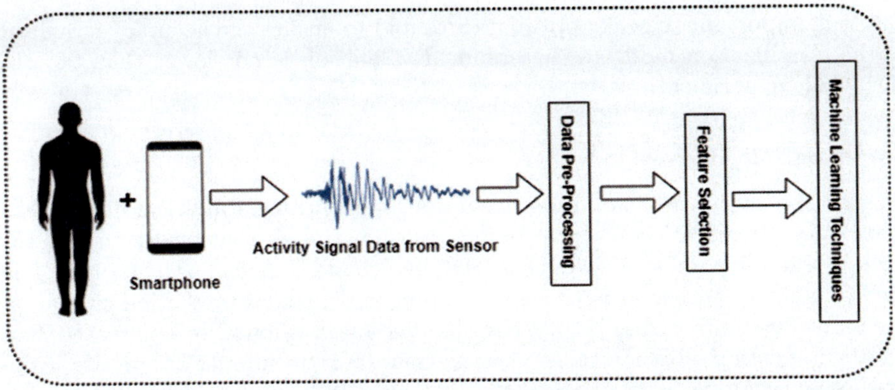

Fig. 1. Workflow diagram of our experiment.

Table 1. No. of samples in the dataset for each class

Activities	No. of samples
Sitting	1777
Standing	1906
Lying	1944
Walking	1722
Walking downstairs	1544
Walking upstairs	1406
Total	**10299**

3.2 Data Pre-processing

Data pre-processing is a holostic process of preparing data to fit in the machine learning models. To get more intuitive and readable data, it is important to pre-process the dataset before deploy into the machine learning models. There are some steps to conduct the pre-processing. In this experiment, authors have checked the missing data, duplicate number of data and imbalanced label of the dataset. And they made the missing data and duplicate data to null.

3.3 Feature Selection

To reduce the feature dimensions, authors have utilized principle component analysis (PCA). PCA is utilized for selecting key features. Authors have used an extensively used technique, variance selection. This Var-PCA technique allows to select the features which are higher than threshold value. This threshold value can be obtained by some machine learning techniques. Low variance features are automatically eliminated by its less importance to identify target features. On the other words, higher variance allows to identify the target sample more easily.

But it is important to select optimal threshold to ensure the accuracy standard of the classification models. The assumed validation is not lower than 96% in this whole experiment.

3.4 Logistic Regression

Logistic Regression is a well recognized machine learning statistical model that is used for classification of data. It focuses on variables that are binary, hence logistic regression only retrieves those data which are classified as 0 or 1 [16]. The idea of logistic regression is to take an input and make a prediction similar to the real valued input class [15]. When the prediction is found to be greater than 0.5 then logistic regression takes yield as class 0 for an input class or else if the prediction results greater or equal than 0.5 then it takes 1 as output class [15]. To predict the probability of output class logistic regression can be expressed with the following equations:

$$P(b = 1|a) = \frac{1}{1 + exp^{-}\theta^{T}a} \equiv \sigma(\theta^{T}a) \tag{1}$$

$$P(b = 0|a) = 1 - P(b = 1|a) \tag{2}$$

$$\sigma(t) = \frac{1}{(1 + e^{-}t)} \tag{3}$$

Equation (3) is known as a sigmoid function which is used to keep the range of between 0 to 1. Then an appropriate value of is estimated so that if a refers to class '1' then probability $P(b = 1|a)$ is high or else its is low if a refers to class '0'.

3.5 Support Vector Machine

Support Vector Machine (SVM) has gained much recognition due to its capability to give accurate results while dealing with large predictors [18]. It is an extensive statistical approach, which is increasingly applied to various applications [14]. Many research studies that requires regression, forecasting and detection uses SVM algorithm. Linear SVM is widely used as it requires very less computational effort for using Non-linear boundaries [17]. Besides it gives a combative performance with other methods in case of predictive analysis. Approximated function of Linear SVM can be expressed as:

$$f = w\phi(x) + b \tag{4}$$

Here, w and b refers to weight vector & threshold w (x) refers to higher dimensional feature space that comes from input vector y. Later using Lagrange multipliers and from optimality requirements the approximated functionality is expressed as a definite form:

$$f(y, x_i, x_i^*) = \sum_{i=1}^{n} (x_i - x_i^*) M(y, y_i) + b \tag{5}$$

3.6 Decision Tree

Decision tree is considered as an expensively used machine learning algorithms for solving problems in the area of machine intelligence. This algorithm solves complex problems and represents them graphically both in human readable and computer readable form [21]. The process of this algorithm is to split a single data set into smaller data sets. The division of this process continues until a small data set is achieved that retains data points which fall under one label. Complex classification problems are thus done with ease using the decision tree algorithm procedure.

3.7 Random Forest

Random Forest Classifier developed by Breiman [20] is a widely used Machine Learning algorithm that generates a set of decision trees imported from a randomly chosen subset of a training set. This classifier gives high prediction accuracy by merging multiple decision trees into a single tree [19]. The performance of RF classifier is evaluated based on four parameters namely Sensitivity, Specificity, Accuracy and Positive predictive value. The only drawback of this classifier is that reducing values of selected features also reduces strength and classifier of the training set.

3.8 Convolutional Neural Network

Several deep learning models have been developed over the years that are being widely used by researchers in the large area of image segmentation, computer-vision, speech identification, machine language processing and other related fields. Convolutional neural network gained much recognition among the deep learning models as complexity of network and required training parameters are greatly reduced in this model. Besides, this model is very easy for training and optimizing any dataset [22]. Recently, CNN models are also implemented in Human computer interface field for face recognition, object detection, image processing and many more. The idea of CNN model is to take the input data and to process and classify it into certain categories. Each dataset is passed through some convolution layers consist of filters (kernels) accommodated with fully connected layers and pooling. The training data undergoes down-testing by pooling layers to reduce the limits inside that incitation. Later after processing through all the networks full convolution is obtained. We have utilized 1D-CNN in this study.

4 Experimental Results and Discussions

In this experiment, the authors have utilized a publicly available smart phone sensor based activity identification dataset. The dataset split into 70% for training and 30% for testing. Machine learing technique are applied to observe the performance in the human activity recognition. Logistic Regression, Support Vector Machine, Random Forest, Decision Tree and 1D-CNN model have been applied in this study. This research has been done in WEKA (Waikato Environment for Knowledge Analysis) environment. The dataset has 10299 records where 7210 records are used to train the models and 3089 records are used to test. However, 1D-CNN model has only 25% dropout, training batch size is 128, training epochs are 50 in this experiment. Which has the best average result among all models. We have evaluated the models in terms of four broadly performance metrics such as accuracy, precision, recall and f1-score. Where accuracy is the correctly identified points out of all points, precision is the genuine positive expectation partitioned by the amount of false positive and genuine positive, recall is the genuine positive forecast isolated by the amount of genuine positive and false negative and f1-score is the convergence of the relation between precision and recall. Table 1 has shows the overall performance of the machine learning and 1D-CNN models. However, the shown performance evaluation is the average performance in each class such as walking, standing, lying, sitting, walking downstairs, walking upstairs. Contrastingly, Fig. 2 has shown the chart-based performance evaluation of each model. It helps to visulize the performance measure in a better way.

In the Kernel-SVM, the value of γ is set to 0.125–1.00 in the interval of 0.125. In logistic regression model, L1 and L2 regularization have been used to validate better outcome. In Random Forest and Decision Tree, cross validation and hyper-parameter tuning have been done to get best possible result. Because before cross validation and hyper-parameter tuning, both models had less than 80% accuracy. We can see that after hyper-parameter tuning and cross-validation the performance has been improved for both Random Forest and Decision Tree.

However, we can see from Table 2, 1D-CNN has on average 97.30% accuracy. Which is higher than other models. Where as Kernel-SVM model has 94.85% accuracy on average, Logistic Regression has 96.00% accuracy on average, Linear-SVM has 93.71% accuracy on average. After hyper-parameter tuning and cross-validation Random Forest has 90.59% accuracy on average. But Decision Tree has 84.46% has accuracy on average, which is lower than other models. Figure 2 has illustrated the performance chart.

Table 2. Performance evaluation of every individual model (%)

Models	Accuracy	Precision	Recall	F1-Score
Decision Tree	84.46	84.60	84.80	89.80
Kernel-SVM	94.85	94.00	94.20	93.00
Linear-SVM	93.71	92.00	93.40	94.50
Logistic Regression	96.00	95.80	97.80	97.60
Random Forest	90.59	90.10	92.70	89.80
1D-CNN	**97.30**	**98.00**	**98.00**	**98.50**

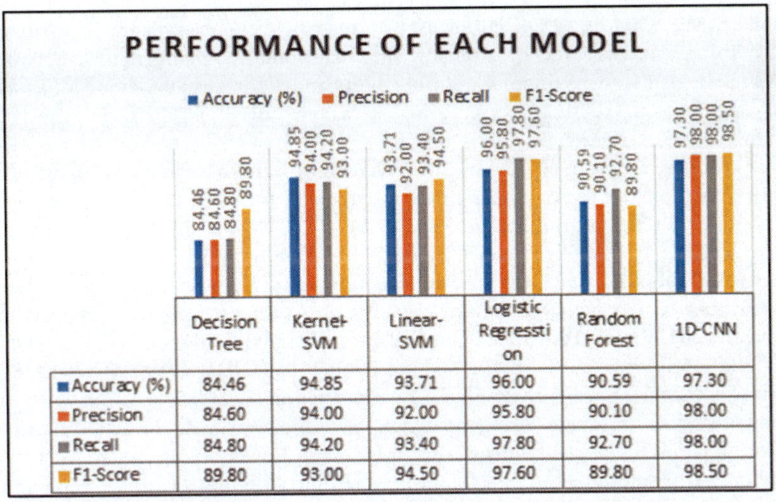

	Decision Tree	Kernel-SVM	Linear-SVM	Logistic Regression	Random Forest	1D-CNN
Accuracy (%)	84.46	94.85	93.71	96.00	90.59	97.30
Precision	84.60	94.00	92.00	95.80	90.10	98.00
Recall	84.80	94.20	93.40	97.80	92.70	98.00
F1-Score	89.80	93.00	94.50	97.60	89.80	98.50

Fig. 2. Performance chart of the machine learning models

5 Conclusion

Human activity identification is a challenging task for low power devices. Activity identification has several applications in various fields. In this study, authors have evaluated several machine learning techniques and convolutional neural network model in activity recognition task. We have trained our model with around 7210 samples out of 10299 samples on the dataset. The main intention of this examination is to determine a robust machine learning technique for human activity identification. In the validation dataset, we have measured average performance of each class. However, the average of each performance has been taken to measure final evaluation for recommendations. In the experiment, 1D-CNN has the best average accuracy of 97.30% with 25% dropout in the model. On the other hand, Logistic Regression, Kernal-SVM, Linear-SVM, Random Forest have a decent average accuracy of 96.00%, 94.85%, 93.71%, 90.59% respectively. But Decision Tree model has 84.46% accuracy, which is slightly deteriorating in accuracy than other models. Although, Random Forest and Decision Tree had hyper-parameter tuning and cross validation to avoid performance decrease. So we recommend 1D-CNN model for further utilization.

References

1. Lee, S.M., Yoon, S.M., Cho, H.: Human activity recognition from accelerometer data using convolutional Neural Network. In: 2017 IEEE International Conference on Big Data and Smart Computing (BigComp), pp. 131–134. Jeju (2017) https://doi.org/10.1109/BIGCOMP.2017.7881728.

2. Babiker, M., Khalifa, O.O., Htike, K.K., Hassan, A., Zaharadeen, M.: Automated daily human activity recognition for video surveillance using neural network. In: 2017 IEEE 4th International Conference on Smart Instrumentation, Measurement and Application (ICSIMA), pp. 1–5. Putrajaya (2017). https://doi.org/10.1109/ICSIMA.2017.8312024.
3. Liu, C., Ying, J., Han, F., Ruan, M.: Abnormal human activity recognition using bayes classifier and convolutional neural network. In: 2018 IEEE 3rd International Conference on Signal and Image Processing (ICSIP), pp. 33–37. Shenzhen (2018). https://doi.org/10.1109/SIPROCESS.2018.8600483.
4. Patel, A.D., Shah, J.H.: Performance Analysis of Supervised Machine Learning Algorithms to Recognize Human Activity in Ambient Assisted Living Environment. In: 2019 IEEE 16th India Council International Conference (INDICON), pp. 1–4. Rajkot, India (2019). https://doi.org/10.1109/INDICON47234.2019.9030353
5. Mekruksavanich, S., Jitpattanakul, A.: Exercise activity recognition with surface electromyography sensor using machine learning approach. In: 2020 Joint International Conference on Digital Arts, Media and Technology with ECTI Northern Section Conference on Electrical, Electronics, Computer and Telecommunications Engineering (ECTI DAMT & NCON), pp. 75–78. Pattaya, Thailand (2020). https://doi.org/10.1109/ECTIDAMTNCON48261.2020.9090711.
6. Nguyen, T.D.T., et al.: Performance analysis of data parallelism technique in machine learning for human activity recognition using LSTM. In: 2019 IEEE International Conference on Cloud Computing Technology and Science (CloudCom), pp. 387–391. Sydney, Australia (2019). https://doi.org/10.1109/CloudCom.2019.00066.
7. Aljarrah, A.A., Ali, A.H.: Human activity recognition using PCA and BiLSTM recurrent neural networks. In: 2019 2nd International Conference on Engineering Technology and its Applications (IICETA), pp. 156–160. Al-Najef, Iraq (2019). https://doi.org/10.1109/IICETA47481.2019.9012979
8. Prabowo, O.M., Mutijarsa, K., Supangkat, S.H.: Missing data handling using machine learning for human activity recognition on mobile device. In: 2016 International Conference on ICT For Smart Society (ICISS), pp. 59–62. Surabaya (2016). https://doi.org/10.1109/ICTSS.2016.7792849
9. Deep, S., Zheng, X.: Leveraging CNN and transfer learning for vision-based human activity recognition. In: 29th International Telecommunication Networks and Applications Conference (ITNAC), pp. 1–4. Auckland, New Zealand (2019). https://doi.org/10.1109/ITNAC46935.2019.9078016
10. Bhattacharjee, S., Kishore, S., Swetapadma, A.: A comparative study of supervised learning techniques for human activity monitoring using smart sensors. In: 2018 Second International Conference on Advances in Electronics, Computers and Communications (ICAECC), pp. 1–4. Bangalore (2018). https://doi.org/10.1109/ICAECC.2018.8479436.
11. Hossain, T., Inoue, S.: A comparative study on missing data handling using machine learning for human activity recognition. In: 2019 Joint 8th International Conference on Informatics, Electronics & Vision (ICIEV) and 2019 3rd International Conference on Imaging, Vision & Pattern Recognition (icIVPR), Spokane, WA, USA, pp. 124–129 (2019). https://doi.org/10.1109/ICIEV.2019.8858520.
12. Singh, D., et al.: Human activity recognition using recurrent neural networks. In: Holzinger A., Kieseberg P., Tjoa A., Weippl E. (eds.) Machine Learning and Knowledge Extraction, CD-MAKE 2017, Lecture Notes in Computer Science, vol. 10410, Springer, Cham (2017). https://doi.org/10.1007/978-3-319-66808-6_18

13. Anguita, D., Ghio, A., Oneto, L., Parra, X., Reyes-Ortiz, J.L.: A public domain dataset for human activity recognition using smartphones. In: 21st European Symposium on Artificial Neural Networks, Computational Intelligence and Machine Learning, ESANN 2013, pp. 24–26. Bruges, Belgium, April 2013
14. Rafi, T.H.: A comparative analysis of hyper-parameter tuned supervised machine learning algorithms on breast cancer prediction. Ann. Eng. **1**(1), 0003 (2020)
15. Jain, H., Khunteta, A., Srivastava, S.: Churn prediction in telecommunication using logistic regression and logit boost. Procedia Comput. Sci. **167**, 101–112 (2020). https://doi.org/10.1016/j.procs.2020.03.187
16. Zhu, C., Idemudia, C.U., Feng, W.: Improved logistic regression model for diabetes prediction by integrating PCA and K-means techniques. Inf. Med. Unlocked **100179**, (2019). https://doi.org/10.1016/j.imu.2019.100179
17. Subasi, A., Ismail Gursoy, M.: EEG signal classification using PCA, ICA, LDA and support vector machines. Expert Syst. Appl. **37**(12), 8659–8666 (2010). https://doi.org/10.1016/j.eswa.2010.06.065
18. Fan, J., Wang, X., Wu, L., Zhou, H., Zhang, F., Yu, X., Xiang, Y.: Comparison of support vector machine and extreme gradient boosting for predicting daily global solar radiation using temperature and precipitation in humid subtropical climates: a case study in China. Energy Convers. Manage. **164**, 102–111 (2018). https://doi.org/10.1016/j.enconman.2018.02.087
19. Subudhi, A., Dash, M., Sabut, S.: Automated segmentation and classification of brain stroke using expectation-maximization and random forest classifier. Biocybernetics Biomed. Eng. (2019). https://doi.org/10.1016/j.bbe.2019.04.004
20. Zhang, T., Chen, W., Li, M.: AR based quadratic feature extraction in the VMD domain for the automated seizure detection of EEG using random forest classifier. Biomed. Signal Process. Control **31**, 550–559 (2017). https://doi.org/10.1016/j.bspc.2016.10.001
21. Trabelsi, A., Elouedi, Z., Lefevre, E.: Decision tree classifiers for evidential attribute values and class labels. Fuzzy Sets Syst. (2018). https://doi.org/10.1016/j.fss.2018.11.006
22. Chen, J.X., Zhang, P.W., Mao, Z.J., Huang, Y.F., Jiang, D.M., Zhang, Y.N.: Accurate EEG-based emotion recognition on combined features using deep convolutional neural networks. IEEE Access **7**, 44317–44328 (2019). https://doi.org/10.1109/ACCESS.2019.2908285
23. Chowdhury, A.I., Ashraf, M., Islam, A., Ahmed, E., Jaman, M.S., Rahman, M.M.: hActNET: an improved neural network based method in recognizing human activities. In: 2020 4th International Symposium on Multidisciplinary Studies and Innovative Technologies (ISMSIT), Istanbul, Turkey, pp. 1–6 (2020). https://doi.org/10.1109/ISMSIT50672.2020.9254992

Wavelet Filtering for Limited-Angle and Low-Dose Computed Tomographic Reconstruction

Thavavel Vaiyapuri(✉)

College of Computer Engineering and Sciences,
Prince Sattam Bin Abdulaziz University, AIKharj, Saudi Arabia
t.thangam@psau.edu.sa

Abstract. Computed tomography (CT) has gained extensive application in nondestructive testing and medical diagnosis. Most of the practical application of CT imaging applications demands the development of more efficient reconstruction techniques for low-dose and limited-angle scanning to avoid the risk of ionizing radiation associated with CT. The widely used reconstruction algorithms, filtered back-projection (FBP), fails to produce satisfactory results because of lacking the projection data. Statistical techniques have been recently introduced for low-dose CT but the success of these algorithms rely on the assumption of noise characteristics and makes clinical application difficult. Aiming to improve the quality of CT reconstruction, this work investigates the incorporation of an alternate filter for FBP reconstruction. To this end, a multi-resolution 1-D Gabor filter is employed due to its band-pass capability and texture segmentation nature. The proposed filter was developed as reusable software components using the object oriented and platform independent features of Java language. The feasibility and validity of Gabor filterfor FBP was evaluated for its potential on projection simulated from Shepp-Logan phantom image. The results demonstrate that the Gabor filter outperformed its counter parts in artifacts reduction, noise suppression, and structure preservation.

Keywords: Tomographic reconstruction · Limited-angle imaging low-dose imaging · Backprojection · Filtered backprojection · Ram-Lak filters · Shepp-Logan filter · Gabor filter

1 Introduction

Computed tomography (CT) is one of the most widely utilized powerful non-invasive medical imaging technique for producing cross sectional images of body organs and tissues from flat X-rays [1,2]. The advent of CT has revolutionized diagnostic radiology. However, the radiation exposure from CT presents significant health hazard. Therefore, the radiation burden is major concern and has come under increased scrutiny in recent years [3,4]. The application of CT for

A. Abraham et al. (Eds.): SoCPaR 2020, AISC 1383, pp. 266–277, 2021.
https://doi.org/10.1007/978-3-030-73689-7_26

screening requires an additional dimension to the controversy. The cost-effective and simplest way to diminish the radiation dose is by reducing current (mAs) in the X-ray tube or by decreasing the amount of X-rays administrated during the data acquisition procedure [5]. But under this condition, the resultant images are severely degraded by excessive X-ray quantum noise which affects and might mislead the diagnostic process. As a result, an efficient noise reduction technique plays a crucial role as a prerequisite for diminishing radiation dose and enabling the low-dose CT in diagnostic process.

Large number of studies has been carried out on this subject and considerable research work has been devoted to formulate alternative reconstruction methods such as algebraic and statistical methods [3,6–11]. Yet, FBP is considered as an industry standard method in clinical context. The reason of its success is three-fold: simple to implement, direct and fast. Nevertheless, for the past decades very few attempts have been made to improve the logic design of FBP but no attempts are made in aspect of filter designing [12–15]. Notably, the spatial resolution of images reconstructed using FBP are solely dependent on the filter function employed during the reconstruction process. Further, the filtering step in FBP plays a key role in circumventing the excessive deblurring, smoothing, and in suppressing aliasing. The most popular filter response functions used for 2D reconstruction of CT images are the Shepp-Logan and Ram-Lak [16–18]. Unfortunately, these conventional filters used for FBP are not adaptable for handling noisy projection data and therefore results in poor reconstructions with severe artifacts. This emphasizes the need for an alternate and efficient filter for FBP that can give best tradeoff between noise suppression and image quality. In this context, the present work investigates the application of 1-D Gabor filter with band-pass characteristic and optimized with localization properties in both spatial and spatial frequency domain for FBP reconstruction.

2 Methodology

2.1 Tomographic Reconstruction

Most of the biomedical imaging modalities acquire data as a set of projections. Consequently, the problem of image reconstruction from acquired projection data arise not only in CT but also in other medical imaging modalities such as EPR, SPECT, PET and also in seismic imaging, radio astronomy and electron microscopy. The process of collecting parallel line of projections at several angles θ around the 2D image $f(x, y)$ is known as Radon transform (RT) or sinogram [19,20]. The continuous version of RT is as follows,

$$P(r, \theta) = R_\theta(f) = \int\int f(x, y)\delta(r - xcos\theta - ysin\theta)dxdy \tag{1}$$

Here r is a spatial coordinate which is represented on x-y plane as $r = xcos(\theta) + ysin(\theta)$. Accordingly, $P(r, \theta)$, the projection data exists for different values of r and varies with image size at different angles θ in the interval of $[0, \pi]$. Now, the process of reconstructing 2D image by combining the set of parallel line of projections collected at different angles is called as image reconstruction or inverse RT. The most widely used FBP comprise two steps [21,22]. The first step, filtering process in frequency domain is defined as follows

$$Q_\theta(\omega) = \widehat{R_\theta} f(\omega)|\omega|W(\omega) \tag{2}$$

Here $\widehat{R_\theta} f(\omega)$ denotes the Fourier transform (FT) of $R_\theta(f)$ and window function is denoted by $W(\omega)$. The second step of FBP, the back-projection process is accomplished as follows,

$$f(x, y) = \int Q_\theta(xcos\theta + ysin\theta)d\theta \tag{3}$$

Owing to the discontinuity of the transfer function at D.C, the impulse response of ramp filter is not compact. As a result of this non-compact characteristics, the application of ramp filter in FBP requires projections from all angles to reconstruct even a small part of an image. Added to, the application of ramp filter to ideal projection data removes 1/r blur effectively. However under the low-dose imaging procedure i.e. when noise is present, the ramp filter amplifies the noise and the reconstructed image are noise degraded. Therefore, in real practice, the RamLak filter which is window approximation of ideal ramp filter is considered.

2.2 Reconstruction Filters of FBP

The Ram-Lak filter used in FBP is a basic filter function derived from ramp function $H(\xi) = |\xi|$ with band limited spatial frequency γ, defined as [17],

$$\begin{aligned} H_{RL}(\xi) &= |\xi|,\ (|\xi| \leq 2\pi\gamma) \\ H_{RL}(\xi) &= 0,\ otherwise \end{aligned} \tag{4}$$

The inverse FT of $H_{RL}(\xi)$ yields the relevant spatial domain filter kernel,

$$h_{RL}(x) = \frac{1}{2\pi} \int_{-2\pi\gamma}^{2\pi\gamma} |\omega| exp(i\omega x) d\omega \tag{5}$$

As the sampled projection data to be processed in the spatial domain are discrete, the filter kernel should also be discrete. Discretizing Eq. (5) at positions $x_k = k\Delta x$, the Ram-Lak filter kernel is obtained as [17],

$$\begin{aligned} h_{RL}(0) &= \frac{1}{4\Delta x^2} \\ h_{RL}(k) &= \frac{-1}{\pi^2 k^2 \Delta x^2}\ forallk = odd \\ h_{RL}(k) &= 0,\ forallk = even \end{aligned} \tag{6}$$

As Ram-Lak filter kernel is sampled in the spatial domain, the filter function in the Fourier domain will necessarily be of periodically repeating band-limited signal of bandwidth $n\gamma$ where n = 1, 2, 3, ... The sharp boundary of Ram-Lak filter produces excellent reconstruction for high dose applications. However, in case of high noise level (low dose) this filter introduces artifacts in the reconstructed image. The most traditional, but yet commonly applied filter to accommodate the Poison noise is the Shepp-Logan filter, which is given as [17],

$$H_{SL}(\xi) = |\xi| sin(\frac{\xi}{4\gamma}), \; (|\xi| < 2\pi\gamma)$$
$$H_{SL}(\xi) = 0, \; otherwise \tag{7}$$

By incorporating suitably, a sine-weighted function into the filter function $|w|$, Shepp-Logan filter effectively removes the ringing artifact. The corresponding filter kernel after sampling at discrete positions is given by

$$h_{SL}(k) = \frac{-8\gamma^2}{\Pi^2(4k^2-1)} \tag{8}$$

In general, the images reconstructed using Shepp-Logan filters are noise-smoothed with less ringing (with good resolution) compared to Ram-Lak reconstructions. Also, both Ram-Lak and Shepp-Logan filters have been referred for deblurring without consideration of noise in the projection data. This necessitates the need for a filter that has a frequency response similar to that of a band-pass filter to demonstrate better reconstruction under noisy and limited-angle projections. Therefore, to achieve reconstructed images with high spatial resolution and acceptable SNR, this work applies a filter function that can retain optimal high frequency content eliminating noise which are at higher frequencies. The section that follows details this concepts introducing a new type of wavelet filter.

2.3 Gabor as Reconstruction Filter

The Gabor filter which has been proven to be optimally localized in both spatial and frequency domain due to its ability to accurately distinguish texture frequencies [23,24]. Also, it poses the characteristics of a band-pass filter. This encouraged to introduce it as a reconstruction filter in FBP to meet the above requirements. The 1D Gabor elementary function (GEF) is represented as [25–27],

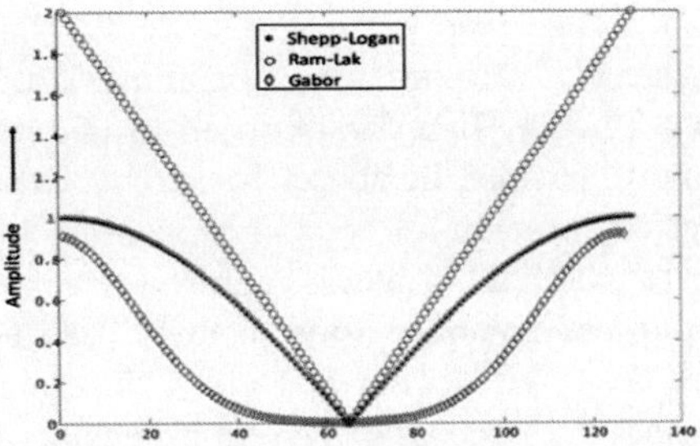

Fig. 1. Frequency response of filter functions

$$h(x, \sigma, \omega) = g(x, \sigma) exp(j\omega x) \tag{9}$$

where $g(x, \sigma)$ is a Gaussian function in 1D with variance σ defined as,

$$g(x, \sigma) = \frac{1}{\sqrt{2\pi\sigma}} exp(\frac{-x^2}{2\sigma^2}) \tag{10}$$

The real and imaginary components of 1D Gabor is given as follows

$$h(x, \sigma, \omega) = g(x, \sigma) cos(\omega x) + jg(x, \sigma) sin(\omega x) \tag{11}$$

The filter responses of Gabor along the traditional reconstruction filter is shown in Fig. 1. From this figure, it can be clearly observed that the Gabor filter only illustrates the response of a filter function with very little high frequency content and band-pass nature.

3 Design and Implementation

3.1 Reconstruction Process

The Gabor filter incorporated image reconstruction system was designed as a platform-independent software using JAVA and was implemented on an Intel (R) core ^{TM}i7-4790 @ 2.53 GHz PC under windows 10 platform. The steps involved are:

(i) Assigns each image pixels to the nearest cluster center according to their Euclidean distance measure
(ii) Read projections from the given data set, $R_\theta f(\omega)$
(iii) Select the filter function, $W(\omega)$ according to Eqs. (5) or (9) or (11).
(iv) Sub-sample the projection data (optional)
(v) For each projection, Apply the filter using Eq. 2
(vi) Back project the filtered projection using Eq. 3.
(vii) Display the resultant reconstructed image.

3.2 Object Oriented Approach

One of the key objectives of this work was to develop an object-oriented library for CT image reconstruction system. The flexibility of this library can be used for other CT reconstruction or analysis techniques being developed in our laboratory. The key benefits of object-oriented approach is that it enables to localize the modifications in object hierarchy to specific given level of object. Also, the object-orientation allows to encapsulate both the object properties and the messages to which an object is required to respond. The encapsulation characteristics of an object not only adds security credits but also allows to predict the object interactions with other objects. Consequently, the object-orientation facilitates and eases the process of system extension and scalability. Here, the concept, class allows to group all methods and variables that are common to a certain kind of software object. Inheritance between objects allows compact codes and the overriding features allow localization of changes. It is this hierarchy that yields an inverted tree structure. Here, the tree at the lower level of hierarchy defines more specific attributes about a particular class of object. The class hierarchy of the CT system is illustrated in Fig. 2.

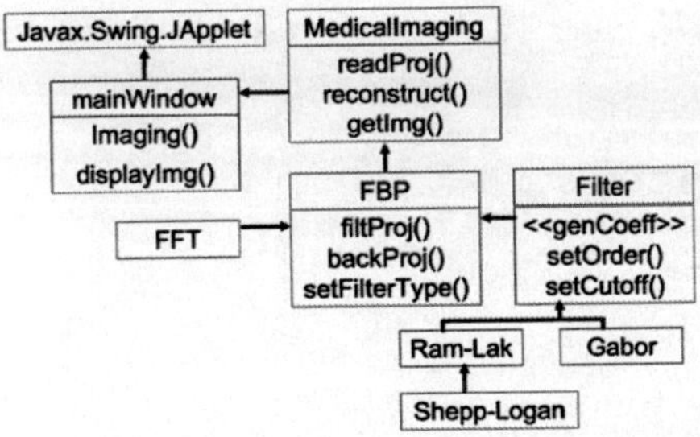

Fig. 2. Class hierarchy of FBP reconstruction

The Java swing provides a large number of GUI controls to quickly and easily change the look and feel of a single component or a group of components. The Java Swing components were employed for GUI design with Document-View architecture. The user-defined classes namely Ram-Lak, Shepp-Logan and Gabor inherits members such as genCoeff and set Order from their parent class namely Filter, to generate filter coefficients using their respective frequency response functions. The class FFT performs the Fourier transformation of projection data. The class Reconstruction performs FBP process using the classes FFT and Filter. Similarly, the class Main Window inherits the JAVA Swing component J Applet to design the container for all GUI controls. Using the class J Dialog, the MainWindow class provides an user-friendly and interactive input environment. The class MainWindow activates CT imaging class to perform the CT reconstruction process.

4 Result and Discussion

In general, phantom studies are conducted in medical imaging to gain insight into the limitations and trade-offs imposed by the experimental parameters. In line to this, the present work employed Shepp-Logan head phantom to demonstrate the feasibility of applying Gabor for FBP reconstruction. In this study, two set of simulation experiments were conducted namely, for limited-angle and for low-dose tomographic reconstruction. Further to prove the performance of Gabor filter method, results were compared with the results of standard Ram-Lak and Shepp-Logan filter. Finally, the performance of the Gabor filter was also quantified with quality metrics to provide evidence for its applicability. The subsection that follows summarizes the results obtained under these two imaging criteria.

4.1 'Limited-Angle' FBP Reconstruction

To demonstrate the viability of Gabor filter for FBP reconstruction with limited angle of projection data, it was tested on Shepp-Logan phantom. The FBP reconstruction images from 18 (Column A), 9 (Column B) and 6 (Column C) projections are illustrated in Fig. 3(a) for visual inspection. In this figure, for the sake of comparison, the images reconstructed using Gabor filter are depicted in row 3, Shepp-Logan filter in row 2 and Ram-Lak filter in row 1.

Observing this figure, it is clear that images reconstructed with Ram-Lak filter comprises streak artifacts and the image reconstruction quality with Ram-Lak is further degraded as the number of projections is reduced. On other hand,

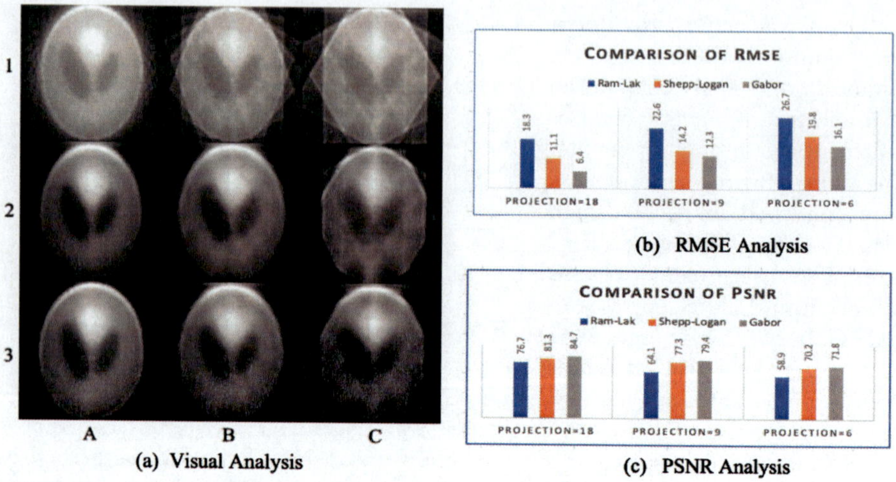

(a) Visual Analysis (b) RMSE Analysis (c) PSNR Analysis

Fig. 3. FBP reconstruction results on Shepp-Logan phantom with 'Limited-Angle' projections

the images reconstructed with Gabor filter displays better quality and are comparably equivalent to Shepp-Logan filter in clearly delineating the ROI without inducing artifacts in the reconstructed image. But observing the resultant images in column C of row 2 and row 3 which are obtained with 6 projects, it can be observed that Gabor filter outperforms Shepp-Logan filter in restoring and delineating the ROI clearly in the reconstructed image.

4.2 Low-Dose 'Noisy' FBP Reconstruction

To evaluate the performance of the Gabor filter for FBP reconstruction for low-dose CT imaging. The Shepp-Logan phantom projections were simulated using Matlab and all the obtained projects were corrupted with various levels of multiplicative noise. Later, the simulated noisy projections were used for FBP reconstruction. Figure 4(a) depicts the images reconstructed from noisy projections. Here, the column A, B and C illustrated the images reconstructed from projections with SNR of 0.5, 0.3 and 0.1 respectively. Observing the results of the Ram-Lak filter shown in Row-1, Shepp-Logan filter in Row-2 and Gabor filter in Row-3, it is clear that the Ram-Lak filter in FBP degrades the quality of the reconstructed image with the increase in noise level. Whilst Gabor filter displays comparably better than Shepp-Logan filter in reconstructing image with good spatial resolution at the same time with minimal artifact even under high noise condition. These obtained results conform the viability of the Gabor filter for low-dose computed tomography.

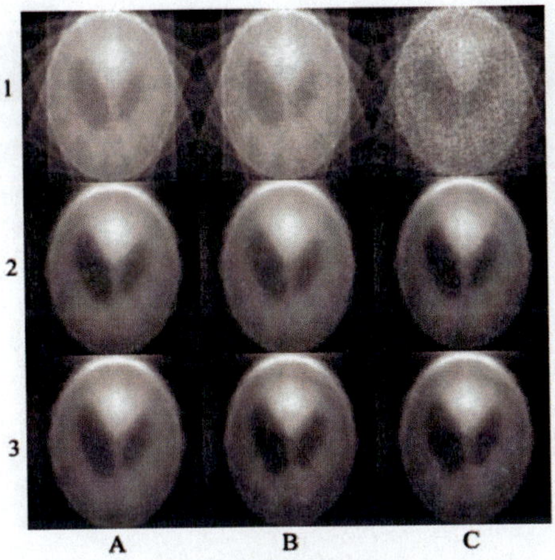

(a) Visual Analysis

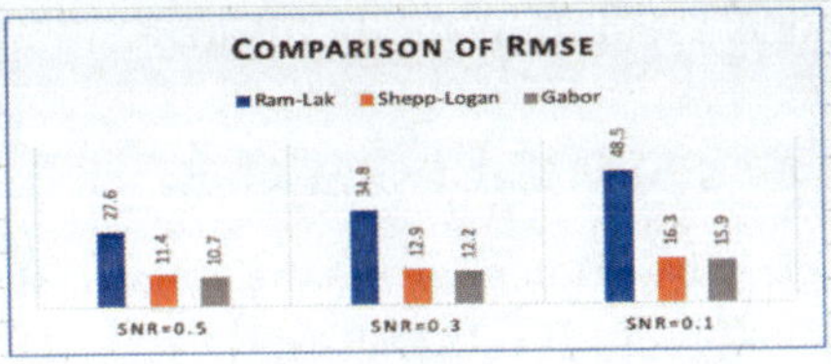

(b) RMSE Analysis

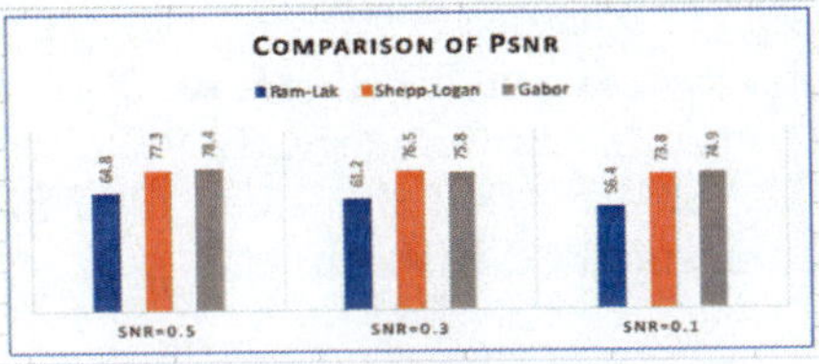

(c) PSNR Analysis

Fig. 4. FBP reconstruction results on Shepp-Logan phantom with 'Noisy' projections

Table 1. QLAQuality metrics RMSE/PSNR for FBP results in Fig. 3(a)

Method	A (p = 18)	B (p = 9)	C (p = 6)
Ram-Lak	18.3/76.7	22.6/64.1	26.7/58.9
Shepp-Logan	11.1/81.3	14.2/77.3	19.8/70.2
Gabor	6.4/84.7	12.3/79.4	16.1/71.8

Table 2. Quality metrics RMSE/PSNR for FBP result in Fig. 4(a)

Method	A (SNR = 0.5)	B (SNR = 0.3)	C (SNR = 0.1)
Ram-Lak	27.6/64.8	34.8/61.2	48.5/56.4
Shepp-Logan	11.4/77.3	12.9/76.5	16.3/73.8
Gabor	10.7/78.4	12.2/75.8	15.9/74.9

4.3 Quantitative Evaluation

For quantitative evaluation of the results, the most commonly used error measurements, viz., RMSE and PSNR were used. These metrics were computed for both limited-angle and low-dose resultant images depicted in Fig. 3(a) and Fig. 4(a) and are collected in Table 1 and Table 2 respectively. Comparing the RMSE and PSNR values given in Table 1 and Table 2, it can be observed that Gabor filter is capable to produce almost similar reconstruction with reduced dose concentration of SNR = 0.3 and SNR = 0.1. Though Shepp-Logan filters show equivalent performance to Gabor filter in SNR, it fails to improve the resolution with dose reduction. In this correspondence, it can be inferred that the Gabor filter can enable low-dose tomographic reconstruction from the usual dosage concentration. Visual Comparison for the quality metrics tabulated in Table 1 and Table 2 are illustrated in Fig. 3 and Fig. 4 respectively.

4.4 Statistical Evaluation

This subsection presents the statistical significance of Gabor filter for FBP reconstruction. In this regard, various statistical analysis was performed for the reconstruction results presented in Table 1 and Table 2. The findings are reported in Table 3. The observation of Table 3 clearly demonstrates that 95% confidence interval obtained for RMSE results of Gabor filter is lower than standard reconstruction filters. Similarly, the 95% confidence interval for PSNR results is higher. This conforms that the Gabor filter displayed a better reconstruction compared to the standard FBP filters.

Table 3. Statistics with ANOVA for quality metrics in Table 1 and Table 2

Metrics	Filters	N	Std. Dev	Std. Error	95%confidence interval	
					Lower bound	Upper bound
RMSE	Ram-Lak	6	10.707	4.371	21.182	38.317
	Shepp-Logan	6	3.312	1.352	11.632	16.934
	Gabor	6	3.599	1.469	9.386	15.146
PSNR	Ram-Lak	6	7.112	2.903	57.992	69.374
	Shepp-Logan	6	3.746	1.529	73.069	79.064
	Gabor	6	4.436	1.811	73.950	81.049

Table 4. ANOVA significance analysis on RMSE and PSNR results

Metrics	ANOVA	Sum square	df	Mean square	F	Sig.
RMSE	Between groups	1097.9	2	548.95	11.88	0.0008
	Within groups	692.86	15	46.19		
	Total	1790.76	17			
PSNR	Between groups	692.6	2	346.302	12.32	0.0007
	Within groups	421.52	15	28.10		
	Total	1114.13	17			

Subsequently, ANOVA significance analysis was conducted using the metrics in Table 1 and Table 2 to demonstrate the performance difference among three reconstruction filters and the findings are shown in Table 4. The box plot analysis for RMSE and PSNR is shown in Fig. 5a and in Fig. 5b respectively. The significance value for RMSE and PSNR in Table 4 which is 0.0008 and 0.007 proves

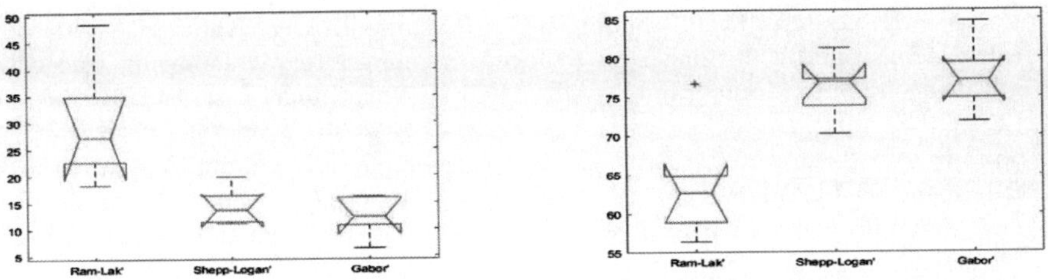

(a) Significance on RMSE Results

(b) Significance on PSNR Results

Fig. 5. Box diagram illustrating the significance of Gabor filter on FBP results

that reconstruction results obtained using three filters differed significantly. The visual inspection of box plot illustrated in Fig. 5a and Fig. 5b conforms that the reconstruction performance of Gabor filter is significant compared to the standard FBP filters.

5 Conclusion

In this filter, Gabor filter is proposed as an alternate filter function for FBP reconstruction under limited angle and low-dose CT imaging procedure. Both the qualitative and quantitative results are the evidences for the enhanced reconstructions of Gabor filter than its counter parts in terms of better discrimination between regions of Shepp-Logan phantom. This demonstrates the feasibility and validity of Gabor Filter for FBP reconstruction. Further, the ANOVA significance analysis results clearly conforms the reconstruction performance of Gabor filter is significantly better compared to the standard FBP filters. Therein, it is evident that the outcome of this work is expected to encourage the applicability of Gabor as an alternate filter for FBP tomographic reconstruction.

References

1. Li, Z., Zhang, J., Liu, D., Du, J.: CT image-guided electrical impedance tomography for medical imaging. IEEE Trans. Med. Imaging **39**(6), 1822–1832 (2019)
2. Dilsizian, V., Pohost, G.M.: Cardiac CT, PET and MR. John Wiley & Sons, New Jersey (2019)
3. Kuanar, S., Athitsos, V., Mahapatra, D., Rao, K., Akhtar, Z., Dasgupta, D.: Low dose abdominal CT image reconstruction: an unsupervised learning based approach. In: 2019 IEEE International Conference on Image Processing (ICIP), pp. 1351–1355. IEEE (2019)
4. Kertesz, H., Beyer, T., Traub-Weidinger, T., Cal-Gonzalez, J., Hacker, M., Kitsos, T., London, K., Kench, P.L., et al.: Reducing paediatric patient radiation exposure during [18F]-PET/CT: the effect of reduced tracer dose and reconstruction methods on image quality (2019)

5. Robbins, H.A., Berg, C.D., Cheung, L.C., Chaturvedi, A.K., Katki, H.A.: Identification of candidates for longer lung cancer screening intervals following a negative low-dose computed tomography result. JNCI: J. Nat. Cancer Inst. **111**(9), 996–999 (2019)
6. Sauter, A., Koehler, T., Brendel, B., Aichele, J., Neumann, J., Noël, P.B., Rummeny, E.J., Muenzel, D.: CT pulmonary angiography: dose reduction via a next generation iterative reconstruction algorithm. Acta Radiol. **60**(4), 478–487 (2019)
7. Ahn, C., Heo, C., Kim, J.H.: Combined low-dose simulation and deep learning for CT denoising: application in ultra-low-dose chest CT. In: International Forum on Medical Imaging in Asia 2019, vol. 11050, p. 110500E. International Society for Optics and Photonics (2019)
8. Nam, H., Guo, M., Yu, H., Lee, K., Li, R., Han, B., Xing, L., Lee, R., Gao, H.: Tensor framelet based iterative image reconstruction algorithm for low-dose multislice helical CT. PloS One **14**(1), e0210410 (2019)
9. Bukreeva, I., Asadchikov, V., Buzmakov, A., Chukalina, M., Ingacheva, A., Palermo, F., Fratini, M., Cedola, A.: Simultaneous iterative reconstruction method for high resolution X-ray phase-contrast tomography. In: Twelfth International Conference on Machine Vision (ICMV 2019), vol. 11433, p. 114331V. International Society for Optics and Photonics (2020)
10. Fu, L., De Man, B.: A hierarchical approach to deep learning and its application to tomographic reconstruction. In: 15th International Meeting on Fully Three-Dimensional Image Reconstruction in Radiology and Nuclear Medicine, vol. 11072, p. 1107202. International Society for Optics and Photonics (2019)
11. Nagarajan, G., Kumar, B.S.: Proficient reconstruction algorithms for low-dose X-ray tomography. In: Hybrid Machine Intelligence for Medical Image Analysis, pp. 237–256. Springer, Singapore (2020)
12. Dietze, M.M., Branderhorst, W., Kunnen, B., Viergever, M.A., de Jong, H.W.: Accelerated SPECT image reconstruction with FBP and an image enhancement convolutional neural network. EJNMMI Phys. **6**(1), 14 (2019)
13. Thavavel, V., Basha, J.J., Krishna, M.C., Murugesan, R.: Heuristic wavelet approach for low-dose EPR tomographic reconstruction: an applicability analysis with phantom and in vivo imaging. Expert Syst. Appl. **39**(5), 5717–5726 (2012)
14. Deák, Z., Grimm, J.M., Treitl, M., Geyer, L.L., Linsenmaier, U., Körner, M., Reiser, M.F., Wirth, S.: Filtered back projection, adaptive statistical iterative reconstruction, and a model-based iterative reconstruction in abdominal CT: an experimental clinical study. Radiology **266**(1), 197–206 (2013)
15. Willemink, M.J., Noël, P.B.: The evolution of image reconstruction for CT–from filtered back projection to artificial intelligence. Eur. Radiol. **29**(5), 2185–2195 (2019)
16. Beckmann, M., Iske, A.: Error estimates for filtered back projection. In: 2015 International Conference on Sampling Theory and Applications (SampTA), pp. 553–557. IEEE (2015)
17. Horbelt, S., Liebling, M., Unser, M.A.: Filter design for filtered back-projection guided by the interpolation model. In: Medical Imaging 2002: Image Processing, vol. 4684, pp. 806–813. International Society for Optics and Photonics (2002)
18. Anam, C., Haryanto, F., Widita, R., Arif, I., Dougherty, G.: An investigation of spatial resolution and noise in reconstructed CT images using iterative reconstruction (IR) and filtered back-projection (FBP). In: Journal of Physics: Conference Series, vol. 1127, p. 012016. IOP Publishing (2019)

19. Dai, X., Shu, H., Luo, L., Han, G.N., Coatrieux, J.L.: Reconstruction of tomographic images from limited range projections using discrete radon transform and Tchebichef moments. Pattern Recogn. **43**(3), 1152–1164 (2010)
20. Ramm, A.G., Katsevich, A.I.: The Radon Transform and Local Tomography. CRC Press, Boca Raton (1996)
21. Lauritsch, G., Härer, W.H.: Theoretical framework for filtered back projection in tomosynthesis. In: Medical Imaging 1998: Image Processing, vol. 3338, pp. 1127–1137. International Society for Optics and Photonics (1998)
22. Pan, X., Sidky, E.Y., Vannier, M.: Why do commercial CT scanners still employ traditional, filtered back-projection for image reconstruction? Inverse Probl. **25**(12), 123009 (2009)
23. Acharya, U.R., Meiburger, K.M., Koh, J.E.W., Vicnesh, J., Ciaccio, E.J., Lih, O.S., Tan, S.K., Aman, R.R.A.R., Molinari, F., Ng, K.H.: Automated plaque classification using computed tomography angiography and Gabor transformations. Artif. Intell. Med. **100**, 101724 (2019)
24. Hu, Q., Hu, S., Zhang, F.: Multi-modality medical image fusion based on separable dictionary learning and Gabor filtering. Sign. Process. Image Commun. **83**, 115758 (2020)
25. Zhang, Y., Li, W., Zhang, L., Ning, X., Sun, L., Lu, Y.: Adaptive learning Gabor filter for finger-vein recognition. IEEE Access **7**, 159821–159830 (2019)
26. Tsai, D.M., Lin, C.P.: Fast defect detection in textured surfaces using 1D Gabor filters. Int. J. Adv. Manufact. Technol. **20**(9), 664–675 (2002)
27. Kumar, T.S., Kanhangad, V.: Gabor filter-based one-Dimensional local phase descriptors for obstructive sleep apnea detection using single-lead ECG. IEEE Sens. Lett. **2**(1), 1–4 (2018)

Handwritten Signature Verification Using Complex Wavelet Transform and Mutual Information Based Kernel

Anushka Desai, Vraj Mashruwala, and Manish Khare(✉)

Dhirubhai Ambani Institute of Information and Communication Technology (DA-IICT), Gandhinagar, India

Abstract. Signature Verification is an essential and challenging task in forensic document analysis. Signature Verification is a process where the authenticity of a given signature is validated; i.e. it is confirmed, after analysis, whether the given signature is forged or not. The objective of the work proposed here is to create a new algorithm that can be used for verification of signatures used in cheques and various legal documents for customer authentication in banking systems. The proposed algorithm is designed using Daubechies complex wavelet transform (DCxWT) which has superior approximate shift invariant and edge representation properties over real valued wavelet transform. This is what motivates the use of DCxWT in the Signature Verification task. For classification of valid and forged signatures we have used Support Vector Machine (SVM) classifier. We have used Mutual Information kernel that is an alternative kernel of the KL divergence kernel in SVM classifier. Proposed method is compared with KL-divergence based kernel and some other standard kernels like Poly, RBF, and Tanh kernel of SVM Classifier. The method we proposed is juxtaposed with different state-of-the-art methods and effectiveness of the algorithm is also measured using different quantitative performance measures.

Keywords: Signature verification · Mutual information kernel · SVM · Convolutional neural network · Daubechies complex wavelet transform

1 Introduction and Related Work

The task of handwritten signature verification deals with classification of genuine or forged signatures in order to verify signatures. Signature verification is used in varieties of applications [1] such as: banking systems, credit card customer authentication, document verification etc. Due to the vital importance of a signature in such applications, signature forgery is carried out by people to perform various misdeeds without the authorized person getting any knowledge of it. The signature verification task can be bifurcated in two parts: On-line and Off-line verification.

Online signature verification is when the signatures which have been obtained with the help of a stylus through which many dynamic attributes of the signer like speed and pressure can be obtained. Whereas what offline signature verification focuses on are

A. Abraham et al. (Eds.): SoCPaR 2020, AISC 1383, pp. 278–288, 2021.
https://doi.org/10.1007/978-3-030-73689-7_27

the static features of the signatures itself which are obtained after scanning the image containing the signatures. Employing an online signature verification system requires complex hardware and scalability of such a system becomes an issue. Thus, offline signature verification has become an extensive area of research, since the signer need not be present at the time of verification which makes it more convenient to be employed in large banking systems. Further distinctions of the signature verification task are the *forgery exposure and writer dependence* tasks. The forgery exposure part of the signature verification task pertains to the type of data which has been used in order to train the classifier; that is if only genuine signatures have been used to train the system or samples of both original and fake signatures are used. For my study, we have used samples original as well as forged signatures to train classifiers. Furthermore, the writer dependence task refers to whether different classifiers have been used for every signer on the system or the same classifier is used. We have used a writer independent model, i.e., a single classifier for all identities.

Various explanations of offline signature verification are explored in different literatures [2–5]. In earlier research offline signature verification is performed using template matching and Hidden Markov Model, these techniques are based on the structure of the signature. These days machine learning based approaches are majorly being used because of the availability of a large amount of training data to carry out unsupervised training. In machine learning based approach, algorithms can answer questions related to a data after learning from a collection of data. Gideon et al. [6] proposed signature verification method based on Convolutional Neural Network. Hafermann et al. [3] proposed signature verification using Deep convolutional neural network. Kruthi et al. [7] proposed a signature verification method based on a support vector machine, in which they used Linear and Poly kernel. Ji and Quan [8] proposed signature verification using Wavelet transform and SVM. Muhamed and Ani [9] proposed signature recognition based on Discrete wavelet transform,

Linear learning machines, like SVM, generate hyperplanes to separate classes. However, there are many such real-life problems where such linear separation is not possible. Kernel method is the method that adds nonlinearity flavor in these machines, and generally projects data into higher dimensions. We can increase the computation power of these machines by choosing an appropriate kernel. In statistical learning, there are two types of classifiers- discriminative model and generative model. Generative models like Hidden Markov Model, Naive Bayes Classifiers use the statistical information of individual data points and deal with variable length sequence data like images. On the other side, discriminative models like Support Vector Machine provide direct decision boundary or mapping function, which often produces better classification results. Some kernels are already existing, which allow us to add advantages of a generative model to Support Vector Machine. Jaakkola et al. [10] provide a theoretical justification for combining these two models. Moreno et al. [11] proposed Kullback-Leibler Divergence based kernel, is an excellent combination of both discriminative and generative models. One more advantage of this KL divergence-based kernel is that it deals with variable length sequence data. In our proposed work we have used a Mutual Information based kernel, which was proposed by Dave et al. [12]. For experimental purposes we have tested and compared various kernels.

Method proposed by Ji and Quan [8] and Muhamed and Ani [9], makes use of real valued wavelet transform for signature verification. However the use of real valued wavelet transform in computer vision applications has various limitations. In order to overcome such difficulties, we can use complex wavelet transforms. In this paper, motivated by the work of Ji and Quan [8] and Muhamed and Ani [9], we have based our proceeding for signature verification using the Daubechies complex wavelet transform (DCxWT). DCxWT, as the name says, is a complex wavelet transform which has advantages over real valued wavelet transform of approximate shift-invariance and better edge representation. For the classification of authentic and forged signatures, we have used the Support Vector Machine (SVM) classifier. We have experimented with different SVM kernels such as: Poly, RBF, Tanh, KLD and Mutual Information kernel. In order to show the effectiveness of DCxWT, our proposed method is juxtaposed with some other methods developed by Gideon et al. [6], Kruthi et al. [7], Ji and Quan [8], Muhamed and Ani [9], and Bhunia et al. [13].

Rest of the paper covers the following: Sect. 2 covers different SVM kernels. Section 3 describes Daubechies complex wavelet transform. Section 4 introduces our proposed method along with steps followed. Section 5 provides the experimental results and observations of comparison between our proposed method with various kernels and other developed methods. The work has been rounded off with a conclusion in Sect. 6.

2 Different SVM Kernels

In the present section we discuss the following distinct SVM kernels: Polynomial, Radial basis Function, Hyperbolic Tangent, KLD, and Mutual Information based kernel.

2.1 Polynomial (Poly)

The polynomial kernel [14] is one of the standard kernels used in SVM, that contains two tunable parameters. Stated below is the equation for Polynomial kernel:

$$k(x, y) = \left\langle x^T y + c \right\rangle^d \ (x, y \in X, d \in \mathbb{N}, c \in \mathbb{R}) \tag{1}$$

In the above equation, d represents the degree of the polynomial equation and c is the intercept constant. When we set c = 0, it becomes homogeneous.

The primary advantage of the polynomial kernel is that it does not use only input features directly, but it allows us conjunction of more than one feature means we can fuse two or more features.

The main disadvantage of the poly kernel is we need to tune c and d carefully. Improper tuning of c and d may lead to numerical instability. If $\left(\left\langle x^T y + c\right\rangle < 1\right)$, then $\left\langle x^T y + c\right\rangle^d$ may lead to zero, and if $\left(\left\langle x^T y + c\right\rangle > 1\right)$, then $\left\langle x^T y + c\right\rangle^d$ may lead to infinity with increasing d. Also if we the value of $x^T y$ is small, there may be a chance that the value of parameter c may vanish the effect of $x^T y$, and it can lead to an underfitting problem.

2.2 Radial Basis Function (RBF)

RBF [14] are the family of kernel functions. Different types of RBF kernels are: Anova kernel, Laplacian kernel, Exponential kernel, Gaussian kernel. Here we are describing the Gaussian kernel. Equation of Gaussian RBF kernel function is

$$k(x, y) = \exp\left(-\frac{\|x - y\|^2}{2\sigma^2}\right)\left(x, y \in X, \sigma^2 \textit{ is variance}\right) \tag{2}$$

It can be depicted as

$$k(x, y) = \exp\left(-\gamma \|x - y\|^2\right)(x, y \in X, \gamma > 0) \tag{3}$$

The Gaussian RBF kernel is one of the standard kernels and the most popular kernel of all available RBF kernels. With popularity, it provides excellent power to the SVM. When the data is distributed in the loop-like shape, the RBF kernel is the most suitable. The RBF kernel can put hypersphere (circular in 2D) boundaries between classes. $\|x - y\|^2$ gives the squared Euclidean distance between two vectors and the range of the RBF function is between 0 and 1. RBF provides the similarity between two vectors. The value of the vector being high (near to 1) implies that the vectors are the same, and if the value is low (near zero), vectors are different.

The RBF kernel has a major disadvantage which is that it is very slow compared to the linear and poly kernel. We also need to tune the γ parameter carefully. If we overestimate γ parameter, it may lead to an overfitting problem and if we underestimate γ parameter, it does not increase computation power(it will act as the linear kernel) of SVM.

2.3 Hyperbolic Tangent (Tanh)

The idea of the Tanh kernel [14], also referred to as the sigmoid kernel, has come from neural networks. Tanh function is a popular nonlinear activation function in neural networks. Equation for Hyperbolic Tangent is given below:

$$k(x, y) = \tanh(\alpha \langle x, y\rangle + \beta), (x, y \in X, \alpha, \beta \in \mathbb{R}) \tag{4}$$

From the above equation, we are able observe that there are two tunable parameters α and β. Generally, we keep α parameter value equal to the *1/N*, N being the dimension. Tanh has a range of between -1 and 1.

The main disadvantage of the tanh kernel is two tunable parameters (α and β) carefully, it may lead to the indefinite kernel function.

2.4 KLD Kernel

KLD kernel [11] works on a set of vectors which are a representation of an individual instance, for example, an image. KLD kernel starts with a statistical model $P\left(x/\theta_i\right)$ of the data which adds advantages of generative models into discriminative models. This kernel estimate parameters θ_i of distribution from a set of vectors. For each instance, this kernel finds Probability Density Function (PDF). For estimating PDF, Gaussian distribution is chosen.

KLD kernel uses the Maximum Likelihood method for estimation of θ_i. Once PDF is computed for each training and testing instance, then apply kernel function. The equation for the KL-Divergence [3] is:

$$KLD\big(P(x/\theta_i)\|(x/\theta_j)\big) = \int_{-\infty}^{\infty} P(x/\theta_i)\log\frac{P(x/\theta_i)}{P(x/\theta_j)}dx \tag{5}$$

KL-Divergence is an asymmetric function. As Mercer Theorem does not allow asymmetric function, so its need to use symmetric KL-Divergence [3].

$$D\big(P(x/\theta_i), P(x/\theta_j)\big) = KLD\big(P(x/\theta_i)\|P(x/\theta_j)\big) + KLD\big(P(x/\theta_j)\|P(x/\theta_i)\big) \tag{6}$$

So,

$$D\big(P(x/\theta_i)\|(x/\theta_j)\big) = \int_{-\infty}^{\infty} P(x/\theta_i)\log\frac{P(x/\theta_i)}{P(x/\theta_j)}dx + \int_{-\infty}^{\infty} P(x/\theta_j)\log\frac{P(x/\theta_j)}{P(x/\theta_i)}dx \tag{7}$$

Still, this does not satisfy Mercer Theorem, because it does not generate PSD gram matrix, so it need to exponentiate the symmetric KL-Divergence with some scale and shift, so modified kernel function is:

$$k(x_i, x_j) = e^{A.D(P(x/\theta_i),(x/\theta_j))+B} \tag{8}$$

here A is for scaling and B for shifting.

2.5 Mutual Information (MI) Kernel

Mutual Information (MI) kernel [12] works on a set of vectors which are a representation of an individual instance, such as an image. MI kernel start with a statistical model $P(x/\theta_i)$ of the data. We estimate parameters θ_i of distribution from a set of vectors. For each object (images or sequence), we find a probability density function (PDF). For computation of PDF, Gaussian distribution function is chosen. For estimation of θ_i, Maximum likelihood method is chosen. Once PDF for each training and testing instance is computed, we apply kernel function with analytical solution for mutual information between two n-variate gaussian distributions. Hence the kernel function is:

$$k(x_i, x_j) = I\big(P(x/\theta_i), P(x/\theta_j)\big) = \frac{1}{2}\log\frac{\det(S_i)\det(S_j)}{\det(S)} \tag{9}$$

Where S_i and S_j are variance-covariance matrices ($n\ x\ n$) of x_i and x_j, and S is full $2n\ x\ 2n$ dimensional matrix covariance matrix. We can get S by concatenating n-dimensional vectors of instance x_i and x_j respectively.

We can observe that there are no tunable parameters as well as there is no use of a mean vector in Eq. 9. MI kernel is a symmetric function, but it does not generate a positive semi-definite (PSD) gram matrix. So it does not satisfy Mercer Theorem. There are many techniques available in the literature for training SVM with an indefinite kernel. As mentioned above in the KLD kernel [11], we can exponentiate the function, but how to choose A and B (shifting and scaling parameters), they do not provide any details. Many of them, project an indefinite gram matrix to the nearest symmetric PSD matrix by vanishing negative eigenvalues. There is another similar approach, shift all the eigenvalues by some constant large enough, so that the smallest eigenvalue becomes positive or flip the negative eigenvalues to be positive [15]. We use this approach and shift eigenvalues to make the gram matrix PSD matrix. It is also possible to directly use an indefinite gram matrix. Ying et al. [16] improve the training algorithm of the SVM.

3 Daubechies Complex Wavelet Transform

For signature verification, we need a feature which is invariant of shift, translation and rotation of signature, considering signatures may be presented in rotated and translated form among different scenes. We have used DCxWT for handwritten signature verification due to the approximate shift-invariance and better edge representation properties that it offers. A description of DCxWT and why it is effectual for the discussed problem is noted below:

A function $f(t)$ can be broken down into complex scaling function and mother wavelet as:

$$f(t) = \sum_{k} c_k^{j_0} \phi_{j_0,k(t)} + \sum_{j=j_0}^{j_{\max}-1} d_k^j \psi_{j,k}(t) \tag{10}$$

where j_0 is the low resolution level already provided, $\left\{c_k^{j_0}\right\}$ and $\left\{d_k^j\right\}$ are approximation coefficients. $[\phi(u) = 2\sum_i a_i \phi(2u - i)]$ and detail coefficients $[\psi(t) = 2\sum_n (-1)^n \overline{a_{1-n}} \phi(2t - n)]$ in which both $\phi(t)$ and $\psi(t)$ allot the identical compact support $[-\mathrm{L}, \mathrm{L}+1]$ and $\sum a_i = 1$ (a_i's being coefficients which can be either real or even complex valued).

One dimensional Daubechies wavelet bases $\{\psi_{j,k}(t)\}$ are interpreted using the above-mentioned scaling function $\phi(u)$ and multiresolution analysis of $L_2(\Re)$ [17]. Complex valued scaling functions are obtained by changing the condition for a_i to be real [17] and calculating the general solution.

DCxWT has various properties [17], from which better edge representation and reduced shift sensitivity properties are of importance for the handwritten signature verification task.

4 The Proposed Method

This section introduces a novel approach for signature verification. In this method, we have used Daubechies Complex Wavelet Transform coefficients as a feature of objects. Figure 1 depicts the Block Diagram of the proposed method.

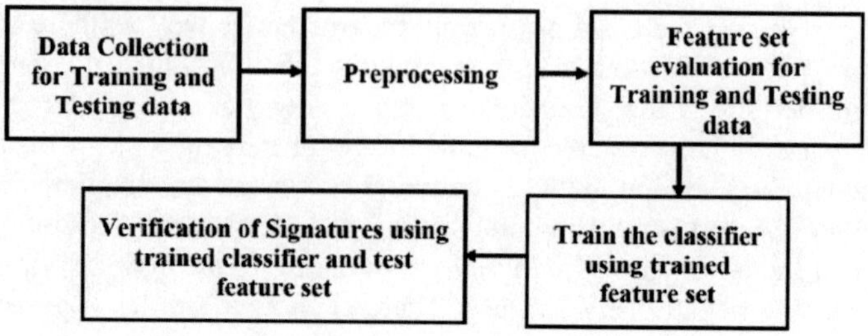

Fig. 1. Block diagram of the proposed method

Following steps describe details of the proposed method:

Step 1: Collecting and representing data for training and testing. The proposed method was tested using different datasets that are publicly available such as: GPDS-300[18], and BHSig[19].

Step 2: Individual images (i.e. signatures) in the selected set can be varied, for example they may not be of the same size or may not have the same color format. Due to this, the images in the dataset need to be normalized in order to maintain uniformity across the dataset while reducing complexity. For our method, we have normalized the collected data to scale to 256 × 256 pixel dimension. The color format of the normalized data after scale normalization on the collected data was converted into gray level format.

Step 3: In the proposed method, DCxWT is used for decomposing the images from the dataset into complex wavelet coefficients for feature set computation. The resulting coefficients after applying DCxWT are in the form of four sub-bands viz. - LL (low-low), HL (high-low), LH (low-high), HH (high-high), wherein LL is the approximation coefficient and LH, HL and HH are the detail coefficients. Each one of these detail coefficient matrices (HL, LH, and HH) is used separately to construct a feature vector. Values of the LL sub-bands are not used for constructing feature vectors. Rather, it is used in further higher-level decomposition due to the multi-resolution property of DCxWT. This means that in order to decompose wavelet transform for level 2, we further decompose approximation coefficients (LL sub-band) of level 1. This would once more split into one approximation coefficient and three detail coefficients matrix. The features of level 1 are then combined to constitute level 2. By carrying out this, we have computed the feature sets.

Step 4: As a final step, we have carried out classification-based verification of signature. We have computed the feature sets for both, test and trained datasets using step 1–3 of the proposed method which are then fed into the SVM classifier. SVM classifier analyzes the test feature set with help of the trained feature set and gives an output in terms of verified and forged signature. The same process has been repeated for all datasets under consideration. For the SVM classifier we have performed the experiment with different kernels: Poly, RBF, Tanh, KLD, and Mutual Information kernel, separately. We have used appropriate kernels as mentioned above during training and testing purposes. We have computed computation speed with different kernels, which is given in Table 1.

Table 1. Computation speed with respect to different kernels for GPDS-300 dataset

SVM classifier with	Computation time (in millisecond)
Poly kernel	0.4156
RBF kernel	0.2276
Tanh kernel	0.2689
KLD kernel	0.2046
MI kernel	0.2001

5 Experimental Results

Experimental results of the method we propose along with the methods developed by Gideon et al. [6], Kruthi et al. [7], Ji and Quan [8], Muhamed and Ani [9], and Bhunia et al. [13], as tabulated with respect to four measures, namely, Precision, Recall, F-Score and Accuracy. The results for all these have been presented in this section for two different datasets: GPDS-300 [18], and BHSig [19].

The GPDS-300 dataset [18] is a collection of 16,200 signatures collected from 300 signers. Each person has signed 24 genuine/valid signatures. For each of the 300 signatures of these people there are 30 skilled forged signatures obtained from 10 different forgers. The BHSig260 dataset [19] is a collection of 6240 genuine signatures and 7800 skilled forged signatures. In this data language signatures are Hindi and Bengali. Some sample signatures of datasets [18, 19] are shown in Table 2.

Table 2. Sample signatures (genuine and forged) from two different dataset

	Genuine		Forgery	
GPDS-300 [18]				
BHSig-260 [19]	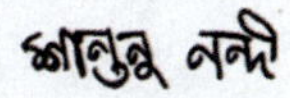	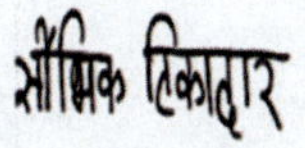	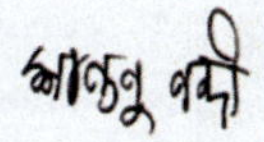	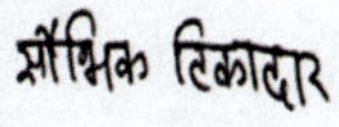

Signature verification results for datasets [18, 19] are shown in Tables 3, and 4 respectively for the proposed method and other state-of-the-art methods [6–9, 13]. For better comparison we have used four different performance metrics [20] namely Precision, Recall, F-Score, and Accuracy. Precision is the average per-class agreement of the data class labels and the classifier labels. Recall is defined as the average per-class effectiveness of the classifier to recognize class labels. F-Score is a measure that conveys the balance between precision and recall. Accuracy is the average per-class effectiveness of a classifier.

Table 3. Performance measure values for GPDS-300 dataset [18]

Method		Precision	Recall	F-Score	Accuracy
The Proposed method with	Poly kernel in SVM classifier	0.8008	0.7933	0.7946	0.9056
	RBF kernel in SVM classifier	0.8488	0.8508	0.8340	0.9274
	Tanh kernel in SVM classifier	0.8204	0.8133	0.8144	0.8912
	KLD kernel in SVM classifier	0.8623	0.8687	0.8525	0.9689
	MI kernel in SVM classifier	**0.8955**	**0.8796**	**0.8898**	**0.9745**
Gideon et al. [6]		0.7787	0.7740	0.7751	0.9095
Kruthi et al. [7]		0.8044	0.7900	0.7927	0.8645
Ji and Quan [8]		0.7558	0.7417	0.7447	0.8873
Muhamed and Ani [9]		0.8325	0.8200	0.8207	0.9210
Bhunia et al. [13]		0.8821	0.8740	0.8483	0.9586

Table 4. Performance measure values for BHSig-260 dataset [19]

Method		Precision	Recall	F-Score	Accuracy
The Proposed method with	Poly kernel in SVM classifier	0.8339	0.8240	0.8257	0.9295
	RBF kernel in SVM classifier	0.8852	0.8780	0.8789	0.9512
	Tanh kernel in SVM classifier	0.8606	0.8520	0.8534	0.9308
	KLD kernel in SVM classifier	0.8981	0.8940	0.8938	0.9572
	MI kernel in SVM classifier	**0.9726**	**0.9720**	**0.9672**	**0.9882**
Gideon et al. [6]		0.8189	0.7729	0.8180	0.9272
Kruthi et al. [7]		0.9510	0.9100	0.8979	0.9347
Ji and Quan [8]		0.9138	0.9100	0.8890	0.9202
Muhamed and Ani [9]		0.9419	0.9440	0.9274	0.9746
Bhunia et al. [13]		0.9604	0.9600	0.9578	0.9800

Table 3 and 4, shows the signature verification results for the proposed method and other state-of-the-art methods [6–9, 13] in terms of four performance evaluation metrics for GDPS-300 dataset [18] and BHSig-260 Dataset [19] respectively. From the table, we see that the proposed method has better precision, recall, F-score and accuracy than other state-of-the-art methods [6–9, 13]. From these tables, one can also see that the MI kernel provides superior results compared to other kernels such as Poly, RBF, Tanh, and KLD kernels with respect to performance measures.

6 Conclusion and Future Work

In this work, we have proposed a new approach for verification of handwritten signatures. The proposed method uses DCxWT coefficients as a feature of signature image. We have compared different SVM Kernels such as Poly, RBF, Tanh, KLD and MI for the signature verification task. We evaluated accuracies and performances of our algorithm and other recognized state-of-the-art methods [6–9, 13] for two different datasets: GDPS-300 [18] and BHSig-260 [19] with respect to four different performance metrics: Precision, Recall F-Score, and Accuracy. Looking at the output for each which has been formulated in tables, we can deduce that the proposed method with MI Kernel performs better in comparison to other SVM kernels such as: Poly, RBF, Tanh, and KLD. Moreover, the proposed method also performs better as compared to other state-of-the-art methods [6–9, 13].

Since signatures are written out in multiple languages with different complexity, our future plan is to probe the usage of a few different scripts for the signature verification task. The accuracy of verification of these can be checked using various SVM kernels.

References

1. Jain, A., Prabhakar, S.: An introduction to biometric recognition. IEEE Trans. Circ. Syst. Video Technol. **14**(1), 4–20 (2004)
2. McCabe, A., Trevathan, J., Read, W.: Neural network based handwritten signature verification. J. Comput. **3**(8), 9–22 (2008)
3. Hafemann, L.G., Sabourin, R., Oliveira, L.S.: Offline handwritten signature verification – literature review. In: 7th International Conference on Image Processing Theory, Tools and Applications (IPTA), pp. 1–8 (2017)
4. Zagoruyko, S., Komodakis, N.: Learning to compare image patches via convolutional neural networks. In: IEEE Conference on Computer Vision and Pattern Recognition, pp. 4353–4361 (2015)
5. Srinivasan, H., Srihari, S.N., Beal, M.J.: Machine learning for signature verification. In: 5TH Indian Conference on Computer Vision Graphics and Image Processing, pp. 761–775 (2006)
6. Jerome, G.S., Kandulna, A., Kujur, A.A., Diana, A., Raimond, K.: Handwritten signature forgery detection using convolutional neural network. In: 8th International Conference on Advances in Computing and Communication (ICACC-2018), pp. 978–987 (2018)
7. Kruthi, C., Deepika, C.S.: Offline signature verification using support vector machine. In: 5th International Conference on Signal and Image Processing, pp. 3–8 (2014)
8. Ji, H.W., Quan, Z.H.: Signature verification using wavelet transform and support vector machine. In: International Conference on Intelligent Computing (ICIC 2005), pp. 671–678 (2005)

9. Muhamed, S.S., Al-Ani, M.S.: Signature recognition based on discrete wavelet transform. UHD J. Sci. Technol. **3**(1), 19–29 (2019)
10. Jaakkola, T., Haussle, D.: Exploiting generative models in discriminative classifiers. In: International Conference on Advances in Neural Information Processing Systems II, pp. 487–493 (1999)
11. Moreno, P.J., Ho, P.P., Vasconcelos, N.: A Kullback-Leibler divergence-based kernel for SVM classification in multimedia applications. Adv. Neural. Inf. Process. Syst. **16**, 1385–1392 (2004)
12. Dave, R., Khare, M., Mitra, S. K.: Mutual information based kernel for image classification. In: 7th IEEE Conference on Soft Computing and Machine Intelligence (2020)
13. Bhunia, A.K., Alaei, A., Roy, P.P.: Signature Verification approach using fusion of hybrid texture features. Neural Comput. Appl. **31**, 8737–8748 (2019)
14. Schölkopf, B., Smola, A.: Learning with Kernels: Support Vector Machines, Regularization, Optimization, and Beyond. MIT Press, Cambridge (2002)
15. Chen, J., Ye, J.: Training SVM with indefinite kernels. In. 25th International Conference on Machine Learning (ICML 2008), pp. 136–143 (2008)
16. Ying, Y., Campbell, C., Girolami, M.: Analysis of svm with indefinite kernels. In: Bengio, Y., Schuurmans, D., Lafferty, J.D., Williams, C.K.I., Culotta, A. (eds.) Advances in Neural Information Processing Systems, pp. 2205–2213 (2009)
17. Clonda, D., Lina, J.M., Goulard, B.: Complex Daubechies wavelets: properties and statistical image modeling. Signal Process. **84**(1), 1–23 (2004)
18. Vargas, F., Ferrer, M.A., Travieso, C.M., Alonso, J.B.: Off-line handwritten signature GPDS-960 corpus. In: 9th International Conference on Document Analysis and Recognition (ICDAR 2007), pp. 764–768 (2007)
19. Pal, S., Alaei, A., Pal, U., Blumenstein, M.: Performance of an off-line signature verification method based on texture features on a large indic-script signature dataset. In: 12th IAPR Workshop on Document Analysis Systems, pp 272–277 (2016)
20. Sololove, M., Lapalme, G.: A systematic analysis of performance measures for classification task. Inf. Process. Manag. **45**, 427–437 (2009)

Multi-Scale Oriented Basic Image Features Column for Handwritten Digit Recognition

Abdeljalil Gattal[1(✉)], Chawki Djeddi[1], Akhtar Jamil[2], and Ameur Bensefia[3]

[1] Department of Mathematics and Computer Science, Larbi Tebessi University, Tebessa, Algeria
{abdeljalil.gattal,mohamed.aouine,c.djeddi}@univ-tebessa.dz
[2] Department of Computer Engineering, Istanbul Sabahattin Zaim University, Istanbul, Turkey
akhtar.jamil@izu.edu.tr
[3] Higher Colleges of Technology CIS Division, Abu Dhabi, UAE
abensefia@hct.ac.ae

Abstract. Handwritten digits recognition is a key research problem in the domain of image analysis and pattern recognition. Specifically, the appearance approaches based on feature extraction have been proposed to solve many research issues. This paper presents a novel way of extending the oriented Basic Image Features column (oBIFs column) to multi-scale features. Moreover, this method is very efficient and robust for hand handwritten digit recognition as the changes in size, slant and shape have little effect on the accuracy. The proposed method is carried in two steps: first, we concatenate oBIF image for scale parameter at two or more scales to provide Multi-scale oBIFs. Second, the two variation Multi-Scale oBIFs are crossed to form columns at each location. Finally, a histogram is created from the occurrences of different patterns. The CVL dataset was considered to evaluate our approach, where the digits used where from different fonts, widths, and directions. We managed to achieve a good recognition of unnormalized digits in almost 96% of the cases, which is comparable with the state-of-the-art approaches.

Keywords: Handwritten digits recognition · Multi-Scale oBIFs Column · CVL dataset

1 Introduction

Automated off-line handwritten digits recognition is not a new research area in classical pattern recognition as many researchers have proposed several methods to solve it. It has applications in several industries such as handwritten postal codes recognition, processing bank check amounts, storing handwritings based on handwritten numeric entries etc. In this context, the main challenges in recognizing handwritten digit arise from different writing styles, varying font sizes, directions, widths, arrangements, and measurements.

The above mentioned problems make the handwritten digit recognition task very challenging and it is an ongoing research field. The recent methods proposed in the literature aim to enhance the performance of recognition which usually employee two most

A. Abraham et al. (Eds.): SoCPaR 2020, AISC 1383, pp. 289–298, 2021.
https://doi.org/10.1007/978-3-030-73689-7_28

commonly used pattern recognition approaches, the feature extraction or/and classification methods [1–6]. Various classification methods have been developed for handwritten digit recognition. These methods have shown superior performance compare to their contemporary methods. For instance, convolutional neural network, Neural Networks [1, 2], Support Vector Machine [3, 5, 7–9], K Nearest Neighbors (KNN) and Hidden Markov Models (HMM) [2–4].

In feature extraction methods for digit recognition, three types of features have been commonly exploited: statistical, structural and textual. The statistical features are directly derived from statistical distribution of information in the input images, e.g. [5, 7, 8]. In structural features, both geometrical and topological properties of the digit are used e.g. [5, 7, 8]. Lastly, the textural feature [8, 9] used to describe the contents of an image or a region in additional to color features [8–10]. In addition, some hybrids methods have also been proposed, e.g. [7, 11], which are based on combination of different features types.

The oriented Basic Image Features column (oBIFs column) has been widely used to derive image texture features. The basic idea of oBIFs is to compose a histogram of local textural patterns which is then used as a texture descriptor. However, the main drawback of this method is that it fails to capture dominant features associated with large structures. These structures are determined using a scale parameter. The oBIFs column is not very robust against local variations in the texture. This problem can be alleviated by using a multi-scale parameter.

The development in the feature extraction stage can improve the recognition rates on CVL dataset [6] have been investigated in this regard. This dataset has been used in the HDRC competition for recognition of handwritten digits organized in conjunction with ICDAR 2013 [6]. The interesting work proposed in this competition is reported by Gattal et al. 2014 [7], which uses both statistical and structural features derived from the whole or different regions from handwritten digit. In another study, Gattal et al. 2016 [9] extracted both oBIFs based histogram and the background concavity features without any size normalization.

In this paper, we present an innovative way of extending the oBIFs column to multiple scale features using the SVM classifiers. We demonstrate that how a concatenate oBIFs image for scale parameter at two or more scales for column scheme can achieve high performance for digit recognition. Furthermore, a quantitative comparative analysis was performed between our system and some recently proposed methods on the CVL Single Digit dataset.

This paper is organized around two main sections: Sect. 2, that presents in detail the proposed approach, where the features and the classifier used are discussed. The performances resulting from the set of experimentations, in addition to a comparative analysis, are presented in Sect. 3.

2 Proposed System

In this section, we first introduce a new Multi-Scale Oriented Basic Image Features Column (Multi-Scale oBIFs Column) to address the limitations of previous oBIFs Column in terms of performance. The Multi-Scale oBIFs Column is able to find more textural information over the multi-scale oBIFs. Therefore, we aim to design the proposed system for recognizing handwritten digits. The schematic diagram of the proposed method is shown in Fig. 1.

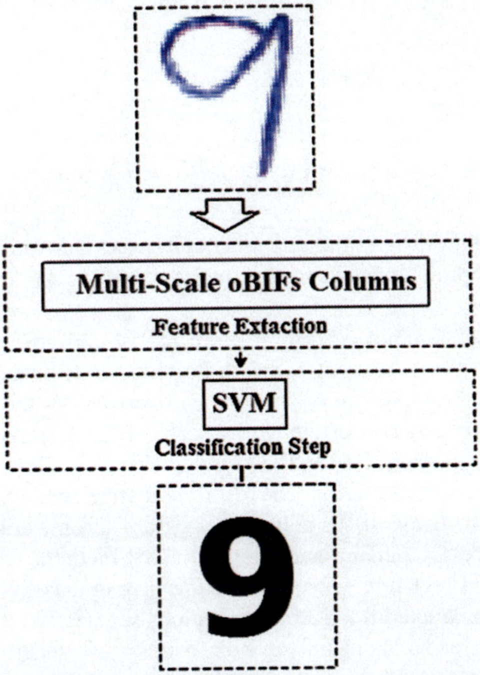

Fig. 1. Workflow of the proposed system

2.1 Multi-Scale Oriented Basic Image Features Column

The classical oBIFs column has been considered in different research problems related to document analysis such as: digit recognition [9], texture classification [12], modern writer identification [13], historical handwritten document recognition [14] and gender classification [15].

The Multi-Scale Oriented Basic Image Features column like classical oBIFs column, is an extension to the Basic Image Features (BIFs) [16, 17]. It catches the textural information in a handwritten document for a discriminatory representation. These representations include flat, dark line on light, and light line slope, dark rotational, light rotational, etc. Based on this local symmetry type, each location in the image is represented in one of these local symmetry classes.

The oBIF features are computed using six different derivatives of Gaussian filters. These filters are parameterized by a scale parameter σ. This is an effective technique for local symmetry class detection. The parameter ε helps identify flat locations from the input image. The oBIF features provide 5n + 3 elements of features dictionary. In our study, 23 features were obtained by setting n = 4 levels.

The oBIF calculation [16, 17] is given using the following equations:

$$\lambda = \sigma^2(s_{20} + s_{02}) \tag{1}$$

$$\gamma = \sigma^2\sqrt{((s_{20} - s_{02})^2 + 4s_{11}^2)} \tag{2}$$

$$s_{mn} = \left\langle \frac{\partial^{m+n}}{\partial x^m \partial y^m} G_\sigma^{(m,n)} | I \right\rangle \tag{3}$$

Where m and n are the differentiation order of Gaussian Derivative function ($G_\sigma^{(m,n)}$), and s_{mn} is the inner product between the image I and a two-dimensional Gaussian Derivative.

To make the oBIFs column descriptor more robust and efficient, we concatenate oBIF image at two or more different scales to provide Multi-Scale oBIFs. In addition, we do the same process to eventually produce two versions of the Multi-Scale oBIFs. Then, we combine two versions of the Multi-Scale oBIFs to provide the Multi-Scale oBIFs column at each location. Hence, the Multi-Scale oBIF column histogram scheme up to $(23)^2$ thus producing 529 entries. The resulting feature vector was also normalized. Different steps of Multi-Scale oBIFs column histogram scheme are shown in Fig. 2.

To further enhance the performance the proposed method, we divided the original image into uniform rectangular blocks of same size and shape. The objective is to combine the features extracted from different regions separately.

The idea is that, instead of using only one oBIF image, we generate two or more oBIF images at different scales, and this is to take advantage of the most pertinent features from these images.

2.2 Classification

For the classification task we selected the Support Vector Machine (SVM) as a classifier [18]. The extracted features from Multi-Scale oBIFs column are used to train our One-Against-All-based multiclass SVM strategies for 10 digit classes. The settings of our SVM classifier are mainly dependent on the soft margin parameter (C) and the Radial Basis Function (RBF) kernel parameter (σ).

An SVM classifier is assigned to the proposed feature vector x of the handwritten digit with a response $f_j(x), j = 1, ..n$; where n is the number of digit classes (10). Equation (4) gives the decision rule of the SVM classifier, where the selected class is the one that maximizes the response:

$$f_{max}(x) = max\{fn_j(x); j = 0, .., n\} \tag{4}$$

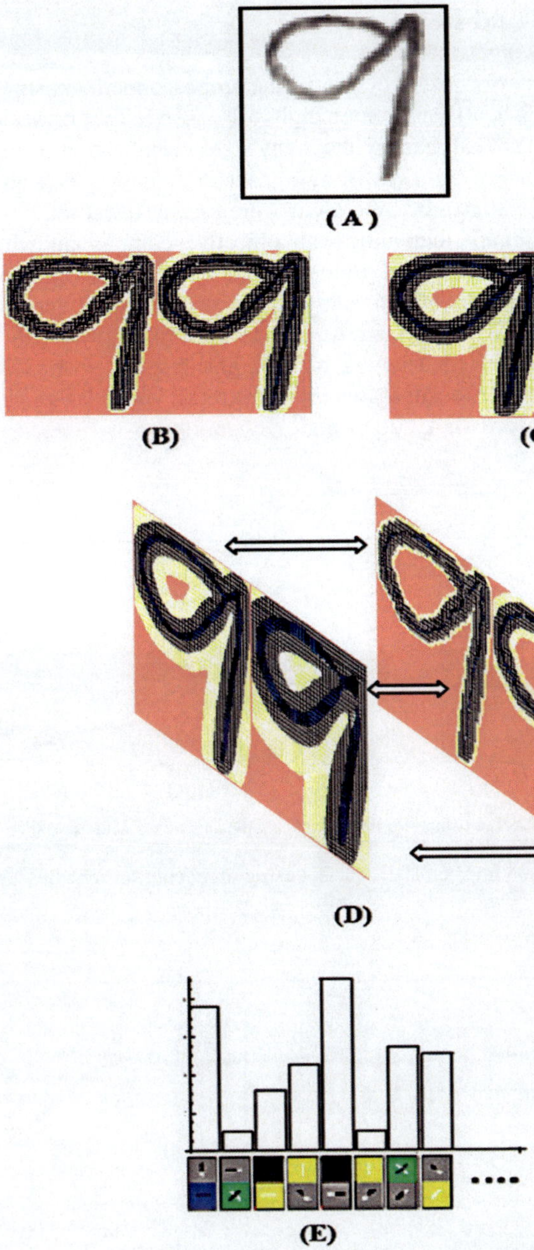

Fig. 2. Multi-Scale oBIFs column histogram method: (A) Input image; (B) oBIFs feture computation: $\sigma = 1$ and $\sigma = 2$, $\varepsilon = 0.0001$ (C) $\sigma = 4$ and $\sigma = 8$, $\varepsilon = 0.0001$ (D) The Multi-scale oBIFs are crossed to form columns at each location (E) Obtained histogram.

3 Experimentations and Results

A set of experiments were performed to evaluated the proposed method using the CVL Single Digit database [6]. First, all handwritten digits are converted to grayscale images. Next, the proposed features are extracted without any size normalization. 14, 000 handwritten digits from the database were used as training set and validation set to parametrize the SVM classifier, and 21, 780 handwritten digits were used as a test set.

To set a neutral benchmark that allows an objective comparison with similar approaches, we adopted the experimental protocol of the ICDAR 2013 competition.

These experiments proved the effectiveness of the texture-based encoding using Multi-Scale oBIFs column for pattern recognition. The proposed feature and the classical oBIFs column feature are generated using two main parameters: scale $\sigma \in \{1,2,4,8\}$ and ε $\varepsilon \in \{0.001\}$. The performance was measured using the recall which is computed using the method proposed in [6] for ICDAR 2013-digit recognition.

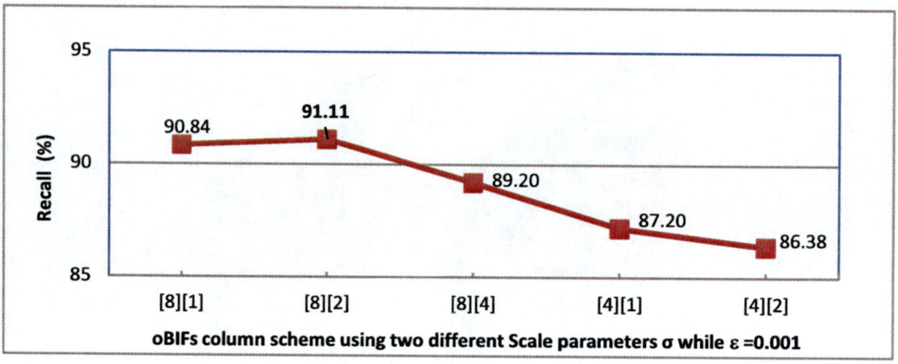

Fig. 3. Recognition rates on CVL dataset using oBIF column scheme.

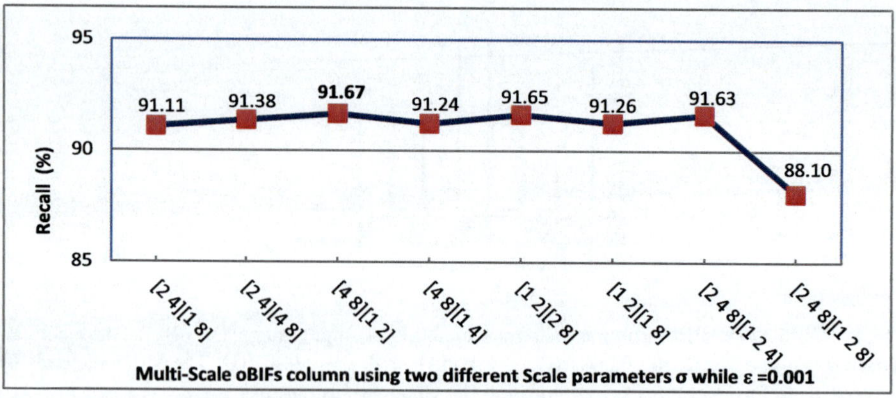

Fig. 4. Recognition rates on CVL dataset using Multi-Scale oBIFs column.

It can be seen from Fig. 3 and Fig. 4 that the proposed Multi-Scale oBIFs column outperforms the oBIFs column. This is due to the cross of best both Multi-Scale oBIFs images. Therefore, the Multi-Scale oBIFs column with $\sigma \in \{1,2\}$ from first Multi-Scale oBIFs image and $\sigma \in \{4,8\}$ from second Multi-Scale oBIFs do better than other Multi-Scale oBIFs column configurations.

In order to improve the system performance, we evaluated Multi-Scale oBIFs column using the best configuration with $\sigma\{[1,2],[4,8]\}$ from different uniform rectangular blocks. Table1 summarizes the recall of combination the extracted feature from different blocks. A highest recall of 94.20% is achieved when the proposed feature extracted from 1×2 blocks making a feature vector of dimension 1,058.

Table 1. Recall with the combination the extracted feature from different blocks

Feature parameters		Blocks	Dim	Recall (%)
F1	$\sigma\{[1,2],[4,8]\}$	1×1	529	91.67
F2	$\sigma\{[1,2],[4,8]\}$	1×2	1,058	89.98
F3	$\sigma\{[1,2],[4,8]\}$	2×1	1,058	94.20
F4	$\sigma\{[1,2],[4,8]\}$	2×2	2,116	93.37
F5	$\sigma\{[1,2],[4,8]\}$	1×3	1,587	89.38
F6	$\sigma\{[1,2],[4,8]\}$	3×1	1,587	94.15
F7	$\sigma\{[1,2],[4,8]\}$	2×3	3,174	91.90

As shown in Table 2, the combined features are used to enhance the performance. The maximum value on the Recall and Precision produced by combination between F1, F3 and F6, which is extracted from the 2×1 blocks, 3×1 blocks as well as from whole image, outperforms all other combined features reporting Recall and Precision of 95.81% and 95.84 respectively.

Table 2. Recall on the combined features

Features combination	Dim	Recall (%)	Precision (%)
F1 + F2	1587	92.47	92.47
F1 + F3	1587	94.76	94.78
F1 + F4	2645	94.94	94.94
F1 + F5	2116	92.83	92.83
F1 + F6	2116	95.43	95.46
F1 + F7	3703	94.63	94.63
F1 + F3 + F6	3174	95.81	95.84

Furthermore, the proposed system was also evaluated for recognition of individual characters. This is especially useful from recognition viewpoint. Table 3 summarizes the results for each class. Generally, the recall values are fairly consistent across different classes. It is worth mentioning that for some digits (3, 8, 9), low recall was obtained. Moreover, some pairs such as "3" and "7", "9" and "8", "7" and "2" etc. are more difficult to distinguish due to high similarity.

A comprehensive comparison of the obtained results and state-of-the-art methods was also performed. Moreover, we compared with two (2) submitted systems (Jadavpur and Tébessa I) to the Digit Recognition Competition (HDRC 2013).

Table 3. Recall on individual classes

Computed classes	Reference classes										
	0	1	2	3	4	5	6	7	8	9	
0	96.51	0.83	0.32	0.05		0.18	0.73		1.01	0.37	
1		99.45	0.28	0.05				0.23			
2	0.55	0.96	97.38	0.28	0.05	0.05		0.28	0.37	0.09	
3	0.09	0.92	0.78	95.09	0.09	0.60	0.05	1.06	0.46	0.87	
4		1.10	0.05	0.05	96.42	0.32	0.55	0.73	0.05	0.73	
5	0.37	0.51	0.09	0.32	0.55	97.47	0.18	0.05	0.09	0.37	
6	0.73	0.78	0.18		0.05	0.69	96.97		0.60		
7	0.09	0.92	1.24	0.23	1.19			95.36	0.32	0.64	
8	0.46	1.15	0.51	0.28	0.09	0.51		0.32	94.31	1.79	
9	0.78	2.20	0.14	2.11	1.65	1.47	0.60	0.32	2.20	89.12	
Recall (%)											**95.81**

It can be seen from Table 4 that our system achieves 2nd highest Precision (95.84%) according to the competition protocol. It is worth mentioning here that we did not employee size normalization. It was also observed that in some cases, our proposed system could not produce desired classification results due to his visual similarity between different classes. In such situation, the proposed method was unable to distinguish between the classes and resulted in miss-classification.

At the opposite of the work presented in Gattal et al. 2014, where the feature space used was high, due to the different types of features used, in this new approach we succeed in reducing the feature space by identifying and using he most robust and significant feature among the set.

Table 4. Comparative analysis of proposed system with other methods

Rank	System	Precision (%)	Normalized digits
1	Gattal et al. 2014 [7]	96.62	No
2	Proposed system	95.84	No
3	Gattal et al. 2016 [9]	95.21	No
4	Jadavpur	94.75	No
5	Tébessa I	77.53	No

4 Conclusion

This paper presented an effective system for isolated handwritten digit recognition that exploits Multi-scale oBIFs column as features. Moreover, the proposed feature is a very efficient for handwritten digit recognition which is robust against size, shape and slant variations. The proposed feature is extracted from the whole image as well as from diverse blocks without any size normalization is investigated. The classification of digits was performed using widely used SVM classifier. Our approach was evaluated under the same conditions of the Digit Recognition Competition (HDRC 2013). The obtained performances were very high comparing the ones reported in the literature.

References

1. Dundar, A., Jin, J., Culurciello, E.: Convolutional clustering for unsupervised learning. In: International Conference on Learning Representations (ICLR), San Juan, Puerto Rico (2016)
2. LeCun, Y., Jackel, L., Bottou, L., Brunot, A., Cortes, C., Denker, J., Drucker, H., Guyon, I., Müller, U., Säckinger, E., Simard, P., Vapnik, V.: Comparison of learning algorithms for handwritten digit recognition. In: Proceedings of the International Conference on Artificial Neural Networks, Nanterre, France, pp. 53–60 (1995)
3. Yamashita, Y., Wakahara, T.: Affine-transformation and 2D-projection invariant k-NN classification of handwritten characters via a new matching measure. Pattern Recogn. **52**, 459–470 (2016)
4. Lauer, F., Suen, C.Y., Bloch, G.: A trainable feature extractor for handwritten digit recognition. Pattern Recogn. **40**(6), 1816–1824 (2007)
5. Gattal, A., Chibani, Y.: SVM-based segmentation-verification of handwritten connected digits using the oriented sliding window. Int. J. Comput. Intell. Appl. (IJCIA) **14**(01), 1550006 (2015)
6. Diem, M., Fiel, S., Garz, A., Keglevic, M., Kleber, F., Sablatnig, R.: ICDAR 2013 competition on handwritten digit recognition (HDRC 2013). In: Proceedings of the 12th International Conference on Document Analysis and Recognition (ICDAR), pp. 1454–1459 (2013)
7. Gattal, A., Chibani, Y., Djeddi, C., Siddiqi, I.: Improving isolated digit recognition using a combination of multiple features. In: Proceedings of 14th International Conference on Frontiers in Handwriting Recognition (ICFHR-2014), Crete Island, Greece, pp. 446–451 (2014)

8. Gattal, A., Chibani, Y.: Segmentation strategy of handwritten connected digits (SSHCD). In: 16th International Conference on Image Analysis and Processing, ICIAP 11, Proceedings, Part II, Ravenna, Italy, September 14–16, pp. 248–254. (2011).
9. Gattal, A., Djeddi, C., Chibani, Y., Siddiqi, I.: Isolated handwritten digit recognition using oBIFs and background features. In: 12th IAPR International Workshop on Document Analysis Systems (DAS-2016), Santorini, Greece, pp. 305–310 (2016)
10. Gattal, A., Djeddi, C., Chibani, Y., Siddiqi, I.: Oriented basic image features column for isolated handwritten digit. In: International Conference on Computing for Engineering and Sciences (ICCES 2017), Istanbul, Turkey, July 22–24, pp. 13–18 (2017)
11. Kimura, Y., Suzuki, A., Odaka, K.: Feature selection for character recognition using genetic algorithm. In: IEEE Fourth International Conference on Innovative Computing, Information and Control (ICICIC), Kaohsiung, December, pp. 401–404 (2009)
12. Newell, A.J., Griffin, L.D.: Natural image character recognition using oriented basic image features. In: Proceedings of the International Conference on Digital Image Computing: Techniques and Applications (DICTA), pp. 191–196 (2011)
13. Newell, A.J., Griffin, L.D.: Writer identification using oriented basic image features and the delta encoding. Pattern Recogn. **47**(6), 2255–2265 (2013)
14. Gattal, A., Djeddi, C., Siddiqi, I., Al-Maadeed, S.: Writer identification on historical documents using oriented basic image features. In: 2018 16th International Conference on Frontiers in Handwriting Recognition (ICFHR), pp. 369–373 (2018)
15. Gattal, A., Chawki, D.I., Siddiqi, C.Y.: Gender classification from offline multi-script handwriting images using oriented basic image features (OBIFs). Exp. Syst. Appl. **99**, 155–167 (2018)
16. Newell, A.J., Griffin, L.D., Morgan, R.M., Bull, P.A.: Texture-based estimation of physical characteristics of sand grains. In: 2010 International Conference on Digital Image Computing: Techniques and Applications, pp. 504–509 (2010)
17. Griffin, L.D., Lillholm, M., Crosier, M., Van Sande, J.: Basic image features (BIFs) arising from approximate symmetry type. In: Lecture Notes in Computer Science (including subseries Lecture Notes in Artificial Intelligence and Lecture Notes in Bioinformatics) (2009). https://doi.org/10.1007/978-3-642-02256-2_29
18. Vapnik, V.N.: The Nature of Statistical Learning Theory. Springer, London (1995)

Delving into Feature Maps: An Explanatory Analysis to Evaluate Weight Initialization

Meenal Narkhede(✉), Prashant P. Bartakke, and Mukul S. Sutaone

College of Engineering, Pune, India
{nmv18.extc,ppb.extc,mssutaone.extc}@coep.ac.in

Abstract. Convolutional neural networks have delivered exceptional performance in various areas of computer vision. There has been growing research to develop deeper architectures with the availability of large datasets. Training such deep networks on large datasets is a tedious process as it involves optimizing a loss function by updating the parameters of the network. Weight initialization is a vital step before training neural networks as the correct choice of network weights ensures that the optimization converges to global minima in the least time. The weight initialization strategies in the literature can be categorized as (1) Initialization without pre-training, and (2) Initialization with pre-training. This paper presents a comparative analysis of the convergence performance of some widely used weight initialization techniques in these categories. This analysis is based on the diversity insights measured in terms of mean standard deviation captured from the feature maps. The experimentation has been carried out by training the AlexNet and VGG16 network on CIFAR-10 and CIFAR-100 datasets. The experimentation results demonstrate that the He initialization technique, which shows the best convergence performance among the others considered for the study, leads the training process such that the diversity of feature maps increases with epochs for both AlexNet and VGG16 network.

Keywords: Weight initialization · Convolutional neural network · Feature map · Diversity · Explanatory analysis

1 Introduction

Convolutional Neural Networks (CNNs) are popularly employed for performing various computer vision applications. The fundamentals of CNNs are based on experiments by Hubel and Wiesel in 1959 [6]. These experiments were carried out to study the response of the visual cortex of cats to the visual stimuli presented in different orientations. It was observed that simple cells responded to a fixed orientation of stimulus in a fixed receptive field, whereas complex cells responded to that orientation irrespective of its position in the receptive field. The authors proposed that the output of complex cells is a result of the sum of outputs of

A. Abraham et al. (Eds.): SoCPaR 2020, AISC 1383, pp. 299–310, 2021.
https://doi.org/10.1007/978-3-030-73689-7_29

simple cells. The notion that complex detectors can be obtained by summing simple detectors is the foundation of CNN.

The first CNN design was the neocognitron presented by Fukushima [1], which performed the visual pattern recognition task on stimulus patterns irrespective of shifts in position and distortions. This network consisted of an input layer followed by a cascade of layers of S-cells and C-cells similar to the simple and complex cells proposed in [6].

The backpropagation algorithm, a popular neural network training algorithm, was presented by Rumelhart et al. [17]. The output for a given set of inputs is predicted in the forward pass of the training process. The error in the predicted and desired output is minimized by updating the network parameters in the backward pass based on the gradient descent rule. A CNN with backpropagation employed for training was proposed by Le Cun et al. [10] for the task of handwritten digit recognition. The proposed network had four hidden layers, followed by an output layer. The hidden layers of the network included feature extractors and subsampling layers. The feature extractors gave an abstraction of the training set while the subsampling layer reduced the feature map resolution. This network performed well in terms of error rate on the test set. The research in CNN has advanced since then and various state-of-the-art CNN architectures like LeNet-5 [11], AlexNet [9], ZFNet [21], GoogleNet [20], VGG16 [19], ResNet [5], etc. have helped to achieved great performance on computer vision tasks. A comprehensive review of advancements and improvements in CNNs related to architecture design, loss functions, regularization, optimization and fast processing is presented in [3].

The trend in the development of CNN shows that networks are going deeper with increasing dataset size, thereby giving better performance accuracy. Deeper networks have a large number of parameters. Training deep networks by backpropagation means optimizing the error function with respect to many parameters, which is a tedious task. The training process must also ensure that the network converges to an acceptable minimum of the loss function. Li et al. [14] have addressed various improvements, such as improving training speed and training accuracy and avoiding false local minima for the backpropagation algorithm. LeCun et al. [12] have discussed that weight initialization plays a vital role in achieving efficient backpropagation. Weight initialization of network is one of the important measures before training a network, as a proper choice of initial weights will ensure convergence of the network to an acceptable minimum and avoid vanishing or exploding gradients. Hence the selected weight initialization strategy should be the one which would lead the network to convergence in less time while maintaining accuracy for any given task.

The contributions of this paper are: A comparative analysis of the convergence performance of CNN initialized with different strategies has been carried out to identify which strategy gives the fastest convergence while maintaining the validation accuracy. This convergence performance has been related and explained by the diversity insights captured from feature maps of the first layer.

This paper is organized as follows: an overview of the weight initialization strategies explored and implemented in this paper is presented in Sect. 2. Section 3 discusses the details of experiments carried out and the results. The conclusions obtained from this study and analysis are presented in Sect. 4.

2 Implemented Weight Initialization Strategies

This section describes the weight initialization strategies which have been explored and considered for the comparative analysis. These initialization strategies belong to one of the following two categories:

- Initialization techniques without pre-training: This broad category explores the initialization strategies where weights are initialized from random distributions and/or later scaled to maintain variance of input and output layers or techniques where weights are derived from the training data.
- Initialization techniques with pre-training: This category includes initialization strategies where the weights of the neural network are obtained by training the network on an unsupervised objective.

2.1 Initialization Techniques Without Pre-training

1. Xavier Initialization:
   Glorot et al. [2] have first considered an initialization technique where weights of a layer have been initialized from a uniform distribution U as shown in Eq. 1.

$$W \sim U\left[\frac{-1}{\sqrt{fan_{in}}}, \frac{1}{\sqrt{fan_{in}}}\right] \tag{1}$$

   where, fan_{in} is the size of previous layer.

   The authors have presented an analysis of the propagation of gradients for the network initialized by this method. It was observed that as the number of layers increased, the backpropagated weight gradients approached to zero. Thus, the authors proposed a normalized initialization technique which derived the scaling factor for weights to be drawn from uniform distribution as shown in Eq. 2.

$$W \sim U\left[\frac{-\sqrt{6}}{\sqrt{fan_{in} + fan_{out}}}, \frac{\sqrt{6}}{\sqrt{fan_{in} + fan_{out}}}\right] \tag{2}$$

   where, fan_{in} is the size of previous layer whereas fan_{out} is the size of next layer.

   The proposed initialization technique maintains the variance of the activations and back propagated gradients.

In this paper, Xavier normal initialization has been implemented where the weights are initialized from a normal distribution with zero mean and standard deviation as given in Eq. 3.

$$sd = \sqrt{\frac{2}{fan_{in} + fan_{out}}} \tag{3}$$

2. He Initialization:
The weight initialization scheme devised in [2] works well for sigmoid and tanh activation functions but has not addressed rectifier nonlinearities. He et al. [4] have proposed an initialization technique by considering ReLU activation function. The authors have selected weights from zero mean Gaussian distribution and derived the standard deviation as given in Eq. 4.

$$W \sim N\left(0, \sqrt{\frac{2}{fan_{in}}}\right) \tag{4}$$

The authors have compared this technique with the one proposed in [2] and the results show that this technique performed better for deep CNNs.
3. Orthogonal Initialization:
Saxe et al. [18] have analyzed the learning dynamics of deep neural networks. This analysis showed that the learning time required for deep neural networks is depth-independent for the networks initialized with weights obtained from unsupervised pre-training. The authors have proposed an initialization technique where the weights from a random normal distribution are decomposed to an orthonormal basis by singular value decomposition (SVD). This proposed method also gave depth-independent learning time and preserved the norm of back propagating gradients as addressed in [2].
4. Layer Sequential Unit Variance (LSUV) Initialization:
Mishkin et al. [16] have proposed an initialization strategy where the variance of each layer output is set equal to one. The algorithm proposed by the authors is as follows:
   - Initialize the network with orthogonal initialization, as proposed in [18].
   - For each layer, compute the output variance by taking mini-batches of training data.
   - Compare the difference of the computed variance and 1 with a pre-defined threshold. If the difference is greater than or equal to a threshold and if the pre-defined number of iterations are not reached, scale weights of that layer as given in Eq. 5.

$$W = \frac{W}{\sqrt{computed\ variance}} \tag{5}$$

5. Initialization based on data statistics:
Koturwar et al. [7] have presented the idea that the weights should be initialized in such a way that the dot product of weights with training data is maximum for relevant information in the training images. For this purpose, the authors have proposed an initialization technique based on the statistics of the data. The proposed algorithm for such initialization is as follows:

- For each layer, draw a batch of training samples from the training data.
- Draw random crops of the filter size of that layer from these images.
- Obtain the mean and covariance of these random crops.
- Initialize the network with weights from a normal distribution with the computed mean and covariance and initialize bias to zero.
- Whiten the initialized weight matrix.
- Scale the whitened weight matrix by the scaling factor as given in [4].

2.2 Initialization Technique with Pre-training

In this type of initialization, the network employed for performing a supervised task is initialized using weights obtained after training the network in an unsupervised way.

Masci et al. [15] have proposed convolutional auto-encoder (CAE) for unsupervised pre-training. The stacked CAE is trained in an unsupervised manner and the weights obtained are used as initial weights for training the CNN on a supervised task. The stacked CAE and CNN have the same topology. The authors have concluded that such an initialization technique helped to capture reasonable features from the data as it learns plausible filters.

3 Experiments and Results

This section discusses the model details, the dataset used, experiments and analysis carried out. The analysis is presented to study the convergence performance and first layer feature maps diversity shown by each of the discussed initialization techniques. The analysis results have also been presented in this section.

3.1 CNN Architectures

The CNN architectures considered for this study are AlexNet [9] and VGG16 [19] as these are two of the standard CNN architectures and have been effective in various image classification tasks. AlexNet architecture has five convolutional layers and three fully connected layers. AlexNet uses a stride of four for the first convolutional layer and a stride of one for the later convolutional layers. VGG16 architecture has thirteen convolutional layers and three fully connected layers. VGG16 uses a stride of one for all convolutional layers.

3.2 Dataset Used

In this paper, CIFAR-10 and CIFAR-100 datasets, which are benchmarking datasets, have been used for the analysis [8]. These datasets are labeled and consist of 60000 images of size 32×32 belonging to 10 and 100 classes respectively. These datasets consists of 50000 training images and 10000 validation images. The CIFAR-10 and CIFAR-100 datasets are first reshaped, such that each image is of size (227,227,3) for AlexNet and (224,224,3) for VGG16. The data is then normalized and augmented using translation and flipping operations.

3.3 Hyperparameter Settings

The activation function in both AlexNet and VGG16 is a rectified linear unit (ReLU). Dropout with a dropping rate of 0.4 has also been employed after every fully connected layer in both the architectures. AlexNet and VGG16 have been trained on the CIFAR-10 and CIFAR-100 dataset with mini-batch stochastic gradient descent optimization for a batch size of 32 and a learning rate of 0.01. Early stopping with the patience of 10 epochs monitoring the validation loss has been employed while training to ensure that the model does not overfit.

3.4 Comparison in Terms of Training and Validation Loss Versus Epochs

The model weights have been initialized with the discussed initialization techniques and the models have been trained to observe the convergence behavior of each of these techniques. Figure 1 and Fig. 2 show the graph of training loss and validation loss versus epochs for training AlexNet on CIFAR-10 and CIFAR-100 respectively. These loss curves show that the model initialized with LSUV and He initialization converges faster as compared to the other techniques.

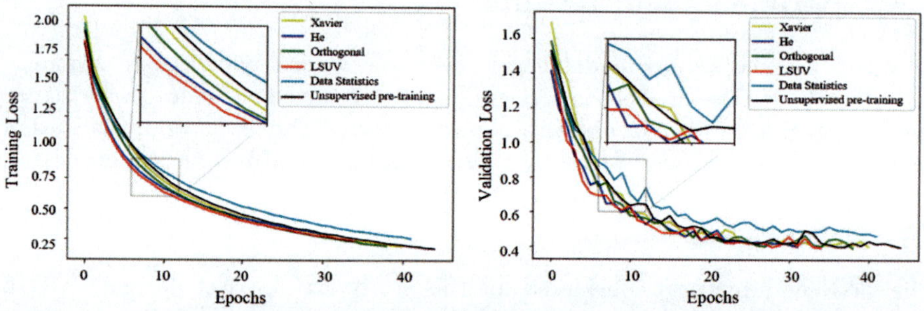

Fig. 1. (a) Training loss versus epochs for AlexNet on CIFAR-10 and (b) Validation loss versus epochs for AlexNet on CIFAR-10 for different weight initialization techniques

The average results after training AlexNet for 10 times with discussed initialization techniques have been shown in Table 1. The training loss (TL), percentage training accuracy (TA), validation loss (VL) and percentage validation accuracy (VA) for epoch numbers (Ep) increasing in powers of two have been mentioned in the table for each initialization technique (Tech). Such increasing epoch numbers have been considered so as to study the convergence performance right from early stages of training. The early stages of training performance are important, as a good performance from the start will lead to fast convergence. The last epoch is the one at which the training was stopped using early stopping.

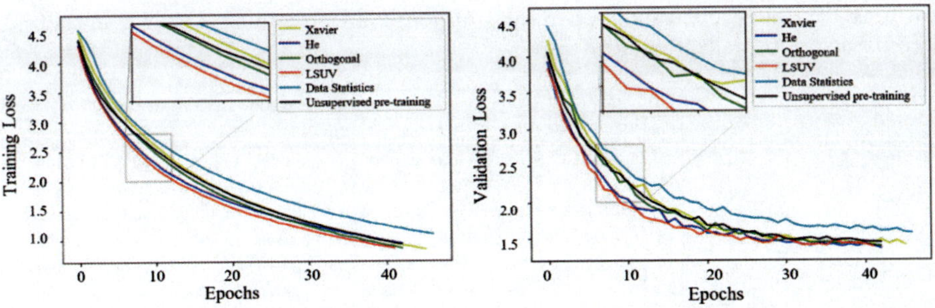

Fig. 2. (a) Training loss versus epochs for AlexNet on CIFAR-100 and (b) Validation loss versus epochs for AlexNet on CIFAR-100 for different weight initialization techniques

It can be observed from Table 1 that the convergence of the He technique is close to that of the LSUV technique in the initial epochs. However, the model with He initialization can reach a low validation loss and validation accuracy in 32 epochs for CIFAR-10 and 43 epochs for CIFAR-100. Thus from the results, it can be concluded that the model initialized with He initialization and LSUV initialization perform better as compared to the other techniques discussed in this paper.

Since the He initialization has shown better performance in terms of convergence as well as diversity of feature maps for AlexNet on both CIFAR-10 and CIFAR-100 datasets, the average training and validation performance of VGG16 network initialized with He initialization technique is shown in Table 2. It can be observed from Table 2 that for convergence VGG16 takes 90 epochs for CIFAR-10 dataset and 128 epochs for CIFAR-100 dataset.

3.5 Analysis of Feature Maps in Terms of Mean Standard Deviation

This subsection presents a brief explanatory analysis of the information variance or diversity in each first layer feature map of a model trained using different initialization techniques. The metric used to quantify such diversity is mean standard deviation, as suggested in [13]. The motivation behind such an analysis of feature maps is to find a quantitative metric to evaluate the features extracted by the learned filters. If meaningful features are extracted from the data in early stages of training, this would help achieve a better performance and faster convergence.

The mean standard deviation (mean-std) of jth feature map in i^{th} layer $F_{i,j}$ where $F_{i,j} \in \mathbb{R}^{w_i \times h_i}$ is given as shown in Eq. 6.

$$Mean\ std(F_{i,j}) = \frac{1}{D} \sum_{d=1}^{D} \sqrt{\frac{\sum_{v=1}^{w_i * h_i} (f_v^d - \bar{f}^d)^2}{w_i h_i - 1}} \tag{6}$$

Table 1. Training and validation results obtained after training AlexNet initialized with discussed initialization strategies

Tech	CIFAR-10					CIFAR-100				
	Ep	TL	TA	VL	VA	Ep	TL	TA	VL	VA
He	1	1.8420	32.3480	1.4109	49.0700	1	4.2926	4.5860	3.9236	10.0199
	2	1.4813	45.9980	1.1947	57.8700	2	3.8614	10.4120	3.4660	18.2699
	4	1.1408	59.2880	0.9210	68.1200	4	3.3049	19.8760	2.9544	27.3400
	8	0.7984	71.9740	0.6458	77.9200	8	2.6388	32.2900	2.3305	40.0099
	16	0.5227	81.7000	0.4996	83.1399	16	1.9015	48.2780	1.7172	53.3599
	32	0.2897	89.8480	0.3999	87.1399	32	1.2073	64.7500	1.4417	60.8099
						43	0.9170	72.5460	1.3607	63.5299
Xavier	1	2.0561	23.5620	1.6874	39.0500	1	4.5027	1.9840	4.2129	5.2499
	2	1.6907	37.9380	1.4511	48.8299	2	4.1418	5.9520	3.8297	11.2400
	4	1.3486	51.2540	1.1553	58.3700	4	3.6210	14.0580	3.3854	18.9899
	8	0.9052	68.1020	0.7719	72.8500	8	2.9511	26.6620	2.6205	33.6800
	16	0.5454	81.0940	0.5425	81.4899	16	2.1149	43.6460	1.9024	49.1200
	32	0.2819	90.1660	0.4133	86.4400	32	1.2555	64.1820	1.4881	60.1300
	41	0.1999	93.0660	0.4003	87.6900	46	0.8201	75.1300	1.4111	62.9999
Orthogonal	1	1.9708	27.3960	1.5791	42.2300	1	4.3841	3.3060	4.0587	7.0500
	2	1.6238	40.3480	1.3957	48.7100	2	4.0046	8.1120	3.6768	13.8999
	4	1.2716	54.2300	1.0298	63.8800	4	3.4951	16.2400	3.3224	19.1799
	8	0.8524	70.0100	0.7668	73.7399	8	2.8314	29.0260	2.5306	35.4799
	16	0.5190	82.0300	0.4878	83.3599	16	2.004	46.2780	1.8513	50.3600
	32	0.2641	90.7940	0.4396	86.5999	32	1.1777	65.9380	1.4610	60.1899
	39	0.2004	93.0440	0.3907	88.0599	43	0.8389	74.7220	1.3903	62.8400
LSUV	1	1.8625	31.5300	1.4921	45.3799	1	4.2575	5.2300	3.8191	11.3200
	2	1.4945	45.9700	1.2631	55.7500	2	3.8060	11.5960	3.5254	16.3699
	4	1.1207	60.1280	0.9538	66.5400	4	3.2654	20.6640	2.9070	27.7700
	8	0.7693	73.1800	0.6943	76.0399	8	2.5600	34.2640	2.1747	42.5900
	16	0.5018	82.5540	0.4785	83.6199	16	1.8212	50.3900	1.7034	53.3500
	32	0.2733	90.5200	0.4025	87.0100	32	1.1089	67.6080	1.4462	60.8399
	35	0.2435	91.4460	0.3949	87.5699	42	0.8394	74.7860	1.3830	63.4800
Data statistics	1	2.0035	26.9800	1.5069	44.8500	1	4.5441	1.8280	4.4105	3.6499
	2	1.5628	43.4080	1.3157	53.8600	2	4.3677	4.6080	4.1870	7.7500
	4	1.2684	54.7500	1.0688	62.2399	4	3.7558	12.5160	3.4188	19.0099
	8	0.9494	66.9900	0.8557	70.7099	8	3.0251	25.4140	2.7329	32.3599
	16	0.6349	78.1200	0.6284	78.6400	16	2.3080	39.7540	2.1348	44.5300
	32	0.3588	87.4980	0.4884	84.3800	32	1.5097	57.8160	1.6958	54.6000
	42	0.2590	91.0020	0.4603	85.8500	47	1.0761	68.4060	1.5704	58.7899
Unsupervised pre-training	1	1.9412	25.5720	1.5271	44.2099	1	4.3570	3.7060	4.0370	8.2299
	2	1.6039	41.5340	1.3830	50.5599	2	3.9401	9.4620	3.6561	14.6500
	4	1.3294	52.2920	1.0986	60.6899	4	3.4046	17.9800	3.1438	23.3600
	8	0.9362	66.9740	0.7763	73.0899	8	2.8369	29.0700	2.6041	34.3800
	16	0.5710	80.0900	0.5764	79.7599	16	2.0840	44.5560	1.8885	49.7799
	32	0.2920	89.8240	0.4192	86.6100	32	1.2475	64.2820	1.4741	60.1000
	45	0.1797	93.7160	0.3965	88.0400	43	0.9057	73.1700	1.4398	61.4799

Where, f_v^d is the vth element in the flattened feature map $F_{i,j} \in \mathbb{R}^{w_i \times h_i}$, $\bar{f}^d$ is the mean of f_v^d and d is the sample index in sample data considered for obtaining feature maps.

The higher is the value of mean-std, the more diversity exists in that feature map, which helps to contribute more to extracting meaningful features.

Table 2. Training and validation results obtained after training VGG16 initialized with He initialization strategy

Tech	CIFAR-10					CIFAR-100				
	Ep	TL	TA	VL	VA	Ep	TL	TA	VL	VA
He	1	2.0003	26.4560	1.7594	36.8400	1	4.5124	2.5260	4.2312	6.5600
	2	1.7734	35.3960	1.8608	35.1099	2	4.2416	5.4380	4.0040	10.5300
	4	1.5549	43.248	1.3376	51.3199	4	3.9434	9.5960	3.6908	14.3800
	8	1.3077	52.8640	1.1285	59.1799	8	3.6222	14.7140	3.3470	20.8700
	16	0.9827	64.8262	0.8360	70.8599	16	3.1720	22.5400	2.8818	28.7800
	32	0.6673	76.5640	0.6156	79.0099	32	2.5341	34.7200	2.2656	40.9799
	64	0.4025	85.8620	0.4030	86.1699	64	1.6638	54.0320	1.6372	55.5599
	90	0.1686	94.1200	0.3257	89.5299	128	1.2282	64.5320	1.3853	61.9000

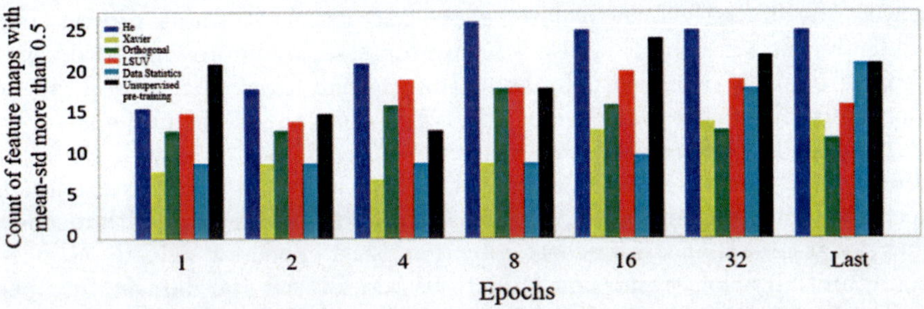

Fig. 3. Count of feature maps with mean standard deviation greater than 0.5 for AlexNet trained on CIFAR-10 dataset

The mean-std values have been obtained for all 96 first layer feature maps of AlexNet and 64 first layer feature maps of VGG16. Hence, there are 96 and 64 mean-std values for AlexNet and VGG16 respectively. These mean-std values have been normalized so that they lie in the range of (0,1). The mean-std values have been found for all the networks initialized with the six discussed techniques for AlexNet. A count of feature maps whose mean-std values are greater than 0.5 for early epochs and last epoch for AlexNet trained on CIFAR-10 and CIFAR-100 have been shown in Fig. 3 and Fig. 4 respectively.

The expected behavior for the count of feature maps with diversity above 0.5 versus epochs should be rapidly increasing with training. It can be observed that for the first epoch, the unsupervised pre-training for CIFAR-10, while the orthogonal initialization for CIFAR-100 gives feature maps with the highest diversity. However, this initialization leads the training process such that diversity does not improve by the last epoch. The He initialization shows the second-highest number of diverse feature maps at the first epoch and leads the model

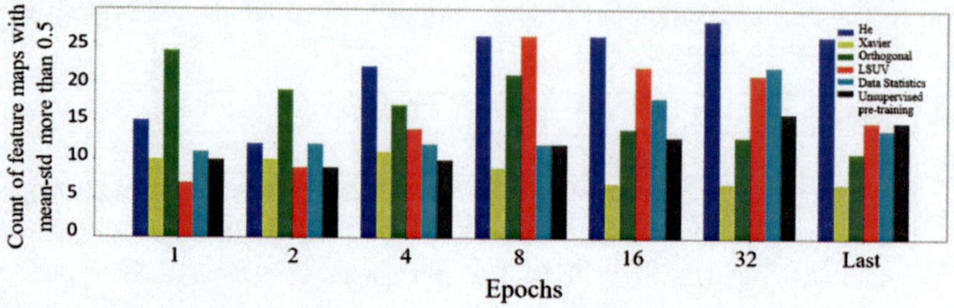

Fig. 4. Count of feature maps with mean standard deviation greater than 0.5 for AlexNet trained on CIFAR-100 dataset

training such that the feature maps' diversity increases for both the datasets. The LSUV initialization shows better and increasing diversity in the early epochs; however, the diversity falls by later epochs. Thus, the training process when weights are initialized by He initialization progresses by improving the diversity in feature maps, which eventually helps this method show a better convergence performance as seen in the previous subsection than the other methods.

The mean-std values of feature maps for VGG16 initialized with He initialization has been obtained as He initialization has shown to obtain diverse feature maps as training progresses for AlexNet. A count of feature maps with mean-std values are greater than 0.5 for early epochs and last epoch for VGG16 initialized with He initialization trained on CIFAR-10 and CIFAR-100 datasets has been shown in Fig. 5. It can be observed from Fig. 5 that He initialization shows close to expected behavior of diversity of feature maps.

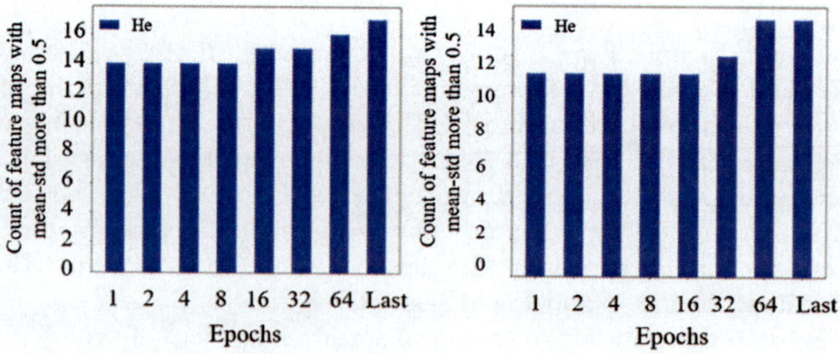

Fig. 5. (a) Count of feature maps with mean standard deviation greater than 0.5 for VGG16 trained on CIFAR-10 (b) Count of feature maps with mean standard deviation greater than 0.5 for VGG16 trained on CIFAR-100

4 Conclusions and Outlook

Weight initialization is a vital step before training neural networks as the correct choice of weights can lead to achieving faster convergence. This paper has presented a study and comparative analysis of some widely adopted techniques of weight initialization. The AlexNet model initialized with the selected techniques has been trained on the benchmark CIFAR-10 and CIFAR-100 datasets. The convergence performance for each initialization technique has been compared in terms of training loss versus epochs and validation loss versus epochs. From the comparison of convergence performance, it has been observed that the He initialization technique and LSUV initialization technique perform closely with each other and better as compared to the others. An explanatory analysis of insights captured from feature maps has also been presented with a view of quantifying the features learned by CNN initialized with different strategies. A mean standard deviation has been obtained from the first layer feature maps at different stages of training. The mean standard deviation analysis of feature maps shows that the He initialization is capable of extracting diverse features from the data, while LSUV initialization improves feature map diversity in early stages after first epoch as the model training progresses. To validate the results, VGG16 network initialized with He initialization was also trained on CIFAR-10 and CIFAR-100 datasets. The He initialization showed increasing count of diverse feature maps as training progresses for VGG16 as well. Thus it can be concluded that the initialization technique which leads to a better convergence and better performance in terms of validation loss and accuracy learns diverse feature maps in the training process.

A weight initialization scheme that would help to reduce the convergence time significantly is required. The random behavior of increasing and decreasing diversity in feature maps through epochs has to be explored and justified. Also, more metrics for quantifying the quality of features learned during the training process would help initialize first layer weights accordingly. We hope that this paper provides an understanding of weight initialization and feature map insights and encourages further research for proposing new schemes to achieve faster convergence.

References

1. Fukushima, K.: Neocognitron: a self-organizing neural network model for a mechanism of pattern recognition unaffected by shift in position. Biol. Cybern. **36**(4), 193–202 (1980)
2. Glorot, X., Bengio, Y.: Understanding the difficulty of training deep feedforward neural networks. In: Proceedings of the Thirteenth International Conference on Artificial Intelligence and Statistics, pp. 249–256 (2010)
3. Gu, J., Wang, Z., Kuen, J., Ma, L., Shahroudy, A., Shuai, B., Liu, T., Wang, X., Wang, G., Cai, J., et al.: Recent advances in convolutional neural networks. Pattern Recogn. **77**, 354–377 (2018)

4. He, K., Zhang, X., Ren, S., Sun, J.: Delving deep into rectifiers: surpassing human-level performance on imagenet classification. In: Proceedings of the IEEE International Conference on Computer Vision, pp. 1026–1034 (2015)
5. He, K., Zhang, X., Ren, S., Sun, J.: Deep residual learning for image recognition. In: Proceedings of the IEEE Conference on Computer Vision and Pattern Recognition, pp. 770–778 (2016)
6. Hubel, D.H., Wiesel, T.N.: Receptive fields of single neurones in the cat's striate cortex. J. Physiol. **148**(3), 574–591 (1959)
7. Koturwar, S., Merchant, S.: Weight initialization of deep neural networks (dnns) using data statistics. arXiv preprint arXiv:1710.10570 (2017)
8. Krizhevsky, A., Nair, V., Hinton, G.: Cifar-10 and cifar-100 datasets, vol. 6 (2009). https://www.cs.toronto.edu/kriz/cifar.html
9. Krizhevsky, A., Sutskever, I., Hinton, G.E.: Imagenet classification with deep convolutional neural networks. In: Advances in Neural Information Processing Systems, pp. 1097–1105 (2012)
10. LeCun, Y., Boser, B.E., Denker, J.S., Henderson, D., Howard, R.E., Hubbard, W.E., Jackel, L.D.: Handwritten digit recognition with a back-propagation network. In: Advances in Neural Information Processing Systems, pp. 396–404 (1990)
11. LeCun, Y., Bottou, L., Bengio, Y., Haffner, P.: Gradient-based learning applied to document recognition. Proc. IEEE **86**(11), 2278–2324 (1998)
12. LeCun, Y.A., Bottou, L., Orr, G.B., Müller, K.R.: Efficient backprop. In: Neural Networks: Tricks of the Trade, pp. 9–48. Springer, Berlin (2012)
13. Li, H., Ma, C., Xu, W., Liu, X.: Feature statistics guided efficient filter pruning. arXiv preprint arXiv:2005.12193 (2020)
14. Li, J., Cheng, J.h., Shi, J.y., Huang, F.: Brief introduction of back propagation (bp) neural network algorithm and its improvement. In: Advances in Computer Science and Information Engineering, pp. 553–558. Springer, Berlin (2012)
15. Masci, J., Meier, U., Cireşan, D., Schmidhuber, J.: Stacked convolutional auto-encoders for hierarchical feature extraction. In: International Conference on Artificial Neural Networks, pp. 52–59. Springer, Berlin (2011)
16. Mishkin, D., Matas, J.: All you need is a good init. arXiv preprint arXiv:1511.06422 (2015)
17. Rumelhart, D.E., Hinton, G.E., Williams, R.J.: Learning internal representations by error propagation. California Univ San Diego La Jolla Inst for Cognitive Science, Technical report (1985)
18. Saxe, A.M., McClelland, J.L., Ganguli, S.: Exact solutions to the nonlinear dynamics of learning in deep linear neural networks. arXiv preprint arXiv:1312.6120 (2013)
19. Simonyan, K., Zisserman, A.: Very deep convolutional networks for large-scale image recognition. arXiv preprint arXiv:1409.1556 (2014)
20. Szegedy, C., Liu, W., Jia, Y., Sermanet, P., Reed, S., Anguelov, D., Erhan, D., Vanhoucke, V., Rabinovich, A.: Going deeper with convolutions. In: Proceedings of the IEEE Conference on Computer Vision and Pattern Recognition, pp. 1–9 (2015)
21. Zeiler, M.D., Fergus, R.: Visualizing and understanding convolutional networks. In: European Conference on Computer Vision, pp. 818–833. Springer, Cham (2014)

Historical Kannada Handwritten Character Recognition Using Machine Learning Algorithm

Parashuram Bannigidad and Chandrashekar Gudada(✉)

Rani Channamma University, Belagavi 591156, KA, India

Abstract. A character is the smallest unit in every line of a script, the development of the script stretches over millions of inscriptions in Kannada, and these inscriptions play a vital role in the reconstruction of Karnataka's history and its culture. The Kannada script has evolved for more than 1500 years and many inventions, and variations have taken place in the scripts used by various dynasties. Due to these variations occurred in the script, it has been improved in text and formed in a new different style. The objective of the present paper is to digitize and restore the historical Kannada handwritten scripts by applying image enhancement techniques and recognize the individual characters by extracting the HOG features. The LDA, K-nearest neighbour (K-NN), and SVM classifiers are used to identify the dynasties of the characters, whether it belongs to the dynasty of the Hoysala or Vijayanagara or the Mysore Wodeyar. The average classification accuracy of the historical Kannada handwritten characters from the different regimes is; the LDA classifier has yielded 68.4%, the K-NN classifier achieved 85.7%, and the SVM classifier yielded 87.5%. Based on the experimentation, it is noted that the overall classification accuracy is relatively improved with the SVM classifier compared to LDA and K-NN classifiers. Further, the results are also verified manually obtained by epigraphists and linguistic experts, which proves the efficacy of the proposed approach.

Keywords: Kannada · Historical documents · Manuscript · HOG · Machine learning · SVM · Character recognition · Document image analysis

1 Introduction

In the state of Karnataka, the people were predominantly spoken Kannada language, which was known as the Dravidian language. The evolution of Kannada script has spread over millions of inscriptions, which play an important role in

The authors are grateful to the authorities of the Rani Channamma University, Belagavi, Karnataka, India for sanctioning minor research project and provided financial assistance to carry out this research work (RCUB/PMEB/2020-21/37 Dt. 15.05.2020).

A. Abraham et al. (Eds.): SoCPaR 2020, AISC 1383, pp. 311–319, 2021.
https://doi.org/10.1007/978-3-030-73689-7_30

reconstructing Indian history and bringing it to the modern generations [5]. The evolution of Kannada's character has undergone many inventions and changes during its usage throughout the centuries of dynasties. In all dynasties, the Kalyana Chalukya script is one of the most decorative forms of ancient Kannada script. The Kalyana Chalukyas script was used by the Hoysala kings in a more decorative and cursive manner. The present script of Kannada is almost like the script of Kalyana Chalukya and Hoysala. During this time soapstones were used to allow such scripts to be cursive in writing. In the 14th and 16th centuries, the rulers of Vijayanagara gave less priority to the writing of documents, as the majority of records were grafted on raw granite stones [6,7]. The Mysore Wodeyars are the last dynasties ruled in 18th to 19th Century AD, during this period, the Vijayanagar scripts were developed, which was almost similar to the Aadhunika Kannada scripts. All these ancient manuscripts were available in different form, namely, Paper (Hastaprati), Palm leaf (Tadole), Metal, Stone carvings, and Cloth inscriptions, etc. In the proposed investigation, only the paper inscriptions of historical Kannada handwritten manuscripts are considered for developing the machine learning algorithms. The various developmental stages of Kannada handwritten manuscripts of different dynasties are shown in Fig. 1.

Most of the research work carried out by various authors in this area are; the Cun-Zhao et al. [1] has proposed an algorithm for recognition of characters from natural scenes using multi-scale graph-matching based kernel. Recognition of Marathi handwritten characters using R-HOG features was investigated by Kamble et al. [9]. Iamsa-at et al. [10] have presented an algorithm for recognition of handwritten characters using HOG features with Deep Learning of Artificial Neural Network. The Surinta et al. [11] investigated the classification of handwritten characters by using HE hotspot features. Recognition of offline handwritten characters using the binarization technique was done by Choudhary et al. [12]. Victor Lavrenko et al. [13] proposed an algorithm for holistic word recognition from historical handwritten documents. The transcriptorium project was investigated with handwritten text recognition for historical documents by Joan Andreu et al. [14]. The feature extraction technique using character geometry for recognition of characters was done by Dileep et al. [2]. Romero et al. [15] proposed the handwritten text identification of historical documents. Classical Mongolian words recognition in the historical document was investigated by Guanglai Gao et al. [16]. Matching the Stone inscription characters of Kannada using SIFT features has been carried out by Mohana et al. [17]. The recent achievements in the offline handwritten recognition system were proposed by Verma et al. [18]. HOG and GLCM feature-based Age-type identification of historical Kannada handwritten manuscript images were presented by Parashuram et al. [3,19]. In [4,8,20], Parashuram et al., have suggested the restoration and binarization methods for degraded Kannada historical handwritten manuscript images and ensures the image quality.

Fig. 1. The various developmental stages of Kannada handwritten manuscripts of different dynasties

2 Proposed Method

The main objective of the proposed research work is to digitize, restore, reconstruct and recognize the Kannada handwritten characters by extracting the HOG features. The LDA, K-nearest neighbour (K-NN) and SVM classifiers are used to recognize the dynasties of the characters using 5-fold experimentation, whether the script belongs to the dynasty of Hoysala or Vijayanagara or the dynasty of Mysore Wodeyar. The historical Kannada handwritten manuscripts were collected from various universities in Karnataka and are obtained in digital format by using digital camera with higher magnification.

2.1 Pre-processing

Generally, the historical handwritten manuscripts contain smear, background uneven illumination, patches due to age and ink bleed-through, normally the writing style of each writer changes from inscription to inscription. Due to this changes, creates the confusion and complexity to recognize the historical document. The stages of pre-processing will play an important role, which includes image enhancement, restoration to improve the quality of the image, and removes the un-wanted area. In the previous work, the new image enhancement technique was carried out for quality enhancement [17–19]. The process of the proposed method is represented graphically in the Fig. 2.

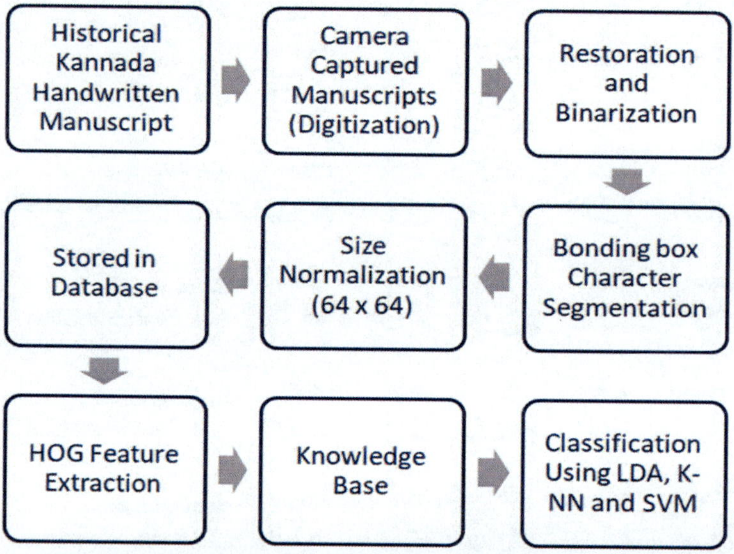

Fig. 2. The process of the proposed method

2.2 Feature Extraction

Obviously, the handwritten manuscript image contain different shapes and styles of character. Hence, the select of dominant features is tedious task. In this paper, the HOG features [1] are used and extracted based on their scale-invariant and rotation invariant for recognition of the Kannada handwritten characters. The detailed process is described in the below algorithm:

Algorithm: Historical Kannada handwritten character recognition using HOG features.

Step 1. Input the camera captured colour image.
Step 2. Apply image enhancement techniques to binarize the image.
Step 3. Then apply the bonding box segmentation method on Step 2 and obtain individual blocks of characters and store them as a individual image in the database.
Step 4. Apply size normalization of size 64×64 to each character.
Step 5. Obtain HOG features from Step 4 and store them for training purpose.
Step 6. Repeat steps between 2 and 4 for all the age-types: namely, Hoysala, Vijayanagar, and Mysore Dynasties
Step 7. Apply LDA, K - nearest neighbor and SVM classifiers to classify and recognize the historical Kannada handwritten characters, whether the script belongs to the dynasty of Hoysala or Vijayanagara or the dynasty of Mysore Wodeyar.

3 Experimental Results and Discussions

In this experimentation, there are 121 historical manuscripts of different dynasties are considered. The bonding box character segmentation method is applied, and HOG features are extracted to recognize the characters of dynasties. The computed results extracted from the manuscripts is shown in Table 1. The experimentation is carried out on Intel Core i5, @ 2.40 GHz system using MATLAB R2015b. The original sample script image of Mysore Wodeyar dynasty (1936 AD) (Fig. 3(a)), is converted into a grayscale image and considered only green channel and apply the combination of local otsu and global otsu to reconstruct the individual character and output the enhanced binarized image (Fig. 3(b)), apply the bonding box character segmentation method on Fig. 3(b) and obtained individual characters (Fig. 3(c)). Apply size normalization of size 64×64 to each individual character (Fig. 3(d)). Extract the HOG features for all the individual characters from Fig. 3(d) and store them in the matrix form for training purpose.

Table 1. The computed results extracted from the manuscripts

Dynasty (Age-type)	No. of Manuscripts used (Written on paper)	Total number of characters extracted from the Manuscripts
Hoysala (1340)	39	12350
Vijayanagara (1400)	30	11467
Vijayanagara (1460)	24	7503
Mysoure Wodeyar (1936)	28	7653

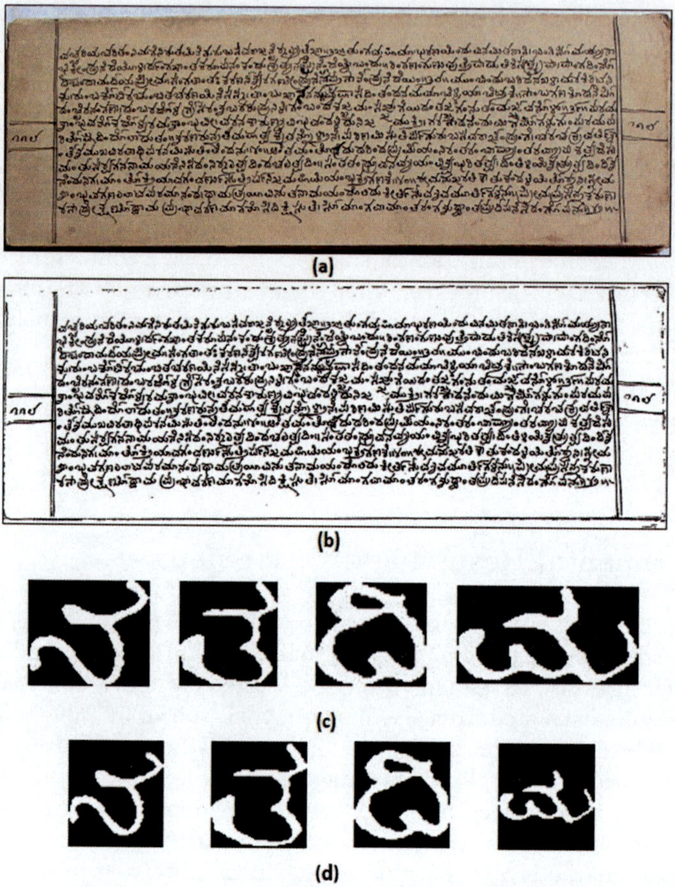

Fig. 3. Sample images of proposed algorithm (a) Original handwritten manuscript image of Mysore Wodeyar dynasty (1936 AD), (b) Binarized image after restoration, (c) Bonding box character segmentation on (b), (d) Size normalized images of size 64×64

Finally, the classification is done by applying LDA, K-NN, and SVM classifiers with 5-fold experimentation. The classification performance of the proposed work with LDA, K-NN and SVM classifiers are given in the Table 2. The average recognition accuracy for different dynasties of historical Kannada handwritten manuscripts are; the LDA classifier has yielded 68.4%, K-NN classifier has 85.7%, and the SVM classifier has achieved 87.5%. The overall results explores that the SVM classifier has an excellent recognition rate for historical Kannada handwritten document images compared with LDA and K-NN classifiers. The results of the confusion matrix of LDA, K-NN, and SVM classifiers are given in the Table 3, Table 4, and Table 5, respectively.

Table 2. The classification performance of the proposed work with LDA, K-NN and SVM classifiers

Dynasty (Age-type)	LDA		K-NN		SVM	
	Recognition rate (%)	Error rate (%)	Recognition rate (%)	Error rate (%)	Recognition rate (%)	Error rate (%)
Hoysala (1340)	72	28	83	17	87	13
Vijayanagara (1400)	66	34	91	9	89	11
Vijayanagara (1460)	76	24	93	7	92	8
Mysoure Wodeyar (1936)	58	42	74	26	82	18
Average accuracy	68.4 (%)		85.7 (%)		87.5 (%)	

Table 3. Confusion matrix for LDA classifier

Dynasty (Age-type)	Hoysala (1340)	Vijayanagara (1400)	Vijayanagara (1460)	Mysoure Wodeyar (1936)	Total no. of characters
Hoysala (1340)	8922	1567	1037	824	12350
Vijayanagara (1400)	1765	7601	655	1446	11467
Vijayanagara (1460)	1018	435	5720	330	7503
Mysoure Wodeyar (1936)	1351	1339	552	4411	7653

Table 4. Confusion matrix for K-NN classifier

Dynasty (Age-type)	Hoysala (1340)	Vijayanagara (1400)	Vijayanagara (1460)	Mysoure Wodeyar (1936)	Total no. of characters
Hoysala (1340)	10310	1107	548	385	12350
Vijayanagara (1400)	619	10463	155	230	11467
Vijayanagara (1460)	321	128	6973	81	7503
Mysoure Wodeyar (1936)	1022	678	292	5661	7653

Table 5. Confusion matrix for SVM classifier

Dynasty (Age-type)	Hoysala (1340)	Vijayanagara (1400)	Vijayanagara (1460)	Mysoure Wodeyar (1936)	Total no. of characters
Hoysala (1340)	10704	611	462	573	12350
Vijayanagara (1400)	703	10239	137	388	11467
Vijayanagara (1460)	385	105	6888	125	7503
Mysoure Wodeyar (1936)	695	498	196	6264	7653

4 Conclusion

In this research work, an algorithm is proposed to recognize handwritten characters from the historical Kannada handwritten manuscripts by applying bonding box segmentation method and extracted the HOG features. The LDA, K-nearest neighbour (K-NN) and SVM classifiers are used to recognize the dynasties of the characters, whether the script belongs to the dynasty of Hoysala or Vijayanagara or the dynasty of Mysore Wodeyar. The average classification accuracy of the historical Kannada handwritten characters from the different dynasties are; the LDA classifier yields 68.4% K-NN classifier has yielded 85.7%, and the SVM classifier has yielded 87.5%. During the implementation, it is noted that, the classification performance of SVM for historical handwritten Kannada documents has been relatively good compared to LDA and K-NN classifiers. However, the same algorithm may be implemented for degraded historical Kannada handwritten document images of other dynasties and the results may be compared with other algorithm in the literature, which will be considered in the future work.

References

1. Cun-Zhao, S., Chun-Heng, W., Bai-Hua, X., Zhang, Y., Song, G.: Multi-scale graph-matching based kernel for character recognition from natural scenes. Acta Automatica Sinica **40**(4), 751–756 (2014)
2. Gaurav, D.D., Ramesh, R.: A feature extraction technique based on character geometry for character recognition (2012). CoRR, vol.abs/1202.3884
3. Bannigidad, P., Gudada, C.: Identification and classification of historical Kannada handwritten document images using GLCM features. Int. J. Adv. Res. Comput. Sci. **9**(1), 686–690 (2018). https://doi.org/10.26483/ijarcs.v9i1.5437
4. Bannigidad, P., Gudada, C.: Restoration of degraded non-uniformally illuminated historical Kannada handwritten document images. Int. J. Comput. Eng. Appl. **XII**(I), 1–13 (2018)
5. Manjunath, M.G., Devarajaswamy, G.K.: Kannada Lipi Vikasa, 1st edn. Jagadhguru Sri Madhvacharya Trust, Sri Raghavendra Swami Matta, Mantralaya (2004)
6. Narasimha Murthy, A.V.: Kannada Lipiya Ugama Mattu Vikasa. Kannada Adhyayana Samsthe, Mysore University, Mysore (1968)
7. Reddy, D.: Lipiya Huttu Mattu Belavanige-Origin and Evolution of Script. Kannada Pustaka Pradhikara (Kannada Book Authority), Bangalore (2011)
8. Bannigidad, P., Gudada, C.: Restoration of degraded historical kannada handwritten document images using image enhancement techniques. In: Proceedings of the Eighth International Conference on Soft Computing and Pattern Recognition (SoCPaR 2016), pp. 498–508. Springer, Cham (2016)
9. Kamble, P.M., Hegadi, R.S.: Handwritten Marathi character recognition using R-HOG feature. Procedia Comput. Sci. **45**, 266–274 (2015). https://doi.org/10.1016/j.procs.2015.03.137
10. Iamsa-at, S., Horata, P.: Handwritten character recognition using histograms of oriented gradient features in deep learning of artificial neural network. In: Proceedings of the International Conference on IT Convergence and Security (ICITCS), pp. 1–5 (2013). https://doi.org/10.1109/ICITCS.2013.6717840

11. Surinta, O., Schomaker, L., Wiering, M.: Handwritten character classification using he hotspot feature extraction technique. In: Proceedings of the 1st International Conference on Pattern Recognition Applications and Methods, pp. 261–264 (2012). https://doi.org/10.5220/0003712002610264
12. Choudharya, A., Rishib, R., Ahlawat, S.: Off-line handwritten character recognition using features extracted from binarization technique. In: AASRI Conference on Intelligent Systems and Control (2013)
13. Lavrenko, V., Manmath, R.R.: Holistic word recognition for handwritten documents. In: Proceedings of the First International Workshop on Document Image Analysis for Libraries (DIAL-04), p. 278 (2004)
14. Sanchez, J.A., Bosch, V., Romero, V., Depuydt, K., de Does, J.: A Handwritten text recognition for historical documents in the tranScriptorium Project. In: DATeCH 2014 Proceedings of the First International Conference on Digital Access to Textual Cultural Heritage, pp. 111–117 (2014)
15. Romero, V., Serrano, N., Toselli, A.H., Sanchez, J.A., Vidal, E.: Handwritten text recognition for historical documents. In: Proceedings of Language Technologies for Digital Heritage and Cultural Heritage Workshop, pp. 90–96 (2011)
16. Gao, G., Xiangdong, S., Wei, H., Gong, Y.: Classical Mongolian words recognition in historical document. In: ICDAR, vol. 1520–5363, no. 11, pp. 692–697. IEEE (2011)
17. Mohana, H.S., Navya, K., Srikanth, P.C., Shivakumar, G.: Stone inscribed Kannada character matching Using SIFT. In: Proceeding of IRF International Conference, pp. 126–131 (2014)
18. Verma, B., Blumenstein, M., Kulkarni, S.: Recent achievements in offline handwriting recognition system. In: International Conference on Computational Intelligence and Multimedia Applications (1998)
19. Bannigidad, P., Gudada, C.: Age-type identification and recognition of historical Kannada handwritten document images using HOG feature descriptors. In: Proceedings of The International Conference on Computing, Communication and Signal Processing (ICCASP) (2018)
20. Bannigidad, P., Gudada, C.: Restoration of degraded Kannada handwritten paper inscriptions (hastaprati) using image enhancement techniques. In: Proceedings of the IEEE International Conference on Computer Communication and Informatics (ICCCI -2017), pp. 1–6 (2017). https://doi.org/10.1109/ICCCI.2017.8117697

Home Health Care Network Management Under Fuzzy Environment Using Meta-heuristic Algorithms

Fariba Goodarzian[1], Aida Goodarzian[1], Ajith Abraham[1](✉), and Sohaib Dastgoshade[2]

[1] Machine Intelligence Research Labs (MIR Labs), Scientific Network for Innovation and Research Excellence, 11, 3rd Street NW, P.O. Box 2259, Auburn, WA 98071, USA
fariba.goodarzian@mirlabs.org, ajith.abraham@ieee.org
[2] Department of Industrial Engineering, Yazd University, Yazd, Iran
Sohaib.dastgoshade@stu.yazd.ac.ir

Abstract. Home Health Care (HHC) Services is an alternative to hospitalization, which plays an important role in reducing hospital costs for patient care costs. Also, resource planning and routing problems are among the most important problems that need to be addressed when planning home health care resources. Home health care scheduling problem is a combination of the generalization of vehicle routing problem with the time window and the employee planning problem, which ultimately aims to identify and allocation tasks to therapists, taking into account time windows for patients and employee preferences while the best routes are identified. In this paper, we presented a bi-objective model in the planning horizon of more than one day in order to reduce system costs and increase employee satisfaction simultaneously. In addition, the routing-allocation-scheduling-planning of home health care problem is of NP-hard problem. Meanwhile, we use a fuzzy approach to cope with this uncertainty due to the uncertain nature of transportation and travel cost parameters in the routing-allocation-planning of HHC problem. To solve this model in different sizes, we used meta-heuristic algorithms.

Keywords: Home health care · Meta-heuristic algorithms · Fuzzy environment

1 Introduction

With the advancement of medical and healthcare in recent years, we often face an increase in the number of elderly people, and since these elderly people need different care, the need to improve the performance of these care centers are increasing day by day. In these situations, more elderly people prefer to live in their homes instead of living in care centers for the elderly and receive the care and treatment they need. Therefore, the older the population grows, the demand for HHC services will also increase. On the one hand, HHC is a unique part of health care that has been little attention in the past, and the industry faces challenges that another industry is not facing and, on the other hand, organizations that provide home care services should be able to optimize their activities in order to meet increasing demand. Recently, we are witnessing the importance of

A. Abraham et al. (Eds.): SoCPaR 2020, AISC 1383, pp. 320–332, 2021.
https://doi.org/10.1007/978-3-030-73689-7_31

planning and determining the route for therapists in HHC centers for therapists, and ignoring them will lead to higher costs, which is of great importance to those involved, such as researchers, doctors, and policymakers [1, 23].

In the HHC operations, therapists are planning in a way and determine the route, which provides a variety of services at the customer home. Because this requires a combination of vehicle routing, synchronization, work-time balance, and time planning approaches, as a result, there will be complex optimization problems. The HHC system can be considered as a health service network that includes patients, people involved in logistics (coordinator for the evaluation of material and human needs, pharmacy), HHC sponsor companies (health insurance), and the HHC group (nurses, doctors, therapists, and others) [25].

This paper is intended to provide a routing-allocation-scheduling-planning model with time windows, synchronization, and work-time balance, in which the travel cost and therapists' travel, which is one of the most important costs of the system, is minimized and the satisfaction of therapists is maximized by allocation and planning. Finally, with optimization aiming to reduce costs and increase the satisfaction of therapists, the model will determine the best routes and the best time schedule for each therapist to provide therapists with satisfaction with planning according to their personal and desirable lifestyle programs, as well as the cost of transportation, which is one of the most important costs of HHC centers. In addition, we use a fuzzy approach to cope with this uncertainty due to the uncertain nature of transportation and travel cost parameters. Therefore, due to the NP-hard and complexity of the model, we used multi-objective meta-heuristic algorithm and the results of this algorithm are compared with other meta-heuristic algorithms for different sizes of the problem.

The structure of the paper is as follows: In the next section of this paper, we first review the review of the literature on HHC and uncertainty approach. In Sects. 3, the problem is described and its definitive model is described and devoted to the approach to uncertainty and the mathematical model in uncertainty. In Sect. 4 the solution methods are described. In Sect. 5, addresses the computational and validation results of the proposed model. Finally, the conclusions and future research fields are presented in Sect. 6.

2 Literature Review

The review of the literature on HHC is very new, the number of papers in this area is estimated to be less than 40, and this number has been moving much forward in recent decades, indicating the need for further review. Table 1 presents an overview of the characteristics of the models used and their solving methods.

In general, according to the literature review, the summary of the results of Table 1 reveals a number of similarities and research gaps:

- Most developed models in this area can be considered as a type of vehicle routing with time windows.
- According to the strategic decision making in the field of health, which is directly related to the lives of patients, there are models which, in addition to the routing, location, scheduling, allocation and scheduling with time windows, synchronization and balance during work time, have not been considered.

Table 1. Summary of literature review

Reference	Multi-objective	Single-objective	Fuzzy	Time windows	Scheduling	Planning/Allocation	Routing	Method
[2]	–	✓	–	✓	✓	–	✓	Exact
[3]	–	✓	–	✓	✓	–	✓	VNS
[4]	–	✓	–	✓	✓	–	✓	Exact
[5]	–	✓	–	✓	✓	–	✓	GA and TS
[6]	–	✓	–	✓	✓	–	✓	Heuristic
[7]	–	✓	–	✓	✓	–	✓	Heuristic
[8]	–	✓	–	✓	✓	–	✓	VNS
[9]	✓	–	–	✓	✓	–	✓	Heuristic
[10]	–	✓	–	✓	✓	–	✓	GA
[11]	–	✓	–	✓	✓	–	✓	GA and HSA
[12]	✓	–	–	–	✓	✓	✓	Heuristic
[13]	✓	–	–	–	✓	–	✓	Ga and HGA
[14]	–	✓	–	–	✓	–	–	Heuristic
[15]	–	✓	–	–	✓	–	✓	VNS
This paper	✓	✓	✓	✓	✓	✓	✓	MOSEO, MOICA. NSGA-II

- Developed methods have generally been based on heuristic and meta-heuristic algorithms.
- Assumptions such as considering different occupations of nurses, the variety of available vehicles and time window constraints have been used in many papers in recent years.

To solve the problems mentioned above and to bring these types of models closer to the real world, this paper aims to develop a new model for home health care problem. Hence, the scheduling-allocation-routing-scheduling problem with time windows, synchronization, and work-time balance were adopted to develop a comprehensive model for this problem. In addition, we use a fuzzy approach to cope with this uncertainty due to the uncertain nature of transportation and travel cost parameters. Another attraction of this paper is the use of different meta-heuristic algorithms.

3 Problem Description and Mathematical Formulation

In this problem, a set of therapists should be given according to the program allocated to them in patient homes on specific days of the week and at the appointed time according to the time windows and with a specific route and service, they will refer and serve. Each customer needs to meet certain schedules in the planning horizon, and they must meet in certain hours (time windows). Each therapist also has definite working hours per day, as well as the number of specific days in each horizon for planning his working days, and, finally, specific days are the closing days where the number of these days is the same for all therapists. In order to increase the satisfaction of the therapist, he/she is given the right to set points prior to the start of the planning horizon, according to his own judgment, for each day, as the day it is more likely to be working on that day, the more points and fewer points to days that, according to their own preferences, tend to be less likely to work. In this regard, the planning horizon is flexible and can last for more than a day.

In this way, the planner considers these preferences before the start scheduling for the planning and allocation of therapists. Also, try to make as many as possible allocations are made in such a way that these privileges and thus the fulfillment of employee preferences, and thus increase their satisfaction to the highest possible level.

In summary, we allocated a set of therapists to a set of patients in the planning horizon of more than one day to maximize the satisfaction of these therapists. In addition, since these therapists need to serve the patients' homes, their best paths should be tailored to the time windows, synchronization and balance at work, and the cost of the system should be minimized.

3.1 Assumption

Modeling assumptions are as follows:

- The number of therapists is fixed and does not change.
- There is no rest (interruption) in working hours.

- The number of working days for therapists is fixed and specific.
- There is a time window for patients' appointment.
- Each patient is only visited by one of the therapists.

3.2 Sets

N : Sets of patients
N_O : Sets of patients and points including point o and $n+1$. A virtual peak or healthcare center that therapists must start every day from a point o and refer to the homes of patients who were allocated to that day. After the completion, all the houses are back to the point $n+1$ and to the same health center will refer.
K : Sets of therapists
T : Sets of the number of planning horizon

3.3 Parameters

$\tilde{M}_{ikt}$: The service time at the patient's house i by the therapist K at the day t
D_{ij} : The distance between node i and node j
t_{ij} : Average time required (per hour) for arcing (travel time from patient's home i to patient's home j)
$\tilde{\lambda}_{kt}$: The privilege by the therapist K to their work day
$\tilde{\alpha}_j$: The earliest start of service at the patient's house j
$\tilde{\beta}_j$: The latest start of service at the patient's house j
S : The number of working days of therapists in the planning horizon T
L_{jt} : If the patient j appointments on request at day t 1, otherwise, then 0 (patient demand in the planning horizon)
V : Maximum permissible hours for each therapist in the per day

3.4 Decision Variable

X_{ijkt} : If the therapist k at the day t goes from home i to home j equal to 1, otherwise 0
U_{kt} : If the therapist k to the day t is allocated to patients equal to 1, otherwise 0
R_{jkt} : The onset time of service at the patient's house j by staff k at the day t

The following model is an integer linear programming (ILP) model:

$$Min\,F1 \sum_{(i,j)\in N_O} \sum_{k\in K} \sum_{t\in T} D_{ij}.X_{ijkt} \tag{1}$$

$$Max\;F2 \sum_{k\in K} \sum_{t\in T} \tilde{\lambda}_{kt}.U_{kt} \tag{2}$$

$$\sum_{i\in N_O} \sum_{k\in K} X_{ijkt} = L_{jt} \;\; \forall j \in N,\; \forall t \in T \tag{3}$$

$$\sum_{j\in N} X_{ijkt} = U_{kt} \;\; \forall k \in K,\; \forall t \in T \tag{4}$$

$$\sum_{i \in N_o} X_{ijkt} - \sum_{i \in N_o} X_{jikt} = 0 \;\; \forall j \in N,\, \forall k \in K,\, \forall t \in T \tag{5}$$

$$X_{ijkt} \leq U_{kt} \;\; \forall (i,j) \in N_o,\, \forall k \in K,\, \forall t \in T \tag{6}$$

$$\sum_{t \in T} U_{kt} \;\; \forall k \in K \tag{7}$$

$$\left(R_{ikt} + \tilde{M}_{ikt} + t_{ij} - R_{jkt}\right) \leq (1 - X_{ijkt}) BigM \quad \forall (i,j) \in N_o,\, \forall k \in K,\, \forall t \in T \tag{8}$$

$$\left(R_{n+1,kt} - R_{okt}\right) \leq V \;\; \forall k \in K,\, \forall t \in T \tag{9}$$

$$\tilde{\alpha}_j \sum_{i \in N_o} X_{ijkt} \leq R_{jkt} \leq \tilde{\beta}_j \sum_{i \in N_o} X_{ijkt} \;\; \forall j \in N,\; \forall k \in K,\, \forall t \in T \tag{10}$$

$$X_{ijkt} \in \{0, 1\} \;\forall (i,j) \in N_o,\;\; \forall k \in K,\;\; \forall t \in T \tag{11}$$

$$U_{kt} \in \{0, 1\} \;\; \forall k \in K,\;\; \forall t \in T \tag{12}$$

$$R_{jkt} \geq 0 \;\forall j \in N_o,\;\; \forall k \in K,\;\; \forall t \in T \tag{13}$$

The objective function (1) shows the minimization of distance travelled by the therapists. The objective function (2) expresses the maximization of employee satisfaction. Constraint (3) ensures that all patients are served according to a pre-determined demand. Constraint (4) shows each therapist leaves the HHC center only once when allocated to the workday. Constraint (5) indicates if a therapist enters a patient's home, will be out of there. Constraint (6) states that if a therapist assigns to a specific day, the therapist will be able to perform the services on that day. Constraint (7) defines the number of working days for each therapist. Constraint (8) indicates the initiation of the therapist's service at the patient's home. The maximum hourly per day for each therapist is determined in Constraint (9). Constraint (10) implies that the patient should be served at the time window set for that patient. Constraints (11)–(13) show the structure of decision variables.

3.5 Fuzzy Model

In this section, the mathematical model presented in this paper is an integer programming model. Since in the real world of uncertainty an inevitable factor is inevitable, most of the parameters used are considered triangular fuzzy numbers because of their uncertain nature. In general, the fuzzy programming problem must first be transformed into a definite equivalent problem and then solved with standard methods and the optimal solution is obtained. As a result, the final solution of the problem is obtained with respect to the fuzzy structure of the problem.

In the following, to solve the model, a two-step approach has been used: In the first step, the proposed model with fuzzy parameters is transformed into a certain auxiliary model by Khimens et al. [16] method. In the second stage, using the Torabi-Hosseini method [17, 24], we solve the multi-objective certain model, which was obtained in the first stage.

3.5.1 Khimens Method

The Khimens presented a method for ranking fuzzy numbers. In this method, defining the fuzzy parameters of the objective functions is calculated based on the concepts of expected distance and expected value for triangular fuzzy numbers $\tilde{c} = (c^p, c^m, c^o)$ according to Eqs. (14) and (15).

$$EI(\tilde{c}) = [E_1^c, E_1^c] = \left[\int_0^1 f_c^{-1}(x)d_x,\ \int_0^1 g_c^{-1}(x)d_x,\right] \\ = \left[\int_0^1 (x(c^m - c^p) + c^p)d_x\right] = \left[\int_0^1 (x(c^o - c^m) + c^o)d_x\right] = \left[\tfrac{1}{2}(c^p + c^m),\ \tfrac{1}{2}(c^m + c^o)\right] \quad (14)$$

$$EV(\tilde{c}) = \frac{E_1^c + E_1^c}{2} = \frac{c^p + 2c^m + c^o}{4} \quad (15)$$

Based on Khaminz's method, Eq. (16) is considered for constraint $\left(\tilde{a}_i X \geq \tilde{b}_i,\ i = 1, 2, \ldots, I\right)$

$$\left(\alpha\frac{a_i^o + a_i^m}{2} + (1-\alpha).\frac{a_i^p + a_i^m}{2}\right)X \geq \left(\alpha\frac{b_i^o + b_i^m}{2} + (1-\alpha).\frac{b_i^p + b_i^m}{2}\right) \quad (16)$$

For equal constraints $\tilde{a}_i X = \tilde{b}_i,\ i = 1, 2, \ldots, I$ converted into the certain equivalent constraints, are represented as Eqs. (17) and (18):

$$\left(\frac{\alpha}{2}\frac{a_i^o + a_i^m}{2} + \left(1 - \frac{\alpha}{2}\right).\frac{a_i^p + a_i^m}{2}\right)X \geq \left(\frac{\alpha}{2}\frac{b_i^o + b_i^m}{2} + \left(1 - \frac{\alpha}{2}\right).\frac{b_i^p + b_i^m}{2}\right) \quad (17)$$

$$\left(\left(1 - \frac{\alpha}{2}\right).\frac{a_i^o + a_i^m}{2} + \frac{\alpha}{2}.\frac{a_i^p + a_i^m}{2}\right)X \geq \left(\left(1 - \frac{\alpha}{2}\right).\frac{b_i^o + b_i^m}{2} + \frac{\alpha}{2}.\frac{b_i^p + b_i^m}{2}\right) \quad (18)$$

After Dfuzzy by the help of Eq. (15), the membership function for the minimization objective function is obtained using the Torabi-Hessian method of Eq. (19).

$$\mu_F = \begin{cases} 1 & if\ Z < Z^{\alpha-PIS} \\ \frac{Z^{\alpha-NIS} - Z}{Z^{\alpha-NIS} - Z^{\alpha-PIS}} & f\ Z^{\alpha-PIS} < Z < Z^{\alpha-NIS} \\ 0 & if\ Z > Z^{\alpha-NIS} \end{cases} \quad (19)$$

And the membership function for the maximization objective function is obtained using Eq. (20).

$$\mu_F = \begin{cases} 1 & if\ Z > Z^{\alpha-PIS} \\ \frac{Z^{\alpha-NIS} - Z}{Z^{\alpha-PIS} - Z^{\alpha-NIS}} & f\ Z^{\alpha-NIS} \leq Z \leq Z^{\alpha-PIS} \\ 0 & if\ Z > Z^{\alpha-NIS} \end{cases} \quad (20)$$

In which the positive ideal solution $(\alpha - PIS)$ and the negative ideal solution $(\alpha - NIS)$ for each objective function and at the level of feasibility (α). According to the above equations, the formulation of a certain auxiliary model is equivalent to the main problem model in Eqs. (21)–(27).

$$Min \sum_{(i,j)\in N_O} \sum_{k\in K} \sum_{t\in T} D_{ij}.X_{ijkt} \tag{21}$$

$$Max \sum_{k\in K} \sum_{t\in T} \left(\frac{\tilde{\lambda}^p_{kt} + 2\tilde{\lambda}^m_{kt} + \tilde{\lambda}^o_{kt}}{4} \right).U_{kt} \tag{22}$$

$$\text{Constraints (3)} - (7) \tag{23}$$

$$\left(R_{ikt} + \left(\left(\frac{\alpha}{2} \frac{\tilde{M}^p_{ikt} + \tilde{M}^m_{ikt}}{2} \right) + \left(1 - \frac{\alpha}{2}\right) \frac{\tilde{M}^o_{ikt} + \tilde{M}^m_{ikt}}{2} \right) + t_{ij} - R_{jkt} \right) \leq \left(1 - X_{ijkt}\right) BigM \quad \forall (i,j) \in N_o,\ \forall k \in K,\ \forall t \in T \tag{24}$$

$$\left(R_{n+1,kt} - R_{okt}\right) \leq V \quad \forall k \in K,\ \forall t \in T \tag{25}$$

$$\left(\left(\frac{\alpha}{2} \frac{\tilde{\alpha}^p_j + \tilde{\alpha}^m_j}{2} \right) + \left(\left(1 - \frac{\alpha}{2}\right) \frac{\tilde{\alpha}^o_j + \tilde{\alpha}^m_j}{2} \right) \right) \sum_{i\in N_o} X_{ijkt} \leq R_{jkt} \leq \left(\left(\frac{\alpha}{2} \frac{\tilde{\beta}^p_j + \tilde{\beta}^m_j}{2} \right) + \left(\left(1 - \frac{\alpha}{2}\right) \frac{\tilde{\beta}^o_j + \tilde{\beta}^m_j}{2} \right) \right) \sum_{i\in N_o} X_{ijkt} \quad \forall j \in N,\ \forall k \in K,\ \forall t \in T \tag{26}$$

$$\text{Constraints (11)} - (13) \tag{27}$$

4 Solution Methods

Multi-objective optimization is one of the most active and highly applied research areas among optimization topics. So far, several methods have been proposed for solving multi-objective optimization problems, among which intelligent optimization methods have a special place. In this paper, the proposed model has been used of meta-heuristic algorithms are including multi-objective imperialist competitive algorithm (MOICA), non-dominated sorting genetic algorithm II algorithm (NSGA-II) and multi-objective social engineering optimization (MOSEO) algorithm.

4.1 Multi-objective Imperialist Competitive Algorithm

Imperialist Competitive Algorithm (ICA) proposed by Atashpaz-Gargari and Lucas [18] is utilized, to solve the proposed problem. MOICA is a type of evolutionary algorithm which maintains a population of solutions to search by an interaction between exploration and exploitation phases, intelligently. This algorithm is greatly welcomed due to its appropriate efficiency, convergence speed, and high ability to optimize in the field of optimization. For more details, scholars can refer to [19, 22].

4.2 Non-dominated Sorting Genetic Algorithm II Algorithm

One of the most efficient and most popular multi-objective optimization algorithms is the Non-Dominated Sorting Genetic Algorithm II (NSGA-II) proposed by Deb [20]. NSGA-II algorithm is one of the fastest and most powerful optimization algorithms that has less complexity than other methods. And by using the non-overcoming and calculating the distance of the nodes, Pareto's optimum points are obtained which have a desirable scope for the purpose of changing functions. In this algorithm, elitism and dispersion are considered, simultaneously. For more details, a reader may refer to [20, 26–28].

4.3 Multi-objective Social Engineering Optimization Algorithm

The social engineering optimization (SEO) algorithm was introduced by Fathollahi-Fard et al. [21]. In this algorithm, each solution is a counterpart to a person and traits of each person such as his/her ability in contents of mathematical, sport, business are the counterparts of all variables of each solution in search space. To start the algorithm, the two random solutions are initialized and better solution is selected as attacker and the other is named as the defender. To simulate training and retraining of the attacker from the defender, some random experiments for each trait of the defender are designed. The attacker tries to evaluate the defender by his/her traits. The counterpart of training and retraining in search space is copying a trait from the attacker to the same trait in defender and computing the rate of retraining of the attacker from the defender, simultaneously. In the next step, spotting a social engineering (SE) attack from the attacker to the defender is the counterpart of changing the position of defender by an intelligent way in feasible space. In responding a SE attack, the fitness of new position of the defender is calculated and the old and current positions of the defender is compared and the best position is selected. If the ability of the defender is better than the attacker, positions of the attacker and the defender are changed. At the end, to strike to the defender, it is annihilated and replaced by a new random solution in search space.

5 Computational Experiments

In this section, the results of the implementation of proposed solution methods on experimental problems are generated and analyzed. In this paper, 30 random tests were designed with the number of patients and the number of different therapists. Hence, the algorithms were implemented for each problem three times, and the average of the three implementations of each problem as the final response variable has been.

To solve numerical examples, at first 15 problems were designed in small, medium and large sizes. The range of sample parameters is shown in Table 2. Also, problems 1 through 30 are illustrated in Table 3.

Also, the total days of the planning horizon of 7 days and the number of working days per therapist is 5 days. The service time for each patient is also 45 min. After these problems are designed, these problems are solved by the algorithms MOICA, NSGA-II, and MOSEO, and the result is shown in Table 4.

Table 2. The range of parameters

T	7 days
S	5 days
$\tilde{M}_{ikt}$	45 min
$\tilde{\lambda}_{kt}$	Rand [1 15]
$\tilde{\alpha}_j$	Rand i([20 250])
$\tilde{\beta}_j$	$\alpha + 10$

Table 3. Numerical example

Problem size	Problem number	Patient number	Therapists number
Small	1	9	2
	2	10	2
	3	12	2
	4	14	3
	5	15	3
Medium	6	22	5
	7	23	5
	8	24	5
	9	25	6
	10	27	6
Large	11	34	8
	12	36	8
	13	37	8
	14	38	9
	15	39	9

In addition, Fig. 1 shows the behaviour of MOICA, NSGA-II, and MOSEO algorithms in terms of solution time. Although, MOICA in all of test problems needed more time, it seems that there is no difference between metaheuristic algorithms in this item. Also, Fig. 2 shows the results of MOICA, NSGA-II, and MOSEO algorithms in the different size. As a result, the MOSEO has a better performance in comparison of the NSGA-II and MOICA algorithms.

6 Conclusion and Future Works

In this paper, an integer linear programming model was employed to formulate a bi-objective home health care routing, allocation, planning and scheduling problem. According to the literature, this paper was for the first time used the fuzzy approach. To solve the model, three metaheuristic algorithms were offered to handle the proposed problem. Finally, results showed the performance of the MOSEO the better than the NSGA-II and MOICA algorithms as well as the efficiency of developed model.

For future works, more comprehensive experiments on the developed model are needed to be probed. In addition, developing other metaheuristic algorithms to compare the results of this study may be required. Also, some new assumptions and innovations can be added into the present model for future studies. For example, considering the

Table 4. The result of MOICA, NSGA-II, and MOSEO algorithms

Problem size	Problem no.	MOICA F1/F2	CPU time	NSGA-II F1/F2	CPU time	MOSEO F1/F2	CPU time
Small	1	3025/4556	21/45	3345/4890	22/35	2334/3244	20/45
	2	3550/4788	23/67	3809/5012	24/07	2455/3440	21/56
	3	3701/4901	29/77	4112/5334	27/68	2677/3778	26
	4	3902/5228	35/45	4453/5899	29/56	2998/3988	32/12
	5	4133/5709	41/12	4667/6233	35/34	3012/4221	34/14
Medium	6	5990/7233	97/33	6553/8556	85/16	4566/6034	82/5
	7	6012/7566	102/45	6990/8991	97/28	4881/6233	92/01
	8	6455/7889	131/45	7233/9344	116/77	5112/6566	112/66
	9	6788/8022	145/78	7880/9880	135/33	5334/6980	127/5
	10	6901/8344	188/76	8112/10455	163/22	5778/7344	140/55
Large	11	8677/9677	356/14	13566/26770	302/33	7811/9233	268/44
	12	9122/9788	387/5	16775/29881	343/13	8122/9533	288/12
	13	9566/9987	422/12	19807/32240	390/12	8344/9788	313/56
	14	9855/10236	467/44	20233/37880	450/78	8766/9877	377/55
	15	10122/1098	500/12	23455/41123	488/23	9012/9909	412/33

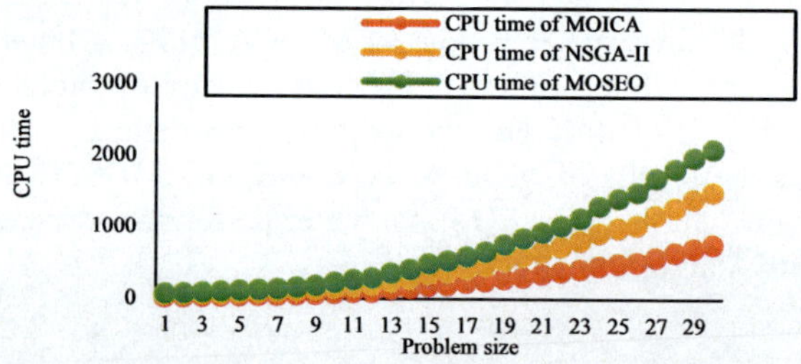

Fig. 1. The CPU time of MOICA, NSGA-II, and MOSEO algorithms

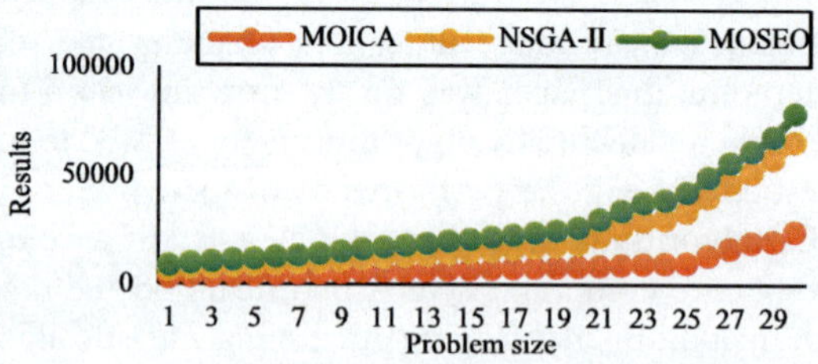

Fig. 2. The results of MOICA, NSGA-II, and MOSEO algorithms

patient's inconvenience, green emission and environmental pollution as different objectives looks interesting. Besides the total cost of transforming nurses and drugs quantity while adding the location decisions for pharmacies and laboratories and proposing multi-period and also multi-products (considering different required drugs for patients) are good ideas for future works.

References

1. Fathollahi-Fard, A.M., Ahmadi, A., Goodarzian, F., Cheikhrouhou, N.: A bi-objective home healthcare routing and scheduling problem considering patients' satisfaction in a fuzzy environment. Appl. Soft Comput. **93**, 106385 (2020)
2. Eveborn, P., Rönnqvist, M., Einarsdóttir, H., Eklund, M., Lidén, K., Almroth, M.: Operations research improves quality and efficiency in home care. Interfaces **39**(1), 18–34 (2009)
3. Trautsamwieser, A., Gronalt, M., Hirsch, P.: Securing home health care in times of natural disasters. OR Spectr. **33**(3), 787–813 (2011)
4. Rasmussen, M.S., Justesen, T., Dohn, A., Larsen, J.: The home care crew scheduling problem: preference-based visit clustering and temporal dependencies. Eur. J. Oper. Res. **219**(3), 598–610 (2012)
5. Liu, R., Xie, X., Augusto, V., Rodriguez, C.: Heuristic algorithms for a vehicle routing problem with simultaneous delivery and pickup and time windows in home health care. Eur. J. Oper. Res. **230**(3), 475–486 (2013)
6. Liu, R., Xie, X., Garaix, T.: Hybridization of tabu search with feasible and infeasible local searches for periodic home health care logistics. Omega **47**, 17–32 (2014)
7. Mankowska, D.S., Meisel, F., Bierwirth, C.: The home health care routing and scheduling problem with interdependent services. Health Care Manag. Sci. **17**(1), 15–30 (2014)
8. Fikar, C., Hirsch, P.: A matheuristic for routing real-world home service transport systems facilitating walking. J. Clean. Prod. **105**, 300–310 (2015)
9. Braekers, K., Hartl, R.F., Parragh, S.N., Tricoire, F.: A bi-objective home care scheduling problem: analyzing the trade-off between costs and client inconvenience. Eur. J. Oper. Res. **248**(2), 428–443 (2016)
10. Shi, Y., Boudouh, T., Grunder, O.: A hybrid genetic algorithm for a home health care routing problem with time window and fuzzy demand. Expert Syst. Appl. **72**, 160–176 (2017)
11. Lin, C.C., Hung, L.P., Liu, W.Y., Tsai, M.C.: Jointly rostering, routing, and re-rostering for home health care services: a harmony search approach with genetic, saturation, inheritance, and immigrant schemes. Comput. Ind. Eng. **115**, 151–166 (2018)
12. Liu, M., Yang, D., Su, Q., Xu, L.: Bi-objective approaches for home healthcare medical team planning and scheduling problem. Comput. Appl. Math. **37**(4), 4443–4474 (2018)
13. Lin, Y.K., Chou, Y.Y.: A hybrid genetic algorithm for operating room scheduling. Health Care Manag. Sci. **23**, 1–15 (2019)
14. Demirbilek, M., Branke, J., Strauss, A.: Dynamically accepting and scheduling patients for home healthcare. Health Care Manag. Sci. **22**(1), 140–155 (2019)
15. Nasir, J.A., Dang, C.: Quantitative thresholds based decision support approach for the home health care scheduling and routing problem. Health Care Manag. Sci. **23**, 1–24 (2019)
16. Jimenez, M., Arenas, M., Bilbao, A., Guez, M.V.: Linear programming with fuzzy parameters: an interactive method resolution. Eur. J. Oper. Res. **177**, 1599–1609 (2007)
17. Torabi, S.A., Hassini, E.: An interactive possibilistic programming approach for multiple objective supply chain master planning. Fuzzy Sets Syst. **159**(2), 193–214 (2008)
18. Atashpaz-Gargari, E., Lucas, C.: Imperialist competitive algorithm: an algorithm for optimization inspired by imperialistic competition. In: Proceedings of IEEE Congress on Evolutionary Computation, Singapore, pp. 4661–4667 (2008)

19. Sahebjamnia, N., Goodarzian, F., Hajiaghaei-Keshteli, M.: Optimization of multi-period three-echelon citrus supply chain problem. J. Optim. Ind. Eng. **13**, 41–50 (2019)
20. Goodarzian, F., Hosseini-Nasab, H.: Applying a fuzzy multi-objective model for a production–distribution network design problem by using a novel self-adoptive evolutionary algorithm. Int. J. Syst. Sci. Oper. Logist. **8**, 1–22 (2019)
21. Fathollahi-Fard, A.M., Hajiaghaei-Keshteli, M., Tavakkoli-Moghaddam, R.: The social engineering optimizer (SEO). Eng. Appl. Artif. Intell. **72**, 267–293 (2018)
22. Fakhrzad, M.B., Goodarzian, F.: A fuzzy multi-objective programming approach to develop a green closed-loop supply chain network design problem under uncertainty: modifications of imperialist competitive algorithm. RAIRO-Oper. Res. **53**(3), 963–990 (2019)
23. Goodarzian, F., Hosseini-Nasab, H., Muñuzuri, J., Fakhrzad, M.B.: A multi-objective pharmaceutical supply chain network based on a robust fuzzy model: A comparison of meta-heuristics. Appl. Soft Comput. **92**, 106331 (2020)
24. Goodarzian, F., Shishebori, D., Nasseri, H., Dadvar, F.: A bi-objective production-distribution problem in a supply chain network under grey flexible conditions. RAIRO Oper. Res. **55**, S1287–S1316 (2021). https://doi.org/10.1051/ro/2020111
25. Goodarzian, F., Hosseini-Nasab, H., Fakhrzad, M.B.: A multi-objective sustainable medicine supply chain network design using a novel hybrid multi-objective metaheuristic algorithm. Int. J. Eng. **33**(10), 1986–1995 (2020)
26. Fakhrzad, M.B., Talebzadeh, P., Goodarzian, F.: Mathematical formulation and solving of green closed-loop supply chain planning problem with production, distribution and transportation reliability. Int. J. Eng. **31**(12), 2059–2067 (2018)
27. Fakhrzad, M.B., Goodarzian, F., Golmohammadi, A.M.: Addressing a fixed charge transportation problem with multi-route and different capacities by novel hybrid meta-heuristics. J. Ind. Syst. Eng. **12**(1), 167–184 (2019)
28. Fakhrzad, M.B., Goodarzian, F.: A new multi-objective mathematical model for a Citrus supply chain network design: metaheuristic algorithms. J. Optim. Ind. Eng. **14**, 127–144 (2020). https://doi.org/10.22094/JOIE.2020.570636.1571

The Lagrangian Relaxation Approach for Home Health Care Problems

Sohaib Dastgoshade[1], Ajith Abraham[2(✉)], and Nazanin Fozooni[3]

[1] Department of Industrial Engineering, Yazd University, Yazd, Iran
Sohaib.dastgoshade@stu.yazd.ac.ir

[2] Machine Intelligence Research Labs (MIR Labs), Scientific Network for Innovation and Research Excellence, 11, 3rd Street NW, P.O. Box 2259, Auburn, WA 98071, USA
Ajith.abraham@ieee.org

[3] Department of Industrial Engineering, K. N. Toosi University of Technology, Tehran, Iran

Abstract. Today, life expectancy and demand for caring for the elderly have been increased in developing countries. However, hospitals and nursing homes have not grown at the same rate. Thus, the home health care problem has been addressed by health academics and practitioners. The home care industry is a unique part of health care with growing demand, which has not been adequately addressed in the past. Based on the assumptions of this context and considering time windows and coordination constraints, a Lagrangian relaxation-based mixed-integer programming model is developed in this paper. To solve this model, since it is NP-hard, two metaheuristic algorithms, including simulated annealing and social engineering optimization, were applied. The results showed that the simulated annealing algorithm outperformed the other algorithm. Besides, a complete sensitivity analysis was performed to investigate the effect of the important parameters on the model outputs. As well as some perspectives on future works will be suggested.

Keywords: Home health care · Routing-scheduling problem · Time window · Lagrangian relaxation approach · Metaheuristic algorithms

1 Introduction

Nowadays, the elderly and patients care is one of the main concerns for the World Health Organization (WHO) [31]. The current care demand is so high that the existing hospitals, health centers, and nursing homes cannot meet it. On the other hand, the elderly prefers to have self-care in their environment rather than be in a hospital. Home care is thus a solution to this problem.

The Homecare industry is a unique part of health care with growing demand, which has not been adequately addressed in the past. It faces technical challenges that other industries do not have [32, 33]. To meet their growing demand, home care organizations should optimize their activities. In this regard, planning and determining a nurse's route to visit patients at home is of particular importance, and ignoring it will lead to increased

A. Abraham et al. (Eds.): SoCPaR 2020, AISC 1383, pp. 333–344, 2021.
https://doi.org/10.1007/978-3-030-73689-7_32

costs, which is of great importance to stakeholders, such as researchers, physicians, and policymakers [34, 35].

In this research, a new multi-purpose location-allocation-routing-scheduling model with a time window is developed, aiming at minimizing the travel cost and travel time of employees. Transportation cost constitutes a major cost of health care centers. To solve this model, the Lagrangian relaxation algorithm was applied to find upper and lower limits that simultaneously satisfy feasibility and optimality conditions. Due to the complexity and NP-hardness of the model, simulated annealing (SA) and social engineering optimizer (SEO) algorithms are used to solve the model. Finally, this model provides the best route and schedule for each employee through which the travel cost is minimized.

Generally, many of the basic models' in-home health care (HHC) problems are vehicle-routing problems. Cheng et al. [1] Investigated a vehicle routing problem with time window (VRPTW) including multiple depots. Considering both full-time and part-time nurses, they have optimized the cost of the working hours. As full-time nurses work overtime and part-time nurses are paid hourly, their objective was to minimize the amount of overtime and part-time work. Fenton et al. [2] studied the allocation and routing of HHC nurses in Germany. They used a two-stage approach to schedule the weekly service. They generated an initial solution using innovative approaches and then the scheduling was improved. Cappanera et al. [3] investigated the impact of politics on production planning and routing decisions in HHC. They used specific scenarios to integrate different decisions. Husebø et al. [4] provided a mixed-integer mathematical model for the routing and scheduling of HHC taking different treatments and time windows into account. Lanzarone et al. [5] proposed a solution to the assignment problem under the care continuity that minimizes nurses' maximum overtime. Yalçındağ et al. [6] considered routing and allocation problems in HHC and used the kernel regression method to estimate the travel time of a vehicle. Braekers et al. [7] addressed a multi-objective routing and scheduling problem in HHC, and to solve it, using a combination of multi-dimensional local search and large neighborhood search. Considering time windows and continuity of care, Valledor et al. [8] used an integer model based on the collection partitioning and neighborhood search. Decerle et al. [9] Used an integer model to incorporate soft time windows and staff competency. [10] Proposed a multi-objective model that minimizes care time and increases the quality of care. They used a memetic algorithm to solve their model. Lin et al. [11] proposed a two-objective model that the first model considered the positions of nurses and vehicles for cost reductions, and the second model expanded the first model considering the occurrence of sudden accidents and used the harmonic search algorithm.

Focusing on environmental pollution problems, Fathollahi-Fard et al. [12] presented a two-objective model and used a modified algorithm to solve the model. They conducted a sensitivity analysis to validate their model. Frazzon et al. [13] provided a flexible model using care prioritization to make the schedule more flexible and increase the number of visited locations. Fard et al. [14] studied an integrated model for basic decisions on allocation problem and used a competitive imperialist algorithm, well suited for large models, to solve the problem. Focusing on economic sustainability and operational efficiency and considering demand and capacity aspects, Nasir et al. [15] considered a MILP model for home care problems. They used a neighborhood search and compared the results with Cplex. Gong et al. [16] used a two-objective model to minimize organizational operations and maximize customer satisfaction. Mohammadi et al. [17] Modeled the home care problem as a covering problem and introduced uncertainty to the nursing network in Shiraz, Iran. Decerle et al. [18] investigated a home care problem and combined ant colony and memetic algorithms to solve it. Grenouilleau et al. [19] looked at home care problems as a perspective robust optimization (RO). Specifically, based on the theory of budget uncertainty, they defined a model with uncertain variables, which should be resolved whenever a new person requested home care. They used taboo search and neighborhood search methods to solve the model and performed a series of experiments to validate the proposed models and algorithms. Erdem et al. [20] investigated a type of home care routing in which a group of healthcare workers performs the required number of jobs using electric vehicles. They explored a combination of genetic and neighborhood variable search algorithms to solve their proposed model [21, 22]. Regarding the literature of home care, a mixed-integer linear programming model based on Lagrangian relaxation is developed in this paper. As the model is NP-hard, simulated annealing (SA), and social engineering optimization (SEO) metaheuristic algorithms are used [23, 24].

This paper is organized as follows. Defining the parameters, variables, and constraints, the problem is modeled in Sect. 2. In Sect. 3 represents the numerical results of the experiments, a description of the test instances, and an analysis of the results. Finally, in Sect. 4 some concluding remarks, as well as some perspectives on the future works, are provided.

2 Mathematical Model

In this section, the mathematical model is formulated based on a routing and scheduling problem in the home health care network. Then, the notations of the proposed model are reported as follows:

n	Set of nodes
o	Set of visits
s	Set of staff members
p	Set of HHC offices
r	Set of job roles
$Sync$	Set of synchronized visits
$\left[i^t, u^t\right]$	Planning period of time
d_{ij}	Distance between the nodes i and j
c_i	Duration of the visit i
$\lfloor \alpha_i, \beta_i \rfloor$	Working time window of the staff member i
$\lfloor a_i, b_i \rfloor$	Availability time window of the patient i
η_i	Qualification of the staff member i
ρ_i	Qualification level required to perform the visit i
y_i^k	Association of the staff member i to the office k
δ_{ij}	Indicates whether i and j are synchronized visits or not

p^t	The coefficient for time windows dissatisfaction
p^s	The coefficient for non-synchronized visits dissatisfaction
π^t_{ijn}	Lagrangian Relaxation Coefficient
λ	The percentage used for the penalties computation
m	A large number
x^k_{ij}	If the employee k travels from i to j, it equals 1, otherwise it is 0
t_i	The start time of visiting i

In this HHC problem, a collection of nurses and a set of required services are programmed such that the transportation cost and dissatisfaction of the patients and nurses are minimized. The working time of each nurse is a fixed value over the week and the home organization should pay them a certain wage [25, 26]. Each nurse is capable of doing some services, which should be considered in the allocation process. In each center, the nurses' time windows are known. Applying the nurses outside of their time windows gives rise to dissatisfaction [27]. They should be scheduled such that their dissatisfaction over the working load is minimized. The patients have also their time windows, where they will be satisfied if they are properly serviced in these time windows. These constraints should be met by the solutions of the model [28].

2.1 Soft Time Window and Synchronization Constraints

In this section, we explain the penalties of not being satisfied with soft time windows and coordination limitations [29]. The penalties are added to the objective function when the nurses are not satisfied with their time windows or the patients are too far apart or when the patients receive care outside their time windows [9, 30]. The penalty for the patient i is denoted by $f_i(t_i)$. The amount of penalty imposed depends on the time the care of the patient i starts. The penalty growth coefficient in the normal range is v^t_i, while it is multiplied by p^t if the care is outside of the normal range. The value of the normal and out of normal time for patient t is given by τ_i and c_i represents the amount of time needed to perform the care (visit time). We consider that $\tau_i = \lambda \times c_i$, where λ is the conformance percentage between the visit time and the time window of a patient. Accordingly, the penalty function is defined as follows:

$$f_i(t_i) = \begin{cases} (\tau_i + (a_i - t_i - \tau_i) \times p^t) \times v^t_i, \textit{ if } t_i < a_i - \tau_i \\ (a_i - t_i) \times v^t_i, \textit{ if } t_i \in [a_i - \tau_i, a_i] \\ 0, \textit{ if } t_i \in [a_i, b_i] \\ (t_i - b_i) \times v^t_i, \textit{ if } t_i \in [b_i, b_i + \tau_i] \\ \left(\tau_i + (t_i - b_i - \tau_i) \times p^t\right) \times v^t_i, \textit{ if } t_i > b_i + \tau_i \end{cases} \tag{1}$$

Also, the penalties imposed on the non-concordance between the two nurses who should take care of a patient is shown by $g_{i,j}(t_i,t_j)$. To separate the normal and non-normal regions of two nurses i and j who have a joint visit, τ_{ij} is used. The rate of the concordance was previously denoted by λ, and accordingly, we have $\tau_{ij} = \lambda \times c_i$.

The penalty for the normal region is shown by is v^s_{ij}. If a nurse arrival time is greater than τ_{ij}, the penalty is multiplied by p^s [9, 36].

$$g_{i,j}(t_i, t_j) = \begin{cases} (\tau_{i,j} + p^s \times (|t_i - t_j| - \tau_{i,j})) \times v^s_{i,j}, \textit{ if } |t_i - t_j| > \tau_{i,j} \\ |t_i - t_j| \times v^s_{i,j}, \textit{ if } |t_i - t_j| \le t_{i,j} \\ 0, \textit{ if } t_i = t_j \end{cases} \quad (2)$$

The HHC routing and scheduling problem is formulated as a Lagrangian relaxation mixed-integer programming model as follows:

$$\min \sum_{i,j\in N} \sum_{K\in s} x^k_{ij} \times d_{ij} + \sum_{i\in o} f_i(t_i) + \sum_{i,j\in o} \delta_{ij} \times g_{ij}(t_i, t_j)$$
$$+\pi^t_{ijn} \left(\begin{array}{l} \left(\sum\sum t_i - t_j + c_i + d_{ij} + \left(\left(\sum x^k_{ij} \right) - 1 \right) \times m \right) + \left(\sum\sum \alpha_k - t_i \left(\left(\sum x^k_{ij} \right) - 1 \right) \right) \\ + \left(\sum\sum\sum t_i - \beta_k + c_i + d_{ij} + (x-1) \right) \times m \end{array} \right) \quad (3)$$

s.t:

$$\sum_{\substack{j\in n \\ i\neq j}} \sum_{k\in s} x^k_{ij} = 1 \quad (4)$$

$$\sum_{j\in o} x^k_{ij} = \sum_{j\in o} x^k_{ji} = \gamma^i_k, \forall i \in p, k \in s \quad (5)$$

$$\sum_{\substack{j\in n \\ i\neq j}} x^k_{ij} = \sum_{\substack{j\in n \\ i\neq j}} x^k_{ji}, \forall i \in n, k \in s \quad (6)$$

$$\sum_{\substack{j\in n \\ i\neq j}} x^k_{ij} = 0, \forall i \in o, k \in s. \eta_k \neq \rho_i \quad (7)$$

$$x^k_{ij} \in \{0, 1\}, \forall i \in N, j \in N, k \in S, t_i \in T, i \in o \quad (8)$$

Regarding the time windows penalties and coordinator constraints, the objective function (3) minimizes the total time of traveling to patient's homes, visiting them in their time windows, and returning to home care center penalties. Constraint (4) ensures that each visit is made by only one person. Constraint (5) examines the staff entry and exit. Constraint (6) regulates the flow of nurses. In Constraint (7), the employees' qualification for visiting a certain patient is evaluated. Constraint (8) specifies the variable domain. Constraints (1) and (2) are used to linearize the model.

3 Numerical Experiments

This section examines the proposed algorithms and compares them with each other. The proposed algorithms were implemented in MATLAB R2016a using a computer with an Intel i7 1.8 GHz processor and 4 GB of RAM under Windows 10 environment.

For both algorithms, 50 iterations were considered. The obtained results are shown in Table 1. As can be seen, for the simulated annealing algorithm, the best solution

Table 1. The result of metaheuristic algorithms (SA-SEO)

	Best cost			Best cost	
Iteration	SA	SEO	Iteration	SA	SEO
1	1030.88	66436.26	26	680.66	7577.97
2	961.70	53899.97	27	680.66	7022.26
3	850.43	42058.14	28	680.66	6899.98
4	753.34	37984.77	29	680.66	6605.31
5	680.66	37592.14	30	680.66	6412.61
6	680.66	34816.33	31	680.66	6412.61
7	680.66	33669.19	32	680.66	6412.61
8	680.66	28756.87	33	680.66	5341.07
9	680.66	28477.25	34	680.66	4846.51
10	680.66	23611.77	35	680.66	4846.51
11	680.66	22890.19	36	680.66	4647.04
12	680.66	20340.99	37	680.66	4635.08
13	680.66	16342.68	38	680.66	4625.01
14	680.66	15008.82	39	680.66	4271.89
15	680.66	14876.09	40	680.66	4028.95
16	680.66	13933.85	41	680.66	3900.29
17	680.66	13353.31	42	680.66	3631.47
18	680.66	12853.43	43	680.66	3462.97
19	680.66	12262.92	44	680.66	3288.80
20	680.66	11641.61	45	680.66	3199.24
21	680.66	11478.63	46	680.66	3199.24
22	680.66	11353.01	47	680.66	3183.90
23	680.66	10071.81	48	680.66	2996.36
24	680.66	7884.06	49	680.66	2802.44
25	680.66	7725.33	50	680.66	2761.59

started at 1030.88, and after 50 replicates reached 680.66. For the social-engineering algorithm, the best solution in the first replicate was 66436.26 and after 50 replicates reached 2761.59. Their corresponding diagrams are shown in Fig. 1. The performance of both algorithms for eight samples with different parameters are presented in Table 2.

According to this table, the SEO algorithm outperforms the SA algorithm. The best solution of the SA algorithm was 680, while the SEO algorithm was able to get to 1095 in replicate 249. The mean and standard deviation of the results are shown in Table 3.

Table 2. Parameter tuning for the SEO and SA algorithms

Algorithms	Parameters	Parameter tuning combinations							
		1	2	3	4	5	6	7	8
SA	Max-iteration	100	200	300	400	600	800	900	1000
	Sub-iteration	50	150	250	350	450	550	650	750
	Mutation	0.2	0.25	0.3	0.35	0.4	0.45	0.5	0.55
	Initial temperature	0.3	0.35	0.45	0.55	0.65	0.75	0.85	0.95
SEO	Max-iteration	200	300	400	600	800	900	1000	200
	Collecting data	0.2	0.25	0.3	0.35	0.4	0.45	0.5	0.55
	Connecting attacker	0.15	0.2	0.25	0.3	0.35	0.4	0.45	0.5
	Connections	50	75	100	125	150	175	200	225

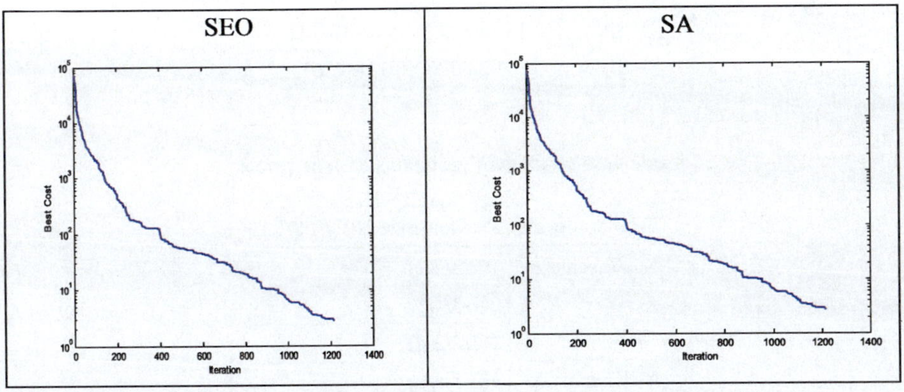

Fig. 1. The objective function values for the proposed SA and SEO algorithms

As evident in Table 3, the solutions obtained by the SA algorithm have a lower mean and standard deviation. The problem has been considered in small, medium, and large scales, the setup of which are given in Table 4.

3.1 Sensitivity Analyses for the HHC Model Developed

To investigate the behavior of the HHC model, a sensitivity analysis was performed on the important parameters of the model. In this regard, a series of changes were introduced to the model parameters, including the number of nurses (s), number of patients (n), and number of HHC centers (p). Each analysis is performed for five samples C1 to C5. The results are reported in Tables 5, 6 and 7 and Fig. 2, 3, and 4.

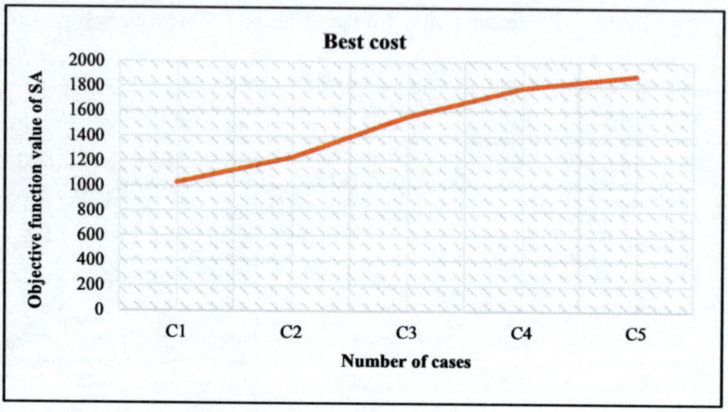

Fig. 2. The objective function values for different number of cases

Table 3. The average and STDEV of the proposed algorithms results in 100 iterations

Algorithm	AVERAGE	STDEV
SA	689.3977	48.002
SEO	7928.525	11861.61

Table 4. The setup of problem sizes and levels.

Problem levels	Problem size N(O/S/P/R)
Small scale	1(2/3/2/2)
	2(2/3/4/1)
	3(1/2/3/4)
Medium scale	4(3/3/5/6)
	5(3/5/6/7)
	6(5/5/7/6)
Large scale	7(6/5/6/7)
	8(7/7/8/5)
	9(9/7/10/9)

Table 5 shows the results of the sensitivity analysis of the number of employed nurses (s). The objective function behavior is shown also in Fig. 2. Table 6 shows the results of the sensitivity analysis on the number of patients served (n). The objective function behavior is clearly displayed in Fig. 3. Finally, Table 7 shows the results of the sensitivity analysis on the number of HHC centers (p), and Fig. 4 illustrates the objective function behavior. The sensitivity analysis of the number of nurses reveals that as the number of nurses increases, the total cost increases. As the number of patients increases, the objective function naturally increases because the number of times patients are visited increases.

Table 5. The sensitivity analyses on the number of employed nurses

Number of cases	S	Best cost
C1	3	1030.88
C2	4	1231.54
C3	5	1560.12
C4	6	1788.22
C5	7	1890.34

Table 6. The sensitivity analyses on the number of patients served

Number of cases	N	Best cost
C1	9	1030.88
C2	11	1567.38
C3	13	1987.25
C4	15	2131.53
C5	17	2431.43

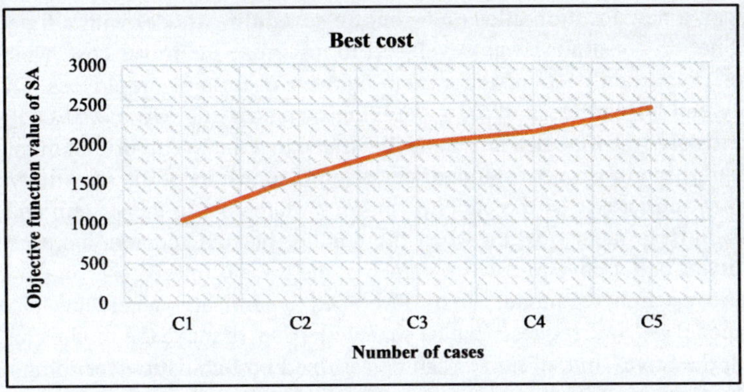

Fig. 3. The objective function values for different number of cases

As the number of home care centers increases, the objective function also increases. By the opening of home care centers, the access and visit time are reduced.

Table 7. The sensitivity analyses on the number of HHC centers

Number of cases	P	Best cost
C1	4	1030.88
C2	6	1621.07
C3	8	1745.21
C4	10	1978.51
C5	12	2014.32

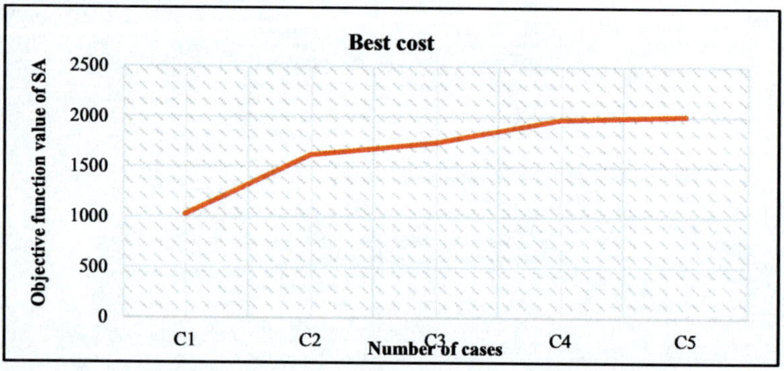

Fig. 4. The objective function values for different number of cases

4 Conclusions

In this paper, a new location-allocation-routing-scheduling model with a time window and coordination constraints was developed to minimize the travel cost, which is one of the most important costs of the system, and travel time of employees. Due to the complexity and NP-hardness of the model, the metaheuristic SA and SEO algorithms were used to solve the model. The results illustrated that the SA algorithm outperformed the SEO algorithm in terms of standard deviation. The results of the sensitivity analysis revealed that an increase of the patients imposed more cost to the system and reduces the accessibility to nurses, while when the number of HHC centers increased, easier accessibility or better coverage was obtained at the price of more cost.

In future research, the model can be developed dynamically, where new patients can enter the model or some requests can be canceled. Also, to make the model closer to the real world, the travel time of nurses can be assumed probabilistic depending on traffic volume. This can increase the efficiency in the system.

References

1. Cheng, E., Rich, J.L.: A home health care routing and scheduling problem (1998)
2. Fenton, J.J., Jerant, A.F., Bertakis, K.D., Franks, P.: The cost of satisfaction: a national study of patient satisfaction, health care utilization, expenditures, and mortality. Arch. Intern. Med. **172**(5), 405–411 (2012)
3. Cappanera, P., Scutellà, M.G.: Home Care optimization: impact of pattern generation policies on scheduling and routing decisions. Electron. Notes Discrete Math. **41**, 53–60 (2013)
4. Husebø, A.M.L., Storm, M.: Virtual visits in home health care for older adults. Sci. World J. **2014**, 1–11 (2014)
5. Lanzarone, E., Matta, A.: Robust nurse-to-patient assignment in home care services to minimize overtimes under continuity of care. Oper. Res. Health Care **3**(2), 48–58 (2014)
6. Yalçındağ, S., Matta, A., Şahin, E., Shanthikumar, J.G.: A two-stage approach for solving assignment and routing problems in home health care services. In: Proceedings of the International Conference on Health Care Systems Engineering, pp. 47–59. Springer, Cham (2014)
7. Braekers, K., Hartl, R.F., Parragh, S.N., Tricoire, F.: A bi-objective home care scheduling problem: analyzing the trade-off between costs and client inconvenience. Eur. J. Oper. Res. **248**(2), 428–443 (2016)
8. Valledor, P., Gomez, A., Priore, P., Puente, J.: Solving multi-objective rescheduling problems in dynamic permutation flow shop environments with disruptions. Int. J. Prod. Res. **56**(19), 6363–6377 (2018)
9. Decerle, J., Grunder, O., El Hassani, A.H., Barakat, O.: A memetic algorithm for a home health care routing and scheduling problem. Oper. Res. Health Care **16**, 59–71 (2018)
10. Allaoua, H., Borne, S., Létocart, L., Calvo, R.W.: A matheuristic approach for solving a home health care problem. Electron. Notes Discrete Math. **41**, 471–478 (2013)
11. Lin, C.C., Hung, L.P., Liu, W.Y., Tsai, M.C.: Jointly rostering, routing, and rerostering for home health care services: a harmony search approach with genetic, saturation, inheritance, and immigrant schemes. Comput. Ind. Eng. **115**, 151–166 (2018)
12. Fathollahi-Fard, A.M., Hajiaghaei-Keshteli, M., Tavakkoli-Moghaddam, R.: A bi-objective green home health care routing problem. J. Clean. Prod. **200**, 423–443 (2018)
13. Frazzon, E.M., Albrecht, A., Pires, M., Israel, E., Kück, M., Freitag, M.: Hybrid approach for the integrated scheduling of production and transport processes along supply chains. Int. J. Prod. Res. **56**(5), 2019–2035 (2018)
14. Fard, A.M.F., Hajaghaei-Keshteli, M.: A tri-level location-allocation model for forward/reverse supply chain. Appl. Soft Comput. **62**, 328–346 (2018)
15. Nasir, J.A., Hussain, S., Dang, C.: An integrated planning approach towards home health care, telehealth and patients group based care. J. Netw. Comput. Appl. **117**, 30–41 (2018)
16. Gong, D., Han, Y., Sun, J.: A novel hybrid multi-objective artificial bee colony algorithm for blocking lot-streaming flow shop scheduling problems. Knowl.-Based Syst. **148**, 115–130 (2018)
17. Mohammadi, M., Dehbari, S., Vahdani, B.: Design of a bi-objective reliable healthcare network with finite capacity queue under service covering uncertainty. Transp. Res. Part E Logist. Transp. Rev. **72**, 15–41 (2014)
18. Decerle, J., Grunder, O., El Hassani, A.H., Barakat, O.: A hybrid memetic-ant colony optimization algorithm for the home health care problem with time window, synchronization and working time balancing. Swarm Evol. Comput. **46**, 171–183 (2019)
19. Grenouilleau, F., Legrain, A., Lahrichi, N., Rousseau, L.M.: A set partitioning heuristic for the home health care routing and scheduling problem. Eur. J. Oper. Res. **275**(1), 295–303 (2019)

20. Erdem, M., Koç, Ç.: Analysis of electric vehicles in home health care routing problem. J. Clean. Prod. **234**, 1471–1483 (2019)
21. Atashpaz-Gargari, E., Lucas, C.: Imperialist competitive algorithm: an algorithm for optimization inspired by imperialistic competition. In: 2007 IEEE Congress on Evolutionary Computation, pp. 4661–4667. IEEE (2007)
22. Bisht, D.C., Srivastava, P.K., Ram, M.: Role of fuzzy logic in flexible manufacturing system. In: Diagnostic Techniques in Industrial Engineering, pp. 233–243. Springer, Cham (2018)
23. Dutta, P., Nagurney, A.: Multitiered blood supply chain network competition: linking blood service organizations, hospitals, and payers. Oper. Res. Health Care **23**, 100230 (2019)
24. Engin, O., Güçlü, A.: A new hybrid ant colony optimization algorithm for solving the no-wait flow shop scheduling problems. Appl. Soft Comput. **72**, 166–176 (2018)
25. Rodriguez-Verjan, C., Augusto, V., Xie, X.: Home health-care network design: location and configuration of home health-care centers. Oper. Res, Health Care **17**, 28–41 (2018)
26. Schønheyder, J.F., Nordby, K.: The use and evolution of design methods in professional design practice. Des. Stud. **58**, 36–62 (2018)
27. Fathollahi-Fard, A.M., Ahmadi, A., Goodarzian, F., Cheikhrouhou, N.: A bi-objective home healthcare routing and scheduling problem considering patients' satisfaction in a fuzzy environment. Appl. Soft Comput. **93**, 106385 (2020)
28. Fakhrzad, M.B., Goodarzian, F.: A fuzzy multi-objective programming approach to develop a green closed-loop supply chain network design problem under uncertainty: modifications of imperialist competitive algorithm. RAIRO-Oper. Res. **53**(3), 963–990 (2019)
29. Sahebjamnia, N., Goodarzian, F., Hajiaghaei-Keshteli, M.: Optimization of multi-period three-echelon citrus supply chain problem. J. Optim. Ind. Eng. **13**(1), 39–53 (2020)
30. Goodarzian, F., Shishebori, D., Nasseri, H., Dadvar, F.: A bi-objective production-distribution problem in a supply chain network under grey flexible conditions. RAIRO-Oper. Res. **55**, 1287–1316 (2021). https://doi.org/10.1051/ro/202011
31. Goodarzian, F., Hosseini-Nasab, H.: Applying a fuzzy multi-objective model for a production–distribution network design problem by using a novel self-adoptive evolutionary algorithm. Int. J. Syst. Sci. Oper. Logist. **8**, 1–22 (2019)
32. Fakhrzad, M.B., Talebzadeh, P., Goodarzian, F.: Mathematical formulation and solving of green closed-loop supply chain planning problem with production, distribution and transportation reliability. Int. J. Eng. **31**(12), 2059–2067 (2018)
33. Goodarzian, F., Hosseini-Nasab, H., Muñuzuri, J., Fakhrzad, M.B.: A multi-objective pharmaceutical supply chain network based on a robust fuzzy model: a comparison of meta-heuristics. Appl. Soft Comput. **92**, 106331 (2020)
34. Fakhrzad, M.B., Goodarzian, F., Golmohammadi, A.M.: Addressing a fixed charge transportation problem with multi-route and different capacities by novel hybrid meta-heuristics. J. Ind. Syst. Eng. **12**(1), 167–184 (2019)
35. Goodarzian, F., Hosseini-Nasab, H., Fakhrzad, M.B.: A multi-objective sustainable medicine supply chain network design using a novel hybrid multi-objective metaheuristic algorithm. Int. J. Eng. **33**(10), 1986–1995 (2020)
36. Fakhrzad, M.B., Goodarzian, F.: A new multi-objective mathematical model for a Citrus supply chain network design: Metaheuristic algorithms. J. Optim. Ind. Eng. **14**, 127–144 (2020). https://doi.org/10.22094/JOIE.2020.570636.1571

Study on Vision System for Disease Detection Using CNN and LabVIEW of an Agricultural Robot

Pradeep Nahak[1], Atanu Jana[1], Dilip Kumar Pratihar[1](✉), Alok Kanti Deb[1], and Hena Ray[2]

[1] Indian Institute of Technology Kharagpur, Kharagpur, India
dkpra@mech.iitkgp.ac.in, alokkanti@ee.iitkgp.ac.in
[2] Centre for Development of Advanced Computing, Kolkata, India
hena.roy@cdac.in

Abstract. Farmers face a lot of problems, while cultivating agricultural plants. They do not get the desired output in agriculture due to the different types of diseases from which the plants suffer. Therefore, it is important to detect the disease(s) of the plant. In this study, a vision-based plant disease detection method has been developed, where convolutional neural network (CNN) model has been used for classification of the plant disease and vision assistant toolbox of LabVIEW software is utilized for disease detection. This vision-based disease detection through a smart camera will be one of the important modules of an intelligent agricultural robot.

Keywords: Vision system · CNN model · Disease detection · Vision assistant toolbox

1 Introduction

In modern era of vastly expanding population, the need for producing the sufficient amount of agricultural products is felt badly for the survival of human-beings. In the recent few years, it has been observed that automation in the field of agriculture plays a vital role in the economy to ensure high productivity. However, the advancement and development of the technology is less in this sector. This could be achieved by using intelligent robot in the agriculture. To fulfil the current demand and need of the agricultural products at low cost, a shift is required from the manual and tractor techniques of cultivation to intelligent agricultural robot. Cultivation and disease detection along with spraying of the necessary pesticides with the help of intelligent agricultural robot can result into a tremendous amount of growth in the field of agriculture.

In agriculture field, most of the plants are found to be suffering from either bacterial infection or fungal infection. To improve the productivity in agricultural industry, one wants to adopt the latest technology. For sustainable agriculture, health monitoring of

A. Abraham et al. (Eds.): SoCPaR 2020, AISC 1383, pp. 345–353, 2021.
https://doi.org/10.1007/978-3-030-73689-7_33

the plant is essential and it cannot be done manually. So, regular monitoring of the plants can be done using vision system. Advancement in the vision system gives an opportunity to enhance the rate of plant protection from the disease. The vision system is mounted on the agricultural robot. So, it can be moved around the plant in the agriculture field to classify the disease and identify the location of the diseased leaf.

To identify the disease(s) present in the given plant, three main parameters can be considered, such as either infected stem or infected fruits or infected leaves. Leaf disease identification is the preliminary criterion to find out the disease present in the plant. In this paper, identification of various types of common diseases present in the plant leaf is done using vision system. This vision system will classify the disease(s) using convolutional neural network (CNN) model and a single image disease detection can be done using vision assistant toolbox of LabVIEW software.

The paper is organized as follows: Sect. 2 gives the details about previous methodologies of disease detection and classification. Section 3 deals with the proposed methods for disease classification and disease detection. Section 4 discusses the output of the proposed method. At last, Sect. 5 describes the conclusion of the paper.

2 Literature Survey

Agricultural automation involves a number of research areas. To provide effective and acceptable result, researchers are focusing on different aspects of the problem. For instance, Gavhale et al. [1] presented an overview on monitoring the health of the plants through disease detection using image processing by utilizing the techniques like K-nearest neighbor (KNN), support vector machine (SVM), back-propagation neural network (BPNN) and spatial gray-level dependence matrices (SGDM). Jhuria et al. [2] used image processing and artificial neural network technique for the plant disease detection. They used two datasets: one for training and another for testing. Back-propagation algorithm was utilized for weight updating of the network. They had used three feature vectors, namely textures, color and morphology. It was found that the morphological features could yield the better results compared to that of the others. Sladojevic et al. [3] developed a deep learning technique to classify and detect infected plant leaves from the images. The developed convolutional neural network model was able to detect the presence of leaf and could differentiate healthy leaves and 13 diseased leaves. Qian et al. [4] developed a computer vision technique to extract the features from tomato leaf's image under different color spaces. To establish the regression and correlation equation between the chlorophyll content and image component, they used statistical analysis method. To conduct these operations, LabVIEW software and IMAQ tool kit had been used. Khirade et al. [5] used image processing technique for plant leaf disease detection. To identify and classify the plant disease textures, color and morphological features are vital parameters. To identify the plant diseases, most common classifiers like SVM and ANN had been used. They discussed about the classification, segmentation and feature extraction of the diseased plant.

Rajmohan et al. [6] introduced a mobile app that used the ML algorithms for disease detection in paddy crops. Deep CNN and SVM classifier were used to detect the disease(s) of infected crop images. They built a database for paddy disease syndrome for the treatment. Their major work concentrated on paddy crop diseases, such as Bacterial leaf blight (BLB), Root-knot nematode, Blast, Sheath blight, Brown spot, False smut, and Whitetip nematode. Here, they used 250 images, out of which 50 images were utilized to train the deep CNN and SVM classifier, and their features were taken for the pattern matching, and the remaining 200 images were used for testing.

Saradhambal et al. [7] developed the k-means clustering method for image processing, to identify the affected portion of the diseased leaves. Image segmentation of the diseased leaves was done by a color-based segmentation model. To identify the disease(s) present in the leaves, they went through various steps as follows: image acquisition, image pre-processing, image segmentation, feature extraction, and classification. They had also used a voice navigation system, so a less skilled person could easily operate. Cruz et al. [8] used convolutional neural networks for image processing. They found the classification precision of 60% in real-time. Sardogan et al. [9] used a CNN model and Learning Vector Quantization (LVQ) algorithm-based techniques to identify the tomato leaf's disease detection and classification. In LVQ algorithm, 500 features vectors had been used for the training and testing. Raveendra et al. [10] described an Arduino control-based spraying and plowing system using LabVIEW software. The main aim of this paper was to monitor the leaves health of the agricultural plants by using herbicide spraying with the proper detection and classification. The presumed datasets of leaves matched with the digitally captured images for health confirmation.

3 Proposed Methodology

The proposed approach uses CNN model and vision assistant toolbox of LabVIEW software for the classification of diseases and disease detection, respectively.

3.1 CNN Model Evaluation

The proposed CNN model performs four different operations, namely data acquisition, data pre-processing, training and testing, and classification. this section deals with a brief discussion on CNN model. The flow-chart is shown in Fig. 1.

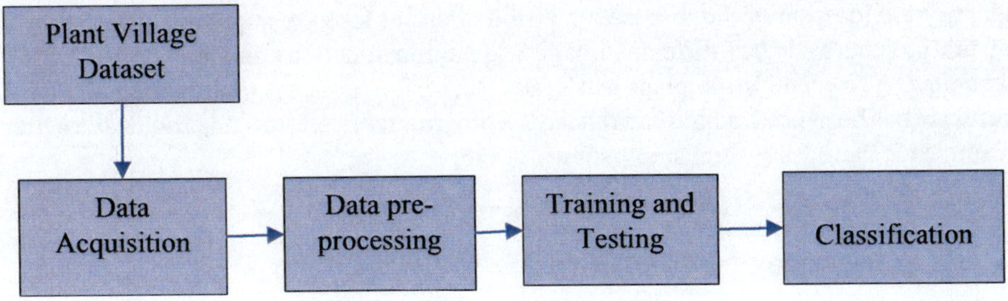

Fig. 1. Proposed methodology of CNN model.

- Data Acquisition

The diseased leaf images are taken from the plant village dataset repository. The dataset has been downloaded from the Kaggle website (link to the dataset: https://www.kaggle.com/abdallahalidev/plantvillage-dataset/version/1).

The images are downloaded using the python algorithm. From the dataset, we have taken 1270 images for two classes. This dataset includes two types of disease(s) present in the apple leaf, such as scab or rot.

- Data pre-processing

This process includes data cleaning, data reduction and data transformation. The major purposes of data preprocessing are to remove noise present in the images and resizing of the images to speed up the training data.

- Training and Testing

For the training and testing of the model, 1200 and 70 images are taken, respectively, from the dataset. While training a model, we do not want overfitting of the model. We use a technique called cross-validation, to avoid the overfitting, where the data are split into two parts: the training set and validation set.

- Classification

The major goal of the classification is to identify the correct class for each data. Multi-class CNN model can predict more than one classes. It requires less preprocessing. The proposed classifier contains two classes. So, after training the model, an evaluation will be done to check whether the proposed CNN model can predict the classes accurately.

3.2 Image Processing Through LabVIEW

After the classification of diseases, image processing can be done through LabVIEW software. A smart camera is compatible with LabVIEW Software. So, image processing will be done through the vision assistant toolbox of LabVIEW software, which helps us to identify the location of the disease(s). Vision assistant toolbox improves the efficiency of the project, as it has different image acquisition functions and image processing techniques. To identify the plant leaf's disease, vision assistant toolbox of LabVIEW software has been used, which can reduce the programming efforts and provide the better result. This includes several steps, shown in Fig. 2:

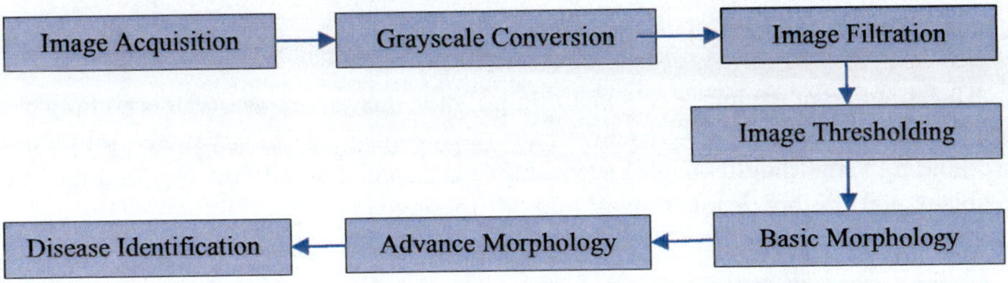

Fig. 2. Flow-chart of image processing in vision assistant.

- Image Acquisition

Acquiring an image from the public domain using a smart camera is the preliminary process of image processing. Acquiring an image acts as input for the process. We can use any format for image acquisition like.jpg,.gif,.jfif as input for image processing. The original image is taken from the site: (https://www.missouribotanicalgarden.org/Portals/0/Gardening/Gardening%20Help/images/Pests/Leaf_Spot_of_Trees669.jpg) for the image processing, shown in Fig. 3.

- Gray scale conversion

It converts the original image of the diseased leaf to a grayscale image which is shown in Fig. 4.

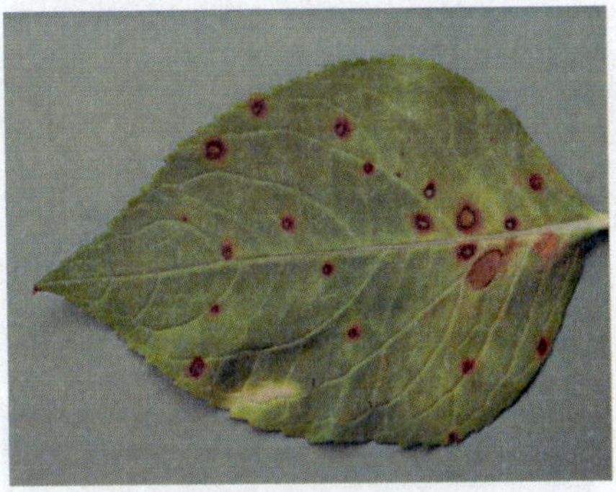

Fig. 3. Original captured image.

Fig. 4. Grayscale image

- Image Filtration

Since the image must be taken from the real field, it contains dust particles and many other impurities as noise. The major purpose to use a filter is to improve the image quality by reducing noise present in the image. Depending on the applications, we use different filters for the image, so that image will appear sharp and fine. The image shown in Fig. 5 is obtained after using a convolution filter.

• Image Thresholding

Thresholding of the image is done to simplify the image complexity. It is nothing but a type of image segmentation. To convert a grayscale image to a binary image, minimum thresholding value should be used according to the application. In the Fig. 6, the white and black portions are denoted by '1' and '0', respectively.

Fig. 5. Image filtration

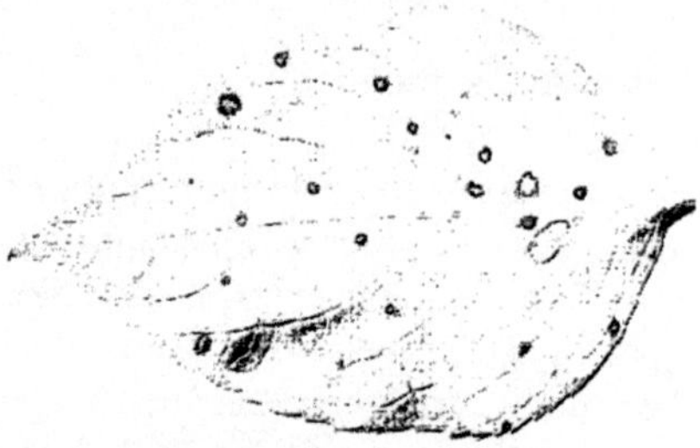

Fig. 6. Image thresholding

• Basic Morphology

In the primary stage of morphology, we are going to remove the boundary of the objects. Now, a major portion of the diseased leaf is reflected in the Image. We can also see dark edge discontinuity of the leaf in Fig. 7. To remove dark edge discontinuity, advanced morphology tool must be used.

• Advanced Morphology

This tool removes the complete background of the diseased image and the diseased portion will be made visible as the spots in Fig. 8. These spots are useful for further, real-time image processing of the plant.

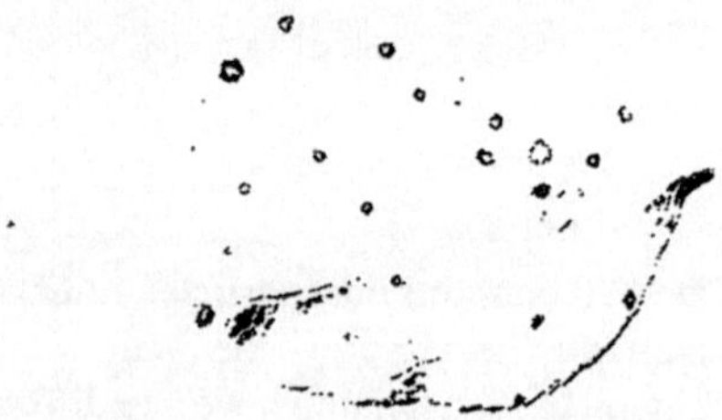

Fig. 7. Basic morphology

Fig. 8. Advanced morphology

4 Results and Discussion

We have used the plant village dataset to train and test our model. The dataset has been downloaded from the kaggle website (link to the dataset: https://www.kaggle.com/abdallahalidev/plantvillage-dataset/version/1).

We have taken the apple leaf samples for the training and testing purpose. We have taken two disease(s) (namely Scab and Rot) in a random manner for the training of the model using 1200 images and its testing utilizing 70 images.

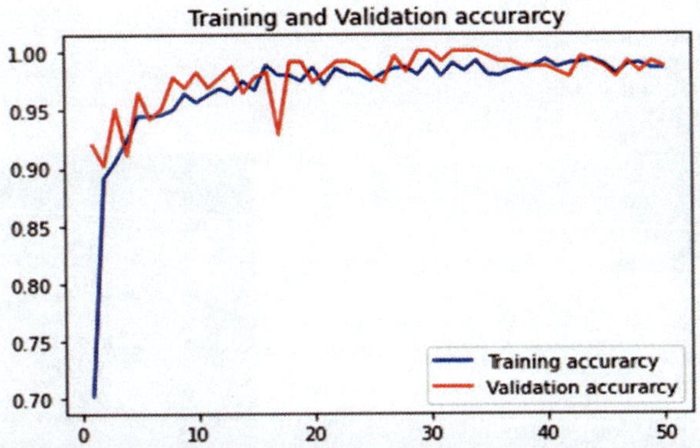

Fig. 9. Training and validation accuracy for epochs 50.

The training set will able to train the CNN model and the testing set will evaluate its performance. The testing set tells about the quality of the model, that is, how well, it can predict the result based on an unseen dataset.

Based on the training and testing data, the accuracy curve (refer to Fig. 9) and loss curve (refer to Fig. 10) have been obtained. The validation accuracy is found to be more than training accuracy, as shown in Fig. 9 and validation loss is seen to be less than training loss, as displayed in Fig. 10. So, this model is found to be good for disease classification.

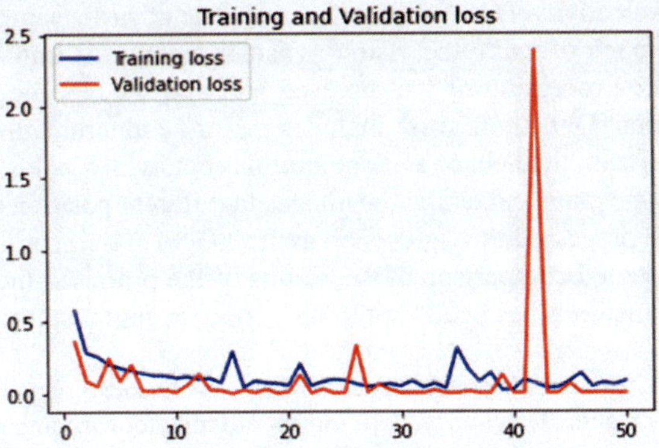

Fig. 10. Training and validation loss for epochs 50.

As the validation loss curve starts decreasing (refer to Fig. 10), the validation accuracy curve starts increasing (refer to Fig. 9). The above observation makes the conclusion that our model is working fine. The results are acceptable for the classification of diseases. The above CNN model is able to classify, whether the disease is scab or rot.

Image processing of diseased leaf using a vision assistant has been done, which follows several steps like image acquisition, Grayscale conversion, image filtration, image thresholding, basic morphology and advanced morphology. After the completion of all the above steps, we will be able to identify the location of the disease present in the leaf (refer to Fig. 11).

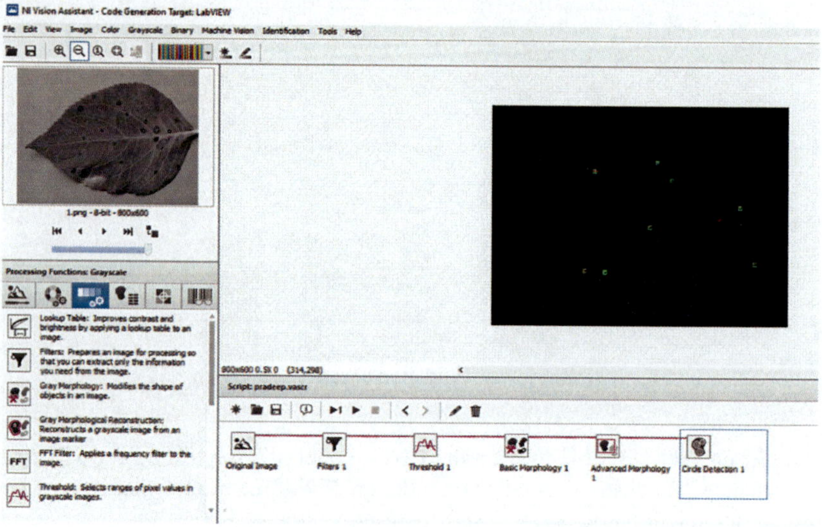

Fig. 11. Image processing using Vision assistant toolbox.

5 Conclusion and Future Work

Nowadays, there have been some development and advancement in the agricultural field to ensure good productivity. However, there is a chance of further improvement. In this paper, a new approach of using the vision assistant toolbox of LabVIEW software is explored for image processing of the diseased leaf and a CNN model is utilized for disease classification. Our future work includes real-time identification of the diseases. This could be done with the help of an agricultural robot, which holds a camera capable of moving around the plant and taking the images in different positions and orientations. This image will be processed through vision assistant and classification will be done by CNN model. A detailed comparison of the results of the proposed method with that of the existing other methods has been kept in the scope for future study.

Acknowledgement. The authors gratefully acknowledge the financial support of the Ministry of Electronics and Information Technology, Government of India, for carrying out this study.

References

1. Gavhale, K.R., Gawande, U.: An overview of the research on plant leaves disease detection using image processing techniques. IOSR J. Comput. Eng. (IOSR-JCE) **16**(1), 10–16 (2014)
2. Jhuria, M., Kumar, A., Borse, R.: Image processing for smart farming: detection of disease and fruit grading. In: IEEE Second International Conference on Image Information Processing (ICIIP) (2013)
3. Sladojevic, S., Arsenovic, M., Anderla, A., Culibrk, D., Stefanovic, D.: Deep neural networks based recognition of plant diseases by leaf image classification. Comput. Intell. Neurosci. **2016**, 1–11 (2016)
4. Qian, J., Zhou, J., Liu, Y.: Labview-based study on the modeling method of chlorophyll content prediction in tomato leaves. AMSE J. **60**, 416–428 (2017)
5. Khirade, S.D., Patil, A.B.: Plant disease detection using image processing. In: 2015 International Conference on Computing Communication Control and Automation, pp. 768–771. IEEE (2015)
6. Rajmohan, R., Pajany, M., Rajesh, R., Raman, D.R., Prabu, U.: Smart paddy crop disease identification and management using deep convolution neural network and SVM classifier. Int. J. Pure Appl. Math. **118**(15), 255–264 (2018)
7. Sheelavantamath, B.: Plant disease detection and its solution using image classification. Int. J. Fut. Res. Dev. **1**(1), 73–79 (2020)
8. de la Cruz, J.E.C., Ramirez, O.J.V.: Convolutional neural networks for the Hass avocado classification using LabVIEW in an agro-industrial plant. In: 2020 IEEE XXVII International Conference on Electronics, Electrical Engineering and Computing (INTERCON), pp. 1–4. IEEE (2020)
9. Sardogan, M., Tuncer, A., Ozen, Y.: Plant leaf disease detection and classification based on CNN with LVQ algorithm. In: 3rd International Conference on Computer Science and Engineering (UBMK), pp. 382–385. IEEE (2018)
10. Raveendra, P., Reddy, V.S., Subbaiah, G.V.: Vision-based weed recognition using LabVIEW environment for agricultural applications. Mater. Today Proc. **23**, 483–489 (2020)

An Autonomous Galactic Swarm Optimization Algorithm Supported by Hidden Markov Model

Mauricio Castillo[1(✉)], Broderick Crawford[1], Ricardo Soto[1], Wenceslao Palma[1], José Lemus-Romani[1], Diego Tapia[1], Felipe Cisternas-Caneo[1], Marcelo Becerra-Rozas[1], Fernando Paredes[2], and Sanjay Misra[3]

[1] Pontificia Universidad Católica de Valparaíso, Valparaíso, Chile
{mauricio.castillo.d,jose.lemus.r,diego.tapia.r, felipe.cisternas.c,marcelo.becerra.r}@mail.pucv.cl
{broderick.crawford,ricardo.soto,wenceslao.palma}@pucv.cl

[2] Universidad Diego Portales, Santiago, Chile
fernando.paredes@udp.cl

[3] Covenant University, Ota, Nigeria
sanjay.misra@covenantuniversity.edu.ng

Abstract. In this work we implemented a version of the Galactic Swarm Optimization metaheuristic algorithm tuned by a hidden Markov model. The Galactic Swarm Optimization algorithm is an abstraction of the motion of stars within galaxies on the first level, and galaxies within a cluster of galaxies on the second level. We address the problem of controlling the metaheuristic parameters by identifying the state of the algorithm at each iteration i, using the Hidden Markov Model framework and updating the Galactic Swarm Optimization parameters accordingly. The results obtained show an improvement compared to the original algorithm using the fixed parameters found in the literature. In addition, the results are compared against other algorithms that use different techniques and hybridizations to solve the same problem, showing an improvement in performance with a similar quality for the solutions obtained.

Keywords: Metaheuristics · Parameter control · Machine learning · Galactic Swarm Optimization · Hidden Markov model

1 Introduction

Metaheuristics (MH) are a group of algorithms designed to solve optimization problems. They are usually inspired by some process observed in nature, such as the movement of flocks of birds looking for food [11] or natural selection [5].

In this work we will use the MH Galactic Swarm Optimization (GSO) algorithm [8], which simulates the movement of stars within galaxies on a first level

A. Abraham et al. (Eds.): SoCPaR 2020, AISC 1383, pp. 354–363, 2021.
https://doi.org/10.1007/978-3-030-73689-7_34

and galaxies within a cluster of galaxies on a second level. The motion of stars depends on the gravity exerted by other stars within the galaxy, accompanied by the inertial motion of the same star.

Each MH has a set of parameters that control its functioning. The definition of parameters is also an optimization problem [3], where the set of parameters chosen represent a solution and the performance of the algorithm corresponds to the fitness. Automatic parameter definition, appears as a method to help researchers create more robust algorithms. Classified by the moment they are made, we can distinguish 2 types of parameter adjustment: the first occurs outside the algorithm and is called offline parameter setting. The second is known as online parameter adjustment, parameter control or autonomous search [2], and is the contribution of this work. In parameter control, parameters are adjusted in relation to the information that the algorithm obtains from the environment in real time, allowing it to react to stimuli it receives from the instance of the specific problem it is solving.

For the parameter control to be effective, it is necessary to determine in the best possible way the stimuli to which the parameter control procedure will react. Machine learning algorithms have proven to be effective in solving pattern recognition problems, so they are a good candidate for building a procedure capable of recognizing patterns that can be used to control GSO parameters. This paper proposes a method to control the parameters of the GSO algorithm by integrating the machine learning (ML) Hidden Markov Model (HMM) technique, which allows to learn to identify the state of each galaxy within GSO, and then to control the parameters of the algorithm autonomously.

This work is organized as follows: in Sect. 2 we review the technique used (GSO), detailing the parameters that control it and showing its movement operators. Section 3 explain the integration with the HMM framework and define the method by which we will update the value of the GSO parameters. In Sect. 4, we show the results obtained and in Sect. 5 we present the final thoughts of this work.

2 Galactic Swam Optimization

This section details the Particle Swarm Optimization (PSO) algorithm, its perturbation operator and its parameters. Then, as a way to extend its operation, the Galactic Swarm Optimization algorithm is detailed, which allows a parallel execution of several instances of the PSO algorithm.

The PSO algorithm [12], simulates the movement of flocks of birds or shoals of fish, which move together to escape from predators or find food sources. Each individual that composes this "swarm" constitutes a solution to the problem we are trying to solve, and its movements are determined by 3 components, the inertial speed, the attraction towards the best historical value of the particle and the attraction towards the particle with the best value of the swarm.

Each of these three components are regulated by 3 coefficients, which are the main parameters of the PSO algorithm. These coefficients are commonly referred to as: inertia coefficient ω, historical coefficient $c1$ and group coefficient $c2$.

The movement operator for each particle within the PSO is defined in the Eq. 1 and Eq. 2.

$$\vec{v}_{t+1} = \omega \otimes \vec{v}_t + c_1 r_1(\vec{p} - \vec{x_t}) + c_2 r_2(\vec{g} - \vec{x_t}) \quad (1)$$

$$\vec{x}_{t+1} = \vec{x}_t + \vec{v}_{t+1} \quad (2)$$

Where $\vec{x_t}$ is the current particle, $\vec{v_t}$ is the particle velocity, (ω) is the coefficient of inertia, $c1$ and $c2$ are the personal and global acceleration coefficients, $r1$ and $r2$ are random numbers in the range [0.1], $\vec{p}$ is the best historical position of the particle and $\vec{g}$ is the best global position.

2.1 Galactic Swarm Optimization GSO

The GSO algorithm, proposed by [8] is a metaheuristic algorithm that simulates the movement of stars and galaxies, which are attracted towards other stars and galaxies with a higher gravity. This MH is a derivative of the PSO algorithms and works on 2 different levels.

The first level runs through the search space by grouping the particles into a defined number of groups, where the particles in each group move in relation to the particle with the best fitness value found within the group and also in relation to the best historical fitness value that the particle has taken, in the same way as the PSO algorithm does. At this level, a particle represents a star and also a solution in the context of the problem. The movement of the stars at this level represents an exploitation of the search space and is the level where the highest number of particles within the algorithm interacts.

The second level goes through the search space in an exploratory way. Each particle on the second level represents the centre of mass of a level 1 group, so it is considered a galaxy. In this level, galaxies are represented only as their center of mass and explore the search space. Each galaxy moves in relation to the particle with the best fitness value obtained so far and in relation to its best historical value.

An important aspect of this algorithm is that level 2 is built from the information obtained in level 1, that is, the centres of mass of each galaxy are defined in the execution of level 1 and are used by level 2. However, once level 2 finishes its execution only the best value obtained by level 2 is kept in memory, in case the best value obtained until that moment is improved. According to the authors [8], experimentally it was proved that this configuration resulted in a better performance of the algorithm.

3 Machine Learning Integration

In the following section, integration with the Machine Learning (ML) Hidden Markov Model (HMM) algorithm is detailed. We use the evolutionary factor f, as the observed states O for the HMM model. In addition, the optimization method of the inertia parameter (ω) is redefined and extended to the optimization of the acceleration coefficients ($c1$ and $c2$).

3.1 PSO Evolutionary Factor f

GSO handles groups of PSOs running in parallel. Therefore, in order to reflect the state of GSO it is necessary to determine the state of each PSO individually. For this, the idea from [13] is taken for the calculation of an evolutionary factor of the PSO. The calculation of the evolutionary factor is done by calculating the average distance from every particle to each other, denoting the maximum and minimum value as d_{min} and d_{max} and the average from the best particle as d_g. The we calculate f in Eq. 3.

$$f = \frac{d_g - d_{min}}{d_{max} - d_{min}} \in [0, 1] \tag{3}$$

Figure 1 shows the relation between the calculated evolutionary state and the position of the particles in the PSO. Part (a) shows a PSO in the exploration state: The particles are far from each other and far from the optimum. Part (b) shows a PSO in the exploitation/convergence state, where the distance between particles is smaller and the optimum is in the centre of the group. Finally, part (c) shows a PSO in the jump-out state, where the particles are in close positions, except for the optimum which has jumped to another zone in the solution space.

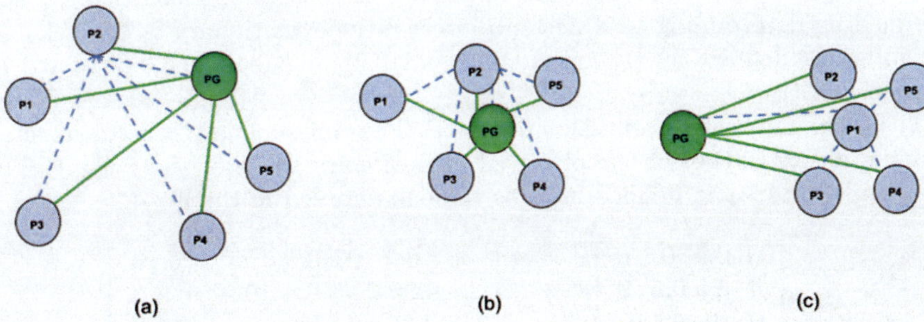

Fig. 1. PSO particle distribution. (a) $d_g \approx d_{pi}$ exploration; (b) $d_g \ll d_{pi}$ exploitation, convergence; (c) $d_g \gg d_{pi}$ jump out

3.2 Modeling the PSO Behavior with Hidden Markov Model

This section defines the Hidden Markov Model, used in this paper to identify the state of each PSO that composes the GSO algorithm. In this section, we show how the states of a PSO can be modeled as a Markov chain, allowing the state to be inferred using the evolutionary factor calculated for a PSO. For more details on hidden markov models refer to [10].

Taking the idea set out in [13] for the PSO algorithm, four evolutionary states are defined: exploration, exploitation, convergence and jump out. These states can be defined as a Markov chain, as detailed in Fig. 2.

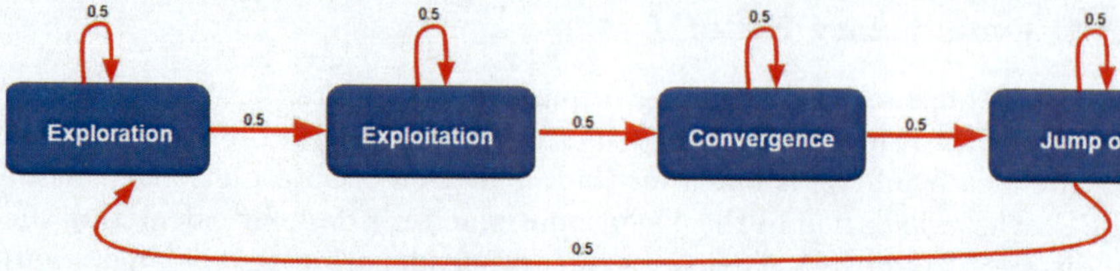

Fig. 2. Evolutionary states defined for the APSO algorithm

This chain is hidden and not directly observable. The observations made to the PSO correspond to a discretization made to the evolutionary factor described in a previous section. The HMM model receives three input parameters, the first is an initial probability vector π, which in this work, as well as in [1], corresponds to a deterministic start in the exploration state. The second parameter of the HMM is a state transition matrix, which describes the probability of passing from one state to another. Figure 2, shows the possible transitions, which start with the same probability of 0.5, and which is represented in matrix A in the Eq. 4.

The observed value for the hidden chain corresponds to a discretization of the evolutionary factor f, calculated in the previous section. This discretization is defined by determining to which sub-interval the evolutionary factor belongs. The intervals defined by [1] are ([0,0.2], [0.2,0.3], [0.3,0.4], [0.4,0.6], [0.6,0.7], [0.7,0.8], [0.8,1]).

The third parameter is emission matrix B. The emission matrix is constructed from the discretization of the observed evolutionary value of the PSO f. The initial emission matrix defined in [1] is the one detailed in the Eq. 4.

$$A = \begin{bmatrix} 0.5 & 0.5 & 0 & 0 \\ 0 & 0.5 & 0.5 & 0 \\ 0 & 0 & 0.5 & 0.5 \\ 0.5 & 0 & 0 & 0.5 \end{bmatrix}, B = \begin{bmatrix} 0 & 0 & 0 & 0.5 & 0.25 & 0.25 & 0 \\ 0 & 0.25 & 0.25 & 0.5 & 0 & 0 & 0 \\ 2/3 & 1/3 & 0 & 0 & 0 & 0 & 0 \\ 0 & 0 & 0 & 0 & 0 & 1/3 & 2/3 \end{bmatrix} \quad (4)$$

With the variables π, A and B we have defined the initial parameters of the HMM. For each iteration the algorithm trains using the Baum-Welch algorithm [9], which allows to adjust the parameters that we defined previously from a set of observations, the evolutionary factors f obtained in each iteration.

Finally, after training, the HMM is able to obtain the hidden state to which the f observations delivered to the model belong, using the Viterbi algorithm [9].

3.3 Parameter Control

Once the state of the algorithm is determined, we can make adjustments to the parameters. For $c1$ and $c2$ respectably: increase and decrease for exploration, slight increase and slight decrease for exploitation, slight increase and slight increase for convergence and decrease and increase for jumping-out.

Through experimental analysis, the percentages of variation for the "Increase and Decrease actions" are determined, increasing or decreasing the current value between 3 and 6%, while the "Slight Increase and Decrease actions" increases or decreases the value between 1 and 3%.

The value of (ω) is updated according to [6], where a different calculation is obtained for each state of the algorithm. The maximum and minimum values for (ω) are 0.9 and 0.1 and where determined by experimental analysis.

In addition, the amount of particles in the PSO is varied for this work, also depending on the state. For the Exploitation and Convergence states new particles are added at a point close to the local optimum, while for the Exploration and Jump out states the particle with the worst fitness value is removed. The minimum and maximum number of particles is 3 and 15, also determined experimentally.

Depending on the state obtained by the HMM, the PSO parameters are updated as explained in the Algorithm 1.

Algorithm 1: Parameter control

Result: Updated inertia, acceleration coefficients and number of particles.

state: The state calculated by HMM;
c1: personal acceleration coefficient;
c2: global acceleration coefficient;
np: particle number;
ω: inertia coefficient;
if state is 'Exploration' **then**
  $c1 = c1 \cdot Rand(1.03, 1.06)$;
  $c2 = c2 \cdot Rand(0.94, 0.97)$;
  $\omega = \omega_{min} + (\omega_{max} - \omega_{min}) \cdot Rand(0, 1)$;
  $np = np - 1$
end if
if state is 'Exploitation' **then**
  $c1 = c1 \cdot Rand(1.01, 1.03)$;
  $c2 = c2 \cdot Rand(0.97, 0.99)$;
  $\omega = \frac{1}{1+1.5\exp^{-2.6f}}$;
  $np = np + 1$
end if
if state is 'Convergence' **then**
  $c1 = c1 \cdot Rand(1.01, 1.03)$;
  $c2 = c2 \cdot Rand(1.01, 1.03)$;
  $\omega = \omega_{min}$;
  $np = np + 1$
end if
if state is 'Jump Out' **then**
  $c1 = c1 \cdot Rand(0.94, 0.97)$;
  $c2 = c2 \cdot Rand(1.03, 1.06)$;
  $\omega = \omega_{max}$;
  $np = np - 1$
end if

4 Comparison of Results

This section compares the results obtained with the GSO prototype using HMM (HGSO) for the adjustment of its parameters. The comparison is made against a GSO instance as described in [8].

4.1 Instances of the SCP Problem

To make the comparison of the results, instances of the Set Covering Problem (SCP) are used. For further reading on the SCP problem, please refer to [4].

4.2 Importance Definition for Self-adaptative Parameters in HGSO

To verify the effectiveness of each optimized parameter is fundamental for the good development of a metaheuristic, therefore it is important to know which parameters of our technique present better performance, so we choose different permutations of parameters to update (the updates are described in Algorithm 1), keeping the others at their initial values. This allows us to find the set of parameters that shows the best performance for a subset of 11 instances of the SCP problem.

The parameters to be evaluated are the four most studied in the literature: the inertia of the set of particles (ω), the personal and overall acceleration coefficients (c_1 and c_2), in addition to the particle size for each group of the 1 np. By allowing all the possible combinations without influencing the order of these, it gives a total of 15 options to be evaluated, which are specified in the Table 1.

Table 1. Parameter permutations

Id	Parameter list
1	c1
2	np
3	c2
4	ω
5	c1, c2

Id	Parameter list
6	ω, c2
7	ω, c1
8	c2, np
9	ω, np
10	c1, np

Id	Parameter list
11	ω, c1, c2
12	c1, c2, np
13	ω, c1, np
14	ω, c2, np
15	ω, c1, c2, np

Experiments are performed for all described parameter permutations. Each combination was tested on several instances of the SCP problem in independent runs 31 times. Then, for each instance the results were stored and the Wilcoxon-Mann-Whitney test was applied [7] to determine whether differences in performance between different sets of parameters are statistically significant. The Table 2 shows the average p-value obtained after performing the test, where the hypothesis to be tested is that the permutation of automated parameters that appears in the rows obtains a lower fitness than the one that appears in the columns. The value "–" represents the case where both samples are the same. Then by adding up the rows (using the exact values) we obtain that the lowest (best) value corresponds to permutation number 9 (ω, np).

4.3 Experimental Results

After having chosen the parameters that will self-adapt as the algorithm progresses, its performance is compared with an instance of the original GSO algorithm proposed in [8].

Table 2. Averages obtained among the p-values for the instances reviewed

P-values average															
	1	2	3	4	5	6	7	8	9	10	11	12	13	14	15	Σ
1	–	1.000	0.306	0.231	0.172	0.086	**0.023**	1.000	1.000	1.000	**0.005**	1.000	1.000	1.000	1.000	8.82387635e+00
2	**0.000**	–	**0.000**	**0.000**	**0.000**	**0.000**	**0.000**	–	–	–	**0.000**	–	–	–	–	3.47374299e–08
3	0.696	1.000	–	0.605	0.500	0.686	0.611	1.000	1.000	1.000	0.400	1.000	1.000	1.000	1.000	1.14983835e+01
4	0.772	1.000	0.396	–	0.520	0.454	0.228	1.000	1.000	1.000	**0.011**	1.000	1.000	1.000	1.000	1.03808462e+01
5	0.829	1.000	0.500	0.481	–	0.365	0.940	1.000	1.000	1.000	0.441	1.000	1.000	1.000	1.000	1.15564287e+01
6	0.916	1.000	0.316	0.549	0.639	–	0.486	1.000	1.000	1.000	0.254	1.000	1.000	1.000	1.000	1.11606288e+01
7	0.978	1.000	0.392	0.774	0.061	0.518	–	1.000	1.000	1.000	0.228	1.000	1.000	1.000	1.000	1.09516240e+01
8	**0.000**	–	**0.000**	**0.000**	**0.000**	**0.000**	**0.000**	–	–	–	**0.000**	–	–	–	–	8.05246105e–08
9	**0.000**	–	**0.000**	**0.000**	**0.000**	**0.000**	**0.000**	–	–	–	**0.000**	–	–	–	–	**6.76330606e–09**
10	**0.000**	–	**0.000**	**0.000**	**0.000**	**0.000**	**0.000**	–	–	–	**0.000**	–	–	–	–	1.02907374e–07
11	0.995	1.000	0.601	0.989	0.564	0.750	0.775	1.000	1.000	1.000	–	1.000	1.000	1.000	1.000	1.26752821e+01
12	**0.000**	–	**0.000**	**0.000**	**0.000**	**0.000**	**0.000**	–	–	–	**0.000**	–	–	–	–	7.58322802e–08
13	**0.000**	–	**0.000**	**0.000**	**0.000**	**0.000**	**0.000**	–	–	–	**0.000**	–	–	–	–	3.74864907e–08
14	**0.000**	–	**0.000**	**0.000**	**0.000**	**0.000**	**0.000**	–	–	–	**0.000**	–	–	–	–	9.77658894e–07
15	**0.000**	–	**0.000**	**0.000**	**0.000**	**0.000**	**0.000**	–	–	–	**0.000**	–	–	–	–	8.77509736e–08

Original GSO Setting. The original GSO instance is set using the values mentioned in [8]. This algorithm performs 50 iterations on level 1 and then 250 iterations on level 2, giving a total of 300 iterations. The Table 3 reflects this configuration.

HGSO Parameter Setting. The modified GSO instance is set differently from the original version. It first performs 100 iterations in total, 50 level 1 and 50 level 2, which are alternated. The parameter settings are detailed in Table 4.

Table 3. Original GSO parameters.

Parameter	Level	Value
ω	1 and 2	$1 - \frac{k}{L+1}$
c1	1 and 2	$2.05 * rand(0,1)$
c2	1 and 2	$2.05 * rand(0,1)$
np	1	50
np	2	5
iter. num.	1	50
iter. num.	2	250

Table 4. HGSO parameters.

Parameter	Level	Value
ω_{min}	1 and 2	0.4
ω_{max}	1 and 2	0.9
np_{min}	1	3 for each PSO
np_{max}	1	15 for each PSO
iter. num.	1	50
iter. num.S	2	50

Comparison of Results. This section shows the results of both algorithms, so that they can be easily compared. The algorithms were tested with 11 instances of the SCP problem. Each problem was run 31 times independently and its results were stored.

First we compare the best results and the average results for each algorithm. The Table 5 summarizes the results obtained.

Table 5. Comparison of results between GSO and HGSO

Instance	Optimum	Best HGSO	Best GSO	HGSO Avg.	GSO Avg.	RPD HGSO	RPD GSO
scp41	429	430	430	430.903	432.419	0.233	0.233
scp51	253	253	255	256.741	260.71	0	0.791
scp61	138	138	140	139.645	140.871	0	1.449
scpa1	253	254	256	255.709	258.097	0.395	1.186
scpb1	69	69	71	69.032	91.129	0	2.899
scpc1	227	227	234	229.741	238.258	0	3.084
scpd1	60	60	79	60.806	123.323	0	31.667
scpnre1	29	29	85	29	106.871	0	193.103
scpnrf1	14	14	39	14	49.29	0	178.571
scpnrg1	176	177	348	179.542	480.839	0.568	97.727
scpnrh1	63	64	277	65.354	349.452	1.587	339.683

5 Conclusions and Future Work

In this work, we solved the SCP problem with an extension of the GSO algorithm, which integrates the hidden markov model framework, for the state classification of each PSO that composes it and finally apply adjustments to their parameters. The integration was done using a metric based on the relationship of the distance between particles, called evolutionary factor. The evolutionary factor was calculated for each PSO independently and used in the HMM framework to deduce the current PSO state: exploration, exploitation, convergence or jumping-out. The state deduced by the HMM was used to adjust four PSO parameters: c_1, c_2, w and np (number of particles). We have performed tests to determine the effectiveness of the adjustments made for all possible parameter combination, where w and np obtained the best performance.

The proposed improvement to the GSO algorithm is shown to be effective when compared to the original algorithm. The optimization of the parameters shows a behavior capable of avoiding the local optimums in which the original algorithm falls, especially in the largest instances. In addition, the statistical analysis confirmed a better performance of the algorithm after applying the improvements. Our algorithm was tested by solving 11 instances of the SCP problem, obtaining the global optimum in 7. In addition, its performance was compared with other similar algorithms, exhibiting very competitive results.

As future work, the parameter modification algorithm will be adjusted, by a method that is closer to the particularity of the problem instance being solved. This method must be able to adjust the parameters according to the fitness obtained and the state deduced. This behavior is the same as that observed in Reinforcement Learning algorithms, where the state corresponds to the environment and the fitness is the reward. Therefore the integration of an RL algorithm for parameter adjustment is considered.

Acknowledgements. Felipe Cisternas-Caneo and Marcelo Becerra-Rozas are supported by Grant DI Investigación Interdisciplinaria del Pregrado/VRIEA/PUCV/039.324/2020. Broderick Crawford is supported by Grant CONICYT/FONDECYT/REGULAR /1171243. Ricardo Soto is supported by Grant CONICYT/FONDECYT/REGULAR /1190129. José Lemus-Romani is supported by National Agency for Research and Development (ANID)/Scholarship Program/DOCTORADO NACIONAL/2019-21191692.

References

1. Aoun, O., Sarhani, M., El Afia, A.: Hidden Markov model classifier for the adaptive particle swarm optimization. In: Recent Developments in Metaheuristics, pp. 1–15. Springer, Heidelberg (2018)
2. Crawford, B., Soto, R., Castro, C., Monfroy, E., et al.: An extensible autonomous search framework for constraint programming. Int. J. Phys. Sci. **6**(14), 3369–3376 (2011)
3. Crawford, B., Soto, R., Monfroy, E., Palma, W., Castro, C., Paredes, F.: Parameter tuning of a choice-function based hyperheuristic using particle swarm optimization. Exp. Syst. Appl. **40**(5), 1690–1695 (2013)
4. Crawford, B., Soto, R., Olivares, R., Embry, G., Flores, D., Palma, W., Castro, C., Paredes, F., Rubio, J.M.: A binary monkey search algorithm variation for solving the set covering problem. Nat. Comput. **19**, 1–17 (2019)
5. Cubillos, C., Rodriguez, N., Crawford, B.: A study on genetic algorithms for the darp problem. In: International Work-Conference on the Interplay Between Natural and Artificial Computation, pp. 498–507. Springer, Heidelberg (2007)
6. El Afia, A., Sarhani, M., Aoun, O.: Hidden markov model control of inertia weight adaptation for particle swarm optimization. IFAC-PapersOnLine **50**(1), 9997–10002 (2017)
7. Mann, H.B., Whitney, D.R.: On a test of whether one of two random variables is stochastically larger than the other. Ann. Math. Stat. **18**, 50–60 (1947)
8. Muthiah-Nakarajan, V., Noel, M.M.: Galactic swarm optimization: a new global optimization metaheuristic inspired by galactic motion. Appl. Soft Comput. **38**, 771–787 (2016)
9. Rabiner, L., Juang, B.: An introduction to hidden Markov models. IEEE ASSP Mag. **3**(1), 4–16 (1986)
10. Rabiner, L.R.: Selected applications in speech recognition. In: Readings in Speech Recognition, p. 267 (1990)
11. Soto, R., Crawford, B., Almonacid, B., Paredes, F.: A migrating birds optimization algorithm for machine-part cell formation problems. In: Mexican International Conference on Artificial Intelligence, pp. 270–281. Springer, Heidelberg (2015)
12. Soto, R., Crawford, B., Toro, J.: Optimal keyboard design by using particle swarm optimization. In: International Conference on Human-Computer Interaction, pp. 281–284. Springer, Heidelberg (2018)
13. Zhan, Z.H., Zhang, J., Li, Y., Chung, H.S.H.: Adaptive particle swarm optimization. IEEE Trans. Syst. Man Cybern. Part B (Cybern) **39**(6), 1362–1381 (2009)

Machine Learning for Breast Cancer Diagnosis and Classification Using Hand-Crafted Features

Dhawa Sang Dong[1(✉)] and Suraj Kandel[2]

[1] Kathmandu Engineering College, Tribhuvan University, Kalimati, Kathmandu, Nepal

[2] Hetauda School of Management and Social Science, Tribhuvan University, Kirtipur, Nepal

Abstract. Analysis of biological processes (signals) is always probabilistic and statistical analysis. Machine Learning stemmed from advanced probabilistic and statistical analysis is the exact tool what the medical workforces are looking for. Research and studies in the area of Machine Learning algorithms enhances its accuracy; its healthcare applications in medical diagnosis can improve the treatment procedure. Healthcare application being directly related to human health, dimensionality reduction is very critical; it sometime triggers to panic losses; therefore, the complete feature involvement in the model development is challenging to enhance the reliability of the Machine Learning algorithm and accuracy of the algorithm. Depending upon the diagnosis area, Regression analysis leading to dimensionality reduction and Principal Component Analysis (PCA) to remove less significant features can be applied. Primarily, in cancer diagnosis, only noise removal is acceptable, not the feature with small significance removal and minimal human intervention is acceptable which could lead to reduced accuracy of diagnosis, so the following treatment procedure.

Keywords: Machine learning · Hand-craft features · Regression analysis · PCA · Accuracy

1 Introduction

1.1 Healthcare

Conventional method of diagnosis involves the visual examination of medical images, and depending up on the experience of physician and his expertise, diagnosis and further clinical treatments will be brought to the patient which is highly subjective. Moreover, diagnosis of features while studying medical images, all the features should be considered as most as we can as suggested in [1] where authors applied Principal Component Analysis (PCA) before classification which removes the very less significant features – features with very low correlation

A. Abraham et al. (Eds.): SoCPaR 2020, AISC 1383, pp. 364–373, 2021.
https://doi.org/10.1007/978-3-030-73689-7_35

coefficient - from the features dimension reducing the computational complexity. Hence, analysis and finding correctly correlated features and corresponding selection of features are very important. However, sometime these could be compromise leading to serious consequences. Rather selecting subset from features, faster algorithm such as Extreme Learning Machine (ELM) or its improved version studied and summarized by Wang *et al.* in [2] could be used; ELM reduces the model training time eliminating gradient based network parameter learning rather using random weights and deviation of input layers.

In traditional method of image analysis, either visual examination by medical personal or image analysis using computer to extract the features of images, there is very little knowledge transfer which directly depends on experience of medical personal, not that of feature extraction algorithm in image computation. However, with the introduction of Machine Learning in modern computer science, knowledge is being transferred with older data and even transfer learning algorithm in feature extraction of medical images.

1.2 Machine Learning

Biological signals are random without well defined mathematical relations between the physical phenomena and bio signals, and classical statistical analysis is used for the analysis of the signals which is limited to calculations of some statistical parameters: variance, mean, median etc. Advanced probabilistic and statistical analysis with larger number of signals or data leading to stochastic signal modeling or random process modeling is one of the advantages of computational mathematics which is ultimately the Machine Learning. Therefore, advanced probabilistic and statistical analysis, with sophisticated term Machine Learning, is one of the prioritized needs in the field of medical and biological science for the modeling of random signals. Mathematically, Machine Learning is modeling of the distribution for given dataset and predict the future value using the model where given dataset is experience from which a computer can learn.

Because of existence of data imbalance [3] in the data especially data from medical healthcare, there is no well defined universal evaluation parameters for all type of Machine Learning algorithm. However, evaluation parameters sensitivity along with specificity does not completely describe the performance of any machine learning model since it excludes the false prediction. Hence, precision and recall (similar to sensitivity) could be the better evaluation parameters the machine learning model which includes the false positive and false negative readings. Higher precision means the diagnostic model has reduced false positive readings; consequently, reduction in physical, economical and psychological cost with unwanted treatments. Similarly, higher value of recall/sensitivity [3,4] means reduced false negative readings which reduces potential higher cost of treatment in the near future which is the primary goal of every Machine Learning model; in overall, there should be reduction in false positive or false negative readings.

Not only does Machine Learning algorithm have applications in the area of classification, but it also can be used to impute missed data as studied [5] where Artificial Neural Network was used for imputing missing data and censored data using multi layer with is also efficient for data redundancy leading to improved classification performance.

2 Related Works

As summarized by Yassin *et al.* in [6], majority of research work for breast cancer is done with digital mammogram (DM) as image model using Support Vector Machine (SVM) – a machine learning technique for malignant or benign cancer stage classification. Some of the biological behaviors such as actin stress fibers in the normal and its deviation in cancerous cell can be used for the analysis and detection of cancer. Optical microscopy images of those actin fibers can be used to classify the cancer cells, but precise detection by human eye is not possible. Use of Convolutional Neural Network (CNN) trained with diverse image datasets – which is known as transfer learning and advantageous for limited data – can be used as studied by Oei *et al.* in [7] which outperformed the human experts. Additionally, there is not proven evidence that image enhancement has not advantage in the classification.

Levine *et al.* [8] studied the current rise of Machine Learning application in medical diagnosis, especially in medical applications. Authors discussed Medical Healthcare sector of radiology and pathology with Machine Learning in tomorrow's prospect where knowledge workers has direct impact and its potential challenges to be faced before entering into the phase when Machine Learning outperforms human expertise. Authors in [9] presented limitation of Naïve Bayes, Support Vector Machine, Recurrent Convolutional Neural Network, and they proposed new modified Neural Network, Deep Neural Network with support Value (DNNS) where their proposed method revised the histogram value, sigmoid function and histo-sigmoid function leading to better performance. Since early and correct detection of breast cancer is crucial and Machine Learning has potential to assist the healthcare workers in the cancer diagnosis, and both study of Machine Learning algorithms and selection of best Machine Learning algorithms is important. Ak in [10] studied different machine learning algorithms: Logistic Regression, K-nearest Neighor, Naïve Bayes, Support Vector Machine, Decision Tree, Random Forest, and Rotation Forest; among them, Logistic Regression was suggested to have best classification accuracy of 98.10%.

Smiti in review work [11] presented some of the Machine Learning application in different medical healthcare type (primary, secondary, and tertiary healthcare) and stated some of the challenges in the application areas: mixed data, misleading outliers. Shaikh and Ali in [12] modified the posterior distribution using discretization filter which changes the numeric properties of Naïve Bayes to reduce both bias and variance leading to higher accuracy compared; different discrete filters such as supervised attribute DISCRETIZE filter (increased to 81.768%), supervised instance RESAMPLE (increased to 85.0829%) and unsupervised attribute Proportional K-Interval Discretization (PKID) (increased to

78.7293%) filter are applied to Naïve Bayes classifier which as accuracy of 73.7569% only. Authors in [13] applied Decision Tree and Logistic Regression model to learn breast cancer, and the study suggests Decision Tree model has higher accuracy of 95.10% compared to Logistic Regression which has 94.40% accuracy. Similarly, Ivaturi, Singh, and Gunanvitha [14] simply brought comparative study among the Machine Learning algorithms SVM, DT, NB, LoR, LDA, and KNN; they studied the parameters precision, recall, ROC, and Accuracy measure, and authors concluded SVM has highest accuracy of 98% compared to other algorithms applied. But study in [15] suggested that SVM has not better performance even compared to DT where only six features are considered and, ANN has better performance of accuracy 84% over KNN, DT and Binary SVM. In the study, authors evaluated the additional parameters Mathew's Correlation Coefficient (MCC) and Jaccard Similarity Coefficient.

Aldhaeebi *et al.* [16] conducted the experiment to study the near-field microwave probe signals to differentiate normal from abnormal breast tissue provided both breast is comprised of symmetric material tissue, and one of the breasts is normal while both phantom size is not to be similar. For the identification of breast tumor, author preprocessed probe signals and analyzed using learning algorithm SVM and DT. Authors in [17] studied histology images using Convolutional Neural Network to predict Invasive Ductal Carcinoma(IDC), a type of breast cancer. Comparative study between two type of CNN using Depthwise Separable Convolution and Standard Convolution suggested that CNN using standard Convolution has better performance compared to Depthwise Separable Convolution whichever the activation function used: ReLU, Sigmoid or tanh. CNN using Standard Convolution has accuracy of 0.8713 using activation function ReLu while that with Depthwise Separable has accuracy of 0.8598; however, latter has less computational complexity and so the memory. Brito-Sarracino *et al.* [18] focused on the interpretability of underlying principle behind the Machine Learning algorithm as physician should be capable of reasoning the diagnosis and classification of highly sophisticated Machine Learning algorithm; they believe better pre-processing and feature selection could lead to the better decision making rather.

Mishra, Singh and Rath [1] carried an experiment where they suggested the feature extraction using SIFT and subsequent classification with Machine Learning has higher accuracy compared to that with SURF a feature extraction tool; additionally, feature extraction and classification after PCA has reduced accuracy of classification in general indicating features have significant role in the classification. As studied in [19], even some of cell characteristics such as groupism and size can be used to classify the cancer if malignant or benign using linear regression which can achieve 84.15% of accuracy. As studied by Ray *et al.* [20], highly correlated signals are to be analyzed, and only some of highly correlated features can be used for the diagnosis and cancer classification. As the author studied higher the number of features selected for machine learning better the accuracy provided features are selected in the rank basis of higher correlation coefficient to lower correlation coefficient. Dhanya *et al.* in [21] studied the three feature selection methods to learn cancer prediction models: two greedy methods to get feature subset with best classification accuracy Sequential Forward Feature Selection and Recursive Feature Elimination, and F-test which calculates F-score and ranked with highest F-score value on top uses higher ranked features in model learning. Study suggests Random Forest model is the best one among Logistic Regression, Naïve Bayes, and Random Forest algorithm.

3 Methodology

In this kind of classification method, human involvement is still very important, and human error can make false classification; therefore, it should be avoided. Input features are selected and its corresponding values by the lab technicians; therefore, probability of false prediction still exists. Machine Learning algorithm can be fed with features either extracted by CNN networks or hand-crafted features as illustrated in Fig. 1 Flow Diagram.

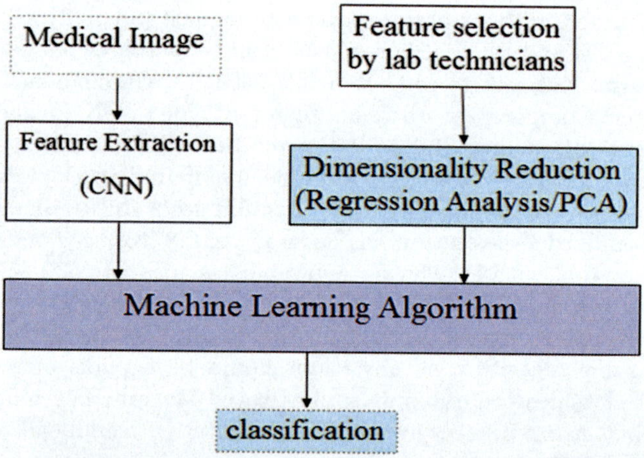

Fig. 1. Flow diagram

3.1 Dimensionality Reduction

Regression analysis was carried to measure contribution of each feature under consideration to study the breast cancer to this data analysis. Definitely, feature removal has impact on diagnosis in terms of clinical decision making which is crucial, but computational complexity could have significant reduction. Mathematically,

$$y = f(x_n) = \beta_0 x_0 + \beta_1 x_1 + \beta_2 x_2 + \beta_3 x_3 .... + \beta_n x_n \quad (1)$$

where $x_0 = 1$ and β are the weight parameters of individual x, features. In regression analysis, weight parameters are evaluated which determines the contribution of corresponding feature to the output $y = f(x_n)$, the class of breast cancer; features associated to very small parameters can be ignored for efficient computation, but accuracy has definitely been reduced. Additionally, in principal component analysis (PCA), major components are included among the list of principal axes reducing computational complexity because of removal of some insignificant data samples considering it as irrelevant to the analysis and diagnosis.

3.2 Data Analysis

Whatever be the machine learning algorithm, the main objective is to increase the prediction accuracy. Alternatively using the following confusion matrix, to increase the accuracy of the algorithm, the machine learning algorithm should decrease value of False Positive (FP) and False Negative (FN) for given data and increase the value of True Positive (TP) and True Negative (TN). The reliability measure depends on the field of application; for example, if the Machine Learning is using in healthcare, there is no compromise for some disease such as cancer. Some dominantly used evaluation parameters are Accuracy (Acc), Sensitivity (Sn), and Specificity (Sp). Accuracy measures true positive rate and true negative rate in totality while Sensitivity and Specificity measure true positive rate and true negative rate respectively.

Confusion Matrix		Actual Observation	
		1	0
Predicted Value	1	True Positive (TP)	False Positive (FP)
	0	False Negative (FN)	True Negative (TN)

4 Results and Discussion

For the model evaluation, different parameters such as Accuracy (Acc), Sensitivity (Sn), and F1 score are studied in model learning and analyzed; however, in overall model, accuracy (Acc) is the major parameter to be evaluated for the model reliability because it measures both true positive and true negative rate in totality. In contrast, for some model learning, both sensitivity and specificity are be analyzed separately – especially in healthcare application, and for cancer data learning, our model should reduce both the false negative and false positive reading as indicated in confusion matrix.

From our study which involves regression analysis (Reg) and principal component analysis (PCA) to reduce dimensionality and correspondingly reduced computational complexity, the dimensionality reduction has suggested the degraded prediction accuracy so the reliability of the learning algorithm with lesser computational complexity.

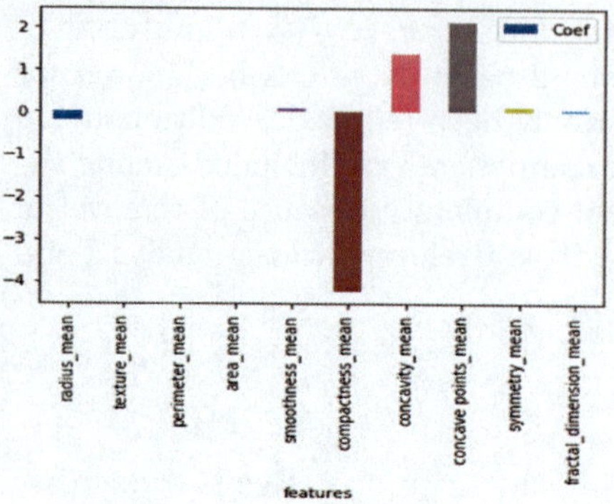

Fig. 2. Feature#1 plot

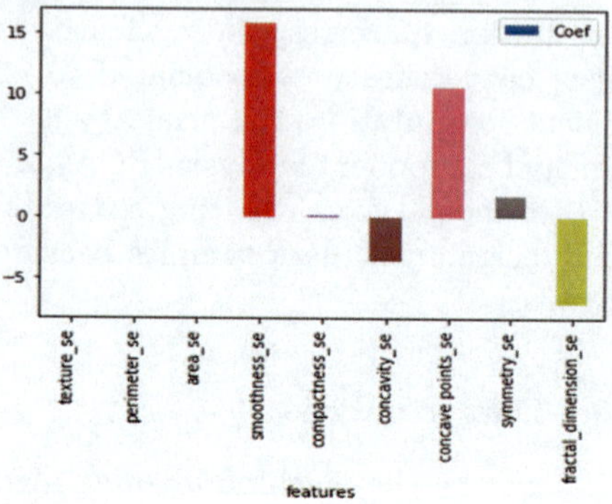

Fig. 3. Feature#2 plot

In this model learning, out of 32 features 21 features (Fig. 2, 3 and 5)were considered using regression analysis which resulted to reduced accuracy from 98.24% to 94.73% in the case of support vector Machine (SVM) while that for multi-layer perceptron (MLP) it has reduction from 98.24% to 96.49%., and degraded model accuracy for Random Forest (RF), and Naïve Bayes except that for Decision Tree (DT) model where Accuracy of 94.74% was increased to 95.61% even for regression analysis. From these results it can be inferred that reduction in dimensionality of feature triggers degraded model accuracy. Additionally, Sensitivity and F1 scores are studied in the simulation, and it is observed parameter values are downgraded for regression analysis. If we observe the comparative accuracy measures for all the algorithms – such as RF, SVM, DT, NB, and MLP – MLP has increased prediction; this is because small error accumulation while moving from one hidden layer to succeeding hidden layer in Neural Network.

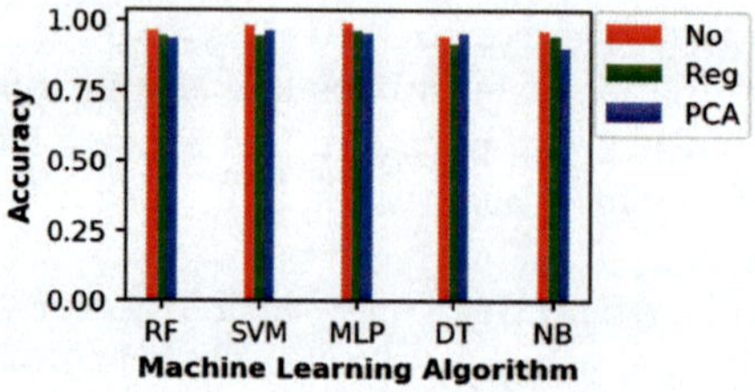

Fig. 4. Accuracy plot

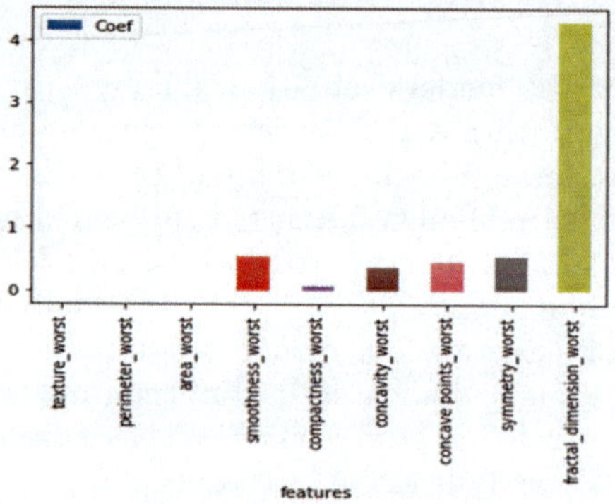

Fig. 5. Feature#3 Plot

Table 1. Parameter readings

Accuracy	Machine Learning Algorithms				
	RF	SVM	MLP	DT	NB
No	0.964912	0.982456	0.991228	0.947368	0.964912
Reg	0.947368	0.947368	0.964912	0.921053	0.947368
PCA	0.938596	0.964912	0.95614	0.95614	0.903509
Sensitivity					
No	0.930233	0.953488	0.976744	0.930233	0.930233
Reg	0.906977	0.906977	0.953488	0.906977	0.906977
PCA	0.906977	0.953488	0.953488	0.953488	0.813953
F1-Score					
No	0.952381	0.97619	0.988235	0.930233	0.952381
Reg	0.928571	0.928571	0.953488	0.896552	0.928571
PCA	0.917647	0.953488	0.942529	0.942529	0.864198

Similar phenomenon (decreased parameter values for Accuracy, Sensitivity and F1-scores) can be observed for Principal Component Analysis (PCA) as demonstrated in Table 1. In this particular data preprocessing, 99% of the variance amounted to the 17 principal components and correspondingly degraded accuracy of the model. Though in our model learning, dataset was of small size, computation was not a big issue, but to the model for which large dataset is available, computational complexity while model learning could be the issue, so the dimensionality reduction and noise removal has significant effect on computational efficiency and very little amount of degradation in model prediction.

When all the available features were considered for model learning, SVM and MLP have better prediction accuracy measure; moreover, algorithm like ELM or its improved version can be used which does not use backward propagation method to learn the model parameters.

5 Conclusion

To summarize, dimensionality reduction is critical in healthcare application of machine learning; moreover, over the cancer diagnosis is feature selection very important. From the computational simulation carried where Regression Analysis and PCA done over the dataset of breast cancer, the corresponding model learning shows the reduced prediction accuracy and other evaluation parameters for the learning model demonstrating features ignorance has unacceptable effects in the learning; therefore, use of high computational device could be preferable rather use of subset of features or use of computationally efficient algorithm like ELM. Convolutional Neural Network (CNN) which learns all of the features of medical images is another promising and challenging technology in machine learning domain. In orthodox method of feature extraction from classical image processing, feature values are evaluated with human knowledge and visual intervention which could result to comparatively lower accuracy compared to machine leaned feature extraction (Convolutional Neural Network involving larger number of convolution filters and input images) followed by Machine Learning for classification. Now the next challenge could be refining Convolutional Neural Network and its corresponding computational complexity.

References

1. Mishra, V., Singh, Y., Rath, S.K.: Breast cancer detection from thermograms using feature extraction and machine learning techniques. In: 2019 IEEE 5th International Conference for Convergence in Technology (I2CT), Bombay, India, pp. 1–5 (2019). https://doi.org/10.1109/I2CT45611.2019.9033713
2. Wang, Z., et al.: Computer-aided diagnosis based on extreme learning machine: a review. IEEE Access **8**, 141657–141673 (2020). https://doi.org/10.1109/ACCESS.2020.3012093
3. Gupta, S., Sedmkar, R.: Machine learning for healthcare: introduction. In: Machine Learning with Health Care Perspective, Learning and Analytics in Intelligent Systems, vol. 13, Springer Nature Switzerland, pp. 1–25 (2020)
4. Rebala, G., Ravi, A., Churiwala, S.: Classification. In: An Introduction to Machine Learning, Springer Nature Switzerland, pp. 57–66 (2019)
5. Ritthipravat, P.: Artificial neural networks in cancer recurrence prediction. In: 2009 International Conference on Computer Engineering and Technology, Singapore, pp. 103–107 (2009). https://doi.org/10.1109/ICCET.2009.84
6. Yassin, N.I., Omran, S., Houby, E.M., Allam, H.: Machine learning techniques for breast cancer computer aided diagnosis using different image modalities: a systematic review. Comput. Methods Programs Biomed. Elsevier **156**, 25–45 (2018)
7. Oei, R.W., et al.: Convolutional Neural Network for cell classification using microscope image of intracellular actin networks. Plos One **14**, e0213626 (2019)
8. Levin, A.B., Schlosser, C., Grewal, J., Coope, R., Jones, S., Yip, S.: Rise of the machines: advances in deep learning for cancer diagnosis. In: Trends in Cancer, Elsevier Inc, vol. 5, no. 3, pp. 157–169, March 2019
9. Vaka, A.R., Soni, B., Reddy, S.: Breast cancer detection by leveraging machine learning. In: The Korean Institute of Communications and Information Science (KICS), pp. 1–5 (2020)
10. Ak, M.F.: A comparative analysis of breast cancer detection and diagnosis using data visualization and machine learning applications. In: Healthcare MDPI (2020)
11. Smiti, A.: When machine learning meets medical world: current status and future challenges. In: Elsevier Inc. Comput. Sci. Rev. **37**, 100280 (2020)
12. Shaikh, T.A., Ali, R.: A CAD tool for breast cancer prediction using Naive Bayes classifier. In: 2020 International Conference on Emerging Smart Computing and Informatics (ESCI), Pune, India, pp. 351–356 (2020). https://doi.org/10.1109/ESCI48226.2020.9167568
13. Sengar, P.P., Gaikwad, M.J., Nagdive, A.S.: Comparative study of machine learning algorithms for breast cancer prediction. In: 2020 Third International Conference on Smart Systems and Inventive Technology (ICSSIT), Tirunelveli, India, pp. 796–801 (2020). https://doi.org/10.1109/ICSSIT48917.2020.9214267

14. Ivaturi, A., Singh, A., Gunanvitha, B., Chethan, K.S.: Soft classification techniques for breast cancer detection and classification. In: 2020 International Conference on Intelligent Engineering and Management (ICIEM), London, United Kingdom, pp. 437–442 (2020). https://doi.org/10.1109/ICIEM48762.2020.9160219
15. Laghmati,S., Tmiri, A., Cherradi, B.: machine learning based system for prediction of breast cancer severity. In: 2019 International Conference on Wireless Networks and Mobile Communications (WINCOM), Fez, Morocco, pp. 1–5 (2019). https://doi.org/10.1109/WINCOM47513.2019.8942575
16. Aldhaeebi, M., Bamatraf, S., Ramahi, O., Binajjaj, S.A.: breast tumor diagnosis using machine learning with microwave probes. In: 2019 First International Conference of Intelligent Computing and Engineering (ICOICE), Hadhramout, Yemen, pp. 1–4 (2019). https://doi.org/10.1109/ICOICE48418.2019.9035150
17. Alghodhaifi, H., Alghodhaifi, A., Alghodhaifi, M.: Predicting invasive ductal carcinoma in breast histology images using convolutional neural network. In: 2019 IEEE National Aerospace and Electronics Conference (NAECON), Dayton, OH, USA, pp. 374–378 (2019). https://doi.org/10.1109/NAECON46414.2019.9057822
18. Brito-Sarracino, et al.: Explainable machine learning for breast cancer diagnosis. In: 2019 8th Brazilian Conference on Intelligent Systems (BRACIS), Salvador, Brazil, pp. 681–686 (2019). https://doi.org/10.1109/BRACIS.2019.00124
19. Murugan, S., Kumar, B.M., Amudha, S.: Classification and prediction of breast cancer using linear regression, decision tree and random forest. In: 2017 International Conference on Current Trends in Computer, Electrical, Electronics and Communication (CTCEEC), Mysore, pp. 763–766 (2017). https://doi.org/10.1109/CTCEEC.2017.8455058
20. Ray, S., AlGhamdi, A., AlGhamdi, A., Alshouiliy, K., Agrawal, D.P.: Selecting features for breast cancer analysis and prediction. In: 2020 International Conference on Advances in Computing and Communication Engineering (ICACCE), Las Vegas, NV, USA, pp. 1–6 (2020). https://doi.org/10.1109/ICACCE49060.2020.9154919
21. Dhanya, R., Paul, I.R., Akula, S.S., Sivakumar, M., Nair, J.J.: A comparative study for breast cancer prediction using machine learning and feature selection. In: 2019 International Conference on Intelligent Computing and Control Systems (ICCS), Madurai, India, pp. 1049–1055 (2019). https://doi.org/10.1109/ICCS45141.2019.9065563

Automatic Seeded Region Growing with Level Set Technique Used for Segmentation of Pancreas

Pradip M. Paithane[1](✉), S. N. Kakarwal[2], and D. V. Kurmude[3]

[1] Dr. B.A.M.U, Aurangabad, India
[2] PES COE, Aurangabad, India
[3] Milind College of Science, Aurangabad, India

Abstract. Image division process is dynamically performed vital character in image investigation. Image division process is called segmentation which applied in many applications like crop disease detection, monitoring of traffic, metal surface crack detection, medical image analysis, and Aerial Application. Top-down approach and bottom-up approaches are running for automatic segmentation of pancreas. Scale-Invariant Feature transforms, Novel Modified Kernel fuzzy c-means clustering (NMKFCM) approaches are compiled for analysis. NMKFCM and SIFT are execute to identify feature of image. In SIFT, ample statistics of native feature of image be located. In bottom-up, Levelset is tested for image segmentation with automatic seeded region growing. Seeded region growing is extracting region from image truly and efficiently. In NMKFC, accuracy is improved but organ not extracted separately. SIFT, more time period to exacted image without single organ segment. LevelSet works on active contour which identifies boundaries of selected organ with neighbor pixel value. Seeded region growing method can discover seeds automatically via Affinity Propagation (AP) method, so it is avoiding user interaction. The speed function is planned to governor drive of the curve; and in diverse application problems, the key is to decide the suitable stopping criteria for the evolution. LevelSet algorithm is providing single organ segmentation with more accuracy than Top-Down approaches Lelel Set method is required minimum time period to calculated image.

Keywords: Abdominal Computed Tomography (ACT) · LevelSet · NMKFCM · Scale Invariant Transform (SIFT) · Seed region growing

1 Introduction

The medicinal imaging plays a progressively noticeable role in the analysis also cure of illnesses, many number of CT, MR, PET, and supplementary images are acquired everyday [1]. Result of input images are challenging for accurate output and fast processing. Automatic or semiautomatic segmentation of various image components are helpful to analyze anatomical organizations, the spatial distribution of their task or movement, and pathological sections. Visualization and diagnosis can be completed by segmentation.

A. Abraham et al. (Eds.): SoCPaR 2020, AISC 1383, pp. 374–382, 2021.
https://doi.org/10.1007/978-3-030-73689-7_36

Subgrouping is a chief subject to various image-processing researches. Image subgrouping of gastric body portion which consist of pancreas, liver and spleen in abdominal computed tomography (C.T.) [2]. NMKFCM assimilate neighborhood pixel data into objective function for progress of result segmentation [3]. The aim of technique remains toward mark simpler or dissimilarity of performance of an input data into more open and informal image to examine. NMKFCM works properly in noisy image but identification of required cluster number is major struggle [4].

In levelset process, structure of speed function is energetic through esteem to conclude consequence. It is calculated to governor for program of the curve also diverse application difficulties, basic point is to decide applicable stopping criteria for the evolution [5].

Relaxation factor be responsible for a technique to discontinue evolving curve next to ventricle borderline even when borderline remains fuzzy. It is decreases bounding state of the alteration ranges from maximum and minimum gradient values of image.

When correlating with dense images, the human optical organism will first of all emphasis on the sections that can effectively express pixels of images, which is entitled Region of Interest (ROI). Extracting ROIs from images with dense background accurately, it can diminish computational complexity efficiently, very significant for analysis and considerate.

2 Top-Down Method

2.1 Novel Modified Kernel Fuzzy C-means Clustering

It is adapting neighbor point value cost in objective function. This system is revised form of K-FCM and which is integrate neighbor point value cost by 3 × 3 or 5 × 5 window and incorporate into objective function [3]. 'α' parameter is educated for governor load of neighbor's point value which is succeeded greater cost through penetrating of noise. Series of α cost lies inside 0 to 1, if ratio of noise is low at that point, elect cost of α ranges 0 and 0.5 and ratio of noise is higher at that point elect cost of α ranges 0.5 and 1.0. After, every iteration, cost of objective function is reduced with neighbor point value [4]. In Eq. 1, announce mask around point value and α parameter.

$$J_{NMKFCM_{obj}}(U, W) = \sum_{y=1}^{Q}\sum_{x=1}^{P} U_{xy}^{m}(1 - K_T(Z_x, W_y))\left(\frac{N_R - \alpha \sum_{k \in N_i} U_{yk}}{N_R}\right) \quad (1)$$

Where: Ni: Collecting Zi point values of neighbors, NR: The cardinality, K_T: Gaussian kernel. Objective function $J_{NMKFCMobj}$ defines a constrained optimization problem. It will convert constrained optimization problem to decision-boundary optimize problematic by relating Lagrange multiplier method [2].

Evaluate membership function U_{ij}:

$$U_{xy} = \left(\frac{((1 - K_T(Z_x, W_y))(\frac{N_R - \alpha \sum_{l \varepsilon N_i} U_{yl}}{N_R}))^{-\frac{1}{m-1}}}{\sum ((1 - K_T(Z_x, W_y))(\frac{N_R - \alpha \sum_{l \varepsilon N_l} U_{kl}}{N_R}))^{-\frac{1}{m-1}}}\right) \quad (2)$$

Reform Cluster epicenter W_j:

$$W_y = \frac{\sum_{y=1}^{Q} U_{xy}^{m} K_T(Z_x, W_y) Z_x}{\sum_{x=1}^{P} U_{xy}^{m}(Z_x, W_y)} \quad (3)$$

This method is beneficial to integrate neighborhood pixel material.

2.2 Scale Invariant Feature Transform(SIFT)

SIFT is designed for identify feature of image and contribute to collection of ample information of local feature of image [2]. Intensity of the input grey-scale image by $P_0(x, y)$, where x, y are pixel coordinate. Intensity of image lies between 0 to1.

For blur images, Gaussian filter used as kernel:

$$B_g(x, y, \sigma) = \frac{1}{2\pi\sigma^2}(-\frac{x^2 + y^2}{2\sigma^2}) \quad (4)$$

Where, σ is blue scale.

With the help of Eq. (5) calculated Blue Image intensity.

$$I_B(x, y, \sigma) = B_G(x, y, \sigma) * P_0(x, y) \quad (5)$$

The points of interest are firmly connected to the native relation of the Laplacian-of-Gaussian function

$$L(x, y, \sigma) = \sigma^2 \Delta I_B(x, y, \sigma) \quad (6)$$

Scale-space pyramid is calculated on the basis of computation of local relation of $D_L(x, y, \sigma)$. $D_L(x, y, \sigma)$ is difference of Gaussian function Eq. (7) and c is constant:

$$D_L(x, y, \sigma) = I_B(x, y, c\sigma) - I_B(x, y, \sigma) \quad (7)$$

Find corrected value of the difference-Gaussian function:

$$D_L(\hat{x}, \hat{y}, \hat{\sigma}) = D_L(x, y, \sigma) + \frac{1}{2}[D_x(x, y, \sigma)dx + D_y(x, y, \sigma)d_y + D_\sigma(x, y, \sigma)d_\sigma] \quad (8)$$

Output image is having low contrast and poor localized edge. To eliminate low contrast edge used threshold value α and β.

$$\left|D_L(\hat{x}, \hat{y}, \hat{\sigma})\right| = \alpha, \frac{D_{xx}D_{yy} - D_{xy}^2}{(D_{xx} + D_{yy})}\rangle\beta \quad (9)$$

SIFT point of objects are extracted from set of reference and stored in dataset.

3 Bottom-Up Method

3.1 LevelSet Algorithm

Levelset technique, introduced by Osher and Sethian, is zero equivalent surface technique [5]. Simple concept is to modification of movable path of a planar curve into movable path of a 3D surface. The main theme stands toward indicate curves or planes as zero level set of an upper dimension hyper-surface. Testing of such practice offers additional perfect algebraic execution also operates topological variation without any problems. The conventional set borderline is explained zero levelset of an implicit symbol ∪ of the evolving curve. In brief, evolve the implicit levelset function $\phi(x, y, t)$ to represent evolution of curve $\tau(t)$ with a speed $F(x, y)$ in normal direction. Time t, zero levelset (x,y)/$\phi(x, y, t) = 0$ defines evolved shape $\tau(t)$.

$$\phi(\tau(t), t) = 0 \tag{10}$$

Differentiate Eq. (10) with respect to t and apply chain rule,

$$\frac{\partial\phi}{\partial t} + \nabla\phi.\frac{\partial\tau}{\partial t} \tag{11}$$

F denotes speed of curve in normal direction, describe below

$$\frac{\partial\tau}{\partial t}.N = F \tag{12}$$

where N denotes outwards normal, and elaborated by Eq. (13)

$$N = \frac{\nabla\phi}{|\nabla\phi|} \tag{13}$$

Substitute Eq. (13) and (12) into (11), and calculate evolution equation for φ as

$$\phi_t + F|\nabla\phi| = 0 \tag{14}$$

This is fundamental equation of the levelset [6], and zero levelset represents item shape curve:

$$\tau(t) = \{(x, y)/\phi(x, y, t) = 0\} \tag{15}$$

Function ϕ is generally determined grounded on signed distance calculation to primary front. Relating to image field purely based on Euclidean distance between the curve and one image point. Describe below:

$$\phi(x, y) = \pm d(x, y) \tag{16}$$

Where: $d(x, y)$ Euclidean distance from point to borderline. Sign: points inside borderline (−) sign and outside (+) sign.

Evolution of borderline is elaborated by partial differential equation on zero levelset of ϕ:

$$\frac{\partial\phi}{\partial t} = -F|\nabla\phi| \tag{17}$$

Where, F: known function, which calculated by the local curvature κ at the zero levelset, i.e.,$F = F(\kappa)$.

Where κ:

$$\kappa = \nabla.\frac{\nabla\phi}{|\nabla\phi|} = \frac{\phi_{xx}\phi_y^2 - 2\phi_x\phi_y\phi_{xy} + \phi_{xy}\phi^2}{(\phi_x^2 + \phi_y^2)^{3/2}} \tag{18}$$

Speed function F is represent as: $F = F(L,G,I)$.

Where *L: local information,* which define by local geometric properties, *G:global property* of front depends on contour and location of front; *I*: *independent properties* that independent of front.

The goal of Medical image is extract an organ from input. Beginning from first contour, moving front to circulate inside similar section and stop at borders.

The propagating task is characterized as:

$$\frac{\partial\phi}{\partial t} = g_I(F_A + F_G)|\nabla\phi| \tag{19}$$

Where, impact of speed function F is divided into two portions, *FA* as well as *FG*. The term *FA*, causing front to consistently multiply or bond with a speed of *FA* dependent on sign. Another word *FG*: portion that relate on geometry of front, such as own local curvature. The curve evolution is joined with image segmentation difficulty by multiplying $(F_A + F_G)$ also stopping word:

$$g_I(x, y) = \frac{1}{1 + |\nabla(G_\sigma \otimes I(x, y))|} \tag{20}$$

Where, $G_\sigma \otimes I$: convolute image I with Gaussian smoothing filter $G\sigma$ with characteristic width of σ. When in uniform sections, $\bigtriangledown(G_\sigma \otimes I\,(x,y))$ will congregate to zero. The boundary filter $g_I\ (x,y)$ falls to zero then performs as a halting criteria for speed function and stops evolving front at preferred section.

General model which assimilate image features into speed function, in result, speed function execute with additional variability and enhanced outcomes.

The form of speed function expressed as [7]: $F = F_A + F_G$. Via calculation control objects or different image force which stated in common but simple form:

$$F = A(F_A + F_G) + B \tag{21}$$

$$A = \frac{1}{(1 + dist(C_{img}, C_{front})} \tag{22}$$

$$B = \overrightarrow{dist}(C_{img}, C_{front}) \tag{23}$$

dist(): appropriate distance function which is used to calculate variance of image features between propagating front and propagated zones. *B:* additional forces that stand up0.325689*/o'; from image content. *C*: image content model, including any of the color, shape, and texture features.

3.2 Automatic Seeded Region Growing Algorithm

Common region growing algorithms decide pixel value manually from every mark zone as seed. The seed point generally consists of great homogenous to adjacent pixels and can represent region [8]. The nature of seed point is very close to cluster center in clustering algorithm. AP clustering algorithm is running to find seed points in an image.

All data points as potential cluster centers, and computes accessibility and accountability data by Eqs. (24)–(26) iteratively between every two data points to find characteristic centers of data points:

$$R(i,k) \leftarrow S(i,k) - \max_{k'\partial t, k' \neq k} \{a(i,k') + S(i,k)\} \tag{24}$$

$$\text{If } i \neq k, ai, k \leftarrow \min\left\{0, R(k,k') + \sum_{i'\partial t. i' \notin i,k} \max\{0, ri', k\}\right\} \tag{25}$$

$$a(k,k) \leftarrow \sum_{i'\partial t. i' \neq k} (0, R(i', k)) \tag{26}$$

To receive the seed *xj* by Eq. (27) and cluster this holds xi:

$$\arg\max_{1 \leq j \leq N} [R(i,k) + a(i,k)] \tag{27}$$

Clustering algorithm only require uniformity between data points rather than number of centers in advance, automatic choice for potential seed pixels in an image.

The simple linear iterative cluster (SLIC) algorithm is running for segmented pixel-level image before applying AP clustering algorithm by merging adjacent pixels with similar characteristics to some irregular pixel blocks.

The region growing is computed by continually assimilation of seed points also surrounding pixels and division of aim objects with background. Lastly design of ROIs is in rectangular boxes according to segmentation results.

Euclidean Distance runs to measure LAB color matches by Eq. (28)

$$S = \sqrt{(L_0 - L_{0j})^2 + (a_0 - a_{0j})^2 + (b_0 - b_{oj})^2} \tag{28}$$

4 Dataset of Pancreas Medical Image

Top-Down and Bottom-Up methods are estimated on a dataset of 80 3D abdominal portal-venous compare enriched CT photographs assimilated from male: 53 and female: 27 patients. The 63 patients are arbitrarily nominated by a radiologist from the Picture Archiving and Communications System (PACS). The CT database is attained from the National Institutes of Health Clinical Center. Patient age series are from 18 to 76 years, with a mean of 46.8 ± 16.7. Scan resolution is pixels including slice thickness, Slice thickness range is from 1.5–2.5 mm on Philips and Siemens MDCT scanners.

5 Experimental Results

Minimum Hardware requisite for proposed method testing is Intel i3 processor M-1.90 GHz, Operating system- Microsoft Windows 10 or Ubuntu 16.6.6, 2 GB RAM and platform is MATLAB 2019a. Matlab Tool Image Processing, Computer Vision (Table 1).

Table 1. Comparison between NMKFCM and level set method using cluster accuracy rate

Image	Cluster accuracy rate	
	NMKFCM	Level set algorithm
Image 1	75.62	58.75
Image 2	46.07	51.87
Image 3	47.82	75.88
Image 4	56.78	58
Image 5	57	80.75

In NMKFCM, $\alpha = 0.5$, $N_R = 8$ and cluster number $c = 6$, value is choose for experiment. From above table, LevelSet algorithm values are more than NMKFCM method. In Image 1 where NMKFCM car value 75.62 is more than Levelset algorithm but Pancreas segmented neatly through LEVELset Algorithm only (Fig. 1).

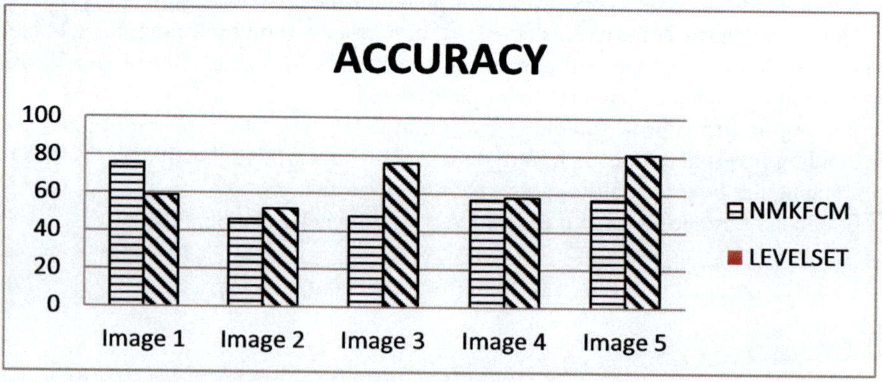

Fig. 1. Comparison between NMKFCM and levelset method using CAR value.

Table 2, describe time period required for exection of each method. NMKFCM method require less time period to satisfy stoping criteria than SIFT algorithm. In Image 1,NMKFCM take very less time period than SIFT and Level set algorithm.In Image 2, 3, 4, 5 NMKFCM take less time period than SIFT but required more time period than Level set algorithm (Fig. 2).

Levelset algorithm easily extracts Pancreas from image but other method failed to single organ segmentation. Fig. 3, segmented Pancreas is shown for every image by

Table 2. Comparison between NMKFCM and level set method using time period

Image	Time period		
	NMKFCM	SIFT	Level set algorithm
Image 1	32	80	73
Image 2	91	140	41
Image 3	78	128	42
Image 4	53	95	45
Image 5	78	110	56

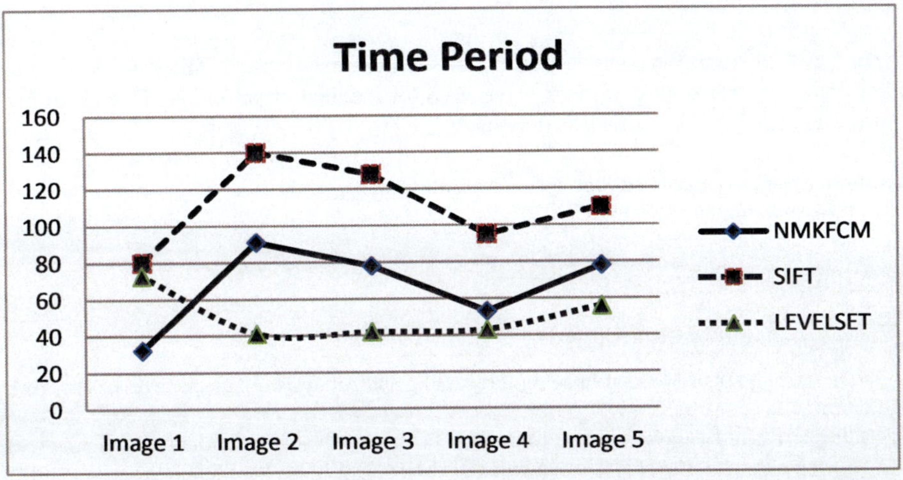

Fig. 2. Comparison between NMKFCM, SIFT and LevelSet method using time period

Levelset algorithm. In Image 1, NMKFCM accuracy is more than other approach but pancreas segmentation not clear. In Image 5, NMKFCM fail to correct segmentation.

6 Conclusion

To abstract seed points, AP clustering algorithm is used in automatic seeded region rowing. The experiment shows that proposed method runs to locate seed points automatically and used manual seeded method algorithm with more robustness. The accuracy of methods is enhanced and extraction effect of ROIs is better after the optimization. Relaxation factor runs a reducing borderline situation, to discontinue the evolving curve at an indistinct or unclear border. Curve evolution is multifaceted problem to solve previously, was fruitfully precise by levelset method. LevelSet algorithm is very advantageous for segmentation of pancreas.

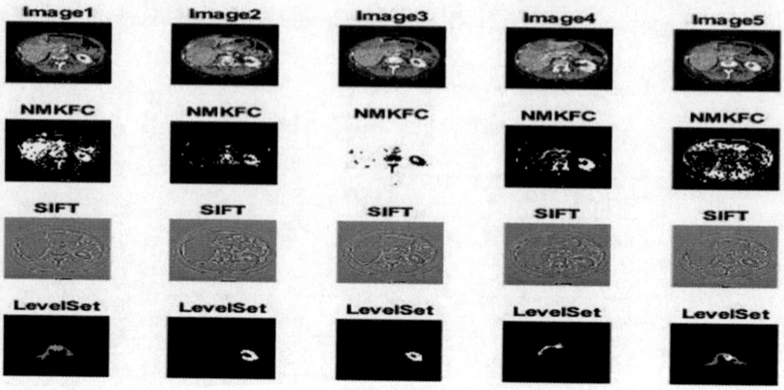

Fig. 3. Segmented images of pancreas medical image

In Levelset algorithm, single organ can segment easily but it failed to determine other organ. Preprocessing of mask is required for medical organ image. Deep learning approach is used to overcome this drawback.

Acknowledgement. Received copyrights certificate from copyright office of government of India with registration number SW-13795/2020.

References

1. IN, B.: Handbook of Medical Imaging: Processing and Analysis. Academic Press, New York (2000)
2. Paithane, P.M., Kakarwal, S.N., Kurmude, D.V.: Top-down method used for pancreas segmentation. Int. J. Innov. Exploring Eng. (IJITEE) **9**(3), 1790–1793 (2020). ISSN 2278-3075
3. Paithane, P.M., Kinariwal, S.A.: Automatic determination number of cluster for NMKFC-means algorithm on image segmentation. IOSR-JCE **17**(1), 12–19 (2015). Ver 2.
4. Paithane, P.M., Kakarwal, S.N.: Automatic determination number of cluster for multi kernel NMKFCM algorithm on image segmentation. In: Intelligent System Design and Applications, pp. 80–89. Springer, Cham, (2018)
5. Qu, Y., Heng, P.A., Wong, T.-T.: Image segmentation using the level set method, pp. 95–122. Springer, New York (2004). ISBN 978-0-387-31204-0
6. Jiang, X., Zhang, R., Nie, S.: Image segmentation based on level set method. In: International Conference on Medical Physics and Biomedical Engineering, pp. 840–845. Elsevier (2012)
7. Rommelse, J.R., Lin, H.-X., Chann, T.F.: A robust level set algorithm for image segmentation nd its parallel implementation (2014)
8. Li, Q., Wei, Z., Zhao, C.: Optimized automatic seeded region growing algorithm with application to ROI extraction. Int. J. Image Graph. **17**(4), 1750024-1-14 (2017)
9. Bandyopadhyay, O., Chanda, B., Bhattacharya, B.B.: Automatic segmentation of bones in X-ray images based on entropy measure. Int. J. Image Graph. **16**(01), 1650001 (2016)

Sustainability Considerations in the Product Design Using System Dynamics and Fuzzy Cognitive Maps

Elahe Mohagheghian[1], Hasan Hosseini-Nasab[1](✉), Ajith Abraham[2], and Mohammad-Bagher Fakhrzad[1]

[1] Department of Industrial Engineering, Yazd University, Yazd, Iran
{hhn,mfakhrzad}@yazd.ac.ir

[2] Scientific Network for Innovation and Research Excellence, Machine Intelligence Research (MIR) Labs, Washington, USA
Ajith.abraham@ieee.org

Abstract. Globally, sustainable product design is an issue that has been receiving more academic attention around the world. Understanding the sustainability-related design decisions seems to be important to ensuring sustainable design. Based on a stance, this study attempted to quantify sustainability factors in the product design through a hybrid method which combines the fuzzy cognitive mapping (FCM) and the system dynamics (SD) approaches. Designer behavior with regard to sustainability criteria was simulated in Vensim and sourced through sessions of two groups who deal with sustainable design. The results provide an *easy*-to-*understand model* of sustainable design to help designers analyze the dynamics of the cause-and-effect relationships between factors of sustainable design and long term scenarios, therefore, different scenarios at the inter and intra-cluster levels were built to determine the impacts of variable changes on the model developed.

Keywords: System Dynamics (SD) · Fuzzy Cognitive Maps (FCM) · Sustainability

1 Introduction

In the last decade, sustainability has become an essential emphasis in product design; companies have been focusing on integrating environmental, (e.g., lifecycle issues and end-of-life recovery factors) and social (e.g., worker safety and health issues) concerns in addition to economic goals during the design process. Due to the critical environmental function of the sustainable design as an intermediary, this design plays an important role as one of the ecology's main pillars. The absence of a sustainable design affects all parts of the society, therefore, considering the issue of sustainable design is crucial for products and the overall society [7]. Indeed, Sustainable design is a complex issue

A. Abraham et al. (Eds.): SoCPaR 2020, AISC 1383, pp. 383–394, 2021.
https://doi.org/10.1007/978-3-030-73689-7_37

with many quantifying factors to assume. The models proposed in the literature on the topic of sustainable product design offer some shortcomings in respect to the definition of variables and their corresponding cause-and-effect relationships. Researchers can not analyze holistically the dynamics of the factors of sustainable design through long-term analyses of the outcomes of changes in these factors.

Given these shortcomings, this study attempted to explore successful factors for sustainable design by using fuzzy cognitive mapping (FCM) and the system dynamics (SD) approach. There have been relatively few studies which examined the proposed hybrid approach in this paper, so the methodology contributes significantly to the existing literature on sustainability and design science.

The variables required for the FCM and the data required for the dynamic analyses were gathered, and their cause-and-effect relationships were determined and mapped, in order to carry out the FCM and SD analyses. To determine the long-term outcomes of changes in the model variables, the mentioned information allowed for the definition of 4 scenarios with 8 simulations.

These methods led to an easy-to-understand model that could aid designers to distinguish the variables that influence sustainable design and their long-term consequences. The rest of this paper is organized as follows. The next section provides a literature review of sustainable design. Section 3 presents the methodology. Section 4 provides the results of the dynamic system created.

2 Literature Review

Sustainability objectives are being designed to measure performance, and to provide the feedback needed to promote improvement of design decisions. This refers to the idea that achieving current goals should be ensured by today's actions, while also achieving that future goals will not be hindered, and it is generally surrounds by three areas: economics, ecology, and society [3]. To enter ecological and social factors into the design decisions, it will be noticed that if the customers value such factors in their purchasing decisions or not. Consumer preferences for socially and environmentally sustainable products have been determined in the recent studies.

One tool for improving these effects is to enhance transparency which could help customers to make smart decisions [10]. The present paper thus attempt to introduce a fully cognitive structure that visibly portrays factors of sustainable design and determines the main factors impacting the implementation of conformant practices. Specifically, through the use of FCM and SD, this study sought to analyze factors of sustainable design in a dynamic manner that ensured a transparent rational analysis of the issue in question.

3 Methodology

In this paper, we show how system dynamics modelling can identify and portray the cause-and-effect relationships of material choice, tolerance and sustainability-related design decisions. Such the cause-and-effect relationships have been shown to affect future decisions and demand for each successive generation of products. By mapping the process, designers could increase understanding of the system level impacts of their decisions.

3.1 Fuzzy Cognitive Mapping

In practice, the phrase cognitive map based on [1], 'relates to models, depicting the structure of causal [...] influences of portrayed situations, or systems'. This description stem from the common method that cognitive maps are an illustration of relationships between variables. These maps are a system of nodes and arrows, in which the direction of the arrows clearly indicated the causality between criteria [11, 14–22].

Kosko [9] was the first to discuss fuzzy cognitive maps (FCMs) according to [2], while also enhancing the power of cognitive maps through fuzzy values. The fuzzy logic of these maps is associated with how cause-and-effect relationships and their concepts can explain real values [12]. The primary difference between cognitive maps and FCMs is the way of demonstration of cause-and-effect relationships. FCMs use a plus (+) or minus (–) sign and a numerical value termed "weight" belong to the range [0.0, 1.0] or [−1.0, +1.0].

Fig. 1, represents a graphical illustration of a FCM. In a simple FCM, as can be seen Fig. 1, there are some nodes that stand for the concepts (C_i), of the modelled system. In addition, there are weighted arcs (W_{ij}) that connect concepts or nodes together, which show the casual relationship between concepts. In respect to the definition of concepts and their *cause-and-effect* relationships in FCMs, Ferreira et al. [4], explained that: 'These relationships can be illustrated by a positive and negative sign, and by a number from –1 to 1 that represents the intensity of the relationships.

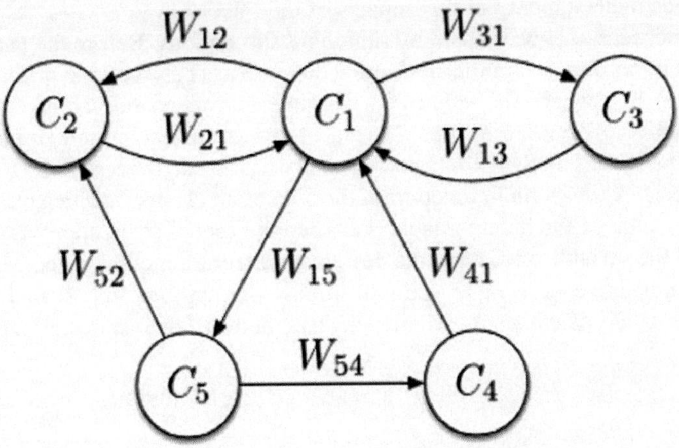

Fig. 1. A simple FCM [13].

More specifically, all the values in the map can be fuzzy and, thus, each concept has a state value A_i that can be a fuzzy value in the range [0, 1] or within $\{0, 1\}$. In addition to, the weights of the arcs can be a fuzzy value within $[-1, 1]$ or within $\{-1, 0, 1\}$'. In respect to the value of the cause-and effect relationships among concepts, one of three situations can be happend. According to [13], when the relationship is negative, $W_{ij} < 0$, when there is no relationship between two concepts, $W_{ij} = 0$, and when it is positive, $W_{ij} > 0$. The mathematical formulation of an FCM is represented by Eq. (1) [8, 13]:

$$A_i^t = f(A_i^{t-1} + \sum_{j \neq i, j=1}^{N} A_j^{t-1}.W_{ji}) \tag{1}$$

in which A_i^t is the activation level of the concept C_i, in the t timeline; $f(x)$ is the activation function; and A_i^{t-1} is the activation level of the concept C_i, in the $t-1$ timeline. Also, A_i^t is the activation level of the concept C_j, in the t timeline, and W_{ji} is the corresponding weight related to the cause-and-effect relationship between concept C_j and C_i.

After each simulation, by using Eq. (1) and defining criteria's activation levels, new state vector with new values are obtained until the system reaches a fixed point that represents the system's equilibrium [4]. The management and modeling of complex dynamic systems can, therefore, advantage from the hybrid application of FCM and SD techniques and tools.

3.2 System Dynamics

SD is a concept was introduced by [5, 6], which was initially related to the basic principles of systems theory. In this paper, the *cause-and-effect* relationships of sustainability-related design decisions have been portrayed through system dynamics modelling. Such relationships have been illustrated to affect future decisions, which decision makers could enhance understanding of the impacts of their decisions.

The proposed model was tested by running 8 simulations. Before the testing phase, different equations utilized in the stock-flow diagram are based on the primary features of FCMs and SD, where the value of a variable considers the value of the sum of all its *cause-and-effect* relationships. In respect to the clusters, an integral aggregation function of the sum of all their concepts' causality was used, divided by 100, to scale the model properly. A logarithmic function of the sum of all clusters' values was applied to stabilize the value of the head variable and keep the model scale appropriate. Finally, in respect to the variable of sustainable design, an integral function of the head variable value was applied to ensure, once again, the model was appropriately scaled. The initial performance values of the head variable, clusters, and sustainable design are located in Table 1.

4 Results

4.1 Dynamic Analyses and Results

The FCM and SD approaches identified three areas of concern, namely: (1) *Design specifications*; (2) *Maintenance requirements* and (3) *External Factors*. Regarding to the cognitive mapping techniques, this procedure was built on sessions with the designers as the decision makers. The first session comprised three steps, namely: (1) deciding factors of sustainable design; (2) creating clusters of factors; and (3) reorganizing the factors by hierarchical order of importance within each cluster. The entire procedure lasted for 4 h.

Notably, all the information which used to create a collective cognitive map are based on the *Decision Explorer* software. This map, is shown in Fig. 2. After the cognitive map was created, a second session was built to quantify the cause-and effect relationships in the cognitive map in order to develop an FCM. At the end of the second session, all the cause-and effect relationships had been determined, and this determination was the beginning point for a dynamic analysis of factors of sustainable design. This analysis allowed for the investigation of the long-term outcomes of cause-and-effect relationships. Therefore, using the Vensim software, a stock-flow diagram according to the FCM, presented in Fig. 3.

In this study, the proposed model was tested by running 8 simulations. Before the testing phase, different equations used in the stock-flow diagram are based on the primary features of FCMs and SD, where the value of a variable considers the value of the sum of all its *cause-and-effect* relationships. In respect to the clusters, an integral aggregation function of the sum of all their concepts' causality was used, divided by 100, to scale the model properly. A logarithmic function of the sum of all clusters' values was used to stabilize the value of the head variable and keep the model scale appropriate. Finally, regarding the variable of sustainable design, an integral function of the value of the head variable was used to ensure, once again, the model was appropriately scaled. The initial performance values of the head variable, clusters, and sustainable design are located in Table 1.

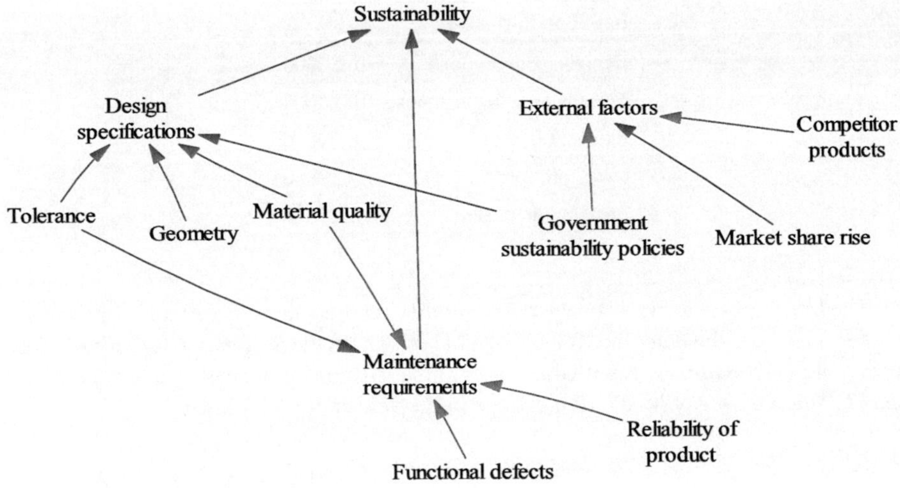

Fig. 2. Collective cognitive map.

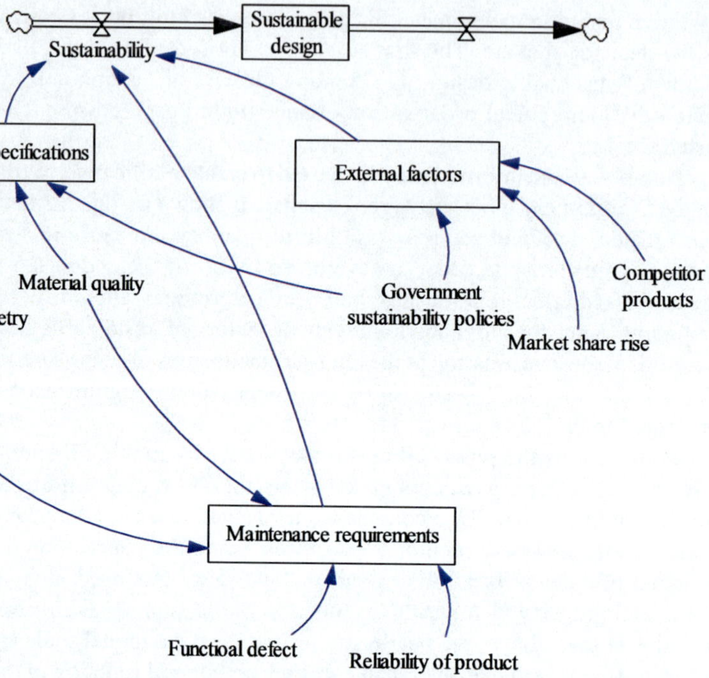

Fig. 3. Stock and flow diagram of dynamics

Table 1. Initial performance values of clusters, head variable and sustainable design.

Initial performance	
Design specifications	16.50000
Maintenance requirement	10.20000
External factors	8.90000
Head variable	4.70366
Sustainable design	3.72018

The inter-cluster analysis consisted of modifying the values of variables of the clusters to understand the outcomes for sustainable design. In the *Scenario 1*, two simulations were built: (1) *Simulation 1*, with the values of the clusters influenced by a − 0.3 change; and (2) *Simulation 2*, with a − 0.6 change in the values of each cluster.

The initial cause-and-effect relationships are determined between the head variable and clusters, and those weights are located in Table 2. The outcomes for the head variable are found in Table 3. As apparent from Table 3, *Simulations 1* and *2* have a negative impact on performance in respect to the sustainability. These results also emphasize that *Simulation 2* has a more negative effect than *Simulation 1* does, and this effect associated with the simulations' initial values. Thus, intra-cluster analysis, presented the alterations in each one of the clusters' criteria, according to the following tables.

Table 2. Weights of cause-and-effect relationships between head variable and clusters.

Determinant	
Design specifications	1
Maintenance requirement	1
External factors	0.9

Table 3. Various values of sustainability in each simulation.

Determinant	Simulation 1	Simulation 2
Sustainability	−0.01738	−0.02113

Table 4. Various values in simulations 3 and 4 for scenario 2.

Determinant	Weight of relationship	Simulation 3	Simulation 4
Tolerance	−0.90	−0.1	−0.1
Geometry	−0.90	−.06	−0.3
Material quality	1	−0.7	−0.5

Table 5. Various values in simulations 5 and 6 for scenario 3.

Determinant	Weight of relationship	Simulation 5	Simulation 6
Functional defects	0.85	−0.4	−0.2
Reliability of product	1	−0.9	−0.6

Table 6. Various values in simulations 7 and 8 for scenario 4.

Determinant	Weight of relationship	Simulation 7	Simulation 8
Government sustainability policies	−0.90	−0.4	−0.1
Market share rise	−0.90	−0.5	−0.3
Competitor products	1	−0.8	−0.4

Tables 4, 5 and 6 related to each of the clusters and their corresponding scenarios. The long-term outcomes for the clusters resulting from the alterations presented above are shown in Table 7. As can be seen from Table 7, *Simulation 3* has a greater negative long-term impact on the performance of the *Design specifications* cluster compared with *Simulation 4* because the simulation values illustrated in Table 4 are more negative. In addition, it can also be concluded that *Simulations 3* and *4* both have negative long-term effects on the performance of the *Maintenance requirement.*

Table 7. Long-term changes in clusters' values for each simulation

Cluster	**Simulation 3**	**Simulation 4**
Design specifications	−4.60	−3.20
Cluster	**Simulation 5**	**Simulation 6**
Maintenance requirement	−4.60	−2.80
Cluster	**Simulation 7**	**Simulation 8**
External factors	−4.60	−2.90

However, *Simulation 5* has a much stronger negative impact. This is because the simulation values shown in Table 5 are also more negative in *Simulation 5*. The results for the performance of the *External Factors* cluster related to changes in the decision criteria, as shown in Table 6, are also clear in Table 7. The long-term outcomes are more negative in *Simulation 7* because, again, the simulation values shown in Table 6 are more negative compared with *Simulation 8*. Fig. 4 and 5 highlight the long-term behavior for all simulations made in this study.

The integrated use of FCM and SD, presented the identification of factors of sustainable design and the ways they correspond with each other. The designers as stakeholders must analyze and understand these changes in order to make more informed decisions. All these factors can impact sustainable product design evolution during the time, so designers require to analyze possible variations of the proposed process and seek sustainable attitudes in their designs.

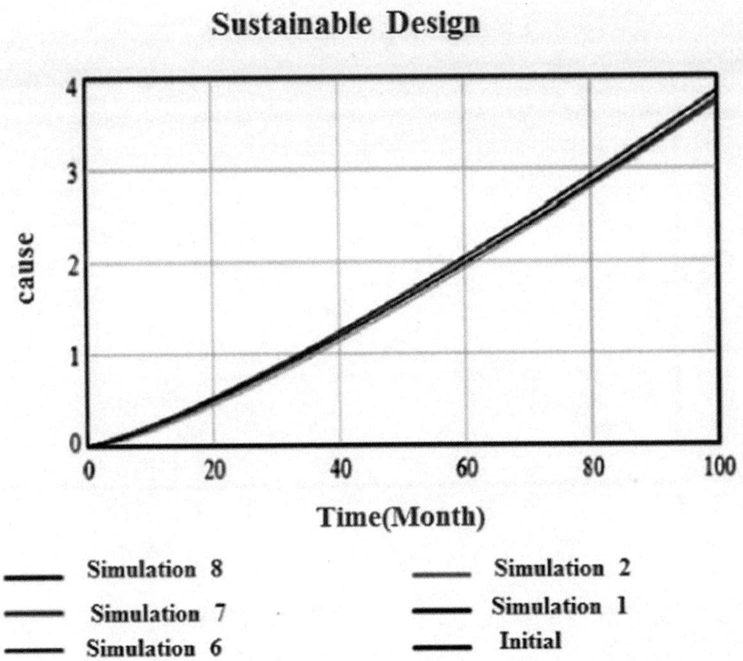

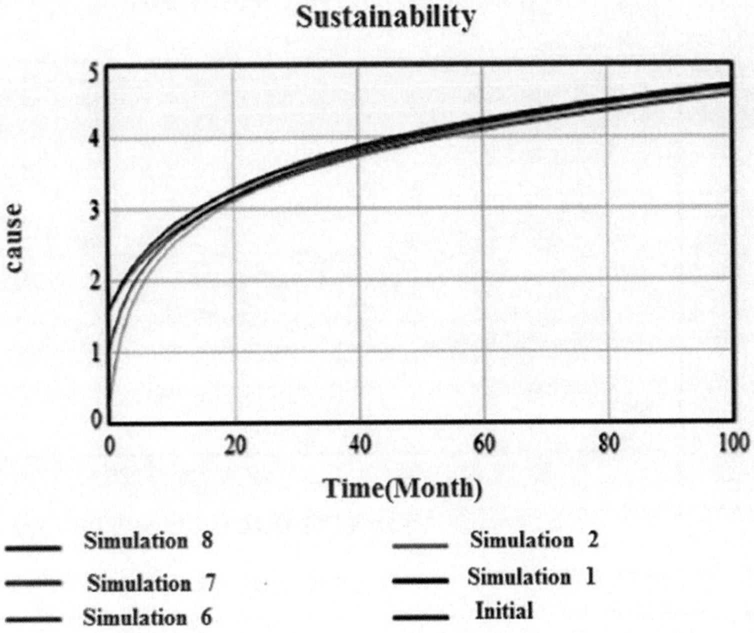

Fig. 4. Long-term performance of sustainable design and head variable for their simulations.

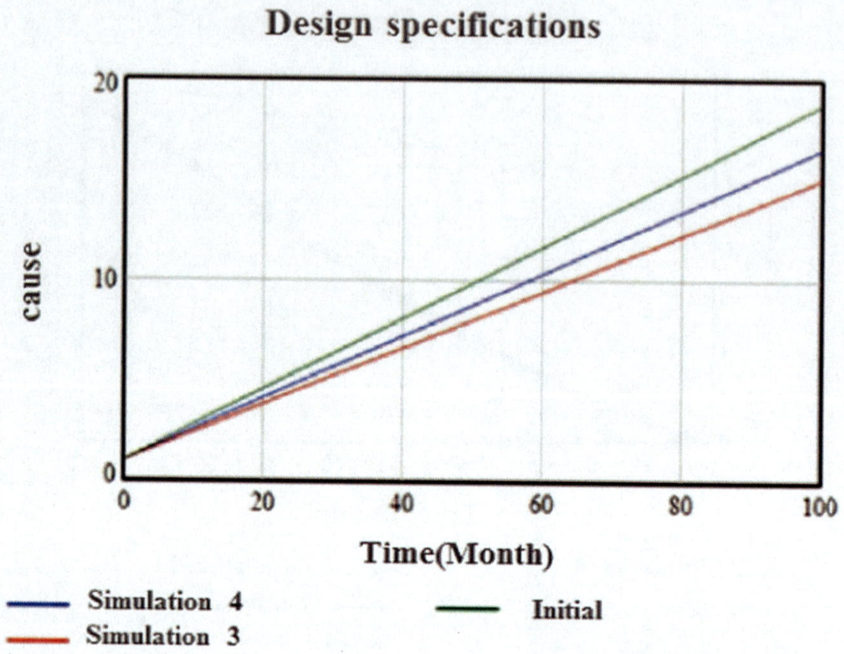

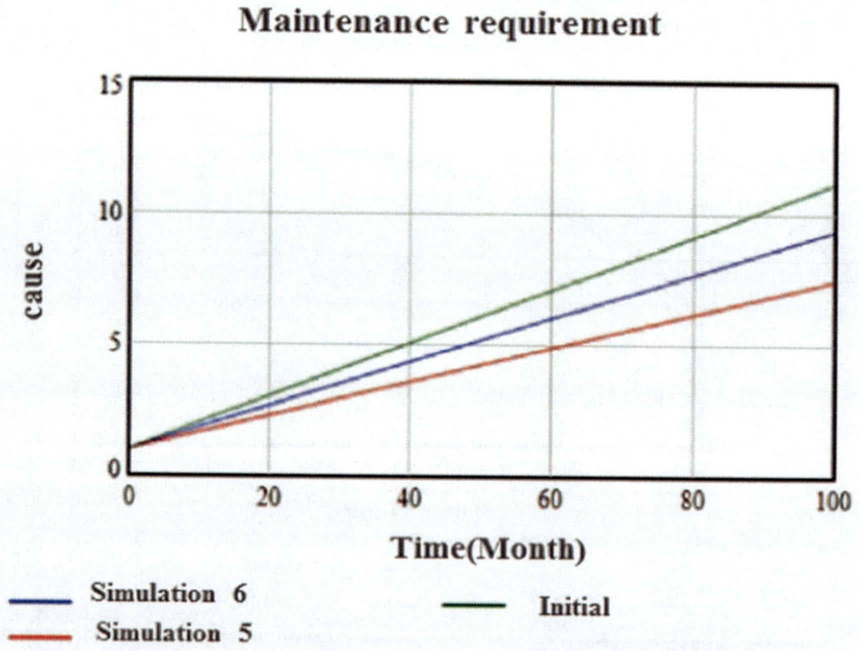

Fig. 5. Long-term performance of clusters for their simulations.

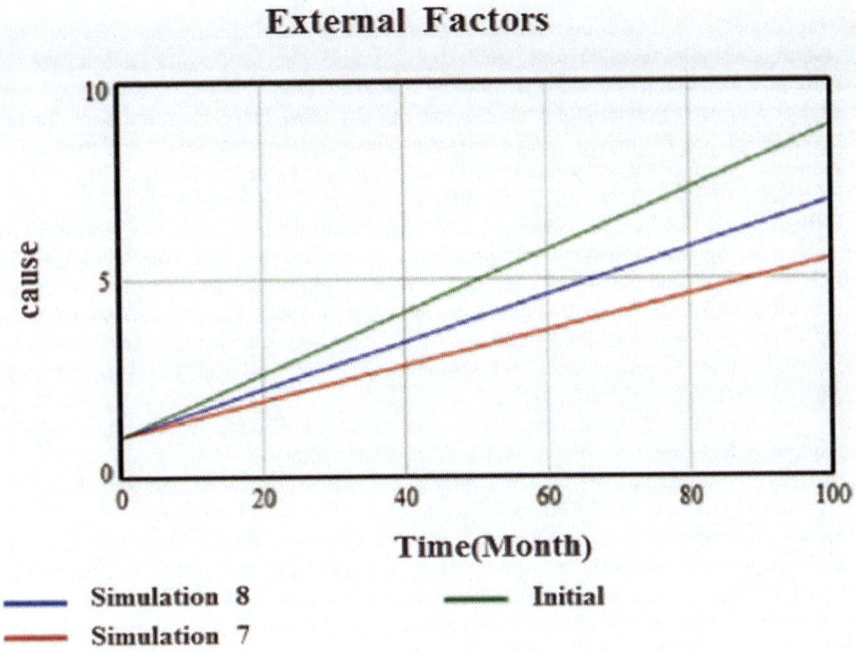

Fig. 5. (*continued*)

5 Conclusion and Future Research

The results of the present paper goes beyond the current approaches because it overcomes some shortcomings by considering a number of variables that influence sustainable product design and the cause-and-effect relationships in the FCM. Additionally, the proposed model provides dynamic analyses through SD, which illustrate and map the behavior of variables over time and the changes that can affect sustainability. Through such a model, designers can change variables of the models to create different scenarios. This approach can help those decision makers make more conscious decisions and understand the magnitude of this trade-off.

For future research, it is recommended to use other ways of investigating the issue of sustainable design, and evaluating tradeoffs among factors while enriching the beneficial for precious insights. Additional studies may also assume performing a greater a cluster's number of factors, the larger number simulations and scenarios because this could lead to another conclusions.

References

1. Abramova, N.: The cognitive approach to the problem of identification validity in cognitive mapping. IFAC-PapersOnLine **49**(12), 586–591 (2016)
2. Azadeh, A., Ziaei, B., Moghaddam, M.: A hybrid fuzzy regression-fuzzy cognitive map algorithm for forecasting and optimization of housing market fluctuations. Expert Syst. Appl. **39**(1), 298–315 (2012). https://doi.org/10.1016/j.eswa.2011.07.020

3. Brundland, G.: World Commission on Environment and Development. Our Common Future Oxford, University Press, Oxford (1987)
4. Ferreira, F, Ferreira, J, Fernandes, C, Meidutė-Kavaliauskienė, I., Jalali, M.: Enhancing knowledge and strategic planning of bank customer loyalty using fuzzy cognitive maps. Technol. Econ. Dev. Econ. **23**(6), 860–876 (2017). https://doi.org/10.3846/20294913.2016.1213200
5. Forrester, J.: Principles of Systems. Wright allen press, Cambridge (1968)
6. Forrester, J.: Industrial Dynamics. The MIT Press, Massachusetts (1961)
7. Hoffenson, S., Dagman, A., Söderberg, R.: Visual quality and sustainability considerations in tolerance optimization: a market-based approach. Int. J. Prod. Econ. **168**, 167–180 (2015)
8. Kok, K.: The potential of fuzzy cognitive maps for semiquantitative scenario development, with an example from Brazil. Global Environ. Change. **19**(1), 122–133 (2019). https://doi.org/10.1016/j.gloenvcha.2008.08.003
9. Kosko, B.: Fuzzy cognitive maps. Int. J. Man. Mach. Stud. **24**(1), 65–75 (1986). https://doi.org/10.1016/S0020-7373(86)80040-2
10. Limnios, E.A.M., et al.: Giving the consumer the choice: a methodology for Product ecological footprint calculation. Ecol. Econ. **68**(10), 2525–2534 (2009)
11. Marques, F.C., Ferreira, F.A., Zopounidis, C., Banaitis, A.: A system dynamics-based approach to factors of family business growth. Ann. Oper. Res. 1–21 (2020)
12. Santos, F., Ferreira, F., Meidutė-Kavaliauskienė, I.: Perceived key factors of payment instrument usage: a fuzzy cognitive mapping-based approach. Technol. Econ. Dev. Econ. **24**(3), 950–968 (2018). https://doi.org/10.3846/20294913.2016.1261374
13. Salmeron, J., Mansouri, T., Reza, M., Moghadam, S., Mardani, A.: Learning fuzzy cognitive maps with modified a sexual reproduction optimization algorithm. Knowl. Based Syst. **163**, 723–735 (2019). https://doi.org/10.1016/j.knosys.2018.09.034
14. Sahebjamnia, N., Goodarzian, F., Hajiaghaei-Keshteli, M.: Optimization of multi-period three-echelon citrus supply chain problem. J. Optim. Ind. Eng. **13**(1), 39–53 (2020)
15. Goodarzian, F., Hosseini-Nasab, H.: Applying a fuzzy multi-objective model for a production–distribution network design problem by using a novel self-adoptive evolutionary algorithm. Int. J. Syst. Sci. Oper. Logistics, 1–22 (2019)
16. Fakhrzad, M.B., Goodarzian, F.: A fuzzy multi-objective programming approach to develop a green closed-loop supply chain network design problem under uncertainty: modifications of imperialist competitive algorithm. RAIRO-Oper. Res. **53**(3), 963–990 (2019)
17. Fakhrzad, M.B., Talebzadeh, P., Goodarzian, F.: Mathematical formulation and solving of green closed-loop supply chain planning problem with production, distribution and transportation reliability. Int. J. Eng. **31**(12), 2059–2067 (2018)
18. Goodarzian, F., Hosseini-Nasab, H., Muñuzuri, J., Fakhrzad, M.B.: A multi-objective pharmaceutical supply chain network based on a robust fuzzy model: a comparison of meta-heuristics. Appl. Soft Comput. **92** 106331 (2020)
19. Fakhrzad, M.B., Goodarzian, F., Golmohammadi, A.M.: Addressing a fixed charge transportation problem with multi-route and different capacities by novel hybrid meta-heuristics. J. Ind. Syst. Eng. **12**(1), 167–184 (2019)
20. Goodarzian, F., Hosseini-Nasab, H., Fakhrzad, M.B.: A multi-objective sustainable medicine supply chain network design using a novel hybrid multi-objective metaheuristic algorithm. Int. J. Eng. **33**(10), 1986–1995 (2020)
21. Fakhrzad, M.B., Goodarzian, F.: A new multi-objective mathematical model for a Citrus supply chain network design: Metaheuristic algorithms. J. Optim. Ind. Eng. (2020). https://doi.org/10.22094/JOIE.2020.570636.1571
22. Fathollahi-Fard, A.M., Ahmadi, A., Goodarzian, F., Cheikhrouhou, N.: A bi-objective home healthcare routing and scheduling problem considering patients' satisfaction in a fuzzy environment. Appl. Soft Comput. **93**, 106385 (2020)

Dominant Voiced Speech Segregation and Noise Reduction Pre-processing Module for Hearing Aids and Speech Processing Applications

Shibani Hamsa[1], Youssef Iraqi[1], Ismail Shahin[2], and Naoufel Werghi[1(✉)]

[1] C2PS, Department of ECE, Khalifa University, Abu Dhabi, UAE
{shibani.koya,Youssef.Iraqi,Naoufel.Werghi}@ku.ac.ae
[2] University of Sharjah, Sharjah, UAE
ismail@sharjah.ac.ae

Abstract. Speech experiences different acoustic obstructions in normal environment, whereas numerous of the applications require a compelling way to partitioned the original dominant speech from the impedance, a perfect hearing framework ought to be able to isolated and recognize sound-related occasions precisely from complex sound-related scenes and in unfavorable conditions. Difficulty in distinguishing a particular speech from a mixture of other unwanted conversations is one of the problems faced by people wearing hearing aid. The possibility of partition of overwhelming discourse from other discourse signals and its enhancement from that point will be accommodating for individuals with hearing disability. The recent literature in the Computational auditory scene analysis (CASA) systems are based on gammatone filter bank and Short time Fourier transform (STFT). But higher computational complexity associated with those models adversely affect the implementation of digital hearing aids. This paper introduces a cochlear model using Wavelet packet transform (WPT) and a novel approach for dominant voiced speech segregation. The experiments confirmed the enhancement of our model in terms of computational complexity and recognition rate when compared to competitive models.

Keywords: Speech signal · Speech segregation · Cochlear modeling

1 Introduction

Understanding biometric modalities through face [1,2], body [3], gesture [4], and speech [5] is nowadays a fundamental tool in many applications spanning healthcare, entertainment, and communication to mention just a few. Emotion recognition of human speech, in particular, has been receiving increased attention, over face and body modalities, for its practicality and lower attain to privacy. In this context, analyzing speech analysis in relation to the human auditory system is an

A. Abraham et al. (Eds.): SoCPaR 2020, AISC 1383, pp. 395–403, 2021.
https://doi.org/10.1007/978-3-030-73689-7_38

essential step towards that goal. Human auditory system is naturally designed to function effectively in complex auditory scene and suitable to sort out and recollect the auditory events precisely. According to audiologists, the challenges being confronted by hearing impeded individuals wearing ear trumpet, namely

- Difficulty in understanding discourse sullied by discourse from other talkers.
- Difficulty in understanding if the speed of the speech is more.

Separation of desired speech from a mixture of speech signals, is a difficult task because features of the target and contaminating signals share obvious similarities. This problem is treated as "Cocktail party effect". The objective of this research is the digital imposition of segregation of prevailing discourse from complex discourse blends utilizing computationally productive approaches.

"Computational Auditory Scene Analysis (CASA)" [6] is a recently coined research filed, aims to deliver the auditory scene analysis (ASA) properties of human auditory system to the machines. The main motive of CASA is dominant speech separation from other noise and interference signals [5]. Bind source separation is another approach used for speech segregation. However, it shows limited success in monaural speech segregation. In fact due to the lack of knowledge about the distribution of the signal source as well as the nature of mixing process, blind source separation is not apt for hearing aid application. The CASA systems can be classified as data-driven or prediction- driven. Prediction driven approaches utilizes classifier models such as GMM (Gaussian mixture model) [7], HMM (Hidden markov model) and NMF (non negative matrix factorization). In data driven approach the cues used for speech segregation is extracted from the input mixture itself, such systems are called bottom up [6].

Computational model for human periferal frequency selectivity can be achieved with filterbanks. The Two common filter banks which mimic cochlea are Gammatone filter bank and short time Fourier transform (STFT) based filter bank. The former is computationally complex and latter has a dilemma in Time—frequency resolution. Hence a new model which is efficient and computationally less complex than the existing cochlear model is designed and implemented in this paper. Inspired by the discriminative power of the wavelet transformation [8–10], we designed a model using Wavelet based filter banks, which closely resemble the critical bands of the human ear.

The rest of the paper is organized as follows: Sect. 2 details the proposed WPT based speech segregation system, Sect. 3 describes the evaluation of the proposed model based on some quantitative and perspective measures. The complexity of the system is also considered in this section and concluded in Sect. 4.

2 System Description

2.1 T-F Decomposition

Acoustic signal properties are difficult to interpret in the time domain, time-frequency analysis allows the technique to study the signal both in time and frequency domain simultaneously. Most of the speech processing algorithms uses auditory filter banks to imitate the resolution of the human auditory system [11] [5]. Gammatone filterbank and STFT filter bank are the commonly used models to mimic the human cochlea. Higher computational complexity of the Gammatone filterbank makes it not suitable for real time applications. STFT have ended up standard apparatuses in signal examination. Be that as it may the STFT has too its impediments, such as the constrain in its time-frequency determination ability, which is due to the uncertainty rule. Low frequencies can be barely portrayed with brief windows, while brief beats can as it were ineffectively be localized in time with large windows.

Table 1. Audio frequency ranges of human cochlea [12]

No.	F_c	BW(Hz)	No.	F_c	BW(Hz)
1	50	0–100	13	1850	1720–2000
2	150	100–200	14	2150	2000–2320
3	250	200–300	15	2500	2320–2700
4	350	300–400	16	2900	2700–3150
5	450	400–510	17	3400	3150–3700
6	570	510–630	18	4000	3700–4400
7	700	630–770	19	4800	4400–5300
8	840	770–920	20	5800	5300–6400
9	1000	920–1080	21	7000	6400–7700
10	1170	1080–1270	22	8500	7700–9500
11	1370	1270–1480	23	10500	9500–12000
12	1600	1480–1720	24	13500	12000–15500

2.2 WPT Cochlear Filter Bank

The details of the frequency band width of the auditory filter created by the human cochlea is given in Table 1. The proposed framework is modelled imitate the human cochlear model and to attain an 18 level decomposition of the input signal. Figure 1 illustrates the decomposition of input speech signal through 18 channels based on the signal frequency. In this work, we have used db4 wavelets for the implementation and evaluation [13]. The input signal is split into short duration of length 20 ms and is allowed to pass through the designed filterbank model. As shown in Fig. 1, the input frame at node (0,0) is filtered using LPF and HPF in the first level of decomposition and forms (1,0) and (1,1) respectively. Then the high frequency and low frequency signals at each node is decimated by a factor of 2 before proceeding with the next level of decomposition.

2.3 Modulation Transform

The WPT representation of the input audio signal $y(n)$ can be represented as,

$$Y(m,k) = \underset{k}{W}(y(n)) \quad (1)$$

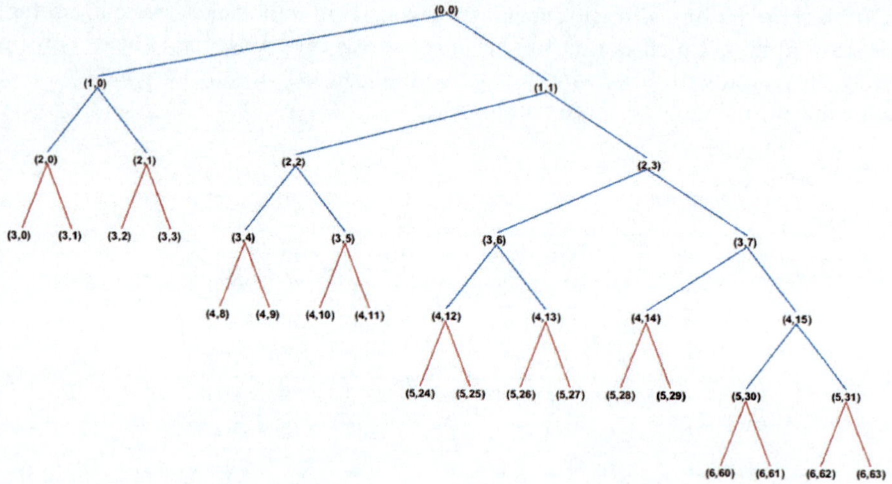

Fig. 1. WPT decomposition model to mimic the human cochlea.

The T-F decomposition of the input speech signal $y(n)$ is represented as Y(m,k), where m and k represents the time and frequency domains respectively. The signal $Y(k)$ taken out during a period of m by the WPT cochlear filter-bank can be represented as the product of message part and the carrier part as described below [14].

$$Y_K = M_K C_K \quad (2)$$

$$M_K = ed[\underset{k}{W}[y(n)]] \quad (3)$$

where ed represents the operation of message signal extraction using Hilbert envelop detection algorithm [15]. The message signal part can be approximated as the magnitude of the signal Y_k and is described as:

$$M_K \cong |Y_K| \quad (4)$$

The frequency-frequency representation of the signal y_k can be obtained by computing the discrete Fourier transform (DFT) of the signal y_k. The relationship between acoustic frequency and modulating frequency are marked with k and i respectively in Eq. (5) and all the transformations are illustrated in Fig. 2

$$Y(k,i) = DF(M_k) \quad (5)$$

$$= \Sigma_{m=0}^{Q-1} M(m,k) e^{(\frac{-j2\pi mi}{Q})}, i = 0, 1, ..., Q-1. \quad (6)$$

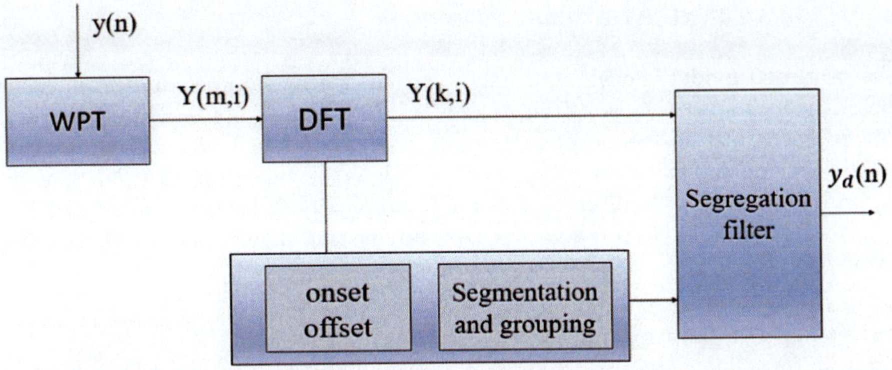

Fig. 2. Block schematic.

2.4 Speech Segregation

The resultant signal $y(k, i)$ in the frequency-frequency domain is filtered using an LPF to get a smooth response with reduced spectral variations. The maximum and minimum frequency variations can be identified by taking its first derivatives. These maximum and minimum points are marked as onset-offset values. The segmentation and grouping using these onset and offset points will helps to identify the pitch range of the dominant signal part over interference [14]. The average modulation spectral energy of the dominant and interference signals can be calculated based on the dominant and interference pitch frequency and are represented as Φ_t and Φ_i respectively.

A dynamic filter is designed based on the spectral energy of the dominant and interfering signals and the filter component is designed as,

$$\Phi_t = \frac{\Sigma(|Y(k,i)|^2)}{(P_h - P_l)}_{dominant} \tag{7}$$

Here P_h and P_l are set as the high pitch and low pitch values respectively.

$$\Phi_i = \frac{\Sigma(|Y(k,i)|^2)}{(P_h - P_l)}_{Interference} \tag{8}$$

The mangitude response of the filter F is designed as,

$$F = \frac{\Phi_t}{\Phi_t + \Phi_i} \tag{9}$$

The filter response in the time domain is:

$$f(m) = \Sigma_{i=0}^{I-1} F e^{(j\psi(i))} e^{\frac{(j2\pi mi)}{I}} \tag{10}$$

Here ψ represents the phase response of the filter and is set as a linear phase, $\psi(\mathrm{i}) = \mathrm{i}$ The convolution of the input speech signal over the filter response $f(m)$ is utilized to separate the dominant voiced speech signals from the noise.

3 Results and Analysis

The proposed model is evaluated using various speech samples and noise types such as voiced speech noise, siren, telephone ring, vehicle noise, white noise etc. The performance of the proposed model is evaluated in terms of various statistical and perceptual parameters such as, signal to noise ration (SNR), mean opinion score (MOS) and computational complexity. The performance of our proposed framework is also examined by integrating it with an emotion recognition system.

3.1 Evaluation in Terms of Signal to Noise Ratio

Signal to noise ratio measures the level of signal strength to noise strength. Higher SNR value ensures less interference and SNR is defined as,

$$SNR = 10log\frac{\Sigma(y(n)^2)}{\Sigma(y(n)^2) - \Sigma(y_s(n)^2)} \tag{11}$$

Where $y(n)$ and $y_s(n)$ are the input speech signal and the segregated speech signals respectively. Evaluation of the proposed model with respect to the STFT based model is given in Table 2. The results shows that the proposed model shows superior performance than the STFT cochlear model based on the evaluation using mixtures of female noise, male noise and female and male noises. The proposed model achieves an SNR of 26.23 over the 17.05 SNR reported by the STFT model using the mixture of male speech and noises in a ratio 2:1.

Table 2. Evaluation in terms of SNR

Type of mixture	Original	STFT [16]	Proposed model
Mixture of male & Female speech	11.0783	11.8667	18.3479
Mixture of male speech	12.4371	17.0505	26.4272
Mixture of female speech	12.8755	15.4600	16.1021

3.2 Mean Opinion Score

Mean Opinion Score (MOS) gives an evaluation score to the perception quality of the reproduced signal after segregation. The MOS assigns a value between 1 and 5 based on the quality of the reconstructed signal. MOS is taken with ten arbitrarily chosen samples and fifty audience members. Table 3 shows the MOS obtained for the STFT model and the proposed model. The results clearly illustrates the superiority of the proposed WPT model and reported an average score of 3 out of 5.

Table 3. Evaluation based on the perceptual quality

Speech	STFT [16]	Proposed
Male & Female voice mixture	2.6	3.5
Mixture of male voices	2.4	2.5
Mixture of female voices	2.5	2.9

3.3 Evaluation Based on Computational Time

In this experiment, we have evaluated the computational complexity of the proposed model with reference to gammatone and STFT filterbanks. Table 4 shows the time taken for speech segregation and the proposed model takes 1.82 s for the same. The time taken by the proposed system is much less than that of gammatone and STFT models.

Table 4. Evaluation based on computational complexity

Cochlear model used	Time required
Gammatone [17]	3.781203 s
STFT [16]	2.080689 s
Proposed	1.818275 s

3.4 Evaluation Based on Application Perspective

Table 5. Evaluation of the proposed model as the pre-processing module of an emotion recognition system

Dataset	Number of emotion classes	Without cochlear filterbank	Gammatone	STFT	Proposed WPT
RAVDESS	8	66.34	83.41	74.31	**86.38**
SUSAS	6	65.9	82.01	72.33	**88.67**
ESD	6	69.03	84.44	75.62	**89.60**

In this evaluation, we have incorporated the proposed WPT model as the pre-processing stage of an emotion recognition algorithm [18] [5] and the results are compared with other filterbank models to gauge the system performance in an application perspective. Table 5 tabulates the impact of Gammatone, STFT and WPT filterbanks in an acoustic emotion recognition system [5]. The experiments are conducted using three different speech datasets. They are RAVDESS and SUSAS in English and ESD in Emirati accented Arabic languages. The results clearly illustrates the performance of the emotion recognition system with the proposed pre-processing module performs superior than the other models. It shows approximately 20% improvements in the recognition rate by the introduction of proposed pre-processing module to the emotion recognition algorithm.

4 Conclusion

We proposed an efficient technique for dominant voiced speech segregation using WPT cochlear filterbank to improve the overall performance of the hearing aids and speech processing systems in real time applications. The performance of the proposed algorithm is evaluated in terms of quantitative and perspective measures and based on computational complexity. The results evidences that the proposed WPT model offers superior performance than that of other techniques based on STFT and Gammatone filter-banks. The reduced computational complexity makes it suitable to implement the pre-processing module of digital hearing aids.

References

1. Hayat, M., Khan, S.H., Werghi, N., Goecke, R.: Joint registration and representation learning for unconstrained face identification. In: IEEE Conference on Computer Vision and Pattern Recognition (CVPR), pp. 1551–1560 (2017). https://doi.org/10.1109/CVPR.2017.169
2. Taha, B., Hayat, M., Berretti, S., Hatzinakos, D., Werghi, N.: Learned 3D shape representations using fused geometrically augmented images: application to facial expression and action unit detection. IEEE Trans. Circuits Syst. Video Technol. **30**(9), 2900–2916 (2020). https://doi.org/10.1109/TCSVT.2020.2984241
3. Xiao, Y., Siebert, P., Werghi, N.: Topological segmentation of discrete human body shapes in various postures based on geodesic distance. In: Proceedings of the 17th International Conference on Pattern Recognition, 2004. ICPR 2004, vol. 3, pp. 131–135 (2004). https://doi.org/10.1109/ICPR.2004.1334486
4. Werghi, N., Xiao, Y.: Recognition of human body posture from a cloud of 3D data points using wavelet transform coefficients. In: Proceedings of Fifth IEEE International Conference on Automatic Face Gesture Recognition, pp. 77–82 (2002). https://doi.org/10.1109/AFGR.2002.1004135
5. Hamsa, S., Shahin, I., Iraqi, Y., Werghi, N.: Emotion recognition from speech using wavelet packet transform cochlear filter bank and random forest classifier. IEEE Access **8**, 96994–97006 (2020). https://doi.org/10.1109/ACCESS.2020.2991811
6. Li, P., Guan, Y., Wang, S., Xu, B., Liu, W.: Monaural speech separation based on MAXVQ and CASA for robust speech recognition. Comput. Speech Lang. **24**(1), 30–44 (2010). https://doi.org/10.1007/s00521-018-3760-2
7. Shahin, I., Nassif, A.B., Hamsa, S.: Novel cascaded gaussian mixture model-deep neural network classifier for speaker identification in emotional talking environments. Neural Comput. Appl. **32**(7), 2575–2587 (2020). https://doi.org/10.1109/ICPR.2002.1044704
8. Werghi, N., Xiao, Y.: Wavelet moments for recognizing human body posture from 3D scans. In: Object Recognition Supported by User Interaction for Service Robots, vol. 1. IEEE, pp. 319–322 (2002). https://doi.org/10.1016/j.patrec.2004.09.018

9. Werghi, N.: A discriminative 3D wavelet-based descriptors: application to the recognition of human body postures. Pattern Recogn. Lett. **26**(5), 663–677 (2005). https://doi.org/10.1016/j.patrec.2004.09.018
10. Taha, B., Dias, J., Werghi, N.: Classification of cervical-cancer using pap-smear images: a convolutional neural network approach. Communications in Computer and Information Science, vol. 723 (2017). https://doi.org/10.1109/ACCESS.2019.2901352
11. Hardcastle, W.J., Laver, J., Gibbon, F.E.: The Handbook of Phonetic Sciences, vol. 116. John Wiley & Sons, Hoboken (2012)
12. Zwicker, E.: Subdivision of the audible frequency range into critical bands. J. Acoust. Soc. Am. **33**(2), 248–248 (1961)
13. Subasi, A., Ercelebi, E.: Classification of EEG signals using neural network and logistic regression. Comput. Methods Programs Biomed. **78**(2), 87–99 (2005)
14. Mahmoodzadeh, A., Abutalebi, H.R., Soltanian-Zadeh, H., Sheikhzadeh, H.: Single channel speech separation with a frame-based pitch range estimation method in modulation frequency. In: 2010 5th International Symposium on Telecommunications. IEEE, pp. 609–613 (2010)
15. Drullman, R., Festen, J.M., Plomp, R.: Effect of temporal envelope smearing on speech reception. J. Acoust. Soc. Am. **95**(2), 1053–1064 (1994)
16. Mahmoodzadeh, A., Abutalebi, H.R., Soltanian-Zadeh, H., Sheikhzadeh, H.: Single channel speech separation in modulation frequency domain based on a novel pitch range estimation method. EURASIP J. Adv. Signal Process. **2012**(1), 67 (2012)
17. Patterson, R.D., Nimmo-Smith, I., Holdsworth, J., Rice, P.: An efficient auditory filterbank based on the gammatone function. In: A Meeting of the IOC Speech Group on Auditory Modelling at RSRE, vol. 2, no. 7 (1987)
18. Shahin, I., Nassif, A.B., Hamsa, S.: Emotion recognition using hybrid Gaussian mixture model and deep neural network. IEEE Access, **7**, 26777–26787 (2019)

Forensic Analysis of Copy-Move Attack with Robust Duplication Detection

Hudan Studiawan(✉), Rahmat Nazali Salimi, and Tohari Ahmad

Department of Informatics, Institut Teknologi Sepuluh Nopember, Surabaya, Indonesia
{hudan,tohari}@if.its.ac.id

Abstract. Copy-move is one type of attack to forge a digital image where the attacker duplicates several areas of the image and paste them in different places to conceal a particular object on the original image. After the forgery, advanced methods such as noise addition and blurring, are often performed in the forged image to make it more challenging to recognize the attack. Therefore, it is required to do a preprocessing before conducting the detection. The preprocessing can be eliminated using a copy-move detection that is more resistant to noise addition and blurring. This paper proposes a new, flexible, and robust method that perform forensic analysis of both regular and advanced copy-move using modification and addition from two methods. The first method is designed to identify a regular copy-move attack, while the second one is effective for an advanced attack. The proposed method combines these two methods, can adapt to the forged image condition, and no preprocessing is required.

Keywords: Copy-move attack · Image forensic · Duplication detection

1 Introduction

Technology advancement nowadays makes digital image manipulation easier. One type of multimedia falsifying attack that involves the manipulation of a digital image is a copy-move attack. This type of attack duplicates a specific digital image region to be pasted into a different place in the same image. It intends to conceal a specific object or particular point of interest in the original image that does not want to be shown. For example, several cases involve a copy-move attack in a newspaper [6].

Luo et al. [5] suggested that people do not stop at making a simple copy-move attack. Some of these attacks are followed by a process called the post region duplication process [5] to further complicate the attacks' detection by doing a certain pixel modification process after the main attack was made. These processes include blurring, sharpening, JPEG compression, and noise addition [8]. The researchers have different strategies for detecting a copy-move attack.

A. Abraham et al. (Eds.): SoCPaR 2020, AISC 1383, pp. 404–413, 2021.
https://doi.org/10.1007/978-3-030-73689-7_39

Several of these approaches are color filter interpolation [9], resampling [9], computing its whole pixel [4] or by divide it into several chunks of data [7]. Every method above has its advantages, but they cannot detect an image attacked by copy-move with post region duplication process.

Post region duplication changes the specific region's pixel value or even the entire image's, so using the method that only compares individual pixel value might not be enough to detect duplicated regions. Another work provides a method that can obtain a particular characteristic of a chunk of an image region that does not significantly change when the post region duplication process is applied [5]. Another approach is to apply a clustering technique to detect copy-move forgery in an image [1].

In this paper, we propose a new variation of the copy-move detection designed to be more robust to an advanced copy-move attack [5] by modifying it with another method that utilizes principal component analysis [8]. We first analyze each method, make improvements from both methods, and the combine them to propose a new technique, namely robust-duplication method.

2 The Proposed Method

The design of the proposed method involves two methods, specifically the duplication detection [8] and the robust detection method [5]. We first discuss both methods and then present the proposed method in the subsequent subsection.

2.1 Duplication Detection Method

Duplication detection method [8] uses principal component analysis (PCA) for determining features of the image block. This method first divides an image $M \times N$ into overlapping blocks with size $L \times L$. The PCA is calculated for each image block and it is considered as the features. The procedure is briefly described as follows.

First, we provide an array $\overrightarrow{x}$ containing the pixel values of an image block with N_b elements. For a grayscale image, $\overrightarrow{x}$ has a size of $L \times L$ containing its pixel values. For a color image, we can 1) compute PCA on each color channel separately, or 2) flatten the image block pixel to a two-dimensional array $3b$, where b is the side of the image block, then compute its features using PCA.

Second, we define the number of dimension N_t so that it satisfies $1 - \epsilon = \frac{\sum_{i=1}^{N_t} \lambda_i}{\sum_{i=1}^{b} \lambda_i}$, where ϵ is a decimal fraction of the variance to be ignored, b is the number of the pixel from an image block, and λ_i is an eigenvalue of the block i. Note that N_t determines the number of principal components from each block.

We can determine principal component of an image block by building a new one-dimensional array $\overrightarrow{x_i} = \sum_{j=1}^{b} a_j \overrightarrow{e_j}$, where $j = 1 \ldots, b$, $\lambda_1 \geq \lambda_2 \geq \ldots \geq \lambda_b$, $\overrightarrow{e_j}$ is the eigenvector $a_j = \overrightarrow{x_i}^T \overrightarrow{e_j}$, and $\overrightarrow{a_i} = (a_1 \ldots a_1)$ are the principal components to represent the image block.

In the next step, the $\overrightarrow{s}$ is then sorted using lexicographical method based on the principal component of each image block. Therefore, the image blocks having the same or similar principal components are located close each other. Subsequently, the procedure creates an array $\overrightarrow{t}$ and fill it with a pair of the coordinate of the image blocks (x_i, y_i) and (x_j, y_j) that satisfies $|i - j| < N_n$, where i and j are the index of image block that stored in $\overrightarrow{s}$, and N_n is the maximum neighbor threshold.

The procedure then computes the offset of each element in $\overrightarrow{t}$. We discard all pair of coordinate in array $\overrightarrow{t}$ having an offset below the offset threshold N_f. We also discard all pair of coordinate in $\overrightarrow{t}$ having offset magnitude below N_d.

2.2 Robust Detection Method

The robust detection method [5] computes the portion of a pixel value in every image block to use as its features. This method is slower with a more complex computation because it redundantly iterates through the pixel. However, this method is more robust in a case where the input is a noised or blurred image because the feature does not significantly change when a post region duplication occurs.

This method starts by dividing the $M \times N$ image into $L \times L$ overlapping image blocks. For each image block, we run a feature extraction. If the color channel of the input image is not grayscale, then the procedure changes the color space to RGB channel to generalize the process. It then computes the first three characteristic features (c_1, c_2, c_3). c_1 represents the sum of a red pixels, c_2 indicates the sum of a green pixels, and c_3 is the sum of a blue ones. If the input image is a grayscale image, then it assigns zero to all of these three features.

Subsequently, the method constructs several types of region border from each image block, specifically vertical, horizontal, and diagonal borders. It then calculates the next four characteristic features (c_4, c_5, c_6, c_7). Each of feature is calculated by $c_i = \frac{\sum(\text{part1}(i))}{\sum(\text{part1}(i)+\text{part2}(i))}, i = 4, 5, 6, 7$, where part1 and part2 are taken from each region. The result then saved in an array $\overrightarrow{s}$ paired with each of the image block's coordinate.

Furthermore, we sort $\overrightarrow{s}$ using lexicographical sorting by its seven features. Therefore, the image blocks that have a similar feature are located close to each other. The procedure then creates an array $\overrightarrow{t}$ containing a pair of image block's coordinate (x_i, y_i) and (x_j, y_j), where i and j is an index of the image block in $\overrightarrow{s}$. Moreover, the procedure calculates the difference of each features and creates a histogram of a set of offsets. It only takes an offset with the greatest frequency and discard all the pair of a coordinate in $\overrightarrow{t}$ that did not have that offset.

2.3 The Proposed Method: Robust-Duplication Method

Analysis of the duplication and robust detection method. Both of aforementioned methods have a similarity in the concept where they both divide the $M \times N$ input image into $L \times L$ overlapping blocks. After the overlapping blocks from the image are obtained, we then do feature extraction to obtain each image block's features. Both methods have a difference in obtaining each image block's features.

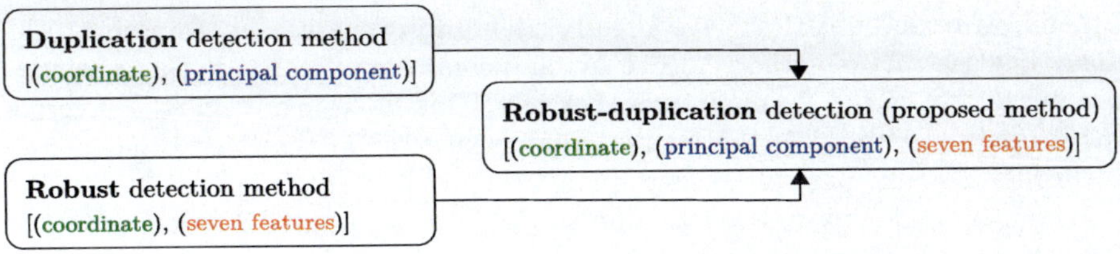

Fig. 1. The proposed method: robust-duplication detection

The duplication detection method proposes the PCA as the image block features. This method is quite fast, but when the input image was noised or blurred, this method does not perform well. This method obtains the absolute value from the offset calculation. A block with a negative distance can be treated to have the same offset. This absolute method is considered to give this method a tolerance. However, this tolerance makes a pair of a block that accidentally have the same offset value to close together, while it is also possible that those pairs are a false positive.

The second method, the robust detection, proposes pixel value comparison to calculate the features. This method is way much slower because calculating the comparison of pixel value involves redundant memory access to obtain a specific pixel value. This method utilizes a histogram to measure the offset frequency, then take the offset with the highest frequency. This procedures makes the method only detect exactly one region of duplication since only one offset is chosen from a set of offsets.

However, this method is robust in handling low-quality input images even without a preprocessing. Some examples of a low-quality input image are an image that contains a considerable amount of noise or a blurring process. This can be achieved because the characteristic features are considered robust towards the image manipulation process. For instance, a blurring process is a method to do an averaging of specific pixels region, so it does not change the region's total pixel. Since the total pixel in the region remains unchanged, the value is identical or at least close to the characteristic features' value because the features are obtained from the comparison of the pixel value. This condition makes the method to be considered robust.

The proposed method. We fix the weaknesses of each method. From the first method, we remove the absolute operation from the offset calculation formula. This procedure makes the offset range from a negative number to a positive number (formerly only zero to infinity because of the absolute process). This offset is accurate because it distinguishes offsets with the same value but different types (i.e., positive or negative). On the other hand, the second method chooses exactly one offset with the highest frequency. We then change it to have a frequency threshold variable that limits the offsets to be chosen. This means that we can take more than one offset, provided all of the taken offset has an offset frequency larger than the threshold.

Using two methods that have already been improved, we created a new technique that uses these methods altogether and creates a tolerance so that a particular method does not dominate another one. This tolerance technique involves round operation towards the characteristic features, so that nearly all of the features could take part in lexicographical sorting [10].

The proposed copy-move detection starts with computing characteristic features from each of the two methods and store it in a single container at once. Thus, we have a container that contains three data, namely the block coordinate, characteristic features from the first method (i.e., the principal components), and characteristic features from the second method (i.e., the pixel value comparison), as shown in Fig. 1.

Each of the characteristic feature obtains a rounding process towards a certain precision. At a glance, this round operation decreases the accuracy of each of the methods. However, a tolerance is obtained from the loss of domination from every method. Note that their features have a decreased precision. The tolerance effect is beneficial when features sorting takes place.

The container is then sorted using a lexicographical sorting based on the features' value from each method. This sorting process draws close the image blocks with similar characteristic features, which means that those blocks are suspected to be identical or at least similar. Knowing that identical or similar image blocks are always close to each other, we create pair sets of an image block from the sorting result close to another within a certain neighboring distance threshold.

A filtering is then performed to the image block sets to remove a pair that does not meet a particular threshold. This filtering leaves only a pair of sets of an image block that are further suspected identical or at least similar. Finally,

(A) Original images (B) Attacked images

Fig. 2. Samples images from the public datasets [3]

(A) Authentic images (B) Regular attack (C) Advanced attack

Fig. 3. Samples of authentic images (A) [3] and copy-move attacks (B-C)

an image of detection result is created based on each coordinate of an image block's remaining pair sets.

3 Experimental Results and Analyses

3.1 Experiment Settings

The dataset used in this experiment is based on a public dataset[1] [3], but most of it did not meet our limitations. Many of them are categorized in multi paste (i.e., the duplicated regions are overlapped). We also realize that no public dataset meet the case. We want to examine how our method identifies an advanced attack because none of them is involved in the post region duplication process. As a solution, we created our dataset consisting of 30 images by modifying the public dataset. It consists of ten images as authentic images, other ten images as a regular copy-move attack images, and the last ten images are for advanced copy-move attack created by blurring the duplicated region's certain edges. Some of the used public datasets are shown in Fig. 2, while some of our datasets are shown in Fig. 3.

We used a predefined threshold value of 2, 50, and 750. There are three experiment scenarios. Scenario 1 is designed to examine how the three methods work to an authentic image and whether they have false positive using ten images from the dataset. Scenario 2 is used to examine how the three methods perform to a regular copy-move attack (i.e., no involvement of post region duplication process). Scenario 3 is performed to investigate how the three methods work to an advanced copy-move attack (i.e., involves post region duplication process).

1 https://www5.cs.fau.de/research/data/image-manipulation/.

Table 1. The experimental results from each scenario

Scenario	Metrics	Method		
		Duplication	Robust	Robust-detection
Authentic image	Accuracy (%)	100.00	100.00	100.00
	MSE	0	0	0
	Similarity (%)	100.00	100.00	100.00
Regular copy-move	Accuracy (%)	83.00	90.00	100.00
	MSE	542.40	1023.98	580.01
	Similarity (%)	99.60	99.37	99.59
Advanced copy-move	Accuracy (%)	75.00	80.00	100.00
	MSE	7504.83	2538.93	2002.91
	Similarity	96.03	98.60	98.86
Regular copy-move from the public dataset	Accuracy (%)	56.00	80.00	80.00
	MSE	6881.08	3772.86	3772.14
	Similarity (%)	96.47	98.07	98.07

(A) Proposed method (B) Robust detection

Fig. 4. Detection of regular copy-move attacks

3.2 Results and Analyses

All of the experimental results of the scenarios are shown in Table 1 with three metrics, namely accuracy, mean squared error (MSE), and similarity. The average MSE of Scenario 1 is zero for all three methods. This value means that all the methods can identify the input image contains no copy-move region (i.e., no

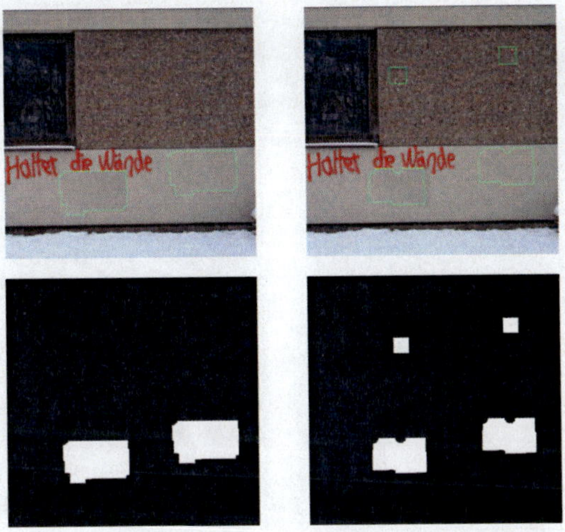

(A) (B)

Fig. 5. Detection of advanced copy-move attacks: (A) the proposed method and (B) the duplication detection method

false-positive condition is raised). Based on the average MSE of Scenario 2, the best method to identify a regular copy-move attack is the duplication detection (542.40), followed by our proposed method (580.01), while the robust detection has the lowest average MSE (1023.98). The utilization of principal components is very accurate for representing the image block's pixel values and far better than the pixel value comparison features. Our proposed method can adapt to the condition and manage to achieve better average MSE than the robust detection with 90% accuracy. The comparison between the methods' detection result is shown in Fig. 4.

Based on the average MSE of Scenario 3, the best method to identify an advanced copy-move attack is our proposed method (robust-duplication detection) with MSE of 2002.91, followed by robust detection (2538.93) and then duplication detection (7504.83). An advanced attack (e.g., involves a blurring operation) is better identified with a robust detection because the pixel comparison features are robust against the change of pixel value.

The duplication detection does not work efficiently because when a pixel value changes, albeit a bit, the principal components of the image block also change and affect the sorting procedure. Our proposed method can manage to adapt, the sorting procedure can tolerate the broken principal components, and it uses the pixel value comparison features instead. The comparison between methods' performance to detect the attack is shown in Fig. 5.

From Scenario 1 and Scenario 2, we understand that our proposed method can adapt to various input images without determining a false positive towards the authentic image. As for Scenario 3, we intend to use public datasets to represent the general experiment result. We only use public dataset images with the duplicated region that did not get overlapped each other with a limitation. We also see that all of the selected public dataset images are considered a regular copy-move attack.

Fig. 6. Detection examples of the proposed robust-duplication method

According to the average MSE of the result images, our proposed method is considered the best way to detect the chosen public dataset images, followed by a robust detection and duplication detection. This proves that our proposed method can be used for a general case on the public dataset. Some results of this scenario using our proposed method are shown in Fig. 6.

As we pick a blurring as the representation of several processes that are considered a post region duplication, we examine how the method reacts to several types of an input image and find that the proposed method can adapt to the type of the input. When a regular attack (i.e., standard copy-move) is performed, the principal components feature become dominant for the attack identification rather than the pixel comparison feature. When an advanced attack (i.e., copy-move that involves post region duplication process) is launched, the pixel comparison feature becomes dominant rather than the principal component feature. The automatic adaptation of these two features could be achieved because the lexicographical sorting is applied to a collection of pairs of these two features that already rounded into a certain precision.

4 Conclusion and Future Work

In this paper, we implement the combination of the robust detection and the duplication detection method, namely a robust-duplication detection method. This method can be used to perform forensic analysis of the digital image copy-move attack. Not only for the regular attack, but the proposed also able to identify a more advanced attack that involves the post region duplication process.

All of these advantages are limited to the fact that this method still could not detect copy-move attack regions that are overlap with each other. In future, we will investigate the application of deep learning technique to copy-move detection that offers more advantages than the traditional methods [2].

References

1. Abdel-Basset, M., Manogaran, G., Fakhry, A.E., El-Henawy, I.: 2-Levels of clustering strategy to detect and locate copy-move forgery in digital images. Multimed. Tools Appl. **79**(7), 5419–5437 (2020)
2. Abidin, A.B.Z., Majid, H.B.A., Samah, A.B.A., Hashim, H.B.: Copy-move image forgery detection using deep learning methods: a review. In: Proceedings of the 6th International Conference on Research and Innovation in Information Systems, pp. 1–6 (2019)
3. Christlein, V., Riess, C., Jordan, J., Riess, C., Angelopoulou, E.: An evaluation of popular copy-move forgery detection approaches. IEEE Trans. Inf. Forensics Secur. **7**(6), 1841–1854 (2012)
4. Fridrich, A.J., Soukal, B.D., Lukáš, A.J.: Detection of copy-move forgery in digital images. In: Proceedings of the Digital Forensic Research Workshop (2003)
5. Luo, W., Huang, J., Qiu, G.: Robust detection of region-duplication forgery in digital image. In: Proceedings of the 18th International Conference on Pattern Recognition, pp. 746–749 (2006)
6. Mirror Board: Sorry. we were hoaxed. Daily Mirror Newspaper (2004)
7. Ng, T.T., Chang, S.F.: A model for image splicing. Proc. Int. Conf. Image Processing. **2**, 1169–1172 (2004)
8. Popescu, A.C., Farid, H.: Exposing digital forgeries by detecting duplicated image regions. Technical. Rep. TR2004-515, Department of Computer Science, Dartmouth College (2004)
9. Popescu, A.C., Farid, H.: Exposing digital forgeries by detecting traces of resampling. IEEE Trans. Signal Process. **53**(2), 758–767 (2005)
10. Vogan, D.: Lexicographic order. Technical report, Department of Mathematics, Massachusetts Institute of Technology. http://www-math.mit.edu/~dav/lex2.pdf

Feature Selection Approach Based on a Novel Variant of Hybrid Differential Evolution and PSO for Sentiment Classification

Abhishek Dixit[1(✉)], Ashish Mani[2], and Rohit Bansal[3]

[1] Department of Computer Science Amity School of Engineering and Technology, Amity University, Noida, Uttar Pradesh, India

[2] Department of EEE Amity School of Engineering and Technology, Amity University, Noida, Uttar Pradesh, India
amani@amity.edu

[3] Department of Management Studies Rajiv, Gandhi Institute of Petroleum Technology, Raebareli, Uttar Pradesh, India

Abstract. Over the last few years, digital text data is increasing with the advancement in use of technologies. Therefore, sentiment analysis is an important step for improving the performance of any classifier. But with the increase in data, this problem has become more complex. To improve the overall performance of classifier and reducing the data-dimensionality, Feature selection is a key sub-stage for the sentiment analysis. In this paper, we propose a feature selection approach, which is based on a novel variant of hybrid differential evolution algorithm and PSO for sentiment classification. Experimental results show that our approach achieves 97.56% and 98.54% of average accuracy by using Naïve Bayes (NB) classifier for the two Twitter datasets, viz., TestData_2015 and twitter sanders, respectively. The results also show that our proposed approach is showing better results with state-of-art algorithms on Twitter datasets. Statistical results also showcase the effectiveness of our proposed model.

Keywords: Feature selection · Differential evolution · PSO · SVM classifier · Sentiment classification

1 Introduction

In the past few years, text mining is becoming crucial with the advancement of digital documents due to the rapid use of digital tools. For text classification, Sentiment analysis is used which divide the data into positive and negative labels. Sentiment analysis is applied usually in three ways viz., document level [1], sentence level [2], and aspect level [3]. There are various machine learning models used for sentiment analysis [4, 5]. These models help various agencies government or private in giving valuable information from the large complex datasets. There are different domains for sentiment analysis like analysis of customer or public reviews, and online content analysis [6].The performance

A. Abraham et al. (Eds.): SoCPaR 2020, AISC 1383, pp. 414–423, 2021.
https://doi.org/10.1007/978-3-030-73689-7_40

of sentiment analysis methods is reduced due to the presence of emoji, shorthand text as these are not included in pre-defined lexicon [7].

Feature selection is a dimensionality reduction method in which redundant and irrelevant features are removed before applying it for classification [8]. There are two categories of feature selection techniques, viz., filter based and wrapper-based technique. In filter approach, feature subsets are evaluated as per the mathematical conditions. On the other hand, wrapper-based technique works on the predictive model. Generally, wrapper-based methods are more popular due the efficiency in producing accurate results based on machine learning model.

Even though feature selection improves the classification accuracy, but this efficiency reduces as the size of feature subset grows. Another problem is the computation time taken by classification model with the increase in size of features. Evolutionary algorithms are used to solve these problems by optimizing the search process efficiently without exploring the entire search space. Many Evolutionary algorithms were proposed like: Genetic Algorithm [9], Ant colony optimization [10], Particle swarm optimization [11], Differential evolution algorithm [12].

Nevertheless, all these feature selection algorithms utilizing evolutionary algorithms still face the issue of getting stuck in local optima and also there is no way to have a good exploration of search space. Therefore, to overcome these issues and balance the exploration and exploitation capability of the algorithm is the main objective of this paper. In this paper we propose the efficient algorithm for the feature selection for sentiment analysis by hybridizing differential evolution algorithm with PSO. In this approach, we have used DE mutation strategy with archival function to explore the large search space. This brings population diversity and avoid particle to stuck in local optima. Lately during the iteration process, we have used PSO mutation to add convergence speed. Additionally, we have used parameter tuning in order to enhance the overall performance. As Naïve Bayes (NB) achieves high performance in sentiment analysis so we have used it as a classification algorithm in our model. We have used twitter datasets: TestData_2015 [13], and twitter sanders [14]. To evaluate the performance of our proposed algorithm and compared it with other state of art algorithms. Twitter is a platform for sharing the text data by user on social media. This raw data is complex, one of the largest databases with millions of messages shared daily [15]. Therefore, in this paper we have used twitter dataset for doing sentiment analysis.

The later section of paper is summarized as: Sect. 2 discusses the differential evolution and PSO algorithm and Naïve Bayes classification model. Section 3 describes the proposed algorithm. Section 4 shows the experiment and results. Section 5 concludes the paper with future scope of study.

2 Background

2.1 Differential Evolution Algorithm

DE proposed by storm and price [16] is a stochastic, population-based algorithm. The algorithm begins by initializing the population and it is further evaluation in successive iteration by the fitness value and updated for each generation. It has mutation, crossover and selection as the three stages.

Mutation: Mutation operating for classical DE is shown as below:

$$v_i(k) = x_{r1}^k + F * \left(x_{r2}^k - x_{r3}^k\right), i = 1, 2, 3 \ldots N \quad (1)$$

where, F is the scaling factor, r_1, r_2, r_3 are the mutually exclusive random values in the range [0, 1], k the generation count.

Crossover: This operation is performed to add the diversity. In this operation x_i, v_i are applied to generate child vector as u_i for the kth generation.

$$u_{ij}^k = \begin{cases} v_{ij}^k \; if \;\; rand \le CR \\ x_{ij}^k \; otherwise \end{cases} \quad (2)$$

CR is the crossover rate and $rand$ is the random value in range [0, 1].

Selection: Greedy selection is applied and best value if selected among parent and child vector.

$$x_i^{k+1} = \begin{cases} v_i^k \; if \;\; f\left(u_i^k\right) \le f\left(x_i^k\right) \\ x_i^k \; otherwise \end{cases} \quad (3)$$

$f(u)$ is the fitness value.

2.2 Particle Swarm Optimization

PSO was proposed by Kennedy and Eberhart [17] is a population based stochastic algorithm and is based on the bird's flocking behavior. Let p is the number of particles in swarm in D dimensional search space and their position and velocity are shown as $x_i = [x_{i1}, x_{i2}, x_{i3}, ..x_{iD}]$, $v_i = [v_{i1}, v_{i2}, v_{i3}, ..v_{iD}]$, where $i = 1, 2, 3 \ldots .p$. In PSO each particle maintains its local position p_{best} and global best for whole swarm is also maintained as g_{best}. These positions changed as per the velocity maintained by particles and is calculated as:

$$\begin{aligned} v_{pD}^i &= wv_{pD}^{i-1} + c_1r_1\left(p_{best,pD}^{i-1} - x_{pD}^{i-1}\right) + c_2r_2\left(g_{best,D}^{i-1} - x_{pD}^{i-1}\right) \\ x_{pD}^i &= x_{pD}^{i-1} + v_{pD}^i \end{aligned} \quad (4)$$

where, i is the iteration number, D is the dimention, w is the inertia weight.

2.3 Naïve Bayes Algorithm

Naïve Bayes (NB) is a simple classifier based on probabilistic model [18]. In case of text classification, any document belong to class c is calculated by Bayes theorem as

$$p\left(\frac{c}{d}\right) = \frac{p(d/c)p(c)}{p(d)} \quad (5)$$

3 Proposed Method

In this section, proposed algorithm is explained in detail. Our model follows the 3-step process as pre-processing, feature extraction, feature selection and classification.

3.1 Pre-processing

In this step, raw data is collected that contains unwanted fuzzy words, stop words. These should be removed before moving to the feature selection stage.

1. Remove urls, email address by regular expression
2. Remove special character as #, /, \, -.
3. Translate all the words to lowercase
4. Change stop words as per the comparison with
5. Remove words that do not start with alphabet.

3.2 Feature Extraction

Applying pre-processing gives dataset that is processed for collecting raw feature set. Total 11 features are extracted from dataset. The collected features are shown in Table 1

Table 1. Extracted feature set

S.No	Name of feature
1	Total characters
2	Positive emoji
3	Negative emoji
4	Neutral emoji
5	Positive exclamation
6	Negative exclamation
7	Negation
8	Positive words
9	Negative words
10	Neutral words
11	Intense words

3.3 Feature Selection and Classification

In this step we have applied our hybrid variant with Naïve Bayes classifier. Irrelevant and redundant features are optimized by our feature selection algorithm by utilizing the exploration and exploitation capabilities of DE and PSO. The optimized feature set and then forwarded to NB classifier for classification. In this step we have first initialized the mandatory parameters and then initialized the velocity of each particles. In the next step DE mutation strategy is applied and fitness value is calculated. In the next step crossover is applied followed by the features forwarded to NB for classification. Updating velocity and position using PSO add diversity and DE mutation strategy helps in improving the local searching capability. The fitness function is defined as:

$$fit = \varphi * \left(\frac{nnz(t)}{length(t)} \right) + (1 - \varphi) * (1 - perf) \tag{6}$$

where, φ is the threshold value between [0,1]. For this experiment we have taken its values as 0.15. t is the temporary value and perf is the NB classifier's outcome.

Algorithm: Feature selection

Input: dataset (samples, classes), Population (NP), Generation, F, CR, a_1, a_2,

repeat
for g = 1: Gen
Iterate over NP
Calculate random integers such that $r1 \neq r2 \neq r3$
For each particle i, Calculate the velocity using as:

$$v_{pD}^{i} = wv_{pD}^{i-1} + c_1 r_1 (p_{best,pD}^{i-1} - x_{pD}^{i-1}) + c_2 r_2 (g_{best,D}^{i-1} - x_{pD}^{i-1})$$

Evaluate the new position and select the best position using as:

$$x_i = \begin{cases} 1 & if\ rand() < f(v_{iD}) \\ 0 & otherwise \end{cases},$$

where $f(v_{iD}) = \frac{1}{1+e^{-v_{iD}}}$

apply DE mutation as:

$$v_i(k) = x_{r1}^{k} + F * (x_{r2}^{k} - x_{r3}^{k}), i = 1, 2, 3 \ldots N$$

Calculate the new fitness value. $f(\gamma_i)$ from above as:

$$\gamma_i = X_{i,g+1} + v_{i,G+1}$$

Apply crossover as

$$C_{i,g+1} = \begin{cases} \gamma_i & if\ rand(0,1) \le CR \\ X_i & otherwise \end{cases}$$

Pass the optimized feature set for Classification
$MultiNB\ (samples, classes, temp)$ and compare with previous feature set.
Choose
End
End

4 Experiment and Results

For this experiment classification accuracy is used as a measurement of fitness. The accuracy is calculated as:

$$acc = \frac{tp + tn}{tp + tn + fp + fn} \tag{7}$$

where, tp *and* tn are the true positive and true negative. fp *and* fn are false positive and false negative. We have applied k-fold cross validation where k = 10 for calculating every accuracy value. In this approach the dataset is divided into equal size section. In this 1 is used for validation and rest for training. This process is repeated 10 times and average value is calculated.

This algorithm is implemented in MATLAB R2018a on computer with configuration Intel® Core™ i7-8665U CPU, 2.11 GHZ and 16.0 GB RAM. The parameters used for the algorithm and other state of art algorithm is shown in Table 2. All the parameters are chosen by trial and error strategy. For this experiment we have used 2 twitter dataset having different topics of tweet. TestData_2015 [13], twitter_sanders [14]. Details of both the datasets are shown in Table 3.

Table 2. Parameter setting

Parameter	DE	PSO
Iteration count	1000	1000
Inertia weight	–	0.9
Crossover rate	0.8	–
Scaling factor	0.6	–
a1, a2	–	2.05,2.05

Table 3. Datasets

Dataset	# instances	# classes	+ ve	−ve	Neutral
TestData_2015	498	3	182	177	139
twitter_sanders	988	3	163	316	509

Table 4. Comparison Results

Dataset	Method	Mean Accuracy	Mean Computational Time	Mean Fitness value
TestData_2015	NB	41.23	319.34	0.2467
	SVM	44.47	324.15	0.2434
	DE-NB	78.56	291.81	0.2876
	PSO-NB	85.65	306.93	
	DEPSO-NB	97.56	289.54	
twitter_sanders	NB	50.53	509.07	0.2455
	SVM	52.27	513.9	0.2423
	DE-NB	82.45	432.77	0.2734
	PSO-NB	87.65	478.48	
	DEPSO-NB	98.54	428.67	

The performance of our proposed method is evaluated on Naïve Bayes, Support vector machine classifiers. Also, we have evaluated the performance of our proposed algorithm on DE-NB, PSO-NB classification model. In order to estimate the accurate

Table 5. Statistical t-test for accuracy

Dataset	Method	Std Error	95% CL	2-tailed P	Significance
TestData_2015	NB	0.0411	40.98–41.54	2.34E-54	Yes
	SVM	0.0401	44.38–44.54	1.10E-55	Yes
	DE-NB	0.0400	58.65–59.41	1.34E-53	Yes
	PSO-NB	0.0410	58.90–59.93	1.72E-53	Yes
twitter_sanders	NB	0.0104	50.08–50.84	2.10E-111	Yes
	SVM	0.0201	52.11–52.35	1.35E-113	Yes
	DE-NB	0.0003	62.84–63.88	1.20E-114	Yes
	PSO-NB	0.0005	61.96–62.40	2.11E-111	Yes

results, values are captured after 30 iterations. The mean value of accuracy, time and fitness value is shown in Table 4. From the results we can see that our proposed model is giving better accuracy as compared with other models. Additionally, our proposed model is giving better results in terms of computation time and fitness values. This shows robustness and computational effectiveness of our proposed approach.

We have also performed statistical t-test to showcase the robustness and efficiency of our proposed model. We have compared our model with other algorithms by performing t-test on 95% confidence level. We have made the null hypothesis as there is no significant difference in the values obtained from all the algorithms after 30 runs. Table 5 shows

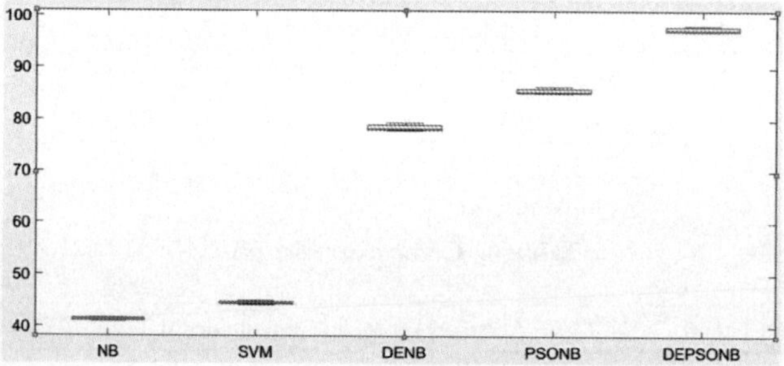

(a)TestData_2015

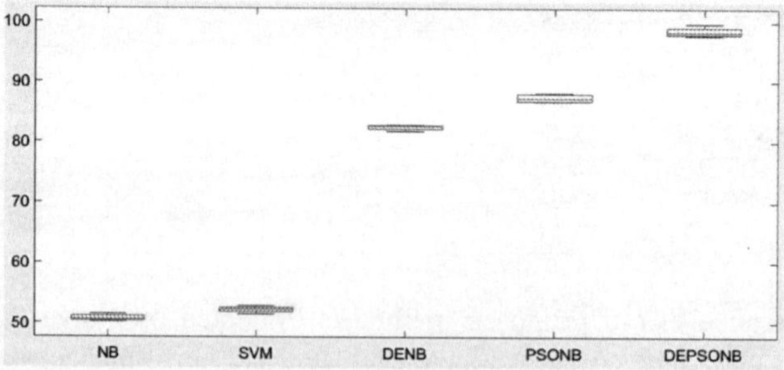

(b) twitter_sanders

Fig. 1. Box plot for NB, SVM, DENB, PSONB, DEPSONB for accuracy.

the t-test results and it shows that our null hypothesis is wrong, and our algorithm is showing significant difference in results. Therefore, null hypothesis is rejected.

Additionally, we have also performed box plot analysis by comparing the performance of all the algorithms. Box plot analysis is done to showcase the distribution of empirical data. Figure 1–3 shows the box plot comparison data for all the algorithms for accuracy, computational time and fitness value. X-axis shows the methods and y-axis shows the accuracy parameter. Figure also shows that our proposed model is giving better results as compared to other algorithms.

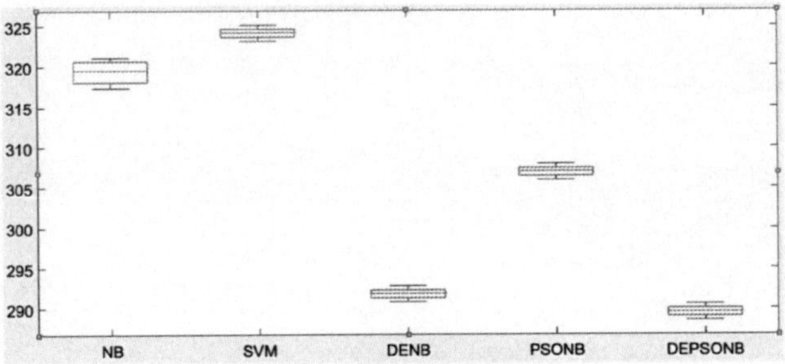

(a)TestData_2015

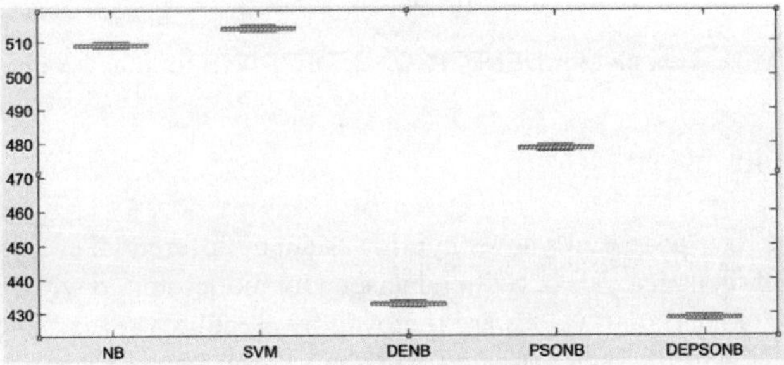

(b) twitter_sanders

Fig. 2. Box plot for NB, SVM, DENB, PSONB, DEPSONB for computational time.

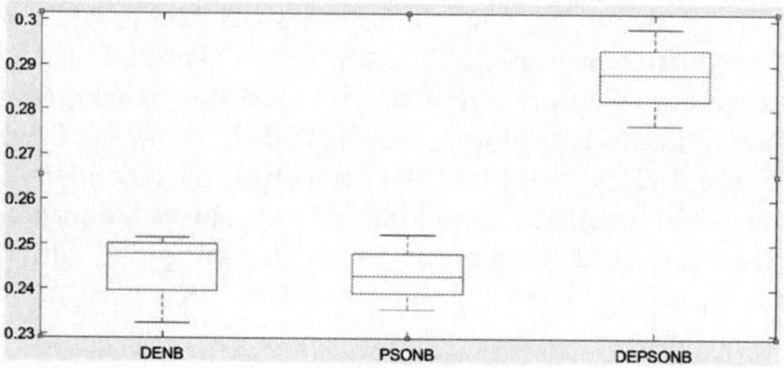

(a)TestData_2015

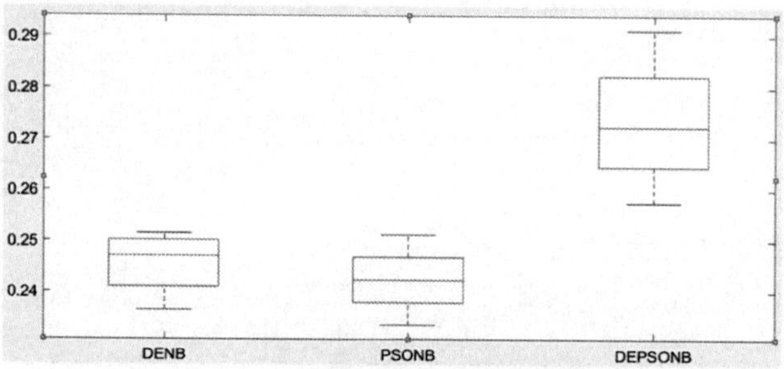

(b) twitter_sanders

Fig. 3. Box plot for DENB, PSONB, DEPSONB for fitness value

5 Conclusion

In this paper, we have proposed a novel hybrid variant of differential evolution algorithm and PSO for sentiment analysis of twitter dataset. Our model started with pre-processing in which unwanted and junk values are removed by creating a regex pattern matching. In the next steps features are extracted and then these features are pushed into our novel feature selection approach for optimizing the feature set. Further the optimized feature set are sent for classification using Naïve Bays classifier. We have compared our proposed approach with NB, SVM, DE-NB, PSO-NB. To showcase the effectiveness of our algorithm, we have also performed statistical t-test. The results show that our proposed model is performing better in comparison to other algorithms.

References

1. Medhat, W., Hassan, A., Korashy, H.: Sentiment analysis algorithms and applications: a survey. Ain Shams Eng. J. **54**, 1093–1113 (2014)
2. Wang, G., Sun, J., Ma, J., Xu, K., Gu, J.: Sentiment classification: the contribution of ensemble learning. Decis. Support Syst. **57**, 77–93 (2014)
3. Wang, G., Zhang, Z., Sun, J., Yang, S., Larson, C.A.: POS-RS: a random subspace method for sentiment classification based on part-of-speech analysis. Inf. Process. Manage. **51**(4), 458–479 (2015)

4. Onan, A., Korukoglu, S., Bulut, H.: A multiobjective weighted voting ensemble classifier based on differential evolution algorithm for text sentiment classification. Expert Syst. Appl. **62**, 1–6 (2016)
5. Gokalp, O., Tasci, E., Ugur, A.: A novel wrapper feature selection algorithm based on iterated greedy metaheuristic for sentiment classification. Expert Syst. Appl. **146**, 113176 (2020)
6. Zhang, C., Zeng, D., Li, J., Wang, F., Zuo, W.: Sentiment analysis of Chinese documents. J. Am. Soc. Inform. Sci. Technol. **60**(12), 2474–2487 (2009)
7. Zhou, Z., Zhang, X., Sanderson, M.: Sentiment analysis on Twitter through topic-based lexicon expansion. In: ADC 2014: Databases Theory and Applications (2014)
8. Xia, R., Zong, C., Li, S.: Ensemble of feature sets and classification algorithms for sentiment classification. Inf. Sci. **181**(6), 1138–1152 (2011)
9. Ghareb, A.S., Bakar, A.A., Hamdan, A.R.: Hybrid feature selection based on enhanced genetic algorithm for text categorization. Expert Syst. Appl. **49**(1), 31–47 (2016)
10. Wan, Y., Wang, M., Ye, Z., Lai, X.: A feature selection method based on modified binary coded ant colony optimization algorithm. Appl. Soft Comput. **49**, 248–258 (2016)
11. Engelbrecht, A.P., Grobler, J., Langeveld, J.: Set based particle swarm optimization for the feature selection problem. Eng. Appl. Artif. Intell. **85**, 324–336 (2019)
12. Elaziz, M.A., Ewees, A.A., Ibrahim, R.A., Lu, S.: Opposition-based moth-flame optimization improved by differential evolution for feature selection. Math. Comput. Simul. **168**, 48–75 (2020)
13. testdata.manual.2009.06.14. https://help.sentiment140.com/for-students/ (2015)
14. Twitter dataset. https://drive.google.com/file/d/0BwPSGZHAP_yoN2pZcVl1Qmp1OEU/view?usp=sharing (2014)
15. Ritter, A., Clark, S., Mausam, Etzioni, O.: Named entity recognition in tweets: an experimental study. In: Proceedings of the Conference on Empirical Methods in Natural Language Processing (2011)
16. Storn, R., Price, K.: Differential evolution – a simple and efficient heuristic for global optimization. J. Global Optim. **11**(4), 341–359 (1997)
17. Kennedy, J., Eberhart, R.: Particle swarm optimization. In: IEEE International Conference on Neural Networks (1995)
18. McCallum, A., Nigam, K.: A comparison of event models for Naive Bayes text classification. In: AAAI-98 Workshop on Learning for Text Categorization (1998)

A Comparative Study for Fisheye Image Classification: SVM or DNN

Zhen Chen(✉) and Anthimos Georgiadis

Leuphana University Lüneburg, Universitätsallee 1, 21335 Lüneburg, Germany
zhen.chen@leuphana.de

Abstract. The comparison between the feature-based method and the learning-based method is conducted in the training time, the accuracy and the generalization capacity, to address the optimisation for the multi-style fisheye imagery classification. We construct an srd-SIFT descriptor based SVM classifier to present the feature-based method for describing the influence of the dataset scale and the visual word scale on the classifier. The SVM classifier achieves 15.98% accuracy on the test set after 162 h training, with the condition that includes 800 images per class in 12 classes and 1500 visual words. For the learning-based method, we propose to expand training samples' style variety, via style transformation, to facilitate the contemporary architecture retraining. Following this approach, we retrain the ResNet-50 by an artificial multi-style fisheye image dataset without complementing new training labels. The performance of the obtained ResNet classifier is evaluated on 6000 images collected in the real-world. The result shows that the retrained classifier has great generalization capacity and reaches 97.19% top-3 accuracy.

Keywords: Fisheye image · Super vector machine · DNN · Style expansion

1 Introduction

In the last few years, cameras with the ultra-wide field of view are integrated into plenty of automation scenes, like automated inspection, unmanned driving vehicle and virtual reality [1]. Owing to the compact optical structure of the fisheye lens, the camera system provides flexible applications in limited workspace. However, the equirectangular image obtained from a fisheye camera suffers from high compression ratios that appear intensely at the edge [2]. The previous research concentrates on omnidirectional image unwrapping algorithms which encourage reusing mature achievements on the perspective image [3]. Due to the extra computing cost from the unwrapping process, research focus is shifting to process the raw fisheye image.

The feature-based method and the learning-based method are two mainstream research aspects of the classification issue. By the feature-based method, a feature point with scale, shift and rotation invariant becomes a unique feature descriptor on image's representation. The classifier can be trained with the aid of the Bag-of-Visual-Words (BoVW) model for restructuring an image's representation [4]. Different to the feature-based method, the learning-based method learns substantial characteristics of training

A. Abraham et al. (Eds.): SoCPaR 2020, AISC 1383, pp. 424–433, 2021.
https://doi.org/10.1007/978-3-030-73689-7_41

samples by layers of artificially constructed neural units. In this paper, we compare two methods for the multi-style fisheye imagery classification. Contrast to the rectilinear image classification task and the regular omnidirectional image processing procedure, our work focuses on samples' geometric models that are an unknown or are mixed with a variety of geometric models. In tests, we adopt the SVM classifier and the Deep Residual Networks (ResNet) as representations of these two methods. The comparison focuses on their performance in terms of the classifier's training speed, the accuracy and the generalization capacity. The artificial fisheye image dataset implemented in experiments is from our previous work [5]. In addition, we give a further test of the achieved model on real-world fisheye images.

2 Related Work

A hand-crafted feature descriptor determines the uniqueness of an image's description, which exhibits great performance for perspective images in the long-term development. Even though feature descriptors SIFT, SURF, BRIEF and ORB are widely used in the image recognition and matching algorithm, their performance on the omnidirectional image is inadequate [6]. For the omnidirectional image matching, Peter Hansen et al. proposed pSIFT that maps the omnidirectional image and spherical Gaussian function to an intermediate stereographic image for approximating the diffusion on the sphere [7]. Cruz-Mota et al. employed a spherical Gaussian filter in spherical SIFT to perform the Gaussian smoothing [8]. In 2012, the srd-SIFT was demonstrated with the radial distortion improvement on the original SIFT framework. Compared to other existing omnidirectional descriptors, the srd-SIFT relies less on camera's calibration parameters [9].

The NN architecture designed for the fisheye image classification is limited. Jeon et al. introduced an active convolution unit to learn position parameters for defining more diverse forms of receptive fields [10]. Coors et al. proposed to dynamically wrap the convolutional kernel sampling pattern around the sphere's surface according to the sampling location [11]. It lessens image distortion by the perspective convolutional neural network (CNN), especially at borders and poles of the equirectangular projection. Khasanova et al. treated each omnidirectional image as a weighted graph signal representation, and extended convolutional neural network on the processing of graphs [12]. However, it was proved just with several specific terms of mapping projections. Above research builds end-to-end training for the omnidirectional image but focuses only on single-style image. Our research targets the classification problem under the multi-style fisheye image. Each object's images achieved from different distances and angles is treated as its non-single representations. If training samples cover variety of representations of an object, existing NN architectures can be reused.

The debate on adopting two different methods continues. KIM et al. showed that SVM classifier outperformed KNN classifier on their Caltech-4-Cropped dataset [13]. Liu et al. had a comparative study between the SVM and the Stacked Auto-Encoders (SAE) on the remote sensing image classification [14]. The comparison showed that SVM takes less time on the small-scale dataset, but SAE can be implemented in parallel clusters. SAE did not show competitive performance on accuracy and noise immunity.

In this paper, a further study between two methods on the performance of the fisheye imagery is conducted.

3 Experimental Framework

3.1 srd-SIFT Descriptor Based SVM Classifier

We implement a three-step process to build a feature-based classification architecture. First, selected representative feature descriptors construct feature vectors as an image's description. Then, feature vectors achieve further dimension scale-down and are also reorganized in the BoVW model. Finally, a classifier is trained with local descriptors from the BoVW model. In the prediction phase, feature vectors from a test image are assigned to the trained classifier for a classification output.

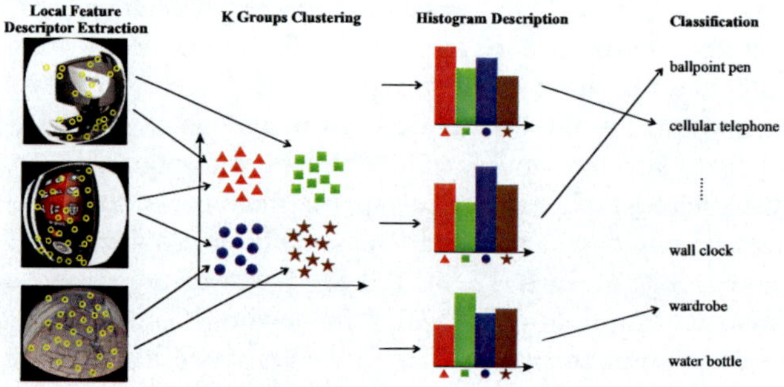

Fig. 1. The framework of the srd-SIFT descriptor based SVM classifier

The srd-SIFT enhances the SIFT algorithm in aspects of the repeatability and the effectiveness on the radial distortion for the omnidirectional image matching. The algorithm processes the original image plane, instead of resampling the image signal. In contrasts to pSIFT and mdBRIEF [15], srd-SIFT has lower accuracy requirement for camera's calibration parameters. Due to high dimensions of achieved srd-SIFT feature descriptors, K-means clustering is adopted to decrease the dimensions. The BoVW is implemented according to replace the concept 'words' by clusters achieved from the K-mean clustering. The SVM solves the multi-class classification problem by simplifying the problem into a set of binary classification problems. We perform one-versus-one SVM strategy by LIBSVM [16].

As shown in Fig. 1, sample images are sent to an srd-SIFT feature extractor. Extracted feature descriptors buildup an Nx128 dimension feature vector. For decreasing the dimension of feature vectors, the vectors are clustered to K classes under the unsupervised learning. To analyse clusters' spatial histogram distribution, feature words are constructed by a BoVW model. Each image can be described by M visual words. Then, we train the SVM classifier with each image's BoVW expression.

3.2 DNN Classifier

DNN architecture is mainly consisted of the convolutional layer, the pooling layer, the fully-connected layer and a series of active functions. The convolutional layer is located on the top layer of a CNN, which extracts features from an input image based on shared weights. It builds the connection between the input volume and the output feature map when the filter constructs convolution operations with the receptive field on each stride. The pooling layer is normally inserted between successive convolutional layers, to reduce the spatial dimension of the representation. It decreases the quantity of parameters in feature maps, but maintains the same depth between the input channel and the output channel. The fully-connected layer is the last output layer of the DNN architecture, which classifies the input image into various classes as pre-set in training dataset. Each neuron in a fully-connected layer has global activations from the previous layer for receiving signals from entire feature maps. High-level features from convolutional layers and pooling layers are reconstructed and activated by the softmax activation function in the output layer.

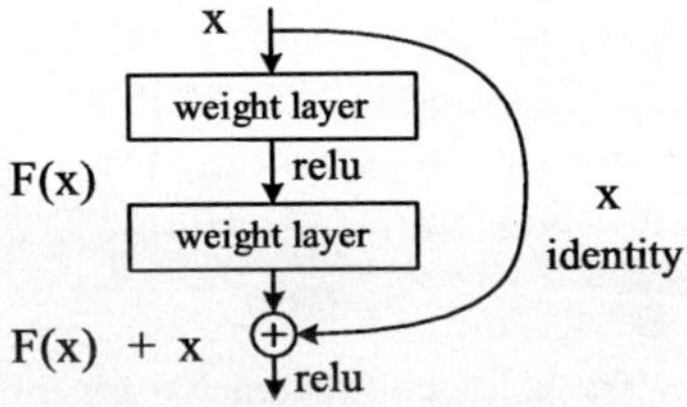

Fig. 2. Residual block building

Residual Block

ResNet lets the stacked layers fit a residual mapping instead of a desired underlying mapping, which is assumed to be easier to optimise than the original mapping [17]. The expression of the short connection in a residual mapping is shown in Fig. 2.

Considering x as the inputs of the first layer and $H(x)$ as an underlying mapping by a few stacked layers, the residual function can be defined as:

$$F(x) = H(x) - x \tag{1}$$

In which the input and the output are the same dimensions. Then a residual unit is defined as:

$$y = q(F(x, w_i) + w_s x) \tag{2}$$

Where y is the output of the residual unit, w_i is a set of weights, w_s is a linear projection matrix for matching dimensions, $F(x, w_i)$ represents the residual mapping to be learned, and q denotes ReLU. For a residual unit that contains two convolutional layers, it is simplified as:

$$F(x, W) = W_2 q(W_1 x) \tag{3}$$

4 Tests and Results

4.1 Training Environment

Tests are executed on Dell Precision T5500 workstation which is equipped with dual Intel Xeon Processor E5645, 23 GB RAM and an MSI Geforce GTX 1080 graphic card. The SVM classifier and the ResNet classifier are trained by MATLAB 2018a on CPU and Deep Learning Library Caffe [18] on GPU separately. The artificial fisheye image dataset includes 167250 synthetic fisheye images in 12 classes, which are transformed by the principle of the equidistant projection.

4.2 SVM Classifier Training and Results

The artificial dataset owns over 10K images on each class. Using the whole dataset to train a 12 classes SVM classifier is extremely expensive. Therefore, we prepare three sample groups which have 300, 500 and 800 random images on each class separately for training and test. Valid images for each sample group reach 3600, 6000 and 9600 where the ratio between the training set and the test set is 8: 2. To understand the influence of the visual word scale on the BoVW model, 600, 1000 and 1500 different quantities of visual words are test for each dataset. Here the visual word scale has the same value as the K value in clustering. Test results are listed as Table 1, 2 and 3.

Table 1. The prediction accuracy for training sets

	The visual word scale		
Img/class	600	1000	1500
300	79.65%	87.56%	93.33%
500	61.75%	78.83%	84.47%
800	53.54%	67.16%	77.73%

Table 1 reflects whether the trained model fits training samples. As shown, the models' accuracy increases when the visual word scale increases, and receives the best achievement at the visual word level 1500. It achieves 93.33% on the training set that includes 300 images per class, with accuracy boosts of 5.77% and 13.68% on the word level 1000 and 600. However, concerning only the word level 1500, the SVM classifier's prediction accuracy drops dramatically to 77.73% when the image quantity increases to 800 images per class.

With the increase in the number of training samples and visual words, training a classifier becomes more and more expensive, rising from 7 h to 162 h (as shown in Table 2). We also give classifiers' test results. As shown in Table 3, classifiers have a slight improvement when a model is trained under a larger dataset at the same word scale level. In contrast, for the same training set, the enlargement of the visual word scale weakens the model's generalization capacity.

Table 2. Time cost for each classifier's training

	The visual word scale		
Img/class	600	1000	1500
300	7 h	12 h	20 h
500	20 h	37 h	59 h
800	59 h	102 h	162 h

Table 3. Prediction accuracy for test sets

	The visual word scale		
Img/class	600	1000	1500
300	15.41%	13.61%	12.5%
500	19.16%	17.16%	15.41%
800	19.89%	18.33%	15.98%

4.3 DNN Classifier Training and Results

We employ 119040 images for the training, 14880 images for the validation and 14880 images for the test separately. Since the fisheye image dataset is originated from the perspective dataset that sharing same objects, weights from the pre-trained model are transferred to new model using transfer learning technology.

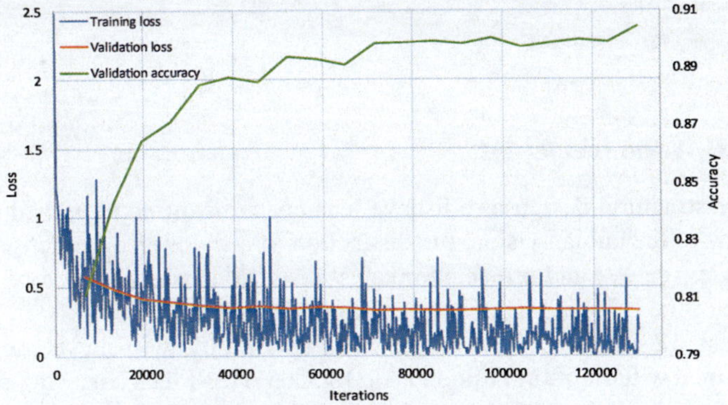

Fig. 3. The fine-tuning model process

In practice, we implement the Stochastic Gradient Descent optimizer with 0.0001 basic learning rate. The multi-step strategy is used with the gamma value 0.5. The training process is shown in Fig. 3. After around 12 h fine-tuning, the new model achieves 90.34% accuracy on the validation set and 89.98% accuracy on the test set, as shown in Table 4. It presents consistent accuracy between the validation set and the test set. The great performance on the test set shows the model's generalization capacity on the unknown data.

Table 4. The prediction accuracy of the CNN classifier on the validation set and the test set

Group number	Validation accuracy	Correct samples	Test accuracy	Correct samples
Ballpoint pen	92.66%	(1149/1240)	92.18%	(1143/1240)
Cellular telephone	88.06%	(1092/1240)	86.77%	(1076/1240)
Desktop computer	85.48%	(1060/1240)	86.13%	(1068/1240)
Espresso maker	92.98%	(1152/1240)	93.63%	(1161/1240)
Printer	86.45%	(1072/1240)	84.84%	(1052/1240)
Projector	85.40%	(1059/1240)	84.35%	(1046/1240)
Shopping cart	92.90%	(1152/1240)	93.95%	(1165/1240)
Stone wall	98.39%	(1220/1240)	97.90%	(1214/1240)
Television	86.53%	(1073/1240)	86.53%	(1073/1240)
Wall clock	92.82%	(1151/1240)	92.02%	(1141/1240)
Wardrobe	94.84%	(1176/1240)	94.44%	(1171/1240)
Water bottle	87.66%	(1087/1240)	87.02%	(1079/1240)
Overall	90.34%	(13443/14880)	89.98%	(13389/14880)

4.4 Physical World Image Test

Although the structural design of a fisheye lens concerns different optical parameters, from the view of the human vision, the distinction of different fisheye lenses reflects in the field of view (FOV) and projected images' distortion status. Therefore, we test the performance of the achieved model on real-world images.

In this test, 12 classes objects images are captured by the fisheye camera system which is equiped with Fujifilm Fujinon F-FE185C057HA-1 lens (focusing range: 0.1 m to infinity; viewing angle: 185 degrees on the 2/3-inch sensor) and Sony Alpha 6000 camera. Two-step strategies are adopted to prepare samples for the evaluation. Firstly, we shoot a video of an object along the vertical, the horizontal and the radial direction on the resolution 1920×1080 with 24 frames per second in AVCHD format. Secondly, we split each video into separate frames and crop out the black border. As a result, images that contain an object's continuous deformation can be obtained as shown in Fig. 4. The resolution of the cropped image is 500×500 pixels.

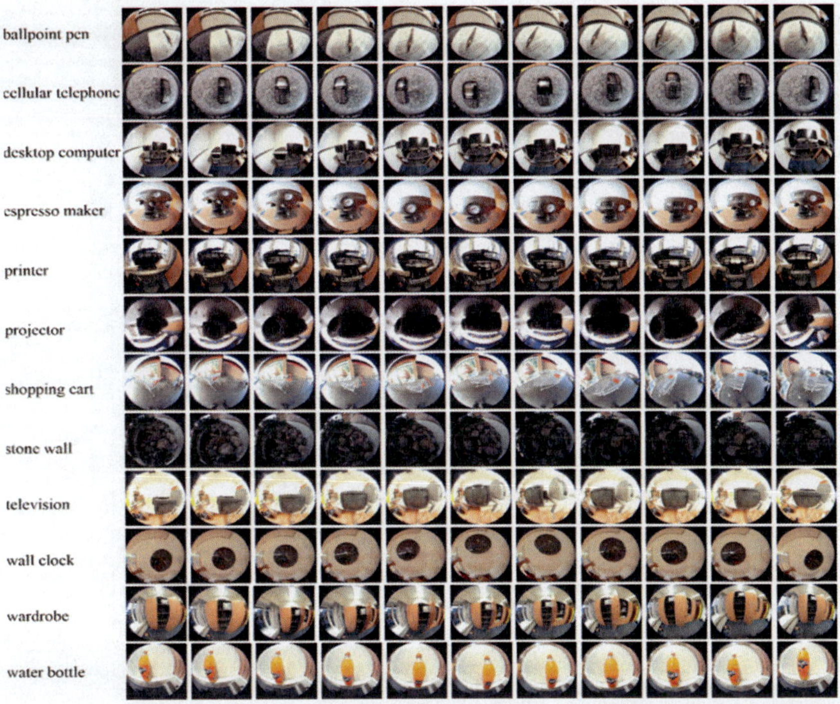

Fig. 4. Parts of collected real-world images for each class

During image collecting procedure, there is no proper scene for the stone wall. Instead, we replace this scene with the cobbled decoration which presents similar morphological feature as the stone wall in this situation. Finally, we combine 500 images for each class to form a new test set (6000 images totally). The dataset covers continuous object deformation on different angles and distances. All obtained images are labelled and sent to the trained classifier. The prediction results are shown in Fig. 5. As illustrated in the confusion matrix, most of the test classes reach 100% prediction accuracy, except ballpoint pen (87.92%), printer (87.17%) and television (92.26%). The overall accuracy of the test set is 97.19%. The result proves that enlarging the style of training set can effectively improve a model's generalization capacity.

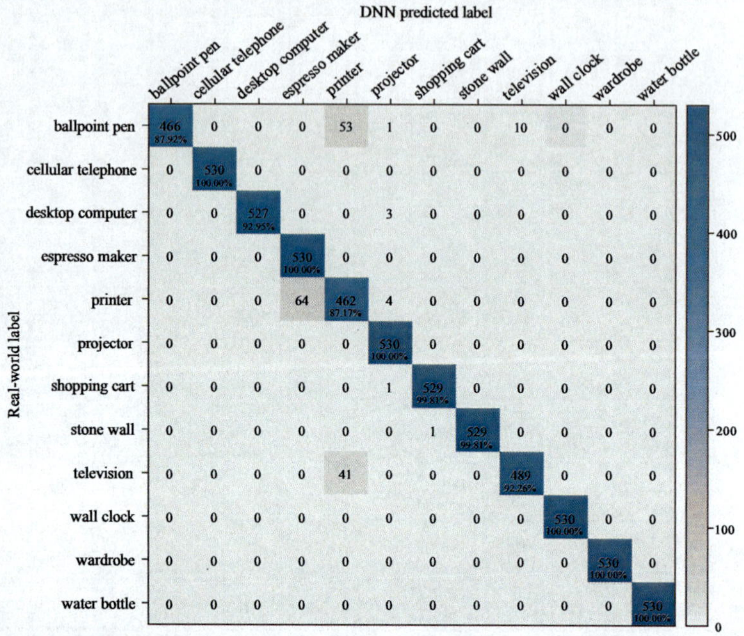

Fig. 5. Confusion matrix for the predictions of the DNN versus the real-world label

5 Conclusion

In this paper, we compare the performance of the srd-SIFT descriptor based SVM classifier and the ResNet on the multi-style fisheye imagery classification. The ResNet shows a great advantage on the training time and the accuracy in the large-scale dataset processing. The visual word scale of the BoVW model deeply influences the SVM classifier's accuracy, especially on the small-scale dataset. From the aspect of the generalization, the diverse simulated camera parameters lead to the fail of the SVM classifiers on the test set. However, the ResNet classifier demonstrates high prediction accuracy on the test set and real-world images.

References

1. Davies, E.R.: Machine Vision: Theory, Algorithms, Practicalities. Elsevier, Amsterdam (2004)
2. Sarkar, M., Brown, M.H.: Graphical fisheye views of graphs. In: Proceedings of the SIGCHI Conference on Human Factors in Computing Systems, pp. 83–91 (1992)
3. Jung, H.G., Kim, D.S., Yoon, P.J., Kim, J.: Structure analysis based parking slot marking recognition for semi-automatic parking system. In: Joint IAPR International Workshops on Statistical Techniques in Pattern Recognition (SPR) and Structural and Syntactic Pattern Recognition (SSPR), pp. 384–393. Springer, Cham (2006)
4. Yang, J., Jiang, Y.-G., Hauptmann, A.G., Ngo, C.-W.: Evaluating bag-of-visual-words representations in scene classification. In: Proceedings of the International Workshop on Multimedia Information Retrieval, pp. 197–206 (2007)

5. Zhen, C., Georgiadis, A.: Parameterized synthetic image data set for fisheye lens. In: 2018 5th International Conference on Information Science and Control Engineering (ICISCE),. pp. 370–374. IEEE (2018)
6. Karami, E., Prasad, S., Shehata, M.: Image matching using SIFT, SURF, BRIEF and ORB: performance comparison for distorted images. arXiv Prepr. arXiv1710.02726 (2017)
7. Hansen, P., Boles, W., Corke, P.: Spherical diffusion for scale-invariant keypoint detection in wide-angle images. In: 2008 Digital Image Computing: Techniques and Applications. pp. 525–532. IEEE (2008)
8. Cruz-Mota, J., Bogdanova, I., Paquier, B., Bierlaire, M., Thiran, J.-P.: Scale invariant feature transform on the sphere: theory and applications. Int. J. Comput. Vis. **98**, 217–241 (2012)
9. Lourenco, M., Barreto, J.P., Vasconcelos, F.: sRD-SIFT: keypoint detection and matching in images with radial distortion. IEEE Trans. Robot. **28**, 752–760 (2012)
10. Jeon, Y., Kim, J.: Active convolution: Learning the shape of convolution for image classification. In: Proceedings of the IEEE Conference on Computer Vision and Pattern Recognition, pp. 4201–4209 (2017)
11. Coors, B., Paul Condurache, A., Geiger, A.: Spherenet: Learning spherical representations for detection and classification in omnidirectional images. In: Proceedings of the European Conference on Computer Vision (ECCV), pp. 518–533 (2018)
12. Khasanova, R., Frossard, P.: Graph-based classification of omnidirectional images. In: Proceedings of the IEEE International Conference on Computer Vision Workshops. pp. 869–878 (2017)
13. KIM, J., Kim, B.S., Savarese, S.: Comparing image classification methods: K-nearest-neighbor and support-vector-machines. In: Proceedings of the 6th WSEAS international conference on Computer Engineering and Applications, and Proceedings of the 2012 American conference on Applied Mathematics, pp. 42122–48109 (2012)
14. Liu, P., Choo, K.-K.R., Wang, L., Huang, F.: SVM or deep learning? a comparative study on remote sensing image classification. Soft Comput. **21**, 7053–7065 (2017)
15. Urban, S., Weinmann, M., Hinz, S.: mdBRIEF-a fast online-adaptable, distorted binary descriptor for real-time applications using calibrated wide-angle or fisheye cameras. Comput. Vis. Image Underst. **162**, 71–86 (2017)
16. Chang, C.-C., Lin, C.-J.: LIBSVM: A library for support vector machines. ACM Trans. Intell. Syst. Technol. **2**, 1–27 (2011)
17. He, K., Zhang, X., Ren, S., Sun, J.: Deep residual learning for image recognition. In: Proceedings of the IEEE Conference on Computer Vision and Pattern Recognition, pp. 770–778 (2016)
18. Jia, Y., Shelhamer, E., Donahue, J., Karayev, S., Long, J., Girshick, R., Guadarrama, S., Darrell, T.: Caffe: convolutional architecture for fast feature embedding. In: Proceedings of the 22nd ACM International Conference on Multimedia, pp. 675–678 (2014)

A Survey on Applications of Modern Deep Learning Techniques in Team Sports Analytics

Gervas Batister Mgaya, Hongbo Liu(✉), and Bo Zhang

Department of Intelligence Science and Technology, College of Information Science and Technology, Dalian Maritime University, Dalian 116026, China
{mgaya,lhb,bzhang}@dlmu.edu.cn

Abstract. The recent advancement of deep learning algorithms has paved the way to many technological applications and research areas. The field of team sport analytics is one of the interesting research areas in which deep neural network techniques are applied. Availability of both tracking and visual data in sports coupled with the recent development in computing process has made sport analytics via deep learning approach possible. However, comprehensive surveys on these deep learning techniques as leveraged in sport analytics are still limited. This paper attempts to provide a survey along with references for further studies on modern state-of-the-art deep learning techniques leveraged to two typical team sport namely soccer and basketball, where multi-agent interactions and coordination are highly observed. We expect this paper to serve as an insightful and handful resource on team sport analytics for both sports practitioners and researchers from machine learning perspective.

Keywords: Behaviour modeling · Multi-agent coordination · Sport analytics · Trajectory generation

1 Introduction

In the last few years, the use of deep learning as a method to solve complex problems has been widespread to computer vision, pattern recognition, computer games, robotics and sports just to mention a few. The sports field which is our main focus is an entertainment at its very core aim and a business venture as well. Sport analytics are a collection of relevant, historical statistics through which investigation and modelling of professional sports performance and contexts using scientific techniques are done to provide competitive advantages to a team or an individual player during pre-match, in-game and post-match performance evaluation. The use of deep learning techniques in sports analytics is becoming increasingly popular. This is due to the availability of sensing technologies that provide high-fidelity data streams extracted from every match that enables variance reduction, players and ball tracking and team behaviour analysis. However, there are very few survey articles on the subject matter. Beal et al.

A. Abraham et al. (Eds.): SoCPaR 2020, AISC 1383, pp. 434–443, 2021.
https://doi.org/10.1007/978-3-030-73689-7_42

[6] provides a broad survey of artificial and machine learning techniques applied in team sport analytics. However, they have little covered modern deep learning techniques.

Our objective in this survey is to provide an overview of modern deep neural techniques as applied in sport analytics from low level perspective (trajectory generation and analysis) to high level perspective (behavioural modeling and coordination modeling). We focus our attention on two typical team sports, namely soccer and basketball. These two team sports are characterised by the use of a goal for scoring, dynamic and tactical features of attack and defence, the containment of space in defence and the dynamic structure (as players change roles) of defence and attack. These characteristics have made the applications of machine learning techniques more appealing. This paper will forgo the basics, graphic illustration of models' architectures and the mathematical formulas associated with surveyed modern deep learning models. For readers seeking a fuller treatment, we refer them to the corresponding reference(s) cited in each surveyed modern deep learning model.

The rest of the paper is structured as follows: Sect. 2 surveys applications of modern deep learning methods in team sport analytics from low to high level perspectives. Finally, Sect. 3 concludes the survey.

2 Deep Learning Methods in Team Sport Analytics

2.1 Low Level Team Sport Analytics

In low level team sport analytics, we focus on survey of how machine learning techniques are leveraged to localize, to track and to generate trajectories of an individual player using spatio-temporal data.

Trajectory Generation: Chen et al. [9] presented a generated adversarial network (GAN) [14], which takes the ball and the offensive players movement as inputs to generate realistic defensive trajectories so as to forecast how the opposing team will react when an offensive strategy is applied. The generated defensive plays by Convolution Neural Networks and Wide Open Penalty CNN+WO (WO means the *wide open penalty*) were evaluated by comparing with real plays in terms of: 1) the degree of realism of players movements (speed and acceleration) and 2) the fulfillment of defense objectives (i.e., to minimise the opportunity of making scores by the opposing team measured by Hausdorff distance). The CNN+WO was compared with other four baseline models namely supervised, Convolution Neural Networks Only (CNNONLY), Recurrent Neural Networks Only (RNNONLY), and Recurrent Neural Networks with Wide Open Penalty (RNN+WO). The defensive plays generated by the proposed model (CNN+WO) is the second best comparing with the mentioned baseline methods.

The same approach was used by [20] where the input of the network is an offensive tactic hand sketched by coaches and the output is a basketball set play, in which the offensive part fulfills the input condition. However, Chen's model and Hsieh's work did not generate offensive strategies and the input sketches for training were so simplified from real players movement trajectories that they could not capture the realistic defensive strategies. Therefore, there is a need to do research in this direction so as to evaluate the effectiveness of defensive strategies. Another direction for research is to use scores or players' ability as a metric to evaluate the effectiveness of offensive and defensive strategies.

Konstantinos et al. [30] estimated the depth map of each player by leveraging Convolutional Neural Networks (CNN) [26] to transform a monocular video of a soccer game into a moving three dimensions (3D) reconstruction. The CNN was trained on three dimensions (3D) player data that were extracted from soccer video games to reconstruct per-player depth maps on the playing field, which then could be viewed interactively on tabletop with an augmented reality device. However, their work had some limitations: Missed detections led to players not appearing in the final reconstruction because players in video clips move quickly and often frames are blurred due to this rapid movement. Errors in the pose estimation could result in incorrect trajectories and segmentation masks (e.g., missing body parts). Lastly, players overlap considerably caused inaccurate depth estimations because of occlusions.

Inspired with the ability of Recurrent Neural Networks (RNN) for sequential data modeling [15], the RNN-based models have been applied in team sport analytics as well. Rajiv et al. [31] applied recurrent neural networks (RNNs) in the form of sequence modeling to predict whether a three-point shot is successful or not. The input of the network is the basketball position data and the game clock. At each time step t, the RNN predicts the probability of a successful shot. However, their model is not fully optimised as evidenced from experimental results. Hence, a more comprehensive search for better performing hyper-parameters is still needed. It is readily apparent that ball trajectories are a much lower level sport analytics problem than other predictive activities. This suggests for more complex prediction tasks with larger datasets will be more beneficial.

Zhao et al. [41] have leveraged deep bidirectional LSTM (BLSTM) [16] and mixture density network architecture namely BLSTM-MDN (that outputs subset of values based on their probability density function) for generating new flying basketball trajectory samples in three dimensions and hit-or-miss classification problem. In both cases the BLSTM-MDN outperformed all the previous state-of-the art models.

Game Action Identification and Classification: Sport games can be described by an ordered sequence of actions happening during the match. These game actions are in most cases, ball related (on-the-ball actions) instances like passes, dribble, tackle, throw-in, free kick, kick-off, goal-kick, etc. Baccouche et al. in [4] developed neural network architectures called *Long Short-Term Memory Recurrent Neural Networks (LSTM-RNN)* that was experimented on the *MICC-Soccer-Actions-4 dataset* [5] to classify four soccer actions (Shot-on-goal, Placed-kick, Throw-in and Goal-kick) using both visual and motion contents. The soccer video sequences were represented by a sequence of descriptors (one descriptor per image) corresponding to a set of features. The LSTM-RNN was trained to categorise each action type based on the temporal evolution of the descriptors. The obtained results were better than the previous state-of-the-arts.

In [2], Agyeman et al. combined LSTM [19] (to model temporal evolution) with three-dimensional Convolutional Neural Networks (for feature extraction) to summarize soccer video (i.e., recognition of actions of interest). The model was trained on UCF101 and soccer-5 datasets, 10 completely new soccer match videos were downloaded from YouTube and summarised using the developed model. Each of the 10 summarised soccer match videos was subjected to human evaluators who gave more Mean Opinion Score (MOS) to this model than the other previous models. A combination of Convolutional Neural Networks and Deep Bidirectional Long Short Term Memory (i.e., CNN + DB-LSTM) has been utilised in [36] to recognise actions from the sequence of frames in basketball and soccer videos. First, CNN features are extracted from the frames of video. Second, the features representing the sequence of action for certain time interval are fed to the DB-LSTM. Last, the final state of each time interval is analysed for final recognition of an action in a video. However, the recognition of actions from salient regions of the frame of video is still an open research problem to address issues of players occlusions (which can be either partial or complete occlusions because of the capturing angle) and the rapid movement of objects in video clips which results in blurred frames.

Game Score and Winning Prediction: Ballan et al. [13] leveraged Mixture Density networks (MDNs) [8] with *team lineup encoding* to predict the final score difference distribution in National Basketball Association (NBA) game in an attempt to address the issues of context and uncertainty in the *win probability model* that attempts to assign a probability to the likelihood of each team winning the game given a current game state. In Ganguly and Frank's work, their mixture density network outperforms their constructed baseline random forest models, by using a larger input vector. However, there are plenty of opportunities for further work in modeling NBA win probability such as to (1) fit mixture density networks that utilise neural networks with a greater number of hidden layers to output the mixture model parameter layer and (2) to utilise the NBA's spatio-temporal player tracking data as inputs to a mixture density network.

Unlike the previous researchers, Fernández et al. [12] have gone further, they have leveraged CNN to develop deep learning architecture namely *SoccerMap* that is capable of estimating full probability surfaces of potential passes in soccer, derived from high-frequency spatio-temporal data. Their model has applications in identification of optimal passing locations, the prediction of optimal positioning for improving pass probability, and the prediction of team-level passing tendencies. Anthony et al. [32] used both a Long-Short Term Memory Network (LSTM) and a Neural Embedding (i.e., embed objects in some latent space such that similar objects are close to each other in this space) to evaluate off-ball events (e.g., screens, cuts away from the ball, etc.) in terms of how they contribute to the success of a possession. Their proposed model (called DeepHoops) processes a dataset composed of spatio-temporal tracking data from NBA games and generates a running stream of predictions on the expected points to be scored as possession progresses within a given temporal window.

2.2 High Level Team Sport Analytics

In high level sport analytics, we focus on how machine learning techniques are leveraged to model movement of an entire team or a set of players through *space* and *time*. Group movement can contain either cooperative or competitive behavior or any combinations of both.

Players and Team Behavioural Modeling: Inspired by the work of [29] and [37] on weak supervision, Eric Zhan et al. [40] used hierarchial VRNN [10] with shared macro intents architecture to effectively model complex interactions between offensive basketball players and generate realistic multi-agent trajectories of basketball game play over long time periods. The concept of macro-intent allowed the researchers to analyse the long-term goals of a player and how they change their objectives during game play. In [24], Lindström leveraged imitation learning to learn an individual player's behaviour and to predict player trajectories in game play sequences of 10, 20, 30, 40 window size policies from Signality soccer dataset [33]. An action (i.e., change in player position) was predicted given *state* and *policy*. A policy was evaluated by rolling it out on a set of sequence over a horizon of t time steps. During the rollout of each sequence, the Euclidian distance d_t (i.e., error or imitation deviation) between player position in a observed sequence and a rolled sequence was calculated for each time step t. *Absolute action* (i.e., player position in the next time step) and *relative action* (i.e., change in player position between current and the next time step) were evaluated to determine which is better in terms of average imitation deviation on rolled out sequences. Training and validation sequences were rolled out with a horizon of 50 time steps. The results showed that relative policy generates a smaller imitation deviation than absolute policy.

Team Coordination Modeling: Another work by Le et al. [23] in *Coordinated Multi-agent Imitation Learning* has leveraged imitation learning method. They presented multi-agent imitation learning algorithm that gradually rolls out sequences with policy during training and cross-update the locations of all players in the consecutive time steps. In their experiment, they aimed at learning multi-agent policies for team soccer defense(excluding the goalkeeper), based on tracking data from real-life professional 45 soccer games [7]. Their coordinated learning approach substantially outperformed conventional imitation learning without structured coordination as shown in Fig. 1 below. Hoang et al. [22] created a ghosting model by learning to mimic multiple players of similar roles in one or more teams. Their methodology involved deep imitation learning [3] to learn the general behaviour of inferred roles, by a process which they called *role-alignment*. They used this process because their data lacked annotations of players' identities in each observed sequence. The roles were inferred by each player's position mean coordinate during a play segment [7]. Their data-driven ghosting method can be applied to various game contexts to better understand

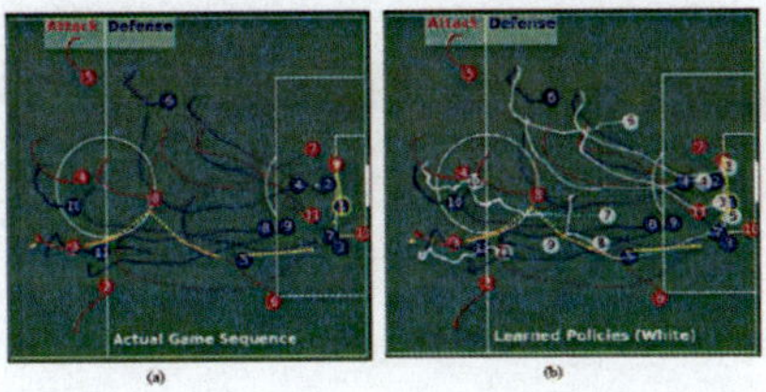

Fig. 1. Coordinating behavior policies for team sports from tracking data. (a) Actual Game sequence, (b) Result of 10 coordinated imitation policies, corresponding with (a). White is the rolled-out imitation policies [23].

and lastly apply the optimal defensive strategies. For instance, Fig. 2 below, Fulham team (red) scores a goal on Swansea team (blue). Ghosts (white) represent where Swansea defenders should have been according to Manchester City Ghost so as to minimise the goal scoring probability.

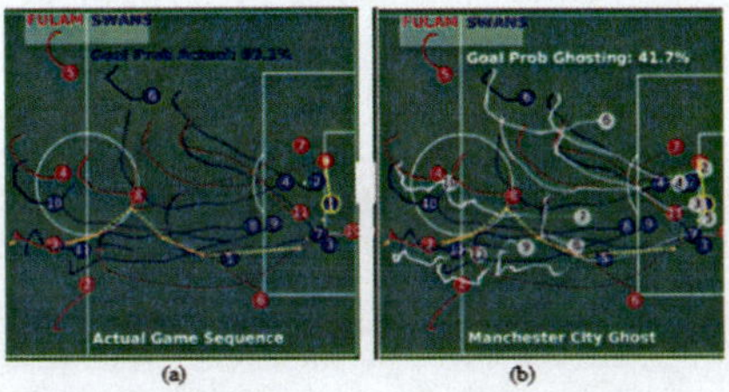

Fig. 2. Application of Ghost Model to improve defensive positioning. (a) Actual game sequences, (b) Learned policies via ghost modeling [22].

Wang et al. [38] used reinforcement learning (RL) framework [21] to determine when and where to effectively double team (i.e., two defensive players focused on one offensive player) another player. Deep RL with a convolutional neural network architecture, called NothingButNet (NBNet) was designed specifically for the task. Over 643,000 possessions from the previous three seasons were entered into the framework. Running the data through the framework several observations were made. First, it was determined that double teaming results in a significantly lower field goal percentage for the offense overall. Second, it is also more likely that the possession will end in a foul. Third, it was also determined that when ball handlers are double teamed they typically tend to pass or dribble the ball.

Automatic extraction of high-level temporal information video sequences to describe their semantic content is one of the major challenges in the field of information systems. Using data from the publicly available SportVu tracking and play-by-play data for 42 Toronto Raptors games played in Fall 2015, Acuna et al. [1] leveraged Variational Auto-Encoders(VAE) [11] (i.e., having ability to regularise the latent space) to synthesise the movement of basketball players with and without conditioning them on the player ID, the trajectories of the ball or both. They firstly trained a vanilla VAE to generate pictorial representation of the movement of basketball players. Then they extended the model and use a Conditional VAE (CVAE) (which helps to learn the latent representation of multi-agent trajectories in context specific manner) to synthesize pictorial representation of the players movements conditioned on ball, player ID and both. Although their approach was able to generate synthetic images that model the trajectories of the basketball players, they noticed that in some cases the generated trajectories were blurred and in other cases they could not see the trajectories at all. This resulted from their simplicity of encoder/decoder architecture.

Table 1. Summary of Modern Deep Neural Network Models applied on team sports analytics

Architecture	Problem addressed	Team Sport type	Dataset(s) used
1. GAN [9]	Generating realistic defensive plays conditioned on ball and offensive term movement	Basketball	NBA SportVu [39]
2. GAN [20]	Simulation of offensive tactic sketched by coaches	Basketball	STATS SportVu [39]
3. RNN [31]	Predict three-point shot	Basketball	NBA SportVu [39]
4. AE [17]	Semantic of player's movement	Basketball	NBA SportVU [25,27]
5. VRNN [40]	Modeling complex multi-agent behaviour	Basketball	NBA SportVU [25,27]
6. CNN+FFN [18]	Predicting shot making	Basketball	NBA SportVU [25,27]
7. LSTM-RNN [4]	classifying four soccer actions	Soccer	MICC-Soccer-Actions-4 datase [5]
8. CNN [34]	sport highlights	Soccer	3 different soccer matches
9. CNN [30]	talent identification	Soccer	soccer videos from YouTube
10. RNN [22]	Ghost modelling	Soccer	STATS
11. CNN +DB-LSTM [36]	Action recognition	Basketball	UCF101 [35]
12. Imitation learning [22]	learn the general behaviour of inferred roles	Soccer	
13. Imitation learning [24]	Predict player trajectories	Soccer	Signality [33]
14. Reinforcement learning [38]	when and where to double team?	Basketball	
15. SoccerMap [12]	Pass probability estimation	Soccer	EPL 2013/2014 and 2014/2015 Season [28]

3 Conclusion

We conducted a survey of modern deep neural network models leveraged on team sports analytics as summarised in Table 1. Several deep neural network approaches were reviewed from low to high level perspectives. Here, the survey clearly shows that modern deep neural networks models are capable of doing efficiently and accurately sport analytics as compared to traditional methods. However, availability of public datasets and the best choice of network model(s) and its internal parameters are the issues that should be addressed in the future works. We believe that our survey can shade light and advance the field of research on team sport analytics.

Acknowledgement. This work is supported by the National Natural Science Foundation of China (Grant No.61702073) and the China Postdoctoral Science Foundation (Grant No. 2019M661079)

References

1. Acuna, D.: Unsupervised modeling of the movement of basketball players using a deep generative model. http://toronto.edu/~david/projects/unsupervised_modeling_using_a_DGM.pdf. Accessed 17 Feb 2020
2. Agyeman, R., Muhammad, R., Choi, G.S.: Soccer video summarization using deep learning. In: Proceedings of IEEE Conference on Multimedia Information Processing and Retrieval, pp. 270–273. IEEE (2019)
3. Attia, A., Dayan, S.: Global overview of imitation learning. arXiv preprint arXiv:1801.06503 (2018)
4. Baccouche, M., Mamalet, F., Wolf, C., Garcia, C., Baskurt, A.: Action classification in soccer videos with long short-term memory recurrent neural networks. In: Proceedings of International Conference on Artificial Neural Networks, pp. 154–159. Springer, Heidelberg (2010)
5. Ballan, L., Bertini, M., Del Bimbo, A., Serra, G.: Action categorization in soccer videos using string kernels. In: Proceedings of Seventh International Workshop on Content-Based Multimedia Indexing, pp. 13–18. IEEE (2009)
6. Beal, R., Norman, T.J., Ramchurn, S.D.: Artificial intelligence for team sports: a survey. Knowl. Eng. Rev. **34**, e28 (2019)
7. Bialkowski, A., Lucey, P., Carr, P., Yue, Y., Sridharan, S., Matthews, I.: Large-scale analysis of soccer matches using spatiotemporal tracking data. In: Proceedings of IEEE International Conference on Data Mining, pp. 725–730. IEEE (2014)
8. Bishop, C.M.: Mixture density networks. IEEE Computer Society (1994)
9. Chen, C.Y., Lai, W., Hsieh, H.Y., Zheng, W.H., Wang, Y.S., Chuang, J.H.: Generating defensive plays in basketball games, pp. 1580–1588. arXiv preprint arXiv:1908.09876 (2018)
10. Chung, J., Kastner, K., Dinh, L., Goel, K., Courville, A.C., Bengio, Y.: A recurrent latent variable model for sequential data. In: Proceedings of Advances in Neural Information Processing Systems, pp. 2980–2988 (2015)
11. Doersch, C.: Tutorial on variational autoencoders. arXiv preprint arXiv:1606.05908 (2016)

12. Fernández, J., Bornn, L.: Soccermap: A deep learning architecture for visually-interpretable analysis in soccer. arXiv preprint arXiv:2010.10202 (2020)
13. Ganguly, S., Frank, N.: The problem with win probability. In: Proceedings of MIT Sloan Sports Analytics Conference (2018)
14. Goodfellow, I.: NIPS 2016 tutorial: Generative adversarial networks. Computer Research Repository (2017). http://arxiv.org/abs/1701.00160
15. Graves, A.: Generating sequences with recurrent neural networks. arXiv preprint arXiv:1308.0850 (2013)
16. Graves, A., Schmidhuber, J.: Framewise phoneme classification with bidirectional LSTM and other neural network architectures. Neural Netw. **18**(5–6), 602–610 (2005)
17. Guttag, J., Nistala, A.: Using deep learning to understand patterns of player movement in the NBA (April 2019). http://www.sloansportsconference.com/content/using-deep-learning-to-understand-patterns-of-player-movement-in-the-nba/. Accessed 03 Jan 2020
18. Harmon, M., Lucey, P., Klabjan, D.: Predicting shot making in basketball learnt from adversarial multiagent trajectories. arXiv preprint arXiv:1609.04849 (2016)
19. Hochreiter, S., Schmidhuber, J.: Long short-term memory. Neural Comput. **9**(8), 1735–1780 (1997)
20. Hsieh, H.Y., Chen, C.Y., Wang, Y.S., Chuang, J.H.: Basketballgan: Generating basketball play simulation through sketching. arXiv preprint arXiv:1909.07088 (2019)
21. Kaelbling, L.P., Littman, M.L., Moore, A.W.: Reinforcement learning: a survey. J. Artif. Intell. Res. **4**, 237–285 (1996)
22. Le, H.M., Carr, P., Yue, Y., Lucey, P.: Data-driven ghosting using deep imitation learning. arXiv preprint arXiv:1609.04849 (2017)
23. Le, H.M., Yue, Y., Carr, P., Lucey, P.: Coordinated multi-agent imitation learning. In: Proceedings of the 34th International Conference on Machine Learning, pp. 1995–2003 (2017)
24. Lindström, P.: Deep imitation learning on spatio-temporal data with multiple adversarial agents applied on soccer (2019)
25. Lino, D.: NBA-player-movements (2020). https://github.com/linouk23/NBA-Player-Movements/tree/master/data/2016.NBA.Raw.SportVU.Game.Logs. Accessed 12 Aug 2020
26. Parekh, M.: A brief guide to convolutional neural network (CNN) (July 2019). https://medium.com/nybles/a-brief-guide-to-convolutional-neural-network-cnn-642f47e88ed4. Accessed 18 Apr 2020
27. Perform, S.: AI-powered basketball player tracking (2020). https://www.statsperform.com/team-performance/basketball/. Accessed 15 May 2020
28. Perform, S.: Better sports predictions and insights (2020). https://www.statsperform.com/. Accessed 25 Sept 2020
29. Ratner, A., Bach, S.H., Ehrenberg, H., Fries, J., Wu, S., Ré, C.: Snorkel: rapid training data creation with weak supervision. Proc. VLDB Endow. **11**(3), 269–282 (2017)
30. Rematas, K., Kemelmacher-Shlizerman, I., Curless, B., Seitz, S.: Soccer on your tabletop. In: Proceedings of the IEEE Conference on Computer Vision and Pattern Recognition, pp. 4738–4747 (2018)
31. Shah, R., Romijnders, R.: Applying deep learning to basketball trajectories. arXiv preprint arXiv:1608.03793 (2016)

32. Sicilia, A., Pelechrinis, K., Goldsberry, K.: Deephoops: Evaluating micro-actions in basketball using deep feature representations of spatio-temporal data. In: Proceedings of ACM International Conference on Knowledge Discovery & Data Mining, KDD 2019, Anchorage, AK, USA, 4–8 Aug 2019, pp. 2096–2104 (2019)
33. Signality: The API for sports (2020). https://www.signality.com/. Accessed 26 Sept 2020
34. Theagarajan, R., Pala, F., Zhang, X., Bhanu, B.: Soccer: Who has the ball? generating visual analytics and player statistics. In: Proceedings of the IEEE Conference on Computer Vision and Pattern Recognition Workshops, pp. 1749–1757 (2018)
35. UCF: Sports web site (2008). http://crcv.ucf.edu/data/UCF_Sports_Action.php. Accessed 8 Jul 2020
36. Ullah, A., Ahmad, J., Muhammad, K., Sajjad, M., Baik, S.W.: Action recognition in video sequences using deep bi-directional LSTM with CNN features. IEEE Access **6**, 1155–1166 (2017)
37. Varma, P., Ré, C.: Snuba: automating weak supervision to label training data. Proc. VLDB Endow. **12**(3), 223–236 (2018)
38. Wang, J., Fox, I., Skaza, J., Linck, N., Singh, S., Wiens, J.: The advantage of doubling: a deep reinforcement learning approach to studying the double team in the NBA. arXiv preprint arXiv:1803.02940 (2018)
39. seal ward: NBA movement (June 2020). https://github.com/sealneaward/nba-movement-data. Accessed 19 Sept 2020
40. Zhan, E., Zheng, S., Yue, Y., Sha, L., Lucey, P.: Generating multi-agent trajectories using programmatic weak supervision. arXiv preprint arXiv:1803.07612 (2018)
41. Zhao, Y., Yang, R., Chevalier, G., Shah, R.C., Romijnders, R.: Applying deep bidirectional LSTM and mixture density network for basketball trajectory prediction. Optik **158**, 266–272 (2018)

Describing Brachiopods Shape Based on Fusion of Curvature and Fourier Descriptors

Youssef Ait Khouya[1(✉)], Mohammed Ait Oussous[2], and Faouzi Ghorbel[3]

[1] Software Engineering and Information Systems Engineering Team, Faculty of Sciences and Techniques Errachidia, Moulay Ismail University, Meknes, Morocco

[2] Qualifying High School Ibn Albana, Ministry of National Education, Rabat, Morocco

[3] GRIFT Research Group, CRISTAL Laboratory, National School of Computer Sciences, University of Manouba, Manouba, Tunisia

faouzi.ghorbel@ensi.rnu.tn

Abstract. The fossils classification has a great importance in palaeontological studies. They make it possible to understand the biodiversity in its morphological dimension and they show the morphological transformations suffered during the biological evolution. In this paper, we propose a descriptor for fossils classification and particularly Brachiopods. The proposed descriptor is a fusion between the curvature and Fourier descriptors. The curvature properties provide an apparently powerful cue to the underlying structure of the curve and captures completely the structure of planar curve. In addition, it is stable and complete. Fourier descriptors are powerful features for the recognition of two-dimensional connected shapes and is supported by the well-developed and well-understood Fourier theory. We used the Fourier descriptors proposed by Ghorbel, this set of invariants is also stable and complete. Experimental results are performed to illustrate the performance of the proposed descriptor.

Keywords: Fossils classification · Complete and stable descriptors · Curvature · Fourier descriptors · CSS

1 Introduction

Shape description is one of the essential task of computer vision and pattern recognition, because the final retrieval or classifications are very dependant of these descriptors. The choice of a description method will depend on the desired application. A good number of shape descriptors, which are prevalent in the literature are largely categorized into two groups: contour based shape descriptors and region-based shape descriptors. Contour-based shape descriptors use only boundary information, they cannot depict shape interior content. Of the contour-based descriptors we cite Fourier descriptors [17,19] which have been

A. Abraham et al. (Eds.): SoCPaR 2020, AISC 1383, pp. 444–455, 2021.
https://doi.org/10.1007/978-3-030-73689-7_43

largely used. In [5] the authors propose a method to define Fourier descriptors for broken shapes, meaning shapes that can have more than one contour. Curvature approaches [15], have also been used. In [15] a shape is described in a scale space by the maximum of the curvature. In [14] the authors presented a pattern description approach based on the multi-scale analysis of the contour of planar objects. Wavelet descriptors can be used to describe a given object shape by wavelet descriptors (WD) [20].

In region based techniques, all pixels surrounded by the shape boundary are taken into consideration to yield the shape descriptor. Frequently referred to methods are based on moment theory to describe shape [1]. These include geometric moments, Legendre moments, Zernike moments and pseudo Zernike moments.

The object descriptions could be also divided into two basic groups: local and global. The first group represents the local feature of the objects. The contour or region of the shape is broken down into sub-sections or sub-regions to generate feature to represent the shape. The second group contains characteristics representing the general qualities of the visible surface.

In this paper, we propose a contour based descriptor, it's a combination between a local and global descriptor. We used the curvature as local descriptor. Local features can consist of a specific shape of a small section of the boundary, and fully capture the structure of a plane curve. Also, it is stable and complete.

We choose the Fourier descriptors, which are considered to be auspicious as global descriptor because they are based on a solid theoretical foundation. The advantages of FD over many other shape descriptors are (i) simple to compute; (ii) each descriptor has a specific physical significance; (iii) simple to do normalization, making shape matching a simple task; (iv) geometrical invariance properties. We used the Fourier descriptors proposed by Ghorbel in [11,12], this set of invariants is stable and complete. Stability criterion means that when two invariant representations have a small difference, the objects they represent should also have a small shape difference. Completeness of the description guaranteed that if two shapes have the same invariant representation then the shapes they represent should also be similar.

For edge detection we have used the Chan and Vese model implemented by the Level set method [2].

We applied the proposed descriptor for classification of fossil species and especially Brachiopods. There are many reasons for fossils classification (i)they are an aid to memory because it is impossible to remember the characteristics of a large number of different things unless we can group them into categories, whose members share many characteristics; (ii) greatly improve our predictive powers;(iii)classification systems improve our ability to explain relationships among things. For biologists, this is especially important when we attempt to reconstruct the evolutionary pathways that have produced the diversity of organisms living today. Several authors tried to classify fossil species [6,7]. In [9] we presented a descriptor for Brachiopods classification it is a fusion between local and global descriptors, and we cite [8], the authors used the Outline Shape Analysis technique for describing Ammonite shape. It aims at approaching the shape by a trigonometric function defined by the sum of sine and cosine terms. This latter is decomposed in either a series of harmonic amplitudes and phase angles or a series of Fourier coefficients, serving as variables for quantitative analysis. This technique was applied to other invertebrate groups such as Trilobites [10], bivalves and Ostracodes [3].

The organization of the paper is as follows. In Sect. 2, we present the Region-based active Contour. Also, we formulate the Chan and Vese model in terms of level set functions and compute the associated Euler-Lagrange equation. The smoothing of the contours and parametrization are recalled in Sect. 3. The Curvature Scale Space and the proposed descriptor are described in Sect. 4 and 5. Finally, experimental results are given in Sect. 6.

2 Overview of Region-Based Active Contour

The region-based active contours model is based on the minimization of a well defined energy to segment the image [13]. Let I an image and Ω his domain. Chan and Vese in [2] proposed to minimize the following energy:

$$\begin{aligned} E(c_1, c_2, \phi) = \mu \int_{\Omega} &\mid \nabla\ H(\phi(x,y)) \mid dxdy \\ + \nu \int_{\Omega} &H(\phi(x,y))dxdy \\ + \lambda_1 \int_{\Omega} &\mid I(x,y) - c_1 \mid^2 H(\phi(x,y))dxdy \\ + \lambda_2 \int_{\Omega} &\mid I(x,y) - c_2 \mid^2 (1 - H(\phi(x,y)))dxdy \end{aligned} \tag{1}$$

where $\mu \geq 0$, $\nu \geq 0$, $\lambda_1 > 0$, $\lambda_2 > 0$ are fixed parameters and H is the Heaviside function. and ϕ is the level set function associated to the contour evolving.The constants c_1 and c_2 are the averages of I in $\phi > 0$ and $\phi < 0$ respectively. The discretized evolution equation is:

$$\begin{aligned} \frac{\phi^{n+1}_{(i,j)} - \phi^n_{(i,j)}}{\Delta t} = \delta_\epsilon(\phi^n_{(i,j)})(\mu div(\frac{\nabla\phi^n_{(i,j)}}{\mid \nabla\phi^n_{(i,j)} \mid}) - \nu \\ -\lambda_1(I(i,j) - c_1(\phi^n_{(i,j)}))^2 + \lambda_2(I(i,j) - c_2(\phi^n_{(i,j)}))^2) \end{aligned} \tag{2}$$

where $\delta_\epsilon(z)$ is a regular form of $\delta(z)$.

3 Smoothing Contours and Parametrization

3.1 Smoothing Contours

The principal cause for smoothing is to remove the noise, which can be product in the contours during edge detection step. The basic idea of smoothing is very simple. We passe through the point by point in the contour. For each data point we yield a new value that is depending of the original value at that point and surrounding data point. There are various methods for smoothing, but here we smooth using a Gaussian kernel. The kernel of smoothing determines the function's form that is used to take the average of the neighboring points. The Gaussian kernel is defined in 1-D as:

$$G_\sigma(t) = \frac{1}{\sigma\sqrt{2\pi}} e^{-\frac{t^2}{2\sigma^2}} \tag{3}$$

with σ determines the width of the Gaussian kernel which controls the degree of smoothing and has to be chosen suitably. A large value of σ will remove all small details of the contour curvature, while a small value will permit false concavities and convexities to remain in the contour.

3.2 Parametrization

The contours Parametrization step provide all the computation results. For a given contour there are many types of parameter. A Parametrization function will be chosen which are suitable for the intended invariance. Here, we are interested in the parametrization invariant under the similarity transformation. To obtain a parametrization invariant under the similarity transformation we used the arc length parametrization, defined by the Euclidean norm which possesses an intrinsic link with the notion of similarity.

Suppose we have the coordinates of the contour $C(t) = (x(t), y(t))$, where t is a real value called the parameter of the contour. To reparametrize the contour by the arc length we need to calculate:

$$s(l) = \frac{1}{L} \int_0^l \mid C(t)' \mid dt \tag{4}$$

where L denotes the curve length.

On the other hand, to obtain a parametrization invariant under affine transformation it's necessary to use affine arc parametrization.

4 Curvature Scale Space

The curvature scale space (CSS) approach was introduced by Mokhtarian et al. in [16] as a shape representation for planar curves. This method is based on finding points of inflection (i.e., curvature zero-crossings) on the curve at varying levels of detail. The curvature k of a planar curve, at a point on the curve, is defined as the instantaneous rate of change of the slope of the tangent at that point with respect to arc length, and it can be expressed as follows:

$$k(s,\sigma) = \frac{X_s(s,\sigma)Y_{ss}(s,\sigma) - X_{ss}(s,\sigma)Y_s(s,\sigma)}{(X_s(s,\sigma)^2 + Y_s(s,\sigma)^2)^{3/2}} \tag{5}$$

where $X_s(s,\sigma)$ and $X_{ss}(s,\sigma)$ correspond to the first and second derivatives of $x(s)$, $Y_s(s,\sigma)$ and $Y_{ss}(s,\sigma)$ correspond to the first and second derivatives of $y(s)$. If we determine the locations of curvature zero crossings of every C_σ during evolution we can display the resulting points in (s,σ) plane where s is the normalized arc length and σ is the width of the Gaussian kernel. The result of this process can represent the CSS image of the curve. CSS is used to describe a contour. It is also an invariant with respect to the similarities of group. CSS is stable with respect to errors due to noise. In addition, it is complete [16]. Mokhtarian et al. used the maxima of CSS to index shapes. This choice is justified by the fact that the maxima of the CSS are the most significant local point on the body of the CSS image. The locations of maxima are not readily available and must be extracted from the CSS image using Algorithm 1 as shown in Fig. 1.

Algorithm 1. Extraction of the CSS maxima

Require: CSS image
  stp is a constant used to increment the variable σ during the construction of the CSS Image
  for $i = \sigma_{max}$ to 1 **do**
    if a zero crossing is found at point (i, s_j) **then**
      check the neighboring points $(i + stp; s_{j-1})$; $(i + stp; s_j)$ et $(i + stp; s_{j+1})$
      if the three neighbors above are not zero-crossings **then**
        (i, s_j) is a candidate for the maxima of the CSS
        find all the other candidates to the maxima as $\sigma = i$
      end if
      For each candidate (i, s_j), check the neighboring candidate
      if (i, s_k) is a neighbor candidate **then**
        if $|j - k| > 5$ **then**
          (i, s_j) is a maxima of CSS
        else
          $(i, (j + k)/2)$ is a maxima of CSS
        end if
      end if
    end if
    $i = i - stp$
  end for

5 The Proposed Descriptor

5.1 Overview of the Fourier Descriptors Methods

Let $x(t)$ and $y(t)$ be the coordinates of the contour, Since boundary is a closed curve and when the two-dimensional plane is considered as a complex plane, the cartesian coordinates of the contour are represented in the complex plane by

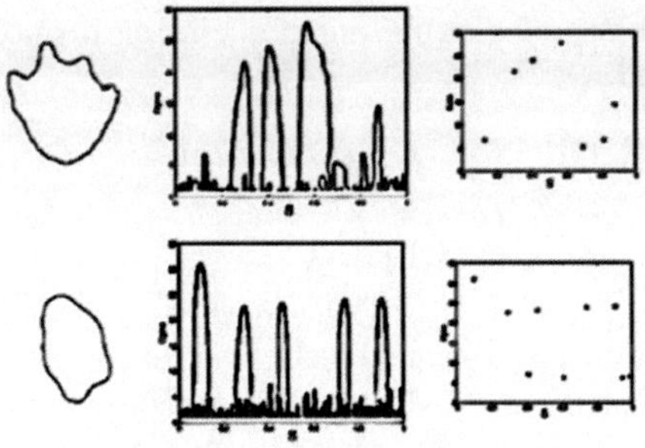

Fig. 1. CSS image and its maxima, left: Contour, middle: normalised CSS image, right: normalised maxima of CSS images.

$z(t) = x(t) + j.y(t)$. The discrete Fourier coefficients of $z(t)$ is defined as follow:

$$c_k = \sum_{k=0}^{N-1} z(t)exp(-j2\pi kt/N) \tag{6}$$

with $t \in [0, N-1]$, and N is a number of points in the contour.

Applying the inverse Fourier transform, we can recreate a contour from its Fourier descriptors. Initially, the first set of invariants was constructed by taking the modulus of Fourier descriptors.

$$\forall k \in [0, N-1], \quad I_k = |c_k|, \tag{7}$$

In [21] the authors used this method in their comparative study. The first coefficient is the only one depends on position of the shape. To obtain translation invariance the coefficient c_0 can be discarded. Scale invariance is achieved by dividing all the Fourier coefficients by the absolute value of the second coefficient. To obtain a rotation and shift invariant descriptor, the phase can completely ignored and the only use the absolute value of the Fourier coefficients.

The acquired Fourier descriptors is used to discriminate between simple-shaped objects [18]. But, this set of invariants is not complete in the sense defined by Crimmins in [4], A set of descriptors is said to be complete if the following property is verified:

two objects have the same shape if and only if they have the same set of invariants.

Completness allows to retrieve shapes from their invariants, up to an Euclidean transformation. In fact, the set of descriptors defined in Eq. (7) is not complete since we can find objects with different shapes but with the same magnitude for their Fourier coefficients.

To resolve the problem of completeness, Crimins proposed in [4] a complete set of invariant, defined as:

$$\begin{aligned} I_{k_0} &= |c_{k_0}|, \quad for \;\; k_0 \;\; such \;\; that \;\; c_{k_0} \neq 0 \\ I_{k_1} &= |c_{k_1}|, \quad for \;\; k_1 \neq k_0 \;\; such \;\; that \;\; c_{k_1} \neq 0 \\ I_k &= c_k^{(k_0-k_1)} \; c_{k_0}^{(k_1-k)} \; c_{k_1}^{(k-k_0)}, \quad \forall k \neq k_0, k_1 \end{aligned} \tag{8}$$

However, this set is not stable, because a slight modification of the invariants can induce a remarkable distortion of shape. To overcome this problem, Ghorbel in [11] presented a complete and stable set of invariant Fourier descriptors, defined as (the reader can consult chapter II page 53–71 in [12] for more details):

$$\begin{aligned} I_{k_0} &= |c_{k_0}|, \quad for \;\; k_0 \;\; such \;\; that \;\; c_{k_0} \neq 0 \\ I_{k_1} &= |c_{k_1}|, \quad for \;\; k_1 \neq k_0 \;\; such \;\; that \;\; c_{k_1} \neq 0 \\ I_k &= \frac{c_k^{(k_0-k_1)} \; c_{k_0}^{(k_1-k)} \; c_{k_1}^{(k-k_0)}}{|c_{k_1}|^{(k-k_0-p)} \; |c_{k_0}|^{(k_1-k-q)}}, \quad \forall k \neq k_0, k_1 \end{aligned} \tag{9}$$

with $p, q > 0$.

5.2 The Proposed Feature Descriptor

The proposed feature descriptors is based on the fusion of local and global feature of the contours of objects. Local features are usually on the boundary of an object or represent a distinguishable small area of a region. Global features can be obtained for either a region by considering all points within a region, or only for those points on the boundary of a region. In each case, the intent is to find descriptors that are obtained by considering all points, their locations, intensity characteristics, and spatial relations.

We used the curvature as a local feature thanks to the arc length parameterization. It has been often noted that curvature properties provide a powerful indicator to the underlying structure of the curve and captures completely the structure of a plane curve. The curvature function is an invariant property of the curve itself. Therefore, it's insensitive to changes in the coordinate system like rotation. When the curve changes size the local relationships between the curvature of different parts of the curve are conserved. This offers that the relative topological structure of the curvature function can achieve a scale invariant. In addition, it is stable and complete. Stable, meaning a slight modification of the invariants can induce a slight distortion of shape. Complete that is two objects have the same shape if and only if they have the same set of invariants. Completeness allows to retrieve shapes from their invariants.

The global feature is based on Fourier descriptors, we used the Fourier descriptor proposed by Ghorbel Eq. (9). This set of invariants is also stable and complete. The proposed feature descriptor has the following form:

$$X = (k(s_1), k(s_2), ..., k(s_N), I_0, I_1, ..., I_n) \tag{10}$$

where $n \in [0, N-1]$ and N is a number of points in the contour.

6 Experimental Results

Following the preprocessing stage, the contour is detected using region based active contours method, specifically chan and vese model implemented with the Level Set method, the parameters are fixed at ($\mu = 1, \nu = 0, \lambda_1 = 1$ *and* $\lambda_2 = 1$), this model provides satisfactory results for the contour detection in an image (see Sect. 2). Every object is represented by the x and y coordinates of its boundary points. The number of these points varies from 600 to 1200 for images in our databases. Due to the boundardiscretey representation and quantization errors, false local concavities and convexities along a contour are formed. This noisy nature of binary contours must be taken into account to obtain reliable calculation of contour curvature. Therefore, a Gaussian filter with ($\sigma = 4$) is used to smooth the contour points to reduce the noise effect. After, we used the arc length for the parametrization of contour.

When the pretreatment step is completed, The curvature for each point on the Gaussian smoothed curve is computed and normalized between -1 and 1. Then, we calculate the set of invariants Fourier descriptors proposed by Ghorbel Eq. (9) also normalized between -1 and 1. This set of invariants is stable and complete. We use the FFT algorithm for reducing the computing time. For efficient shape description, Zhang and Lu in [21] have found that 10 Fourier descriptors features are sufficient to represent shape. In our study, the number of invariants Ghorbel descriptors is calculated in advance based on experimentation reconstruction performed on the entire database of Brachiopods. We proceed by an inversion invariants over the entire database of Brachiopods using Eq. (11) and compare the original contour with the contour reconstructed in the sense of the norm 2 with truncation (see [12] for more details).

$$c_k = I_k^{\frac{1}{k_0-k_1}} I_{k_0}^{\frac{-p}{(k_0-k_1)(k_0-k_1+p)}} I_{k_1}^{\frac{-q}{(k_0-k_1)(k_1-k_0+q)}} \\ exp(\frac{2i\pi}{k_0-k_1}(k(\theta_0-\theta_1)-k_1\theta_0+k_0\theta_1)) \tag{11}$$

$$C_r(t) = FFT^{-1}(c_k)$$

where k_0 and k_1 are constants fixed when computing Ghorbel descriptors I_k, c_k is the Fourier coefficients after inversion of Ghorbel invariants, θ_0 and θ_1 are the arguments of c_{k_0} and c_{k_1} respectively. FFT^{-1} is the inverse discrete Fourier transform, $C_r(t)$ is the reconstructed Contour. The number p of invariants Ghorbel descriptors retained will be one that allows a reconstruction with an error less than 5%. After several tests on our Brachiopods database, we found that p = 16 is sufficient to represent shape with an error less than 5% as show in Fig. 2.

Once p is found, we sample equidistantly the p arc length of the curvature function. We take the values of normalized curvature at these p points of the curve. This reduces the number of local descriptors to $p < N$. We choose the point that has the highest value of the curvature as a starting point.

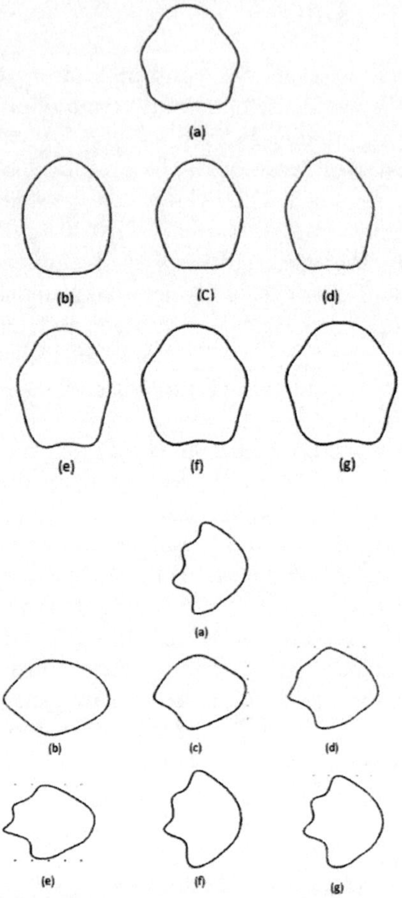

Fig. 2. (a) original Contour; (b), (c), (d), (e), (f), and (g) are reconstructed contour with 8, 10, 12, 14, 16, and 18 of Ghorbel Coefficients respectively.

For evaluation of the proposed descriptor, we realized a comparative study with Curvature, Ghorbel and CSS descriptors for classification of Brachiopods. We dispose a database of Brachiopods that contains 800 binary images, 20 are original others are distorted copies of the image database, which they are subject to rotations and scaling, as shown in Fig. 3. The search process in the system using CSS as a descriptor begins with a pre-selection phase during which we rejected all shape not having the same number of maximum points of the CSS that the request shape. The normalization relative to the arc length ensures invariance in translation and scale. The rotational invariance is also provided by a lateral shift of the maximum of CSS. For two shapes represented by their descriptors, the similarity between the two shapes is measured by the city-block distance between the two feature vectors of the shapes. Therefore, the online

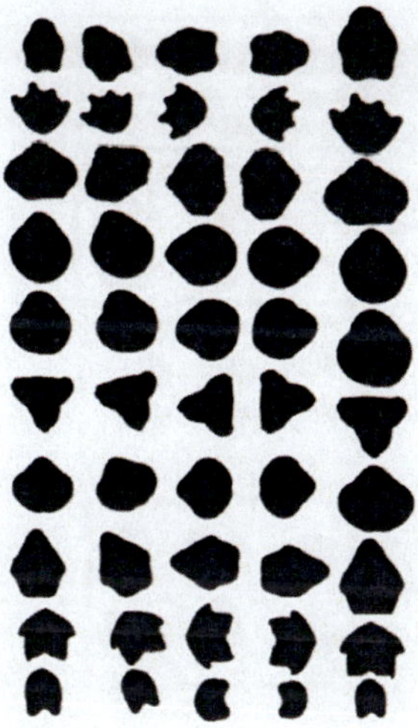

Fig. 3. Example of Brachiopods database.

matching is efficient and simple. We used the recognition rate to compare the discrimination of the two descriptors.

We applied request the seven groups each group contains 100 images of Brachiopods. The first two groups contain images of the database. Others contain images that do not belong to the database, they are the images of database that have met a change as rotation and scale.

The results of this comparison for the classification of Brachiopods show that the system using the proposed descriptor gives the greatest recognition rate comparable to rates obtained by the system using CSS descriptor Fig. 4. In addition, the proposed descriptor is fast to compute compared to CSS descriptor. The two methods are executed on a processor machine Intel Core 2 Duo 2 GHz with 2 GB of RAM. The proposed descriptor requires a computing time equal to 0.078 s, opposed to the CSS method that requires 3.844 s (Fig. 5).

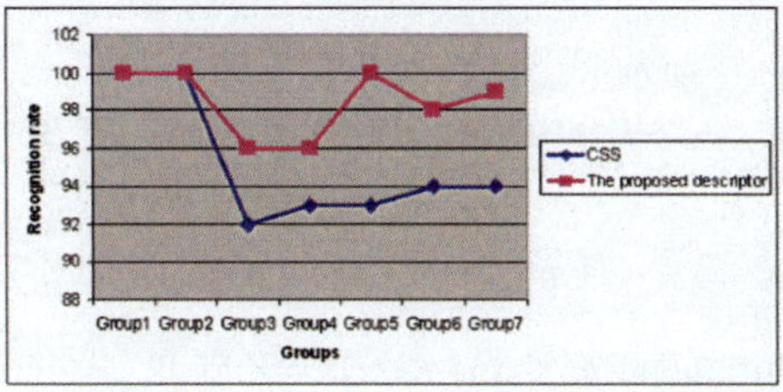

Fig. 4. Recognition rate for Groups.

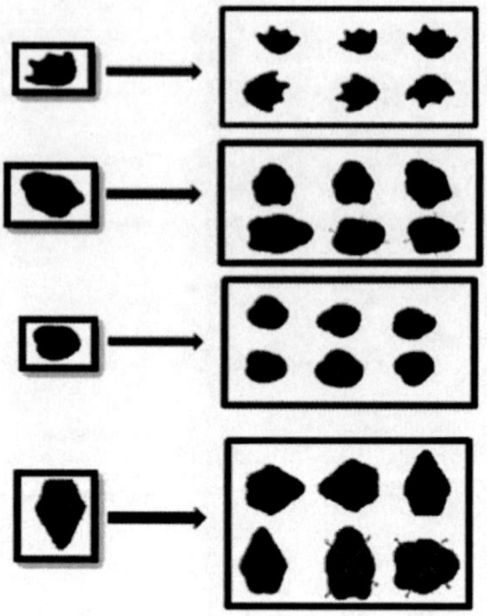

Fig. 5. Retrieval result of arbitrary queries on our database using the proposed descriptor.

7 Conclusions

This paper present a contour based descriptor for Brachiopod classification. The proposed descriptor is based on the fusion of curvature and Fourier descriptors. The curvature presents the local information of the curve and provides a powerful cue to underlying the structure of the shape. In addition, it is stable and complete. To enhance the ability of the proposed descriptor to capture global shape information, We used the stable and complete set of invariants Fourier descriptors proposed by Ghorbel as a global information of the contour. The performance of the proposed descriptor is then compared with that of Curvature Scale Space. The results of experimental show that the proposed descriptor is a reliable choice for Fossils classification. The combination may increase the performance of pattern recognition and the image retrieval systems.

References

1. Bailey, R., Srinath, M.: Orthogonal moment features for use with parametric and non-parametric classifiers. IEEE Trans. Pattern Anal. Mach. Intell. **18**(4), 389–399 (1996)
2. Chan, T., Vese, L.: Active contours without edges. IEEE Trans. Image Process. **10**(2), 266–277 (2001)
3. Crampton, J.: Elliptic Fourier shape analysis of fossil bivalves: some practical considerations. Lethaia **28**(2), 179–186 (1995)
4. Crimmins, T.R.: A complete set of Fourier descriptors for two dimensional shapes. IEEE Trans. Syst. Man Cybern. **12**, 848–855 (1982)
5. Dalitz, C., Brandt, C., Goebbels, S., Kolanus, D.: Fourier descriptors for broken shapes. EURASIP J. Adv. Sig. Process. **2013**, 161 (2013)

6. Ait Khouya, Y., Ghorbel, F.: Brachiopods classification based on fusion of contour and region based descriptors. Int. J. Adv. Intell. Paradigms (IJAIP) **14**(1/2), 140–156 (2019)
7. Ait Khouya, Y., Ghorbel, F.: Description shape based on B-spline approximation and contour based shape descriptors: application to Brachiopods classification. J. Netw. Innov. Comput. **7**, 025–031 (2019)
8. El Hariri, K., Bachnou, A.: Describing ammonite shape using Fourier analysis. J. Afr. Earth Sci. **39**, 347–352 (2004)
9. Ait khouya, Y., Ghorbel, F., Alaa, N.: Brachiopods classification based on fusion of global and local complete and stable descriptors. In:12th ACS/IEEE International Conference on Computer Systems and Applications, Marrakech, Morocco, 17–20 November 2015 (2015)
10. Foote, M.: Perimeter-based Fourier analysis: a new morphometric method applied to the trolobite cranidium. J. Paleontol. **63**(6), 880–885 (1989)
11. Ghorbel, F.: Stability of invariant Fourier descriptors and its inference in the shape classification. In: 11th International Conference in Pattern Recognition, The Hague, The Netherlands, 30 August–3 September 1992 (1992)
12. Ghorbel, F.: Invariants de formes et de mouvement Onze cas du 1D au 4D et de l'euclidien aux projectifs, Arts Pi, Tunis (2013)
13. Jehan-Besson, S., Barlaud, M., Aubert, G.: DREAM2S: deformable regions driven by an Eulerian accurate minimization method for image and video segmentation. Int. J. Comput. Vision **53**(1), 45–70 (2003)
14. Kpalma, K., Ronsin, J.: Multiscale contour description for pattern recognition. Pattern Recogn. Lett. **27**(13), 1545–1559 (2006)
15. Mokhtarian, F., Abbasi, S.: Shape similarity retrieval under affine transforms. Pattern Recogn. **10**(2), 31–41 (2002)
16. Mokhtarian, F., Mackworth, A.K.: A theory of multiscale, curvature based shape representation for planar curves. IEEE PAMI **14**, 789–805 (1992)
17. Persoon, E., Fu, K.: Shape discrimination using Fourier descriptors. IEEE Trans. Syst. Man Cybern. **7**(3), 170–179 (1977)
18. Persoon, E., Fu, K.: Shape discrimination using Fourier descriptors. IEEE Trans. Pattern Anal. Mach. Intell. **8**(3), 388–397 (1986)
19. Rui, Y., She, A.: A modified Fourier descriptor for shape matching in MARS. Image Databases Multimedia Search **8**, 165–180 (1998)
20. Tieng, Q.M., Boles, W.W.: Recognition of 2D object contours using the wavelet transform zero crossing representation. IEEE Trans. PAMI **19**(8), 910–916 (1997)
21. Zhang, D.S., Lu, G.: A comparative study of Fourier descriptors for shape representation and retrieval. In: Proceedings of the Fifth Asian Conference on Computer Vision (ACCV02), Melbourne, Australia, 22–25 January 2002, pp. 646–651 (2002)

Early Diagnosis of Parkinson's Disease Using LSTM: A Deep Learning Approach

Amit Gawade[1(✉)], Rohit Pandharkar[1], Subodh Deolekar[2], and Uday Salunkhe[2]

[1] REDx Innovation Lab, Prin. L. N. Welingkar Institute of Management Development and Research, Mumbai, India
amit.gawade@welingkarmail.org

[2] Prin. L. N. Welingkar Institute of Management Development and Research, Mumbai, India

Abstract. Parkinson's disease (PD) is one of the most rapidly growing neurodegenerative diseases in the world. Due to motor symptoms, it affects the normal life of a person. There is a severe need to identify PD in its early stage to avoid it getting worse and to control its symptoms easily. The advancements in Artificial Intelligence (AI) and the Internet of Things (IoT) open up new avenues for the analysis of various data points such as the gait of a person for early-stage detection. In this paper, we propose a methodology based on the use of Long Short-Term Memory (LSTM) architecture for PD diagnosis. We have used time series analysis to find the gait patterns and deep learning techniques to extract the features and to build a classifier model. The proposed model is predicting the PD disease with 85% testing accuracy and with an F1 score of 0.90. The validation is performed using Cohen's Kappa statistical method and obtained a score of 0.631.

Keywords: Parkinson's disease · Deep learning · LSTM · RNN · Gait analysis · IoT

1 Introduction

Common movement disorders like Parkinson's and Huntington's disease are caused due to illnesses that involve the death of certain parts of the brain. Almost 30% of people across the globe are facing this neurodegenerative incurable disease.

The very first information on Parkinson's disease (PD) was given by James Parkinson in the early 19th century [1]. It is one of the primary public health diseases in the globe that is severely rising day by day and has its effect on many countries. Symptoms are categorized into Motor symptoms and non-Motor Symptoms. In Motor symptoms, voluntary actions are involved. It indicates movement-related disorders such as tremor, rigidity, freezing, or any voluntary muscle movement. Non-motor symptoms include disorders of mood and affect apathy, cognitive dysfunction like memory loss as well as complex behavioral disorders [2].

The use of Artificial Intelligence (AI) to find symptoms of PD in its early stages is of current interest for many researchers [3]. Generally, medical doctors screen various questionnaires to detect PD symptoms and check the severity of the same [4]. It

A. Abraham et al. (Eds.): SoCPaR 2020, AISC 1383, pp. 456–465, 2021.
https://doi.org/10.1007/978-3-030-73689-7_44

is observed that there were major differences between responses from individuals in two groups, normal and affected. This makes the overall process time-consuming with discrepancies of questionnaires.

The main goal of this paper is to explore the use of different machine learning and deep learning algorithms that process raw data obtained from various sensors placed in the shoes' sole and to predict parkinsonism by analyzing the gait data. We propose a methodology that uses the Recurrent Neural Network (RNN) Long Short-Term Memory (LSTM), a deep learning algorithm for the early detection of PD.

2 Related Work

Researchers have been using various machine learning algorithms such as regression, clustering and classification to predict PD. Most of them have used Non-motor symptoms such as Rapid Eye Movement (REM) and Sleep Behavior Disorder (RBD). Additionally, olfactory loss and machine learning models have been developed, which can help in predicting the disease and can play a vital role in early prediction. In [5], automated diagnostic models using Multilayer Perceptron, BayesNet, Random Forest, and Boosted Logistic Regression are proposed to diagnose PD. It has been observed that Boosted Logistic Regression achieved an accuracy of 90.159% and the area under the ROC curve was 86.9%. The authors have used Parkinson's Progression Markers Initiative (PPMI) database for this experiment.

Based on the handwriting markers, PD patients can be differentiated from healthy ones. In this experiment, the patient has been asked to draw a straight line, as doing this involves certain muscular movements that were tracked and used for prediction. Naive Bayes algorithm has been used to build a classification model that gives 90% accurate results [6].

Using UCI's Parkinson's Telemonitoring Voice dataset of patients and Deep Neural Networks, the author [7] proposed a methodology for the prediction of PD severity. The accuracy values obtained by the mentioned method are better as compared to the accuracy obtained in previous research work, 81% accuracy has been achieved using this method. The author also tried regression and decision tree, but the accuracy is less in comparison to neural networks.

Attention-based bi-directional RNN is being used to analyze the gait data. Flagg et al. [8] have used the GaitPDB dataset that has an equal number of control subjects and PD was selected at random to provide a balance of control and abnormal subjects as training input. The First 80% of data is selected as a training set and the remaining 20% is used as a validation set. For the testing set, 16 patients with PD and control patients were selected at random from the files that are not included in the training set. The model achieved 88% training accuracy.

LSTM and one-dimensional Convolutional Neural Networks are being used for clinical PD diagnosing and performed well in terms of accuracy. These algorithms can be used to analyze gait patterns using motion data obtained from low-cost RGB-D cameras. The classifier performed well and had given 83% prediction accuracy with 83.4% recall and 83.5% precision [9].

A deep time series-based approach is used to detect abnormal walking patterns in the gait style of elderly people by analyzing the acceleration values of movements. The results show a training accuracy and testing accuracy of over 90% [10].

The manual feature engineering approach is currently being used with various machine learning approaches to predict PD.

3 Proposed Methodology

In our proposed system, we have used the LSTMnetwork, a special type of RNN for the prediction of PD. LSTM learns from raw data coming from sensors and finds the important features directly. It is responsible for detecting deep features with respect to the input. It also handles multiple inputs parallelly.

3.1 The Architecture of the Proposed System

The proposed methodology for predicting PD using deep learning is shown in Fig. 1. In the first step, the gait data of various subjects have been collected using sensors. There were 8 sensors placed underneath each foot and generated gait data stored in.csv file. The collected data is standardized by removing the mean and scaling to unit variance. By considering the sequences of data, time steps have been generated. In the next step, RNN is designed with an input layer, a hidden layer with 50 LSTM cells, and an output layer. The number of neurons in the input layer is fixed as the number of attributes in the input data. One sample is processed at a time, at time step 1, all 50 cells would get the first-time step of data with all 18 features. The output layer contains two neurons corresponding to the two classes – "normal" and "PD". The standardized data is given into the formed RNN for training and validation.

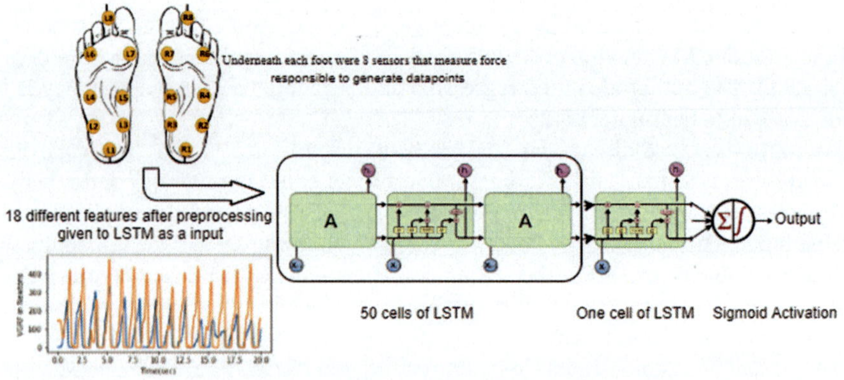

Fig. 1. Architecture of proposed system.

The system shown in Fig. 1 takes 18 different features as input to RNN-LSTM with hidden layers. We applied a multivariate sliding windowing method of time series analysis to find features that were further given as input to LSTM for analysis. RNN-LSTM takes preprocessed data and analyzes different patterns in the same. Combining different results in a single vector using a fully connected layer as the output will tell us the status of subject 0 as normal and 1 as a PD patient.

RNN is a type of neural network in which the output of a layer is fed back to the input layer multiple times in order to learn from the past data.

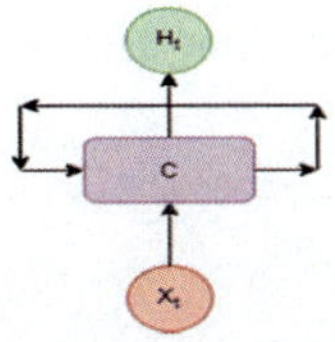

Fig. 2. Recurrent neural network structure

In Fig. 2, C is a cell looking for input Xt and gives the output as Ht. A close loop allows data to be passed from one cell to another. Here, each cell transfers a message or data to a successor. LSTM [11] can learn long-term dependencies in data, which is useful to find patterns.

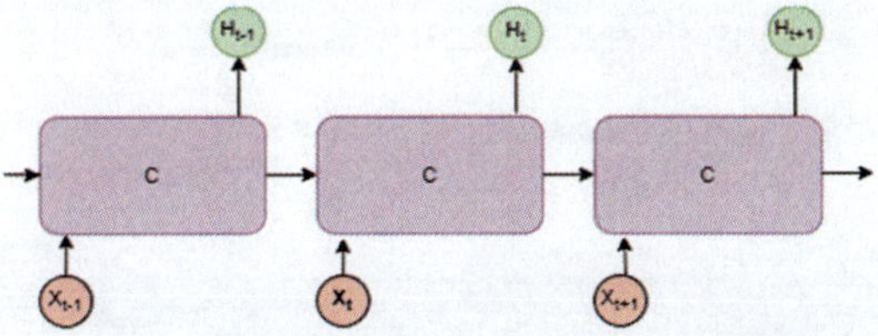

Fig. 3. Long short-term memory networks

In Fig. 3, each arrow carries an entire vector from the input of one cell 'C' to the output of another. Every LSTM cell has its cell state that carries filtered information from one cell to another. LSTM can add or remove data to the cell state using gates. Each gate consists of a sigmoid neural net layer, which gives the output as 0 or 1 based on the information. In the next stage, the new information that has to be added to the cell state is decided. In this step, the tanh layer creates a vector of new values. In the third stage, we combine both stages to update the cell state. Next, we update the old cell state to the new state. In the end, we decide the output based on the combined cell state.

We have also tried machine learning classification algorithms such as Logistic Regression, ANN, Decision Tree and Naive Bayes algorithms. Logistic regression is a classification algorithm based on probability, generally used for binary classification. It is based on sigmoid cost function [12]. ANN is the network of functions (units) connected with weight to analyze the new data and detect the patterns that are used for classification [13]. By learning decision rules inferred from the data features Decision

Tree classify the data in different labels [14]. Naive Bayes algorithm uses Bayes theorem that works on conditional probability. It is a highly extensive algorithm used for classification [15].

4 Experimental Setup

The proposed model is implemented on a system with Intel Core i3 1080Ti GPU at 2.20 GHz and 8 GB RAM. The Python libraries and TensorFlow are used to implement the RNN with LSTM.

4.1 Flowgraph

A workflow of the system is shown in Fig. 4. We have extracted the data from the source [12]. In Pre-processing and data cleaning stage, null values, outliers, and distribution of data are checked. Feature Selection stage comprises the filtration of various features that may include or exclude the features as per their importance decided in exploratory data analysis stage. Various machine learning algorithms such as Logistic Regression, Decision Tree and Naive Bayes have been used along with deep learning algorithms like ANN and LSTM. In the last stage accuracies of these algorithms have been compared.

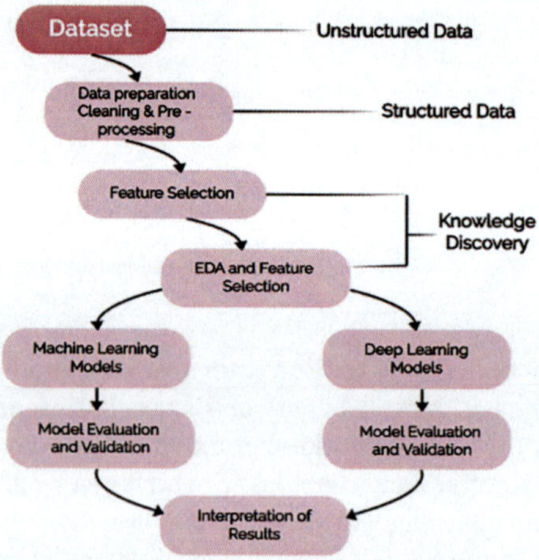

Fig. 4. The flow of the entire experiment.

4.2 Dataset

The dataset contains measures of gait from 93 patients with PD and 73 healthy subjects. The dataset includes the vertical ground reaction force records of subjects as they walked for 2 min on level ground. Under each foot, there were 8 sensors that measure the force in Newtons as a function of time. The output of each of these 16 sensors has been recorded at 100 data points per second and the records also include two signals that have the sum of the 8 sensor outputs for each foot [16]. Dataset contains around 70% PD subjects and 30% normal subjects.

4.3 Time Series Windowing Method

We have selected an interval of 20 s to separate the original data into several consecutive data sets to find patterns that will help our model to understand the data and will predict the result.

Considering only one sensor, here are some plots shown in Fig. 5 that depict the patterns of control and PD subjects. These plots helped us while preparing input to LSTM. Input to every LSTM layer must be in three-dimensions i.e. Samples, Time Steps & Features. Sample is the number of data points; time steps is how many previous sequences and features is the number of columns.

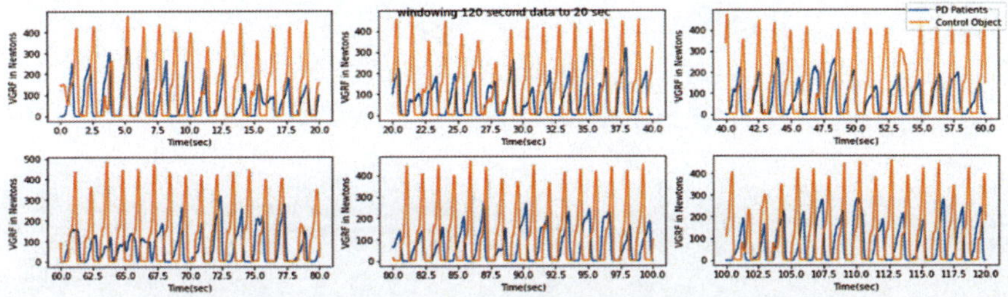

Fig. 5. Left side 7th sensor data divided into different windows

5 Performance Measures

5.1 Accuracy

The classification accuracy obtained is 92% and 85% for the train and test dataset respectively. With precision being 0.854167, Recall: 0.953488 and F1 score: 0.901099. The model accuracy graph has been shown in Fig. 6. The graph represents the model performance during training and testing with respect to the number of epochs.

5.2 Area Under the Curve (AUC) Receiver Operating Characteristics (ROC)

To evaluate the model's classification performance, we have used AUC that represents a measure of separability, and ROC which is a probability curve. It showcases how much a model is capable of distinguishing between classes. We have obtained 80% ROC AUC confidence, which shows that the classifier is capable to predict classes more accurately (Fig. 7).

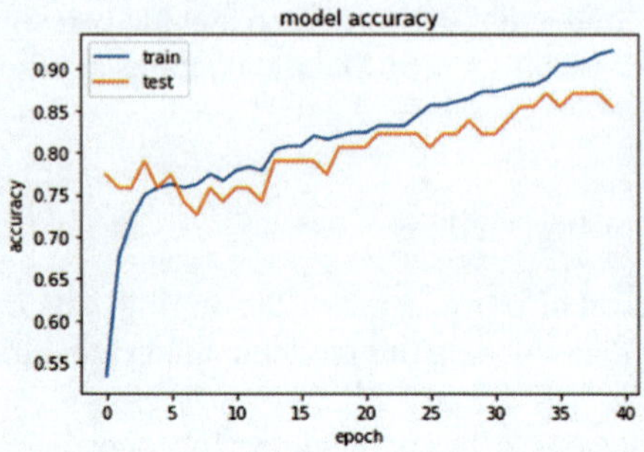

Fig. 6. Performance of model during training.

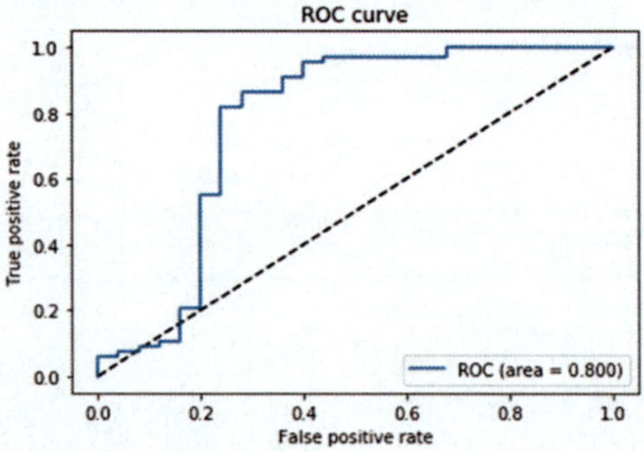

Fig. 7. Receiver Operating Characteristic (ROC) curves of PD patients and for Normal patients obtained when predicting classes using LSTM.

5.3 Cohen's Kappa Statistic

In order to measure inter-rater reliability for quantitative variables, we have used Cohen's Kappa Statistic. It can be defined as,

$$k = (Ao - Ae)/(1 - Ae) = 1 - (1 - Ao)/(1 - Ae) \quad (1)$$

where k is Cohen's Kappa coefficient, Ao is the observed value and Ae is the expected value.

As per Landis and Koch machinate [17], the coefficient k is always less than or equal to 1. If the value of k is less than 0 it shows no use of a classifier, 0 to 0.20 is inconsiderable, 0.21 to 0.40 can be considered. Whereas, 0.41 to 0.60 is modest, 0.61 to 0.80 is substantial and 0.81 to 1 is a perfect classifier.

Since the dataset is not balanced, Cohen's Kappa statistic method can be used to evaluate classifiers that handle imbalanced data. The proposed method has given 0.631440 Cohen's Kappa value. As per the values set by Landis and Koch, the classifier performs well.

5.4 Parkinson's Disease and Control Subject Classification

To measure the effectiveness and performance of trained classifiers, we used a confusion matrix measure. It describes the combination of four different cases i.e. true positive, true negative, false positive and false negative. True positive means PD patients were properly classified as PD, true negative means normal subjects were identified as normal. Furthermore, false positive means number of normal subjects improperly identified as PD patients, false negative means number of PD patients improperly identified as normal. A graphical representation of the confusion matrix is shown in Fig. 8.

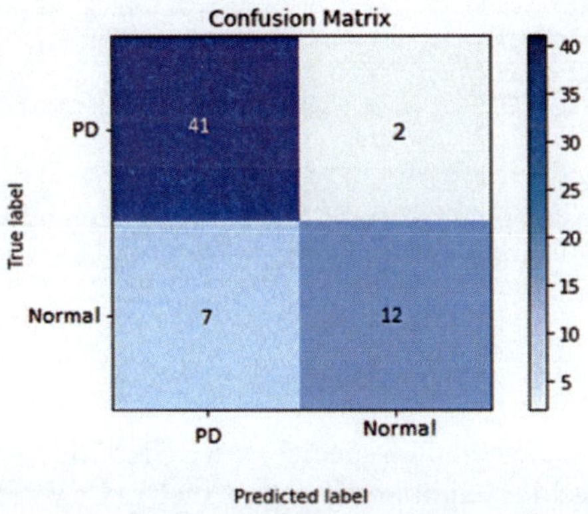

Fig. 8. Confusion matrix.

5.5 A Learning Curve

A learning curve is a plot of model learning performance over time. The number of epochs will be on x-axis and loss will be on the y-axis. Figure 9 shows that the model is trained correctly and periodically reduces loss after every epoch.

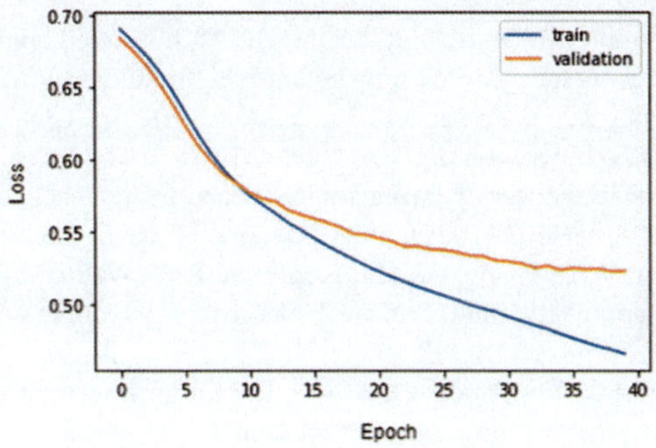

Fig. 9. Learning curve epoch vs loss.

5.6 Performance Measure Comparison

Table 1. Comparativeresult of different machine learning algorithms

Sr. No	Algorithm	Precision	Recall	F1 Score	Accuracy
1	Logistic Regression	0.76	0.69	0.57	0.69
2	Decision Tree	0.70	0.69	0.60	0.80
3	ANN	–	–	–	0.76
4	Naive Bayes	0.58	0.55	0.55	0.68
5	**RNN-LSTM**	**0.85**	**0.95**	**0.90**	**0.85**

We have tried various machine learning algorithms and compared their results shown in Table 1. The proposed method performs better than the other methods.

6 Conclusion

In this paper, we have implemented a special type RNN, LSTM and its performance is compared with other machine learning algorithms such as Logistic Regression, Decision Tree, Artificial Neural Network and Naive Bayes. The proposed LSTM model achieved 92% training accuracy and 85% testing accuracy. To validate the model, we have used various performance measures like confusion matrix, ROC and Cohen's Kappa coefficient. The results indicate that our proposed model performs better and is able to detect Parkinson's disease in an early stage in an efficient way.

References

1. Pearce, J.M.S.: Aspects of the history of Parkinson's disease. J. Neurol. Neurosurg. Psychiatry (1989). https://doi.org/10.1136/jnnp.52.Suppl.6
2. Challa, K.N.R., Pagolu, V.S., Panda, G., Majhi, B.: An improved approach for prediction of Parkinson's disease using machine learning techniques. In: International Conference on Signal Processing, Communication, Power and Embedded System, SCOPES 2016, Proceedings, pp. 1446–1451 (2017). https://doi.org/10.1109/SCOPES.2016.7955679
3. Senturk, Z.K.: Early diagnosis of Parkinson's disease using machine learning algorithms. Med. Hypotheses **138**, 109603 (2020). https://doi.org/10.1016/j.mehy.2020.109603
4. Chan, D.K.Y., Hung, W.T., Wong, A., Hu, E., Beran, R.G.: Validating a screening questionnaire for parkinsonism in Australia. J. Neurol. Neurosurg. Psychiatry **69**(1), 117–120 (2000). https://doi.org/10.1136/jnnp.69.1.117
5. Challa, K.N.R., Pagolu, V.S., Panda, G., Majhi, B.: An improved approach for prediction of Parkinson's disease using machine learning techniques. In: International Conference on Signal Processing, Communication, Power and Embedded System, SCOPES 2016 - Proceedings, pp. 1446–1451 (2017). https://doi.org/10.1109/SCOPES.2016.7955679

6. Kotsavasiloglou, C., Kostikis, N., Hristu-Varsakelis, D., Arnaoutoglou, M.: Machine learning-based classification of simple drawing movements in Parkinson's disease. Biomed. Signal Process. Control (2017). https://doi.org/10.1016/j.bspc.2016.08.003
7. Grover, S., Bhartia, S., Yadav, A., Seeja, K.R.: Predicting severity of Parkinson's disease using deep learning. Procedia Comput. Sci. **132**(ICCIDS), 1788–1794 (2018). https://doi.org/10.1016/j.procs.2018.05.154
8. Flagg, C., Frieder, O., MacAvaney, S., Motamedi, G.: Streaming gait assessment for Parkinson's disease. In: CEUR Workshop Proceedings, vol. 2551, pp. 34–42 (2020)
9. Reyes, J.F., Steven Montealegre, J., Castano, Y.J., Urcuqui, C., Navarro, A.: LSTM and Convolution Networks exploration for Parkinson's Diagnosis. In: IEEE Colombian Conference on Communications and Computing, COLCOM 2019, Proceedings, pp. 2–5 (2019). https://doi.org/10.1109/ColComCon.2019.8809160
10. Paragliola, G., Coronato, A.: Gait anomaly detection of subjects with Parkinson's disease using a deep time series-based approach. IEEE Access (2018). https://doi.org/10.1109/ACCESS.2018.2882245
11. Hochreiter, S., Schmidhuber, J.: Long short-term memory. Neural Comput. **9**(8), 1735–1780 (1997). https://doi.org/10.1162/neco.1997.9.8.1735
12. Peng, C.Y.J., Lee, K.L., Ingersoll, G.M.: An introduction to logistic regression analysis and reporting. J. Educ. Res. **96**(1), 3–14 (2002). https://doi.org/10.1080/00220670209598786
13. Agatonovic-Kustrin, S., Beresford, R.: Basic concepts of artificial neural network (ANN) modeling and its application in pharmaceutical research. J. Pharm. Biomed. Anal. **22**(5), 717–727 (2000). https://doi.org/10.1016/S0731-7085(99)00272-1. ISSN 0731-7085
14. Song, Y.Y., Lu, Y.: Decision tree methods: applications for classification and prediction. Shanghai Arch. Psychiatry **27**(2), 130–135 (2015). https://doi.org/10.11919/j.issn.1002-0829.215044
15. Salmi, N., Rustam, Z.: Naïve Bayes classifier models for predicting the colon cancer. IOP Conf. Ser. Mater. Sci. Eng. **546**(5) (2019). https://doi.org/10.1088/1757-899X/546/5/052068
16. Goldberger, A.L.: PhysioBank, PhysioToolkit, and PhysioNet: components of a new research resource for complex physiologic signals. Circulation (2000). https://doi.org/10.1161/01.cir.101.23.e215
17. Performance Measures: Cohen's Kappa statistic - The Data Scientist, 2 September 2020. https://thedatascientist.com/performance-measures-cohens-kappa-statistic

Segmentation of Foetus Brain MRI Based on K-nn Algorithm, A Case Study

Paulo Salgado[1] and Teresa-Paula Azevedo Perdicoúlis[2](✉)

[1] CITAB-Dep. de Engenharias, ECT, UTAD, Quinta de Prados, 5000-801 Vila Real, Portugal
psal@utad.pt

[2] Institute of Systems and Robotics, Department of Electrical and Computer Engineering, University of Coimbra, Rua Silvio Lima-Polo II, 3030-290 Coimbra, Portugal
tazevedo@utad.pt

Abstract. The segmentation of magnetic resonance images of the fetal brain has been emerging as a clinical tool to detect abnormalities during the development of the foetus. Since the brain is still in development, a mixture of regions of white matter, grey matter and transition structures that are related to brain growth are still associated with it. In this work, two versions of the K-nearest neighbour algorithm are proposed as the core method for the recognition of different regions of an image; the first one is a refinement of the standard algorithm since some statistics are associated with the pixel, and the second a reinforcing iterative version of the same method. Both versions are used to identify 3 *apriori* selected regions—the brain, the intracranial region and the remaining part of the foetus. The effectiveness of the method has been demonstrated in a magnetic resonance image segmentation that was first pre-processed with digital filters for feature extraction. The present study illustrates the capabilities of this type of method to support obstetricians and general practitioners to assess the foetus and, in particular, its brain.

Keywords: Image segmentation · K-nn algorithm · Ultrasound image processing · Magnetic resonance images

1 Introduction

In the image processing realm, image segmentation (IS) refers to the process of dividing a digital image into multiple regions (sets of pixels), to simplify its representation in order to facilitate the analysis. As a result, a set of regions or contours is extracted from the image, where every pixel in the same region has similar characteristics, such as color, intensity, texture, or continuity.

In biomedical engineering, the combination of magnetic resonance imaging (MRI) with computational techniques warrants the development of high-interest tools for analysis of brain imaging to assist neuroscientists and general practitioners in the early diagnosis. Therefore, MRI of the human foetus is emerging

A. Abraham et al. (Eds.): SoCPaR 2020, AISC 1383, pp. 466–475, 2021.
https://doi.org/10.1007/978-3-030-73689-7_45

as a clinical tool for early detection of brain abnormalities, due to its ability to evaluate morphometric measurements of a brain in development, and promises a range of new quantitative biomarkers to be used in the clinical evaluation of pregnancy. It is factual that MRI of foetus may be corrupted by noise or blurred by the movement of the foetus during the examination [5].

However, there are methods of image reconstruction that have been developed to correct fetal movement and permitting to obtain 3D images of the fetal brain [8]. As reported also in [8], the anatomy of the developing fetal brain is significantly idiosyncratic in terms of both geometric and underlying tissue morphology, as it consists of a mixture of white matter, gray matter, and transitory structures related to brain growth. The segmentation of the foetus has been extensively studied by a number of authors over the years and several methods are described for this type of study. In [2,9], the construction of an atlas with 10 isotropic images is proposed as a first method. A second method is the construction of a simple atlas based on a segmentation that aims to find the transformation between the atlas and the target low-resolution images. Following a similar methodology, [4] put labels on the atlas to specify different structures of the brain, since these are usually used as a paradigm for automatic segmentation algorithms.

In this paper, the fetal brain is segmented through neighbouring characteristics using the K-nn algorithm [10] in order to attribute every image pixel to one of the pre-defined regions. Two versions of the K-nearest neighbour algorithm, K-nn, are proposed as the core method for the recognition of different regions of an image. The first one is a refinement of the standard K-nn, RK-nn, and the second an iterative version, K-iterative nearest neighbour algorithm, KI-nn. As a result, a clear distinction between different regions is automatically obtained as well as between the different structures that are present in the acquired MR images. In our case study, these regions are the foetus brain, the intracranial

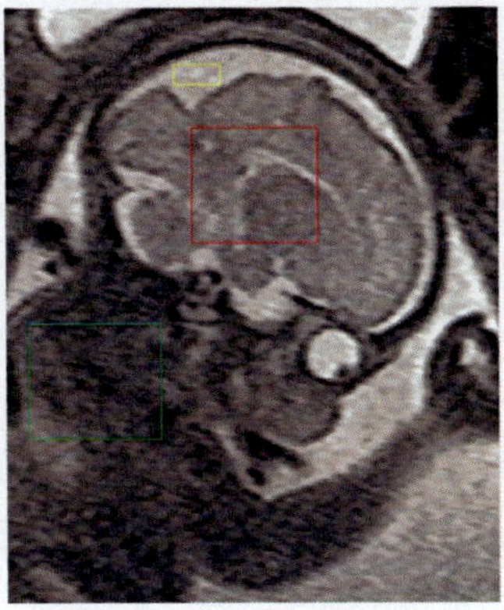

Fig. 1. Selected atlas image. The coloured boxes outline the local regions: brain, intracranial space and partial fetal body.

region and partially the body. The remainder of the work is organised as follows: Image acquisition and its pre-processing are explained in Sect. 2. The K-nn Algorithm is described in Sect. 3. Our two versions of the K-nn algorithm are explained in Sect. 4. Section 5 contains the results and the analysis of the image segmentation task realised by our algorithms. Section 6 concludes the paper and withdraws some directions for future work.

2 Image Acquisition and Pre-processing

Digital image processing (DIP) refers to the manipulation of digital images through processing methods that are able to make them more clear, either by removing noise and/or highlighting certain features such as contours or textures [5].

Test images can be are downloaded from [6], where one can find several kinds of foetus MRI similar to the image shown in Fig. 1, the one that was selected as a case study for this work.

Once the image is uploaded, the first step is it to be normalised, since this will improve the performance of the different methods to be applied next. Thence, three square regions of the image are marked as sample sub-regions—the brain, the intracranial area and the part of the foetus body (see the rectangles in Fig. 1)—whose respective pixels are taken as training data—the training set (TS). These sub-regions are to be used to define the classes. DIP algorithms are used to provide some local features of the image to the modified K-nn algorithm, which is trying to recognise regions of the image with similar characteristics.

The algorithm input is the data features of a circular region around every pixel that can be obtained either with (1) low-pass filters: the mean, the median and the variance; or (2) edge or contours operators: *Sobel, Canny, Prewitt* and the *Laplacian of the Gaussian Operator*. The expected output is the classification of every image pixel into one of the pre-defined regions.

In the processing of the image of Fig. 1, low-pass filters were used to assess some statistical properties of a local region of the image, since they allow for the smoothing of the image. In particular, median filters can reduce image noise without blurring the image contours. This type of filter is especially suited for removing impulsive noise that appears in limited regions of the image [3]. Additionally, variance filters provide the information variation contained in the local image region, in order to understand what is common in the vicinity of every pixel, which is important for distinguishing textures and some kind of patterns. The variance is calculated around each pixel using the following function:

$$Var_{ij} = \left(\frac{1}{n} \sum_{(r,s)\in R_{ij}} I_{rs}^2 \right) - \bar{I}_{ij}^2, \qquad (1)$$

where I_{rs} is the intensity of pixel–(r, s) of the image contained in the vicinity R_{ij}—the local region centred in pixel–(i, j) and with cardinality n—and $\overline{I}_{ij}$ is the mean value of the intensity of all the pixels in R_{ij}. Var_{ij} is the variance of the circular region R_{ij}. The result of the application of the different filters to the image is shown in Fig. 2.

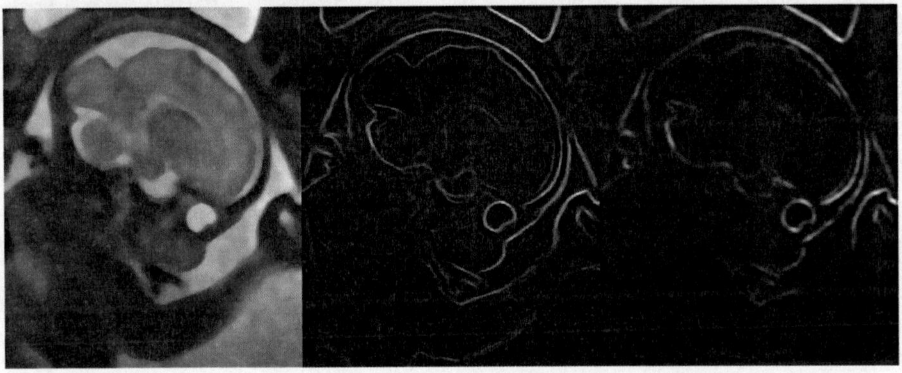

Fig. 2. Image features extracted by filter: (a) Median; (b) Gradient magnitude; (c) Variance;

The image treated by the median and variance filters are the components of the input vector of the K-nn algorithms. These features components are sufficient to find the regions of interest (ROI) identified by both versions. Other components can be added to improve the results of the algorithms if they provide more rich and discriminant features.

For validation purposes, contours filters were used to identify the borders of ROI. Contour filters, such as the operator *Sobel*, *Canny*, *Prewitt* or the *Laplacian of the Gaussian Operator* may also be applied to the same image to assess the results obtained with the segmentation procedure proposed in this work, that has shown better performance than other traditional methods [1]. Canny filter presents the best results since the contours of the brain, intracranial space, and the fetal body are all clearly expressed and highlighted. This outcome is perfectly expected since the detection of contours by the Canny filter is one of the best-defined methods to provide reliable detection [7]. However, all contour filters are used for comparison purposes, to check which of the two versions of the algorithm best represents our image and validate the results.

3 The K-nn Algorithm

The K-nn algorithm is a non-parametric method for classification and regression whose input is, in both situations, a set of training examples, of K-elements, in the feature space. The output of the algorithm depends on whether is used for classification or regression. When being used for classification, that is the case here, it consists in the association of every element to one or several pre-defined classes, i.e., an object is classified by a plural vote of its K-neighbours and then assigned to the most common class among them. Neighbours are taken from a set of objects for which the class is known. This can be considered the TS, although no explicit training step is required.

The training examples are vectors in a multidimensional feature space, each with a class label. The training phase of the algorithm consists only in storing the feature vectors and class labels of the training samples.

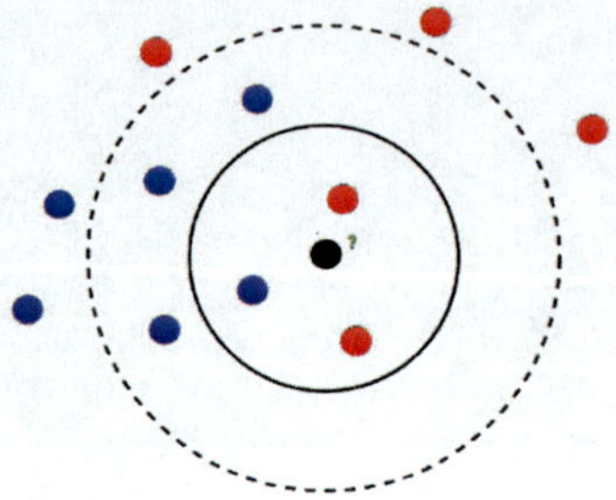

Fig. 3. K-nn algorithm. (Image taken from Wikipédia)

In the example of Fig. 3, the K-nn algorithm is illustrated in its simplest form. We have two different classes: the red points and the blue points, and the black point is meant to be classified. If $K = 3$, it means that we consider 3 nearest-neighbours. In this case, the majority is red and, thus, the point is classified as red. If $K = 6$, it means that we consider 6 nearest-neighbours. Thus, the majority is blue, and the point is classified as blue.

A peculiarity of the K-nn algorithm is being sensitive to the local structure of the data. The algorithm approximates the functions locally, and the entire calculation is deferred until the function is evaluated. Since this algorithm depends on the distance used in the classification task, the normalisation of the training data can significantly improve its accuracy. A useful technique is to assign weights to the neighbour contributions, which gives a larger participation of the closest neighbours to the average. In the K-nn algorithm, K is a user-defined parameter—the number of nearest neighbours. When $K = 1$ the algorithm is known as the nearest-neighbour algorithm, and this is the most intuitive version [10].

Thence, the image is a set of n pixels, everyone with a d-dimensional feature vector, $X_i \in \mathbb{R}^d$, $i = 1, \cdots, n$, e.g. the mean or the median value and with the variance, as defined in (1), or the gradient (with the last one being strongly correlated to the variance). Some of these pixels belong to the training set, $X_i \in TS \subset \mathbb{R}^d$. These are classified and partitioned into n_w–classes, where Y_i is the class label value of the ith sample, $Y_i \in \mathrm{N}_1^{n_w} = \{1, 2, \cdots, n_w\}$. An unlabelled vector is classified by assigning to it the label of the most frequent class among its K nearest neighbours belonging to the TS.

Consider some norm $\|\cdot\|$ on $\mathbb{R}^d$—e.g. the Euclidean distance, Hamming distance, Manhattan distance or Minkowski distance—and a non-classified pixel of index ℓ with feature vector $X_\ell \in US$ (Unlabelled Set). Hence, consider a sequence of ordered pixels of the TS: $\{X_{I(k)}\}_{k\in I}$, where I is the list of indices of the K nearest neighbours training pixels, such that

$$\left\|X_{I(1)} - X_\ell\right\| \leq \left\|X_{I(2)} - X_\ell\right\| \leq \cdots \leq \left\|X_{I(K)} - X_\ell\right\|. \tag{2}$$

The classification of point-ℓ is done by the majority vote of its K neighbours. The score value is

$$V_{j\ell} = \sum_{k=1}^{K} u_{j,I(k)} w_{I(k),\ell}, \quad j = 1, \ldots, n_w \tag{3}$$

for the ℓ-point to the jth class. u_{jr} is the membership value of the r training point that belongs to the jth class (e.g. $u_{jr} = 1$ if $Y_r = j$ and $u_{jr} = 0$ otherwise). $w_{I(k),\ell}$ is the weight of every k^{th} nearest-neighbour for the point-ℓ. Finally, the class with most votes is taken as the prediction, i.e., Y_i when $V_{i\ell} > V_{j\ell}, \forall i \neq j$. The quality of the classification of the unmarked samples, X_ℓ-point, is better when the result of the vote is concentrated in a certain class, by expressive clear majority voting. For this we define a measure that we call the *certainty belief factor* (CBF):

$$S_{i\ell} = \frac{V_{i\ell}^{\alpha}}{\sum\limits_{j=1}^{n_w} V_{j\ell}^{\alpha}}, \tag{4}$$

where α is a shape parameter. This measure can be seen as a confidence value of the classification process. Values of S close to 1 reveal high level of confidence and express goodness of the classification.

4 Description of the *K*-nn Algorithm Versions

Before defining the algorithm used in the segmentation procedure, some notation is in place. Every pixel of the image is represented by p_i. We consider two disjoint sets of pixels of the image: P and $\bar{P}$, where P is the set of labelled pixels (TS) and $\bar{P}$ is the set of unlabelled pixels (US). Every pixel $p_i \in P \cup \bar{P}$ has associated a vector $X_i \in \mathbb{R}^d$ of features. In this work, X_i stores the mean value, the variance and the median value of every pixel, considering a circular neighbourhood of radius δ for every pixel. Therefore, we can write $(p_i, X_i), \forall p_i \in P \cup \bar{P} =$ image. We define n_w classes, the pre-defined regions: $Y_j, j = 1, \ldots, n_w$. Only $p_i \in P$ have a class associated to it. Therefore, we can write $(p_i, X_i, Y_j), j = 1, \ldots, n_w, \forall p_i \in P$.

Next, we outline the steps of the first version of the K-nn algorithm— RK-nn:

Step 1 Upload the image.
Step 2 Pre-process the image. For instance to remove noise.
  2.1 Define a circular neighbourhood for every $p_i \in \bar{P}$ of radius $\delta : \mathscr{V}_{i(p_i,\delta)}$.

2.2 Calculate X_i for every $p_i \in P \cup \bar{P}$ in $\mathscr{V}_{i(p_i,\delta)}$.

Step 3 Define the classes: brain, intracranial space, part of the foetus body. $n_w = 3$

Step 4 Mark the three regions of the image that correspond to the defined classes: $Y_j, j = 1, 2, \ldots, n_w$. At this step the points get classified into P and $\bar{P}$. Every $p_i \in P$ belongs to one class $Y_j, j = 1, \ldots, n_w$.

Step 5 Choose parameter K, the number of nearest neighbours to be considered.

Step 6 The K-nn algorithm follows to classify every $p_i \in \bar{P}$. I.e., to attribute a class Y_j to every unlabelled point.

Hence, for every $p_i \in \bar{P}$, we do:

6.1 Use (2) to calculate the similarity of point p_i with every labelled point, $p_i \in P$.

6.2 Select the K most similar ones.

6.3 Use (3) to calculate the similarity degree of point $p_i \in \bar{P}$ to every class $Y_j, j = 1, \ldots, n_w$.

6.4 Classify $p_i \in \bar{P}$ with the class Y_j that obtains the highest score in **Step 6.3.**

For the iterative version of the K-nn algorithm—KI-nn, steps 1–5 are equal. Step 6 has some extra steps. Fundamentally, P is no longer a static set. It will grow with the points best classified in the current iteration (e.g. the CBF value is above a certain threshold) to be used in the next iteration. Hence:

Step 1–5 Equal to the first version.

Step 6 The RK-nn algorithm follows to classify every $p_i \in \bar{P}$. I.e., to attribute a class Y_j to every unlabelled point.

Hence, for every $p_i \in \bar{P}$, we do:

6.1 Use (2) to determine the similarity of point p_i with every labelled point, $p_i \in \bar{P}$.

6.2 Select the K most similar ones.

6.3 Use (3) to calculate the similarity degree of point $p_i \in \bar{P}$ to every class $Y_j, j = 1, \ldots, n_w$.

6.4 Classify $p_i \in \bar{P}$ with the class Y_j that obtains the highest score in **Step 6.3**.

6.5 Calculate S_{ij}-value for $p_i \in P$ using (4).

6.6 Define a thresholds: tol_M.

6.7 If $p_i \in \bar{P}$ with S_{ij}-value> tol_M then it goes to P.

Step 7 If there were new points added to P in **6.7** or the maximum number of iterations has not been reached then go to **Step 6**. Else STOP.

5 Application of the *K*-nn Algorithms and Results

Two distinct versions of the K-nn algorithm are implemented. First, the segmentation of the brain image is done in a non-iterative way by the RK-nn algorithm described in the previous section.

Both algorithms were tested on the same image and training data, with $K = 200$. The components of the input vector X are the mean and variance intensity of a circular region centred in each pixel and the value of the contour filter. For the RK-nn, the threshold value for inclusion into the TS was around 0.9.

Fig. 4. Simple K-nn algorithm. Classes: (1) brain; (2) intracranial space; (3) partial fetal body.

The results obtained with the first version are shown in Fig. 4, where the three segmentation classes are fully distinguished: brain (1), intracranial space (2) and partial fetal body (3). All pixels of the image are classified into these three classes. However, pixels that belong to the border regions or edges of the structure show low levels of discrimination. The respective CBF is represented in Fig. 5.

In a second version—KI-nn, an iterative version of the algorithm reinforces the classification process. In every iteration, the pixels with lower CBF are rejected and not classified, and pixels with higher S-value are added to the TS for the next iteration. Also, TS pixels with lower mean weight values are discarded from the TS. Then, at the new iteration, the non-classified points are subject to a new classification process.

In the second version, the method terminates after 2 iterations. The method starts with 10.7% of the image points belonging to the pre-defined classes. For the specified threshold value, 49.2% of the pixels were well classified during the first iteration and 26.5% in the second and the remaining pixels, 13.6%, are not classified. These results are shown in Fig. 6.

The results of kI-nn are shown in Fig. 7, where white zones represent a high membership value of the image region to the class. Black regions represent no pixels in the class. With this version, the confidence results of the classification process improved around 8% for the well-classified pixels when compared with RK-nn algorithm results.

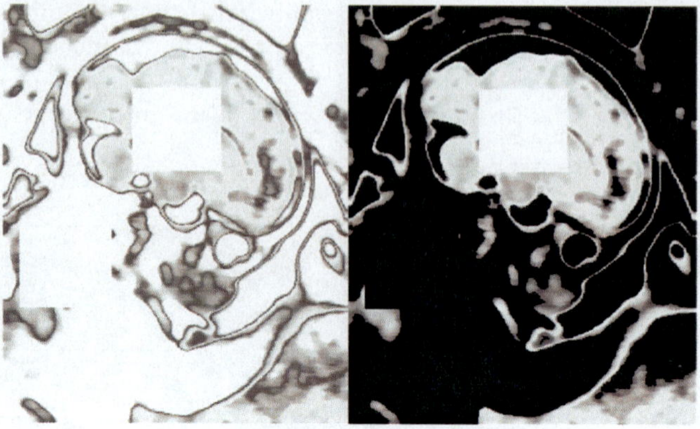

Fig. 5. Total sensitivity—credibility belief factor—of the points classified can be seen on the left and the sensitivity for class–1, the brain, can be seen on the right.

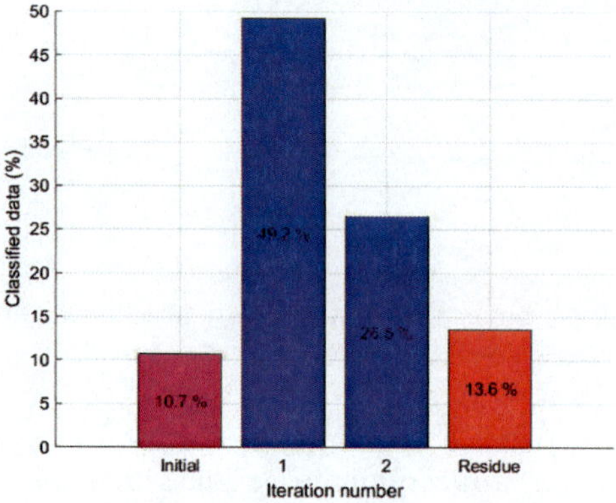

Fig. 6. Classified data (%) in each iteration.

Fig. 7. *K*I-nn algorithm. Classes: (1) brain; (2) intracranial space; (3) partial fetal body.

6 Conclusion and Future Work

In this work, two versions of the K-nn algorithm were used for the segmentation of a foetus brain MRI. The first version is a modified version of the standard algorithm and the second is an iterative version of the first one. Both algorithms have produced good results to segment the image into sub-regions, although the second one shows a higher confidence value. Besides the good results already obtained, the performance of both versions of the method can be improved by taking other features of the images as the input of the K-nn classifier. The presented study illustrates the capabilities of this type of method to support obstetricians and general practitioners to assess the foetus and, in particular, its brain.

References

1. Attallah, O., Sharkas, M., Gadelkarim, H.: Fetal brain abnormality classification from MRI images of different gestational age. Brain Sci. **9**(9), 231 (2019)
2. Gholipour, A., et al.: A normative spatiotemporal MRI atlas of the fetal brain for automatic segmentation and analysis of early brain growth. Sci. Rep. **7**(1), 476 (2017)
3. da Silva, A.M.M., Patrocínio, A.C., Schiabel, H.: Processamento e análise de imagens médicas. Revista Brasileira de Física Médica **3**(1), 34–48 (2019)
4. Makropoulos, A., Counsell, S., Rueckert, D.: A review on automatic fetal and neonatal brain MRI segmentation. NeuroImage **170**, 231–248 (2017)
5. Lucas, C.F.: Técnicas de Segmentação e Classificação de Imagens em Exames de Ressonância Magnética Nuclear. Ph.D. thesis, Pontifícia Universidade Católica do Rio Grande do Sul, Faculdade de Engenharia e Faculdade de Informática, Porto Alegre, Brasil (2014)
6. Levine, D., et al.: Compedium of fetal MRI @ONLINE (2002). http://radnet.bidmc.harvard.edu/fetalatlas/atlas.html
7. Canny, J.: A computational approach to edge detection. IEEE Trans. Pattern Anal. Mach. Intell. **8**(6), 679–698 (1986)
8. Habas, P., et al.: Atlas-based segmentation of the germinal matrix from in utero clinical MRI of the fetal brain, vol. 11, pp. 351–358, February 2008
9. Tourbier, S., et al.: Automatic brain extraction in fetal MRI using multi-atlas-based segmentation. In: Ourselin, S., Styner, M.A. (eds.) Medical Imaging 2015: Image Processing, vol. 9413, pp. 248–254. International Society for Optics and Photonics, SPIE (2015)
10. Cover, T., Hart, P.: Nearest neighbour pattern classification. IEEE Trans. Inf. Theory **IT–13**(1), 21–27 (1967)

Design of a Closed Supply Chain Under Uncertainty with Regards to Social and Environmental Impacts

Omid Abdolazimi[1], Mitra Salehi Esfandarani[2], and Ajith Abraham[3](✉)

[1] Department of Economy, Kharazmi University, Tehran, Iran
[2] Department of Civil Engineering, University of Memphis, Memphis, TN, USA
mslhsfnd@memphis.edu
[3] Scientific Network for Innovation and Research Excellence, Machine Intelligence Research (MIR) Labs, Washington, USA
Ajith.abrham@ieee.org

Abstract. In recent years, the decrease of resources and the increase of environmental concerns regarding the burial of industrial waste have led to developing the integrated green supply chain. In this study a multi-objective, multi-echelon, and multi-product closed-loop supply chain system is formulated in a way that considers the total profit, social responsibility, and environmental impacts. To overcome the innate uncertainty in some parameters such as transportation cost, operational costs, and customer demand, a robust counterpart based on the Bental-Nimrovski approach is developed. Finally, the performance of the deterministic and robust models evaluated by implementing the different experiments. The result shows the superiority of the deterministic model based on the mean of all objective functions and the robust model's superiority based on standard deviations.

Keywords: Social and environmental supply chain · Uncertainty · Multi-objective model

1 Introduction

Designing the supply chain system is considered one of the most critical decisions in the last few decades. On the other hand, environmental concerns related to the supply chain network is growing, which leads researchers to focus more on reusing the defective products and prevent further waste to reduce environmental pollution and achieve profitability by considering social and commercial considerations [1–3, 14]. Another concept that has been highlighted by researchers is considering corporate social responsibility throughout the supply chain [4, 19–25]. This paper considers these aspects for a closed-loop supply chain (CLSC) through three target functions: total profit, social responsibility, and environmental effects. The most important novelty of this paper is (I) considering the three aspects of economic, social, and environmental impacts simultaneously, (II) considering the robust approach (Bental-Nimrovski) to design a CLSC, (III) considering all levels of

A. Abraham et al. (Eds.): SoCPaR 2020, AISC 1383, pp. 476–488, 2021.
https://doi.org/10.1007/978-3-030-73689-7_46

forward and reverse supply chain, and (IV) using the goal-achieving technique to solve the problem as a single-objective.

The rest of this paper is as follows: the literature review is presented precisely in Sect. 2. The mathematical model is described and formulated in Sect. 3. The numerical example and comparisons based on two criteria are presented in Sect. 4. Finally, future research and conclusions are demonstrated in Sect. 5.

2 Literature Review

According to studies in the literature, designing the reverse and CLSC system based on mixed-integer programming is a widely used approach [5, 27]. Over the past decade, many reverse supply chain network design models have been developed as closed-loop supply chains. Wang and Hsu proposed a nonlinear model for designing the CLSC system. In this model, the potential locations of factories, distributors, and disposal centers are determined while the total cost was minimized [6, 15]. Uster et al., designed a multi-product CLSC network that contains the remanufacturing and collection facilities in a forward and backward loop. They used the Benders approach to minimize the fixed cost of locating facilities as well as transportation costs and processing costs [7, 16]. Shih developed a mixed-integer programming (MIP) model for the reverse logistics network. In this model, the transportation costs, fixed construction costs, and disposal costs of non-recoverable products were minimized [8]. Data uncertainty is an inevitable phenomenon in the supply chain field that needs to be addressed [9, 28]. A probabilistic approach has been used in many studies to deal with this uncertainty [10, 11]. Lickens et al., demonstrated a mixed-integer nonlinear programming model that considered the returned product rate as an uncertain parameter. They used the genetic algorithm to solve the model based on the differential evolution technique [12]. Recently, the green supply chain concepts and social responsibility in designing the supply chain network are highlighted. For example, Millet evaluated the criteria to achieve a robust supply chain that considered the economic, social, and environmental impacts simultaneously [13]. In the following section, the proposed mathematical model is presented.

3 Proposed Mathematical Model

The designed CLSC network in this study is a multi-objective, multi-level, multi-product system that maximizes the total profit, social effects, and minimizes the environmental impacts simultaneously. In the following subsections, the sets, parameters, and decision variables are described.

3.1 Sets

v	The fixed locations for the supplier
c	The fixed locations for the first market

(*continued*)

(*continued*)

i	The fixed location for the second market
j	The fixed locations for disposal center
k	The potential locations for opening the factory
l	The potential locations for opening the collection center
s	The potential locations for opening the repair center
r	The potential locations for opening the redistribution center
p	The potential locations for opening the recycling center
e	Products
h	Capacity levels for potential locations

3.2 Parameter

Pr_{cp}	The purchasing cost of product p in the first market c
Prr_{qp}	The purchasing cost of product p in the second market q
DE_{cp}	The demand quantity of product p in the first market c
DDE_{qp}	The demand quantity of product p in the second market q
FX_{ih}	The fixed cost of opening factory i with level h
FY_{jh}	The fixed cost of opening distribution center j with level h
FH_{kh}	The fixed cost of opening inspection center k with level h
FW_{lh}	The fixed cost of opening repair center l with level h
FO_{sh}	The fixed cost of opening re-distribution center s with level h
FU_{rh}	The fixed cost of opening recycling center r with level h
SC_{ve}	The purchasing cost of raw material e from supplier v
MC_{ip}	The production cost of product p in factory i
RMC_{ip}	The re-production cost of product p in factory i
DC_{jp}	The holding cost of product p in the distribution center j
CC_{kp}	The collection cost of product p in collection center j
DPC_{dp}	The disposal cost of product p in disposal center d
RPC_{lp}	The repair cost of product p in repair center l
RDC_{sp}	The re-distribution cost of product p in re-distribution center s
RC_{rp}	The recycling cost of product p in recycling center r
TVI_{vie}	The transportation cost of raw material e between supplier v and factory i
TIJ_{ijp}	The transportation cost of product p between factory i and distribution center j
TJC_{jcp}	The transportation cost of product p between distribution center j and first market c
TCK_{ckp}	The transportation cost of product p between first market c and inspection center k

(*continued*)

(*continued*)

TKI_{kip}	The transportation cost of product p between inspection center k and factory i
TKD_{kdp}	The transportation cost of product p between inspection center k and disposal center d
TKL_{klp}	The transportation cost of product p between inspection center k and repair center l
TKR_{krp}	The transportation cost of product p between inspection center k and recycling center i
TLS_{lsp}	The transportation cost of product p between repair center l and recycling center s
TSQ_{sqp}	The transportation cost of product p between re-distribution center s and second market q
TRV_{rve}	The transportation cost of raw material e between recycling center r and the supplier v
$CapV_{v}$	The maximum capacity for supplier v
$CapI_{ih}$	The maximum capacity for factory i with level h
$CapJ_{jh}$	The maximum capacity for distribution center j with level h
$CapK_{kh}$	The maximum capacity for inspection center k with level h
$CapL_{lh}$	The maximum capacity for repair center l with level h
$CapS_{sh}$	The maximum capacity for redistribution center s with level h
$CapR_{rh}$	The maximum capacity for recycling center r with level h
RT	The rate of returned products from the first market to the collection center
RM	The re-production rate in factory
RS	The recycling rate in recycling centers
RD	The disposal rate in disposal center
RR	The repair rate in repair centers
θ_{JO}	The normalized weighted for the created job opportunities in facility
θ_{LD}	The normalized weight for the total number of lost working days due to workplace injury at the facility
JOI_{ih}	The quantity of created job opportunity in case of opening factory i with level h
JOJ_{jh}	The quantity of created job opportunity in case of opening distribution center j with level h
JOK_{kh}	The quantity of created job opportunity in case of opening inspection center k with level h
JOL_{lh}	The quantity of created job opportunity in case of opening repair center i with level h
JOS_{sh}	The quantity of created job opportunity in case of opening reproduction center s with level h
JOR_{rh}	The quantity of created job opportunity in case of opening recycling center r with level h

(*continued*)

(*continued*)

LDI_{ih}	Average lost working days due to the work injury in case of opening the factory *i* with level *h*
LDL_{lh}	Average lost working days due to workplace injury in case of opening the repair center l with level h
LDR_{rh}	Average lost working days due to the workplace injury in case of opening the recycling center *r* with level *h*
$ETVI_{vie}$	The carbon emission rate for transporting the raw material *e* between supplier *v* and factory *i*
$ETIJ_{ijp}$	The amount of carbon emission for transporting product *p* between factory *i* and distribution center *j*
$ETJC_{jcp}$	The amount of carbon emission for transporting product *p* between the distribution center*j* and first market *c*
$ETCK_{ckp}$	The amount of carbon emission for transporting product *p* between the first market *c* and inspection center *k*
$ETKI_{kip}$	The amount of carbon emission for transporting product *p* between the first market *c* and factory *i*
$ETKD_{kdp}$	The amount of carbon emission for transporting product *p* between the inspection center *k* and disposal center *d*
$ETKL_{klp}$	The amount of carbon emission for transporting product *p* between the inspection center *k* and repair center *l*
$ETKR_{krp}$	The amount of carbon emission for transporting product *p* between the inspection center *k* and recycling center *r*
$ETLS_{lsp}$	The amount of carbon emission for transporting product *p* between the repair center *l* and re-distribution center *s*
$ETSQ_{sqp}$	The amount of carbon emission for transporting product *p* between redistribution centers *s* and the second market *q*
$ETRV_{rvp}$	The amount of carbon emission for transporting the raw material *e* between recycling center*r* and the supplier *v*
EM_{ip}	The amount of carbon emission for producing product *p* in factory *i*
ERM_{ip}	The amount of carbon emission for reproducing product *p* in factory *i*
ERP_{lp}	The amount of carbon emission for repairing product *p* in repair center *l*
ERC_{rp}	The amount of carbon emission for recycling product *p* in recycling center *l*
NX	The maximum number of factories
NY	The maximum number of distribution center
NH	The maximum number of collection center
NW	The maximum number of repair center
NO	The maximum number of re-distribution center
NU	The maximum number of recycling center

3.3 Decision Variable

QVI_{vie}	The amount of raw material *e* transported between supplier *v* and factory *i*
QIJ_{ijp}	The amount of product *p* transported between factory *i* and distribution center *j*
QJC_{jcp}	The amount of product *p* transported between distribution center *j* and first market *c*
QCK_{ckp}	The amount of product *p* transported between the first market *c* and inspection center *k*
QKI_{kip}	The amount of product *p* transported between the inspection center *k* and factory *i*
QKD_{kdp}	The amount of product *p* transported between the inspection center *k* and disposal center *d*
QKL_{klp}	The amount of product *p* transported between the inspection center *k* and repair center *l*
QKR_{krp}	The amount of product *p* transported between the inspection center *k* and recycling center *r*
QLS_{lsp}	The amount of product *p* transported between the repair center*l* and re-distribution center *s*
QSQ_{sqp}	The amount of product *p* transported between the re-distribution center*s* and the second market *q*
QRV_{rve}	The amount of product *p* transported between the recycling center*r* and supplier *v*
X_{ih}	If factory *i* with level *h* is opened, otherwise 0
Y_{jh}	If distribution center *j* with level *h* is opened, otherwise 0
T_{kh}	If inspection center *k* with level *h* is opened, otherwise 0
W_{lh}	If repair center *l* with level *h* is opened, otherwise 0
O_{sh}	If re-distribution center *s* with level *h* is opened, otherwise 0
U_{rh}	If recycling center *r* with level *h* is opened, otherwise 0

$$
\begin{aligned}
Max\, Z_1 = & \left(\sum_j \sum_c \sum_p \Pr_{cp} QJC_{jcp} + \sum_s \sum_q \sum_p \Pr r_{qp} QSQ_{sqp} \right) \\
& - \left(\begin{array}{l} \sum_i \sum_h FX_{ih} X_{ih} + \sum_j \sum_h FY_{jh} Y_{jh} + \sum_k \sum_h FH_{kh} T_{kh} + \\ \sum_l \sum_h FW_{lh} W_{lh} + \sum_s \sum_h FO_{sh} O_{sh} + \sum_r \sum_h FU_{rh} U_{rh} \end{array} \right) \\
& - \left(\begin{array}{l} \sum_v \sum_i \sum_e TVI_{vie} QVI_{vie} + \sum_i \sum_j \sum_p TIJ_{ijp} QIJ_{ijp} + \sum_j \sum_c \sum_p TJC_{jcp} QJC_{jcp} + \\ \sum_c \sum_k \sum_p TCK_{ckp} QCK_{ckp} + \sum_k \sum_i \sum_p TKI_{kip} QKI_{kip} + \sum_k \sum_d \sum_p TKD_{kdp} QKD_{kdp} + \\ \sum_k \sum_l \sum_p TKL_{klp} QKL_{klp} + \sum_k \sum_r \sum_p TKR_{krp} QKR_{krp} + \sum_l \sum_s \sum_p TLS_{lsp} QLS_{lsp} + \\ \sum_s \sum_q \sum_p TSQ_{sqp} QSQ_{sqp} + \sum_r \sum_v \sum_e TRV_{rve} QRV_{rve} \end{array} \right)
\end{aligned}
$$

$$-\begin{pmatrix} \sum_v\sum_i\sum_e SC_{ve}QVI_{vie} + \sum_i\sum_j\sum_p MC_{ip}QIJ_{ijp} + \sum_j\sum_c\sum_p DC_{jp}QJC_{jcp} + \\ \sum_c\sum_k\sum_p CC_{kp}QCK_{ckp} + \sum_k\sum_i\sum_p RMC_{ip}QKI_{kip} + \sum_k\sum_d\sum_p DPC_{dp}QKD_{kdp} + \\ \sum_k\sum_l\sum_p RPC_{lp}QKL_{klp} + \sum_i\sum_j\sum_p RC_{rp}QKR_{krp} + \sum_l\sum_s\sum_p RDC_{sp}QLS_{lsp} \end{pmatrix} \quad (1)$$

$$Max\,Z_2 = \theta_{JO}\begin{pmatrix} \sum_i\sum_h JOI_{ih}X_{ih} + \sum_j\sum_h JOJ_{jh}Y_{jh} + \sum_k\sum_h JOK_{kh}T_{kh} + \\ \sum_l\sum_h JOL_{lh}W_{lh} + \sum_s\sum_h JOS_{sh}O_{sh} + \sum_r\sum_h JOR_{rh}U_{rh} \end{pmatrix}$$

$$-\,\theta_{LD}\left(\sum_i\sum_h JDI_{ih}X_{ih} + \sum_l\sum_h LDL_{lh}W_{lh} + \sum_r\sum_h LDR_{rh}T_{rh}\right) \quad (2)$$

$$\begin{aligned} Min\,Z_3 = {} & \sum_v\sum_i\sum_e ETVI_{vie}QVI_{vie} + \sum_i\sum_j\sum_p (ETIJ_{ijp} + EM_{ip})QIJ_{ijp} + \sum_j\sum_c\sum_p ETJC_{jcp}QJC_{jcp} \\ & + \sum_c\sum_k\sum_p ETCK_{ckp}QCK_{ckp} + \sum_k\sum_i\sum_p (ETKI_{kip} + ERM_{ip})QKI_{kip} + \sum_k\sum_d\sum_p ETKD_{kdp}QKD_{kdp} \\ & + \sum_k\sum_l\sum_p (ETKL_{klp} + ERP_{lp})QKL_{klp} + \sum_k\sum_r\sum_p (ETKR_{krp} + ERC_{rp})QKR_{krp} + \sum_l\sum_s\sum_p ETLS_{lsp}QLS_{lsp} \\ & + \sum_s\sum_q\sum_p ETSQ_{sqp}QSQ_{sqp} + \sum_r\sum_v\sum_p ETRV_{rve}QRV_{rve} \end{aligned} \quad (3)$$

The first objective function (Eq. 1) maximizes the profits (total revenue minus total costs) from establishing the facilities in four parts. In the first part, the revenue obtained from the product's sale in the first and second market is computed. The fixed costs of establishing factory, distribution, collection, repair, redistribution, and recycling center are calculated in the second part. In the third part, the transportation cost between facilities is shown. Finally, in the fourth part, the operating cost of the supply chain is presented. The second objective function (Eq. 2) maximizes the social effects of the supply chain network by maximizing the created job opportunities as well as minimizing the average lost days due to workplace injury. The third objective function (Eq. 3) minimizes the environmental impacts, CO_2 emission resulted from transportation, production, reproduction, repair, and recycling throughout the supply chain.

Constraint

$$\sum_j\sum_p QIJ_{ijp} = \sum_v\sum_e QVI_{vie} + \sum_k\sum_p QKI_{kip} \qquad \forall i \quad (4)$$

$$\sum_i QIJ_{ijp} = \sum_c QJC_{jcp} \qquad \forall j, p \quad (5)$$

$$\sum_j QJC_{jcp} \geq DE_{cp} \qquad \forall c, p \quad (6)$$

$$\sum_k QCK_{ckp} = DE_{cp} \times RT \qquad \forall c, p \quad (7)$$

$$\sum_d QKD_{kdp} = \sum_c QCK_{ckp} \times RD \qquad \forall k, p \quad (8)$$

$$\sum_{l} QKL_{klp} = \sum_{c} QCK_{ckp} \times RR \quad \forall k, p \tag{9}$$

$$\sum_{r} QKR_{krp} = \sum_{c} QCK_{ckp} \times RS \quad \forall k, p \tag{10}$$

$$\sum_{i} QKI_{kip} = \sum_{c} QCK_{ckp} \times RM \quad \forall k, p \tag{11}$$

$$\sum_{c} QCK_{ckp} = \sum_{d} QKD_{kdp} + \sum_{i} QKI_{kip} + \sum_{l} QKL_{klp} + \sum_{r} QKR_{krp} \quad \forall k, p \tag{12}$$

$$\sum_{k} QKL_{ckp} = \sum_{s} QLS_{lsp} \quad \forall l, p \tag{13}$$

$$\sum_{l} QLS_{lsp} = \sum_{q} QSQ_{sqp} \quad \forall s, p \tag{14}$$

$$\sum_{s} QSQ_{sqp} \geq DDE_{qp} \quad \forall q, p \tag{15}$$

$$\sum_{k}\sum_{p} QKR_{krp} = \sum_{v}\sum_{e} QRV_{rve} \quad \forall r \tag{16}$$

$$\sum_{i}\sum_{e} QVI_{vie} \leq CapV_{v} \quad \forall v \tag{17}$$

$$\sum_{j}\sum_{p} QIJ_{ijp} \leq \sum_{h} CapI_{ih} X_{ih} \quad \forall i \tag{18}$$

$$\sum_{c}\sum_{p} QJC_{jcp} \leq \sum_{h} CapJ_{jh} Y_{jh} \quad \forall j \tag{19}$$

$$\sum_{c}\sum_{p} QCK_{ckp} \leq \sum_{h} CapK_{kh} T_{kh} \quad \forall k \tag{20}$$

$$\sum_{k}\sum_{p} QKL_{klp} \leq \sum_{h} CapL_{lh} W_{lh} \quad \forall l \tag{21}$$

$$\sum_{l}\sum_{p} QLS_{lsp} \leq \sum_{h} CapL_{sh} W_{sh} \quad \forall s \tag{22}$$

$$\sum_{k}\sum_{p} QKR_{krp} \leq \sum_{h} CapR_{rh} U_{rh} \quad \forall r \tag{23}$$

$$\sum_{h} X_{ih} \leq 1 \quad \forall i \tag{24}$$

$$\sum_{h} Y_{jh} \leq 1 \quad \forall j \tag{25}$$

$$\sum_{h} T_{kh} \leq 1 \quad \forall k \tag{26}$$

$$\sum_h W_{lh} \leq 1 \quad \forall l \tag{27}$$

$$\sum_h O_{sh} \leq 1 \quad \forall s \tag{28}$$

$$\sum_h U_{rh} \leq 1 \quad \forall r \tag{29}$$

$$\sum_h X_{ih} \leq NX \quad \forall i \tag{30}$$

$$\sum_h Y_{jh} \leq NY \quad \forall j \tag{31}$$

$$\sum_h T_{kh} \leq NH \quad \forall k \tag{32}$$

$$\sum_h W_{lh} \leq NW \quad \forall l \tag{33}$$

$$\sum_h O_{sh} \leq NO \quad \forall s \tag{34}$$

$$\sum_h U_{rh} \leq NU \quad \forall r \tag{35}$$

$$X_{ih}, Y_{jh}, T_{kh}, W_{lh}, O_{sh}, U_{rh} \in \{0, 1\} \tag{36}$$

$$QVI_{vie}, QIJ_{ijp}, QJC_{jcp}, QCK_{ckp}, QKI_{kip}, QKD_{kdp}, QKL_{klp}, QKR_{krp}, QLS_{lsp}, QSQ_{sqp}, QRV_{rve} \geq 0 \tag{37}$$

Constraint 4 guarantees that the total inflows and outflows from all suppliers and collection centers are equal for each factory. Constraint 5 ensures that the summation of the inflows to each distribution center from all production centers is equal to the summation of the outflows from these distribution centers. Constraint 6 guarantees that the demand in the first market area is satisfied for all customers. Constraint 7 indicates the relationship between customer demand in first market areas and the returned products to collection centers. Constraint 8 shows that the outflows from the collection centers to all disposal centers are equal to the inflows to each collection center from all first market customers. Constraint 9 depicts that the outflows from the collection centers to all repair centers are equal to the inflows to each collection center from all first market customers. Constraint 10 illustrates that for each product, the outflows from the collection centers to all recycling centers are equal to the inflows to each collection center from all first market customers. Constraint 11 demonstrates that the outflows from the collection centers to all factories are equal to the inflows to each collection center from all first market customers. Constraint 12 shows that the inflows to each collection center from all the first market customers are equal to the total outflows from this center to the repair, reproduction, recycling, and disposal centers. Constraint 13 ensures that the total the inflows to each repair center from all collection centers is equal to the summation of the outflows from these repair centers. Constraint 15 guarantees that the demand in the second market is satisfied for all customers. Constraint 16 ensures that for each raw material, the sum of inflows to each recycling center from all collection centers is equal to the sum of the outflows from these recycling centers. Constraint 17 ensures that the total outflows from each supplier to all factories does not exceed the capacity of supplier for each raw material. Constraint 18 ensures that the total output from each factory to all distribution centers does not exceed the capacity of factories for each product. Constraint 19 ensures that the total the outflows from each distribution center to all first market customers does not exceed the distribution center's capacity for each product. Constraint 20 guarantees that the total inflows to the collection centers from all customers in the first market do not exceed the collection centers' capacity. Constraint 21 guarantees that the total inflows to the repair centers from all collection centers do not exceed the capacity's repair centers. Constraint 22 ensures that the total inflows to the redistribution centers from all collection centers do not exceed the capacity of redistribution centers. Constraint 23 guarantees that the total inflows to recycling centers from collection centers do not exceed recycling centers' capacity. Constraints 24 to 29 ensure that the factories, distribution, collection, repair, redistribution, and recycling centers are built with at most one capacity level, respectively. Constraints 30 to 35 limit the maximum number of facilities, including factories, distribution, collection, repair, redistribution, and recycling centers for the establishment, respectively. Constraints 36 and 37 show the binary and positive variables of the model, respectively.

4 Numerical Examples

In this section, five different scenarios (test) are defined to evaluate the proposed deterministic model. To solve the model, GAMS, solver CPLEX with a system of Cori7 and RAM 16 GIG is used. In addition, to overcome the innate uncertainty, the Bental-Nimrovski approach is implemented. To compare the obtained results from deterministic and robust models, two criteria are defined, standard deviation and mean of solutions which are summarized in Table 1 [17, 18]. As was expected, the results demonstrate the superiority of the deterministic model in terms of the first criteria (mean) in most cases, which is due to implementing the Bental Nimrovski approach, which provides the worst optimized solutions. On the other hand, the robust model shows a better solution than the deterministic model in terms of standard deviation [26].

Table 1. First and second criteria of the objective functions under different numerical examples

Test	Level of uncertainty	Mean of the objective functions under tests						Standard deviation of the objective functions under tests					
		Certain			Robust			Certain			Robust		
		Z_1	Z_2	Z_3	Z_1	Z_2	Z_3	Z_1	Z_2	Z_3	Z_1	Z_2	Z_3
1	0.20	1471389	1425	4979	1443491	1395	6391	23280	14	105	12827	6	66
	0.50	1438239	1429	4971	1410900	1376	6559	53807	42	142	20108	3	25
	1	1468308	1437	4908	1372803	1365	6615	33885	24	63	18554	2	19
2	0.20	1802385	1778	13362	1950852	1777	17169	95885	137	157	41117	20	102
	0.50	1897217	1798	13571	1873418	1671	17376	157140	103	328	37141	37	46
	1	1848915	1744	13663	1757935	1618	17871	125590	130	110	43065	12	72
3	0.20	2530859	2297	13003	2522298	2436	17383	70324	161	513	15913	8	24
	0.50	2491791	2342	12954	2484593	2401	17561	37530	220	798	23407	33	105
	1	2469014	2357	12825	2411694	2107	17849	40128	109	1030	16067	73	94
4	0.20	2894264	2500	13692	2719786	2575	18318	78356	107	133	31941	28	22
	0.50	2902641	2560	13834	2629353	2465	18472	81214	110	273	18249	14	150
	1	2854546	2475	13437	2548031	2415	18788	65352	111	256	33258	31	91
5	0.20	2729494	3208	16491	3613648	3005	14937	334308	226	2262	108021	4	937
	0.50	2333410	2846	15638	3414474	2933	16726	252544	136	1354	162274	46	399
	1	3284102	3020	16285	2956161	2757	18772	211693	139	1825	105276	24	705

5 Conclusion

In this paper, a multi-objective, multi-product robust optimization model for the CLCS was presented. According to the multidimensional concept of sustainability, the goals of maximizing profits and social effects as well as minimizing the environmental impacts were considered. To design a CLSC network, a mixed-integer linear programming model was formulated, then to extend the proposed model under uncertainty, some model parameters, such as first and second market customer demand, transportation costs between facilities, as well as operating costs related to each facility, were considered uncertain. Then, based on the Bental-Nimrovski approach, the robust counterparts presented. Finally, five different scenarios (test) were defined to evaluate the deterministic and robust model's performance based on two criteria. The results demonstrated the superiority of the deterministic model in most cases in terms of objective function's means, while the robust method had better performance in terms of standard deviation. In conjunction with the potential factories, distribution, collection, repair, redistribution, and recycling centers, as well as the number of products transported between facilities, were determined. Since the proposed robust model is NP-hard, it requires considerable solving time for the large dimension problems, so as future studies implementing the heuristic methods are recommended.

References

1. Goodarzian, F., Hosseini-Nasab, H., Fakhrzad, M.B.: A multi-objective sustainable medicine supply chain network design using a novel hybrid multi-objective metaheuristic algorithm. Int. J. Eng. **33**(10), 1986–1995 (2020)
2. Fakhrzad, M.B., Goodarzian, F.: A new multi-objective mathematical model for a Citrus supply chain network design: metaheuristic algorithms. J. Optim. Ind. Eng. (2020). https://doi.org/10.22094/JOIE.2020.570636.1571
3. Fathollahi-Fard, A.M., Ahmadi, A., Goodarzian, F., Cheikhrouhou, N.: A bi-objective home healthcare routing and scheduling problem considering patients' satisfaction in a fuzzy environment. Appl. Soft Comput. **93**, 106385 (2020)
4. Chan, H.L., Wei, X., Guo, S., Leung, W.H.: Corporate social responsibility (CSR) in fashion supply chains: a multi-methodological study. Transp. Res. Part E Logist. Transp. Rev. **142**, 102063 (2020)
5. Abdolazimi, O., Esfandarani, M.S., Shishebori, D.: Design of a supply chain network for determining the optimal number of items at the inventory groups based on ABC analysis: a comparison of exact and meta-heuristic methods. Neural Comput. Appl., 1–16 (2020). https://doi.org/10.1007/s00521-020-05428-y
6. Wang, H.-F., Hsu, H.-W.: A closed-loop logistic model with a spanning-tree based genetic algorithm. Comput. Oper. Res. **37**, 376–389 (2010)
7. Üster, H., Easwaran, G., Akçali, E., Çetinkaya, S.: Benders decomposition with alternative multiple cuts for a multi-product closed-loop supply chain network design model. Nav. Res. Logist. (NRL) **54**, 890–907 (2007)
8. Shih, L.-H.: Reverse logistics system planning for recycling electrical appliances and computers in Taiwan. Resour. Conserv. Recycl. **32**, 55–72 (2001)
9. Pishvaee, M.S., Kianfar, K., Karimi, B.: Reverse logistics network design using simulated annealing. Int. J. Adv. Manuf. Technol. **47**(2), 69–81 (2010)
10. El-Sayed, M., Afia, N., El-Kharbotly, A.: A stochastic model for forward–reverse logistics network design under risk. Comput. Ind. Eng. **58**, 423–431 (2010)
11. Pishvaee, M.S., Farahani, R.Z., Dullaert, W.: A memetic algorithm for bi-objective integrated forward/reverse logistics network design. Comput. Oper. Res. **37**, 1–2 (2010)
12. Lieckens, K., Vandaele, N.: Reverse logistics network design with stochastic lead times. Comput. Oper. Res. **34**, 395–416 (2007)
13. Millet, D.: Designing a sustainable reverse logistics channel: the 18 generic structures framework. J. Clean. Prod. **19**, 588–597 (2011)
14. Abdolazimi, O., Esfandarani, M.S., Salehi, M., Shishebori, D.: Robust design of a multi-objective closed-loop supply chain by integrating on-time delivery, cost, and environmental aspects, case study of a Tire Factory. J. Clean. Prod. **264**, 121566 (2020)
15. Shakhsi-Niaei, M., Esfandarani, M.S.: Multi-objective deterministic and robust models for selecting optimal pipe materials in water distribution system planning under cost, health, and environmental perspectives. J. Clean. Prod. **207**, 951–960 (2019)
16. Ben-Tal, A., Nemirovski, A.: Robust optimization–methodology and applications. Math. Program. **92**(3), 453–480 (2002)
17. Pishvaee, M.S., Rabbani, M.: A graph theoretic-based heuristic algorithm for responsive supply chain network design with direct and indirect shipment. Adv. Eng. Softw. **42**, 57–63 (2011)
18. Vahdani, B., Tavakkoli-Moghaddam, R., Jolai, F.: Reliable design of a logistics network under uncertainty: a fuzzy possibilistic-queuing model. Appl. Math. Model. **37**(5), 3254–3268 (2013)

19. Goodarzian, F., Shishebori, D., Nasseri, H., Dadvar, F.: A bi-objective production-distribution problem in a supply chain network under grey flexible conditions. https://doi.org/10.1051/ro/202011
20. Sahebjamnia, N., Goodarzian, F., Hajiaghaei-Keshteli, M.: Optimization of multi-period three-echelon Citrus supply chain problem. J. Optim. Ind. Eng. **13**(1), 39–53 (2020)
21. Goodarzian, F., Hosseini-Nasab, H.: Applying a fuzzy multi-objective model for a production–distribution network design problem by using a novel self-adoptive evolutionary algorithm. Int. J. Syst. Sci.: Oper. Logist. **8**(1), 1–22 (2021). https://doi.org/10.1080/23302674.2019.1607621
22. Fakhrzad, M.B., Goodarzian, F.: A fuzzy multi-objective programming approach to develop a green closed-loop supply chain network design problem under uncertainty: modifications of imperialist competitive algorithm. RAIRO-Oper. Res. **53**(3), 963–990 (2019)
23. Fakhrzad, M.B., Talebzadeh, P., Goodarzian, F.: Mathematical formulation and solving of green closed-loop supply chain planning problem with production, distribution and transportation reliability. Int. J. Eng. **31**(12), 2059–2067 (2018)
24. Goodarzian, F., Hosseini-Nasab, H., Muñuzuri, J., Fakhrzad, M.B.: A multi-objective pharmaceutical supply chain network based on a robust fuzzy model: a comparison of meta-heuristics. Appl. Soft Comput. **92**, 106331 (2020)
25. Fakhrzad, M.B., Goodarzian, F., Golmohammadi, A.M.: Addressing a fixed charge transportation problem with multi-route and different capacities by novel hybrid meta-heuristics. J. Ind. Syst. Eng. **12**(1), 167–184 (2019)
26. Ramezani, M., Bashiri, M., Tavakkoli-Moghaddam, R.: A robust design for a closed-loop supply chain network under an uncertain environment. Int. J. Adv. Manuf. Technol. **66**(5–8), 825–843 (2013)
27. Keshavarz Ghorabaee, M., Amiri, M., Olfat, L., Khatami Firouzabadi, S.A.: Designing a multi-product multi-period supply chain network with reverse logistics and multiple objectives under uncertainty. Technol. Econ. Dev. Econ. **23**(3), 520–548 (2017)
28. Ramezani, M., Bashiri, M., Tavakkoli-Moghaddam, R.: A new multi-objective stochastic model for a forward/reverse logistic network design with responsiveness and quality level. Appl. Math. Model. **37**(1–2), 328–344 (2013)

Contour Method for Ultrasound Images

Paulo Salgado[1] and Paulo Afonso[2(✉)]

[1] Universidade de Trás-os-Montes e Alto Douro, 5000-801 Vila Real, Portugal
psal@utad.pt

[2] IT-Instituto de Telecomunicações/ESTGA, Universidade de Aveiro, 3754-909 Águeda, Portugal
pafnaa@ua.pt

Abstract. Medical ultrasonography is widely used because it is a safe, non-invasive and non-ionizing diagnosis method. For these reasons, medical ultrasound is recommended for monitoring parts of body. This paper proposes a new contour methodology for ultrasound images, based on the Dijkstra algorithm. It will is used to detect structures or boundaries regions. It belongs to Live wire segmentation techniques group, for extracting quickly and accurately regions of interests in images. This can be done through a set of seed points, generated by an automatic process or selected manually by the user. Experimental results show that this method is robust and efficient in segmentation of fetal ultrasound images. It is less sensible to noise and, shadows and to imprecise boundary regions.

Keywords: Segmentation of image · Contour algorithm · Dijkstra algorithm

1 Introduction

Medical ultrasonography is the use of medical ultrasonic equipment and imaging techniques to visualize internal organs or tissues. The images must capture their size, structure and any pathological lesions with the most clarity possible. It is a diagnostic technique that is safe for the patient and operators, as well as being cheaper than others medical imaging. Due these facts it has become one of the most preferred imaging techniques. The function of these exams are to measure abnormalities in internal organs or tissues [1]. Usually these measurements are obtained manually by the doctor from ultrasound images which they have poor quality, low contrast and with high level of noise, that makes it difficult to interpret an image. Incomplete or erroneous measures frequently are the cause of a bad diagnostic. An automatic system to support processing, analysis and extraction of accuracy measurements and features from the images can be useful to imaging task support, namely to improving the work-flow and shorten the time of each examination [2].

Unfortunately, ultrasound images (UI) generally reveal poor detail of organs and tissues, partly due to the low level contrast of images. They are usually

A. Abraham et al. (Eds.): SoCPaR 2020, AISC 1383, pp. 489–495, 2021.
https://doi.org/10.1007/978-3-030-73689-7_47

strongly degraded by a blurred, distorted shape and high dotted noise (speckle) which are due of physical limitations process of the acquisition system, from the sensor resolution to the limitations of waves sound behaviour and its non-ideal wave propagations conditions inside the body [3]. Image processing techniques need to be used to make the image better and clearer for diagnostic purposes. Standard processing algorithms, which are been used with advantage in many other application domains, such as hard edge or contour detection, they frequently fail when applied to ultrasound image processing task [4].

Three kinds of interactive image segmentation methods are frequently used in edge detection [5]: (1) Active contour model and level set methods. These methods reveal strong limitations when applied to UI. (2) random walk [6] and graph cut methods [7], which are part of Contour based methods with initial target points and background points. The performance of these methods are affected by the location and the number of the initial target points and background points. (3) Live Wire (LW) method, where is included the Dijkstra algorithms and its variants, starting from marked points on the edge. These methods have less manual intervention and have high-precision results when in presence of blur boundary of UI.

An Live-Wire Image Segmentation method based on Dijkstra Algorithm (DA) in here presented and tested in contour detection of the foetus body in an womb ultrasound image context. Experimental results show that this method can run image segmentation successfully, with precise contour close line. More details about the method will be found in [8].

This paper is organized as follows. First it is present an introduction about the edge and contour image detection problem with focus in the ultrasound image context. In second section are presents methods, particularly the Sigmoid Edge Filter (SEF) and DA for cutting regions of interest (DA-ROI). Next, the proposed Dijkstra Contour Algorithm (DCA) is described. Test results of the proposed algorithm with ultrasound image is presented in Sect. 3. Last section presents the main conclusions.

2 Method

The proposed method for segmentation foetus UI, namely the contour detection, is described summary in the diagram block of Fig. 1.

The segmentation method has the following steps:

1. *Filtering*: The SEF filter is applied to the early imported image. It try do approximate each region around each pixel by a sigmoid surface, which is characterized by three parameters: magnitude, homogeneity (also designed as offset) and orientation. Let A_{ij} a circular region of image centred on pixel (i, j) and diameter $2 \times n + 1$ which will be approximated by sigmoid S of magnitude m_{ij}, offset h_{ij} and orientation angle θ_{ij}. As results of filtering process three matrix are done which represents: magnitude, M, and potential orientation of the edges, Θ, and the level of homogeneity over each pixel, H. Their values

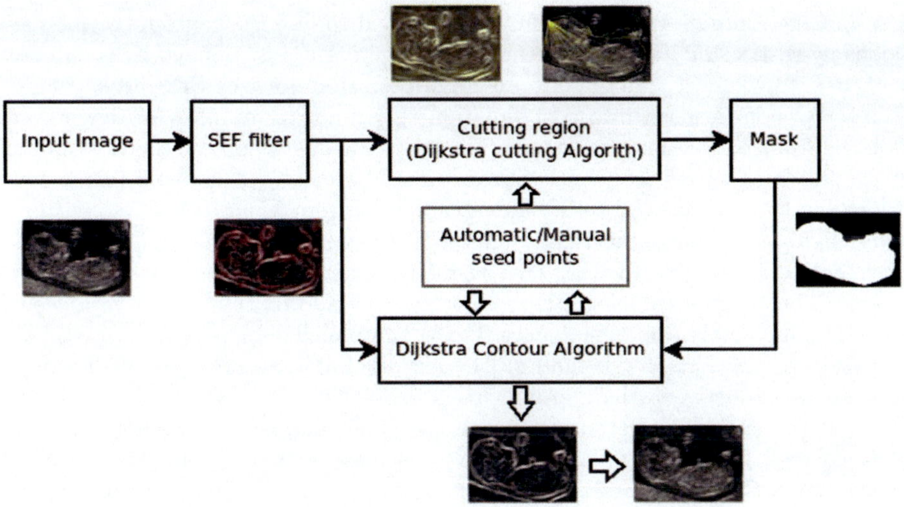

Fig. 1. Method steps

are $m_{ij} = M(i,j)$, $h_{ij} = H(i,j)$ and $\theta_{ij} = \Theta(i,j)$. From this information, the pseudo-image of the image regions is generate: $C = M \cdot (1 - H)$

2. *Foetal region cutting*: Typically the region of the fetus in the image is surrounded by a darker and more homogeneous region of the uterus. Delimiting this region of interest is a first task of the present method, which consists in determining a closed line of fountain through the pixels resulting in a lower cumulative value of its intensity per unit of defined area. For that, a grid of points is placed on image, which are the nodes of a weight graph. They cover all image region. Neighbouring nodes are linked by straight segment with width value w (i.e., it is a box with width $2 \times w + 1$). When $w = 0$ the box is a straight segment between the two nodes. Each segment is assigned a length that it is computed by adding the intensities of the pixels that are inside box, i.e., $(i,j) \in Box(r,s,w)$. Finally, based on this metric distance, the DA is used to found a closed path that joins a order set of given nodes trough low intensity paths. The results is a close polygon line on amniotic sac region, which is usually a region on image with low intensity pixels.
3. *Image Mask*: DA algorithm result is a closed path formed by a linked chain of straight segments, which draw is polygon usually non-convex. An algorithm was developed in this work to iteratively split polygon area into a set of disjoint triangles. In each iteration three consecutive nodes, not yet used, are chosen to extract a triangle from the polygon in a sequence in which the convexity of the erased region is preserved. At the end of this process, the polygon is full decomposed into a sum of disjoint triangles, which are convex objects. A mask, $Mask$, is now made by the area of triangles that cover the polygon region of image.

4. *Dijkstra Contour Algorithm* (DCA): It is used to find the contours of the foetus by restricting its search within the region of interest defined by the mask, $Mask$. Optionally, the direction edge information done by SEF filter (i.e., Θ matrix) can be used into cost function to penalizing paths what have non-concordance direction with edges. The input image C can be see as a weighted graph which nodes are pixels within region. Let $C_{r,3\times3} \in C$ a sub-image centred on the rth pixel (absolute index value). Each pixel r is connected with its 8-neighbours pixels (generically refereed as pixel s), which links or edge have associated a weighted value. This segment make an angle β_{rs} with horizontal line. The weights values are here computed by adding the intensity values of neighbour pixels (i.e., the square 3×3) those elements are placed behind a dividing line. This line is hold on the rth pixel and orientation $\frac{\pi}{2} + \beta_{rs}$. Distance value between the r and s pixels is done by $\rho_{rs} = \Phi_{rs} * C_{r,3\times3}$, where d_{rs} is the euclidean distance between the pixels and $C_{r,3\times3}$ is a sub-image of C centred in the rth pixel, and $\Phi(\beta)$ the mask of angle β for the convolution operator $*$.The result of this algorithm is a closed path over the contour region of the foetus with detail accuracy.

3 DijkStra Contour Algorithm

In this paper two closed directional graphs are used to segment image of Fig. 2)(a). The first case is used as a cutting graph method to extract the region of interest (ROI) in the image (see Fig. 2)(b). Its edges, in the form of line segments, are the edges of a non-convex polygon, separates the graph into two parts, inside and outside regions. In the second task the graph serves to extracts contours of objects in ROI through a closed line (see Fig. 2)(c). Both methods are based in Dijkstra algorithm (DA). DA try to finding the shortest paths between two nodes in a graph [9]. In graph-image context these edge weights are typically chosen to reflect the image content in some way and may be based on, e.g., local differences in intensity, gradient or other features, between adjacent image elements. In this paper, graph try solve the problems to finding minimum cost close path through seeds nodes, on in a graph. This may be interpreted as the distance or degree of connectedness between pairs of vertices.

4 Results

Edge detection methods that aim at identifying points ultrasound image filter a digital image at which the image brightness changes sharply which can organized into a set of straight edge or curved line segments termed edges. Due the reasons presented before, the Edge detection in an ultrasound images is a hard task, but extremely necessary into feature detection and feature extraction [3]. A set of ten grey-scale images, with full size 600×800 pixels, was collected from an ALOKA ultrasound equipment (Hospital of São Marcos;year, 2009). They were used to test proposed methods. The effective work region of image is a region size with 400×560 pixels. All images are filtered by SEF filter, without any additional precessing task, for edges detection. The results on one of these test image are show in Fig. 2. The first sub-figure display the magnitude of sigmoid approximation, while the second sub-figure display the correspondent sigmoid offset value. The third sub-figure is the edge output image used by algorithms. Orientation of relevant sigmoid are presents in last sub-figure by versors.

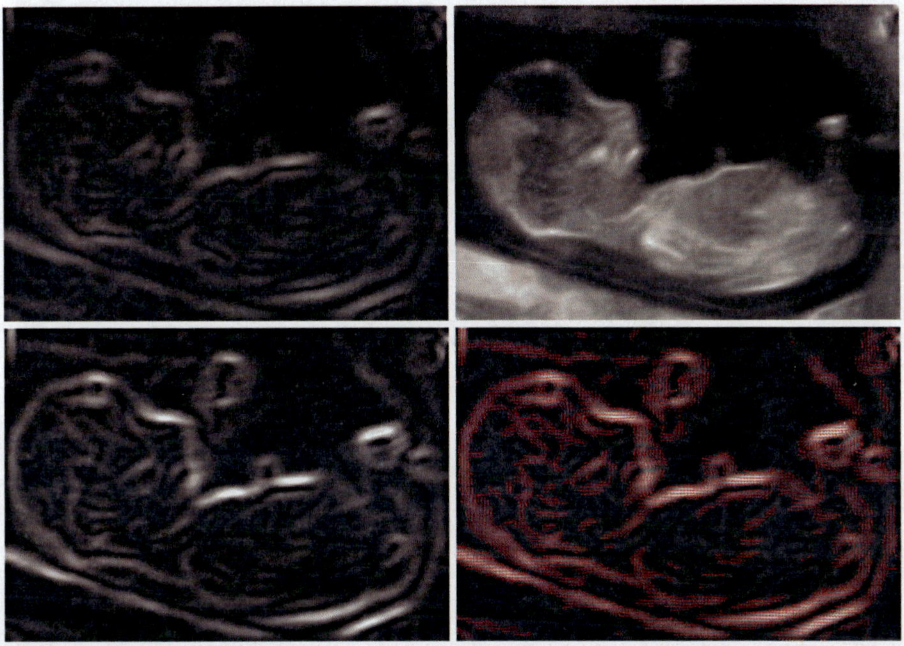

Fig. 2. SEF results: (a) Intensity, M.; (b) Homogeneity, H.; (c) Edge, C; (d) Orientation, Θ(versors).

Finally, the Fig. 3 display the results of DCA algorithm. The left sub-figure is display the close path results of the use o DCA algorithm in edge image C while the right sub-figure show the contour path in the original image. These results show us the goodness of the proposed method as well as a accuracy contour of foetus.

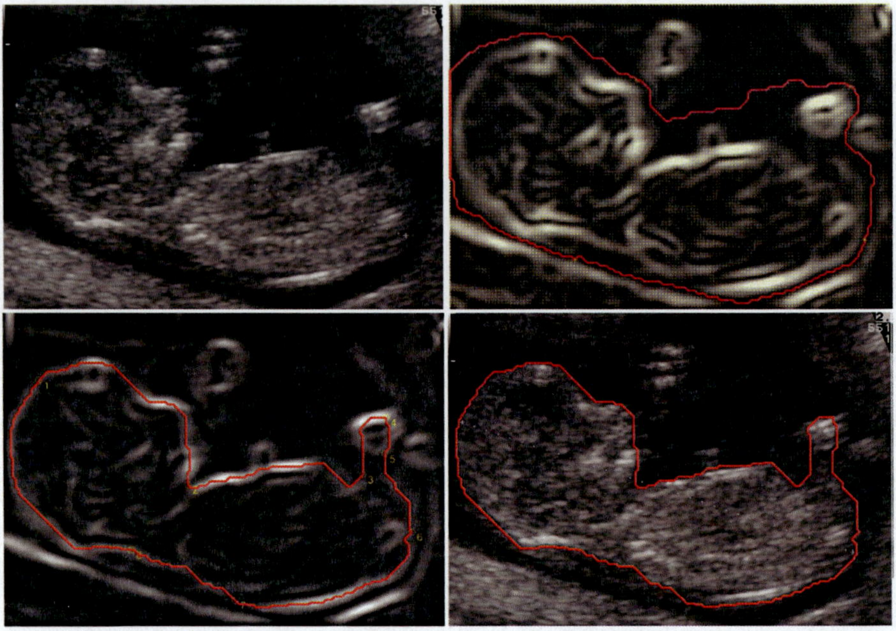

Fig. 3. DA algorithms: (a) Original image, (b) Foetal region cutting, (c) DCA result and (d) Segmented image.

5 Conclusions

In this paper, two algorithms are presented for cutting image regions and for contour (foetus) edges in an ultrasound image. Both are based on Dijkstra algorithm, that by using predefined edge weight and interactively labeled seeds, a shortest path algorithm for image segmentation is realized. However, for a best performance of theses algorithms was necessary to present a new edge sigmoid filter, which approximate edges by orientated sigmoid dependent. Moreover, the proposed shortest path based DA algorithm for image segmentation performs quite well in degraded noisy, blurred and distorted images. Lastly, owing to its high efficiency, these segmentation and contour algorithms can be used for real-time interactive process.

References

1. Edward, L., Hui, L.: First and second trimester screening for fetal structural anomalies. Semin. Fetal Neonatal Med. **23**(2), 102–111 (2018). https://doi.org/10.1016/j.siny.2017.11.005
2. Abinader, R., Steven, L.W.: Benefits and pitfalls of ultrasound in obstetrics and gynecology. Obst. Gynec. Clin. N. Am. **46**(2), 367–378 (2019). https://doi.org/10.1016/j.ogc.2019.01.011

3. Szabo. T.L.: Imaging Systems and Applications, Diagnostic Ultrasound Imaging: Inside Out, 2 Edn., pp. 365-430. Academic Press (2014). https://doi.org/10.1016/B978-0-12-396487-8.00010-0. (Chapter 10)
4. Tay, P.C., Garson, C.D., Acton, S.T., Hossack, J.A.: Ultrasound despeckling for contrast enhancement. IEEE Trans. Image Proc. **19**(7), 1847–1860 (2010)
5. He, J., Kim, C.-S., Jay Kuo, C.-C.: Interactive Segmentation Techniques Algorithms and Performance Evaluation - Algorithms and Performance Evaluation. Springer (2014). ISBN 978-981-4451-59-8 https://doi.org/10.1007/978-981-4451-60-4
6. Grady, L., Walks, R.: Image segmentation. IEEE Trans. Pattern Anal. Mach. Intell. **28**(11), 1768–1783 (2006). https://doi.org/10.1109/TPAMI.2006.233
7. Yi, F., Moon, I.: Image segmentation: a survey of graph-cut methods. In: 2012 International Conference on System and Informatics, pp. 1936–1941 (2012)
8. Paulo S., et al.: Iterative Contour Method for Fetal Ultrasound Images, Submitted to a Journal (2020)
9. Dijkstra, E.W.: A note on two problems in connexion with graphs. Numer. Mathem. **1**, 269–271 (1959)

Unsupervised Learning from Data and Learners

Yohan Foucade(✉) and Younès Bennani

LIPN UMR 7030 CNRS, Université Sorbonne Paris Nord, LaMSN,
La Maison des Sciences Numériques, 93430 Villetaneuse, France
{yohan.foucade,younes.bennani}@sorbonne-paris-nord.fr

Abstract. In the collaborative clustering framework, the hope is that by combining several clustering solutions, each one with its own bias and imperfections, one will get a better overall solution. The goal is that each local computation, quite possibly applied to distinct data sets, benefits from the work done by the other collaborators. This article is dedicated to collaborative clustering based on the Learning Using Privileged Information paradigm. Local algorithms weight incoming information at the level of each observation, depending on the confidence level of the classification of that observation. A comparison between our algorithm and state of the art implementations shows improvement of the collaboration process using the proposed approach.

Keywords: Unsupervised learning · Learning from learners · Collaborative clustering · Distributed learning · Federated learning

1 Introduction

Cluster analysis is a widely used technique in data analytics. Its goal is the organization of a collection of patterns into homogenous (small intra-group variability) and distinct (large inter-group variability) clusters. There is a large variety of methods for clustering and most require all the data to be available on the same place and at the same time [1]. However, technical, legal, confidentiality or computational performance reasons might prevent us from gathering the data or exchanging sensitive information.

Our work is based on an advanced learning paradigm proposed by Vapnik *et al.* called Learning Using Privileged Information (LUPI) [2,3]. In addition to examples, learners can be provided with hidden information that exists in the form of explanations, results, or comparisons, etc. The important question which Vapnik asks in the supervised case is: can the generalization performance be improved using the privileged information? Vapnik also showed this is true in the case of SVM.

The idea of LUPI is the following: suppose we want to establish a decision rule to determine a label based on certain characteristics X, but during the

The authors thank Facebook France for financially supporting this research.

A. Abraham et al. (Eds.): SoCPaR 2020, AISC 1383, pp. 496–507, 2021.
https://doi.org/10.1007/978-3-030-73689-7_48

learning phase, in addition to X, we also receive additional information, referred to as "Privileged Information" X^* which will not be available during the test phase. In such a case, how can we use X^* to improve learning? In this article, we propose a new approach based on the LUPI paradigm for collaborative clustering based on knowledge sharing among a set of learners.

In collaborative clustering [4,5], we assume that we have a dataset distributed among different nodes of a network (also called sites, or experts). The goal of each expert is to obtain a clustering of its local data using findings of its remote counterparts without exchanging the data themselves. This is often done in two phases. First, a clustering algorithm is applied locally and independently on each database. This is called the local phase. Then, during the collaboration phase, data sites exchange their findings in order to try to improve their results.

The kind of information algorithms share will differ depending on what kind of data distribution one is dealing with. When objects in different data sites lie in the same feature space, then experts can share information that can be represented in the feature space itself. This task in known as vertical collaborative clustering. In horizontal collaborative clustering, on the other hand, data sites agree on the identity of the objects but these latter are described in different feature spaces, then they can no longer share that kind of information. But they can still compare the clusters they created, using *e.g.* there partition matrices. A partition matrix is a $N \times C$ matrix where N is the number of objects in the dataset and C is the number of clusters looked for. The k-th row of this matrix represents the membership of each object to the k-th cluster.

This paper is dedicated to horizontal collaborative clustering, and our main contributions are:

- An implementation of an intuitive idea for collaboration: the more doubts we have about our findings, the more weight we will give to remote findings.
- Collaboration with every experts at the same time. Thus there is no more need to choose which site to collaborate with at each step.
- Collaboration at the level of each observation. This allows algorithms to fine-tune the way they collaborate with others.
- Detection of examples that are difficult to cluster.
- Improvement of efficiency by executing in parallel the base-learning processes on subsets of the training data set.
- Advantage of learning from small subsets of data that fit in main memory.
- Improves predictive performance by combining different learning systems each having different inductive bias.

It is organized as follows: Sect. 2 discusses related work. Then our main contribution –Collaborative Learning Using Privileged Information– is introduced in Sect. 3. Next, we present the results obtained with different datasets. Finally, we discuss the results in Sect. 5.

2 Related Work

Research in collaborative clustering has been introduced by Pedrycz in 2002 [4]. The author proposed a collaborative version of the Fuzzy C-Means algorithm

[6]. For this purpose, the objective function of the fuzzy c-means algorithm is extended with a second term which forces a clustering based on a subset to be "aware" of other partitions.

The SAMARAH method introduced by Wemmert in [7] can deal with heterogeneous hard clustering algorithms. This method attempts to increase similarity between clusterings through conflicts evaluation and resolution. In order to compare partitions with possibly different number of clusters, a confusion matrix is computed between each pair of algorithms. Then, a consensus is found by applying a voting algorithm to these results.

In 2015, Sublime *et al.* proposed an approach lifting some limits of previous works. Indeed, they introduce a framework for collaboration with heterogeneous algorithms. This method has the advantage not to require any global exchange confidence parameter, thus the impact of a distant algorithm on the local results might be different for each observation.

However, using hard clustering results for the construction of the confusion matrices leads to some loss of information. And every algorithm has the same weight in this process. As it makes no difference between a data site with well separated clusters and one with overlapping clusters, it allows low-performing algorithms to impede the results of the better ones.

After having proposed collaborative versions of Self Organizing Maps (SOM) and Generative Topographic Mapping (GTM), Ghassany *et al.* combined Fuzzy C-Means and GTM algorithms to obtain a collaborative clustering algorithm.

In [9], Sublime *et al.* describe a collaboration framework for model-based clustering algorithms. They define a global likelihood function and try to optimize a proxy for this function. This process requires a weight parameter between local and external information to be fixed.

More recently, an automated method for optimizing exchange confidence has been proposed in 2018 by Sublime *et al.*, and the collaboration involved 4 different views. The authors achieved successful results in detecting noisy views, but the method tends to favor collaboration between very similar views.

For a more comprehensive survey of the collaborative clustering literature, the reader may refer to [5].

3 Theoretical Framework and Learning Algorithm

In this section, we describe the general framework in which our method applies and we describe our algorithm.

3.1 Setting

The goal of collaborative learning is to learn from local data and from a set of learners with data stored across a large number of sites.

More formally, let the global data set X distributed among P sites: $X^{[1]}, \ldots, X^{[p]}, \ldots, X^{[P]}$, where $X^{[p]} = \{x_i^{[p]}\}_{i=1}^{N}$ is a set of N objects and each object $x_i^{[p]} \in \mathbb{R}^d$ is characterized by d features (i.e. is a vector in an d-dimensional feature space).

Our study is centered on the horizontal collaborative clustering problem. Thus, each data site has access to different $d^{[p]}$ features describing the same individuals. On each data site, there is an algorithm $\mathcal{A}^{[p]}$, $p = 1, \ldots, P$ trying to come up with a clustering of its data $X^{[p]}$. The process has two steps: a local step, and a collaborative step.

3.2 Underlying Framework

Local Step. During the local step, each algorithm $\mathcal{A}^{[p]}$ works on its local data set $X^{[p]}$ and fits its parameters as it would do in a non-collaborative framework. One requirement is that all algorithms have to be probabilistic ones. In particular, one of the output is a responsibility matrix $R^{[p]}$. This matrix contains the contributions of each component to each observation, $R_{i,k}^{[p]} = \mathbb{P}(Z_i = k|X_i, \theta)$, where K is the number of clusters, $Z_i \in \{1, \ldots, K\}$ are the components of the model and the elements of θ are the distributions parameters.

Once every model has been trained, we want them to exchange information in order to improve their performances. This is done in the collaborative step.

Collaborative Step. In the collaborative steps, the different algorithms $\mathcal{A}^{[p]}$ will exchange information in order to try to improve their respective classification. In this context, the idea of LUPI is the following: suppose we want to establish a clustering based on certain characteristics $X^{[p]}$, but during the learning phase, in addition to $X^{[p]}$, we also receive additional information, $X^{[p]^*}$. This additional information is presented in the form of partition matrices $R^{[p]}$. The local learner therefore has access to $X^{[p]}$ and $X^{[p]^*}$, instead of only $X^{[p]}$, to do its clustering. In our case, for each algorithm $\mathcal{A}^{[p]}$, $X^{[p]^*} \equiv R^{-[p]}$, where:

$$R^{-[p]} = \{R^{[q]} : q \in \{1, \ldots, P\} \setminus \{p\}\} \quad (1)$$

The set $R^{-[p]}$ contains all partition matrices of site p's counterparts. In this case, how can we use $X^{[p]}$ and $X^{[p]^*}$ to improve each local learners $\mathcal{A}^{[p]}$?

For the sake of simplicity, we consider two learners (sites): $P = 2$. On the local step, they yield two partition matrices: $R^{[1]}(t)$ and $R^{[2]}(t)$. In particular, for sample x_i, we have $R_{i,\cdot}^{[1]}$ and $R_{i,\cdot}^{[2]}$. Thus, the update rule for learner 1 for sample x_i is:

$$R_{i,\cdot}^{[1]}(t+1) \longleftarrow f\left(R_{i,\cdot}^{[1]}(t), R_{i,\cdot}^{[2]}(t)\right) \quad (2)$$

We want this update rule to depend on the level of certainty data site 1 has about its opinion. The higher this level, the lower the difference between $R_{i,\cdot}^{[1]}(t+1)$ and $R_{i,\cdot}^{[1]}(t)$. Conversely, the lower it is, the more weight will be put on $R_{i,\cdot}^{[2]}(t)$. Moreover, the higher (resp. the lower) site 2's certainty, the more (resp. less) it will influence $R_{i,\cdot}^{[1]}(t+1)$.

It turns out that the probabilistic framework is endowed with a measure that can be interpreted as the amount of uncertainty in a distribution, namely the entropy. It is defined as follows:

$$H(X) = -\sum_{i=1}^{d} p(x_i)\log_2 p(x_i) \tag{3}$$

Where X is a random variable with possible values $\{x_1,\ldots,x_d\}$. It follows that the uncertainty for a distribution R_i is:

$$H(R_{i,.}) = -\sum_{k=1}^{K} R_{i,k}\log_2 R_{i,k} \tag{4}$$

And its normalized version, using the fact that it is positive and maximized when X is uniformly distributed:

$$\mathcal{H}(R_{i,.}) = \frac{H(R_{i,.})}{log_2(K)} \tag{5}$$

Then $0 \le \mathcal{H}(R_{i,.}) \le 1$, and

- $\mathcal{H}(R_{i,.}) \approx 0 \Rightarrow R_{i,.} \approx \mathbb{1}_{Z_i=k}$ for some $k \in 1,\ldots,K$, i.e. we have high confidence about membership of observation i.
- $\mathcal{H}(R_{i,.}) \approx 1 \Rightarrow R_{i,.} \approx \frac{1}{K}\ \forall k \in 1,\ldots,K$, i.e. we have high uncertainty about membership of observation i.

Equation 6 gives the update rule for $R^{[ii]}$ when P algorithms are involved in the collaboration.

$$R^{[p]}(t+1) \longleftarrow \alpha^{[p]} \cdot R^{[p]}(t) + \sum_{R^{[q]}(t)\in R^{-[p]}} \beta^{[p]}_{[q]} \cdot R^{[q]}(t) \tag{6}$$

where

$$\begin{cases} \alpha^{[p]} = \left(\frac{1}{P-1} \sum_{R^{[q]}(t)\in R^{-[p]}} \mathcal{H}(R^{[q]}(t))\right)\cdot(1-\mathcal{H}(R^{[p]}(t))) \\ \beta^{[p]}_{[q]} = \mathcal{H}\left(R^{[p]}(t)\right)\cdot\left(1-\mathcal{H}\left(R^{[q]}(t)\right)\right) \end{cases} \tag{7}$$

In these equations, $\alpha^{[p]}$ is a vector of size N. Each element i of $\alpha^{[p]}$ is a weight associated with the local classification of the i-th observation. This weight depends negatively on the amount of uncertainty carried by the local classification, and positively on the average remote classification uncertainty.

Similarly, $\beta^{[p]}_{[q]}$ are vectors of size N. Each element i of $\beta^{[p]}_{[q]}$ is a weight associated with the classification of the i-th observation by algorithm $\mathcal{A}^{[q]}$. This weight depends positively on the amount of uncertainty carried by the local classification, and negatively on the uncertainty of the remote classification.

Note that the data sites only share there partition matrices. Therefore, no matter what the underlying algorithms are, provided that they look for the same number of components, the collaboration algorithm is still relevant. However, the data sites need to agree on the identity of the clusters. Our implementation uses the Hungarian algorithm to reorder the clusters on each data site. Next, we show how the values of α and β can be used to visualize information flow in the collaboration process, and describe the CoLUPI algorithm.

Collaborative Learning Algorithm. Based on the theoretical formalism developed and presented in the previous section, we can design a learning algorithm to establish exchanges between the different sites through a collaborative process. This algorithm therefore uses the Eq. 6 for updating the parameters of the sites in collaborative interaction.

Algorithm 1. CoLUPI algorithm

Local step:
for all algorithms $\mathcal{A}^{[p]}$ **do**
  train $\mathcal{A}^{[p]}$ on the data $X^{[p]}$
  get local parameters $\theta^{[p]}$ and partition matrix $R^{[p]}$
end for
Collaborative step:
repeat
  for all algorithms $\mathcal{A}^{[p]}$ **do**
    $R^{[p]}(t+1) \leftarrow f\left(R^{[p]}(t), R^{-[p]}(t)\right)$
    train model starting from new partition $R^{[p]}(t+1)$ using data $X^{[p]}$
    get $\theta^{[p]}(t+1)$ and $R^{[p]}(t+1)$
    if quality$\left(R^{[p]}(t+1)\right)$ is better than quality$\left(R^{[p]}(t)\right)$ **then**
      Accept collaboration
    else
      $\theta^{[p]}(t+1) \leftarrow \theta^{[p]}(t)$
      $R^{[p]}(t+1) \leftarrow R^{[p]}(t)$
    end if
  end for
until no algorithm improves for its criterion

3.3 Visualization of the Collaboration Process

In this section, we are interested in visualizing the information flow during collaboration. As it has been mentioned above, each collaborator assigns one weight to each source data site (including itself), for each observation. The mean of these weights for each remote site gives us the average weight assigned to that site and can be regarded as a confidence coefficient, except that it is not uniform across observations. If there are P sites, then this gives us a $P \times P$ matrix as defined in 8 which can be regarded as a confidence matrix. We show this collaboration heatmap for the Wdbc dataset in Sect. 4.

$$C_{(P\times P)} = \begin{pmatrix} \overline{\alpha^{[1]}} & \overline{\beta^{[1]}_{[2]}} & \cdots & \overline{\beta^{[1]}_{[P]}} \\ \overline{\beta^{[2]}_{[1]}} & \overline{\alpha^{[2]}} & \cdots & \overline{\beta^{[2]}_{[P]}} \\ \vdots & \vdots & \ddots & \vdots \\ \overline{\beta^{[P]}_{[1]}} & \overline{\beta^{[P]}_{[2]}} & \cdots & \overline{\alpha^{[P]}} \end{pmatrix} \tag{8}$$

Where

$$\overline{\alpha^{[p]}} = \frac{1}{N}\sum_{i=1}^{N} \alpha_i^{[p]}, \tag{9}$$

and

$$\overline{\beta^{[p]}_{[q]}} = \frac{1}{N}\sum_{i=1}^{N} \beta_i{}^{[p]}_{[q]}. \tag{10}$$

4 Experimental Validation

In this section, we present the results obtained after running our algorithm on several datasets. First, we describe the datasets that has been used in the experiments, then we present two kinds of results: performance assessment and visualization of the collaboration process. Finally, we discuss the execution time of our algorithm with different input values.

4.1 Data Sets

- The Breast Cancer Wisconsin (Diagnostic) (WDBC) dataset consists in 569 digitized images of a breast mass. There are 30 real-valued input variables describing the cell nuclei present in each image. Each observation is labelled as benign or malignant.
- The Spambase data set consists in 57 attributes describing a collection of 4601 spam and non-spam emails.
- The Battalia3 data set is an artificial data set describing 2000 generated exoplanets with 27 numerical attributes.
- The MV2 dataset features 2000 data points, each described by 6 features. They have been randomly generated from a mixture of one noise and four Gaussian components.
- The Isolated letters (Isolet) data set has 617 input variables describing 7797 voice recordings of individuals who spoke the name of each letter of the alphabet.
- The Madelon data set is an artificial data set containing 4400 data points grouped in 32 clusters placed on the vertices of a five dimensional hypercube. Then 15 redundant features and 480 useless features (random probes) were added, for a total of 500 attributes.

4.2 Experimental Protocol

In our experiments, we split the databases in order to obtain a horizontal collaborative clustering setting – i.e. each data site has access to different input variables for the same set of observations.

4.3 Results and Analysis

Co-LUPI Using GTM Models. The Co-LUPI algorithm has been applied to each of the 6 data sets mentioned in Sect. 4.1 using Generative topographic mappings. The criterion for acceptance of collaboration was the improvement of the Davies-Bouldin index. This is an internal index, thus it does not require prior knowledge on the data structure. A second version of the algorithm, RCo-LUPI, has been implemented. It features a new, random, initialization of the responsibility matrix at each step, along with the collaboration matrix. This technique is often used in unsupervised learning, it is meant to reduce dependence on the initial parameters. Table 1 shows that in most cases, RCo-LUPI performed slightly better than Co-LUPI.

In order to visualize the dynamic of this process, one can look at the successive confidence matrices of the collaboration step. Figure 2 represents such data. The Co-LUPI algorithm has been applied to the WDBC data set, split among 18 data sites. Obviously, not all algorithms benefited from the collaboration at each step. In particular, the algorithm running on data site number one did not improve its results until the fifth iteration in the collaboration step. Furthermore, on iteration number 7, only the second algorithm did improve. While this can be interpreted as a sign of an imminent end of the process, 3 other algorithms benefited from these new findings on the next iteration. The process did not end before 4 more iterations.

Table 1. Experimental results, Davies-Bouldin index - Co-EM: [9], Co-MV: [10], Co-GTM: [8], Co-SOM: [11], Co-LUPI, RCo-LUPI

Name	Co-EM	Co-MV	Co-GTM	Co-SOM	Co-LUPI	RCo-LUPI
Wdbc	0.85	0.97	0.9	0.84	0.78	**0.69**
Spam base	0.94	1.27	0.92	0.87	**0.42**	0.59
Battalia3	2.43	2.83	2.68	2.51	1.47	**1.37**
MV2	1.34	1.34	1.61	1.44	0.86	**0.85**
Isolet	–	–	–	–	1.33	**1.31**
Madelon	–	–	–	–	0.87	**0.82**

4.4 Comparison with Other Collaborative Approaches

The Co-LUPI and RCo-LUPI algorithms were empirically compared to four recent implementations of collaborative clustering algorithms. The optimization process behind the Co-EM algorithm is based on variational EM. It optimizes a collaborative term which is equivalent to the entropy [9]. The same principle is used in Co-EM with the difference that it is based on prototypes, while the Co-EM is based on partitions [10]. In the Co-SOM and Co-GTM methods, the SOM and GTM loss functions where modified in order to penalize the difference between local and remote parameters [8,11].

In order to assess the performance of our approaches, we use the Friedman test and Nemenyi test recommended in [12]. First, algorithms are ranked according to their performance on each dataset. There are as many rankings as their are datasets. Then, the Friedman test is conducted to test the null-hypothesis under which all approaches are equivalent, and in this case their average ranks should be equal. If the null hypothesis is rejected, then the Nemenyi test will be performed. If the average ranks of two approaches differ by at least the critical difference(CD), then it can be concluded that their performances are significantly different. In the Friedman test, we set the significance level $\alpha = 0.05$. The Fig. 1 shows a critical diagram representing a projection of average ranks of the algorithms on enumerated axis. The methods are ordered from left (the best) to right (the worst) and a thick line connects the groups of algorithms that are not significantly different (for the significance level $\alpha = 5\%$). As shown in Fig. 1, Co-LUPI and RCo-LUPI seem to achieve some improvement over the other proposed techniques. But the results are not sufficient to conclude to a statistically significant improvement. This result can be explained by the small number of datasets and by the fact that the Nemenyi test only considers algorithms performances through their ranks and is blind to the actual value of the performance index.

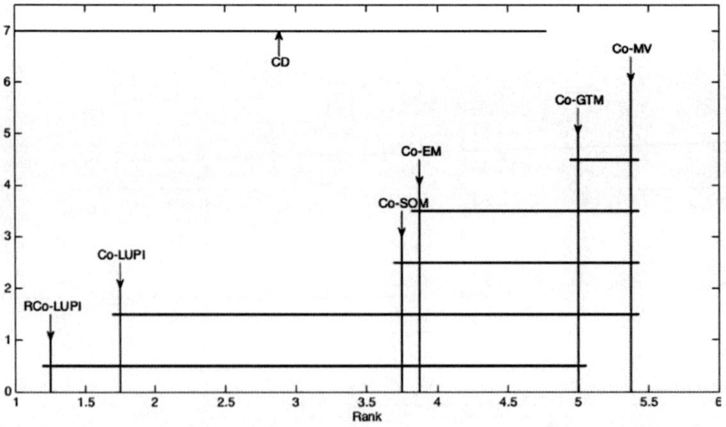

Fig. 1. Friedman and Nemenyi test for comparing multiple approaches over multiple data sets: Approaches are ordered from left (the best) to right (the worst)

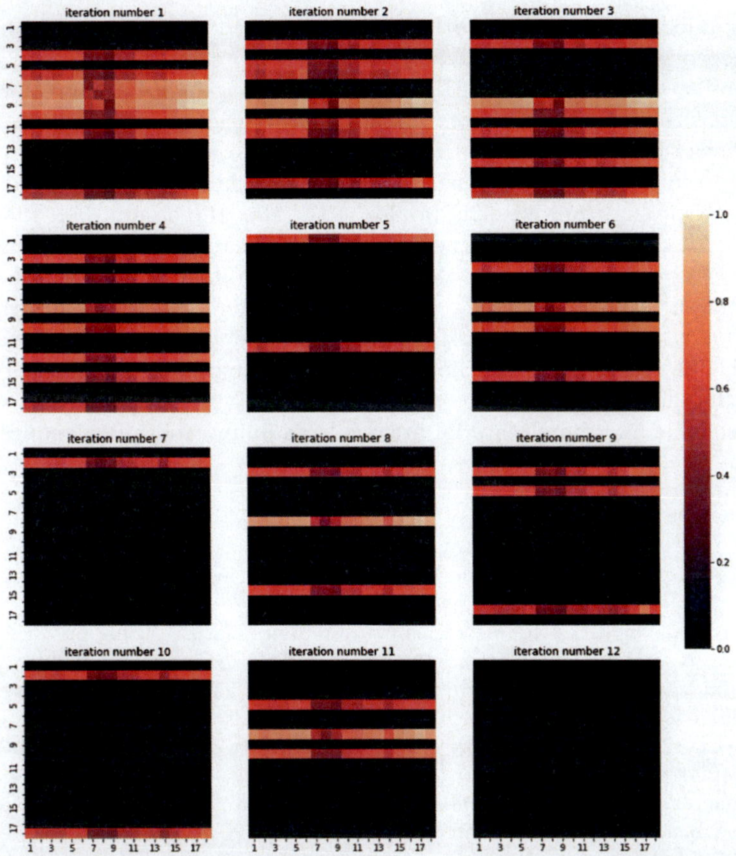

Fig. 2. Information flow during collaboration for the Wdbc data set. Each row represents a data site and each column the total weight assigned to that source.

5 Conclusion and Future Work

We introduced Co-LUPI and RCo-LUPI, algorithms based on usage of privileged information and probability for collaborative clustering. This allows local algorithms to fine-tune collaboration depending on their (un)certainty –and their remote counterparts' (un)certainty– about the classification of each data point, as measured by the entropy. This update rule is straightforward and the collaboration is done with every remote site at the same time. This avoids the classical problem of choosing which site to collaborate with at each step.

We tested our approach on several data sets in the horizontal collaboration setting, but it is also applicable in the hybrid setting. The results showed improvement over state of the art. The framework also provides a way to visualize information flow during the process. It exhibited interesting behaviors, as algorithms with lower initial performance tended to make heavier use of incoming information than the other ones. Moreover, even the algorithms with the best results after the local step were able to improve in the process. Indeed, the flexibility brought by Co-LUPI in the weighting of incoming information allows algorithms to benefit from globally less efficient counterparts, because the latter can be locally more efficient.

These results are very promising and the algorithm can be improved in several ways. One big improvement would be to relax the assumption that every algorithm look for the same number of clusters. This might be challenging as the algorithms also have to agree on the identity of the different clusters. This problem is related to the generalized assignment problem. We will also investigate how is the CoLUPI algorithm performing when used with some of the most representative algorithms to solve the problem, e.g. monarch butterfly optimization (MBO) algorithm [13].

Source code and more material are available online and under request [14].

References

1. Jain, A.K., Murty, M.N., Flynn, P.J.: Data clustering: a review. ACM Comput. Surv. **31**(3), 264–323 (1999)
2. Vapnik, V., Vashist, A.: A new learning paradigm: learning using privileged information. Neural Netw. **22**(5–6), 544–557 (2009)
3. Vapnik, V., Izmailov, R.: Learning using privileged information: similarity control and knowledge transfer. J. Mach. Learn. Res. **16**, 2023–2049 (2015)
4. Pedrycz, W.: Collaborative fuzzy clustering. Pattern Recogn. Lett. **23**(14), 1675–1686 (2002)
5. Cornuéjols, A., Wemmert, C., Gançarski, P., Bennani, Y.: Collaborative clustering: Why, when, what and how. Inf. Fusion **39**, 81–95 (2018)
6. Bezdek, J.C.: Pattern Recognition with Fuzzy Objective Function Algorithms. Springer, Kluwer Academic Publishers, New York (1981). https://doi.org/10.1007/978-1-4757-0450-1
7. Wemmert, C., Gançarski, P., Korczak, J.J.: A collaborative approach to combine multiple learning methods. Int. J. Artif. Intell. Tools **09**(01), 59–78 (2000)
8. Ghassany, M., Grozavu, N., Bennani, Y.: Collaborative generative topographic mapping. In: Huang, T., Zeng, Z., Li, C., Leung, C. (eds.) Neural Information Processing - 19th International Conference, ICONIP 2012, Doha, Qatar, November 12–15 (2012) Proceedings, Part II. Lecture Notes in Computer Science, vol. 7664, pp. 591–598. Springer, Heidelberg (2012)
9. Sublime, J., Matei, B., Cabanes, G., Grozavu, N., Bennani, Y., Cornuéjols, A.: Entropy based probabilistic collaborative clustering. Pattern Recogn. **72**, 144–157 (2017)
10. Ghassany, M., Grozavu, N., Bennani, Y.: Collaborative multi-view clustering. In: The 2013 International Joint Conference on Neural Networks, IJCNN 2013, Dallas, TX, USA, August 4–9, 2013, pp. 1–8. IEEE (2013)
11. Grozavu, N., Bennani, Y.: Topological collaborative clustering. Aust. J. Intell. Inf. Process. Syst. **12**(3), 1–6 (2010)
12. Demsar, J.: Statistical comparisons of classifiers over multiple data sets. J. Mach. Learn. Res. **7**, 1–30 (2006)

13. Wang, G.-G., Deb, S., Cui, Z.: Monarch butterfly optimization. Neural Comput. Appl. **31**(7), 1995–2014 (2019)
14. Foucade, Y.: CoLUPI: learning from data and learners [Source code and more material] https://github.com/yfoucade/colupi

Using Microsoft Kinect V2 for Custom Upper-Limb Rehabilitation Exercises

Shourav Bin Rabbani, Amin Ahsan Ali, and M. Ashraful Amin(✉)

Agency Lab, Department of Computer Science and Engineering,
Independent University, Bangladesh, Dhaka, Bangladesh
{1531162,aminali,aminmdashraful}@iub.edu.bd

Abstract. Motor disability due to stroke, accident, or other illnesses in many cases can be partially or fully recovered with guided and/or prescribed physiotherapy. Moreover, due to the nature of the problem, in most of the case, the treatment is needed to be continued for years. Thus, one of the main challenges in this is the availability of an expert physiotherapist to guide the treatment. Most of the time, after initial guided physiotherapy sessions, patients are prescribed exercises to practice at home by themselves. Patients may be able to perform the task at home; however, the expert is present during home sessions, hence, no performance assessment. In this paper, we present a prototype system to resolve this problem using 3D sensor technology: Microsoft Kinect v2. An expert can record an exercise by performing it in front of the 3D camera and provide the prescription (how many times and when the exercise should be performed) on the system. At home, the patient performs the exercise in front of our Kinect-powered system which guides the patient through visual cues on the monitor. In this initial study, five occupational arthritis exercises on the right arm were recorded. We matched these exercises with 15 healthy volunteers (10 male, 5 female) standing 1.5 ms from the system. To overcome different body shapes, we used geometric coordinates from recorded frames and recreated the visual cues. Our system matched the exercises with uncertainty of 10px (2% of depth resolution) and volunteers adapted to screen instructions with minimum supervision.

Keywords: Rehabilitation · Microsoft kinect · Therapy · Telehealth · Human-computer interaction · Human activity

1 Introduction

Strokes and post-recovery motor disabilities are every common for older population [3] which can be recovered through exercises. Unfortunately, such programs are expensive [12], so research to reduce cost and create better method of rehabilitation is a hot topic. Recently the use of Microsoft Kinect, a motion capture

Supported by a grant from the ICT Division of the People's Republic of Bangladesh Government.

A. Abraham et al. (Eds.): SoCPaR 2020, AISC 1383, pp. 508–519, 2021.
https://doi.org/10.1007/978-3-030-73689-7_49

tool for video game is gaining popularity in the field of Rehabilitation Engineering. As the Kinect is able to get body locations through its depth sensor, it was used in several "telehealth" systems to assist expert instructors and make therapy more accessible to the patients.

Firstly, Kinect provide convenient and affordable access. For example, one of the first use of Kinect in rehabilitation were to recover balance for stroke patients by Lai et al. [6]. And Chang et al. [2] used a Kinect based system to visualize and get more correct motions from the participants (intervention phase) than by a therapist telling what to do (baseline phase). Both studies concluded positive results and discussed about the Kinect's affordability and support.

Secondly, Kinect provides accurate representation of the human body and can be used in several rehabilitation systems. For example, Tupa et al. [8] started to use Kinect for observing stroke recovery. They selected a few exercises and use the angles between limbs to access the motion from three different angles to get the complete picture of the patient. Exercise movement was scored ranging from 0 to 6, where 0 means the subject was unable to do the exercise and 6 means the subject completed the exercise successfully. Or for computing the Fugl-Meyer Assessment (FMA) [11] as done by Lee et al. [7] using Microsoft Kinect V2. The proposed system automated FMA with a very high correlation ($r2 = 0.982$) with the in-person FMA.

Thirdly, rehabilitation with Kinect was used with virtual reality. For example, Demers et al.[4] used the Kinect v1 to score patients doing mimic real life tasks such as making toasts, doing dishes etc. and found improvement over time. Similarly, a Kinect based virtual reality based rehabilitation impact and acceptability study was conducted by Adams et al. [1]. He received patients' feedback via questionnaires and interviews and reported that such systems are fun with some minor eye fatigue for a few individuals. Furthermore, Moldovan et al. [9] developed video games for virtual rehabilitation using MIRA platform (analytic platform), Microsoft Kinect and Leap sensors. He reported that his patient's score improved from 46 to 52.

However, most of these studies are done with a few preset exercises or tasks. Therapist often customize the exercise based on patients as different patient might require different range of motions for rehabilitation. So we propose a system, as a pilot study, that allows recording and comparing exercises done by patients. We limited the system to the right arm to reduce system and study complexity. Our aim is to make rehabilitation more affordable and deployment to community clinics and personal home computers.

The paper layout is as followed: (i) System design describes how the main modules of the proposed system; (ii) Study design tests the proposed system by recording 5 different exercises and matching them on 15 volunteers; (iii) Results show the system performance on different exercises and person of different height, weight and gender; (iv) Discussion provides study limitations and possible improvement of the proposed system; (v) Conclusion and acknowledgement.

2 System Design

Microsoft Kinect v2 has two cameras, which provides the image frames and metadata. The image can be broken into the color frame (1920 × 1080) and depth frame (512 × 424). The metadata is timestamped and contains the coordinates of the 24 logical joints of up to 6 human bodies. Figure 1 shows the joint locations overlaid on a human body and their indices [5]. Kinect provides joint data as 2D for screen usage and 3D for realistic spatial processing. Such convenient access to joint data makes the Kinect a good tool for analysing human motion.

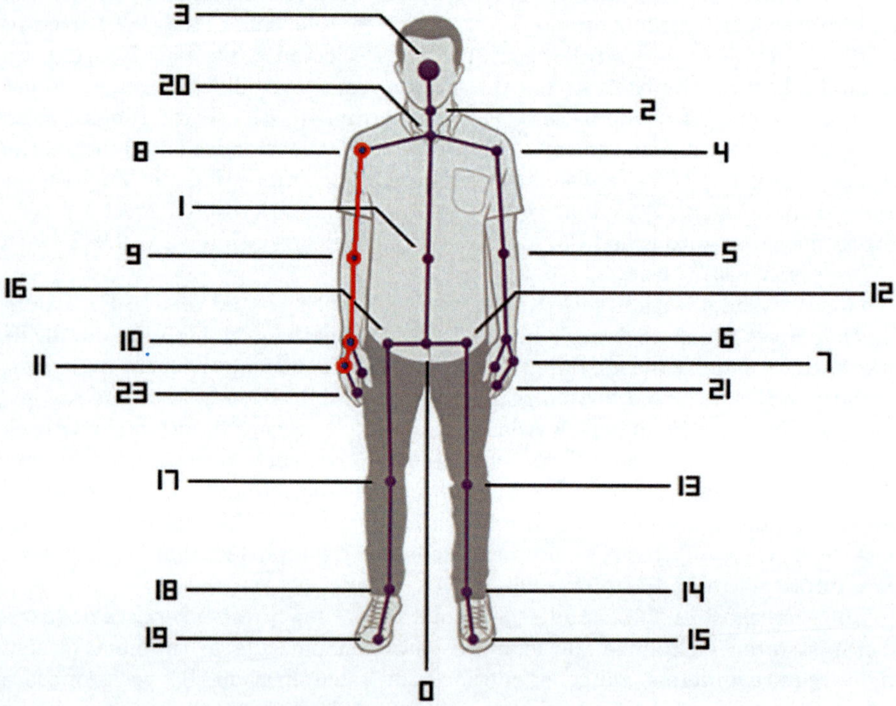

Fig. 1. Microsoft Kinect v2 provides the 3D locations of 24 joints in real time, which makes it a convenient tool for therapy. The figure shows all the joints and their indices. For the study, only the right arm, 4 joints (shown in red) were used.

Our proposed system (Fig. 2) will use in the Kinect image data and metadata for recording custom and matching custom exercises with different persons. The system can be broken in 4 modules: (i) master module deals with data input-output, communicates data with other modules and deals with storage; (ii) drawing module displays the image frames and other overlaying graphics to the screen; (iii) Recording and matching modules are the main functional modules and are the discussed in the following subsections.

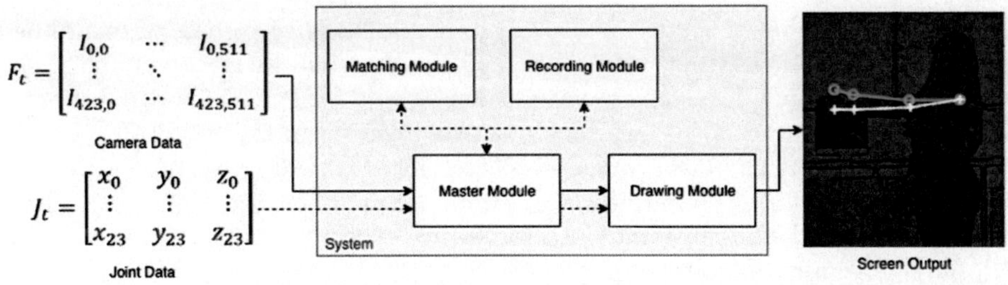

Fig. 2. The system acquires the image data (512 × 424), F_t, and joint data (3 × 24), J_t, every frame. The system will use the joint data to record exercises and match them to different users. Finally it will display the image and overlaid joint data, in white, while also showing the correct arm location to move in green. In this case, the screen is instructing the user to move up her arm.

2.1 Recording Module

A doctor or an expert can record any exercise as needed for the patient with a press of a button on the screen interface. The system will start acquiring a stream of frame data and joint data and show the exercise to the Kinect. Let the recording be for a total of N frames, then on each frame the recording module will store the following data: (i) 3D locations of the joints with timestamp; (ii) Calculate angles between adjacent joints. These angles are used to overcome the difference in joint distances which varies from person to person.

2.2 Matching Module

For matching, a user can select the exercise and follow the on screen instructions to do the exercise effectively. The system loads the recorded data files and angles between two adjacent joints are used to recreate the exercise overlay instruction for the current user.

Afterwards, for each frame from frame 1 to N (as stored during recording session), the system does the following: (i) acquire user's 3D locations of the joints with timestamp. (white line markers); (ii) Use the joint angles to re-create the destination joint locations for the current user's body shape on screen as shown with green line marker in Figs. 3 and 5; (iii) Calculate error distances of the joints in pixel units; (iv) Accepts and increases frame number by 1 when error distances are within the threshold parameter; (v) Store all data for later use. This way the system ensures that the user follows the recorded motion trajectory. The threshold parameter is kept at 10px i.e. 2% of the depth resolution for the study design. This can be adjusted as per needed, e.g. lower values will cause stricter matching.

3 Study Design

The study is limited to right arm joints which are shoulder, elbow, wrist and palm and for one body at a time. This is to reduce study complexity and empirically test the proposed system modules. To check the effectiveness of recording and matching modules, first, a few different exercises will be recorded and then matched with different persons of various height and weight. Since real patients are hard find and special permissions must be taken from the hospital, healthy volunteers from the campus ground is selected for this study. Lastly, the recording user and matching volunteers will stand 1.5 ms from the Kinect to allow the system record the full joint data for future study. For this distance, 10px threshold gave a good match.

3.1 Recording Exercises

Five arthritis exercises about occupational exercises are selected from an online source (see Fig. 3) [10] as they contain a good range of motions for the arm. The exercises are recording by standing 1.5 ms away from the Kinect and turned to give best view of the joints (see Fig. 3 and Fig. 7). The five exercises are as follows: (i) Flexion (ii) Extension (iii) Abduction (iv) Adduction (v) Circumduction.

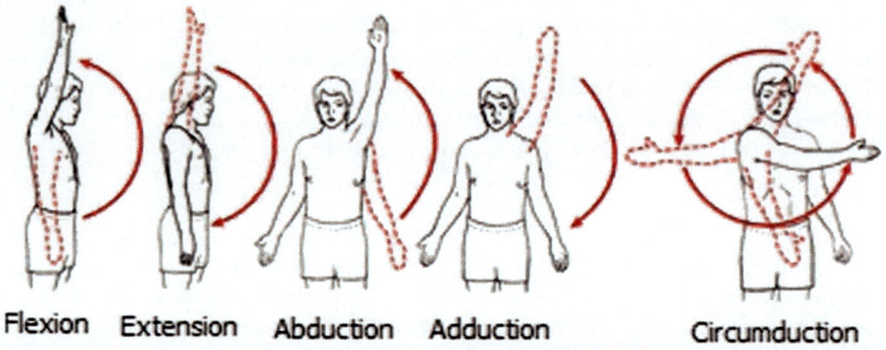

Fig. 3. Five exercises selected to be recorded which contain a good range of different motion to test.

3.2 Matching the Exercises for Different Persons

We match the 5 recorded exercises done by 15 healthy volunteers (5 female) for the study. The volunteers are briefed about the setup (Fig. 4). All volunteers will stand at 1.5 m from the Kinect's field of view and try to imitate the exercise as shown from the on-screen display. They will enter the room one by one and do the tests and were asked to refrain from discussing the test with other volunteers after they were done. After the briefing, a demonstration was shown for a simple arm motion. A computer operator is present to confirm the data storage and monitor any error messages.

Fig. 4. Experiment setup, from left to right, the volunteer standing 1.5 m from the Kinect, the Screen Display which instructs how to move the right arm as per the exercise selection and a computer operator is overseeing the data storage procedure.

We record the volunteers' physical attributes like age, height and height individually before sending them to perform the exercises. For each exercise, 10 iterations were completed. As well as the raw data from the Kinect, the error distances for different joint lengths were recorded for each exercise and iteration. The total time taken for each iteration were also recorded. Finally, after the test they were asked about the test, their performance score and if they faced any difficulty with the screen instructions.

As the exercises were recording, the users had to turn around as can be seen in Fig. 3 and Fig. 5 (standing at the same position) to get the best view of the arm and the motion. The motion range are as follows: (i) the flexion is the raising of the arm from waist height to above shoulders by rotating the hand from the front of the body. (Fig. 5 block 1) (ii) the extension is the opposite of the flexion. It starts by moving the arm from above the head to the waist position (Fig. 5 block 2) (iii) abduction starts by opening the palm and extending out toward the waist height, then raise up from sideways (not front like the flexion) and up till its above of the shoulder (Fig. 5 block 3) (iv) adduction is the reverse of the exercise 3. It starts from top of the shoulders, palms descending to the waist height (Fig. 5 block 4). (iv) right arm circumduction starts from the mid part of the vertical abduction, 360° arm rotation counter-clockwise wise around the shoulder and back to the original position. (Fig. 5 block 5)

4 Results

First, the volunteers are all young individuals who had a mean age, height and weight of 21.9 years, 1.66 m and 65.9 kg respectively. Their data is shown in

Table 1. They did not face any difficulties following the on-screen instructions and recreated the exercises correctly.

Second, the error distances of all the exercises and iterations are averaged for each joints and the standard deviation was calculated. The summary was shown in Fig. 6. The mean error distances under 10px and has a standard deviation under 1px. Having mean error under 10px means the system is following the threshold parameter and a low standard deviation shows that the system had no difficulty matching different bodies.

Finally, the time taken to complete each iteration are also analysed. It can be seen that eventually all volunteers learn to do the exercise and performs faster with each iteration. The exercise 5 takes about 5.67 s during the recording phase and the volunteers are reaching duration as close the base value. Table 2 show the data, here each volunteers (row-wise) shows over decrease in time with iterations (column-wise). Few volunteers show increased time at last iteration as their arms get tired and they slow down. This shows that the volunteers indeed understood the screen instructions and imitating exercise was a good method to convey information. Figure 7 shows 4 random volunteers' time taken to complete the 5th exercise.

5 Discussion

Our study had three limitations to reduce complexity and prevent adding error sources. First, we focused on the right arm only as most physiotherapy exercise target parts of the body rather than the whole but if needed the full body exercise can also be done similarly. Second, we limited tracking to only one body, but the Kinect is capable of tracking up to 6 bodies which makes it a great tool to be deployed in remote clinics. Our system can also be modified to accommodate this with some tweaking as discussed next. Third, all study was done standing 1.5 m away from the Kinect to prevent changing error distances. However, to accommodate up to 6 patients, our system requires additional module that will auto-select the matching threshold based on the distances.

Moreover, we calculated error distances are in pixel units with a fairly simple matching algorithm and got satisfactory results. Further development and study will be conducted with advanced techniques, depending on the patient's ailment (weakness in particular joints or range of motion), we will use weighted error calculation method and more sophisticated matching methods that may give better results. The human-computer interactions should also be considered as our actual audience (older patients) may have difficulty using technology unlike the younger generation. The exercises we used was only for testing the system, in reality, doctors may prescribe very different exercises but we think the recording module will work well regardless.

Furthermore, we studied the system using healthy volunteers in their 20s but our result might change when tested on patients who might be over 50 year old and are suffering through multiple complications which may change how the system is used.

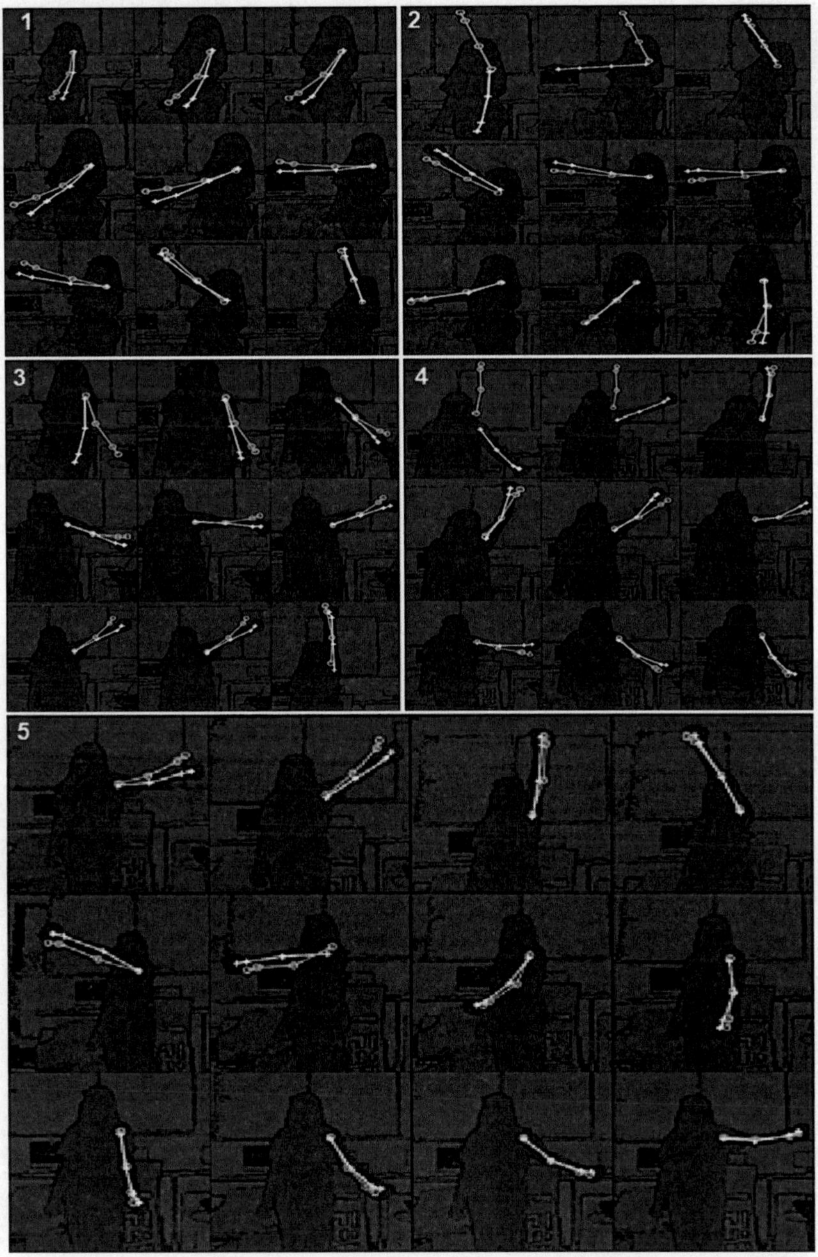

Fig. 5. The figure is set as numbered blocks, each showing range of motion done by each exercise. Each exercise starts from a rest position (top-left of each block) and ends at another rest position (bottom-right of each block). The green line tells the user where to move the hand as shown as the white line, each of the joints are also shown by different markers.

Table 1. Volunteers' age, height and weight

#	Gender	Age (yrs)	Height (m)	Weight (kg)
1	F	18	1.60	69
2	F	20	1.55	60
3	F	23	1.57	48
4	F	23	1.57	58
5	F	23	1.55	40
6	M	24	1.72	84
7	M	23	1.72	72
8	M	20	1.78	58
9	M	23	1.65	55
10	M	20	1.67	70
11	M	23	1.67	59
12	M	20	1.65	65
13	M	21	1.72	80
14	M	25	1.69	80
15	M	23	1.80	90
Mean		21.9	1.66	65.9

Table 2. Time taken in seconds by volunteers (rows) on each iteration (columns) for exercise 5

	Iterations									
#	1	2	3	4	5	6	7	8	9	10
1	8.4	6.4	6.4	7.2	8.4	10.2	8.2	8.2	7.6	11.4
2	6.0	9.4	7.0	7.2	7.8	6.8	6.2	6.2	6.1	5.8
3	5.4	5.6	5.7	5.6	5.5	5.4	5.5	5.5	5.5	5.7
4	7.4	5.5	6.4	7.2	8.4	10.2	8.2	8.2	7.6	11.4
5	8.5	5.6	7.0	7.2	7.8	6.8	6.2	6.2	6.1	5.8
6	5.4	5.7	5.8	5.7	5.6	5.4	7.4	5.8	5.8	6.1
7	6.9	6.1	6.0	5.8	5.5	9.0	6.2	5.7	6.8	6.0
8	11.4	7.9	5.9	5.6	6.1	6.0	5.9	5.7	6.4	6.0
9	5.9	6.1	5.9	6.5	5.9	6.1	5.8	5.9	6.0	7.5
10	7.0	8.0	8.3	6.4	5.6	5.6	6.0	5.5	5.5	5.5
11	6.0	5.7	5.5	5.9	5.4	5.7	6.0	5.6	5.4	5.6
12	6.0	5.4	6.0	5.4	5.6	5.4	5.4	5.4	5.4	5.4
13	11.0	6.5	7.1	9.7	6.5	6.6	10.8	7.9	7.2	6.7
14	8.1	8.0	11.1	6.7	5.9	5.7	5.9	7.1	7.3	7.0
15	6.6	5.7	19.5	5.4	5.9	5.9	5.4	5.9	5.7	6.4

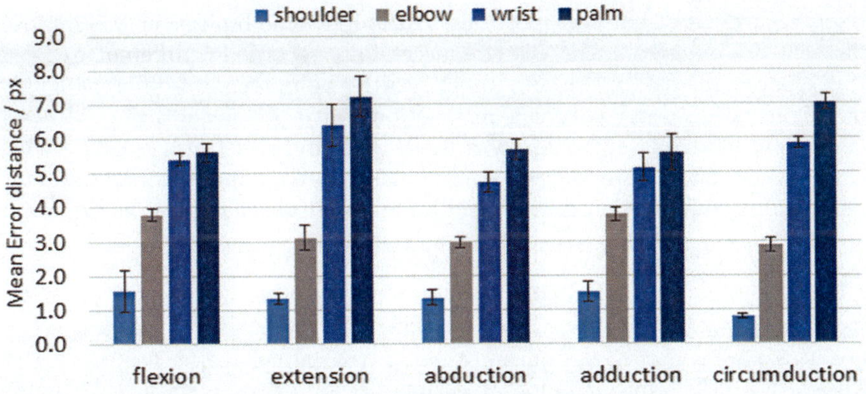

Fig. 6. Shows the mean error distances in pixel units of all users and their standard deviation. The joints that are further away from shoulders have more error distance. The low standard deviation shows that the matching module had no problem with different persons.

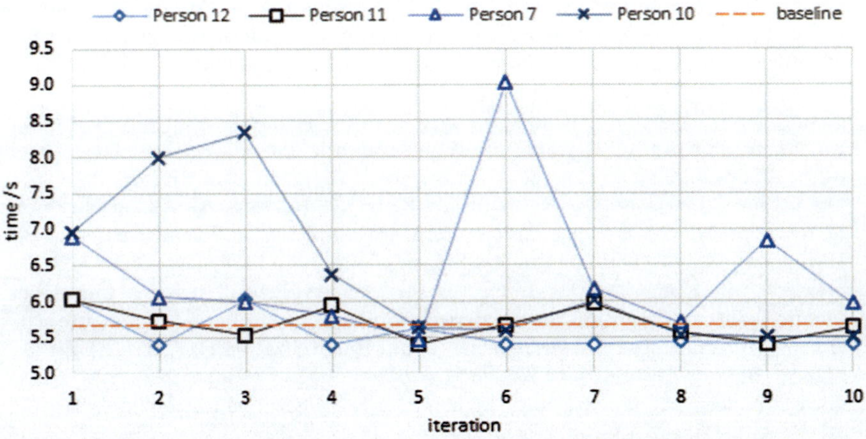

Fig. 7. Time taken to complete the 5th exercise by four volunteers whereas the original time taken was 5.67 s (baseline). Time eventually decreases showing the volunteers learns the exercise with iterations and performs faster. Person 7 made a mistake at 6th iteration and took a longer time.

6 Conclusion

Motor disability is a common problem and most patient can recover fully given they get proper exercise and therapy. But rehabilitation is not affordable and available in remote locations. So, a system was proposed which will use the Microsoft Kinect v2, a motion sensor that allows us to conveniently access human joint data for easy spatial and motion processing. The proposed system is able to record motion for different exercises and match it with different persons regardless of body proportions. Although some limitation was imposed to prevent adding errors, the system successfully recorded 5 different exercises and compared the said exercise with 15 healthy volunteers, of different heights and weight, using only on-screen instructions with 2% uncertainty. Although fairly simple matching algorithm was used, we found satisfactory results which we can fine tune for more focused study in future. We hope to collaborate with experts from rehabilitation centers and build a digital rehabilitation aid system that will connect patients and doctors at remote clinics or homes at affordable rate.

Acknowledgement. We thank all the 15 volunteers who donated their time for this study. This work is part of the Human Activity Recognition project for partially disabled patients seeking rehabilitation. It is supported by a grant from the ICT Division of the People's Republic of Bangladesh Government.

References

1. Adams, R.J., Lichter, M.D., Krepkovich, E.T., Ellington, A., White, M., Diamond, P.T.: Assessing upper extremity motor function in practice of virtual activities of daily living. IEEE Trans. Neural Syst. Rehabil. Eng. **23**(2), 287–296 (2015). https://doi.org/10.1109/tnsre.2014.2360149
2. Chang, Y.J., Chen, S.F., Huang, J.D.: A kinect-based system for physical rehabilitation: a pilot study for young adults with motor disabilities. Res. Dev. Disabil. **32**(6), 2566–2570 (2011). https://doi.org/10.1016/j.ridd.2011.07.002
3. Chowdhury, R., Hasan, A., Rahman, Y., Khan, S., Hussain, A., Ahsan, S.: Pattern of neurological disease seen among patients admitted in tertiary care hospital. BMC. Res. Notes **7**(1), 202 (2014). https://doi.org/10.1186/1756-0500-7-202
4. Demers, M., Kong, D.C.C., Levin, M.F.: Acceptability of using a kinect-based virtual reality intervention to remediate arm motor impairments in subacute stroke. In: 2017 International Conference on Virtual Rehabilitation (ICVR). IEEE (2017). https://doi.org/10.1109/icvr.2017.8007496
5. Joreg: Kinect documentation. https://vvvv.org/documentation/kinect (2017). Accessed 03 Nov 2018
6. Lai, C.L., Huang, Y.L., Liao, T.K., Tseng, C.M., Chen, Y.F., Erdenetsogt, D.: A microsoft kinect-based virtual rehabilitation system to train balance ability for stroke patients. In: 2015 International Conference on Cyberworlds (CW). IEEE (2015). https://doi.org/10.1109/cw.2015.44
7. Lee, S., Lee, Y.S., Kim, J.: Automated evaluation of upper-limb motor function impairment using Fugl-Meyer assessment. IEEE Trans. Neural Syst. Rehabil. Eng. **26**(1), 125–134 (2018). https://doi.org/10.1109/tnsre.2017.2755667
8. Lin, S., Mann, J., Mansfield, A., Wang, R.H., Harris, J.E., Taati, B.: Investigating the feasibility and acceptability of real-time visual feedback in reducing compensatory motions during self-administered stroke rehabilitation exercises: a pilot study with chronic stroke survivors. J. Rehabil. Assist. Technol. Eng. **6**, 205566831983163 (2019). https://doi.org/10.1177/2055668319831631

9. Moldovan, I.M., Tric, L., Ursu, R., Podar, A., Calin, A.D., Cantea, A.C., Dascalu, L.A., Mihaiu, C.A.: Virtual rehabilitation programme using the MIRA platform, kinect and leap motion sensors in an 81 years old patient with ischemic stroke. In: 2017 E-Health and Bioengineering Conference (EHB). IEEE (2017). https://doi.org/10.1109/ehb.2017.7995427
10. Nisabel: Occupational therapy fieldwork toolki. http://speakingofot.blogspot.com/2012/04/toolkit-of-all-toolkits.html (2012). Accessed 03 Nov 2018
11. Sanford, J., Moreland, J., Swanson, L.R., Stratford, P.W., Gowland, C.: Reliability of the Fugl-Meyer assessment for testing motor performance in patients following stroke. Phys. Therapy **73**(7), 447–454 (1993). https://doi.org/10.1093/ptj/73.7.447
12. Urmi, U.K.: Barriers of stroke patients during receiving treatment at CRP. http://dspace.crp-bangladesh.org:8080/xmlui/handle/123456789/322 (2017). Accessed 03 Nov 2018

Customer Churn Prediction Using Deep Learning

Omer Faruk Seymen[1], Onur Dogan[2(✉)], and Abdulkadir Hiziroglu[2]

[1] Sakarya University, Esentepe Campus, Üniversite Caddesi, 54050 Sakarya, Turkey
ofseymen@sakarya.edu.tr

[2] İzmir Bakırçay University, Gazi Mustafa Kemal Mahallesi, Kaynaklar Caddesi Seyrek, 35665 Izmir, Turkey
{onur.dogan,kadir.hiziroglu}@bakircay.edu.tr

Abstract. Churn studies have been used for years to achieve profitability and to establish a sustainable customer-company relationship. Deep learning is one of the contemporary methods used in churn analysis due to its ability to process huge amounts of customer data. In this study, a deep learning model is proposed to predict whether customers in the retail industry will churn in the future. The model was compared with logistic regression and artificial neural network models, which are also frequently used in the churn prediction studies. The results of the models were compared with accuracy classification tools, which are precision, recall and AUC. The results showed that the deep learning model achieved better classification and prediction success than other compared models.

Keywords: Predictive analytics · Customer churn modelling · Deep learning · RFM

1 Introduction

In today's competitive world in retail industry, customer churn is one of the main concerns. In the retail domain, churn is described which customers end doing transactions and leave a company [21]. This results in a possible loss of income for the company. Furthermore, it has become a compelling issue to maintain customers. Therefore, it is essential to recognize the customers who may potentially leave the company in a short period. This process is described as churn prediction.

The concept of churn is the application area of analytical customer relationship management in the widest perspective. In fact, this application area is a sub-part of customer behavior modeling in customer analytics. From the marketing perspective, the concept of churn is associated with customer loyalty and customer value concepts that are related to each other. When looking at the basic variable sets of customer value, they are an indication of the investment or cost associated with corresponding behavioral transactions for a customer

A. Abraham et al. (Eds.): SoCPaR 2020, AISC 1383, pp. 520–529, 2021.
https://doi.org/10.1007/978-3-030-73689-7_50

within his/her lifetime. There are different variables that are essential in behavioral customer modeling. Recency, frequency and monetary (RFM) is one of the most common variable sets that are utilized in behavioral customer modeling.

Recency, frequency and monetary (RFM) analysis is a robust and popular method for many domains such as the retail industry [4], banking [9], e-Commerce [6,13]. It is generally applied to measure the customer value according to the past shopping records [10]. RFM was firstly introduced by Hughes [11] as he describes the important customers as those concurrently possessing low recency, high frequency and high monetary values. Recency denotes the period between the time of the latest and current events, while frequency describes the event repetitions in the past transaction, and monetary expresses the total expense amount made for the company.

Many studies combined successfully RFM analysis with data mining process in recent years [5,9]. Some studies first segment customers using RFM analysis and then apply data mining methods to create patterns for a set of important customers [15,16]. Also, data mining methods can be considered as efficient in churn prediction according to the research performed in the last years [21]. Development of the practical churn prediction models is a vital duty that includes many examinations from identifying the best predictor variables to the choice of a powerful predictive method fitting for these variables. Retail industries collect huge amount of data about customers and using the past shopping data, the possibility of a customer's churn rate can be calculated to decide whether they will leave or stay. Due to the necessity of processing huge amount of data, deep learning methods among up-to-date machine learning techniques, have an important place for customer churn modeling.

The performance of a churn prediction model relies extremely on the variables selected from the dataset [12]. The traditional approaches include two significant obstacles; i) Feature extraction procedure among thousands of customer characteristics is mainly tedious and time-consuming and frequently applied by an expert. ii) The churn prediction model is often developed for a particular dataset. Because deep learning methods can automatically find out useful features, a deep learning algorithm can be applied for customer churn prediction in retail domain. Especially in studies with a lot of input attributes, it is important to use current ML methods such as DL to produce better predictive results. In this study, a deep neural network structure was developed and applied to the churn analysis using a supermarket transaction dataset. The empirical outcomes present that deep-learning-based models yield solutions as good as traditional models, without using the manual interaction for feature extraction. The aim of this study is to analyze customer churns using a deep learning algorithm in retail domain and compare its performance with other well-known churn modeling approaches.

Within this framework, the study first expresses a detailed literature review to show the scientific gap in this domain in Sect. 2. Then the proposed deep learning algorithm is explained in Sect. 3. Section 4 compares the results of proposed methodology and various algorithms considering some performance metrics. The results show that the proposed deep learning algorithm surpasses the selective comparative algorithms. Section 5 concludes the study and Sect. 6 discusses the results.

2 Background and Related Work

Deep learning is a sub-domain in artificial neural networks (ANNs), which consists of many hidden layers. Each layer of neurons in deep learning algorithms is trained on the data features of the previous layer. Therefore, a feature hierarchy of increasing abstraction can be created, which is good at discovering patterns within a dataset [21]. The use of past transaction data is broadly used in former studies [3,14]. Using transaction data is regarded to be an advantage that allows individualizing a special offer to enhance sales performance. However, because the customers' choices may change in time, such a procedure may cause an information loss or incorrect performance results.

There have been applications of churn modeling in retail domain, only the studies that are of the domain and with RFM focus as well as deep learning applications in churn context will be reviewed. Churn studies mostly based on contractual setting sectors such as telecommunication, banking, insurance, etc. where a customer has to sign a contract to get the service from the provider. Thus, a customer who terminates the contract labeled as a churner by the company whereas the customer who continues to get service accepted as a non-churner. For a non-contractual setting such as the retail sector, churn studies needed to determine who is churner or not. Churn prediction studies in the retail sector have been handled differently in the literature using different types of features and models. Churn prediction studies in the retail sector using deep learning algorithms, logistic regression and neural networks are presented in Table 1. Buckinx and Van den Poel [2] suggested a partial-churn definition for non-contractual settings. The customers who do not shop for 3 consecutive months or shops under a predefined spending limit, will be taken as churners. They used an RFM model with transactional data and demographic variables for predicting churn and used logistic regression, neural networks, and random forests to evaluate the model. Migueis et al. [19], developed an RFM based model to predict customer churn by using the succession of product category purchasing history. Purchase history handled as a basket analysis method and but only the purchased products used for the model. The logistic regression technique was used to evaluate the RFM model. Migueis et al. [18] developed an RFM model via using prediction techniques, Multivariate Adaptive Regression Splines and Logistic regression. These studies mostly based on RFM data and lack the information on purchasing under promotional data. Dingli et al. [7] compared two deep learning algorithms, Convolutional Neural Network and Restricted Boltzmann Machine, for predicting churn in their retail sector study. The dataset and features were unknown but the RFM was declared to be used. Bashir et al. [20] used a Feature Learning Deep Learning model for classification in a large-Scale Business (B2B). Transactional data (RFM) was used as a dataset with 6000 customers. Martines et al. [17] proposed a customer prediction with a new set of customer-relevant features updated every month from previous purchases and used state of the art machine learning algorithms. They found that the gradient tree boosting method was superior to others with high accuracy and classification performance.

Table 1. Churn prediction studies with deep learning and comparison with the current work.

Study	Sector	Models/Techniques	Features	Data	Goal
[2]	Supermarket	RFM, LR, NN, RF	Transactional data, Demographic variables	5 months, 32,371 customers	Predicting partial defection by behaviourally loyal clients
[7]	Grocery	CNN, Restricted Boltzmann	Transactional data	Unknown	Implementing DL algorithms to predict customer churn
[18]	Supermarket	LR, MARS	Transactional data, Demographic variables	24 months, 130284 customers	Compare the performance of MARS with LR in modeling customer attrition
[19]	Supermarket	Forward and backward model, LR	Transactional data, Forward and backward categories succession	24 months, 74,607 customers	Predicting partial churn by considering a variable length of the first-category purchase sequence for each customer
[20]	B2B	Classification, Feature Engineering	Transactional data	28 months, 6000 customers	A generic representation of feature engineering steps that could be applied to any non-subscription based business for predicting churn
[1]	Telecom	Statistical model, RFM, LSTM, CNN	Daily call features	5 months, 20,000 customers	Investigating daily based churn prediction comparison to monthly based churn prediction
[17]	B2B	LL, ELM, GTB	Transactional data	25 months, 10,000 customers	Developing machine learning tools that predict future customer behavior in the non-contractual setting
[21]	Telecom	NN, CNN	75 features	40,000 customers	Investigating the deep learning algorithms and their applications for customer churn prediction problem
Current work	Supermarket	Classification, DL, LR, NN	Transactional Data, Basket Analysis	27 months, 10,340 customers	A DL model for churn prediction using basket analysis and RFM variables especially the effect of promotion

LR: Logistic regression, NN: Neural networks, RF: Random forests, MARS: Multivariate adaptive regression splines, LSTM: Long short-term memory, CNN: Convolution neural network, B2B: Business to Business, ELM: Extreme learning machine, GTB: Gradient tree boosting, LL: Logistic Lasso

Even though some of those studies are not related to non-contractual setting, they were included in this literature review because they analyze churn prediction with deep learning methods. Alboukaey et al. [1] proposed a daily based churn prediction model and suggested daily behaviors as a multivariate time to predict churn. A statistical model, an RFM model, A Long Short-Term Memory (LSTM) model and a Convolution Neural Network (CNN) model were applied on a mobile telecom dataset. They found that daily based churn prediction was better than the monthly based predictions. Umayaparvathi and Iyakutti [21] emphasize on feature selection of the churn prediction models. They suggested that deep learning methods were as successful as the traditional methods without picking or extracting features from datasets like the others.

As it can be seen in previous churn prediction studies in the retail domain, transactional data with RFM data was mostly used in prediction models. In this study, a partial churn model for non-contractual setting was taken into account and the data of purchasing under promotion sales were also added to the model in addition to the RFM variables. The aim of adding this feature is to measure the effect of promotion on churn studies by using a deep learning model.

3 Methodology

3.1 Data Description

In this study, a national supermarket's 27-month retail scanner dataset was used. The dataset include the following features: customer's identification number, transaction dates, and spending per each product category if the customer made transactions when promotion applied. The database was created with eight major product categories consist of spices, vegetables, and fruits; appetizers; dairy products and breakfast food; canned food; deli food; sauces and beverages; hygiene products; margarine, butter, and cooking oils.

Table 2. Features in the dataset.

Feature	Explanation	Number of features
Recency	The last day of shopping for a month	24
Frequency	The number of how many times customers shopped for a month	24
Monetary	The monthly total spend of a customer	24
Category spending	monthly spend by each product category	192
Promotion	Whether a customer buys if there is a promotion in that product category that month	192

The dataset consists of a total of 10000 customers' transactions from two major cities of Turkey, Istanbul, and Ankara. Customers from Istanbul are 6000 and the rest are from Ankara. Table 2 gives features (totally 456) for each month in the dataset. The number of customers and features are enough to train for a deep learning neural network. The dataset contains 27 months of data but only 24 months were used to train the model. The last three-month data were excluded because this part of data shows whether a customer is a churned one or not. The data first was cleaned by excluding missing values and then normalized to ensure data integrity and to eliminate data redundancy. Recency, frequency, monetary and product category spending values were ranged between 0–1 after the normalization.

3.2 Proposed DL Algorithm

This study uses a deep learning algorithm for future prediction to reveal the effect of promotion buying on customer churn.

Definition of Churn in This Model. For a non-contractual setting, a customer who has no transaction for three consecutive months or periods considered a churned customer. However, Buckinx and Van den Poel [2] and Migues [19] also considered customers who spent less than a certain spending threshold for three consecutive months as churners. Nevertheless, in this study, only customers who did not shop at all were considered churners. Starting from the 27th month, backward, a customer who does not shop for three or more consecutive months is considered as churner.

Implementation of the Model. A deep learning model, a logistic regression model and a neural network model were implemented using the same data with the same classes of churn or non-churn. The first major city customers were used in the training set (5967) and the other city customers were used in the test set (3978). So, there is a 60%–40% apportionment and all the prediction models used this proportion for the validation.

Deep Learning Model. We use Rapidminer Studio software H2O artificial intelligence library to conduct our prediction models including DL, LR and NN. Our deep learning model uses RFM variables, product category purchases and promotion data for future predictions. The effect of promotion buying on churn behavior was revealed. We used three hidden layers with 200, 200, 200 neurons respectively in deep learning model with a learning rate of 0.01, different from the standard neural network architecture which uses only one hidden layer. Figure 1 shows the proposed the churn prediction model used in the study. Customer transactions and customer demographics data were extracted from the customer retail scanner data which provided by the company. These data were cleaned, transformed and normalised in pre-processing phase to prepare for the models.

Feature extraction was applied to data before performing deep learning model to derive informative and non-redundant values. Later, the DL model was first applied to the training data for the activation of the deep learning model, Rectifier Linear Unit was selected and dataset was iterated for 10 times. The number of training data rows to be processed per iteration was set as −2. The classifiers resulting from this application were also applied on the test data. Then results were evaluated with the performance metrics such as AUC, Accuracy, Recall and Precision.

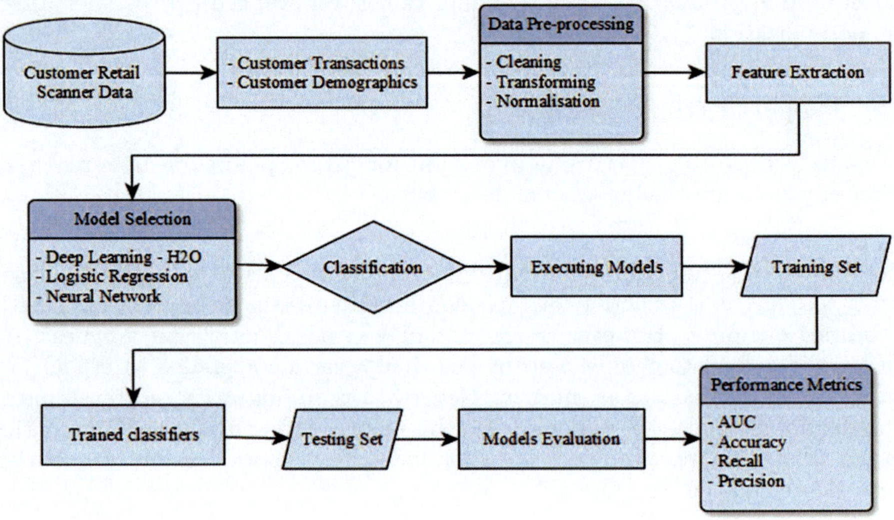

Fig. 1. Deep learning models.

3.3 Performance Metrics

Accuracy, precision, recall, and area under the curve (AUC) were used to evaluate the performance of the models. For extracting the accuracy, precision and recall ratios, a confusion matrix was used. A confusion matrix is a simple table to reveal ratios of the performance of a classification/prediction model on which the values are known. Also, it helps to visualize the performance of the model using Receiver Operations Curve. For the explanations and the corresponding calculations (equations regarding those performance metrics) the readers can consult with [8].

4 Results and Evaluation

After executing the models on RapidMiner Studio with two different datasets (with purchasing under a promotion data) evaluation metrics for the comparative models are obtained in Table 3. The term PD stands for "including the promotion data". Deep Learning PD has an accuracy of 0.908 while other models have 0.893, 0.847, 0.847, 0.832, 0.843 accuracy, respectively. The AUC performances of the models where Deep Learning PD was superior to the others are 0.961, 0.950, 0.957, 0.911, 0.901 and 0.903, respectively.

Table 3. Evaluation metrics for models.

Evaluation metric	Deep learning PD	Logistic regression PD	Neural network PD	Deep learning	Logistic regression	Neural network
Accuracy	0.908	0.893	0.897	0.847	0.832	0.843
Recall	0.921	0.904	0.912	0.839	0.885	0.886
Precision	0.935	0.918	0.924	0.832	0.896	0.898
AUC	0.961	0.950	0.957	0.911	0.901	0.903

Evaluating the results performed by these models showed that the Deep Learning PD model has better evaluation metrics and values compared to Logistic Regression and Neural Network models with and without purchasing under promotion sales data scenarios. These results showed that Deep Learning is a very successful technique in churn prediction, and the purchase under promotion sales variable is a powerful predictor that empowers the predictions. Including additional basket analysis features such as purchase under promotion sales data empowers the prediction under the same dataset. Based on these findings, it could be worthy of investigation that different variable combinations from basket analysis features may have a particular effect on customer churn prediction results.

5 Conclusion

So far, in the relationship marketing context, customer churn prediction studies were conducted to reveal which customers are probably to leave the company. Several models and prediction techniques have been used. The deep learning methods, as one of the recent developments within the scope of artificial neural networks and are frequently used in image processing and image definition, can also be utilized in churn studies. In this study, a deep learning model in customer churn prediction with basket analysis features was presented. Apart from RFM variables and demographic factors, the effect of purchases during the promotion sales, one of the basket analysis variables, on churn prediction was examined. The data was derived from a supermarket retail chain in Turkey which includes two major cities' customers' transactions. In this study, six different prediction models were compared. The first three models use deep learning, logistic regression, and neural network techniques and include promotional data as well as transactional data. The other models executed with the mentioned comparative techniques and include only transactional data.

RFM variables and promotional data were used both in the deep learning model and in logistic regression and artificial neural networks models, which are frequently used in the literature. Models' performances were measured with the confusion matrix outputs including accuracy, recall, precision, and AUC. According to these performance metrics, the DL model with the inclusion of promotional data revealed higher accuracy of churn prediction compared to the other models.

6 Discussion

This churn prediction study has two major contributions in churn context. First, from practitioners' points of view, via including additional information from the basket analysis in the prediction models showed that it can provide better prediction performances in terms of i) achieving higher customer profitability, ii) ensuring to attain loyal customers which may lead to increase in customer value in the long-run. Second, from academic point of view, it has been observed that the deep learning technique shows better predictive performance compared to other models whether promotional data is used or not.

This study has also limitations. The data size obtained from the supermarket chain is relatively less than other churn studies in the literature. For a future study, it can be considered that a data set with more transactions, especially attributes that can be extracted from the customers' basket analysis may provide better predictive performance. Also, the length of the data was the other limitation of the study. We assume that the results of longer period of transaction data than the 27-month limited data may be different. Therefore, future work investigating the effect of the different transactional periods could be worthy of consideration. In addition to this, researchers can also make a predictive analytics study by creating customer segments that contain RFM or other behavioral variables, examining customers' transitions between these segments over time. Last but not least, the study needs further investigation of the proposed model in the same industry with different datasets or in other application domains.

References

1. Alboukaey, N., Joukhadar, A., Ghneim, N.: Dynamic behavior based churn prediction in mobile telecom. Expert Syst. Appl. **162**, 113779 (2020)
2. Buckinx, W., Van den Poel, D.: Customer base analysis: partial defection of behaviourally loyal clients in a non-contractual FMCG retail setting. Eur. J. Oper. Res. **164**(1), 252–268 (2005)
3. Changchien, S.W., Lee, C.F., Hsu, Y.J.: On-line personalized sales promotion in electronic commerce. Expert Syst. Appl. **27**(1), 35–52 (2004)
4. Chen, Y.L., Kuo, M.H., Wu, S.Y., Tang, K.: Discovering recency, frequency, and monetary (RFM) sequential patterns from customers' purchasing data. Electron. Commer. Res. Appl. **8**(5), 241–251 (2009)
5. Cheng, C.H., Chen, Y.S.: Classifying the segmentation of customer value via RFM model and RS theory. Expert Syst. Appl. **36**(3), 4176–4184 (2009)

6. Chiang, W.Y.: To mine association rules of customer values via a data mining procedure with improved model: an empirical case study. Expert Syst. Appl. **38**(3), 1716–1722 (2011)
7. Dingli, A., Marmara, V., Fournier, N.S.: Comparison of deep learning algorithms to predict customer churn within a local retail industry. Int. J. Mach. Learn. Comput. **7**(5), 128–132 (2017)
8. Dogan, O., Oztaysi, B.: Genders prediction from indoor customer paths by Levenshtein-based fuzzy KNN. Expert Syst. Appl. **136**, 42–49 (2019)
9. Hsieh, N.C.: An integrated data mining and behavioral scoring model for analyzing bank customers. Expert Syst. Appl. **27**(4), 623–633 (2004)
10. Hu, Y.H., Yeh, T.W.: Discovering valuable frequent patterns based on RFM analysis without customer identification information. Knowl.-Based Syst. **61**, 76–88 (2014)
11. Hughes, A.M.: Strategic Database Marketing: The Masterplan for Starting and Managing a Profitable, Customer-Based Marketing Program, vol. 12. McGraw-Hill, New York (2000)
12. Ładyżyński, P., Żbikowski, K., Gawrysiak, P.: Direct marketing campaigns in retail banking with the use of deep learning and random forests. Expert Syst. Appl. **134**, 28–35 (2019)
13. Li, Y.M., Lin, C.H., Lai, C.Y.: Identifying influential reviewers for word-of-mouth marketing. Electron. Commer. Res. Appl. **9**(4), 294–304 (2010)
14. Liao, S.H., Chen, Y.J.: Mining customer knowledge for electronic catalog marketing. Expert Syst. Appl. **27**(4), 521–532 (2004)
15. Liu, D.R., Lai, C.H., Lee, W.J.: A hybrid of sequential rules and collaborative filtering for product recommendation. Inf. Sci. **179**(20), 3505–3519 (2009)
16. Liu, D.R., Shih, Y.Y.: Integrating AHP and data mining for product recommendation based on customer lifetime value. Inf. Manag. **42**(3), 387–400 (2005)
17. Martínez, A., Schmuck, C., Pereverzyev Jr., S., Pirker, C., Haltmeier, M.: A machine learning framework for customer purchase prediction in the non-contractual setting. Eur. J. Oper. Res. **281**(3), 588–596 (2020)
18. Miguéis, V.L., Camanho, A., e Cunha, J.F.: Customer attrition in retailing: an application of multivariate adaptive regression splines. Expert Syst. Appl. **40**(16), 6225–6232 (2013)
19. Miguéis, V.L., Van den Poel, D., Camanho, A.S., e Cunha, J.F.: Modeling partial customer churn: on the value of first product-category purchase sequences. Expert Syst. Appl. **39**(12), 11250–11256 (2012)
20. Shah, M., Adiga, D., Bhat, S., Vyeth, V.: Prediction and causality analysis of churn using deep learning. Comput. Sci. Inf. Technol. **9**(13), 153–165 (2019)
21. Umayaparvathi, V., Iyakutti, K.: Automated feature selection and churn prediction using deep learning models. Int. Res. J. Eng. Technol. (IRJET) **4**(3), 1846–1854 (2017)

The Fusion of Features for Detection of Cotton Wool Spots in Digital Fundus Images

Parashuram Bannigidad and Asmita Deshpande(✉)

Department of Computer Science, Rani Channamma University, Belagavi, India

Abstract. The Cotton Wool Spots are lesions seen on the surface of the retina among diabetic patients. It indicates a pre-proliferative Diabetic Retinopathy state and needs to be treated. This paper describes a method based on morphological operations and thresholding for detection of Cotton Wool Spots. Feature level fusion is a technique used in image processing and pattern recognition that can improve the accuracy of object recognition and computer vision. Hence, in this experiment, LBP and HOG features have been combined as it improves the ability to detect affected areas in a retinal image considerably. Also, Decision tree based classification method is used to classify and distinguish between healthy and diseased images. The proposed method yielded Precision 0.92, Recall 0.88 and F-Score 0.90 for IDRiD database. For the Kaggle database, Precision, Recall and F-score were computed as 0.95 each respectively. The performance evaluation measures comprising Precision, Recall and F-Score for STARE database were computed as 0.98, 0.93 and 0.96 respectively.

Keywords: Cotton Wool Spots · Soft exudates · Thresholding · Morphological · Operations · Decision tree classifier

1 Introduction

Healthcare professionals in the field of ophthalmology rely on a non-mydriatic fundus camera to capture digital fundus images that aid in the diagnosis of various ophthalmological disorders like Diabetic Retinopathy, Macular Edema, Diabetic Maculopathy, Glaucoma etc. Diabetic Retinopathy is an eye disorder in which different lesions such as Microaneurysms, Hemorrhages, hard and soft exudates can be commonly observed on the surface of the retina. These lesions appear when blood sugar levels increase and affect the blood vessels and capillaries found on the retina. Microaneurysms and Hemorrhages are small red dots that appear during the early stage of Diabetic retinopathy whereas hard and soft exudates are seen as the disease progresses. Hard exudates are yellowish-white deposits formed due to leakage of lipid proteins. Soft Exudates are also known as cotton wool spots and are generally greyish-white patches that have fluffy edges and are formed due to discolouration in the nerve fibre. A few cotton wool spots do not require any medical attention; however multiple such lesions indicate a pre-proliferative Diabetic Retinopathy state and need to be treated. Figure 1(a) shows a normal fundus image and Fig. 1(b) depicts a fundus image with Cotton Wool Spots.

A. Abraham et al. (Eds.): SoCPaR 2020, AISC 1383, pp. 530–538, 2021.
https://doi.org/10.1007/978-3-030-73689-7_51

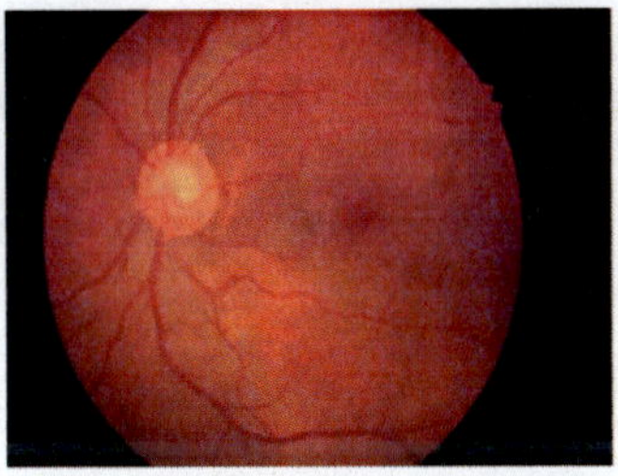

(a)

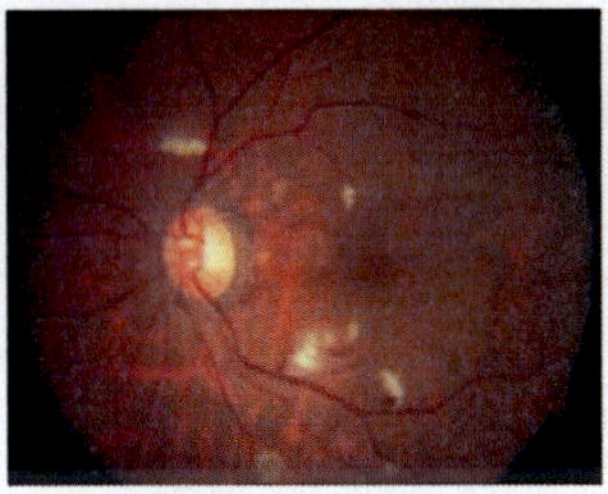

(b)

Fig. 1. (a) Normal fundus image (b) Fundus image with Cotton Wool Spots

From the literature, it can be observed that different methodologies have been proposed for segmentation and classification of lesions. Vasanti Satyananda et al. [1] proposed an approach based on embedded systems for detection of exudates from digital fundus images with an accuracy of 90%. Yogesh Rajput et al. [2] developed a methodology for the extraction of cotton wool spots in fundus images with the help of multiresolution analysis and application of symlet wavelet that yielded 92% accuracy. Giancardo et al. [3] implemented Kirsch's edges and stationary wavelets for detection of exudates. They focused on computational performance and found that the lesion segmentation using wavelet approach consumed ~2.4 s per image and Kirsch's edges consumed ~1.9 s per image. Shengchun Long et al. [4] have used dynamic thresholding and fuzzy C-Means clustering along with SVM classifier. The proposed method yielded an average accuracy of 97.7%. Balint Borsos et al. [5] extracted homogenous superpixels in a fundus image using SLIC algorithm along with ANN classifier to detect hard and soft exudates with 94% precision. Samra Irshad et al. [6] presented a method using Gabor filtering and global thresholding to detect cotton wool spots. They achieved 82.2% PPV. An automatic cotton wool spot detection method using thresholding, ant colony optimization and SVM classification was proposed by Syna Sreng et al. [7]. This method yielded 96.96% accuracy. Bannigidad and Deshpande [8] explored a method based on morphological operations and thresholding for identification of exudates from digital fundus images that yielded 96.6%. This paper focuses on the extraction of cotton wool spots from digital fundus images and exploring the use of a fusion of HOG and LBP features to distinguish between diseased and healthy images.

2 Materials and Methods

A number of fundus databases are available online. However, for the purpose of implementing the proposed work 60 fundus images were selected from Indian Diabetic Retinopathy image Dataset (IDRiD) [9]. These images were captured at a 50° field of view using Kowa-VX-10 alpha digital fundus camera having 4288 × 2848 pixels resolution and jpg format. This database also provides ground truth images are that graded by medical experts. The proposed method was also tested on 20 images from Kaggle.com [10]. These images are provided by Joint Shantou International Eye Centre (JSIEC), China and freely available for research purposes. The STARE [11] database also has exclusive fundus images containing Cotton Wool Spots. 54 images from the STARE database have been considered for this experiment. Among these 47 are diseased images sand 7 are healthy images. These images have been captured by a TopCon TRV-50 fundus camera with 24-bit gray scale resolution and spatial resolution of 700 × 605 pixels.

3 Proposed Methodology

The proposed methodology for detection of Cotton Wool Spots is based on image processing techniques that include fundus image preprocessing, segmentation, feature extraction and classification. The flow of steps in the proposed method is presented in the flow diagram in Fig. 2.

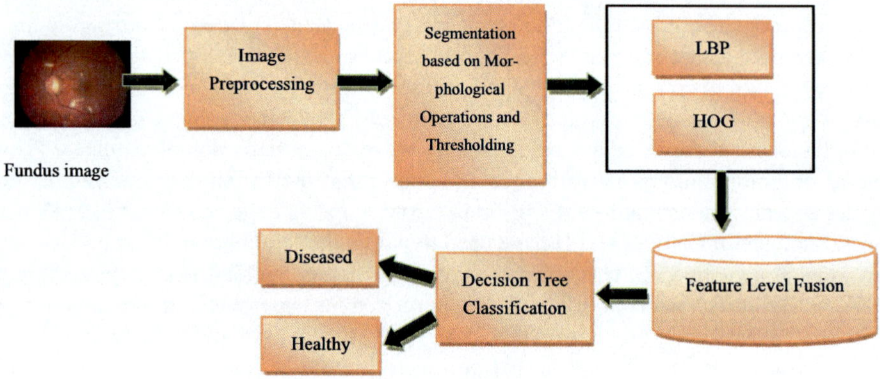

Fig. 2. Flow diagram of the proposed method

3.1 Preprocessing

The initial step in most image processing applications is preprocessing. It is particularly essential for fundus images because of differences in illumination, field-of-view, resolution and quality of images. Hence, during this step, each of the fundus images from IDRiD and Kaggle datasets is normalized for accurate segmentation and feature extraction. In the proposed technique, the fundus images are resized to 576 x 720 pixels and the image is converted from RGB to grayscale. Further, Contrast Limited Adaptive Histogram Equalization is applied to highlight the cotton wool spots present on retina surface [12]. The original fundus image is shown in Fig. 3(a) and image after preprocessing is shown in Fig. 3(b).

3.2 Segmentation

Proposed method for segmentation of cotton wool spots from digital fundus images is based on morphological operations and thresholding [17]. However, to accurately

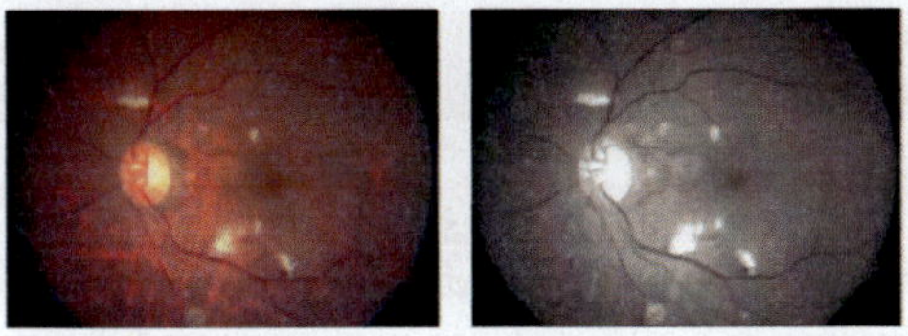

(a) (b)

Fig. 3. (a) Original fundus image (b) Image after preprocessing

detect these lesions, it is necessary to detect and eliminate the vasculature comprising blood vessels and capillaries as well as the optic disc present on the retinal surface [12]. The blood vessels are removed from the fundus image initially by applying canny edge detection followed by morphological opening and closing operations, whereas, the optic disc is eliminated by computing a circular mask and subtracting it from the segmented image. Cotton wool spots are present on the surface of the retinal image [13]. Since digital fundus images exhibit dissimilarities in intensity and texture, segmentation based on local thresholding [14] is proposed in this paper.

$$f(x, y) = 0 \text{ if } g(x, y) > T \\ 1 \text{ if } g(x, y) <= T \tag{1}$$

where f(x,y) is the fundus image, g(x,y) represents the intensity of a pixel and T indicates the threshold. Figure 4(a) shows canny edge detection, Fig. 4(b) presents the image after the optic disc is eliminated and Fig. 4(c) shows the segmented image.

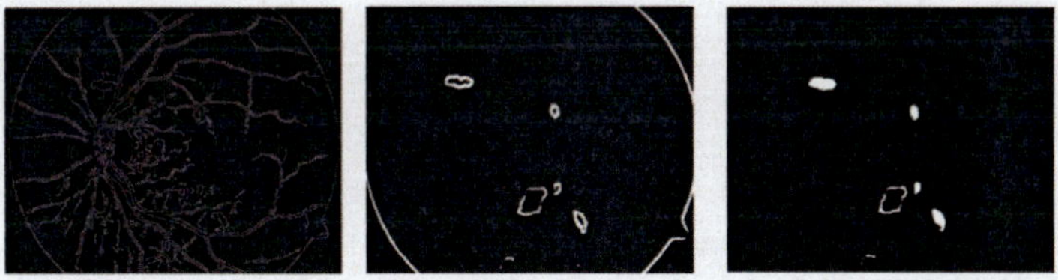

(a) (b) (c)

Fig. 4. (a) Canny edge detection (b) Optic disc eliminated (c) segmented image

4 Feature Extraction

4.1 LBP Features and Classification

This work is based upon extracting two important features set namely, Local Binary Pattern (LBP) and Histogram of Oriented Gradients (HOG). LBP feature set was first introduced in 1994 by Ojala et al. [15]. They are texture-based features used commonly for object recognition in computer vision. LBP operates by dividing an image into cells.

For instance, each cell can be 3 × 3 size. Each pixel has an intensity between 0–255. Now, the central value of the matrix under consideration is treated as a threshold. Each of the neighbouring pixels is set 1 if its intensity is greater than or equal to the threshold, else it is set to 0. The matrix hence produced contains only binary values. This binary pattern is now converted into its corresponding decimal value. Once all the new gray values have thus been computed, a histogram is constructed that represents the feature set. The process of LBP feature computation is shown in Fig. 5.

$$\mathrm{LBP}(x_{p,c}) = \sum\nolimits_{p=0}^{p=7} (u(xp - xc)2p \tag{2}$$

where x_p represents the sample, x_c represents the central pixel and p denotes the neighbouring pixels.

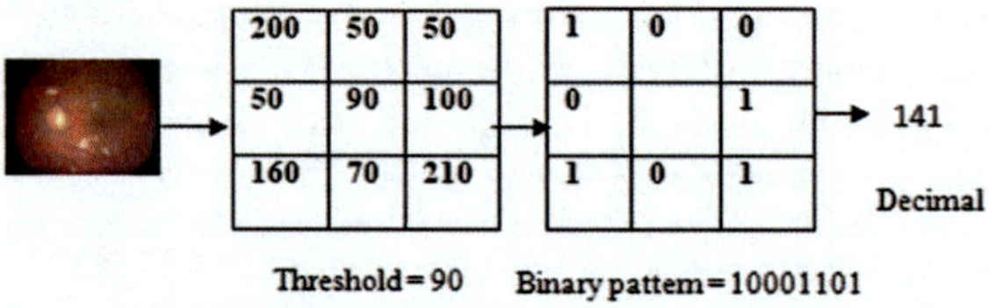

Fig. 5. Process of LBP feature computation

4.2 HOG Features

Image processing applications and computer vision algorithms often use Histogram of Oriented Gradients (HOG) [16] features for object detection. The main concept in HOG revolves around the idea that the appearance and shape of an object can be represented by understanding the gradients of intensity or directions of edges. It divides an image into square regions containing cells and calculates Histogram of Oriented Gradients for every cell. This computation is normalized with the help of a block-wise pattern that generates a descriptor for every cell. Every pixel gradient has magnitude and direction. These are computed as follows:

$$g = \sqrt{g^2x + g^2y} \tag{3}$$

$$\theta = \arctan \frac{gy}{gx} \tag{4}$$

The x-gradient works on vertical lines and y-gradient on horizontal lines. The magnitude of gradient triggers when there is any sudden change in intensity and it remains the same when the region is uniform. The HOG descriptor represents an image patch that extracts necessary information and disposes unwanted information. By looking at the gradient image it is possible to visualize the object being segmented. Figure 6(a) shows the segmented image and Fig. 6(b) depicts its HOG visualization.

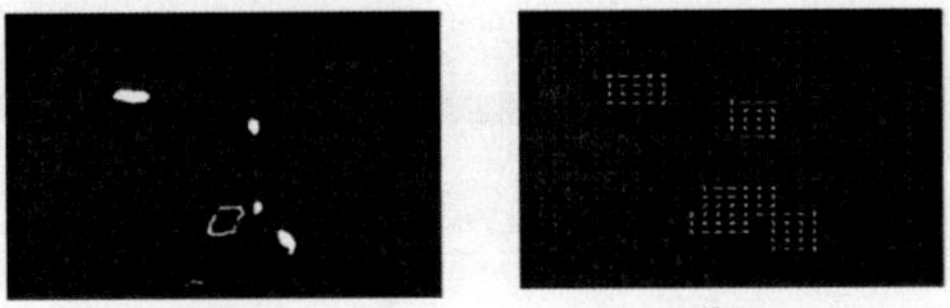

(a) (b)

Fig. 6. (a) Segmented image (b) HOG visualization

4.3 Fusion Features

Feature level fusion is a technique used in image processing and pattern recognition that can improve the accuracy of object recognition and computer vision. One of the advantages of fusion of features is that it can build a compact set of correlated features that can greatly aid the process of classification. Fusion at feature level can be implemented by simply concatenating the features obtained from different sources. Let $A = \{a_1, a_2, a_3, \ldots\ldots, a_n\}$ and $B = \{b_1, b_2, b_3, \ldots\ldots, b_n\}$ represent feature vectors that have been extracted from two different sources. The aim is to combine these sets into a new set C that can represent the image characteristics in a better manner. In this experiment, LBP and Hog features have been combined as it improves the ability to detect affected areas in a retinal image considerably. Further, a database of this feature set is used to train and test the system.

4.4 Classification

This experiment was conducted on 134 digital fundus image belonging to IDRiD, Kaggle and STARE databases. Various classifiers such as k-NN classifier, SVM classifier, Decision Tree classifier were tested during this experiment. However, Decision tree yielded the best results of classification. Decision tree classifier is a supervised learning algorithm and analysis technique applicable to a number of classification problems. It uses a tree-like model with the root at the top and leaves at the bottom. It is constructed by splitting a dataset based on various conditions. The conditions are represented as internal nodes and the decisions are represented as leaf nodes. As the decision tree grows, the conditions are tested based on some feature and every edge leads to possible solutions to the condition. It is a recursive process and repeated for every subtree of the tree. Figure 7(a) shows the classification tree with 10 fold cross-validation with a maximum number of splits = 7.

5 Experimental Results and Discussion

The proposed algorithm is coded in MATLAB R2015b with Intel i3 processor and 4 GB RAM. The confusion matrix obtained after classification of the images is shown in Table 1.

From the above table, it can be observed that for IDRiD database, out of 40 images comprising soft exudates, 34 were correctly classified and 6 were misclassified and 20 healthy images were correctly classified. For Kaggle database out of total 20 images, 10 were diseased and 10 images were healthy. All the diseased images were correctly

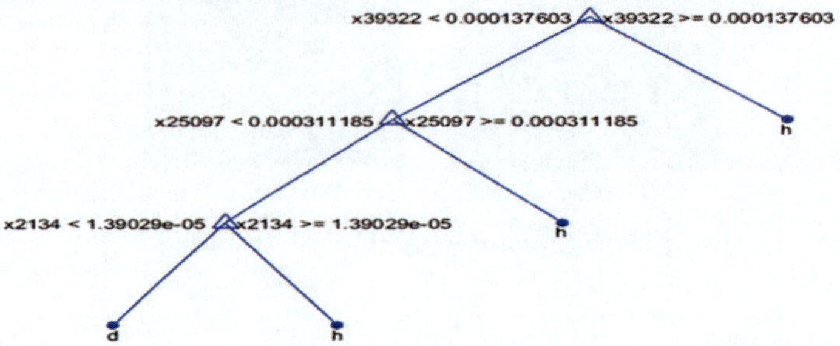

Fig. 7. (a) Classification tree

Table 1. Confusion matrix

Database	Classification results			
	TP	FP	FN	TN
IDRiD	34	06	00	20
Kaggle	10	00	01	09
STARE	46	01	00	07

classified and among healthy images, only 1 was misclassified whereas 9 were correctly classified. Form the STARE out of 54 images, 47 were diseased and 7 were healthy images. 46 of the diseased images were correctly classified and only one was misclassified and all the healthy images were correctly classified. The algorithm was tested for its performance using various performance evaluation measures namely, precision, recall and F-score. Precision provides the proportion of results that are relevant while recall represents the total relevant results correctly classified by a classification method. F-Score is the measure of the accuracy of a test. It is the harmonic mean of Precision and Recall. The performance evaluation measures computed for each of the databases are depicted in Table 2.

Table 2. Performance evaluation measures

Database	Precision	Recall	F-Score
IDRiD	0.92	0.88	0.90
Kaggle	0.95	0.95	0.95
STARE	0.98	0.93	0.96

The proposed method yielded Precision 0.92, Recall 0.88 and F-Score 0.90 for IDRiD Database. For the Kaggle database, Precision, Recall and F-score were computed as

0.95 respectively. The performance evaluation measures comprising Precision, Recall and F-Score for STARE database were computed as 0.98, 0.93 and 0.96 respectively.

6 Conclusion

Cotton Wool Spots are lesions seen on the surface of the retina among diabetic patients. They indicate a pre-proliferative Diabetic Retinopathy state and needs to be treated. This paper describes a method based on morphological operations and thresholding for detection of Cotton Wool Spots. As a part of the feature extraction stage, feature level fusion comprising LBP and HOG features is explored as it yields better accuracy in detection of affected areas in fundus images. The Decision tree based classification method is used to classify and distinguish between healthy and diseased images. The proposed method yielded Precision 0.92, Recall 0.88 and F-Score 0.90 for IDRiD database. For the Kaggle database, Precision, Recall and F-score were computed as 0.95 respectively. The performance evaluation measures comprising Precision, Recall and F-Score for STARE database were computed as 0.98, 0.93 and 0.96 respectively.

Acknowledgements. The authors wish to thank the developers of IDRid, Kaggle and STARE databases for making available digital fundus images publicly and freely to the research community.

References

1. Satyananda, V., Karibasappa, N.K.V.: FPGA and MATLAB based solution for retinal exudate detection. Int. J. Recent Technol. Eng. (IJRTE) **8**(6), 727–234 (2020). ISSN: 2277-3878
2. Rajput, Y.M., Manza, R.R., Patwari, M.B.: Extraction of cotton wool spot using multi resolution analysis and classification using K-means clustering. Int. J. Comput. (2015). National Conference on Digital Image and Signal Processing, vol. DISP 2015, no. 1, pp. 6–10
3. Giancardo, L., Meriaudeau, F., Karnowski, T.P., Yi, L., Tobin, K.: Automatic retina exudates segmentation without a manually labelled training set. In: International Symposium on Biomedical Imaging, pp. 1–6 (2011)
4. Long, S., Huang, X., Chen, Z., Pardhan, S., Zheng, D.: Automatic detection of hard exudates in color retinal images using dynamic threshold and SVM classification: algorithm development and evaluation. BioMed Res. Int. **2019**, 1–3 (2019)
5. Borsos, B., Nagy, L., Iclănzan, D., Szilágyi, L.: Automatic detection of hard and soft exudates from retinal fundus images. Acta Univ. Sapientiae Informatica **11**(1), 65–79 (2019)
6. Irshad, S., Usman Akram, M., Salman, M., Yasin, U.: Automated detection of cotton wool spots for the diagnosis of hypertensive retinopathy. In: 7th Cairo International Biomedical Engineering Conference Cairo, Egypt, pp. 121–124 (2014)
7. Sreng, S., Maneerat, N., Hamamoto, K., Panjaphongse, R.: Cotton wool spots detection in diabetic retinopathy based on adaptive thresholding and ant colony optimization coupling support vector machine. IEEEJ Trans. Electr. Electron. Eng. **14**, 884–893 (2019)
8. Bannigidad, P., Deshpande, A.: Exudates detection from digital fundus images using GLCM features and decision tree classifier. In: Santosh, K.C., Hegadi, R.S. (eds.) Recent Trends in Image Processing, pp. 245–257. Springer (2019)

9. Porwal, P., Pachade, S., Kamble, R., Kokare, M., Deshmukh, G., Sahasrabuddhe, V., Meriaudeau, F.: Indian Diabetic Retinopathy Image Dataset (IDRiD), IEEE Dataport. Dataset IDRiD (2018). https://doi.org/10.21227/H25W98. Accessed 02 Aug 2020
10. Dataset Kaggle: https://www.kaggle.com/linchundan/fundusimage1000
11. Dataset STARE: STARE. https://cecas.clemson.edu/~ahoover/stare
12. Bannigidad, P., Deshpande, A.: A hybrid approach for digital fundus images using image enhancement techniques. Int. J. Comput. Eng. Appl. **XII**(I), 122–131 (2018)
13. Bannigidad, P., Deshpande, A.: Automated method for optic disc detection and elimination in digital fundus images. Int. J. Recent Technol. Eng. (IJRTE) **8**(4), 12558–12563 (2019)
14. Gonzalez, R.C., Woods, R.E.: Digital Image Processing, 3rd edn. Pearson Education (2014)
15. Ojala, T., Pietikainen, M., Harwood, D.: A comparative study of texture measures with classification based on feature distributions. Pattern Recogn. **29**(1), 51–59 (1996)
16. Dalal, N., Triggs, B.: Histograms of oriented gradients for human detection. In: Proceedings of the 2005 IEEE Computer Society Conference on Computer Vision and Pattern Recognition (CVPR 2005), vol. 1, pp. 886–893 (2005)
17. Shilpa, B.V., Nagabhushan, T.N.: An ensemble approach to detect exudates in digital fundus images. In: 2016 Second International Conference on Cognitive Computing and Information Processing (CCIP), Mysore, pp. 1–6 (2016)

Thyroid Nodule Cell Classification in Cytology Images Using Transfer Learning Approach

Ahsan Baidar Bakht[1], Sajid Javed[1], Roberto Dina[2], Hasan Almarzouqi[1], Ahsan Khandoker[1], and Naoufel Werghi[1](✉)

[1] Khalifa University, KUCARS, Abu Dhabi, UAE
naoufel.werghi@ku.ac.ae
[2] Cleveland Hospital, Abu Dhabi, UAE

Abstract. Analysis of fine needle aspiration cytology FNAC slides of thyroid nodules is a very crucial test before the preoperative diagnosis of thyroid malignancy. Cytology slides may be composed of different types of cells. Differentiating between cancerous cells and healthy cells plays an important role in the treatment. However, the conventional visual inspection is very time consuming and the process might endure inaccuracies because of the subject-level assessment. To the best of our knowledge, no work has been done for the multi-class cell level classification of thyroid nodules. In this paper, we propose a method for classification of cytology images at the cell level by using fine-tuned VGG-19 and AlexNet models, exploiting the transfer learning approach to better fit the model for classification of our dataset. Model evaluations are done by calculating the precision, recall, F1-score, and accuracy. Although the data is highly imbalanced, but both model have shown very good performance by achieving an accuracy of 93.05% and 92.88% by VGG-19 and AlexNet respectively.

Keywords: Thyroid cancer detection · FNAC · Deep learning

1 Introduction

Cancer is a leading cause of death worldwide, accounting for an estimated 9.6 million deaths in 2018. Early and accurate cancer detection will allow clinicians to initiate early intervention and appropriate treatment at earlier stages, thus decreasing the mortality rate. In this context several computer-aided diagnosis system have been proposed for the early screening of cancer using different imaging modalities. These systems covered lung [1,2], colon [3,4],cervix [5], prostate [6–8], colon [9], and retina [10,11] In this work, we propose a novel framework for the early detection of thyroid cancer from cytology image modalities.

Thyroid gland is a vital organ of the human body which regulates several body functions by releasing thyroid hormones [12]. Thyroid cancer occurs when old thyroid cells do not die and instead shows an abnormal growth of cells,

A. Abraham et al. (Eds.): SoCPaR 2020, AISC 1383, pp. 539–549, 2021.
https://doi.org/10.1007/978-3-030-73689-7_52

this can form a mass of tissues commonly known as thyroid nodule. In most cases thyroid nodule is benign and is not fatal. According to the results of ultrasound-guided fine-needle aspiration, 9–15% of nodules are malignant [13]. So it is necessary to differentiate between benign and malign thyroid nodules to avoid unnecessary surgeries for the patients having benign nodule. American Cancer Society has estimated that 52,890 new cases and 2,180 deaths has been reported for thyroid cancer in 2020.

Fine needle cytological aspiration (FNAC) is a fast, inexpensive, minimally invasive and readily accessible method of examining cells obtained from patients that can classify the majority of lesions [14]. In practice, a cytopathologist signed the FNAC pathology report after microscopic analysis of the slices/smears. Routine inspection of cytology slides is therefore a very tedious task. In order to increase the detection sensitivity and minimize the time consumption, it is important to build computer-aided diagnostic systems (CADs). Due to the fast computing technologies, it provides an advantage for the processing of large quantities of medical images in a short span of time.

Detection of cancer at early stage plays an important role in the treatment of cancer patients. Researchers are striving to develop near human level accurate automated systems for the detection of cancer at early stages. With the advancement of technology in high computing devices, deep learning has drawn a lot of attention. Deep learning is mainly used for image processing and analysis problems.

Several research studies have been done on the development of computer aided diagnostic systems for cancer detection. Atsushi Teramoto et al. have done research on the classification of benign and malign cells in lungs cytology images. Using fine-tuned VGG-16 model they achieved an accuracy of 87% [15]. Moumita Dholey et al. have proposed a two-step method which includes nucleus segmentation by using k-means clustering method and classification using feature-based ANN and SVM [16]. Sanaullah khan et al. have developed a method for detection and classification of breast cancer in breast cytology images utilizing the transfer learning technique. They combined features from three pre-trained models (VGGNet, GoogLeNet, ResNet) and fed to a fully connected layers for classifying it as either benign or malign [17]. Amartaya Ranjan Saikia et al. have done a comparative analysis between different available CNN models for the classification of FNAC breast cytology images. Their study have shown that GoogLeNet V3 outperformed all other models [18]. Miao Wu et al. have proposed a method for classifying between four different types of ovarian cancer. In their method they used Alexnet model for performing their classification task [19].

In this study, our research is focused on the detection and classification of thyroid cancer. Several computer aided diagnostic systems have been developed by researchers for the detection of thyroid cancer [20–22]. Daskalakis et al. proposed multiple techniques for discriminating the H&E stained slides of thyroid nodule either as benign or malign. Their study utilized Probabilistic Neural Network (PNN) as single classifier and k-NN/PNN/Bayesian as multiple classifiers giving result based on the majority vote [20]. Guan Q et al. have used two DCNN models for diagnosing the papillary thyroid carcinoma (PTC) in cytology images. Training and testing has been done on VGG-16 and Inception-v3 models. Results have shown that VGG-16 has performed better than the inception-v3 for differentiating between PTC and benign cytology images [21]. Gopinath B et al. have developed an automated system for identifying thyroid cancer in Fine Needle Aspiration Cytology (FNAC) images. Their proposed method used gabor features as input to the SVM to classify between benign and malign cases of thyroid nodules [22]. In another research, Parikshit et al. have developed ANN to classify between papillary thyroid carcinoma and non-papillary thyroid carcinoma [23].

Till now, all the research work done in thyroid cancer cytology image classification has focused on distinguishing between either malign or benign. Cytology slides are made up of different cells which includes Abnormal epithelial (Cancerous cell), Normal epithelial, Oncocytic epithelial, Lymphocyte and Macrophages. In this research, Our main contributions are. (1) Annotation and extraction of cell level image data for classifying between different cells in thyroid cancer cytology slides. (2) Application of transfer learning approach using VGG-19 and AlexNet models for the classification of different cells.

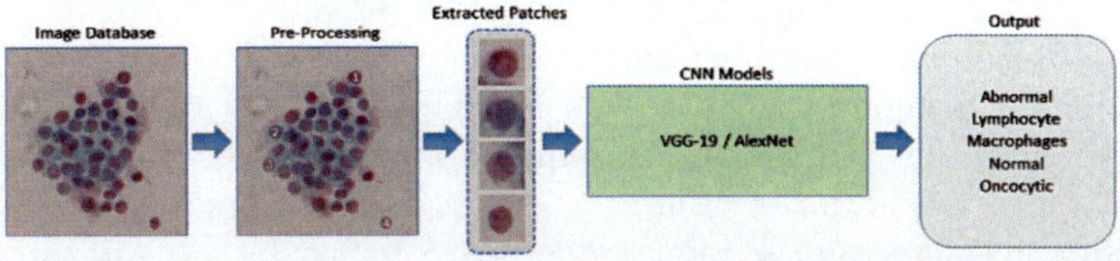

Fig. 1. Method block diagram: Firstly, cytology images are obtained. After that preprocessing step involves identification of center co-ordinates of cells in cytology images. Patches are extracted based on these co-ordinates. These patches are fed to CNN models for training and later testing for generating output labels

2 Methodology

In this study, transfer learning approach has been applied to classify five types of cells present in cytology images of thyroid cancer. As shown in Fig. 1, after acquiring cytology images of patients, patch level dataset consisting of five classes is made by extracting patches from the cytology image. Using transfer learning approach VGG-19 and AlexNet models are trained and then tested on the testing dataset to generate the output labels.

2.1 Dataset

For this study, 11 cases of thyroid cancer were collected using fine needle aspiration cytology (FNAC). The collected cytology images contains 4 benign cases and

10 malign cases. All these cases are annotated by the expert pathologist identifying the cell type present in each cytology image of patients. Cytology image of thyroid nodule sample is composed of multiple cells named as Abnormal epithelial, Normal Epithelial, Oncocytic Epithelial, Lymphocyte and Macrophages. The digital slide may contain only one type of cell or it can be made up of multiple type of cells. Abnormal epithelial is a cell that corresponds to the tumor in the thyroid nodule whereas normal epithelial shows the healthy cells. Oncocytic epithelial shows a type of cells which are not clear to be declared as Abnormal epithelial or Normal epithelial. Lymphocyte and Macrophages counts as immune system cells.

2.2 Class Imbalance

Training deep neural network requires a large amount of data for better performance. In many cases, distribution of data is uneven which causes the network to make biased decisions. Same is the problem in our case as shown in Table 1. Abnormal epithelial makes 69.55% of the whole dataset while Normal Epithelial, Oncocytic Epithelial, Lymphocyte and Macrophages contains 29.28%, 8.93%, 0.8% and 0.96% respectively. To overcome class imbalance data problem there are two major types of techniques named as data focused and algorithm focused. In data focused technique, up sampling of the weak classes has been done by accumulating new data or down sampling of major class is done by dropping redundant data. While in algorithm focused technique, modification is done in the classification algorithm so that model learn better features to do unbiased classification. In this research, algorithm focused technique is applied to overcome the class imbalance problem.

Table 1. Dataset

Class	Samples
Abnormal epithelial	6405
Lymphocyte	74
Macrophages	89
Normal epithelial	1818
Oncocytic epithelial	823

2.3 Data Augmentation

High quality and abundance of data is the essential to the successful implementation of various deep learning models. Small dataset may cause overfitting problem in deep convolutional neural networks. To avoid overfitting, DCNNs are usually trained on large amount of dataset. But for every classification problem acquiring large dataset is itself a difficult task. To overcome this problem, data augmentation techniques are used to partially solve the problem, which transforms the current dataset images by performing rotation, translation, shearing, scaling and reflection operations and hence increases the number of samples in the dataset. Due to the acquisition and annotation process, the classification problem presented in this study lacks a sufficient quantity of data to fed into deep learning architecture and exploit the performance of models under study. Thus, in order to enhance the accuracy we extended the existing data by applying 3 different augmentation techniques. The employed augmentation techniques are rotation, translation and scaling. Augmentation is done by applying rotation operation to images randomly between [90°,270°], and applying translation operation randomly between [−10 10] for both X and Y axis. And scaling is done by modifying image between [0.8,1.5] times. Thus increasing our dataset 3 times where each sample has 3 different variants. A sample of data augmented image is shown in Fig. 2. Although data augmentation increases the dataset size still it cannot fully cover up the shortage of originally acquired dataset. For this purpose, transfer learning approach is used in addition to data augmentation which is discussed in the next sub-section.

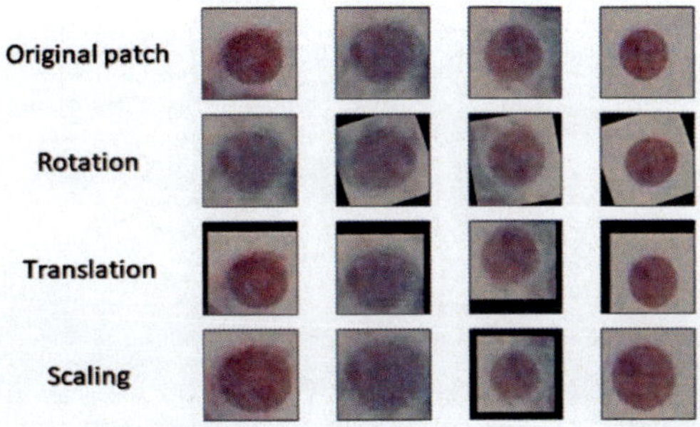

Fig. 2. Sample patches of the original and augmented data.

2.4 Transfer Learning

While the CNN framework functions very well on large data sets, it struggles to produce meaningful results on small data sets. The concept of transfer learning can be utilized to enhance the efficiency of individual CNN frameworks by integrating their information gained from previous dataset with new data sets under consideration in order to gain greater recognition precision and minimize computational costs. The most common challenge in medical image analysis and processing is that the volume of usable information for research purposes is minimal and limited. Training an entirely deep network framework such as CNN with a limited amount of data can therefore subject to overfitting, which is typically the reason for poor performance and generalization capacity [24].

To resolve these problems, CNN models can be re-trained by fine-tuning, and this technique is called transfer learning. Transfer learning is very easy and quick way to do representation learning for classification task. Fine tuning of a model can be done in multiple ways depending on the problem in hand. For instance, training the network only on the fully connected layers of the model while freezing all the convolutional layers [25]. On the other hand one can utilize the potential of convolution layers to better learn the features although it cost more time. Our cytology dataset images contains only cells which are hard to differentiate. Figure 3 shows the two models namely VGG-19 and AlexNet and that has been used in this research, we fine tuned the model by freezing some of the convolutions layers and including some deep convolutional layers in the training part to better learn the representation learning. Transfer learning has been widely used in the area of medical imaging and has demonstrated considerable success in terms of accuracy, training time and error rates.

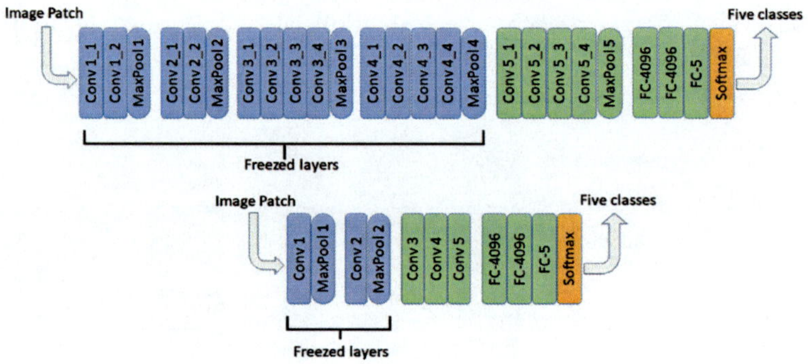

Fig. 3. DCNN Models: Top VGG-19, bottom: AlexNet

2.5 CNN Model

CNN models have shown near human level performance in classification tasks but designing CNN from scratch has its own advantages and disadvantages. One main problem is it require a very large database. There are many DCNNs developed by researchers especially for image classification tasks. Out of those models, VGG-19 and AlexNet has been used in this study. VGG-19 and AlexNet model has shown very promising results in several image classification problems. In the proposed network, these two models architectures are fine-tuned for cell classification of thyroid nodules in cytology images. As shown in Fig. 3 (a), VGG-19 model contains 16 convolutional layers and 3 fully connected layers. In VGG-19 architecture, first two blocks contain two convolutional layers each followed by maxpool layer. In next 3 blocks, each block contains 4 convolutional layers followed by a maxpool layer. Last three layers are fully connected layers, while final layer is softmax layer which acts as classification layer. Also as shown in Fig. 3(b). AlexNet model contains less number of layers having 5 convolutional layers and 3 fully connected layers. In AlexNet model, first two convolutional layers are followed by maxpool layer while the rest three layers are connected directly. Fine tuning has been done by freezing some layers in the selected models and by modifying the output layer. As shown in Fig. 3(a), first four blocks of VGG-19 model has been freezed while keeping the fifth convolutional block open for training on new dataset. Although it cost some more time to train but it makes the model learn better feature for classification. For our five class dataset, VGG-19 model is fine tuned by replacing the last fully connected layers having 1000 neurons with a fully connected layer having 5 neurons that acts as input to the softmax layer. Similarly as shown in Fig. 3(b), first two layers are freezed while keeping the rest three convolutional layers open for training. classification layer of the AlexNet model has been replaced by 5 neuron layer to classify five different classes.

2.6 Weighted Classification Layer

Imbalance data causes the network to be biased and tend to only predict the majority class data. The attributes of the minority class are viewed as noise and are mostly overlooked. Because of that there is a high probability of misclassification of the minority class relative to the dominant class. As the generated dataset is also highly imbalanced. So, to cater for the negative effect of imbalance data on our models output, we modified the classification layer of VGG-19 and AlexNet models and replace it with the weighted classification layer instead of initializing with random weights. In this modified classification layer, weights are initialized for every class based on the formula:

$$w_i = 1 - \frac{samples\ of\ class\ A}{samples\ of\ all\ classes}$$

Where w_i shows the initialized weights of each class based on their number of samples in the dataset. This initialization gives more weights to the classes with less number of samples which makes the algorithm pay more attention to these classes during the training session.

3 Experimental Setup

In this section, the experimental environment is explained followed by the performance metrics used to evaluate the performance of the model.

3.1 Implementation

First of all cell patches are extracted from the cytology images of patients. Each patch has a size of 61×61 with cell as the center of patch. Overall dataset is composed of 9209 samples with per class distribution as 6405, 54, 82, 1818, 523 samples for abnormal epithelial, lymphocyte, macrophages, normal epithelial and oncocytic epithelial respectively. Dataset is randomly divided into 70% for training and 30% for testing containing 6446 and 2773 collective samples respectively. As shown in Fig. 3(a), initial 4 blocks of VGG-19 model are freezed to preserve the features of the pre-trained network and also it reduces the time cost for the training. Similarly as shown in Fig. 3(b) first two convolution blocks of alexnet model has been freezed. So, firstly these fine-tuned models are trained using the training dataset to learn the features using convolution layers and then fed to the fully connected layers for the classification. Both models have been trained on 70% of the dataset and performance measures are done on the remaining 30% of the dataset. All the experimentation has been done on matlab 2020a. Learning rate is set to 0.0003, and sgdm is used as the optimizer for these models. Models are trained on 40 epochs with a batch size of 20.

3.2 Evaluation Metrics

Due to the imbalance data, relying only on the accuracy does not show the true ability of the model. Therefore, confusion matrix is computed. Confusion matrix also known as error matrix or contingency table consists of four terms namely true positives (TP), false positives (FP), true negative (TN) and false negatives (FN). Based on these four terms different performance measures are calculated which includes accuracy, precision, recall and F1-score.

3.3 Results

Five class thyroid nodule cytology cell data has been classified using fine-tuned VGG-19 and AlexNet models. To the best of our knowledge, no research has been done in classifying cytology images at multi-class cell levels. However, some researches has been done in binary classification of images whose summary is given in Table 2-a. Classification report consisting of accuracy, precision, recall and f1-score is given in Table 2-b. It can be seen that although data is highly imbalanced but still both models manage to achieve a handsome results. Both VGG-19 and AlexNet models have achieved 93.05% and 92.88% accuracy respectively.

Table 2. Top: Summary of previous methods performing binary classification. Bottome: Performance of our method for the 5-class classification.

Method	Features	Model	Dataset	Accuracy
[21]	Deep features	VGG-16, Inception-v3	279 cytology images	97.6%, 92.7%
[22]	Gabor features	SVM	110 images	96.7%
[23]	CNN features	Customised ANN	544 images	85.1%

	VGG-19			AlexNet		
	Precision	Recall	F1-Score	Precision	Recall	F1-score
Abnormal epithelial	0.947	0.963	0.955	0.957	0.953	0.955
Lymphocyte	0.692	0.409	0.515	0.631	0.545	0.585
Macrophages	0.533	0.296	0.381	0.395	0.629	0.485
Normal epithelial	0.888	0.946	0.916	0.879	0.924	0.901
Oncocytic epithelial	0.930	0.753	0.832	0.934	0.805	0.865
Weighted average	0.927	0.929	0.928	0.931	0.927	0.929

4 Conclusion

Analysis of FNAC slides is very important step in thyroid cancer treatment. FNAC slides are composed of multiple type of cells and their analysis plays an important role in the detection of thyroid cancer. In this study, two deep learning models has been used for the automatic analysis of cytology slides. VGG-19 and AlexNet models have been tested and compared to show their performance on the cytology dataset. It has been observed that VGG-19 and AlexNet models have shown very good results by achieving an accuracy of 92.88% and 93.05% respectively.

References

1. Taher, F., et al.: Bayesian classification and artificial neural network methods for lung cancer early diagnosis. In: 19th IEEE International Conference on Electronics, Circuits, and Systems, ICECS, pp. 773–776 (2012). https://doi.org/10.1109/ICECS.2012.6463545
2. Taher, F., et al.: Extraction and segmentation of sputum cells for lung cancer early diagnosis. Algorithms **6**(3), 512–531 (2013). https://doi.org/10.3390/a6030512
3. El Khatib, A., et al.: Automatic polyp detection: a comparative study. In: Proceedings of the Annual International Conference of the IEEE Engineering in Medicine and Biology Society, EMBS, vol. 2015-November, pp. 2669–2672 (2015). https://doi.org/10.1109/EMBC.2015.7318941
4. Taha, B., et al.: Automatic polyp detection in endoscopy videos: a survey. In: Proceedings of the 13th IASTED International Conference on Biomedical Engineering, BioMed, pp. 233–240 (2017). https://doi.org/10.2316/P.2017.852-031
5. Taha, B., et al.: Classification of cervical-cancer using pap-smear images: a convolutional neural network approach. In: Communications in Computer and Information Science, vol. 723 (2017). https://doi.org/10.1007/978-3-319-60964-5_23

6. Reda, I., et al.: Computer-aided diagnostic tool for early detection of prostate cancer. In: Proceedings - International Conference on Image Processing, ICIP, vol. 2016-August, pp. 2668–2672 (2016). https://doi.org/10.1109/ICIP.2016.7532843
7. Reda, I., et al.: A comprehensive non-invasive framework for diagnosing prostate cancer. Comput. Biol. Med. **81**, 148–158 (2017). https://doi.org/10.1016/j.compbiomed.2016.12.010
8. Alkadi, R., et al.: A deep learning-based approach for the detection and localization of prostate cancer in t2 magnetic resonance images. J. Digital Imaging **32**(5), 793–807 (2019). https://doi.org/10.1007/s10278-018-0160-1
9. Javed, S., Mahmood, A., Werghi, N., et al.: Multiplex cellular communities in multi-gigapixel colorectal cancer histology images for tissue phenotyping. IEEE Trans. Image Process. **29**, 9204–9219 (2020). https://doi.org/10.1109/TIP.2020.3023795
10. Raja, H., Hassan, T., Akram, M.U., et al.: Clinically verified hybrid deep learning system for retinal ganglion cells aware grading of glaucomatous progression. IEEE Trans. Biomed. Eng. (2020). https://doi.org/10.1109/TBME.2020.3030085
11. Hassan, T., Akram, M.U., Werghi, N., et al.: RAG-FW: a hybrid convolutional framework for the automated extraction of retinal lesions and lesion-influenced grading of human retinal pathology. IEEE J. Biomed. Health Inf. **25**, 108–120 (2020). https://doi.org/10.1109/JBHI.2020.2982914
12. Keles, A.: Keles : a ESTDD: expert system for thyroid diseases diagnosis. Expert Syst. Appl. **34**, 242–246 (2008)
13. Nam-Goong, I.S., et al.: Ultrasonography-guided fine-needle aspiration of thyroid incidentaloma: correlation with pathological findings. Clin. Endocrinol. **60**, 21–28 (2004)
14. Meinkoth, J.H., Cowell, R.L.: Sample collection and preparation in cytology: increasing diagnostic yield. Vet. Clin. North Am. Small Anim. Pract. **32**(6), 1187–1207 (2002)
15. Teramoto, A., et al.: Automated classification of benign and malignant cells from lung cytological images using deep convolutional neural network. Inf. Med. **16**, 100205 (2019). https://doi.org/10.1016/j.imu.2019.100205
16. Dholey, M. et al.: A Computer Vision Approach for Lung Cancer Classification Using FNAC-Based Cytological Images (2018)
17. Su, K., et al.: A novel deep learning based framework for the detection and classification of breast cancer using transfer learning. Pattern Recogn. Lett. **125**, 1–6 (2019)
18. Saikia, A., et al.: Comparative assessment of CNN architectures for classification of breast FNAC images. Tissue Cell **57**, 8–14 (2019). https://doi.org/10.1016/j.tice.2019.02.001
19. Wu, M., et al.: Automatic classification of ovarian cancer types from cytological images using deep convolutional neural networks. Biosci. Rep. **38**, BSR20180289 (2018). https://doi.org/10.1042/BSR20180289
20. Daskalakis, A., et al.: Design of a multi-classifier system for discriminating benign from malignant thyroid nodules using routinely H&E-stained cytological images. Comput. Biol. Med. **38**(2), 196–203 (2008). https://doi.org/10.1016/j.compbiomed.2007.09.005
21. Guan, Q., et al.: Deep convolutional neural network VGG-16 model for differential diagnosing of papillary thyroid carcinomas in cytological images: a pilot study. J. Cancer **10**(20), 4876–4882 (2019). https://doi.org/10.7150/jca.28769

22. Gopinath, B., Shanthi, N.: Support Vector Machine based diagnostic system for thyroid cancer using statistical texture features. Asian Pac. J. Cancer Prev. APJCP **14**(1), 97–102 (2013). https://doi.org/10.7314/apjcp.2013.14.1
23. Sanyal, P., et al.: Artificial intelligence in cytopathology: a neural network to identify papillary carcinoma on thyroid fine-needle aspiration cytology smears. J. Pathol. Inf **9**, 43 (2018)
24. Long, M., et al.: Learning transferable features with deep adaptation networks. In: Proceedings of the 32nd International Conference on Machine Learning, Lille, France, pp. 97–105 (2015)
25. Hosny, K.M., et al.: Skin cancer classification using deep learning and transfer learning. In: 9th Cairo International Biomedical Engineering Conference (CIBEC), Cairo, Egypt, pp. 90-93 (2018). https://doi.org/10.1109/CIBEC.2018.8641762

Deep Fusion Driven Semantic Segmentation for the Automatic Recognition of Concealed Contraband Items

Muhammad Shafay(✉), Taimur Hassan, Divya Velayudhan, Ernesto Damiani, and Naoufel Werghi

C2PS, Khalifa University, Abu Dhabi, UAE
100057573@ku.ac.ae

Abstract. Automatic detection of prohibited items in passenger baggage is a challenging task, especially in cluttered and occluded concealment scenarios. In this paper, we present a deep fusion driven semantic segmentation network that leverages multi-scale feature representations (extracted via CNN backbone) to generate highly accurate segmentation masks of the suspicious items irrespective of the clutter and concealment. Assessed with the public GDXray, SIXray and OPIXray datasets our proposed architecture reached a mean IoU performance of 0.7768, 0.6263, and 0.6713 respectively, outperforming the leading frameworks.

Keywords: Aviation security · Convolutional neural networks · Object recognition · Semantic segmentation · X-ray baggage imagery

1 Introduction

X-ray imagery, due to its cost-effectiveness, is a common medium for investigating luggage at the airports, malls, and cargoes [1]. Manuel detection is tiring, subjective and resource demanding To address these shortcomings, many researchers have developed autonomous frameworks capable of recognizing baggage threats on-the-fly. Yet, the detection of extremely cluttered and occluded contraband items remain a serious challenge. Although several sophisticated occlusion-aware detectors work well on RGB images, their performance deteriorates on the X-ray scans owing to the intrinsic disparities in RGB images and X-ray images. One way to deal with occlusion is to utilize semantic segmentation models due to their pixel-wise recognition capacity. These models have the potential to recognize extremely cluttered threatening items [2].

This paper proposes a unique semantic segmentation architecture that encompasses a deep fusion of different combinational blocks from backbone architecture, inspired from the RefineNet [3,4], to generate the sparse latent representation enabling accurate pixel-wise recognition of the concealed contraband items.

A. Abraham et al. (Eds.): SoCPaR 2020, AISC 1383, pp. 550–559, 2021.
https://doi.org/10.1007/978-3-030-73689-7_53

The rest of the paper is structured as follows: A detailed review of related work is presented in Sect. 2 while Sect. 3 focuses on the proposed framework. Section 4 discusses the experimental setup. The outcomes of the tested model The findings of the research work and the assessment of the proposed approach with the leading methods are presented in Sect. 5 followed by the conclusion in Sect. 6.

2 Related Work

Initial methods for detection of contraband items employed classical machine learning techniques including a Bag of Visual Words (BOVW) [10] coupled with adaptive feature vectors [11]. Moreover, Franzel et al. [12] proposed a filter-based recognition via HOG features and Support Vector Machine (SVM) model. Similarly, SIFT and SURF features are also used in parallel with the SVM to apprehend potential forbidden objects [13]. Also, Dai et al. [14] detected prohibited items using K-NN classifier.

More recently, with the emergence in computer vision and deep learning and its successful usage in a variety of applications [5–9], researchers have used deep learning for detecting baggage threats due to its robustness. The early wave of deep learning-driven frameworks involved supervised classification and detection strategies where AlexNet [15] coupled with SVM was used for the baggage threat classification [16]. Similarly, Xu et al. [17] suggested convolutional network, centred around the attention mechanism, for the detection of contraband objects. Griffin et al. [18] combined InceptionV2 [19] along with Gaussian distribution for the threat detection. Miao et al. [20] proposed a class-balanced hierarchical refinement (CHR) to deal with the imbalance contraband items. GANs have also proved useful for detecting baggage threats as anomalies. GANomaly [21] and Skip-GAN [22] architectures pioneered this by generating distinct latent space representations from the X-ray scans to detect baggage threats in an adversarial manner. Apart from this, Hassan et al. recently proposed Cascaded Structure Tensors (CST) [1], Trainable Structure Tensors (TST) [23], and dual tensor-shot detection networks (TSD) [24] frameworks that exploit the contours of the concealed contraband items to effectively recognize them. Wei et al. [25] recently suggested DOAM, otherwise known as the De-Occlusion Attention Module to deal with the occluded contraband items where they have thoroughly tested their framework on the publicly available OPIXray dataset (introduced in the same paper). Although, researchers have addressed the cluttered and imbalanced nature of the threatening items to some extent [20–25]. However, all of these frameworks lack on one thing. Either they are validated on an exclusive dataset [20,25], or a lot of parameter refinement is required to accommodate for different kind of scans from different scanners [1].

In this paper, we present a single-staged semantic segmentation model that employs deep fusion blocks to generate sparse representation of the concealed contraband items irrespective of the scanner characteristics as it equally performs well on all three publicly available datasets, including OPIXray [25] which is the latest addition to the X-ray datasets and to the best knowledge of the authors, this is first of its kind research work, to conduct extensive experimentation regarding semantic segmentation on this new dataset.

3 Methodology

The proposed deep fusion driven semantic segmentation framework, inspired by the RefineNet [3] and Lightweight RefineNet (LWRN) [4], leverages the multi-scale feature representations of the input scan to generate sparse latent representation which enables effective segmentation of concealed contraband items in baggage X-ray scans. Typical semantic segmentation model suffers from limited resolution as multiple stages of pooling and variable stride rates in convolution operation decrease the dimensionality of the feature maps and thereby limiting themselves in utilizing the features representation from the earlier convolutional blocks. In the proposed framework, multi-resolution feature maps are extracted at different levels from the ResNet [32] backbone, and are incorporated at the decoder end. The decoder, consisting of Residual Convolutional Units (RCU), Fusion blocks and Chained Residual Pooling (CRP), aids then in combining the finer feature representations with the coarsest ones (extracted from the stacked convolutional layers).

The backbone network in the proposed framework, is divided into N disjoint blocks (with index $i = 0, ..., N-1$) as shown in Fig. 1. Similarly, the decoder also has N blocks (with multiple sub-blocks as well). Here, the decoder block at index i is fused with the ith block within the backbone network (via addition) to generate multi-scale feature representations. Afterwards, this fused representation (at the ith level) is merged with the $i-1$ decoder block and this process is repeated till $i - (N-1)$(where i has reached $N-1$ since it is the index of the blocks) level to extract the concealed contraband items from the candidate scan. Unlike the conventional segmentation models, the decoder within the proposed framework utilizes fused feature representations generated via its in-built Residual Convolutional Units (RCU), Fusion Blocks and Chained Residual Pooling (CRP) (across various decomposition scales) to effectively segment the concealed and cluttered contraband items. The detailed description of each of these modules are described next.

3.1 Residual Convolutional Units

RCUs consist of a convolutional block which is a toned-down variant of ResNet without batch-normalization layers. It is responsible for extracting the coarser and finer features from lower and upper levels of the CNN Backbone and forward the maps towards Fusion Block.

3.2 Chained Residual Pooling

CRP combines the features from multiple decomposition scales and adds them together using its adaptable weights. Each unit includes a max-pool followed by

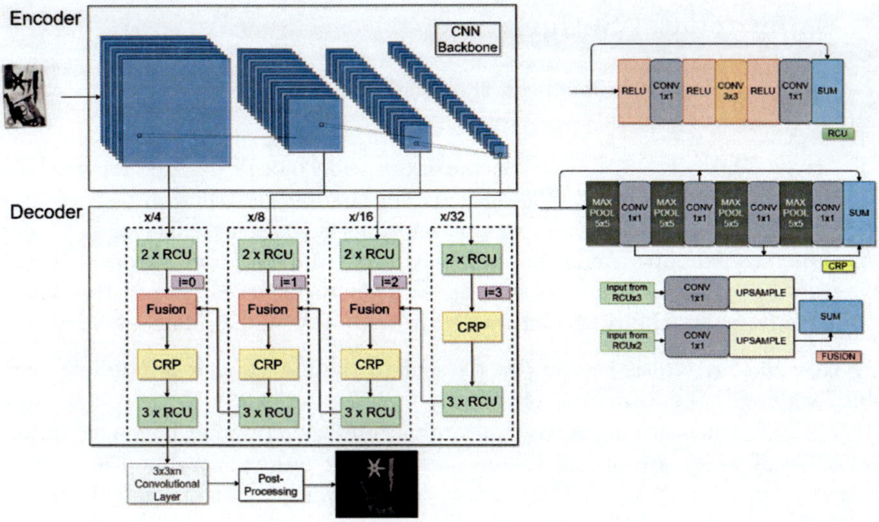

Fig. 1. Schematic diagram of the proposed framework. CNN (ResNet) backbone extracts the features from X-ray scans and RCU acquires the features of different resolutions from the backbone. The Deep Fusion Block is responsible for integrating the features from a lower resolution to the high-resolution feature maps. On the right-hand side, from top to bottom, there is Residual Convolution Unit (RCU), a series of convolutions and activation functions, Chained Residual Pooling (CRP) is a series of pooling function and convolutions & Deep Fusion Block is responsible for up-sampling the inputs and creates an integrated map.

a convolution layer that serves as the summation operation. During training, the convolutional layers learn to adapt to the pooling framework. Each pooling block adds to the input and summation is performed in the final unit of CRP. Consequently, it obtains a vast field without using large pooling kernels, reducing thus the loss of information.

The CRP blocks are responsible for capturing the global data through multiple pooling and convolutional layers. CRP blocks then extract the global information, and Fusion blocks integrate the information of lower and higher resolution, so the combination of both produces robust extraction of the cluttered baggage threats.

3.3 Fusion Blocks

By summing up residual connections, the resulting feature maps of the CRP are merged with the input feature map. As depicted in Fig. 1, lower resolution features from CRP and RCU, are combined with the consecutive higher resolution features from the RCU (which takes the input directly from the backbone network).

4 Experimental Setup

The proposed framework has been thoroughly assessed on three public datasets described below:

GDXray: GDXray [26] dataset is the most widely used dataset for screening prohibited items. It contains 19,407 scans for the non-destruction testing. However, for screening baggage threats, it contains 8,150 grayscale baggage scans containing contraband items like *guns*, *knives*, *shuriken* and *razors*. We used 400 scans for training and the 400 for evaluation purposes, as per the dataset standard, to assess the proposed model on the GDXray [26] dataset.

SIXray: SIXray [20] is the largest dataset to the date and it is publicly accessible for detecting contraband objects. It includes 1,059,231 X-ray color images, 1,050,302 of which are negative (containing only the common baggage content) and 8,929 of which are positive (containing one or more contraband objects like *guns*, *knives*, *wrenches*, *scissors*, *pliers*, and *hammers*). For evaluating the proposed model, we need the test scans to have some mask ground truths so that we can analyze the model performance. For that purpose, we have divided 8,929 scans for testing and training purposes as well. As per the dataset standard, 80% of the scans (7128 scans for training) should be used for the training and the rest of 20% (1777 scans for evaluation) should be used for the testing purposes.

OPIXray: OPIXray [25] is a recently introduced dataset for the detecting occluded contraband items like *folding knives*, *straight knives*, *scissors*, *utility knives*, and *multi-tool knives*. As per the dataset standard, we used 8,885 scans for training and 1,776 scans for the testing purposes.

4.1 Implementation Details

The proposed model is executed using PyTorch 1.5.1 on Anaconda with Python 3.8, CUDA 10.2, cuDNN v7.6. Some utility functions (for the evaluation and comparison purposes) have also been written on MATLAB R2020b. The machine specifications are: Intel Core i7-9750H@2.6 Ghz, 16 GB RAM, NVIDIA GeForce GTX 1650. The training for conducted for 60 epochs (where the number of iterations in each epoch varied as per each dataset). Stochastic gradient descent is employed as an optimizer, the momentum is set to 0.9 and learning rate of 0.0005.

4.2 Evaluation Metrics

We used performance indicators such as IoU otherwise known as intersection-over-union, accuracy, recall and F1 value to test the proposed system, where true positives (TP), false positives (FP), false negatives (FN), and true negatives (TN) were computed pixel-wise in each scan.

5 Results

Here, we put forward a meticulous assessment of the suggested method on the three public baggage X-ray datasets. We also report its comparative evaluation with the state-of-the-art segmentation models. First of all, the proposed framework is evaluated on GDXray datasets. Afterwards, to further test its capacity to recognized cluttered baggage threats, we evaluated on the SIXray datasets. Lastly, the performance of the proposed model is evaluated on the most recent publicly available baggage threat detection dataset i.e., the OPIXray. Moreover, the qualitative evaluation of the suggested architecture and its comparative analysis with the state-of-the-art segmentation models (such as SegNet [27], U-Net [29], PSPNet [28], FCN-8 and FCN-32 [30], and RAGNet [31]) on all three datasets is presented in Fig. 2.

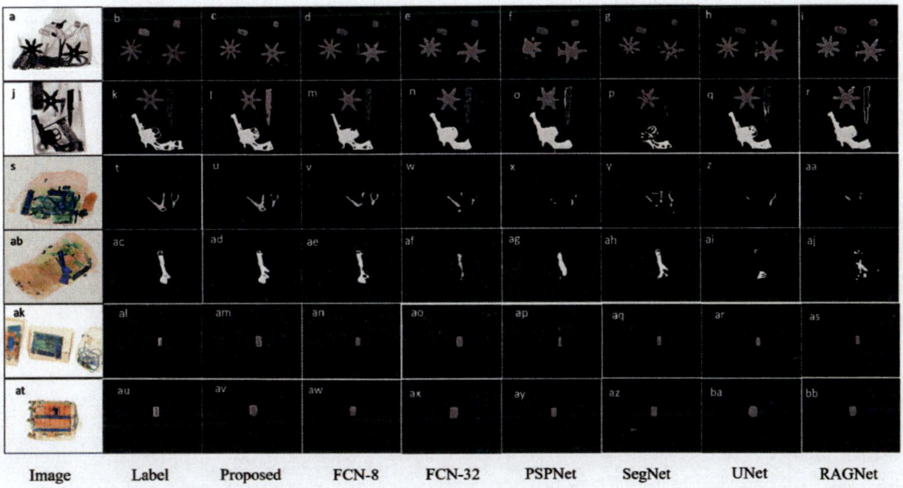

Fig. 2. A Qualitative performance comparison of the proposed architecture with other leading models. Here, Column 1 shows the original scans from GDXray, SIXray and OPIXray datasets, respectively. Column 2 shows the ground truth labels, Column 3 shows the results of the proposed framework, Column 4 to 9 showcase the results of the FCN-8, FCN-32, PSPNet, SegNet, UNet, and RAGNet, respectively. Please zoom-in for better visualization.

5.1 Evaluations on the GDXray Dataset

The assessment of the suggested method on GDXray dataset is reported in Table 1. In terms of IoU, F1 and Recall, the proposed framework outperforms other segmentation networks by 8.16%, 4.95%, and 2.69%, respectively except for CST. Here, it should be noticed that GDXray is a highly complex dataset containing gray-scale X-ray scans and due to the texture-less nature of the scans, detecting contraband items robustly from GDXray dataset is a challenging task.

Table 1. Performance comparison of the suggested method with leading semantic segmentation networks on GDXray [26] dataset. Numbers in bold show the leading results meanwhile the second-best scores are underlined.

Method	Precision	Recall	mIoU	F1 score
Proposed	0.9169	<u>0.8367</u>	<u>0.7768</u>	<u>0.8631</u>
SegNet [27]	0.8659	0.3804	0.3617	0.5286
PSPNet [28]	0.8391	0.7590	0.6645	0.7970
UNet [29]	0.8586	0.7681	0.6952	0.8108
FCN-8 [30]	<u>0.9170</u>	0.7103	0.6639	0.8005
FCN-32 [30]	0.8709	0.7483	0.6722	0.8049
RAGNet [31]	0.8271	0.8098	0.6879	0.8136
CST [1]	**0.9714**	**0.9969**	**0.9644**	**0.9836**

5.2 Evaluations on SIXray Dataset

The assessment of the suggested method on SIXray is shown in Table 2. Here, it can be noticed that in all the metrics, the proposed system surpasses state-of-the-art solutions. To the most of our comprehension, SIXray is the most demanding and imbalanced dataset till date and we believe that the output of the proposed model on SIXray dataset is significant, particularly because it surpassed the leading model by 6.92% regarding the F1 (where F1 possesses the capacity to fairly evaluate the frameworks on the highly imbalanced data).

Table 2. Performance comparison of the suggested method with leading semantic segmentation networks on SIXray [20] dataset. Numbers in bold show the leading results meanwhile the second-best scores are shown with an underline. Also, '-' means that the corresponding metric is not evaluated.

Method	Precision	Recall	mIoU	F1 score
Proposed	**0.8001**	**0.7225**	<u>0.6263</u>	**0.6780**
SegNet [27]	0.6449	0.3661	0.3121	0.4671
PSPNet [28]	0.7037	<u>0.5364</u>	0.4532	<u>0.6088</u>
UNet [29]	0.7069	0.4587	0.3936	0.5564
FCN-8 [30]	<u>0.7284</u>	0.5167	0.4415	0.6052
FCN-32 [30]	0.7091	0.5288	0.4503	0.6058
RAGNet [31]	0.5828	0.2769	0.2565	0.3754
CST [1]	–	–	**0.9689**	–

5.3 Evaluations on OPIXray Dataset

The assessment of the suggested method on the OPIXray [25] is reported in Table 3, where it can be seen that the suggested framework surpasses the leading solutions across all the reported metrics. Nevertheless, the performance of the second-best, FCN-8 [30], is quiet appreciable as in the terms of mean IoU, it only lags from the proposed system by 4.86%.

Table 3. Performance comparison of the suggested method with leading semantic segmentation networks on OPIXray [25] dataset. Numbers in bold show the leading results meanwhile the second-best scores are shown with an underline.

Method	Precision	Recall	mIoU	F1 score
Proposed	**0.7878**	**0.8311**	**0.6713**	**0.7883**
SegNet [27]	0.7352	0.5850	0.5179	0.6516
PSPNet [28]	0.7767	0.6991	0.5861	0.7359
UNet [29]	0.7447	0.6963	0.5677	0.7197
FCN-8 [30]	<u>0.7842</u>	<u>0.7399</u>	<u>0.6227</u>	<u>0.7614</u>
FCN-32 [30]	0.7687	0.7384	0.6073	0.7533
RAGNet [31]	0.7114	0.7221	0.5673	0.7167

The robustness of the proposed architecture (on all the three datasets) stems from the fact that it fuses multi-resolution representations of the candidate scan to create the sparse latent vectors, which are further regularized in generating the contraband items masks regardless of the baggage clutter. One thing to notice here is that CST significantly outperforms the proposed method in terms of mIoU scores. This is because the CST framework exploits the contours of the concealed contraband items to recognize them effectively. However, CST requires extensive parameter tuning since it is a non-configurable. Therefore, it does not possess the capacity to be well-generalized across diverse ranging datasets containing sparse scanner specifications.

6 Conclusion

This paper introduces a unique semantic segmentation method that employs deep fusion, residual convolutional, and pooling blocks for autonomously recognizing the concealed contraband items in the baggage X-ray images. The proposed technique is rigorously assessed on three publicly available datasets. It can be seen in the results section that it outperforms leading baggage threat detection and popular semantic segmentation models against various metrics. For future work, the capacity of the proposed model to extract cluttered objects can be evaluated on the RGB scans as well. Also, it can be availed in segmenting the 3D printed and dismantled contraband objects within the baggage X-ray scans.

Acknowledgement. This research work is enabled by the research fund from ADEK Ref: AARE19-156 and a Khalifa University: Ref: CIRA-2019-047.

References

1. Hassan, T., Bettayeb, M., Akçay, S., et al.: Detecting prohibited items in X-ray images: a contour proposal learning approach. In: 2020 IEEE International Conference on Image Processing (ICIP), pp. 2016–2020. Abu Dhabi, United Arab Emirates (2020). https://doi.org/10.1109/ICIP40778.2020.9190711
2. Purkait, P., Zach, C., Reid, I.: Seeing Behind Things: Extending Semantic Segmentation to Occluded Regions. arXiv:1906.02885v2 (2019)
3. Lin, G., Milan, A., Shen, C., et al.: RefineNet: multi-path refinement networks for high-resolution semantic segmentation. In: CVPR (2017)
4. Nekrasov, V., Shen, C., Reid, I.: Light-weight RefineNet for real-time semantic segmentation. In: BMVC (2018)
5. Taha, B., Werghi, N., Dias, J.: Automatic polyp detection in endoscopy videos: a survey. In: Proceedings of the 13th IASTED International Conference on Biomedical Engineering, BioMed, pp. 233–240 (2017). https://doi.org/10.2316/P.2017.852-031
6. Taha, B., Dias, J., Werghi, N.: Classification of cervical-cancer using pap-smear images: a convolutional neural network approach. Communications in Computer and Information Science (2017). https://ieeexplore.ieee.org/abstract/document/7893296
7. Reda, I., Shalaby, A., Elmogy, M., et al.: A comprehensive non-invasive framework for diagnosing prostate cancer. Comput. Biol. Med. **81**, 148–158 (2017)
8. Alkadi, R., Taher, F., El-baz, A., et al.: A deep learning-based approach for the detection and localization of prostate cancer in T2 magnetic resonance images. J. Digit. Imaging **32**, 793–807, (2019)
9. Hayat, M., Khan, S.H., Werghi, N., et al.: Joint registration and representation learning for unconstrained face identification. In: IEEE Conference on Computer Vision and Pattern Recognition, CVPR, pp. 1551–1560 (2017). https://doi.org/10.1109/CVPR.2017.169
10. Turcsany, D., Mouton, D.A., Breckon, T.P.: Improving feature-based object recognition for X-ray baggage security screening using primed visual words. In: IEEE International Conference on Industrial Technology, pp. 1140–1145 (2013)
11. Bastan, M., Yousefi, M.R., Breue, T.M.: Visual Words on Baggage X-Ray Images. Springer, Heidelberg, pp. 360–368 (2011)
12. Franzel, T., Schmidt, U., Roth, S.: Object detection in multi-view X-ray images. In: Pinz, A., Pock, T., Bischof, H., Leberl, F. (eds.) Pattern Recognition, DAGM, OAGM. Lecture Notes in Computer Science, vol. 7476, p. 2012. Springer, Heidelberg (2012)
13. Oquab, M., Bottou, L., Laptev, I., et al.: Learning and transferring mid-level image representations using convolutional neural networks. In: CVPR, vol. 2, pp. 1717–1724. IEEE (2014)
14. Dai, J., Li, Y., He, K., et al.: R-FCN: Object detection via region-based fully convolutional networks. CoRR 06409 2(3) (2016). arXiv:16054
15. Krizhevsky, A., Sutskever, I., Hinton, G.E.: ImageNet classification with deep convolutional neural networks. Adv. Neural Inf. Process. Syst. **25**, 1097–1105 (2012)

16. Akçay, S., Kundegorski, M.E., Willcocks, C.G., Breckon, T.P.: Using deep convolutional neural network architectures for object classification and detection within X-ray baggage security imagery. IEEE Trans. Inf. Forens. Secur. **13**(9), 2203–2215 (2018)
17. Xu, M., et al.: Prohibited item detection in airport x-ray security images via attention mechanism based CNN. In: Chinese Conference on Pattern Recognition and Computer Vision, pp. 429–439 (2018)
18. Griffin, L.D., Caldwell, M., Andrews, J.T.A., et al.: Unexpected Item in the Bagging Area: Anomaly Detection in X-ray Security Images, pp. 1539–1553 (2019)
19. Szegedy, C., Liu, W., Jia, Y., et al.: Going Deeper with Convolutions (2015)
20. Miao, C., Xie, L., Wan, F., et al.: SIXray: a large-scale security inspection X-ray benchmark for prohibited item discovery in overlapping images. In: IEEE International Conference on Computer Vision and Pattern Recognition (CVPR), pp. 2119–2128 (2019)
21. Akçay, S., Abarghouei, A.A., Breckon, T.P.: GANomaly: semisupervised anomaly detection via adversarial training. In: Asian Conference on Computer Vision (2018)
22. Akçay, S., Abarghouei, A.A., Breckon, T.P.: Skip-GANomaly: skip connected and adversarially trained encoder-decoder anomaly detection. In: International Joint Conference on Neural Networks (2019)
23. Hassan, T., Akçay, S., Bennamoun, M., et al.: Trainable structure tensors for autonomous baggage threat detection under extreme occlusion. In: Asian Conference on Computer Vision (2020). https://arxiv.org/abs/2009.13158
24. Hassan, T., Shafay, M., Akçay, S., et al.: Meta-transfer learning driven tensor-shot detector for the autonomous localization and recognition of concealed baggage threats. MDPI Sensors **20**, 6450 (2020). https://doi.org/10.3390/s20226450
25. Wei, Y., Tao, R., Wu, Z., et al.: Occluded Prohibited Items Detection: an X-ray Security Inspection Benchmark and De-occlusion Attention Module (2020)
26. Mery, D., Riffo, V., Zscherpel, U., et al.: GDXray: the database of X-ray images for nondestructive testing. J. Nondestr. Eval. **34**(4), 1–12 (2015)
27. Badrinarayanan, V., Kendall, A., Cipolla, R.: SegNet: a deep convolutional encoder-decoder architecture for image segmentation. IEEE Trans. Pattern Anal. Mach. Intell. **39**(12), 2481–2495 (2017)
28. Zhao, H., Shi, J., Qi, X., et al.: Pyramid scene parsing network. In: IEEE International Conference on Computer Vision and Pattern Recognition (CVPR), pp. 2881–2890 (2017)
29. Ronneberger, O., Fischer, P., Brox, T.: U-Net: Convolutional Networks for Biomedical Image Segmentation. arXiv:1505.04597 (2015)
30. Long, J., Shelhamer, E., Darrell, T.: Fully convolutional networks for semantic segmentation. In: IEEE International Conference on Computer Vision and Pattern Recognition (CVPR), pp. 3431–3440 (2015)
31. Hassan, T., Akram, M.U., Werghi, N., et al.: RAG-FW: a hybrid convolutional framework for the automated extraction of retinal lesions and lesion-influenced grading of human retinal pathology. IEEE J. Biomed. Health Inf. **25**, 108–120 (2020). https://doi.org/10.1109/JBHI.2020.2982914
32. He, K., Zhang, X., Ren, S., Sun, J.: Deep residual learning for image recognition. In: IEEE Conference on Computer Vision and Pattern Recognition (2016)

Check for updates

Project Comparison Method Supporting Participatory Budget Process - Poznan Case Study

Aleksandra Rutkowska[1](✉), Dariusz Walczak[2], and Agata Filipowska[1]

[1] Poznan University of Economics and Business, al. Niepodleglosci 10, 61-875 Poznań, Poland
{aleksandra.rutkowska,agata.filipowska}@ue.poznan.pl
[2] The Bit By Bit Sp. z o.o., ul. Za Bramka 9/10, 61-842 Poznan, Poland
dariusz.walczak@thebitbybit.com

Abstract. Participatory Budget (PB) is a process that distributes part of the city budgets among projects submitted and selected by the dwellers. The key challenge for IT-supported e-PB is the projects' comparison and ranking. This paper focuses on an empirical test of a hybrid method for comparing BP projects. In this study, we investigate two difficult to measure dimensions: beneficiaries and categories. We use ontology to describe and map distances between concept and then generate ranking based on the fuzzy TOPSIS method. The method is validated through experiments with annotators and working methods. The results surpass semantic measure, but also show the space for further development.

Keywords: Participatory budget · Similarity measure · Fuzzy TOPSIS

1 Introduction

Citizen participation management is a new trend in public administration aimed at activating society, and Participatory Budget (further PB) is one of the most popular forms of participation. The process splits a fraction of the city budget among projects submitted and selected by the dwellers [3,7,9,14]. The first PB process was carried out in the Brazilian city of Porto Alegre in 1989 [10]. Since then, PB has become a phenomenon and regularly gains followers. In Poland, in 2017, all large cities (above 100k citizens), 2/3 of mid-size cities (20k–40k citizens), and 1/4 of small cities were splitting budget in a participatory manner. According to the newest data from 2019, Poland is the European leader with the highest number of processes, namely from 1840 to 1860, a situation to which the creation of national legislation to encourage the development of these processes at the municipal level contributed to a decisive way [1]. More than half of the cities let their citizens participate via the Internet - which makes it a special kind of web economy. The E-PB systems became essential when the number

A. Abraham et al. (Eds.): SoCPaR 2020, AISC 1383, pp. 560–571, 2021.
https://doi.org/10.1007/978-3-030-73689-7_54

of users and projects increased significantly. For different types of supporting systems, regardless of whether we are talking about a recommendation system or a system for processing projects by the city hall, the key element is to build a ranking and compare projects. This paper focuses on an empirical test of methods of comparing PB projects, as it guarantees the correct operation of the whole system. A similarity measure that allows comparing project descriptions should take into account multiple description dimensions and obtain results at least as good as the basic measures. Despite that, the applied measure should work according to human intuition or in a similar way. The method must accept project descriptions expressed in natural language. The main novelty of this paper lies in the hybrid approach to similarity measures. With this study, we also expand and test empirically fuzzy TOPSIS, which was proposed in [15]. We also aim to best validation of similarity measures, so an experiment with a group of respondents was conducted. The rest of the paper proceeds as follows. In the next section, we introduce methods for comparison projects. After that, we present the experiment on real data from Poznan PB. First, we introduce data, then the results of algorithmic comparison with validation made by a test group invited to the experiment.

2 Methods

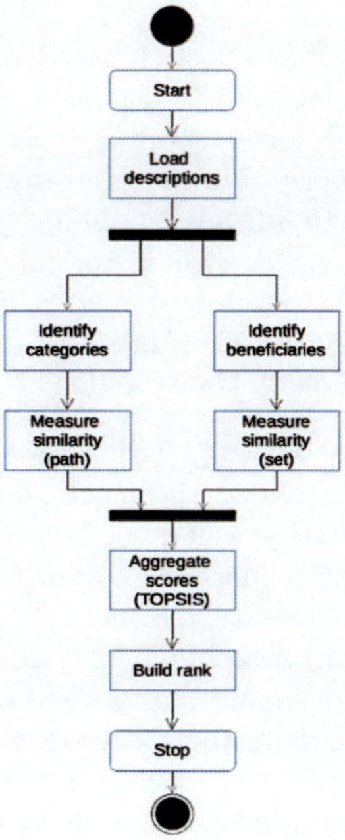

Fig. 1. The algorithm of measure similarity of PB project.

Due to the multidimensional nature of BP project descriptions, the applied measure will be used to determine similarity in different ways for different dimensions. Within our research we identify around 10 dimensions used in Poland to describe PB, these are: title, description, cost, location, motivation, municipality comment, categories, beneficiaries, future costs, citizen comments. Some of them are obvious to compare: cost, future cost, category or location. In this paper we focus on two of them with major importance: category and beneficiary. First one says what is the subject of the project. The descriptions were load into the system and indexed based on keywords. The beneficiary dimension seems to be important, because we assume that decision maker maximise his/her own utility and selects/compares projects that he/she can benefit from. Both selected dimensions are hard to compare as both use plain text.

Before similarity calculation the cleaning and indexing has been done. The following steps has been done:

- single letters, number or special characters has been removed
- the text has been sliced into words
- for each slice we've determined the lemma of a word
- words has been checked against stop word lists

At the end the bag of words has been calculated for each project. The keywords were selected based on existence of the corresponding term in the proposed ontology (for both, category and beneficiaries dimension). The ontology has been build in following manner:

- The top level terms has been selected based on the law defining scope of responsibilities for municipalities in Poland.
- For each of the top level terms more specific words has been added to the ontology based on Polish Wordnet Database. This operation has been repeated recursively for 3 to 8 levels.
- Each ontology has been tested with historical projects descriptions. If the term was not present in the ontology we've tried to connect it via it's generalisations (disovered via Wordnet). If there was a link between the term and the ontology terms the whole link was added.

First two steps gave us generic ontology within the scope of municipality responsibilities while the last one extended the most popular areas for PB. However the task of measure the similarity was done in a different manner for category and beneficiary dimension. After mapping to proper concepts the projects were compared on each dimension separately. Finally the results were aggregated with TOPSIS method and final rank was built. The whole algorithm is presented in Fig. 1. The proposed algorithm should be classified as a hybrid solution, although even among measures of this type there are no solutions using different techniques depending on the type of relationship between concepts. For dimensions with hierarchy structure, a measure using edge counting is used, for dimensions based on labels the common part of the set.

2.1 Comparing Projects According to the Category

The category is a feature with a hierarchy structure, which means higher-level concepts are more generic while their children are more specific. For example, when considering an element of urban infrastructure, it can be placed in a hierarchy such as the following:

1. City infrastructure
   (a) Roads
      i. residential roads
      ii. ring roads
      iii. bicycle paths
      iv. ...
   (b) Bridges
   (c) Buildings
      i. Operas
      ii. Offices
      iii. Warehouses
      iv. ...

For the dimension with hierarchy structure, project category, the starting point were the works of [2,5,12,13] and the assumption that the similarity of the words describing PB projects categories is inversely proportional to the distance that separates the concepts associated with the words in an ontology. Using Rada's work [12] you can define the function of the distance between words a and b, which is represented by the concepts c_a and c_b in ontology with the relations *is type*:

$$Distance(a, b) = \text{min. no. separating edge } c_a \text{ i } c_b \tag{1}$$

We use one of the existing implementations of the Dijkstra algorithm [6] - the algorithm to find the shortest path in the graph. The distance between sets of concepts representing project descriptions will be calculated as an average of the minimum distances between individual pairs of concepts using the function:

$$\begin{aligned} Distance(A, B) &= Distance(a_1 \wedge ... \wedge a_k, b_1 \wedge ... \wedge b_m) \\ &= \tfrac{1}{k} \textstyle\sum_{i=1}^{k} \min_{j \in 1,...,m} Distance(a_i, b_j) \end{aligned} \tag{2}$$

Originally, [12] uses an average distance between all words. This converges the results of all comparisons to the average distance between concepts in an ontology. This situation is visible with large discrepancies in the number of identified concepts. If one project has a wide range and the other is a subset of it, then in the case of the original solution their similarity will be low. If we use only minimal distance we get two results, the general project will be less similar to the specific than the specific to the general. This is in line with the intuition, as the general project will also meet the detailed objectives, while the detailed project will be fully consistent with the general objectives.

2.2 Comparison of Projects According to Beneficiaries

The types of beneficiaries, unlike the categories, have not any hierarchical structure. The distance between the various concepts does not reflect their similarity. Beneficiaries can be divided into many subcategories according to their age, interests, social role, and the way they move around the city. It would be very difficult to take into account the different types and the placement of concepts in a hierarchical ontology, and it might be impossible to do it objectively. The beneficiary's dimension comes down to a set of labels representing various characteristics of social groups. The similarity can be calculated as the intersection of these collections. Therefore we use Jaccard's [11] similarity measure (3). The distance is obtained by subtracting the obtained result from 1 (4):

$$J(X,Y) = \frac{|X \cap Y|}{|X \cup Y|} \tag{3}$$

$$J_d(X,Y) = 1 - J(X,Y) = \frac{|X \cup Y| - |X \cap Y|}{|X \cup Y|} \tag{4}$$

An important step preceding the calculation is to determine an appropriate set of concepts describing beneficiaries on the basis of thematic categories. This is done by searching for links in the prepared ontology. If a given concept is not directly related to specific beneficiaries its ancestors are checked.

2.3 Fuzzy Topsis

To aggregate different dimension of comparison, we use the fuzzy TOPSIS method proposed in [15]. Let us assume that there are k decision makers, n possible alternatives called $A = A_1, A_2, \ldots, A_n$, which are evaluated against m criteria $C = C_1, C_2, \ldots, C_m$. For every decision maker, the decision matrix $D_{n \times m}$ contains performance ratings for each alternative A_i with respect to criteria C_j, which are denoted as x_{ij}. The primary modification presented of the TOPSIS method is that we do not seek the perfect solution among the maximum and minimum values of the criteria because the PIS must be chosen by a participant. For every criterion i, the NIS_i value is that with the greatest distance from the PIS: $NIS_i = \max_j (d_{jPIS})$. The algorithm contains the following steps:

1. Determine the PIS, as a project chosen by participant $PIS = A_j$.
2. Determine distance from the PIS.
3. Create the distance measures matrix.
4. Determine the NIS.
5. Normalize the decision matrix.
6. Weight the normalized decision matrix with linguistic weights.
7. Calculate distance from the NIS.
8. Calculate the closeness coefficient for each project.
9. Rank the projects.

In the fuzzy extension of TOPSIS, the ratings of alternative A_i with respect to criterion C_j and the weights of the criteria w_j are fuzzy numbers or linguistic variables. Thus, the decision matrix is a fuzzy decision matrix $\tilde{D}$ and the weight vector is a fuzzy vector $\tilde{W} = (\tilde{w_1}, \tilde{w_2}, \ldots, \tilde{w_n})$. The importance weight for each criterion can be obtained by direct assignment or based on pairwise comparisons. In the case of PB, the weights are subjective values and can be different for each citizen, so decision support system should allow individual decision-makers to determine the weights. Note that steps 3–6 are the same as in standard algorithm, but the operations are performed on fuzzy numbers. Following [4] we use the vertex method to calculate the distance between two triangular fuzzy numbers.

2.4 Verification and Validation

To verify the conformity of the annotators, Cohen's kappa statistic was used, which measures the interrater reliability (further IRR) [8]. The idea of the measure is based on the assumption that the observed value of a given variable (X) consists of the actual value (T) and the error (E). Moreover, the observed variance is equal to the variance of the real value and the error [8]: $Var(X) = Var(T) + Var(E)$. The coefficient determines the value of the observed variance, after removing the error variance, in relation to the actual variance of [8]:

$$Reliability = \frac{Var(T)}{Var(X)} = \frac{Var(X) - Var(E)}{Var(X)} = \frac{Var(T)}{Var(T) + Var(E)} \tag{5}$$

The designed method will be evaluated in the context of its application, i.e. comparing projects. In practice, it is done at the stage of submitting proposals and selecting initiatives in the voting process. The measure is supposed to indicate what other projects the user should get familiar with before making a decision. It is important to demonstrate to the user all projects that are potentially important to him/her. The standard in the assessment of the effectiveness of information search systems is the precision (6) & recall (7).

$$P = \frac{tp}{tp + fp} \tag{6}$$

$$R = \frac{tp}{tp + fn} \tag{7}$$

where tp - true positive value, tn - true negative values, fp - false positive value, fn - false negative value In order to obtain a result in the form of a single value, the measure F is used:

$$F = \frac{1}{\frac{P}{+}(1-\alpha)\frac{1}{R}} = \frac{(\beta^2 + 1)PR}{\beta^2 P + R}, \tag{8}$$

where $\beta^2 = \frac{1-\alpha}{\alpha}$. For $\beta < 1$ precision is more important, for $\beta > 1$ sensitivity. For ranking evaluation Mean Average Precision (further MAP) is used. MAP is the average precision for the most important k observations additionally averaged by the number of all relevant elements. For query/informational need $q_j \in Q$ the relevant documents are $d_1, \ldots, d_{m_j}$ and the search result is a ranking of R_{jk}:

$$MAP(Q) = \frac{1}{|Q|} \sum_{j=1}^{|Q|} \frac{1}{m_j} \sum_{k=1}^{m_j} Precision(R_{jk}) \tag{9}$$

As benchmarks two popular measures cosine (cosine of the angle between two vectors and determines whether two vectors are pointing in roughly the same direction) and Rada (1) were chosen.

Before the evaluation of different types of similarity measures, we establish the similarity of project descriptions by the annotators. For this purpose, two groups of students from the Faculty of Computer Science and Electronic Economy of the Poznań University of Economics were selected, 32 people in total. At the beginning of the experiment, the participants were instructed on technical and substantive issues of the evaluation process. Each participant of the experiment using a specially prepared application had to evaluate two sets of 105 pairs of project descriptions. In the next step, a one-sided statistical test was used to determine which pairs of project descriptions could be treated as similar, on the basis of data collected from the annotators from the sample of project descriptions. For both sets, the average cross-compliance of annotators was examined. Cohen's Kappa [8] ratio was used. The average value of the Kappa coefficient for the A set is 0.3018 and for the B set 0.3710. It shows that according to the Landis and Koch scales, in both cases a reliable result was obtained.

3 Empirical Data

The source of data in the experiment is the Poznań Civic Budget for 2017 (PBO2017) - the set contains descriptions of 127 proposals, each of which is described with the necessary attributes. Since the comparison of all possible pairs of projects was not possible for experiment with annotators, we choose a two random sample of 15 projects. Project descriptions contained for set A an average of 49 words with a standard deviation of 28.3685 and 47 words for B with a standard deviation of 26.1683. The shortest description in the first sample has 17 words and the longest 99, while in the set B 15 and 101 accordingly. As a result of cleaning and indexing, the projects were assigned to 5 concepts

Table 1. Characteristics of descriptions of selected sets after indexing

Set	Avg. no. of categories	Std. for categories	Avg. no. of beneficiaries	Std. no. of beneficiaries
A	4.6666	2.0548	7.5333	2.1868
B	6.7333	4.5088	6.6667	1.9889

representing the category in set A and 7 in B and 2 and 4 concepts describing the beneficiaries respectively (Table 1).

In TOPSIS step we use the linguistic variables proposed by Chen [4] and used also by [15] represented by following fuzzy triangular numbers:

- Very low: (0.00;0.00;0.10)
- Low: (0.00;0.10;0.30)
- Medium Low: (0.10;0.30;0.50)
- Medium: (0.30;0.50;0.70)
- Medium high: (0.50;0.70;0.90)
- High: (0.70;0.90;1.00)
- Very high: (0.90;1.00;1.00)

4 Results

Table 2. The value of the MAP measure for rankings from the set A and B.*

Set A				Set B			
Project	TOPSIS	$Cos(X,Y)$	Rada	Project	TOPSIS	$Cos(X,Y)$	Rada
P2017.122	1	0.25	0.3333	P2017.15	0.3278	0.1991	0.1927
P2017.23	1	0.25	0.5	P2017.2	0.2083	0.1548	0.7
P2017.118	0.7503	0.5382	0.6780	P2017.22	0.2978	0.4515	0.3390
P2017.119	0.1111	0.1667	0.0769	P2017.28	0.25	0.25	0.2
P2017.25	0.75	0.6771	0.6104	P2017.35	0.1111	0.0714	0.5
P2017.32	0.125	0.3333	0.0714	P2017.60	0.6786	0.8690	0.4837
P2017.42	0.3735	0.3854	0.3833	P2017.61	0.8714	0.9667	0.4130
P2017.5	0.5	0.2728	0.65	P2017.65	0.7361	0.6905	0.3603
P2017.6	0.8333	0.5	0.325	P2017.72	0.2897	0.3131	0.2209
P2017.62	0.5833	0.75	0.1714	P2017.75	0.3354	0.6556	0.2991
P2017.65	0.95	0.3930	0.95	P2017.95	0.225	0.2111	0.2429
P2017.86	0.1429	0.375	0.1325				
P2017.97	0.6667	0.7875	0.95				
Avg.	0.5989	0.4368	0.4486	Avg.	0.3937	0.4393	0.3592

*from set A: P2017.92 and P2017.1; from set B: P2017.86, P2017.16, P2017.3 and P2017.52 are exclude due to lack of similar documents

The calculation results are presented in the Table 2. In two cases, the TOPSIS algorithm obtained the maximum result i.e., all descriptions indicated by the annotators as similar were included in the ranking at the beginning. In these two cases, the set of essential elements was single-element. In the same rankings, the cosine measure placed the important project on the 4th position and the Rada on the 3 and 2. For project no. 65, the TOPSIS and Rada measures obtained the same MAP value, although the rankings are not identical:

- TOPSIS: projects no.: 97, 118, 25, 92, 5,
- Rada: projects no.: 97, 5, 25, 92, 118,

The rankings are different, but the MAP result is identical (0.95). In many applications, this behaviour may be negative, but in the case of social participation, with a large divergence of opinion among endorsers, it is acceptable. It is impossible to establish the correct order.

The participants in their assessments were consistent with the RADA ranking, which proves that the category is incorrectly related to the dimension of beneficiaries in the ontology. The high position of project no. 92 shows an not obvious relationship in the hierarchy of categories. In total, in set A, in 6 cases the TOPSIS measure obtained the best result, in 5 better than the cosine measure or Rada. In one case it was significantly worse.

For two cases of the A set, the ranking could not be established. The participants of the survey according to (<20% for similarity) described pairs with these projects as not similar. In 5 cases we cannot speak of unanimity, more than 40% were responsible for the similarity. It is not possible to clearly resolve the similarity of these projects. Therefore, the issue of including them in the set of documents is debatable.

Set B turned out to be much more difficult for annotators. Projects drawn in this set were extremely different or similar in many ways. They were partially overlapping in scope, thematic links and beneficiaries were difficult to recognise. Probably this is the reason why no similar project was selected in 4 cases. The order in which comparisons were made was also important. Most participants started with set A. The task was not easy, and the tiring of subsequent comparisons may have resulted in less inquisitiveness about the similarity. Such a situation may have promote a measure based on a word - the most easily identifiable form of similarity. In this set, until in 3 cases, the cosine measure reached the highest score of the MAP measure. In the case of the project P2017.22 (title: "On foot after Poznań - Urban Planner for the inhabitants"), all measures received a low score, TOPSIS the worst. This is the only case where the TOPSIS measure received a significantly weaker result. The cosine measure gained an advantage by choosing only one project better than TOPSIS. For both sets the Kappa coefficient was calculated between the results of TOPSIS baseline measurements and all participants. The result is presented in Table 3. All the measurements were given a low result. The reason for that can be a small number of similar projects - the annotators unanimously agreed that only one pair is similar. Low compliance of annotators occurred despite clear guidelines for comparing projects. The guidelines stated that it was necessary to identify beneficiaries, but they did not affect the way of applying. This dimension was much more likely to affect the negative outcome of the TOPSIS measure. This indicates the need for an attribute describing potential project beneficiaries - participants interpret this factor in different ways. In order to identify the right reference point for measures, participants were grouped based on their responses. The K-Means algorithm was used for this, assuming a different number of possible k's, from 2–10. In Table 4 we present results for set A with $k = 4$ and set B with $k = 5$, which were the best partition.

Table 3. Avg. value of Cohen's Kappa coeff. for baseline measures, TOPSIS and all study participants for set A i B.

Measure	Set A	Set B
$Cos(X,Y)$	0.1108	0.2069
Rada	0.1125	−0.0039
TOPSIS	0.1756	0.1292

In the case of set A, the best result for a single group of participants was 0.5709, unfortunately, the group consisted of only 3 people and was one of the 9 groups (k = 9). For k 4, one of the groups is much larger than the others, with the highest IRR (Inter-Rater Reliability) value. In group 1 the average IRR value is 0.3971, the average IRR value for each of the participants and measuring $Cos(X,Y)$ 0.0793, Rada 0.0853, TOPSIS 0.1535.

For set B, the best result for a single group was 0.6407, the result was also achieved for three annotators. For k 5, the most numerous group did not obtain the best IRR score, but due to its size, it was verified for consistency with the results obtained using the designed and basic measures. With an average annotator of 0.4134, $Cos(X,Y)$ received a compliance of 0.2547, Rada −0.0214, and TOPSIS 0.1155.

Table 4. The result of grouping the study participants according to the answers.

Group Id	av. IRR	No. of people	Group Id	av. IRR	No. of people
Set A			Set B		
0	0.377416	3	0	0.603809	4
1	0.39714	15	1	0.413378	12
2	0.275815	8	2	0.424432	2
3	0.384475	6	3	0.474043	9

Despite an attempt to find social groups with greater than average compliance, this was achieved to a limited extent. The obtained IRR results are higher than in the full group, however, their absolute value is still low, which indicates a large variation in the perception of similarity between projects among annotators.

For the first set of evaluated projects, the multi-criteria semantic measure of project similarity obtains better results than traditional dictionary methods and semantic measures using a hierarchy of types for both project equalisation and ranking purposes. In the case of the second set, the designed measure has also been more consistent with annotations than the semantic base measure. Unfortunately, both of them obtained a lower score than the dictionary measure. The most common reason for lowering the score was a misidentified similarity or lack of similarity for the beneficiary dimension. By analysis particular cases, the following factors were identified that affected the result established by the TOPSIS and are also potential directions for further studies as follows. We do not include a connection between dimensions. A method of discovering relations between them should be developed. There is a need to place additional concepts in the ontology and to extend it with additional relationships. Moreover the links between subject categories and beneficiaries were defined subjectively by the authors of the work - they may not be consistent with the assessment of the annotators.

5 Conclusion

The paper presents an application of a new method for multicriteria semantic comparison of projects descriptions. The method combines the features of path length measures developed by Rada et al. [12] and a part of Jaccard's common collection. According to the current knowledge of the authors, it is probably the first hybrid measure that applies various methods of measuring similarity depending on the nature of the feature being studied. The dimensions are aggregated with the modified TOPSIS method, in which the positive ideal solution is an actual project and not an artificial, ideal case. Such an approach allows to easily extend the measure by further dimensions, e.g. cost, location, using traditional distance measures.

Empirical studies proved possibility to rank PB projects by fuzzy TOPSIS with different similarities measures per dimension.

Within the presented results, weights for aggregation were defined for whole population while the agreement within smaller groups were more consistent. These leads to the conclusion that comparison of multidimensional information requires individual approach. In order to support decisions with BP one have to take into consideration individual weight of the project features.

References

1. The world atlas of participatory budgeting. https://www.pbatlas.net/index.html. Accessed 31 Oct 2010
2. Alani, H., Brewster, C.: Metrics for ranking ontologies (2006). 10.1.1.61.5749
3. Alfaro, C., Gómez, J., Ríos, J.: From participatory to e-participatory budgets. In: e-Democracy, pp. 283–299. Springer (2010)
4. Chen, C.T.: Extensions of the TOPSIS for group decision-making under fuzzy environment. Fuzzy Sets Syst. **114**(1), 1–9 (2000)
5. Corley, C., Mihalcea, R.: Measuring the semantic similarity of texts. In: Proceedings of the ACL Workshop on Empirical Modeling of Semantic Equivalence and Entailment, June 2005, pp. 13–18 (2005)
6. Dijkstra, E.W.: A note on two problems in connexion with graphs. Numer. Math. **1**(1), 269–271 (1959)
7. Gomez, J., Rios Insua, D., Alfaro, C.: A participatory budget model under uncertainty. Eur. J. Oper. Res. **249**(1), 351–358 (2016)

8. Hallgren, K.: Computing inter-rater reliability for observational data: an overview and tutorial. Tutor. Quant. Methods Psychol. **8**(1), 23–34 (2012)
9. Kębłowski, W.: Budżet partycypacyjny: krótka instrukcja obsługi. Instytut Obywatelski (2013)
10. Marquetti, A., da Silva, C.E., Campbell, A.: Participatory economic democracy in action: participatory budgeting in Porto Alegre, 1989–2004. Rev. Radic Polit. Econ. **44**(1), 62–81 (2012). https://doi.org/10.1177/0486613411418055
11. Niwattanakul, S., Singthongchai, J., Naenudorn, E., Wanapu, S.: Using of Jaccard coefficient for keywords similarity. In: International MultiConference of Engineers and Computer Scientists, vol. I, Hong Kong (2013). ISBN 978-988-19251-8-3
12. Rada, R., Mili, H., Bicknell, E., Blettner, M.: Development and application of a metric on semantic nets. IEEE Trans. Syst. Man Cybern. **19**(1), 17–30 (1989). https://doi.org/10.1109/21.24528
13. Resnik, P.: Semantic similarity in a taxonomy: an information-based measure and its application to problems of ambiguity in natural language. J. Artif. Intell. Res. **11**(3398), 95–130 (1999). https://doi.org/10.1613/jair.514
14. Rios, J., Rios Insua, D.: A Framework for Participatory Budget Elaboration support (2008). https://doi.org/10.1057/palgrave.jors.2602501
15. Walczak, D., Rutkowska, A.: Project rankings for participatory budget based on the fuzzy TOPSIS method. Eur. J. Oper. Res. **260**(2), 706–714 (2017)

Seeding Initial Population, in Genetic Algorithm for Features Selection

Marc Chevallier(✉), Nicoleta Rogovschi, Faouzi Boufarès, Nistor Grozavu, and Charly Clairmont

LIPN Laboratory, Sorbonne Paris Nord University, Villetaneuse, France
{mchevallier,nicoleta.rogovschi,faouzi.boufares, Nistor.Grozavu}@lipn.univ-paris13.fr, charly.clairmont@synaltic.fr

Abstract. The feature selection process is a difficult task that can be tackled by various algorithms. Our work uses a subclass of metaheuristic algorithms called genetic algorithms (GA) to select the best subset of features that has given, for a machine learning algorithm, the best results (based on accuracy). GA are easy to implement and understand, and their results are readily explainable. However, they don't ensure to find the absolute best solution for a given problem, but only the best solution found. In order to improve the performance of GA, we introduce two seeding methods for the initial population of the GA that rely on the use of a Random Forest algorithm. The two methods are applied on two different GA using Bayesian networks as classifier to evaluate accuracy. The tests are done on five data-sets, and the two methods are compared to other dimensional reduction techniques. Our results show a better convergence of the genetic algorithms when they are seeded.

Keywords: Machine learning · Meta-heuristic · Random forest · Bayesian networks · Features selection · Genetic algorithm

1 Introduction

In a world where more and more data are collected, ensuring the quality of the data has become a huge problem. Several studies on data quality have been carried out in the literature for several years [10], particularly those dealing with duplicates in the data (anomalies between rows) [10], and those dealing with anomalies caused by the non-respect of functional dependencies that may exist between columns [12]. A way to analyse the quality of structured data is to discover the functional dependencies they contain (relationships between columns). In this domain, we explore a new path to discover those dependencies. Our current algorithm relies on machine learning and extracting numerous features from the data. However, we need a way to select the best features in order to accomplish our goal.

Supported by organization Synaltic.

A. Abraham et al. (Eds.): SoCPaR 2020, AISC 1383, pp. 572–582, 2021.
https://doi.org/10.1007/978-3-030-73689-7_55

Features selection is a well-known problem in machine learning. The goal is to select the best subset of features for a given machine learning problem. The domain is composed of three types of methods: embedded, wrappers and filters methods [9]. Wrappers and embedded methods require the use of a classifier to select the best features, common algorithm in this area are: LASSO [16], random selection [4], Recursive feature Elimination [7], genetic algorithms [13]. On the other hand, filters methods do not need a classifier, example of algorithms of this kind are: Correlation-based Feature Selection, Information Gain, ReliefF [8].

We decided to use genetic algorithms for two reason: optimise the feature selection process for a specific classifier, and because such algorithms are easy to implement.

2 Genetic Algorithms

2.1 Overview of Genetic Algorithms

Genetic algorithms are a class of metaheuristic algorithms inspired by genetics. The goal of these algorithms is to find the best parameters that optimize a function called fitness function. An initial population is generated, each individual of the population contains a set of parameters called chromosomes. Each parameter in a chromosome is called gene and is encoded in various ways. Algorithms rely on two mechanisms, mutation and crossover. Mutation randomly modifies genes in some individuals, and crossover takes place in a reproduction phase where two individuals exchange parts of their chromosomes in order to create new individuals called offsprings. After each generation each individual is evaluated with the fitness function and the best are kept for the next generation.

GA are used for a large range of problems. In some cases, seeding the initial population with non-random individuals lead to better results [5,15].

2.2 Eclectic GA (EGA)

We will first introduce the genetic algorithm called Eclectic GA (EGA). EGA uses annular crossover, uniform mutation and full elitism [2]. If we denote G the number of generations, n the number of individuals, *B2M* the number of genes to mutate, *I(n)* the *n-th* individual, L the chromosome's length, P_c the probability of crossover and P_m the probability of mutation. EGA [2], is described in Algorithm 1.

2.3 GAAM

The genetic algorithm with aggressive mutation (GAAM) is a GA with some particularities [11,14]. If we denote L the dimension of the features set, K the number of individuals, V the number of genes per chromosome (the expected maximum of features kept) and G the number of generations. GAAM uses an integer encoding system for genes, each chromosome contains a list of values between *0* and L. GAAM is described in Algorithm 2.

Algorithm 1. EGA

step 0 $B2M \leftarrow [nLxP_m]$
step 1 $i \leftarrow 1$
step 2 Generate a random population
step 3 Evaluate population
step 4 Duplicate population
for j = 1 to n **do**
  I(n+j)←I(j)
  fitness(n+j)←fitness(j)
end for
step 5 Deterministic Selection Annular Crossover
for j = 1 to n/2 **do**
  Generate a uniform random number $0 \leq \rho < 1$
  if If $\rho \leq P_c$ **then**
    Generate a random number $1 \leq \rho <$ L/2
    Interchange the semi-ring starting at locus ρ between I(j) and I(n-j-1)
  end if
end for
step 6 Mutation
for j to B2M **do**
  Generate uniform random numbers $0 < \rho_1, \rho_2 < 1$ Mutate Bit $\lfloor\rho_2 L\rfloor$ of $I(\lfloor\rho_1 n\rfloor)$
end for
step 7 Selection
Sort the 2n individuals by their fitness, ascending
step 8 $i \leftarrow i + 1$
if If i = G **then**
  return I(1) and stop
else
  go to step 3
end if

3 Experiment

3.1 Experiment Setups

All our tests are run on a google Colab instance, the CPU is an Intel Xeon 2.30 GHz with 4 cores and 12Go of RAM.

For the experiment, we have chosen five datasets described in Table 1.

We focus our analysis on two datasets Semeion and Madelon. We compare different proportions of seeding of the initial population: 0%, 10%, 50% and 100%. For EGA, we have selected an initial population of 100 individuals, and set the maximum number of iterations to 100. The probability of crossover and the probability of mutation are dynamically calculated following the ILM/DHC strategy [1]:

Gi denotes the current generation number, *Gm* the maximal number of generations.

$$P_c = 1 - \frac{Gi}{Gm} \tag{1}$$

Algorithm 2. GAAM

INPUT : *G, K, V, P*
$g = 0$
1 Build K individuals with V genes randomly picked in {0,1,2..L} in order to produce the initial population *Ip*
2 Aggressive mutation : Create *Mp* the mutate population
for j=1 to K **do**
  for x=1 to V **do**
    pick a random value m in {*0,1,2..L*}
    Assign to D a copy of **Ip(k)**
    $D(x)=m$
    add D to *Mp*
  end for
end for
3 Crossover : Apply a classical holland crossover on each individuals
4 Create *Tp=Ip+Mp+Cp* the total population evaluate each individual with the fitness function and rank them according to their fitness.
5 Drop the *K+K*V* individuals with the lowest fitness from *Tp* and replace *Ip* by the remaining,*g+=1* If $g = G$ return *Ip(0)* else back to step 2

Table 1. Description of dataset

Dataset	Number of features	Number of lines	Number of classes
Madelon	500	2600	2
Semeion	256	1593	10
Har	561	10299	6
Parkinson	753	754	2
Hill-Valley	100	1212	2

$$P_m = \frac{Gi}{Gm} \tag{2}$$

The classifier used is a Gaussian Naive Bayes [18]. The result is measured with the means of the accuracy calculated over a 5 cross-validation that we note *ma*. The fitness function is defined by 1 - *ma*. For the GAAM algorithm, we decided to keep 20 features, an initial population of 20 individuals, 100 iterations and the same fitness function.

4 Seeding of the Initial Population

For the seeding part of the process, we have developed two seeding methods denoted SSM (standard seeding method) and ESM (elitism seeding method). Our model for the generation of the chromosomes needed for the seeding process relies on the use of a random forest classifier [3]. We first train a random forest classifier on the data-set with the following parameters: 100 estimators, maximum depth

8. The importance of every feature is evaluated using the gini impurity measure and normalized (the sum of the score of all features is 1). Then, we keep the feature importance ranking and use the importance score as the probability of being select by our algorithm.

For SSM, we then randomly pick the number of desired features for each individual needed according to the probability distribution.

For ESM, we generated individuals in the same way as for SSM but we generated 5 times more individuals than desired. For each we train a Gaussian Naive Bayes classifier to evaluate their fitness and rank them, we then only pick the top n individuals we needed.

Chromosomes generated this way can have duplicated values. GAAM natively handles duplicated features by keeping one of them and set others to 0 (i.e.: no features). For EGA, when the integer representation of features is transformed into a binary representation, duplicate features are deleted.

To ensure the validity of the methods, we generated 1000 examples (on the Semeion data-sets, using a bayesian classifier, the accuracy is the mean of the accuracy calculated over a 5 cross-validation) with each of the methods and analysed the results for each case (Fig. 1).

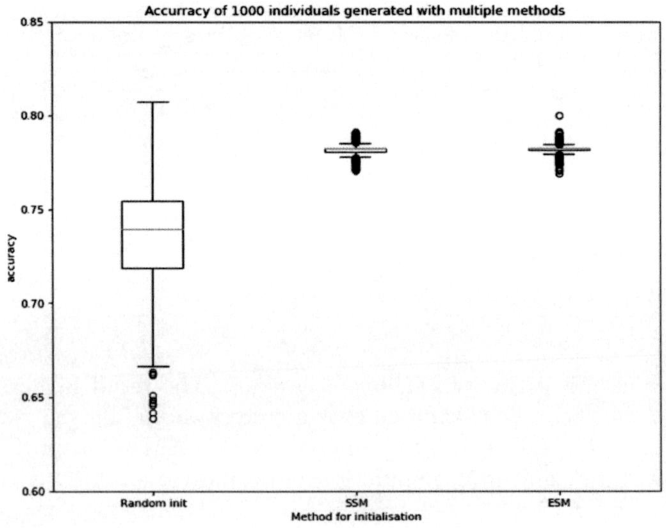

Fig. 1. Boxplot of accuracy for 1000 individuals generated with each method

The median of accuracy is higher in the SSM and the ESM generated groups. Results prove that, on average, the two methods created better individuals than with random generation. In addition, individuals generated with SSM and ESM present far less variability (so we can expect more stability in the results) but just enough to preserve diversity.

5 Results

Due to the probabilistic nature of GA, each simulation is run ten times and the mean value of the all simulations altogether is kept. We can draw some conclusions from these results.

On the 60 experiments run with an initialisation technique, 50 perform better than those without initialization. For the 10 remaining results, they are near to identical to those with no seeding. This confirms the usefulness of the approach. This conclusion is logical because individuals that are seeded go through an optimisation process before the start of the GA algorithm.

In the following section, we will analyse more deeply the results on Semeion and Madelon.

5.1 Results with EGA Algorithm

For EGA, results are particularly interesting because seeding the initial population leads most of the time to better results in accuracy and to more stability in the results (lower standard deviation) Fig. 2 and Fig. 3 show the results of EGA with SSM seeding technique.

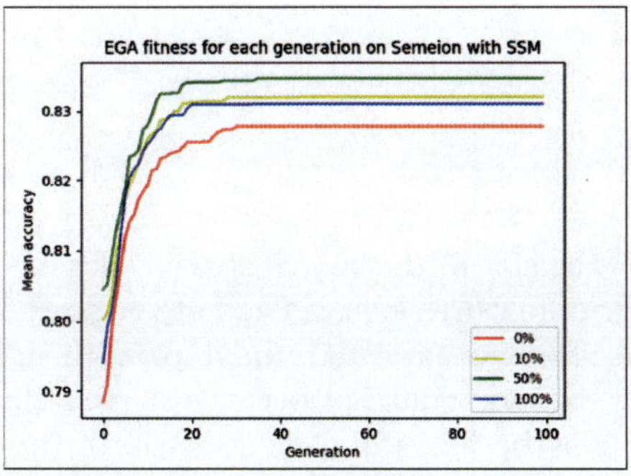

Fig. 2. Evolution of the mean of accuracy calculated on ten runs for each with 100 generations for SSM seeding on Semeion data-set

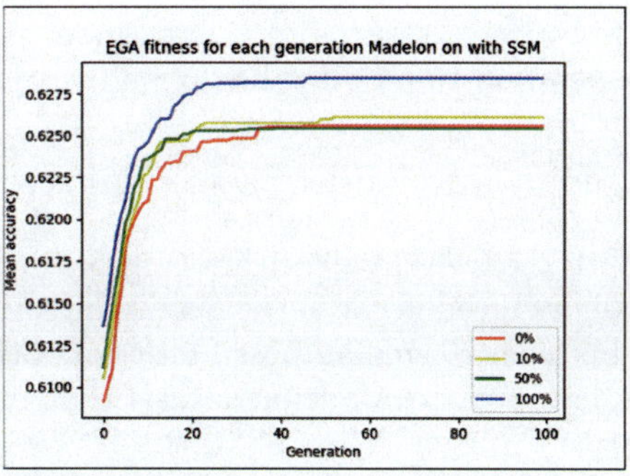

Fig. 3. Evolution of the mean of accuracy calculated on ten runs for each with 100 generations for SSM seeding on Madelon data-set

Results indicate here that seeding the initial population with SSM leads to better results on average than random seeding, and - in the worst case scenario - to results as good as random seeding. Due to the optimisation of the initial population, results are always better for seeded populations after 20 generations. EGA converges at the same speed with or without seeding (and stabilise around 50 iterations in each scenario) but the optimum result is better with seeding. In some cases, like 50% seeding, the results show no improvement. This is due to the probabilistic nature of the GA and the seeding process.

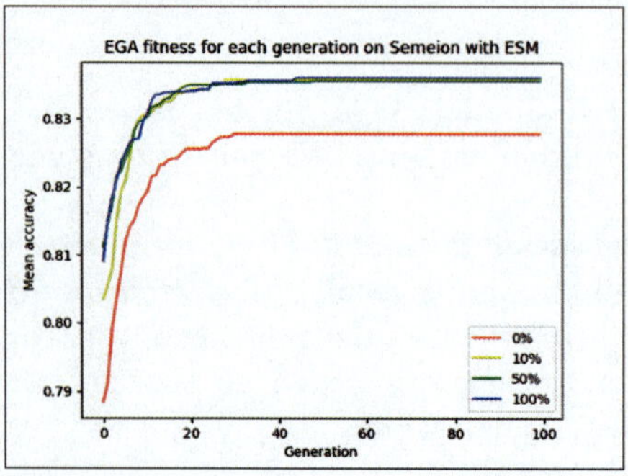

Fig. 4. Evolution of the mean of accuracy calculated on ten runs for each with 100 generations for ESM seeding on Semeion data-set

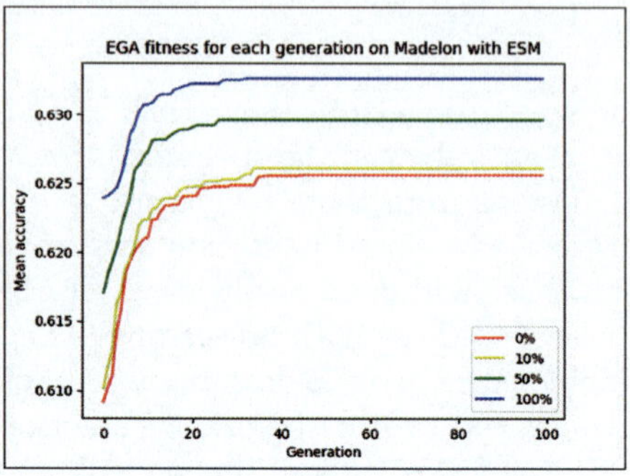

Fig. 5. Evolution of the mean of accuracy calculated on ten runs for each with 100 generations for ESM seeding on Madelon data-set

With the ESM technique (Fig. 4 and 5), the seeding method always outperforms the random initialisation. On the Semeion dataset, which only has 256 features, all the seeding proportions converge to the same result. On the Madelon datasets, results improve following the initial seeding level. We can infer that ESM always returns to a very good initial group of individuals that leads the population to a better final result.

5.2 Results with GAAM Algorithm

With the GAAM algorithm (Fig. 6, 7, 8, 9), the size of the search space is drastically reduced because the maximum number of features kept is fixed at the start. This leads to smaller possibilities of optimisation. As expected, the improvement is better on the Madelon dataset than on the Semeion dataset because the search space is bigger. With SSM, the 10% seeding setup leads in each scenario to the best result. This effect can be explained by the way GAAM works: during the aggressive mutation step, each gene of each chromosome is mutated one by one. If there are too many similarities in the initial population due to the seeding process, good genes stay but the proximity between the first individuals leads to less possibilities of exploration.

We summarise the results of all our experiments in Table 2 and Table 3. We add the standard deviation calculated for the 10 run in each case in order to evaluate the dispersion of the results for each scenario.

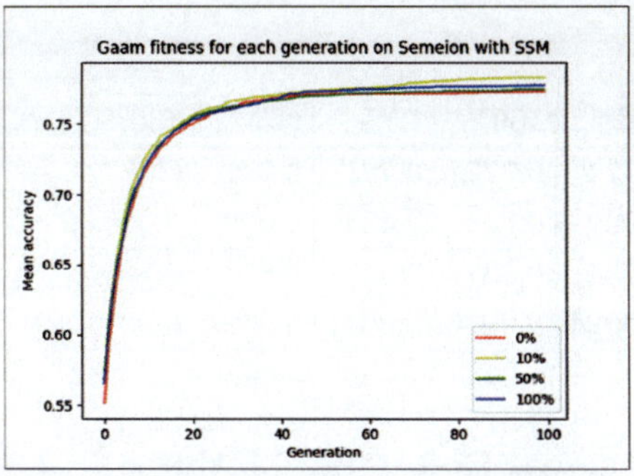

Fig. 6. Evolution of the mean of accuracy calculated on ten runs for each with 100 generations for SSM seeding on Semeion data-set

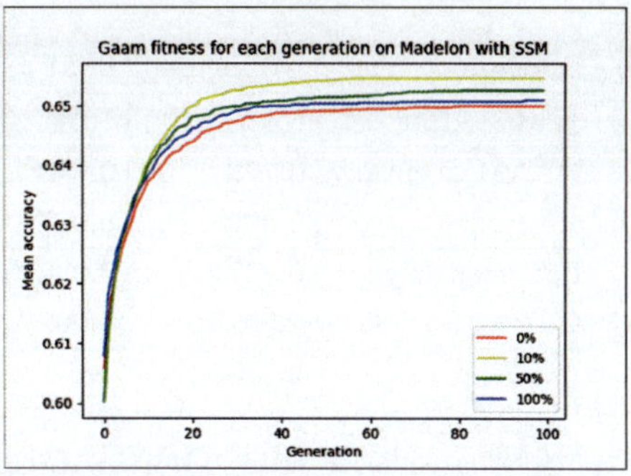

Fig. 7. Evolution of the mean of accuracy calculated on ten runs for each with 100 generations for SSM seeding on Madelon data-set

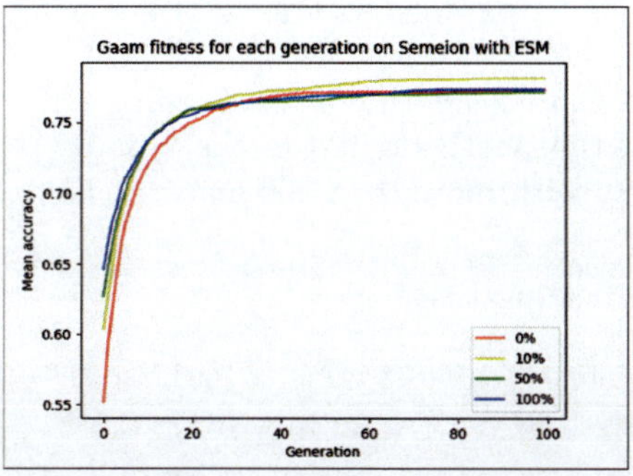

Fig. 8. Evolution of the mean of accuracy calculated on ten runs for each with 100 generations for ESM seeding on Semeion data-set

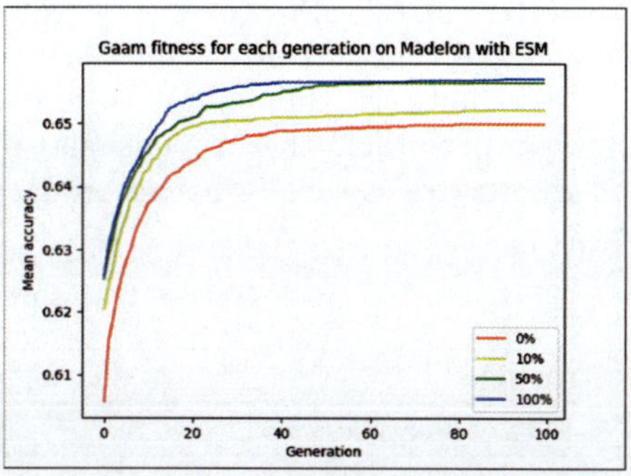

Fig. 9. Evolution of the mean of accuracy calculated on ten runs for each with 100 generations for ESM seeding on Madelon data-set

Table 2. Experimental results for Madelon dataset

Madelon	No seeding	SSM			ESM		
% of seeding	0%	10%	50%	100%	10%	50%	100%
GAAM accuracy	0.649	0.655	0.653	0.651	0.652	0.656	0.657
std	0.007	0.004	0.005	0.002	0.007	0.008	0.004
Ega accuracy	0.625	0.626	0.625	0.628	0.626	0.630	0.632
std	0.002	0.001	0.001	0.001	0.002	0.0018	0.001

Table 3. Experimental results for Semeion dataset

Semeion	No seeding	SSM			ESM		
% of seeding	0%	10%	50%	100%	10%	50%	100%
GAAM accuracy	0.773	0.783	0.774	0.777	0.781	0.771	0.773
std	0.002	0.007	0.009	0.008	0.004	0.009	0.006
Ega accuracy	0.827	0.832	0.834	0.831	0.835	0.835	0.836
std	0.0028	0.0027	0.0024	0.0026	0.0041	0.0027	0.0013

5.3 Comparison with Others Dimensional Reduction Methods

We have compared the results of our methods on five different datasets described in Table 4 versus two classical methods of dimensional reduction: PCA [6] and Auto-encoder [17]. The tests are run with 2 settings. The first is a reduction to 20 dimensions in order to compare with GAAM. The second is a reduction of the feature numbers to half the initial size in order to compare with EGA. This is because classical genetic algorithms led by accuracy tend to converge near half size of the number of features [11]. Auto encoder parameters are: 22 layers using sigmoid and selu activation functions.

The results (in Table 4 and 5) indicated that GA with our new seeding technique outperforms other dimensional reduction methods in 4 of the 5 datasets and results are especially better on datasets with more than 500 features.

Table 4. Results with 20 features

Methods	Madelon	Semeion	Hill-Valley	Har	Parkinson
PCA 20 features	0.627	**0.867**	0.511	0.794	0.725
Auto-encoder 20 features	0.550	0.784	0.511	0.779	0.737
GAAM SSM best result	0.654	0.783	0.518	**0.927**	0.871
GAAM ESM best result	**0.657**	0.781	**0.519**	0.927	**0.873**

Table 5. Results on half number of features

Methods	Madelon	Semeion	Hill-Valley	Har	Parkinson
PCA half number of features	0.582	**0.883**	0.504	0.675	0.681
Auto-encoder half number of features	0.545	0.819	0.510	0.779	0.7369
EGA SSM best result	0.628	0.835	0.518	0.878	0.805
EGA ESM best result	**0.632**	0.836	**0.519**	**0.881**	**0.807**

6 Conclusion

In this paper, we studied the influence of using seeding techniques in genetic algorithms applied to feature selection. The experiments were conducted with two genetic algorithms, two seeding techniques and five datasets. For each scenario, three levels of seeding were compared and the two techniques were compared with other dimensional reduction methods. The results indicate that seeding techniques are a powerful tool to optimise the performance of GA for dimensional reduction. We have showed that GA can surpass other dimensional reduction methods and that they are the best choice when it comes down to obtain the best results on high dimensional datasets. However, further studies have to be done to come up with a reliable way to identify the best level of seeding needed in order to obtain an optimal effect.

References

1. Hassanat, A., Almohammadi, K., Alkafaween, E., Abunawas, E., Hammouri, A., Surya Prasath, V.B.: Choosing mutation and crossover ratios for genetic algorithms-a review with a new dynamic approach. Information **10**(12), 390 (2019)
2. Kuri-Morales, A., Aldana-Bobadilla, E.: The best genetic Algorithm I. In: Castro, F., Gelbukh, A., González, M. (eds.) Advances in Soft Computing and Its Applications, MICAI 2013. Lecture Notes in Computer Science, vol. 8266. Springer, Heidelberg (2013). https://doi.org/10.1007/978-3-642-45111-9_1
3. Breiman, L.: Random forests. Mach. Learn. **45**, 5–32 (2001). https://doi.org/10.1023/A:1010933404324
4. Burduk, R.: Recognition task with feature selection and weighted majority voting based on interval-valued fuzzy sets. In: Nguyen, N.T., Hoang, K., Jędrzejowicz, P. (eds.) Computational Collective Intelligence. Technologies and Applications, ICCCI 2012. Lecture Notes in Computer Science, vol. 7653. Springer, Heidelberg (2012). https://doi.org/10.1007/978-3-642-34630-9_21
5. Osaba, E., Carballedo, R., Diaz, F., Onieva, E., Lopez, P., Perallos, A.: On the influence of using initialization functions on genetic algorithms solving combinatorial optimization problems: a first study on the TSP. In: 2014 IEEE Conference on Evolving and Adaptive Intelligent Systems (EAIS), Linz, Austria, pp. 1–6 (2014). https://doi.org/10.1109/EAIS.2014.6867465
6. Karl Pearson, F.R.S.: LIII. On lines and planes of closest fit to systems of points in space. Lond. Edinb. Dubl. Phil. Mag. J. Sci. **2**(11), 559–572 (1901). https://doi.org/10.1080/14786440109462720
7. Guyon, I., Weston, J., Barnhill, S., Vapnik, V.: Gene selection for cancer classification using support vector machines. Mach. Learn. **46**(1), 389–422 (2002)
8. Hall, M.: Correlation-based feature selection for machine learning. Department of Computer Science, 19 (2000)
9. Liu, H., Motoda, H.: Feature Selection for Knowledge Discovery and Data Mining. Kluwer Academic Publishers, Norwell (1998)
10. Ilyas, I.F., Chu, X.: Data Cleaning. Association for Computing Machinery, New York (2019)
11. Rejer, I., Lorenz, K.: Classic genetic algorithm vs. genetic algorithm with aggressive mutation for feature selection for a brain-computer interface. Przeglad Elektrotechniczny **1**, 100–104 (2015). https://doi.org/10.15199/48.2015.02.24

12. Berti-Equille, L., Harmouch, H., Naumann, F., Novelli, N., Thirumuruganathan, S.: Discovery of genuine functional dependencies from relational data with missing values. Proc. VLDB Endow. **11** (2018). https://doi.org/10.14778/3204028.3204032
13. Reeves, C.R.: Genetic Algorithms. Springer, Boston (2010)
14. Rejer, I.: Genetic algorithm with aggressive mutation for feature selection in BCI feature space. Pattern Anal. Appl. **18**(3), 485–492 (2015). https://doi.org/10.1007/s10044-014-0425-3
15. Bryant, A.J.: Seeding the population: improved performance in a genetic algorithm for the rectilinear Steiner problem. In: Proceedings of the 1994 ACM Symposium on Applied Computing (SAC 1994), pp. 222–226. Association for Computing Machinery, New York (1994). https://doi.org/10.1145/326619.326728
16. Tibshirani, R.: Regression shrinkage and selection via the Lasso. J. R. Stat. Soc.: Ser. B (Methodol.) **58**, 267–288 (1996). https://doi.org/10.1111/j.2517-6161.1996.tb02080.x
17. Wang, Y., Yao, H., Zhao, S.: Auto-encoder based dimensionality reduction. Neurocomputing **184**, 232–242 (2016). https://doi.org/10.1016/j.neucom.2015.08.104, http://www.sciencedirect.com/science/article/pii/S0925231215017671, roLoD: Robust Local Descriptors for Computer Vision 2014
18. Zhang, H.: The optimality of Naive Bayes. In: Proceedings of the Seventeenth International Florida Artificial Intelligence Research Society Conference, FLAIRS 2004, vol. 2 (2004)

An Incremental Fuzzy Learning Approach for Online Classification of Data Streams

Vladyslav Yavtukhovskyi, Rami Abukhader, Nils Tillaeus, and Ning Xiong(✉)

School of Innovation, Design and Engineering, Mälardalen University, Västerås, Sweden
{vyi19001,rar19002,nts14001}@student.mdh.se, ning.xiong@mdh.se

Abstract. Online learning concerns analyzing a continuous stream of transient data and progressively updating the knowledge model without revisiting previously encountered examples. This paper proposes a new incremental fuzzy learning approach for online classification of data streams. It enables an existing fuzzy rule set to be efficiently updated based on a new training example without learning from scratch. The proposed algorithm can not only incrementally construct new fuzzy classification rules but also update the content with old rules to assimilate information from new data. The efficacy of the proposed incremental fuzzy learning method has been demonstrated in a set of simulation tests where the benchmark data sets were treated as data streams for learning.

Keywords: Fuzzy systems · Incremental fuzzy learning · Online classification · Data stream

1 Introduction

In recent years, fuzzy systems have achieved widespread applications around the globe in many industrial and technical fields, including process modeling, pattern recognition and automatic control [1]. These fuzzy rule-based systems are more suitable for complex system problems where it is very difficult to describe the systems mathematically. Fuzzy system design is a non-trivial task of creating linguistic fuzzy rules and specifying the associated fuzzy set membership functions. Machine learning is gaining high attention to alleviate the knowledge acquisition bottleneck in the development of fuzzy systems.

So far, many learning methods of fuzzy systems have been developed, whereas they mainly work in a batch-based and offline learning manner. A major issue with offline learning is that, if one gets a set of new examples in addition to previous ones, it requires restarting the whole learning process with all data available. Hence it can lead to an inefficient learning process with slow adaptation to assimilate knowledge from new data. Particularly, offline learning is incapable to handle data streams in an online environment, in which fresh data are generated continuously with open ends.

This paper is devoted to online learning for data stream classification. We propose a new incremental fuzzy learning approach to modifying an existing fuzzy rule set when a new training example is available, without requiring to learn with all examples anew. The proposed algorithm can not only incrementally construct new fuzzy rules but

A. Abraham et al. (Eds.): SoCPaR 2020, AISC 1383, pp. 583–592, 2021.
https://doi.org/10.1007/978-3-030-73689-7_56

also update the content with old fuzzy rules in accordance with information from new arrived examples. This enables the fuzzy rule set to self-refine to accommodate new information while not forgetting useful knowledge acquired before. The efficacy of the proposed incremental learning method has been demonstrated in a set of simulation tests where the benchmark data sets were treated as data streams for learning.

The incremental learning method proposed in this paper would have strong potential for computational intelligent systems such as autonomous agents interacting with the real-world. Whenever a new piece of experience is obtained from interaction, it can be used immediately to tune and augment the current knowledge model, which will in return support the agent with better strategy in future problem solving. The life-long learning capability for agents can thereby be cultivated when they continually evolve with streams of experiences over time.

The remainder of the paper is organized as follows. Section 2 reviews the related work. The standard fuzzy classification system is described in Sect. 3. Subsequently we present the proposed incremental fuzzy learning approach in Sect. 4. The results of the simulation tests are given in Sect. 5. Finally, the paper concludes with Sect. 6.

2 Related Work

Indeed, incremental learning of fuzzy systems for online classification is a relatively new research area. The number of papers addressing this topic seems very limited. In this section we attempt to give a review of several relevant papers found in the literature.

Reignier [2] considered the framework of reinforcement learning of robots and advocated incremental learning of fuzzy rules to improve the estimation of the optimal values of robot actions in various states. Based on the impact analysis of individual rules, three ways of rule modification were suggested in that paper: 1) creation of a new rule; 2) adaptation of an existing rule; 3) generalization of a selected rule. However, the author did not present a complete algorithm for rule learning, nor the performance evaluation in the robot control system.

Incremental learning for fuzzy pattern matching was addressed in [3], which uses probability-possibility transformation for creating membership functions to decide the class of a new sample. The possibility densities are then refined based on each new classified sample for improving future performance. A similar work was presented in [4], which relies on data frequency to update fuzzy sets of classes in the online learning of a fuzzy classifier. The difference of [3] and [4] from this paper is that they don't handle the learning and revision of fuzzy classification rules.

The series of works in [5, 6] and [7] investigated the incremental methods to extract fuzzy classification rules from big data. The basic idea is to split the initial training examples into multiple subsets, with each corresponding to a learning episode. In each episode, the fuzzy knowledge acquired from the preceding stage is updated and refined based on the new examples that are allocated to the current episode. However, the adaptation of fuzzy rules in each of the episodes is still an off-line learning procedure that resorts to a genetic algorithm. Evidently these methods cannot be applied in an online learning scenario for data stream classification.

3 Standard Fuzzy Classification System

This section briefly introduces a fuzzy rule-based classification system in the basic form. It should be noted that the ideas of measuring a fuzzy rule with weight and how these weights can affect the final decision, as depicted in the following, are also valid for other forms of fuzzy classifiers.

Let us consider a pattern classification problem with n dimensionality and M possible classes. Basically, a fuzzy if-then rule R_i for classification is of the form:

$$R_i : \textit{If } x_1 \textit{ is } A_{i1} \textit{ and } \cdots \textit{ and } x_n \textit{ is } A_{in}$$

$$\textit{Then Class } C_i \textit{ with wight } w_i \quad i = 1, \cdots, N$$

where (A_{i1}, ..., A_{in}) represents a set of antecedent fuzzy sets, C_i is the consequent class of the rule, w_i is the weight measuring the confidence of the rule, and N is the total number of fuzzy rules in the rule set.

Given a new sample with its features represented by $(x_1, \cdots, x_n)$, the fuzzy reasoning procedure is as follows. First, we evaluate the firing strength of each rule, which is defined as the compatibility of this feature vector with respect to the condition of the rule. Hence the firing strength for rule R_i is calculated by:

$$t_i = \prod_{j=1}^{n} \mu_{A_{ij}}(x_j) \tag{1}$$

The next step is to calculate the voting score for each candidate class h (=1 $\cdots M$) as:

$$\beta_h = max_{C_i = h}(t_i \bullet w_i) \tag{2}$$

Finally, the pattern of the new sample is classified as class C that receives the highest voting score, i.e.:

$$C = \text{arg}max_h(\beta_h) \tag{3}$$

4 The Proposal of Incremental Fuzzy Learning

The proposed incremental learning method is contingent upon pre-specification of the fuzzy sets and their membership functions for each of the features. In this section we first describe the designed fuzzy sets, as well as the structure of the fuzzy rules that are adopted in our learning system. We then discuss the mechanisms of how to update these fuzzy rules in an incremental manner.

4.1 Representation of Fuzzy Sets and Rules

Each feature is normalized into the interval [0, 1] beforehand. We use triangular membership functions, as shown in Fig. 1, to depict the fuzzy sets of each feature. Furthermore, for simplicity, each feature is equally partitioned with L fuzzy sets, with the case of $L = 7$ being illustrated in Fig. 2. It follows that the membership function of the kth fuzzy set of a feature can be expressed by:

$$\mu_k(x) = max\left(1 - \frac{|x - x_k|}{z}, 0\right), k = 1 \cdots L \tag{4}$$

where

$$x_k = \frac{k-1}{L-1}, k = 1, \cdots L \tag{5}$$

$$z = \frac{1}{L-1} \tag{6}$$

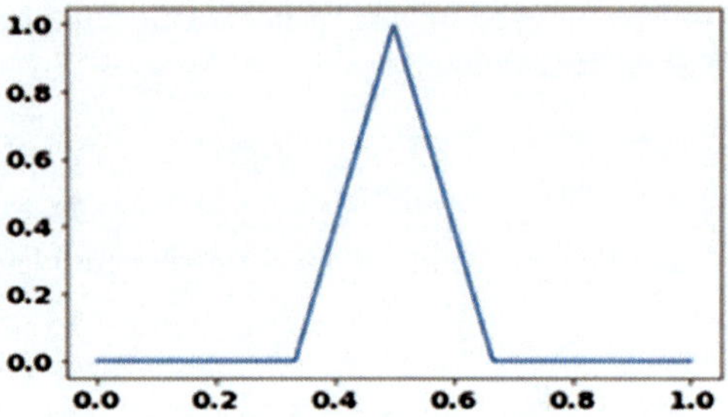

Fig. 1. A triangular membership function.

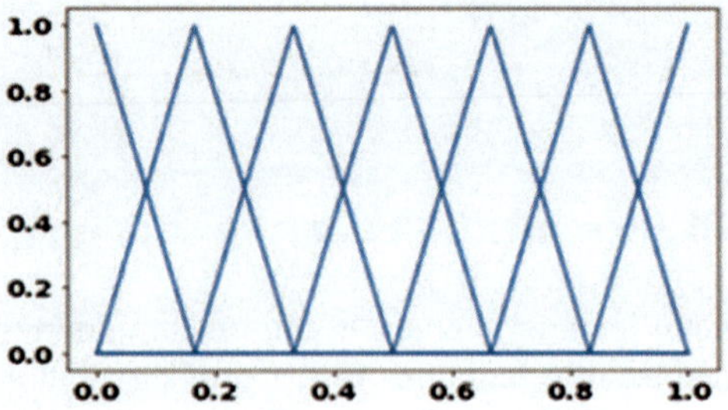

Fig. 2. Equal partition of a feature with seven fuzzy sets.

Regarding fuzzy rules, we adopt a general premise structure allowing for OR connection of fuzzy sets in rule antecedents [8]. Hence a fuzzy rule considered for learning in this paper is formulated in the following form:

$$R : If\ x_1\ is \bigcup_{k \in D(1)} B_{1k}\ and\ \cdots\ and\ x_n\ is \bigcup_{k \in D(n)} B_{nk}$$

$$\textit{Then Class } C_R \textit{ with wight } w_R \tag{7}$$

where (B_{i1}, ..., B_{iL}) represents the L fuzzy sets depicting the ith feature, and $D(i) \subseteq \{1, \cdots, L\}$ indicates which of them are included in the OR connection of the antecedent.

An important merit of the rules with general premise structure is that they can cover a group of elementary rules that use AND connections of single fuzzy sets in rule premises. For instance, the rule "If (x_1 = NZ ∪ PZ) and (x_2 = NZ ∪ PZ) then class C" has the premise as illustrated in Fig. 3, and it covers the following four elementary rules:

1) If (x_1 = NZ) and (x_2 = NZ) Then Class C
2) If (x_1 = NZ) and (x_2 = PZ) Then Class C
3) If (x_1 = PZ) and (x_2 = NZ) Then Class C
4) 4) If (x_1 = PZ) and (x_2 = PZ) Then Class C

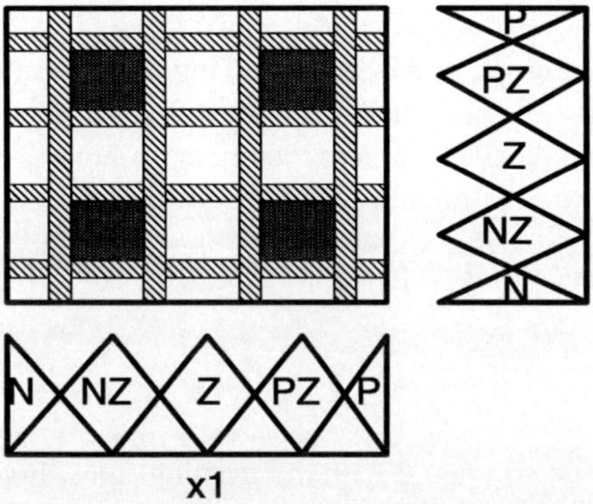

Fig. 3. Rule premise with OR-connection of fuzzy sets

4.2 Incremental Updating of Rule Weights

A new training example can be considered as positive/negative evidence when it is matched with the antecedent of a rule. It provides positive support with which to increase the confidence of the rule if the class of the examples is consistent with the consequent of the rule. Otherwise, it is considered as a counter-example decreasing the confidence of the rule. The extent of change of confidence depends on the degree of compatibility between the feature vector and the rule antecedent, i.e. the firing strength of the rule.

Now suppose the arrival of a new training example $E = (x_1, \cdots, x_n, C)$, with $x_1, \cdots, x_n$ being the feature values, and C as the associated class. Considering the OR connections in the antecedent, the firing strength of rule R in (7) can be yielded by

$$t_R(E) = \prod_{j=1}^{n} \max_{k \in D(j)} \mu_{B_{jk}}(x_j) \tag{8}$$

With the firing strength calculated above, we can incrementally update the positive and negative scores of rule *R* by:

$$S_R^+ = \begin{cases} S_R^+ + t_R(E), if C_R = C \\ S_R^+, if C_R \neq C \end{cases} \quad (9)$$

$$S_R^- = \begin{cases} S_R^- + t_R(E), if C_R \neq C \\ S_R^-, if C_R = C \end{cases} \quad (10)$$

Finally, the weight of the rule is calculated based on the updated positive and negative scores. Hence, we have:

$$w_R = \frac{S_R^+ - S_R^-}{S_R^+ + S_R^-} \quad (11)$$

4.3 Creation of a New Rule

It is also possible to generalize the new training example $E = (x_1, \cdots, x_n, C)$ into a fuzzy rule to cover the fuzzy region that surrounds it. This fuzzy rule can be created by including the fuzzy sets that receive non-zero membership degrees in the OR connections of the rule antecedent. Formally this fuzzy rule can be created as follows:

$$R' : \text{ If } x_1 \text{ is } \bigcup_{\mu_{B_{1k}}(x_1)>0} B_{1k} \text{ and} \ldots \text{and } x_n \text{ is } \bigcup_{\mu_{B_{nk}}(x_n)>0} B_{nk}$$

$$\textit{Then Class } C \quad (12)$$

The rule *R*' will then be compared with the existing rules. If no matching is found, the rule will be added into the rule set, and its weight and positive and negative scores are initialized as:

$$w_{R'} = 1, \; S_{R'}^+ = t_{R'}(E), \; S_{R'}^- = 0$$

given that this rule is newly created without having encountered counter-examples.

4.4 The Incremental Learning Algorithm

Given a new training example, the existing fuzzy rule set can be incrementally updated with the following three steps:

Step 1: Calculate the matching degrees (firing strengths) between this new example and each existing rule using Eq. (8). Subsequently use the matching degrees to update the positive and negative scores and weights of all rules according to Eqs. (9), (10), and (11).
Step 2: Generalize the new training example into a rule as defined in (12) and compare this rule with all rules in the fuzzy rule set.
Step 3: If no identical rule is found from the existing ones, add this new rule to the fuzzy rule set and initialize the weight and scores of this new added rule.

The overall algorithm for incremental learning of fuzzy rules for data stream classification is given in **Algorithm 1**.

Algorithm 1: Incremental learning of fuzzy rules

Input: a data stream of labelled examples, *stream*(·)

Output: a set of fuzzy classification rules, *RuleSet*

$RuleSet = \emptyset;\ t = 0$

while $(strteam(t) \neq \emptyset)$ **do**

  $E = stream(t)$

  for R $RuleSet$ **do**

    calculate the rule firing strength $t_R(E)$

    $S_R^+ = \begin{cases} S_R^+ + t_R(E), & if\ C_R = C \\ S_R^+, & if\ C_R \neq C \end{cases}$ //C: the class of E

    //C_R: the conclusion of R

    $S_R^- = \begin{cases} S_R^- + t_R(E), & if\ C_R \neq C \\ S_R^-, & if\ C_R = C \end{cases}$

    $w_R = \dfrac{S_R^+ - S_R^-}{S_R^+ + S_R^-}$

  end for

  generate rule *R'* according to (12)

  if $(R' \notin RuleSet)$ **then**

    calculate the rule firing strength $t_{R'}(E)$

    $w_{R'} = 1,\ S_{R'}^+ = t_{R'}(E),\ S_{R'}^- = 0$

    $RuleSet = RuleSet \cup \{R'\}$

  end if

  $t = t + 1$

end while

---

5 Results and Evaluation

To examine the efficacy of the proposed incremental method for learning of fuzzy rules, we conducted simulation tests on a set of benchmark data sets taken from the UCI machine learning repository [9]. The properties of the 10 data sets that have been used in our evaluation are presented in Table 1.

In the simulation tests for evaluation, we used the *K*-fold cross-validation with *K* being set to 6. This means that each benchmark data set was split into 6 equal parts, where one part was used as test data and the others as the training data in all the 6 trials. In each trial, the training data was treated as a data stream for learning of the fuzzy rules and the test data was used to assess the quality of the fuzzy rule sets acquired during the learning procedure.

In addition to the rules with a general premise structure, we also built and evaluated the fuzzy rules of the basic form where AND connections of single fuzzy sets were used as rule antecedents. The single fuzzy sets were picked such that they received the highest membership degrees with respect to the individual features of the new training example, as described in the Wang-Mendel model [10]. The mechanisms of adding new rules as well as updating old rule weights still follow Eqs. (9–11) and what is specified in **Algorithm 1**.

Table 1. The 10 data sets used for evaluation

Data	# of examples	# of attributes	# of classes
Caesarian	80	5	2
Diabetes	768	9	2
Haberman's	306	2	2
Wheat seeds	210	7	2
Trains	10	32	2
Iris	150	4	3
Breast cancer wisconsin	569	10	2
Breast cancer coimbra	116	10	2
Sonar	208	60	2
Banknote	1372	5	2

We evaluated the quality of the generated fuzzy rule sets by applying them to the test data. The classification accuracy on the test data sets when 50% and 100% of the training examples had been visited are presented in Table 2. From this table we see that learning of rules with general premises can result in higher test accuracy in the mid-stage of the learning on 9 of the 10 benchmarks. This is owing to the fact that the general premises of rules cover larger regions of the feature space, yield higher generalization ability, and therefore accelerate the learning process. Regarding the results when all the training examples were used, the rule sets with general premises achieved significantly better accuracy than the basic rules on 3 benchmarks, similar performance (slightly better or infetrior) compared to the counterpart on 6 benchmarks. Only on the benchmark date set "Bannote", the final accuracy with the basic fuzzy rules was obviously superior. This can be explained by their relatively longer stream of the "Banknote" data, which provides sufficient examples to create basic rules of more specific nature so that the generalization capability of fuzzy rules becomes less important.

Since there is no other related work reported concerning online learning of fuzzy rules for classification problems, our incremental learning method is assessed here in comparison with some other offline learning techniques on the same benchmarks. In Table 3 we compare the test accuracy of our incremental learning algorithm when the training samples had been visiterd once with that of offline learning methods in which the training examples can be visited a number of times. It can be seen that, on 8 of the 10 benchmarks, our algorithm is quite compertitive or similar in performance to the offline learning methods. Only on the "Banknote" and "Breast Cancer Wisconsin" data, the offline learning performed obviously better. But our algorithm still obtained satisfactory accuracy of over 80% on these two benchmarks. Further more there will be room to improve the accuracy of our incremental learning method by adaptation of the parameters of the fuzzy set membership functions.

Table 2. Comparison of test accuracy for basdic and general rules

Data set	Basic rules		Rules with general premises	
	50%	100%	50%	100%
Caesarian	0.1621	0.2738	0.2775	0.4002
Diabetes	0.51114	0.5763	0.6414	0.6766
Haberman's	0.4314	0.6569	0.5719	0.6438
Wheat seeds	0.5576	0.8538	0.6127	0.8494
Trains	0.6701	0.7250	0.7306	0.7430
Iris	0.64	0.9067	0.66	0.92
Breast cancer wisconsin	0.5062	0.5941	0.8101	0.8506
Breast cancer coimbra	0.4632	0.5846	0.4404	0.5943
Sonar	0.4148	0.5905	0.4790	0.6185
Banknote	0.5543	0.9818	0.5551	0.8133

Table 3. Comparison with offline learning methods

Dataset	Accuracy of our algorithm	Offline learning methods	Offline accuracy
Caesarian	0.4002	AdaboostM1 [11]	0.42
Diabetes	0.6766	Genetic-decision trees [12]	0.651
Haberman's	0.6438	CART [13]	0.6402
Wheat seeds	0.8494	CART [15]	0.8684
Trains	0.7430	RELAGGS [14]	0.75
Iris	0.92	Logit-boost NB [19]	0.9487
Breast cancer wisconsin	0.8506	Model trees [20]	0.9654
Breast cancer coimbra	0.5943	kNN [17]	0.5814
Sonar	0.6185	Naïve Bayes [16]	0.6711
Banknote	0.8133	MLP [18]	0.9883

6 Conclusions

In this paper we propose a novel incremental approach for online learning of fuzzy classifiers with data streams. The results of the simulation tests, made on a set of benchmark data sets, have shown that the proposed method can acquire quite competitive results in comparison with some offline learning algorithms. One interesting direction of future work will be enhancing our current method to deal with ill-balanced and more complex big data streams.

References

1. Murphey, Y., Chen, T.: Incremental learning in a fuzzy intelligent system. In: Proceedings of the 16th International Joint Conference on Artificial Intelligence, IJCAI, vol. 2, pp. 1376–1384. Morgan Kaufmann Publishers Inc. (1999)
2. Reignier, P.: Supervised incremental learning of fuzzy rules. Robot. Auton. Syst. **16**(1), 57–71 (1995)
3. Mouchaweh, M.S., Devillez, A., Lecolier, G.V., Billaudel, P.: Incremental learning in fuzzy pattern matching. Fuzzy Sets Syst. **132**(1), 49–62 (2002)
4. Visa, S., Ralescu, A.: Towards online learning of a fuzzy classifier. In: Meeting of the North American Fuzzy Information Processing Society, Cincinnati, USA (2005)
5. Orriols-Puig, A., Casillas, J.: Fuzzy knowledge representation study for incremental learning in data streams and classification problems. Soft. Comput. **15**(12), 2389–2414 (2011)
6. Gámez, J.C., García, D., González, A., Pérez, R.: On the use of an incremental approach to learn fuzzy classification rules for big data problems. In: IEEE International Conference on Fuzzy Systems (FUZZ-IEEE), pp. 1413–1420. IEEE (2016)
7. Romero-Zaliz, R., González, A., Pérez, R.: Incremental fuzzy learning algorithms in big data problems. a study on the size of learning subsets. In: 2017 IEEE International Conference on Fuzzy Systems (FUZZ-IEEE), pp. 1–6. IEEE (2017)
8. Xiong, N.: Evolutionary learning of rule premises for fuzzy modelling. Int. J. Syst. Sci. **32**(9), 1109–1118 (2001)
9. UCI machine learning repository. https://archive.ics.uci.edu/ml/index.php. Accessed 31 Oct 2020
10. Wang, L.X., Mendel, J.M.: Generating fuzzy rules by learning from examples. IEEE Trans. Syst. Man Cybern. **22**(6), 1414–1427 (1992)
11. Ayyappan, G.: Meta classifications for caesarian section classification dataset data set. Indian J. Comput. Sci. Eng. **9**(6), 148–150 (2019). Chennai, India
12. Kwedlo, W, Krętowski, M.: Discovery of decision rules from databases: An Evolutionary Approach. Technical University of Bialystok, Białystok, Poland (2006)
13. Mansourifar, H., Shi, W.: Toward efficient breast cancer diagnosis and survival prediction using L-Perceptron. University of Houston Houston, Texas (2018)
14. Krogel, M.A., Rawles, S., Železný, F., Flach, P.A., Lavrač, N., Wrobel, S.: Comparative evaluation of approaches to propositionalization. In: Proceedings of the 13th International Conference on ILP, Szeged (2003)
15. Agrawal, D., Dahiya, P.: Comparisons of classification algorithms on seeds dataset using machine learning algorithm. An Int. J. Adv. Comput. Technol. **7**, 2760–2765 (2018)
16. Džeroski, S., Ženko, B.: Stacking with multi-response model trees. In: International Workshop on Multiple Classifier Systems. Springer, Heidelberg (2002)
17. Hall, M.A.: Feature selection for discrete and numeric class machine learning (1999)
18. Austria, Y.D., Jay-ar, P.L., Maria, L.B.S., Jr., Goh, J.E.E., Goh, M.L.I., Vicente, H.N.: Comparison of machine learning algorithms in breast cancer prediction using the coimbra dataset. Cancer **7**, 10 (2019)
19. Kumar, C., Dudyala, A. K.: Bank note authentication using decision tree rules and machine learning techniques. In: International Conference on Advances in Computer Engineering and Applications, pp. 310–314 (2015)
20. Kotsiantis, S.B.: Logitboost of simple Bayesian classifier. Informatica **29**(1), 53–59 (2005)

Depth Image Completion Using Anisotropic Operators

Vanel Lazcano[1(✉)], Felipe Calderero[2], and Coloma Ballester[3]

[1] Universidad Mayor, 08544 Santiago, Chile
vanel.lazcano@umayor.cl
[2] Glovo, 08005 Barcelona, Spain
felipe.calderero@glovoapp.com
[3] Universitat Pompeu Fabra, 08018 Barcelona, Spain
coloma.ballester@upf.edu

Abstract. Our work tackles the problem of anisotropic data interpolation in the context of depth computation in large regions of an image. Usually, depth data present areas without depth data due to low confidence in the data. We consider the image endowed with an anisotropic metric g_{ij} taking spatial and the geometry information. Our proposal is a numerical implementation based on "eikonal" operators to compute the solution of the biased Infinity Laplacian or biased Absolutely Minimizing Lipschitz Extension (bAMLE). The infinity Laplacian operator interpolates creating cones. The biased infinity Laplacian interpolates creating exponential cones, making it more adapted to real scenes that, in general, present smooth surfaces. Thus, the biased Infinity Laplacian properties make this interpolator able of completing depth data to large image areas. We performed an experimental evaluation comparing our numerical model comparing Infinity Laplacian, biased Infinity Laplacian, and bilateral filter in two databases: Middlebury2014 and KITTI (for depth completion). In Middlebury2014 biased infinity Laplacian outperforms Infinity Laplacian in the up-sampling depth data task, and in KITTI dataset, completing depth images, biased Infinity Laplacian exceeds both bilateral and Infinity Laplacian. The simplicity of our practical implementation makes biased Infinity Laplacian a low-cost-implementation alternative for many computer vision community interpolation problems.

Keywords: Biased Absolutely Minimizing Lipschitz Extension · Bilateral filter · Depth completion

1 Introduction

Nowadays, the use of depth images is common in applications such that: autonomous vehicles, 3D reconstruction, or video games. Frequently the depth images present holes or low confidence levels in the data. The lack of depth data region could occur by occlusions or sensor misinterpretation in the case of Time-Of-Flight camera, Lidar sensor or Kinect sensor. Our work studies the

A. Abraham et al. (Eds.): SoCPaR 2020, AISC 1383, pp. 593–604, 2021.
https://doi.org/10.1007/978-3-030-73689-7_57

problem of anisotropic depth data completion applied to depth interpolation in large areas. The problem of depth completion consists of interpolating the empty depth data area in a depth map, taking into account the information presented in the self depth image and in a color reference image.

We present an empirical study of an interpolation operator to complete depth maps. This considered interpolation operator is the biased Absolute Minimizing Lipschitz Extension or biased Infinity Laplacian that appeared in the axiomatic approach propose in [6] and [5]. The AMLE interpolator was presented in [2] and [3] as a completion tool from a theoretical point of view. As explained in [5] and [6], Infinity Laplacian operator emerges as a operator very simple that satisfies a small set of mathematical axioms. In this work, we proposed a numerical scheme to solve the biased Infinity Laplacian. The numerical method uses the eikonal operator. This scheme generates a simple implementation of the biased Infinity Laplacian, consisting only of a weighted average. Making this interpolate easy and fast to implement.

1.1 Related Works

In [14], the depth interpolation has been addressed to synthesize several different perspectives of the same scene from a multiview-video-plus-depth representation. New datasets have been created to evaluate in a systematic procedure the performance of depth completion methods such that KITTI (Depth Completion Evaluation [20]).

Nowadays, new approaches based on convolutional networks (CNN) have been utilized to the depth completion problem. In [11] and [9], the authors assess convolutional networks in a database that contains depth images obtained by a Kinect sensor. The authors sampled these depth images every other 8 pixels square; then, the authors used CNN to complete the data. In [12] using CNN authors simultaneously reconstructed the depth map and also the reference image. This approach let the CNN net understand better the context and the objects in the scene. In [7] is presented an approach to identify disturbed depth data by learning an input confidence estimator based on the normalized convolutional neural networks (NCNNs). Obtained results in KITTI dataset shows that the proposal outperforms Bayesian Deep Learning approaches. In [1] AMLE was applied to interpolate elevation data between level lines. Authors have created an elevation reconstruction model from a scanned map. Authors have at their disposal few data contained in few iso-level curves. In [5] the Infinity Laplacian operator was used for completing surfaces embedded in $\mathbb{R}^3$ given iso-level curves.

Recently in [17] the interpolations properties of the AMLE operator has been applied to the flow densification and large hole completion in optical flow.

In Sect. 2 we present the biased infinity Laplacian (bAMLE) we used to complete sparse disparities. In Sect. 3 we explain our numeric implementation of the bAMLE and in Sect. 4 is presented the performance of our proposal in KITTI [19] and in Middlebury2014 [4]. Finally, Sect. 4.3 presents our conclusions.

2 Method

The bAMLE (biased infinity Laplacian) was mentioned in the analysis of interpolators in manifolds [5], although it was not empirically studied there (it was unnoticed in [6]). AMLE operator is incredibly efficient in completing sparse depth maps. This operator is beneficial in cases where a large depth map presents an empty area.

We take into account the problem,

$$\Delta_{\infty,g} u + \beta \left|\nabla u\right|_{\xi} = 0 \in \Omega, \tag{1}$$

jointly with this boundary conditions $u|_{\partial\Omega} = \theta$, $\Omega \subset \mathbb{R}^2$ is the problem domain, g_{ij} is the metric and, ξ direction of the gradient, $\beta \in \mathbb{R}^+$, ∇u gradient of the depth map, $\Delta_{\infty,g} u$ depth map infinity Laplacian given the metric g_{ij}. The metric g_{ij} measures the geodesic spatial and photometric similarity between pixels of a reference image. The Eq. (1) in the literature is called the biased AMLE or biased infinity Laplacian.

For convenience of notation, let us write,

$$\Delta_{\infty,g} u := D^2_{\mathcal{M}} u \left(\frac{\nabla u}{\left|\nabla u\right|_{\xi}}, \frac{\nabla u}{\left|\nabla u\right|_{\xi}} \right), \tag{2}$$

where $\mathcal{M} = (\Omega, g)$ is the manifold and also,

$$D^2_{\mathcal{M},ij} u(\mathbf{x}) = \frac{\partial^2 u}{\partial x^i \partial x^j}(\mathbf{x}) - \Gamma^k_{ij}(\mathbf{x}) \frac{\partial u}{\partial x^k}(\mathbf{x}).$$

where Γ^k_{ij} is the Christoffel symbol in the coordinates system and the indices i. j, $k = 1, 2$. Given this notation write again Eq. (1):

$$\frac{1}{|\nabla u(\mathbf{x})|^2_{\mathbf{x}}} D^2_{\mathcal{M},ij} u(\mathbf{x}) g^{i\alpha}(\mathbf{x}) \frac{\partial u}{\partial x^{\alpha}}(\mathbf{x}) g^{j\gamma}(\mathbf{x}) \frac{\partial u}{\partial x^{\gamma}}(\mathbf{x}) + \beta |\nabla u(\mathbf{x})|_{\mathbf{x}} = 0, \tag{3}$$

where

$$|\nabla u(\mathbf{x})|^2_{\mathbf{x}} = g^{ij}(\mathbf{x}) \frac{\partial u}{\partial x^i}(\mathbf{x}) \frac{\partial u}{\partial x^j}(\mathbf{x}).$$

Let $I : \Omega \subset \mathbb{R}^2 \to \mathbb{R}^3$ a color reference image and $u(\mathbf{x}) : \Omega \to \mathbb{R}$ is the scene depth data. Supposing that the available data is located in a region $O \subset \Omega$ with a boundary ∂O. Assuming that $g_{ij}(\mathbf{x}) = h(\mathbf{x})\delta_{ij}$ i.e. $h(\mathbf{x}) = (\kappa_1 + \kappa_2 |\nabla I(\mathbf{x})|^2)^p$. Thus, we can write Eq. (3):

$$\frac{1}{|\nabla u(\mathbf{x})|^2} D^2_{\mathcal{M},ij} u(\mathbf{x}) \frac{\partial u}{\partial x^i}(\mathbf{x}) \frac{\partial u}{\partial x^j}(\mathbf{x}) + \beta |\nabla u(\mathbf{x})|_{\mathbf{x}} = 0. \tag{4}$$

In this case, $h(\mathbf{x})$ was canceled in the numerator and the denominator in the first term of the Eq. (4). This biased AMLE exactly fits the available data, and it can interpolate depth data on isolated points. This feature makes this interpolator able of extending known data to large empty areas.

2.1 A Numerical Implementation for the Biased Infinity Laplacian

Let us consider as a graph the discrete grid. Taking pair of points $\mathbf{x}, \mathbf{y}$ in the grid, let $d_{\mathbf{xy}}$ be its distance. If the considered $\mathbf{x}, \mathbf{y}$ are neighbor in the grid, $d_{\mathbf{xy}}$ is computed by,

$$d_{\mathbf{xy}} = \sum_{i=R,G,B} \kappa_c |I_i(\mathbf{x}) - I_i(\mathbf{y})|^2 + \kappa_x \|\mathbf{x} - \mathbf{y}\|^2, \tag{5}$$

where κ_c, κ_x, are constant $\in \mathbb{R}$, $I(\mathbf{x})$ is the reference image with three color components R, G, B.

Considering a curve $\{\gamma(i)\}_{i=0}^{m}$ in the grid, its length γ is given by

$$L_g(\gamma) = \sum_{i=0}^{m} d_{\gamma(i),\gamma(i+1)}. \tag{6}$$

Given a pair of points $\mathbf{x}$ and $\mathbf{y}$ in the grid, the distance d could be defined by:

$$d = \inf\{L_g(\gamma) : \gamma \text{ is a curve from } \mathbf{x} \text{ joining to } \mathbf{y}\}. \tag{7}$$

The distance d should be computed using Dijkstra's algorithm. Given a point $\mathbf{x}$ in the grid, let $\mathcal{N}(\mathbf{x})$ be a neighborhood around $\mathbf{x}$. The positive eikonal operator on a graph is defined by [13] as:

$$\|\nabla u(\mathbf{x})\|_{\mathbf{x}}^{+} = \max_{\mathbf{y} \in \mathcal{N}(\mathbf{x})} \frac{u(\mathbf{y}) - u(\mathbf{x})}{d_{\mathbf{xy}}}. \tag{8}$$

The negative eikonal operator on a graph is defined by,

$$\|\nabla u(\mathbf{x})\|_{\mathbf{x}}^{-} = \min_{\mathbf{y} \in \mathcal{N}(\mathbf{x})} \frac{u(\mathbf{y}) - u(\mathbf{x})}{d_{\mathbf{xy}}}. \tag{9}$$

The discrete Absolutely Minimizing Lipschitz Extension following [15] is given:

$$\Delta_{\infty,g} u(\mathbf{x}) = \frac{\|\nabla u(\mathbf{x})\|_{\mathbf{x}}^{+} + \|\nabla u(\mathbf{x})\|_{\mathbf{x}}^{-}}{2}. \tag{10}$$

The discretized version of the biased Infinity Laplacian corresponds to:

$$\frac{\|\nabla u(\mathbf{x})\|_{\mathbf{x}}^{+} + \|\nabla u(\mathbf{x})\|_{\mathbf{x}}^{-}}{2} + \beta \left| \, \|\nabla u(x)\|_{\mathbf{x}}^{+} \right| = 0, \tag{11}$$

with $\beta > 0$ the expression (11) depends on the sign of $\beta \|\nabla u(\mathbf{x})\|_{\mathbf{x}}^{+}$. If the eikonal is positive we obtain:

$$\frac{\|\nabla u(\mathbf{x})\|_{\mathbf{x}}^{+} + \|\nabla u(\mathbf{x})\|_{\mathbf{x}}^{-}}{2} + \beta \|\nabla u(\mathbf{x})\|_{\mathbf{x}}^{+} = 0, \tag{12}$$

in the other hand, if the eikonal operator is negative we have:

$$\frac{\|\nabla u(x)\|_{\mathbf{x}}^{+} + \|\nabla u(\mathbf{x})\|_{\mathbf{x}}^{-}}{2} - \beta \|\nabla u(\mathbf{x})\|_{\mathbf{x}}^{+} = 0. \tag{13}$$

We can write both expressions (12) and (13) as:

$$\frac{\|\nabla u(\mathbf{x})\|_{\mathbf{x}}^{+} + \|\nabla u(\mathbf{x})\|_{\mathbf{x}}^{-}}{2} + \beta \text{sgn}(\|\nabla u(\mathbf{x})\|_{\mathbf{x}}^{+})\|\nabla u(\mathbf{x})\|_{\mathbf{x}}^{+} = 0. \tag{14}$$

The discretized biased AMLE is given by:

$$\beta_{+}\|\nabla u(\mathbf{x})\|_{\mathbf{x}}^{+} + \beta_{-}\|\nabla u(\mathbf{x})\|_{\mathbf{x}}^{-} = 0, \tag{15}$$

where $\beta_{+} = \frac{1}{2} + \beta \text{sgn}(\|\nabla u(\mathbf{x})\|_{\mathbf{x}}^{+})$ and $\beta_{-} = \frac{1}{2}$. Given $\mathbf{x}$, let $\mathbf{y}, \mathbf{z} \in \mathcal{N}(\mathbf{x})$ be such that:

$$\|\nabla u(\mathbf{x})\|_{\mathbf{x}}^{+} = \frac{u(\mathbf{y}) - u(\mathbf{x})}{d_{\mathbf{xy}}}, \quad \text{and} \quad \|\nabla u(\mathbf{x})\|_{\mathbf{x}}^{-} = \frac{u(\mathbf{z}) - u(\mathbf{x})}{d_{\mathbf{xz}}}.$$

By introducing these expressions into (15),

$$u(x) = \frac{\beta_{+} d_{\mathbf{xz}} u(\mathbf{y}) + \beta_{-} d_{\mathbf{xy}} u(\mathbf{z})}{\beta_{+} d_{\mathbf{xz}} + \beta_{-} d_{\mathbf{xy}}}. \tag{16}$$

The numerical implementation for the iterative discretized biased Infinity Laplacian is:

$$u^{k+1}(\mathbf{x}) = \frac{\beta_{+} d_{\mathbf{xz}} u^{k}(\mathbf{y}) + \beta_{-} d_{\mathbf{xy}} u^{k}(\mathbf{z})}{\beta_{+} d_{\mathbf{xz}} + \beta_{-} d_{\mathbf{xy}}}, \qquad k = 0, 1, ... \tag{17}$$

2.2 Distance Approximation

In practice, we have approximated the length of the geodesic path (L_g) (joining two points $\mathbf{x}$ and $\mathbf{y}$) by the distance $d_{\mathbf{xy}}$ defined in (5). Different neighborhoods size can be taken, for example, the 4−connected pixels of $\mathbf{x}$, or the 8−connected pixels of $\mathbf{x}$. Taking a larger neighborhood size we use the approximated value of the distance $L_g \approx d_{\mathbf{xy}}$.

2.3 Algorithm to Estimate Parameters

We have estimated the parameters of AMLE (neighborhood size (r), κ_c, κ_x, and iteration number) using the PSO (Particle swarm optimization) algorithm [8]. This algorithm was used to estimate an optical flow model's parameters, which is described in [10]. The success of that previous work motivates us to use PSO optimization algorithm in our proposal.

PSO Algorithm. This algorithm optimizes a function by iteratively improving many candidate solutions. The algorithm performs each iteration using the process to optimize updating those candidates. The updating of these candidates' solutions is performed according to dynamic positions and velocities of those candidates' answers. In our case, those candidate solutions are different model parameters and the function to optimize is the depth estimation error.

Let $f : \mathbb{R}^n :\rightarrow \mathbb{R}$ be the depth estimation error obtained by each candidate solution and n is the parameters number to estimate. Let μ_i be a candidate solution that minimizes the depth estimation error with $i = 1, ..., n$. In each iteration, the best candidate solution is stored in μ_b, and also, the best candidate found in all performed iterations is stored in μ_g. The μ_b and μ_g are called the best candidate and best global candidate, respectively. Each candidate solution (μ_i) are updated according to the evolution equation given by,

$$\nu_i^{t+1} = \omega \nu_i^t + \varphi_g(\mu_i^t - \mu_g) + \varphi_b(\mu_i^t - \mu_b) \tag{18}$$

and,

$$\mu_i^{t+1} = \mu_i^t + \nu_i^t, \tag{19}$$

where ω is the evolution parameter for each solution candidate μ_i, φ_g and φ_b are positive weight parameters, ν_i is the velocity for each candidate solution. A saturation for ν_i is usually incorporated. We used $\nu_{max} = 2$ and $\nu_{min} = -2$.

3 Experiments and Dataset

KITTI depth completion suite contains RGB reference images and its corresponding ground-truth. The ground-truth consists of semi-dense depth obtained by raw LiDaR scans. The considered dataset consists of 1000 reference images and its corresponding sparse depth map.

Middlebury2014 consists of 22 color stereo pair with its corresponding depth ground-truth for each stereo image. These images present size of 2994 × 2000 pixels. The ground-truth is dense and was obtained with different illumination conditions, as explained in [18]. In Fig. 1(a) and Fig. 1(b) we show two reference images from and (d) and (e) images of the sparse depth data. The depth was color-coded using jet colormap of Matlab. In (c) and (f), we show an example of one reference image of the Middlebury2014 [18] dataset and its corresponding depth, respectively.

3.1 Training Model

To estimate parameters (or train the model), we used a subset of KITTI. Each subset element is a pair of images, one is the incomplete depth data, and the other is the reference color image. We trained the model applying the PSO algorithm. We give ten initial values of the model parameters (μ_i). For each parameter value in the complete KITTI subset ($MAE(\mu_i)$ and $(MSE(\mu_i))$, we estimate Mean Absolute Error (MAE) and Mean Square Error (MSE). We updated the values of the parameter following the evolution equations of the PSO algorithm. We followed the steps: i) Initialize μ_i with random values for each individual. ii) Compute MAE y MSE for each μ_i. iii) Compare the performance of each μ_i. iii) Update μ_i according the evaluation equation of PSO Eq. (18) and Eq. (19). iv) goto to ii) until reach 30 iterations.

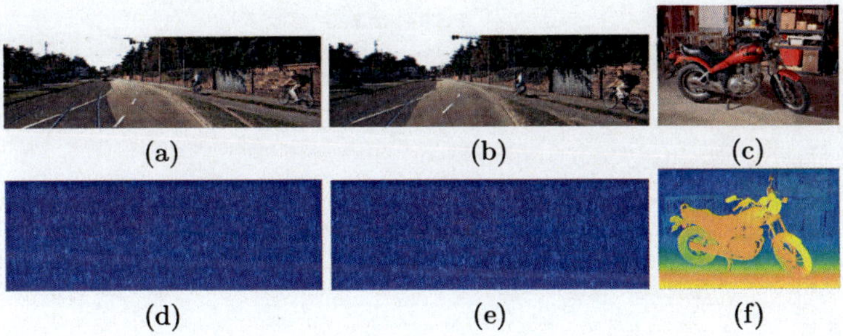

Fig. 1. Example of RGB reference and depth images of the KITTI dataset and Middlebury2014. (a) left camera images and (b) right camera images. (c) Color reference image of the sequence Motorbike of Middlebury2014. (d) y (e) depth images of reference RGB images (a) and (b), respectively. and (f) depth of the scene in (c).

Model Parameter Estimation. We set $\omega = 0.5$, $\varphi_g = 0.5$ and $\varphi_b = 1.0$. We used ten individuals, stop criteria thirty iterations, three depth and reference images. The functional we minimized was the MSE + MAE of our selected depth and reference images:

$$J(r_j, \kappa_{xj}, \kappa_{cj}) = \sum_{i=1}^{NI} MSE_i(r_j, \kappa_{xj}, \kappa_{cj}) + MAE_i(r_j, \kappa_{xj}, \kappa_{cj}). \quad (20)$$

where NI is the size of the training dataset.

Training Dataset

We selected three reference images and its corresponding depth from the validation dataset to determine best model parameters, shown in Fig. 2.

Fig. 2. Images of the training set.

Estimation of r (Neighborhood Size), κ_c and κ_x. We estimated the parameter of the AMLE operator. In Fig. 3 we show the evolution of the PSO algorithm during the iterations. We plot in different colors the change of the performance of the ten candidate solutions. In this figure, we have shown that the performance improves as iterations go. In Fig. 3 we show that the PSO algorithm always minimizes the MSE+MAE error in Eq. (20). The obtained best parameters set were $n_{iter} = 40$, $r = 1$, $\kappa_c = 310.97$, and $\kappa_x = 280.62$.

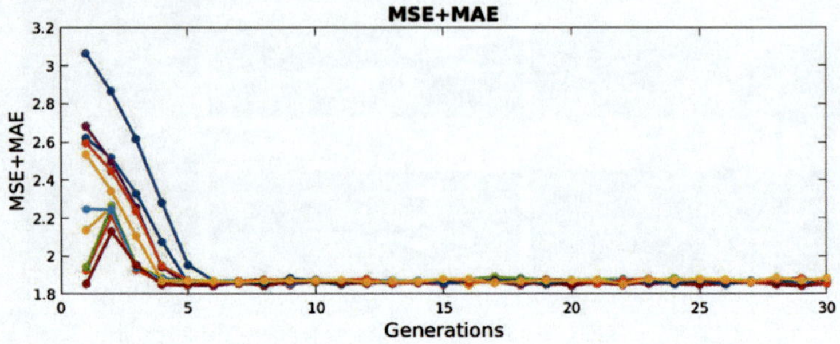

Fig. 3. Evolution of ten individuals of the PSO algorithm during iterations.

3.2 Estimation of β Parameter

The β value span $(0.1 \leq \beta \leq 1.5)$ was defined based on the shape of the exponential cone as we show in Fig. 4.

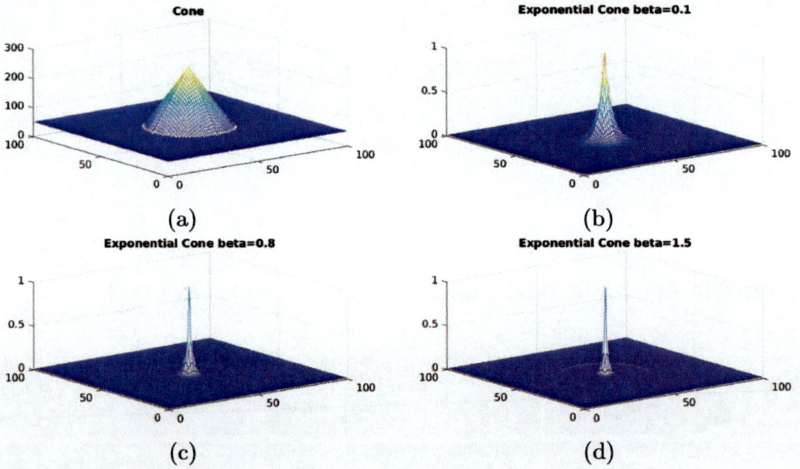

Fig. 4. Different shapes for exponential cones. (a) Shape for $\beta = 0$. (b) $\beta = 0.1$. (c) $\beta = 0.8$. (d) $\beta = 1.5$

In Fig. 4, we show the shape of exponential cones for values of β parameter. It is shown that given shape of the exponential cone for values larger that $\beta = 1.5$ (shape is very thin), the interpolation of large region should take more iterations to cover it.

To estimate the parameters of the bAMLE we used parameters estimated in the above section but the β parameter was estimated varying its value from 0.1 to 1.5 in the training set, as we explained above. We show in Fig. 5 the performance of the bAMLE in the training set: We observed in Fig. 5 that the best performance in the training set was obtained for $\beta = 0.7$.

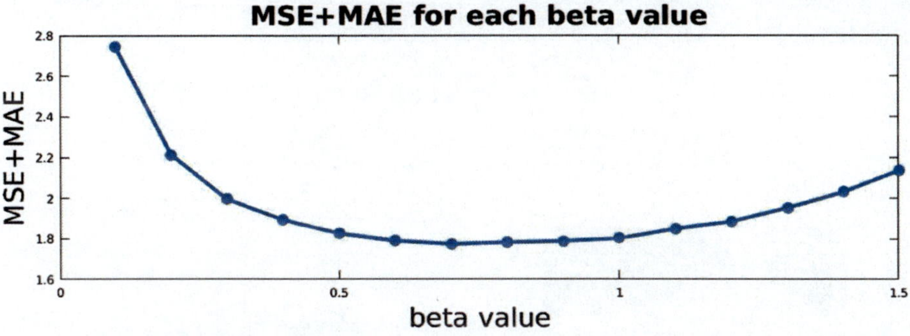

Fig. 5. Performance of bAMLE using different β value in training set. Best performance for $\beta = 0.7$.

3.3 Sampling Middlebury Depth Images

We sampled the depth images of the Middlebbury on a regular spatial grid using a Δ size. We used $\Delta = 2$, $\Delta = 4$, $\Delta = 8$ and $\Delta = 16$. If $\Delta = 2$ we sampled every 4 other pixels square, for $\Delta = 4$ we sampled every 16 other pixels square and so and on for all Δ values.

4 Results and Discussion

4.1 Results of Depth Completion on KITTI Dataset

We present obtained results in the complete validation dataset. In Table 1 we show obtained results by AMLE and bAMLE. As we show in Table 1 bAMLE outperforms the AMLE and both outperform bilateral filter (reported in [16]). Contemporaneous methods obtain better results than bAMLE.

In Fig. 6 we present qualitative evaluations of the bAMLE in the depth completion task.

4.2 Results Up-Sampling Middlebury2014

We up-sampled our dataset using AMLE and bAMLE. We computed the MSE and MAE for each up-sampled data. We show in Table 2 the obtained results by each method. In Table 2 the last row, the difference in performance between these two models is more notorious for the low sub-sample rate. For large sample rates $\Delta = 8$ or $\Delta = 16$ the effect of β is less important in the bAMLE. In Fig. 7, we show an example of an up-sampled image and the original ground-truth. We observe that the result is still acceptable but presents many error in fine details or in small objects.

Table 1. Results obtained by different methods in KITTI depth completion validation set.

Method	MSE	MAE
DeepLidar [16]	0.687	0.215
bAMLE	1.786	0.440
AMLE	1.801	0.471
CNN [20]	2.100	0.680
Bilateral	2.989	1.201

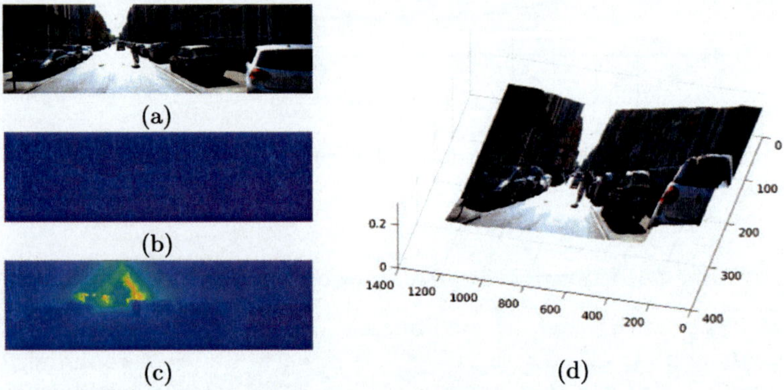

Fig. 6. Obtained results: (a) Color reference image. (b) Incomplete depth image. (c) Completed depth image. (d) 3D model of the scene using the completed depth map.

Table 2. Results obtained by AMLE and bAMLE in up-sampling task.

	$\Delta = 16$		$\Delta = 8$		$\Delta = 4$		$\Delta = 2$	
Method	MSE	MAE	MSE	MAE	MSE	MAE	MSE	MAE
AMLE	35.22	8.73	34.78	8.30	34.43	7.80	34.35	7.47
bAMLE	34.82	8.98	34.19	8.56	33.63	7.84	33.42	7.32
Difference	0.40	−0.25	0.59	−0.26	0.80	−0.04	0.93	0.15

4.3 Discussion

As we see in Table 1 AMLE obtained $MSE = 1.801$, which performs better than bilateral filter and than CCN [19] but cannot defeat bAMLE and neither more contemporary methods. Given sparse data, AMLE is a cone interpolator. KITTI dataset contains scenes where spherical or rounded objects are presents. The cone interpolator seems to be less suitable to this kind of surface than the bAMLE (which is an exponential cone interpolator). In Table 2 AMLE performs similar than the bAMLE. For $\Delta = 16$, $\Delta = 8$ and $\Delta = 4$ the MAE of the AMLE is inferior to the MAE associated to bAMLE. MAE means "Mean Absolute

(a) (b)

Fig. 7. Depth ground-truth and upsampled data. (a) ground truth. (b) Up-sampled depth with $\Delta = 16$.

Error" which is a metric that is less sensitive to outlier showing that AMLE is more sensitive to outliers than bAMLE as is shown in MSE values.

5 Conclusions

We have implemented the biased AMLE or biased infinity Laplacian endowing the depth image with an anisotropic metric g_{ij}. On the one hand, biased AMLE does not perform like a contemporaneous methods. Still, on the other hand, the simplicity of its implementation based on exponential cones makes the AMLE operator a low-implementation-cost solution in the problem of data completion. In our evaluation, the implemented model outperforms Infinity Laplacian and bilateral in depth interpolation in KITTI dataset. The upsampling task using Middelbury2014 outperforms the AMLE model for sampling rates such that $\Delta = 2$, $\Delta = 4$. In future work, we will test different metrics to obtain better performance and instead of using an approximated geodesic distance between two pixels, we will consider the exact length using Dijkstra's algorithm. Additionally, we will test different algorithm to train our model. We will evaluate the performance of methods such that: i) krill herd optimization algorithm, ii) monarch butterfly optimization, iii) earthworm optimization or iv) elephant herding optimization in parameter estimation task, etc. We will consider validate the model in other dataset such that Middelbury2005-2006 and/or NYU-v2 dataset.

References

1. Almansa, A., Cao, F., Gousseau, Y., Rougé, B.: Interpolation of digital elevation models using AMLE and related methods. IEEE Trans. Geosci. Remote Sens. **40**(2), 314–325 (2002)
2. Aronsson, G.: Extension of functions satisfying Lipschitz conditions. Aktiv fuer Mathematik **6**(6), 551–561 (1967)
3. Aronsson, G.: On the partial differential equation $u_x^2 u_{xx} + 2u_x u_y u_{xy} + u_y^2 u_{yy} = 0$. Aktive fuer Mathematik **7**(5), 395–425 (1968)
4. Baker, S., Scharstein, D., Lewis, J., Roth, S., Black, M., Szelensky, R.: A database and evaluation methodology for optical flow. Int. J. Comput. Vision **92**, 1–31 (2011)
5. Caselles, V., Igual, L., Sander, O.: An axiomatic approach to scalar data interpolation on surfaces. Numer. Math. **102**(3), 383–411 (2006)

6. Caselles, V., Morel, J.M., Sbert, C.: An axiomatic approach to image interpolation. IEEE Trans. Image Process. **7**(2), 376–386 (1998)
7. Eldesokey, A., Felsberg, M., Holmquist, K., Persson, M.: Uncertainty-aware CNNs for depth completion: uncertainty from beginning to end. In: Proceedings of the IEEE/CVF Conference on Computer Vision and Pattern Recognition (CVPR), June 2020
8. Kennedy, J., Eberhart, R.: Particle swarm optimization. In: Proceedings of IEEE International Conference on Neural Networks, IV, pp. 1942–1948 (1995)
9. Lai, W.S., Huang, J.B., Ahuja, N., Yang, M.H.: Fast and accurate image super-resolution with deep Laplacian pyramid network. Trans. Pattern Anal. Mach. Intell. **41**(11), 2599–2613 (2019)
10. Lazcano, V.: An empirical study of exhaustive matching for improving motion field estimation. Information **9**, 320 (2018)
11. Li, Y., Huang, J.B., Ahuja, N., Yang, M.H.: Joint image filtering with depth convolutional networks. Trans. Pattern Anal. Mach. Intell. **41**(8), 1909–1923 (2019). https://doi.org/10.1109/TPAMI.2018.2890623
12. Lu, K., Barnes, N., Anwar, S., Zheng, L.: From depth what can you see? Depth completion via auxiliary image reconstruction. In: Proceedings of the IEEE/CVF Conference on Computer Vision and Pattern Recognition (CVPR), June 2020
13. Manfredi, J., Oberman, A.M., Sviridov, A.P.: Nonlinear elliptic partial differential equation and p-harmonic functions on graphs. Diff. Integr. Equ. **28**(1/2), 79–102 (2015)
14. Ndjiki-Nya, P., Koeppel, M., Doshkov, D., Lakshman, H., Merkle, P., Mueller, K., Wiegand, T.: Depth image based rendering with advanced texture synthesis. In: Proceedings of the IEEE International Conference on Multimedia and Expo, pp. 424–429 (2010)
15. Oberman, A.M.: A convergent difference scheme for the infinity Laplacian: construction of absolutely minimizing Lipschitz extensions. Math. Comput. **74**(251), 1217–1230 (2005)
16. Qiu, J., Cui, Z., Zhang, Y., Zhang, X., Liu, S., Zeng, B., Pollefeys, M.: DeepLiDAR: deep surface normal guided depth prediction for outdoor scene from sparse lidar data and single color image. In: The IEEE Conference on CVPR), June 2019
17. Raad, L., Oliver, M., Ballester, C., Haro, G., Meinhardt, E.: On anisotropic optical flow inpainting algorithms. Image Process. Online **10**, 78–104. https://doi.org/10.5201/ipol.2020.281
18. Scharstein, D., Hirschmüller, H., Krathwohl, G., Kitajima, Y., Nešić, N., Wang, X., Westling, P.: High-resolution stereo datasets with subpixel-accurate ground truth. In: GCPR 2014. Lecture Notes in Computer Science, vol. 8753. Springer, Cham (2014)
19. Schneider, N., Schneider, L., Pinggera, P., Franke, U., Pollefeys, M., Stiller, C.: Semantically guided depth upsampling. In: German Conference on Pattern Recognition, pp. 37–48. Springer (2016)
20. Uhrig, J., Schneider, N., Schneider, L., Franke, U., Brox, T., Geiger, A.: Sparsity invariant CNNs. In: International Conference on 3D Vision (3DV) (2017)

Hyperparameter Tuning Based Performance Analysis of Machine Learning Approaches for Prediction of Cardiac Complications

Shital Patil[1,2](✉) and Surendra Bhosale[1]

[1] Department of Electrical Engineering, VJTI, Mumbai, India
snpatil_p17@ee.vjti.ac.in, shital.patil@rait.ac.in,
sjbhosale@vjti.org.in

[2] Department of Instrumentation Engineering, RAIT, Navi Mumbai, India

Abstract. Nowadays, cardiac disorders are the biggest causes for morbidity and mortality in the world causing huge number of deaths over the last few decades and have emerged as the most life-threatening disease. So, this calls for immediate need of reliable, accurate and feasible system to diagnose and predict such diseases for timely diagnosis. This paper emphasizes on different approaches using classical machine learning algorithms with their hybrid approaches for the predictions of heart attacks, so as to compare their performance. Here, classical and hybrid machine learning algorithms and techniques have been applied to the standard medical data sets to come out with lab-based prototype in an attempt to automate the analysis of large and complex patient data attributes. Comparative performance analysis of different scalar has been carried out to use standard scalar for feature scaling method. It initially deals with approach pertaining to classical models based on supervised learning algorithms such as Support Vector Machines, K-Nearest Neighbor, Random Forest and Deep learning models. Hybrid approach using Fuzzy C means based Neural Network gives classification accuracy of 98.76% which shows that this proposed optimized technique outperforms much better over the simple classifiers in terms of prediction accuracy. Using huge medical data in health care systems with validated and appropriate data splitting for training as well as testing, attempts have been made to compare performance of different algorithms for different ratios of data used for training and testing to fine tune accuracy requirement to understand the impact of these ratios. This work envisages GridsearchCV to find the best suitable values for the hyper parameters and its impact on improvising accuracy of algorithmic models leading to more accurate prediction of heart complications.

Keywords: K-Nearest neighbor · Support vector machine · Random forest · Neural networks · Hyperparameter

1 Introduction

Heart, being one of the most important organs of the human body, needs proper care for the life of human beings. The heart runs the circulatory systems of the body by

A. Abraham et al. (Eds.): SoCPaR 2020, AISC 1383, pp. 605–617, 2021.
https://doi.org/10.1007/978-3-030-73689-7_58

pumping blood throughout the body. So, functioning of the heart is an utmost important for survival else it could even lead to death. Heart disease describes a range of conditions that affect one's heart. The term "heart disease" is often used interchangeably with the term "cardiovascular disease". Cardiovascular diseases are the conditions that causes narrowed or blocked blood vessels that can lead to a heart attack, chest pain angina or stroke [1]. Using this huge amount of data, a disease can be detected, predicted or used by cardiac physician to address emergency needs of the patient [1].

In today's world death because of the heart diseases has become a major problem and approximately one person dies per minute due to heart disease. This problem can also start in the early age group and predicting the cause and disease is a major challenge nowadays. There is a huge amount of data being generated by various applications related to heart patients that is needed to be stored in data storage in an arranged manner and this data can be used to predict the heart attacks by using various ML methods [2].

1.1 Condition of Heart Attack

A heart attack occurs when a part of the heart muscle is damaged or dies because blood flow is reduced or completely blocked. A heart attack, also called a myocardial infarction, occurs when a part of the heart muscle doesn't receive enough blood flow. The more time that passes without treatment to restore blood flow, the greater the damage to the heart muscle. A heart attack occurs when the blood flow that brings oxygen to the heart muscle is severely reduced or cut off completely. This happens because coronary arteries that supply the heart muscle with blood flow can become narrowed from a build-up of fat, cholesterol and other substances that together are called plaque [1]. Coronary artery disease (CAD) is the main cause of heart attack. If someone knows the signs and symptoms of a heart attack and recognize that he or she is having a heart attack, should seek immediate proper treatment. If the treatment is delayed which pertains to further damage to the heart muscle [2].

1.2 Identification of Factors for Cardiac Prevention

Cardiac diseases should be handled by following healthy lifestyles in regard to heart. Cardiac physicians have identified few factors causing heart diseases like, Family history, Age, Sex etc. helping us to explore plenty ways to reduce the risk [1]. Here are the tips for boosting healthy heart are:

- Quit Smoking: Chemicals in tobacco is prone to mouth cancer as well as lung cancer due to smoke in cigarette. This reduces blood supply to the blood increasing blood pressure and hence is very important data attribute is to be considered in such diseases.
- Physical Exercise Sports: Daily at least 30 min of workout or physical exercise, puts off the burden from heart. Not only it keeps blood pressure in control but also Cholesterol and type-2 diabetes. So, heart rate, blood cholesterol and sugar level of the blood are considered to be data attributes for such prediction.
- Healthy Diet: Healthy diet keeps healthy heart. Avoid intake of processed foods, trans fat, saturated fats.

- Obesity: Weight plays crucial role in maintaining good health. Increase in weight, leads to not only bad cholesterol but also blood pressure. This highlights the body weight as one of the most important data attributes for this study.
- Quality Sleep and Managing Stress: For Adults, seven hours' sound sleep is recommended for healthy mind. Inadequate sleep leads to stress and hypertension. Such people are at higher risk of developing obesity and blood pressure.
- Family History: Family history plays a vital role in cardiac disorder as it has hereditary features.
- Regular Health Check-ups: In addition to annual health screenings parameters like High blood pressure, cholesterol, sugar level, Electrocardiogram (ECG), cardiac risk markers with its analytical parameters, other physiological parameters like angiographic study-based parameters like number of vessels where calcium deposits were observed, in consultation with senior cardiologist in Mumbai, India.

2 Proposed Methodology

Problem Statement: Comparative performance analysis of various classical machine learning algorithmic hybrid approaches adopted for predicting heart diseases and further improving these accuracy using data optimization algorithms using grid search CV method, hyper parameter based tuning approach.

The block diagram in Fig. 1 shows the basic steps followed for the proposed data classification.

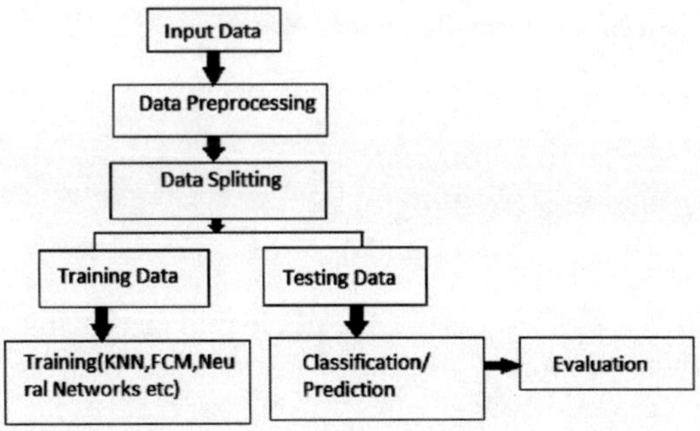

Fig. 1. Basic steps for data classification system

A. Input Dataset

Heart dataset which has been used here, is taken from the UCI Machine Learning Repository. This dataset has 76 attributes, but for our work we have considered only 14 attributes for significance of this study in initial stage of this research [2]. Some of the attributes presented similar information while some attributes were not related with the target attribute. So, redundant attributes have been discarded. Only relevant and unique attributes which can help in predicting our target feature are considered in this dataset. Table 1 gives the information about the attributes used for our work.

Table 1. Attribute information of the heart dataset

Sr No.	Data attributes	Description
1	Age	Age in years
2	Sex	Gender, 1 = Male, 0 = Female
3	Chest Pain (Cp)	Types of chest pain (0: typical angina, 1: Atypical angina, 2: Non-angina, 3: Asymptomatic.)
4	Trestbps	Resting blood pressure (in mm Hg on admission to the hospital)
5	Chol	Serum cholesterol in mg/d
6	FBS	Fasting blood sugar > 120 mg/dl) (1 = true, 0 = false)
7	Restecg	Resting electro cardiographic results (0): normal, 1: ST-T wave abnormality 2: probable or definite left ventricular hypertrophy by Estes' criteria)
8	Thalach	Maximum heart rate achieved
9	Exang	Exercise induced angina (1 = yes, 0 = no)
10	Oldpeak	ST depression induced by exercise relative to rest
11	Slope	Slope of the peak exercise (0: up sloping, 1: flat, 2: down sloping)
12	Ca	Number of major vessels (0–3) colored by fluoroscopy
13	Thal	3 = normal; 6 = fixed defect; 7 = reversible defect
14	Num (Target)	Diagnosis of heart disease 0: Healthy, 1: UnHealthy

B. Data Pre-Processing and Splitting

1. *Data Cleaning*: In case of data cleaning, all the null or missing values are either be dropped or imputed. However, in case of this dataset, there is no null or missing value and none of the column is single valued in our dataset. So, no specific need for data cleaning is required and hence we can move further with data preprocessing.
2. *Feature Scaling*: Machine learning algorithm just sees numbers, if there is a vast difference in the data range say few ranging in thousands and few ranging in the tens, it makes the underlying assumption that higher ranging numbers have superiority of some sort. So, these more significant number starts playing a more decisive role while training the model. So, feature scaling is generally required when we are dealing with attributes of different scale, otherwise, it may lead to a dilution in effectiveness of an equally important attribute [3]. As dataset is diverse on the scale, it becomes prudent to consider feature scaling whose three types are as below

- Standard Scalar: This standardizes the feature by sub-tracing the mean and then scaling to unit variance. It transforms the data in such a manner that it has mean as 0 and standard deviation as 1. It arranges the data in a standard normal distribution and is more useful in classification than regression.

- Min Max Scalar: Transforms feature by scaling each feature in a given range. It preserves the shape of the original distribution. It scales and translates each feature individually with the help of the formula.

$$X_{scaled} = \frac{X - X_{min}}{X_{max} - X_{min}} \quad (1)$$

where X is a feature, Xmin refers to the minimum value of that feature, Xmax refers to the maximum value of the feature and Scaled feature is represented by Xscaled.

- Robust Scalar: It is used mostly when we want to reduce the effect of the outliers, relative to that of the min-max scalar. It does not scale the data between the pre-determined intervals like what happens in the min max scalar.

All the above three scaling methods were applied to a machine learning algorithm to see which method suits best for our dataset. Prediction accuracy were noted while using each method at a time. Accuracy with respect to each of these methods have been presented in the Table 2.

Table 2. Comparative performance analysis of the different scalers

Scalar	Accuracy (%)
Standard scalar	86.79
Min max scalar	84.91
Robust scalar	85.85

As can be seen from Table 2, best result was obtained with the help of standard scalar. So, we have used standard scalar for feature scaling. The Standard Scalar assumes that the data is normally distributed within each feature and will scale them such that the distribution is centered around zero with standard deviation of 1.

Basis for Training and Testing Data Split: Splitting ratio for the training and testing dataset is the foremost requirement for any further processing of the machine learning algorithms. For e.g. It is not a good choice to take 50:50 as the split ratio because if we take 50:50 then we will have only half the samples for training and half for testing.

For any algorithm, it is recommended that the algorithm gets maximum data for its training. In context of this dataset, we will have only 150 sample if we take 50:50 samples for testing which is not a good choice. In this paper, we chose 4 different split ratios, and applied it on different algorithms like SVM, Random Forest, neural networks and noted the accuracy in each set of ratios. The Table 3, shows comparative analysis of training accuracy (%) because of using different split ratio. It was found that the best result was obtained with 70:30 split. Hence split ratio of 70:30 is used as the train-test split ratio for our rest of the algorithms in this paper.

Table 3. Accuracy of different algorithms due to of various Split Ratios

Algorithm	60:40	70:30	75:25	80:20
SVM	86.7	86.79	85.02	84.30
Neural network	84.53	84.91	84.14	82.64
Random forest	86.74	92.45	90.31	85.12

C. Data Visualization

Data Visualization affects user's decision making. With visualization it becomes easier to communicate findings in a constructive way. Figure 2 indicates basically the graphical representation of the dataset for its better understanding. It's a histogram plot where on x axis, we have different attributes of dataset and y axis represents the count of the respective attributes [4].

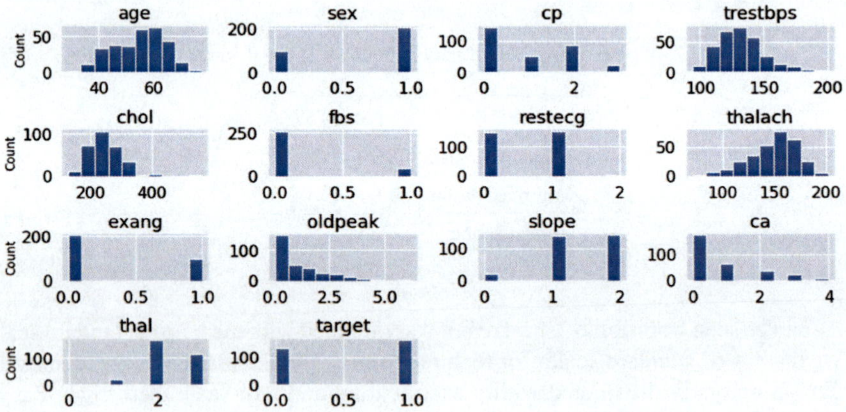

Fig. 2. Histogram showing count of heart patient with different attributes

Figure 2 shows distribution of people with different parameters as mentioned in the Table 1 viz. age, sex, Cpetc. with heart diseases. It can be clearly inferred from the histograms which attributes are categorical variables and which one are numerical (integer type) [4]. Some of the key findings which can be interpreted from the histogram plots are mentioned here [5].

- Age: It is observed from the Fig. 2 that people with age groups around 60 are more prone to heart attacks.

- Sex: Heart diseases in Men (value 1) are way higher than women (value 0) which means that male are more prone to heart diseases than female.
- Cp: Majority of people are suffering with type 0 chest pain, while least number of people are suffering with type 3 chest pain.
- Trestbps: Histogram shows that people with resting BP of around 140 are mostly prone to heart diseases with count around 50.
- Chol: People with Cholesterol level of 250 have more heart disease, while there are some outliers with cholesterol level of 500.
- FBS: People with FBS > 120 are represented with 1 and are more in count for heart disease than the people with fbs < 120. This means, people suffering with heart diseases have Fasting Blood sugar > 120 mg/dl.
- Restecg: in case of ECG parameters, majority of people are having ST-T wave abnormality (value 1) while least people are showing probable or definite left ventricular hypertrophy.
- Thalach: Majority of people have maximum heart rate above 150.
- Exang: People with exang value of 0 are more in count. So, people with heart disease will not have exercise induced angina.
- Oldpeak: People with value between 0 and 1 have more chances of heart disease.
- Slope: People with value 0 have least heart disease while majority of people have value 1 and 2.
- Ca: People with no major vessels colored by fluoroscopy are more in count in terms of heart diseases.
- Target: Majority of people in the given dataset are Un-healthy with value = 1.

D. Different Algorithms for Prediction of Heart Diseases

1. ***Support Vector Machine:*** Support Vector Machine (SVM) classifier is a linear classifier where the separating hyper plane is chosen to minimize the expected classification error of the unseen test patterns. SVM is a strong classifier which can identify two classes. SVM classifies the test image to the class which has the maximum distance to the closest point in the training. For linearly non-separable data the input is mapped to high-dimensional feature space where they can be separated by a hyperplane. This projection into high-dimensional feature space is efficiently performed by using kernels. SVM aims to find the best separating hyperplane given by the equation that maximizes the distance between the two classes [6].

$$Y = Wx + B \tag{2}$$

2. ***Neural Networks:*** Artificial Neural networks (ANN) try to simulate the decision-making process of the human brain. ANN can realize and study patterns for obtaining knowledge. It displays a link between an input neuron and an output neuron. Neurons have some specified weights. Output is calculated by multiplying the input with the

specific neuron weight and then comparing it with the threshold value. If it's above the given threshold then it is contemplated as the output. In general, we cannot analytically calculate the number of layers or the number of nodes to use per layer in an artificial neural network to address a specific real-world predictive modeling problem. It must be discovered using careful experimentation to see what works best for the specific dataset. So, after trying different values for the size of hidden layer best results were obtained with the hidden layer size = 100 [8].

3. ***Random Forest:*** Random Forest (RF) constructs many individual decision trees at training. Predictions from all trees are pooled to make the final prediction. The importance for each feature on a decision tree is then calculated as

$$fi_i = \frac{\sum_{j:node\ j\ splits\ on\ feature\ i} ni_j}{\sum_{k \in all\ nodes} ni_k} \tag{3}$$

Where fi_i is the importance of feature i, ni_j is the importance of node j. The final feature importance, at the Random Forest level, is its average over all the trees. The sum of the feature's importance value on each tree is calculated and divided by the total number of trees [6].

$$RFfi_i = \frac{\sum_{j \in all\ trees} norm\ fi_{ij}}{T} \tag{4}$$

$RF\ fi_i$ is the importance of feature i calculated from all trees in the Random Forest model, $norm\ _{fi}ij$. is the normalized feature importance for i in the tree.

4. ***K-Nearest Neighbours:*** K-Nearest Neighbours (KNN) is the simplest one amongst all other machine learning algorithms. On the basis of the majority votes of the neighbors, an object can be classified. The pattern space for k training samples which are nearest to unknown samples is searched by KNN classifier in the case when an unknown sample is given. The Euclidean distance helps in defining the closeness of samples. Thus, amongst two given points X = (x1, x2,..., xm) and Y = (y1, y2,..., ym) the Euclidean distance can be defined by the Eq. 5.

$$d(x, y) = \sqrt{\sum_{i=1}^{m} (x_i - y_i)^2} \tag{5}$$

5. ***K-Means Based SVM:*** The prediction analysis requires two phases. In the first phase, the k-mean clustering is applied which will cluster the similar and dissimilar type of data. In the second phase, the SVM classifier is applied which will classify the data.

6. ***K-Means Based Neural Network:*** The prediction analysis requires two phases. In the first phase, the k-mean clustering is applied which will cluster the similar and dissimilar type of data. In the second phase, the Neural Network classifier is applied to classify the data. Initial steps remain same as K means based SVM.

7. ***Fuzzy C-Means Based Neural Network:*** Fuzzy C-means clustering is a soft clustering technique which allows partial belonging ness of pixels into different clusters. The sum of all membership degrees for any given data point is equal to 1. This method has better applicability to segmentation applications than hard clustering algorithms. The algorithm finds 'c' clusters by minimizing the objective function given by the Eq. 6.

$$J_{FCM} = \sum_{k=1}^{n} \sum_{i=1}^{c} (u_a)^q d^2(x_k, v_i) \ldots \quad (6)$$

In above equation, $x_k = \{x1, x2, x3 \ldots xn\}$ are the data points, n is the number of data items, c means no.of clusters, the degree of membership of $_{xk}$ in the ith cluster is represented as u_{ik}, q is the weighting exponent on each fuzzy membership, vi represents the centre of cluster i, d^2 (x_k, v_i) is the distance between data point x_k and cluster centre v.

In this hybrid approach the Fuzzy C-means output is given as input to the Neural network for better classification based on the fuzzy c-partitioned matrix obtained.

It has been seen that after implementing this Hybrid method there is an increase in the accuracy of classification rather than using only Neural Network.

8. ***Fuzzy C-Means Based SVM***: Classification based on the fuzzy c-partitioned matrix obtained

- The first step, Fuzzy c means the algorithm is applied on the dataset available. Finally, a Fuzzy c-partitioned matrix is obtained.
- Output from FCM i.e. Fuzzy partition matrix and is combined with the feature set already available and will be given as input to the SVM classifier for the classification we saw that after implementing this Hybrid method there is an increase in the accuracy of classification rather than using only SVM.

2.1 Result and Analysis

Comparative analysis of all the 8 algorithms in terms of their prediction accuracy in terms of training and testing is given the Table 4.

Data Optimization Techniques There are multiple ways to optimize the data for applying it to the algorithm.

1. *Data Imputation*: Most of the models are unable to fit and predict the values when we feed them with missing values.

Table 4. Prediction accuracy of different algorithms prior to tuning

Algorithm	Training accuracy	Testing accuracy
SVM	86.22	80.22
Neural network	90.21	82.41
Random forest	87.74	75.82
KNN	84.91	82.42
Algorithm	Training accuracy	Testing accuracy
KNN Means based SVM	80.06	84.62
KNN Means based NN	91.98	83.52
FCC Means based SVM	87.6	86.89
FCC Means based NN	95.04	88.52

- Drop Missing Values: This is the fastest and easiest step to handle missing values. However, it is not generally advised. This method reduces the quality of the model as it reduces sample size. It works by deleting all the observations where any of the variable is missing.
- Fill Missing Values by Test Statistics: This is the most common method of handling missing values. This is a process whereby missing values are replaced with a test statistic like mean, median or mode of the feature column
- Predict Missing Values with a Machine learning Algorithm: This is an efficient method for handling missing data. Depending on the class of data that is missing, one can either use a regression model or classification to predict missing data. This works by turning missing features to labels and then using columns without missing values to predict columns with missing values.

2. *Handling Categorical Values*: Before we feed the data set into our machine learning model, we need to transform categorical values into numerical values because many models do not work with textual values. Solution is to use encoders to assign a unique numerical value to each textual value by using different encoders available.
3. *Data Standardization*: Z-score is one of the most popular methods to standardize data, and can be done by subtracting the mean and dividing by the standard deviation for each value of each feature.
4. *Using Machine Learning by Various Optimization Algorithms:* Thissuggests method to select best hyperparameters required for different algorithms.

GridsearchCV: This technique is used to find the optimal parameters to use with an algorithm called as hyperparameters. It is a library function that is a member of sklearn's model selection package. It helps to loop through predefined hyperparameters and fit the estimator (model) on the training set. So, we can select the best parameters and also additionally we can specify the number of times for the cross-validation for each set of hyperparameters. Here GridsearchCV is applied to all the 8 algorithms, to find the best suitable values for the hyperparameters. With the new values of these hyperparameters, we again performed 8 machine learning algorithms. Improvement in the prediction accuracy for all the 8 algorithms has been observed as specified in Table 5 as well as in Fig. 3 for Comparison of Training Accuracy before and after implementing GridsearchCV.

Table 5. Prediction accuracy of models after implementing grid search CV

Algorithms	Training accuracy	Testing accuracy
SVM	86.32	83.52
Neural network	93.87	83.52
Random forest	86.32	77.24
KNN	87.26	83.54
KNN Means based SVM	80.06	84.66
KNN Means based NN	93.4	84.42
FCC Means based SVM	87.19	88.52
FCC Means based NN	98.76	86.89

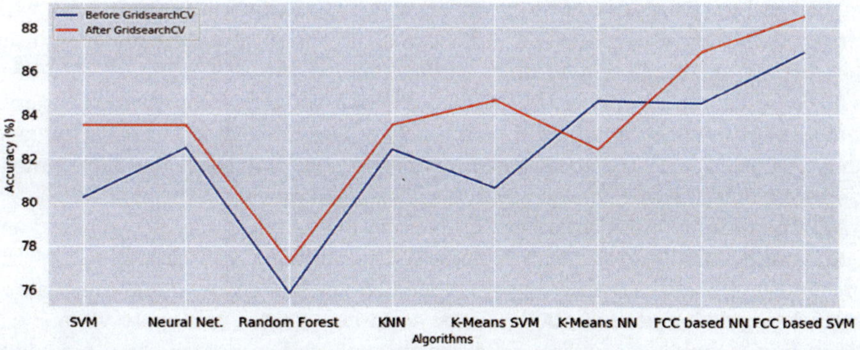

Fig. 3. For comparison of training accuracy before and after implementing GridsearchCV

Figure 4 for Comparison of Testing Accuracy before and after implementing GridsearchCV.

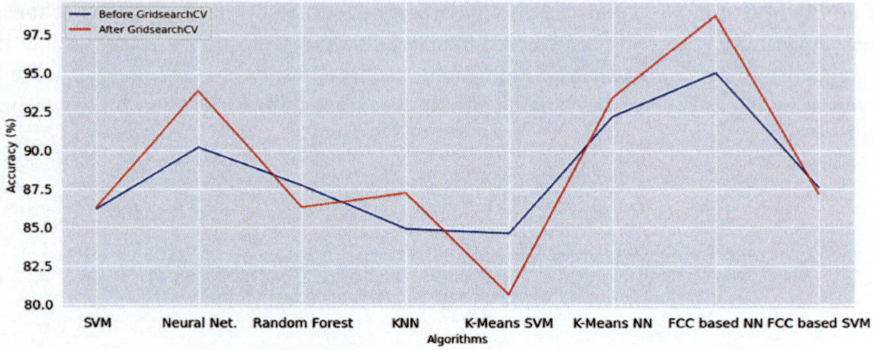

Fig. 4. For comparison of testing accuracy before and after implementing GridsearchCV

3 Conclusion and Future Work

The purpose of this study is to predict heart disease problem using different classical machine learning algorithms at the same time comparing them the hybrid models. For this research work, hybrid models like K Means Neural network and Fuzzy C-means clustering based Neural network performed best in terms of training accuracy K

Means Neural Network achieved 91.98% prediction accuracy which further improved in Fuzzy C-Means with Neural network i.e. FCC based NN 95.05% when no hyper tuning parameter is used. After careful selection of hyperparameters using GridsearchCV, the prediction accuracy further improved to 93.4% and 98.76% for K means based neural network and FCM based neural network respectively. Clustering techniques in combination with the Neural networks and SVM and their performance is thus compared with statistical classifiers and it was found that the hybrid model performed better in terms of accuracy than the classical models.

After consulting with cardiologist and cardio thoracic surgeon, it is planned as a future work, to add diversified physiological data attributes like cardiac risk marker elements, thyroid values, family history related parameters etc. with real time data set for larger population around 5000 patients in Indian context and apply various data optimization technique to improve the accuracy of cardiac complication prediction. With these large and diversified datasets, it looks prudent to see impact of hyper tuning on improvising the accuracy of prediction. Also, an approach will be formalized to use anomaly detection technique for cardiac complication prediction with better accuracy.

References

1. M. C. Staff: Strategies to prevent heart disease. https://www.mayoclinic.org/diseases-conditions/heart-disease/indepth/heart-disease-prevention/art-20046502
2. Nasrabadi, A., Haddadnia, J.: Predicting heart attacks in patients using artificial intelligence methods. Mod. Appl. Sci. **10**(3), 66 (2016)
3. Ramalingam, V., Dandapath, A., Raja, M.K.: Heart disease prediction using machine learning techniques: a survey. Int. J. Eng. Technol. **7**(2.8), 684–687 (2018)

4. Zriqat, I.A., Altamimi, A.M., Azzeh, M.: A comparative study for predicting heart diseases using data mining classification methods. arXiv preprint arXiv:1704.02799 (2017)
5. Aljanabi, M., Qutqut, H., Hijjawi, M.: Machine learning classification techniques for heart disease prediction: a review. Int. J. Eng. Technol. **7**(4), 5373–5379 (2018)
6. Gavhane, A., Kokkula, G., Pandya, I., Devadkar, K.: Prediction of heart disease using machine learning. In: 2018 Second International Conference on Electronics, Communication and Aerospace Technology (ICECA), pp. 1275–1278. IEEE (2018)
7. Khourdifi, Y., Bahaj, M.: Heart disease prediction and classification using machine learning algorithms optimized by particle swarm optimization and ant colony optimization. Int. J. Intell. Eng. Syst. **12**(1), 242–252 (2019)
8. Ozbay, Y., Ceylan, R., Karlik, B.: A fuzzy clustering neural network ¨ architecture for classification of ecg arrhythmias. Comput. Biol. Med. **36**(4), 376–388 (2006)
9. Patil, P.M., Deshmukh, M.P., Mahajan, P.: A novel fuzzy clustering neural network. In: Proceedings 2005 IEEE International Joint Conference on Neural Networks 2005, vol. 3, pp. 1989–1994. IEEE (2005)
10. M. H. Asvinth, A.: A computational model for prediction of heart disease based on logistic regression with gridsearchcv. International Journal of Scientific Technology Research, vol. 9 (2020)
11. Guzm´an, J.C., Miramontes, I., Melin, P., Prado-Arechiga, G.: Optimal genetic design of type-1 and interval type-2 fuzzy systems for blood pressure level classification. Axioms **8**(1), 8 (2019)
12. Ramirez, E., Melin, P., Prado-Arechiga, G.: Hybrid model based on neural networks, type-1 and type-2 fuzzy systems for 2-lead cardiac arrhythmia classification. Expert Syst. Appl. **126**, 295–307 (2019)
13. Miramontes, I., Guzman, J.C., Melin, P., Prado-Arechiga, G.: Optimal design of interval type-2 fuzzy heart rate level classification systems using the bird swarm algorithm. Algorithms **11**(12), 206 (2018)
14. Melin, P., Miramontes, I., Prado-Arechiga, G.: A hybrid model based on modular neural networks and fuzzy systems for classification of blood pressure and hypertension risk diagnosis. Expert Syst. Appl. **107**, 146–164 (2018)
15. Santhana Krishnan, J., Geetha, S.: Prediction of heart disease using machine learning algorithms. In: 2019 1st International Conference on Innovations in Information and Communication Technology (ICIICT), pp. 1–5 (2019)

Acousto-Prosodic Delineation and Classification of Speech Disfluencies in Bilingual Children

Sheena Christabel Pravin(✉) and M. Palanivelan

Department of ECE, Rajalakshmi Engineering College, Chennai, India
{sheena.s,palanivelan.m}@rajalakshmi.edu.in

Abstract. Speech disfluencies in spontaneous speech are valuable cues indicating language planning and speech execution problems in bi-lingual children during the secondary-language acquisition phase which is often confused with stuttering disfluencies. In this work, unique speech features are evaluated to set clinical benchmarks towards distinguishing linguistic impairment from stuttering disorder in bilingual Tamil-English (TE) speaking children whose primary language is Tamil (L1) and their secondary-acquired language (L2) is English. Acoustic and prosodic speech features extracted from disfluent segments of speech in the self-annotated disfluent speech corpus reveal divergence in spectral energy, pitch, formants and speech rate in Children with Stuttering Disorder (CSD) compared amid the Children with Language Impairment (CLI). The speech feature-based classification of CSD from CLI was verified with the implementation of linear and non-linear Support Vector Machine (SVM) models. The execution of the classifying models was evaluated based on their classification accuracy, F1-score, time complexities in discriminating CLI from CSD. The non-linear SVM was observed to exhibit the best classification accuracy and F1-score when the 'polynomial' kernel was chosen though its time complexity was higher than the linear SVM.

Keywords: Bilingual children · Children with Stuttering Disorder · Children with linguistic impairment · Speech features · Support Vector Machine

1 Introduction

Fluent speech is symbolized by lack of interruptions at a normal speaking rate with adequate prosody intonation, rendered with a minimal mental effort. Disfluency, on the other hand, signifies any disorder in the fluency constraints. Children during their language development phase, tend to exhibit disfluencies either due to developmental stuttering or other language impairments arising from linguistic complexity of the language learnt. Essentially, the second language learners, called as the bilingual children tend to be more disfluent than their monolingual peers. These disfluencies are inevitable in distinguishing developmental stuttering from language impairment. Typical disfluencies occur most likely in second language utterance other than the mother tongue as the speakers present a limited proficiency in second language. The analysis of disfluencies can shed light on

A. Abraham et al. (Eds.): SoCPaR 2020, AISC 1383, pp. 618–628, 2021.
https://doi.org/10.1007/978-3-030-73689-7_59

the process of second language acquisition due to the problems in the language-specific use of silent and filled pauses. Conventionally, the Clinicians judge a child to be at the risk of continuing as a stutterer if a child expresses three or more intra-word disfluencies per 100 words of conversational speech [1]. The peril is more for children who are open to two languages before joining school compared with mono-lingual children.

Repetitions of phrases, interjections and revisions were considered to be non-stuttering-like disfluencies [2]. However, monosyllabic word repetitions, parts of word repetition, sound prolongations are commonly considered to be indicators of stuttering [2]. The frequency and types of stuttering-like and non-stuttering-like speech disfluencies produced by bilingual children can be considered as a differential diagnosis of stuttering [3]. Physiological exertions including tension during speech rendition were observed in stuttering children. Prosodic analysis of hesitations in a cross-language study reveal that disfluencies carried language-specific information and that the timbre or voice quality of vocalic hesitations varied with respect to the positions of first and second formant frequencies and pitch [4]. The stand-alone speech features-based disfluency type detection is enhanced with the subjugation of conventional machine learning [5, 6] and deep learning models [7, 8]. An end-to-end deep neural network [7] distinguishes different stutter disfluencies with a moderate accuracy while the same authors report an improved accuracy of classification in [8] using a Bi-directional Long Short Term Memory model hybridized with a residual network.

The key contributions of this article are:

(1) This investigation presents the acoustic and prosodic speech feature exploration of disfluent segments of spontaneous speech by bilingual Tamil–English (TE) speaking children. The characteristics of disfluency is inspected by observing the patterns of pitch, speech formants, duration and energy which distinguish one disfluency from another.
(2) False diagnosis of stuttering disorder in children when language impairment is the issue was comparably curtailed by introducing non-linear Support Vector Machine (SVM) model for categorizing speech segments of Children with Stuttering Disorder (CSD) from Children with Language Impairment (CLI).

Section 2 illustrates the disfluent speech corpus. Section 3 briefs on the difference between linguistic disfluency due to bilingualism and stuttering disfluency. Section 4 elaborates a detailed acoustic analysis of disfluencies. Section 5 elaborates the prosodic characterization of disfluencies in bilingual children and a machine learning based classification in Sect. 6. Section 7 presents our conclusions and future work.

2 Disfluent Speech Corpus

The disfluent speech data corpus was built by collecting speech samples from 27 bilingual children, ranging in age between 4 and 7 years of age, out of which, 14 were boys and 13 were girls. Boys pre-dominantly exhibited more disfluencies than girls in conversational speech. A signed consent was taken from the parents for each of the participants. The recordings were carried out at the comfort of the subjects' home in order to ensure that they felt at ease. After warming up, Tamil-English speaking bilingual children (speakers) were asked to perform the following tasks, namely, (1) produce a few minutes of spontaneous speech on a topic they chose from the visual stimuli. (2) Phonation of different vowels were recorded for at least 5 s each, from children for phonation analysis. The sampling rate was set to 16 kHz. The narrative speech or spontaneous speech recorded from children was effective in distinguishing Children with Stuttering Disorder (CSD) from Children with Language Impairment (CLI). The speech signals were annotated on three tiers in the Text Grid associated with each audio file using the PRAAT tool [9], and disfluency intervals were forced-aligned together with the start and end of the associated word intervals. The frequency of the subjects' individual disfluencies was observed and the percentage calculated. The proficiency of the candidate in his second language is described in Table 1.

Table 1. Proficiency of the subjects in second language

Proficiency in second language	Repetitions	Revisions	Prolongation	Blocks	Filled pauses
High	18%	8%	18%	20%	53%
Medium	23%	25%	27%	26%	45%
Low	48%	35%	38%	39%	32%

3 Linguistic Disfluency Due to Bilingualism (vs) Stuttering Disfluency

Distinguishing linguistic disfluencies from stuttering is the crucial part of assessment, particularly for early school-going children as it is essential to discriminate stuttering disorder from language impairments like the reading disorders, word finding problems, organization of speech, intelligibility of speech. Table 2 presents a quick comparison between linguistic disfluency and stuttering disfluency.

Bilingual children are usually evaluated equally in both languages as stuttering bilingual children exhibit stutters in both languages [10]. Children with Stuttering Disorder (CSD) principally exhibit disfluencies like the word-medial repetitions ("I **w-w-w-**want"), prolonged sounds – ("**Sssssssam** is good"). On the other hand, Children with Language Impairment (CLI) exhibit disfluencies as well but they show prominence over interjections – ("I ***um*** need to go home."), word repetitions with more than one syllable ("My **puppy-puppy-puppy**"), phrase repetition – ("***He is-he is*** three years old."), reviewing the words, called revision ("**He came; They came**") yet, their disfluencies are not probable to encompass prolongations, long pauses and other secondary behaviors such as physical tension.

Table 2. Linguistic disfluency (vs) stuttering disfluency

	Linguistic disfluency	Stuttering disfluency
Causes	Bi/Multilingualism	Genetics, Speech/language disorder, fear
Primary traits	Difficulties with utterance formulation or word finding, Inter-word disfluencies, Silent Pauses, Revisions Interjections	Part-word and Mono-Syllabic reiterations, Sound prolongations, Blocks (Unfilled Pause)
Secondary traits		Jaw tremors, Muscle tightening, Struggle to Speak, Disordered Breathing Patterns, Excess Anxiety, Head movements, Bodily Gyration

4 Acoustic Analysis of Disfluencies in Bilingual Children

The phonatory competence of every child was investigated by extracting acoustic features [11] presented in Table 3, related to trepidation measures such as Shimmer, Jitter, Pitch Perturbation Quotient (PPQ) and Amplitude Perturbation Quotient (APQ) using the SciPy library [12] in Python.

Amplitude Perturbation Quotient (APQ) is the mean absolute dissimilarity between the amplitude of one period of the speech signal and the mean of the signal magnitude with its four adjoining neighbors, normalized by the mean amplitude. Likewise, the Pitch Perturbation Quotient (PPQ) is the absolute assessment of the period-to-period

Table 3. Extracted acoustic features

Name of the feature	Description
d(F0), mean(d(F0), std(d(F0)), skew(d(F0)), kurtosis(d(F0)), dd(F0), mean(dd(F0), std(dd(F0)), skew(dd(F0)), kurtosis(dd(F0))	First and second derivatives of the fundamental frequency and its statistical functionals
J, mean(J), std(J), skew(J), kurtosis(J)	Jitter and its statistical functions
Sh, mean(Sh), std(Sh), skew(Sh), kurtosis(Sh)	Shimmer and its statistical functions
APQ, mean(APQ), std(APQ), skew(APQ), kurtosis(APQ)	Amplitude Perturbation Quotient and its statistical functions
PPQ, mean(PPQ), std(PPQ), skew(PPQ), kurtosis(PPQ)	Pitch Perturbation Quotient and its statistical functions
Log(E), mean(Log(E)), std(Log(E)), skew(Log(E)), kurtosis(Log(E))	Logarithmic Energy and its statistical functions
DUV	Degree of Unvoiced sounds
F1, DF1, DDF1, F2, DF2, DDF2	First and second formant frequencies and their derivatives

inconsistency of the pitch within the analyzed speech sample with a normalizing factor of 5 periods. APQ and PPQ are higher for the CSD than the CLI as shown in Fig. 1(b). Hoarse and/or breathy voices in stutterers lead to an increased PPQ. Jitter and Shimmer are phonation perturbation parameters that state the phonatory capabilities of the child. Jitter describes the temporal trepidation of the pitch while Shimmer defines the amplitude trepidation of the pitch. Significant differences between stutterers and children with language impairment were observed for jitter and shimmer measures as shown in Fig. 1(a). Thus, jitter and shimmer analysis of sustained phonation using the Scikit learn Python library [13] reveal that the stutterers exhibit less stability in neuro-muscular co-ordination and regulated respiration. Jitter and shimmer were found to be higher in children who stuttered compared to the children with language impairment.

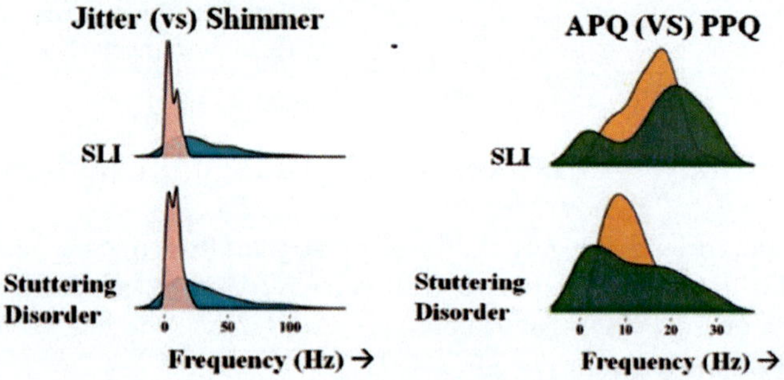

Fig. 1. Relation between (a) Jitter and Shimmer (b) APQ nd PPQ for CSD, CLI

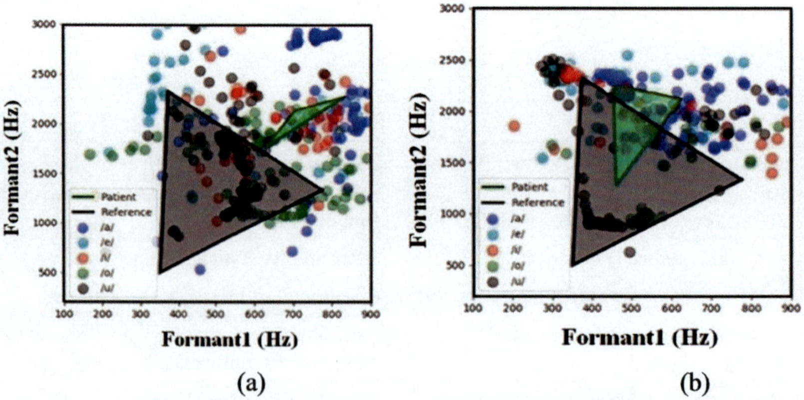

Fig. 2. Formant vocal triangle for (a) CSD (b) CLI

In Fig. 2, a reduction in the vocal triangle's area for the speech event from the CLI, is observed in contrast to the area acquired for CSD speakers that shows a drop in the articulatory proficiency of the CSD when compared with the CLI. The reference vowel triangle was fixed as in [14].

The CLI presented a better vocal triangle as compared against the CSD indicating well-defined vowel rendition with relatively constant amplitude and sustenance. The relationship between the resonant frequencies viz. Formant 1 and Formant 2 yields the characteristics of the oral cavity and the trachea's shape of the individual. Tables 4–6 summarize the average acoustic feature values measured during the task of sustained vowel analysis of the children under test.

Table 4. Sustained Tamil vowel analysis: /ஆ/

	d(F0)	Jitter	Shimmer	APQ	PPQ	Log(E)
Fluent	40	33	39	44	43	45
CLI	65	36	36	24	30	36
CSD	92	88	78	96	79	65

Table 5. Sustained English vowel analysis:/u/

	d(F0)	Jitter	Shimmer	APQ	PPQ	Log(E)
Fluent Subjects	44	32	38	43	42	44
CLI	39	27	33	38	37	39
CSD	76	64	70	75	74	55

Table 6. Sustained English vowel analysis:/e/

	d(F0)	Jitter	Shimmer	APQ	PPQ	Log(E)
Fluent	31	19	25	30	29	31
CLI	25	13	19	24	23	25
CSD	80	68	74	79	78	80

The formant frequencies or the speech spectral peaks for each vowel were spanned out more for the CSD than for the CLI speakers, demonstrating a deficit in the regulation of the tongue movement and velum positions during sustained vocalization of vowels as shown in Fig. 3.

5 Prosodic Characterization of Disfluencies in Bilingual Children

Prosody of speech expounds features such as duration, loudness, intonation and rhythm during the rendition of speech which play a vital role in intelligible speech production. In this study, prosody was analyzed in addition to the acoustical analysis with quantities

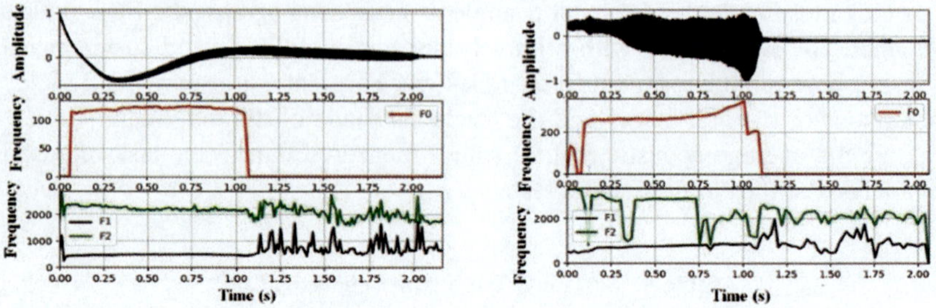

(a) (b)

Fig. 3. Formant analysis of (a) CSD (b) CLI

derived from the pitch, energy and duration of utterances. 13 static prosody features were computed from continuous speech based on duration, fundamental frequency and energy as shown in Table 7. With the objective of modelling intonation precedents of the speakers, the contour of their pitch was computed along with their statistical functionals such as the mean, standard deviation, peak values to assess the speech pattern of the speaker and the highest pitch each speaker can attain. Analogous to the pitch contour, the energy contour of speech was computed together with its statistical functionals.

Table 7. Prosodic features

Name of the feature	Description
avgF0, stdF0, VarF0, maxF0	Mean, Standard Deviation, Variance, maximum of pitch in voiced segments
avgE, stdE, maxE	Mean, Standard Deviation, Variance, maximum of energy in dB
Vrate, avgV, stdV	Rate, mean duration, Standard deviation of Voiced speech segments
Pr, avgP, stdP	Number of pauses per second, mean and standard deviation of Pauses

The average energy and pitch of a CSD is plotted in Fig. 4 (a) while the same is plotted for a CLI in Fig. 4(b). It is evident that the stuttered utterances exhibit lower spectral energy and fluctuating fundamental frequency which are evident from the absence of spectral spikes essentially at the commencement of the utterance while the utterances from CLI display steady pitch with appreciable energy. Additionally, the duration features such as the count of voiced speech segments per second measured as voiced rate were observed along with the mean and standard deviation of the voiced segment duration. The pause rate of the speech utterances from CSD and CLI were measured as the number of silence sections per second and the statistical values of the duration of mute sections were observed which revealed that the CSD exhibited high pause rate than the CLI. The peak, mean and maximum value of pitch along with statistical functionals

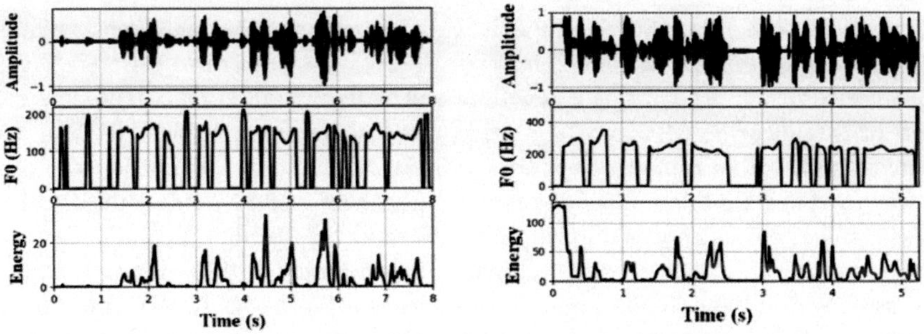

(a) (b)

Fig. 4. Prosody analysis of (a) CSD (b) CLI

calculated on the energy contour reveal the reduced spectral energy and highly varying pitch at the start of the utterance by the CSD.

6 Machine-Learning Based Classification

The Support Vector Machine (SVM) model [15] was built to verify the importance of the speech features extracted to classify CSD from CLI. The training set comprising of features $(x_1, y_1), (x_2, y_2), \ldots\ldots..(x_n, y_n)$ where input,$x \in R^N$, N being the input feature set dimension and y being the target output indicating either of the two classes viz. $y \in \{0, 1\}$: {CSD, CLI}. The sigmoid activation function was chosen as the decision function shown in Eq. (1).

$$f_{v,b} = sgn((v.x) + b) \tag{1}$$

where, (v. x) characterizes the inner product of v and x; v being the vector representing weight and 'b', the bias signal. It is imperative to obtain a decision function $f_{v,b}$ comprising of the properties shown in Eqs. (2) and (3).

$$y_i((v, x_i), b) \geq 1, i = 1, ..n \tag{2}$$

The slack variables,ε_i were presented as $\xi_i \geq 0, i = 1, \ldots n$ to get the function as shown in Eq. (3).

$$y_i((vx_i) + b) \geq 1 - \xi_i \tag{3}$$

The sample point x_i when misclassified yields an error ξ_i. The classification error which has to be diminished during SVM training is given in Eq. (4) subject to the constraints given in Eqs. (2) and (3).

$$\phi(v, \xi) = (v.v) + \mathrm{C} \sum\nolimits_{i=1}^{n} \xi_i \tag{4}$$

where, C is the parameter for regularization that regulates the balance linking the extension of the hyperplane margin and the drop in classification error.

Four SVM-based classification models were implemented independently including the linear Support Vector Machine and non- linear SVM with three different kernels namely the 'Linear', 'Radial Basis Function' and 'Polynomial' kernels. The time complexity of the linear SVM was of the order: $\boldsymbol{O}(\boldsymbol{n})$ while the non-linear SVM models rendered a higher time complexity of $\boldsymbol{O}(\boldsymbol{n}^2)$; '*n*' being the input dimension. 80% of the data was used to train the models of which 20% was utilized to validate the classifiers and the residual 20% data was utilized to test the models. The non-linear SVM with the 'Polynomial' kernel exhibited better classification accuracy of 93% during the testing phase as shown in Table 8 and Fig. 5. The non-linear SVM was observed to filch the highest run time than the other models but was found to yield the highest F1-score which is the harmonical mean of precision and recall.

Table 8. Evaluation of SVM models with $\xi = 1.0$; $C = 1$

Model	Kernel	Validation accuracy	Test accuracy	Precision	Recall	F1-score
SVM (Linear)	-	89	81	78	83	81
SVM (Non-Linear)	RBF	100	89	83	87	85
	Linear	89	81	79	77	78
	Polynomial	**100**	**93**	92	90	91

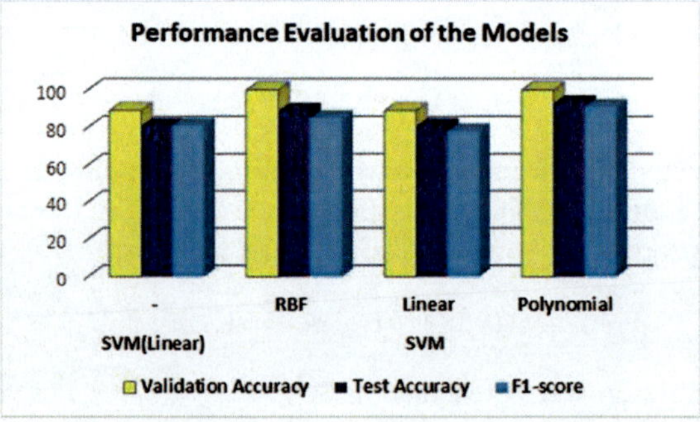

Fig. 5. Performance evaluation of the models

The SVM-based classification of the CSD and CLI classes were compared with the state-of-the-art models in the literature to detect stuttering disfluencies and to demarcate them from other common disfluent events in spontaneous speech in the recent literature as presented in Table 9.

Table 9. Comparison of SVM with State-of-the-Art models

Model	Year	Missed rate (%)	Classification accuracy (%)
Non-Linear SVM (Polynomial Kernel)	-	-	**93**
FluentNet [9]	2020	13.03	86.70
Deep residual network [10]	2020	10.03	91.15
Multi-layer perceptron neural network [8]	2018	13	82.6
Hidden markov model [8]	2018	15.2	70.0

7 Conclusion

In this paper, a detailed acoustic and prosodic investigation of the speech disfluencies pertinent to stuttering disorder and language impairment is presented to distinguish stuttering speech disorder from speech language impairment which is difficult to decipher in children below the age of 7 years. The feature-based investigation was verified by implementing Support Vector Machine-based classification with different 'kernel' functions and the 'polynomial' kernel was found best for classification between CSD and CLI with an accuracy of **93%**. Speech analysis in this study revealed unique variations in spectral energy, pitch, formants and speech rate in CSD which have been concluded as valuable speech features that aid in distinguishing natural disfluencies in bilingual children due to their lack of proficiency in second language from the disfluencies uttered by a stuttering child. These discoveries carry evidence that the spoken language theatres a significant role affecting vocal traits of bilingual children when they render speech in diverse languages. In future, the authors wish to take forward the investigation on acoustic and prosodic characterization of stutter and common disfluencies to develop a self-assessment tool that can render fluency score of an individual. Further, stuttered speech transformation and resynthesis using generative models is also focused as a prospective future work.

8 Data Availability Statement

The disfluent speech dataset and code generated during the current study is made available in the link: https://github.com/SheenaPravin/Acousto-Prosodic-Disfluent-Speech-Corpus.

Acknowledgment. This project is partly aided by AICTE, India (Research Progress Scheme - Grant Ref. No. 8–40/RIFD/RPS/Policy-1/2017–18) 15th March 2019. The Authors are the Investigators of the project.

Conflict of Interest. The authors declare no conflict of interest.

References

1. Howell, P., Davis, S.R., Williams, R.: The effects of bilingualism on stuttering during late childhood. Arch. Dis. Child. **94**(1), 42–46 (2009)
2. Howell, P., Lu, C.: Assessing risk for stuttering in children. J. Fluency Disor. **38**(1), 63–65 (2013)
3. Watson, J., Byrd, C.T., Carlo, E.: Effects of length, complexity, and grammatical correctness on stuttering in Spanish-speaking preschool children. Am. J. Speech Lang. Pathol. **20**(1), 209–220 (2011)
4. Wang, H., Van Heuven, V.J.: Acoustical analysis of English vowels produced by Chinese, Dutch and American speakers. Linguist. Neth. **23**(1), 237–248 (2006)
5. Mahesha, P., Vinod, D.S.: Automatic segmentation and classification of dysfluencies in stuttering speech. In: Proceedings of the Second International Conference on Information and Communication Technology for Competitive Strategies, pp. 1–6 (2016)
6. Pravin, S.C., Palanivelan, M.: Analysis of liguistic disfluencies in billigual children's discourse. Int. J. Cogn. Lang. Sci. **12**(6), 710–714 (2018)
7. Kourkounakis, T., Hajavi, A., Etemad, A.: Fluentnet: End-to-end detection of speech disfluency with deep learning (2020). arXiv preprint arXiv:2009.11394
8. Kourkounakis, T., Hajavi, A., Etemad, A.: Detecting multiple speech disfluencies using a deep residual network with bidirectional long short-term memory. In: Proceedings of IEEE International Conference on Prosodics, Speech and Signal Processing (ICASSP), pp. 6089–609, Barcelona, Spain (2020)
9. Boersma, P., Weenink, D.: Praat: Doing phonetics by computer [Computer program]. Version 5.3.51. https://www.praat.org/. Accessed 20 July 2020
10. Van Borsel, J., Maes, E., Foulon, S.: Stuttering and bilingualism-a review. J. Fluency Disor. **26**(1), 179–206 (2001)
11. Orozco-Arroyave, J.R., Vásquez-Correa, J.C., et al.: NeuroSpeech: an open-source software for Parkinson's speech analysis. Digit. Speech Process. **77**(1), 207–221 (2018)
12. SciPy. https://www.scipy.org/. Accessed Aug 2020
13. Pedregosa, F., Varoquaux, G., Gramfort, A., et al.: Scikit-learn: machine learning in Python. J. Mach. Learn. **12**(1), 2825–2830 (2012)
14. Shenker, R.C.: Multilingual children who stutter: clinical issues. J. Fluency Disor. **36**(3), 186–193 (2011)
15. Support Vector Machines. https://scikit-learn.org/stable/modules/svm.html. Accessed 5 Oct 2020

Real-Time Traffic Sign Recognition and Classification Using Deep Learning

Ilaiah Kavati(✉), E. Suresh Babu, and Ramalingaswamy Cheruku

National Institute of Technology Warangal, Hanamkonda 506004, India
{ilaiahkavati,esbabu,rmlswamy}@nitw.ac.in

Abstract. Traffic signs convey information, an instruction, or a warning to the driver. Recognizing these traffic signs and classifying them into their appropriate classes is an imperative task for autonomous driving assistant systems. Each country has its own traffic signs which vary much in their physical appearance, thus it is much more difficult to design classification systems that succeed. Since it is a real time problem, along with the recognition accuracy of the algorithm, its real-time performance is also much desirable. In this work, the Belgium traffic sign dataset has been used in which the traffic signs have been segmented into 62 classes. It is a subset of the European traffic sign dataset that includes traffic signs from 6 countries, namely, Germany, Belgium, France, Croatia, Netherlands, and Sweden with 162 classes. We have used the Keras library (provided by TensorFlow) to build the neural network. The model uses Softmax activation along with ReLu function. A dropout of 15% has been used to avoid over fitting. A fully connected neural network with five layers which employs Adam optimizer and cross entropy has been used in the model to train the given images. Our experimental trials, by varying different parameters, have resulted an accuracy of 91.35%.

Keywords: Traffic signs · Autonomous driving · Deep learning

1 Introduction

An important part of the road infrastructure is traffic signs. To the drivers as well as pedestrians, traffic signs provide interpretative information and credible enforcement. Based on the information perceived from the signs, the driver can then change his behavior, with respect to driving, to abide by the recommendation that the traffic sign has imposed. Since these signs give critical assessment about various conditions like the speed limit for the driver to safely go, drivable lanes, temporary situations like ongoing work, restricted areas and directions, sharp turns, accident-prone zones, school zones etc. There are presumable chances that there could be more road accidents than those that currently happens now, if there was no use of traffic signs [1].

Traffic sign recognition and classification is a demanding real-world artificial intelligence problem because of the various complicated scenarios like illumination changes, weather conditions, perspectives, fading, and human errors, etc. [2].

A. Abraham et al. (Eds.). SoCPaR 2020, AISC 1383, pp. 629–636, 2021.
https://doi.org/10.1007/978-3-030-73689-7_60

Every year, around 1.3 million people die on the roads in our modern age. If there were no road signs, the number has much more possibility of being higher. Since autonomous vehicles must also be as safe as a human driver, taking into account all the possible inputs from the road, including the traffic signs, autonomous vehicles should also be trained to observe and follow the rules enforced by these laws of the road. Therefore they must learn to recognize and understand traffic signs to help reduce the number of traffic accidents. Traffic signs are more difficult to recognize at night or in bad weather conditions, and drivers are easily affected by the headlights of incoming vehicles. These situations can lead to accidents in the traffic and serious injuries [3]. In order to avoid traffic hazards, a system of recognition is desirable to catch the attention of a driver.

The main objective of this work is to design a system which recognizes and classifies the traffic signs through images captured and recorded from the camera installed in the Automated Driver Assistance System (ADAS) by utilizing neural networks combined with data augmentation to improve the accuracy.

2 Related Work

Conventionally, standard methods of computer vision were used to recognize and classify traffic signs. Most of these techniques are based on template matching [4], Support Vector Machines (SVM) [5,6] and recently CNNs [7,8]. On the classification of traffic signs, CNN's outperformed human performance. However, their architectures differ substantially from one another. Aghdam et al. [8] proposed and designed CNN architecture and proved that it is transferable. All the previous papers till then only focused much on efficiency. This method was among the first few ones to take into account not only efficiency but also transferability. Y. Yang et al. [5] proposed a method based on Support Vector Machines to classify the detected signs into their subclasses within each superclass. G. Piccioli et al. [4] uses cross-correlation techniques in its recognition stage. Only after the sign is validated, is classified based on the database of the signs.

Apart from CNNs, random forests were used to obtain a competitive result. F. Zauklota et al. [9] proposed random forests of K-Decision trees for traffic sign recognition and classification. G.Wang et al. [10] proposed hierarchical SVMs which tries to reduce the number of parameters by 22% and achieves a result that is slightly better than MCDNN (Multi-Column Deep Neural Network). Recently, CNNs with three stages were trained on hinge loss stochastic gradient descent (HLSGD) [11]. Their CNN consists of three stages (70-110-80) consisting of 11,62,284 parameters used to train the network. However, these approaches classify signs that have already been detected, which is not practical in real applications [12].

3 Data Collection

We worked on Belgium Traffic Sign Dataset [13]. It was chosen since it is sufficiently large to train on, and also sufficiently small to be handy to work with.

The dataset consists of two parts namely, training (5000 images) and testing (1700 images). Each part has 62 classes each representing a unique traffic sign such as a stop sign, yield, speed limit, etc. Sample image of each label is shown in Fig. 1.

Fig. 1. Sample image from each label of the dataset

4 Methodology

4.1 Data Preprocessing

This stage can be divided into three parts: Image rescaling, Normalization and Data Augmentation.

Image Rescaling: The aspect ratio of the dataset images are different from one other but all are square shaped. Neural networks require images of fixed size. Typical sizes used are 28×28, 32×32, 16×16, 20×20 etc. Our dataset has images whose shapes over around 128×128. 16×16 and 20×20 are very small and may cause data loss so we resized to 32×32. This reduces the size of the network model and is easy to identify. Images in our dataset are colored. The complexity of a model will be increased due to color. Thus grayscaled images are preferred as the inherent complexity of gray level images is lower than that of color images. Hence, by converting RGB to grayscale, we have three times less of the data to process which strongly influences the training time.

Normalization: Data normalization is a very important step that ensures that every input parameter (pi el, during this case) includes a similar knowledge distribution. This makes convergence quicker during training the network. Data normalization is finished by subtracting the mean from every pixel and dividing the result by the standard deviation. The distribution of such data would look like a Gaussian curve centered at zero. For image inputs, pi el values should be positive, therefore it is preferred to scale the normalized data within the vary [0,1] or [0, 255].

Data Augmentation: Data Augmentation is a method using which more data is created from the existing data by making little changes like flips, zoom in, zoom out, rotate, and translate brightness variations. Neural networks assume that this data is different. In actuality, we might have a dataset of pictures taken in a restricted set of conditions. But, our target application might exist in an exceedingly kind of conditions, like a completely different orientation, location, scale, brightness, etc. we have a tendency to account for these things by feeding our neural network with further synthetically generated data. A histogram of a number of images in each label in our dataset is shown in Fig. 2.

We can see that few labels have a large number of images while others have very few images. Thus we require data augmentation. We opted Keras library for this. The main function is ImageDataGenerator. This takes several parameters like rescale factor, height shift range, width shift range, rotation range, horizontal flip, vertical flip, etc. Each parameter can be adjusted as per the requirements and dataset to produce synthetic data.

4.2 Network Architecture

Our neural network has an input layer, an output layer, a hidden layer, and two dropout layers (Fig. 3). All layers are fully connected i.e. each node takes input from all nodes in its previous layer. The output layer consists of 62 nodes each node representing each label. Each node in the input layer takes a pi el value thus forming $32 \times 32 \times$ number of images nodes. The hidden layer has 128 nodes. Drop out layer is used to avoid overfitting. It nullifies (makes zero) a fraction of inputs taken from the previous layer.

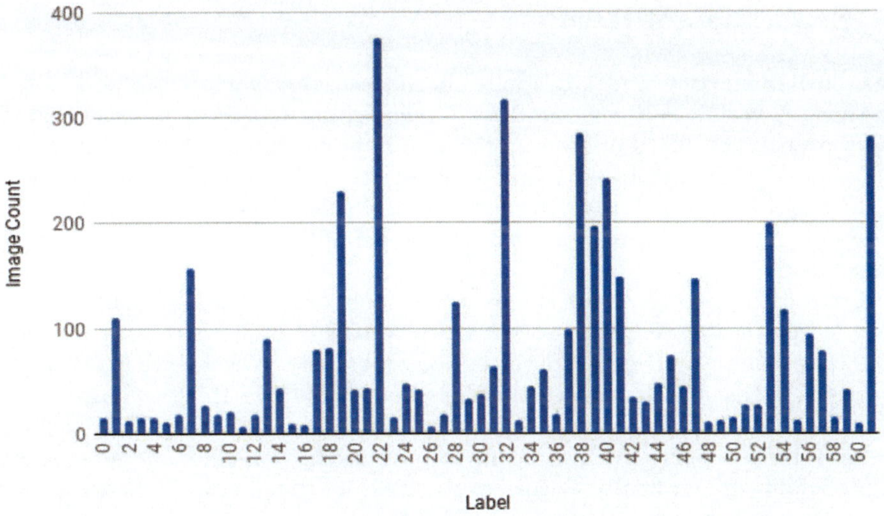

Fig. 2. Histogram showing the variation in image counts for each label

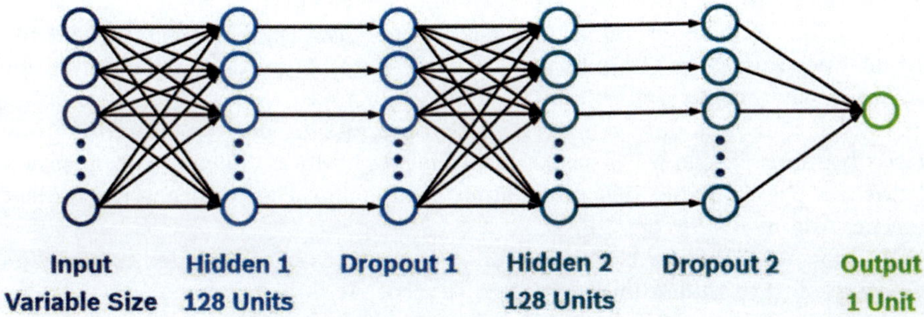

Fig. 3. Proposed neural network model

We used Tensorflow Keras library to build the neural network. It has inbuilt functions to build a neural network. We add each layer sequentially. We used ReLu function as the activation function because the inputs are pixel values and cannot be non-negative. The output of our model assigns a value [0, 1] to each class. These are not probability distribution i.e, it doesn't sum up to 1 but they denote the probability that the image belongs to respective class. To calculate accuracy, we need to convert the output into probability distribution so as to compute the class it belongs to which can be verified with its actual label. To achieve this, softmax cross entropy is used. Several optimization techniques are available among which we chose Adam Optimizer to obtain better results.

5 Experimental Results

The softmax function is mostly applied in the last layer of neural network based classifier. A vector of K real numbers is taken as input then it is standardized into a probability distribution comprising of K probabilities. Besides, the larger probabilities occur due to large input values. For mapping the normalized yield of a network to a probability distribution over predicted output classes in neural networks, generally, softmax is used. Experimentally, K (number of classes) is chosen as 62.

Cross-entropy is the most common loss function for classification tasks. It is a measure of the difference between two vectors of probabilities. So we need to convert labels and logits to vectors of probability. The function takes the generated logits and ground truth labels and does three things: converts the label indexes of shape [None] to logits of shape [None, 62] (one-hot encoded vectors), then it runs softmax to convert both prediction logits and label logits to probabilities, and finally calculates the cross-entropy between the two. This generates loss value.

Optimization algorithm allows us to maximize or minimize an Error function, $E(\ )$ which is a mathematical function reliant on the internal learnable parameters of the model, that are necessary for calculating the output values (Y) from the set of input values (X) used in the network. The Bias (B) and the Weight (W) values are the internal learnable parameters that are required for evaluating the output values and to get an optimized solution i.e., decreasing the Loss through training process of network and thus play an important role in the training procedure of the Neural Network Model. Several optimization algorithms are available like mini batch gradient descent, stochastic gradient descent, Adam, RMSProp [14].

We have used the ADAM optimizer because it has been revealed to converge at a quicker rate than simple gradient descent. It is an extension to stochastic gradient descent which has seen a greater implementation for deep learning applications in recent past in computer vision and natural language processing. Adam is nothing like classical stochastic gradient descent. Stochastic gradient descent maintains a single learning rate for all weight updates which does not change during training. In the training loop, we iteratively train the model to minimize the loss function. We start the training loop and run the train operation repeatedly. We run the loss operation as well to print its values and monitor the progress of the training. This is implemented through one of the arguments while compiling the Keras network model. Sample predictions of the proposed model is shown in Fig. 4. It can be observed that, our model performs accurately in most of the cases (represented with green color labels in Fig. 4). The best accuracy achieved is 91.35%.

Fig. 4. Truth v/s prediction outputs of the approach

6 Conclusions

We covered how deep learning can be used to classify traffic signs, employing a variety of pre-processing and regularization techniques (e.g. dropout), and trying different model architectures. We presented a network architecture using Keras library with Data Augmentation. This architecture reached till 91.35% accuracy using 32 * 32 grayscale data using a five layered fully connected neural network. Furthermore, we found that some traffic signs are miss-recognized when we apply this method in the unmanned ground vehicle. Future work can focus on expanding the data set by seeking out more images of traffic signs, especially the images at night. Then, the speed can be accelerated by optimizing the algorithm for real-time application in vehicles.

References

1. Liu, C., Li, S., Chang, F., Wang, Y.: Machine vision based traffic sign detection methods: review, analyses and perspectives. IEEE Access **7**, 86578–86596 (2019)
2. Zhou, S., Deng, C., Piao, Z., Zhao, B.: Few-shot traffic sign recognition with clustering inductive bias and random neural network. Pattern Recogn. **100**, 107160 (2020)
3. Liang, Z., Shao, J., Zhang, D., Gao, L.: Traffic sign detection and recognition based on pyramidal convolutional networks. Neural Comput. Appl. **32**, 1–11 (2019)
4. Piccioli, G., De Micheli, E., Parodi, P., Campani, M.: Robust method for road sign detection and recognition. Image Vision Comput. **14**(3), 209–223 (1996)
5. Yang, Y., Luo, H., Xu, H., Wu, F.: Towards real-time traffic sign detection and classification. IEEE Trans. Intell. Transp. Syst. **17**(7), 2022–2031 (2015)
6. Bascón, S.M., Rodríguez, J.A., Arroyo, S.L., Caballero, A.F., López-Ferreras, F.: An optimization on pictogram identification for the road-sign recognition task using svms. Comput. Vision Image Underst. **114**(3), 373–383 (2010)
7. Stallkamp, J., Schlipsing, M., Salmen, J., Igel, C.: Man vs computer: benchmarking machine learning algorithms for traffic sign recognition. Neural Netw. **32**, 323–332 (2012)
8. Aghdam, H.H., Heravi, E.J., Puig, D.: A practical and highly optimized convolutional neural network for classifying traffic signs in real-time. Int. J. Comput. Vision **122**(2), 246–269 (2017)
9. Zaklouta, F., Stanciulescu, B., Hamdoun, O.: Traffic sign classification using kd trees and random forests. In: The 2011 International Joint Conference on Neural Networks, pp. 2151–2155. IEEE, 2011
10. Wang, G., Ren, G., Wu, Z., Zhao, Y., Jiang, L.: A hierarchical method for traffic sign classification with support vector machines. In: The 2013 International Joint Conference on Neural Networks (IJCNN), pp. 1–6. IEEE (2013)
11. Jin, J., Fu, K., Zhang, C.: Traffic sign recognition with hinge loss trained convolutional neural networks. IEEE Trans. Intell. Transp. Syst. **15**(5), 1991–2000 (2014)
12. Zhu, Z., Liang, D., Zhang, S., Huang, X., Li, B., Hu, S.: Traffic-sign detection and classification in the wild. In: Proceedings of the IEEE Conference on Computer Vision and Pattern Recognition, pp. 2110–2118 (2016)
13. Mathias, M., Timofte, R., Benenson, R., Van Gool, L.: Traffic sign recognition—how far are we from the solution? In: The 2013 International Joint Conference on Neural networks (IJCNN), pp. 1–8. IEEE (2013)
14. Ayachi, R., Afif, M., Said, Y., Atri, M.: Traffic signs detection for real-world application of an advanced driving assisting system using deep learning. Neural Process. Lett. **51**(1), 837–851 (2020)

Merits of Mixotrophic Cultivation for Microalgal Biomass Production

Geetanjali Udawat[1], Niketa Gandhi[2], and Sanjay Deshmukh[2](✉)

[1] Department of Environmental Sciences Ratnagiri Sub-Centre, University of Mumbai, Ratnagiri, India
[2] Department of Life Sciences, University of Mumbai, Mumbai, India
niketa@ieee.org, sanjaydeshmukh@mu.ac.in

Abstract. Renewable sources of energy and chemicals are a viable solution in current situation where world is facing extreme energy crisis. Achieving carbon neutral options for energy supply and high productive options for bulk chemicals by renewable source. Microalgal Biomass is a promising option and has the potential to provide renewable energy and products for future grown in minimum inputs and gives higher output. In this review paper three cultivation techniques are discussed such as autotrophic, heterotrophic, and mixotrophic. Mixotrophic cultivation involves combination of both auto and heterotrophic modes where both light reaction and dark reaction are combined for maximum results. Cost effective options of bioenergy and valuable co products are efficiently produced from microalgae in mixotrophic cultivation conditions. Bioprospection of microalgal options for bioenergy and cultivating those strains in mixotrophic conditions combining with wastewater treatment and CO_2 fixation from atmosphere to produce biofuel and valuable co products in the main theme of this review. Concept of biorefinery is prescribed for future options.

Keywords: Mixotrophic cultivation · Bioprospection · Bioenergy

1 Introduction

Population growth, high paced growth of manufacturing industries, over demand of non-renewable sources of energy are crucial reasons of growing global warming and energy crisis. In last decades industries with scientific community started identifying new renewable energy options which can help future generations to fulfil their energy demand [1, 2]. But there is a great challenge in replacing non renewable sources of energy and bulk chemicals produced by these sources to renewable fuels and bioactive compounds [2–4]. Bioenergy production from raw material has major concerns like accessibility, cost effective production and processing for future use [1, 3, 5, 6]. Microalgae is third generation bioenergy resource, and it provides high growth in low area and higher output and converted into biofuels and bioactive compounds [1, 7, 8]. It can grow on land, water, and extreme harsh conditions too [1, 4]. These are unicellular organism which grow rapidly using available carbon source by photosynthesis and converts it into oxygen and biomass. They require less water and nutrients [4].

A. Abraham et al. (Eds.): SoCPaR 2020, AISC 1383, pp. 637–647, 2021.
https://doi.org/10.1007/978-3-030-73689-7_61

Microalgae can grow in wastewater and uses its nutrients and makes it useable. Since they use carbon source to grow, help in sequester the extra carbon di oxide emitted by industries and other man-made processes [9–13].

Microalgae is sustainable feedstock for biofuels, bioactive compounds and biorefining. Algal technology can grow by undertaking biochemical pathways of different species under various conditions to achieve best and efficient production of biomass and by-products [5, 14–21].

2 Microalgae as Biomass

Microalgae are abundant unicellular autotrophic organism found in water bodies such as freshwater, seawater, hypersaline and considerable quantities in desert and cold arctic area. There are two kinds of microalgae, eukaryotic and prokaryotic. Due to simple structure and photosynthetic mechanism they are efficient in consuming carbon di oxide and produce high quality biomass [6, 9]. Many have tendency to switch from photo auto trophic to heterotrophic mode and becomes feedstocks with high nutrient content and high lipid content [21, 23].

Screening strains depends on the area of production, output and its doubling rate. Genetically engineered mechanisms can target production of biofuel and bioactive compounds [23, 24].

3 Microalgae Cultivation

Biomass cultivation is an important step in upstream process and is crucial for maximum production and high-quality biomass [25]. This involves three process depending on culture system and they are batch, semi batch and continuous. The bulk production depends on factors like physical (light temperature, pH, salinity, carbon di oxide, oxygen, stress etc.) and biological. Cultivation methods involve four types [2, 7, 26].

3.1 Autotrophic Cultivation

Autotrophic cultivation is primitive method. Microalgae are high in photosynthetic rate and growth than green plants. Open ponds and enclosed photobioreactors are two methods (systems under natural ponds, lagoons, or artificial ponds or photobioreactors which are closed containers). Carbon di oxide requirement is completed by air on surface or artificial submerged aerators [27, 28, 31].

3.1.1. **Open pond system** –Maximum production and simplest method for microalgae is open system. Size, shape, and type of agitators determine types e.g. Raceway ponds (paddle wheel, extensive shallow, circular ponds, cascade system with thin layer slope). Low energy inputs make this system very efficient, easily scalable. Disadvantages lies when microalgae is prone to contamination from air and ground [29–31,33].

3.1.2. **Closed photobioreactor system** –Plate photobioreactor, tubular photobioreactor, vertical photobioreactor, main photobioreactor are made with transparent materials. Optimum supply of growth requirement such as light, nutrient and carbon di oxide speeds up growth. Column Photobioreactor are very efficient because of gas transfer ratios, controlled growth, and agitation.

3.1.3. **Flat panel photobioreactor** are thin layer of dense culture with clear panel where maximum growth occurs. These produce high biomass with less contamination, and growth can be controlled. But because of high cost in overall process compensate high growth production. In some proposed system Photobioreactor and open pond system are combined. Seed culture is grown in Photobioreactor and scaled up to open ponds by transfer [30–38].

3.2 Heterotrophic Cultivation

Under autotrophic cultivation high cell density and photoinhibition leads to slow growth. Introduction of heterotrophic cultivation where microalgae utilize carbon source is a good option to maximize growth. This system does not need light and surface to volume ratio is also nonimportant. E coli contamination is a concern. Heterotrophic growth has many advantages like high growth rate, high biomass, high lipid content and wastewater reuse [37–39].

3.3 Mixotrophic Cultivation

Autotrophic and heterotrophic cultivation when occurs together in some microalgae is called mixotrophic cultivation. Here microalgae are not light dependent or carbon source dependent e.g. *Spirulina, Chlamydomonas.* They produce biomass either by photosynthesis or undergo aerobic respiration where they depend on carbon source is always media dependent.

Light Reaction: In this reaction light is captured and energy saved in form of NADPH and ATP. This absorption is done by light harvesting photosystem or pigment protein complexes which result in activation of cells by positive and negative charge distribution.

Dark Reaction: Calvin cycle where ATP and NADPH help in fixation of carbon di oxide by RUBISCO where 3- phospho- glyceraldehyde reduce to Glycero-3 – phosphate by NADPH reduction and cycles to it again to ribulose1-5, biphosphate. Carbon metabolism is species specific in microalgae and also depends on carbon source if pathway is mixotrophic, C-6 sugars are consumed through glycolysis and C5 by pentose phosphate pathway further these metabolites moves to TCA cycle for energy and converted to amino acids, carotenoids and fatty acids.

Mixotrophic cultivation is advantageous because here limitations of photosynthesis are bypassed, and carbon is utilized, and channels carbon dioxide produced by respiration and oxygen produced by photosynthesis are involved in building cell components. Glycerol, malonate, ribose has showed positive effect on biomass, doubling rate and production volumetrically, enhanced lipids and carbohydrates. Organic carbon eliminated photoinhibition of growth at high light intensity [40–46] (Fig. 1 and Table 1).

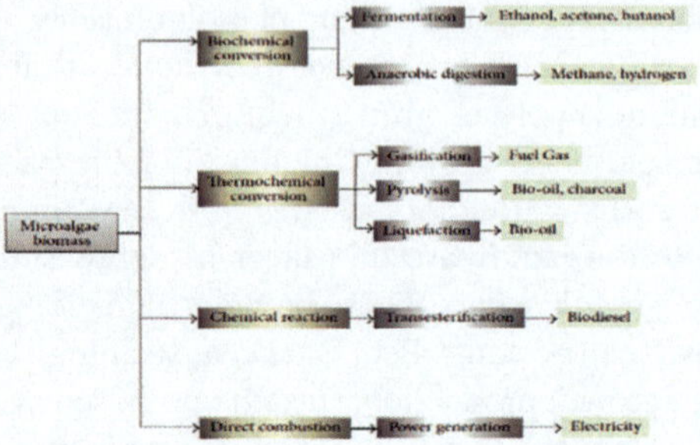

Fig. 1. Diagram of biomass conversion technology

Table 1. Overall difference between three cultivation modes.

Cultivation mode	Carbon source	Energy supply	Advantages	Disadvantages
Autotrophic	CO_2	Light	1. Low cost 2. High cost of pigmentation and phytochemicals	1. Low growth rate, low biomass and lipid accumulation 2. Need special bioreactors or a higher dependence of weather conditions
Heterotrophic	Organic carbon source (Glucose, glycerol, acetate etc.)	Organic carbon source	1. Higher growth rate, higher biomass and lipid accumulation 2. Bioreactors with little constraints	1. Higher cost 2. Need for sterile medium and east to be contaminated 3. Production of CO_2 4. Dark conditions diminish the pigmentation and production of phytochemicals
Mixotrophic	$CO_2$and organic carbon source	Light and organic carbon source	1. Higher growth rate, higher biomass and lipid accumulation 2. Sustain of pigmentation and phytochemicals production 3. Decreased production of CO_2	1. Higher cost 2. Need for sterile medium and easy to be contaminated 3. Reduced conversion efficiency

4 Uses of Microalgae Biomass Produced by Mixotrophic Cultivation

Many microalgae species such as *Chlorella vulgaris, Nannochloropsis, Neochlorisoleobundens* showed higher growth of biomass under mixotrophic condition. As increased cell growth biomass is used in various forms.

4.1 Bioenergy

Higher biomass harvested is converted to bioenergy. Bioenergy from biofuels is categorized in many forms.

4.1.1 Biofuel

Mixotrophic cultivation of algae increases lipid accumulation which are dominant molecules stored by reduction of carbon after photosynthesis or metabolism under stress. Lipid accumulation is much high in microalgae compared to oil producing plants Ch*lorella* species has shown positive results in lipid accumulation when grown in mixotrophic condition (Fig. 2).

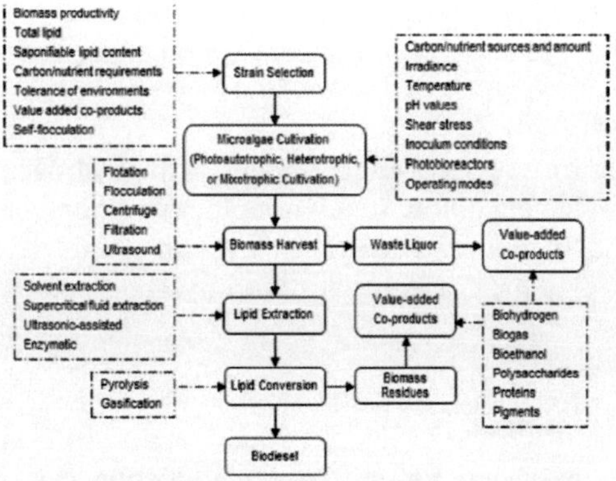

Fig. 2. Biodiesel production from microalgae

Lipid production or conversion is dependent on carbon source comparison is s shown in Table 2. Glucose fed in mixotrophic condition does not affect the lipid content, but it has more unsaturated fatty acid. Other characteristics like seter content, cetane number, iodine value, degree of unsaturation and cold filter plugging point from biofuel produced by *Chlorella vulgaris* met the standard requirement (European standard EN 14214, US Standard ASTM 6751, Australian Standard and Brazilian Standard Petroleum ANP -255) [54–57].

4.1.1.1 Bioethanol

In mixotrophic condition there is increase carbohydrate content or reducing sugar content

Table 2. Lipid content of different algal strains under mixotrophic condition with various organic carbon sources.

Microalgae	Carbon species	Lipid content (% dry weight)	Reference
Chlorella sp.	Technical glycerol	15.9	28
Chlorella sp. Y8-1	Sucrose	35.5	52
Chlorella sorokiniana CCTCC M209220	Glucose	51.0	28
Chlorella vulgaris	Glucose	13.8	60
Chlorella vulgaris	Glycerol	40.1	64
Chlorella vulgaris	Glucose-glycerol (4:6)	26.0	64
Chlorella vulgaris	Acetate	14.2	56
Desmodesmus sp. PCH19	Glycerol	28.2	29
Chlamydomonas sp. BTA 9032	Cheese whey permeate	38.2	56
Haematococcus sp.	Technical glycerol	16.2	53
Marine Chlorella sp.	Glucose	25.4	52
Nannochloris sp.	Technical glycerol	16.2	53
Nannochloropsis sp.	Glucose	25.3	62
Scenedesmus sp.	Technical glycerol	15.1	61
Scenedesmus obliquus	Glucose	11.8	33

in microalgae. In microalgae cellulose structure and high carbohydrate is converted to bioethanol which remain after lipid extraction. Their cell wall have inner and outer layer. Outer layer contains pectin, agar and alginate and inner layer contain cellulose, hemicellulose etc. which can be easily fermented to bioethanol [56–62].

4.1.1.2 Biohydrogen

Chlamydomansreinhardtii produces high amount of biohydrogen under mixotrophic condition than autotrophic condition. In mixotrophic condition, dark fermentation with mixotrophic condition it produced hydrogen by two phase incubation one is dark fermentation and other is light reaction leads to hydrogen production and evolution [57–60].

4.1.2 Wastewater Treatment

Mixotrophic cultivation system for bioenergy production cost of substrate can be decreased by using waste material. Mixotrophic cultivation system technology is less studied in wastewater because of contamination issue and high cost. Some microalgae showed maximum growth in sewage as carbon source and co culture of microalgae and bacteria has removed DOC in extremely high percentage from wastewater [63–73].

Solid waste with high carbohydrate and less nitrogen can increase biohydrogen production. Combining CO_2 fixation from flue gas and nutrient removal from wastewater is a good study to understand carbon capture strategies [74, 75].

4.1.3 Value Added Coproducts

High value products of nutraceutical, pharmaceutical and cosmetic industry can be produced from microalgal biomass. Value products like lutein, astaxanthin (antioxidant against UV radiation), beta carotene (antioxidant, Vitamin A supplement, Anticancer, antiaging properties), zeaxanthin (cosmetic and food industry), DHA, EPA,

Lycopene(antioxidant, anticarcinogenic), phycobiliprotein(dyes, fluorescent agents). By exploiting mixotrophic cultivation this production of valuable compounds is a future prospect [77–83].

5 Biorefining

The process of conversion of biomass to bioenergy and value-added products combined which is more economical for future. This setup can be used if we focus on biomass production, good strain with best and useful biochemical combination and value product extraction [76].

6 Conclusion

Cultivation of microalgae in mixotrophic condition can give a cost effective alternative for production of sustainable and renewable energy. Integration of biorefining, exploitation of wastewater, carbon di oxide utilization can reduce the carbon source cost and results high biomass production.It is highly advantageous over autotrophic and heterotrophic for high biomass, high lipid, and less dependence on light.

So in future Optimization and comprehensive study of growth factors will lead to increase production and further Strain selection, carbon di oxide mitigation, scaling up and valuable product extraction can increase feasibility of microalgae biomass growth by mixotrophy.

References

1. Maiti, S., Maiti, D.C., Verma, M., Brar, S.K.: Biobutanol—"a renewable green alternative of liquid fuel" from algae. In: Soccol, C.R., Brar, S.K., Faulds, C., Ramos, L.P. (eds.) Green Fuels Technology, pp. 445–465. Springer, Cham (2016)
2. Feng, Q., Lin, Y.: Integrated processes of anaerobic digestion and pyrolysis for higher bioenergy recovery from lingo cellulosic biomass: a brief review. Renew. Sustain. Energy Rev. **77**, 1272–1287 (2017)
3. Lyman, R.: Why renewable energy cannot replace fossil fuels by 2050. Energy (2016). https://www.mysearch.org.uk/website2/pdf/2016-Lyman.pdf
4. Goh, B.H.H., Ong, H.C., Cheah, M.Y., Chen, W.-H., Yu, K.L., Mahlia, T.M.I.: Sustainability of direct biodiesel synthesis from microalgae biomass: a critical review. Renew. Sustain. Energy Rev. **107**, 59–74 (2019)
5. Ilic, D., Williams, K., Farnish, R., Webb, E., Liu, G.: On the challenges facing the handling of solid biomass feedstocks: on the challenges facing the handling of solid biomass feedstocks. Biofuels Bioprod. Biorefin. **12**(2), 187–202 (2018). https://doi.org/10.1002/bbb.1851
6. Ullah, K., Ahmad, M., Sharma, V.K., Lu, P., Harvey, A., Zafar, M., et al.: Assessing the potential of algal biomass opportunities for bioenergy industry: a review. Fuel **143**, 414–423 (2015)
7. Thrän, D., Billig, E., Brosowski, A., Klemm, M., Seitz, S.B., Witt, J.: Bioenergy carriers-from smoothly treated biomass towards solid and gaseous biofuels. Chem. Ing. Tech. **90**, 68–84 (2018)

8. Alaswad, A., Dassisti, M., Prescott, T., Olabi, A.: Technologies and developments of third generation biofuel production. Renew. Sustain. Energy Rev. **51**, 1446–1460 (2015)
9. Aslan, S., Kapdan, I.K.: Batch kinetics of nitrogen and phosphorus removal from synthetic waste water by algae. Ecol. Eng. **28**, 64–70 (2006)
10. Markou, G., Georgakakis, D.: Cultivation of filamentous cyanobacteria (blue-green algae) in agro-industrial wastes and wastewaters: a review. Appl. Energy **88**, 3389–3401 (2011)
11. Voltolina, D., Gómez-Villa, H., Correa, G.: Nitrogen removal and recycling by Scenedesmus obliquus in semicontinuous cultures using artificial wastewater and a simulated light and temperature cycle. Bioresour. Technol. **96**, 359–362 (2005)
12. Chisti, Y.: Biodiesel from microalgae. Biotechnol. Adv. **25**, 294–306 (2007)
13. Shi, J., Podola, B., Melkonian, M.: Removal of nitrogen and phosphorus from waste water using microalgae immobilized on twin layers: an experimental study. J. Appl. Phycol. **19**, 417–423 (2007)
14. Abinandan, S., Shanthakumar, S.: Challenges and opportunities in application of microalgae (Chlorophyta) for wastewater treatment: a review. Renew. Sust. Energ. Rev. **52**, 123–132 (2015)
15. Karthikeyan, S.: A critical review: microalgae as a renewable source for biofuel production. Int. J. Eng. Res. Technol. **1**, 1–6 (2012)
16. Li, Y., Horsman, M., Wu, N., Lan, C.Q., Dubois-Calero, N.: Biofuels from microalgae. Biotechnol. Prog. **24**(4), 815–820 (2008)
17. Wang, B., Li, Y., Wu, N., Lan, C.Q.: CO_2 bio-mitigation using microalgae. Appl. Microbiol. Biotechnol. **79**(5), 707–718 (2008)
18. Acién Fernández, F.G., Fernández Sevilla, J.M., Molina Grima, E.: Costs analysis of microalgae production, 2nd edn. In: Pandey, A., Chang, J.-S., Soccol, C.R., Lee, D.-J., Chisti, Y. (eds.) Biofuels from Algae, pp. 551–566. Elsevier (2019). https://doi.org/10.1016/B978-0-444-64192-2.00021-4
19. Ahmed, A.B.A., Adel, M., Karimi, P., Peidayesh, M.: Pharmaceutical, cosmeceutical, and traditional applications of marine carbohydrates, chap. 10. In: Kim, S.-K. (ed.) Advances in Food and Nutrition Research, Volume 73 of Marine Carbohydrates: Fundamentals and Applications, Part B, pp.197–220. Academic Press (2014)
20. Arad, S.M., Yaron, A.: Natural pigments from red microalgae for use in foods and cosmetics. Trends Food Sci. Technol. **3**, 92–97 (1992). https://doi.org/10.1016/09242244(92)90145-M. JanISSN0924-2244
21. Adarme-Vega, T.C., Lim, D.K.Y., Timmins, M., Vernen, F., Li, Y., Schenk, P.M.: Microalgal biofactories: a promising approach towards sustainable omega-3 fatty acid production. Microb. Cell Fact. **11**, 96 (2012). https://doi.org/10.1186/1475-2859-11-96
22. Becker, E.W.: Micro-algae as a source of protein. Biotechnol. Adv. **250**(2), 207–210 (2007). https://doi.org/10.1016/j.biotechadv.2006.11.002
23. Liang, Y., Sarkany, N., Cui, Y.: Biomass and lipid productivities of Chlorella vulgaris under autotrophic, heterotrophic and mixotrophic growth conditions. Biotech. Lett. **31**(7), 1043–1049 (2009)
24. Lum, K.K., Kim, J., Lei, X.: Dual potential of microalgae as a sustainable biofuel feedstock and animal feed. J. Anim. Sci. Biotechnol. **4**(1), 53 (2013)
25. Andrade, M.R., Costa, J.A.V.: Mixotrophic cultivation of microalga Spirulina platens is using molasses as organic substrate. Aquaculture **264**, 130–134 (2007)
26. Bharathiraja, B., Jayamuthunagai, J., Chakravarthy, M., Ranjith Kumar, R., Yogendran, D., Praveenkumar, R.: Algae: promising future feedstock for biofuels. In: Singh, B., Bauddh, K., Bux, F. (eds.) Algae and Environmental Sustainability, pp. 1–8. Springer India, New Delhi (2015). https://doi.org/10.1007/978-81-322-2641-3_1
27. Borowitzka, M.A.: Culturing microalgae in outdoor ponds. In: Anderson, R.A. (ed.) Algal Culturing Techniques, pp. 205–218. Elsevier, London, UK (2005)

28. Borowitzka, M.A.: Commercial production of microalgae: ponds, tanks, tubes and fermenters. J. Biotechnol. **70**(1–3), 313–321 (1999)
29. Brennan, L., Owende, P.: Biofuels from microalgae-a review of technologies for production, processing, and extractions of biofuels and co-products. Ren. Sus. Energy Rev. **14**, 557–577 (2010)
30. Carvalho, A.P., Meireles, L.A., Malcata, F.X.: Microalgal reactors: a review of enclosed system designs and performances. Biotechnol. Prog. **22**, 1490–1506 (2006)
31. Bhatnagar, A., Chinnasamy, S., Singh, M., Das, K.C.: Renewable biomass production by mixotrophic algae in the presence of various carbon sources and wastewaters. Appl. Energy **88**, 3425–3431 (2011)
32. Morita, M., Watanabe, Y., Saiki, H.: Photosynthetic productivity of conical helical tubular photobioreactor incorporating Chlorella sorokiniana under field conditions. Biotechnol. Bioeng. **77**, 155–162 (2002)
33. Hall, D.O.; Fernández, F.A.; Guerrero, E.C.; Rao, K.K.; Grima, E.M. Outdoor helical tubular photobioreactors for microalgal production: modeling of fluid-dynamics and mass transfer and assessment of biomass productivity. Biotechnol. Bioeng. **82**, 62–73 (2003). [PubMed]
34. Fernandez, F., Camacho, F.G., Perez, J., Sevilla, J., Grima, E.M.: Modeling of biomass productivity in tubular photobioreactors for microalgal cultures: effects of dilution rate, tube diameter, and solar irradiance. Biotechnol. Bioeng. **58**, 605–616 (1998)
35. Rodolfi, L., Zittelli, G.C., Bassi, N., Padovani, G., Biondi, N., Bonini, G., Tredici, M.R.: Microalgae for oil: strain selection, induction of lipid synthesis and outdoor mass cultivation in a low-cost photobioreactor. Biotechnol. Bioeng. **102**(1), 100–112 (2009). https://doi.org/10.1002/bit.22033
36. Perez-Garcia, O., Escalante, F.M.E., Luz, E., de-Bashan, Yoav Bashan, : Heterotrophic cultures of microalgae: metabolism and potential products. Water Res. **45**(1), 11–36 (2011). https://doi.org/10.1016/j.watres.2010.08.037
37. Mohammad Mirzaie, M.A., Kalbasi, M., Mousavi, S.M., Ghobadian, B.: Investigation of mixotrophic, heterotrophic, and autotrophic growth of Chlorella vulgaris under agricultural waste medium. Prep BiochemBiotechnol **46**, 150–156 (2016)
38. Bell, G.: Experimental evolution of heterotrophy in a green alga. Evolution **67**, 468–476 (2013)
39. Andruleviciute, V., Makareviciene, V., Skorupskaite, V., Gumbyte, M.: Biomass and oil content of Chlorella sp., Haematococcus sp., Nannochloris sp. and Scenedesmus sp. under mixotrophic growth conditions in the presence of technical glycerol. J. Appl. Phycol. **26**(1), 83–90 (2014). https://doi.org/10.1007/s10811-013-0048-x
40. Orosa, M., Franqueira, D., Cid, A., Abalde, J.: Carotenoid accumulation in Haematococcus pluvialis in mixotrophic growth. Biotechnol. Lett. **23**, 373–378 (2001)
41. Sun, H., Kong, Q., Geng, Z., Duan, L., Yang, M., Guan, B.: Enhancement of cell biomass and cell activity of astaxanthin-rich Haematococcus pluvialis. Bioresour. Technol. **186**, 67–73 (2015). https://doi.org/10.1016/j.biortech.2015.02.101
42. Kong, W.B., Yang, H., Cao, Y.T., Song, H., Hua, S.F., Xia, C.G.: Effect of glycerol and glucose on the enhancement of biomass, lipid and soluble carbohydrate production by Chlorella vulgaris in mixotrophic culture. Food Technol. Biotechnol. **51**, 62–99 (2013)
43. Wan, M., Liu, P., Xia, J., Rosenberg, J.N., Oyler, G.A., Betenbaugh, M.J., Nie, Z., Qiu, G.: The effect of mixotrophy on microalgal growth, lipid content, and expression levels of three pathway genes in Chlorella sorokiniana. Appl. Microbiol. Biotechnol. **91**(3), 835–844 (2011)
44. Cheirsilp, B., Torpee, S.: Enhanced growth and lipid production of microalgae under mixotrophic culture condition: effect of light intensity, glucose concentration and fed-batch cultivation. Biores. Technol. **110**, 510–516 (2012). https://doi.org/10.1016/j.biortech.2012.01.125

45. Müller, P., Li, X.-P., Niyogi, K.K.: Non-photochemical quenching. a response to excess light energy. Plant Physiol. **125**(4), 1558–1566 (2001)
46. Srirangan, K., Pyne, M.E., Perry Chou, C.: Biochemical and genetic engineering strategies to enhance hydrogen production in photosynthetic algae and cyanobacteria. Bioresour. Technol. **102**(18), 8589–8604 (2011)
47. Hervás, M., Navarro, J.A., De la Rosa, M.A.: Electron transfer between membrane complexes and soluble proteins in photosynthesis. Acc. Chem. Res. **36**, 798–805 (2003)
48. Buchanan, B.B.: The carbon (formerly dark) reactions of photosynthesis. Photosynth. Res. **128**, 215–217 (2016)
49. Chae, S.R., Hwang, E.J., Shin, H.S.: Single cell protein production of Euglena gracilis and carbon dioxide fixation in an innovative photo-bioreactor. Bioresour. Technol. **97**(2), 322–329 (2006)
50. Ren, H., Tuo, J., Addy, M.M., Zhang, R., Qian, L., Anderson, E., Chen, P., Ruan, R.: Cultivation of Chlorella vulgaris in a pilot-scale photobioreactor using real centrate wastewater with waste glycerol for improving microalgae biomass production and wastewater nutrients removal. Bioresour. Technol. **245**, 1130–1138 (2017)
51. Mostafa, M., EL-Sheekh, Mohamed Y Bedaiwy, Mohamed E Osman, Mona M Ismail, : Mixotrophic and heterotrophic growth of some microalgae using extract of fungal-treated wheat bran. Int. J. Recycl. Org. Waste Agric. **1**(1), 12 (2012)
52. Li, X., Li, W., Zhai, J., Wei, H.: Effect of nitrogen limitation on biochemical composition and photosynthetic performance for fed-batch mixotrophic cultivation of microalga Spirulina platens is. Bioresour. Technol. **263**, 555–561 (2018)
53. Sharma, Y., Singh, B.: Development of biodiesel: current scenario. Renew. Sustain. Energy Rev. **13**, 1646–1651 (2009)
54. Zhu, L.: Microalgal culture strategies for biofuel production: a review. Biofuels Bioprod. Biorefin. **9**, 801–814 (2015)
55. Gouveia, L., Oliveira, A.C.: Microalgae as a raw material for biofuels production. J. Ind. Microbiol. Biotechnol. **36**, 269–274 (2009)
56. Zhu, L.D., Hiltunen, E., Antila, E., Zhong, J.J., Yuan, Z.H., Wang, Z.M.: Microalgal biofuels: Flexible bioenergies for sustainable development. Renew. Sustain. Energy Rev. **30**, 1035–1046 (2014)
57. Chen, C.Y., Chang, H.Y., Chang, J.S.: Producing carbohydrate-rich microalgal biomass grown under mixotrophic conditions as feedstock for biohydrogen production. Int. J. Hydrog. Energy **41**, 4413–4420 (2016)
58. Leite, G.B., Abdelaziz, A.E., Hallenbeck, P.C.: Algal biofuels: challenges and opportunities. Bioresour. Technol. **145**, 134–141 (2013)
59. Chen, J., Li, J., Dong, W., Zhang, X., Tyagi, R.D., Drogui, P., et al.: The potential of microalgaein biodiesel production. Renew. Sustain. Energy Rev. **90**, 336–346 (2018)
60. Pittman, J.K., Dean, A.P., Osundeko, O.: The potential of sustainable algal biofuel production using wastewater resources. Bioresour. Technol. **102**(1), 17–25 (2011)
61. Li, Y., Horsman, M., Wang, B., Nan, W., Lan, C.Q.: Effects of nitrogen sources on cell growth and lipid accumulation of green alga Neochloris oleoabundans. Appl. Microbiol. Biotechnol. **81**(4), 629–636 (2008)
62. Castro, Y.A., Ellis, J.T., Miller, C.D., Sims, R.C.: Optimization of wastewater microalgae saccharification using dilute acid hydrolysis for acetone, butanol, and ethanol fermentation. Appl. Energy **140**, 14–19 (2015)
63. Cheng, H.H., Whang, L.M., Chan, K.C., Chung, M.C., Wu, S.H., Liu, C.P., Tien, S.Y., Chen, S.Y., Chang, J.S., Lee, W.J.: Biological butanol production from microalgae-based biodiesel residues by Clostridiumacetobutylicum. Bioresour. Technol. **184**, 379–385 (2015)

64. Melis, A., Zhang, L., Forestier, M., Ghirardi, M.L., Seibert, M.: Sustained photobiological hydrogen gas production upon reversible inactivation of oxygen evolution in the green alga Chlamydomonas reinhardtii. Plant Physiol. **122**, 127–136 (2000)
65. Eroglu, E., Melis, A.: Microalgal hydrogen production research. Int. J. Hydrog. Energy **41**, 12772–12798 (2016)
66. Wang, Y., Ho, S.H., Cheng, C.L., Guo, W.Q., Nagarajan, D., Ren, N.Q., Lee, D.J., Chang, J.S.: Perspectives on the feasibility of using microalgae for industrial wastewater treatment. Bioresour. Technol. **222**, 485–497 (2016)
67. Markou, G., Angelidaki, I., Nerantzis, E., Georgakakis, D.: Bioethanol production by carbohydrate-enriched biomass of Arthrospira (Spirulina) platens is. Energies **6**, 3937–3950 (2013)
68. Lakaniemi, A.M., Tuovinen, O.H., Puhakka, J.A.: Anaerobic conversion of microalgal biomass to sustainable energy carriers—a review. Bioresour. Technol. **135**, 222–231 (2013)
69. Appels, L., Baeyens, J., Degrève, J., Dewil, R.: Principles and potential of the anaerobic digestion of waste-activated sludge. Prog. Energy Combust. Sci. **34**, 755–781 (2008)
70. Gonzalez-Fernandez, C., Mandy, A., Ballesteros, I., Ballesteros, M.: Impact of temperature and photoperiod on anaerobic biodegradability of microalgae grown in urban wastewater. Int. Biodeterior. Biodegrad. **106**, 16–23 (2016)
71. Hernandez, D., Riano, B., Coca, M., Solana, M., Bertucco, A., Garcia-Gonzalez, M.C.: Microalgae cultivation in high rate algal ponds using slaughterhouse wastewater for biofuel applications. Chem. Eng. J. **285**, 449–458 (2016)
72. Hirano, A., Ueda, R., Hirayama, S.: Ogushi Y (1997) CO2 fixation and ethanol production with microalgal photosynthesis and intracellular anaerobic fermentation. Energy **22**, 137–142 (1997)
73. Raeesossadati, M.J., Ahmadzadeh, H., McHenry, M.P., Moheimani, N.R.: CO_2 bioremediation by microalgaein photobioreactors: impacts of biomass and CO_2 concentrations, light, and temperature. Algal Res. **6**, 78–85 (2014)
74. Yen, H.W., Hu, I.C., Chen, C.Y., Ho, S.H., Lee, D.J., Chang, J.S.: Microalgae-based biorefinery—from biofuels to natural products. Bioresour. Technol. **135**, 166–174 (2013)
75. Barrow, C.; Shahidi, F.: Marine Nutraceuticals and Functional Foods. CRC Press, Boca Raton (2007)
76. Wijesekara, I., Pangestuti, R., Kim, S.K.: Biological activities and potential health benefits of sulphated polysaccharides derived from marine algae. Carbohydr. Polym. **84**, 14–21 (2011)
77. Da Silva Vaz, B., Moreira, J.B., de Morais, M.G., Costa, J.A.V.: Microalgae as a new source of bioactive compounds in food supplements. Curr. Opin. Food Sci. **7**, 73–77 (2016)
78. Adarme-Vega, T.C., Thomas-Hall, S.R., Schenk, P.M.: Towards sustainable sources for omega-3 fatty acids production. Curr. Opin. Biotechnol. **26**, 14–18 (2014)
79. Wang, H.M., Chen, C.C., Huynh, P., Chang, J.S.: Exploring the potential of using algae in cosmetics. Bioresour. Technol. **184**, 355–362 (**2015**)
80. Brown, M., Jeffrey, S., Volkman, J., Dunstan, G.: Nutritional properties of microalgae for mariculture. Aquaculture **151**, 315–331 (1997)
81. Marris, E.: Putting the carbon back: black is the new green. Nature **442**, 624–626 (2006)
82. Guedes, A.C., Amaro, H.M., Malcata, F.X.: Microalgae as sources of high added-value compounds-a brief review of recent work. Biotechnol. Prog. **27**, 597–613 (2011)
83. Becker, E.: Micro-algae as a source of protein. Biotechnol. Adv. **25**, 207–210 (2007)

Deep Neural Network Based Phishing Classification on a High-Risk URL Dataset

Mehmet Korkmaz[1](✉), Emre Kocyigit[1], Ozgur Koray Sahingoz[2], and Banu Diri[1]

[1] Yildiz Technical University, Davutpasa, Esenler, Istanbul, Turkey
{mehmet.korkmaz.501,f0119017}@std.yildiz.edu.tr, diri@yildiz.edu.tr

[2] Biruni University, Topkapi, 34100 Istanbul, Turkey
osahingoz@biruni.edu.tr

Abstract. Due to the growing trend of Internetization, the number of connected computers has been increasing day by day. Almost all companies are transferring their main operations from the real world to the cyberworld. Although this increases the marketplace of the firms, it also brings lots of vulnerabilities, such as cyber-attacks, especially with the anonymous structure of the Internet. Phishing is one of the popular attack types which exploits the vulnerabilities to user unawareness. There are some works in the literature that gets help from the rule-based detection systems as a static preventions mechanism, and machine learning-based systems as dynamic prevention mechanisms. In this work, we implemented a deep neural network (DNN) based phishing detection system by analyzing the URL of the suspicious websites. Although in almost all previous researches the used datasets are collected by different resources in which legitimate and phishing websites are clear, in this research, we firstly create a high-risk dataset, which contains only the suspicious websites which are reported to PhishTank website. Experimental research showed that the proposed system gives a very good level of efficiency both in accuracy and execution time manner.

Keywords: High-Risk phishing dataset · Phishing detection · Deep neural network · Website classification

1 Introduction

Communication is indispensable for human beings. People can express themselves, socialize, share information, get educated, and trade etc. with communicating. Similarly, many vital and commercial activities can be done by communicating. Communication has four main elements: Source, Recipient, Message, and Channel. Nowadays, e-mail and instant messaging applications have been accepted as communication channels with the increase of phone and internet usage in general. However, the source and message of this communication channel started to lose its reliability over time. Therefore, Recipients have become insecure communication from unreliable sources in the

A. Abraham et al. (Eds.): SoCPaR 2020, AISC 1383, pp. 648–657, 2021.
https://doi.org/10.1007/978-3-030-73689-7_62

communication channel they consider reliable. At the same time, questions about the reliability of messages have also occupied experts for a while. That is why, in recent years, academic and commercial studies have been carried out in order not to affect people (victims) of an unreliable source or message. Among the primary examples of the mentioned studies, the reliable source approach of companies providing communication services was included. Today, different tools and certificates are used in order to be included in the trusted list of the source. Apart from this, the firewall or antivirus software that can be used is also intended to prevent possible threats. Despite these precautions, there are different studies to detect possible threats to users. In particular, attacks that can be easily created by e-mail as a communication channel are among the high-impact cyber-attacks of recent years. Cyber-attacks carried out by this type of e-mail and aiming to harm users are called ***Phishing Attacks.***

There is striking information in the report prepared by Proofpoint [1] regarding phishing attacks and it showed that the unreliable source (attacker) in communication affects the receiver (victim) with a high rate of deception. Therefore, phishing attacks are one of the cyber-attacks frequently preferred by attackers. According to the 2020s report of APWG [2], which is one of the worldwide organizations working within the scope of phishing attacks to support making communication reliable, SAAS/WebMail has been the most targeted sector as mentioned in Fig. 1.

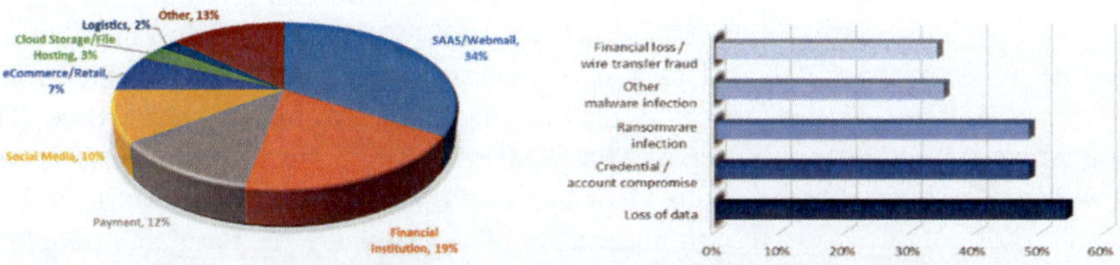

a) most targeted industry sectors b) impact of successful phishing attack

Fig. 1. Phishing statistics

In this research we proposed a DNN based phishing detection system by analyzing the URL of the websites. In the literature there are many approaches for this type protection. Due to the dynamicity of the system, machine learning based models are mostly preferred especially for the detection of the zero-day attacks. However, these systems are generally used a well categorizable dataset in their implementation. These systems use the legitimate sites from the ranking engines which are not the main target of the phishing query. In this research, we firstly generate a High-Risk URL dataset [3] which is collected from the PhishTank which contains the website URL lists that are asked to PhishTank and labeled as suspicious in the first time. Therefore, it is thought as a real dataset for Phishing Attacks. Because standard website queries such as "https://www.db.com/index?language_id=1" or "https://www.amazon.com/" can easily be categorized by a rule based system. However the suspicious websites such as "https://amaaon1.amazon-rjp.xyz/pc/" are risky and detection systems must be focus on them.

2 Background

2.1 Computer Security and Phishing Attacks

Computer security is the mechanism to protect it from all electronic threats. Today, security problems are especially caused by attacks over the network. Any attack, organized against another device on the network, can be described as a cyber-attack. Many efforts are made to minimize security vulnerabilities in individual or corporate terms and security support is received. Firewall, VPN (Virtual Private Networks), Antivirus software, IDS (Intrusion Detection Systems), Web filtering, and Antiphishing mechanisms are the main methods used.

Antiphishing is a method based on preventing unauthorized access to secure and sensitive information as a protection service against phishing attacks. Phishing is a type of cyber-attack in which the source, which appears to be reliable through different communication channels, tries to obtain personal information by deceiving the victim. This personal information is used later to access important accounts, resulting in identity theft or financial loss.

As seen in Fig. 2, phishing attacks operate based on the fake website prepared by the attacker. The victim is asked to take an action on a fake website with a design very similar to the original. The goal is to watch the keyboard movements unnoticed by the victim. According to the working logic of this attack the most important job is a phishing kit. This kit allows information received from the victim to be transmitted to the attacker. Here, the hardest step for the attacker is to make the victim use the website with a convincing message. In fact, the name of the attack comes precisely from here: The victim's free will access to this prepared mechanism and enter his information. The next step will result in the attacker's stealing using this data. What is stolen can be money, data, personal information. Every month, thousands of websites are prepared and people or organizations are targeted using all credible factors. Despite the measures taken against these attacks, the attack may not be detected. Because each detection causes different attack behaviors. Thus, existing systems must be constantly improved.

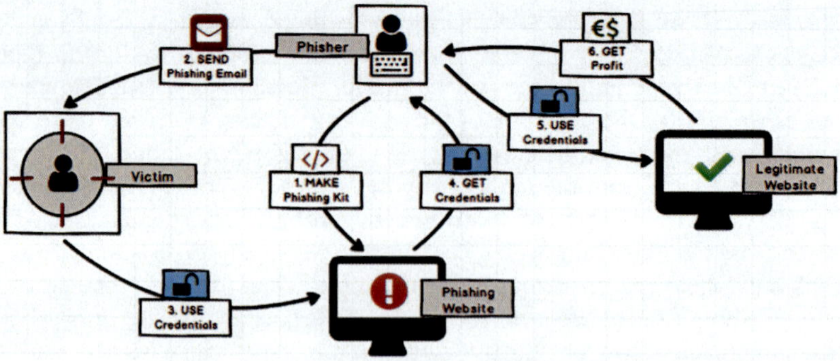

Fig. 2. Life cycle of a phishing attack

These protection systems detect phishing attacks according to different approaches. These approaches have been developed by looking at the content of the website, analyzing its URL, or hybrid. Good results have been obtained in studies that detect attacks by looking at the content of the website. To increase the real runtime and accuracy, studies were carried out with URL analysis. As mentioned earlier in URL analysis, different detection systems have been developed according to attack behavior.

In *list-based systems*, the blacklist in which the phishing attacks are kept and the white list of suitable URLs to be used, the URL is searched in this database according to the preferred approach and the attack is detected. In rule-based systems, the URL is tagged according to pre-determined rules. In *similarity-based systems,* an attack is detected by comparing the visual similarity between the fake and the original of the website. Attack detection is made with a plug-in or application in 3rd party-based systems. In *machine learning-based systems,* detection is made according to a dataset trained with machine learning algorithms based on URL properties, website contents or e-mail contents. In deep learning-based systems, they are systems based on the detection of the model trained according to deep learning algorithms with more data.

2.2 Related Works

There are many works focused on the detection of phishing attacks by using different machine learning approaches. The research about the string properties of the URL were removed without the need for third-party support and prior knowledge [4]. Then, they combined the statistical properties of the URL, website content properties and website text properties. 1,021,758 phishing extracted features using the Convolutional Neural Networks (CNN) - Long Short Term Memory (LSTM) method in their experiments in the dataset they created with 989,021 legal URLs and obtained results based on machine learning by combining them with statistical URL features. In the tests, the accuracy was 98.99% and the FP rate was 0.59%.

To detect phishing websites, a deep learning-based framework was developed [5]. Two types of features have been designed as Original features and Interaction features. Later, a model based on Deep Belief Networks (DBN) is presented. They obtained the dataset from the real-time ISS (Internet Service Provider). In their tests on this dataset, they found that the detection model based on DBN achieved a TP ratio of approximately 90% and an FP rate of 0.6%.

In another research, URLNet has been proposed [6] to overcome difficulties such as not capturing semantic meaning and sequential patterns in URL strings, requiring manual feature engineering, not processing non-preliminary features, and not generalizing for data testing. They created the framework for nonlinear URL embedding. For this, they applied CNN. Thus, they have captured many types of semantic information. In 2017, the dataset they created with 15 million data from VirusTotal site, 5 million training and 10 million were tested. According to the test results, they obtained 98.58% TP rate and 0.1% FP rate.

Adversarial Generating Artificial Neural Networks (GANs) were used to classify URLs [7]. Experiments on real data have shown that the synthetic URLs produced are realistic and an accuracy rate of close to 99% has been achieved. Again, in the research using the GAN method [8], it was aimed to increase the number of phishing URLs in order to adjust the class distribution in the dataset. As a result of the tests performed with a dataset consisting of 127,881 data belonging to Ebay, PayPal, Bank of America, and Sorio sites, an accuracy rate of 70%, 81%, 78% and 97% respectively was obtained. In another research, a deep learning-based approach to phishing attacks was proposed [9]. It is aimed to make the URLs of the websites and the characters in these URLs more controlled with word2vec based embedding. For the reviews, the dataset they created includes 587,668 phishing URLs and 584,909 legitimate URLs. As a result of the tests, they reached an accuracy of 99.69% with 0.40% FP and 99.79% TP. They also detected each URL in an average of 32 ms, with the recommended approach much faster than current approaches.

In [10], authors focused on the selection of the features that will be used in the classification process. In a similar research, authors tried to use the Natural Language Processing techniques to identify the phishing attacks [11].

In another research [12], experiments were performed on two different datasets to investigate the efficiency of the proposed three-layer deep neuron network. One of these datasets is obtained from Kaggle and the other is the newly created dataset using 6 million phishing and 6 million legitimate URLs. The methods used for phishing URL detection are a combination of CNN, LSTM, and CNN-LSTM. Experimental results found that LSTM had 99.99% and 97.82% precision, CNN 99.84% and 96.76%, CNN-LSTM 99.90% and 96.97% precision. In [13], they proposed a tool to prevent phishing URL attacks, distributed as a chrome extension. This tool analyzes URLs and classifies them in two different neural networks using a specific type of Recurrent Neural Network (RNN), Artificial Neural Network (ANN) and LSTM networks. As a result of the tests performed on a dataset consisting of 2 million data, RNN was achieved with an accuracy of 95.57% and LSTM with an accuracy of 96.89%.

3 Proposed System

In this research, it is aimed to detect phishing URLs by using the DNN method. By empirically examining our dataset which is newly created, we hope to produce a more complete understanding of the DNN approach. That's why in this section we will focus on the dataset we have created the features we use, and our learning model.

3.1 Dataset

As mentioned before regarding cyber-attacks, different protection methods are used. For this purpose, there are organizations operating around the world. One of these organizations is PhishTank. PhishTank website can be used by everyone. And the PhishTank URL list is used by virus programs such as McAfee, Kaspersky and Avira, as well as applications such as APWG, YahooMail, Mozilla and Opera. The working logic of PhishTank, which also serves as an API, is as follows: Users add the URL of the website they consider phishing to the list to be checked. All users can view this list and rate it as a phishing or non-phishing website. According to the result of this voting, the URL is added to the relevant list.

Approximately 7 million URLs have been entered into the system, which has been operating since 2006. Some of these URLs were tagged as phishing, some were never tagged, and some were tagged as non-phishing. Until the beginning of 2020, approximately 113,000 non-phishing URLs were entered into the system. This data is the most important part of our dataset, which we consider as high risk. The reason it is considered high risk is that users see the URL as phishing, but it is not tagged as phishing after voting. There are about 2.5 million URLs tagged as phishing until this date. Within the scope of this research, the Phishing part of our dataset (with the same size as legitimate part) was chosen *randomly* among these data. Therefore, it is planned to work on a dataset that is difficult to detect a phishing attack which contains 226.378 data that are equally distributed as *Phishing* and *Legitimate* [3].

3.2 Used Features

Today, the importance of the features used in machine learning has been accepted. In this context, there are many publications in the literature. There are literature studies in the field of determination of statistically significant features, the order of importance of features and feature reduction. When looking at the studies conducted for the detection of phishing attacks, it can be seen that different categories are studied. These studies were made by looking at the content, visuality, meta-information or URL of the website. As a real-time application, especially the studies on URL analysis are remarkable in terms of less processing and acceptable time. Therefore, only the URL analysis will be applied in this research. Within the scope of this subject, it is thought that it will be more effective to work with features that do not need a 3rd party services in the light of the information obtained from the publication that [14] handled with a detailed literature review in 2020. And among the features in this research, 73 features were selected and coded.

3.3 Used Learning Models

Broadly speaking, I agree with the positive effects of machine learning on detecting phishing web sites because previous studies show that it achieves better results than other approaches. The future work sections of the research published on this subject generally contain the following purposes: Can training and test time be improved? Can a higher rate of accuracy be achieved? In this context, the use of deep learning method has been considered to answer these questions with more data, even high-risk data. The first answer we were looking for was to investigate the consequences of deep learning in this dataset. So in this research, results were obtained with the basic architecture, DNN. To compare with machine learning methods, results from Logistic Regression (LR), K-Nearest Neighborhood (KNN), Decision Tree (DT), Naive Bayes (NB), XGBoost (XG), Random Forest (RF) and ANN algorithms will be given. First step in these algorithms, cross validation was not used. In the second step, the result was retested with 5 fold cross validation for the RF which has the highest accuracy rate in first step.

In our research, we experimented on a structure consisting of 73 inputs and 1 output. Results were obtained in 2 to 5 hidden layers. On some models, we used dropout as a regularization technique to reduce overfitting. We tried different dropout values in different models. We share them in detail below. We have used various activation functions for nonlinear transformation operations. Since we got the best values using *relu* and *sigmoid*, we decided to use these two activation functions. Again, in order to achieve better results, we used *adam* optimizer in all our models, from which we get the best results from the algorithms we tried in optimization. Similarly, we used *binary_crossentropy* as the loss function. We created six models. These models are given in Table 1. It is shown as Dense D and Units U in the table.

Table 1. Detailed information of models used

Name	Information of layers hidden layers/#Of nodes dropout rate activation function	#Of hidden layers/Epochs	Batch size
DNN M1	D/U: 1024–A: relu D/U: 1–A: sigmoid	2/100	128
DNN M2	D/U: 1024–A: relu D/U: 512–A: relu D/U: 1–A: sigmoid	3/100	128
DNN M3	D/U: 1024–A: relu D/U: 512–A: relu D/U: 128–A: relu D/U: 1–A: sigmoid	4/100	128
DNN M4	D/U: 1024–A: sigmoid D/U: 512–A: sigmoid D/U: 128–A: sigmoid D/U: 1–A: sigmoid	4/100	128
DNN M5	D/U: 1024–A: relu DR/R: 0.5 D/U: 512–A: relu DR/R: 0.5 D/U: 128–A: relu DR/R: 0.5 D/U: 32–A: relu DR/R: 0.5 D/U: 1–A: sigmoid	5/100	128
DNN M6	D/U: 1530–A: relu DR/R: 0.4 D/U: 512–A: relu DR/R: 0.2 D/U: 128–A: relu DR/R: 0.3 D/U: 32–A: relu DR/R: 0.1 D/U: 1–A: sigmoid	5/100	1000

4 Experimental Results

Machine learning is a good technique for detecting phishing attacks especially for zero-day attacks. The fact that virus programs also use this method is a great indicator of this.

So, in particular, what consequences can machine learning and deep learning approaches get in a high-risk dataset? What will be the effects of deep learning compared to machine learning? What results can be achieved with deep learning? To find out, experiments will be done with those selected from machine learning and deep learning models and the results will be compared. In [14], model training was performed in less time compared to studies based on content analysis or image processing. This was performed using the URL analysis method. A high rate of accuracy was achieved with this.

In this research, the machine learning algorithms NB, LR, XG, DT, KNN, ANN and RF algorithms used in the above mentioned study were used. RF5 is the result from Random Forest algorithm with 5 Fold cross validation. On the other hand six different DNN models which discussed in detail in the previous section were used to compare with the results of these algorithms. In this way, the effect of a simple deep learning approach can be observed. The results of all these experiments are shown in Table 2.

Table 2. Results on high-risk dataset

Classifier	TPR (%)	FNR (%)	TNR (%)	FPR (%)	Precision (%)	Recall (%)	F1 score (%)	Accuracy (%)	Time (sec.)
NB	65.12	34.88	67.28	32.72	70.41	65.12	67.66	66.1	0.25
LR	75.88	24.12	76.19	23.81	76.84	75.88	76.36	76.04	5.43
XG	79.94	20.06	80.58	19.42	81.16	79.94	80.54	80.25	35.18
DT	81.25	18.75	81.09	18.91	81.4	81.25	81.32	81.17	6.51
KNN	81.74	18.26	82.82	17.18	83.42	81.74	82.57	82.26	56.93
ANN	85.42	14.58	84.27	15.73	84.29	85.42	84.85	84.84	995.08
RF	88.07	11.93	86.17	13.83	86.05	88.07	87.05	87.1	384.82
RF 5	87.23	12.77	86.02	13.98	85.79	87.23	86.51	86.62	586.28
DNN M1	85.67	14.33	85.23	14.77	85.4	85.67	85.53	85.45	427.59
DNN M2	84.68	15.32	85.47	14.53	85.89	84.68	85.28	85.07	441.83
DNN M3	84.77	15.23	84.88	15.12	85.17	84.77	84.97	84.83	455.89
DNN M4	86.07	13.93	84.06	15.94	83.88	86.07	84.96	85.05	460.48
DNN M5	87.35	12.65	85.51	14.49	85.4	87.35	86.36	86.41	507.33
DNN M6	87.24	12.76	86.04	13.96	86.06	87.24	86.64	**86.64**	**82.11**

In the research [15] showed that there is no big difference in terms of machine learning algorithms between test times. However, there are differences between the training times of the models. If we look at the machine learning approaches in this study, we can see that RF gives the best results with 87.10% accuracy rate. The RF algorithm, with the highest accuracy rate, was re-run with 5 fold cross validation (RF5) and an accuracy rate of 86.62% was achieved. Thus, the algorithm with the highest accuracy rate among machine learning algorithms was found. RF's training time was recorded as 384.82 s, while RF5's was recorded as 586.28. Although the best time as training time is NB with 0.25 s, the accuracy rate is very low with 66.10%. DT or KNN can be used for both an acceptable accuracy rate and a good training time.

Among the six different DNNs, the DNN M6 model achieved the highest accuracy rate with 86.64%. It cannot be said to make a huge difference compared to other results. However, the DNN M6 was the model that completed the training in the shortest time with 82.11 s. This is a pretty good result. Although other models received similar accuracy rates to DNN M6, training times took longer. Therefore, DNN M6 can be expressed as the most ideal model according to the deep learning approach.

From our point of view, the most important result of the research is the accuracy rate obtained in a dataset that is very difficult to predict, regardless of machine learning or deep learning. Current research has shown that a legitimate URL with a high probability of being tagged as phishing is predicted 86.64% correctly. When we look at the research as a whole, on the one hand, high accuracy rate was obtained. On the other hand, a basic deep learning approach, according to the machine learning approach, completed the training in a very short time.

The results obtained here prove the positive effects of a simple DNN model. With more different deep learning architectures to work on, this research will be a prelude to future work, especially for deep learning.

5 Conclusions

In this research it is aimed to draw attention to the effects of phishing attacks. We have expressed the approaches to the detection of phishing attacks applied until today. Especially in recent years, the success of studies on AI is undeniable for almost all cyber world applications. Some of the research conducted with a deep learning approach on phishing attacks has been examined and written in the part of related works. The foregoing discussion has attempted to determine the level of success of the learning method in detecting phishing attacks, as the accuracy rate and training time of the model. Based on the questions we expressed previously, the results obtained with a DNN approach on a high-risk dataset would be the answer we are looking for. In this context, a dataset consisting of 226.378 URLs which equally distributed as phishing and legitimate, was created by collecting data from the PhishTank.

On our dataset, results were obtained by using 73 features [14] and different parameters in seven different machine learning algorithms. The results obtained with machine learning algorithms indicated the highest accuracy rate of 86.62% and high training time 586.28 s in RF5. On the other hand, models were designed with different hidden layers, dropout, 100 epochs and different batch size considering what kind of a result a simple DNN model would produce. With an accuracy rate of 86.64%, the DNN M6 (5 hidden layers, 4 losses, 100 periods and 1000 batch sizes) has almost the same accuracy as the RF5, but is slightly better. However, with a very low training time of 82.11 s, a very good result was obtained within the scope of this research.

References

1. Proofpoint Report. https://www.proofpoint.com/au/resources/threat-reports/state-of-phish Accessed 31 Oct 2020
2. APWG 2020 2nd Quarter Report. https://docs.apwg.org/reports/apwg_trends_report_q2_2020.pdf. Accessed 31 Oct 2020
3. High Risk URL Dataset. www.kaggle.com/dataset/a38e161d237f30579f2e90d8da6bb7dc50717eff63642abdd7adfda0d02218b9 (2020)
4. Yang, P., Zhao, G., Zeng, P.: Phishing website detection based on multidimensional features driven by deep learning. IEEE Access https://doi.org/10.1109/access.2019.2892066 (2019)
5. Yi, P., Guan, Y., Zou, F., Yao, Y., Wang, W., ZhuWeb, T.: Phishing Detection Using a Deep Learning Framework, Hindawi Wireless Communications and Mobile Computing Volume 2018, Article ID 4678746, (2018)
6. Le, H., Pham, Q., Sahoo, D., Hoi, S.: URLNet: learning a URL representation with deep learning for malicious url detection. In: Proceedings of ACM Conference, Washington, DC, USA, July 2017 (Conference 2017) (2018)
7. Yuan, H., Yang, Z., Chen, X., Li, Y., Liu, W.: URL2Vec: URL modeling with character embeddings for fast and accurate phishing website detection. In: IEEE (2018)
8. Trevisan, M., Drago, I.: Robust URL Classification with Generative Adversarial Networks, Workshop on AI in Networks (WAIN) 2018, Toulouse (2018)
9. Anand, A., Gorde, K., Moniz, A., Park, N., Chakraborty, T., Chu, B.: Phishing URL detection with oversampling based on text generative adversarial networks. In: 2018 IEEE International Conference on Big Data (2018)
10. Buber E., Demir O., Sahingoz O.K., Feature selections for the machine learning based detection of phishing websites. In: International Artificial Intelligence and Data Processing Symposium (IDAP), Malatya, pp. 1–5 (2017)
11. Buber, E., Diri, B., Sahingoz, O.K.: Detecting phishing attacks from URL by using NLP techniques. In: 2017 International Conference on Computer Science and Engineering (UBMK), Antalya, pp. 337–342 (2017)
12. Pham, T.T.T., Hoang, V.N., Ha, T.C.: Exploring efficiency of character-level convolution neuron network and long short term memory on malicious URL detection. In: Proceedings of the 2018 VII International Conference on Network, Communication and Computing (ICNCC 2018), ACM, New York, pp. 82–86 (2018)
13. Shivangi, S., Debnath, P., Sajeevan, K.: Chrome extension for malicious URLs detection in social media applications using artificial neural networks and long short term memory networks. In: 2018 International Conference on Advances in Computing, Communications and Informatics (ICACCI), Bangalore, pp. 1993–1997 (2018)
14. Korkmaz, M., Sahingoz, O.K., Diri, B.: Feature selections for the classification of webpages to detect phishing attacks: a survey. In: 2020 International Congress on Human-Computer Interaction, Optimization and Robotic Applications (HORA), Ankara (2020)
15. Korkmaz, M., Sahingoz, O.K., Diri B.: Detection of phishing websites by using machine learning-based URL analysis. In: 11th International Conference on Computing, Communication and Networking Technologies (ICCCNT), India, pp. 1–7 (2020)

A Secure Bayesian Compressive Spectrum Sensing Technique Based Chaotic Matrix for Cognitive Radio Networks

Salma Benazzouza[1](✉), Mohammed Ridouani[1], Fatima Salahdine[2], and Aawatif Hayar[1]

[1] RITM Laboratory, CED Engineering Sciences, Hassan II University, Casablanca, Morocco
salma.benazzouza@ensem.ac.ma, {MOHAMMED.RIDOUANI, AAWATIF.HAYAR}@etu.univh2c.ma

[2] Department of Electrical and Computer Engineering, The University of North Carolina at Charlotte, Charlotte, USA
fsalahdi@uncc.edu

Abstract. Compressive spectrum sensing is an advanced technique to improve the detection process in cognitive radio networks. It aims at reconstructing sparse signals from only few measurements and then achieving spectrum sensing process using only the estimated signal. In cooperative cognitive radio networks, security, time, and uncertainty issues can affect the detection performance when external malicious users try to join the network. Thus, enhanced compressive spectrum sensing techniques are extremely needed. We proposed a novel combination based on the efficiency of the Bayesian recovery inference and the strengths of the Chaotic sensing matrices to accelerate the spectrum sensing, deal with uncertainty in measurements, and provide inherent security of the system. The proposed technique is implemented and tested using several performance metrics. The simulation results prove that our proposed technique is highly efficient over the acquisition and the recovery processes.

Keywords: Compressive spectrum sensing · Cognitive radio networks · Chaotic matrices · Bayesian models

1 Introduction

Wideband spectrum sensing is one of the most attractive solutions of cognitive radio systems. It refers to sensing any activity of a licensed/primary user (PU) by an unlicensed/secondary user (SU) to detect spectrum holes over a wide spectral bandwidth. To ensure the detection process, cooperative spectrum sensing techniques [1] have been proposed as efficient solutions in which SUs share sensing data and collaborate with each other to make the final decision about the spectrum status (occupied/vacant). Cooperative spectrum sensing techniques can be classified into two main schemes: centralized [2], distributed [3]. In centralized scheme, SUs report their sensing information to a fusion center, which collects all sensing results to decide about the spectrum occupancy. On the

A. Abraham et al. (Eds.): SoCPaR 2020, AISC 1383, pp. 658–668, 2021.
https://doi.org/10.1007/978-3-030-73689-7_63

contrary, in the distributed scheme, the spectrum sensing is performed independently by SUs without the fusion center intervening.

The performances of the network using these architectures can be affected when external malicious users try to hack the system. As a result, there are several security challenges that must be treated [4, 5], including confidentiality of the sensing information, sensing time, and uncertainty in the sensing measurements. To handle these different issues, numerous solutions based on compressive spectrum sensing (CSS) technique are proposed in literature. CSS technique involves three main processes: Sub-Nyquist sampling, signal recovery, and spectrum detection [6, 7]. In the first process, a sparse signal is multiplied with a sensing matrix to extract only the main information of the signal. In the second process, a recovery algorithm is used to reconstruct the original signal from the compressed measurements. In the last process, one of the spectrum sensing techniques can be used to make the sensing decision about the primary signal presence.

For Sub-Nyquist sampling operation, two main categories of sensing matrices have been proposed in the literature, including random matrices [8] and deterministic matrices [9]. These matrices are easy to construct and satisfy the Restricted Isometry Property (RIP) [10]. In this paper, we focus on other types of sampling matrices called chaotic sensing matrices. Through comparing chaotic matrices to other matrices, chaotic structure produces pseudo-random matrix in deterministic approach and satisfies the RIP, similar to random matrices. Furthermore, it enhances the network security by making it very difficult for malicious users to generate the original signal without having appropriate information of the parameters used to reconstruct the sensing matrix before the signal sampling.

For the recovery process, there are three categories of reconstruction algorithm: convex optimization [11], Greedy [12] and Bayesian approaches [13]. Convex algorithms are accurate but complex, slower, uncertain, and difficult to implement for large size applications. Greedy algorithms are fast but require more measurements than convex algorithms and are uncertain. Bayesian models combine the advantages of both categories. They are fast, require less measurement for a high recovery rate, and can deal with uncertainty [14].

Recently, a few works have been proposed to reconstruct a chaotic matrix [15–18]. For instance, the authors in [15] generated a Toeplitzed structurally chaotic matrix (TSCM) based on two steps. Firstly, they multiplied an orthonormal matrix with a chaotic-based Toeplitz matrix, which is constructed by chaotic sequences. Secondly, they subsampled the resultant matrix to obtain the structural one. In [16], the authors proposed an Incoherence Rotated Chaotic (IRC) matrix based on QR decomposition. They choose first an appropriate chaotic system and remove the first 1000 chaotic elements. Then, the first row of the sensing matrix is fixed and the rest rows are generated by a circular way and finally, a QR decomposition of the sensing matrix is used to construct IRC. In [17], the authors represent chaotic stream sensing matrices based on topologically conjugate chaotic systems that produce a large class of chaotic sensing matrices for compressive sensing. In [18], the authors proposed a new method to fill up a chaotic sensing matrix to increase the randomness of the matrix, by using different initial value for every column of the sensing matrix.

However, all these methods are based on convex or Greedy recovery algorithms with more complexity of the system and do not deal with uncertainty in measurements. To address these limitations, we propose a novel CSS technique based on a robust combination that exploits the efficiency of Bayesian recovery algorithms and the strengths of chaotic matrices. The rest of this work is arranged as follows: Sect. 2 represents system model and the methodology of the proposed technique. In Sect. 3, the simulation results of the proposed technique are compared and discussed based on different metrics. Last, a conclusion is given.

2 Methodology

Figure 1 represents the general model of the proposed CSS. The aim of this technique is to recover the original sparse signal, x, from only few measurements, y, and then perform spectrum sensing using the recovered signal x'. The sensing decision is performed by selecting between binary hypotheses, H1 (occupied) and H0 (vacant). A detection threshold, denoted λ, is chosen according to the energy detection technique [19] and then compared with the output signal.

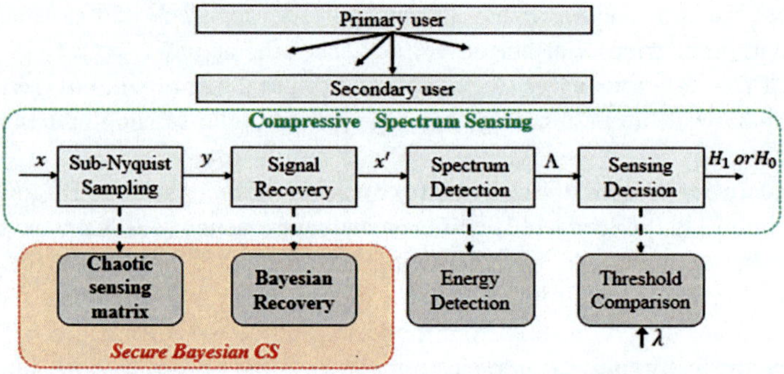

Fig. 1. Secure Bayesian compressive spectrum sensing diagram.

Figure 2 illustrates the main blocs of the proposed system. Assuming the signal sparsity, SU received the signal x with N samples, its sparsity level D satisfies D << N. Based on central limit theorem, x can be approximated as a Gaussian variable with x = $(x_1, x_2..., x_n)$. During acquisition process x is multiplied with the chaotic sensing matrix C [M × N], to generate the compressive vector y, which selects only M samples from x (N, 1). A noise, w, is added to the signal measurements y, including the noise measurements and the sparse representation error and can be approximated as a zero mean Gaussian noise with unknown variance δ_w. Finally, the input signal is reconstructed from the noisy measurements y_w by using the Bayesian compressive sensing recovery. The system output is the recovered signal x' with N samples and D sparsity.

2.1 Chaotic Sensing Matrix

The Chaotic matrices are constructed by using chaotic sequences known as the chaotic map. In the literature, there are several types of maps: the logistic map [20], Tent map

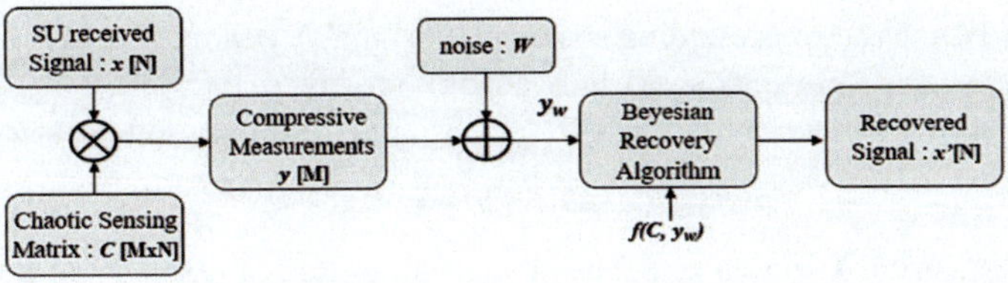

Fig. 2. General model of secure Bayesian compressive spectrum sensing.

[21], quadratic map [22], Chebychev map [23] and many other types. Each map has an initial value and parameters that manages its behavior. Thus, any slight modification in these parameters leads to different sequences, which can affects the performances of the generated chaotic sensing matrix.

To construct a chaotic matrix, let first define a chaotic map as

$$V_{n+1} = g(V_n) \quad (1)$$

where the function g is determined by a chosen chaotic sequence (logistic, Tent, Chebyshev, …) which can be described as follows $V(h, d, v_0) = \{v_n, v_{n+d}, \ldots, v_{n+hd}\}$ with h = 0, 1, 2 …, a sampling distance d and initial condition v_0.

To guarantee the randomness of the chaotic map used, the first 1000 terms of the sequence are ignored [18], as result we denote the regularization of $V(h, d, v_0)$ as

$$C_h = V_{1000+hd} \quad (2)$$

where $C_h \in C(h, d, c_0)$ present the new sequence that used to fill up the sensing matrix column by column such that

$$C = \sqrt{\frac{2}{M}} \begin{pmatrix} C_0 & C_M & \cdots & C_{M(N-1)} \\ C_1 & C_{M+1} & \cdots & C_{M(N-1)+1} \\ \vdots & \vdots & \vdots & \vdots \\ C_{M-1} & C_{2M-1} & \cdots & C_{MN-1} \end{pmatrix}$$

where the constant $\sqrt{2/M}$ is for normalization.

The sensitivity to the initial conditions present the key of security of each chaotic matrix, that makes this type of matrices a suitable solution to enhance the security in case of cooperative CRNs; only by sharing the accurate initial parameters of the chaotic map can any user in the network perfectly reconstruct the received signal.

2.2 Bayesian Recovery

The recovery process aims at finding the unique and sparse solution of the following underdetermined system

$$y = Cx + w \quad (3)$$

where x is an N × 1 vector presenting sparse input signal, y denotes the M × 1 compressive measurements, C presents an M × N chaotic sensing matrix and w is the additive Gaussian noise. This process can be expressed as an l_1-minimisation problem

$$min\,||x||_1 \;\; subject\, to\, y = Cx + w \tag{4}$$

where$\|.\|_1$ is the norm 1, which calculates the absolute sum of elements of a vector.

To solve this problem, we apply the Bayesian recovery algorithm based on Laplace prior technique [24] to handle uncertainty in measurements. Bayesian algorithm is a statistical technique that defines all unknown variables through probability distributions. In the proposed model, compressive measurements y, original signal x, chaotic sensing matrix C, and noise w are the main parameters in the system. The objective is to provide a full posterior distribution for the values of x based on prior belief including the sparsity assumption of x and the relationship between the noisy measurements y and the received signal x. To realize this, the Bayesian inference requires the definition of a joint distribution P (x, α, β, y) of all unknown parameters by introducing new variables α and β called hyperparameters. To simplify the analysis of the joint distribution the following factorization is used [24]

$$P(x,\ \alpha,\ \beta,\ y) \ = \ P(y|x,\ \beta)\ P(x|\alpha)\ P(\beta)\ P(\alpha) \tag{5}$$

Where P (x| α) is a prior distribution which defines our knowledge on the nature of x and P(y|x, β) is the conditional distribution with $\beta = 1/\sigma^2$ is the inverse noise variance.

According to the central limit theorem, when the number of measurements is small compared to length of the received signal (N ≫ M), thus the measurements vector y can be approximated as Gaussian distribution with zero mean and a variance β^{-1}. Hence, the conditional distribution of y given the sparse signal x can be expressed as

$$P(y|x,\ \beta) \sim N\left(y|C_x,\ \beta^{-1}\right) \tag{6}$$

To simplify the analysis, the Gamma distribution is used to define the hyper parameter β as follows

$$P\left(\beta|c^{\beta},\, d^{\beta}\right) \sim \Gamma\left(\beta|c^{\beta}\,,\, d^{\beta}\right) \tag{7}$$

Where c^{β}, d^{β} are the scale and the shape parameter, respectively, assigned to β. They are called hyperpriors.

To exploit the sparsity of the received signal x, a Laplacian density function is used to describe the prior distribution P (x| α) as follows

$$P(x|\alpha) = (\alpha/2)^N exp\,(-\,\alpha_1\,\|\boldsymbol{x}\|_1) \tag{8}$$

Finally, we maximize the joint distribution P (x, α, β, y) or its logarithm, to estimate the hyper parameters β, α and then we consider the posterior distribution for x as a multivariate Gaussian distribution with a mean μ and covariance Σ given by

$$\mu = \beta \sum C^T y$$
$$\sum = \left[\beta C^T C + A\right]^{-1} with\, A = diag(1/\alpha_i) \tag{9}$$

and subsequently we can solve the l_1-minimisation problem given by (4).

3 Simulation Results

To evaluate the efficiency of the proposed technique we consider two main aspect using different metrics: i) the system performance of signal reconstruction process in order to evaluate the strength of the Bayesian model compared to the two other techniques convex and greedy algorithms. ii) the system performance of signal acquisition process to evaluate the reliability and sensitivity of the system to the initial parameters of the chaotic matrices, and to show the robustness of the proposed matrices when a hacker tries to reconstruct the original signal.

In this paper, we consider a basic model with one PU and one SU to simplify the analysis and to confirm the efficiency of the proposed solution. First, the system performance using the proposed technique is compared to two other recovery techniques, namely, basis pursuit (BP) [25] and matching pursuit (MP) [11] algorithms. These algorithms were used because they provide the best performance in two key categories of recovery processes: greedy and convex.

For simulation purposes, the same sensing chaotic matrix is used and generated via logistic map with the following recurrence formula

$$V_{n+1} = r V_n(1 - V_n) \tag{10}$$

Where V_n denotes the nth term in the chaotic sequence with $V_0 = 0.5$ presents the initial value and $r = 3.9$ is the bifurcation parameter.

The following performance metrics are used: recovery error [26], spikes error rate (SER), recovery time and sampling time. The recovery error is defined as the distance between the estimated signal and the original signal x. It is formulated as follows

$$R_e = \frac{\|\hat{x} - x\|}{\|x\|} \tag{11}$$

The spikes error rate SER measures the error rate of the recovered signal spikes. It is the sum of the false and missed spikes divided by the signal's length. It is given by

$$SER = \frac{N_{fs} + N_{ms}}{N} \tag{12}$$

where N_{fs} and N_{ms} are the number of false and missed recovered spikes, respectively.

Recovery time is the elapsed time during the reconstruction process. Sampling time is the time elapsed to compress the acquired signal by a particular sensing matrix.

Figure 3 and 4 illustrate a comparison results between three recovery algorithms: Bayesian, basis pursuit, and matching pursuit algorithms with the chaotic matrix.

Figure 3A shows the input signal with $N = 200$ samples and $D = 15$ spikes. Figure 3B–D presents the output signal using the chaotic matrix according to (10) with the Bayesian recovery, BP and MP algorithms, respectively. As we can see in Fig. 3B, the Bayesian recovery, is more efficient in combination with chaotic sensing matrix, and gives a more similar output signal to the input one. However, the estimated signal using the BP and MP algorithms reveals some variations, as shown in Fig. 3 C, D. As result, our proposed combination is efficient and reliable.

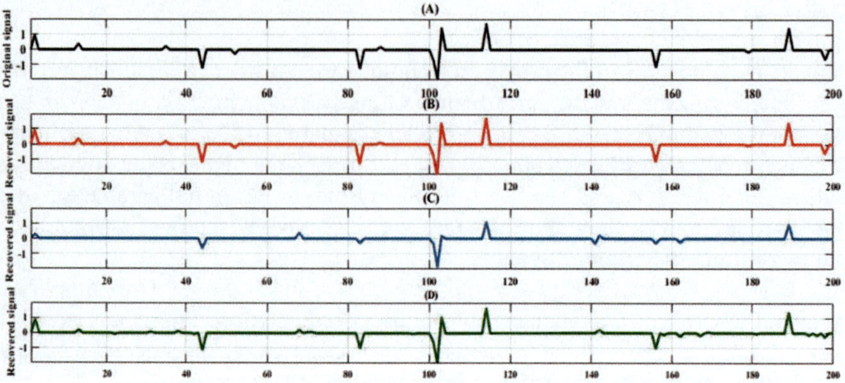

Fig. 3. Signal output comparison of three recovery techniques with Chaotic sensing matrix: A, Input signal. B, Output signal of Bayesian recovery. C, Output signal of matching pursuit technique. D, Output signal of basis pursuit technique.

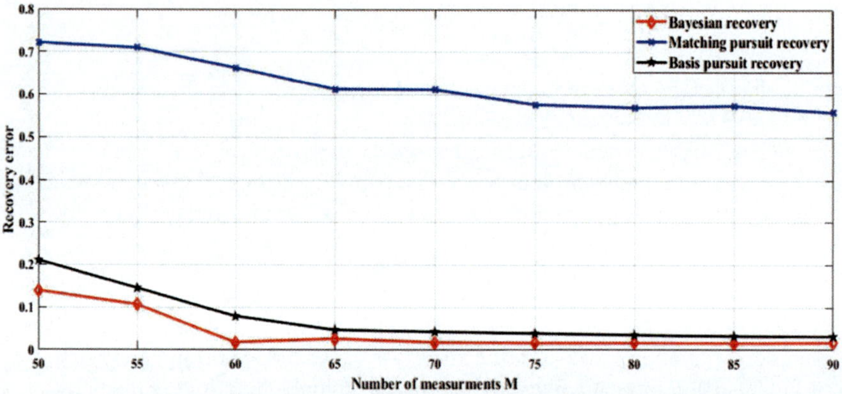

Fig. 4. Comparison of recovery error metric for multiple value of compressive measurements M for Bayesian, matching pursuit and basis pursuit algorithms.

Figure 4 shows the comparison results based on the recovery error metric for multiple values of the compressive measurements M. As we can see, the three recovery algorithms minimize the recovery error with the growth of number measurements M. In addition, the Bayesian model gives a better recovery error compared to the other recovery techniques.

For the recovery time, Table 1 presents the result of each algorithm, where the Bayesian method requires lower value of recovery time than MP and BP method.

Next, the system performance using the chaotic sensing matrix is tested by comparing 3 different sensing matrices: the random matrix [8], the circulant matrix [27] and two chaotic matrices based on two different maps: the logistic map according to (10) and the quadratic map [22] defined as follows

$$V_{n+1} = r - V_n^2 \tag{13}$$

Table 1. Recovery time comparison.

Recovery algorithm	Bayesian	Matching pursuit	Basis pursuit
Recovery time (ms)	4.90	19.52	68.37

With the initial value $V_0 = 0.2$ and the bifurcation parameter $r = 1.2$.

Figure 5 shows the comparison results based on the SER metric for multiple values of the compressive measurements M. It is clear that chaotic matrices, in some points are similar to the random and circulant matrices. Thus, the small variations existing between chaotic matrices and the other types can be considered insignificant. Furthermore, the chaotic matrices have more advantages comparing to the other sensing matrices used, including the security issues. Thus, to verify the sensitivity of the system to the initial parameters of the chaotic matrix, we used two different matrices: one for the acquisition process and another for recovery process. The sensing matrix generated in reconstruction process used different initial parameters (different bifurcation parameter r and initial value V_0).

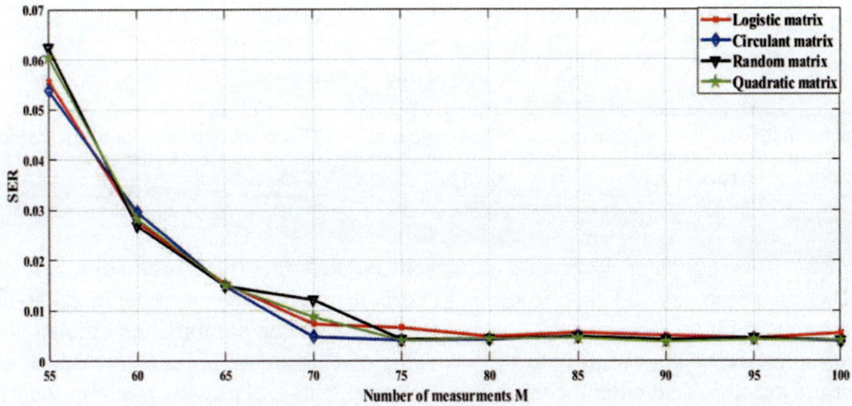

Fig. 5. Comparison of SER metric for multiple value of compressive measurements M for logistic matrix, circulant matrix, random matrix and quadratic matrix.

Figure 6 shows an example of the comparison results in case of using a logistic map according to (10) with the same parameters ($r_0 = 3.9$, $V_0 = 0.3$) in both acquisition and reconstruction step and two other matrices with different initial parameters ($r_1 = 3.7$, $V_0 = 0.3$) and ($r_0 = 3.9$, $V_{0+0.2} = 0.5$), respectively, in reconstruction step. As we can see, any variations in the initial chaotic parameters produce different sequence for the sensing matrix, which generate different recovered signal with high error rate. This means that it is very difficult for an outside user to use the network without having prior information about the initial parameters used to generate the sensing matrix.

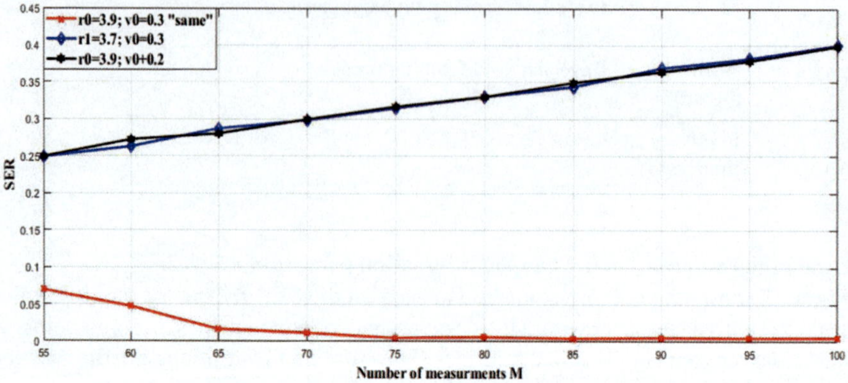

Fig. 6. Sensitivity of the system to changing the initial parameters of the chaotic sensing matrix based on logistic map according to (10).

Sampling time is another metric used to evaluate the performance of the system using chaotic matrices in term of the speed of signal acquisition. Examples of the comparison results for original signal with N = 400 samples and D = 20 spikes are presented in Table 2. The simulation results show that chaotic matrices require less sampling time to acquire the input signal.

Table 2. Sampling time comparison.

Sensing matrices	Logistic matrix	Quadratic matrix	Circulant matrix	Random matrix
Sampling time (ms)	1.17	0.78	2.35	4.79

These illustrations of different comparisons confirm that the combination between the Bayesian recovery and chaotic sensing matrix is reliable, robust, faster, and handles uncertainty in measurements. Thus, our proposed technique combine the efficiency of Bayesian recovery algorithm with the strengths of Chaotic matrices to accelerate the spectrum sensing, deal with uncertainty in measurements and provide inherent security of the system.

4 Conclusion

Compressive spectrum sensing techniques are an emerging solution in cooperative cognitive radio networks to improve the detection performance. In this paper, we aim at improving the acquisition and recovery processes taking into consideration the network security, time, and uncertainty issues. Our proposed solution provides a robust combination using chaotic sensing matrices and Bayesian recovery algorithm. Chaotic matrices provide integral security of the system by making it difficult to malicious users to generate the sensing matrix and integrate the cooperative network. Then, the Bayesian recovery based on Laplace prior technique is used to handle uncertainty in measurements.

To evaluate the proposed technique, we opt for several performance metrics including, recovery error, spikes error rate, recovery time, and sampling time. Through analyzing the simulation results, we can conclude that the Bayesian algorithm with the chaotic sensing matrix is more efficient than the Bayesian algorithm with circulant and random matrices, as well as the basis pursuit and matching pursuit algorithm with the chaotic matrices. In further works, we will extend the scenario to multiple PUs and SUs.

References

1. Salahdine, F., Ghribi, E., Kaabouch, N.: A cooperative spectrum sensing scheme based on compressive sensing for cognitive radio networks. Int. J. of Digit. Inf. Wirel. Commun. 124–136 (2019)
2. Gao, Z., Zhu, H., Li, S., Du, S., Li, X.: Security and privacy of collaborative spectrum sensing in cognitive radio networks. IEEE Wirel. Commun. **19**(6), 106–112 (2012)
3. Akyildiz, I.F., Lo, B.F., Balakrishnan, R.: Cooperative spectrum sensing in cognitive radio networks: a survey. Phys. Commun. **4**(1), 40–62 (2011)
4. Salahdine, F., Kaabouch, N.: Security threats, detection, and countermeasures for physical layer in cognitive radio networks: Surv. Phys. Commun. **39**, 101001 (2020)
5. Arabia-Obedoza, M.R., Rodriguez, G., Johnston, A., Salahdine, F., Kaabouch, N.: Social engineering attacks: a reconnaissance synthesis analysis. In: IEEE Annual Ubiquitous Computing, Electronics & Mobile Communication Conference, pp. 1–6 (2020)
6. Ridouani, M., Hayar, A., Haqiq, A.: Continuous transmit in cognitive radio systems: outage performance of selection decode-and-forward relay networks over Nakagami-m fading channels. J. Wirel. Com. Netw. **2015**, 102 (2015)
7. Ridouani, M., Hayar, A., Haqiq, A.: A novel power control based on a relaxed constraint in cognitive system. Trans. Emerg. Telecommun. Technol. **27**(5), 745–758 (2016)
8. Arjoune, A., Kaabouch, N., El ghazi, H., Tamtaoui, A.: A performance comparison of measurement matrices in compressive sensing. Int. J. Commun. Syst. **31**(10), e3576 (2018)
9. Thu, L.N., Shin, Y.: Deterministic sensing matrices in compressive sensing: a survey. Sci. World J. (2013)
10. Baraniuk, R., Davenport, M., DeVore, R., Wakin, M.: A simple proof of the restricted isometry property for random matrices. Constr. Approx. **28**, 253–263 (2008)
11. Meenu, R., Dhok, S.B., Deshmukh, R.B.: A Systematic review of compressive sensing: concepts, implementations and applications. IEEE Acces, **6**, 4875–4894 (2018)
12. Salahdine, F., Kaabouch, N., El Ghazi, H.: A survey on compressive sensing techniques for cognitive radio networks. Phys. Comm. **20**, 61–73 (2016)
13. Benazzouza, S., Ridouani, M., Salahdine, F., Hayar, A.: A survey on compressive spectrum sensing for cognitive radio networks. In: 2019 IEEE International Smart Cities Conference (ISC2), pp. 1–6 (2019)
14. Salahdine, F., Kaabouch, N., El Ghazi, H.: A Bayesian recovery technique with Toeplitz matrix for compressive spectrum sensing in cognitive radio networks. Int. J. of Commun. Syst. **30**(15), e3314 (2017)
15. Zeng, L., Zhang, X., Chen, L., Cao, T., Yang, J.: Deterministic construction of Toeplitzed structurally chaotic matrix for compressed sensing. Circ. Syst. Sign. Process. **34**, 797–813 (2015)
16. Yao, S., Wang, T., Shen, W., Shaoming, P.: Research of incoherence rotated chaotic measurement matrix in compressed sensing. Multimed. Tools Appl. **76**, 17699–17717 (2017)
17. Gan, H., Xiao, S., Zhao, Y.: A large class of chaotic sensing matrices for compressed sensing. Sign. Process. **149**, 193–203 (2018)

18. Kamel, S.H., Abdel Malek, M.B., El-Khamy, S.E.: Compressive spectrum sensing using chaotic matrices for cognitive radio networks. Int. J. Commun. Syst. **32**(6), e3899 (2019)
19. Ridouani, M., Hayar, A., Haqiq, A.: Perform sensing and transmission in parallel in cognitive radio systems: spectrum and energy efficiency. Digit. Sign. Process. **62**, 65–80 (2017)
20. Suneel, M.: Electronic circuit realization of the logistic map. Sadhana **31**, 69–78 (2006)
21. Li, C., Luo, G., Qin, K., Li, C.: An image encryption scheme based on chaotic tent map. Nonlinear Dyn. **87**, 127–133 (2017)
22. Ramadan, N., Ahmed, H.H., Elkhamy, S.E., Abd-El-Samie, F.E.: Chaos-based image encryption using an improved quadratic chaotic map. Am. J. Sign. Process. **6**(1), 1–13 (2016)
23. Gan, H., Li, Z., Li, J., Wang, X., Cheng, Z.: Compressive sensing using chaotic sequence based on Chebyshev map. Nonlinear Dyn. **78**(4), 2429–2438 (2014)
24. Babacan, S.D., Molina, R., Katsaggelos, A.K.: Bayesian compressive sensing using laplace priors. IEEE Trans. Image Process. **19**, 53–63 (2010)
25. Abo-Zahhad, M.M., Hussein, A.I., Mohamed, A.M.: Compressive sensing algorithms for signal processing applications: a survey. Int. J. Commun. Netw. Syst. Sci. **8**, 197–216 (2015)
26. Salahdine, F., Ghribi, E., Kaabouch, N.: Metrics for evaluating the efficiency of compressing sensing techniques. In: 2020 International Conference on Information Networking (ICOIN), pp. 562–567 (2020)
27. Salahdine, F., Kaabouch, N., Ghazi, H.E.: Bayesian compressive sensing with circulant matrix for spectrum sensing in cognitive radio networks. In: IEEE 7th Annual Ubiquitous Computing, Electronics & Mobile Communication Conference (UEMCON), pp. 1–6 (2016)

Optimized Machine Learning Techniques for IoT 6LoWPAN Cyber Attacks Detection

Yassine Maleh[1(✉)], Abdelkbir Sahid[2], and Mustapha Belaissaoui[2]

[1] LaSTI Laboratory, ENSAK, University Sultan Moulay Slimane, Beni Mellal, Morocco
yassine.maleh@ieee.org
[2] ENCG, University of Hassan, 1, B.P.: 577, 26000 Settat, Morocco

Abstract. Particle swarm optimization (PSO) is a versatile and easy-to-use optimization technique. It was used to develop and enhance the efficiency of machine learning techniques. In this paper, we compare various machine learning techniques for IoT intrusion detection using PSO. A major issue is the lack of data on IoT 6lowPAN attacks. The Cooja IoT simulator has been used in our research to generate attacks. We provide a performance overview of different machine learning models optimized by a PSO algorithm for the detection of IoT routing attacks with high precision and accuracy. The effective network architecture for all system models is chosen based on a comparison of the performance of different network topologies and network scenarios.sThe experimental results offer a brief idea of the cost of each Machine Learning Model for intrusion detection, especially in terms of precision and accuracy.

Keywords: Cyber-attacks · Swarm optimization machine learning · Internet of Things · 6lowpan · Intrusion detection

1 Introduction

Today, machine learning is one of the most common subjects for identifying cyber threats on the Internet of Things IoT. So machine intelligence approaches can provide a comprehensive mechanism for advanced attacks [1]. On the opposite, the main challenge in IoT security analysis is the lack of accessible and modified datasets.

Traditional machine learning techniques, such as Bayesian Belief Networks (BBN) [2], Support Vector Machines (SVM) [3–5] have been applied for cybersecurity. However, the large-scale data produced in IoT demands an effective machine-based learning system that can be tailored to the IoT requirements.

In this research, we have demonstrated the feasibility of our approach with simulations of up to 100 nodes relative to established studies., such as [6]. We used data generated by real-life analogous simulations due to a lack of public IoT attack datasets. Cooja simulation produces raw packet capture files that are first translated to Comma Separated Values (CSV) files for text-based processing. The CSV files are then entered into the pre-processing module of our system.

A. Abraham et al. (Eds.): SoCPaR 2020, AISC 1383, pp. 669–677, 2021.
https://doi.org/10.1007/978-3-030-73689-7_64

Firstly, we have identified 24 features as an initial candidate feature set. Afterwards, feature normalization is applied to all datasets to reduce the negative effects of marginal values. As a result of this analysis, some of the features are abandoned in the pre-functional selection process. We have minimized the number of features to 16 main features. After feature pre-processing, the datasets corresponding to each scenario is labelled and mixed to produce a pre-processed dataset, consisting of a mixture of attack and benign data. These datasets are fed into the machine learning algorithms. We perform a classification based on different classification algorithms such as Support Vector Machine, Naïve Bayes, K-Nearest Neighbour, Random Forest and a false multilayer perceptron neural network optimized by particle swarm optimization (PSO) algorithm.

The findings show the efficacy and robustness of the suggested hybrid approach in the analysis of multiple types of data for the detection of IoT attacks. It can be inferred that K-Nearest Neighbor and SVM are the best algorithms for classifying intrusions in the IoT setting. The.contributions of this paper are:

- Generate a new dataset for IoT routing attacks.
- Delete obsolete, unnecessary functions.
- Optimizations using the PSO Particle Swarm Optimization algorithm.
- Comparison of various IoT attacks application data mining algorithms.
- Identification of the best IoT intrusion performance-based algorithm.

This paper discusses the related work in Sect. 2 and a detailed summary of the litterature review in Sect. 3. Section 4 outlines the experiments using the models suggested. Section 5 points out findings and future directions.

2 Literature Review

Machine learning is now an emerging field due to the increasing amount of data. Machine learning makes it possible to acquire knowledge from a massive amount of data, which is very heavy for humans and sometimes impossible [7]. There is no clear programming in ML; it can be named 'training' because of the learning process. The ML method does not do much except to investigate the mathematical nature of the results.

There are several types of models [8–11]. In addition to the fact that they can be differentiated by their supervised or unsupervised side, they can also be differentiated by their classification or regression side.

The first classifies the data and the second predicts values for each data. Since the IDS problem appears to be a classification problem, the following is a non-exhaustive list of classification models that could be considered for the implementation of the IDS.

k-nearest neighbour belongs to the family of Instance-Based Learning [12]. It looks for the k-nearest neighbour of the example to be predicted. The majority class of these neighbours is returned as the answer for this example. The distance can be calculated using a Euclidean or other distance. The learning process is simple since you only have to memorize examples. But the prediction can be very slow since one has to look for the nearest k neighbours, in a potentially very high dimensional space.

Naive Bayes uses Bayes rules [13]. One assumption is that each class of examples is independent. The learning phase calculates each of their probability and that of each attribute knowing it. This is done by looking in the training examples at the frequency of appearance of the class and the value of the attribute for each class. The phase the greatest conditional probability to determine that a particular instance of enforcement is unknown belongs to a known class. One of the strong points is the need for a small amount of information for the learning phase.

Learning Decision Trees is equivalent to Learning Decision Rules [14]. The acyclic tree contains nodes labelled with attributes. Each stop is labelled with a predicate that applies to the parent node. Each leaf is labelled by a class which is the target attribute. The learner builds a tree recursively by choosing the attribute that best fits the examples in their classes, creating child nodes for each value of the chosen attribute. Iteratively, the algorithm removes those chosen by creating child nodes. The algorithms for creating a decision tree are ID3, C4.5, CART. To classify an example, you have to start from the root of the tree and go down to a leaf respecting the predicates.

Support Vector Machine (SVM) are generalizations of linear classifiers [15]. The goal is to separate space using a hyperplane. In the case where a hyperplane separator does not exist 10, we pass the whole to a higher dimension.

3 Materials and Methods

The suggested intrusion detection architecture is a hybrid IDS that implements both the anomaly-based IDS and the signature-based IDS in the system. The proposed IDS uses the 6LoWPAN compression header to understand and identify the form of attacks as a function of the machine learning algorithm. The rule or signature generated by the machine learning algorithm is then put at 6BR. Over time, as a new signature for routing attacks is available, 6BR will be modified with the new rule or signature created by the new features. The proposed IDS structure is divided into three layers. The first layer consists of a sensor that collects network data using the Cooja traffic analyzer. The collected data is then processed and refined using a second layer approach that only extracts distinct characteristics that differentiate normal and irregular network operations. At this stage, the data is labelled as usual, hello, hello, abyss or a wormhole attack. The suggested IDS structure will be seen in Fig. 1.

3.1 Data Generation

We have created various attack scenarios in a wide range of IoT nodes, up to 100, with different percentages (10%, 20%, etc.) of malicious nodes for simulating routing attacks such as hello-blood attacks, wormhole attacks and Sinkhole attacks. The scenarios are seen in Table 1. We simulated these scenarios with the Cooja network simulation, preventing the creation of a virtual dataset, as Cooja allows the real RPL code to be executed on the simulated nodes.

3.2 Data Preprocessing

Machine learning algorithms require certain learning data attributes that are extracted from the extraction of a function. The datasets are generated as PCAP files after scenarios

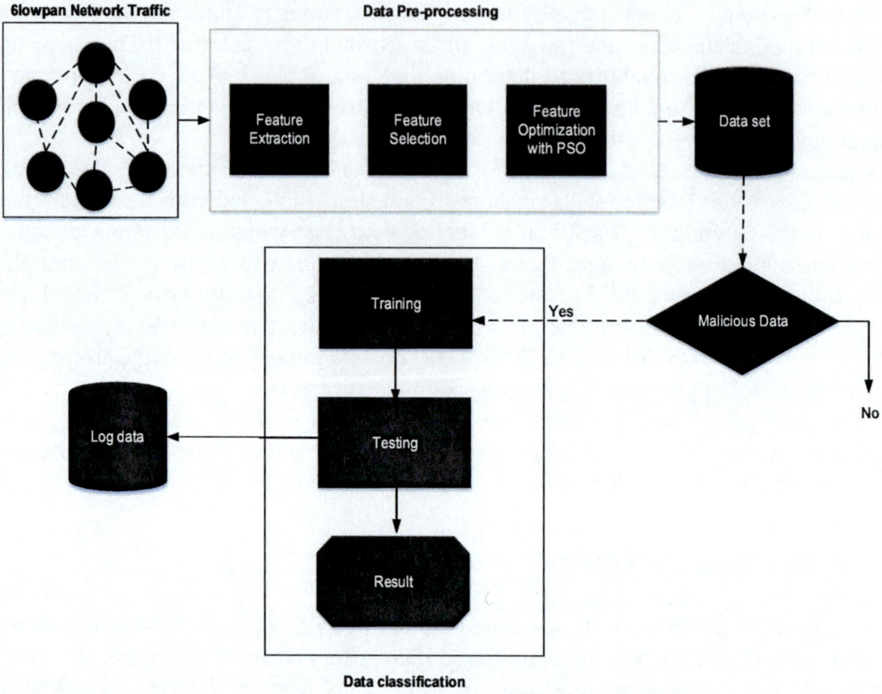

Fig. 1. The proposed architecture

are simulated. We dissected the PCAP file to CSV using Wireshark, and these CSV files are fed to the pre-processing section. The pre-processing data phase requires extracting valuable data features to avoid over-fitting and to acquire problem-oriented attributes. We simulated various situations of different network topologies and sizes for each type of attack. We get the raw datasets as a result of the simulations. After the simulation is done, Cooja exports PCAP and CSV files. However, raw data files are not adequate to be input to learning algorithms since the raw dataset contains information such as the address of the source/destination nodes and the length of the packet, which creates noise and over-fitting in the learning algorithm. For these purposes, we have built a function extraction algorithm using Python 3.7. These libraries make the mathematical operations required to remove the functionality simpler. To cope with a large number of nodes, we also create a dictionary structure. We decided not to calculate over the overall time of simulation or the overall number of packets, since this calculation could minimize the impact of the key features extracted. We have therefore separated all simulation into time frames or windows with a length of 5000 ms. The extracted features are presented in Table 2.

In our datasets, attack packets are labelled 1 and benign packets are labelled 0. We labelled datasets with malicious node 1 and labelled benign datasets 0.

Table 1. Dataset

	Malicious				Normal		
Datasets	Scenarios	No. of nodes	No. of M./N. nodes	Total packet count	Scenarios	No. of nodes	Total packet count
Hello Flood	HF_10	10	2/8	212.134	Normal_10	10	176.286
	HF_50	50	8/50	328.465	Normal_50	50	218.724
	HF_100	100	16/84	416.274	Normal_100	100	310.187
	HF_500	500	50/500	675.765	Normal_500	500	501.075
Wormhole	WH_10	10	2/8	121.126	Normal_10	10	118.172
	WH_50	50	8/50	147.465	Normal_50	50	129.557
	WH_100	100	16/84	317.673	Normal_100	100	238.933
	WH_500	500	50/500	719.764	Normal_500	500	450.193
Sinkhole	SK_10	10	2/8	121.361	Normal_10	10	117.213
	SK_50	50	8/50	227.186	Normal_50	50	165.763
	SK_100	100	16/84	301.392	Normal_100	100	216.031
	SK_500	500	50/500	815.534	Normal_500	500	721.554

3.3 Features Optimization Using PSO

The PSO method is a meta-heuristic method that enables the optimum of a function to be found in reasonable processing time, except for large instances where scientific computing is intensive, which requires considerable computing time. The use of appropriate parallel models reduces computation time and gives more satisfactory results than sequential models [16–18].

In particle swarm optimization, each population is called a particle. In standard PSO, after initialization of the population, each particle updates its velocity and location in each iteration on the basis of its own experience (pbest) and the best experience of all particles (gbest) as seen in Eq. 1 and Eq. 2. The efficiency of all particles will be measured by predefined cost functions at the end of each iteration.

$$v^i[\mathrm{t}+1] = \mathrm{w}.v^i[t] + c_1 r_1 (p^{i,best}[t] - p^i[t] + c_2 r_2 (p^{g,best}[t] - p^i[t]) \tag{1}$$

$$p^i[\mathrm{t}+1] = p^i[t] + v^i[t+1] \tag{2}$$

Where, $i = 1, 2, \ldots, N$, N is the number of swarm populations. $v^i[t]$ is the velocity vector in $[t]th$ iteration. $p^i[t]$ Represent the is the current position of the ith particle. $p^{i,best}[t]$ is the previous best position of ith particle and $p^{g,best}[t]$ is the previous best position of the whole particle. To control the pressure of local and global search, w has been used.c_1 And c_2 are positive acceleration coefficients which respectively called cognitive parameter and social parameter. r_1 And r_2 are random numbers between 0 and 1.

Table 2. Extracted features

Num	Abbreviation	Description
1	Num	Packet sequence number
2	Time	Simulation time
3	Src	Source node IP
4	Des	Destination node IP
5	TR	Transmission rate
6	RR	Reception rate
7	TAT	Transmission average time
8	RAT	Reception average time
9	TPC	Transmitted packet count
10	RPC	Received packet count
11	TTT	Total transmission time
12	TRT	Total reception time
15	DIO	DIO packet count
13	DAO	DAO packet count
14	DIS	DIS packet count
16	Tag	Normal/malicious label

4 Implementation and Evaluation

4.1 Experimental Setup

The three IoT routing attacks, namely hello flood, sinkhole and wormhole attacks, are launched in the Contiki network simulator known as Cooja. Contiki has proved to be a strong toolbox for developing complicated wireless networks and has produced a practical outcome in the actual network [19–21]. Furthermore, all the data used in the simulation are from the real network environment. In the simulation, Tmote Sky is used as client node and Cooja mote is used as a 6BR or sink node.

As seen in Table 1, we evaluate the different ML models across 4 network configurations (10 nodes, 50 nodes, 100 nodes, and 500 nodes). In the experimental section, we will discuss the findings of our 500-node experiment to demonstrate the feasibility of the model proposed in large networks.

4.2 Experimental Results and Evaluations

In this study, the best algorithm to classify the routing attacks is obtained by analyzing five machine-learning algorithms, i.e. K-Nearest Neighbour (K-NN), Support Vector Machine (SVM), Naïve Bayes (NB), Random Forest (RF) and Multilayer Perceptron (MLP).

• Hello Flood Attack

We evaluated the various machine learning approaches for the Hello Flood Attack Detection Model for the proposed IoT intrusion dataset. Performance indicators are presented in Table 3.

The result reveals that RF PSO and K-NN PSO have the highest TP rating and manage to detect Hello Flood at 99.5% and 99.14% respectively, as seen in Table 4. RF PSO achieved the best results with 99.8% precision, 99.5% recall and 99.8% accuracy when checked on the Hello Flood dataset. The K-NN PSO algorithm is ranked second with a TP rate of 99.14%, a precision of 98.6% and a precision of 99.2%.

Table 3. Hello flood attack detection

Evaluation criteria						
Classifiers optimized by PSO		Precision	Recall	Accuracy	TP rate	FP rate
	K-NN_PSO	98,6%	99,3%	99,2%	99,14%	0,019%
	SVM_PSO	96,5%	97,3%	99,9%	95,30%	0,057%
	RF_PSO	99,8%	99,5%	99,8%	99,50%	0,017%
	NB_PSO	97,3%	97,1%	96,8%	97,50%	0,027%
	MLP _PSO	96,9%	97,9%	97,8%	97,10%	0,031%

• **Wormhole Attack**

For the Wormhole attack, RF PSO and NB PSO attain 100% precision, recall and precision, while among other algorithms, SVM PSO achieves 98.7% accuracy, 98.2% recall and 98.5% accuracy.

Table 4. Wormhloe attack detection

Evaluation criteria						
Classifiers optimized by PSO		Precision	Recall	Accuracy	TP rate	FP rate
	K-NN_PSO	99,1%	99,4%	99,3%	98,15%	0,041%
	SVM_PSO	98,7%	98,2%	98,5%	96,85%	0,054%
	RF_PSO	100%	100%	100%	98,40%	0,014%
	NB_PSO	100%	100%	100%	98,14%	0,016%
	MLP _PSO	99,5%	98,8%	99,3%	97,25%	0,035%

- **Sinkhole Attack**

In the Sinkhole attack, RF PSO again achieves the highest detection efficiency as seen in Table 5, with 99.6% precision, 99.4 recall and 99.7 precision, 99.71 TP and 0.002 FP.

Table 5. Sinkhole attack detection

Evaluation criteria						
Classifiers without optimization		Precision	Recall	Accuracy	TP rate	FP rate
	K-NN_PSO	95,4%	95,7%	95,5%	95,25%	0,065%
	SVM_PSO	94,7%	94,2%	94,2%	94,65%	0,079%
	RF_PSO	99,6%	99,4%	99,7%	99,71%	0,002%
	NB_PSO	94,2%	94,1%	93,9%	94,15%	0,071%
	MLP _PSO	95,3%	95,7%	96,1%	96,12%	0,055%

5 Conclusion and Future Works

IoT-6LoWPAN nodes are vulnerable to multiple attack attacks. The proposed ML detection models accurately identify IoT routing attacks (hello-flood attack, wormhole attack, sinkhole attack). This paper also addresses a weakness in detecting IoT routing attacks. The key concern in this field is the lack of datasets, and current databases like KDDcup 99 are too old. We generated real data from IoT-6LoWPAN network traffic recorded by network sniffers like Wireshark. We also developed a detection model based on machine learning techniques powered by realistic PSO swarm optimization, trained them with routing data sets generated and created different models for attack detection. We evaluated five streamlined algorithms, i.e. MLP, SVM, Naïve Bayes, Logistic and Random Forest use five tools to track and identify routing attacks. The Random Forest algorithm is selected for the proposed IDS as RF provides the best results of all algorithms in detecting routing attacks. The proposed ML PSO IDS detected both individual and new anomaly attacks generated by mixing routing attacks.

Further experiments will be conducted for multiple scenarios with more attacks and usual node values to equate the efficacy of the dataset used in this paper with other literature intrusion detection datasets. We intend to enhance our IoT dataset with new routing attacks. Our goal is to improve predictive output for three routing attacks and include further routing attacks.

References

1. Hajiheidari, S., Wakil, K., Badri, M., Navimipour, N.J.: Intrusion detection systems in the Internet of things: a comprehensive investigation. Comput. Netw. **160**, 165–191 (2019). https://doi.org/10.1016/j.comnet.2019.05.014

2. Janakiram, D., Reddy, V.A., Kumar, A.V.U.P.: Outlier detection in wireless sensor networks using Bayesian belief networks. In: 2006 1st International Conference on Communication Systems Software & Middleware, pp. 1–6 (2006). https://doi.org/10.1109/COMSWA.2006.1665221.
3. Zhang, Y., Meratnia, N., Havinga, P.J.M.: Distributed online outlier detection in wireless sensor networks using ellipsoidal support vector machine. Ad Hoc Netw. **11**(3), 1062–1074 (2013). https://doi.org/10.1016/j.adhoc.2012.11.001
4. Jha, J., Ragha, L.: Intrusion detection system using support vector machine. Int. J. Appl. Inf. Syst. (ICWAC), 25–30 (2013). https://doi.org/10.5120/758-993
5. Maleh, Y., Ezzati, A.: Lightweight intrusion detection scheme for wireless sensor networks. IAENG Int. J. Comput. Sci. **42**(4) (2015)
6. Nait-Abdesselam, F., Bensaou, B., Taleb, T.: Detecting and avoiding wormhole attacks in wireless ad hoc networks. IEEE Commun. Mag. **46**(4), 127–133 (2008). https://doi.org/10.1109/MCOM.2008.4481351
7. Alpaydin, E.: Introduction to Machine Learning. MIT Press, Cambridge (2014)
8. Verma, A., Ranga, V.: Evaluation of network intrusion detection systems for RPL based 6LoWPAN networks in IoT. Wirel. Pers. Commun. **108**(3), 1571–1594 (2019). https://doi.org/10.1007/s11277-019-06485-w
9. Verma, A., Ranga, V.: Machine learning based intrusion detection systems for IoT applications. Wirel. Pers. Commun. **111**(4), 2287–2310 (2020). https://doi.org/10.1007/s11277-019-06986-8
10. Liu, J., Kantarci, B., Adams, C.: Machine Learning-driven intrusion detection for Contiki-NG-based IoT networks exposed to NSL-KDD Dataset. In : Proceedings of the 2nd ACM Workshop on Wireless Security and Machine Learning, pp. 25–30 (2020). https://doi.org/10.1145/3395352.3402621
11. Maleh, Y.: Machine learning techniques for IoT intrusions detection in aerospace cyber-physical systems. Mach. Learn. Data Min. Aerosp. Technol. (2020). https://doi.org/10.1007/978-3-030-20212-5_11
12. Laaksonen, J., Oja, E.: Classification with learning k-nearest neighbors 1996. In: IEEE International Conference on Neural Networks, 1996, vol. 3, pp. 1480–1483 (1996). https://doi.org/10.1109/ICNN.1996.549118.
13. Rish, I.: An empirical study of the naive Bayes classifier. In : IJCAI 2001 Workshop on Empirical Methods in Artificial Intelligence, vol. 3, no. 22, pp. 41–46 (2001)
14. Quinlan, J.R.: Induction of decision trees. Mach. Learn. **1**(1), 81–106 (1986). https://doi.org/10.1023/A:1022643204877
15. Jing, R., Zhang, Y.: A view of support vector machines algorithm on classification problems. In: 2010 International Conference on Multimedia Communications, pp. 13–16 (2010). https://doi.org/10.1109/MEDIACOM.2010.21
16. Kennedy, J., Eberhart, R.C.: Particle swarm optimization. Encycl. Mach. Learn.**4**, 760–766 (2010)
17. Wang, D., Tan, D., Liu, L.: Particle swarm optimization algorithm: An overview. Soft Comput. **22**(2), 387–408 (2018). https://doi.org/10.1007/s00500-016-2474-6
18. Shi, Y.: Particle swarm optimization. IEEE Connect. **2**(1), 8–13 (2004)
19. Pongle, P., Chavan, G.: Real time intrusion and wormhole attack detection in internet of things. Int. J. Comput. Appl. **121**(9) (2015)
20. Raza, S., Wallgren, L., Voigt, T.: SVELTE: real-time intrusion detection in the Internet of Things. Ad Hoc Netw. **11**(8), 2661–2674 (2013). https://doi.org/10.1016/j.adhoc.2013.04.014
21. Napiah, M.N., Idris, M.Y.I.B., Ramli, R., Ahmedy, I.: Compression header analyzer intrusion detection system (CHA - IDS) for 6LoWPAN communication protocol. IEEE Access **6**, 16623–16638 (2018). https://doi.org/10.1109/ACCESS.2018.2798626.

Inadequacy of IT Approaches to Manage Cyber Security in ICS Context

Omar El Idrissi(✉), Abdellatif Mezrioui, and Abdelhamid Belmekki

STRS Lab, RAISS Team, INPT, Rabat, Morocco
{elidrissi.omar,mezrioui,belmekki}@inpt.ac.ma

Abstract. The paper "Inadequacy of IT approaches to manage cyber security in ICS context" gives an overview of the application of different IT solutions to deal with cyber security for industrial control systems (ICSs). The real time aspect is important for ICSs to monitor and control effectively different processes. On the other hand, some traditional IT solutions for cyber security do not comply with it. After a careful analysis, it is concluded that there is no ISO 27001 requirement to meet the "non interference" cyber security goal, used by the working group PCSRF in the System Protection Profile for ICSs to ensure that the security functions are implemented in a non-interfering way and do not affect the performance constraints of ICSs. Therefore, a new component, called "Cyber Safety", is proposed to be added to the existing ones (confidentiality, integrity and availability). Its role would be to check the safety requirements (controls or measures) before their implementations. It is proposed a new model for the management of the cyber security risks that reuses existing results issued from the safety risk analysis and may add safety requirements.

Keywords: Industrial Control System · Critical infrastructure · Cyber security · Risk management

1 Introduction

Industrial Control Systems (ICS) are commonly used to monitor and control industrial infrastructures providing vital services like electricity, water, transportation, manufacturing, etc. Initially, ICS were designed to operate in isolated and autonomous mode. They were designed without any security requirements in mind, while safety and physical security have been considered as the most crucial concern in industrial control system. As the industry opens up to the outside world, with increased connectivity of external systems, industrial cyber security is becoming one of the most complex and delicate topics in industrial environments. Indeed, the convergence of IT and Operational Technology (OT) in recent years makes industrial control systems a prime target for hackers and cybercriminals [1, 2]. The number of cyber incidents targeting ICS/SCADA systems has increased dramatically in recent years, and the same goes for the number of ICS vulnerabilities which in turn has increased considerably [3]. While, the cyber security threats can lead to the same dangerous and disastrous phenomenon as a safety incident [4], we note

A. Abraham et al. (Eds.): SoCPaR 2020, AISC 1383, pp. 678–689, 2021.
https://doi.org/10.1007/978-3-030-73689-7_65

on the other hand, some limitations of management methods, untrained and unskilled developers in business management and organizational areas, various classifications and non-unique taxonomies of vulnerabilities [5]. 77% of industrial organizations are concerned about cyber security incidents, but half have no appropriate response program [6]. Contributing to this challenge is that the methodologies and approaches available today to handle cyber risk in operating industrial environments have been developed for traditional IT domains, and when applied to engineering environments, there are unfit for this purpose because factors such as exposure, threats, and consequences are different [7]. In this regard, implementing IT security solutions and tools (proxies, Firewalls, IDS…) has a negative impact on the performance of the ICS, including real-time requirements and system availability and can lead to communication latency between the different industrial instruments [8, 9], therefore and to maintain system availability, IT security solutions are rarely applied to protect ICS from cyber risks [10]. To face this problem, a significant number of research works have been published in the field of cyber security risk management related to the industrial control system with two different perspectives. Some researchers try to develop new methodologies exclusive to industrial environments, for example IEC-62443 which is a series of standards including technical reports to secure Industrial Automation and Control Systems (IACS), and it provides a systematic and pragmatic approach to industrial systems cyber security. Every stage and aspect of industrial cyber security is covered, from risk assessment through operations, while others prefer not to reinvent the wheel and work to adapt available IT approaches to be applicable in the ICS domains. In Park and Lee (2014), the authors discuss the need to update and adapt international security standards such as NIST SP 800-53 and ISO 27001 in order to consider and address the specifics of the ICS. For that, in this paper, we will analyze and study the second perspective and verify whether we can adapt IT approaches and find a way to bring them back useful in the industrial context. This paper is structured as follows:

We clarify in the next section, the meanings of the terms safety and security used in the context of this paper. In Sect. 3, we provide an overview of the ICS architecture and we introduce the concepts of safety requirements for industrial control systems. In Sect. 4, we present some constraints related to ICSs cyber security management. In Sect. 5, we highlight the global idea behind this study and we undertake an ICS cyber security management analysis. Section 6 proposes a process for managing ICS cyber security using IT mechanisms. Finally, the paper overview and perspectives are summarized in Sect. 7.

2 Terminology

The definitions of the terms safety and security vary widely in different contexts and technical communities [11]. The goal of cyber security is to protect or defend the use of cyberspace from cyber-attacks [12]. Siwar Kriaa et al., [13] studied the interdependencies between safety and security and explain the similarities and differences between safety and security. In our context we relate the cyber security risks to the risks of malicious origin, while we associate the safety with the risks of accidental nature such as natural disasters, environmental, mechanical failure, and inadvertent actions of an authorized user [14].

3 Background of ICSs

3.1 Industrial Control System Operation

ICSs are designed to support industrial processes. It aims to monitor and to control in real time a large number of processes and operations of complex infrastructures. ICS are used in different domain such as gas and electricity distribution (conventional and nuclear), water treatment, oil refining and rail transportation etc. They consist of numerous control loops, human-machine interfaces, diagnostic and maintenance tools that communicate with each other across an industrial network. The figure below (Fig. 1) illustrates the basic operating process of the ICSs.

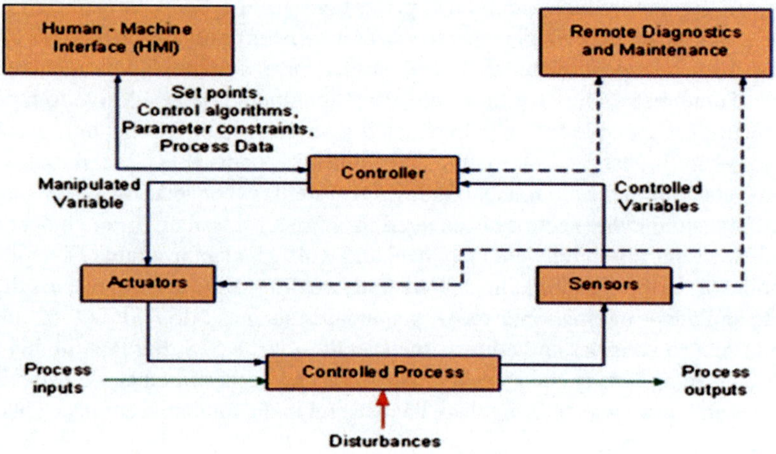

Fig. 1. ICS operation [15]

There are also different types of ICS (SCADA, DCS, PLC and SIS) that will be presented below.

SCADA Systems (Supervisory Control and Data Acquisition): SCADA systems are used to control geographically dispersed assets. These systems are used in distribution systems such as water and wastewater systems, oil and gas pipelines, electricity grid, rail and other transportation systems.

DCS system (Distributed Control Systems): A DCS is a control system of industrial processes located geographically in the same area, such as oil refineries, water and wastewater treatment, power plants, power plants manufacture of chemicals, etc.

PLC (Programmable Logic Controller): PLCs are used as a primary controller in smaller control system configurations to provide operational control of processes.

SIS (Safety Instrumented Systems): SISs are designed to bring the process to a safe state when process conditions that threaten safety are detected. SIS is used to perform safety functions. For example it can quickly shutdown a process and isolate it completely to prevent dangerous situations from occurring or getting worse. It may order a shutdown of the entire factory, unit or equipment if necessary.

3.2 Safety Requirement for Industrial Control Systems

Physical safety is the most crucial and the main requirement in industrial control [1]. It is a prime concern of industrial activities that directly affect the engineering and operational decisions of an industrial control system. Indeed, ICS systems are designed according to IEC 61508 and IEC 61511 to ensure certain prerequisites and requirements related to safety of the industrial infrastructure.

Table 1. IEC 61511 (Requirements)

IEC 61511 (Requirements)
Management of functional safety
Safety life-cycle requirements
Verification
Process hazard and risk analysis
Allocation of safety functions to protection layers
SIS safety requirements specification
SIS design and engineering
Requirements for application software including selection criteria for utility software
Factory acceptance testing
SIS installation and commissioning
SIS safety validation
SIS operation and maintenance
SIS modification
SIS decommissioning
Information and documentation requirements

IEC 61508 is a generic and complete standard that covers the complete safety life-cycle including the analysis, realization and operation phases and common to several industries, from which different industries derived their own standards (e.g., IEC 61513 for the nuclear industry, IEC 61511 for industrial processes, and ISO 26262 for the automotive industry) [16]. The safety requirements of IEC 61511 are composed of 15 domains and the total number of controls is 215 pieces, divided into five safety parts and the safety parts consist of development, allocation, design, installation, commissioning, validation, operation, modification, and decommissioning for an ICS [17]. Safety requirements are met partly by the safety-related technology other than safety instrumented systems (SIS), such as relief valves, alarms, and other specific-safety devices [18]. Any safety strategy must therefore consider not only all the elements within an individual system (for example sensors, controlling devices and actuators) but also all the safety-related systems making up the total combination of safety-related systems [19]. The Table 1 shows the safety requirements according to IEC 61511.

4 Industrial Control Systems Cyber Security Management Constraints and Difficulties

In addition to its primary purpose of controlling and supervising the industrial process, ICS also aims to protect industrial infrastructure, users and environment from physical accidents in general. However, in recent years, ICS general safety concern may arise in particular from malicious cyber threat factors that attempt to disrupt an industrial process such as interfering with its specific operations (e.g. to create a power outage) or to negatively impact the environment and/or personal safety (e.g. exploding a fuel tank or destabilizing chemical process to free noxious gases) [20]. Unfortunately, facing this type of threat suffers from many limitations and obstacles, we provide some examples below.

4.1 Insecure by Design

Initially, the ICS were designed to operate, with specific protocols and hardware, in isolated and autonomous mode. They were designed without any cyber security requirements in mind. Most components of ICSs (PLCs, protocol converters or data acquisition servers) lack even basic authentication and accept generally any properly formatted command [21]. On the other hand, the communication protocols used by ICS environments to control field devices are proprietary protocols that are designed to ensure efficiency, reliability and accuracy in real-time operation. This means that any other functions of the protocol have been omitted [22].

4.2 Resources Constraints

ICS operating systems (OS) and applications may not have the computing resources to tolerate typical security practices including upgrade these systems with current security capabilities and desired features such as encryption capabilities, error logging and password protection [23].

4.3 Interaction with Physical Environment

ICS have been designed to meet high performance and reliability requirements, it have different priorities and involve risks that are much broader in scope and impact. Cyber security issues in ICS systems can disrupt system functionality and interfere with functional requirements that have a potential physical impact including significant risks to the safety of human lives and severe damage to the environment [24, 25].

4.4 Management Skills

In addition to all constraints we have mentioned above, there is a very critical point that further complicates the situation, as the ICS management team generally consists of automation professionals and their knowledge of the norms and standards related to cyber security is often very limited. Unfortunately and despite the criticality of industrial cyber security, in the European Union, limited resources are dedicated specifically to this area and the most efforts in cyber security risk management are limited to the IT field [26].

5 ICS Cyber Security Management Analysis

For a longtime, only accidental component failures and software errors were traditionally addressed in industrial environment. Today, cyber-attacks and security breaches can also compromise the safety of the system [13] and can lead to various risks affecting the critical infrastructure business continuity, including degradation of production and performance, unavailability of critical services, and violation of the regulation [27]. Therefore, great attention should be paid to cyber security issues and their potential impacts on critical infrastructure systems, but as we have seen in the previous section, securing ICS is a huge challenge due especially to the fact that most IT security approaches are inadequate and not suitable for ICS that have specific requirements. In this regard, we will extract the points of incompatibility and inadequacy that exist between traditional cyber security mechanisms and industrial environments by studying and analyzing de gap between ICS cyber security objectives and ISO 27001 controls and requirements.

5.1 Cyber Security Objectives for Industrial Control Systems

We chose to work with the objectives instead of the requirements because working directly with the requirements of all ICS applications is very complicated due to the differences between processes and networks deployed across different industries. On the other hand, common cyber security objectives are sufficient for every ICS, regardless of the specific application. For that and to identify the cyber security objectives for ICSs, we will use the System Protection Profile for ICSs (SPP-ICS) document produced by the working group PCSRF that initiated by NIST. The goal of PCSRF group is to assess vulnerabilities and to design and document a set of security specifications using the Common Criteria for Information Technology Security Evaluation, also known as ISO/IEC 15408 [19]. The Table 2 below shows the cyber security objectives for industrial control systems according to SPP-ICS document.

Table 2. Cyber security objectives for ICS (SPP-ICS) [20]

Cyber security objectives for ICS (SPP-ICS)
O.1 Physical
O.2 Risk
O.3 Non_interference
O.4 Interconnectivity
O.5 Data_Backup
O.6 Data_authentication
O.7 Continuity
O.8 Management
O.9 Migration

(continued)

Table 2. (*continued*)

Cyber security objectives for ICS (SPP-ICS)
O.10 Compliance
O.11 3Rdparty
O.12 Remote
O.13 Acess_control
O.14 Secure_comms
O.15 Data_integrity
O.16 Confidentiality
O.17 Availability
O.18 System_integrity
O.19 System_diagnostics
O.20 Monitoring
O.21 Audit
O.22 IDS

5.2 Security Requirement for Information Technology

The main cyber security concern according to the Standard "Information Security Management System" (ISMS) is to protect information from harmful threats, as ISMS aims to continuously control, monitor and maintain IT security whenever required. ISO 27001 as an IT security management system offers a set of controls to protect assets and reduce risk to an acceptable level (Table 3).

Table 3. ISO 27001 controls and requirements

ISO 27001 (Requirements + Controls)
R.4 Context of the organization
R.5 Leadership
R.6 Planning
R.7 Support
R.8 Operation
R.9 Performance evaluation
R.10 Improvement
A.5 Information security policies
A.6 Organization of information security

(*continued*)

Table 3. (*continued*)

ISO 27001 (Requirements + Controls)
A.7 Human resources security
A.8 Asset management
A.9 Access control
A.10 Cryptography
A.11 Physical and environmental security
A.12 Operations security
A.13 Communications security
A.14 System acquisition, development and maintenance
A.15 Supplier relationships
A.16 Information security incident management
A.17 Information security aspects of business continuity management
A.18 Compliance

5.3 Mapping Cyber Security Objectives for ICS (SPP-ICS) with ISO 27001 Controls

The aim of this section is to extract the points of incompatibility and inadequacy that exist between traditional standards and industrial environments. For that we compare and analyze the cyber security objectives for ICS in accordance to SPP-ICS with ISO 27001 controls and requirements. Our comparison method is as follows: For each ICS cyber security objective identified in the System Protection Profile for Industrial Control Systems (SPP-ICS) document, we examine a similar security topic (requirements and controls) in ISO/IEC 27001 (Table 4).

Table 4. Mapping cyber security objectives for ICS (SPP-ICS) with ISO 27001 controls

Cyber security objectives for ICS	ISO/IEC 27001 (Requirements & Controls)
O.1 Physical	A.11
O.2 Risk	R.6, R.8, A.12
O.3 Non_interference	**None**
O.4 Interconnectivity	A.9
O.5 Data_Backup	A.12
O.6 Data_authentication	A.14, A.9, A.10, A.12
O.7 Continuity	A.17
O.8 Management	R.4, R.5, A.5, A.6

(*continued*)

Table 4. (*continued*)

Cyber security objectives for ICS	ISO/IEC 27001 (Requirements & Controls)
O.9 Migration	R.10, A.12,
O.10 Compliance	A.18
O.11 3Rdparty	A.7, A.9, A.14
O.12 Remote	A.6,
O.13 Acess_control	A.9
O.14 Secure_comms	R.9, A.10, A.13
O.15 Data_integrity	A.9, A.10, A.12
O.16 Confidentiality	A.9 A.10
O.17 Availability	A.17
O.18 System_integrity	R.9, A.9, A.11, A.12
O.19 System_diagnostics	R.9
O.20 Monitoring	R.9, A.12
O.21 Audit	A.12, A.16, A.18
O.22 Ids	A.12, A.14, A.18

5.4 ICS Risk Management Requirements

Through our mapping analysis, we conclude that there is no ISO 27001 requirement or control to meet the following cyber security objective: "O.3 NON_INTERFERENCE: *The ICS security functions shall be implemented in a noninterfering manner such behavior of the ICS functions and safety functions are able to meet their performance constraints*" [20]. This objective makes it possible to verify and assess the impact of cyber security measures and controls on the operational safety part. Since this objective is not covered by ISO 27001, the use of this kind of ISMS to manage ICS cyber security based on Confidentiality, Integrity and Availability (CIA) triad never achieves mutually exclusive security policy for an ICS and must to be improved by adding some components to manage interference of IT controls with functional safety [28]. For that we associate the objective "O.3 NON_INTERFERENCE", with a new component other than CIA (confidentiality, Integrity and availability): "Cyber-Safety". Therefore cyber security for industrial control systems should be based on the Cyber-Safety like as CIA. This new principle "Cyber-Safety" consists of verifying the effect of each security control or measure on safety requirements before it is implemented.

6 ICS Cyber Security Risk Management Process

The application of security controls in the industrial control system can have an impact on the safety and the functioning of the industrial system, as we have seen in the above section, it is necessary to integrate the principle "Cyber-safety" in the process of risk management to evaluate and verify the impact after the implementation of a security measure (for example a firewall blocks or delays the sending of alarms) on the functional safety. In the following we propose a new process to manage cyber security risk in ICS environment that integrate the "Cyber-Safety" principle as recommended in Sect. 4 (Fig. 2).

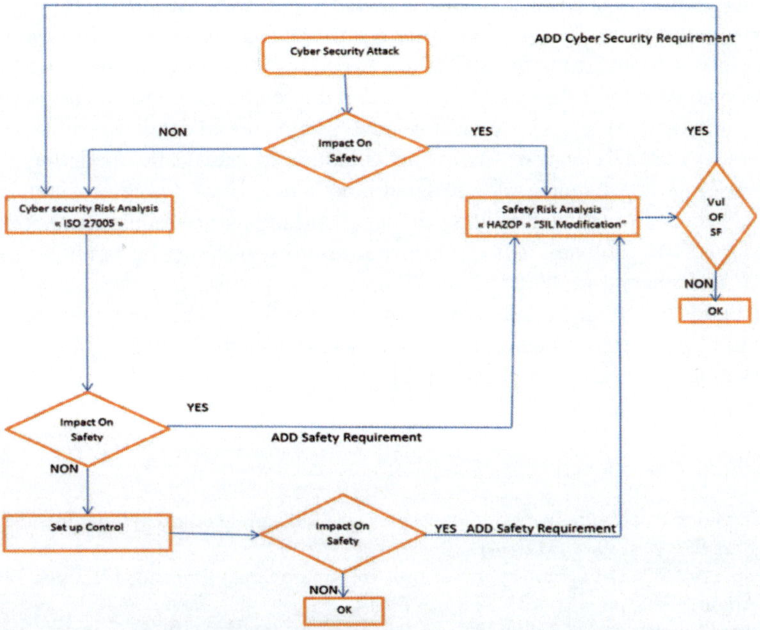

Fig. 2. ICS cyber security risk management process

The added value of our process is to simplify cyber security risk management by reusing existing results from safety risk analysis, and to update the safety risk management plan by adding safety requirements.

7 Conclusion and Perspectives

ICS systems are exposed to potential cyber security risks of different natures and patterns that could have a serious impact to the health, safety of human lives and serious damage to the environment [15] and can lead to the same dangerous phenomenon as a safety incident [4]. On the other hand, the simple implementation of some standard security measures also may compromise the functional safety requirements and affect negatively the proper functioning of the system; we have cited some scenarios to illustrate these kinds of situations. Through our mapping analysis, we concluded that "non_interference" objective is obligatory to manage cyber security in ICS in order to evaluate and avoid negative impact on industrial safety aspect, we have proposed a preliminary process which meet the Cyber-safety requirements. However, we believe that the challenge is more complex than we have seen, given the complex interference and mutual influence between safety and cyber security fields. Since the requirements and objectives related to security and safety domains converge and can interact with each other, the introduction of cyber security risks in the industrial environment makes the risk analysis related to functional safety already undertaken not credible and requires redoing it from the beginning (did not consider all possible risks, especially, cyber security risk). Therefore, ICS cyber security risk management system must be redesigned in a coherent manner that takes into account the cyber security and functional safety requirements (IEC 61508 and derived standards) at the same time. For that the relationship and interdependencies between safety and cyber security must be analyzed in careful detail. Scientific research in this area will need to focus on finding new methods to manage the modeling of safety and security interdependencies and respond how to deal if safety requirements conflict with good security practice, including the legal and regulatory requirements related to the activity of the company. Also, ICS cyber security plans and programs should be consistent and integrated with current IT security expertise, programs, and practices as they should be explained and clarified the method for coordination and cooperation between the cyber security team and the safety team which is essential for the success of the cyber security risk management project.

References

1. Xu, X.: Global and initiative safety mechanism in industrial control system. Int. J. Comput. Sci. Eng. **9**(1–2), 139–146 (2014)
2. Fovino, I.N.: SCADA system cyber security. In: Secure Smart Embedded Devices, Platforms and Applications, pp. 451–471. Springer (2014)
3. ENISA: Communication network dependencies for ICS/SCADA Systems, December 2016
4. Abdo, H., Kaouk, M., Flaus, J.-M., Masse, F.: A safety/security risk analysis approach of industrial control systems: a cyber bowtie - combining new version of attack tree with bowtie analysis. Comput. Secur. **72**, 175–195 (2017)
5. Hentea, M.: Improving security for SCADA control systems (2008)
6. Kaspersky: https://www.developpez.com/actu/213156/Kaspersky-77-pourcent-des-organisations-industrielles-s-inquietent-des-incidents-de-cybersecurite-mais-la-moitie-n-a-aucun-programme-de-reponse-approprie/
7. Mohr, R.: Evaluating cyber risk in engineering environments: a proposed framework and methodology (2016)
8. Stouffer, K.A., et al.: Roadmap for measurement of security technology impacts for industrial control systems (2014)
9. Falco, J., Gilsinn, J., Stouffer, K.: IT security for industrial control systems: requirements specification and performance testing (2004)
10. Hashimoto, Y., et al.: Safety securing approach against cyber-attacks for process control system. Comput. Chem. Eng. **57**, 181–186 (2013)
11. Burns, A., McDermid, J., Dobson, J.: On the meaning of safety and security. Comput. J. **35**, 3–15 (1992)
12. Kissel, R.: Glossary of key information security terms. U.S. Department of Commerce, National Institute of Standards and Technology (2013)

13. Kriaa, S., Pietre-Cambacedes, L., Bouissou, M., Halgand, Y.: A survey of approaches combining safety and security for industrial control systems. Reliab. Eng. Syst. Saf. **139**, 156–178 (2015)
14. Piètre-Cambacédès, L.: Des relations entre sûreté et sécurité. Cryptographie et sécurité [cs.CR]. Télécom ParisTech (2010)
15. NIST: Guide to Industrial Control Systems (ICS) Security: NIST Special Publication 800-82 R2 (2014)
16. Kriaa, S.: Joint safety and security modeling for risk assessment in cyber physical systems. Other. Université Paris Saclay (COmUE) (2016)
17. International Electrotechnical Commission, IEC 61511-1 (2003)
18. Marcin, Ś, Emilian, P.: Integrated approach for functional safety and cyber security management in maritime critical infrastructures. J. Pol. Saf. Reliab. Assoc. Summer Saf. Reliab. Semin. **10**, 137–148 (2019)
19. IEC 61508:2010: Functional safety of electrical/electronic/programmable electronic safety-related systems (2010)
20. NIST: System Protection Profile--Industrial Control Systems Version 1.0. (2004)
21. ICS-CERT: Overview of Cyber Vulnerabilities. https://us-cert.cisa.gov/ics/content/overview-cyber-vulnerabilities
22. Drias, Z., Serhrouchni, A., Vogel, O.: Analysis of cyber security for industrial control systems. In: 2015 International Conference on Cyber Security of Smart Cities, Industrial Control System and Communications (SSIC), Shanghai, pp. 1–8 (2015)
23. Macaulay, T., Singer, B.: Cybersecurity for Industrial Control Systems: SCADA, DCS, PLC, HMI, and SIS, p. 204. Auerbach Publications (2011). ISBN 9781439801963
24. El Idrissi, O., Mezrioui, A., Belmekki, A.: A lightweight risk analysis of a critical infrastructure based ICSs. In: 2019 1st International Conference on Smart Systems and Data Science (ICSSD), Rabat, Morocco, pp. 1–8 (2019)
25. Lyu, X., Ding, Y., Yang, S.: Safety and security risk assessment in cyber-physical systems. IET Cyber-Phys. Syst. Theory Appl. **4**(3), 221–232 (2019)
26. ENISA: Analysis of ICS-SCADA Cyber Security Maturity Levels in Critical Sectors (2015)
27. Kure, H.I., Islam, S., Razzaque, M.A.: An integrated cyber security risk management approach for a cyber-physical system. Appl. Sci. (Switzerland) **8**(6), 898 (2018)
28. Park, S., Lee, K.: Advanced approach to information security management system model for industrial control system. Sci. World J. **2014** (2014). Article no. 348305, 13 pages

Trust-Based Intrusion Detection for Multi Path OLSR Protocol

Hicham Zougagh[1](✉), Noureddine Idboufker[2], and Youssef Saadi[1]

[1] Computer Sciences Department, Sultan Moulay Slimane University, Beni Mellal, Morocco
[2] National School of Applied Sciences, Cadi Ayyad University, Marrakesh, Morocco

Abstract. In this paper, we have focused on the implementation of a mechanism for detecting malicious nodes in the MP-OLSR routing protocol through the integration of trust reasoning. The study carried out has enabled the emergence of trust rules between nodes based on local information and messages exchanged during the neighborhood discovery process and network topology to make the MP-OLSR protocol more trusted. The correlation between the received control messages allows nodes to validate the neighborhood at one hop and two hops. When the information received is consistent with the MP-OLSR protocol specifications, this reasoning allows the creation of a trust relationship. Otherwise, it allows nodes to be wary of the sources of its inconsistencies. When an attack is detected, the countermeasure proposed allows the attacking node to be isolated from the network and to guarantee its non-participation in the routing mechanism by simply integrating this node in the HELLO message with a modified Link Code, so that it is declared by all the nodes of the network as an attacker.

Keywords: Trust · Ad hoc networks · Security · Multipath OLSR routing protocol

1 Introduction

An ad hoc network is a Distributed System, composed of multiple autonomous entities capable of communicating via a radio medium, among them without the need for a centralized infrastructure [1]. In addition, cooperation between the different entities of the ad hoc network allows the continuity of services to be maintained.

Routing takes an important place in an Ad-hoc network in that it allows logical paths to be established and maintained so that any entity can communicate with any other entity, even if one is outside the transmission range of the other [2].

Ad hoc mobile networks are confronted to many problems [3] due to their intrinsic characteristics making security solutions developed for wired or infrastructure-based networks generally not adapted to the context of Ad hoc mobile networks. Indeed, security in Ad-hoc networks is a rather complex problem due to the conjunction of several factors that increase vulnerability to various types of attacks that can be launched in a relatively simple way [4, 5]. Indeed, the use of wireless connections facilitates eavesdropping, allowing the analysis of network traffic and subsequently making it possible to launch active types of attacks.

A. Abraham et al. (Eds.): SoCPaR 2020, AISC 1383, pp. 690–705, 2021.
https://doi.org/10.1007/978-3-030-73689-7_66

Due to the fact that the functioning of the ad hoc networks depends entirely on cooperation between its entities, significant degradations can be expected when entities attempt to divert operations from the bases for which they are responsible. Unfortunately, most ad hoc routing protocols do not incorporate security control mechanisms, as they assume honest behavior between collaborating entities. However, the reality can be very different in the presence of malicious entities capable of performing intrusions by exploiting vulnerabilities in the ad hoc infrastructure [6].

Our paper presents a trust-based solution that allows, based on individual observations and control messages received, to check the behavior of nodes and detect anomalies in their behavior. We will detail prevention and countermeasure techniques to overcome some of the vulnerabilities of MP-OLSR to stop attacks by integrating some mistrust rules [7–9] deduced from the MP-OLSR [10] protocol specifications.

The paper is organized as follows. In Sect. 2, related works on security in ad hoc networks are summarized. In Sect. 3, an overview of MP-OLSR is presented. In Sect. 4 we introduce the concept of trust management and trust specification language. In Sect. 5, we introduce the analysis of implicit trust in MP-OLSR, then we present trust reasoning developed to secure MP- OLSR. In Sect. 6. Countermeasures concerning the attacks against MP-OLSR. Finally, we conclude this paper by presenting simulation results.

2 Related Works

In this section, we propose the different Multi-Path routing strategies for MANET networks that exist in the literature. Most of the work concerning Multi-Path routing focuses on reactive routing protocols.

In [11] a summary of the different Multi-Path routing protocols derived from DSR and AODV is presented. These contributions classify the different Multi-Path routing strategies according to their goal (reachability, QOS,...). In [12] the authors propose a Multi-Path Routing Protocol (MPR-LB) to reduce congestion in MANETs. This protocol is based on the reactive approach. It defines two other phases: data transmission and overhead balancing.

According to [13], Multi-Path reactive routing protocols increase the number of control messages. As a result, redundancy of routing packets is introduced in networks. Moreover, the search for disjoint node routes makes the discovery of routes very long compared to DSR or AODV.

In addition, several Multi-Path approaches take advantage of the proactive behavior and flooding mechanism by MPR nodes proposed in the OLSR protocol. In [14], the author proposed a different version of multi-path OLSR using source routing, based on the dijkstra algorithm, nodes compute disjoint paths. In addition, the authors introduced an algorithm that allows the distribution of the network load over the different calculated paths.

In a work [15], the author have developed another multi-path approach for the OLSR protocol, taking into account the QoS constraints, i.e. bandwidth and throughput (QOLSR). This work also includes a comparative study between single-path QOLSR and multi-path QOLSR. In order to secure the routing mechanism.

In [16], the authors propose a solution to remedy this problem. Zhang et al. proposed in [10] the protocol (SR-MPOLSR) which uses source routing and the dijkstra algorithm in the calculation of multiple paths.

Not all of the proposed strategies take into consideration the security constraint. Both types of proactive and reactive routing protocols are vulnerable to multiple attacks. Especially at the level of control message routing. Therefore, in order to guarantee a very high level of security of the data exchanged and to ensure an optimal level of the rate of data packets delivered in a MANET network, we have adopted the MP-OLSR routing protocol [13] which is an extension of the OLSR protocol. To guarantee the desired optimal level of delivered data packets, and in order to limit attacks against this protocol, each node of the network must build multiple and disjoint paths to the other nodes of the network with more than two hops.

As MP-OLSR inherits the OLSR protocol. Its mechanisms for discovering neighborhoods and network topology. So, they have the same vulnerabilities. In order to solve the security problems in MP-OLSR, we present in this chapter a trust-based reasoning to check the behavior of nodes and detect behavioral anomalies, based on individual observations and control messages received. Next, we detail prevention and countermeasure techniques to overcome some of the vulnerabilities of MP-OLSR and to stop attacks by integrating some mistrust rules deduced from the MP-OLSR protocol specifications.

3 Multipath OLSR Routing Protocol

Multipath Optimized Link State Routing (MP-OLSR) [6] is an extension of OLSR. multipath that can be considered as a hybrid routing method. It is based on the periodic sending of HELLO messages and TC messages to find out the network topology, exactly like OLSR. The difference between the two is that MPOLSR does not always keep a routing table. It only calculates routes when there are of the data packets to be transmitted. The MP-OLSR specification has two parts main: the knowledge of topology and the calculation of routes.

3.1 Topology Detection

This detection (or discovery) consists in obtaining the network topology, the knowledge of existing radio links and discovery of neighbors. As in OLSR, MPOLSR benefits from the optimization due to the use of MPR (Multi Point Relay) reducing the traffic inherent to this detection phases, avoiding a complete flooding of the network. The number of message retransmissions is then limited to these Multipoint relays. By sending of the control messages (TC and HELLO), each node can know the complete network topology. This topology base is stored in three tables: topology table, 1-hop neighbor table and 2-hop neighbor table.

3.2 Route Calculation

In OLSR, routes are recalculated by the nodes whenever they detect a change in the local neighborhood or network topology. Routes to all network destinations are stored in the routing table. On the other hand, MPOLSR uses an on-demand route calculation scheme, i.e. the node only starts the route calculation procedure if it has data to transmit to the destination. In this part, we propose the assumptions followed by the proposed algorithm for route calculation.

4 Implicit Trust Analysis in MP-OLSR

The concept of trust varies according to the field in which it is to be applied (authentication [17], electronic commerce [18, 19], P2P [20]. The authors in [21] sees that it is more judicious to be interested in the behavior of trust than in trust itself, thus removing the need to adhere to a specific definition. On the other hand, he pays particular attention to the many facets of trust, from biology to trust between agents that interact in a distributed mode.

4.1 Notations

We use the language proposed in [17] for the modeling of trust rules. The notion of trust underlying this language is expressed by the fact that if an entity X trusts an entity Y about a certain aspect, it means that X believes that Y will behave in a certain way realizing (or not) a certain action of the specific circumstances.

There are two types of trust relationships: direct and derived. Derived trust relationships are obtained from the recommendations of other units. To better exploit the notion of trust in the ad hoc context, we distinguish different classes of trust. An entity can trust another for one of the following actions:

- Provide the identity of an entity to another entity (identification class: id).
- Do not interfere in activities, provide correct data, and do not modify the data before retransmitting it (non-interference class: ni).
- Forward its packets (forwarding class: fw).
- Correctly declare all LS neighbors (class declare link state information, dlsi).
- Provide recommendations on the trust characteristics of other entities (rec).
- Trust granted by an entity for access to its resources or services by other entities (***class access trust: at***).
- Trust delegated to another entity to act and make decisions on behalf of an entity (class delegation ***trust: dt***).

The trust clauses are expressed with the following notation:

- The expression X trusts_{cc}(Y) means that X trusts Y to perform the action cc.

X trusts (SET) means that X trusts all entities of the SET to perform the action cc, SET being defined as a set of entities for which a certain predicate is valid;

- X trusts_{cc}-C(Y) means that X trusts Y to perform the cc action for entity C, but not necessarily for other entities;

- X *trusts.rec*$_{cc}$(Y) *when.path* [*SET_1*]*when.target*[*SET_2*] means that X trusts Y's recommendations about the ability of other units to be trusted to perform the cc action. The when clauses are used to define the constraints of these recommendations, with the path specifying a set of SET_1 entities intermediate between entity Y and the target entities of the SET_2 recommendation specified in the target.
- X trusts.rec*cc(Y) represents the recommendations derived from the previous recommendations;
- X¬ trusts (Y) means that X does not trust Y (exact distrust);
- X¬trusts(SET) means that X does not trust the nodes belonging to the ENS set (partial distrust);

In our design we use the following sets to represent node topology information under Multipath_OLSR:

- MANET is the set of nodes of the MANET network.
- NS: is the set of symmetrical neighbours of node X
- 1HNSx: the set of immediate neighbours of node X.
- 2HNSx: the set of two-hop neighbors of node X.
- MPRSx: the set of MPR nodes of X.
- MPRSSx): the set of nodes with symmetrical neighborhood that have chosen node X as MPR.
- TSx: the set of information concerning the topology of X.
- Degree (X, Y): the number of nodes in the neighborhood two hops away from X, such that $\{2HNx \cap 1HNy \neq \emptyset / Y \in 1HNx\}$.
- Reachability (X,Y): the number of nodes in the neighborhood at two hops of X which are not yet covered by any one node belonging to the MPRSx set and which are neighbors to a hop of Y.
- Poorly_set: the subset of nodes belonging to 2HNx that is covered by less than K nodes in 1HNx.
- $x \xleftarrow{helloy} y$ The reception from x of the hello message sent by its neighbor y.
- $x \xleftarrow{TCy} y$ The reception by x of a TC message sent by its MPR y.
- $x \not\xleftarrow{helloy} y$ The absence of an incoming hello message from y.
- $x \xrightarrow{TCx} *$, $x \xrightarrow{DATAx} *$ The broadcast by x of a TC message or a data packet to be retransmitted by its MPRs.
- RTX: is the routing table of X.
- Route x → y: is the sequence of nodes that make up the route (shortest path) between x and y, this sequence has the form of the following predicate:

$$\text{Routey}1 \rightarrow \text{yn} = \text{y}1, \ldots \text{yn}\,,\ \text{avec yi} + 1 \in \text{MPRSy}_\text{i}$$

- Length (Route x → y) the length of the road between x and y.

4.2 Analysis of the Confidence Implicit in the MP_OLSR Protocol

In this section we present the different actions of the MP_OLSR protocol, namely the neighborhood detection, the calculation of the MPRs nodes and the calculation of the routing table, using the language described in the previous section. We show that the "explicit" trust can be a way of reasoning for the node about the information it receives [22, 23].

Neighborhood Discovery

Initially the nodes are generally confident [21], since they know nothing about their environment and believe that all information received from others is correct without any verification.

First, node x broadcasts a HELLO message to its neighborhood including node y, after this reception node y proceeds in the same way by broadcasting the HELLO control message to x. As a result, a relationship of trust has just been established, since x considers y as its symmetrical neighbor and the symmetrical neighbors of y as two-hop neighbors.

$$\begin{array}{lll} x \xleftarrow{HELLOy} y,\ (x, asym) \in LS_y & \Rightarrow & x\ trusts_{idUni}(y) \\ \text{Ou} \quad (x, sym) \in LS_y & & LS_x \leftarrow LS_x \cup (y,\ sym) \\ & & \forall z \in 1HNS_y,\ z \notin 1HNS_x,\ z \neq x : \\ & & 2HNS_x \leftarrow 2HNS_x \cup \{z\} \end{array} \tag{1}$$

The relationship is also seen as symmetrical, since y should behave in the same way as the relationship (2).

$$\begin{array}{lll} y \xleftarrow{HELLOx} x,\ (y,\ asym) \in LS_x & \Rightarrow & y\ trusts_{idUni}(x) \\ \text{Ou}(y,\ sym) \in LS_x & & LS_y \leftarrow LS_y \cup (x,\ sym) \\ & & \forall z \in 1HNS_x \setminus 1HNS_y \setminus \{y\} : \\ & & 2HNS_y \leftarrow 2HNS_y \cup \{z\} \end{array} \tag{2}$$

These two relationships allow the optimized selection of the MPRs nodes that will be in charge of broadcasting TC messages, so that each node calculates the routes to all the nodes of the network on demand.

Selection of MPRs by The K_Coverage Algorithm

In OLSR, the MPR node selection algorithm optimizes the overhead by selecting the smallest set of MPR nodes. In contrast, MP_OLSR uses the principle of K_coverage to allow each node to calculate multiple paths. In MP_OLSR, the two criteria for selecting MPRs for a node x are: degree, denoted Degree (y), and reachability, denoted reachability(y):

$$\text{Degree}(x,\ y) \leftarrow \left|1HNS_y \setminus 1HNS_x \setminus \{x, y\}\right|$$

$$\text{Reachability}(x,\ y) \leftarrow \left|\{F / F \in 2HNS_x \cap 1HNS_y \text{ and } MPRS_x \cap 1HNS_F = \emptyset\}\right|$$

Initially, the MP_OLSR node selects nodes with a Willigness field equal to Will_always as MPR.

$$MPRS_x = MPRS_x \cup \{y / y \in 1HNS_x,\ y\_Willigness = Will\_always\}$$

Then it proceeds by selecting as MPR the nodes covering the elements of the set poorly_set_x.

$$\text{MPRS}_x = \text{MPRS}_x \cup \{1\text{HN}_x \cap 1\text{HN}_y \,/\, y \in Poorly\_set_x\}$$

As long as there are nodes at two hops not yet covered by at least K nodes of the MPRS set, do:

- Calculate the reachability of each neighbor at one hop that does not belong to the MPRSx set.
- Add the node with the highest reachability and degree to MPRSx:

$$\text{MPRS}_x = \text{MPRS}_x \cup \{y \in \text{NS}_x \setminus \text{MPRS}_x : \text{reachability}(x, y) = \text{Max}\{\text{reachability}\ (x, Z) \,/\, \forall Z \in \text{NS}_x \setminus \text{MPRS}_x\}\} \text{ and}$$
$$\text{MPRS}_x = \text{MPRS}_x \cup \{y \in \text{NS}_x \setminus \text{MPRS}_x : \text{Degree}(x, y) = \text{Max}\{\text{Degree}(x, Z) \,/\, \forall \in Z\,\text{NS}_x \setminus \text{MPRS}_x\}$$

- Eliminate all two-hop nodes that are covered by at least K nodes in the MPRSx set.
- These last steps are repeated until all two-hop nodes are accessible.
- When x selects neighbors as MPR, it means that x trusts them to be its routers:

$$\forall y \in \text{MPRS}_x : x\ \textit{trusts}_{fw}(y) \tag{3}$$

By recurrence, the neighbors selected as MPRs nodes also choose their MPRs and so on. Thus, they recommend paths to remote nodes to the neighbors who selected them as MPRs.

So, $\forall Z \in MANET : x\ \textit{trusts.rec}_{fw}(y)\ \textit{when.path}[MPRS_y]\ \textit{when.target}[Z]$.

The path from x to Z is formed by a chain of MPRs nodes:

$\text{route}_{x \to Z} = y_1, \ldots y_n \text{avec}\, y_{i+1} \in \text{MPRS}_{yi}$, which allows the extension of the above expression.

$$\forall Z \in MANET : x\ \textit{trusts.rec}_{fw}(y)\ \textit{when.path}[\text{route}_{x \to Z}]\ \textit{when.target}[Z]. \tag{4}$$

Each node selected there as MPR must declare in the TC messages broadcast at a regular time interval all the neighboring nodes which have selected it as MPR. The calculation of the MPRSSy set implies the following trust relationships:

✓ y trusts x to use its routing resources without causing harm:

$$y\ \textit{trusts}_{at}(x) \tag{5}$$

✓ y trusts x to announce that y is an MPR:

$$y\ \textit{trusts}_{dt}(x) \tag{6}$$

he procedure for selecting the MPRs of neighboring nodes and accepting to be chosen as MPRs by these neighbors presents the basic trust relationships for routing in MP_OLSR.

Routing Table Calculation

The routing table represents the final result of the MP-OLSR protocol, on which the routing is based. Any change in the sets: LS, NS, 2HNS, TS and MPRS induces an immediate change in the routing table of the MP-OLSR node.

When a node wants to transmit data to a destination in the MANET network, it first starts by searching for different possible paths to reach the destination. To do this, it proceeds as follows:

✓ In case no route connecting source node s to destination node d exists in the RTs routing table, it starts the route search procedure, once found, it adds them to the RTs routing table:

$$\text{Add}\,(route_{s\to d}^{i:1\to N},\ \text{RT}_s)$$

✓ If the route to destination d exists but with an updated Flagd = False, it restarts the path calculation and updates the entry for destination node d:

$$\text{Update}\left(\text{Entry}_d,\ \text{RT}_s\right) = \text{Vraie}.$$

✓ Otherwise (i.e. the route exists with an updatedFlagd = True it uses the list of already existing routes in the RT_s routing table).

The routing table is calculated in such a way that there are several disjoint routes to each destination.

$$\forall \text{x},\ \text{Z} \in \text{MANET et Z} \notin \text{NS}_\text{x} \Rightarrow \exists\, route_{y_i\to Z}^{i:1\to card\,(MPRS_x)}\ \text{telle que}\ \text{y}_\text{i} \in \text{MPRS}_\text{x}.$$

The index i varies between one and one is the value card(MPRSx), this is explained by the fact that in order to route the data it is necessary to go through another MPR different from the previous one.

The basic idea of the routing table calculation mechanism is to iterate the Dijkstra algorithm N times. But in the case of low density networks one cannot find the N disjoint routes especially in the case of N very large ones. Therefore the choice of the value of N must be optimized in order to minimize redundancy in the network.

Let i be the number of iterations in progress:

$$\text{Length}(route_{y_i\to Z}) \leq \text{Length}(route_{y_{i+1}\to Z}) \tag{7}$$

5 Intrusion Detection Using the MP-OLSR Protocol's Implicit Trust Reasoning

Since, some attacks against the MP-OLSR protocol exploit the vulnerability resulting from the lack of validation of the existing chain of trust between MPRs nodes. Nodes must have a certain degree of distrust regarding the information used in the calculation of the different paths to all destinations.

In this paper, we focus on the detection of malicious behavior based on node observations. This anomaly detection includes checking the consistency of MP-OLSR (Hello, TC) messages and reasoning about the trust that can be made by each node in the network.

5.1 Verifying the Consistency of Basic Information

One Hop Neighbor Validation

In order to know their neighbors, MANET nodes exchange HELLO messages at two-second intervals. In addition, any MPR node in the network must participate in the transmission and retransmission of TC messages at a time interval equal to five seconds. Therefore the possibilities of an attacker to abuse these two time properties, allows the construction of these rules of distrust:

- If a node y is declared as a symmetric neighbour of x, but y does not send a hello message for duration of two seconds. Then its behavior does not meet the MP-OLSR specifications. Therefore, x must be wary of y. The problem for nodes under the MP-OLSR protocol is that even if they do not receive HELLO messages from these jump neighbors that have been declared as symmetric, they will not be removed from the 1HNSx set until after a time duration equal to 3*HELLO_refresh_Interval = 6 s.

$$x \overset{hello_y}{\not\longleftarrow} y \;, y \in 1HNS_x \text{ avec } \quad T > 2s \quad \Rightarrow x \neg trusts(y) \tag{8}$$

T: duration of time after the last reception of the $hello_y$ message.

- If x includes y in its MPRS set, then x must receive a TC message from y every five seconds. Otherwise the behaviour of y does not respect with the MP-OLSR protocol specifications and shall automatically beware of it. The vulnerability exploited in this case is that even if x does not receive TC messages from its MPR y, it is still declared as an MPR for a duration equal to 3*TC_Refresh_Interval = 30 s in its MPRS set.

$$x \overset{TCy}{\longleftrightarrow} y,\ y \in MPRS_x \text{ avec } \ T > 5s \ \Rightarrow x \neg trusts(y) \tag{9}$$

T: time duration after the last reception of the TC_y message.

- Another vulnerability of the MP-OLSR protocol is that a node y sends a TC message to its selector MPR x with a TTL (Time To Live) field different from 255 in order to limit the broadcast of TC messages throughout the network. Therefore x has to beware of the neighboring node y.

$$x \overset{TCy}{\longleftarrow} y,\ y \in MPRS_x \text{ avecTTL} \neq 255 \ \Rightarrow x \neg trusts(y) \tag{10}$$

Other vulnerabilities have been studied by [24] and [22]. Using the notion of trust, we describe these vulnerabilities in order to extract the rules of mistrust that can be used by a node to ensure its self-protection against neighboring nodes with malicious behaviors:

- Any node announcing its selector MPRs in TC messages must also announce them as symmetric neighbour in HELLO messages. Otherwise, all its neighbours must mistrust of it:

$$x \xleftarrow{helloy} y,\ x \xleftarrow{TCy} y,\ \mathrm{MPRSS_y} \not\subseteq \mathrm{NS_y} \Rightarrow x \neg \textit{\textbf{trusts}}(y) \tag{11}$$

- If x detects that a non-neighboring node y is announcing it in its TC messages, then it should be distrusted because its behavior does not meet the specifications of the MP-OLSR protocol:

$$x \xleftarrow{TCy} *,\ \mathrm{x} \in \mathrm{MPRSS_y}, \mathrm{y} \notin \mathrm{x} \Rightarrow x \neg \textit{\textbf{trusts}}(y) \tag{12}$$

- If a node y neighbour of x, not chosen as MPR by x, announces x in its TC message, then x should be distrusted because its behaviour does not respect the MP-OLSR specifications.

$$x \xleftarrow{TCy} *, \mathrm{x} \in \mathrm{MPRSS_y}, \mathrm{y} \notin \mathrm{MPRS_x} \Rightarrow x \neg \textit{\textbf{trusts}}(y) \tag{13}$$

- If x receives the same TC message (same sequence number and source address) from two different nodes, and detects that the two messages are different, then it must be mistrust of both senders:

$$\forall \mathrm{z, w} \in \mathrm{MANET},\ x \xleftarrow{(TCy)z} *,\ x \xleftarrow{(TCy)w} *,\ \left(\mathrm{TC_y}\right)_\mathrm{z} \neq \left(\mathrm{TC_y}\right)_\mathrm{w} \Rightarrow x \neg \textit{\textbf{trusts}}(\{z, w\}) \tag{14}$$

Two-hops Neighbors Validation

Due to the properties of Ad Hoc networks, each node can listen to the messages sent by its neighbors at a hop. As a result, it can decide on the consistency of the information exchanged by its neighbors during the neighborhood discovery process. The above properties can be extended to allow nodes to check the local view of their neighbors about the network:

- Let z be a neighbor at one hop of x. if a node y announces z as selector MPR in its TC messages received by node x, but y has not been selected as MPR by z. Therefore, x must be distrust of y.

$$x \xleftarrow{helloz} z,\ x \xleftarrow{TCy} *,\ \mathrm{z} \in \mathrm{NS_x} \cap \mathrm{MPRSS_y}, \mathrm{y} \notin \mathrm{MPRS_z} \Rightarrow x \neg \textit{\textbf{trusts}}(y) \tag{15}$$

- Two nodes in a MANET network that are electromagnetically symmetrical must declare their symmetry in their HELLO messages to their neighbourhood.

$$x \xleftarrow{helloy} y,\ x \xleftarrow{helloz} z,\ \mathrm{z} \in \mathrm{NS_y},\quad \mathrm{y} \notin \mathrm{NS_z} \Rightarrow x\neg trusts(\{y, z\}) \tag{16}$$

We can note that the reasoning on distrust, does not always allow us to detect the node that caused the anomaly, but it allows us to simply detect the group of nodes containing the attacker.

Verification of Consistency in the Selection of MPRs Nodes

The selection of MPR nodes is of great importance in the mechanism of the MP-OLSR routing protocol, such that each node x must select two MPR nodes, in order to reach a node z in its two-hop neighborhood based on the number of neighbors covered at two hops by each neighboring one-hop node.

The vulnerability of an MPR selection with K-Coverage based on the degree of accessibility comes from the fact that any node at one hop can give false information. The latter cannot be verified; for example by declaring non-existent nodes or simply by declaring nodes more than two jumps away from the target.

The two rules of mistrust that can be extracted from the MP-OLSR protocol specifications are as follows:

- Let y be an MPR node of x. if y does not generate TC messages correctly or generates these messages without including its selector x. so x must be wary of y.

$$\mathrm{y} \in \mathrm{MPRS_x},\ (x \xleftarrow{TCy} y)\ \mathrm{ou}\ (x \xleftarrow{TCy} \!\!\!\!/\ \ y,\ \mathrm{x} \notin \mathrm{MPRSS_y}) \Rightarrow x\neg trusts(y) \tag{17}$$

- If a node y selected as MPR does not retransmit data packets (DATAx) or messages (TCx) sent by a node x that has selected it as MPR, then y does not comply with the MP-OLSR specification and x must be wary of it:

$$\mathrm{y} \in \mathrm{MPRS_x},\ (x \xrightarrow{TCx} *,\ x \xleftarrow{(TCx)y} \!\!\!\!/\ \ y)\ \mathrm{ou}\ (x \xrightarrow{DATAx} *,\ x \xleftarrow{(DATAx)y} y) \Rightarrow x\neg trusts(y) \tag{18}$$

6 Prevention and Countermeasures

Intrusion detection based on trust reasoning allows each node to check the consistency of the information received. In this section, we propose countermeasures to secure the MP-OLSR protocol using trust reasoning.

Considering that the MP-OLSR protocol is an extension of the OLSR protocol, it is proposed to apply the work proposed in [25] to ensure the integrity and authenticity of the nodes. This work ensures the exchange of messages and packets through the use of a message signing system (M-OLSR) to prevent malicious nodes from injecting incorrect information into the network, a signature is generated by the generator of each control message and transmitted with it. In addition, a timestamp is associated with each signature to ensure the freshness of the messages (Fig. 1).

Sig. Method	Reserved	MSN Referrer
mestamp		
Signature		

Fig. 1. Signature format in SOLSR

Generally, signing and verification of messages in [21] can be done with a private and public key pair or with a shared secret key.

The system we propose to deal with malicious nodes is based on the following two approaches:

Firstly, each time a situation of inconsistency is detected between a node and another malicious node, the target node must necessarily break the symmetrical link connecting it to the attacker, and then reject all control messages including the original address and the attacker's address. The target node must then reselect its MPR nodes based on the K-Coverge method. Since the K-Coverge method allows nodes to have K MPRs nodes to reach a neighbor at two hops, the K-Coverge method is a good way to reach a neighbor at two hops. Therefore, the elimination of malicious MPR neighbors will not affect the optimization of the network. Subsequently this isolated node will be registered in a blacklist. At each route request to a new destination, the node detecting the attack, executes the Dijksta algorithm N times in order to search for the various existing disjointed routes that do not include the malicious node.

Sharing information about the attack is accomplished by broadcasting an alert to inform all nodes in the network to be vigilant about the attack. To avoid overloading the network by adding control messages to propagate the alert. We suggest using the length of the Link code field is eight bits, but the protocol specification uses only the first four bits (maximum value 15 in decimal), this field contains both information about links to neighboring nodes and the type of the links.

We propose to include malicious nodes by simply extending the Link code field by setting the first five bits to $10000 = 16$ to indicate that the nodes following the field are malicious (Fig. 2):

In this way, each node receiving a HELLO message from its neighbor checks the Link Code field: if it is strictly less than 16, it proceeds with a normal execution of the link state calculation algorithm. On the other hand, if the Link Code value is equal to 16, it records the malicious nodes included in the HELLO message in its Blacklist set and recalculates its MPRS set without taking into account the nodes belonging to the Blacklist set.

The two messages HELLO and TC received from **B** allow Node **A** to change its mental state by detecting an inconsistency produced by the conflicting information contained

Reserved		HTime	Willigness
Link code	Reserved	Link Message Size	
Neighbor Interface Adress			
Neighbor Interface Adress			
....			
Link code = 16	Reserved	Link Message Size	
Nœud malicieux_1			
Nœud malicieux_2			
....			

Fig. 2. Hello message with alert declaration

in the two messages (Fig. 3), because Node **B** declares that the link with its neighbor **M** is asymmetrical while it declares **M** as an MPR selector in its TC message. Therefore, A broadcasts an alert to its neighbors in the HELLO message with a Time To Live equal to 1, and a Link Code equal to 16 to indicate the malicious node. Each neighboring node receiving the alert adds the malicious node to its Blacklist, removes all these entries from its local database, rejects all messages received from the attacker, and in turn broadcasts an alert to its neighbors at one hop via the HELLO message.

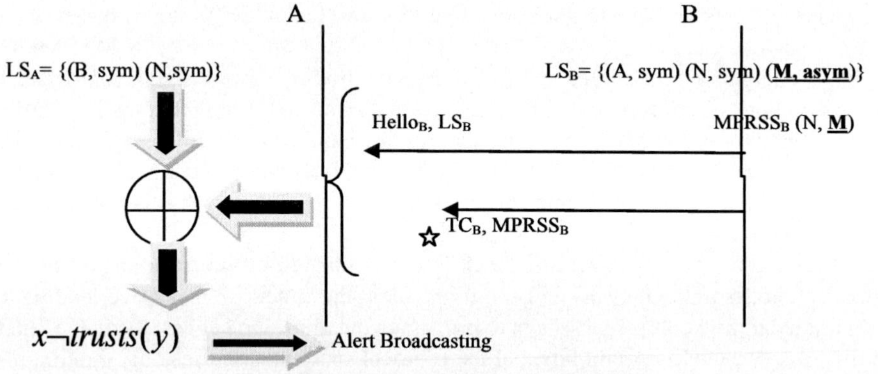

Fig. 3. Sample for inconsistency detection

7 Simulation

We have used the NS2 Simulator and the MP-OLSR patch. We have added to this patch a module implementing trust rules, and several attack scenarios. In our simulations, ad hoc networks are composed of 40 nodes which are placed randomly. Moreover, the attackers are selected randomly, and each one selects an attack scenario, as well as a set of targets according the selected attack. However, since the ad hoc networks have to be stabilized to allow the attacker to perform the attacks, we have considered that nodes are not mobile. The following attack was implemented: generation and modification of HELLO messages, generation and modification of TC messages.

In this part we will consider the symptoms of the following attacks:

1. If a node y symmetrically neighbors a node x so that y does not send a Hello message for 2 s, then node x will consider y an attacker.
2. If node x including node y in its MPRS set and if node y does not transmit the TC message before 5 s, node x will take y as an Attacker.
3. If a node y sends a TC message to its selector MPR x with a TTL (Time To Live) field different from 255 in order to limit the diffusion of TC messages throughout the network, node x will consider y as an attacker.

Figure 4, illustrates the detection rate as a function of density. This rate varies between 65% and 85%. During the MPRs selection phase, if a node has only one router node in its MPR set table, it is required to not delete it, even if it shows symptoms of the attack.

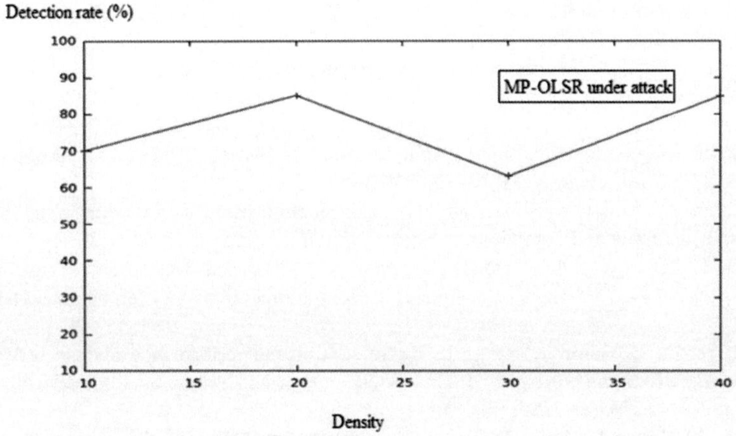

Fig. 4. Detection rate of MP-OLSR under attack vs density

8 Conclusion

In this paper, we have focused on the implementation of a mechanism for detecting malicious nodes in the MP-OLSR routing protocol through the integration of trust reasoning. The study carried out has enabled the emergence of trust rules between nodes based on local information and messages exchanged during the neighborhood discovery process and network topology to make the MP-OLSR protocol more trusted. Our approach allows an exchange of information concerning the attack without having to resort to other control messages that will overload the network and limit the optimization brought to the protocol by the use of the Multi Point Relay principle.

These results motivate extending the approach for establishing more example of simulation and to check degree of automation of reasoning about trust in order to measure the impact of these solutions on Qos of the protocol, while preserving the self-organization and the dynamic environment of ad hoc network.

References

1. Russia, S., Anita, R.: Joint cost and secured node disjoint energy efficient multipath routing in mobile ad hoc network. Wirel. Netw. **23**, 2307–2316 (2017)
2. Kok, G., Chow, C., Xu, Y., et al.: EAOMDV-MIMC: a multipath routing protocol for multi-interface multi-channel mobile ad-hoc networks. Wirel. Pers Commun. **73**, 477–504 (2013). https://doi.org/10.1007/s11277-013-1198-5
3. Kim, S.: An ant-based multipath routing algorithm for Qos aware mobile ad-hoc networks. Wirel. Pers. Commun. **66**, 739–749 (2012)
4. Zhang, Y., Yan, T., Tian, J., Hua, Q., Wangb, G., Li, Z.: TOHIP: a topology-hiding multipath routing protocol in mobile ad hoc networks. Ad Netw. **21,** 109–122. ISSN: 1570-8705
5. Zhang, Y., Wang, G., Hu, Q., Li, Z., Tian, J.: Design and performance study of a topology-hiding multipath routing protocol for mobile ad hoc networks. In: Proceedings of the IEEE INFOCOM (2012)
6. Hothefa, S., Raed, A.: Multipath routing algorithm using an electromagnetic-like mechanism with threshold acceptance for mobile Ad hoc networks. Wirel. Person. Commun. **81**(2), 607–630 (2015)
7. Marchang, N., Datta, R.: Light-weight trust-based routing protocol for mobile ad hoc networks, IET Inf. Secur. **6**(2), 77–83 (2012)
8. Mawloud, O., Challal, Y., Bouabdallah, A.: Certification-based trust models in mobile ad hoc networks: a survey and taxonomy. J. Netw. Comput. Appl. **35**, 268–286 (2012)
9. Huang, J., Woungang, I., Chao, H., Obaidat, M.: Multi-path trust-based secure AOMDV routing in Ad Hoc Networks. In: Global Telecommunications Conference (GLOBECOM) (2012)
10. Yi, J., Cizeron, E., Hamma, S., Parrein, B.: Simulation and performance analysis of MP-OLSR for mobile ad hoc networks. In: IEEE WCNC: Wireless Communications and Networking Conference, Las Vegas, USA (2008)
11. Tarique, M., Tepe, K., Adibi, S., Erfani, S.: Survey of multipath routing protocols for mobile ad hoc networks. J. Netw. Comput. Appl. **32**(6), 1125–1143 (2009)
12. Pham, P.: Multi-path routing with load balance and cross-layer design in MANETs. PhD thesis, University of South Australia, Australia, August 2004
13. Yi, J., Adnane, A., David, S., Parrein, B.: Multipath optimized link state routing for mobile ad hoc networks. Ad Hoc Netw. **9**(1), 28–47 (2011)
14. Kun, M., Jingdong, Y., Zhi, R.: The research and simulation of multipath-OLSR for mobile ad hoc network. In: IEEE International Symposium on Communications and Information Technology (ISCIT), pp. 540–543, October 2005
15. Badis, H., Al Agha, K.: QOLSR multi-path routing for mobile ad hoc networks based on multiple metrics: bandwidth and delay. In: Vehicular Technology Conference, vol. 4, pp. 2181–2184. IEEE, May 2004
16. Aiache, H., Haettel, F., Lebrun, L., Tavernier, C.: Improving security and performance of an ad hoc network through a multipath routing strategy. J. Comput. Virol. **4**, 267–278 (2008)
17. Yahalom, R., Klein, B., Beth, T.: Trust relationships in secure systems a distributed authentication perspective. In: Proceeding of the IEEE Symposium on Security and Privacy (1993)

18. Grandison, T., Sloman, M.: A survey of trust in internet applications. In: IEEE Communications Surveys and Tutorials (2000)
19. Viljanen, L.: Towards an ontology of trust. In: Trust, Privacy and Security in Digital Business. Springer (2005)
20. Aberer, K., Despotovic, Z.: Managing trust in P2P information system. In: Proceeding of the Tenth International Conference on Information and Knowledge Management (CIKM01). ACM Press (2001)
21. Marsh, S.: Formalising Trust as a Computational Concept. PHD thesis, University of Stirling, 148 (1994)
22. Adnane, A., Timoteo, R., Bidan, C., Mé, L.: Analysis of the implicit trust within the OLSR protocol. In: IFITM 07: Joint iTrust and PST Conference on Privacy, Trust Management and Security, volume 238 of IFIP International Federation for Information Processing. Springer, July 2007. ISBN: 978-0-387-73654-9
23. Timoteo, R., Adnane, A., Bidan, C., Mé, L.: Analyse de la confiance implicite requise dans le outage adhoc OLSR. In: SARSSI 2007: The 2nd Joint Conference on Security in Network Architectures and Information Systems, June 2007. ISBN: 2-916377-49-2
24. Wang, M., Lamont, L., Mason, P., Gorlatova, M.: An effective intrusion detection approach for OLSR MANET protocol. In: 1st IEEE ICNP Workshop on Secure Network Protocols (NPSec 2005), pp. 55–60 (2005)
25. Adjih, C., Clausen, T., Jaquet, P., Muhlethaler, P., Laouiti, A., Raffo, D.: Securing the OLSR protocol. In: Med Hoc-Net, pp. 25–27, June 2003

Bitcoin Technology and Its Security Challenges: A Research Survey

Aman Bhargava, Kartik Shankhavaram, Priyanshu Agarwal, and Jayaprakash Kar(✉)

Centre for Cryptography, Cyber Security and Digital Forensics, Department of Computer Science and Engineering, The LNM Institute of Information Technology, Jaipur, India
jayaprakashkar@lnmiit.ac.in

Abstract. In this paper, we discuss the bitcoin technology such as structuring of bitcoin, building of blockchain used in bitcoin, mining, transaction and bitcoin security. The various attacks and its countermeasures have been presented. Bitcoin is a peer-to-peer decentralised electronic cash mechanism which does not depend on any form of central monetary issuing authority or validation and settlement of transactions. For security aspect Blockchain is used in bitcoin in the form of distributed computation. In particular decentralized computation relies on nodes in its network to process, verify and store the information disseminating inside the network. Each node acts as both a client and a server and so the load of catering to requests from clients is spread out and not concentrated on one system. In the case of blockchain, a single data structure is shared across all the nodes, and any updates made to that data structure are circulated across the network.

Keywords: Blockchain · Transaction · Decentralize · Proof-of-concept · Cryptocurrency

1 Introduction

Virtual currencies play a major role in digital money ecosystem. It can be used in a same way as conventional currency to buy and sell goods and services and send money to people. As of November 2019, there are more than 18 million Bitcoins is in circulation, valued at around 7,341.38 USD [1]. And there are many retailers who are accepting bitcoins for making payment like expedia.com and overstock.com and also some of start-up companies have developed services to allow users of bitcoin to shop at the retailers that have not yet started to take virtual currency. While many other virtual monetary enjoy no more than marginal use, Bitcoin has supported for a long time and has gathered the consideration of the standard money related world.

It has features like encryption and digital signatures that ensure the security of the bitcoin network. The process from which Bitcoin is generated is called

A. Abraham et al. (Eds.): SoCPaR 2020, AISC 1383, pp. 706–715, 2021.
https://doi.org/10.1007/978-3-030-73689-7_67

mining which involves looking for a solution to a difficult problem and there is a miner which could be any participant in bitcoin network who is using their system to find a solution to this problem. There are four key innovations that have been brought together in a very effective and powerful combination with over decades of research in cryptography and distributed systems which constitute bitcoin. It consists of (i) the bitcoin protocol, (ii) the blockchain, (iii) distributed mining, (iv) transaction script.

1.1 The History of Bitcoin

Bitcoin was proposed by Satoshi Nakamoto in 2008 with the publication of a paper "Bitcoin: A Peer-to-Peer Electronic Cash System" [10]. His main aim was to build a truly decentralised electronic cash mechanism which does not depend on any form of central monetary issuing authority or validation and settlement of transactions. Previously in digital currency, the double spend problem (where a single currency unit can be spent twice) was a weak link and was fixed by authenticating all transactions through a centrally-regulated authority but after the invention of distributed computation system we don't want to do that. Since it is a popular cryptocurrency, we have many implementations of bitcoin client software. The three major forms are:

1. **Full Client**: This client stores the entire blockchain locally and fully verifies all transactions against the bitcoin rules. If this node finds an invalid transaction, it is rejected from the network.
2. **Light Client**: These clients do not download the entire blockchain as full clients do. Instead, they obtain the headers of the block and then use a merkel tree path to verify the transaction.
3. **Web Client**: These run a light or full client in the back-end, while presenting the user with a web interface which is accessible from anywhere.

In full client user have the maximum control but on the contrary it puts load of security and backups on the user. Where as web client is easy to use but there is too much risk involved because both the owner and the user of web service shared the security and control that means if service will be compromised at any cost that will directly results in loss of all the funds of given user.

1.2 Concept of Bitcoin

It is quite different from our traditional banking system, there is no central authority in bitcoin it is completely a part of decentralized but trusted mechanism where trust can be achieved by interactions between various users in bitcoin system [2]. The idea driving the formation of Bitcoin is the simple transfer of the money without paying a lot of transaction charges. It is evident that Bitcoin does not have physical evidence. It is demonstrated uniquely as balance in the record of the client of the Bitcoin profile. The balance is preserved, considering all accounting, like the ledger or the balance sheet. At the hour of transaction, a sort of check is done as to reveal that the exchange of the money has been managed with no sort of issue or misusing. The coin is obtainable in different denominations too, for example, micro bitcoin and mill bitcoin and also Bitcoins aren't printed, similar to Euros or Dollars they're produced by individuals. It very well may be utilized to purchase things electronically. In that sense, it resembles regular dollars, euros, or yen, which are additionally exchanged carefully. User can also track transactions made through bitcoin network and blockchain. It is better to use blockchain explorer a web application to visualize each step of transaction and can see is there any connection or flows between them. These explorers uses a search function for finding data on bitcoin network which takes either address, transaction hash or block number to search.

Transactions in Bitcoin. A transaction happens when the owner of bitcoin send some of those bitcoin to another owner in authorized manner. Transactions are basically entries in double-entry bookkeeping ledger where every transaction has one or more inputs those are debits from bitcoin account and one or more output those are credits added to the account on other side of transaction. Remember it's not compulsory that sum of inputs and outputs end up with same result because we have transaction fee also which is collected by miner who includes those transactions in ledger. We have three forms of Transactions which includes:

1. **Common Transaction**: Here transaction is simply a payment from one address to another address with some change that original owner gets back after transaction. Basically we have here one input and two output with us.
2. **Aggregating Transaction**: Another form of transaction that combines several input to a individual output. We can see it as adding payments given by customer to pay a large amount to supplier.
3. **Distributing Transaction**: This form of transaction includes distribution of a individual input to several outputs. This transaction can be seen when a company distributes payroll payments to the workers [2].

Constructing a Transaction. So to build a transaction user should have appropriate inputs and outputs. Like a user must have some inputs with him/her to pay someone and after paying user will have transaction output as a script which tells that whoever with a signature corresponding to payer address will have that output. And payer will also have a second output as well that is the remaining balance in his/her account after subtracting the transaction fee and that paid amount. Now this transaction presents in bitcoin network (peer-to-peer network where each bitcoin are connected to each other) will added to the blockchain. But before going to the ledger, all transaction goes through a process called mining which have two major role in bitcoin and these are, it generates new bitcoin in each block and it also generates trust which increases with every adding block. Finally the transaction has been added in blockchain as a part of block and will be visible to all bitcoin applications.

1.3 Bitcoin's Impact as a Global Currency

USD as a world currency has survived for quite a long time but because of the rise of several cryptocurrencies like bitcoin which may represent a genuine risk to USD as it could be considered as an alternative to existing global currency including USD because of it's features. A research has also been done on this and from there we got a model from which four independent factors have been recognised and these are: 1) A technology that guarantees the confidentiality and anonymity of financial transactions, and is very cost-effective at the very same time. 2) Laws around Bitcoin worldwide or absence of it. 3) Bitcoin as a Currency and 4) Bitcoin's ever-increasing economy; all these four have a direct impact on the variable that depends on it which is "Impact of Bitcoin on USD, a world currency". At the end after keeping all the stats and other things about bitcoin in mind we can conclude that bitcoin has an immense potential, yet it can't in its present form, influence the USD however it can harm it essentially, if its conversion scale versus USD increments drastically.

2 Bitcoin and the Blockchain

Blockchain is the medium in which bitcoin operates. All transactions involving bitcoin are stored in the blockchain, which is essentially a shared public database. It relies on nodes in its network to process, verify and store the information disseminating inside the network. Each node acts as both a client and a server and so the load of catering to requests from clients is spread out and not concentrated on one system. In the case of blockchain, a single data structure is shared across all the nodes, and any updates made to that data structure are circulated across the network. Blockchain has many uses. It can be used in currencies, finance, contracts, IoT, and much more [5].

The blockchain consists of *blocks*. This chain can be visualized as a stack, with the genesis block at the lowest position and new blocks being added dynamically. A block is identified by the hash of its header. The contents of the header are discussed below in another section. A block contains a huge number of transactions. Thus, the blockchain forms a huge data structure containing all the transactions performed, like a public ledger. And since it is not stored at a single locations, it is truly decentralised. A node can fetch the blockchain from any of the peers it is connected to. This increases reliability and removes a single point of failure and regulation. A block can be identified either by the double SHA-256 hash of its header (unique for each block) or by the height (sequence number of the block starting at the genesis block) at which it appears in the blockchain. Some key characteristics of blockchain are:

1. Decentralization: No single controller.
2. Persistence: Increased difficulty to change as more blocks are added on top.
3. Anonymity: No proof of identity required.
4. Auditability: All transactions are publicly visible.

Since bitcoin is built upon peer-to-peer network architecture, data is requested from all peers instead of a single one (server). The node starts by connecting to seed nodes, which are long running stable nodes that act as beacons for any new node to connect. Through them, the node discovers new peers, and then disconnects from the seed nodes. After this, the process of "syncing" starts. If the node is a "full" node (it maintains a full copy of the blockchain), it starts by requesting the blocks it does not have from its peers. It uses the highest block it currently has as a base to request more blocks. If the node does not have any blocks, it starts from the first block, also known as the "genesis block", the first block ever mined.

2.1 Security Issues and Challenges

Blockchain, by virtue of being decentralized, does not have a single failure point. Hence, no attacker can control the network through a single point. Thus, issues here concerns multiple nodes, or exploitation of singular nodes. Some issues and challenges that arise with the current implementation of blockchain are [3]:

1. **Size of the block**. A larger size enables us to store more transactions, but can put strain on the nodes and the network. Smaller sizes are easier to transmit and maintain, but have a high overhead-to-data ratio.
2. **The 51% attack**. When an attacker controls more than half of the processing power in a network, they can decide which blocks to add to the blockchain and which to discard. This gives them the power to manipulate transactions. They can also cripple the network.
3. **Compromised cryptography mechanisms**. Blockchain relies on a lot of cryptographic algorithms and mechanisms to function, such as public key cryptography, hashing, etc. A user's entire wallet is denoted by a user-generated cryptographic private key. If that key is lost, then access to that wallet and the assets linked to it is lost permanently. Also, if there is a vulnerability in certain cryptographic algorithms or their implementation, the blockchain may not be as secure as intended.

3 Bitcoin Security

In this section, we describes the security challenges in digital-asset maintenance and transaction in bitcoin. Bitcoins and satoshis (smallest unit of BTC) do not have unique identities. Digital signatures, digital keys and bitcoin addresses are used for declaring bitcoin's ownership. Security in Bitcoin relies on the **public key cryptography** paradigm [12]. Public key cryptography places it's leverages the fact that some mathematical functions are easy to solve one way but retrieving the original values from the end values is practically infeasible [8]. For security, algorithms like SHA-256 are used, which can convert a bitstring of any length less than 2^{64} into 256 bits. The collision resistant property provides its security [7,8].

3.1 Wallet Key Generation

1. Private Key Generation: Private key generation from random number generators.
2. Public key generation: Creating a public key with ECDSA.
3. Encryption: Encrypting the key pair with SHA-256 and RIPEMD-160
4. Verification: Verification of the key pair by calculating the checksum with double SHA-256
5. Key Compression: Encoding the key with Base58 for easier representation.

3.2 Digital Identity for Transactions

In a conventional banking mechanism, the receiver communicates their *account details* to the sender who in turn forwards these details to their bank along with a proof of authority of the sender to make the transaction. Similar to this, the bitcoin system uses **bitcoin addresses** in place of the *account details* of the receiver. A bitcoin address is a series of alphanumeric characters which can be shared with any number of senders in order to receive BTC [16]. The digital identity is generated for the transactions. Similar to how a public key is derived from a private key, the *bitcoin address* is derived from the user's public key in a similar fashion ensuring it's a one-way link only, that is a user's public key cannot be derived from the bitcoin address [4]. The wallet address A is derived from the public key K in the operations: $A = RIPEMD160(SHA256(K))$ *i.e.* the public key is double hashed successively with first *SHA256* algorithm and then *RIPEMD160* algorithm. This value produces a 160 digest. The obtained hash value is then encoded using *Base58CheckEncoding* which helps human readability, adds checksum capability for protecting against errors during multiple phases of transactions.

3.3 Safe Storage of the Digital Identity

The digital public-private key pairs are stored by the end-users in encrypted files or, in most cases, in a database service known as **wallets** either on their own computer or on an online wallet service [4]. There are two kinds of wallets, deterministic and non-deterministic wallets.

- **Non-deterministic Wallets**: Private keys are generated randomly without any correlation with each other and each is used only once for a transaction. Hence, wallet management gets burdensome and clunky as soon as transaction volumes increase. Keys are lost *irrevocably* in case of a wallet data corruption.
- **Deterministic Wallets**: These wallets are **seeded** with an initial value and after that all successive private keys can be generated from the initial key using a *one-way hash function* eliminating the need to store all private keys. This makes wallet backup and import/export much more easier. Unlike non-deterministic wallets, deterministic wallets can use mnemonic code words as *seeds* and generate the initial key and hence recover the private keys in case of an inaccessible wallet, hence the user's digital assets are not lost.

3.4 Preventing Theft

Since the ownership of bitcoins is determined by access to the private key, the only possible way for an attacker to gain illegitimate access to bitcoins is through gaining access to the private key. This can happen in two ways as follows:

1. **Gain direct access to private key**: A hacker is able to hack into a user's wallet or system and retrieve the private key or an attacker is able to gain access to a user's private key stored physically someplace.
2. **Using only information in the blockchain ecosystem**: Assuming that the attacker is somehow able to gain access to a user's public key for example through a challenge script (Multisig or P2PK), the problem of computing the private key from the public key again forms the ECDLP problem as discussed earlier, which is computationally unfeasible to solve.

3.5 Preventing Transaction Tampering

Consider a transaction between Mary and John where Mary is the sender. The potential tamper scenarios are as follows:

1. Another node who is acting as an accomplice of Mary might attempt to modify the transaction communicated by Mary to the miner node, and attempt to put its own challenge script in place of Mary's in the transaction outputs.
2. John could attempt to change the transaction and send more coins to himself than what Mary intended.
3. John could initiate replay attacks in the event that he can modify the output location in the transaction input, thereby being able to unlock other UTXOs containing bitcoins which were not intended to be given to John through the modified transaction.
4. In the eventuality that Mary could change the transaction after successfully coming in possession of some services or goods from John, she can attempt to modify the amount of bitcoin being sent to John and modify the transaction to change her UTXO.

The first three scenarios are prevented using digital signatures as Mary's private key is not known to anyone and hence it is not possible to modify transactions created by her. In the fourth scenario, Mary is able to update the transaction that till current time not appeared on the blockchain **(Unconfirmed Transaction)** by re-generating an authentic digital signature for the modified transaction. However, once the transaction has been confirmed by enough nodes, Mary will be unable to update/delete it until and unless she controls the majority of the overall network hash rate which is mining blocks (Goldfinger Attack).

3.6 Spam Transactions

An attacker may try to launch the cryptocurrency equivalent of a DDoS attack by trying to spam the blockchain network with transactions that do not actually hold any business value but are only playing the role of increasing the network fee and thus increasing the transaction fee for other users in the network. For this to work the attacker would have to control a significant amount of the network hash rate if he does not want to lose the transaction fee received after mining the transaction to other honest miners [16].

4 Attack Vectors

Over the past few years, a few vulnerabilities have come to the light, some of them have been exploited while others have not.

4.1 Selfish Mining Attack

In selfish mining [13], an oligopoly of users decide to not release to the public blockchain the newly discovered blocks thus creating a fork. When the fork is larger than the public chain, it is released and because of the virtue of being the largest chain, honest users will consider this to be the true chain and the original chain and underlying data will hence be discarded [15]. This wastes the mining power of honest users thus becoming a security vulnerability.

4.2 Eclipse Attack on Overlay Networks

An overlay network is a network that is built on top of another network. An example of an overlay network is a distributed network like a P2P network, since all the nodes in the P2P network run on top of the internet. In the "Eclipse" attack vector, an honest node is made to connect with a group of coordinating, malicious nodes in an overlay network. On successful attack penetration, the attacker can intercept/modify the communication from the victim. Hence, the attacker can utilize the victim's computing power to conduct their own malicious acts and reap rewards [14]. Heilman et al. showed that Bitcoin's peer-to-peer network peer-discovery-mechanism is vulnerable to eclipse attacks [6] and thus preventing a node from having complete information about the blockchain network. Heilman et al. also demonstrated how eclipse attacks can be leveraged as an instrument to amplify the efficacy of the double spend attack vector on a merchant. First, the attacker obstructs the merchant's view of the blockchain using the overlay layer, then, he proceeds to communicate to the vendor a supposedly erudite transaction $\boldsymbol{H}$, which contains the payment for a good. Next, the attacker sends the miners an erroneous transaction $\boldsymbol{F}$ that shifts the funds to a different location. Next, he creates and sends a series of $\boldsymbol{Z}$ blocks to the victim merchant such that $\boldsymbol{H}$ is part of the first block. Finally, the eclipse attack progresses till the real blockchain has grown to at least $\boldsymbol{Z+1}$ blocks. After that, the attacker can retain both the goods and the blockchain.

4.3 Balance Attack

This attack is aimed at PoW-based blockchain, in which communication between subgroups with comparable hashing powers can be momentarily disrupted by an attacker with low hashing power. This attack was proposed by Christopher et al. [11]. First, blockchain is represented as Directed Acyclic Graph tree in which all nodes (or blocks) are connected with a directed edge. A delay is introduced between the different subgroups, then, transactions are issued in one sub group, while blocks are mined in the other subgroup by the attacker. This is done so that the attacker can mine blocks with all transactions but his own. The motive of the attacker is to make the chain of the subgroup in which he is mining longer, so that it can replace he other chains and all nodes accept it as the one true chain. Thus, this attack allows double spending, as the attacker's transactions are no longer included in the blockchain.

To execute this attack successfully, the attacker needs to identify the subgroup to which the merchant belongs and issue transactions to purchase goods from this merchant to the subgroup. Now, the attacker can propagate the mined blocks to the rest of the nodes. The attacker stops delaying messages when the merchant has shipped the purchased items [9]. There is a high probability that the directed acyclic graph tree seen by the merchant is outweighed by another tree [11], and thus the attacker, using the same coins could successfully reissue another transaction. This attack proves that PoW-based blockchain is block oblivious, i.e. there is a certain chance that the attacker can modify or delete blocks containing his own transactions and make the malicious tree the new default.

5 Conclusion

Bitcoin is a cryptocurrency, invented by Satoshi Nakamoto, as a replacement for fiat currency which does not require a central institution regulating it. Instead, it uses a decentralized mechanism to process and validate transactions, and a global ledger to store those transactions. Since there is no central regulating authority, the authenticity of the network depends upon the ratio of honest nodes and malicious nodes. As long as that ratio is greater than 0.5, all the malicious transactions will be discarded and not added to the blockchain. The blockchain is the global ledger that stores all the transactions. It is the one source of truth for the entire bitcoin network. It consists of blocks, which contain transactions, chained to one another. The only way to find a specific block is to find the block that is above it. Hence, if the chain is broken, there is no way to find the blocks below the fracture. This secures the blockchain against many malicious attacks. Since everything in the bitcoin network happens in the public domain, it needs to be able to handle many different types of attacks. We have mentioned some of the attacks above.

References

1. Bitcoin charts and graphs - blockchain. https://www.blockchain.com/charts?. Accessed 1 Dec 2019
2. Antonopoulos, A.M.: Mastering Bitcoin: Unlocking Digital Crypto-Currencies, 1st edn. O'Reilly Media, Inc., Sebastopol (2014)
3. Dai, F., Shi, Y., Meng, N., Wei, L., Ye, Z.: From bitcoin to cybersecurity: a comparative study of blockchain application and security issues. In: 2017 4th International Conference on Systems and Informatics (ICSAI), pp. 975–979, November 2017. https://doi.org/10.1109/ICSAI.2017.8248427
4. Er-Rajy, L., El Kiram My, A., El Ghazouani, M., Achbarou, O.: Blockchain: bitcoin wallet cryptography security, challenges and countermeasures. J. Internet Bank. Commer. **22**(3), 1–29 (2017)
5. Foroglou, G., Tsilidou, A.L.: Further applications of the blockchain. In: 12th Student Conference on Managerial Science and Technology (2015)
6. Heilman, E., Kendler, A., Zohar, A., Goldberg, S.: Eclipse attacks on bitcoin's peer-to-peer network. In: 24th USENIX Security Symposium (USENIX Security 2015), pp. 129–144 (2015)
7. Kar, J., Naik, K., Abdelkader, T.: An efficient and lightweight deniably authenticated encryption scheme for e-mail security. IEEE Access **7**, 184207–184220 (2019)
8. Kar, J., Naik, K., Abdelkader, T.: A secure and lightweight protocol for message authentication in wireless sensor networks. IEEE Syst. J. (2020)
9. Li, X., Jiang, P., Chen, T., Luo, X., Wen, Q.: A survey on the security of blockchain systems (2018)
10. Nakamoto, S.: Bitcoin: a peer-to-peer electronic cash system (2009). http://www.bitcoin.org/bitcoin.pdf
11. Natoli, C., Gramoli, V.: The balance attack against proof-of-work blockchains: the R3 testbed as an example (2016)
12. Nisha, S., Farik, M.: RSA public key cryptography algorithm - a review. Int. J. Sci. Technol. Res. **6**, 187–191 (2017)
13. Sapirshtein, A., Sompolinsky, Y., Zohar, A.: Optimal selfish mining strategies in bitcoin. In: Grossklags, J., Preneel, B. (eds.) Financial Cryptography and Data Security, pp. 515–532. Springer, Heidelberg (2017)
14. Singh, A., et al.: Eclipse attacks on overlay networks: threats and defenses. In: IEEE INFOCOM. Citeseer (2006)
15. Solat, S., Potop-Butucaru, M.: Zeroblock: preventing selfish mining in bitcoin. arXiv preprint arXiv:1605.02435 (2016)
16. Vijayakumaran, S.: The security of the bitcoin protocol, May 2019

Zero Trust User Access and Identity Security in Smart Grid Based SCADA Systems

Abdul Wahid Mir(✉) and Ketti Ramachandran Ram Kumar

Institute of Engineering and Technology, Chitkara University, Punjab 140401, India
k.ramkumar@chitkara.edu.in

Abstract. Smart grid based systems are part of modernized critical infrastructures that countries use to operate in various critical sectors ranging from electricity services, nuclear stations, transportation systems, hospital services, waste management, water services, etc. The modern smart grid based systems comprises of operational technology (OT) components and the information technology (IT) components and the convergence of both has introduced new security challenges for the operations of the systems in terms of its safety, reliability, efficiency and stability. The security perimeter is completely redrawn with the OT and IT convergence. The need of the hour is the new approaches towards the security that are innovative and will ensure the security of critical infrastructures. The traditional approach of perimeter-based security defense is obsolete with the convergence of OT and IT. The Zero Trust model for user access and identity is based on concept of "Never Trust but Always Verify" that enable organizations to secure the critical infrastructures by ensuring only trustworthy and validated components are allowed into the network. The perimeter-centric security architecture is replaced by Zero Trust model to ensure and enable security and access decisions for devices, identity and user context to be enforced dynamically. It also ensures that only authorized and authenticated users and devices can have access to the network, systems, applications and data. In this paper we are proposing Zero Trust User Access & Identity Security model that can be implemented in smart grid based SCADA systems.

Keywords: Zero Trust · Critical infrastructure (CI) · Industrial control systems (ICS) · Smart grid · SCADA · Operational technology (OT) · User access · User identity

1 Introduction

The critical infrastructure (CI) based industrial control systems (ICS) are deployed all over the world to manage the vital operations with Supervisory Control and Data Acquisition (SCADA) systems being an important part of it [1]. The smart grid based SCADA systems are typically implemented by electricity companies to enhance the control and monitoring of the operations. It enables them to achieve efficacy in the operations as well as efficiency in the cost. The following are some of the key bene-fits of the smart grid based systems:

A. Abraham et al. (Eds.): SoCPaR 2020, AISC 1383, pp. 716–726, 2021.
https://doi.org/10.1007/978-3-030-73689-7_68

- It enables improved and efficient transmission of electricity.
- The outages are addressed on time and the restoration of electricity is faster.
- It helps to reduce the operations and management cost resulting in lower electricity cost for the customers.
- It enables to manage the load and reduce peak demand resulting in lower electricity rates.
- It enables improved integration of large-scale renewable energy systems including the hybrid models.
- It enables improved security in terms of the systems and the assets within the grid.

The security threat landscape for the smart grid is very huge as it includes large number of components that include electricity generation plants, transmission networks, distribution networks, communication networks, consumers endpoint devices and networks, etc. The security of critical infrastructure based smart grid is very crucial both for countries as well as organizations running this business as any security incident can be catastrophic [2] (Fig. 1).

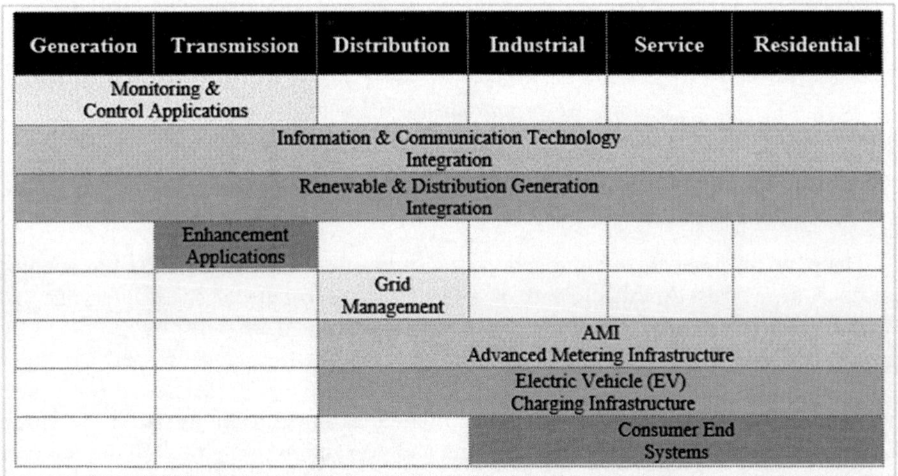

Fig. 1. Various system deployment across smart grid

The convergence of OT and IT leads to opening of previously "air gapped" ICS thus leading to potential threats of security incidents. The traditional security approach based architectures like perimeter model is obsolete with emerging smart grid technologies. In today's security context, all the companies are using firewalls to implement perimeter security. Nevertheless, this security model fails as soon as the perimeter is breached, and the attacker has easy access to the company's internal systems and data. With respect to smart grid architecture, it is really difficult to rely merely on the perimeter level security and its enforcement.

The major objectives of the attackers are loss, denial or manipulation of the systems and applications with the sole motive of bringing harm to the organization. The risk appetite in critical infrastructures is very low as any security incident can have disastrous consequences Table 1.

Table 1. List of attackers objectives

#	Attacker objective	Attack target
1	Loss	• Loss of view • Loss of control • Loss of monitoring • Loss of service
2	Denial	• Denial of view • Denial of control • Denial of monitoring • Denial of safety • Denial of service
3	Manipulation	• Manipulation of view • Manipulation of control • Manipulation of sensors & instruments • Manipulation of safety • Manipulation of service

The new and innovative approach to security i.e. Zero Trust model is the need of the hour. Zero Trust Architecture is based on "Trust No One, Suspect Everyone" or "Never Trust but Always Verify" concept [3]. The term "Zero Trust" was coined by Forrester in 2010. The concept of Zero Trust has been further developed by major market vendors including Google (BeyondCorp, 2014), Microsoft (Microsoft Advocates "Assume Breach", 2014), etc. It ensures no one is assumed to be trusted even within the perimeter and instead everyone, anything and everything trying to avail the services must be verified prior to granting access to any organizational resources.

2 Zero Trust Architecture Principles

The principles of Zero Trust are fundamentally built on notion of not trusting users, devices, networks, systems, applications, locations, etc. The access to all the critical resources are based on explicit verification of each of the identity type and their related data points. The main principles of Zero Trust are defined below (*Microsoft Zero Trust model*) [4]:

A. **Verify explicitly:** All the access requests from anyone and anywhere needs to be authenticated and authorized. The requests will be accessed based on multiple data points that include user identity, endpoint hardware type, geo-location, firmware versions, device health, operating system versions, service or workload, applications installed, data classification, patch levels, user logins, security or incident detections, anomalies, etc.

B. **Least Privilege access:** All the privileges must be least as a default policy with provisioning of Just-In-Time (JIT) and Just-Enough Access (JEA) methodology along with risk-based adaptive polices.

C. **Assume breach:** The security strategy should be of "assume breach" concept. The attack surface should be minimized for breaches and lateral movement should be prevented with the help of network segmentation by various categories that include access by devices, network, user, application awareness, etc. All the sessions should be protected by end to end encryption. Use of user and entity behavior analytics (UEBA) technology to get full visibility of users and their actions (Fig. 2).

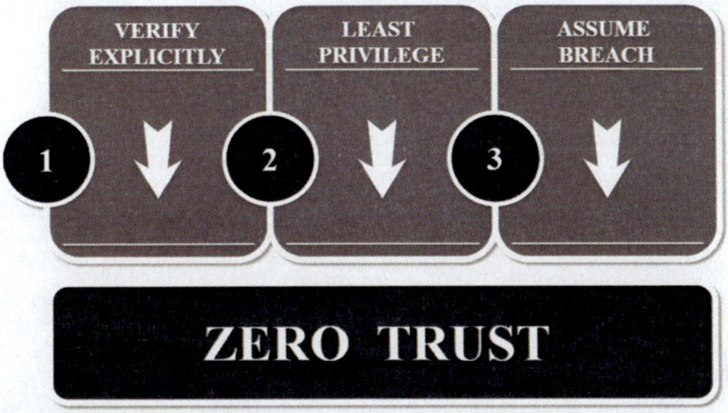

Fig. 2. Zero trust principles

3 Zero Trust – User Access and Identity Centric

The user access and identity is essentially core of the security model and identity management being at the heart of Zero Trust architecture. Zero Trust model can be defined by the identities of an organizational employees, the systems deployed in an organization and the data management practices in an organization. It is very effective model for organizations to control user access and identity management for their applications, systems, networks and data. It is mandatory requirement in Zero Trust implementation to ensure that all access requests are scrutinized continuously before granting them access to the organizational assets. The Zero Trust policies profoundly depend on real-time user behavior analytics that include the following user attributes (Table 2):

The main tools that are core requirement for implementing Zero Trust include Single Sign-On (SSO), Multi-Factor Authentication (MFA), Fast Provisioning Systems and Device Protection [5] (Fig. 3).

Table 2. List of attribute types

#	Attribute type
1	User identity
2	Endpoint hardware type
3	Geo-location
4	Firmware versions
5	Device health
6	Operating system versions
7	Service or workload
8	Applications installed
9	Data classification
10	Patch levels
11	User logins
12	Security or incident detections
13	Anomalies

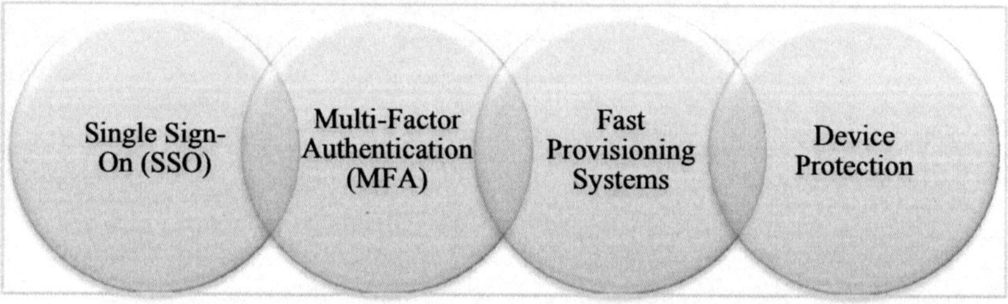

Fig. 3. Tools for implementing zero trust model

4 Implementation

The goal of the Zero Trust implementation should be that it is deployed across all the organization i.e. end to end strategy. The model should serve as an integrated security philosophy for an organization. The execution can be done by implementing controls and technologies across foundational elements that include identities, devices, applications, data, infrastructure, and networks. Every element comprises of the data attributes for the control plane for decision making whether to grant the access or defend the potential attack. All the elements are very critical for the successful implementation of the Zero Trust model [6]. The following are the details of the each of them:

A. **Identities**:
They can be users, applications, services, devices, etc. For every access request, the verification of the identity needs to be ensured using strong authentication methods in order to ensure that access is granted to legitimate user or device. The least access privilege policy needs to be applied all the time irrespective to the requester type of request type.

B. **Devices**
The data transfer takes place across the various devices after the identity has been allowed and granted access to the requesting resource. Since the attack surface is huge, the continuous monitoring is very important function that needs to be active so that device health and compliance requirements are met.

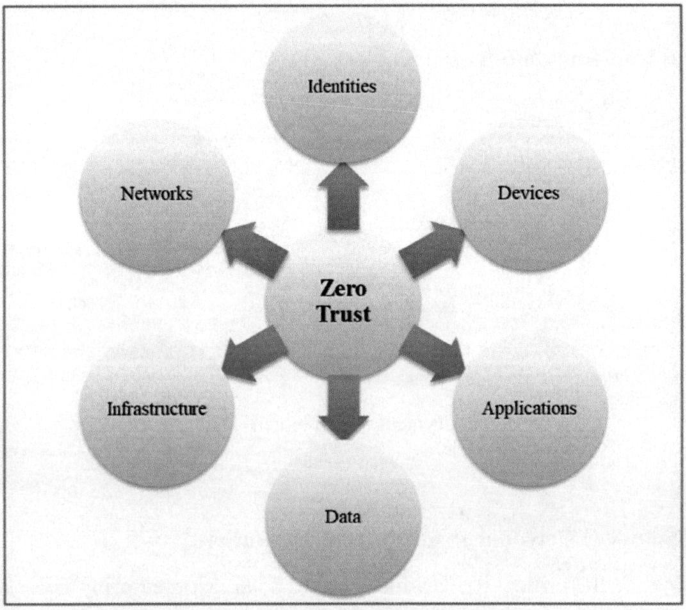

Fig. 4. Elements of Zero Trust model

C. **Applications**
Data is usually consumed by applications and API's as they provide the interface for the same. There is a lot of room for security breaches in this area owing to on-premises legacy systems and other related scenarios. The need is to have an appropriate controls in place that include permission for applications (in-app), real-time analytics based gate access, monitoring of abnormal user behavior, control and validation of user's actions.

D. **Data**
The ultimate focus of security is to protect data from unauthorized access. The provisions and methodology should be in place that ensures even in case the data leaves the organization, it remains safe and secure. The proper data classification should be applied along with data encryption.

E. **Infrastructure**
The infrastructure element is very critical irrespective of its type i.e. on-premises servers, virtual machines, containers, cloud services, infrastructure as a service, etc. This should be focused area with appropriate policies, controls, access and monitoring in place. The continuous monitoring for detecting any attacks, anomalies, risky behavior should be handled appropriately.

F. **Networks**
Network infrastructure is the core element that is used to access all the data. Proper management of network will help in preventing attackers to traverse through the network thus reducing the effect of any security incident. Network segmentation is key mechanism which can be implemented to handle the real-time threats and encryption.

4.1 Typical Implementation Steps

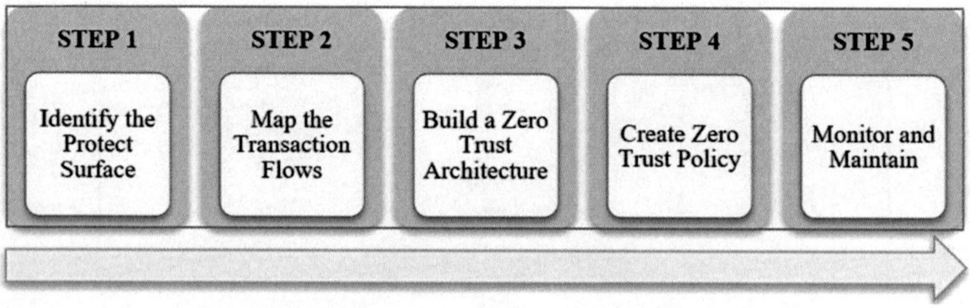

Fig. 5. Typical implementation steps

4.2 Key Factors to Consider Prior to Implementation

The following are the critical items which should be considered prior to the implementation of the Zero Trust model [7–10]:

a) The strong authentication processes must be implemented i.e. multi-factor authentication (MFA) technologies. There should be continuous session risk monitoring and detection as a part of access strategy in order to minimize identity compromise risk.
b) The mechanism of policy-based adaptive access must be implemented. The policies should be well defined for all the components and pertinently enforced.
c) The network forms the backbone for of the infrastructure and services. It should be appropriately segmented in order to reduce the attack surface and prevent any eventuality.
d) It is nearly impossible to manually manage security operations. The automation of alerts and remediation actions helps to decrease the mean time to respond (MTTR) to attacks. The user events and behavior analytics tool can help to automate this process.

e) Threat Intelligence is important component against the against attacks especially for real time attacks and zero day attacks.
f) It is really important to establish an organizational vide data classification policy. The policy should cover all the aspects of data available within an organization in terms of creation, storage, access, disposal, etc.

5 Advantages of Zero Trust Model

The primary objective of Zero Trust model implementation is to protect an organization externally as well as internally in particular. The biggest drawback of traditional security models like defense in depth has been its limitation and focus on organizational network perimeter. The internal organizational security was assumed to be covered while many of the breaches are ironically happening within inside only. Many cases of data exfiltration are coming to the surface as the within the network security is not the area of focus for the security teams [11]. Zero Trust employs different take where everyone and everything needs to be authenticated and authorized thus minimizing the breaches whether internal or external. Since the focus of security is not limited to perimeter only, organizational data that resides in cloud is still under the protection as it can be protected using identity verification method. This enables organizations to store and protect data anywhere irrespective of the location.

6 Testing Scenarios

The Threat Intelligence system is one of the core component of the Zero Trust model. The test scenarios are defined with some of the key data points that include Geo-Location, User Requests & Actions and User Events. The test data was used from AlienVault® USM Anywhere [12] application for the following set of events and alarms for the simulation and testing scenarios as shown in Fig. 4 to Fig. 6:

a) User Activities Events test data containing 1,700 + records.
b) Events by Authentication test data containing 50,000 records.
c) Events by Intrusion Detection test data containing 50,000 records.
d) ISO 27001 A.12.2.1 Controls against malware and location based test data containing 50,000 records (Fig. 5, Fig. 7, Fig. 8).

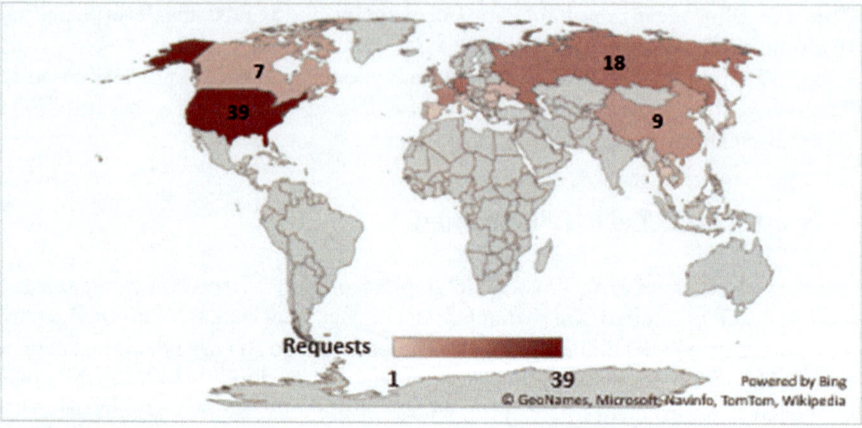

Fig. 6. Test scenario 1: country wise access requests map depicting the access requests from various countries

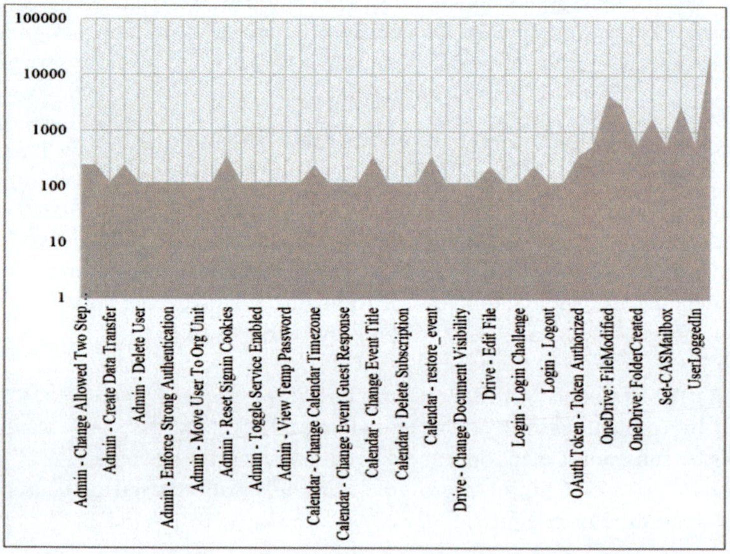

Fig. 7. Test scenario 2: user requests and user actions (logarithmic)

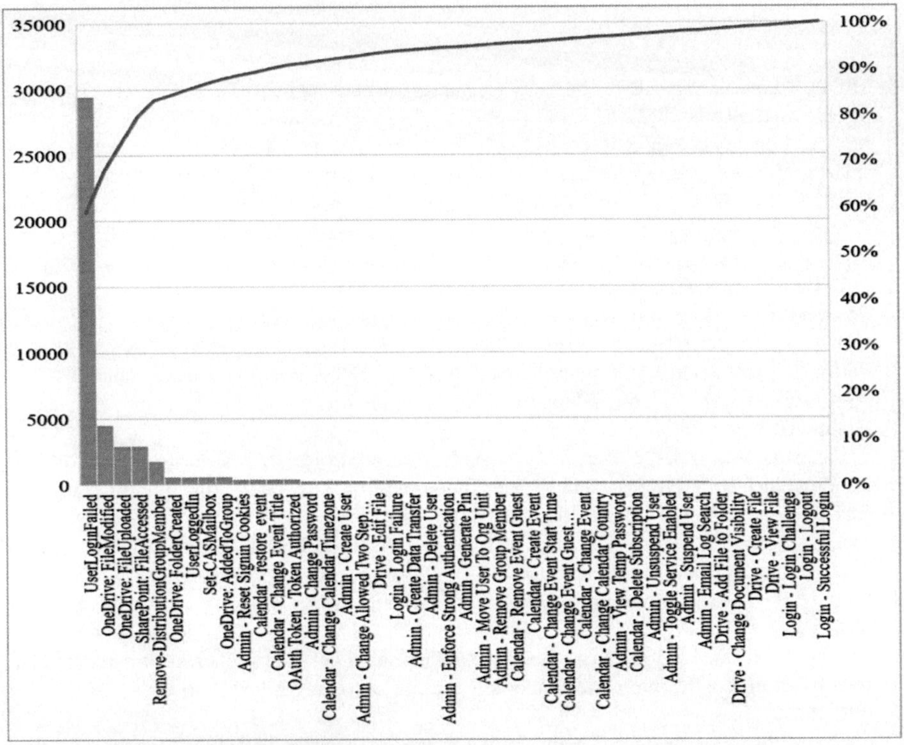

Fig. 8. Test scenario 3: pareto chart - user events & resulting actions

7 Conclusion

Smart grid based critical infrastructures rely profoundly on SCADA systems to monitor, control and operate the vital services. The security of smart grid is very critical and needs to be addressed as a top critical priority. Zero Trust model offers the next level of security architecture which can be deployed to protect critical infrastructures such as electricity utility based smart grid. The Zero Trust model offers identity centric approach that transforms legacy security models in future ready model. Although Zero Trust is not easy to implement, but with the resources and technologies at our disposal, it is achievable. The Zero Trust model is best solution to build security not only from outside in but from the inside out approach. Whereas no security model or architecture is offering 100% security assurance, the Zero Trust model is offering best chance to protect critical infrastructures from security incidents prior to any disastrous incident takes place. With the organizations like Microsoft and Google implementing Zero Trust models, it is evident that this is best in the class approach at the moment.

References

1. Mir, A., Ramachandran, R.K.: Security gaps assessment of smart grid-based SCADA systems. Inf. Comput. Secur. **27**(3), 434–452 (2019)
2. Mir, A., Ramachandran, R.K.: A survey on security challenges and research opportunities in smart grid based SCADA Systems. Int. J. Comput. Sci. Eng. **7**(3), 734–755 (2019)
3. Rose, S., Borchert, O., Mitchell, S., Connelly, S.: Zero-Trust Architecture NIST Special Publication (SP) 800–207. National Institute of Standards and Technology (2019)
4. Microsoft Homepage. https://www.microsoft.com/en-us/security/business/zero-trust. Accessed 01 July 2020
5. Gordon, S.: A matter of trust. Netw. Secur. **2019**(5), 9–11 (2019)
6. Keeriyattil, S.: Zero-Trust Networks with VMware NSX. Apress, Berkeley, CA (2019)
7. Zero-Trust architecture design principles. National Cyber Security Center, United Kingdom. https://www.ncsc.gov.uk/blog-post/zero-trust-architecture-design-principles. Accessed 10 Jan 2020
8. Berhe, A.B., Kim, K.H., Tizazu, G.A.: Industrial control system security framework for Ethiopia. In: 2017 Ninth International Conference on Ubiquitous and Future Networks (ICUFN), pp. 814–817. IEEE (2017)
9. Yan, X., Wang, H.: Survey on zero-trust network security. In: International Conference on Artificial Intelligence and Security, pp. 50–60. Springer, Singapore (2020)
10. Embrey, B.: The top three factors driving zero trust adoption. Comput. Fraud Secur. **9**, 13–15 (2020)
11. Ahmed, I., Nahar, T., Urmi, S.S., Taher, K.A.: Protection of sensitive data in zero trust model. In: Proceedings of the International Conference on Computing Advancements, pp. 1–5. ACM (2020)
12. Cybersecurity Response Anywhere, Unified Security Management (USM) Anywhere. https://cybersecurity.att.com/products/usm-anywhere. Accessed 01 June 2020

Man-in-the-Browser Attacks Against IoT Devices: A Study of Smart Homes

Sampsa Rauti(✉), Samuli Laato, and Tinja Pitkämäki

University of Turku, Turku, Finland
{sjprau,sadala,tievpi}@utu.fi

Abstract. Smart environments such as smart homes are a collection of IoT devices, sensors, artificial intelligence and remote control systems. These technologies come with many possible benefits, such as improved energy-efficiency, easier maintenance and increased living comfort. On the other hand, ubiquitous use of these technologies raise cybersecurity and privacy concerns. Understanding the vulnerabilities, attack vectors and potential exploits of these systems is crucial for enabling precise and effective countermeasures. In this study, we investigate the potential of man-in-the-browser attacks for targeting the remote control systems of smart homes. We implement a malicious browser extension to the Chrome web browser that can alter the user input in a smart home management console. We empirically demonstrate that these browser extensions can manipulate IoT devices in a smart home. We discuss countermeasures for securing the remote control systems of smart homes against man-in-the-browser attacks.

Keywords: IoT security · Browser security · Cybersecurity · Man-in-the-browser · Smart home · Smart environment

1 Introduction

The Internet of Things (IoT) consists of billions of interconnected devices and sensors – smart home appliances, industrial control systems, smart traffic systems, medical devices, and other networked devices all exchanging data. The fact that IoT makes everyday life easier cannot be denied. However, the security is often poor and technology is prone to advanced attacks. These remaining challenges degrade users' trust towards connected devices. The cybersecurity risks are present at every step along the journey when transferring data in a distributed IoT system, and there are many adversaries that wish to take advantage of a system's weaknesses.

One particular threat in today's internet is the man-in-the-middle attack (MitM) [1], in which the adversary stealthily relays and possibly modifies the data transmitted between two communicating parties. A malicious piece of code can alter or spy on the data in outgoing and incoming messages, while the

A. Abraham et al. (Eds.): SoCPaR 2020, AISC 1383, pp. 727–737, 2021.
https://doi.org/10.1007/978-3-030-73689-7_69

communicating parties (usually the user and the server) think they are directly exchanging messages with each other and do not notice anything suspicious.

As cryptography and endpoint authentication have made MitM attacks between endpoints more difficult, adversaries are increasingly targeting the endpoints themselves. A man-in-the-browser (MitB) attack is a special subtype of MitM attack taking place at the communication endpoint. As the name suggests, this attack intercepts and possibly modifies the data inside a web browser [7].

There are several reasons why we think MitB attacks are a significant threat for IoT systems. First, the trend of migrating from desktop applications to web environment also affects the IoT environment. IoT devices and smart systems are increasingly being monitored and controlled using web-based user interfaces, which exposes them to MitB attacks. Second, the huge growth in the number of IoT devices is making them common and interesting targets for cyber attacks. Finally, the increasing interconnection between the real world and cyberspace makes the potential effects of a MitB attack much more serious, as altering data can have immediate consequences in the physical world.

The aim of this study is to investigate MitB against IoT devices in the context of smart environments, more specifically, smart homes. To this end, we create a proof-of-concept implementation for a JavaScript-based MitB attack and empirically test it against a web interface for smart home devices, Mozilla WebThings[1]. We also discuss potential MitB attacks scenarios in the smart home environment and assess their consequences. Based on these tests and previous work, we suggest countermeasures for smart device designers, software vendors and smart home residents to guard their systems against MitB attacks.

2 Basics of a MitB Attack

Because of point-to-point encryption, modifying or eavesdropping traffic in a network is challenging. Instead, it is much easier for the adversary to target an endpoint of communication – that is, the user's device – where more attack vectors exist and the user can make mistakes that compromise cybersecurity. Infecting the user's web browser and performing a MitM attack before the data is encrypted is a tempting option. The fact that software developers and device manufacturers often do not consider MitB attacks a serious threat [2] is not making the situation any better.

As applications are increasingly being moved to the web environment, adversaries build malicious browser extensions that intercept data inside the web browser. As discussed, this kind of deceitful proxy inside the browser is called a MitB attack. The malicious browser extension aims to spy on or alter the data transmitted between the user and the server [7,16]. More specifically, the following functionalities are possible:

- capturing and storing sensitive data and delivering it to the adversary's command and control server [4]

[1] https://iot.mozilla.org/framework/.

- tampering with the data in incoming or outgoing messages [7]
- modifying the contents of web pages before they are rendered
- issuing additional malicious HTTP requests that are not initiated by the user

A MitB malware can contain one or several of the above functionalities. The malicious extension usually does its job silently, without the user or the server noticing its existence. Figure 1 depicts a MitB attack in the smart home environment.

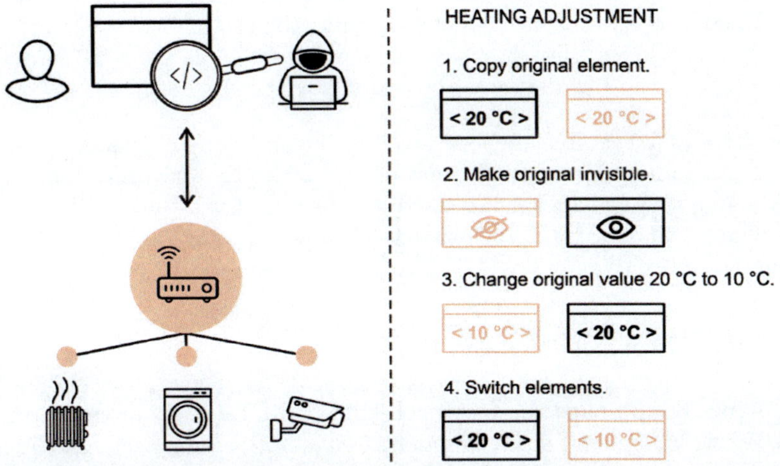

Fig. 1. A MiTB attack in the smart home environment (left) and the phases of an attack adjusting the room temperature (right).

Next, we describe an example scenario clarifying how a MitB attack can be used to stealthily modify the data given by the user and sent to the server, as well as the data later returned by the server. The attack alters the data on the user interface level and proceeds as follows:

1. The user's client device is infected with the malware. Implemented as a browser extension, the malware hides and executes inside the browser.
2. The malware contains a list of web addresses that activate the MitB functionality. In this case, opening the web page (user interface) for monitoring and controlling IoT devices launches the MitB attack.
3. The malware waits until the user causes the web application to make a HTTP request – for instance, the user adjusts the room temperature through a smart home user interface.
4. Before the user's request is transmitted to the server, the malware modifies it according to the adversary's wishes. One way of doing this is to make use of the DOM (Document Object Model) interface to manipulate values of HTML elements. For example, the adversary could change the room temperature set by the user.

5. After intervening with the HTTP request and changing values submitted by the user, the MitB extension allows the browser to proceed with sending the data to the server.
6. The tampered HTTP request with the modified values are transmitted to the server. Naturally, the server does not know the request has been modified by a malicious program and accepts the values as a valid input, trusting it reflects the users real intent.
7. When the user checks the status of an IoT device or sensor, for example the room temperature, the malware changes the value displayed in the user interface to deceive the user. In our example, the user thinks the temperature corresponds to the value they have set previously, while the real temperature controlled by the IoT device is the one set by the adversary.

The user and the IoT devices involved in the data exchange have been deceived. At some point the user may notice that the real room temperature is not the one they have set using the web interface. Even then, realizing where the problem is, finding the malicious extension and getting rid of it can be quite challenging.

3 Study Context: Smart Homes

Smart homes are perhaps the most well-known sub-category of smart environments. We define smart homes as *"houses enhanced with sensors and remotely controllable devices which utilize intelligence to increase the level of automation"*. This definition aligns well with that of Jiang et al. [8] who provide three defining characteristics which must exist within a dwelling for it to be called a smart home: (1) internal network; (2) home automation; and (3) intelligent control. Based on these characterizations, a smart home contains at least two or more smart devices or sensors that are connected to each other, either directly or via a central control system. Because of the strong link between smart home technologies and the real world, stakeholders place high cybersecurity requirements on these systems. As such, a single security solution is not enough and scholars emphasize the importance of implementing multi-layered security measures in smart home technologies [12]. Security of smart homes can also be improved by providing access to remote monitoring, which can further be supplemented with AI solutions to also monitor the behavior of IoT devices [18].

Forcing smart home devices and sensors to connect to the internet only through a central control system can enhance the cybersecurity of a smart home. Lin and Bergmann call this the gateway architecture [10]. Having a central control hub enables a more effective monitoring of online traffic within the house, and offers house residents a way to monitor and control their home through sensor data and connected devices [15]. The sensor data needs to be secured to ensure residents' privacy and the device control needs to be secured to ensure residents' security. However, house residents do not typically directly connect to the control hub with their smart devices, instead, a cloud server is implemented in most commercial solutions between the hub and the user, which is controlled by the hub provider [15]. This solution compromises user's privacy as their data is shared to a third party. On the other hand, it enables the hub provider to gather training data for smart home AI solutions which can then be used to further optimize, for example, the energy consumption of smart homes.

In a gateway smart home architecture [10], securing the connection between the hub and the cloud server, and the cloud server and the user, is thus paramount for maintaining user's privacy and security. These connections are password encrypted and cannot be interpreted or modified by a malicious agent, but as previously discussed, the MitB attack may be harnessed to interfere with the input of the user as they are accessing their home's control hub [13]. This way, even though the communication between the endpoints is encrypted, the malicious modifications to the user input (conducted by the MitB attack) are already included in the communication on the client device. In the next section, we describe a proof-of-concept implementation of how this kind of an attack can be carried out against smart home control interfaces.

4 The Proof-of-Concept Implementation

In this study, a MitB attack is implemented as a Chrome web browser extension. This extension manipulates the data the user has filled in and the settings they have made on a web site before they are sent to an IoT device. We also carry out experiments with the extension by testing it against popular smart home interfaces.

A MitB attack can be implemented on many layers [13]. For example, the adversary can modify the data in the HTTP requests or use DOM to manipulate data before it is sent to the server. Because intercepting HTTP traffic has been made difficult in Chrome by restricting this functionality for extensions [11], we opt for UI-level DOM modification, which is an easier and browser agnostic method. When manipulating the UI and replacing the user's input with modified data using DOM, the adversary has to do this silently without the user noticing anything out of ordinary. Simply changing the value of an HTML element in the UI which is easily seen by the user, does not work, for example. There are many ways to modify the input data stealthily, our implementation uses the following one:

1. Find the UI element O containing the value that will be modified.
2. Make a copy C of the original element O.
3. Make O invisible.
4. Replace the value of the invisible element O with a modified value.
5. Insert the copied element C in the place of O.

The user will only see the copy of the original element and interact with it. However, only the value of the original, hidden element, modifiable by the malware but unreachable to the user, is going to be sent to the server. The same technique can be used to change values of multiple elements. In a smart home management console, several settings can be changed and several devices controlled with one attack.

There is one more thing we have to implement: the malicious extension is also supposed to deceive the user when information about IoT devices is displayed. On pages that display status information about the smart home, the extension just searches the elements with the data that was sent to the server earlier (e.g. the real room temperature) and replaces it with the original data the user filled in. This can be achieved by storing the data input by the user, and displaying it in the appropriate element instead of the modified value that was actually sent to the server. The whole modification functionality of the extension can be written in just a couple of lines of basic JavaScript making use of the DOM API. Full implementation of the MitB extension is available upon request.

In the Chrome browser, extensions are normally only installed from the Chrome Web Store. However, users can also experiment with an extensions by turning on Chrome's developer mode. If the adversary cannot slip the extension into Chrome Web Store (which actively tries to screen new extensions and remove the malicious ones, but sometimes fails to detect dubious extensions), they can use social engineering techniques to convince the user to use developer mode to launch the extension. Because our extension uses JavaScript and DOM, the same code could also be used in a Firefox extension with very little effort. In this sense, our implementation of the MitB attack is browser and platform agnostic.

5 Attack Scenarios

There exists countless of IoT devices, and in addition, basic home functionality such as doors, light switches, heating and air conditioning can be connected to a control hub and controlled remotely. The exact composition of a smart home varies between houses. Due to this, in the following attack scenarios we focus on some of the more common IoT devices and sensors that are currently popular in smart homes, such as smart plugs, thermostats or fridges.

Scenario 1: Smart Plug. A smart plug is essentially a switch that can be used to turn on and off all kinds of electronic devices connected to it. If a MitB attack turns the smart plug on, this can lead to potentially dangerous situations and at the very least, increase energy consumption. One example could be turning on a radiator plugged into smart switch and causing a fire. A fridge could be turned off so that the food in the fridge goes bad or stereos could be turned on with a high volume. Many smart plugs can also be timed and this timing can be programmed through a web interface. The problem with smart plugs is that many devices that were never supposed to be remotely controlled can now be connected to the internet. Had these devices been designed for remote use, many of them would have more sophisticated safety mechanisms.

Scenario 2: Spying on the Smart Home. Malicious extensions with MitB functionality do not always aim to modify data. Often the mission is to spy on the user and report the information back to the adversary's server. The information could be sold to third parties or used for criminal purposes (e.g. for finding out when the user is not at home). As more and more devices are added to smart homes, there is lots of data on people's personal lives available for malicious adversaries. Spying can be done very stealthily and cannot be easily detected. When the data is being sent back to the adversary, an intrusion detection system might notice a suspicious connection in some cases depending on how the data is delivered.

Scenario 3: Smart Thermostat. If a smart thermostat was allowed to be turned off, a MitB attack could cause the indoors temperature to become freezing. This can of course be prevented by designing the device correctly and setting limits for the room temperature.

Scenario 4: Removing a Smart Device. A MitB attack can easily modify the user interface so that a certain IoT device is no longer displayed, essentially causing a denial of service. The user can no longer control the device (through the web interface) but the MitB malware can keep controlling the device as long as it is connected to the web. Another option would be to leave the device in the web interface but prevent any changes when the user tries to control it. In both cases, the user would probably think the device is malfunctioning and finding the real culprit would be very challenging.

We tested the feasibility of a MitB attack by building an extension to Google Chrome. The extension was tested against Mozilla WebThings, a piece of software for smart homes allowing users to monitor and control their smart devices over the web. By manipulating the user input, we successfully controlled a smart plug so that it was always on.

6 Suggestions for Mitigating MitB Attacks

Most scenarios against IoT devices covered in this study are dependent on the actions that the IoT devices offer the user, and consequently, a adversary through the MitB attack. There are several countermeasures that IoT device developers and software vendors can implement to mitigate the threat. Based on these findings, we suggest countermeasures against the MitB attack for (1) IoT device and smart home hub developers; (2) browser vendors and virus protection providers; and (3) individuals using IoT devices.

6.1 Suggestions for IoT Device and Control Hub Developers

Smart home device and control hub designers and manufacturers have the obvious burden of creating systems which cannot be exploited by adversaries. In doing so, a solid architecture and design are important. Following the guidelines of Lin and Bergmann, smart homes should not allow individual IoT devices to freely connect to and be visible on the internet, but should rather operate and be accessed through a control hub- a gateway [10]. The cybersecurity considerations of any networked device apply both to the control hub and the individual IoT devices nonetheless. Furthermore, situations where a user owns only a single IoT device but not a control hub need to be considered, as denying IoT devices from using the internet without a hub in between can limit the usability of these devices and can therefore be a sub-optimal solution. In addition to these considerations, we list the following guidelines for IoT device designers and manufacturers who wish to secure their devices against MitB attacks.

- *Input Validation.* The user input should always be validated. The developer has to ensure that the values sent to IoT device are feasible and cannot cause damage, such as allowing the user to set the room temperature to freezing. The user input should never be trusted, users can make mistakes or a malicious program can modify the input on the fly.
- *Out-of-Band Verification.* Use out-of-band (OOB) verification to verify input that differs from the user's usual behavioral patterns or is potentially harmful (such as shutting down a fridge). In OOB verification, the transaction initiated by the user is verified using a second channel other than the client device [5,19]. For example, a mobile application can be used to verify a command sent to an IoT device. However, if this kind of verification is used all the time, it becomes a routine and users may just allow any transaction without really thinking. OOB verification also degrades usability, as users may feel it adds unnecessary complexity. Therefore, OOB verification should be invoked sparingly.
- *Anomaly Detection.* Use an intrusion detection system and anomaly detection [6,18] to find unusual patterns in the network traffic between the client and the control hub. For example, the number of commands sent to IoT devices may increase substantially due to a MitB attack or their contents may be unusual.
- *Multi-layered Security.* Enhance the multi-layered security of IoT devices by applying proactive cybersecurity measures to their operating systems such as internal interface diversification [12].

6.2 Suggestions for Browser Developers and Antivirus Software Vendors

In many ways, web browsers are becoming more like operating systems, because they are now platforms for running web applications, and there is the trend of migrating many desktop applications to the modern web. The security of the browser environment, however, is not fully keeping up with this development. Just as programs in an operating system, extensions and applications running inside a web browser should be better monitored.

- *Ensuring DOM Integrity.* Inside browsers, the functionality of extensions and flow of data could be better monitored. For example, an extension making critical input fields invisible is suspicious. Also, there are solutions to cryptographically ensure the integrity of web pages [17] so that DOM cannot be manipulated.

- *Hardened Browser.* Hardening limits the attack surface by pruning off functionality and implementing other measures that thwart attacks. Ronchi and Zakhidov suggest loading a web browser from an external tamper-proof device [14]. A hardened browser would not allow installing extensions. Therefore, realizing a man-in-the-browser attack becomes much more difficult for the adversary. Using an external device is a solution that degrades usability and probably is not a good fit for an ordinary home computer. However, browser vendors should still consider minimizing the attack surface. Extensions could also be disabled for certain web pages.
- *Improved Malware Detection.* The detection rate for malicious browser extensions is still low [3]. Therefore, anti-virus software developers and browser vendors should co-operate in order to monitor and detect malicious extensions.

6.3 Individual's Precautions

A MitB attack is invisible to the user when it is happening. However, as the MitB malware used in this work was delivered as a browser extension, individuals need to be careful with the extensions they install. This is a problem that goes beyond the IoT environment, as previous work has shown MitB attacks to be effective against, for example, online banking systems [11]. In addition, users should be encouraged to educate themselves on the aspects of cybersecurity to increase their ability to identify and respond to cybersecurity risks within smart homes [9]. Users should also gain a better understanding about the meaning of permissions given to browser extensions and learn to avoid social engineering schemes that are often used to spread malware.

7 Conclusions and Future Work

In this work, we looked at a JavaScript-based MitB attack and empirically observed how it can be used to exploit currently available browser-based IoT device control interfaces. We have seen that MitB attacks can pose a significant cyber security and privacy treat in smart home environment. Still, it seems MitB attacks are not considered a serious threat by software and device vendors.

Naturally, it is important to understand that IoT devices have numerous other vulnerabilities that were not covered in this study. In addition, there are several smart home solutions and control interfaces we did not test. As such, our results are not aimed to provide a comprehensive and precise description of attacks against individual systems, but rather prove the magnitude and potential of MitB attacks in the context of smart homes.

As IoT devices are becoming popular for smart environments such as smart homes, and real and virtual worlds are getting increasingly tangled together, it is clear that cybersecurity research needs to keep up with the changes. Consequently, further work is needed to devise effective counter-measures against concrete MitB attacks as well as solutions for detecting them.

References

1. Bhushan, B., Sahoo, G., Rai, A.K.: Man-in-the-middle attack in wireless and computer networking—a review. In: 2017 3rd International Conference on Advances in Computing, Communication & Automation (ICACCA) (Fall), pp. 1–6. IEEE (2017)
2. Blom, A., de Koning Gans, G., Poll, E., De Ruiter, J., Verdult, R.: Designed to fail: a USB-connected reader for online banking. In: Nordic Conference on Secure IT Systems, pp. 1–16. Springer (2012)
3. DeKoven, L.F., Savage, S., Voelker, G.M., Leontiadis, N.: Malicious browser extensions at scale: bridging the observability gap between web site and browser. In: 10th USENIX Workshop on Cyber Security Experimentation and Test (CSET 17), Vancouver, BC, USENIX Association (2017)
4. Dougan, T., Curran, K.: Man in the browser attacks. Int. J. Ambient Comput. Intell. (IJACI) **4**(1), 29–39 (2012)
5. Entrust: defeating man-in-the-browser malware – how to prevent the latest malware attacks against consumer and corporate banking. White paper (2014)
6. Fernandes, G., Rodrigues, J.J., Carvalho, L.F., Al-Muhtadi, J.F., Proença, M.L.: A comprehensive survey on network anomaly detection. Telecommun. Syst. **70**(3), 447–489 (2019)
7. Gühring, P.: Concepts against man-in-the-browser attacks. Technical report (2006)
8. Jiang, L., Liu, D.Y., Yang, B.: Smart home research. In: Proceedings of 2004 International Conference on Machine Learning and Cybernetics (IEEE Cat. No. 04EX826), vol. 2, pp. 659–663. IEEE (2004)
9. Laato, S., Farooq, A., Tenhunen, H., Pitkamaki, T., Hakkala, A., Airola, A.: AI in cybersecurity education-a systematic literature review of studies on cybersecurity MOOCs. In: 2020 IEEE 20th International Conference on Advanced Learning Technologies (ICALT), pp. 6–10. IEEE (2020)
10. Lin, H., Bergmann, N.W.: IoT privacy and security challenges for smart home environments. Information **7**(3), 44 (2016)
11. Rauti, S.: Man-in-the-browser attack: a case study on malicious browser extensions. In: International Symposium on Security in Computing and Communication, pp. 60–71. Springer (2019)
12. Rauti, S., Laurén, S., Mäki, P., Uitto, J., Laato, S., Leppänen, V.: Internal interface diversification as a method against malware. J. Cyber Secur. Technol. **5**, 1–26 (2020)
13. Rauti, S., Leppänen, V.: Man-in-the-browser attacks in modern web browsers. In: Emerging Trends in ICT Security, pp. 469–480. Elsevier (2014)
14. Ronchi, C., Zakhidov, S.: Hardened client platforms for secure internet banking. In: ISSE 2008 Securing Electronic Business Processes, pp. 367–379. Springer (2009)
15. Stojkoska, B.L.R., Trivodaliev, K.V.: A review of internet of things for smart home: challenges and solutions. J. Clean. Prod. **140**, 1454–1464 (2017)
16. Ståhlberg, M.: The trojan money spinner. In: Virus Bulletin Conference, vol. 4 (2007)

17. Toreini, E., Shahandashti, S.F., Mehrnezhad, M., Hao, F.: DOMtegrity: ensuring web page integrity against malicious browser extensions. Inte. J. Inf. Secur. **18**, 1–14 (2019)
18. Zainab, A., Refaat, S.S., Bouhali, O.: Ensemble-based spam detection in smart home IoT devices time series data using machine learning techniques. Information **11**(7), 344 (2020)
19. Zhang, P., He, Y., Chow, K.: Fraud track on secure electronic check system. Int. J. Digit. Crime Forensic **10**(2), 137–144 (2018)

Scambaiting as a Form of Online Video Entertainment: An Exploratory Study

Samuli Laato(✉) and Sampsa Rauti

University of Turku, Turku, Finland
{sadala,sjprau}@utu.fi

Abstract. Popular scambaiting content creators and streamers have turned scambaiting into a form of video entertainment, which has accumulated hundreds of millions of views on YouTube alone. In this explorative study, we followed four popular scambaiters and their viewers to gain insight into the phenomena. Four levels of scambaiting were observed: (1) wasting scammer's time; (2) annoying scammers;(3) making fun of or ridiculing scammers; and (4) collecting assets from the scammers for authorities. While the first level could be regarded as a harmless activity, the rest require some ethical justification. To this end, scambaiters were found to gain sympathy by highlighting and showing the ruthlessness of the scammers. Scambait video viewers regularly shared experiences of their relatives being scammed. While recent work has suggested scambaiting could enhance racial stereotypes and hatred towards certain groups, such behavior could not be seen in the analyzed YouTube comments.

Keywords: Scambaiting · Technical support scam · Scammer · Entertainment · Video

1 Introduction

Scambaiting as a form of entertainment, that also serves the purpose of educating the masses about the harmful activity of scamming, started to emerge on online video platforms in the 2000s. It has been profiled as a controversial activity, sparking ethical discussions regarding both the activity itself and its viewership [13–16]. One of the initial ideas behind scambaiting aimed at wasting online scammers' time as much as possible, with the aim of making their business less profitable. However, recently it has also received aspects of entertainment at the expense of scammers [13]. YouTube videos where scammers are laughed at have accumulated hundreds of millions of views to date. The rapid increase of scambaiting channels and streamers during the past few years have contributed to scambaiting becoming its own genre of online video entertainment.

S. Laato and S. Rauti—The authors contributed equally to this work.

A. Abraham et al. (Eds.): SoCPaR 2020, AISC 1383, pp. 738–748, 2021.
https://doi.org/10.1007/978-3-030-73689-7_70

Scambaiting as a form of video entertainment can be decomposed in two ways. First, we can consider the characteristics of the scam. Previous work has discussed several different types of scams from advance fee frauds [6] to technical support scams [10]. Another way to categorize scams is to look at whether they occur in real time synchronously, or asynchronously as a correspondence [8]. Asynchronous scamming such as typical advance fee frauds [3] are more difficult to turn into online entertainment. Still, examples of this also exist. On January 8th, TED uploaded a video on YouTube titled: *"This is what happens when you reply to spam email"* which quickly accumulated millions of views. In the video, comedian James Veitch presents a humorous reconstructed scenario of his conversation with a scammer.

A pioneer of synchronous scambaiting videos, JimBrowning, uploaded his first videos the first half of 2014. Today, thousands of channels dedicated to synchronous scambaiting content exist on Twitch and YouTube. Most typically, scambaiter streamers engage in phone conversations with scammers, wasting their time and trying to do so in a humorous fashion. However, scambaiters have also exposed the real names and faces of scammers, initiating a discussion of the lawfulness and ethics of the activity. This is especially true for sites such as 419eater.com where scammers are publicly shamed [14]. Furthermore, in their pursuit of making the scambaiting as entertaining as possible, scambaiters have relied on ridiculing and purposefully enraging scammers. While stopping people from scamming others can be regarded as a noble activity, questions arise of whether the scambaiters themselves are also wrongdoers [11,13]. Furthermore, even ethical scambaiters could unintentionally encourage viewers to partake in morally questionable online vigilantism.

Scambaiting literature currently primarily focuses on addressing it from the perspective of online vigilantism [11] or "digilantism" [13]. These studies observe the phenomena from an ethical perspective. In addition, there is a strong research tradition in understanding phenomena specifically related to the so called 419 scams or asynchronous scams initially originating from Nigeria (e.g. [1,14]). We depart from these veins of research and observe scambaiting as a form of popular video entertainment. We study what the central phenomena related to video entertainment scambaiting are, how popular videos are constructed and how the viewing community responds to them. As the current research is exploratory in nature, we invoke the ethnographic research methodology [7]. The rest of this study is structured as follows. Next, we present our research methodology followed by the findings of our ethnographic observations and thematic clustering analysis. We then summarize our main findings, discuss the implications of our research and present the limitations of our study.

2 Method

2.1 Netnographic Observations and Analysis

We decided to carry out exploratory content-driven research [4], as the study topic was still fresh with only a few academic studies published beforehand.

Consequently, to explore scambaiting as a form of entertainment, we conduct netnographic observations [7] on popular scambaiter channels on the streaming service Twitch and the video sharing platform YouTube. As a form of participant observation research, we observed both the video content and viewers' reactions to the content as has been done in previous work on Twitch streamers [2]. We chose four popular YouTube channels for this analysis: (1) Jim Browning; (2) Kitboga; (3) ScammerRevolts; and (4) Scammer Hotline. The second channel, Kitboga, was also followed on the streaming platform Twitch.

For analyzing the video and text data, we utilized the applied thematic analysis method [4] by clustering observed phenomena into clusters based on their theme. In doing so, we initially identified several clusters which we started to combine together. The thematic framework was iterated between two authors until a consensus on a framework that describes the data was reached. To report our findings, we dedicate a subsection for each identified thematic cluster, describe main phenomena within the cluster and also include comments from both scambaiter video creators as well as chat participants.

2.2 Information About Scambaiter YouTube Channels Followed in This Study

Jim Browning[1] is a YouTube channel created in 2014 with currently 132 million views across 75 videos and 2.13 million subscribers. The channel contains various kinds of scambaiting videos where scammers are lured into revealing information about themselves, or even giving the scambaiter access to their system.

Kitboga[2] is a channel created in 2017 with 168 million views across 347 videos and 1.28 million subscribers. The channel differs from Jim Browning in that the scambaiting is mostly done live, is uncut and less technical. The channel description includes a justification of scambaiting as follows: *"Everyday there are scammers taking advantage of people. I call them to waste their time, walk people through their "script" and lies, report info when I can, and otherwise make light of a dark situation."*

ScammerRevolts[3] currently has 71 million views across 505 videos and 849k subscribers. The channel description reads as follows: *"I troll and expose scammers and do my best to make sure they can't scam any more innocent people!"* However, some of the channel videos go beyond trolling and involve doing harm to the scammers' PC. Similarly to Kitboga, ScammerRevolts also streams his scambaiting activities.

Scammer Hotline[4] is a channel founded in 2016 that has 10 million views across 31 videos and 144k subscribers. The channel does not aim to get any information out of the scammers like Jim Browning and ScammerRevolts, but instead focuses simply on having fun and entertaining conversations with scammers to waste their time. In this regard the channel is similar to Kitboga. The

[1] https://www.youtube.com/c/JimBrowning/.
[2] https://www.youtube.com/c/KitbogaShow/.
[3] https://www.youtube.com/c/ScammerRevolts/.
[4] https://www.youtube.com/c/ScammerHotline/.

channel description describes the content as follows: *"I annoy telephone scammers and upload the recordings here on YouTube for your enjoyment! Check it out!"*

3 Findings

3.1 Scambaiting Videos Are Primarily Watched for Entertainment

With hundreds of millions of views across several channels on YouTube alone, scambaiting has become a popular form of online video entertainment. This has had several consequences. First of all, the scambaiters have brought public attention to the scam baiting business. The risen awareness and education concerning the issue has likely had beneficial impacts on reducing the number of victims. In addition, scambaiters have motivated others to start scambaiting as a hobby, marketing it as a fun and rewarding activity. Because of these issues and the increased number of scambaiters, scammers have become more paranoid of who they are scamming and are resorting to measures for verifying that they really are scamming clueless people instead of scambaiters. The YouTuber Jim Browning explicitly states in his video that almost all scammers are currently aware of his channel and his voice, and accordingly he has been forced to use voice modulation. The popularity of scambaiter has ignited an arms race between scammers and scambaiters where scammers are coming up with new measures of finding real victims, and ways to detect scambaiters. Scambaiters are responding by coming up with new ways for getting into contact with scammers and remaining undetected.

3.2 Scambaiters Have Several Ways to Gain Sympathy and Justify Their Activity

An important factor in justifying scambaiting is the demonstration of the unethical, criminal conduct of the scammers. A good example of this comes from Kitboga. He regularly uses a voice modulator and acts as an old clueless grandma, "Edna". He says things like *"My daughter Nina goes there to get my prescriptions"*, *"I told the previous technician that... [implying she has been scammed before]"* and *"I am 81 years old, I can hardly move out of this couch"*. The viewers then can listen to the scammer forcefully continue their scam despite obviously realizing their target is defenseless and clueless. Besides voice modulation and appearing clueless, scambaiters can demonstrate their total lack of technical know-how by having a large number of useless or even malicious software and add-ons in their system. Furthermore, when scammers attempt to invade the user's privacy, further anger is stirred. As an example, YouTuber Monkey Wrench Moment writes: *"This one gives me high anxiety, I feel like KitBoga was truly angry when they tried to access his camera. It made me mad too. It felt intrusive and scary."*

With this type of a setup, the viewers realize that the scammer is prepared to attack clueless victims, and not only go against their money, but also have complete disregard towards their privacy. In the chat, some commentators shared personal experiences of their own relatives falling victim to these types of scams. YouTuber Brandon Barber writes: *"My grandmother has dementia and has had contact with these scammers before. I want to personally thank you for messing with these scumbags."* Taken together, scambaiters gain sympathy by showing the ruthless nature of the scammers, and people further draw parallels to their own helpless relatives. This serves as a justification for scambaiters' activities, and might also explain why people enjoy watching scambaiting videos.

3.3 Confrontations with the Scammers Initiates an Ethical Debate

An important part of scambaiting is the confrontation of the scammers. This typically occurs towards the end of video clips displayed on YouTube. If using voice modulators, the scambaiters switch them off revealing to the scammers that they were in fact not what they pretended to be. The confrontation often involves phrases such as *"You go through a script and take advantage of innocent old ladies"* (Kitboga). This initiates an conversation with the scammer where they either admit to scamming, switch topics or hang up. Here another form to gain sympathy and acceptance for scambaiting might come if the scammers admit to malicious activity, and justify it by calling people stupid, weak, fools or other names.

Arguments by scammers who admitted to being scammers were, for example, the following:

"You people are just fools. People who live in the USA. That's why we are doing this."

"I am earning a lot of money from you people. That's why we are doing this."

Arguments by scammers denying they were scammers were, for example:

"No no sir, we are not scamming you. This is technical support, you called us."

"This is not a scam sir. I can show you our reviews."

However, upon increased pressure by the scambaiters, the initial deniers sometimes later on admitted to scamming. The most popular YouTube clips rarely include examples of scammers hanging up after realizing they are baited, but in the Twitch stream of Kitboga such cases could be regularly seen.

3.4 Scambaiting Has Parallels to Bullying

As noted by several authors, scambaiting features aspects of ridiculing the scammers, which sometimes may turn towards ridiculing an entire nation of people as has arguably happened in the case of the 419 scams [14,16]. This aspect could be seen in particular on the videos of Kitboga and ScammerRevolts. Highlights of Kitboga's stream started appearing on YouTube with titles such as: *"WHY DID YOU REDEEM IT? Full video. Scamming scammers"*[5] that showcases a scammer being mad, and which the viewers laugh at. Comments such as nedra

[5] https://www.youtube.com/watch?v=8r-cEirkuTI.

lemieux's: *"Lmmfao! I'm crying laughing over here"* and Asha Greig's *"This is the best entertainment!"* were commonplace on the particular video and others like it. This behavior of laughing at someone being mad, or laughing at an intentional ridicule of someone, draws a parallel to bullying [12]. ScammerRevolts at times even went as calling scammers names and demanding money from them as follows: *"You dumb bitch. You need to give me iTunes gift cards, then I will give you the password."*

As several of the tech support scammers are poor and have no other income, it needs to be questioned whether they should be ridiculed publicly despite scamming being wrong. On the other hand, before ridiculing a scammer, the scambaiters typically verify that the scammers are aware of the fowl nature of their activity. In some instances, in particular on several videos of Jim Browning, the scambaiting activity involves accessing the scammer's PC, or in the case of ScammerRevolts even removing some of their files. In these cases we can talk about doing harm to the scammer beyond wasting their time, which further emphasises the bullying aspect of scambaiting. The illegal aspect of scambaiters' activities is also at times brought up by the scammers. For example, in the Kitboga YouTube video "Scammer Lost his Mind After Failed SYSKEY"[6] the scammer replies: *"Even you were taking advantage [of me]"*.

3.5 While Scambaiting Might Bring Forth Racial Stereotypes, It Does Not Show in the YouTube Comments

The racial stereotype problem is acknowledged in the comments. As an example, YouTube user carlotapuig writes: *"This scammer embarrassed his whole country and almost 5 million people watched this. If I were from his country I would be very very mad. He brought a lot of shame to India."* YouTube user Eshita Goel writes the following: *"I live in Delhi, where we have a lot of scam call centers. such a disgrace, I feel so embarrassed and ashamed. We all aren't like this, I promise."* These comments received hundreds of likes and were commonplace across videos across YouTube channels. This suggests that while prior research has alerted the scientific community to the possibility of scambaiting leading to the exploitation and ridicule of certain groups of people [14], such activity is not representative of the typical Youtube scambaiting audience.

A particularly interesting example related to this theme is the video *"SAD SCAMMER SAYS HE WANTS TO DIE???"*[7] from the Scammer Hotline. The video features a scammer who seems genuinely depressed. Upon asked about his dreams, the scammer replies *"my dream is to die"*. The YouTube comment section response to the video shows overwhelming support for the particular scammer. For example, YouTuber Nami Zou wrote: *"Poor Adam he sounded like he wanted to cry...I'm so sad for him my heart ached hearing his sad voice. And he sweetly says everybody has dreams...then.." my dream is to die "omg I'll pray for Adam."* YouTuber Princess Games echoed these thoughts: *"I feel bad*

[6] https://www.youtube.com/watch?v=698Rm2FV6ik.
[7] https://www.youtube.com/watch?v=dMRX_qU37f4.

for "Adam", he is doing this brutal job, he's forced to scam people, and the fact that he doesn't have a big dream other than dying, is so sad, and so heartbroken. These scammers need to stop forcing innocent people, money gets you nowhere, only hard work and dedication gets you somewhere, stay strong "Adam"".

Taken together, these examples demonstrate that YouTube viewers seem to be motivated to gang up and enjoy the ridicule of only those scammers who are arrogant and ruthless. In fact, during our research sample we found no counterexamples to this besides very few comments.

3.6 Scambaiting Video Creators Have Multi-layered Motivation for Their Activity

In addition to motivation related to the scambaiting activity itself, such as wasting scammer's time and gratifications derived from psychologically dominating the scammer, scambaiting video creators have several other reasons to create and post videos online. Here we discuss three of them.

Entertainment Value. Scambaiting has entertainment value for both the scambaiter and their potential audience on video sharing platforms, as they laugh at the expense of obnoxious scammers getting scammed. Not only is the content enjoyable and humorous to many, the community also enjoys the aspect of inflicting vengeance on criminals. In addition to entertaining others, scambaiters seemed entertained by the activity themselves, as evident by, for example, Kitboga laughing several times during his stream.

Educating Internet Users. Since scambaiting has become something people want to share on social media and video platforms, educating the public about this threat has also become an important justification for the activity. As technical ways to prevent scams based on social engineering are limited, disseminating knowledge about the modus operandi of scams has value for the large scale presentation of scams. Even scambaiting videos made solely for entertainment purposes often provide good lessons for internet users, and teach them to be wary of different kinds of scams rampant on the web. Through watching scambaiting videos, users can also learn general IT skills. Most scambait videos include some educational elements, and not only about how to avoid being scammed, but also on how to scambait. For example, Scammer Hotline videos include the warning *"NEVER USE YOUR REAL TELEPHONE NUMBER TO CALL SCAMMERS!!!!!!!"* as the first thing in their video description.

Scambaiting as a Source of Income. With over a million cumulative views on YouTube, advertisements running during them and additional income from Patreon and other sources, the observed scambaiting channels are making a lot of money from viewers supporting their activities. The rationale for many viewers to donate to scambaiters is that rather they see scambaiters have the money than actual scammers who could target their relatives unless stopped. Some viewers seemed to also be motivated to support scambaiters simply because their content was entertaining to them. From the scambait video creators' perspective, video sharing platforms enable them to make a living out of scambaiting and gives them the time and money to come up with ever more complex and elaborate scambait strategies. On the other hand, the popularity of their channels can backfire as scammers become aware of them and can quickly learn to avoid their counter-schemes.

4 Discussion and Conclusion

4.1 The Four Levels of Scambaiting

Based on our findings, we can derive four distinct levels of scambaiting ranging from wasting scammer's time to collecting assets from the scammer. Hypothetical levels beyond these four could already be considered criminal, and are thus not included here. These levels are described below.

1. Wasting Scammers' Time and Money. Scambait video creators are likely motivated by several factors. Perhaps the most popular, and initial reason for baiting scammers is to waste their time, energy, money and other resources. This way, scammers will have less time to deceive more gullible victims. Scambaiters have different strategies to prolong phone calls, in some cases even taking several hours of a scammer's time during a single session. Losing time and money this way is also a significant problem for call canters where scammers work. Most commentators on YouTube seemed to be motivated to waste scammer's time due to personal prior experiences with scammers. From this, we arrive to our next subtheme.

2. Annoying Scammers. Many scambaiters want to pay back to scammers for annoying innocent users with scam calls and deceptive messages. There are many strategies used to vex scammers, such as intentionally misunderstanding them, forcing them to diverge from the script and making all kinds of excuses for dragging out the call. Continuously prank calling scammers and asking them stupid questions was popular in the observed content creators (e.g. Scammer Hotline's and KitBoga's videos). Sometimes calls escalate into shouting and yelling insults, especially from the scammers part. This can be seen as morally questionable, but also adds entertainment value for viewers. Instances where scammers lose their temper also often found their way into the titles of posted YouTube videos.

3. Ridiculing Scammers Publicly. Scammers are laughed at and ridiculed publicly by scambaiters and their audience, although faces or identities usually cannot be shown on the video (because of YouTube's terms of service). However, more detailed information is often available on scambaiter's Patreon. Scambaiting sites such as 419eater.com also publish photos of scammers (called trophies by scambaiters).

4. Collecting Information or Assets from Scammers. In some cases, scambaiters are able to reverse the remote connection so that they can take a look at the scammer's machine and gather information on their identities, victims and operations. For example, a call center owner was arrested after Jim Browning hacked into the scammers' system and collected information on identities of the scammers[8]. Browning also managed to obtain 70,000 recordings in which innocent victims were being ripped off by the scammers.

4.2 Main Contributions

The identified and above described four levels of scambaiting can be considered one of the main contributions of this work. Scammer Hotline and mostly Kitboga operated on the first level, with the main proportion of their content focusing on having funny harmless and long conversations with scammers. Kitboga also included aspects of the second level by ridiculing and making fun of scammers. ScammerRevolts was the only content creator who was found to aggressively call the scammers with names, ridiculing them. On the fourth level, Jim Browning and ScammerRevolts seemed to take slightly different approaches. While ScammerRevolts "fought fire with fire", Jim Browning tried to take a more legal route in his activity. He informed viewers in his videos that he has reported the scammer's activities to the service providers they use, and in the case of one video, even managed to get some people arrested through contacting the local police force.

Previous work has discussed scambaiting as a form of vigilantism [11,13], but it can also be seen as a form of citizen activism that assist law enforcement rather than acts as one itself. The English philosopher and political theorist Michael Oakeshott stated that people continuously seek to balance their environment and strive to change the current state of affairs into a more appealing one [9, pp. 247–285]. Our study revealed that scambaiters and their viewers strongly feel that authorities are not doing enough to catch scammers from other countries. There are barriers here involving international law and other factors. For these reasons, and judged by the rising popularity of scambaiting, there is definitely a gap to be filled with activism in order to satisfy the basic human need for justice with regards to online scams.

4.3 Limitations and Future Work

This study has the following limitations. The netnographic observations were carried out by two authors independently and focused primarily on four YouTube content creators. While we believe this sample to be representative of the currently popular scambaiting channels, we acknowledge that there is a large scambaiting community beyond the observed platforms, as also discussed in other recent work [11,14]. Furthermore, there is always some subjectivity involved in ethnographic work, which is why several studies supplement observations with

[8] https://www.bbc.com/news/technology-51740214.

interviews [5]. We identify this as an important venue for future work on the topic. Furthermore, our analysis was limited in depth. We went through several videos from each content creator, but not all of them. We read through the most popular and controversial comments in these videos, but not all of them. A more systematic approach to content analysis could improve the robustness of the findings.

This study was one of the first to examine scambaiting as a form of online video entertainment. Its aim is to take the first steps towards understanding why scambaiting has grown to become a popular form of online video entertainment. Future work on this research area should involve both quantitative and qualitative data from scambaiters and those viewing scambaiting content to verify the findings of our work on why people watch scambaiters and what makes the activity appealing. The educational potential of scambaiting also needs be studied in further detail. Finally, studies on the implications of the popularity of scambaiting on scammer's activities should be assessed to establish the impact it has on scamming as a business.

References

1. Brody, R.G., Kern, S., Ogunade, K.: An insider's look at the rise of Nigerian 419 scams. J. Finan. Crime (2020)
2. Consalvo, M., Phelps, A.: Performing game development live on twitch. In: Proceedings of the 52nd Hawaii International Conference on System Sciences (2019)
3. Edwards, M., Peersman, C., Rashid, A.: Scamming the scammers: towards automatic detection of persuasion in advance fee frauds. In: Proceedings of the 26th International Conference on World Wide Web Companion, pp. 1291–1299 (2017)
4. Guest, G., MacQueen, K.M., Namey, E.E.: Applied Thematic Analysis. Sage Publications, Thousand Oaks (2011)
5. Hammersley, M.: Ethnography. In: The Blackwell encyclopedia of sociology (2007)
6. Holt, T.J., Graves, D.C.: A qualitative analysis of advance fee fraud e-mail schemes. Int. J. Cyber Criminol. **1**(1), 137–154 (2007)
7. Kozinets, R.V.: Netnography. In: The International Encyclopedia of Digital Communication and Society, pp. 1–8 (2015)
8. Laato, S., Murtonen, M.: Improving synchrony in small group asynchronous online discussions. In: World Conference on Information Systems and Technologies, pp. 215–224. Springer (2020)
9. Oakeshott, M.: Experience and Its Modes. Cambridge University Press, Cambridge (1933)
10. Rauti, S., Leppänen, V.: "You have a Potential Hacker's Infection": a study on technical support scams. In: 2017 IEEE International Conference on Computer and Information Technology (CIT), pp. 197–203. IEEE (2017)
11. Smallridge, J., Wagner, P.: The rise of online vigilantism. In: The Palgrave Handbook of International Cybercrime and Cyberdeviance, pp. 1307–1331 (2020)
12. Søndergaard, D.M.: The thrill of bullying. Bullying, humour and the making of community. J. Theor. Soc. Behav. **48**(1), 48–65 (2018)
13. Sorell, T.: Scambaiting on the spectrum of digilantism. Crim. Justice Ethics **38**(3), 153–175 (2019)

14. Yékú, J.: Anti-afropolitan ethics and the performative politics of online scambaiting. Soc. Dyn. **46**, 1–19 (2020)
15. Zingerle, A.: Towards a categorization of scambaiting strategies against online advance fee fraud. Int. J. Art Cult. Des. Technol. (IJACDT) **4**(2), 39–50 (2014)
16. Zingerle, A., Kronman, L.: Humiliating entertainment or social activism? Analyzing scambaiting strategies against online advance fee fraud. In: 2013 International Conference on Cyberworlds, pp. 352–355. IEEE (2013)

Create a Hybrid Algorithm by Combining Hill and Advanced Encryption Standard Algorithms to Enhance Efficiency of RGB Image Encryption

Rania A. Tabeidi(✉), Hanaa F. Morse, Samia M. Masaad, Reem H. Al-shammari, and Dalia M. Alsaffar

Computer Science Department, College of Science and Humanities, Imam Abdulrahman Bin Faisal University, 31961, Jubail, Kingdom of Saudi Arabia
ratabeidi@iau.edu.sa

Abstract. The greatest challenge of this century is the protection of stored and transmitted data over the network. This paper provides a new hybrid algorithm designed based on combination algorithms, in the proposed algorithm we combine the Hill and the Advanced Encryption Standard Algorithms, to increase the efficiency of color image encryption and increase the sensitivity of the key to protect the RGB image from Keyes attackers. The proposed algorithm has proven its efficiency in encryption of color images with high security and countering attacks. The strength and efficiency of combination the Hill Chipper and Advanced Encryption Standard Algorithm is tested by statical analysis for RGB images histogram and correlation of RGB images before and after encryption using hill cipher and proposed algorithm and also analysis of the secret key and key space to protect the RGB image from brute force attack. The result of combining Hill and Advanced Encryption Standard Algorithm achieve strength in encrypting images rather than encrypting images by using the single Hill algorithm.

Keywords: Image · Encryption · Algorithm · AES · Hill

1 Introduction

Recently, with the development of digital devices and the need to send images via the Internet, the importance of information security has emerged [1]. Cryptography is one mechanism used in securing information. There are two types of cryptography, one key encryption, and two key encryptions. The symmetric key encryption is called a private-key scheme, where the secret key not transmitted by the sender and receiver [2, 3]. Asymmetric key encryption is called a public-key scheme with two different keys. One of the keys, the public key is available for anyone, the other key is known as the private key [4]. The symmetric cryptosystems security, such as AES, RC4, DES depends on the key size, however, the asymmetric cryptosystems, such as RSA, DSA, and Elliptic Curves depend on the difficulty of the mathematical problem [5]. In a new schema, we propose an algorithm that is a combination of the Advanced Encryption Standard (AES) algorithm and Hill chipper algorithm. AES was adopted by the U.S. government, which

A. Abraham et al. (Eds.): SoCPaR 2020, AISC 1383, pp. 749–757, 2021.
https://doi.org/10.1007/978-3-030-73689-7_71

is a symmetric key that has a 128-bit block size [6]. The key length of AES cipher is 128, 192, and 256 bits. [7]. Hill cipher was originating in 1929 by the mathematician Lester Hill, it is a multi-letter cipher. Hill cipher is used the secret key that uses matrix size n x n as the key for encryption and decryption [8]. The paper consists of the following: Next section presents the related work on image encryption. Section 3 describes the methodology used to encrypt the images. Section 4 presents the results of the proposed algorithm. Finally, Sect. 5 represents the conclusion.

2 Related Work

In [9], the researchers used the Advanced Encryption Standard (AES) Algorithm in the process of encoding and decoding color images by using MATLAB. The result indicates that no information about the image can be obtained after encrypting and the histogram is different between the images before and after encryption. Therefore, the images were successfully encrypted using the mentioned method.

In [10], the researchers combined encryption by using the Advanced Encryption Standard (AES) algorithm and the steganography which is a way to embed secret information. Combining encryption and steganography gives more efficiency when transmission images, especially in the unsecured networks.

In [11], the researchers improved the Hill cipher algorithm to overcome the disadvantages of the algorithm, which consists of not encrypting all the image features with a homogeneous background. The results showed that the proposed algorithm encrypt images with high security and cope the weakness of the original algorithm.

In [12], the researchers combine two methods for encryption namely, Hill Cipher and Caesar Cipher method. They used the proposed method to encrypt the data on exam questions. The method the researchers used relies on one key in the encryption process, which made it simple and easy to understand.

In [13], the proposed algorithm about image encryption based on a secure variant of Hill Cipher (HC) and three improved one-dimensional (1D) chaotic maps. This suggestion algorithm encrypts pixel by pixel in the images. The result showed that the encrypted result by using the proposed algorithm was effective and the time of encryption was short.

In [14], the researchers proposed a new approach to increase the security on the Hill cipher algorithm. They proposed a hybrid encryption algorithm between the elliptic curve cryptosystem and hill cipher (ECCHC). The methodology they proposed converting Hill Cipher from a symmetric technique (private key) to asymmetric one (public key) and increase the security level and efficiency.

In [15], the researchers proposed a framework to provide security in the cloud. They used two encryption algorithms namely Extended hill cipher and Homomorphic encryption. The first algorithm was applied to encrypt dynamic data while the second algorithm was applied to encrypt static data. The result showed that the proposed algorithm ensures the transmission of information between the cloud and the customer is efficient and safe.

3 Methodology

We proposed a hybrid algorithm by combined the Hill cipher algorithm and Advanced Encryption Standard (AES) cipher algorithm, the Hill cipher inputs are RGB image I_1

and secret key (k_1) contains the secret matrix ($K_{2 \times 2}$) with two numerical values ($C_{2 \times 2}$), the output is encrypted RGB image I_2. The RGB image I_2 becomes one of the inputs of the AES cipher with secret key K_2. Figure 1. shows the encryption and decryption of the proposed scheme of hybrid algorithms, in the decryption method, the output encrypted RGB image (I_3) used as the input in Fig. 1(b).

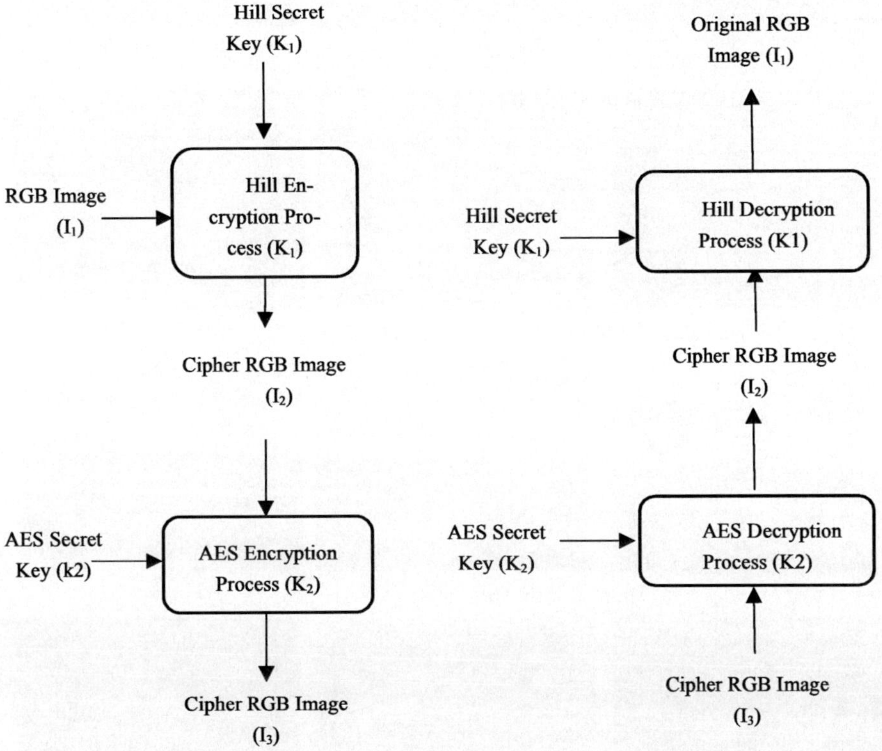

(a) The encryption of Proposal Scheme (b) The decryption of Proposal Scheme

Fig. 1. Proposal scheme of hybrid algorithms

4 Results

4.1 Security and Performance Analysis

We used MATLAB 2020 to analyze several properties of the image before and after encoding, four images were used for encryption and decryption.

(a) Flower (b) Hill Cipher Flower (c) Cipher Flower

(d) Fruits (e) Hill Cipher Fruits (f) Cipher Fruits

(g) Horse (h) Hill Cipher Horse (i) Cipher Horse

(j) Pabon (k) Hill Cipher Pabon (l) Cipher Pabon

Fig. 2. The plain RGB images and their ciphered RGB image using hybrid algorithm and Hill cipher

4.1.1 Analysis of Key Space

Try to find the secret key used to decrypt the message, as it depends on the number of attempts, which is directly proportional to a key space. In the proposed algorithm, along the key (256 bits, 2^{256} area with 2^{256} Attempts) was used sufficiently to protect the color image from this type of attack.

4.1.2 Analysis of Secret Key

The key sensitivity means the change ratio of the cipher-image pixels if a slight bit of the key is changed. Also, shows the effectiveness of changing only one bit key, or choosing the wrong key lead to a different result for obtaining the corresponding plain image. Figure 3 shows the differences between the two ciphered images for two identical images with only one difference bit. It is shown there is a full difference between two ciphered images.

(a) Original image before encryption

(b) Image after decryption using the proper keys

(b) Cipher Image using Hybrid algorithm

(d) Image after decryption using different bit key

Fig. 3. Image after decryption using the correct secret key and secret key with a difference bit

4.2 Statistical Analysis

Most cipher algorithms are broken using statistical analyzes. Analyzes are used to find a linkage between the cipher and the original images to gain access to the secret key, to overcome statistical attacks the publishing and disturbance features of the cipher algorithm must be available. The following statistical analyzes were performed for the encrypted image:

4.2.1 Histogram Analysis

It indicates how the values of the RGB image's pixels are distributed by drawing the number of times values are repeated different from the histogram of the plan image. Figure 4 shows the histogram of the flower plain red, green, and blue images and the corresponding ciphered images are shown in Fig. 5 histograms of the cipher RGB image are different from the plain image histogram and uniform.

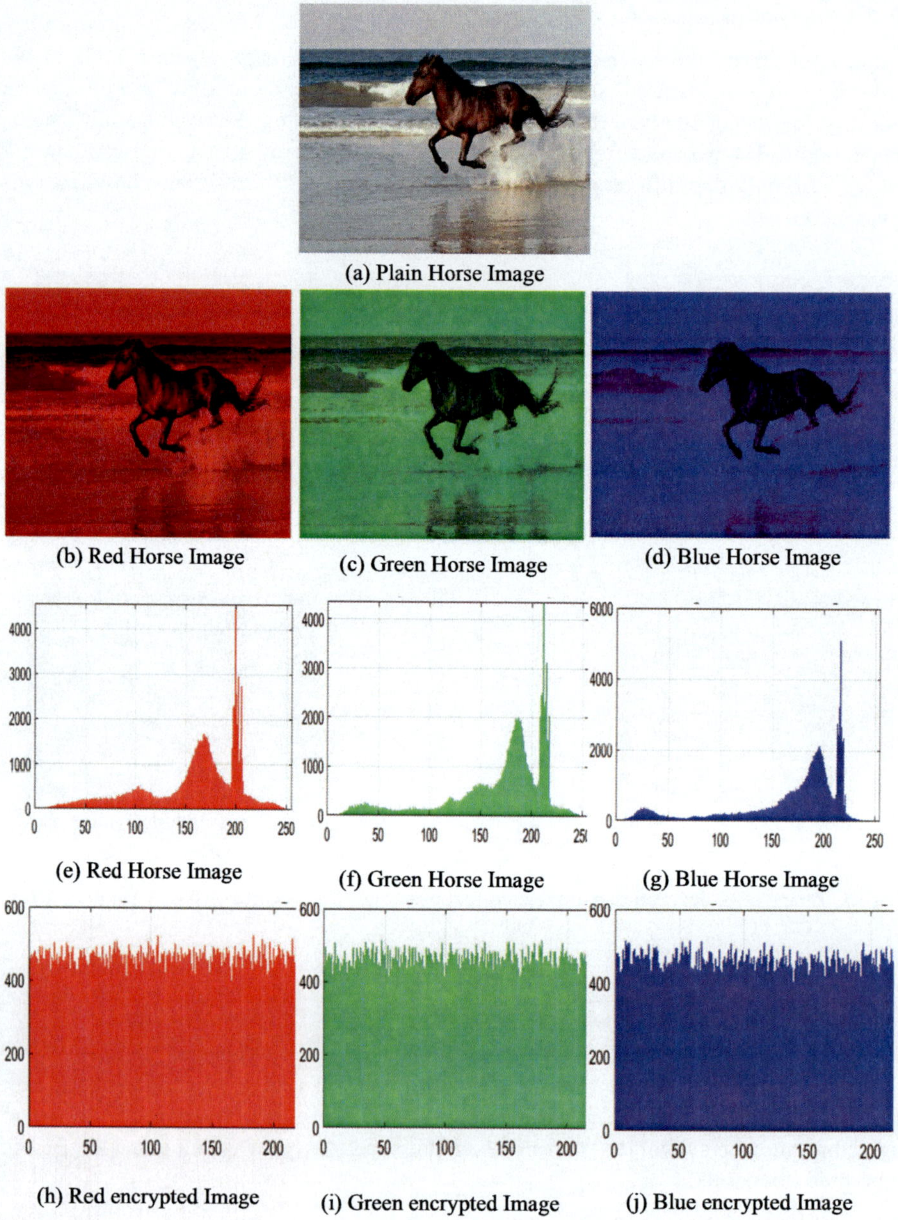

(a) Plain Horse Image
(b) Red Horse Image (c) Green Horse Image (d) Blue Horse Image
(e) Red Horse Image (f) Green Horse Image (g) Blue Horse Image
(h) Red encrypted Image (i) Green encrypted Image (j) Blue encrypted Image

Fig. 4. (e-g) Histogram of RGB plain image (h-j) Histogram RGB corresponding cipher image.

4.2.2 Correlation Between Original and Cipher Images

If the coefficient of correlation is close to zero, the correlation between the image after and before encryption decreases, and there's difficulty extract the original image [16].

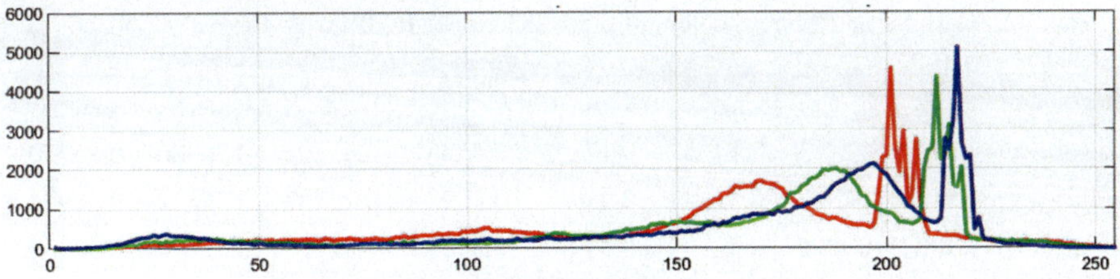

(a) RGB image

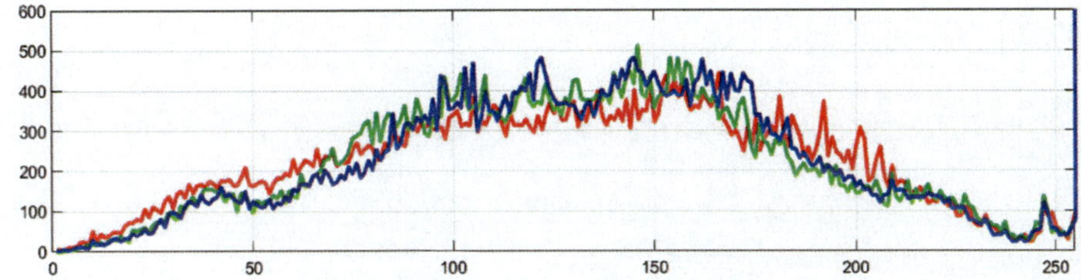

(b) Original Hill RGB Image Histogram

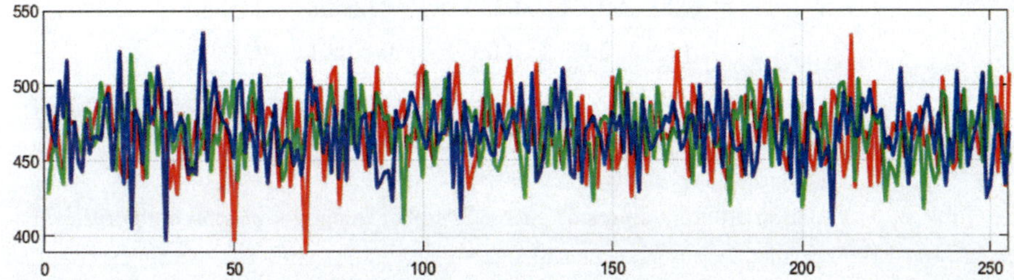

(c) Proposed Cipher RGB Image Histogram

Fig. 5. The histogram of (a) RGB image, (b) Original Hill RGB image, and (c) Proposed cipher RGB image

When comparing the correlation coefficient using the Hill algorithm as in Table 1 and Table 2 and the proposed algorithm in Table 3 and Table 4 notice the strength of the proposed schema.

Table 1. Correlation same colors of plain and cipher RGB image for Hill algorithm

Image name	Size	C_{RR}	C_{GG}	C_{BB}
Flower	392 * 512 * 3	0.19420	0.45205	0.27087
Fruits	512 * 512 * 3	0.37708	0.28527	0.36954
Horse	300 * 400 * 3	0.55843	0.51348	0.50823
Baboon	512 * 512 * 3	0.47476	0.53106	0.40563

Table 2. Correlation different colors of plain and cipher RGB image for Hill algorithm

Image name	Size	C_{RG}	C_{RB}	C_{GR}	C_{GB}	C_{BR}	C_{BG}
Flower	392 * 512 * 3	0.14352	0.17845	0.45951	0.45262	0.26960	0.25770
Fruits	512 * 512 * 3	0.31644	0.28964	0.30927	0.25707	0.35839	0.38944
Horse	300 * 400 * 3	0.52559	0.53971	0.54613	0.52117	0.53494	0.50559
Baboon	512 * 512 * 3	0.46167	0.45781	0.51660	0.52012	0.40598	0.42377

Table 3. Correlation same colors of plain and cipher RGB image for proposed algorithm

Image name	Size	C_{RR}	C_{GG}	C_{BB}
Flower	392 * 512 * 3	−0.00034	−0.00226	−0.00003
Fruits	512 * 512 * 3	0.00036	−0.00118	−0.00137
Horse	300 * 400 * 3	0.00189	0.00506	0.00051
Baboon	512 * 512 * 3	0.00303	0.00601	0.00399

Table 4. Correlation different colors of plain and cipher image for proposed algorithm

Image name	Size	C_{RG}	C_{RB}	C_{GR}	C_{GB}	C_{BR}	C_{BG}
Flower	392 * 512 * 3	−0.00330	−0.00062	−0.00193	0.00133	−0.00122	−0.00149
Fruits	512 * 512 * 3	−0.00252	−0.00231	−0.00142	0.00077	0.00091	−0.00076
Horse	300 * 400 * 3	0.00453	0.00183	0.00135	0.00191	0.00161	0.00382
Baboon	512 * 512 * 3	0.00396	−0.00146	0.0045	−0.00046	0.00204	0.0074

5 Conclusion

We proposed an asymmetric hybrid algorithm; Hill cipher and Advanced Encryption Standard Algorithms were joined to propose a new encryption algorithm to encrypt color images. The results showed that the proposed system was able to encode color images with high efficiency against attacks. The encryption key was used to encrypt the images, therefore increasing the key strength to counter attacks and statistical attacks. The results showed that encrypting the images using the proposed schema is more secure and effective than the original one. The combination of both algorithms has been successfully applied to RGB images of different size.

References

1. Wang, X., Zhang, H., Madhuviswanatham, A.V., Krishna, N.: A color image encryption with heterogeneous bit-permutation and correlated chaos Elsevier (2014)
2. Li, Y., Zhang, F., Li, Y., et al.: Asymmetric multiple-image encryption based on the cascaded fractional Fourier transform. Opt. Lasers Eng. **72**, 18–25 (2015)
3. Daemen, J., Rijmen, V.: The Design of Rijndael: AES -The Advanced Encryption Standard J. Springer-Verlag, pp. 55–56 (2002)
4. Kartalopoulos, S.: Security of Information and Communication Networks. Wiley-IEEE Press (2009)
5. Patil, P., Narayankar, P., Narayan, D.G: Meena, S.Md.: A comprehensive evaluation of cryptographic algorithms: DES, 3DES, AES, RSA, Blowfish, **78**, 617–624 (2016)
6. Liu, H., Kadir, A., Li, Y.: Asymmetric color image encryption scheme using 2D discrete-time map, 104–112 (2015)
7. Kocarev, L., Lian,, S.: editors. Chaos based cryptography. Springer (2011)
8. Reddy, K.A., Vishnuvardhan, B., Madhuviswanatham, V., Krishna, A.V.N.: A modified hill cipher based on circulant matrices Procedia Technol. **4**, 114–118 (2012)
9. Zhang, Q., Ding, Q.: Digital image encryption based on advanced encryption standard (AES). In: 2015 Fifth International Conference on Instrumentation and Measurement, Computer, Communication and Control (IMCCC), pp. 1218–1221. IEEE, September 2015
10. Reddy, M.I.S., Kumar, A.S.: Secured data transmission using wavelet based steganography and cryptography by using AES algorithm. In: Elsevier, International Conference on Computational Modeling and Security (CMS), Procedia Computer Science, vol. 85, pp. 62–69, January 2016
11. Hraoui, S., Gmira, F., Abbou, M.F., Oulidi, A.J., Jarjar, A.: A new cryptosystem of color image using a dynamic-chaos Hill Cipher algorithm. Procedia Comput. Sci. **148**, 399–408 (2019)
12. Agung, A., Heryana, N., Solehudin, A.: Combination of hill cipher algorithm and Caesar cipher algorithm for exam data security. Buana Inf. Technol. Comput. Sci. (BIT and CS) **1**(2), 42–45 (2020)
13. Essaid, M., Akharraz, I., Saaidi, A.: Image encryption scheme based on a new secure variant of hill Cipher and 1D chaotic maps. J. Inf. Secur. Appl. **47**, 173–187 (2019)
14. Almaiah, M.A., Dawahdeh, Z., Almomani, O., Alsaaidah, A., Al-khasawneh, A., Khawatreh, S.: A new hybrid text encryption approach over mobile ad hoc network. Int. J. Electr. Comput. Eng. (IJECE) **10**(6), 6461–6471 (2020)
15. Das, B.K., Garg, R.: Security of cloud storage based on extended hill cipher and homomorphic encryption. In: 2019 International Conference on Communication and Electronics Systems (ICCES), pp. 515–520. IEEE, July 2019
16. El-Fishawy, N., Zaid, O.: Quality of encryption measurementof bitmap images with rc6, mrc6, and rijndael block cipher algorithms. Int. J. Netw. Secur. 5
17. Arab, A., Rostami, M.J., Ghavami, B.: An image encryption method based on chaos system and AES algorithm. J. Supercomput. **75**(10), 6663–6682 (2019). https://doi.org/10.1007/s11227-019-02878-7

An Improved Smart Wheat Health Monitor for Smart Farming Application Using Deep Residual Learning Framework

Mohammed Elidrissi(✉), Omar Elbeqqali, and Jamal Riffi

Sidi Mohamed Ben Abdellah University, B.P. 1796, Fez-Atlas 30003, Morocco

Abstract. In this paper, we propose a classification model called SWHM (Smart Wheat Health Monitor) that aims to help farmers in wheat plants health monitoring task using residual learning framework. Training a very deep network of CNN (Convolutional Neural Network) creates a vanishing gradient problem, which is interpreted as a loss of information during training process, because learned features in the later layers vanish while being sent back to the early layers, in some other words, when updating the weights by backpropagation in a deep neural network, the multiple derivation of the loss function (gradient) may make the gradient disappear. We chose residual learning because it allows us to go deeper in the training process by extending the number of layers to achieve a few hundreds of layers without being worry about vanishing gradient or model complexity issues. proposed amelioration allows us to achieve 83.33% of accuracy, and minimize error rate to 16.67%, within a reasonable time of training.

Keywords: Smart farming · Big data · Deep learning · Residual learning

1 Introduction

Smart agriculture is one of the fields that have taken advantages from the wide use of new technologies in most of their processes, furthermore, the diversity of bio-aggressors, and the need to improve crops productivity, make crops health monitoring tasks using traditional methods very difficult for farmers. Starting from this issue, decision support systems have been emerged thanks to new methods of image classification, and significant power processing provided by new hardware and software resources. Decision support systems aim to help end users in making decision process based on image analysis, particularly with the decrease in capacity of humans to extract intrinsic information inside huge amount of data, on the other side, the increase in capacity of machine to develop hypothesis intelligently, and to detect and construct data models by applying known rules, these models are used to: categorize and classify objects or humans, predict actions, detect patterns and unknown relations, or to identify anomalies and unexpected behaviors. In this context we propose an enhanced model for plant health monitoring, using residual neural network techniques as a based deep learning technique to distinguish between infected and healthy wheat plants, and thus to improve other parameters,

A. Abraham et al. (Eds.): SoCPaR 2020, AISC 1383, pp. 758–767, 2021.
https://doi.org/10.1007/978-3-030-73689-7_72

such as increasing model size without being worry about vanishing or exploding gradient problem, that can be occurred while going deeply in the neural network model, accuracy also is one of things that has been improved, because pertinent information, that are used to enhance accuracy, are extracted by early layers and forwarded to the later layers using shortcuts. The algorithm implemented in this work is proposed by K. He et al. [1], it is a deep residual learning algorithm for image recognition that won the 1st place on the ILSVRC 2015, it achieved 3.57% top-5 error on the ImageNet test set, the Table 1 compares between plain neural models and residual learning models in terms of several characteristics:

Table 1. Comparison between plain and residual learning models

Characteristics	Shallow plain neural model	Deep plain neural model	Deep residual learning model
Number of layers	Few number	Big number	Big number
Accuracy	Important	Minimal	Maximal
Error rate	Important	Maximal	Minimal
Architecture flow	Contains only one main path	Contains only one main path	Contains two paths, main and shortcut, and two types of building blocks (identity, bottleneck)
Training time	Training time is acceptable	Training time is important	Requires a huge amount of time
Complexity	Minimal	Maximal	Minimal

We first open this paper with a short look of residual learning framework in Sect. 2, we then build the enhanced ResNet plant health monitor in Sect. 3, model training results and evaluation are discussed in Sect. 4, Sect. 5 is conclusion and perspectives.

2 Residual Learning Framework

Some recent researches have noticed that the depth of neural network is an important key factor for accuracy improvement [3, 7]. Piloted by this concern, how can the network architecture have many layers without encountering the gradient vanishing or exploding issues, as well as decreasing the global complexity of the network? Many solutions have been proposed to fix this problem such as alternate weight initialization schemes [4], unsupervised pre-training [5], layer-wise training [6], and the Adaptive optimization approach [8]. Many algorithms have been proposed in the last approach like Adaptive gradient algorithm (ADAGRAD) [8], ADADELTA [9] and Adaptive moment estimation (ADAM) [10] in order to achieve a rapid training process, based on calculating the adaptive learning rate hyper-parameter instead of extensively adjusting it. Each one of the adaptive previous methods has some points of weakness, for instance: in the

ADAGRAD method, the decay of the learning rate (LR) is considered as a problem in case of training deep neural networks. To overcome the LR decay, the ADADELTA algorithm is introduced an extension of ADAGRAD that limits the LR values in a moving window of gradient updates, such method ensures a continual LR during the learning process. ADAM optimization algorithm is proposed to combine the advantages of both Root Mean Square Propagation (RmsProp) [11] and ADAGRAD with momentum, by calculating the squared gradients to scale the learning rate.

ResNet [12] is the acronym of Residual Network, this type of neural networks introduces a new machine learning concept, it is making direct connection between the input of a given layer and the nth layer, this method of training allows to go deeply in learning process, increasing number of layers with a minimum of complexity, and learn features at many levels of abstraction from edge to very deep layers. Important results of ResNet is minimizing Top 5 errors to 3.57%, thus learning precision is improved. This family of neural networks contains many types, for instance ResNet50 with 50 residual layers, ResNet101 composed of 101 residual layers, and ResNet152 with 152 residual layers. According to [1, 12], the principal component of a residual model is the residual block (Fig. 1), which also has two types: identity block, and convolutional block, depending on whether the input and output have the same dimensions or not, both of these block types make skip connection between the input of one or more layers and their output, in order to make activation of the previous layer contribute to the activation of the next layers, which is formally denoted by H(x) = F(x) + x, with x is the signal used to activate the previous block, named identity, and F(x) is the output of the last layer of the residual building block, named residual function.

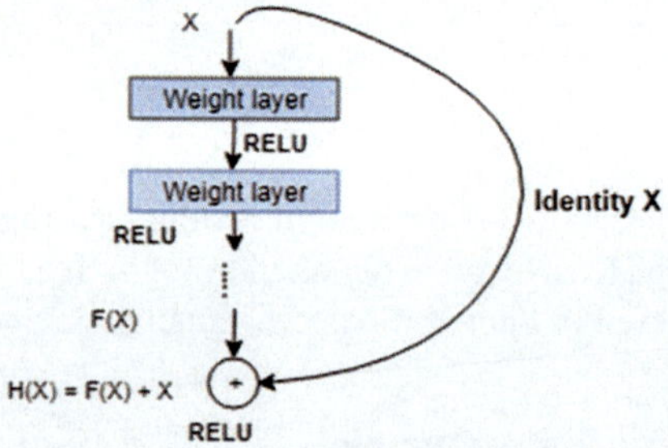

Fig. 1. ResNet building block

2.1 Identity Block

This type is the default block used in ResNets, its input is supposed to have the same dimension as its output, and a direct shortcut path is made to skip one or more layers of the main path, each skipped block may consist of two (Fig. 2 a) or three CONV2D-BatchNorm-Relu triplet layers (Fig. 2 b), the output of the main path is added to the value of the shortcut, then it passed through a ReLu activation function. This kind of blocks will be denoted Iden_BL in the rest of this paper.

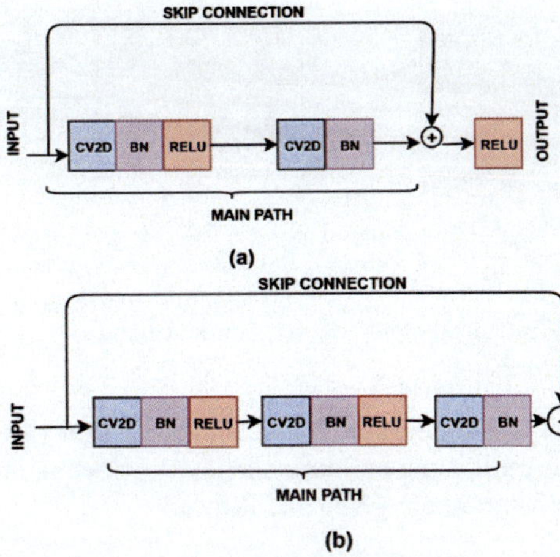

Fig. 2. Identity block

2.2 Convolutional Block

This type of residual bocks is normally used when input and output dimensions are different, its main function is to compensate this difference by reducing the input dimension to be adequate for the next steps, without using any non-linear activation function. This task is achieved using a CONV2D-BatchNorm block in the shortcut path. The values of the main path and the skip connection will be added together, then the output passes through a Relu activation function. This type of blocks will be denoted Conv_BL in the rest of this document.

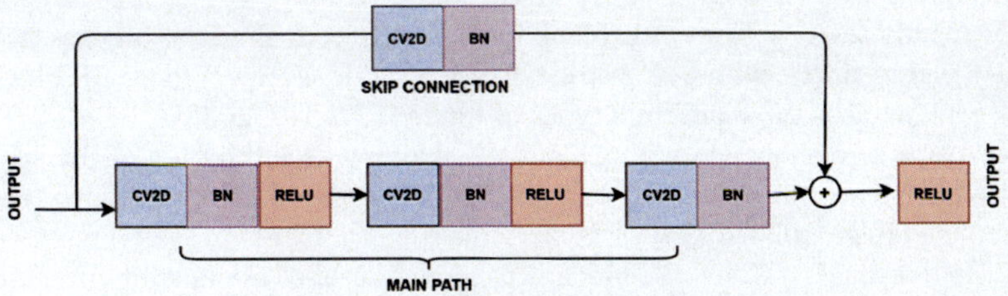

Fig. 3. Convolutional residual block

3 Building ResNet Model for Plant Health Monitoring

After this short introduction of ResNet terminology, and after giving its fundamental building blocks, now Fig. 4 illustrates the logical structure of enhanced plant health monitor built using ResNet framework, collected data are split into three repositories, one part is reserved for training process, another part is used for model pertinence validation, the third one is used for final test purpose. Communication between data repositories and ResNet model is performed using 64 bits batch size.

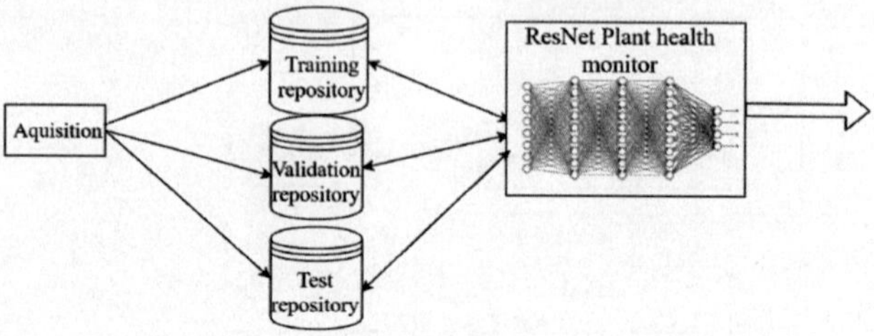

Fig. 4. Enhanced plant health monitor structure

The structure of plant health monitoring model is composed of seven stages as described below, each stage is built using building blocks described in the Sect. 2 above, details of interconnected architecture are given as follows:

Stage 1: The first Conv2D function used in this stage has 64 (7×7) kernels, with a stride of 2, and activated with Relu activation function. Next BatchNorm function is used in the three channels axis of the input image. The stage is closed by a Maxpooling function using a (3×3) window, and uses a stride of (2×2).
Stage 2: The stage 2 is opened with a Conv_BL block that uses a set of filters with different sizes [64, 64, 256], stride used in this block is 1. Next, two Iden_BL blocks are used with a three set of filters [64, 64, 256].
Stage 3: This stage is also opened with a Conv_BL block with a three set of filters [128, 128, 512], and a stride of 2. Then a series of three (3) Iden_BL blocks, using the same set of kernels with the same sizes [128, 128, 512] as convolutional block, are stacked.
Stage 4: Here, a Conv_BL block is used in the header of this stage, this convolutional block uses [256, 256, 1024] set of filters, with 2×2 stride. Next, a series of five (5) Iden_BL blocks with [256, 256, 1024] are stacked.
Stage 5: In this stage, a Conv_BL block is used, this convolutional block uses [512, 512, 2048] set of filters, with 2×2 stride. Then, a set of seven (7) Iden_BL blocks are applied, each one has three different types of filters [512, 512, 2048].
Stage 6: This stage is opened with a Conv_BL block with a three set of filters [1024, 1024, 4096], and a stride of 2. Then a series of nine (9) Iden_BL blocks, using the same set of kernels [1024, 1024, 4096].
Stage 7: In this stage, a Conv_BL block is used with a set of filters with different sizes [2048, 2048, 8196], stride used here is 1. Next, two Iden_BL blocks are used with the three set of filters used in Conv_BL.

4 Experiment

ResNet model is implemented in python using Tensorflow and Keras, and executed on a virtual machine hosted on a cluster composed of eight (8) cisco unified computing system nodes, each node has an Intel(R) Xeon(R) CPU E5-2630 v4 @ 2.20 GHz, and 256 gigabytes of RAM. The CGIAR Computer Vision for Crop Disease [14] dataset is used in the experiment. Training process starts with data preparation and augmentation strategies. The algorithm described below, monitors cost and accuracy functions, and plot their curves at the end of the training process. The model is trained in 100 epochs, and the number of steps by epoch depends on the learning rate and the number of samples in each operation type (training or validation), optimization function used in training process is RMSprop (Root Mean Square Propagation) since we use mini batches, and we want to freely vary our learning rate to look for the best accuracy value. The process begins with training and validation data preprocessing operations, such as flipping, rotation, and rescaling. Next, identity block function is defined followed by definition of convolutional block function, the mapping of identity blocks and convolutional blocks is then performed, to construct ResNet model architecture described earlier in the previous section, learning process is configured using RMSProp with 10^{-3} of learning rate, followed by model compilation by means of all predefined parameters, and the learning process get started.

```
Input: RGB image dataset; Output: Pre-Trained Model
STEPS =>100; BATCH SIZE => 64; RATIO => 0.1; TARGET SIZE => 200;
NTR = > Number of Training Samples; NVAL => Number of Validation Samples;
TARGET WIDTH; TARGET HIGH => TARGET SIZE; TARGET SIZE;
Definition of Data Preparation Strategy (TRAIN DATAGEN, VAL DATAGEN);
Definition of Data Augmentation Strategy (TRAIN GENERATOR, VALIDATION GENERATOR);
Iden Block function definition; Conv Block function definition;
Mapping of ResNet Model; Learning Process Configuration; Model Compilation;
while STEPS ≤ 100 do
STEP PER TRAINING EPOCH => NTR * (1 − RATIO) / BATCH SIZE;
STEP PER V ALIDATION EPOCH => NVAL * RATIO / BATCH SIZE;
   if accuracy > 0 then
      TRAINING PROCESS;
   else
      Wait for Two Iterations then execute Early Stopping Function;
   end
end
```

4.1 Training and Validation of the Enhanced Model

After 100 iterations of training and validation, we obtained accuracy (Fig. 5 (a)) and loss (Fig. 5 (b)) curves for training and validation processes, training accuracy value starts form 55% and continues increasing till epoch 91 where it achieves its maximal value (95%), between epochs 40 and 91, accuracy value variation basically flattens.

In order to more evaluate our trained model, we plot confusion matrix figured in Fig. 6 of eighteen (18) new samples, nine (9) of them are healthy plants, and nine (9) plants are infected with a variety of diseases.

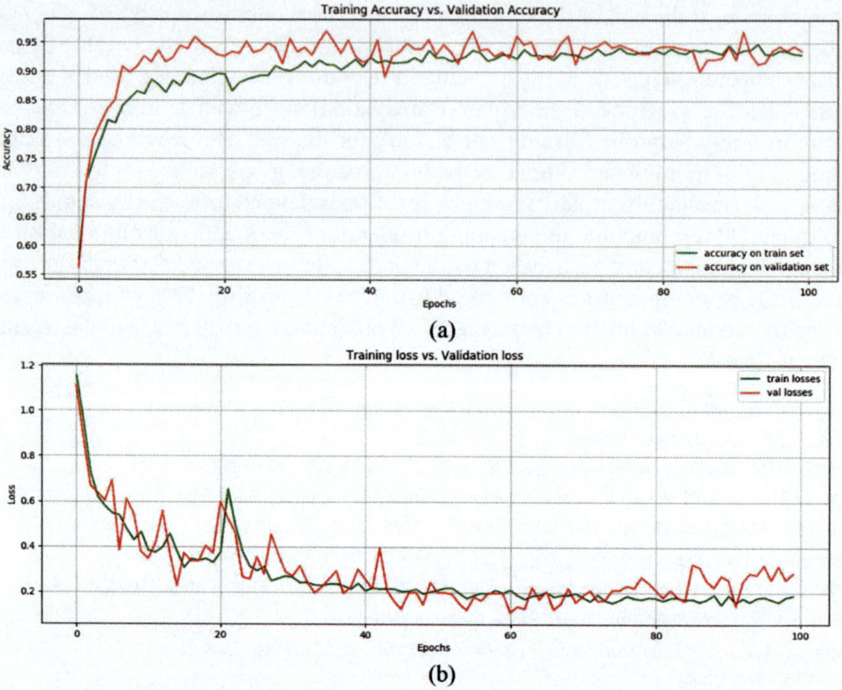

Fig. 5. Accuracy and cost function curves of training and validation processes

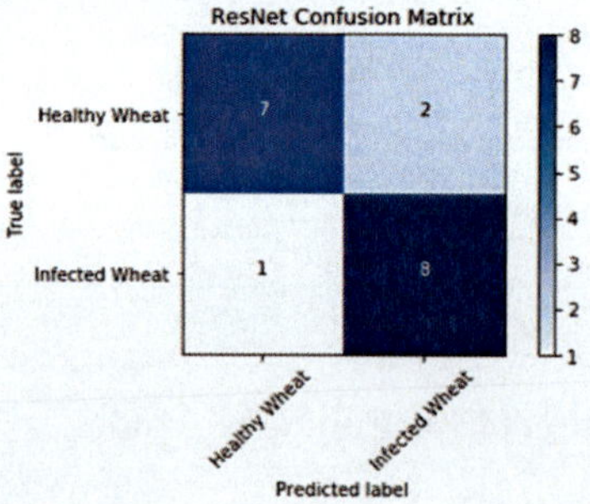

Fig. 6. Confusion matrix of ResNet enhanced plant health monitor

From confusion matrix we can extract accuracy parameter denoted AC, which gives us an idea about how good is the model to detect infected and healthy wheat plants, accuracy is calculated using the following formula:

$$AC = (TN + TP) / (TN + FP + FN + TP) = 83.33\% \quad (1)$$

With:

TN: Healthy wheat crops identified as healthy.

TP: Infected wheat crops detected as infected.

FN: Infected wheat crops incorrectly identified as healthy.

FP: Healthy wheat crops incorrectly detected as infected.

Another important parameter that can help us to evaluate our model is True Positive Rate, called also sensitivity, and is denoted TPR, it refers to the probability to correctly detect infected wheat plants as infected, a system 100% sensitive can correctly identify all infected plants as infected, but this value of sensitivity is rarely obtained. From confusion matrix TPR can be calculated as follows:

$$TPR = TP/(FN + TP) = 88.88\% \tag{2}$$

On the other side, True Negative Rate called also specificity, denoted TNR, refers to the capability of the model to correctly detect healthy wheat plants as healthy, it gives us a clear idea on how good the model is to predict crops as healthy. From confusion matrix, TNR is obtained using the following formula:

$$TNR = TN/(TN + FP) = 77.77\% \tag{3}$$

Separability of the model is proved by the Area Under Curve (AUC) value and ROC (Receiver Operation Characteristic) [13] curve illustrated in Fig. 7, this characteristic denotes how good the model is to do a binary classification of the different available classes. The AUC obtained value is 93%, the next table summarizes the enhanced parameters:

Table 2. Summary of improved parameters

Evaluation parameter	Enhanced value
Precision (accuracy)	83.33%
Error rate	16.67%
Sensitivity (TPR)	88.88%
Specificity (TNR)	77.77%
AUC-ROC	93%

The ROC curve shows the great ability of the enhanced plant monitor model to distinguish between infected and healthy wheat crops, using 93% of AUROC calculated value.

4.2 Evaluation and Field Test

In the field test, we injected dozens of wheat crops images in the input of the ResNet pre-trained model, some of the results that we obtained are shown in Fig. 8.

A real implementation of this enhanced model can be concretized using a drone or a CCTV system to take real time images of the crops, and send them to a fog computing local platform for decision making, then notify made decisions to the farmer in order to have an idea about his crops, and intervenes if it is required, the following Fig. 9 illustrates this process.

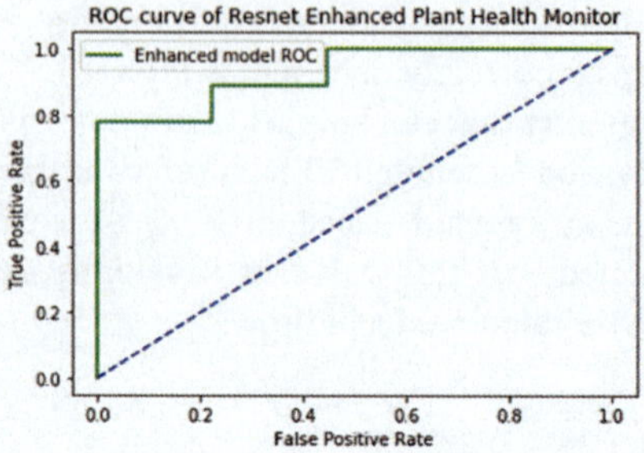

Fig. 7. ROC curve of enhanced plant health monitor

Fig. 8. Some results obtained in the field test

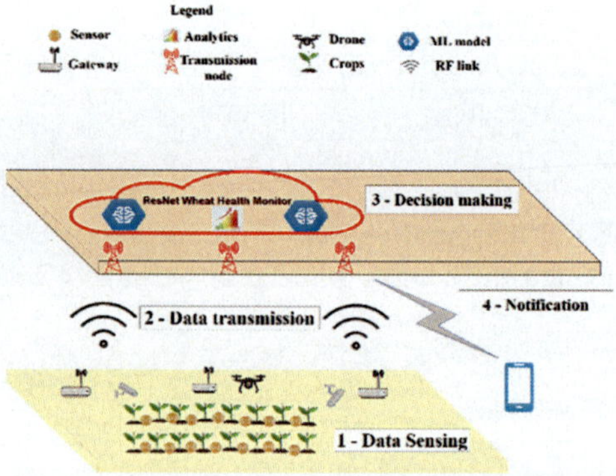

Fig. 9. Plant health monitoring process

5 Conclusion and Perspectives

As we can see, the implementation of residual learning algorithm is a good choice that helped us to enhance performances of plant health monitoring model, especially accuracy of detection which was improved to 83.33%, and error rate which was decreased to 16.67%. This work represents one of the three pillars of a smart farming system based on internet of things, fog computing and cloud computing layers, which basically has three main machine learning modules, respectively used to predict the exact amount of nutrient that the crops need, to forecast crops amount of irrigation water need, and to monitor wheat plant health. In this paper we present an enhanced version of a machine learning model to help farmers to monitor their crops, by injecting an image of their crops in the input of a pre trained ResNet model. Next works will be focused on the other pillars required to build the smart farming system.

References

1. He, K., et al.: Deep residual learning for image recognition. In: 2016 IEEE Conference on Computer Vision and Pattern Recognition (CVPR), Las Vegas, pp. 770–778 (2016)
2. Simonyan, K., Zisserman, A.: Very deep convolutional networks for large-scale image recognition. In: 3rd International Conference on Learning Representations (ICLR). San Diego, CA, USA (2015). https://arxiv.org/pdf/1409.1556.pdf. Accessed 25 June 2020
3. Szegedy, C., et al.: Going deeper with convolutions. In: 2015 IEEE Conference on Computer Vision and Pattern Recognition (CVPR), Boston, MA, pp. 1–9 (2015)
4. Adjusting for Dropout Variance in Batch Normalization and Weight Initialization. https://arxiv.org/pdf/1607.02488v1.pdf. Accessed 3 July 2020
5. Dumitru, E., et al.: Why does unsupervised pre-training help deep learning? J. Mach. Learn. Res. **11**, 625–660 (2010)
6. Yoshua, B., et al.: Greedy layer-wise training of deep networks. In: 19th International Conference on Neural Information Processing Systems (NIPS 2006), MIT Press, Cambridge, MA, USA, pp. 153–160 (2006)
7. Tan, H.H., Lim, K.H.: Vanishing gradient mitigation with deep learning neural network optimization. In: 7th International Conference on Smart Computing & Communications (ICSCC), Sarawak, Malaysia, Malaysia, pp. 1–4 (2019)
8. Duchi, J., Hazan, E., Singer, Y.: Adaptive subgradient methods for online learning and stochastic optimization. J. Mach. Learn. Res. **12**, 2121–2159 (2011)
9. ADADELTA: an adaptive learning rate method. https://arxiv.org/pdf/1212.5701.pdf. Accessed 14 June 2020
10. Adam: a method for stochastic optimization. https://arxiv.org/pdf/1412.6980.pdf. Accessed 20 June 2020
11. Neural Networks for Machine Learning Lecture 6e, rmsprop: Divide the gradient by a ru ning average of its recent magnitude. http://www.cs.toronto.edu/~tijmen/csc321/slides/lecture_slides_lec6.pdf. Accessed 23 July 2020
12. Zhang, K., et al.: Residual networks of residual networks: multilevel residual networks. IEEE Trans. Circ. Syst. Video Technol. **28**(6), 1303–1314 (2018)
13. Powers, W.: ROC-ConCert: ROC-Based Measurement of Consistency and Certainty. In: Spring Congress on Engineering and Technology, Xian, pp. 1–4. (2012)
14. CGIAR Computer Vision for Crop Disease. https://www.kaggle.com/shadabhussain/cgiar-computer-vision-for-crop-disease. Accessed 20 May 2020

A New Electronic Voting System Based on Elliptic Curve in a Cloud Computing Environment

Meryem Cherkaoui Semmouni[1(✉)], Ali Azougaghe[1], Abderrahmane Nitaj[2], and Mostafa Belkasmi[1]

[1] ICES Team, ENSIAS, Mohammed V University in Rabat, Rabat, Morocco
meryem.semmouni@um5s.net.ma, mostafa.belkasmi@um5.ac.ma
[2] LMNO, University of Caen Normandie, Caen, France
abderrahmane.nitaj@unicaen.fr

Abstract. The process of democratization is developing and promoting due to electronic voting. However, the increase of frauds and the increasing amount of attacks launched by hackers, gave birth to privacy and authentication problems. Cryptography offers multiple solutions to overcome the sensitive data protection issues in e-voting. In this paper, we study the application of elliptic curve cryptography, and use the homomorphic encryption properties to present a new electronic voting system. Our new scheme is based on the homomorphic cryptosystem EC-ELGAMAL [7], and zero knowledge algorithm of Schnorr algorithm for identification and authentication.

Keywords: Elliptic curve · Homomorphic encryption · Zero knowledge · Electronic voting

1 Introduction

New technologies make more accessible the relations between Human and the machines in many evolutive domains such as electronic voting and remote voting. An election can be carried out without requiring the participants to go to the polling place. To construct an electronic voting system, we propose to use cloud computing for data processing without trusted third party. This needs various cryptographic tools for security and efficient calculators for processing the votes. Homomorphic encryption is a fundamental tool for processing Data in a protected way. It uses specific operations to manipulate encrypted Data without decrypting it. The decrypted result corresponds to the result of the same operations on plain Data. On the other hand, zero knowledge proof is a good and secure algorithm for identification and authentication.

Various schemes for voting systems has been done in the literature. In 2009, Tohari et al. [1] proposed a mobile voting schemebased on the elliptic curve cryptography. In 2014, Zhao et al. [15] presented an electronic voting system

A. Abraham et al. (Eds.): SoCPaR 2020, AISC 1383, pp. 768–778, 2021.
https://doi.org/10.1007/978-3-030-73689-7_73

based on homomorphic encryption to ensure anonymity, privacy, and reliability in the voting, but there are still some unattended issues in their work, such as how to supervise the voting process to prevent conspiracy. At the same time, Cervero, M. A. et al. [2] presented an efficient homomorphic e-voting system over elliptic curves.

In this paper, we propose an electronic voting system based on additive homomorphic encryption and Modified Schnorr's digital signature algorithm that achieves the following security requirements: Eligibility, Authentication, Integrity, Privacy, Fairness, and Verifiability (see [9] for more details).

The rest of this paper is organized as follows. In Sect. 2, we recall the main properties of the elliptic curve cryptography. This includes the elliptic curve secp256k1, and the elliptic curve discreet logarithm problem. In Sect. 3, we briefly describe five schemes that will be used in this paper, namely homomorphic encryption, EC-ELGAMAL cryptosystem, Zero knowledge proof, and Schnorr's digital signature scheme. In Sect. 4, we present our new electronic voting scheme. In Sect. 5, we present an analysis of the security of the new scheme. We conclude the paper in Sect. 6.

2 Elliptic Curve Cryptography

Invented independently by Koblitz [10] and Miller [8] in 1985, the elliptic curve Cryptography is specifically suitable for machines with limited computational resources such as mobile phones and cloud computing. With significantly smaller parameters, it provides fast computation, small secret keys and guarantees the same level of security than RSA and the schemes based on the discrete logarithm problem (see [13] for more details). In the rest of this section, we briefly review the elliptic curve Secp256k1, and the elliptic curve discrete logarithm problem.

2.1 The Secp256k1 Elliptic Curve

Definition 1. *Let $p \geq 5$ be a prime number and $\mathbb{F}_p$ be the field of p elements. Let a and b be two elements of $\mathbb{F}_p$ such that $4a^3 + 27b^2 \neq 0$. An elliptic curve E $\mathbb{F}_p$ is the set $\mathbb{F}_p(E)$ of the solutions of the short affine Weierstrass form equation*

$$E : y^2 = x^3 + ax + b, \quad (a, b) \in \mathbb{F}_p^2,$$

together with a special point $\mathcal{O}$, called the point at infinity which serves as the neutral element for the addition operation in $\mathbb{F}_p(E)$.

A typical elliptic curve is secp256k1, defined by the Standards for Efficient Cryptography Group (SECG) [3]. It is deployed in the Bitcoin system [11] and has the following properties:

- the prime number: $p = 2^{256} - 2^{32} - 2^9 - 2^8 - 2^7 - 2^6 - 2^4 - 1$,
- the equation: $y^2 \equiv x^3 + 7 \pmod{p}$,
- the base point: $P = (55066263022277343669578718895168534326250603453$
  $777594175500187360389116729240,$
  $32670510020758816978083085130507043184471273380659243$
  $27593890433575733748242 4)$,
- the order n_0 of P: $n_0 = 2^{256} - 432420386565659656852420866394968145599$.

With the point at infinity $\mathcal{O}$, secp256k1 has n solutions. The discrete logarithm problem in secp256k1 is secure and supply approximately 128 bits of security.

2.2 The Elliptic Curve Discrete Logarithm Problem

The Elliptic Curve Discrete Logarithm Problem (ECDLP) is a hard problem in the theory of elliptic curves: Given an elliptic curve E over a finite fields $\mathbb{F}_p$ and two points $(P, Q) \in \mathbb{F}_p(E)^2$ such that $Q = dP$, the ECDLP consists in finding the integer d. The security of several schemes is based on the hardness of ECDLP (see [6] for a detailed analysis of ECDLP).

3 Homomorphic Encryption and Zero Knowledge

In cryptography, a cryptosystem is said to be homomorphic when computations, essentially addition and multiplication can be made on the encrypted data without decrypting it. The results on the encrypted data is similar than the results of the same operations on the decrypted one.

3.1 Homomorphic Encryption

As cloud computing is increasingly growing, the need of protecting the privacy of sensitive data stored in a cloud server or a distant machine becomes a prevalent task. The traditional encryption schemes such as RSA, AES, and the elliptic curve cryptography, can provide privacy of data but allow only partial operations on encrypted data. An encryption scheme allowing all the operations is called a fully homomorphic encryption (FHE) scheme. The problem of constructing a fully homomorphic encryption scheme supporting arbitrary operations was first solved by the breakthrough work of Gentry [4] in 2009.

Definition 2. *Let E be an encryption algorithm. E is partially homomorphic if it satisfies one of the following properties:*

- *Additively homomorphic: $E(m_1 + m_2) = E(m_1) + E(m_2)$.*
- *Multiplicatively homomorphic: $E(m_1 m_2) = E(m_1)E(m_2)$.*

3.2 Class Function on Elliptic Curves

Let $p \geq 5$ be a prime number and $a, b \in \mathbb{F}_p$ be two integers such that $4a^3 + 27b^2 \neq 0 \pmod{p}$. We consider elliptic curves defined over the ring $\mathbb{Z}/p^2\mathbb{Z}$ with equations of the form $y^2 = x^3 + ax + b$. The points at infinity come up with the projective form

$$Y^2 Z = X^3 + aXZ^2 + bZ^3$$

.

Let $E\left(\mathbb{Z}/p^2\mathbb{Z}\right)$ be the set of points on this elliptic curve over $\mathbb{Z}/p^2\mathbb{Z}$. Similarly, let $E_1(\mathbb{Z}/p^2\mathbb{Z})$ be the subset of points that reduce to $\mathcal{O} = (0:1:0)$ modulo p. We have (see [7])

$$E_1(\mathbb{Z}/p^2\mathbb{Z}) = \{(\alpha p : 1 : 0) | 0 \leq \alpha \leq p-1\}.$$

The theory of formal groups [7] implies that $E_1(\mathbb{Z}/p^2\mathbb{Z})$ is a group isomorphic to the additive group $(\mathbb{Z}/p\mathbb{Z})^+$. We have:

$$\lambda : E_1(\mathbb{Z}/p^2\mathbb{Z}) \longrightarrow (\mathbb{Z}/p\mathbb{Z})^+, (\alpha p : 1 : 0) \longrightarrow \alpha.$$

Given a finite point $P = (x, y) \in E(\mathbb{F}_p)$, with $y \neq 0$, we define

- $\Delta(P) = \dfrac{x^3 + ax + b - y^2}{p} \pmod{p^2}$.
- $\varpi(P) = \dfrac{\Delta(P)}{2y} \mod p$.

This gives rise to the map:

$$\omega : E(\mathbb{F}_p) \longrightarrow E(\mathbb{Z}/p^2\mathbb{Z}), \begin{cases} \mathbf{O} \longrightarrow \mathbf{O} \\ (x, y) \longrightarrow (x, y + \varpi(P)p) \end{cases}$$

- $\beta = \widehat{Q} = \dfrac{\lambda(nQ)}{n} \mod p$. where $Q \in E(\mathbb{Z}/p^2\mathbb{Z})$, $Q = (\beta p : 1 : 0)$, and $n = ord_E(P)$

3.3 ElGamal-Type Encryption Schemes on Elliptic Curves Cryptosystem (EC-ELGAMAL)

The **Encoding-Free ElGamal-Type Encryption Schemes on Elliptic Curves** [7] scheme named EC-ELGAMAL is a typically additive homomorphic cryptosystem. Its ellipstic curve version is based on the following three algorithms:

1. **Key generation:**
   - Choose a large prime number p.
   - Choose two integers $a, b \in \mathbb{F}_p$ that define an elliptic curve E over $\mathbb{F}_p$.
   - Choose a point $P \in \mathbb{F}_p(E)$ with a large order n.
   - Choose a random integer d such that $2 \leq d \leq n-1$.
   - Compute $Q = dP$ on the elliptic curve E.
     (a) **Private key**: The private key is the integer d.
     (b) **Public key**: The public key is the tuple (p, a, b, P, Q, n).
2. **Encryption:**
   - Choose a random integer r such that $2 \leq r \leq n-1$.
   - Compute $C_1 = rP$ on the elliptic curve E.
   - Compute $C_2 = rQ$ on the elliptic curve E.
   - Compute $\beta = \widehat{\omega(C_2)}$.

- Define $c_2 = m + \beta \pmod p$
- Output the ciphertext $C = (C_1, c_2)$.

3. **Decryption:**
   - Compute $m = c_2 - \widehat{\omega(dC_1)} \pmod p$

The Encoding-Free ElGamal-Type Encryption Schemes on Elliptic Curves is an additive homomorphic cryptosystem [7].

3.4 Zero Knowledge Interactive Proof and Schnorr Scheme

In cryptography, Zero-Knowledge systems have numerous applications such as secure communication, authentication, and privacy (see [14] for more applications). A proof of knowledge is a protocol that enables one party to convince another of the validity of a statement, without revealing all the details. One classic example of zero-knowledge proof is the Schnorr protocol, also known as the Σ-protocol (see [12]). The Schnorr protocol can be based on elliptic curves [5] as follows.

Let E an elliptic curve over a prime finite field $\mathbb{F}_p$ with a point P of order n. First, the prover chooses a random private integer k with $2 \leq k \leq n$ and computes the public key $Q = kP$ on the elliptic curve. Then, the protocol works in three interactions:

1. The prover chooses a random integer v with $2 \leq v \leq n-1$ and computes $V = vP$. He sends V to the verifier.
2. The verifier chooses a random integer c with $2 \leq c \leq n-1$, and sends c to the prover.
3. The prover computes $b = v - kc \pmod p$ and sends b to the verifier.
4. The verifier checks whether Q is on the curve and whether $V = bP + cQ$. If any check fails, then the verification is not correct since we should have

$$bP + cQ = vP - ckP + ckP = vP = V.$$

3.5 Modified Schnorr Scheme

A modified Schnorr Scheme could be used to prove that an encrypted message is sent by the correct source, and to avoid a man in the middle attacks.

Let E an elliptic curve over a prime finite field $\mathbb{F}_p$ with a point P of order n. First, the prover chooses a random private integer k with $2 \leq k \leq n$ and computes the public key $Q = kP$ on the elliptic curve, and the encrypted value that wants to send (C_1, c_2), where C_1 is a point on the curve. Then, the protocol works in three interactions:

1. The prover chooses a random integer v with $2 \leq v \leq n-1$ and computes $V = vP + C_1$. He sends V to the verifier.
2. The verifier chooses a random integer c with $2 \leq c \leq n-1$, and sends c to the prover.
3. The prover computes $b = v - kc \pmod p$ and sends b to the verifier.

4. The verifier checks whether Q is on the curve and whether $V = bP + cQ + C_1$. If any check fails, then the verification is not correct since we should have

$$bP + cQ + C_1 = vP - ckP + ckP + C_1 = vP + C_1 = V.$$

4 The New Electronic Voting Scheme

In this section we present our new electronic voting scheme. It is based on the following schemes: the EC-ELGAMAL [7] cryptosystem, the Schnoor scheme, and the Zero Knowledge Interactive Proof.

The arithmetical operations of the new scheme are based on the elliptic curve secp256k1 with the prime number p, the generator point P with order n. The main participating parties of the new scheme are the Voters, the Cloud, and the Trusted Party.

4.1 Participating Parties

1. **Trusted Party:** The trusted party plays the role of authorities. It generates and stores the election private key and publishes the election public key on the cloud. It also decrypt the encrypted votes of the voters, calculate the vote results, and sends the final results to the Cloud.
2. **Voters:** Voter's actions during the electronic voting are kept at minimum. To vote for a candidate, the voter first interacts with the Trusted Party in order to be authenticated. The vote is then encrypted and sent to the Cloud.
3. **Cloud:** The role of the cloud is to publish the public keys as lists of voters, to publish the list of the candidates, to collect the encrypted votes, to calculates the encrypted results using homomorphic properties, and finally to publish the final result of the elections.

4.2 Scenario of the New Scheme

To set up the new scheme, we list the parameters as follows.

- v is the number of the voters,
- m is the number of the candidates,
- u is the vote basis number which satisfies $u > v$,
- E is an elliptic curve over a prime finite field $\mathbb{F}_p$,
- P is a point on the elliptic curve E with order larger vu^m,
- E_k is the EC-Elgamal encryption scheme with private key k and public key Q.

The main steps of the new scheme are listed as follows.

1. If the voter j wants to vote for the candidate i, he has to perform the following two operations.
   - He computes $R_i = u^i$.

- He computes $C_j = E_k(u^i) = (A, B)$ using the EC-ElGamal scheme.

2. Before voting, each voter j has to be identified and authenticated by the Trusted Party. Moreover, a secret key k_j will be allowed to the voter so that the public key of the voter j is $Q_j = k_j P$. This is achieved by the following interaction.
   - Voter j chooses a number v_j uniformly at random from $\{1, ..., p-1\}$ and computes $V_j = v_j P + A$. He sends V_j to the Trusted Party.
   - The Trusted party chooses a challenge number c_j uniformly at random from $\{1, ..., p-1\}$, and sends it to the Voter j.
   - The Voter j computes $b_j = v_j - k_j c_j \pmod p$ and sends it to the Trusted Party.
   - The Trusted Party checks weather the following equality holds: $V_j = b_j P + c_j Q_j + A$. If it holds, then the voter j is authenticated.
3. After being authenticated by the Trusted Party, the voter j is allowed to vote.
4. Trusted Party sends C_j to the Cloud.
5. After recording the encrypted vote C_j, the Cloud drops the public Q_j of the voter j from the system.
6. The system will do this process until the deadline of the vote.
7. The Cloud uses the properties of homomorphic encryption of the Ec-ElGamal scheme to compute sum of all the votes the encryption of voting result as follows:

$$C = \sum_{j=1}^{v} C_j = \sum_{j=1}^{v} E_k\left(u^i\right) = \sum_{i=1}^{m} E_k\left(x_i u^i\right) = E_k\left(\sum_{i=1}^{m} x_i u^i\right),$$

   where x_i is the number of votes for the candidate i. Observe that $x_i \le v < u$.
8. The cloud sends C to the Trusted Party.
9. The Trusted Party uses the private key k inside the EC-ElGamal scheme to compute the value $s = \sum_{i=1}^{m} x_i u^i$.
10. The Trusted Party solves the super-increasing equation $\sum_{i=1}^{m} x_i u^i = s$ using the solver Φ to obtain all the number of votes $x_1, \ldots, x_m$ of the candidates.

Algorithm 1. Φ Algorithm

Require: The sum $s = \sum_{k=1}^{m} x_k u^k$, the number m of candidates, and u is the vote basis number satisfying $u > v$.
Ensure: The numbers $x_1, x_2, ..., x_m$ of votes of the candidates.

Set $s_1 = s$.
Compute $x_1 = \frac{s}{u} \pmod u$.
for $i = 2$ to m **do**
    Compute $s_i = s_{i-1} - x_{i-1} u^{i-1}$.
    Compute $x_i = \frac{s_i}{u^i} \pmod u$.
end for
Return $x_1, x_2, ..., x_m$.

Lemma 1 (Correctness) .
If the sum $s = \sum_{k=1}^{m} x_k u^k$ *is final result of all votes,* m *the number of candidates,* u *is the vote basis number satisfying* $u > v$, $s_1 = s$, *and* $x_1 = \frac{s}{u} \pmod{u}$. *then*

$$\forall i \in \{1, ..., m\} : x_i = \frac{s_i}{u^i} \pmod{u}.$$

Where $s_i = s_{i-1} - x_{i-1}u^{i-1}$. *and* x_i *is the number of votes of candidate* i.

Proof. First we start by proofing by induction on i that $s_i = s - \sum_1^{i-1} x_k u^k$. then we will proof the lemma.

- **Base case:** for $i = 1$: $s_1 = s$
- **Inductive step:**
  Suppose that $s_i = s - \sum_1^{i-1} x_k u^k$ is correcte.
  so:
  $s_{i+1} = s_i - x_i u^i = s - \sum_1^{i} x_k u^k$

Hence,
$\forall\, i \in \{2, ..., m\}$: $s_i = s - \sum_1^{i-1} x_k u^k$
Then we have

$$\begin{aligned}
\frac{s_i}{u^i} \pmod{u} &= \frac{s - \sum_1^{i-1} x_k u^k}{u^i} \pmod{u} \\
&= \frac{\sum_{k=1}^{m} x_k u^k - \sum_1^{i-1} x_k u^k}{u^i} \pmod{u} \\
&= \frac{\sum_{k=i}^{m} x_k u^k}{u^i} \pmod{u} \\
&= \sum_{k=i}^{m} x_k u^{k-i} \pmod{u} \\
&= x_i + \sum_{k=i+1}^{m} x_k u^{k-i} \pmod{u} \\
&= x_i \qquad (u > v \Rightarrow u > x_i)
\end{aligned}$$

This terminates the proof.

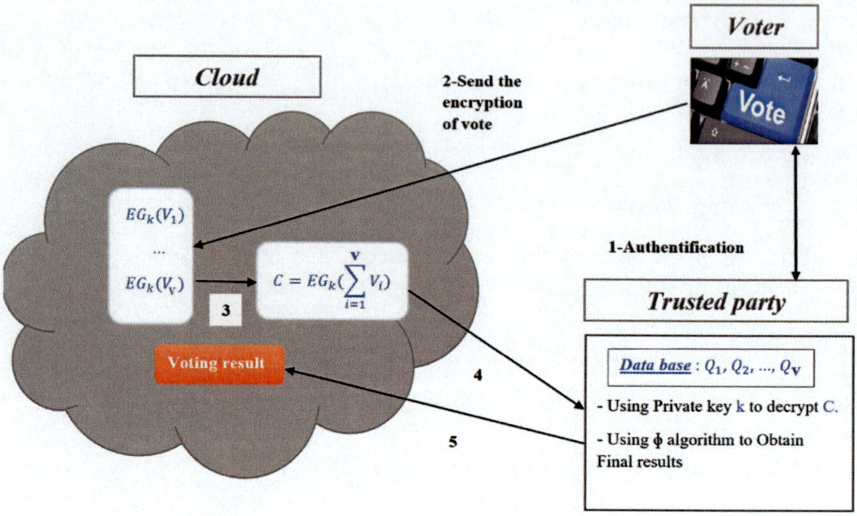

Fig. 1. Scenario of a new voting system based on elliptic curve

5 Security Analysis

In this section we show how the proposed system achieves the security requirements:

5.1 Authentication

Only people in the electoral roll can vote.

1. Each voter authenticate with trusted party with zero knowledge proof, and trusted party check that he correspends of one of the Data base public keys.

5.2 Unicity

Every participant can vote once at most.

1. Voter sends the encryption of vote to a cloud, and after his public key is dropped from the "data base" for unicity.

5.3 Privacy

Votes can not be related to voter identities

1. All the votes are encrypted using the EC-ElGamal cryptosystem, so that no information can be obtained from an encrypted result.
2. Only trusted party which acts honestly can decrypt and has got the result of vote.
3. The algorithm used for authentication is zero knowledge. Hence, no information leaks from it.

5.4 Fairness

No partial results can be revealed before the end of the voting period.

1. No vote is decrypted before the ending of the voting period, and only trusted party can decrypt.

5.5 Verifiability

Correctness of the process can be checked

1. The calculation of result (ϕ algorithm) is public and anyone can verify his correctness,
2. The correctness of decryption is proved.

6 Conclusion

We presented a new secure electronic voting scheme based on EC-ELGAMAL [7] which is an additive homomorphic cryptosystem to guarantee Privacy, and modified schnorr scheme for authentication and identification. We conclude that the correctness and the security of our scheme is verified, that shows that the scheme is practical, and could be realized in reality for elections or any form of voting process.

References

1. Ahmad, T., Hu, J., Han, S.: An efficient mobile voting system security scheme based on elliptic curve cryptography. In: 2009 Third International Conference on Network and System Security, pp. 474–479. IEEE (2009)
2. Cervero, M.A., Mateu, V., Miret, J.M., Sebe, F., Valera, J.: An efficient homomorphic E-voting system over elliptic curves. In: International Conference on Electronic Government and the Information Systems Perspective, pp. 41–53. Springer, Cham (2014)
3. Certicom Research: Standards for efficient cryptography 2: Recommended elliptic curve domain parameters. Standard SEC2, Certicom (2000)
4. Gentry, C.: Fully homomorphic encryption using ideal lattices. In: Proceedings of STOC, pp. 169–178. ACM (2009)
5. Hao, F.: Schnorr non-interactive zero-knowledge proof. RFC **8235**, 1–13 (2017)
6. Galbraith, S.D., Gaudry, P.: Recent progress on the elliptic curve discrete logarithm problem. Des. Codes Crypt. **78**(1), 51–72 (2016)
7. Joye, M., Libert, B.: Encoding-free ElGamal-type encryption schemes on elliptic curves. In: Cryptographers Track at the RSA Conference, pp. 19–35. Springer, Cham (2017)
8. Miller, V.S.: Use of elliptic curves in cryptography. In: Williams, H.C. (ed.) Advances in Cryptology - CRYPTO'85. Lecture Notes in Computer Science, vol. 218, Springer, pp. 417–426 (1986)

9. Mursi, M.F.M., Assassa, G.M.R., Abdelhafez, A., Abo Samra, K.M.: On the development of electronic voting: a survey. Int. J. Comput. Appl. **61**(16), 0975–8887 (2013)
10. Koblitz, N.: Elliptic curve cryptosystems. Math. Comput. **177**(1987), 203–209 (1987)
11. Nakamoto, S.: Bitcoin: a peer-to-peer digital cash system, 24 May (2009). https://bitcoin.org/bitcoin.pdf
12. Cramer, R., Damgård, I., Schoenmakers, B.: Proofs of partial knowledge and simplified design of witness hiding protocols. In: Proceedings of the 14th Annual International Cryptology Conference (CRYPTO'94), Santa Barbara, California, USA. LNCS, vol. 963, pp. 174–187. Springer (1994)
13. Afreen, R., Mehrotra, S.C.: A review on elliptic curve cryptography for embedded systems. Int. J. Comput. Sci. Inf. Technol. (IJCSIT) **3**(3), 84–103 (2011)
14. Goldwasser, S., Micali, S., Rackoff, C.: The knowledge complexity of interactive proof systems. SIAM J. Comput. **18**(1), 186–208 (1989)
15. Zhao, Y., Pan, Y., Wang, S., et al.: An anonymous voting system based on homomorphic encryption. In: 2014 10th International Conference on Communications (COMM), pp. 1–4. IEEE (2014)

Investigation of Stereo Camera Calibration Based on Python

Elmehdi Adil[1] (✉), Mohammed Mikou[1], and Ahmed Mouhsen[2]

[1] Faculty of Sciences and Techniques, MISI Laboratory, Hassan First University of Settat, 26000 Settat, Morocco

[2] Faculty of Sciences and Techniques, IMII Laboratory, Hassan First University of Settat, 26000 Settat, Morocco

Abstract. Camera calibration is a crucial process in the field of computer vision. In this paper, describing the pinhole camera model is introduced. A calibration process of two similar cameras using a 2D pattern is then described. A Python program is developed to extract the cameras intrinsic and extrinsic parameters, and save them for a further use. The OpenCV library based the proposed program takes in consideration the radial and tangential distortion of the camera. To evaluate the performance of the proposed program, real data experiment has been performed. The results show the effectiveness of the proposed program to calibrate the cameras and form the basis for the next stage of the research. The proposed program can be used as first step in real time applications such as obstacle detection and navigation systems.

Keywords: Camera calibration · 2D pattern · Python · OpenCV

1 Introduction

Stereo vision is a process of depth reconstruction from two images of the same scene taken by two or more cameras. This process is modelled on the human perception of relief thanks to the two planar images that are perceived from each eye. The first step of every computer vision application is calibration which is a process of extracting the intrinsic and extrinsic parameters. The camera calibration that has an important role in the computer vision affects the precision of vision measurement [1, 2]. It has been widely studied because it's one of the most important techniques in the field of vision measurement.

Hall et al. [3] presented the first linear calibration method. This method computes the 3×4 transformation matrix based on the pinhole model. Next, the nonlinear calibration methods have been developed, in which lens distortion is introduced [4–9]. Zhang's method [10] is one of the most used camera calibration methods, because it's based on a simple and fast procedure for cameras calibration. He proposed a flexible technique based on the observation of a 2D pattern from different positions. Accurately calibrating a camera can not only improve the measurement precision but also provide a good basis for stereo image matching and 3D reconstruction [11–13].

A. Abraham et al. (Eds.): SoCPaR 2020, AISC 1383, pp. 779–789, 2021.
https://doi.org/10.1007/978-3-030-73689-7_74

OpenCV (Open Computer Vision) is a free graphics library, initially developed by Intel, specialized in real-time image processing. In camera calibration, it uses the Zhang's calibration method. The OpenCV camera calibration module supports both operating systems Windows and Linux, enhancing development efficiency. The module's execution speed is fast, and it has good cross-platform portability. Several previous research works were based on this library [14–18].

In this work, two similar cameras were fixed on a rigid support, a 2D chessboard pattern was used during the calibration process, and then calibration image pairs at multiple positions were captured. The real data experiments show that this proposed Python program for extracting the intrinsic and extrinsic parameters of the cameras gives accurate results. This program is based on the OpenCV library.

This paper is organized as follows. Section 2 gives a description of the camera model. In Sect. 3, a real data is used to verify the proposed Python program. The paper ends with a conclusion in the last section.

2 Camera Model and Lens Distortion

2.1 Camera Model

A camera model describes a projection from 3D to 2D. The most commonly used camera model is the pinhole camera where any 3D point Q in space is "projected" onto a 2D point q, obtained by intersecting the line (CQ) with the image plane. C is a fixed point related to the camera, called the optical center of the camera.

The pinhole camera is a simple model to approximate the imaging process. In the physical camera, image is formed behind the camera center. To make the math easier, the image and the center of the camera are swapped. The difference is now the image of the object appears right side up. This is shown in Fig. 1, where Z is the distance between the camera and the object, X is the length of the object, x is the length of the object's image on the image plane, and f is the focal length of the camera. x that is obtained by using similar triangles is expressed by the following equation:

$$x = f\frac{\mathrm{X}}{\mathrm{Z}} \tag{1}$$

The camera center C is reinterpreted as the center of projection. Optical axis is also called principal axis. The intersection between principal axis and the image plane is the principal point.

Camera projection. The utilized coordinate systems of object position in 3D are: World coordinate system (U, V, W) and Camera coordinate system is (X, Y, Z). Figure 2 shows a forward projection onto image plan. In other words, the projection of a point Q in 3D (U, V, W) is a point q in 2D pixel coordinates (u, v).

- Perspective projection:
  $(\mathrm{X_Q}, \mathrm{Y_Q}, \mathrm{Z_Q})$ are the coordinates of a 3D point Q in the space. The projection of Q onto the image plane is a 2D point q located at (x_q, y_q) as illustrated in Fig. 3

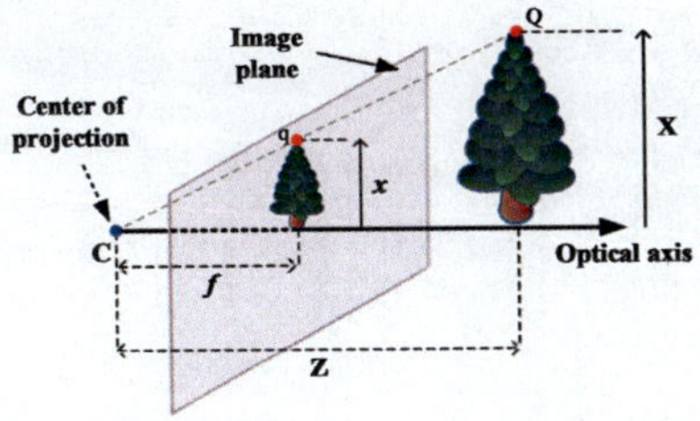

Fig. 1. New arrangement of the pinhole camera

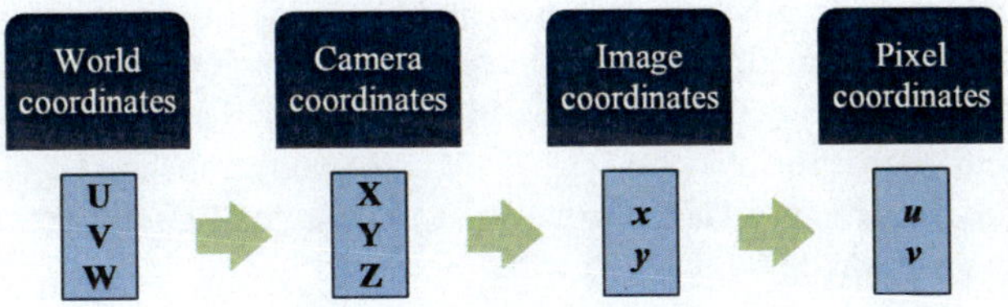

Fig. 2. Forward projection

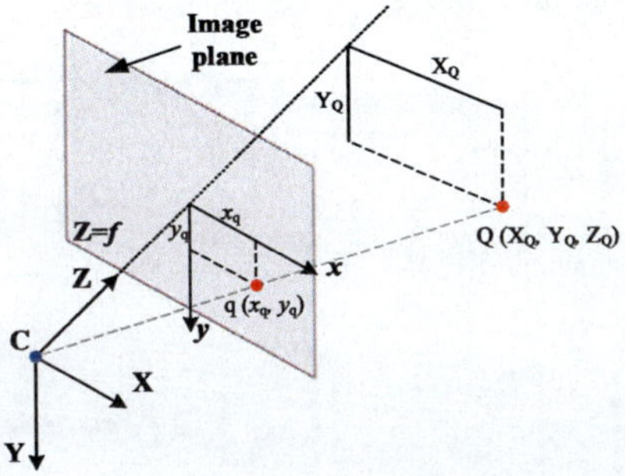

Fig. 3. Perspective projection

Using Thales theorem, the relationship between the world point $\left(X_Q, Y_Q, Z_Q\right)$ and the corresponding image point $\left(x_q, y_q\right)$ is given in the following equation:

$$x_q = f\frac{X_Q}{Z_Q}\ and\ \ y_q = f\frac{Y_Q}{Z_Q} \tag{2}$$

These equations are nonlinear. Homogeneous coordinates are used. The 2D point (x, y) can be recovered in function of the coordinates $(\tilde{x},\ \tilde{y},\ \tilde{z})$ using the equations bellow.

$$x = \frac{\tilde{x}}{\tilde{z}}and\ \ y = \frac{\tilde{y}}{\tilde{z}} \tag{3}$$

Equation (3) can be expressed as:

$$\begin{pmatrix}\tilde{x}\\ \tilde{y}\\ \tilde{z}\end{pmatrix} \sim \begin{pmatrix}x\\ y\\ 1\end{pmatrix} = \begin{bmatrix} f\,0\,0\,0\\ 0\,f\,0\,0\\ 0\,0\,1\,0\end{bmatrix}\begin{pmatrix}X\\ Y\\ Z\\ 1\end{pmatrix} = \mathbf{M_{projection}}\begin{pmatrix}X\\ Y\\ Z\\ 1\end{pmatrix} \tag{4}$$

where f is the focal length.

- Extrinsic parameters:
  Extrinsic parameters define the orientation and position of the camera with respect to the frame of external world. These parameters may change with respect to the world. The Q_{cam} and Q_{wor} that are represented in Fig. 4 describe the same physical point, just in two different coordinate systems. (U, W, V) are firstly translated by −C; therefore, the two origins are aligned. Then, a rotation is applied to align the axes. These points Q_{cam} and Q_{wor} are linked by the following relation:

$$Q_{cam} = R(Q_{wor} - C) \tag{5}$$

Using the homogeneous coordinates, the Eq. (5) is rewritten in Eq. (6).

$$\begin{pmatrix}X\\ Y\\ Z\\ 1\end{pmatrix} = \begin{bmatrix} r_{11} & r_{12} & r_{13} & 0\\ r_{21} & r_{22} & r_{23} & 0\\ r_{31} & r_{32} & r_{33} & 0\\ 0 & 0 & 0 & 1\end{bmatrix}\begin{bmatrix} 1 & 0 & 0 & -C_X\\ 0 & 1 & 0 & -C_Y\\ 0 & 0 & 1 & -C_Z\\ 0 & 0 & 0 & 1\end{bmatrix}\begin{pmatrix}U\\ V\\ W\\ 1\end{pmatrix}$$

$$\begin{pmatrix}X\\ Y\\ Z\\ 1\end{pmatrix} = \begin{bmatrix} \mathbf{R}_{3x3} & \mathbf{T}_{3x1}\\ 0_{1x3} & 1\end{bmatrix}\begin{pmatrix}U\\ V\\ W\\ 1\end{pmatrix} = \mathbf{M_{extrinsic}}\begin{pmatrix}U\\ V\\ W\\ 1\end{pmatrix} \tag{6}$$

where **R** is a 3 × 3 rotation matrix and **T** is a 3 × 1 translation vector, both link the world coordinate system to the camera's coordinate system.

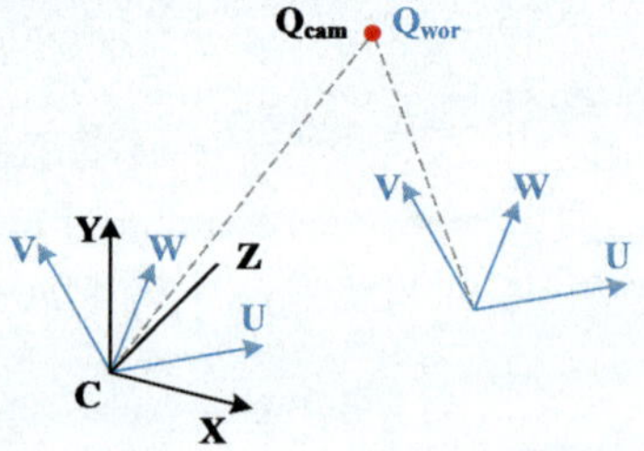

Fig. 4. A representation of a point in both world coordinates and camera coordinates

- Affine transformation:
  The optical measurements (x, y) are converted to pixel coordinates (u, v) by using the affine transformation as shown in Fig. 5. Two factors are considered. The first factor, the image is represented in the computer as a matrix; therefore, the origin (0, 0) of the matrix is usually at the top left corner of the image. Whereas in the optical image the center of the image plane p is (0, 0) where x is pointing to the right and y is pointing down. The second factor is the scale factor. A simple linear transformation expressed by two equations is used to shift from the principal point to the top left corner followed by a scale factor Eq. (7).

$$u = s_x x + p_x \ \textit{and} \ v = s_y y + p_y \tag{7}$$

  And it can be written down as a matrix transformation:

$$\begin{pmatrix} u \\ v \\ 1 \end{pmatrix} = \begin{bmatrix} \mathbf{s_x} & 0 & \mathbf{p_x} \\ 0 & \mathbf{s_y} & \mathbf{p_y} \\ 0 & 0 & 1 \end{bmatrix} \begin{pmatrix} x \\ y \\ 1 \end{pmatrix} = \mathbf{M_{affine}} \begin{pmatrix} x \\ y \\ 1 \end{pmatrix} \tag{8}$$

- Intrinsic parameters:
  Intrinsic parameters that are expressed in the Eq. (9) are specific to each camera and allow the mapping between camera coordinates and pixel coordinates in the image frame.

$$\begin{pmatrix} u \\ v \\ 1 \end{pmatrix} = \begin{bmatrix} s_x & 0 & p_x \\ 0 & s_y & p_y \\ 0 & 0 & 1 \end{bmatrix} \begin{bmatrix} f & 0 & 0 & 0 \\ 0 & f & 0 & 0 \\ 0 & 0 & 1 & 0 \end{bmatrix} \begin{pmatrix} X \\ Y \\ Z \\ 1 \end{pmatrix}$$

$$\begin{pmatrix} u \\ v \\ 1 \end{pmatrix} = \begin{bmatrix} f_x & 0 & \mathbf{p_x} & 0 \\ 0 & f_y & \mathbf{p_y} & 0 \\ 0 & 0 & 1 & 0 \end{bmatrix} \begin{pmatrix} X \\ Y \\ Z \\ 1 \end{pmatrix} = \mathbf{K} \begin{pmatrix} X \\ Y \\ Z \\ 1 \end{pmatrix} \tag{9}$$

  where $\mathbf{K}$ is the intrinsic parameters matrix, $\left(p_x, p_y\right)^T$ is the coordinates that model a possible displacement, away from the optical axis, of the center of the image plane. f_x and f_y are the product of the physical focal length of the lens and the size s_x and s_y of the pixel, respectively.

2.2 Lens Distortion

A lens that do not introduce any distortion can theoretically be modeled. In practice, for reasons of manufacturing, no lens is perfect. Based on the work of Brown [19] and Fryer and Brown [20], the two main lens distortions are represented and modeled. The radial distortion results because of the shape of the lens, while tangential distortion is the result of the assembly process of the camera as a whole.

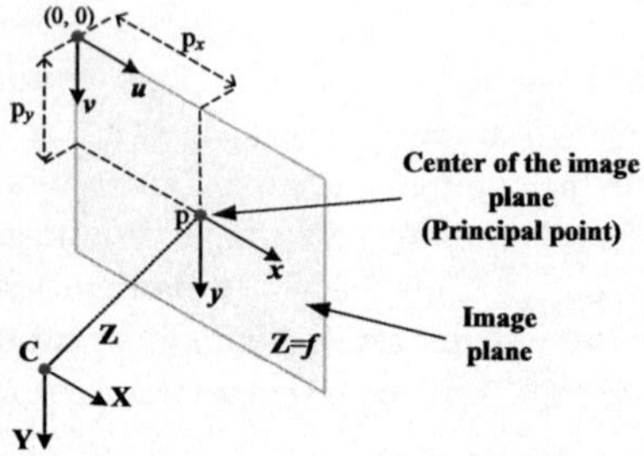

Fig. 5. From 2D image to 2D pixel array

For radial distortion, the distortion is 0 at the optical center of the imager and increase as getting closer to the periphery. In practice, the original location of the distorted point on the imager is considered (x, y), the radial location of this point is rescaled by using the following equation.

$$\begin{pmatrix} x_{cor} \\ y_{cor} \end{pmatrix} = \begin{pmatrix} x\left(1 + k_1 r^2 + k_2 r^4 + k_3 r^6\right) \\ y\left(1 + k_1 r^2 + k_2 r^4 + k_3 r^6\right) \end{pmatrix} \tag{10}$$

where (x_{cor}, y_{cor}) is the new location resulting from the correction.

Tangential distortion is characterized by two parameters p_1 and p_2:

$$\begin{pmatrix} x_{cor} \\ y_{cor} \end{pmatrix} = \begin{pmatrix} x + \left(2p_1xy + p_2\left(r^2 + 2x^2\right)\right) \\ y + \left(p_1\left(r^2 + 2y^2\right) + 2p_2xy\right) \end{pmatrix} \tag{11}$$

OpenCV gathers the five distortion coefficients into a 1×5 distortion vector in the following order k_1, k_2, p_1, p_2 and k_3.

3 Experimental Results

The proposed program has been tested on real data. The two similar cameras that have been used consists of a CMOS (OV2710) and lens with 8 mm focal length. The image resolution is 640×480 pixels. The calibration target is a 2D chessboard pattern with 10×7 corner points evenly distributed. To mimic the human binocular vision, the two cameras are fixed on a rigid support. The chessboard is moved and rotated so that the captured images cover most of the camera's field of view. For good results, at least 10 chessboard images are needed. The image pairs are taken from 45 different orientations. Some examples of image pairs used for calibration are shown in Fig. 6. The calibration steps of the two cameras are presented Fig. 7 in the form of a flowchart.

Zhang's method and Hong's method using MATLAB software were implemented on a laptop with Windows 10, 2,60 GHz Intel Pentium Dual-Core and 2 GB RAM. The proposed Python program is implemented on a 7th generation laptop with Windows 10, 2,70 GHz 2,90 GHz Intel® Core™ i7 CPU and 8 GB RAM. To evaluate the performance with respect to the number of patterns, five groups of experiments are carried out. The position and rotation of the first seventeen images are the same during the hole

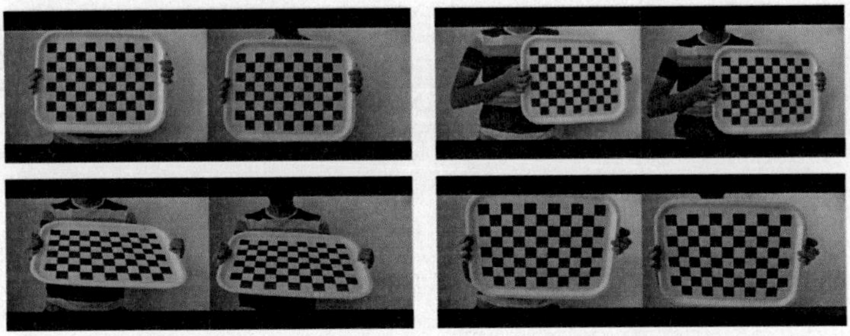

Fig. 6. Examples of image pairs used for calibration

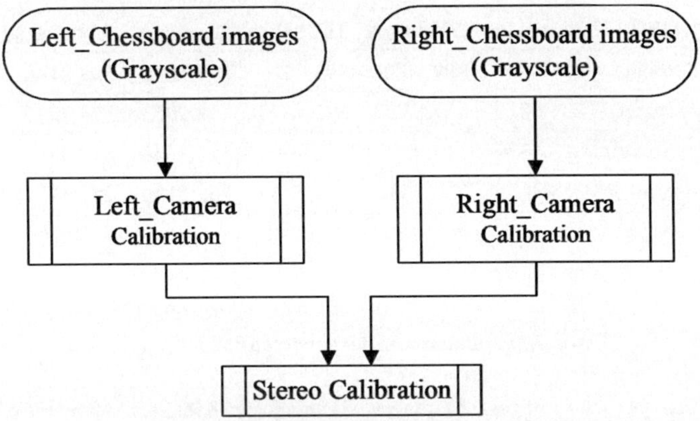

Fig. 7. The calibration process flowchart

experiments. From the second group, the number of pattern images are increased by seven.

First step of the process is to take pattern image pairs from each camera, simultaneously. Next step is to calibrate the left and right camera by finding the chessboard corners position, and for that a corner detection method with sub-pixel accuracy are used. The extracted parameters for the left and right camera are displayed in Table 1 and Table 2, respectively. After applying stereo calibration, two 3 by 4 projection matrices in the new coordinate systems, P_{Left} for the left camera and P_{Right} for the right camera are obtained. Note that the first three columns of P_{Left} and P_{Right} will be the new rectified camera matrices. The obtained left and right camera parameters are listed in Table 3 and Table 4, respectively. The relative error for p_x and p_y is shown in Fig. 8. The error decreases when the number of pattern images are increased. This error decrease significantly from 24 to 31 images.

To further examine the validity of the suggested code a mean reprojection error is computed for each camera. The error distributions for each experiments are displayed in Fig. 9.

Table 1. Left camera parameters

Groups	17 images	24 images	31 images	38 images	45 images
f_x (pixel)	899.7239	900.6110	898.0849	896.9915	897.4030
f_y (pixel)	897.3411	899.7707	896.9851	895.8734	896.3139
p_x (pixel)	305.2764	301.2389	302.1301	304.0177	304.5951
p_y (pixel)	226.2922	224.1501	224.1750	226.2341	227.0204

Table 2. Right camera parameters

Groups	17 images	24 images	31 images	38 images	45 images
f_x (pixel)	896.8507	897.7450	895.6470	894.5016	894.8146
f_y (pixel)	894.6519	897.5522	895.1675	893.9063	894.2277
p_x (pixel)	315.4324	312.9390	312.9690	313.3348	313.2862
p_y (pixel)	253.3629	250.7038	250.5745	252.5190	253.3451

Table 3. Parameters of the new rectified left matrix

Groups	17 images	24 images	31 images	38 images	45 images
f_x (pixel)	321.3056	407.0584	602.5674	715.4424	736.4040
f_y (pixel)	321.3056	407.0584	602.5674	715.4424	736.4040
p_x (pixel)	275.5169	389.2623	354.16391	339.5778	334.6583
p_y (pixel)	205.5450	209.6384	244.9582	238.7447	240.3073

Table 4. Parameters of the new rectified right matrix

Groups	17 images	24 images	31 images	38 images	45 images
f_x (pixel)	321.3056	407.0584	602.5674	715.4424	736.4040
f_y (pixel)	321.3056	407.0584	602.5674	715.4424	736.4040
p_x (pixel)	275.5169	389.2623	354.1639	339.5778	334.6583
p_y (pixel)	205.5450	209.6384	244.9582	238.7447	240.3073

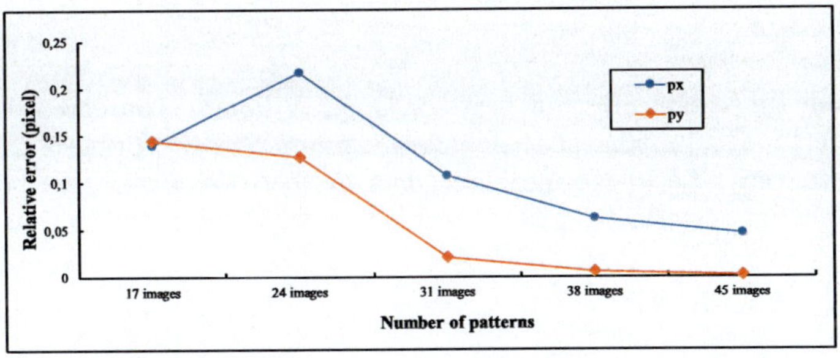

Fig. 8. Relative error with respect to the number of patterns

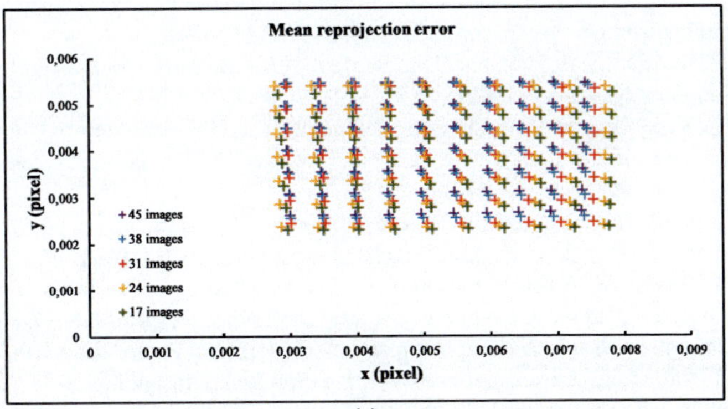

(a)

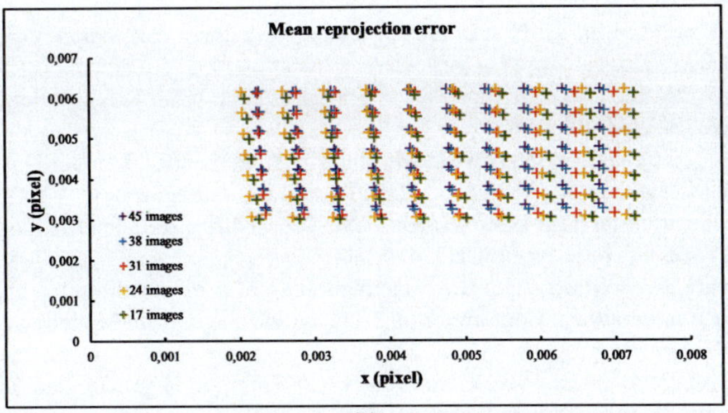

(b)

Fig. 9. Reprojection error distribution for the five experiments. (a) Left camera; (b) Right camera. Each "+ " represents the mean reprojection error of the corresponding chessboard corner in the test images.

4 Conclusion

A Python code for stereo calibration based on OpenCV library is proposed. This library gives consideration to the radial and tangential distortion. The effectiveness and accuracy of this code are verified on the results of real data. Furthermore, the extracted parameters will be used in future work.

References

1. Cui, Y., Zhou, F., Wang, Y., Liu, L., Gao, H.: Precise calibration of binocular vision system used for vision measurement. Opt. Express **22**, 9134–9149 (2014)
2. Usamentiaga, R., Garcia, D.F.: Multi-camera calibration for accurate geometric measurements in industrial environments. Measurement **134**, 345–358 (2019)
3. Hall, E.L., Tio, J.B.K., McPherson, C.A., Sadjadi, F.A.: Measuring curved surfaces for robot vision. Computer **15**, 42–54 (1982)
4. Tsai, R.: A versatile camera calibration technique for high-accuracy 3D machine vision metrology using off-the-shelf TV cameras and lenses. IEEE J. Robot. Autom. **3**, 323–344 (1987)
5. Weng, J., Cohen, P., Herniou, M.: Camera calibration with distortion models and accuracy evaluation. IEEE Trans. Pattern Anal. Mach. Intell. **14**, 965–980 (1992)
6. Heikkila, J., Silven, O.: A four-step camera calibration procedure with implicit image correction. In: Proceedings of IEEE Computer Society Conference on Computer Vision and Pattern Recognition, pp. 1106–1112. IEEE Computer Soc, San Juan, Puerto Rico (1997)
7. Heikkila, J.: Geometric camera calibration using circular control points. IEEE Trans. Pattern Anal. Mach. Intell. **22**, 1066–1077 (2000)
8. Wang, J., Shi, F., Zhang, J., Liu, Y.: A new calibration model of camera lens distortion. Pattern Recogn. **41**, 607–615 (2008)
9. Hong, Y., Ren, G., Liu, E.: Non-iterative method for camera calibration. Opt. Express **23**, 23992–24003 (2015)
10. Zhang, Z.: A flexible new technique for camera calibration. IEEE Trans. Pattern Anal. Mach. Intell. **22**, 1330–1334 (2000)
11. Rodriguez, T., Sturm, P., Wilczkowiak, M., Bartoli, A., Personnaz, M., Guilbert, N., Kahl, F., Johansson, M., Heyden, A., Menendez, J.M., Ronda, J.I., Jaureguizar, F.: VISIRE: photorealistic 3D reconstruction from video sequences. In: Proceedings 2003 International Conference on Image Processing (Cat. No. 03CH37429), Barcelona, Spain, pp. III-705–708. IEEE (2003)
12. Sundareswara, R., Schrater, P.R.: Bayesian modelling of camera calibration and reconstruction. In: Fifth International Conference on 3-D Digital Imaging and Modeling (3DIM 2005), Ottawa, ON, Canada, pp. 394–401. IEEE (2005)
13. Perez, A.J., Perez-Cortes, J.-C., Guardiola, J.-L.: Simple and precise multi-view camera calibration for 3D reconstruction. Comput. Ind. 7 (2020)
14. Wang, Y.M., Li, Y., Zheng, J.B.: A camera calibration technique based on OpenCV. In: The 3rd International Conference on Information Sciences and Interaction Sciences, Chengdu, China, pp. 403–406. IEEE (2010)
15. Xu, G.Y., Chen, L.P., Gao, F.: Study on binocular stereo camera calibration method. In: 2011 International Conference on Image Analysis and Signal Processing, Wuhan, Hubei, China, pp. 133–137. IEEE (2011)
16. Peng, H.X., Zou, X.J., Xiong, J.T., Wu, D.Z., Shen, J.H., She, Z.X.: Picking robot camera calibration system based on openCV. Adv. Mater. Res. **430–432**, 1963–1966 (2012)

17. Qiao, Y.J., Liu, Z.H., Hu, D.X., Xu, J.W.: Camera calibration method based on openCV. Appl. Mech. Mater. **330**, 517–520 (2013)
18. Wu, Y., Hua, Y., Chao, X.: The camera calibration of binocular vision measurement based on OpenCV. In: Proceedings 2013 International Conference on Mechatronic Sciences, Electric Engineering and Computer (MEC), Shengyang, China, pp. 1835–1839. IEEE (2013)
19. Brown, D.C.: Close-range camera calibration. Photogramm. Eng. 37(8), 855–866 (1971)
20. Fryer, J.G., Brown, D.C.: Lens Distortion for Close-Range Photogrammetry. Photogramm. Eng. Remote Sens. **52**, 51–58 (1986)

Study on the Different Types of Neural Networks to Improve the Classification of Ransomwares

Houria Madani(✉), Noura Ouerdi, Ahmed Boumesaoud, and Abdelmalek Azizi

Faculty of Sciences, Mohammed Premier University, Oujda, Morocco

Abstract. Among the diversity of malware, we mention "Ransomware". Its main objective is to take a user's data hostage, preventing and blocking access to the data and even to the computer (depending on the type of attack). The data is released once the ransom is paid. Despite the efforts of developers and the research community, this scourge remains a viable security threat that is why there is competition between IT community (developers and researchers) to realize and create a tool and method to detect malware, among the methods and techniques used are Artificial Intelligence and more precisely machine learning using neural networks. In this paper, we will discuss the different types of neural networks, the related work of each type, aiming at the classification of malware in general and ransomware in particular. After this study, we will talk about the methodology followed for the implementation of our neural network model (multilayer perceptron). We tested this model, firstly, with the binary classification if it is a malware or a goodware, and secondly, with the classification of the nine families of Ransomware by taking the vector of our previous work and we will make a comparison on the accuracy rate of the instance that are correctly classified.

Keywords: Classification · Malware · Ransomware · Artificial neural network · Multilayer perceptron

1 Introduction

Malware has been a problem from the first uses of computers and the Internet, and it is increasing and has become one of the serious problems in terms of computer security. After the exponential growth of Internet use, malware is affecting the IT community and the world in general, and any Internet user can be affected by this "epidemic". Many researches are underway to find a solution to this problem. Malware can easily sneak into our computers without knowing how or when we are attacked.

Among the malware that is so grave is Ransomware. Ransomware comes from the name «Ransom-Malware», it is a malware that attacks a user to take his data and divulge the functioning of his system under a hostage, if he pays the ransom, the data and the operating system will be released. This type of virus, in addition to its maliciousness, can also be considered lucrative. The issue of paying the ransom is under discussion, because if the user pays, then it will open a vulnerability for the attackers, they can threaten him and attack him again.

A. Abraham et al. (Eds.): SoCPaR 2020, AISC 1383, pp. 790–798, 2021.
https://doi.org/10.1007/978-3-030-73689-7_75

This paper presents a general study of artificial neural network types and an implementation of a multilayer perceptron to classify and deduce the recognition rate of the classes identified by Ransomware. The main challenge of classifying and identifying the correct Ransomware classes (families) is to find the correct rate between the learning set and the test set.

The first section will discuss the state of art with concepts and the related works to get an idea about similar works and to be update with the new methods used. The second section, we will detail our methodology. Results and discussion will be treated in the 3rd part. At the end, we will close our paper with a general conclusion.

2 State of Art

2.1 Concepts

Classification
Classification is part of supervised learning technique, it assigns a class to data whose category or class is unknown. To do this, a classifier (classification algorithm) is trained on a dataset often called a training set (or learning set), and this training set will be applied to another test set with similar data that we want to classify and to make prediction, if the data are correctly classified or not.

Artificial Neural Network (ANN)
Today, there are many definitions of artificial neural networks, but most agree that it is a developed computer system; its functioning is inspired by neurons of the human brain. ANN can be described as system composed of at least 2 layers, an input and output layers, and this system can also contains an intermediate layer which is called a hidden layer, each layer is characterized by a number of neurons. Neural networks can have as many layers depending on the complexity of the problem to be solved.

Convolutional Neural Network (CNN)
One of the types of Neural Network is the «Convolutional Neural Network or CNN». They are often used and well suited for image recognition and processing. This specific type is composed of several layers (it is multilayer) the layers are: Input layer, Convolution layer, Pooling layer, Fully Connected layer.

Recurrent Neural Network (RNN)
RNN is one of the types of neural networks that is widely used in Artificial Intelligence, especially in deep learning and dedicated to the discipline of Automatic Natural Language Processing (NLP). In the case of RNN, contextual information is given by a loop connection. This connection makes it possible to take into account at the current step one or more information predicted in a previous step.

2.2 Related Works

Recently, malware classification and detection has become very competitive among researchers [1], each one use a method to prove the effectiveness of its results.

We often see the use of machine learning techniques and neural networks with its different algorithms because they are able to analyze in depth the ransomware behavior [2]. Among the proposed ransomware classification methods, the author [3] suggested an approach using machine learning algorithms which have been used for binary classification of ransomware using static analysis of opcodes transformed into n-gram, they achieved an accuracy of 91.43%.

Classification of Malware Using ANN
Rad et al. [4] have collected different types of malware from a malicious repository shared by Naiyarah Hussain [5]. The authors used as many samples as possible (both benign and malicious), the samples are placed in folders as follows:

Malicious samples are placed in the "malicious_samples" folder and the benign samples are placed in the "benign_samples" folder. With this way, they can dispense with manually labeling each obtained file. And to ensure that the samples obtained are usable and properly labelled (malicious or benign), they proposed criteria to be followed:

- An executable file (PE for Portable Executable) is valid by checking the first 2 bytes of the binary file if it contains "MZ".
- Exclude packaged malware from the dataset so as not to damage the classification model by verifying the signature of each piece of software to see if it is packaged by a packer (such as UPX) or not.
- The collected samples are compared with anti-viruses. If the malicious sample is detected (even at a low rate), it will be removed. For the benign sample, if the antivirus has not detected anything (with a rate of 0%), the sample will be in the list of benign samples.

The authors required the relevance and effectiveness of criteria aforementioned in order to remove any sample that might affect the model that they want to realize. They proposed a neural network who do a binary classification to classify an invisible PE (Portable Executable) file as a begnin or malicious, developed with the Python library «TensorFlow-GPU». In addition, they used «scikit-learn» to do the cross-validation method. The dataset contained 4000 samples (1000 benign and 3000 malicious), each class is represented by vector indicating (0, 1) if the sample is malicious and (1, 0) if it is benign. The input size (Windows library function calls) was 34087 after pre-processing of the dataset. They validated the model using the cross-validation method, they have assumed that the model can perform on an unknown dataset, and will operate based on the results that have been achieved. With this model, they obtained an average accuracy of 97.8%, with 97.6% precision, and 96.6% recall.

Malware comes in several types; our goal is to focus on ransomware. The paper [6] presents our previous work, which was a comparative study of a proposed ransomware dataset on work [7] with a new proposed dataset. We changed the learning files but we did not have a test dataset, so we took 20% of the learning database file to build our test database. The results were high compared to that in paper [7] but the problem was that the 20% files taken, was not removed (but copied) from the learning database, we copied the test part in question to another file. We obtained approximately a rate of 70% for correctly classified instances, but were not as surprising and relevant as expected.

We also compared the two algorithms Artificial Neural Network (ANN) and Bayesian Network (BN), to deduce the one that will give us the best results in our work, and we got the right results with the ANN.

Classification of Malware Using CNN

Many researchers use convolutional neural networks to classify and detect malware. Kabanga et al. [8] proposed a model of convolutional neural networks to extract features from image at a time, this images extracted from a sample of malware will be classified. The model has images of a size 128 * 128 * 1 (1 being the channel width), they used the image library from the PIL (Python Image Library) package of Python to generate vectors of images and after the generation of the vectors, processing was performed on these vectors. Then, they designed a three-layer deep convolutional neural network to do the classification task. The output layer has 25 neurons that correspond to the 25 families of malware available in the dataset used. The goal of this neural network is to have a single class containing the malware.

The Malimg dataset [9] that was used, has 25 malware families (with several grayscale images), which 90% are used for training and 10% for testing. They obtained in the results an accuracy of 98%.

Using this technique, they have proven that it is the best compared to other traditional methods of classification. However, using images to classify malware can lead to erroneous results due to poor image extraction.

Classification of Malware Using RNN

Pascanu et al. [10] try to learn the language of malware in order to detect them, by forming a recurrent model to predict the next API call, using the hidden state of the model (which encodes the history of past events) as a fixed length vector that is given to a classifier (by the logistic regression or Multi-layer Perceptron "MLP" algorithm). And to improve their recurrent model, they introduce Max-Pooling on the values of the hidden units in the input sequence, and they propose the Half-Frame model that increases the memory capacity of the final representation by including the mid-state information of the sequence of events in the final state. The authors suppose that the memory window of the standard ESN (Echo State Network) and RNN models is not adequate, to solve that, a Leaky-Units architecture [11] has been added to the recurrent architectures, which allows the increase of the system memory in the long term. The authors also added a bidirectional mode [12] that combines two distinct models, one learning by treating events in the forward direction and the other learning by processing the events in the reverse direction.

The results show that the combination of a recurrent model with a standard classifier can improve the classification of malware, and on the other hand, ESN models outperform RNN models in the majority of experiments, but the use of recurrent neural networks is a bit complicated especially when we want to do malware detection. We cannot assume that this is the right method to do malware classification and detection because of the need to learn the language of malware, which is a bit difficult considering that malware varies.

Ransomware detection systems can have performance by having the ability to capture repetitive behavior and general sequence learning. That's why, Agrawal et al. [13] worked on the attention mechanisms while processing executable sequences for ransomware detection. They proposed an implementation of an improved neural cell to integrate the attention mechanism in learning named ARI (Attended Recent Inputs); this ARI is used by LSTM (Long Short-Term Memory) networks, named ARI-LSTM. This cell allows to perform a detailed analysis of ransomware executable, i.e. it processes input sequences by taking the attention weights for each recent input.

In the learning phase, they took the LSTM model (improved by incorporating the ARI cell) and Max Pooling (LaMP). The goal of the learning is to classify the input sequences by labeling the ransomware by "1" and the benign by "0". They used a set of sequence data containing benign and ransomware executable of a Windows operating system captured from users' computers.

The results show that the accuracy of the ARI-LSTM is better than an LSTM, proving the effectiveness of attention mechanisms in the learning phase.

3 Methodology

The objective of this work is to compare the results of the classification of "ransomwares" with the old ones based on the previous work [6]; this classification will be done with an artificial neural network (MLP) that has been implemented.

3.1 Implementation of the Artificial Neural Network

We took the same files (dataset of ransomware Fig. 1), features and parameters in the previous work [6].

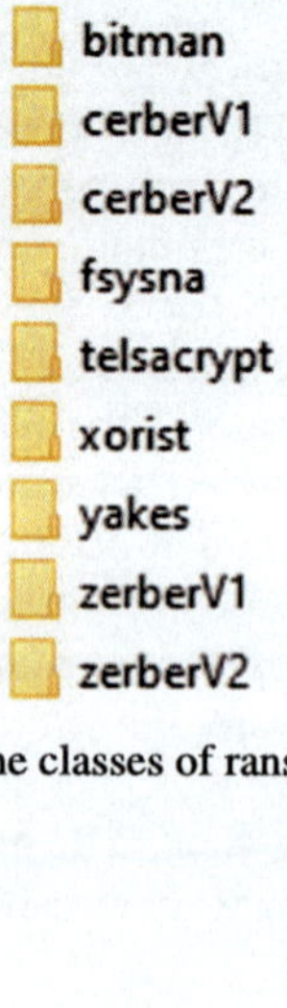

Fig. 1. The classes of ransomwares.

We suggested some deep models for artificial neural network to solve the classification problem. The implementations were made by the universal programming language Python.

We used TensorFlow to do the machine learning and the Keras API. Keras allow to build deep learning models, adjust them, evaluate them and use them to make predictions. For proposed model, we will follow the following steps: prepare the data, then define, compile, adjust, and evaluate the model of network, and at the end doing predictions.

4 Results and Discussion

4.1 Classification of Ransomware

After the implementation of the multilayer perceptron, we decided to test the model to do the Malware classification (precisely the 9 families of ransomware).

Based on the previous work [6], we took the vector V1 with the size 1089, with the extracted instances which were 437, we formed a multilayer perceptron on 80% of these data (349 samples), and tested it on the remaining data.

```
Train on 349 samples, validate on 88 samples
Epoch 1/3
349/349 [==============================] - 9s 27ms/step - loss: 0.7317 - accuracy: 0.7794
- val_loss: 0.5866 - val_accuracy: 0.8750
Epoch 2/3
349/349 [==============================] - 9s 25ms/step - loss: 0.3680 - accuracy: 0.8854
- val_loss: 0.4282 - val_accuracy: 0.8523
Epoch 3/3
349/349 [==============================] - 9s 26ms/step - loss: 0.2553 - accuracy: 0.9140
- val_loss: 0.2077 - val_accuracy: 0.9091
time (s):  28.69094157218933
88/88 [==============================] - 0s 458us/step
Accuracy:  90.90909361839294 %
loss:  20.770819214257326 %
```

Fig. 2. Evaluation of the multilayer perceptron on the different files ransomware

- Model Definition: The model is defined in 3 layers (an input layer, a hidden layer and an output layer), knowing that the output layer has 9 neurons (corresponds to 9 ransomware categories) with the softmax activation function, for the purpose of making a classification of 9 classes. We have compiled this model using the Adam optimizer and the multi-class log loss function (categorical).
- Evaluation: The model is evaluated in 3 epochs with a batch size of 5 lines, we got the following results (Fig. 2):
- Classification report:
  The accuracy of the model is 91%. If we compare with the results of the previous work [6] we notice an increase in accuracy, from 70% to 91%. This is due to the good methodology followed and it also means that the developed model works well by perfectly classifying 4 classes without any error, except 2 classes (ZerberV1, ZerberV2) with the recall rate equal to 0 (Fig. 3).

	precision	recall	f1-score	support
bitman	1.00	1.00	1.00	15
cerberV1	0.50	0.83	0.62	6
cerberV2	0.93	0.72	0.81	18
fsysna	0.00	0.00	0.00	1
telsacrypt	1.00	1.00	1.00	23
xorist	1.00	1.00	1.00	21
yakes	1.00	1.00	1.00	3
zerberV1	0.00	0.00	0.00	0
zerberV2	0.00	0.00	0.00	1
accuracy			0.91	88
macro avg	0.60	0.62	0.60	88
weighted avg	0.93	0.91	0.91	88

Fig. 3. The metrics of the multilayer perceptron

Estimated Classes

Real Classes	Bitman	cerberV1	cerberV2	Fsysna	Teslacrypt	xorist	yakes	zerberV1	zerberV2
Bitman	15	0	0	0	0	0	0	0	0
cerberV1	0	5	0	0	0	0	0	1	0
cerberV2	0	5	13	0	0	0	0	0	0
Fsysna	0	0	0	0	0	0	0	1	0
Teslacrypt	0	0	0	0	23	0	0	0	0
Xorist	0	0	0	0	0	21	0	0	0
Yakes	0	0	0	0	0	0	3	0	0
zerberV1	0	0	0	0	0	0	0	0	0
zerberV2	0	0	1	0	0	0	0	0	0

Fig. 4. Confusion matrix

- Confusion matrix:
  It is true that some classes are not correctly classified, but the model worked very well for the rest. We will soon see other classification methods and we will try to find the real instances for the classes that are incorrectly classified (Fig. 4).

4.2 Comparison Between the Different Neural Network

The table below shows a comparison of several classification methods between the different types of neural networks (Table 1).

Table 1. Comparison of the classification between the different types of neural networks.

Type	Our recent result	Related work	Methods	Results
ANN	The binary classification (malware and goodware) gave us an accuracy of 100%	[4]	The authors proposed a Neural Network who do a binary classification (begnin or malicious) for invisible PE files	The accuracy for the classification of malware is 97.8%
	The accuracy for classification of ransomware is 91%	[6]	Comparison between the old and the new dataset to classify ransomware, this new dataset gave us better results, we also tested Bayesian networks	The accuracy for the classification of ransomware: - using ANN is ≈70% - using BN is ≈49%
CNN		[8]	The authors proposed a model of convolutional neural networks to extract features from image at a time, this images extracted from a sample of malware will be classified	They obtained in the results an accuracy of 98%
RNN		[10]	The authors try to learn the language of malware in order to detect them, by forming a recurrent model to predict the next API call, using the hidden state of the model as a fixed length vector that is given to a classifier	No accuracy value
		[13]	The authors worked on the attention mechanisms while processing executable sequences for ransomware detection	- ARI-LSTM (L = 5) = 0.93 (93%) - ARI-LSTM (L = 8) = 0.91 (91%)

5 Conclusion

In this paper, we have implemented a neural network (MLP) by dealing the binary classification method (malware/goodware) as well as ransomware classification. To perform this classification, we used data sets (goodware data set and another dataset malware/ransomware used in [6]), we took 80% for the training set and 20% for testing, in order to evaluate if our neural network model detects instances correctly or not. The result obtained shows that our model detects correctly and perfectly the instances of the binary classification (goodwares/malwares) with a rate of 100%, and also a 91% accuracy rate for the ransomware classification, which means that there is an improvement on the results compared to those found in the work [6].

References

1. Idika, N., Mathur, A.P.: A survey of malware detection techniques, Technical report, Purdue University, p. 48, February 2007
2. Kok, S., Abdullah, A., Jhanjhi, N., Supramaniam, M.: Ransomware, threat and detection techniques: a review. Int. J. Comput. Sci. Netw. Secur. (IJCSNS) **19**(2), 136–146 (2019)
3. Zhang, H., Xiao, X., Mercaldo, F., Ni, S., Martinelli, F., Sangaiah, A.K.: Classification of ransomware families with machine learning based on N-gram of opcodes. Fut. Gener. Comput. Syst. **90**, 211–221 (2019). https://doi.org/10.1016/j.future.2018.07.052
4. Rad, B.B., Nejad, M.K.H., Shahpasand, M.: Malware classification and detection using artificial neural network. J. Eng. Sci. Technol. **13**, 14–23 (2018). Special Issue on ICCSIT 2018
5. GitHub - naisofly/Static-Malware-Analysis: Static Feature Extraction & Selection (used in conjunction with the MASTIFF framework). https://github.com/naisofly/Static-Malware-Analysis. Accessed 23 Jul 2020
6. Madani, H., Ouerdi, N., Palisse, A., Lanet, J.-L., Azizi, A.: Classification of ransomwares using artificial neural networks and Bayesian networks. In: 2019 3rd International Conference on Intelligent Computing in Data Sciences (ICDS), Marrakech, Morocco, October 2019, pp. 1–6 (2019). https://doi.org/10.1109/ICDS47004.2019.8942294
7. Ouerdi, N., Hajji, T., Palisse, A., Lanet, J.-L., Azizi, A.: Classification of ransomware based on artificial neural networks. In: Rocha, Á., Serrhini, M. (eds.) Information Systems and Technologies to Support Learning: Proceedings of EMENA-ISTL 2018, pp. 384–392. Springer, Cham (2019)
8. Kabanga, E.K., Kim, C.H.: Malware images classification using convolutional neural network. JCC **6**(01), 153–158 (2018). https://doi.org/10.4236/jcc.2018.61016
9. https://old.vision.ece.ucsb.edu/spam/malimg.shtml
10. Pascanu, R., Stokes, J.W., Sanossian, H., Marinescu, M., Thomas, A.: Malware classification with recurrent networks. In: 2015 IEEE International Conference on Acoustics, Speech and Signal Processing (ICASSP), South Brisbane, Queensland, Australia, pp. 1916–1920 (2015). https://doi.org/10.1109/ICASSP.2015.7178304
11. Bengio, Y., Boulanger-Lewandowski, N., Pascanu, R.: Advances in Optimizing Recurrent Networks. arXiv:1212.0901 [cs], December 2012. Accessed 9 Sept 2020
12. Schuster, M., Paliwal, K.K.: Bidirectional recurrent neural networks. IEEE Trans. Sig. Process. **45**(11), 2673–2681 (1997). https://doi.org/10.1109/78.650093
13. Agrawal, R., Stokes, J.W., Selvaraj, K., Marinescu, M.: Attention in recurrent neural networks for ransomware detection. In: 2019 IEEE International Conference on Acoustics, Speech and Signal Processing (ICASSP), ICASSP 2019, May 2019, pp. 3222–3226 (2019). https://doi.org/10.1109/ICASSP.2019.8682899

Quality of Service Evaluation with DSR (Dynamic Source Routing) Protocol in the Classroom Ad Hoc Network of the New Generation of Digital Open Universities (DOUNG)

Mahamadou Issoufou Tiado(✉), Ibrahim Ganaou Noura, Chaibou Dan Inna Hussein, Col. Harouna Gazobi Souleymane, and Hamani Mounkaila Mahamadou

Department of Mathematics and Computer Science, Research Team on Network and Telecommunication, University of Abdou Moumouni, BP 10662 Niamey, Niger

Abstract. Many learner devices with multimedia capacities such as computers, laptops and telephone cells derive from the nowadays technology evolution. The wide range of these electronic devices brings a real opportunity to build an ad hoc network in the DOUNG [1, 2] classroom with real time option of following courses. In addition of previous works that define significant Quality of Service (QoS) parameters for following courses alive in the DOUNG model, this paper aims to contribute by using the learner devices to extend the DOUNG network architecture. The idea is to use an ad hoc network in the classroom after a Wi-Fi antenna instead of duplicating this antenna when the signal is low. Thus, with an access mode that brings the network resource consumption to reach a critical threshold, we measure some QoS parameters by simulation in this new architecture using SCTP (Stream Control Transport Protocol) at the transport layer and DSR (Dynamic Source Routing) at the network layer. Curves are produced and the results are discussed.

Keywords: Distance education · DOUNG · Classroom ad hoc network · QoS · SCTP · DSR

1 Introduction

Ad hoc network derives from recent technology evolution that allows mobiles nodes to use radio links to communicate without infrastructure. A node in this network communicates with others in its range of emission/reception. It maintains its autonomy and independence through mobility, function and routing information in the network.

The great density of electronics devices such as computers, laptop and telephone cells bring the opportunity to build a classroom ad hoc network as an extension of the DOUNG architecture. Learners are thus able to follow courses in real time with their

A. Abraham et al. (Eds.): SoCPaR 2020, AISC 1383, pp. 799–809, 2021.
https://doi.org/10.1007/978-3-030-73689-7_76

own means. But this way of improving the DOUNG architecture brings to the need to determine the level of the reachable QoS. For this purpose, this paper presents at first the goals of using the ad hoc network in the classroom before the description of the attenuation factors of that network limitation. In addition, the architecture of the wireless classroom network is described with the QoS parameters deriving from the communicating vessels model [3] and summarized in the IIMSG (Intelligent Interface of Monitoring the Service Guarantee). The description of the simulation scenarios is realized before presenting and interpreting the obtained results.

2 The Using of an Ad Hoc Classroom Network

Wireless environments give opportunity to extend the DOUNG initial architecture defined in [1] for the synchronous mode of following courses. As a reminder, the DOUNG architecture includes m-learning [4–6] and cloud computing option. But in spite of many advantages, this solution brings to the great challenge of measuring the performance of the system when the network is subjected to the constraint of conveying multimedia stream. Particularly in this paper, the main problematic is to determine the QoS parameters that allow to follow a course in real time. For this purpose, the values of the required QoS parameters are evaluated by simulation to establish the balance between the advantage and the inconvenient of using this solution. Before that, we first analyze the nature of the DOUNG wireless network classroom having many weakness attenuation factors.

In general case, wireless networks are time characterized by (1) large–scale variation of the link quality that can change quickly and causes bursts of errors. In addition, (2) the wireless channel is subjected to the influence of obstacles and interferences and (3) the dynamic topology of the ad hoc networks brings to the problematic of routing information. Also, (4) the limitation of their available bandwidth, (5) their average channel state depending on user mobility, are other additional weakness that need to be taken into account in the DOUNG extension. We then determine the intensity of these weakness negative influences on the QoS that can compromised the session of a course alive.

3 Attenuation Factors of the Ad Hoc Network Limitations

The DOUNG closed or open classroom has main characteristics that facilitate the using of ad hoc network with attenuation of the precedent weakness. For the poor wireless link quality (limitation 1), we identify three first attenuation factors related to the DOUNG studied geographic position, the limited size of the open or closed classroom and the low range of the wireless signal. When joined together, they allow to avoid the signal degradation. For example in this case, pirates frequency blocking action can be easily detected and will then find an appropriate solution. These three factors have the same attenuation effect on the influence of obstacles and interferences (limitation 2), on the low available bandwidth (limitations 4) and on the average channel state depending on user position because of the mobility (limitations 5). An additional factor considers the mobility speed at the human scale and the topology change that impact (3) the routing problematic is widely dependent on the learner's action of starting or stopping devices, or if they enter or give up the classroom. Normally during the course period, they are mostly concentrated on the teacher message delivery and make the classroom ad hoc network topology stable.

4 Classroom Ad Hoc Network Architecture

Nowadays, the spontaneous concentration of learner's mobile devices provides wireless interconnection inside a classroom. Then the DOUNG can easily extend its wired Ethernet network with an ad hoc network after a Wi-Fi antenna. As this rapid deployment of the wireless network is easy, with low coast, by just making available a Wi-Fi antenna, thus the QoS of the synchronous access mode of course becomes interesting to study. A recent work [7] over the comparative analysis of infrastructure and ad hoc networks makes interesting the using of this solution. The first extension covers the even reduced zone of the antenna. The second considers the signal propagation between neighbors by using the resource of learner device. Thus we complete the DOUNG initial architecture defined in [1] as shown in the figure below subdivided in four zones. The first zone is given by the DOUNG server available in the Ethernet network that offers a Wi-Fi antenna being the second zone. The antenna access for the learners of the front line (that is the third zone) is realized by using the means of a wireless network with infrastructure. At the end in the fourth zone, the learners can follow a course alive by using the means of ad hoc network (Fig. 1).

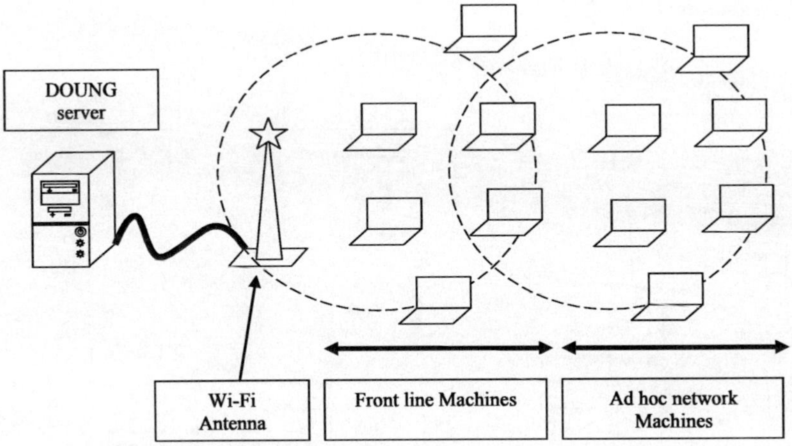

Fig. 1. DOUNG architecture extension with ad hoc network

When analyzing the previous figure, the problematic of QoS evaluation derives mostly from the need to establish the limit that makes better to use the ad hoc network instead of the traditional reflex of Wi-Fi antenna multiplication.

5 Overview of the Communicating Vessels Model Parameters

The DOUNG QoS parameters are defined in [3] through the communicating vessels model (CVM). This real time system of course delivery considers first the camera multimedia stream temporally saved in the transmission buffer called Server Buffer (SB). At the opposite side, the learner's application manages an anticipation window that brings also to the management of a local buffer called Client Buffer (CB). Then, the CVM model illustrates the moving of the content from SB to CB. The same δ amount of the content subtracts to SB is theoretically equal to the amount that reaches CB.

In that architecture, the network is built with materials and protocols used for the system operation. The figure below presents a summary of the IIMSG parameters showing the network extension. The goal is to evaluate the performance gains through the QoS level that can be reached according to the attendance of learner. It is supposed that these parameters get their best values in a fair used of the DOUNG Ethernet network. And then the interest to study the wireless extension of the network derives from the limited size of that wired network that can't be indefinitely extended. In addition, the Wi-Fi antenna is to be used in an efficient manner by minimizing the architecture coasts. Under these consideration, the using of the MANET (Mobile Ad hoc NETwork) allows to overcome this limitation barrier but with the main question of the balance to be established with the security and the low bandwidth. It becomes then interesting to measure the influence of the mobiles density in the wireless extension and to deduce the QoS rank that verifies the theory of the better choice of using this architecture instead of the multiplication of the Wi-Fi antenna (Fig. 2).

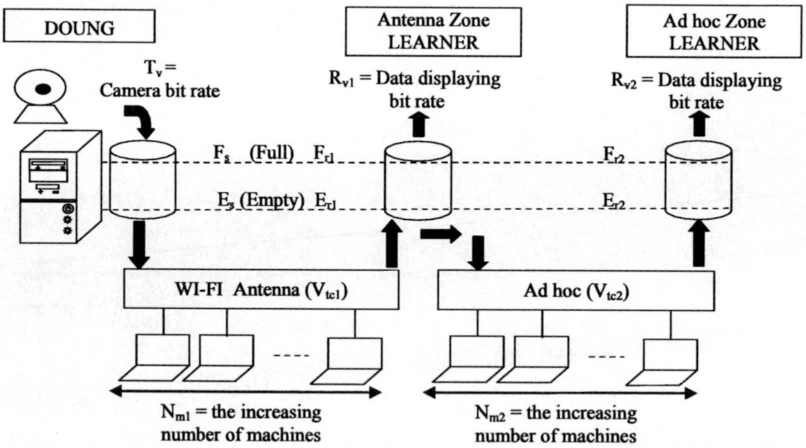

Fig. 2. The communicating vessels model in the DOUNG Extended architecture

In the last figure, T_v is the camera bit rate (in bits per second). E_s and F_s (in byte) are respectively the empty and the full level of the SB while E_{s1} and F_{s1} are the same level for the mobile in the Wi-Fi antenna zone. Their equivalent E_{s3} and F_{s3} are used for the ad hoc network zone. R_{v1} and R_{v2} are data displaying bit rate (in bit per second) used by the learner application to follow the course in real time. V_{tc1} is the theoretical Wi-Fi antenna bit rate (in bit per second) and its equivalent for the ad hoc network is V_{tc2}.

6 Summary of the IIMSG Parameters

6.1 Parameters Description

In this paper, simulations are run to evaluate some QoS parameters deriving from the IIMSG interface. We consider the two zones of the classroom network and measure the influence of the attendance of learners in each zone over the synchronization delay between the teacher and the learner. By the same way, we are evaluating the influence of the same attendance on the buffers SB and CB size that allow the learner to follow the course alive.

We consider ST_s as the DOUNG initial time of starting the session. This parameter is associated to ST_{v1} and ST_{v2}, the real time synchronization delays of the learners. ST_{v1} is for the Wi-Fi antenna zone and ST_{v2} for the ad hoc zone. They are defined as the time used by the DOUNG system to convey the traffic from SB to CB including the crossing of the network. We measure the difference between these three parameters under the network constraints, notably, the attendance of learners in each zone. Particularly when initiating a CBR traffic, a QoS level is reached according to the shift between the time T of the camera multimedia stream registration in the SB and the same stream reception time by a learner (registration in the CB). The two delays ST_{v1} and ST_{v2}, depend normally to many other factors that are not studied and not modified in our simulations as for example T_v the camera bit rate or the time used by the network to transfer δ bits in CB before the application starts playing the video.

6.2 Definition of Mathematic Model

This first inequality is established between the previous parameters from an intuitive approach:

$$ST_s < ST_{v1} < ST_{v2} \tag{1}$$

Let consider TTL_{max} as the maximal duration of a multimedia stream in SB and T_δ the synchronization extension parameter at the learner side during the transfer of δ bits. Two global delays DG_1 and DG_2 can be calculated as follow:

$$DG_1 = ST_{v1} + T_\delta \text{ for the learners in the Wi - Fi antenna zone} \tag{2}$$

$$DG_2 = ST_{v2} + T_\delta \text{ for the learners in the ad hoc network zone} \tag{3}$$

From the previous, the QoS level becomes a minimal value called S_{DOUNG} that is set by the DOUNG. It gives the tolerable threshold of the precedent equalities (2) and (3). The value of S_{DOUNG} must imperatively be set in the following interval:

$$S_{DOUNG} < TTL_{max} \tag{4}$$

And the following inequality is established for a good level of the far learners QoS value,

$$DG_2 >= S_{DOUNG} \quad (5)$$

Let assume Q_s as the amount of information available in the SB at the time CT. Q_{r1} is the same amount in the learner's CB of the Wi-Fi antenna area and Q_{r2} is for the ad hoc area. The value of Q_s depends of the camera bit rate and the multimedia time to live in the SB. It is given as follow:

$$Q_s = T_v * TTL_{max} \quad (6)$$

In the best theoretical case with no slowing factor, Q_{r1} and Q_{r2} can get values near than Q_s. But as this case is special, it is better to consider the values of these parameters as given by the following inequality by taken into account the propagation and network additional delays:

$$Q_{r2} < Q_{r1} < Q_s \quad (7)$$

7 QoS Parameters Evaluation by Simulation

In this section, the previous QoS parameters are evaluated by simulation in ns-2 environment. We use a protocol stack with SCTP at transport layer [8]. The DSR (Dynamic Source Routing) [9] and IP protocols are at the network layer. For lower layers, we use the 802.11 protocols. The scenario is run with a traffic source representing the Wi-Fi antenna. Reception nodes are both in the antenna and in the ad hoc areas. To simplify our model, a Constant Bit Rate (CBR) traffic is initiated. It helps to measure the influence of the learners attendance in the two zones. The inter-arrival time of messages during peak is set to 0.1 s and 1500 bytes are used for the SCTP segments size.

7.1 The SCTP Protocol of the Transport Layer

SCTP protocol is used at the transport layer as in our case mainly for multimedia stream transmission in the wireless environment. It provides a reliable and orderly transmission of data. It is based on the stream and the association concepts. The stream is a sequence of messages that are to be transmitted in order. The association concept is a group of streams in which each stream endpoint provides a list of its transport addresses (@IP + port). An association is established between the SCTP sender and receiver before transferring data.

SCTP is a message oriented protocol used over connectionless protocol of the network layer. It offers many services as the data transfer with acknowledgement and no duplication and data fragmentation according to the network constraints. It delivers messages in order within streams with individual message delivery. The transferred streams are independent to avoid that the failure of one impacts the others. SCTP offers also error management through the multi domiciliation. The congestion avoidance, the path failure management within a timer, the flooding and attacks overcoming are its other additional mechanisms.

7.2 The DSR Protocol of the Network Layer

DSR protocol at the network layer is used for routing information in the ad hoc network. It offers a source routing service that allow a node to reach a particular destination through packet propagation in the network. When receiving data from the IP protocol, a DSR node uses many routing structures and mechanisms with also special messages. Mainly, it performs the route discovery and the route maintenance. The route discovery function is based on the route request and the route response messages. The route maintenance function is ensured through transmission control (with acknowledgement), route error management, packet salvaging and packet segmentation to adjust its size to a path.

7.3 Description of Scenarios

The goal is to measure the influence of the learner's attendance over the synchronization delay and over the size of SB and CB. We evaluate the DG_1 parameter for the synchronization of the Wi-Fi learners and DG_2 for the ad hoc zone. We evaluate also the values of Q_s, Q_{r1} and Q_{r2} parameters that show the amount of data available at the time T respectively in the buffers of the DOUNG server, the antenna and the ad hoc zones.

7.4 Results Interpretation Basis

The curves are product by crossing one of the previous parameters with the increasing number of nodes in the network. For the attendance, we choose to increase the number of nodes within the interval growing from 10, 15, 20, 25, 30, 35, 40, 45, to 50. It is assumed that in the real classroom, the promiscuous mode brings one simulated node to represent many others. Simulations are run for each case and the results are average values extracted from the traces file. These results are analyzed and explained by highlighting the QoS parameters evolution. We choose a low mobility speed [1 m/s, 10 m/s] with 25 ms of break time and the nodes moving in their respective area. The global simulation area dimension is set to 1500×300. The area is subdivided in two zones. The antenna zone is set to 1/3 of the global area and begins from the left border. The ad hoc network takes the remaining 2/3 of the area. Each zone is having the same number of nodes. The simulation duration is set to 150 s.

7.5 Influence of the Attendance Over the Synchronization Delay

The two curves in the figure below show the influence of the learners increasing attendance on their synchronization time according to their zone in the classroom wireless network. For each attendance, reception in the antenna area is immediate, resulting in a zero value of the curve. These results confirm the previous expectations and mathematical models as well as the deductions from the intuitive approach with the following inequality:

$$\begin{aligned}
&ST_{v1} < ST_{v2} && \text{by inequality (1)}\\
&ST_{v1} + T_d < ST_{v2} + T_d && \text{by (2) and (3)}\\
&DG_1 < DG_2.
\end{aligned}$$

When considering the average value of DG_2, the curve of the classroom ad hoc network reflects the influence of the DSR protocol because the antenna can't reach immediately all the nodes in this area. Indeed, to join the ad hoc nodes, it is necessary to carry out the routing of the packets step by step. If the less distant nodes can synchronize quickly, those distant have their synchronization time extended because of the propagation and execution of the routing primitives. The observed values are therefore explained through the behavior of the DSR.

By operating in the ad hoc area for routing needs, the DSR adds an incompressible delay in the reception of packets by the nodes and in the routing of these packets to the other nodes of the network. Thus, step by step, the cumulative delay increases and negatively impacts the average value of the DG_2 ratio. Therefore, between the output of the packets from the SB buffer of the server and their entry into the CB buffer of the ad hoc zone far learner, a considerable time elapses consumed by the routing primitives including the DSR route discovery. In the case of the antenna area with direct transmission, route discovery does not exist. In contrast, each node in the ad hoc area is influenced by the degree of dispersal of other nodes in the network and with distance; some receive packets slower than others.

It is possible to note that the curve of the ad hoc zone admits an evolution not far from that of the immediate transmission. Its appearance suggests a non-ascending progression for the low attendance rates of learners. However, from the value 40, the synchronization delay undergoes a significant evolution. The explanation for this evolution is due to the density of exchanges and the success of the DSR route discovery in the ad hoc area when the number of nodes increases. This configuration gives many available routes than the less values of learner attendance. In addition as favorable factor, the DSR protocol does not carry out residual periodic activities. This factor is expected to be further accentuated in the case of protocols that add more overhead to the network. The comparative study of the contributions of the routing protocols is therefore necessary to find the most suitable for the context of the ad hoc network of the classroom before determining the maximum average offset allowing the DOUNG to find a solution for determining the S_{DOUNG} threshold. The cross analyze allows also to suggest a model of the content of the course which facilitates the synchronization of learners in the ad hoc area, without letting any disruption to becomes detrimental to the understanding of the course despite the shift indicating by the DG_2 ratio (Fig. 3).

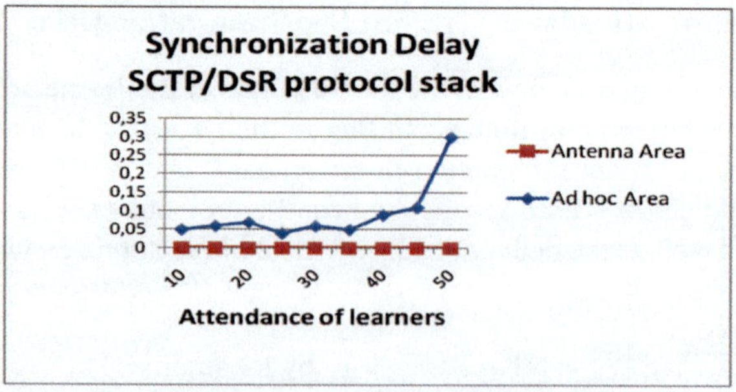

Fig. 3. Influence of the learner attendance over the synchronization delay

7.6 Influence of the Attendance Over the Learner Buffers Size

The three curves in the figure below show the evolution of the buffers size, that of the DOUNG server with Q_s, the antenna zone learner with Q_{r1} and the ad hoc zone learner with Q_{r2}. The graph is drawn to reflect the influence of the increasing number of learners on these three QoS parameters according to the learner area in the classroom wireless network. For each value of the learners attendance, there is no big difference in the evolution of the size of the buffers even if the antenna zone has the best values close to those of the DOUNG server and higher then those in the ad hoc area. These results therefore confirm the observations and the previous mathematical results as well as the deductions from the intuitive approach with:

$Q_{r2} < Q_{r1} < Q_s$ by inequality (7).

The case is particular of the learner's attendance close to the threshold of 10 with: $Qr2 = Qr1$.

The explanation of the average values reflected by the curves below comes as previously from the influence of the routing time and the propagation delay which make the ad hoc zone buffers to fill less quickly than those of the antenna area. Indeed, the ad hoc area is handicapped by the propagation, by the scattering and by the weight of the number of nodes which constitute delay parameters. They have the effect of slowing down the evolution of the size of the learner buffer in the ad hoc zone in relation to the source (DOUNG server) and in relation to the nodes in the intermediate zone of the antenna.

In fact, when considering the average value of Q_{r2}, the result calculated reflects the influence of the DSR protocol. As previously specified, the need to maintain the connectivity with the other nodes in this part of the classroom network requires packet routing. With this principle, the least distant nodes can receive information more quickly in comparison with distant nodes which see their delay being extended: the packets recorded in the buffer of intermediate nodes at the calculation instant CT (Current Time) may not have yet reached the distant nodes. Thus, these nodes bear the brunt of the effect of the propagation and execution of the primitives of the DSR protocol. As a reminder, even by limiting the overhead in the network, this protocol initiates route discovery and route maintenance mechanisms to guarantee communication between nodes. While

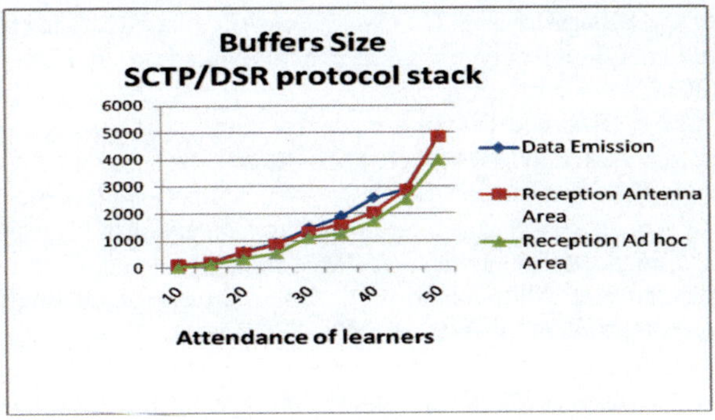

Fig. 4. Influence of the attendance over the buffers size

network stability reduces the impact of route maintenance, route discovery remains active despite the management of local temporary routing tables. The solution for DOUNG is to avoid taking a low TTL_{max} to allow learners to follow the course without breaking the sequence, compromising thus the understanding of the teacher's message. As a result, the end time of the course for the most distant learner may coincide with part of the time spent on questions and answers between learners and teacher (Fig. 4).

8 Conclusion

The DOUNG is subjected to the rigor of control and the achievement of a significant level of QoS in its activities. This challenge is more reinforced since the evolution of technology constantly poses the problem of its extension in order to cover ever larger geographic areas while minimizing deployment costs. One of this paper contribution consists of extending its network architecture by taking advantage of the generalization of laptops and mobile phones with multimedia capacities. This extension, initiated with a Wi-Fi antenna leads to an ad hoc network and still poses acutely the problem of the QoS guarantee.

In order to assess the reachable QoS gains from the juxtaposition of the two wireless networks; we study by intuitive approach, by mathematical model and by simulation, the influence of the increasing number of learners over their synchronization period according to their area in the classroom. Likewise, the ratio of the number of learners to the quantity of information available in the buffers was assessed to measure the impact of the chosen models on the synchronous access to the courses. The goal is to avoid a break in sequence prejudicial to the understanding of conveyed messages. The results obtained consolidate the expected observations and differentiate the level of contributions of the two wireless extensions.

It emerges from the present study that the DSR routing protocol used in the ad hoc area has a significant influence on the QoS parameters. As such, the study of the routing protocols contribution is a complementary perspective allowing to consolidate the assets of the present works and to lead the DOUNG to choose the most adapted to its context.

References

1. Issoufou Tiado, M., Saliah-Hassane, H.: Cloud-computing based architecture for the advent of a new generation of digital open universities in m-learning. In : ICEER13 Proceedings, pp 572–579, July 2013. www.labader.org
2. Tiado, M.I., Idrissa, A., Karimou, D.: Improved text reading system for digital open universities. Int. J. Adv. Res. Artif. Intell. (IJARAI) **4**(10), 29–34 (2015)
3. Noura, I.G., Tiado, M.I., Mahamadou, H.M.: Communicating vessels model for the Intelligent Monitoring System of the service guarantee in the New Generation of Digital Open Universities (NG-DOU). Adv. Wirel. Mob. Commun. (AWMC) (2020)
4. Dolgova, N., Larionova, J., Shirokolobova, A.: Engineering students English teaching in e-learning environment. MATEC Web Conf. **297**. ISPCIME-2019. https://doi.org/10.1051/matecconf/201929706007
5. Basak, S.K., Wotto, M., Bélanger, P.: E-learning, M-learning and D-learning: conceptual definition and comparative analysis. E-Learn. Digit. Med. **15**(4), 191–216 (2018). Universite du Quebec à Montréal (UQAM), Canada. https://doi.org/10.1177/2042753018785180

6. VeeraManickam, M.R.M., Mohanapriya, M.: Research study on centralized e-learning architecture model for educational institutes in INDIA: teaching & learning process. MATEC Web Conf. **57**, 02017 (2016) https://doi.org/10.1051/matecconf/20165702017
7. Li, S.: Comparative Analysis of Infrastructure and Ad-Hoc Wireless Networks. ITM Web Conf. **25**, 01009 (2019). ICICCI 2018. https://doi.org/10.1051/itmconf/2019250100
8. Tahir, H.M., Said, A., Seman, M.A.A., Shelena, S.N., Selan, M.S., Abdi, S.H.: Performance comparison of SCTP and UDP over mobile Ad Hoc networks. Int. J. Comput. Sci. Issues (IJCSI) **9**(4), 443–448 (2012). ISSN 1694-0814. www.IJCSI.org
9. Khabir, K.M., Siraj, S., Habib, A., Hossain, T., Ahad, A.R.: A study on DSR routing protocol in Ad Hoc network for daily activities of elderly living. In: 7th International Conference on Informatics, Electronics and Vision, Fukuoka, Japan (2018)

Detection of Vulnerabilities Related to Permissions Requests for Android Apps Using Machine Learning Techniques

Didier Bassolé[1(✉)], Yaya Traoré[1], Gouayon Koala[1], Franklin Tchakounté[2], and Oumarou Sié[1]

[1] Laboratoire de Mathématiques et d'Informatique, Université Joseph Ki-Zerbo, Ouagadougou, Burkina Faso

[2] Faculty of Science, Department of Mathematics and Computer Science, University of Ngaoundéré, Ngaoundéré, Cameroon

http://www.univ-ouaga.bf, http://www.univ-ndere.cm/

Abstract. Android application vulnerabilities are weaknesses in the applications or their design that allow an attacker to execute commands, access unauthorised data and carry out denial of service attacks. In this paper, we try to improve the detection of vulnerabilities, especially those based on permissions in Android applications. We propose a model that uses machine learning techniques for the automatic detection of vulnerabilities in the permissions system on Android devices. This model allows the user to identify the risk of vulnerabilities associated with an application via a permission or combination of permissions. We applied eight (8) machine learning techniques to analyse the use of permissions for malware detection. Our goal is to enable users to assess the vulnerabilities of applications that may access their personal and other sensitive data.

Keywords: Vulnerability · Permission-based · Machine learning · Android applications

1 Introduction

Smartphones and tablets are used nowadays like computers. In addition to the high storage capacity of these devices, their applications have access to sensitive resources (camera, microphone or GPS), confidential user data (mail or contact list), etc. Thus, data protection becomes imperative. Android is the most widespread mobile operating system in the world and therefore the preferred target of attacks [3]. Despite Android patching efforts, the number of vulnerabilities in Android applications continues to grow [1,2,9].

These vulnerabilities exist in applications from official and third party sites [4]. Also, Android's security model is mainly based on permission management. To improve data security on Android, the researchers have analysed vulnerabilities and made proposals as contributions [3,8,10,21–24] and [25]. Permission-based malware detection using machine learning techniques is one of the methods

A. Abraham et al. (Eds.): SoCPaR 2020, AISC 1383, pp. 810–822, 2021.
https://doi.org/10.1007/978-3-030-73689-7_77

of static analysis [5]. Several studies have proposed alternatives for mobile data protection, using machine learning techniques. In this study, we propose a model that uses machine learning techniques to automatically detect vulnerabilities in the permissions system of Android applications. This model aims to perform vulnerability assessments while improving the detection of those related to permissions in order to propose mechanisms for protecting data stored on Android. The rest of this document is structured as follows: the Sect. 2 deals with related work. The Sect. 3 presents our process of building a database of permissions related vulnerabilities and the classification of machine learning techniques used to predict the type of vulnerability. The Sect. 4 illustrates the description of our dataset, the experimental set-up and discusses the results of our analysis. We conclude this work in the Sect. 5.

2 Related Works

Studies on data security on Android have revealed that its permission-based security model does not help the majority of users to make security decisions in the face of threats and attacks [3,27,28]. These studies have shown that permission management remains one of the drawbacks of the Android security model. In [29], Felt et al. presented Stowaway, a permissions analysis tool that detects permissions abuse in Android applications. Au et al. in [30], proposed PScout, a tool for obtaining API permission to specify Android OS source code through static analysis. Several researchers have used machine learning to improve the accuracy and efficiency of vulnerability detection. In [11], Quiao et al. propose an approach to detect malware by exploiting permission models and API function calls acquired and used by Android applications. Cen et al. [14] proposed a model based on API calls and permissions that uses regularised logistic regression (RLR) to detect malware in Android. DroidMat [15] classified malware and benign apps according to intentions, permissions and API calls. The author used the K-Nearest Neighbors algorithm (KNN) and k-means clustering to extract static characteristics. Arora et al. [31] propose PermPair, a detection model that constructs and compares the graphs for malware and normal samples by extracting the permissions pairs from the manifest file of an application. In [16], an algorithm based on API calls, permissions and network access was proposed and an SVM classifier was developed for malware detection on devices. Munoz et al. [12] considered the metadata, obtained from PlayStore, as the characteristics that would allow a machine learning approach to be applied to identify the highest predictive characteristics and detect malware. Similarly, Peiravian et al. [13] propose a feature-based learning framework that focuses on requested permissions and API call behaviour and applies Support Vector Machine (SVM), Decision Tree and Bagging algorithms to the same end. Zhou et al. [6] proposed a permission-based behavioural approach to malware detection. Yerima et al. [17,23] used Random Forest for a set learning models to detect malware, which were based on permissions, API calls, embedded commands and intentions. Wang et al. [18], applied SVM, Decision Trees and Random Forest to analyse the use of vulnerable permissions for malware detection.

Varsha et al. [19] extracted static features from the application's manifest and executable files; their localisation strategy yielded the SVM, Rotation Forest and Random Forest over three datasets. Dapasa [20] used sensitive sub-graphics to construct five(5) features describing invocation patterns, and the Random Forest machine learning algorithm achieved the best detection performance.

3 Methodology

In this section, we present our methodology (Fig. 1) to examine the relationship between authorisation request and Android's apps vulnerabilities in order to identify malware using machine learning techniques. Our approach develops risk criteria based on the impact on data access via Android authorisations. In this approach, we derive characteristics from a set of benign and malicious data for malware detection. The security rules are then constructed based on the permission request templates in the dataset. Our goal is twofold. Firstly, the construction of a database with permission rules based on a set of risky permissions of Android applications. Secondly, the construction of a model that uses machine learning algorithms for the detection of vulnerabilities in Android applications.

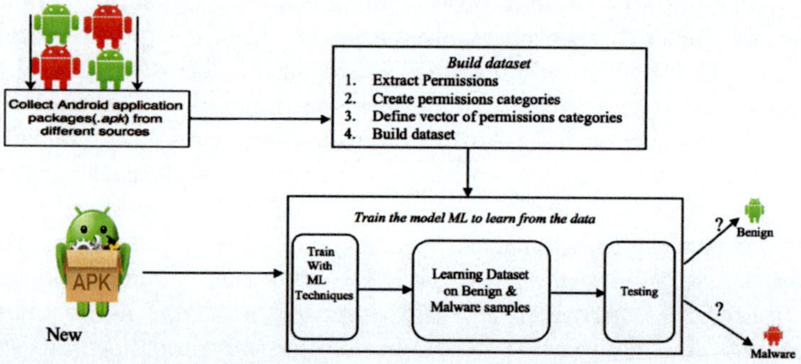

Fig. 1. Malware detection process with machine learning techniques

3.1 Permissions-Based Vulnerability Database Building Process

Permissions Type. Depending on the level of authorization protection in an application, Android defines four types of permissions that characterize the level of sensitivity implicit in the requested authorizations. The values associated with these types are Signature, SignatureOrSystem, Normal and Dangerous. In this study, we are interested at the impact of Normal and Dangerous authorisations on the protection of personal data. Normal permissions are low-risk permissions that give applications access to isolated functionality at the application level with minimal risk to other applications, the system or the user. Dangerous permissions are high-risk permissions that give an application access to sensitive user data or device control that could have a negative impact on the user. Most dangerous permissions are grouped together. The request for these permissions is always made through the group. Authorising a group of permissions means to authorize all the permissions in that group.

The Specific Permissions Attribute. For the security of personal data, each application, depending on its functionalities, must ask for permissions. To avoid burdening users, Android organizes permissions into groups related to the basic functionality of a device. Thus, permission requests are processed at the group level, which can correspond to several permission statements in the application. This affects the user experience because the request does not describe the specific permission within the group. As a result, the permission granted to the group of permissions does not allow the user to better understand the potential risk of his or her choices when using the applications. In addition, if an application has already received at least one authorization in a permission group, any other request for authorization in that group is granted without interaction with the user.

The Criticality of Permissions. Declaring that an application uses a permission allows its users to spot abuse and malicious use. Considering that each specific permission gives access to a specific resource and the associative relationships of the resources in the Table 1, we have grouped the permissions into categories. In this table, we have listed the semantically associative resources. This allowed us to identify permissions that may increase the vulnerability of applications with respect to the confidentiality, integrity and availability of sensitive and personal data. Based on the types of permissions and the sensitivity of the resources, we have developed a table of four(4) permissions categories (PCs) according to the risk induced by the permission in Android applications. Thus, the level of impact of permissions allowed us to determine two(2) categories for dangerous permissions (PC1 and PC2) and two(2) categories for normal permissions (PC3 and PC4). Semantically associative resources constitute exploitable vulnerabilities for attacks on the confidentiality, integrity and availability of data. An application with a permission that accesses a given resource can potentially access other resources on the device. For example, the relationship between location and contact resources allows an application that accesses contacts to be able to access the location. PC1 category permissions are permissions that have severe (or high) criticality because they are dangerous permissions and further expose user data. Their criticality is described as catastrophic because these permissions activate other permissions in the same group as them (Fig. 2) or access resources through resource relationships. The other dangerous permissions are grouped in the PC2 category which is qualified as critical. Normal permissions have a light criticality depending on the sensitivity of the resource they access. The category PC3 groups normal permissions with marginal criticality because their uses require the user's opinion. The fourth category (PC4) consists of normal permissions with negligible criticality. Most permissions for this criticality are granted without the intervention of the application user.

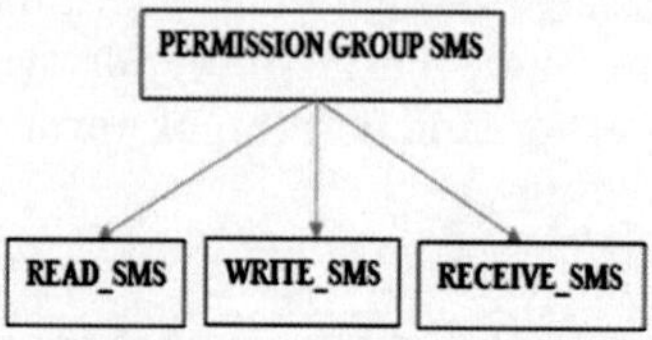

Fig. 2. Permission group

Table 1. Semantically associative resources

	Calendar	Camera	Contacts	Location	Phone	Sensors	SMS	Storage	Microphone
Calendar	X				X			X	X
Camera		X	X		X			X	X
Contacts			X	X	X		X	X	
Location			X	X				X	
Phone			X		X			X	X
Sensors						X		X	
SMS			X				X	X	
Storage		X	X		X		X	X	
Microphone				X				X	X

Table 2. Distribution of permissions for android applications

Permissions type	Permissions Category (PC)	Specific Permissions (SP)
Dangerous	Catastrophic (PC1)	READ_PHONE_STATE
		CALL_PHONE
		READ_CALL_LOG
		WRITE_CALL_LOG
		ADD_VOICEMAIL
		USE_SIP
		PROCESS_OUTGOING_CALLS
		SEND_SMS
		RECEIVE_SMS
		READ_SMS
		RECEIVE_WAP_PUSH
		RECEIVE_MMS
		ANSWER_PHONE_CALLS
		READ_PHONE_NUMBERS
		READ_CALENDAR
		WRITE_CALENDAR
		CAMERA

(*continued*)

Table 2. (*continued*)

Permissions type	Permissions Category (PC)	Specific Permissions (SP)
Dangerous	Critical (PC2)	READ_CONTACTS
		WRITE_CONTACTS
		GET_ACCOUNTS
		ACCESS_FINE_LOCATION
		ACCESS_COARSE_LOCATION
		BODY_SENSORS
		READ_EXTERNAL_STORAGE
		WRITE_EXTERNAL_STORAGE
		ACCEPT_HANDOVER
		RECORD_AUDIO
Normal	Marginal (PC3)	ACCESS_LOCATION_EXTRA_COMMANDS
		ACCESS_NETWORK_STATE
		ACCESS_WIFI_STATE
		BLUETOOTH
		BLUETOOTH_ADMIN
		CHANGE_NETWORK_STATE
		CHANGE_WIFI_STATE
		DISABLE_KEYGUARD
		EXPAND_STATUS_BAR
		GET_PACKAGE_SIZE
		INSTALL_SHORTCUT
		UNINSTALL_SHORTCUT
		INTERNET
		KILL_BACKGROUND_PROCESSES
		NFC
		NFC_TRANSACTION_EVENT
		READ_SYNC_SETTINGS
		RECEIVE_BOOT_COMPLETED
		REORDER_TASKS
		SET_ALARM
		SET_WALLPAPER
		SET_WALLPAPER_HINTS
		USE_BIOMETRIC
		VIBRATE
		WAKE_LOCK

(*continued*)

Table 2. (*continued*)

Permissions type	Permissions Category (PC)	Specific Permissions (SP)
Normal	Negligible (PC4)	ACCESS_NOTIFICATION_POLICY
		BROADCAST_STICKY
		CHANGE_WIFI_MULTICAST_STATE
		FOREGROUND_SERVICE
		MANAGE_OWN_CALLS
		MODIFY_AUDIO_SETTINGS
		READ_SYNC_STATS
		REQUEST_COMPANION_RUN_IN_BACKGROUND
		REQUEST_COMPANION_USE_DATA_IN_BACKGROUND
		REQUEST_DELETE_PACKAGES
		REQUEST_IGNORE_BATTERY_OPTIMIZATIONS
		REQUEST_INSTALL_PACKAGES
		SET_TIME_ZONE
		TRANSMIT_IR
		USE_FINGERPRINT
		WRITE_SYNC_SETTINGS

Build Dataset. Table 2 gives an overview of each specific permission with its functionality, associated with a permission category, grouped by permission type according to the sensitivity of the resources. This illustration shows that the behaviour of an application is characterised by risk and an index of permission sensitivity. This makes it possible to study the vulnerability of each application. The authorisation sensitivity index is a measure of the sensitivity of an application based on the risk of the authorisations it requests. Thus, we collect Android apps from different sources. Then, we extract the authorisations of each Android application. Each extracted permission is associated with one of four permission categories (PC1, PC2, PC3, PC4) and we calculate the number of permissions for PC1, PC2, PC3 and PC4.
For example, consider two Android apps:

- an Android app A we are extracted 10 specific permissions distinct of PC1, 0 for PC2, 0 for PC3 and 0 for PC4: : PCA = [10,0,0,0],
- and for Android app B, we are extracted 10 specific permissions distinct of PC1, 0 for PC2, 0 for PC3 and 14 for PC4: PCB = [10,0,0,14].

Finaly we assigned the type of vulnerability detected by malware or benign for every Android app used. The Algorithm 1 is used to calculate the value of each permission category.

Algorithm 1: Vecteur_PC

```
Entry :
        Android app : .apk
        PC1, PC2, PC3, PC4
Output: SP
N1 = N2 = N3 = N4 = 0
SP = extract(Android SP)
foreach SPi ∈ SP do
    if SPi ∈ PC1 then
        N1 ← N1 + 1
    end
    else
        if SPi ∈ PC2 then
            N2 ← N2 + 1
        end
        else
            if SPi ∈ PC3 then
                N3 ← N3 + 1
            end
            else
                N4 ← N4 + 1
            end
        end
    end
end
return N1,N2,N3,N4
```

3.2 Model Based on a Machine Learning Algorithm

The Algorithm 2 is used to learn from the data. The classification basis of machine learning (i.e. LR, LDA, MLP, KNN, DTREE-CART, RForest, SVM, NB) is used to predict the type of vulnerability. If it is a vulnerability, the predicted class is 'Malware', else 'Benign'. The predicted 'Malware' class means that there is a high impact of the vulnerability. The 'Benign' class means that there is a low impact of the vulnerability.

Algorithm 2: Machine Learning algorithm to detect malware or benign

Entry :
X = Android Dataset for model
A = newAndroid
Output: vulnerability predicted
$DTra$ = TrainingPartition(X)
$DTst$ = TestPartition(X)
foreach *instance* $Xi \in DTst$ **do**
  Li = labels of Xi //Li in (Malware,Benign)
end
//model construction
$Clas$ = ObtainClassifier($DTra$) //use classifier base (ie. : SVM, NB, KNN, CART, MLP,...)
// test and evaluation of model
foreach *instance*$Xi \in DTst$ **do**
  Pi = Clas(Xi) //Obtain classifier's prediction
  Evaluate(Pi, Li) //obtain precision, recall, f1-score for each class
end
Evaluate(P) //obtain Accuracy, macro precision, macro recall, macro f1-score
//for a new Android app, use the model clas to detect malware et benign apps
vulnerability predicted=clas(A) // use clas for new Android app

4 Experiments

4.1 Dataset and Experimental Setup

Dataset. To validate our approach we collected data based on the permissions granted by users of Android phones from Contagio repository[1], the Koodous repository[2] and VirusShare[3]. Thus we were able to build dataset with 18095 Android apps classified Malware or Benign.

Evaluation Metric. The metrics precision (P), recall (R) and F-measure (F1), Accuracy are proposed to evaluate our method. The precision, recall and F1 for example are defined as:

$$P(precision) = \frac{TP}{TP + FP} \qquad R(recall) = \frac{TP}{TP + FN}$$

$$F1 = \frac{2 * P * R}{P + R} \qquad Accuracy = \frac{TP + TN}{TP + FP + FN + TN}$$

[1] http://contagiomobile.deependresearch.org/index.html.
[2] https://koodous.com/.
[3] https://virusshare.com.

Where:

- TP (True Positive): when the actual class and the predicted class are all yes.
- TN (True Negative): when the actual class and the predicted class are all no.
- FP (False Positive): when the actual class is no and the predicted class is yes.
- FN (False Negative): when the actual class is yes and the predicted class is no.

4.2 Results and Analysis

To analyze the performance of our model, we have implemented the proposed approach using eight (08) machine learning algorithms namely: Logistic Regression (LR), Linear Discriminant Analysis (LDA), K-Nearest Neighbors Classifier (KNN), Decision Tree Classifier (DTREE-CART), Naive Bayes (NB), MLP Classifier (MLP), Random Forest Classifier (RFOREST), Support Vector Machine (SVM) that were selected based on results of several preliminary experiments realized.

Table 3. Comparison of de machine learning algorithms

	Malware			Benign apps		
	Precision	Recall	F1-score	Precision	Recall	F1-score
LR	0.93	0.99	0.96	0.54	0.15	0.24
LDA	0.93	0.99	0.96	0.58	0.19	0.28
KNN	0.98	1.00	0.99	0.96	0.75	0.84
TREE	0.98	0.99	0.99	0.93	0.80	0.86
NB	0.94	0.96	0.95	0.31	0.24	0.27
MLP	0.98	1.00	0.99	0.96	0.77	0.85
SVM	0.98	1.00	0.99	0.96	0.77	0.85
RFOREST	0.98	1.00	0.99	0.94	0.81	0.87

Table 3 presents a comparison of performance of proposed approach with the metrics of precision, recall and F1-score for Android vulnerability (Malware or Benign) detection. The strategy gives a good performance for all machine learning algorithms used. The best performance is given by the Random Forest algorithm with is 0.99 F-score for Malware detection and the 0.87 F-score for Benign detection.

Table 4 presents the results of implementation for all machine learning algorithms with Accuracy and Macro F1-score. The best performance algorithm achieved 0.98 Accuracy and 0.93 Macro f1-score.

Figure 3 shows that the Random Forest algorithm is better than the others. These good performances are linked to our strategy to categorize the specific permission by permission categories. Thus, we can conclude that our method proposed improves the performance of the vulnerability detection.

Table 4. Implementation for all machine learning algorithms

	Macro precision	Macro recall	Macro F1-score	Accuracy
LR	0.74	0.57	0.60	0.93
LDA	0.76	0.59	0.62	0.93
KNN	0.97	0.87	0.91	0.98
TREE	0.96	0.90	0.93	0.98
NB	0.63	0.60	0.61	0.91
MLP	0.96	0.89	0.92	0.98
SVM	0.97	0.88	0.92	0.98
RFOREST	0.97	0.90	0.93	0.98

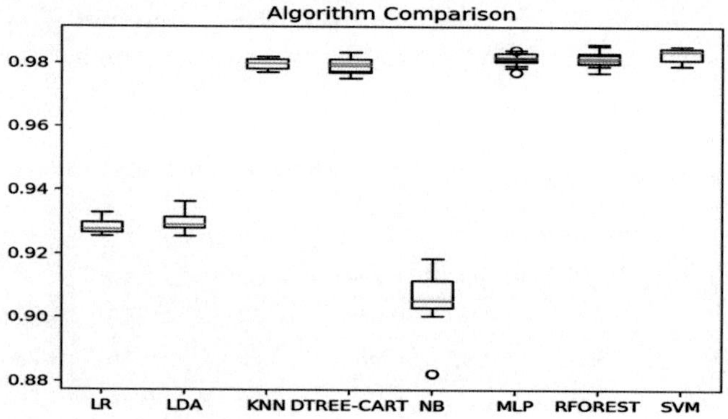

Fig. 3. Result of algorithm comparison

5 Conclusion

We have propose in this paper a model that uses Machine Learning techniques to automatically detect vulnerabilities of Android apps permissions system. This model allows the user to identify the risks of vulnerabilities associated with a permission or combination of permissions. Our work focused on analysing the potential risks associated with the use of these applications. It aims to conduct vulnerability assessments, facilitate an informed assessment of the data security and privacy risks that Android applications incorporate, and raise awareness among the research and practitioner communities.

References

1. Shewale, H., Patil, S., Deshmukh, V., Singh, P.: Analysis of android vulnerabilities and modern exploitation techniques. ICTACT J. Commun. Technol. **5**(1), 863–867 (2014)

2. Jimenez, M., Papadakis, M., Bissyandé, T.F., Klein, J.: Profiling Android Vulnerabilities. In: IEEE International Conference on Software Quality, Reliability and Security (QRS), Vienna, 2016, pp. 222–229 (2016)
3. Koala, G., Bassolé, D., Zerbo/Sabané, A., Bissyandé, T.F., Sié, O.: (2020) Analysis of the impact of permissions on the vulnerability of mobile applications. In: Zitouni, R., Agueh, M., Houngue, P., Soude, H. (eds.) e-Infrastructure and e-Services for Developing Countries. AFRICOMM 2019. Lecture Notes of the Institute for Computer Sciences, Social Informatics and Telecommunications Engineering, vol 311. Springer, Cham. https://doi.org/10.1007/978-3-030-41593-8_1
4. Vulnerabilities and threats in mobile applications, 19 June 2019. https://www.ptsecurity.com/ww-en/analytics/mobile-application-security-threats-and-vulnerabilities-2019/
5. Mahindru, A., Singh, P.: Dynamic permissions based android malware detection using machine learning techniques. In: Proceedings of the 10th Innovations in Software Engineering Conference (ISEC'17), 05–07 Feb 2017, Jaipur, India, pp. 202–210. https://doi.org/10.1145/3021460.3021485
6. Zhou, Y., Zhang, X., Jiang, X., Freeh, V.W.: Taming information - stealing smartphone applications (on android). In: Proceedings of the 4th International Conference on Trust and Trustworthy Computing, TRUST'11 (2011)
7. Zhao, M., Ge, F., Zhang, T., Yuan, Z.: AntiMalDroid: an efficient SVM-based malware detection framework for android. In: Liu, C., Chang, J., Yang, A. (eds.) Information Computing and Applications (ICICA). Communications in Computer and Information Science, vol. 243. Springer, Berlin, Heidelberg (2011)
8. Tam, K., Feizollah, A., Anuar, N.B., Salleh, R., Cavallaro, L.: The evolution of android malware and android analysis techniques. ACM Comput. Surv. **49**(4), 1–41 (2017)
9. Li, L., et al.: Static analysis of android apps: a systematic literature review. Inf. Softw. Technol. **88**, 67–95 (2017)
10. Nix, R., Zhang, J.: Classification of android apps and malware using deep neural networks. In: 2017 International Joint Conference on Neural Networks (IJCNN), Anchorage, AK, pp. 1871–1878 (2017)
11. Qiao, M., Sung, A.H., Liu, Q.: Merging permission and API features for android malware detection. In: 5th IIAI International Congress on Advanced Applied Informatics (IIAI-AAI), Kumamoto 2016, pp. 566–571 (2016)
12. Muñoz, A., Martín, I., Guzmán, A., Hernández, J.A.: Android malware detection from Google Play meta-data: selection of important features. In: 2015 IEEE Conference on Communications and Network Security (CNS), Florence, pp. 701–702 (2015)
13. Peiravian, N., Zhu, X.: Machine learning for android malware detection using permission and API calls. In: 2013 IEEE 25th International Conference on Tools with Artificial Intelligence, Herndon, VA, pp. 300–305 (2013)
14. Cen, L., Gates, C.S., Si, L., Li, N.: A probabilistic discriminative model for android malware detection with decompiled source code. In: IEEE Transactions on Dependable and Secure Computing, vol. 12, no. 4, pp. 400–412, 1 July–Aug 2015
15. Wu, D., Mao, C., Wei, T., Lee, H., Wu, K.: DroidMat: android malware detection through manifest and API calls tracing. In: 2012 Seventh Asia Joint Conference on Information Security, Tokyo, pp. 62–69 (2012)
16. Arp, D., Spreitzenbarth, M., Hubner, M., Gascon, H., Rieck, K.: DREBIN: effective and explainable detection of android malware in your pocket. In: Proceedings of the 2014 Network and Distributed System Security Symposium, San Diego, CA, USA, 23–26 Feb 2014

17. Yerima, S.Y., Sezer, S., Muttik, I.: High accuracy android malware detection using ensemble learning. In: IET Information Security, vol. 9, no. 6, pp. 313–320 (2015)
18. Wang, X., Wang, W., He, Y., Liu, J., Han, Z., Zhang, X.: Characterizing Android apps' behavior for effective detection of malapps at large scale. Future Gener. Comput. Syst. **75**, 30–45 (2017)
19. Varsha, M.V., Vinod, P., Dhanya, K.A.: Identification of malicious android app using manifest and opcode features. J. Comput. Virol. Hack. Tech. **13**, 125–138 (2017). https://doi.org/10.1007/s11416-016-0277-z
20. Fan, M., Liu, J., Wang, W., Li, H., Tian, Z., Liu, T.: DAPASA: Detecting Android Piggybacked Apps Through Sensitive Subgraph Analysis. IEEE Trans. Inf. Forensics Secur. **12**(8), 1772–1785 (2017)
21. Liu, K., Xu, S., Xu, G., Zhang, M., Sun, D., Liu, H.: A review of android malware detection approaches based on machine learning. In: IEEE Acess, vol. 8, pp. 124579–124607 (2020). https://doi.org/10.1109/ACCESS.2020.3006143
22. Alqahtani, E.J., Zagrouba, R., Almuhaideb, A.: A survey on android malware detection techniques using machine learning algorithms. In: 2019 Sixth International Conference on Software Defined Systems(SDS), Rome, Italy, 2019, pp. 110–117. https://doi.org/10.1109/SDS.2019.8768729
23. Yerima, S.Y., Sezer, S., Muttik, I.: Android malware detection using parallel machine learning classifiers. In: 2014 Eighth International Conference on Next Generation Mobile Apps, Services and Technologies, Oxford, pp. 37–42 (2014)
24. Wei, L., Luo, W., Weng, J., Zhong, Y., Zhang, X., Yan, Z.: Machine learning-based malicious application detection of android. IEEE Access **5**, 25591–25601 (2017)
25. Ma, Z., Ge, H., Liu, Y., Zhao, M., Ma, J.: A combination method for android malware detection based on control flow graphs and machine learning algorithms. IEEE Access **7**, 21235–21245 (2019)
26. Li, L., et al.: Understanding android app piggybacking: a systematic study of malicious code grafting (2016)
27. Bassolé, D., Koala, G., Traoré, Y., Sié, O.: Vulnerability analysis in mobile banking and payment applications on android in African Countries. In: Thorn, J., Gueye, A., Hejnowicz, A. (eds.) Innovations and Interdisciplinary Solutions for Underserved Areas. Intersol 2020. Lecture Notes of the Institute for Computer Sciences, Social Informatics and Telecommunications Engineering, vol. 321. Springer, Cham (2020). https://doi.org/10.1007/978-3-030-51051-0_12
28. Frank, M., Dong, B., Felt, A.P., Song, D.: Mining permission request patterns from android and facebook applications. In: IEEE 12th International Conference on Data Mining (ICDM), pp. 870–875 (2012)
29. Felt, A.P., Chin, E., Hanna, S., Song, D., Wagner, D.: Android permissions demystified. In: Proceedings of the 18th ACM Conference on Computer and Communications Security, pp. 627–638 (2011)
30. Au, K.W.Y., Zhou, Y.F., Huang, Z., Lie, D.: Pscout: analyzing the android permission specification. In: Proceedings of the 2012 ACM Conference on Computer and Communications Security, pp. 217–228 (2012)
31. Arora, A., Peddoju, S.K., Conti, M.: PermPair: android malware detection using permission pairs. In: IEEE Trans. Inf. Forensics Secur. **15**, 1968–1982 (2020). https://doi.org/10.1109/TIFS.2019.2950134

A New Approach to Detecting Bot Attack Activity Scenario

Dandy Pramana Hostiadi[1,2], Tohari Ahmad[1(✉)], and Waskitho Wibisono[1]

[1] Department of Informatics, Institut Teknologi Sepuluh Nopember, Surabaya, Indonesia
dandy.17051@mhs.its.ac.id, {tohari,waswib}@if.its.ac.id

[2] Department of Informatics, Institut Teknologi dan Bisnis STIKOM Bali, Bali, Indonesia

Abstract. Botnets are a dangerous threat to computer networks. A botnet consists of a bot-master and bot-client connected and communicated through command and control (C&C) servers. When the bot attacks and infects the target computer, it performs several activities. Nevertheless, the introduced model may not detect the connection between one activity and others as a whole botnet attack scenario. The connection between activities is required to get the attack step carried out by each bot. This paper proposes a new approach to detect linkages between bot activities by analyzing the network traffic flows and obtaining a bot attack scenario. The analysis is carried out by finding the frequency of each activity that is sequentially connected. The results show that the proposed model successfully detects interrelated bot activity scenarios based on its pattern.

Keywords: Bot activity · Bot network · Intrusion detection system · Network security

1 Introduction

Nowadays, data and information security are an issue that needs concern. Botnet is a threat that needs to be addressed seriously [1]. Botnet uses malicious software [2] to carrying out malicious activities, such as phishing, theft of credential information and data, spamming, Distributed Denial of Service (DDoS) attacks to illegal communication access [3]. A botnet has a communication structure that consists of bot-master and bot-client. Bot-client is often referred to as zombie computers [4], carrying out some malicious activities on the bot-master's instructions. The bot-client receives instructions from the bot-master via Control and Command (C&C) communications [1, 5].

The malicious activity carried out by botnets is different from the types of attacks that commonly occur. The botnet has a life cycle scheme [6] and is currently developing into a decentralized [7]. In this decentralized model, the botnet network topology is dynamic. The bot network topology is updated when the C&C service is blocked, and the botnet network will form a new C&C service. Each bot that will infect the target computer performs several stages of activity. Generally, the attack begins with spamming the target, finding vulnerabilities from the target computer, and infecting the target with

A. Abraham et al. (Eds.): SoCPaR 2020, AISC 1383, pp. 823–835, 2021.
https://doi.org/10.1007/978-3-030-73689-7_78

malicious software to carry out attack activities such as DDoS. In an intrusion detection model, they are detected as different attack activities.

Several studies to detect botnet activity have been conducted. In [5] botnet activity detection was carried out based on the peer-to-peer (P2P) botnet traffic activity analysis. In other research [8, 9] conducted the detection of botnet activity based on DNS query access analysis on botnet C&C communications and analyzed using a traffic segmentation model. Research in [10] detects botnet activity by analyzing bot behavior, representing the bot as a node with incoming and outgoing communications as a flow activity. In [5, 11, 12], the detection of botnet activity using classification methods is introduced. Some related botnet activity detection models show high detection accuracy for detecting the existence of botnets until detecting botnet activity on computer networks. However, the previous research model has not shown the connection between each activity, which is a continuous collection of bot activity. It is only can detect bot activity as a single or spreading activity. The interrelated activity is required to get a botnet attack scenario.

This paper proposes a new approach to detect the connection between botnet activities based on traffic flow analysis. This proposed model uses a bot detection model to label botnet activities, classify them based on the time, calculate the frequency of each activity, and analyze bot scenarios. The proposed scheme has the renewability to model the relationship between bot activities, thus collecting scenarios for attacking bot activities as a stage of attack activity. The purpose is to get an activity scenario performed by bots when performing infection to attack activities, making it easier to take preventive steps against botnet attack activities on computer networks.

This paper consists of several sections, namely Sect. 2 describes the related research, Sect. 3 explains the proposed model, Sect. 4 the analysis of the experimental results in this study, and finally, the conclusion of this research is drawn in Sect. 5.

2 Related Research

Bots can have several different attack scenarios [13], according to the attacks' purpose. Several models of botnet activity detection have been introduced in the literature. In previous studies was introduced the botnet activity detection using classification methods such as k-Nearest Neighbor (k-NN) [14], Decision Tree [5], Random Forest and Naïve Bayes [11]. The detection model analyzes the Domain Name Service (DNS) query data. It is to detect the existence and activity of bot in network traffic. Similar research to the classification model was conducted in [5]; however, the proposed model has the advantage of detecting the Peer-to-Peer (P2P) activity of botnets. It also combines two bot activity detection techniques, namely traffic behavior and flow similarity, which has high accuracy results. The classification model can show reasonably high accuracy in detecting bot activity, but it does not yet declare a set of interrelated bot activity as an attack scenario.

A detection model with a clustering approach is proposed in [10] besides the classification model. In this scheme, bot activity is analyzed based on traffic behavior and represented as a vertex connected through edges. Features extraction of out-flows, in-flows, in-degree flows, out-degree flows, node betweenness centrality, clustering coefficient, and eigenvector centrality are obtained from connected vertices. The Self-Organizing Map (SOM) method is used to form clusters, which results in group activities with the same feature group. The detection accuracy shows that the model can work well, but the activities detected in the cluster cannot be described as a scenario with a continuous sequence of activities.

Botnet activity detection with a time-based data segmentation approach is proposed in [8, 9]. Time segmentation is carried out to obtain a set of activities called bot group activities contained in each segment, which are measured for similarity in different segments using the Cosine, Jaccard, and Kulczynski similarity methods. The result of this similarity detection is an intense, periodic, and consistent bot group activity. The study results indicate that their proposed model can detect known bot and new types of bots. However, in that study, each similar bot group activity does not show mutually sustainable activities. Besides, the segmentation model is performed statically, affecting the loss of information on a series of activities in the segmentation transition.

Several botnet activity detection models that have been introduced have good detection performance and accuracy, capable of detecting every bot activity in detail. In contrast to previous studies, this study analyzes the relationship between bot activity in the form of a bot activity scenario in the network. The relationship between activities is used as the basis for showing the bot activity stage carried out by the bot's attack or infection process to the target computer. Besides, it used to prevent and anticipate a series of other bot attack activities.

3 Proposed Model

This section describes the proposed model, where the proposed model is a new approach to detect related bot activity. The proposed model consists of two main parts, namely bot detection, to obtain a set of bot activity based on activity time and analysis of bot activity scenarios. The proposed model is shown in Fig. 1.

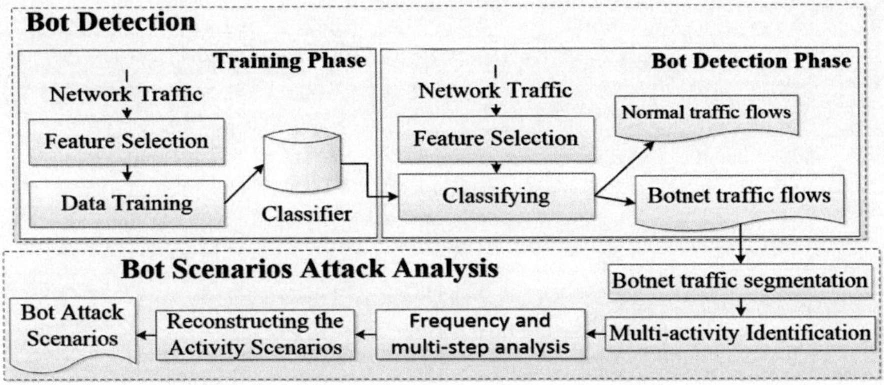

Fig. 1. Proposed model for Bot scenario attack detection.

3.1 CTU-13 Dataset

The CTU public dataset is a dataset from the Stratosphere IPS Laboratory project of Czech Technical University. The dataset contains network traffic with malware activities at Czech Technical University in 2011. The purpose of this dataset is to obtain an actual record of botnet malware. Several studies have used this dataset to model the detection of botnet activities [5].

The CTU-13 dataset consists of thirteen datasets that have different types of bot activity. Each dataset contains a different number of bots, duration of activities and number of traffic records. The description of the CTU-13 dataset is shown in Table 1.

Table 1. CTU-13 dataset description.

ID	Duration (hours)	Bots count	Bots name	Total netflows	Botnet flows	Normal flows	Background flows
1	6.15	1	Neris	2.824.637	39.933 (1.41%)	30.387 (1.07%)	2.753.290 (97.47%)
2	4.21	1	Neris	1.808.123	18.839 (1.04%)	9.120 (0.5%)	1.778.061 (98.33%)
3	66.85	1	Rbot	4.710.639	26.759 (0.56%)	116.887 (2.48%)	4.566.929 (96.94%)
4	4.21	1	Rbot	1.121.077	1.719 (0.15%)	25.268 (2.25%)	1.094.040 (97.58%)
5	11.63	1	Virut	129.833	695 (0.53%)	4.679 (3.6%)	124.252 (95.7%)
6	2.18	1	Menti	558.920	4.431 (0.79%)	7.494 (1.34%)	546.795 (97.83%)
7	0.38	1	Sogou	114.078	37 (0.03%)	1.677 (1.47%)	112.337 (98.47%)
8	19.5	1	Murlo	2.954.231	5.052 (0.17%)	72.822 (2.46%)	2.875.282 (97.32%)
9	5.18	10	Neris	2.753.885	179.880 (6.5%)	43.340 (1.57%)	2.525.565 (91.7%)
10	4.75	10	Rbot	1.309.792	106.315 (8.11%)	15.847 (1.2%)	1.187.592 (90.67%)
11	0.26	3	Rbot	107.252	8.161 (7.6%)	2.718 (2.53%)	96.369 (89.85%)
12	1.21	3	NSIS.ay	325.472	2.143 (0.65%)	7.628 (2.34%)	315.675 (96.99%)
13	16.36	1	Virut	1.925.150	38.791 (2.01%)	31.939 (1.65%)	1.853.217 (96.26%)

3.2 Bot Detection

Bot activity detection is the initial process for detecting scenarios of bot attack activity. In this paper, the classification model used to detect bot activity that has been implemented in [5, 11] and improved in the feature selection process. The bot activity detection model employs several classification methods, such as the Decision Tree, k-Nearest Neighbor (k-NN), Naïve Bayes, Logistic Regression, and Support Vector Machine (SVM). The proposed model selects the basic features of network traffic based on the study [15] and added two features to network traffic, namely source IP address, destination IP address, start_time, duration_time, protocol, destination port address, source port address, total_packets, and total_bytes. An example of network traffic is shown in Fig. 2.

```
2011-08-17 15:55:53,0.000596,udp,147.32.84.192,3337,  <->,147.32.80.9,53,CON,0.0,0.0,2,512
2011-08-17 15:55:53,0.00035800000000000003,udp,147.32.84.192,3338,  <->,147.32.80.9,53,CON,0.0,0.0,2,384
2011-08-17 15:55:53,3351.880859,udp,77.110.41.177,62334,  <->,147.32.86.165,12114,CON,0.0,0.0,5,349
2011-08-17 15:55:53,25.580938,udp,66.184.194.249,33853,  <->,147.32.86.165,12114,CON,0.0,0.0,4,272
2011-08-17 15:55:53,3126.79126,udp,147.32.84.59,60276,  <->,208.88.186.4,34015,CON,0.0,0.0,20,2693
2011-08-17 15:55:53,2.917608,tcp,147.32.84.192,3324,   ->,189.212.138.56,6667,S_,0.0,,2,124
2011-08-17 15:55:53,8.580406,udp,164.111.132.38,22057,  <->,147.32.84.229,13363,CON,0.0,0.0,4,456
2011-08-17 15:55:53,2.168745,udp,123.21.153.28,54829,  <->,147.32.86.165,12114,CON,0.0,0.0,4,280
2011-08-17 15:55:53,0.000328,udp,147.32.84.189,58001,  <->,147.32.80.9,53,CON,0.0,0.0,2,304
2011-08-17 15:55:53,0.000319,udp,147.32.84.189,58001,  <->,147.32.80.9,53,CON,0.0,0.0,2,216
2011-08-17 15:55:53,2562.404053,udp,147.32.84.229,13363,   ->,95.153.104.73,31272,INT,0.0,,12,2208
2011-08-17 15:55:53,0.33223800000000003,udp,147.32.84.192,3340,  <->,147.32.80.9,53,CON,0.0,0.0,2,200
2011-08-17 15:55:53,0.0,tcp,147.32.84.207,1143,   ->,64.18.5.10,25,S_,0.0,,1,62
2011-08-17 15:55:53,2.91129,tcp,147.32.84.193,2489,   ->,115.178.24.226,6667,S_,0.0,,2,124
2011-08-17 15:55:53,0.04996000000000004,udp,147.32.84.229,13363,  <->,94.26.133.73,65466,CON,0.0,0.0,2,129
2011-08-17 15:55:53,3130.97168,udp,190.51.199.98,52415,  <->,147.32.86.165,12114,CON,0.0,0.0,5,353
2011-08-17 15:55:53,0.000251,udp,147.32.84.138,46237,  <->,147.32.80.9,53,CON,0.0,0.0,2,214
2011-08-17 15:55:53,0.00031299999999999996,udp,147.32.84.138,37989,  <->,147.32.80.9,53,CON,0.0,0.0,2,214
```

Fig. 2. Network traffic example in CTU-13 dataset 1.

If the network traffic defined as *NetT*, which has a set of activity records denoted as *r* then expressed as:

$$NetT = f(r_1, r_2, \ldots, r_k);\ k = 1, 2, 3, \ldots, n, \tag{1}$$

and each activity record has a network header information called a feature in the detection model, denoted as *F*, then expressed in the detection model as:

$$r_k = f\left(F_1, F_2, \ldots, F_j\right);\ k = 1, 2, 3, \ldots, n;\ j = 1, 2, 3, \ldots, m \tag{2}$$

The classification results of several methods show different accuracy values, and it is used in the scenario bot detection process. To optimize the detection of bot attack activity scenarios, a classifier is explored.

3.3 Bot Scenarios Attack Detection

The scenario of bot activity detection is a process to analyze bot activity attack scenarios. In some bot activity detection models, the bot activity can be detected repeatedly and randomly detected. Thus, in this paper, the detection of bot activity scenarios is carried out to obtain bot attack stages to analyze a set of bot activity detected in the bot detection model.

The initial process in detecting scenario bot activity is the segmentation of bot traffic data. This process is used to analyze the chain of bot activity and detect periodic activity by performing a sliding window (Fig. 3).

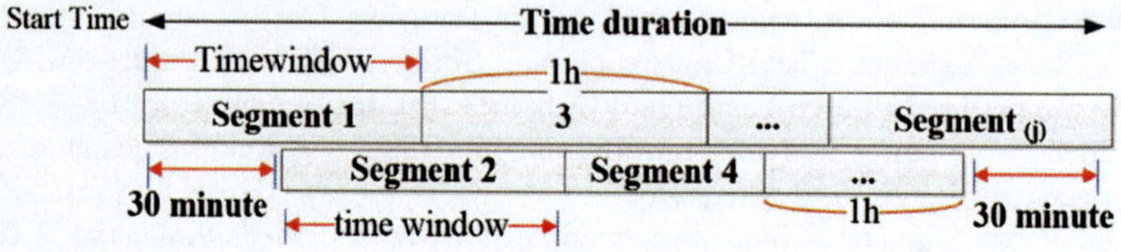

Fig. 3. Bot flows traffic segmentation.

Bot traffic activity has a total duration of time denoted as (*T*), which is segmented for 1 h with a sliding process at the initial time of 30 min. The segmentation process with a time window duration (τ) produces a smaller scale bot traffic segment, denoted as (*Net-B*$_j$), where j is the index segment with the value $j = 1, 2, \ldots$, n. Thus, $\tau j =$ activity duration in *Net-B*$_j$ segment. The bot activity in each segment (*Net-B*$_j$) consists of different bot group activities (a_m). Then it is defined as:

$$Net{-}B_j = f(a_1, a_2, \ldots, a_m), \tag{3}$$

with the total activity duration, it is defined as:

$$T = \left(\sum\nolimits_{j=1}^{j=n} \tau_j\right) - 1 \tag{4}$$

Bot activities in different segments relate to activities in other segments as a bot attack scenario. In the process of identifying multi bot activity, the model traces periodic activities in a segment. The model classifies several activities in a multi-activity group, whose process is shown in Fig. 4.

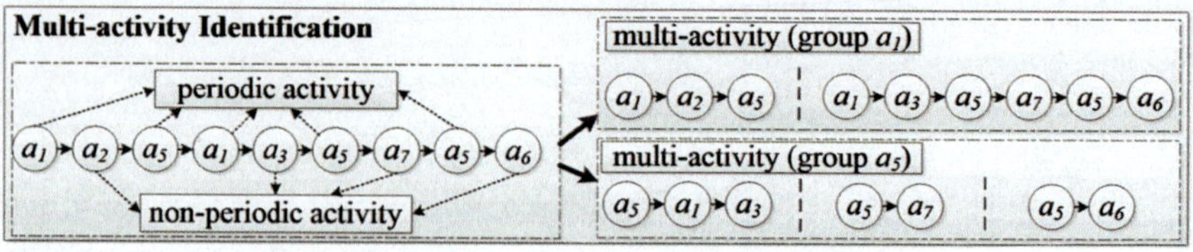

Fig. 4. Bot multi-activity identification.

The multi-activity identification process generates a set of different scenarios based on the type of bot activity. The model then calculates the frequency of each repetitive activity and regrouping them based on their destination IP address information. These two steps are to obtain patterns of bot attack scenarios against different attack targets, resulting in possible interrelated activity patterns.

Activity scenarios in the form of multi-activity in one segment can occur in other segments. The similarity of activities in different segments can result in overlapping activity scenarios and become independent scenarios. To avoid this case, the model reconstructs each multi-activity in a different segment. The process of activity scenario reconstruction obtained from multi-activity uses intersection analysis. Here, the intersection sought is from multi-activity in different segments with the number of activities intersect with at least two types of bot activity. After getting the slices, the two multi-activities in the two segments are merged into one new multi-activity collection, which is called the botnet activity scenario. If *Net-*B_j is a set of different bot activities and a multi-activity, then the intersection tracing between the two segments *Net-*B_x and *Net-*B_y is written as:

$$Net-B_x,\ Net-B_y;\ x,y \in j \begin{cases} Net-B_x \cap Net-B_y,\ then\ Net-B_x \cup Net-B_y\ as\ bot\ attack\ scenario \\ Net-B_x;\ Net-B_y\ as\ bot\ spread\ activity \end{cases}, \quad (5)$$

where $Net - B_x = f(a_1, a_2, a_b, \ldots, a_z)$ and $Net - B_y = f(a_1, a_2, a_\beta, \ldots, a_\gamma)$. The process of reconstructing bot activity scenarios produces several patterns of bot attack activity scenarios. This scenario pattern consists of sequential and gradual botches of bot activity based on the bot attacking activity.

4 Implementation

In this section, the discussion covers experiments that have been carried out, including discussion of bot detection, detection of bot attack activity scenarios, and the comparisons of the proposed model. In this paper, we test the CTU dataset using the CTU-1 database scenario. This dataset has 2,824,637 traffic network flows and consists of 39,933 bot traffic flow with activity duration while recording for 6.15 h.

4.1 Bot Detection

In the bot detection process, the proposed model develops a classification machine learning model. In the training phase, the model uses 70% traffic data in which there are normal activities and bot activities along with activity labels as training data. Meanwhile, for testing data, the model uses the CTU-13 dataset 1 traffic data whose activity labels have been removed. To perform classification, the network header is specified as Feature (F), namely F_1 = start_time, F_2 = duration_time, F_3 = source IP address, F_4 = destination IP address, F_5 = protocol, F_6 = source port address, F_7 = destination port address, F_8 = total_packets and F_9 = total_bytes. The results of the classification process show different bots detection accuracy shown in Table 2.

Table 2 shows that the method producing the highest detection accuracy is k-NN, and it is used in the bot scenario attack process. The k-NN generates 14,500 bot activity records with 23 different types of activities, shown in Table 3.

Some bot activities detected in the classification process have the nature of repetitive activities with dynamic activity occurrences. Not only repetitive, but bot activity also spreads. However, this spreading activity cannot show the stages as a set of a bot attack scenario. This underlies that the analysis process of the bot detection results in the classification model is required to analyze to get bot attack scenario activities.

Table 2. Bot detection result using classification methods.

Algorithm	Accuracy (%)
Decision Tree	89.43
Support Vector Machine (SVM)	93.69
k-Nearest Neighbor (k-NN)	94.20
Logistic Regresion	89.61
Naïve Bayes	90.27

Table 3. Bot activity labels.

No.	Label of bot activity
1	flow = Botnet-UDP-DNS
2	flow = Botnet-TCP-HTTP-Google-Net-Established
3	flow = Botnet-TCP-Established
4	flow = Botnet-TCP-Established-HTTP-Binary-Download
5	flow = Botnet-TCP-Established-HTTP-Binary-Download-Custom-Port
6	flow = Botnet-TCP-CC-Plain-HTTP-Encrypted-Data
7	flow = Botnet-TCP-Established-HTTP-Ad
8	flow = Botnet-TCP-WEB-Established
9	flow = Botnet-TCP-Attempt
10	flow = Botnet-TCP-CC-Custom-Encryption
11	flow = Botnet-TCP-CC-HTTP-Not-Encrypted
12	flow = Botnet-TCP-Attempt-SPAM
13	flow = Botnet-TCP-Not-Encrypted-SMTP-Private-Proxy
14	flow = Botnet-UDP-Established
15	flow = Botnet-TCP-Established-HTTP-Adobe
16	flow = Botnet-TCP-Established-SPAM
17	flow = Botnet-TCP-Established-HTTP-Ad-yieldmanager
18	flow = Botnet-TCP-HTTP-Not-Encrypted-Down
19	flow = Botnet-TCP-WEB-Established-SSL
20	flow = Botnet-TCP-Established-Custom-Encryption
21	flow = Botnet-TCP-Established-SSL-To-Microsoft
22	flow = Botnet-ICMP
23	flow = Botnet-UDP-Attempt-DNS

4.2 Bot Scenarios Attack

A bot attack activity scenario is a set of interrelated activities. This attack scenario shows the stages of infection attack by bots to the target computer. During attacking and infecting a new computer, the bot will repeatedly carry out several activity types to find vulnerabilities or weaknesses from the target computer until it can be fully infected. The segmentation is performed to get a collection chain of bot activities to get the repetitive bot scenario activity. From the bot detection process, the model gets bot activity with T for 3 h. The model then segmented with a time window duration (τ) for 1 h, with an initial sliding time for 30 min, and produced five segments with different bot traffic records in each segment. The model also analyzes the number of different bot activities in each segment, which may have different activity types. With the F_1 = start_time feature, the model performs a group analysis of activities sorted by bot activity time called a multi-activity. The results of the segmentation process and multi-activity tracing are shown in Table 4.

Table 4. Bot traffic segmentation result.

Segment	Start_Time	End-Time	Bot_Flows	Bot_activity_Label	Multi-activity
1	11:00:00	12:00:00	2,825	22	77
2	11:30:00	12:30:00	3,478	20	97
3	12:00:00	13:00:00	5,165	19	99
4	12:30:00	13:30:00	9,868	14	109
5	13:00:00	14:00:00	6,508	10	57

Multi-activities in each segment have a relationship as a chain of bot activity scenarios. From the multi-activity tracing results for each segment, the model analyzes the multi-activity slice between the two segments, namely *Net-*B_x and *Net-*B_y, and performs multi-activity merging to get bot attack activity scenarios. The model gets the overall bot activity scenario pattern and its frequencies from the tracing results of bot attack activity scenarios. Scenario frequency is a pattern of attack scenarios by a bot carried out repeatedly against the target computer. Each activity scenario has several activity stages referred to as the bot activity step. The proposed model can obtain a pattern of bot attack activity scenarios whose frequency and stages of each pattern are shown in Table 5.

Based on the experimental results, it is found that the proposed model can detect 231 patterns of bot attack activity scenarios with different stages of attack activity. The lowest number of activity stages is 3, and the highest number of activity stages is 7, with examples of activity scenario patterns are shown in Fig. 5.

We compared the results of scenario analysis from the proposed model's results with the CTU-13 dataset 1 scenario descriptions manually. The dataset explained that bots generally perform three main activities; namely, the bot sent spam, connected to an HTTP CC, and use HTTP to do some ClickFraud [16]. For each main activity, it

Table 5. Bot Scenarios Pattern results.

No	Scenario pattern	Frequency	Bot-activity-step
1	"flow = Botnet-TCP-Attempt-SPAM", "flow = Botnet-TCP-Attempt", "flow = Botnet-TCP-Established-HTTP-Ad"	58	3
2	"flow = Botnet-TCP-Attempt-SPAM", "flow = Botnet-UDP-DNS", "flow = Botnet-TCP-Established-HTTP-Ad"	57	3
3	"flow = Botnet-TCP-Established-HTTP-Ad", "flow = Botnet-TCP-Attempt-SPAM", "flow = Botnet-UDP-DNS"	40	3
4	"flow = Botnet-TCP-Attempt", "flow = Botnet-TCP-Attempt-SPAM", "flow = Botnet-TCP-Established-HTTP-Ad", "flow = Botnet-TCP-Attempt-SPAM"	32	4
5	"flow = Botnet-TCP-Established-HTTP-Ad", "flow = Botnet-TCP-Attempt-SPAM", "flow = Botnet-TCP-Attempt", "flow = Botnet-TCP-Attempt-SPAM", "flow = Botnet-TCP-Attempt"	11	5
…	…	…	…

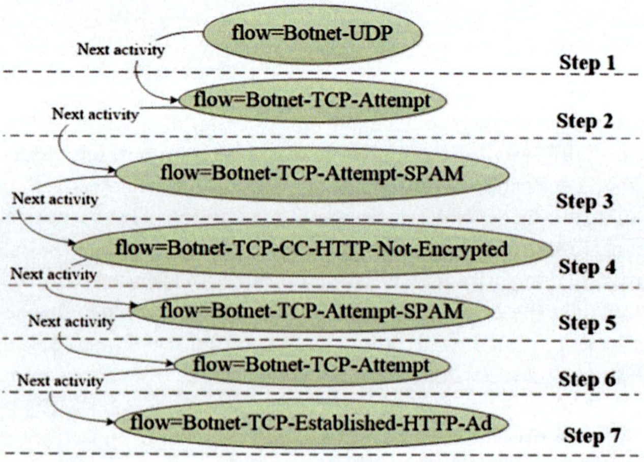

Fig. 5. The result of bot scenarios attack Detection.

consists of several bot flows (shown in Table 3). For example, a SPAM activity consists of a bot flow labeled "flow = Botnet-TCP-Attempt-SPAM" and "flow = Botnet-TCP-Established-SPAM". Thus, from the proposed model results shown in Fig. 5, the spam-sending process is detected in step 3 and repeated in step 5, CC connections are shown in step 4, and used HTTP to do some ClickFraud is shown in step 7. This indicates that the proposed model can detect bot activity, the relation between bot activity and obtain bot attack scenarios.

4.3 Comparison with Another Algorithm

Detection of bot attack activity scenarios that is done by analyzing bot traffic flows is the main' objective of this proposed paper. Here, activities within the activity group are interrelated. This paper compared the bot attack scenario's detection results from the proposed approach with the apriori algorithm, which is generally used to analyze the sequential relationship between two events, such as analyzing intrusion activity or detecting anomalous activity [17, 18]. Our proposed model shows better detection results for bot's attack activity scenarios than this Apriori algorithm. The comparison results are shown in Table 6.

Table 6. Comparing result with apriori algorithm.

Model	Number of scenario detection	Bot-activity-count					Sequence
		(1)	(2)	(3)	(4)	(>5)	
Apriori method	51	11	22	15	3	-	Yes
Proposed model	231	-	8	64	79	80	Yes

As shown in Table 6, the proposed model can detect patterns of bot attack activity scenarios better than the Apriori algorithm. Besides the number of activity scenario patterns that can be detected, the proposed model can also detect more activity stages than the Apriori algorithm. The highest number of stages that can be detected by the Apriori algorithm is four. In comparison, the proposed model can detect more than five, where the highest is seven stages. It shows that the proposed model is better in detecting bot attack scenarios and detecting detailed activity stages in network traffic.

5 Conclusion

This paper proposed a new approach to detect linkages between bot activities by analyzing network traffic flows and obtaining bot attack activity scenarios. The proposed model consists of two processes: bot detection to get the bot label activity and detection of bot attack activity scenarios to get the bot's attack stages. In bot activity detection, the model succeeds in obtaining activity labels using the classification machine learning approach. The classification method used is the method with the highest detection accuracy results, namely, the k-NN method with an accuracy of detection is 94.20%, and the processing time is 0.441501 s.

In the process of bot activity scenario detection, the proposed model has recognized interrelated stages of bot attack activity, which is depicted by successfully detecting 231 activity scenario patterns. There is a repeated scenario among these detected patterns, represented by the frequency and stage of attack activities. The lowest number of activity stages is three, and the highest number of activities is seven. Compared with the Apriori algorithm, the proposed model shows a better activity scenario detection rate and can detect more detailed activities, more than three activity stages.

In future research, we will conduct further analysis of the pattern of bot attack activity scenarios. This research was carried out to detect a new variant type of bot attack pattern. Thus, it can be used as a knowledge base for bot activity detection and early warning models for bot attack activity.

References

1. Aleksieva, Y., Valchanov, H., Aleksieva, V.: An approach for host based botnet detection system. In: Proceedings of the 2019 16th Conference on Electrical Machines, Drives and Power Systems, ELMA 2019, June, pp. 6–8 (2019)
2. Álvarez Cid-Fuentes, J., Szabo, C., Falkner, K.: An adaptive framework for the detection of novel botnets. Comput. Secur. **79**, 148–161 (2018)
3. Wang, C.Y., et al.: BotCluster: a session-based P2P botnet clustering system on NetFlow. Comput. Netw. **145**, 175–189 (2018)
4. Mathur, L., Raheja, M., Ahlawat, P.: Botnet Detection via mining of network traffic flow. Procedia Comput. Sci. **132**, 1668–1677 (2018)
5. Khan, R.U., Zhang, X., Kumar, R., Sharif, A., Golilarz, N.A., Alazab, M.: An adaptive multi-layer botnet detection technique using machine learning classifiers. Appl. Sci. **9**(11), 2375 (2019). https://doi.org/10.3390/app9112375
6. Eslahi, M.: Bots and botnets: an overview of characteristics , detection and challenges, pp. 23–25, 2012.
7. Al-Hakbani, M.M., Dahshan, M.H.: Avoiding honeypot detection in peer-to-peer botnets. In: 2015 IEEE International Conference on Engineering and Technology, ICETECH 2015, March 2015
8. Choi, H., Lee, H., Lee, H., Kim, H.: Botnet detection by monitoring group activities in DNS traffic. In: 7th IEEE International Conference on Computer and Information Technology, CIT 2007, pp. 715–720 (2007)
9. Choi, H., Lee, H., Kim, H.: BotGAD: detecting botnets by capturing group activities in network traffic. In: Proceedings of the 4th IEEE International Conference on Software Testing, Verification and Validation (ICST), Middlew, pp. 1–8 (2009)
10. Chowdhury, S., Khanzadeh, M., Akula, R., Zhang, F., Zhang, S., Medal, H., Marufuzzaman, M., Bian, L.: Botnet detection using graph-based feature clustering. J. Big Data **4**(1), 14 (2017). https://doi.org/10.1186/s40537-017-0074-7
11. Hoang, X., Nguyen, Q.: Botnet detection based on machine learning techniques using DNS query data. Future Internet **10**(5), 43 (2018)
12. Hung, C., Sun, H.: A botnet detection system based on machine-learning using flow-based features. In: The 12th International Conference on Emerging Security Information, Systems and Technologies A, SECURWARE 2018, The Twelft, pp. 122–127 (2018)
13. Garcia, S., Zunino, A., Campo, M.: Survey on network-based botnet detection methods. Sec. Commun. Netw. **7**(5), 878–903 (2014)
14. Dollah, R.F.M., Faizal, M.A., Arif, F., Mas'ud, M.Z., Xin, L.K.: Machine learning for HTTP botnet detection using classifier algorithms. J. Telecommun. Electron. Comput. Eng. **10**(1–7), 27–30 (2018)

15. Beigi, E.B., Jazi, H.H., Stakhanova, N., Ghorbani, A.A.: Towards effective feature selection in machine learning-based botnet detection approaches. In: 2014 IEEE Conference on Communications and Network Security, pp. 247–255 (2014)
16. Garc, S., Grill, M., Stiborek, J., Zunino, A.: An empirical comparison of botnet detection methods. Comput. Secur. **45**, 100–123 (2014)
17. Hanguang, L., Yu, N.: Intrusion detection technology research based on apriori algorithm. Phys. Procedia **24**, 1615–1620 (2012)
18. Ohrui, M., Kikuchi, H., Rosyid, N.R., Terada, M.: Mining botnet coordinated attacks using Apriori-Prefixspan hybrid algorithm. J. Inf. Process. **21**(4), 607–616 (2013)

A Log-Based Method to Detect and Resolve Efficiently Conflicts in Access Control Policies

Maryem Ait El Hadj[1(✉)], Ahmed Khoumsi[2], Yahya Benkaouz[3], and Mohammed Erradi[1]

[1] ENSIAS, Mohammed V University in Rabat, Rabat, Morocco
maryem_aitelhadj@um5.ac.ma, mohamed-erradi@um5.net.ma
[2] Department of Electrical Engineering and Computer Engineering, University of Sherbrooke, Sherbrooke, Canada
ahmed.khoumsi@usherbrooke.ca
[3] LCS, FSR, Mohammed V University in Rabat, Rabat, Morocco
yahya.benkaouz@um5.ac.ma

Abstract. Typically, a security audit is conducted to detect and track inappropriate activities, such as security policy misconfigurations and attacks. Practically, an audit can be done through the analysis and assessment of data in logs registering traces of queries according to predefined policies. In this paper, we present an auditing approach that detects and resolves efficiently conflicting rules of a security policy. Such efficiency translates into a reduction in the time it takes to detect and resolve conflicts. Such efficiency is a consequence of the fact that conflict detection is executed only among suspicious pairs of rules, instead of all pairs of rules. The idea of using suspicious pairs of rules has recently been applied to reduce the execution time of previous detection methods. The present study goes further by applying the idea not only for conflict detection, but also for reducing the *resolution* time of the detected conflicts. We present experimental results that illustrate the efficiency of the suggested method.

Keywords: Access control policy · Access log clustering and analysis · Contradictory access records · Suspicious pairs of rules · Log-based conflict detection and resolution

1 Introduction

A security audit is a technical assessment to analyze the security of an organization's information system by evaluating if the deployed security policy is adequate or not. Auditing requires the logging of all requests and activities performed within an organization systems and networks for later analysis. These activities are stored in a log file that provides pieces of evidence of the sequence of events, operations and procedures carried out. Therefore, a log plays a key

A. Abraham et al. (Eds.): SoCPaR 2020, AISC 1383, pp. 836–846, 2021.
https://doi.org/10.1007/978-3-030-73689-7_79

role in detecting, tracking and preventing inappropriate activities. For instance, when a service fails, logs are examined and analyzed to gain insights into the failure and potential problems (i.e. identify the problem source). Among various categories of logs, we consider particularly access logs, that keep track of access requests to existing resources and corresponding decisions of the security policy to accept or reject these requests. Usually, recorded data must be in compliance with the considered security policy. However, due to inappropriate activities (e.g. conflicting rules in security policy, intrusions or attacks), log analysis may show contradictions that do not comply with the security policy. Hence, ensuring that the data in logs are conform with the security policy deployed is crucial. In this paper, we only consider the case of conflicting rules, while intrusions and attacks detection are beyond the scope of this paper.

Researchers have devoted special attention to some aspects of log analysis such as anomaly detection [2,5,7], intrusion detection [6,8] and performance monitoring [9]. The aspect that has received less attention is the policy compliance verification based on log analysis. In this direction, this paper suggests a method to correct the security policy in order to prevent inappropriate decisions. The present work is based on [1,4]. In [1], a method is proposed to detect (but not resolve) conflicting rules among *suspicious* pairs of rules, instead of all policy pairs of rules. Those suspicious pairs of rules are retrieved from analyzing the access log that records the history of access requests and the corresponding decisions made by the security policy. In [4], a method is developed to detect and resolve conflicts, but without using logs and the notion of suspicious pairs of rules. Conflicts are searched among all rules, instead of suspicious pairs of rules. In this paper, the log-based approach of [1] is applied to the method of [4], which results in a reduction of the detection and resolution time. In a nutshell, the log-based security analysis helps to narrow down the sets of rules to be verified, especially in the case where the security policy is dynamic (i.e. a policy to which rules are added from time to time or modified in response to new situations). Moreover, an important motivation behind this paper is to overcome the manual intervention of the security administrator and contribute in automating as much as possible the conflict resolution process.

The remainder of this paper is structured as follows: Sects. 2 and 3 present respectively the methods of [1,4], on which is based our new proposal. In Sect. 4, we propose how to combine the methods of [4] and [1] to obtain an efficient access log-based method to detect and resolve conflicts. Section 5 reports and discusses experimental results. Finally, the conclusion and expected future work are presented in Sect. 6.

2 Conflict Detection and Resolution Method

We first recall some definitions related to security rules and conflicts, and then we present an overview of the conflict detection and resolution procedure of [4].

2.1 Security Policy: Rules and Conflicts

A policy P is a non-empty set of rules: $P = \{r_1, r_2, ..., r_n\}$. Each rule $r_i \in P$ is specified by a condition and an access decision. The condition of a rule is specified by one or several assignments $att \in V_{att}$, where att is an attribute name and V_{att} is a set of possible values of att. An access decision of a rule is noted X_{act}, where X is the decision *Permit* or *Deny*, and act is a set of actions (e.g. *read*, *write*). A rule will be expressed as follows:

$$r_i : X_{act}((att_1, att_2 ..., att_m) \in V_{att_1} \times V_{att_2} \times ... \times V_{att_m}) \tag{1}$$

The set $V_{att_1} \times V_{att_2} \times ... \times V_{att_m}$ is called the *access domain* of r_i and noted AD_{r_i}. Hence, a rule can be expressed more briefly in the form $r_i = X_{act}((att_1, ..., att_m) \in AD_{r_i})$ and yet more simply $r_i = X_{act}(AD_{r_i})$.

A policy P contains a *Conflict*, when P includes two or more rules that generate contradictory decisions on an access request. Formally, $r_i = X_a(AD_{r_i})$ and $r_j = Y_b(AD_{r_j})$ are conflicting iff:

$$\begin{cases} AD_{r_i} \cap AD_{r_j} \neq \emptyset \\ X \neq Y, and \\ a \cap b \neq \emptyset \end{cases} \tag{2}$$

2.2 Conflict Detection and Resolution

The method of [4] detects and resolves conflicts in a security policy P by constructing a graph (N, E), where N is a set of nodes, and E is a set of edges, where each edge is defined by a pair of nodes, i.e. $E \subseteq N \times N$. Each node represents a rule r_i of P, and each edge (r_i, r_j) means that we have to verify if there is a conflict between two rules r_i and r_j of P, and resolve the conflict, if any. Initially, the graph contains the nodes representing all rules, and an edge between each pair of rules. This initial graph is the input of the procedure which will verify and modify iteratively the graph until we obtain a graph without edges, which represents a conflict-free set of rules. At each iteration, the conflict detection and resolution procedure is applied to an edge (r_i, r_j) of E as explained below, where $r_i = X_a(AD_{r_i})$ and $r_j = Y_b(AD_{r_j})$:

1. If no conflict is detected between r_i and r_j, the resolution removes the link between r_i and r_j.
2. If a conflict between r_i and r_j is detected, it is resolved by one of the following two strategies (permissive or restrictive). Since $X \neq Y$ (from (2)), let us take $X = Permit$ and $Y = Deny$.

- Permissive resolution: allows the access requests for which the two rules disagree. This is realized by not modifying r_i and replacing r_j by:
  $r'_j = Deny_b(AD_{r_j} \setminus AD_{r_i})$
  $r''_j = Deny_{b \setminus a}(AD_{r_i} \cap AD_{r_j})$

- Restrictive resolution: refuses the access requests for which the two rules disagree. This is realized by not modifying r_j and replacing r_i by the following two rules:
$r'_i = Permit_a(AD_{r_i} \setminus AD_{r_j})$
$r''_i = Permit_{a \setminus b}(AD_{r_i} \cap AD_{r_j})$

Then, the two new nodes (for any strategy) are linked to all the nodes of the graph, except the unmodified node. The complete detection and resolution procedure is realized by verifying all edges until the graph has no edge, i.e. until E is empty. Such complete execution is illustrated in the example below.

Example 1. *Consider a policy P with three rules:*

- r_1: $Permit_{\{read,write\}}$ (IpSRC ∈ {10.1.1.122, 10.1.2.122, 10.1.3.122}, IpDST ∈ {10.1.1.120, 10.1.2.120, 10.1.3.120}, Dept ∈ {Production, Marketing, Development})
- r_2: $Deny_{\{read\}}$ (IpSRC ∈ {10.1.1.122, 10.1.3.122}, IpDST ∈ {10.1.1.120, 10.1.3.120}, Dept ∈ {Production, Development})
- r_3: $Deny_{\{read\}}$ (IpSRC ∈ {10.1.2.122}, IpDST ∈ {10.1.2.120}, Dept ∈ {Marketing})

Since P consists of three rules, we get the 3-nodes and 3-edge as shown in iteration 0 of Fig. 1. The conflict detection and resolution procedure proceeds iteratively as follows (assuming the restrictive resolution strategy is applied):

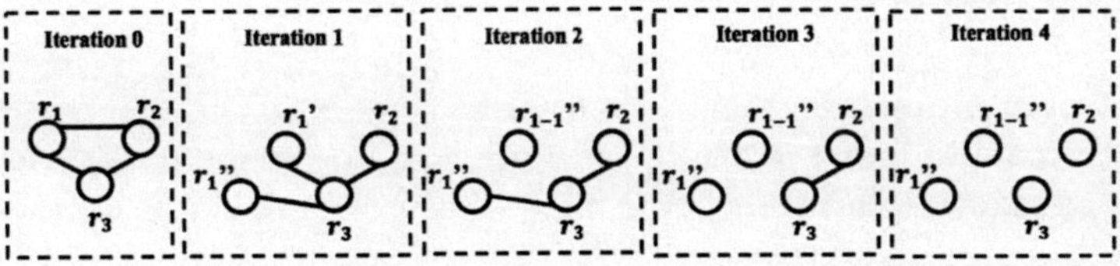

Fig. 1. Constructed graphs from Example 1

- **First iteration**: Consider the pair (r_1, r_2). Since $AD_{r_1} \cap AD_{r_2} =$ {10.1.1.122, 10.1.3.122} × {10.1.1.120, 10.1.3.120} × {Production, Development} $\neq \emptyset$, $\{read, write\} \cap \{read\} = \{read\} \neq \emptyset$, and with different decisions (*Permit* $\neq$ *Deny*), a conflict is detected between r_1 and r_2. We keep r_2 and r_1 is replaced by the following two rules, we then obtain the graph in iteration 1 of Fig. 1:
  - $r'_1 = Permit_{\{read,write\}}$ (IpSRC ∈ {10.1.2.122}, IpDST ∈ {10.1.2.120}, Dept ∈ {Marketing})
  - $r''_1 = Permit_{\{write\}}$ (IpSRC ∈{10.1.1.122, 10.1.3.122}, IpDST ∈{10.1.1.120, 10.1.3.120}, Dept ∈ {Production, Development})

- **Second iteration**: Consider the pair (r_1', r_3). Since $AD_{r_1'} \cap AD_{r_3} = \{10.1.2.122\} \times \{10.1.2.120\} \times \{Marketing\} \neq \emptyset$, $\{read, write\} \cap \{read\} = \{read\} \neq \emptyset$, and with different decisions ($Permit \neq Deny$), a conflict is detected between r_1' and r_3. We keep r_3 and r_1' is replaced by the following two rules, we then obtain the graph in iteration 2 of Fig. 1:
  - r_{1-1}' $= Permit_{\{read,write\}}$ ($\emptyset$) (r_{1-1}' is removed since its access domain is empty)
  - r_{1-1}" $= Permit_{\{write\}}$ (IpSRC $\in$ {10.1.2.122}, IpDST $\in$ {10.1.2.120}, Dept $\in$ {Marketing})
- **Iterations 3 and 4**: Either the intersection of sets of actions or the intersection of access domains is empty, for the rest of connected pairs in the graph in iteration 2 (the pairs: (r_1", r_3) and (r_2, r_3)). Therefore, their links are removed through the two iterations 3 and 4. We obtain the graph in iteration 4 of Fig. 1, which has no link. Therefore, the procedure terminates and we get a conflict-free set of rules.

3 Access Log-Based Conflict Detection Method

The idea behind the conflict detection method in [1] is that potential conflicting rules are detected among suspicious pairs of rules, instead of all policy rules. Suspicious rules are retrieved from access log analysis. In what follows, we first recall some definitions related to access records and contradictions, and then we present an overview of the access log-based conflict detection method of [1].

3.1 Access Log: Access Records and Contradictions

An access record ρ_i is formally specified by an action α (e.g. read, write), a decision X (Permit or Deny), and one or more expressions "$att_k = v_{att_k}$", where att_k is an attribute name and v_{att_k} is a value of att_k. The access record ρ_i is expressed as follows, where m_i is the number of attributes in ρ_i:

$$\rho_i : X_\alpha(att_1 = v_{att_1}, \cdots, att_{m_i} = v_{att_{m_i}}) \quad (3)$$

Intuitively, (3) means that the action α has been requested with the specified attribute values and the policy has made the decision X for such request.

An access log is defined as a finite set of access records: $L = \{\rho_1, \rho_2, \cdots, \rho_l\}$, where ρ_i is an access record, and l is the number of access records.

A *Contradiction* occurs in an access log L, when L contains two or more access records that have recorded the same attribute values on different decisions for the same action. Formally, consider $\rho_i : X_\alpha(att_1 = u_1, \cdots, att_{m_i} = u_{m_i})$ and $\rho_j : Y_\beta(att_1' = v_1, \cdots, att_{m_j}' = v_{m_j})$. Let $m_{i,j}$ be the number of attributes that are common to ρ_i and ρ_j and $\mu_{i,j}$ be the number of common attributes that have the same value in ρ_i and ρ_j. ρ_i and ρ_j are contradictory iff:

$$\begin{cases} X \neq Y \\ \alpha = \beta, and \\ \frac{\mu_{i,j}}{m_i + m_j - m_{i,j}} = 1 \end{cases} \quad (4)$$

3.2 Access Log-Based Conflict Detection

In [1], conflicts are searched (detected, but not resolved) among suspicious pairs of rules of a security policy P, instead of all the rules of P. Suspicious pairs of rules are determined by analyzing L. The suggested access log-based approach consists of [1]:

1. *Access Log Preprocessing*: Local access logs of the various components of a system are collected into a single global access log. Then the access records are extracted and transformed into a structured representation.
2. *Access Log Clustering*: The clustering aims to regroup similar access records into the same cluster based on computer similarity measures.
3. *Access Log Contradiction Detection*: We aim by this step to detect access record contradictions within each cluster.

For each pair of detected contradictory access records (ρ_i, ρ_j), we search the corresponding sets R_{ρ_i} and R_{ρ_j} of rules, i.e. the set of rules that match ρ_i and ρ_j respectively. We say that a rule r matches an access record ρ, if ρ has recorded the same decision as r, the recorded action in ρ belongs to the set of actions of r, and the recorded attribute values in ρ satisfy the attributes conditions of r. Every pair of rules in $R_{\rho_i} \times R_{\rho_j}$ is said to be suspicious, in the sense that there exist potential conflicts between pairs of rules of $R_{\rho_i} \times R_{\rho_j}$. The principle of [1] is to verify the presence of conflicts only between those suspicious pairs of rules.

4 Proposed Efficient Method for Detecting and Resolving Conflicts

4.1 Proposed Method Overview

Given a security policy P and its generated access log L, the objective is to detect and resolve conflicts in P based on L. We propose a new method that generalizes [1], since not only it detects conflicts, but it also resolves the detected conflicts. Conflict resolution is done as in [4], but in a more efficient way, since we consider only suspicious pairs of rules instead of all pairs of rules of the policy.

An important motivation behind this paper is to overcome the manual intervention of the security administrator that costs a lot of time [3]. It is intended that the automatic identification of suspicious pairs of rules will assist the security analyst in quickly determining possible conflicts that are worth correction. Therefore, the proposed approach contributes in automating as much as possible the conflict resolution process without the intervention of the security administrator.

In order to illustrate the benefit of taking into account suspicious pairs of rules, we compare in the example below:

- *The method of* [4]: conflicts are verified between all the rules. In this case, the initial graph contains all policy rules and an edge between each pair of rules.

– *The proposed log-based method*: conflicts are verified in suspicious pairs of rules. In this case, the initial graph contains all policy rules and edges ONLY for suspicious pairs of rules.

Consider the policy P in Example 1, and suppose that the following three access records are part of the access log generated from the execution of P after numerous requests.

– ρ_1 : $Permit_{read}$ (IpSRC = 10.1.1.122, IpDST = 10.1.1.120, Dept = Production)
– ρ_2 : $Deny_{read}$ (IpSRC = 10.1.1.122, IpDST = 10.1.1.120, Dept = Production)
– ρ_3 : $Deny_{read}$ (IpSRC = 10.1.2.122, IpDST = 10.1.2.120, Dept = Marketing)

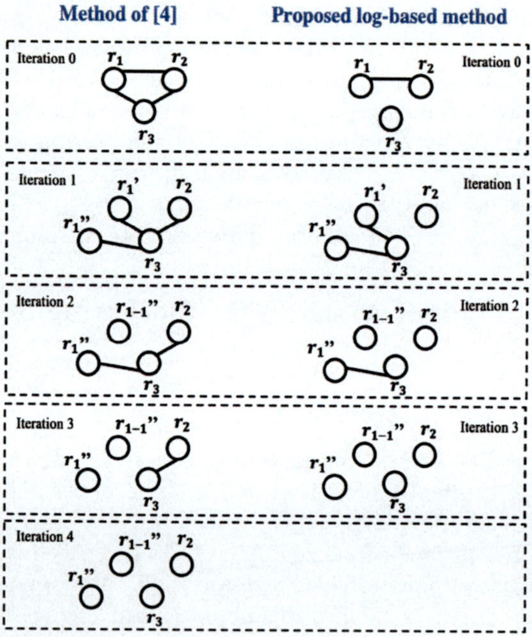

Fig. 2. Constructed graphs from Example 1: method of [4] vs. proposed log-based method

By analyzing the above access log, we deduce that ρ_1 is in contradiction with ρ_2 (i.e., they have the same attribute values, but different decisions). Since r_1 and r_2 match ρ_1 and ρ_2 respectively, therefore the pair (r_1, r_2) is qualified as suspicious. Figure 2 presents the necessary iterations to obtain a graph without edges (which models a conflict-free set of rules), for the method of [4] and the suggested method. In iteration 0, the method of [4] links ALL nodes (r_1, r_2 and r_3), while the suggested method links ONLY suspicious pair (r_1, r_2). From Fig. 2, we can obviously notice that [4] requires more iterations than the new method. These results can be explained by the fact that [4] links all nodes in the first iteration, while the suggested method takes into account only suspicious pairs of rules. We have saved one iteration with the 3-rule policy of Example 1. The gain can be further noticeable with a very large number of rules, which is confirmed by experiment results in the next section.

4.2 Discussion

With the suggested method, the initial graph contains only the links corresponding to suspicious pairs of rules, but the new rules generated by the resolution procedure are linked with *all* rules. By doing so, we increase the probability of detecting conflicts in non-suspicious pairs of rules. To illustrate this point, consider the iterations of the proposed method in Fig. 2. The pair (r_1, r_3) is not suspicious, although r_1 and r_3 are conflicting. When r_1 is replaced by r_1' and r_1'' (in iteration 1), the conflict of (r_1, r_3) is transferred to (r_1', r_3). Linking r_1' to all rules (and hence to r_3) implies that the existence of conflict between r_1' and r_3 will be verified, and hence detected (in iteration 2). If r_1' was linked only to the rules that were linked to r_1, there will be no link between r_1' and r_3, and hence the existence of conflict between r_1' and r_3 is not verified and hence not detected.

We have explained in the previous paragraph how to detect conflicts which do not correspond to suspicious pairs, by linking the newly generated rules to *all* rules. However, this approach does not guarantee to detect *all* conflicts which do not correspond to suspicious pairs. Another aspect that influences the capacity of detecting conflicts is that the more rules that have many matching records in the log, the more chance there is of detecting conflicts. To illustrate this point, consider Example 1 where we have seen that r_3 is not in a suspicious pair, although r_1 and r_3 are conflicting. Assume now that the new access record $\rho : Permit_{read}$ (IpSRC = 10.1.2.122, IpDST = 10.1.2.120, Dept = Marketing) is added to the log. Since ρ matches r_1 and is in contradiction with ρ_3 which matches r_3, we deduce that the addition of ρ has made the pair (r_1, r_3) suspicious. Hence the initial graph contains a link between r_1 and r_3. Therefore, the conflict between r_1 and r_3 is detected in the first iteration, even if the method links the new generated rules r_1' and r_1'' only to the rules that were linked to r_1 (instead of all rules). However, the detection and resolution method connects the new generated rules to all rules, so that it remains applicable also in cases of a relatively "poor" log, i.e. such that some rules of the policy do not match enough access records in the log.

5 Experimental Setup and Tests

The proposed method is implemented using Java programming language and the experiments were made using a laptop with a 2.7 GHz Intel Core i5 CPU, 8 Gb RAM. For testing purposes, we have applied our method to access logs generated by monitoring a system traffic controlled by a Cisco *ASA* firewall, with four network interfaces containing the external network (Internet) on the first network interface, the internal network (Intranet) on the second interface, the zones DMZ 1 and DMZ 2 on the third and fourth interface respectively. The incoming and outgoing network traffic is filtered based on a set of defined rules as described in the Access Control Lists (ACLs) in the matrix flow shown in Table 1. The network traffic that traverses the firewall is matched against these rules to determine if it should be allowed through or not. For each generated log, we have obtained a set of suspicious pairs of rules and then applied the detection and resolution procedure, by constructing a graph with only edges corresponding to suspicious pairs of rules. To analyze our results, we have considered the running time of the clustering, the conflict detection and resolution and the overall execution of the method of [4] and the proposed log-based method.

Table 1. Firewall matrix flow

↱	Internet	Intranet	DMZ 1	DMZ 2
Internet	Permit	Deny	Permit	Deny
Intranet	Permit	Permit	Permit	Permit
DMZ 1	Permit	Deny	Permit	Deny
DMZ 2	Permit	Deny	Deny	Permit

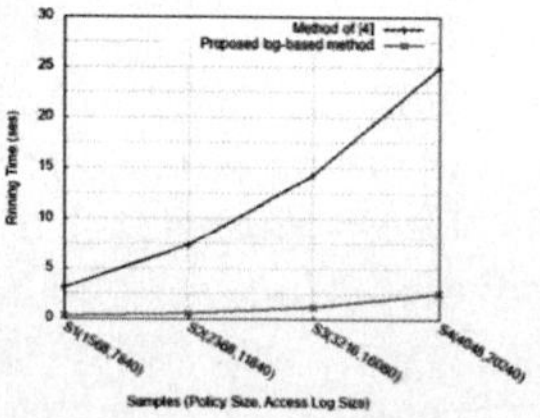

Fig. 3. Clustering running time

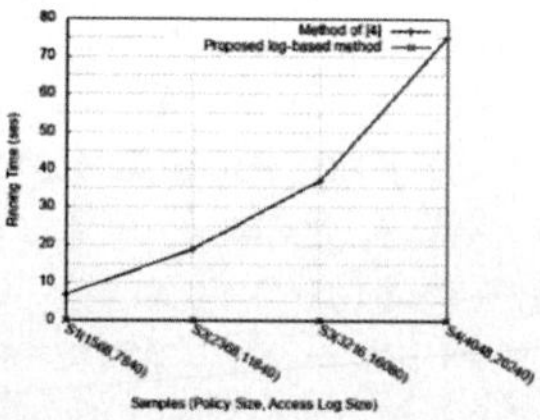

Fig. 4. Conflict detection and resolution running time

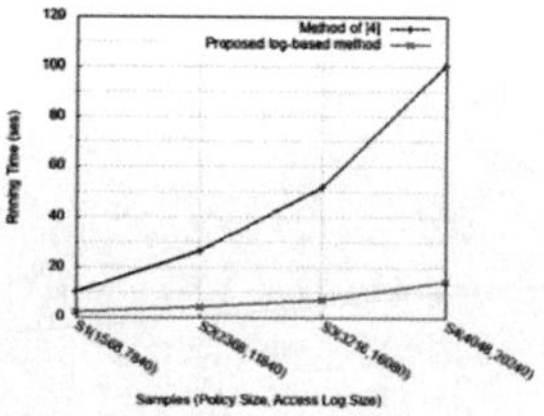

Fig. 5. Overall running time

In Figs. 3 and 4, we compare the two methods based on the clustering, the detection and the resolution durations using policies and access logs of different sizes. Figure 5 presents the overall time, that sums the times of Figs. 3 and 4 plus the preprocessing time. The figures show that the execution time of the log-based suggested method is significantly smaller than the execution time obtained in [4]. As shown in Fig. 4, the detection and resolution times undergo a significant reduction when we replace the method of [4] with the proposed log-based method. However, this improvement arises the following question:

- Is it convincing to analyze only a small part of a policy instead of the whole policy, if on the other hand, it is necessary to add the analysis of an access log?

Our answer is Yes, from the fact that on average a rule of a security policy contains significantly more attributes than an access record (especially in ABAC rules that are fine-grained by considering a lot of attributes, unlike in access log that records just the attributes that have been satisfied in requests). Therefore, the time required to analyze a rule is significantly more than the time required to analyze an access record. This fact is proved by the clustering running time, since the number of attributes influences the time required for computing the similarity scores, which in turn influences the execution time of clustering.

In order to demonstrate that the detection and resolution method remains applicable and efficient in case of a relatively "poor" log (as explained in Sect. 4.2), we have evaluated the *recall* measurement $\frac{TP}{TP+FN}$, which reflects the percentage of conflicts that are detected. *True positive* (TP) is the number of conflicts that are detected, while *False negative* (FN) is the number of conflicts that are not detected. Experimental results show that the proposed method identifies security conflicts with an average recall of 95.65% on all datasets.

6 Conclusion

In this paper, we have presented a method to improve the efficiency of security conflict detection and resolution time. The suggested method is made efficient by detecting and resolving conflicts in suspicious pairs of rules, instead of all pairs of rules. Those suspicious pairs of rules are determined by analyzing the access log that records the access requests and the corresponding decisions taken by the considered security policy. Moreover, the proposed method allows us to detect and resolve some conflicts in non-suspicious pairs of rules. A perspective to enhance this work could be to realize a parallel implementation of the method based on *MapReduce*, to support distributed storage and enable processing of large data sets in a parallel way, which makes it suitable to Big Data systems.

References

1. Ait El Hadj, M., Khoumsi, A., Benkaouz, Y., Erradi, M.: Efficient security policy management using suspicious rules through access log analysis. In: International Conference on Networked Systems, pp. 250–266. Springer (2019)
2. Du, M., Li, F., Zheng, G., Srikumar, V.: DeepLog: anomaly detection and diagnosis from system logs through deep learning. In: Proceedings of the 2017 ACM SIGSAC Conference on Computer and Communications Security. ACM (2017)
3. Lin, Q., Zhang, H., Lou, J.-G., Zhang, Y., Chen, X.: Log clustering based problem identification for online service systems. In: Proceedings of the 38th International Conference on Software Engineering Companion, pp. 102–111. ACM (2016)

4. Ait El Hadj, M., Khoumsi, A., Benkaouz, Y., Erradi, M.: Formal approach to detect and resolve anomalies while clustering ABAC policies. EAI Endorsed Trans. Secur. Saf. **5**(16) (2018)
5. Nguyen, Q.P., Lim, K.W., Divakaran, D.M., Low, K.H., Chan, M.C.: GEE: a gradient-based explainable variational autoencoder for network anomaly detection. In: 2019 IEEE Conference on Communications and Network Security (CNS), pp. 91–99. IEEE (2019)
6. Patcha, A., Park, J.-M.: An overview of anomaly detection techniques: Existing solutions and latest technological trends. Comput. Netw. **51**(12), 3448–3470 (2007)
7. Yagoub, I., Khan, M.A., Jiyun, L.: IT equipment monitoring and analyzing system for forecasting and detecting anomalies in log files utilizing machine learning techniques. In: 2018 International Conference on Advances in Big Data, Computing and Data Communication Systems (icABCD), pp. 1–6. IEEE (2018)
8. Yang, H., Cheng, L., Chuah, M.C.: Deep-learning-based network intrusion detection for SCADA systems. In: 2019 IEEE Conference on Communications and Network Security (CNS), pp. 1–7. IEEE (2019)
9. Yuan, D., Park, S., Huang, P., Liu, Y., Lee, M.M.-J., Tang, X., Zhou, Y., Savage, S.: Be conservative: enhancing failure diagnosis with proactive logging. In: OSDI, vol. 12, pp. 293–306 (2012)

A Hybrid Policy Engineering Approach for Attribute-Based Access Control (ABAC)

Manar Alohaly[1(✉)] and Daniel Takabi[2]

[1] Princess Nourah Bint Abdulrahman University, Riyadh, Kingdom of Saudi Arabia
mfalohaly@pnu.edu.sa
[2] Georgia State University, Atlanta, USA
takabi@gsu.edu

Abstract. The development of an attribute-based access control (ABAC) system requires a substantial amount of manual effort to analyze requirements documents and derive a set of appropriate machine-readable rules. In the literature, two primary approaches have been proposed to reduce the development cost, namely bottom-up and automated top-down policy engineering. Both approaches have their own benefits and shortcomings. In this work, we propose a hybrid ABAC policy engineering approach combining elements from bottom-up and top-down approaches in a systematic framework. The results show that our approach enables reliable construction of machine-readable ABAC rules with a significant improvement rate of 77.5% over the bottom-up policy engineering approach.

Keywords: Access control · ABAC · Policy engineering · Natural language policy

1 Introduction

Attribute-based Access Control (ABAC) provides an unprecedented amount of flexibility and security to withstand today's dynamic business environments [8]. Due to its capabilities, ABAC was recognized as a recommended access control model [8]. Despite the promising potential, the manual development of an initial policy of an ABAC-based system has stood out as a costly effort [8].

An ABAC policy consists of one or more ABAC rules against which access requests are evaluated. The rules are defined using attributes which describe properties of various entities involved in authorization decisions. Since manually crafting these rules can be difficult [4] and certainly expensive task [8], a major challenge is to effectively automate this process and reduce costs associated with the manual execution. Two prime approaches have been studied to help with cost reduction goal, namely automated top-down and bottom-up ABAC policy engineering.

A. Abraham et al. (Eds.): SoCPaR 2020, AISC 1383, pp. 847–857, 2021.
https://doi.org/10.1007/978-3-030-73689-7_80

Top-down approaches focus on the analysis of the organizational business processes, system's use-cases and authorization requirements, which are often written in natural language, to work them into machine-readable ABAC policies [1–3,12–17]. However, the difficulty of the problem and the limitations of current natural language processing techniques restrict this approach.

The bottom-up approaches utilize information of existing access control model and attribute data to partially automate the construction of ABAC policy. While this is a cost-effective approach, the practical usefulness of the resulting policy is not clear. This is to say the produced policy can be irrelevant to the actual organization policy and unintuitive to security architects, who have to post-process them [6,9–11,18].

Since 100% automation of ABAC policy development is beyond the current state of the art, it is worthwhile to study how to aggregate the benefits of both approaches. This requirement has motivated our research on hybrid ABAC policy engineering. The goal is to automate -to the extent possible- the construction of ABAC rules and to aid administrators in processing the mined rules. **Hence, the main research contributions of this work are**: (1) Designing a hybrid ABAC policy engineering framework; At the core of our framework is a natural language policy analysis component interacting with a bottom-up policy miner for ABAC rules generation and refinements; (2) Designing a similarity metric that matches mined rules with natural language policy sentences they implement.

2 Our Proposed Framework

The core idea of our proposed framework centers around using the organizational policy as a point of reference to determine the proper input to the miner and to enhance its output. Particularly, as illustrated in Fig. 1, the proposed process begins with analyzing organizational policy along with the attribute data from existing databases to determine attributes necessary for the construction of ABAC rules. After identifying the set of relevant attributes, it is fed to the miner to produce the first draft of formal ABAC rules. Next, the process proceeds to improve the mined rules to be as close as possible to the actual organizational policy. This entails mapping mined rules to policy sentences they likely implement, comparing both representation and refining the mined rules accordingly.

As input, our hybrid ABAC policy engineering framework accepts the tuple $<U, R, O, A_u, A_r, d_u$ and $d_r>$, where U, R, O, A_u, A_r, d_u and d_r denotes the set of users, resources, operations, user attributes, resource attributes, user attribute data, and resource attribute data, respectively. Besides, it also requires the data dictionary DD defining the meanings of attribute data, access log L and the organizational policy $NLPs$ in its natural language form. The goal is to produce ABAC rules that conform with the organizational policy using a combination of low-level information of the existing system and a high-level organizational policy.

As illustrated in Fig. 1, the pipeline of our framework consists of five modules. It begins with a preprocessing step to establish the link between the low-level and high-level information. Then, it goes through four other tasks of identifying relevant attributes, mining ABAC rules, mapping mined rules to natural language policy and refining the mined rules.

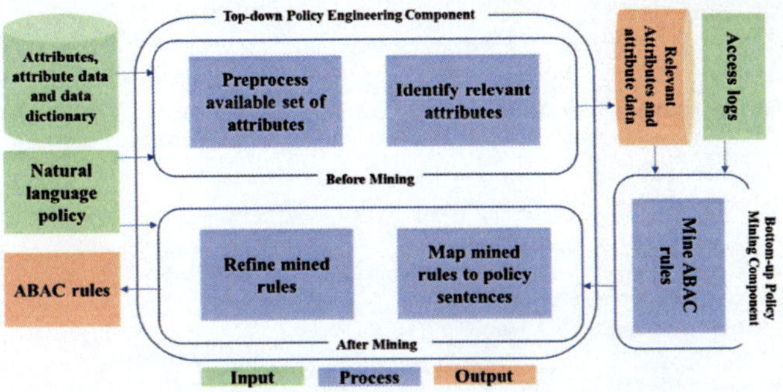

Fig. 1. High level overview of the proposed hybrid ABAC engineering framework

2.1 Module 1: Preprocess Available Set of Attributes

To enable the integration of different levels of information, a link needs to be established between low-level information such as access logs and attribute data, on one hand, and the high-level natural language policies, on the other. To provide such a link, we use the data dictionary that defines attributes names and their allowable values in natural language vocabulary similar to that used in the original policy. In particular, we replace symbols and abbreviations encoding the key-value pairs of attributes in existing database with their equivalent semantic expressions using data dictionary. For instance, consider a university system has defined an attribute named *position* with *1*, *2* and *3* as possible values. Suppose that their data dictionary indicates that these values are defined as *professor*, *lecturer* and *student*, respectively. Then, this preprocessing step provides *position* as an attribute with *professor*, *lecturer* and *student* as the possible values.

2.2 Module 2: Identify Relevant Attributes

Existing data sources are likely to contain data that is not relevant to the construction of ABAC rules. Simply using all available data is not a solution as shown in Sect. 3. The goal of this module is to determine the set of attributes that are relevant to authorization decisions. By definition, ABAC rule is a conjunction of attribute expressions and constraints that together determine sets of users and resources to which the rule applies [18]. Hence, we categorize relevant attribute data into two potentially overlapping sets: (1) set of attributes used to specify attribute expressions and (2) the set used for constraints specification.

To identify attributes to be used in attribute expressions, the rule of thumb is to adhere to the natural language policy that determines the attributes required

Table 1. Resource attributes data

rID	Type	Title	Resource department	Prerequisites
MthCrs1	Course	Applied statistics	Mathematics	Null
MthCrs2	Course	Numerical analysis	Mathematics	Null
csCrs1	Course	Introduction to programming 1	Computer science	Null
csCrs2	Course	Machine learning	Computer science	csCrs1
eeCrs1	Course	IntroductionToBatteryManagement	Electrical engineering	Null

Table 2. User attributes data

uID	Position	Ethnic group	User department	Region	Gender	Disability	Certification	Age band	Course taken
csStu1	Student	African American	Computer science	Albany	Female	No	A certificate or equivalent	20–25	csCrs1
eeStu1	Student	White	Electrical engineering	Amsterdam	Male	No	B certificate or equivalent	20–25	eeCrs1
eeStu2	Student	American Indian	Computer science	Albany	Male	No	Null	20–25	csCrs1,csCrs2

for access approval [5,8]. To achieve this, we exploit the semantic similarity between the key-value pairs of attributes as obtained from Module 1 and the given policy $NLPs$. In particular, this module chunks each sentence in $NLPs$ into noun phrases. Then, it examines the semantic similarity between the resulting noun phrases and key-value pairs obtained from Module 1. The semantic similarity is computed using spaCy which computes, among other things, the semantic similarity between words, sentences, and documents using the state-of-the-art statistical NLP. If the similarity score between an attribute and a noun phrase exceeds a predefined threshold, that attribute is deemed relevant to the construction of ABAC rules. To identify attributes that can potentially be used in constraints expressions, we locate user and resource attributes whose values are derived from the same domain. For illustration, consider the simplified resource and user attributes data presented in Table 1 and Table 2, respectively, along with the following policy sentences: (1) Computer science students may take courses in applied statistics and numerical analysis; (2) Students must be able to access the system to add or drop courses. The result of chunking the above sentences are: computer science students, courses, applied statistics, numerical analysis, students, the system and courses. Matching against these noun phrases suggest that the attributes *user department*, *resource department*, *position*, *type* and *title* are relevant and can potentially be used in attributes expressions. Additionally, comparing the values of these attributes suggest that *user department*, *resource department*, *prerequisites*, *course taken* and *rID* are also relevant and can potentially be used in constraints expressions.

2.3 Module 3: Mine ABAC Rules

Once the list of relevant attributes is identified, it is fed along with the access log to the bottom-up miner to produce the first draft of ABAC rules. To ensure reliable performance and the ease of use of the mining module we analyze and compare several mining approaches based on their performance, generality, and the required human resources to operate each miner. Xu's algorithm [18] better fits our selection criteria and hence we used it. The analysis was removed due to the space limit.

To begin the mining process, the miner selects a log entry, denoted as $<u, r, a>$; here u, r and a are elements in U, R, and O, respectively. Then, it computes the set CC of candidate constraints. In this context, constraints are interpreted as relationships that hold between the attributes of subject and resource in the selected tuple. Then, it examines the set $U' \subseteq U$ of users who are subject to the constraints in CC and are granted the same permission $<r, a>$ in order to infer user-attribute expressions e_u of a candidate rule. Heuristics producing e_u are biased to attributes which have known values characterizing all users in U as opposed to unique identifiers. It also examines the set $R' \subseteq R$ of resources that are subject to a by u for the same motive of inferring the resource-attribute expressions e_r of the rule. Next, the miner attempts to generalize the candidate rule by eliminating unnecessary conjuncts in e_u and/or e_r to increase the rule coverage. Then, it proceeds to improve the overall policy by merging and simplifying candidate rules. Finally, it selects, according to some quality measures, the highest-quality candidate rules for inclusion in the generated policy.

The output of this module is a set of ABAC rules constructed based on attribute data available at the operation time and consistent with the given access log. The following rules are examples of the output of this module from our experiment: <position = student, department = computerScience; type = course; drop; > and <; type = course; add; preRequisites ⊆ crsTaken>.

2.4 Module 4: Map Mined Rules to Policy Sentences

Once the set of rules produced by Module 3 are made available, the pressing question becomes whether these rules align with the original organizational policies. To ensure the conformance of rules-to-NLPs, an essential step is to map each rule to its potential natural language counterpart. To enable the mapping, this module introduces a policy similarity measure to quantify a rule-to-policy alignment and carries out the mapping task accordingly. As illustrated in Fig. 1, the inputs to this module are the set of mined rules, MRs, and the natural language policies, $NLPs$. To begin, each single rule in MRs is compared with each policy sentence in $NLPs$, and a similarity score between the rule and policy sentence is computed, as discussed later in this section. The similarity scores are then sought to find the many-to-many mappings, denoted as Φ. The mapping Φ determines for each policy sentence in $NLPs$, the most similar rules in MRs.

Thus we define a Φ mapping as follows:

$$\Phi(nlp_i) = \{mr_j | S(nlp_i, mr_j) \geq S(nlp_k, mr_j) \forall nlp_k \in NLPs\} \quad (1)$$

Where $mr_j \in MRs$ and S is the proposed similarity score. Intuitively, a rule and a policy sentence are similar if the attributes along with the semantics of the conjunction of these attributes are similar in both representations. We, therefore, design our similarity measure, S to capture this intuition. In particular, we define the similarity function S between a mined rule and policy sentence as follows:

$$S_{(nlp_i, mr_j)} = S_A(nlp_i, mr_j) + S_{rule}(nlp_i, mr_j) \quad (2)$$

S_A is a function which computes a similarity score of attributes in common in each pair, i.e., mr_j and nlp_i, as defined in Eq. 3. S_{rule}, on the other hand, is a function computing the semantic similarity between the mined rule mr_j and the policy sentence nlp_i. While S_A provides a rough evidence of how close a rule to a policy sentence, it is not enough to reason about their similarity. This is because a policy in its natural language form may express variety of rules using the same set of attributes. It also may contain more than one rule in a sentence. The rationale of introducing the second term, i.e., S_{rule}, to the right side of Eq. 2 is to capture the context of the overlapping attributes in the policy sentence nlp_i and its semantic similarity to mr_j.

$$S_A(nlp_i, mr_j) = |A_{mr_j} \cap A_{nlp_i}| \quad (3)$$

As indicated in Eq. 3, the computation of S_A involves comparing the values of attributes in mr_j with the policy sentence nlp_i. For this, we expand the set of values in mr_j to include their synonyms. The value contributes to the score if it or any of its synonyms occurs in nlp_i.

$$S_{rule}(nlp_i, mr_j) = Max_{p_k \in nlp_i}\{Sim(R_{Pred_k}, R_{mr_j})\} + S_{Args} \quad (4)$$

$$S_{Args} = |A_u \cap Arg_0| + |A_r \cap Arg_{1-3}| + |action \cap Pred| \quad (5)$$

S_{rule} is computed by comparing the arguments of each predicate in nlp_i against the mined rule mr_j to capture the semantic distance between a pair of mined rule and policy sentence. This process involves a total of 4 steps. First, we extract all predicates and their arguments from nlp_i using semantic role labeling [14]. Second, we concatenate each predicate with its respective arguments, on one hand, and the values of attributes of the mined rule, on the other. The intuition is to create a semi-sentence structures of both representations, denoted as R_{Pred} and R_{mr} in Eq. 4. Here, the semi-sentence is a sentence incomplete in its structure but communicates a meaningful concept. Third, we compute the semantic similarity of both structure using spaCy. Fourth, we repeat the third step for each predicate and assign the maximum value of similarity to the first term of in the right side of Eq. 4.

S_{Args} ensures the proper correspondence between subject, object and action attributes of the mined rule and the subject, object and the predicates elements of the matching predicate argument structure. This term is necessary to account for cases when subject attributes are also valid object attributes. We note that, in this process, we disregard attribute values that have no lexical meaning, e.g. unique identifiers. Following the above-mentioned steps, this module successfully maps the rules: <position = student, department = computerScience; type = course; drop; > and <; type = course; add; preRequisites ⊆ crsTaken> to the policy sentence "Students must be able to access the system to add or drop courses." to enable the forward and backward tractability between the formal and natural language expressions of a rule.

2.5 Module 5: Refine Mined Rules

Mapping rules to their natural language counterpart is essential to evaluating the conformance between rules and sentences they implement. Consider, for instance, the rule <position = student, department = computerScience; type = course; drop;> and the policy sentence "Students must be able to access the system to add or drop courses." This rule is produced by the miner in Module 3 and mapped by Module 4 to the corresponding policy sentence. As can be seen, the mined rule does not fully conform with the policy sentence it intends to represent. That is, the policy sentence indicates granting drop requests to individuals identified as *students* without imposing any restriction on an individual's department. This is to say that the correct ABAC rule for the given sentence is to be defined as <position = student; type = course; drop; >.

This module exploits the forward and backward tractability between the mined rule and its natural language counterpart, achieved by Module 4, to overcome the above-mentioned mismatch issue. The refinement process begins with parsing the policy sentence with the semantic parser to extract predicates and its arguments -if any- from the sentence. Chances are, more than one predicate may appear in the sentence. Each set of predicate and its arguments can potentially represent different rule. We identify the set that is most similar to the mined rule in-hand using Eq. 4. This set is then used as a point-of-reference to update the rule by: (1) removing subject and object attributes that are in the mined rule but not in the sentence; and (2) adding subject and object attributes that are in the sentence but not in the rule.

For this, we begin with the subject and object attributes that correspond to subject and object policy elements, for instance, *position* and *type*. If the similarity between a value of these attributes and the subject and object arguments of the predicate that best matches the mined rule exceeds a threshold, then these values will be added to the rule (if not already in it). After identifying the subject and object elements, the goal is to extract their attributes, if any. These attributes will be added to the rule (if not already in it). To address this task, we use the dependency tree patterns, identified in Alohaly's et al. study of ABAC attributes extraction [1,2], as patterns encoding subject-attribute and object-attribute relations. Next is to remove attributes that are in the mined rule but are irrelevant to the policy sentence. In order to do so, we use the similarity between the noun phrases in the sentence and attribute values in the rule. If the score is less than a predefined threshold, the attribute is removed from the rule.

3 Experimental Results and Performance Evaluation

Data: We need data of access control system for which we have: (1) its policy documents written in natural language, (2) attributes of different entities in the system, (3) attributes data representing instances of each type of user and resource in the system, and (4) log of access requests. For evaluation, we also have to have ABAC rules representing access control sentences in policy documents. Unfortunately, we know of no data that is readily available for our purpose. Therefore, we created a sample dataset using policy-related data and data generation method existing in the literature as detailed in the followings. For natural language policies, we adopted authorization requirements of the IBM Course Registration System, IBMApp for short. This dataset was developed to study the automated translation of natural language policies to machine-readable form [1,13–17]. Now that we have the policies in natural language, next step is to identify the attributes necessary to express these policies. For this, we read through the policy sentences and break them down into attributes. This set of attributes is to be used for the bottom-up mining process. However, this set as-is would not suffice to capture the real sense/behaviour of bottom-up mining. That is, in a real-world scenario, attributes to be supplied to the miner are likely not designed specifically for ABAC rules. We, therefore, extended this set with attributes that would normally be used in the development of systems in a similar domain. Next, we define number instances of each type of user and resource in IBMApp system: ten students and professors, a few staff members, number of courses and course offerings, etc. Using the set of attributes, we define ABAC rule(s) corresponding each policy sentence. For the log of access requests, we use the algorithm and the implementation of Xu et al. discussed in [18]. In total the data contains 82 policy sentences, 29 attributes with 334 distinct attributes values and 62 rules. Of the 29 attributes, only 14 attributes are relevant to the expression of rules. Our policy is written in a syntax similar to that of Xu et al. to enable the use of their mining and log generation algorithms [18]. For instance, userAttrib $(uid, a1 = v1, a2 = v2, \ldots)$ provides user attribute data for a user whose "uid" attribute equals uid and whose attributes $a1, a2, \ldots$ equal $v1, v2, \ldots$, respectively. Resource attributes statement is defined in similar terms.

Experimental Results: Here, we present the experiments that we conducted to assess the effectiveness of the proposed approach. Particularly, we address the following research questions: **(RQ1)** How well does the proposed approach perform in selecting relevant attributes, i.e. Module 2?; **(RQ2)** To what extent does attributes selection step, discussed in Module 2, improve the results of the bottom-up miner?; and **(RQ3)** How well are the rules, generated by our hybrid approach, in terms of their similarity to the truth?

To answer RQ1, we evaluated Module 2 with IBMApp policy document, the set of available attributes and their values, discussed in the Data subsection.

Based on data inspection, we set a similarity score of 0.7 as a threshold above which an attribute value and a noun phrase obtained from policy sentence are flagged as similar indicating relevant attribute. Our attributes selection approach showed reasonable performance with precision, recall and F_1-score values of 0.77, 1.0 and 0.87, respectively. A higher threshold would improve the precision but at the cost of lower recall as we are more likely to miss relevant attributes.

To answer RQ2, we evaluate the output of the bottom-up miner considering two cases: (1) all available attributes are given to the miner as part of the input, and (2) only those that are identified by Module 2 as relevant. The comparison was made with respect to Jaccard similarity, discussed in [18], between mined rules and the truth and the execution time in both cases. Jaccard Similarity score falls in [0–1]; 0 indicates completely dissimilar while 1 indicates completely similar rules. To compute the execution time, we repeated the experiment three times and reported the average of them. The experiments were performed on a 64-bit Windows 10 machine having 16 GB RAM and Intel Core i7-4720HQ processor. Table 3 provides summary of the results. Several observations can be made from the table: (1) The similarity scores in both cases are relatively low. However, mining with the set of relevant attributes does improve the similarity between mined ABAC rules and the truth; (2) The execution time of the bottom-up mining process decreases approximately by a factor of 2.5 given the set of relevant attributes. These observations suggest that attributes selection is essential not only for better mining results but also for reduced computation complexity. In addition, the obtained similarity score confirms the fact that bottom-up miners are likely to produce semantically irrelevant rules. This justifies the need for rules refinement step.

Table 3. Comparison between mining with all versus relevant attributes

	All	Relevant
Jaccard similarity	0.34	0.40
Execution time	75 min 36 s	29 min 1 s

To answer RQ3, we map mined rules, produced by Module 3, to policy sentences they potentially implement and perform the refinements steps as detailed in Module 4 and 5. We, then, evaluate the similarity between the refined rules and the truth. Our hybrid approach achieves a similarity score of 0.71. While this is not ideal, it is a significant improvement over the 0.40 achieved using pure bottom-up miner. In addition to improved similarity, our hybrid approach establishes traceability between machine-readable rules and sentences they implement. Such a link is essential for further refinements and/or future maintenance for policy sentences and their implementations.

4 Related Work

Das et al. proposed HyPE, a heuristic-based ABAC hybrid engineering framework [7]. HyPE begins with generating candidate rules by combining the attributes of users and resources according to their entropy and frequency. The access decision of each candidate rule with respect to any operation is then resolved by individuals from the security team. This procedure continues iteratively until a complete ABAC policy is constructed. With this design, a rule produced by HyPE does not capture constraint expressions defining relationships between the targeted user and resource. In addition, it heavily relies on an input of the end-user, i.e. security architect, for authorization decisions. Unlike HyPE, our proposed framework has a bottom-up mining component that produces rules consisting of both attributes expressions and constraints necessary for rule definition. Besides, our work has the top-down component that highly automates attributes selection and rule refinement steps. This in turn reduces the burden on security personal and decreases potential human errors. It is also worth mentioning that Das's work used attributes entropy and frequency as rough estimate of authorization-relevance of the attributes. In contrary, our work uses the actual organizational policy, which is the recommended source of information for selecting relevant attributes [5,8].

5 Conclusions

To the best of our knowledge, we present results of the first study on the effectiveness of combining low-level information of access control and high-level natural language organization policies in one hybrid ABAC engineering framework. The experimental results show significant improvement over pure bottom-up miner in terms of the similarity between produced rules and the ground truth. Besides, we designed a similarity measure that provides links between rules and sentences they implement to facilitates further refinements of policies.

Acknowledgements. Daniel Takabi's work is supported in part by the National Science Foundation under Grant No. 2006329. Any Opinions, findings and conclusions or recommendations expressed in this material are those of the author(s) and do not necessarily reflect those of the National Science Foundation.

References

1. Alohaly, M., Takabi, H., Blanco, E.: A deep learning approach for extracting attributes of ABAC policies. In: Proceedings of the 23rd ACM Symposium on Access Control Models and Technologies (2018)
2. Alohaly, M., Takabi, H., Blanco, E.: Automated extraction of attributes from natural language attribute-based access control (ABAC) policies. Cybersecurity **2**(1), 2 (2019)

3. Alohaly, M., Takabi, H., Blanco, E.: Towards an automated extraction of ABAC constraints from natural language policies. In: IFIP International Information Security Conference (2019)
4. Beckerle, M., Martucci, L.A.: Formal definitions for usable access control rule sets from goals to metrics. In: Proceedings of the Ninth Symposium on Usable Privacy and Security, p. 2. ACM (2013)
5. Cavoukian, A., Chibba, M., Williamson, G., Ferguson, A.: The importance of ABAC: attribute-based access control to big data: privacy and context (2015)
6. Cotrini, C., Weghorn, T., Basin, D.: Mining ABAC rules from sparse logs. In: 2018 IEEE European Symposium on Security and Privacy (EuroS&P). IEEE (2018)
7. Das, S., Sural, S., Vaidya, J., Atluri, V.: HyPE: a hybrid approach toward policy engineering in attribute-based access control. IEEE lett. Comput. Soc. **1**(2), 25–29 (2018)
8. Hu, V.C., Ferraiolo, D., Kuhn, R., Friedman, A.R., Lang, A.J., Cogdell, M.M., Schnitzer, A., Sandlin, K., Miller, R., Scarfone, K., et al.: Guide to attribute based access control (ABAC) definition and considerations. NIST Special Publication 800–162 (2013)
9. Iyer, P., Masoumzadeh, A.: Mining positive and negative attribute-based access control policy rules. In: Proceedings of the 23rd ACM on Symposium on Access Control Models and Technologies. ACM (2018)
10. Karimi, L., Joshi, J.: An unsupervised learning based approach for mining attribute based access control policies. In: 2018 IEEE International Conference on Big Data (Big Data), pp. 1427–1436. IEEE (2018)
11. Mocanu, D., Turkmen, F., Liotta, A., et al.: Towards ABAC policy mining from logs with deep learning. In: Proceedings of the 18th International Multiconference, IS 2015, pp. 124–128 (2015)
12. Narouei, M., Takabi, H., Nielsen, R.: Automatic extraction of access control policies from natural language documents. IEEE Trans. Dependable Secur. Comput., 1 (2018). https://doi.org/10.1109/TDSC.2018.2818708
13. Narouei, M., Khanpour, H., Takabi, H., Parde, N., Nielsen, R.: Towards a top-down policy engineering framework for attribute-based access control. In: Proceedings of the 22nd ACM on Symposium on Access Control Models and Technologies (2017)
14. Narouei, M., Takabi, H.: Towards an automatic top-down role engineering approach using natural language processing techniques. In: Proceedings of the 20th ACM Symposium on Access Control Models and Technologies, pp. 157–160. ACM (2015)
15. Slankas, J., Williams, L.: Access control policy identification and extraction from project documentation. Academy of Science and Engineering Science (2013)
16. Slankas, J., Xiao, X., Williams, L., Xie, T.: Relation extraction for inferring access control rules from natural language artifacts. In: Proceedings of the 30th Annual Computer Security Applications Conference, pp. 366–375. ACM (2014)
17. Xiao, X., Paradkar, A., Thummalapenta, S., Xie, T.: Automated extraction of security policies from natural-language software documents. In: Proceedings of the ACM SIGSOFT 20th International Symposium on the Foundations of Software Engineering, p. 12. ACM (2012)
18. Xu, Z., Stoller, S.D.: Mining attribute-based access control policies from logs. In: IFIP Annual Conference on Data and Applications Security and Privacy, pp. 276–291. Springer (2014)

Anomaly Detection for Internet of Things (IoT) Using an Artificial Immune System

Noe Elisa[1], Longzhi Yang[1(✉)], Fei Chao[2], and Nitin Naik[3]

[1] Department of Computer and Information Sciences, Northumbria University, Newcastle upon Tyne NE1 8ST, UK
{noe.elisa,longzhi.yang}@northumbria.ac.uk

[2] Department of AI, Xiamen University, Xiamen, China
fchao@xmu.edu.cn

[3] School of Informatics and Digital Engineering, Aston University, Birmingham B4 7ET, UK
n.naik1@aston.ac.uk

Abstract. Internet of Things (IoT) have demonstrated significant impact on all aspects of human daily lives due to their pervasive applications in areas such as telehealth, home appliances, surveillance, and wearable devices. The number of IoT devices and sensors connected to the Internet across the world is expected to reach over 50 billion by the end of 2020. The connection of such rapidly increasing number of IoT devices to the Internet leads to concerns in cyber-attacks such as malware, worms, denial of service attack (DoS) and distributed DoS attack (DDoS). To prevent these attacks from compromising the performance of IoT devices, various approaches for detecting and mitigating cyber security threats have been developed. This paper reports an IoT attack and anomaly detection approach by using the dendritic cell algorithm (DCA). In particular, DCA is an artificial immune system (AIS), which is developed from the inspiration of the working principles and characteristic behaviours of the human immune system (HIS), specifically for the purpose of detecting anomalies in networked systems. The performance of the DCA on detecting IoT attacks is evaluated using publicly available IoT datasets, including DoS, DDoS, Reconnaissance, Keylogging, and Data exfiltration. The experimental results show that, the DCA achieved a comparable detection performance to some of the commonly used classifiers, such as decision trees, random forests, support vector machines, artificial neural network and naïve Bayes, but with reasonably high computational efficiency.

Keywords: IoT · Dendritic cell algorithm · Anomaly detection · Artificial immune systems · Cyber-attacks

This work has been supported by the Commonwealth Scholarship Commission (CSC-TZCS-2017-717), the Royal Academy of Engineering (IAPP1\100077), and Mr. Aminu Abulmalik who contributed to data processing and part of the experimentation under the Royal Academy of Engineering project.

A. Abraham et al. (Eds.): SoCPaR 2020, AISC 1383, pp. 858–867, 2021.
https://doi.org/10.1007/978-3-030-73689-7_81

1 Introduction

The demand and deployment of IoT automated networks have been increasing significantly in the past years [1,2]. IoT sensors and actuators are deployed in various places, such as in industry, health monitoring systems, weather, transportation system etc., for monitoring, reporting, and activating different events for timely and informed decision making [2]. IoT devices can misbehave due to cyber-attacks or even due to breakdown of the system itself. As IoT networks expand, attacks and anomalies in IoT networks increase significantly [3]. Typical cyber attacks to business websites, e-Government [4], and internet devices [5], such as malware, DoS, keyloggers, network scan, spying, DDoS, Ramsonware, are more and more commonly appeared in IoT networks, which can cause serious damages to IoT services and applications.

Many machine learning approaches ranging from supervised, unsupervised to semi-supervised algorithms have been well exploited to develop IoT intrusion detection models with promising performance generated [1,6]. One of the methods that was developed for detecting intrusions in networked systems is AIS algorithms. This study develops an intrusion detection and mitigation approach for IoT networks by using AIS. One of the robust, effective and recently proposed AIS algorithm for anomaly detection is the DCA algorithm [7], and this algorithm has been used in this study to develop the anomaly detection approach for IoT networks. The development of AIS algorithms is achieved by mimicking the characteristics of natural HIS.

The DCA follows four phases to detect anomalies from the dataset including feature selection and signal categorisation, context determination of data samples, assignment of context to data samples, and finally classification of data samples [7]. Briefly, the DCA applies features selection process to a training dataset to select the most informative attributes. Then, the selected attributes are assigned into either PAMP signal, danger signal (DS) or safe signal (SS), depending on their characteristic behaviour abstracted from the natural immune signals. Then, the DCA initialises a number of artificial DCs which use a weighted function to determine the context of each data sample from the three signals. Since each DC process multiple data samples, during the context assignment phase, each DC assigns all its data samples to the context it has found either as normal or anomaly. During the classification phase, the final class of each data sample is determined from a consensus decision reached by multiples DCs which have sampled the data sample.

In this study, particles swarm optimisation algorithm (PSO) [8] was used to generate and optimise the weights required by the DCs to process the signals of data samples during the context determination phase of the DCA. The PSO was selected since it can easily achieve a good balance between exploration and exploitation of search space, implying faster performance compared to other optimisation methods, such as genetic algorithm, which has also been applied to optimise the DCA [9,10].

The performance of the proposed approach was evaluated by using publicly available dataset named BoT_IoT [6] which is comprised of five attacks including

DoS, DDoS, Reconnaissance and Keylogging and Data_Exfiltration. The performance of the DCA method was compared with five state-of-the-art machine learning techniques including Support Vector Machine (SVM), Artificial Neural Network (ANN), Naïve Bayes (NB), Decision Tree (DT) and Random Forest (RF). The result shows that, the DCA method achieved better performance compared to SVM, ANN and NB; whilst producing the comparable performance to that of DT and RF.

The rest of this work is organised as follows. Section 2 introduces the theoretical underpinnings of the biological DCs and the DCA algorithm. Section 3 presents the DCA-based approach for detecting attacks in IoT. Section 4 details the experimentation process and finally, Sect. 5 summarises this study and suggests feasible future works.

2 Background

This section presents the theoretical underpinnings of biological DCs and the DCA algorithm.

2.1 Biological DCs

In natural HIS, the DCs are available in the body tissues such as skin, nose and lung where they act as the first line of defense against foreign invaders [11]. DCs are responsible for capturing antigens such as bacteria or anything identified as harmful and present them to HIS for a specialised immune response. On their own cell surface, the DCs express co-stimulatory molecules for limiting the amount of time spent while gathering antigens in the tissue. The activation of DCs depends on the concentrations of the following three signals in HIS [11]:

- **PAMP** are chemical substances which are produced by foreign invaders such as virus or bacteria and can easily cause immune deletion.
- **DS** are generated by stressed cells within a disrupted tissue to indicate an anomalous situation but with lower score than the PAMP.
- **SS** are generated by normal cell death process in the tissue, which is an indicator of normal cell behavior.

Generally, DCs exist in three states in HIS [11] depending on the concentrations of SS, PAMP or DS signals in the tissue as follows:

1. **Immature DCs (iDCs):** reside in tissue in their initial sampling state, where they absorb both normal and harmful antigens but they cannot initiate any immune response. The amount of the signals of the absorbed antigens causes iDCs to differentiate to either full-mature or semi-mature state.
2. **Full-mature DCs (mDCs):** iDCs differentiate into mDCs when they sample antigens with significant larger quantity of either PAMP or DS than SS which causes immune deletion.
3. **Semi-mature DCs (smDCs):** iDCS differentiate into smDCs when they collect antigens with greater concentrations of SS than PAMP and DS which causes immune tolerance.

2.2 Dendritic Cell Algorithm

DCA is a population based attacks mitigation system where a population of artificial DCs is initialised in a sampling pool, then, a pre-selected number of DCs are taken to perform signal and data sampling, context analysis and classification [7]. DCs in the sampling pool collect data samples included in the data source and their corresponding current signal values [12]. Each DC can collect and process multiple data samples and therefore, during the labelling phase, a consensus context value from multiple DCs for a particular data sample is computed which is used to classify a data sample as normal or anomalous.

Normally, the DCA first applies a feature selection process to select the most significant features from the dataset. Then, the selected attributed are manually or automatically assigned into either "PAMP", "DS" or "SS" signal. Briefly, PAMP indicates an definite abnormality associated with a particular feature. DS represents an abnormality associated with a feature but with lower score than PAMP; and SS indicates a normality behaviour associated with a feature. Commonly, there are two signal generation and categorisation approaches used with the DCA including the manual technique based on the expert knowledge [7,13], and the automatic approaches such as principal component analysis [14], fuzzy-rough data pre-processing [15], genetic algorithm [16] or fuzzy interpolation systems [17,18]. In addition, the input features can also be aggregated into the three signals using inference approaches [19].

Once the signal generation and categorisation phase is done, a population of artificial DCs is initialised by the DCA in a pool (often 100) [7]. Then, the three input signals are processed by a pre-selected number of DCs (often 10) in order to get three output signals namely co-stimulatory signal (csm), mature signal (mDC) and semi-mature signal ($smDC$) by applying:

$$[csm, smDC, mDC] = \sum_{s=1}^{m} \frac{\sum_{k,l=1,1}^{3}(x_k * w_i^l)}{\sum_{k,l=1,1}^{3} w_k^l}, \tag{1}$$

where s represent a data sample being processed, $x_k (k = 1, 2, 3)$, represent its SS, DS and PAMP values respectively; and $w_k^l (k, l = 1, 2, 3)$ are the parameters of weight regarding csm, $smDC$ and mDC corresponding to SS, DS and PAMP, respectively. Conventionally, the parameters of weight are either empirically derived from the data or a user manually pre-defined them [15]. Note that, a random migration threshold value is assigned to each DC in order to determine its lifespan while sampling data samples and signals from the data source. Generally, DCs keep accumulating the values of mDC, $smDC$ and csm overtime for multiple data samples they sample to obtain the cumulative values.

Once the cumulative csm value surpasses the migration threshold assigned to a DC, the DC ceases sampling, matures and moves to the context detection and assignment phases. Then, the DC compares the cumulative values of $smDC$ and mDC to determine the nature of the sampled data samples. If cumulative mDC is greater than cumulative $smDC$, the DC differentiate to full-mature context, implying that the data samples collected are potentially anomalous.

Otherwise, the DCs goes to semi-mature context, which indicates that the data samples are under normal condition.

Note that, since one data sample is usually sampled by multiple DCs, based on the context values attached to it, the ratio of the number of abnormal context attachments to the total number of attachments of the data item can thus be calculated, which therefore determines the likelihood of the data time belongs to the abnormal class. Based on the training dataset, the percentage of the abnormal data items within the entire dataset can be calculated, which is usually used as a threshold. If the likelihood of the data belongs to the abnormal class is greater than the threshold, the data item will be labelled as abnormal; otherwise normal.

3 Anomaly Detection in IoT Networks Using DCA

The proposed system for detecting anomalies and attacks in IoT networks by using the DCA algorithm is illustrated in Fig. 1. Firstly, the most informative attributes from the dataset were selected by applying a feature selection technique. Secondly, the selected attributes were categorised into either PAMP, DS or SS signal. Then, the PSO algorithm was exploited to generate and optimise the weights used by the DCA in its context detection phase. The last two phases in this study are indistinguishable from that of the conventional DCA, and thus the rest of this section only focuses on the context determination and detection phase of the DCA.

3.1 Weights Generation and Optimisation Using PSO

PSO is a stochastic optimisation approach inspired by the behaviour of bird flocking or fish schooling [8]. Usually, PSO begins with an initial population of random solutions (i.e. particles) and explore the search space for an optimal solution by updating the velocity and position of each particle while iterating over a search space over a number of iterations. In PSO, diversification of potential solutions depends on the movement of particles (i.e.; velocity), particle's direction towards the best regions and the best regions within the search space.

Typically, the initialised particles in PSO algorithm follow the direction and velocity of the best particle in the current iteration to navigate through the search space. Hence, the are two parameter values which are exploited to update the direction and location of particles in each generation. Firstly is the fitness value of a particle that it has attained so far, called personal best (*pbest*). Secondly is the fitness value of the best particle in the current generation called global best (*gbest*).

After evaluating the best values, each particle update its velocity ($V_j(t+1)$) and position ($Z_j(t+1)$) by using:

$$V_j(t+1) = \omega V_j(t) + c_1 r_1 [pbest - Z_j(t)] + c_2 r_2 [gbest - Z_i(t)], \tag{2}$$

$$Z_j(t+1) = Z_j(t) + V_j(t+1), \tag{3}$$

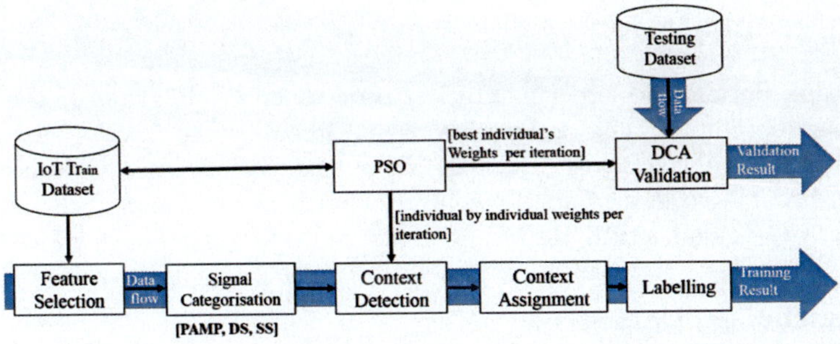

Fig. 1. The proposed DCA-based anomaly detection system in IoT networks

where j is the identity of a particle; ω is the coefficient regarding a particle's inertia; c_1 and c_2 are particle's acceleration coefficients $(0 \leq c_1, c_2 \leq 2)$; r_1 and r_2 are random values $(0 \leq r_1, r_2 \leq 1)$ which are changed every time the velocity of a particle is updated; $V_j(t)$ is the particle's velocity at time t; $Z_j(t)$ is the particle's location at time t; *pbest* is the best position of a particle at time t; *gbest* is the position of the best particle in the current iteration at time t.

Each parameter of the PSO exploited in this study is detailed in the following main steps.

1) Particle size and depiction: In this study, each particle (P) within a population of swam is a solution that consists of all the weights of DCA's Eqs. 1. Therefore, an individual is represented as $I = \{w_1^{smDC}, w_2^{smDC}, w_3^{smDC}, w_1^{mDC}, w_2^{mDC}, w_3^{mDC}, w_1^{csm}, w_2^{csm}, w_3^{csm}\}$, where 1,2,3 represent the three signals (i.e., SS, DS and PAMP).

2) Initialisation of each solution: The population of swarm $\mathbb{P} = \{P_1, P_2, ..., P_T\}$ is randomly initialised from a Gaussian function with a standard deviation of 10 and a mean of 1. Where, T is the maximum number of solutions (i.e.; particles) in the population of swam, with widely used values being 10–50 [8].

3) Objective function: In this study, the classification accuracy computed by the DCA for each particle was taken as the objective function of the PSO.

4) Coefficients of acceleration: The acceleration coefficients c_1 and c_2, and the random values r_1 and r_2, control the stochastic influence and collaboration among particles in the swarm. Lower values for c_1 and c_2 allow the particles to visit the search space far from the best regions before being pulled back towards the best particles [8].

5) Inertia component: The ω is used to keep the stable movement of particles to the desirable regions within the search space. Smaller value of ω influence faster convergence and larger value encourages exploration of the search space, and is normally set between 0.8 and 1.2 [8].

6) Termination and iteration: In this study, once the classification accuracy of the DCA surpasses the pre-determined optimum accuracy or the pre-specified maximum number of generations, the PSO terminates the search process.

4 Experimentation

This section details the experimentation process and validation of the results as well as the comparative evaluation. Experiments were developed by using JAVA language in NetBeans IDE 8.2. Then, the performance validation were accomplished by exploiting an HP computer with Intel® Xeon™ E5-16030 v4 CPU @3.70 GHz and 32 GB RAM.

4.1 Benchmark Datasets

The dataset named BoT_IoT [6] from Cyber Range Lab of the University of New South Wales, Australia was used to analyse the performance of the DCA-based approach in comparison to other methods. The BoT_IoT dataset was created by designing a realistic IoT network environment while incorporating legitimate and simulated IoT network traffic, along with five types of attacks including DDoS, DoS, Reconnaissance, Keylogging and Data exfiltration attacks (i.e., Information Theft) [6]. The dataset contains 7336090 samples and 17 features.

Pre-processing: Feature selection process was accomplished by using the Information Gain technique [20]. Then, signal assignment was performed by opimising the feature-to-class mutual information [20]. Any selected feature with a substantial larger mutual information with the normal class and significant lesser mutual information with the abnormal class, it is assigned to SS. Similarly, any feature with significant greater mutual information with the abnormal class but substantial lesser mutual information with the normal class, it is assigned to PAMP. Otherwise, the feature is assigned to DS.

DCs Initialisation: The size of sampling DCs population in the pool was set to 100 and the size of the mature pool was set to 10 DCs. The DC's migration threshold values were initialised by using a Gaussian distribution function with a standard deviation of 1.0 and a mean of 5.0.

PSO Parameters: The parameters of the PSO used in this study include the initial velocity $V_0 = 0$, maximum velocity $V_{max} = 10$, $c_1 = c_2 = 2.0$, 20 particles in a swarm, 250 number of generations, $0 \leq r_1, r_2 \leq 1$, and $\omega = 0.95$, since they are being widely used [8].

Evaluation Metrics. The performance of the DCA-based approach was evaluated in terms of testing accuracy and attack's detection rate.

Accuracy and detection rate are defined as follows:

$$
\begin{aligned}
Accuracy &= \frac{TP+TN}{TP+TN+FP+FN}, \\
Detection\_Rate &= \frac{TP}{TP+FN},
\end{aligned}
\tag{4}
$$

where TP, FP, TN, and FN refer respectively to true positive, false positive, true negative and false negative.

4.2 Results and Analysis

The testing results on accuracy and detection rate for each attack type are summarised in Table 1 regarding the proposed DCA-based approach and four widely used classifiers including SVM, ANN, DT, RF and NB. The best performance amongst these approaches on overall accuracy and each attack type is highlighted in bold.

Table 1. Performance results and comparison

Approach	Overall accuracy (%)	Attacks detection rate (%)			
		DoS	DDoS	Reconn	Info theft
DCA	97.96	**99.38**	**98.24**	97.73	73.74
SVM	72.65	90.49	37.12	92.84	25.32
DT	95.92	96.94	95.95	93.94	89.67
RF	**97.97**	97.95	96.98	**100.00**	97.47
ANN	89.15	96.67	85.42	84.24	0.0
NB	66.10	92.38	95.18	2.68	**98.73**

From the results presented in Table 1, it can be noticed that, the DCA-based approach has produced higher detection rates on DoS and DDoS attacks compared to the state-of-the-art methods. The classification accuracy of the DCA is comparable to that of the best performing accuracy of RF, while it has best overall detection performance for each attack type compared to other classification methods. For instance, RF has better detection rate on reconnaissance attack but significant less effective on DoS and DDoS and information theft when compared to the DCA. Same remark is noticed for NB classifier where it has produced better detection rate on information theft attack but it has poor detection performance on other attacks and the overall classification accuracy. Ultimately, it can be concluded that, the DCA can be effective on detecting anomalies and attacks in IoT networks with better detection performances compared to some best state-of-the-art machine learning methods.

Furthermore, the performance comparison in terms of computation time for the DCA and the compared classifiers is illustrated in Fig. 2. Although DCA takes much more time to train in comparison to RF and DT, taking its detection performances in account, DCA is more appropriate and reasonable for mitigating anomalies in IoT networks. Generally, the main goal is to make sure that, when the DCA is deployed to detect attacks in IoT networks, it is able to produce higher and effective detection results compared to the referenced commonly used classification techniques.

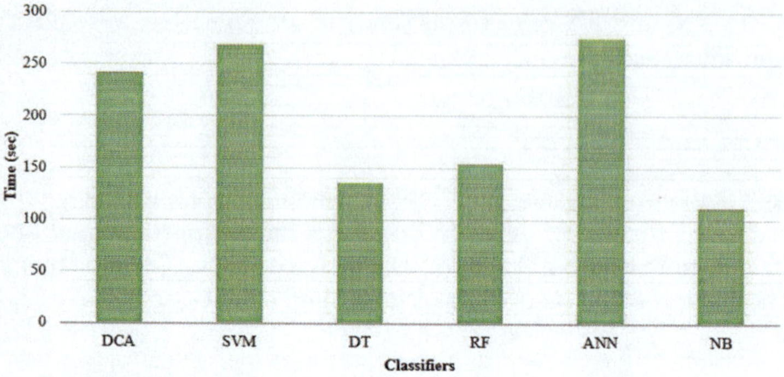

Fig. 2. Running time (in seconds) for classifiers

5 Conclusions

This work presented an approach for detecting anomalies and attacks in IoT networks by using the DCA algorithm. The performance of the DCA-based approach was evaluated by using publicly available IoT datasets which include five attacks DoS, DDoS, Reconnaissance and Keylogging and Data_Exfiltration. The DCA-based approach achieved a comparable performance to some of the commonly used techniques, such as SVM, DT, RF, ANN and NB. Although promising, future work is to further explore how the DCA-based approach can be integrated with privacy-preserving technologies such as the blockchain technology and cryptography to develop a privacy-preserving system for IoT networks in order to reinforce information privacy and security.

References

1. Zarpelão, B.B., Miani, R.S., Kawakani, C.T., de Alvarenga, S.C.: A survey of intrusion detection in internet of things. J. Netw. Comput. Appl. **84**, 25–37 (2017)
2. Ray, P.P.: A survey on internet of things architectures. J. King Saud Univ. Comput. Inf. Sci. **30**(3), 291–319 (2018)
3. Oracevic, A., Dilek, S., Ozdemir, S.: Security in internet of things: a survey. In: 2017 International Symposium on Networks, Computers and Communications (ISNCC), pp. 1–6. IEEE (2017)
4. Yang, L., Elisa, N., Eliot, N.: Privacy and security aspects of e-government in smart cities. In: Smart Cities Cybersecurity and Privacy, pp. 89–102. Elsevier (2019)
5. Naik, N., Jenkins, P., Kerby, B., Sloane, J., Yang, L.: Fuzzy logic aided intelligent threat detection in cisco adaptive security appliance 5500 series firewalls. In: 2018 IEEE International Conference on Fuzzy Systems (FUZZ-IEEE), pp. 1–8 (2018)
6. Koroniotis, N., Moustafa, N., Sitnikova, E., Turnbull, B.: Towards the development of realistic botnet dataset in the internet of things for network forensic analytics: Bot-IoT dataset. Future Gener. Comput. Syst. **100**, 779–796 (2019)
7. Greensmith, J., Aickelin, U., Cayzer, S.: Introducing dendritic cells as a novel immune-inspired algorithm for anomaly detection. In: International Conference on Artificial Immune Systems, pp. 153–167. Springer, Heidelberg (2005)

8. Engelbrecht, A.P., Cleghorn, C.W.: Recent advances in particle swarm optimization analysis and understanding. In: Proceedings of the 2020 Genetic and Evolutionary Computation Conference Companion, pp. 747–774 (2020)
9. Elisa, N., Yang, L., Chao, F., Naik, N.: A comparative study of genetic algorithm and particle swarm optimisation for dendritic cell algorithm. In: 2020 IEEE Congress on Evolutionary Computation (CEC), pp. 1–8 (2020)
10. Elisa, N., Yang, L., Naik, N.: Dendritic cell algorithm with optimised parameters using genetic algorithm. In: 2018 IEEE Congress on Evolutionary Computation (CEC), pp. 1–8. IEEE (2018)
11. Matzinger, P.: Essay 1: the danger model in its historical context. Scand. J. Immunol. **54**(1–2), 4–9 (2001)
12. Chelly, Z., Elouedi, Z.: A survey of the dendritic cell algorithm. Knowl. Inf. Syst. **48**(3), 505–535 (2016)
13. Elisa, N., Yang, L., Qu, Y., Chao, F.: A revised dendritic cell algorithm using k-means clustering. In: 2018 IEEE 20th International Conference on High Performance Computing and Communications; IEEE 16th International Conference on Smart City; IEEE 4th International Conference on Data Science and Systems (HPCC/SmartCity/DSS), pp. 1547–1554. IEEE (2018)
14. Gu, F.: Theoretical and empirical extensions of the dendritic cell algorithm. PhD thesis, University of Nottingham (2011)
15. Chelly, Z., Elouedi, Z.: Hybridization schemes of the fuzzy dendritic cell immune binary classifier based on different fuzzy clustering techniques. New Gener. Comput. **33**(1), 1–31 (2015)
16. Elisa, N., Yang, L., Chao, F.: Signal categorisation for dendritic cell algorithm using GA with partial shuffle mutation. In: UK Workshop on Computational Intelligence, pp. 529–540. Springer, Cham (2019)
17. Yang, L., Chao, F., Shen, Q.: Generalised adaptive fuzzy rule interpolation. IEEE Trans. Fuzzy Syst. **25**(4), 839–853 (2017)
18. Yang, L., Shen, Q.: Closed form fuzzy interpolation. Fuzzy Sets and Syst. **225**, 1–22 (2013)
19. Elisa, N., Li, J., Zuo, Z., Yang, L.: Dendritic cell algorithm with fuzzy inference system for input signal generation. In: UK Workshop on Computational Intelligence, pp. 203–214. Springer, Cham (2018)
20. Witten, I.H., Frank, E., Hall, M.A., Pal, C.J.: Data Mining: Practical Machine Learning Tools and Techniques. Morgan Kaufmann, Burlington (2016)

Agreement-Broker: Performance Analysis Using KNN, SVM, and ANN Classifiers

Zineb Bakraouy(✉), Amine Baina, and Mostafa Bellafkih

STRS Laboratory, National Institute of Posts and Telecommunications, 2, Av. Allal El Fassi, Madinat Al Irfane, Rabat, Morocco
{bakraouy,baina,bellafkih}@inpt.ac.ma

Abstract. For several years now, thanks to the emergence of new information and communication technologies, information has become increasingly available and accessible. The IT field today has a very large amount of data allowing the search for any information. However, exploiting this large amount of data makes finding and classifying precise data complex and time consuming. This difficulty has motivated the development of new adapted data classification tools. Classification is first used to refer to the division of a set of individuals into classes so that every individual belongs to one class and only one. But the term classification is also used to designate nested systems of classes so we can say that it is a statistical operation which consists in grouping objects (individuals or variables or observations) into a limited number of groups (classes, segments), and to classify individuals according to some of their characteristics. There are different types of classification, but one of the most intuitive and widely used is supervised classification. The overall objective of the classification is to identify the classes to which objects belong based on descriptive features (attributes, characteristics). Data classification is a delicate problem that arises in many sciences such as data mining analysis. The broker's primary goal is to reduce the overhead costs associated with managing web service level agreements, while maintaining compatibility with existing infrastructure and vendors.

Keywords: SLA · Negotiation · QOS · Availability · Web services · Service broker · Classification · ANN · KNN · SVM

1 Introduction

Artificial intelligence [1] is a subfield of computing. In recent times, the expression "artificial intelligence" is frequently used in the public because it is a constantly evolving field, in particular thanks to the progress of computer technologies and, among other things, thanks to the ever greater capacities of machines to perform tasks. Calculations. Machine learning is the fact that a machine with artificial intelligence can be able to manage itself but above all to be self-taught. Thus through a systematic process, Machine learning is able to conceptualize, develop and assimilate tasks of different natures. This consists of the implementation of algorithms aimed at obtaining predictive analysis from data, for a specific purpose. Classification [2] is a basic method of machine learning; the

A. Abraham et al. (Eds.): SoCPaR 2020, AISC 1383, pp. 868–879, 2021.
https://doi.org/10.1007/978-3-030-73689-7_82

objective goes beyond the strictly exploratory framework. It is the search for a typology, or segmentation, that is to say a partition, or distribution of individuals into classes, or categories. This is done by optimizing a criterion aimed at grouping individuals into classes, each as homogeneous as possible and, between them, as distinct as possible. In general, this process is associated with the possibility of measuring in some sense the quality and accuracy of the results. It is possible to consider several different learning techniques: supervised, semi-supervised, unsupervised transfer and reinforcement learning within the framework of this dissertation, we will mainly talk about supervised learning. In particular, supervised learning involves the modeling of an input-output relationship based solely on the observation of input-output pairs resulting from this relationship. Statistical learning is today confronted with data whose nature is more and more complex (curves, images) and which take values in spaces whose dimension is always higher. In particular, these data can be in the form of random functions, or curves. Nevertheless, the perimeter of use of traditional statistical methods is often limited to the case where the space of observations is of finite dimension. In this new context, the challenge is then to propose methods for processing this functional data. In this paper, we provide an overview of the machine learning and approaches adopted in the QoS Classification on the Agreement-Broker platform to expose their performance analysis. The paper is organized as follows: Sect. 2 is dedicated to related work, supervised and unsupervised classification, in Sect. 3, we present the algorithms comparison study and the Sect. 4 presents the results, eventually in Sect. 5 we draw our conclusions.

2 Background

There are several types of machine learning for classification: supervised, unsupervised, semi-supervised and active. Supervised learning allows users to specify the desired classes according to the specific needs of the application. It uses a learning database in which each piece of data is manually labeled with its respective class. Obtaining a successful classifier requires manual labeling of enough data for each of the classes, which turns out to be costly. Unsupervised learning does not require labeling and instead tries to discover clusters (groups of data that are close to each other) in an unlabeled data set. But the clusters discovered do not always correspond to the specific classes desired by the user. Semi-supervised learning allows users to specify classes without requiring a lot of labeling. This is because semi-supervised learning techniques can learn using both labeled and unlabeled data, reducing the cost of labeling. Finally, active learning allows an operator to tag a few appropriately selected "important" data, which dramatically reduces the cost of tagging. The main problem to be solved for active learning is how to select the data to label. This is not a trivial problem because an ideal solution must allow both (i) to label as much data as necessary to have a powerful classifier, and (ii) to minimize the cost of labeling, which can be estimated by the number of data to be labeled. Machine learning is a concept that is gaining more and more attention in the computer world, and relates to the field of artificial intelligence. Also called "statistical learning", this term refers to a process of development, analysis and implementation leading to the establishment of systematic processes. Simply put, it is a kind of program that allows a computer or machine to learn automatically, so that a number of very complex operations can be performed. The goal is to make the machine capable of providing solutions to complicated problems by processing amount of information. This thus offers an opportunity to analyze and highlight the correlations that exist between two or more given situations, and to predict their different implications. Machine learning is broken down into 2 stages: a training phase (we learn about part of the data) and a verification phase (we test on the second part of the data). We will therefore have 3 phases: representation, evaluation, optimization. The representation phase consists in finding the most suitable mathematical model. There is a large number of modeling. The assessment measures the gap between the model and the reality of the test data. Finally, optimization aims to narrow this gap. We can count 3 basic methods:

- Classification: modeling of several groups of data in existing classes. For example: the classification of types of orchids, the tendency of a political party,
- Clustering: looks like the classification but they are not known classes,
- Regression: the data are linked to other numerical data by a correlation (a line, a curve, a trend).

The Support Vector Machine (SVM) is a supervised learning algorithm that produces a data classifier. Wide-margin separators are classifiers based on two key ideas, which allow the classification problem to be reformulated as a quadratic optimization problem. The first key idea is the notion of maximum margin which is only the maximum distance between the separation boundary and the closest samples. These are called support vectors. Hence the name of the method. The second key idea of SVMs is to transform the representation space of input data that is not linearly separable into a larger dimensional space, in which there is likely to be a linear separator. The idea of maximum margin hyperplanes was explored as early as 1963 by [3], and in 1973 by [4]. The usefulness of kernel functions in the context of machine learning was shown as early as 1964 by [5]. It was not until 1992, however, that these ideas were fully understood and brought together by [6] in an article, which is the founding article of the wide margin separators. From 1995, which corresponds to the publication of [7], SVMs gained in popularity and were used in many applications. To imitate and duplicate human intelligence is undoubtedly the ultimate dream of researchers in the field of artificial intelligence. The first work on Artificial Neural Networks (ARNs) was carried out by [8] in 1943. They show that it is possible to build systems satisfying Turing's definition for calculating machines and therefore capable of calculating logical functions. In 1949, [9] established the law or theory of assemblies of neurons. This is a rule for learning artificial neural networks in the context of studying neural assemblies. This rule now known as 'Hebb's rule" suggests that when two neurons are excited together, a bond between them is created or strengthened. It describes how cells learn to modify the weights of the connections between them. It is used both as a hypothesis in neuroscience and as a concept in neural networks in mathematics. In 1957, the integration of Hebb's law while taking into account the error observed in the output allowed the success of Franck Rosenblatt's perceptron [10]. This model of perceptron constitutes the first artificial system capable of learning by experience. In the 1950s, researchers began to develop not only software but also hardware models to simulate low level functions of the nervous system [11]. Despite all these efforts, the triumph of ANR fell in the 1970s. A serious blow was dealt to the scientific community revolving around neural networks. Indeed, in 1969, [12] highlighting some theoretical limitations of Perceptron Rosenblatt, and more generally of linear classifiers, notably the impossibility of dealing with nonlinear or connectedness problems. But at the start of the 1980s, several researchers made it possible to relaunch research in this field. In 1982, [13] breathed new life into neural networks by publishing an article introducing a new model of neural network which is completely recurrent. This model suffers from the inability to deal with non-linear problems. A new generation of neural networks, capable of successfully processing non-linear phenomena and not possessing the defects highlighted by Marvin Minsky, was initially proposed for the first time by [14, 15] which is the first to study the back-propagation of the error. This multilayer Perceptron simultaneously introduced in detail by [16, 17] in 1986 is based on the back-propagation of the error gradient in its different layers. Neural networks subsequently experienced a considerable boom. Currently the use of neural networks is becoming a reality and research continues. Several articles have been published on the reciprocal contributions of neural networks [18–20]. Artificial neural networks are therefore an attempt to model biological neural networks from which implicit associations in the brain emerge. According to researchers and artificial intelligence experts, translating their structure into computer systems would allow programs to acquire implicit knowledge on their own. A good understanding of the functioning of biological neurons allows us to better understand that of their computer analogues.

3 Research Methodology

Over the past few years, a solid technical foundation on cloud computing has been developed, ranging from allocating resources on the Infrastructure-as-a-Service (IaaS) layer to performing tasks and applications on the layer. PaaS (Platform-as-a-Service), Use of software and services on the SaaS (Software-as-a-Service) layer. Cloud computing services have emerged as a result of efforts to organize and commission IT resources. Typically, these services are owned by service providers that allow consumers to use cloud resources on a paid basis: the consumer only pays for the resources that were actually used to solve their problem (for example: bandwidth, storage space, etc.). Managing current SLAs in the cloud market is often non-transparent and inefficient for both customer and vendor. However, SLAs are essential to build trust among all participants, especially for businesses whose success depends on purchased cloud service with appropriate quality of service. The way legal contracts and tier agreements of service (SLA) between cloud computing service providers and consumers of cloud computing services are established and managed is currently less than ideal from a customer perspective [21, 22]. This subject naturally obtains crucial importance for customers who are firms whose success depends partially or totally on the advertised quality of service (QoS) provided by the service suppliers. Three facts documented by [23–25] illustrate in an exemplary manner the gap that exists between service level contracts in the cloud market: inflexible pricing and conditions, non-transparent verification of the advertised quality of service and non-rigid expression of service level agreements to establish a network of trusted vendors as a global marketplace for cloud services, where all participants can come together, showcase their product offering and reach a wider customer community.

This thesis introduces a new concept of inter-cloud agreement broker. The contract brokers of this global network of suppliers act as neutral and autonomous entities in the cloud market and build trust between independent suppliers and between suppliers and customers (Fig. 1).

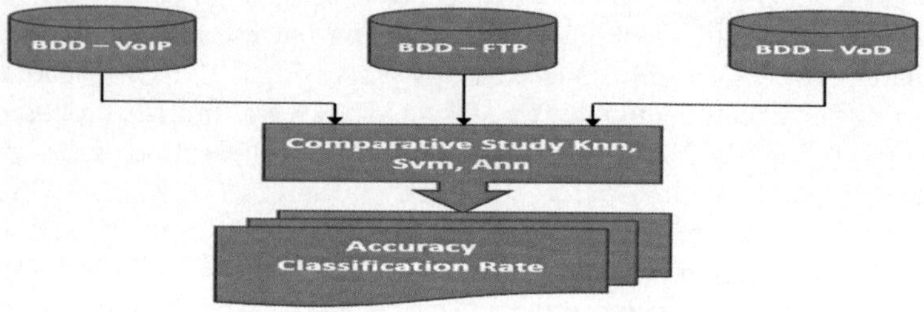

Fig. 1. synoptic diagram of our study

3.1 Experimental Study

3.1.1 Dataset

The quality of service requirements of our framework are based on the minimum accepted values of each service, as described in the table below (Table 1):

Table 1. QOS requirements

Service	Data rate	Delay	Loss rate
VoIP	64 Kbps	<150 ms	<0,1%
FTP	11,8 Kbps	~10 s	0
Video	4–60 Mbps	<150 ms	<0,0001%

Our dataset is based on a set of offers proposed by several service providers and which are approximate values of that cited in the table.

3.1.2 Algorithm Description

SVM

To solve classification problems, this method relies on the existence of a linear classifier in an appropriate space, and on the use of kernel functions that allow optimal data separation [26]. The goal of the svm is to find the optimal separator hyperplane that maximizes the margin between classes in a large space. Let's define the margin of a hyperplane as the distance between the hyperplane and the nearest datum. SVM allows you to use a number of cores in your Support Vector Machines models. These kernels can be of several types linear, polynomial, radial Gaussian function (RBF) and sigmoid.

Algorithm 1: Classification of provider's offers by SVM

---

1. database construction
2. Reading the Learning Matrix, and the Test Matrix
3. SVM training
4. Calculation of the decision function f (x) with the chosen kernel
5. Prediction of the elements of the test matrix by f (x)
6. Calculate the recognition rate;
7. Calculate the accuracy;

ANN

The number of layers, the number of neurons per layer and the interconnections between the different units of the network define the architecture. Multilayer neural networks are usually built according to the "normalized" model and include 3 or 4 layers in all (so 1 or 2 hidden layers). Most multilayer neural networks are, in practice, multilayer perceptron (PMC) [27].

Algorithm 2: Classification of provider's offers by ANN

---

1. Attribute to all the linkages a Random weights to start the algorithm.
2. Find the activation rate of Hidden Nodes Using the inputs and the (Input => Hidden node) linkages
3. Find the activation rate of Output Nodes Using the activation rate of Hidden Nodes and linkages to Output.
4. Find at the output node the error rate and recalibrate all the linkages between Output Nodes and hidden nodes.
5. Throught the error and Weights found at Output node, cascade down the error to Hidden Nodes
6. Recalibrate the weights between hidden node and the input nodes.
7. Repeat or iterate the process till the convergence criterion is met.
8. Using the final linkage weights score note the activation rate of the output nodes.
9. Calculate the recognition rate;
10. Calculate the precision;

KNN

The k-nearest neighbor algorithm is an intuitive, easily configurable algorithm to deal with a classification problem with any number of labels [28]. The k-nearest neighbor algorithm is one of the simplest supervised machine learning classification algorithms. The only tool we need is a distance between the elements we want to classify. If we represent these elements by coordinate vectors, there are generally quite a few possible choices for these distances, starting from the simple usual distance (Euclidean).

Algorithm 3: Classification of provider's offers by KNN
D: the set of learning vectors of the QOS such that D = {(x ', c), c ∈ C};
x: QOS offers from providers whose membership class is to be determined;
Exit: The membership of the offer x;
1.for each ((x ', c) ∈ D) do
2. Calculate the distance d (x, x ');
3.end
4.for each {x ' ∈ knn (x)} do
5. count the number of occurrences of each class;
6.end
7. Assign the most frequent class to x;
8. Calculate the recognition rate;
9. Calculate the precision;

3.1.3 Environment

Table 2 details the key artifacts characterizing our experimental environment.

Table 2. Experimental environment

Component	Configuration
OS	Windows 10
CPU	Intel Core i7–7500 270 GHz
RAM	8 Go
GPU	GeForce GTX 1080
Python	3.5

Our experimental study was performed on Matlab tool: Matlab is a commercial interactive calculation software. It allows to perform numerical simulations based on numerical analysis algorithms. It can therefore be used for the approximate resolution of differential equations, partial differential equations or linear systems. Matlab is composed of a suite of instructions and dedicated toolbox to perform vector or matrix operations and statistical calculations, it is increasingly used in industry and banks to develop software prototypes and test new algorithms.

4 Result and Discussion

The results of the comparative study between the three data classification algorithms (RN, SVM, KNN) (Fig. 2 and Table 3):i

SVMs represent several advantages, notably that different kernel functions can be specified, they have a solid theoretical basis and they are efficient in large spaces. Despite their performance, SVMs also represent weaknesses. They use complex mathematical functions for classification which require an enormous amount of time during the test

Table 3. Assessment test of KNN.

Assessment test	Value of K	VoIP	VoD	FTP
Classification rate (%)	K = 1	0,6719	0,6763	0,6998
	K = 5	0,7252	0,7053	0,7311
	K = 9	0,7278	0,7303	0,7442
	K = 13	0,7462	0,7382	0,7429
	K = 17	**0,7447**	**0,7645**	**0,752**
Accuracy (%)	K = 1	0,5493	0,7494	0,7537
	K = 5	0,5228	0,7514	0,7754
	K = 9	**0,8381**	0,7656	0,7827
	K = 13	0,5361	**0,8098**	**0,7827**
	K = 17	0,5218	0,7975	0,6805

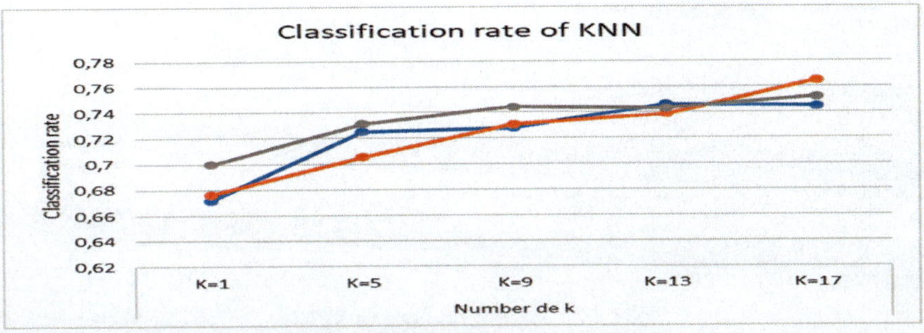

Fig. 2. Classification rate of KNN

phases. The results of this experiment for the three databases are better (Fig. 3 and Table 4).

ANN are flexible and generic. They deal with unstructured problems on which no information is available in advance even they work on incomplete or noisy data. This information gap can be supplemented by adding more neurons to the hidden layer. Neural networks also have drawbacks, the slowness of learning and the difficulty of choosing initial values of the connection weights. The results of this experiment for the three databases are satisfactory, that the classification rate values for the three databases with all the tools are satisfactory with a disturbance sometimes the rate increases and sometimes it decreases by a small difference when changes the number of neurons which indicates that the change in PMC architecture may influence the results (Fig. 4 and Table 5).

The KNN algorithm is robust against noisy data and the k nearest neighbors method is efficient if the data is large and incomplete, in addition KNN is one of the simplest of all machine learning algorithms. On the other hand, we always need to determine the

Table 4. Assessment test of SVM

Assessment test	Polynomial degree	VoIP	VoD	FTP
Classification rate (%)	P = 1	**0,8654**	**0,88**	**0,837**
	P = 5	0,8365	0,83	0,7704
	P = 7	0,827	0,82	0,8037
	P = 9	0,827	0,82	0,8037
	P = 10	0,8173	0,79	0,7407
Accuracy (%)	P = 1	0,7272	**0,75**	0,81
	P = 5	**1**	0,5217	0,8173
	P = 7	**1**	0,5	0,8301
	P = 9	0	0,4444	0,7574
	P = 10	0	0,4137	**0,8425**

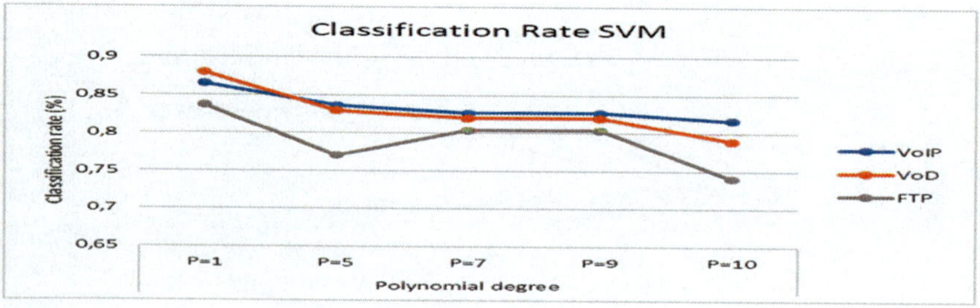

Fig. 3. Classification rate of SVM

value of the number of closest neighbors (the parameter k) so the prediction time is very long since we must calculate the distance of all the examples and finally This method is greedy in memory space because it uses large storage capacity for processing. The experiment was carried out on bases of different size. The classification rate is optimistic and the results very close for the different values of k. from an execution time point of view, we observe that the nearest neighbor k classifier classified the data well, as an important remark the more the number of K increases the more the classification rate increases, this shows that the choice of K is important and affects the results, when the K is large means that the KNN classifier exploits more information about the neighborhood which can improve the classification results.

For KNN classifier, the highest accuracy 83.81%, was achieved at k = 9. For SVM classifier, the highest accuracy 100%, was obtained at Polynomial degree 5 and 7. For ANN classifier, the highest accuracy 84.56%, was obtained at N = 15. Classification time is more useful than training time due too training that can be executed offline and is only needed to be executed once. Accordingly, the accuracy and classification rate of three types of classifiers are compared in this paper.

Table 5. Assessment test of ANN

Assessment test	Number of neurons	VoIP	VoD	FTP
Classification rate (%)	N = 5	0,8333	0,8112	0,8141
	N = 10	0,8296	0,8186	0,8111
	N = 15	0,8407	0,8075	**0,8259**
	N = 20	0,8444	**0,8449**	0,8259
	N = 25	**0,8481**	0,8038	0,8037
Accuracy (%)	N = 5	0,8004	0,7948	0,8129
	N = 10	0,7852	**0,8141**	0,8235
	N = 15	0,789	0,7881	**0,8456**
	N = 20	**0,8153**	0,7966	0,8322
	N = 25	0,7993	0,7815	0,8212

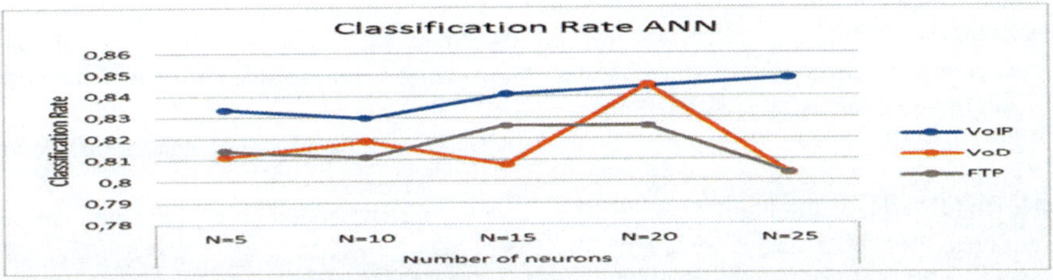

Fig. 4. Classification rate of ANN

5 Conclusion and Future Work

In our work, we are interested in the problem of classification, this paper led us to the development of a comparative study using the three algorithms RN, SVM and Knn applied to the three databases. The results of this study led us to choose the SVM classifier as an algorithm for classifying web service providers' offers for our broker, because the SVM classifier achieved the fastest classification rate and the highest accuracy (88%). SVM and ANN are the most effective classification methods. Our work is a new contribution to the many studies on classification. For a more complete assessment, and from what we have seen through this work we could consider some perspectives in order to note what we can improve in this one, it would be relevant to test other types of classifiers, especially those that support the unsupervised model and use other software like Python or Keel to see the behavior of these algorithms.

References

1. Marcon, E., Chaabane, S., Sallez, Y., Bonte, T., Trentesaux, D.: A multi-agent system based on reactive decision rules for solving the caregiver routing problem in home health care. Simul. Model. Pract. Theor. **74**, 134–151 (2017)

2. Vapnik, V., Lerner, A.: Pattern Recognition using Generalized Portrait Method. Autom. Remote Control **24**, 774–780 (1963)
3. Dash, M., Liu, H.: Feature selection for classification. Intell. Data Anal. **1**, 131–156 (1997)
4. Shakhnarovich, G., Darrell, T., Indyk, P.; Nearest-Neighbor Methods in Learning and Vision: Theory and Practice. The MIT Press, Cambridge (2006)
5. McQueen, J.: Some methods for classification and analysis of multivariate observations. In: Proceedings of the 5th Berkeley Symposium on Mathematical Statistics and Probability, vol. 1, pp. 281–297 (1967)
6. Aizerman, M., Braverman, E., Rozonoer, L.: Theoretical foundations of the potential function method in pattern recognition learning. Autom. Remote Control. **25**, 821–837 (1964)
7. Bernhard, E., Boser, I., Guyon, M., Vapnik, V.: A training algorithm for optimal margin classifiers. In: 5th Annual Workshop on Computational Learning Theory, Pittsburgh, pp. 144–152. ACM (1992)
8. Vapnik, V.N.: The Nature of Statistical Learning theory. Springer, New York (1995)
9. McCulloch, W.S., Pitts, W.: A logical Calculus of the ideas immanent in nervous activity. Bull. Math. Biophys. **5**, 115–133 (1943)
10. Hebb, D.O.: The Organization of Behavior. Wiley, New York (1949)
11. Rosenblatt, F.: The perceptron: a probabilistic model for information storage and organization in the brain. Psychol. Rev. **65**(6), 386–408 (1958)
12. Riviere, D., Mangin, J., Papadopoulos, D., Martinez, J., Frouin, V., Regis, J.: Automatic recognition of cortical sulci of the human brain using a congregation of neural networks. Med. Image Anal. **6**(2), 77–92 (2002)
13. Minsky, M., Papert, S.: Perceptrons: an introduction to computational geometry. Inf. Control **17**, 501–522 (1972)
14. Hopfield, J.: Neural networks and physical systems with emergent collective computational abilities. Proc. Natl. Acad. Sci. **79**, 2554–2558 (1982)
15. Werbos, P.J.: Beyond Regression: New Tools for Prediction and Analysis in the Behavioral Sciences. Ph.D. thesis, Harvard University (1974)
16. Werbos, P.J.: Back propagation through time: what it does and how to do it. Proc. IEEE **78**(10), 1550–1560 (1990)
17. Rumelhart, D.E., Hinton, G.E., Williams, R.J.: Learning internal representations by error propagation. In: Parallel Distributed Processing: Explorations in the Microstructure of Cognition: Foundations, pp. 318–362. MIT Press (1987)
18. LeCun, Y.: A learning scheme for asymmetric threshold networks. In: Proceedings of Cognitive, Paris, France, vol. 85, pp. 599–604 (1985)
19. Zouari, H.: Contribution à l'évaluation des méthodes de combinaison parallèle de classifieurs par simulation. Université de Rouen, Thèse de Doctorat (2004)
20. Ripley, B.D.: Pattern Recognition and Neural Networks. Cambridge University Press, Cambridge (1996)
21. Rasheed, H., Rumpl, A., Waldrich, O., Ziegler, W.: A standards-based approach for negotiating service QoS with cloud infrastructure providers. In: eChallenges e-2012 Conference Proceedings, pp. 1–9 (2012)
22. Bakraouy, Z., Abbass, W., Baina, A., Bellafkih, M.: The IT infrastructure's industrialization and mastering. J. Commun. **14**(10), 884–891 (2019)
23. Hon, W.K., Millard, C., Walden, I.: Negotiating cloud contracts: looking at clouds from both sides now. Stanford Technol. Law Rev. **16**(1), 79–129 (2013)
24. Bakraouy, Z., Abbass, W., Baina, A., Bellafkih, M.: MAS for services availability in cloud of things network: monitoring and reactivity. In Proceedings of the 12th International Conference on Intelligent Systems: Theories and Applications, vol. 10, pp. 1–6 (2018)

25. Bakraouy, Z., Baina, A., Bellafkih, M.: System multi agents for automatic negotiation of SLA in cloud computing. In: Abraham, A., Haqiq, A., Muda, A.K., Gandhi, N. (eds.) Innovations in Bio-Inspired Computing and Applications. Advances in Intelligent Systems and Computing, vol. 735, pp. 234–244. Springer, Cham (2018)
26. Mohamadally, H., Fomani B.: SVM : Machines à Vecteurs de Support ou Separateurs a Vastes Marges. BD Web, ISTY3, Versailles St Quentin, pp. 1–20, 16 January 2006
27. Krogh, A.: What are artificial neural networks? Nat. Biotechnol. **26**, 195–197 (2008)
28. Mathieu-Dupas, E.: Algorithme des k plus proches voisins pondères et application en diagnostic. In: Actes des 42èmes Journées de Statistique, France, pp. 2–4 (2010)

Secret Disclosure Attacks on a Recent Ultralightweight Mutual RFID Authentication Protocol for Blockchain-Enabled Supply Chains

Paolo D'Arco and Zahra Ebadi Ansaroudi(✉)

Dipartimento di Informatica, Università degli Studi di Salerno, Fisciano, Italy
{pdarco,zebadiansaroudi}@unisa.it

Abstract. Radio Frequency Identification (RFID) technology has emerged as a suitable technology for various applications of the Internet of Things (IoT). Two types of components, tiny labels called tags and small devices called readers, enable them to associate identifying information to objects (through the tags), which can be automatically read and identified (through the readers). Hence, RFID authentication protocols, needed by each entity to be sure of the identity of the other entity with whom it is interacting, play a crucial role in the deployment of secure IoT applications. Many efforts have been devoted in recent years to the design of efficient and secure protocols. However, these protocols usually need a server to maintain a database of sensitive information for all the tags used in the application, making such a server more vulnerable to security attacks. Several blockchain-based authentication protocols have been developed to take advantage of some blockchain capabilities, e.g. decentralization and immutability, to address this issue and design secure authentication protocols. In this paper, we focus on one of these: we analyze the security vulnerabilities of a recent ultra-lightweight mutual RFID authentication protocol for blockchain-enabled supply chains. Despite the detailed formal security analysis provided by the authors, carried out by using the Gong, Needham and Yahalom logic, and by using automatic validation security tools, we present two secret disclosure attacks against the protocol. The first one is an active attack, while the second is a fully passive attack.

Keywords: Ultra-lightweight · Authentication · RFID · Blockchain

1 Introduction

The Internet of Things (IoT) notion has evolved over the last decades. It refers to a network of connected objects. Such objects may be cellular phones, household appliances, industrial machineries, vehicles, sensors, to name a few, which collect and share data over the Internet. The data are information about objects' identity, location, and so on.

A. Abraham et al. (Eds.): SoCPaR 2020, AISC 1383, pp. 880–890, 2021.
https://doi.org/10.1007/978-3-030-73689-7_83

Radio Frequency Identification (RFID) as an automatic identification method plays a major role in IoT realization. It uses radio waves to detect, track, and identify objects and thereby manage them. Typically, an RFID system includes one or more RFID readers and tags. An RFID tag attached to an object collects and transmits data to an RFID reader. The RFID reader, which is connected to the internet, then reads them. Eventually, the data are processed to suit the needs of a particular IoT application. RFID tags may fall into two groups: active and passive. The classification depends on how powered they are. The active one is powered by a source such as a battery and has powerful computation capacity, while the passive one is battery-free and performs very simple operations.

Over the years, RFID technology has been widely developed but privacy and security issues continue to be a serious obstacle to its fully deployment. That is due to the ease of malicious attacks on wireless communications including eavesdropping, traffic analysis, denial of service, de-synchronization, impersonation, and several other attacks. Identifying the legitimate tag/reader and ensuring that an adversarial party can not trace a tag are, for example, two main properties that we would like to get. Other researchers have addressed these properties over the years and proposed various solutions, such as authentication [4,12,21].

Bearing in mind that RFID tags are limited in resource, computationally intensive cryptographic algorithms can not be used to secure the wireless channel between the RFID reader and tags. It thus makes sense investigating the design of ultralightweight authentication protocols that only use simple bit-wise operations such as XOR, OR, AND, and rotation.

So far, a large number of research papers on RFID ultralightweight authentication protocols have appeared in the literature since 2006, with the works of Peris-Lopez et al. [18–20], and one year later, Chien [7] as a starting point for the diffusion of the approach. However, almost all of them presented security weaknesses and privacy flaws; for example see [3,5,6,8,11] and recently [9,10] for details and an overview of the field.

Blockchain is a tamper-proof digital distributed ledger based on consensus rules and without a central entity. It allows a community of entities to record their transactions within that community in a shared ledger, to transfer value among themselves, such that transactions are unalterable [28]. Consequently, it is believed to be a suitable tool for addressing security challenges in IoT [17] in light of the blockchain decentralized and immutability capabilities. Accordingly, over the last few years, an increasing number of research studies have reported on blockchain-based decentralized RFID authentication. Just to name a few: Wang et al. [27] proposed a new blockchain-based mutual authentication protocol for distributed RFID systems. Using this protocol, RFID tags are authenticated using an immutable transaction on the blockchain. Hammi et al. [14] proposed a decentralized blockchain-based approach, named bubbles of trust, to provide authentication and identification of IoT devices. The bubbles of trust are secure virtual zones (bubbles), and IoT devices inside them can identify and trust each other. Lin et al. [16] proposed a novel framework that takes into account the advantages of blockchain along with cryptographic primitives like attribute-based signatures, multi-receiver encryption, and message authentication code to ensure secure anonymous authentication, auditability, and confidentiality. Jangirala et al. [15] proposed a new efficient lightweight blockchain-enabled RFID-based authentication protocol for supply chains in 5G mobile edge computing environment, which is better off in terms of security and communication and computational costs than the existing state of art protocols.

Very recently, Sidorov et al. [24] proposed an ultralightweight mutual authentication RFID protocol. It utilizes a decentralized database to set up a secure blockchain-enabled supply chain management system. The authors have also proven the security of the protocol by using the Gong, Needham, and Yahalom logic [13], and AVISPA tool [2]. Although, it was broken a few months later by Safkhani and Bagheri [23]. They proposed a full disclosure attack on it. In this paper, we present two other scenarios of full disclosure attacks: passive and active. Our proposed passive attack is more efficient than the active attack proposed in [23] as it only needs to eavesdrop two sessions of the protocol.

The rest of this paper is organized as follows: Sect. 2 gives an overview of the above-mentioned protocol. Section 3 presents our methods of attack on the protocol with a toy example. Section 4 reviews the previously known attack on the protocol. Finally, Sect. 5 provides some final comments and closes the paper.

2 Protocol Overview

In this section, we present a brief overview of the protocol and particularly the authentication phase of it (for details, please refer to [24]).

An ultralightweight mutual RFID authentication protocol for blockchain-enabled supply chains is a three-round authentication protocol between an RFID tag and a supply chain node, which want to be sure of each other. The supply chain node interacts with the tag through an intermediate reader. In Fig. 1 the protocol we look at is shown. In a nutshell, each tag has two secrets, IDS and K, stored in the tag as well as in a blockchain. Each supply chain node has a database stored in a blockchain with all the tag secrets. The secrets are updated after any successful authentication session. An authentication session works as follows: the reader first queries the tag with a 'Hello' message and a 96-bit random number T. Then the tag chooses a 96-bit random n and replies with two messages A and B, computed by using n. The maximum number of attempts the tag can send messages A and B to a given number T has been set by the protocol to five. Once it is exceeded, the reader sends a 'KILL command' and terminates the tag. If the supply chain node verifies A and B, it replays with two other messages, C and D, and ends the protocol. Finally, the tag checks C and D, and, if they are correct, it updates its secrets. Figure 1 displays relevant message structures and their computations. The notations are summarized in Table 1.

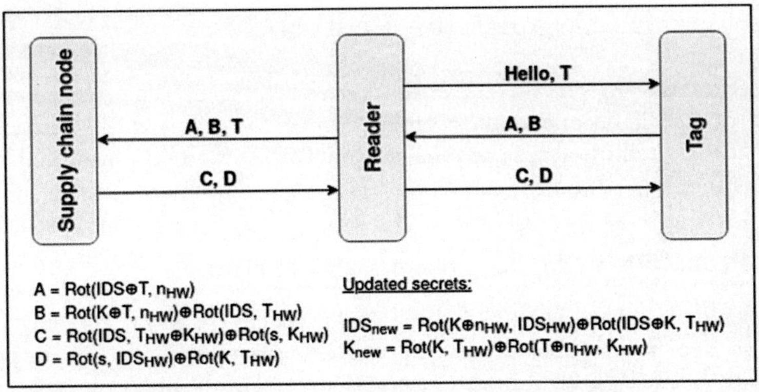

Fig. 1. Authentication part of the protocol

Remark 1. In [24] n_{HW} is not specified in IDS_{new} and K_{new} formulas, however, *the computational cost for the data update phase of the RFID tag* provided in Section VII of that paper indicates that it is the Hamming weight of the generated random number n. In the following, we use the same notation to denote the Hamming weight of a string.

Table 1. Protocol notations

Notation	Meaning
T, n, s	Random numbers generated by reader, tag and supply chain node, respectively
$\oplus$	Bit-wise exclusive or
K, K_{new}	Current key and updated key, respectively
IDS, IDS_{new}	Current and updated index pseudonym, respectively
$Rot(X, Z_{HW})$	Left rotation of X by the Hamming weight of Z
$RRot(X, Z_{HW})$	Right rotation of X by the Hamming weight of Z

3 Our Attacks Against the Protocol

In this section, we propose two secret disclosure attacks on the protocol in detail. They are probabilistic and deterministic, respectively. The probabilistic one is an active attack and discloses the tags' secrets with probability around 0.33. The deterministic one is a passive attack and is always successful.

3.1 A Probabilistic Attack Against the Protocol

Our probabilistic attack is an attack on the protocol based on guessing the tag secrets. To accomplish this attack, an attacker eavesdrops on an authentication session between a reader and tag and uses the obtained messages to get a successful guess. Figure 2 shows the attack procedure. In brief, the attacker follows Algorithm 1 to find secrets. First, it eavesdrops an authentication session messages T_1, A_1 and B_1. It then attempts to extract the session secrets from messages A_1 and B_1. It does this by *guessing* the Hamming weight of the random number generated by the tag n_{HW} and computing five *tentative* updated values for IDS_{new} and K_{new}. It tries to verify his guesses, later, when the reader starts the next successive authentication session and sends a 'Hello' message and a random number T_2. At this point, the attacker blocks the real tag response messages and sends his messages A_2 and B_2 for a maximum of five times (as stated earlier, the tag is allowed to respond only five times). The messages are computed using the tentative updated secrets. If the response is accepted, then the secrets have been revealed. Notice that, in the pseudocode, with $IDS_{1,i_{HW}}$ and $k_{1,i_{HW}}$ we denote the Hamming weight of $IDS_{1,i}$ and $k_{1,i}$, respectively. Same notation for the Hamming weights of the other values and messages.

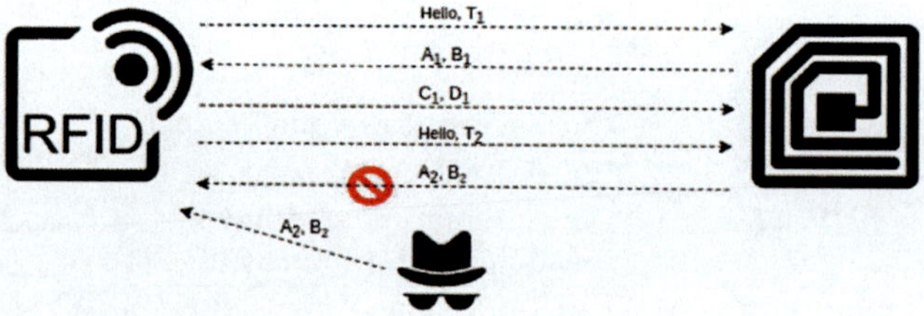

Fig. 2. Probabilistic attack on the protocol

We succeed with high probability in up to five trials. Indeed, the Hamming weight of a 96-bit number has a binomial distribution in the set $\{0,\ldots,96\}$, instead of a uniform distribution. That is why, considering $K = \{0,1\}^{96}$ as a set of 96-bit strings, the probability of choosing a string that gives us Hamming weight d in the set $\{0,\ldots,96\}$ would be $\frac{1}{2^{96}}\binom{96}{d}$. On the other hand, if d is chosen uniformly at random, the probability would be $\frac{1}{96}$. We have shown the Hamming weight distribution for 1,000,000 binary numbers in Fig. 3. So we

Algorithm 1: Find secrets - probabilistic way

Input: A_1, B_1, T_1, T_2
for $i \leftarrow 1$ *to* 5 **do**
  $n_{HW} \leftarrow$ choose a random number from (43,53)
  $IDS_{1,i} \leftarrow RRot(A_1, n_{HW}) \oplus T_1$
  $k_{1,i} \leftarrow RRot((Rot(IDS_{1,i}, T_{1_{HW}}) \oplus B_1), n_{HW}) \oplus T_1$
  $IDS_{1_{new_i}} \leftarrow Rot(k_{1,i} \oplus n_{HW}, IDS_{1,i_{HW}}) \oplus Rot(IDS_{1,i} \oplus k_{1,i}, T_{1_{HW}})$
  $K_{1_{new_i}} = Rot(k_{1,i}, T_{1_{HW}}) \oplus Rot(T_1 \oplus n_{HW}, k_{1,i_{HW}})$
end
The tag receives a new 'Hello' message with T_2
Block the tag response
for $i \leftarrow 1$ *to* 5 **do**
  $n_{HW} \leftarrow$ choose a random number from (0,95)
  $A_{2,i} \leftarrow Rot(IDS_{1_{new_i}} \oplus T_2, n_{HW})$
  $B_{2,i} \leftarrow Rot(K_{1_{new_i}} \oplus T_2, n_{HW}) \oplus Rot(IDS_{1_{new_i}}, T_{2_{HW}})$
  send $A_{2,i}$ and $B_{2,i}$ to the reader
  if *accepted* **then**
    $IDS \leftarrow IDS_{1_{new_i}}$
    $k \leftarrow K_{1_{new_i}}$
    break
  end
end
return IDS, K

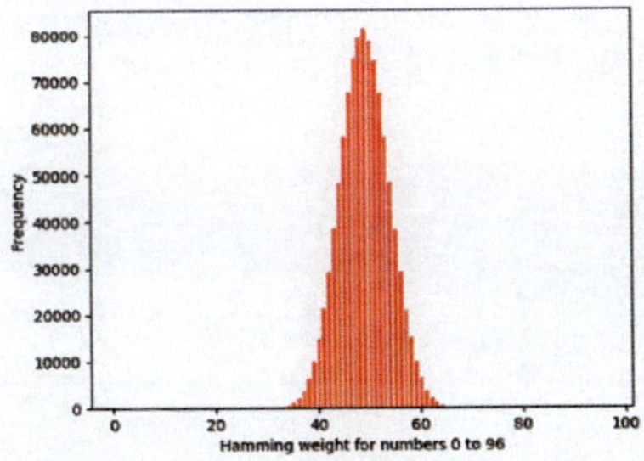

Fig. 3. Hamming weight distribution

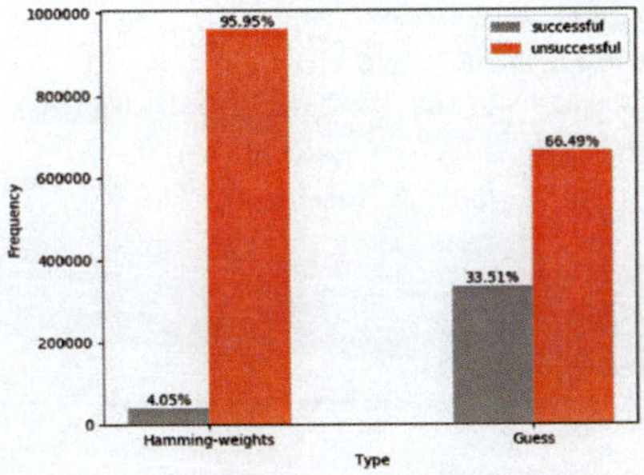

Fig. 4. Prediction of IDS using Guess and Hamming-weights sets

would expect the subset Guess = $\{43, \ldots, 53\}$ to cover almost all the cases. We run Algorithm 1 with n_{HW} values selected from Guess set. The attack efficiency is measured by comparing it with the case where n_{HW} is drawn according to the uniform distribution on Hamming-weights = $\{0, \ldots, 96\}$. We run 1,000,000 tests. As shown in Fig. 4, the attack is successful in roughly 33% of the cases.

3.2 A Deterministic Attack Against the Protocol

Our deterministic attack is a passive attack on the protocol where the attacker only eavesdrops on two consecutive authentication sessions. It finds the secrets in several steps, as follows: first, it sets n_{HW} (the hamming weight of a random number generated by the tag) to be equal to v, for each value, v in $\{0, \ldots, 96\}$ and, from the messages A_1, B_1 and T_1 eavesdropped in the first session, it extracts 97 tentative values for the secrets IDS and K. Then, for each choice of v, it computes the corresponding updated secret IDS_{new}. Similarly, using the messages A_2, B_2 and T_2 eavesdropped in the second session, it extracts 97 tentative values of IDS. Since these are consecutive sessions, the computed IDS_{new} from the first session should be equal to IDS of the second session. So a simple search between the two lists yields IDS and subsequently K values (see Algorithm 2 for details).

Algorithm 2: Find secrets - deterministic way

Input: $A_1, B_1, T_1, A_2, B_2, T_2$
for $i \leftarrow 0$ *to* 96 **do**
  $IDS_{1,i} \leftarrow RRot(A_1, i) \oplus T_1$
  $k_{1,i} \leftarrow RRot((Rot(IDS_{1,i}, T_{1_{HW}}) \oplus B_1), i) \oplus T_1$
  $IDS_{1_{new_i}} \leftarrow Rot(k_{1,i} \oplus i, IDS_{1,i_{HW}}) \oplus Rot(IDS_{1,i} \oplus k_{1,i}, T_{1_{HW}})$
  $IDS_{2,i} \leftarrow RRot(A_2, i) \oplus T_2$
end
for $i \leftarrow 0$ *to* 96 **do**
  if $(IDS_{1_{new_i}} == IDS_{2,i})$ **then**
    $IDS \leftarrow IDS_{2,i}$
    $k \leftarrow RRot((Rot(IDS, T_{2_{HW}}) \oplus B_2), i) \oplus T_2$
    break
  end
end
return IDS, K

Toy Example: A toy example with six-bit length strings is given below to show how the attack works. We follow the procedure used by Algorithm 2 and compute IDS and K using the values provided in Table 2.

$$T_1 = 010010, A_1 = 111100, B_1 = 110111$$
$$IDS = 110101, IDS_{1_{new}} = 011011, K = 010110, n_{HW} = 3$$
$$T_2 = 111010, A_2 = 110000, B_2 = 011010$$
$$IDS_2 = 011011, K_2 = 001110, n_{HW} = 5$$

Table 2. $IDS1_{new}$ and IDS_2 sets

i	0	1	2	3	4	5	6
$IDS1_{new_i}$	110000	110100	110010	011011	010110	110101	010001
$IDS2_{2,i}$	001010	100010	110110	111100	111001	011011	001010

$IDS = IDS1_{new} \cap IDS2 = 011011$, $n_{HW} = 5$
$K = RRot((Rot(IDS, 4) \oplus 011010), 5) \oplus 111010 = 001110$

4 Related Works

As noted above, the Hamming weight function follows a binomial distribution that makes the use of it inappropriate in the design of security protocols [3]. There is much in the literature that confirms this fact. The various security attacks on RFID authentication protocols explained in [1,22,23,25,26], are a few examples. These attacks use the Hamming weight to estimate all possible values of the security parameters used in the protocol and eventually identify the right one. In [23], a full disclosure attack is presented by Safkhani and Bagheri on the protocol discussed in this paper. They use an adversary model where an adversary can eavesdrop and block messages transmitted via reader-tag communications. Their attack scenario as shown in Fig. 5 includes the following steps: first, the adversary eavesdrops all transmitted messages in an authentication session, that are T_1, A_1, B_1, C_1, and D_1. The adversary then blocks the reader reply C_1 and D_1 and prevents updating the tag current IDS and K. Afterward, the adversary waits until the reader initiates a new session and sends a 'Hello' message along with a newly generated random number T_2 to the tag. At this point, the adversary once again eavesdrops messages A_2, B_2, C_2, and D_2 transmitted over the session. Finally, it processes the collected messages, as shown in the following box, using nested for loops and outputs IDS and K. It is further explained below with a toy example.

$for\ i \in 0, 1, ..., 95$
$for\ j \in 0, 1, ..., 95$
$if(RRot(A_1, i) \oplus RRot(A_2, j) == T_1 \oplus T_2)$
$IDS = RRot(A_1, i) \oplus T_1$
$K = RRot((Rot(IDS, T1_{HW}) \oplus B_1), i) \oplus T_1$
$return\ IDS,\ and\ K$

Toy Example: Six-bit strings are provided here. We use the nested for loop and calculate IDS. Knowing IDS, it is straight-forward to calculate K.

$T_1 = 010010, A_1 = 111100, B_1 = 110111$
$IDS = 110101, K = 010110, n_{HW} = 3$
$T_2 = 111010, A_2 = 100111, B_2 = 001011$
$IDS = 110101, K = 010110, n_{HW} = 5$
$T_1 \oplus T_2 = 101000$

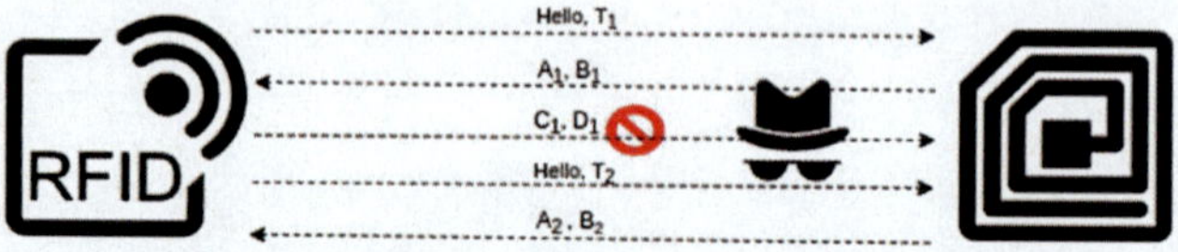

Fig. 5. Proposed attack on the protocol in [23]

$RRot(A_1, i) \oplus RRot(A_2, j)$ for $i, j \in \{0, 1, ..., 5\}$ is shown in gray cells in Table 3. Two cells have value of $T_1 \oplus T_2$. The IDS formula (i.e., $IDS = RRot(A_1, i) \oplus T_1$) is used to calculate IDS as follows:

1) $i = 2, j = 0 \Rightarrow IDS = 001111 \oplus 010010 = 011101$ ✗
2) $i = 3, j = 5 \Rightarrow IDS = 100111 \oplus 010010 = 110101$ ✓
$\rightarrow K = RRot((Rot(IDS, T_{1_{HW}}) \oplus B_1), i) \oplus T_1 = 010110$

Table 3. $RRot(A_1, i) \oplus RRot(A_2, j)$ for $i, j \in \{0, 1, ..., 5\}$

⊕	i	0	1	2	3	4	5
j	$RRot(A_1, i)$ / $RRot(A_2, j)$	111100	011110	001111	100111	110011	111001
0	100111	011011	111001	101000	000000	010100	011110
1	110011	001111	101101	111100	010100	000000	001010
2	111001	000101	100111	110110	011110	001010	000000
3	111100	000000	100010	110011	011011	001100	000101
4	011110	100010	000000	010001	111001	101101	100111
5	001111	110011	010001	000000	101000	111100	110110

5 Conclusion

We addressed the security vulnerabilities of a recent ultra-lightweight mutual RFID authentication protocol for blockchain-enabled supply chains. The main weakness of the protocol is the use of the Hamming weight function to compute authentication messages. We have shown how an adversary could take advantage of this weakness designing a passive and an active secret disclosure attacks. Both attacks are easy to implement, and the active attack is very efficient.

References

1. Ahmadian, Z., Salmasizadeh, M., Aref, M.R.: Desynchronization attack on RAPP ultralightweight authentication protocol. Inf. Process. Lett. **113**(7), 205–209 (2013)

2. Armando, A., Basin, D., Boichut, Y., Chevalier, Y., Compagna, L., Cuéllar, J., Drielsma, P.H., Héam, P.C., Kouchnarenko, O., Mantovani, J., et al.: The AVISPA tool for the automated validation of internet security protocols and applications. In: International Conference on Computer Aided Verification, pp. 281–285. Springer (2005)
3. Avoine, G., Carpent, X., Hernandez-Castro, J.: Pitfalls in ultralightweight authentication protocol designs. IEEE Trans. Mob. Comput. **15**(9), 2317–2332 (2015)
4. Ayoade, J.: Roadmap to solving security and privacy concerns in RFID systems. Comput. Law Secur. Rev. **23**(6), 555–561 (2007)
5. Bilal, Z., Martin, K.: Ultra-lightweight mutual authentication protocols: weaknesses and countermeasures. In: 2013 International Conference on Availability, Reliability and Security, pp. 304–309. IEEE (2013)
6. Bilal, Z., Masood, A., Kausar, F.: Security analysis of ultra-lightweight cryptographic protocol for low-cost RFID tags: Gossamer protocol. In: 2009 International Conference on Network-Based Information Systems, pp. 260–267. IEEE (2009)
7. Chien, H.Y.: SASI: a new ultralightweight RFID authentication protocol providing strong authentication and strong integrity. IEEE Trans. Dependable Secure Comput. **4**(4), 337–340 (2007)
8. Chien, H.Y., Huang, C.W.: Security of ultra-lightweight RFID authentication protocols and its improvements. ACM SIGOPS Oper. Syst. Rev. **41**(4), 83–86 (2007)
9. D'Arco, P.: Ultralightweight cryptography. In: International Conference on Security for Information Technology and Communications, pp. 1–16. Springer (2018)
10. D'Arco, P., De Prisco, R.: Design weaknesses in recent ultralightweight RFID authentication protocols. In: IFIP International Conference on ICT Systems Security and Privacy Protection, pp. 3–17. Springer (2018)
11. D'Arco, P., De Santis, A.: On ultralightweight RFID authentication protocols. IEEE Trans. Dependable Secure Comput. **8**(4), 548–563 (2010)
12. Garfinkel, S.L., Juels, A., Pappu, R.: RFID privacy: an overview of problems and proposed solutions. IEEE Secur. Privacy **3**(3), 34–43 (2005)
13. Gong, L., Needham, R.M., Yahalom, R.: Reasoning about belief in cryptographic protocols. In: IEEE Symposium on Security and Privacy, pp. 234–248. Citeseer (1990)
14. Hammi, M.T., Hammi, B., Bellot, P., Serhrouchni, A.: Bubbles of trust: a decentralized blockchain-based authentication system for IoT. Comput. Secur. **78**, 126–142 (2018)
15. Jangirala, S., Das, A.K., Vasilakos, A.V.: Designing secure lightweight blockchain-enabled RFID-based authentication protocol for supply chains in 5G mobile edge computing environment. IEEE Trans. Industr. Inf. **16**(11), 7081–7093 (2019)
16. Lin, C., He, D., Huang, X., Choo, K.K.R., Vasilakos, A.V.: BSeIn: a blockchain-based secure mutual authentication with fine-grained access control system for industry 4.0. J. Netw. Comput. Appl. **116**, 42–52 (2018)
17. Panarello, A., Tapas, N., Merlino, G., Longo, F., Puliafito, A.: Blockchain and IoT integration: a systematic survey. Sensors **18**(8), 2575 (2018)
18. Peris-Lopez, P., Hernandez-Castro, J.C., Estevez-Tapiador, J.M., Ribagorda, A.: EMAP: an efficient mutual-authentication protocol for low-cost RFID tags. In: OTM Confederated International Conferences "On the Move to Meaningful Internet Systems", pp. 352–361. Springer (2006)
19. Peris-Lopez, P., Hernandez-Castro, J.C., Estévez-Tapiador, J.M., Ribagorda, A.: LMAP: a real lightweight mutual authentication protocol for low-cost RFID tags. In: Proceedings of 2nd Workshop on RFID Security, vol. 6 (2006)

20. Peris-Lopez, P., Hernandez-Castro, J.C., Estevez-Tapiador, J.M., Ribagorda, A.: M^2AP: a minimalist mutual-authentication protocol for low-cost RFID tags. In: International Conference on Ubiquitous Intelligence and Computing, pp. 912–923. Springer (2006)
21. Peris-Lopez, P., Hernandez-Castro, J.C., Estevez-Tapiador, J.M., Ribagorda, A.: RFID systems: a survey on security threats and proposed solutions. In: IFIP International Conference on Personal Wireless Communications, pp. 159–170. Springer (2006)
22. Peris-Lopez, P., Hernandez-Castro, J.C., Phan, R.C.W., Tapiador, J.M., Li, T.: Quasi-linear cryptanalysis of a secure RFID ultralightweight authentication protocol. In: International Conference on Information Security and Cryptology, pp. 427–442. Springer (2010)
23. Safkhani, M., Bagheri, N.: Cryptanalysis of two recently proposed ultralightweight authentication protocol for IoT. arXiv preprint arXiv:1907.11322 (2019)
24. Sidorov, M., Ong, M.T., Sridharan, R.V., Nakamura, J., Ohmura, R., Khor, J.H.: Ultralightweight mutual authentication RFID protocol for blockchain enabled supply chains. IEEE Access **7**, 7273–7285 (2019)
25. Sun, H.M., Ting, W.C., Wang, K.H.: On the security of Chien's ultralightweight RFID authentication protocol. IEEE Trans. Dependable Secure Comput. **8**(2), 315–317 (2009)
26. Wang, K.H., Chen, C.M., Fang, W., Wu, T.Y.: On the security of a new ultralightweight authentication protocol in IoT environment for RFID tags. J. Supercomput. **74**(1), 65–70 (2018)
27. Wang, S., Zhu, S., Zhang, Y.: Blockchain-based mutual authentication security protocol for distributed RFID systems. In: 2018 IEEE Symposium on Computers and Communications (ISCC), pp. 00074–00077. IEEE (2018)
28. Yaga, D., Mell, P., Roby, N., Scarfone, K.: Blockchain technology overview. arXiv preprint arXiv:1906.11078 (2019)

A Novel Image Encryption Technique Using Multiple One Dimensional Chaotic Map

K. U. Shahna(✉) and Anuj Mohamed

School of Computer Sciences, Mahatma Gandhi University, Kottayam, India

Abstract. In cryptography, chaotic systems are commonly used for image encryption due to the complex properties of chaos such as sensitivity to initial conditions, unpredictability, and ergodicity. An image encryption method using one-dimensional chaotic maps and Josephus permutation sequence is depicted in this paper. The keystreams required for the encryption process are generated from one-dimensional chaotic maps. Double scrambling operation is proposed in the pixel permutation phase to enhance security. Josephus permutation sequence is used in the first level scrambling. For the second level of scrambling, a novel method is implemented using the keystream generated from a chaotic tent map. Also, the keystream generated from the logistic map is used in the substitution phase. Experimental results and security analysis shows that the proposed encryption scheme can resist various attacks and ensures high security.

Keywords: Image encryption · Tent map · Logistic map · Josephus scrambling method · Security analysis

1 Introduction

As a result of advancements in communication technology, an enormous volume of information is transmitted through various communication networks. Cryptographic techniques such as encryption and decryption are widely used to ensure security while transmitting digital images. When encrypting images we have to consider the larger image size and higher pixel correlation of images. The strong correlation and high similarity between the adjacent pixels of images result in data redundancy. So it is necessary to consider these special properties of digital images while developing encryption methods. Chaos-based encryption techniques can address these issues in a better way than conventional encryption techniques. Chaos theory [1] was introduced by Edward Lorenz in 1963. Chaos is a phenomenon that exists in deterministic nonlinear systems. They have some complex properties like unpredictability and sensitivity to initial conditions.

Recently, researchers have proposed several techniques for image encryption using diverse chaotic maps [2–7] as chaos has excellent properties like speed and security. A one-dimensional chaotic map is employed for image encryption in [2, 8–10]. Two-dimensional chaotic maps are used for ensuring the confusion and diffusion properties of image encryption techniques in [11, 12]. In [13–15], hyper-chaos ensures image security

A. Abraham et al. (Eds.): SoCPaR 2020, AISC 1383, pp. 891–903, 2021.
https://doi.org/10.1007/978-3-030-73689-7_84

with improved performance and multifaceted structure. Multiple one dimensional or higher-dimensional chaotic maps are used to generate strong keys for encryption.

Josephus permutation sequence is used for performing image scrambling in [16–19]. In [16], the encryption is performed only through the pixel permutation process using Josephus sequence. This scheme is susceptible to known plain/ciphertext attack [20] because it simply scrambles the image pixels without altering pixel values. A better security system with two phases of pixel permutation was introduced in [21]. In that method, the first phase scrambling is performed with Josephus permutation sequence and the notion of matrix diagonal is used for the second phase scrambling. In the second phase scrambling, the initially scrambled matrix is diagonally scanned from the bottom right corner to the top left. Similarly, the upper and lower half portions of the diagonals are scanned. Based on the diagonals of this new matrix they performed a swap of the upper and lower half segments. But this can be easily attacked through a brute force attack by scanning the matrix in the above manner.

Motivated by the above discussions and to overcome the drawbacks in [21], we have introduced novel image encryption using double scrambling operation and 1-D chaotic maps. A new method for second round scrambling is proposed in this paper. Here, the second round of scrambling is performed using a keystream generated from a chaotic tent map to ensure enhanced security in the scrambling process. In the proposed encryption, the double scrambling based permutation process modifies the pixel positions effectively. Also, the pixel values are modified by an efficient substitution process. Using this method the pixel values and adjacent pixel correlations of an image are simultaneously broken. Thus the permutation-substitution process ensures the diffusion-confusion principle effectively. A new method is also proposed to generate the secret keystreams from chaotic maps. Since the values generated from the chaotic maps are unpredictable, the proposed system can resist any attacks and the method has high security and sensitivity. All these build up the security of the proposed cryptosystem. Performance analysis proves that the proposed encryption method can resist various attacks and ensures improved security.

The remaining sections of the paper are organized as follows. Section 2 introduces the chaotic logistic map, tent map, and the fundamental idea about the sequence generation using the Josephus permutation problem. Section 3 describes the proposed image encryption method. The security issues of the proposed system are analyzed in Sect. 4. Section 5 presents the concluding remarks.

2 Preliminary Concepts

2.1 Chaotic Logistic Map and Tent Map

Tent Map

Tent map is a one-dimensional discrete dynamic system that shows chaotic behavior. It is described as follows:

$$x_{n+1} = \begin{cases} \beta\, x_n : if\ x_n < 0.5 \\ \beta\,(1 - x_n) : if\ x_n \geq 0.5 \end{cases} \tag{1}$$

where, $\beta \in [1, 2]$.

Logistic Map
The logistic map is a discrete one-dimensional dynamic system and is defined as follows:

$$y_{k+1} = \delta\, y_k(1 - y_k) \tag{2}$$

where $\delta \in [0, 4]$. The map becomes chaotic when the parameter δ takes the value within 3.57 and 4.

2.2 Josephus Permutation Problem

In Mathematics, Josephus problem is a well-known problem. It is a counting game with three parameters (*n, start, j*) representing the total number of persons in the game, the starting point, and the counting sequence respectively. Based on these parameters we can define the problem as follows:

Let us assume *n* persons are standing in a circle. We can start counting from any particular person represented as the starting position *start.* A counting period *j* is also set, which indicates that the jth person is detached from the circle and continues the process with a new circle. Finally, one person remains and he will become the winner. Josephus permutation sequence [21] is the series generated based on the persons who are detached from the circle. The Josephus scrambling sequence can be symbolized as J_Sequence(*n*,*start*,*j*).

For example, J_Sequence (15,1,7) = [7,14,6,15,9,3,13,11,10,12,2,8,1,4,5]

J_Sequence (12, 3, 6) = [8, 2, 9, 4, 12, 10, 7, 11, 3, 1, 6, 5]

From the above examples, one can see that the Josephus permutation creates a random number sequence equal to *n.* The major drawbacks in Josephus permutation sequence are:

1. If the starting counting position (*start*) is 1 then the permutation sequence first number will indicate the counting period (*start*).
2. If the parameters *n* and *j* are the same then the randomness property of the sequence tends to degrade. E.g: consider the sequences, J_Sequence (15, 5, 5) and J_Sequence (15, 7, 5). The difference between these two sequences corresponding position values reveals the initial counting position (*start*) difference if we cyclically consider the sequence.

The local random scrambling in the spatial domain of Josephus permutation method results in a lack of efficiency. To overcome that, the proposed method calculates the parameters like counting sequence and starting position from a randomly generated secret key. Thus our method produces an unpredictable and efficient scrambling sequence.

3 The Proposed Method

The symmetric cryptosystem for grayscale images is detailed in this section. The architecture of the proposed cryptosystem is shown in Fig. 1. It consists of 3 phases, viz.;

key generation phase, image pixel permutation phase, and substitution phase. For the keystream generation, a new methodology is adopted. In the pixel permutation phase, two rounds of scrambling are performed. Josephus permutation sequence is used for initial round scrambling. The pixel permutation process is enhanced with a novel second level scrambling process. The random matrix obtained from the chaotic tent map is used for the second level scrambling process. The pixel values are completely altered in the substitution phase with the secret keystream obtained from the logistic map. Performance analysis illustrates that the proposed method stands firm against various attacks.

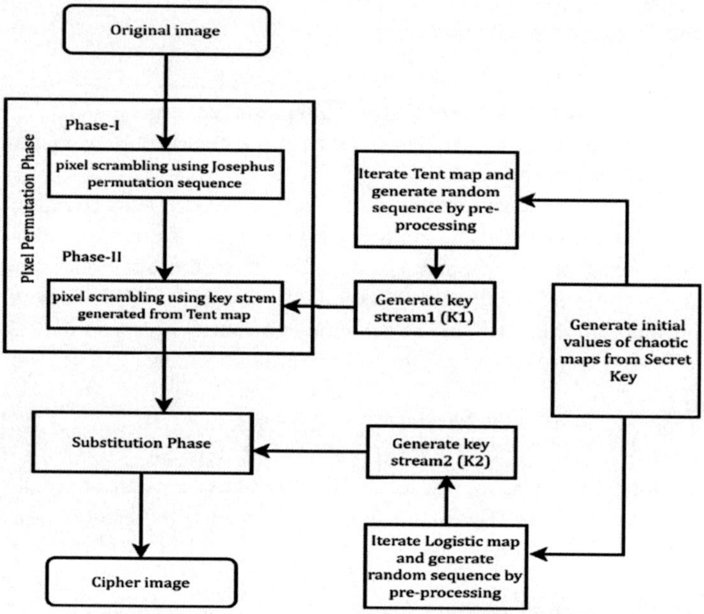

Fig. 1. Architecture diagram of proposed system

3.1 New Method for Secret Keystream Generation

a. Let the randomly selected 128-bit key can be considered as 32-hexadecimal key values $K = K_1\ K_2K_3\ldots\ldots\ K_{32}$
b. The initial values of the chaotic tent map (x_0) and logistic map (y_0) are calculated from the key K as follows:

$$S1 = \sum_{i=1}^{32} K(i) + \left(\sum_{i=1}^{16} K(i) \oplus \sum_{i=17}^{32} K(i) \right) \tag{3}$$

$$Sum1 = S1 \bmod 256 \tag{4}$$

$$Sum2 = (S1 * S1) \bmod (M * N) \tag{5}$$

$$\text{Now, } x_0 = Sum1/(M * N) \tag{6}$$

Using the initial value x_0 iterate the tent map (MN) + (2M) times and discard the first 2M values to obtain improved randomness.

$$y_0 = \frac{\sqrt{S1} * 10^2}{Sum2} * (x_0/2) \tag{7}$$

Similarly, using the initial value y_0 iterate the logistic map (MN) + (2M) times, and to get better randomness simply discard the first 2M values.

c. The parameter λ is calculated as follows:

$$\lambda = floor\ (S1/(x_0 * y_0)) \tag{8}$$

d. The value of μ is obtained as follows:

$$\mu = floor\left(S1/(\sqrt{(Sum1 * Sum2)}\right) * M * N \tag{9}$$

The parameters λ and μ are used in the substitution phase to change the pixel values. The parameter λ is also used in the pre-processing operation of chaotic random sequences. Now, Pre-process the random values x_i and y_i using the following Equations.

$$Val1(i) = \left\lfloor (abs\ (\lfloor x_i * x_i * \lambda \rfloor) * 10^{15})\ \text{mod}\ (M * N) \right\rfloor \tag{10}$$

$$Val2(i) = \left\lfloor (abs(\lfloor y_i * y_i * \lambda \rfloor) * 10^{15})\ \text{mod}\ (M * N) \right\rfloor \tag{11}$$

Now construct two key streams K1 and K2 each of size M × N from the above two values Val1 and Val2 respectively. The keystream K1 is used in the permutation phase and K2 in the substitution phase.

3.2 Pixel Permutation Phase

Pixel scrambling is done in two phases. Initial scrambling is performed with Josephus permutation sequence. The second round of scrambling introduces a novel approach with the help of a tent map. The two scrambling phases enhance the pixel permutation process. The first-round permutation with Josephus traversal is described below:

Translate plain image O_IMG into one dimension array of size M × N. The Josephus scrambling series permutes the array and then converts the scrambled array again into a matrix. In J_Sequence(n, start, j) the parameter values are assigned as follows:

(i) The parameter n is assigned with the size of the plain image M × N.
(ii) The value Sum2 obtained from Eq. (5) is assigned to start.
(iii) The value j representing the count period is calculated using:

$$j = \left(S1 + \sqrt{(Sum1 * Sum2)}\right)\ \text{mod}\ (M * N) + 1 \tag{12}$$

The scrambled image is denoted as SC_IMG1. Figure 2 shows the first level permutation of pixels with Josephus scrambling.

The second round pixel permutation randomly shuffles the image pixel positions according to the keystream K1. This phase scrambling is carried out as follows.

a. Take the scrambled image SC_IMG1 and the chaotic matrix K1, then the next phase scrambling is performed as follows:

   i. Take the ith row of SC_IMG1 and i^{th} column of K1 where $i = 1, 2 \ldots M$.
   ii. Find the smallest element in the ith column of K1 and notify its position (pos). In a new matrix S_Mat, towards this position (pos) place the first element in the ith row of SC_IMG1.
   iii. Now find the second smallest element in the ith column of K1 and notify its position. Towards this position (pos) in SC_IMG place the second element in the ith row of SC_IMG1.
   iv. Continue this process for $i = 1, 2 \ldots M$. That is, the entire rows in the matrix SC_IMG1are shuffled and stored in SC_IMG.

The final scrambled image is denoted as SC_IMG. The scrambling sequence cannot be predicted since the permutation is based on keystream K1 generated from the chaotic map. From Fig. 3 one can monitor that the pixels in the image are shuffled satisfactorily after this round. The double scrambling decreases the correlation of the neighboring pixels in the plain image.

3.3 Substitution Phase of the Proposed Scheme

In the substitution phase, pixel values are modified successively by performing xor operation with the keystream K2 that is generated by a one-dimensional chaotic logistic map. The cipher image E_IMG is attained through the substitution phase is as follows:

$$E\_IMG(i,j) = \begin{cases} (E\_IMG(i-1,j-1) + SC\_IMG(i,j) + \lambda)\ mod\ 256 \oplus K2(i,j) : for\, i,\ j > 1 \\ (E\_IMG(i,j-1) + SC\_IMG(i,j) + \lambda)\ mod\ 256\ \oplus K2(i,j) : for\, i = 1,\ j \neq 1 \\ (E\_IMG(i-1,j) + SC\_IMG(i,j) + \lambda)\ mod\ 256 \oplus K2(i,j) : for\, j = 1,\ i \neq 1 \end{cases} \tag{13}$$

$$E\_IMG(i,j) = (SC\_IMG(i,j) + \lambda + Sum2 + \mu)\ mod\ 256 \oplus K2(i,j);\ for\, i = j = 1 \tag{14}$$

Here, *i* and *j* can take the values 1, 2, 3, …, 256. After the successful execution of the permutation and substitution phases, the unintelligible cipher image is generated from the original image. Figure 4 shows the original, double scrambled, and encrypted images of pepper, Lena, and baboon. The decryption method involves the inverse process of the above steps and is defined as:

$$SC\_IMG(i,j) = \begin{cases} (E\_IMG(i,j) \oplus K2(i,j) - E\_IMG(i-1,j-1) - \lambda)\ mod\ 256 : for\, i,j > 1 \\ (E\_IMG(i,j) \oplus K2(i,j) - E\_IMG(i,j-1) - \lambda)\ mod\ 256;\ for\, i = 1, j \neq 1 \\ (E\_IMG(i,j) \oplus K2(i,j) - E\_IMG(i-1,j) - \lambda)\ mod\ 256;\ for\, j = 1, i \neq 1 \end{cases} \tag{15}$$

$$SC\_IMG(i,j) = (E\_IMG(i,j) \oplus K2(i,j) - \lambda - Sum2 - \mu)\ mod\ 256;\ for\, i = j = 1 \tag{16}$$

1	2	3	4	5	6
7	8	9	10	11	12
13	14	15	16	17	18
19	20	21	22	23	24
25	26	27	28	29	30
31	32	33	34	35	36

(a)

12	17	22	27	32	1
6	11	18	24	30	36
7	14	21	29	2	9
19	28	3	13	25	35
15	31	8	26	10	34
23	20	33	5	16	4

(b)

Fig. 2. Phase-1 scrambling (a) Plain image matrix (b) Scrambled matrix with J_Sequence (36, 8, 5)

1	2	3	4	5	6
7	8	9	10	11	12
13	14	15	16	17	18
19	20	21	22	23	24
25	26	27	28	29	30
31	32	33	34	35	36

(a) SC_IMG1

122	131	7	8	100	13
201	224	18	101	111	234
179	6	184	251	33	79
2	11	87	136	214	114
98	82	253	142	1	137
14	210	98	75	68	120

(b)K1

4	10	13	19	28	31
6	12	14	21	29	36
5	7	17	24	26	32
1	8	15	22	30	33
3	9	18	23	25	35
2	11	16	20	27	34

(c) SC IMG

Fig. 3. Phase-II permutation (a) plain image matrix (b) key stream (c) Scrambled matrix based on phase-II

4 Results and Discussion

In this section, experiments are carried out and the results are compared. Performance analyses include various statistical analyses such as correlation analysis, entropy analysis, histogram analysis, and differential attack analysis. Here experiments are conducted on a personal computer with Intel(R) Core(TM) i3-6100U CPU@2.30 GHz, 4G RAM running on Windows 7 using MATLAB 9.0 (R2016a).

4.1 Statistical Analysis

Different categories of statistical analysis are described below.

Correlation Coefficient Analysis

Correlation analysis says whether the adjacent pixels of an image are diagonally, vertically, or horizontally correlated or not. This analysis was conducted by randomly selecting adjacent pixel pairs from plain and cipher images. It can be obtained using the formula:

$$\mathrm{r}(\mathrm{x_n}, \mathrm{y_n}) = \frac{\mathrm{cov}(\mathrm{x_n}, \mathrm{y_n})}{\sqrt{\mathrm{D}(\mathrm{x_n})}\sqrt{\mathrm{D}(\mathrm{y_n})}} \tag{17}$$

$$\text{Here, } \mathrm{cov}\left(\mathrm{x_n}, \mathrm{y_n}\right) = \frac{1}{N}\sum\nolimits_{i=1}^{N} ((x_{ni} - E(x_n))(y_{ni} - E(y_n))) \tag{18}$$

$$E(x_n) = \frac{1}{N}\sum_{i=1}^{N} x_{ni} \tag{19}$$

$$E(y_n) = \frac{1}{N}\sum_{i=1}^{N} (x_{ni} - E(x_n))^2 \tag{20}$$

Generally, the correlation between adjacent pixels of the plain image is nearly 1(high correlation) and that of cipher image is 0 (less correlation). Table 1 shows the experimental values obtained in our scheme. The correlation coefficients of both (plain and cipher) images are shown in Fig. 5.

Histogram Analysis
One can easily understand the pixel values distribution from an image histogram. If encryption is performed efficiently then the histogram of the cipher image represents homogeneous distribution. An original image histogram has large spikes and has no uniform distribution. Figure 6 shows the histogram analysis of both plain and cipher images.

4.2 Differential Attack Analysis

It shows how minor modifications in the plain image will affect the encrypted image. The two measures to oppose the assaults are the Number of Pixels Change Rate (NPCR) and Unified Average Changing Intensity (UACI).

$$NPCR = \frac{\sum_{\mathrm{i}_1,\mathrm{j}_1} f(\mathrm{i}_1,\mathrm{j}_1)}{M \times N} \times 100\% \tag{21}$$

$$\text{Where, } f(\mathrm{i}_1,\mathrm{j}_1) = \begin{cases} 0, & \mathrm{E}_1(\mathrm{i}_1,\mathrm{j}_1) = \mathrm{E}_2(\mathrm{i}_1,\mathrm{j}_1) \\ 1, & \mathrm{E}_1(\mathrm{i}_1,\mathrm{j}_1) \neq \mathrm{E}_2(\mathrm{i}_1,\mathrm{j}_1) \end{cases}$$

$$\mathrm{UACI} = \frac{1}{\mathrm{M} \times \mathrm{N}}\left[\sum_{\mathrm{i}_1,\mathrm{j}_1} \frac{\mathrm{E}_1(\mathrm{i}_1,\mathrm{j}_1) - \mathrm{E}_2(\mathrm{i}_1,\mathrm{j}_1)}{255}\right] \times 100\% \tag{22}$$

The height and width of the encrypted images are symbolized using N and M. The two cipher images that are generated from the plain images, with a single bit deviation, are symbolized using E1 and E2. The expected values of NPCR and UACI for an 8-bit image are 99.6094% and 33.4635% respectively. The NPCR and UACIresults of encrypted images generated from the proposed method are shown in Table 2. From the table, one can see that the NPCR and UACI values of the proposed method approximate to the expected values which indicate that our proposed algorithm ensures better security.

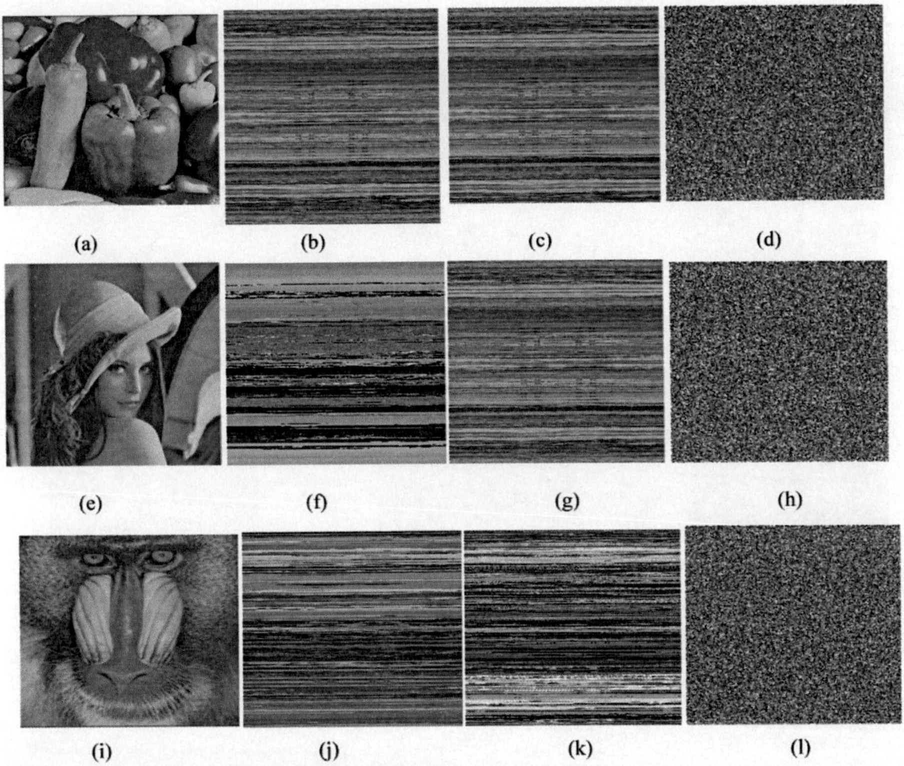

Fig. 4. (a) Original Pepper image. (b) Round1-permuted Pepper. (c) Round2-permuted Pepper (d) Encrypted Pepper (e) Original Lena image. (f) Round1-permutedLena. (g) Round2-permuted Lena (h) Encrypted Lena (i) Original Baboon image. (j) Round1-permuted Baboon. (k) Round2-permuted Baboon (l) Encrypted Baboon.

4.3 Information Entropy

The entropy value can be calculated as follows

$$H(y) = \sum_{j=0}^{2^n-1} p(y_j) log_2 \frac{1}{p(y_j)} \tag{23}$$

Table 3 indicates the entropy values of different images. The theoretically calculated entropy value is 8 and our values are nearer to that value. The random distribution nature of the encrypted image can be easily understood from these values.

Table 4 shows the performance comparison of the proposed scheme and other state-of-the-art methods for the Lena (256 × 256) image. From the results, one can see that our proposed algorithm outperforms other algorithms in [22, 23], and [25] for entropy, NPCR, and UACI measures.

4.4 Speed Analysis

The time taken by the proposed method for performing encryption of different images is shown in Table 5. Fast execution is an indication of an ideal encryption method.

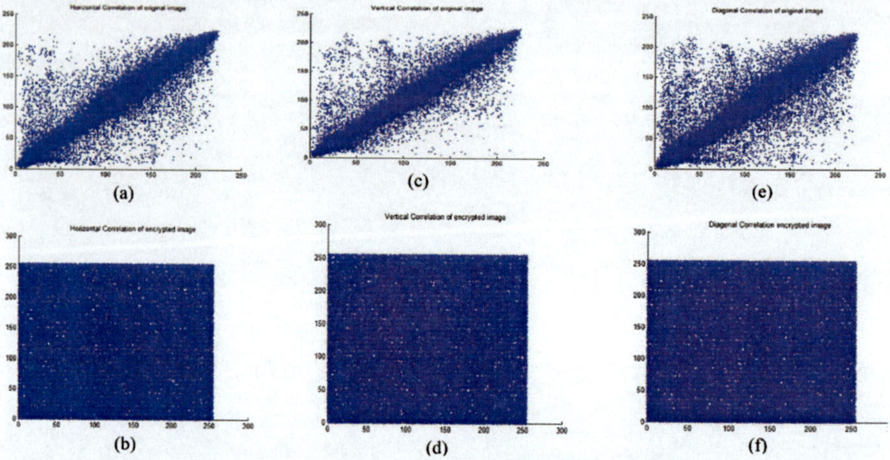

Fig. 5. Correlation distribution - (a), (c), (e) represents plain image-horizontal, vertical and diagonal correlation - (b), (d), (f) represents encrypted image-horizontal, vertical and diagonal correlation

Table 1. Values of correlation coefficient

Correlation	Lena plain image	Lena Cipher image	Baboon plain image	Baboon Cipher image	Pepper plain image	Pepper Cipher image
Horizontal correlation	0.9684	0.0004	0.9499	0.0003	0.9469	0.0023
Vertical correlation	0.9860	−0.0013	0.9395	−0.0012	0.9546	0.0034
Diagonal correlation	0.9516	0.0041	0.9061	−0.0043	0.9138	0.0027

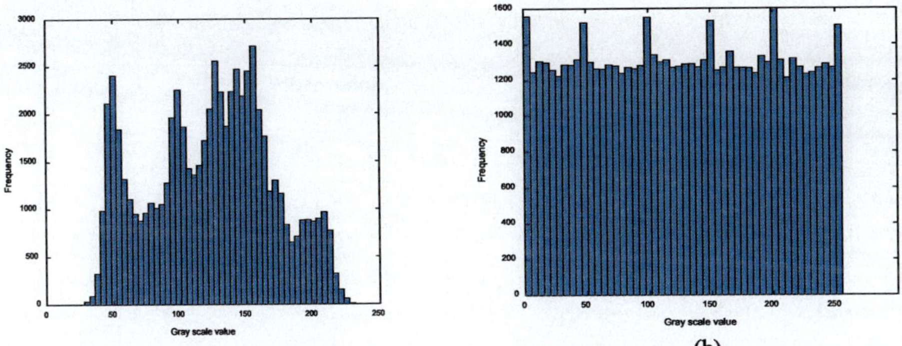

Fig. 6. Histogram of (a) plain image-Pepper (b) cipher image-Pepper

Table 2. NPCR, UACI values

Image name	NPCR	UACI
Lena	99.5881	33.4031
Baboon	99.5901	33.4113
Pepper	99.5863	33.3979

Table 3. Information entropy values

Image name	Entropy values	
	Original	Cipher
Lena	7.3879	7.9978
Baboon	7.1287	7.9879
Pepper	7.5931	7.9775

Table 4. Comparison with other schemes based on Lena image

Schemes	Correlation coefficient			Information entropy	NPCR	UACI
	HC	VC	DC			
[22]	−0.0063	0.0095	0.0089	7.9974	99.4602	37.638
[23]	−0.0074	0.0069	−0.0191	7.9963	99.504	31.6551
[24]	0.0004	0.0013	−0.0023	7.9988	99.59	33.5
[25]	0.0018	0.0012	0.0015	7.9028	99.5875	33.4605
Proposed method	0.0004	−0.0013	0.0041	7.9978	99.5881	33.4031

Table 5. Encryption speed (sec)

Image name	Speed (sec)
Lena	0.321016
Baboon	0.341734
Pepper	0.332471

5 Conclusion

A novel grayscale image encryption system with a double round scrambling operation in the permutation phase has been proposed here. Initial level scrambling is carried out using the Josephus scrambling method. The second level scrambling is performed using the keystream generated from the chaotic tent map. In the substitution phase, the values of the pixels are changed using the keystream generated from the logistic map. Statistical analysis shows that the security of the proposed system is better in terms of correlation coefficient analysis, differential attack analysis, and Entropy values. In the future, we want to implement the concept of hyper-chaos to develop a cryptosystem for image encryption.

Acknowledgment. Authors acknowledge the support from DST-PURSE (Phase - II), Government of India.

References

1. Lorenz, E.N., Haman, K.: The essence of chaos. Pure Appl. Geophys. **147**(3), 598–599 (1996)
2. Wu, X., Zhu, B., Hu, Y., Ran, Y.: A novel color image encryption scheme using rectangular transform-enhanced chaotic tent maps. IEEE Access **5**, 6429–6436 (2017)
3. Wang, X., Wang, S., Wei, N., Zhang, Y.: A novel chaotic image encryption scheme based on hash function and cyclic shift. IETE Tech. Rev. **36**(1), 39–48 (2019)

4. Shahna, K.U., Mohamed, A.: A novel image encryption scheme using both pixel level and bit level permutation with chaotic map. Appl. Soft Comput. **90**, 106162 (2020)
5. Kar, M., Kumar, A., Nandi, D., Mandal, M.K.: Image encryption using DNA coding and hyper chaotic system. IETE Tech. Rev. **37**, 1–12 (2018)
6. Fu, X.Q., Liu, B.C., Xie, Y.Y., Li, W., Liu, Y.: Image encryption-then-transmission using DNA encryption algorithm and the double chaos. IEEE Photon. J. **10**(3), 1–15 (2018)
7. Ge, R., Yang, G., Wu, J., Chen, Y., Coatrieux, G., Luo, L.: A novel chaos-based symmetric image encryption using bit-pair level process. IEEE Access **7**, 99470–99480 (2019)
8. Xu, L., Li, Z., Li, J., Hua, W.: A novel bit-level image encryption algorithm based on chaotic maps. Opt. Lasers Eng. **78**, 17–25 (2016)
9. Ye, G., Huang, X.: An image encryption algorithm based on autoblocking and electrocardiography. IEEE Multimedia **23**(2), 64–71 (2016)
10. Shahna, K., Mohamed, A.: An image encryption technique using logistic map and Z-order curve. In: 2018 International Conference on Emerging Trends and Innovations in Engineering and Technological Research (ICETIETR), pp. 1–6. IEEE (2018)
11. Zhang, W., Wong, K.W., Yu, H., Zhu, Z.L.: An image encryption scheme using reverse 2-dimensional chaotic map and dependent diffusion. Commun. Nonlinear Sci. Numer. Simul. **18**(8), 2066–2080 (2013)
12. Hua, Z., Zhou, Y.: Image encryption using 2D Logistic-adjusted-Sine map. Inf. Sci. **339**, 237–253 (2016)
13. Niyat, A.Y., Moattar, M.H., Torshiz, M.N.: Color image encryption based on hybrid hyper-chaotic system and cellular automata. Opt. Lasers Eng. **90**, 225–237 (2017)
14. Norouzi, B., Mirzakuchaki, S.: A fast color image encryption algorithm based on hyper-chaotic systems. Nonlinear Dyn. **78**(2), 995–1015 (2014)
15. Zhu, C.: A novel image encryption scheme based on improved hyperchaotic sequences. Opt. Commun. **285**(1), 29–37 (2012)
16. Desheng, X., Yueshan, X.: Digital image scrambling based on Josephus traversing. Comput. Eng. Appl. **10**, 44–46 (2005)
17. Ye, G., Huang, X., Zhu, C.: December. Image encryption algorithm of double scrambling based on ASCII code of matrix element. In: 2007 International Conference on Computational Intelligence and Security (CIS 2007), pp. 843–847. IEEE (2007)
18. Yang, G., Jin, H., Bai, N.: Image encryption using the chaotic Josephus matrix. Math. Probl. Eng. **2014**, 632060 (2014)
19. Wang, X., Zhu, X., Zhang, Y.: An image encryption algorithm based on Josephus traversing and mixed chaotic map. IEEE Access **6**, 23733–23746 (2018)
20. Furht, B., Kirovski, D. (eds.): Multimedia Security Handbook. CRC Press, Boca Raton (2004)
21. Shahna, K., Mohamed, A.: An image encryption method using Henon map and Josephus traversal. In: 2018 World Congress on Information and Communication Technologies (WICT). Advances in Intelligent Systems and Computing, Springer (2018)
22. Bakhshandeh, A., Eslami, Z.: An authenticated image encryption scheme based on chaotic maps and memory cellular automata. Opt. Lasers Eng. **51**(6), 665–673 (2013)
23. Guo, Y., Shao, L.P., Yang, L.: Bit-level image encryption algorithm based on Josephus and Henon chaotic map. Appl. Res. Comput. **32**(4), 1131–1137 (2015)
24. Wang, S., Wang, C., Xu, C.: An image encryption algorithm based on a hidden attractor chaos system and the Knuth-Durstenfeld algorithm. Opt. Lasers Eng. **128**, 105995 (2020)
25. Wang, M., Wang, X., Zhang, Y., Gao, Z.: A novel chaotic encryption scheme based on image segmentation and multiple diffusion models. Opt. Laser Technol. **108**, 558–573 (2018)

Use of Cryptography for Monetary Issuance and Securing Financial Transactions

Mamadou Mbaye(✉) and Cheikhou Thiam

Thies University, Thies, Senegal
m.mbaye@univ-thies.sn

Abstract. During the financial subprime crisis (2007/2008), the banking system showed its limits and at the same time lost the privilege and monopoly of trading in money. The immediate consequence of this banking commoditization was the decompartmentalization of markets, leading in its wake to the dematerialization of the borders of the economical system. This deregulation movement is characterized by the emergence of complex innovations combdigitqapointing financial theory and mathematical applications. Among them we have cryptography applied to monetary creation (crypto currency) which continues to push the limits of financial engineering. Despite a hesitant penetration into the financial ecosystem, it is revolutionizing the conception of money and third-party holding. In this article we analyze the use of cryptography tosecure monetary issuance and financial transactions in the block chain technology.

Keywords: Cryptography · Cryptocurrency · Digital currency · Blockchain

JEL Classification: E22 · E55 · G11 · G22

1 Introduction

Cryptocurrency is a concrete response to the major changes that have continued, since the 1980s, to disrupt the financial system. The first change follows the economic liberalization policies carried out, initially in developed countries, then in so-called poor and emerging countries. Governments and regulatory authorities had decided to promote deregulation policies with regard to economic agents based on the belief in the self-regulation of markets and the ability of financial actors to self-control.

They have thus favored the development of disintermediated, opaque and difficult to control markets, characterized by a virtual absence of prudential rules common to all players [1]. The second upheaval concerns the modalities of banking intermediation. Indeed, subject to the double pressure of the performance requirements of shareholders (owners of financial structures), and increasingly restrictive prudential supervision standards, the financial system has sought to increase and diversify its supply of currencies [2].

A. Abraham et al. (Eds.): SoCPaR 2020, AISC 1383, pp. 904–910, 2021.
https://doi.org/10.1007/978-3-030-73689-7_85

It has developed an innovative form of intermediation, characterized by intense and aggressive activity in the capital markets, increasingly robotization of processes and dematerialization of liquidity.

All these changes have fostered the emergence of a new paradigm based on questioning the conception of money and, in the broad sense, the notion of money. The system as a whole is subtly and continuously undergoing profound changes leading to an increasingly accentuated disengagement of the human being as an actor, in favor of the machine, for the issuance and securing of currency.

This robotization reduces the probability of failure linked to the human factor while favoring the safety and speed of operations. Thus, we find ourselves in an inclusive "bankless", disintermediated and decompartmentalized finance focused on a currency generated from cryptographic protocols [3]. In this "high trading frequency" liquidity with all the attributes of a classical currency is created by the logarithm; it is digital but also decentralized. Trust is no longer placed in humans through central banks but in cryptography and blockchain; in addition, the central authority for issuing and regulating the money supply is non-existent. Indeed, we are in an innovative finance at the cutting edge of technology, deeply rooted in the theory of free trade and secured by cryptographic keys.

In this article, our main objective is to show the importance of cryptography in the issuance of digital currency, and in the blockchain. Our main assumption is that it allows inclusive and participatory finance in a context of optimal security. This hypothesis is justified by the generalization of the use of cryptocurrencies in the financial system.

2 Conceptual Analysis of Crypto Currency

Crypto currency is a virtual liquid good that circulates on the internet. It is based on cryptographic technology for its issuance and validation of its transactions. Popularized by [4], it frees the agent from the need for financing of traders, national and supranational institutional intermediaries. It eliminates commissions and brokerage fees, while optimizing peer-to-peer exchanges. In a dynamic of participatory management, it has become a product analogous to cash, in a secure environment without central authority.

The advent of cryptography for the issuance of currency is greeted as a radical but above all positive development for the financial system. It's a new paradigm of digital assets that have no intrinsic cost let alone centralized regulators and issuers, and are not backed by any tangible good. They circulate in a distributed network called blockchain which allows participatory control thus avoiding any risk of fraud, redundancies and duplication of transactional activities. The concept of decentralized digital currency and its dependencies such as property directories have been the subject of several studies for decades. The anonymous electronic crypto-currency protocols of the 1980s were mainly derived from a cryptographic function of the Chaumian Blind type [5]. This function guaranteed digital currency a high degree of protection and anonymity. However, this model could not be popularized due to its dependence on a centralized intermediary.

Dai's "b-money" cryptocurrency (1998) was the first experiment in issuing money by solving cryptographic primitives in a context of decentralization. However, she was very ambiguous about the process of applying the distributed consensus.

A decentralized currency was first implemented by [5] by combining primitives, using asymmetric cryptography and the consensus algorithm. This process made it possible to introduce the notion of "proof of work" to archive in an indelible and tamper-proof manner the trace of possession.

Cryptocurrency today enjoys a certain maturity characterized by effective and generalized disintermediation of all operations, by permanent traceability, and finally by distributed consensus. Trust is no longer entrusted to a moral institution but to all users who validate or deny transactions. Today, there are about thirty cryptocurrencies in circulation. In terms of market dynamism, bitcoin leads the way with a market cap of $ 19.4 billion in April 2017, followed by Ethereum valued at $ 3.9 billion, then ripple at $ 1.2 billion. dollars still for the same period [6].

At the macroeconomic level, it is diluted in the M1 aggregate. Unlike central money, cryptocurrency has a "because of use" price even though more and more economic operators accept it as a means of payment and debt extinction. In fact, with it, operations are possible only between economic agents who accept it as an intermediary without any legal constraints. Compared to other assets whose mass is regulated and the value framed by central banks, cryptocurrency is stateless. Its price fluctuates only because of supply and demand; which sometimes makes it a speculative currency, sometimes very volatile [7].

Some digital currencies favor the total anonymity of agents in transaction, where it is only pseudonymity with many others. Most digital platforms use standard asymmetric cryptography concepts, including private key and public key principles. A private key can only be known to its owner who uses it as a signature to validate its transactions and as a code for decryption of messages sent to it. Therefore, any cryptocurrency transaction between two agents induces an individual digital address of the type: 35sv1egv71wmzD1GePte0dtzVZdca5wK1e generated randomly when an account is produced on a cryptocurrency monetization platform. Anyone who owns it can send cryptocurrency to its owner.

On the other hand, a public key is an address that can be found by analogy in a "directory" and which functions like a post office box. During a transaction, each injected token implements its origin, in order to allow network nodes to trace ownership.

Thus, considering that all transactions are published in a distributed ledger, each node will be able to track the movements of the encrypted tokens. The balance of a digital address is comparable to a net accounting between the inputs and outputs of money at this address. Once validated, the transaction leaves indelible marks, becomes irrevocable and irreversible thanks to a decentralized architecture [8]. It is added to the chain of blocks and will be permanently kept there. As we can see, the underlying contribution of cryptography to the modern economy is digital finance with cryptocurrency and blockchain technology as instruments. The latter is a tool of distributed consensus, guarantor of the essential fiduciary aspect, even inseparable from the concept of money.

3 Blockchain Description

The blockchain is a register of transactions that allows information to be stored in a transparent, secure manner, operating without a central control body. It is distributed across multiple computers through a peer-to-peer network. Blocks are added and chained to previous blocks using a cryptographic hash function, resulting in a continuous increase in the size of the chain [9]. The registry is validated and kept by a network node (user). Each node maintains a complete copy of the entire ledger. This database stores the history of all exchanges made since its creation between its users. It is secure and distributed. Indeed, it is shared by its autonomous users, thus allowing everyone to check the validity of the chain. The first goal of the blockchain was to solve the existing problems in the Bitcoin cryptocurrency. In order to optimize the security of the issuance of cryptocurrency and financial transactions, the blockchain will have to meet the challenge of energy consumption and inclusive use.

3.1 Cryptographic Applications in the Issuance of Digital Currency

The process of introducing a new block into the blockchain is called mining. To check if the coin (cryptocurrency) is legitimate or if it has not already been spent, each node uses the blockchain. Before transaction records are added to the blockchain, more participants come to an agreement. The data mining process is a resource intensive task, which makes it difficult for an attacker to approve a transaction that is invalid. Every time we extract a block, we check whether it has valid proof of work or proof of stake. The following scenario is the most common in cryptocurrency: (i) a generated address (public key) is available for a user who has a wallet (ii) a private key is assigned to the wallet. It is used to sign the transaction and prove ownership (iii) the payer sends the coin to the payee using a given address and signs it using the payer's private key, and finally (iv) the transaction is committed through the data mining process.

3.2 Progress of a Transaction

Transactions between network users are grouped into blocks. Each block is validated by the miner's node, according to the type of blockchain. In Bitcoin this technique is called "Proof-of-Work", and consists of solving algorithmic problems.

As soon as the block is validated, it is added to the chain after having time-stamped it. The receiver and the entire network can see the transaction. This transaction takes place in three phases:

✓ Phase 1

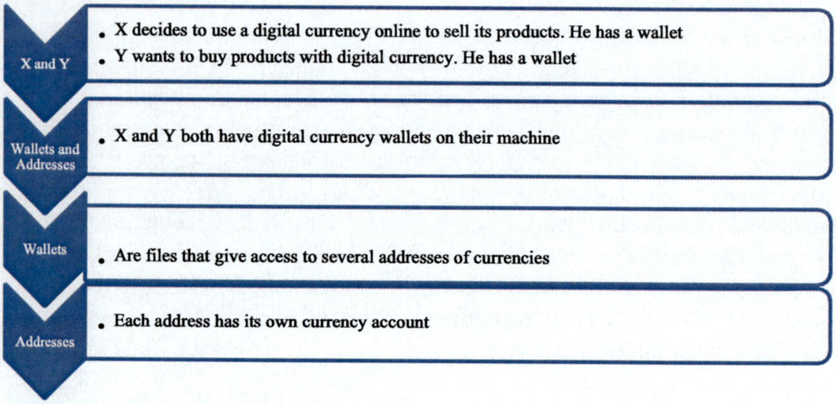

✓ Phase 2 : Creation and submission for transaction

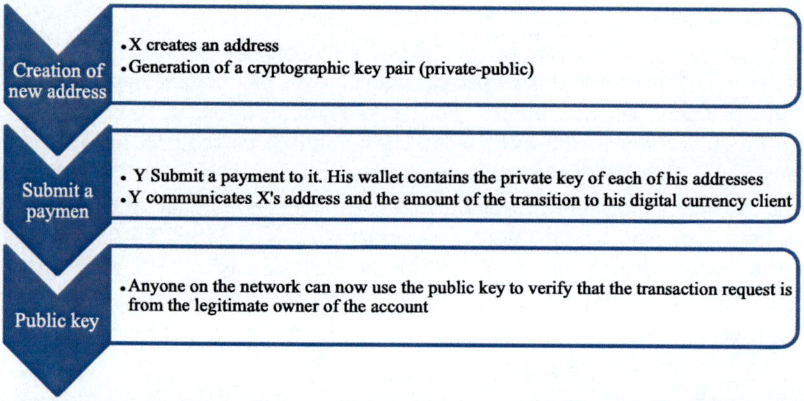

✓ Phase 3: Security and finalization

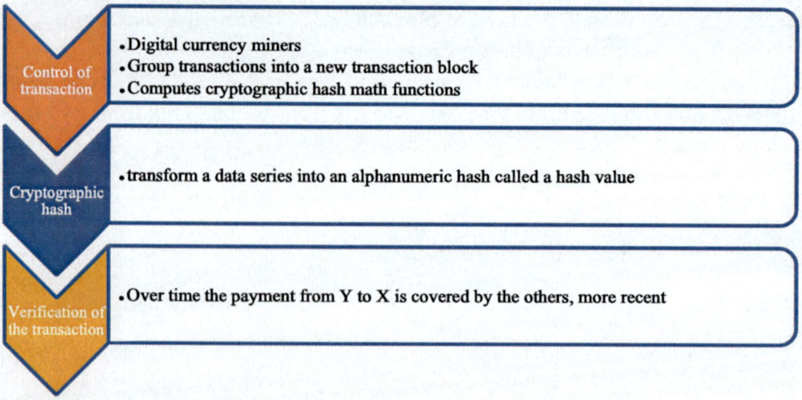

This is the use of asymmetric cryptography which works thanks to a public key and a private key. The public key is the encryption key that allows anyone who has it to encrypt a transaction. The issuer of a transaction signs it with his private key, and once broadcast in the blockchain network, anyone can verify that the signature is valid using the issuer's public key. This approach is much more secure than that of Dai's "b-money" cryptocurrency (1998). In this method any node of the network can verify the validity of the solution. It can calculate the block header hash very quickly and verify that the target is reached. Once the proposed block header is suitable for the entire network, the mined block is permanently added to the chain. This makes the transaction log immutable.

3.3 Mining

Mining is a distributed consensus system. It is used to confirm pending transactions, including them in the chronologically numbered blockchain. This numbering protects network neutrality and allows computers on the network to cooperate on the state of the system. Confirmation of transactions requires their inclusion in a block and must correspond to very strict cryptographic rules which are then verified by the network [10]. This creates a competitive lottery that prevents any agent from adding additional blocks to the blockchain. This process prevents anyone from changing what is included in the blockchain or replacing parts of it to reverse their own transactions. The first block contains the first transaction. The hash of the first block is passed to the miner, who employs it and generates a hash for the second block. Similarly, the third block creates a hash that includes the first two blocks, etc. All the blocks that make up the blockchain can be traced. Cryptography has been widely used for digital currency financial transactions. Nowadays, cryptocurrencies have emerged as leading software systems.

4 Conclusion a Private

In financial system, the introduction of cryptography is an important innovation that allows economic agents to conduct secure, verifiable and authenticated transactionsmini. It allows the control of uncertainty and induces a general change in the perception of asymmetry linked to the monetary system. For the economic actor, risk aversion is thus reduced to its lowest expression.

However, mastering cryptocurrency is undoubtedly a major challenge for the regulator because it contains specific risks. Its interconnection with traditional money (central as well as scriptural) could generate systemic crises that are difficult to contain. Nonetheless, it is useful and should be encouraged because it creates competition in a financial sector dominated by a few large banks under the authority of financial institutions which very often obey political interests. In its most successful version, it fulfills all the functions of general equivalent and all the roles devolved on conventional money. Cryptocurrency is a financial reality that fulfills a function necessary for a strong and sustainable economy.

References

1. Gutmann, R., Plihon, D.: The consumer debt at the center of finance-led capitalism. Int. Rev. Appl. Econ. **24**(3), 269–283 (2010)
2. Pluchart, J.J.: Shadow banking: a new financing lever for emerging economies. Eyrolles Edition (2015)
3. Yahanpath, N., Wilton, Z.: Virtual money: betting on bitcoin. Auckland Univ. Bus. Rev. **17**(1), 36 (2014)
4. Nakamoto, S.: Bitcoin: cryptomoney Cash System. Satoshi Nakamoto Institute (2008)
5. Chaum, D.L.: Untraceable electronic mail, return addresses, and digital pseudonyms. Commun. ACM **24**(2), 84–90 (1981)
6. Haber, S., Stornetta, W.S.: How to time-stamp a digital document. In: Journal of Cryptology, pp. 87–1100 (1991)
7. Varriale, G.: What is the best way to regulate Bitcoin? International Financial Law Review vol. 32, no. 7 (2013)
8. Nellen, A.: Taxation and today's digital economy. Journal of Tax Practice and Procedure, vol. 17, no. 2 (2015)
9. Bayer, D., Haber, S., et Stornetta, W.S.: Une Amélioration de l'efficacité et de la fiabilité du systéme d'horodatage numérique, pp. 319–354 (1993)
10. Merkle, R.C.: Protocols for public key cryptosystems. In: Proceedings of the 1980 Symposium on Security and Privacy, pp. 122–133. IEEE Computer Society, April 1980

MAAMSIC: Multimodal Authentication and Authorization Model for Security of IoT Communication via GSM Messaging in Sub-Saharan Africa

Joan Nabusoba(✉), Calvins Otieno, and Wilson Cheruiyot

Jomo Kenyatta University of Agriculture and Technology, P.O. Box 81310-80100, Mombasa, Kenya

Abstract. Internet of Things (IoT) which consists of heterogeneous devices is an enabling technology that can greatly improve the quality of lives in Sub-Saharan Africa. For instance, soil humidity and irrigation for e-agriculture, energy consumption, or even health data. As with technology, however, IoT has introduced security and privacy challenges. IoT devices create, transfer, process, and store sensitive data that must be protected from unauthorized access. Similarly, the devices and infrastructure linking with IoT and the IoT devices themselves are assets that must be protected. Though IoT devices are being adopted, there isn't wide access to GPRS in rural parts of Africa; hence users need to have access to technology that is seamless, viable, and easy to use. This paper introduces a Multimodal Authentication and Authorization Model for the Security of IoT Communication via GSM Messaging (MAAMSIC). The MAAMSIC introduces a secure server that will encrypt devices' data and the AT Commands from the IoT. The model is illustrated with a case study of a soil and moisture control system. The system will have a secure server to store data and send feedback to the user. In case of any anomalies, mitigation protocols will be executed. The system will be available for both internet capable phones and simple phones that can't access the internet. In smartphones, a mobile app will be used to send and receive messages from the server. JSON Web Encryption strategy will be used to encrypt and decrypt messages to prevent third party apps from reading them.

Keywords: Internet of Things · GSM · JSON Web Encryption

1 Introduction

Internet of things (IoT) is a new paradigm that relies on widely spread connected objects that cooperate to automate many everyday actions (Rak et al. 2020). Innovations in IoT are improving lives in rural Africa. For example, to solve water problems, Ingram and Memon (2020) are utilizing IoT technology to scale water supply systems in sub-Saharan Africa for sustainable water supply.

A. Abraham et al. (Eds.): SoCPaR 2020, AISC 1383, pp. 911–920, 2021.
https://doi.org/10.1007/978-3-030-73689-7_86

According to BRCK, about 800 million Africans do not have internet access making internet coverage, especially in rural areas, low (brck.com). As much as the acquisition of phones is growing in Africa, smartphone adoption is modest. Common types of mobile devices owned are basic phones. Worldwide, sub-Saharan Africans report the lowest ownership rate of smartphones compared to any other geographic region. Additionally according to M-Kopa (https://m-kopa.com/impact/), one of the leading solar home system states that 75% of sub-Saharan Africa remains unconnected to the internet. Sending IoT data can use general packet radio service (GPRS), which is an advancement of GSM. However, since GPRS is not dedicated for transmission of IoT data due to weaknesses in power efficiency and coverage, GSM is preferred (Bima et al. 2020).

There is explosive development of IoT, spurring a massive demand for many smart devices to access the wireless networks concurrently (Priyanka et al. 2020). It is predicted that over 20 billion devices will connect to wireless internet by the end of 2020 (Xu et al. 2020).

In addition to low network coverage, half of Sub-Saharan Africa does not have access to electricity, which consequently hampers GPRS equipment (Bakibinga-Gaswaga et al. 2020). The latter highlights how unstable GPRS is as a choice of IoT infrastructure, affecting its adoption.

Owuor et al. (2020) state that there is rapid increase in IoT applications in almost every area of our society. However, according to Franklin (2020), these increases of IoT startups fail to profit from their IoT innovations. This is due to security related problems during IoT development, which are often underestimated or overlooked (Selgert 2020). Even with inadequate coverage of the internet in rural parts of Africa, smart farming innovations are still being implemented using IoT and wireless devices (Oliveira-Jr et al. 2020).

The use of technology still presents security issues to the African continent. Despite network coverage challenges, Ndubueze (2020) suggested that the number of cybercrime and victimization incidents is on the rise. However, the African continent has been slow in responding to crime and disorder in cyberspace and is still establishing cybercrime legislations. This makes Africans vulnerable in cyberspace, without the assailants facing the consequences (Ndubueze 2020).

The objective of this paper was to construct a multimodal technique for securely sending and receiving data from IoT systems using GSM in sub-Saharan Africa (MAAMSIC).

We run exhaustive experiments using both real and simulated data to evaluate whether the MAAMSIC can send and receive data securely. The experimental analysis shows that MAAMSIC is capable of sending and receiving data from verified users to the IoT. In case of unauthorized access, the MAAMSIC system will reject and flag these incidences for further action.

The paper is organized as follows: related work, requirement and system design, evaluation, conclusions, and further work.

2 Related Work

Recently, Waghmare et al. (2020) integrated IoT and GSM with a transformer to monitor and protect it. Their system uses a microcontroller and is monitored using GSM technology. The microcontroller gets initialized, and C code is compiled and uploaded to the Arduino board from a USB port. The transformer values are then compared and checked against the preset values that are fed in the Arduino. Examples of these values include voltage where, if an overvoltage occurs, its relay starts to operate. Its values are displayed on an LCD screen, and messaging is done using GSM technology. In the presence of any abnormality, transformer details are updated on a webpage, and an alert text message is sent. Although the system utilizes a GSM communication network with low investment and operation costs, no data protection or security of the transformer is proposed. If any, it is undisclosed.

Antonio et al. (2020) propose an IoT platform for the benefit of rural African farmers. The platform has an IoT sensing mechanism that provides the state of the soil information like moisture, light, air temperature, color, and texture. The system's hardware is housed in a specially designed casing for easy assemblage, protection, deployment and transport. While the solution targets rural farmers, there isn't any security protocols implemented to ensure that the user data is not acquired maliciously.

Sigu et al. (2020), in collaboration with the International Cancer Institute, are utilizing IoT to support oncology clinics in the rural areas of Kenya. Cancer patients get the information needed from the physician without going to the clinic facilities. Patients are monitored, and can access information in real-time. This reduces costs and improves the patient outcome of treatment. Patients are also trained through an online module. Remote monitoring is beneficial to prevent lengthy hospital durations and readmissions. The use of technology has proved to have a significant impact on cancer patients in rural sub-Saharan Africa. The application of this technology, however, lays minimum emphasis on the security of sensitive data of patients. It is not apparent whether any security strategies have been put in place to protect patients' data.

With the era of the use of informational renovation to enhance performance in water pipelines, Priyanca et al. (2020) proposed integration of Internet of Things to monitor pressure, viscosity, pumping station parameters and other external parameters. The paper shows how the Internet of Things has progressed the industrial element of monitoring and intellectual control of the pipeline systems. The IoT system proposed is a smart module to enhance efficient data communication. The system module however, isn't clear on security of its data, if any.

3 System Architecture

The system comprises three major modules:

An IoT sensing platform, with a soil moisture sensor, detects the moisture content and sends signals to initialize crop irrigation. Signals from the device are sent through the GSM network. A water level sensor is added to monitor the amount of water in the irrigation tank. The water level sensor is connected to the server. This is to ensure that users can make requests through their phones to monitor the water level through the SMS gateway.

The server software is written in PHP and is hosted on a server. The server will be connected to an SMS gateway provider to receive messages from the device and send them to the user. The server will ensure that the user is registered and has a verified phone number before sending and receiving commands. The server also ensures that users are authorized so that messages are sent and received securely. Additionally, to mitigate issues arising, the server will notify the user if the device is offline. Since the users' phone numbers are verified and linked to a specific device ID that only the server knows, SIM card replacement fraud and impersonation are prevented (Fig. 1).

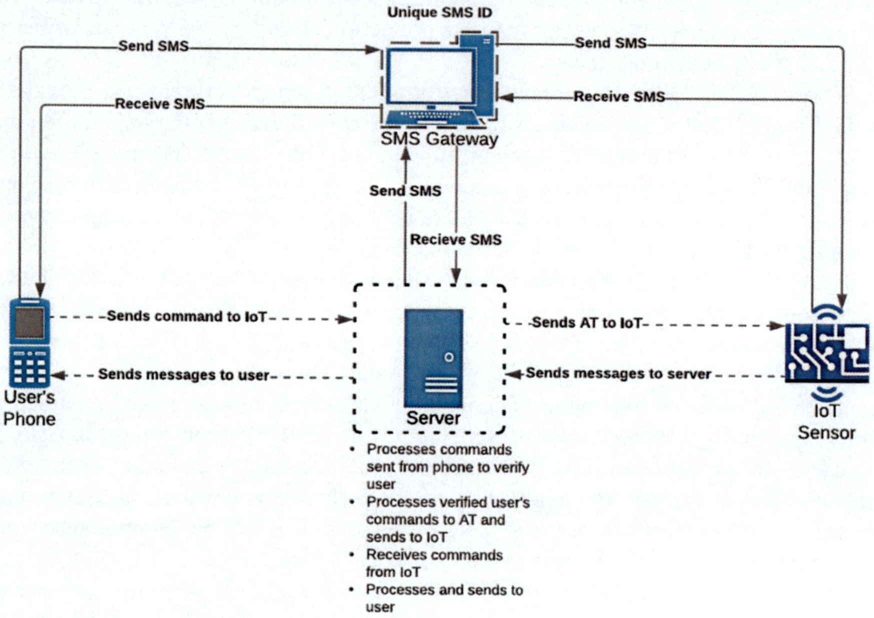

Fig. 1. MAAMSIC architecture

The users are connected to the device via the server through their verified mobile phones. They can join either via smartphones or the simple phones that do not have any access to the internet. For smartphones, the user uses a Flutter app whose data is encrypted with JSON Web Encryption (JWE) to prevent third-party apps from reading user information. For simple phones, the user will send SMS to a specific SMS code, which will concurrently send commands to the server. The server then verifies the user and relays the command to the device. The server also sends feedback to the user from the device.

3.1 Server App

The server contains a database module that stores data on user details and device details. It also has a web app that comprises of five main modules: authentication, message processing module, SMS gateway, and IoT connector module.

The authentication module ensures that the user is registered and verified; their phone numbers that will be used to send SMS through commands are verified and are linked to the device(s) ID. When a user sends a message, the authentication module immediately checks whether the user is verified and connected to a device or devices in the system. The authentication module sends a message to the user in case there is a malicious attempt to send commands to their device. These attempts are logged on the server for future reference (Fig. 2).

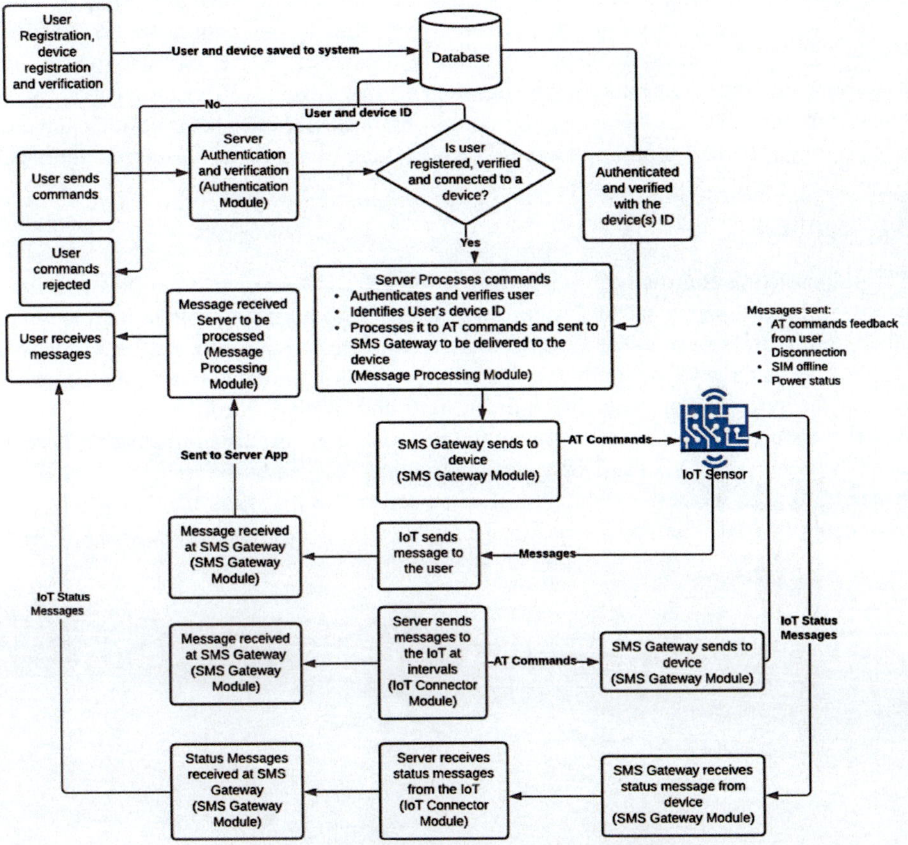

Fig. 2. Software architecture block diagram

The message processing module handles received and sent messages. When a user sends a command to the device, it is verified through the auth server. After successful verification, their message is processed into an AT command and sent to the device through the SMS gateway module. When an IoT device sends feedback or information to the user, the message is sent as SMS to the message processing module through the SMS gateway. The message is processed and the user connected to the device identified. The message processing module then sends a feedback SMS to the user through the SMS gateway.

The SMS gateway module is in charge of sending and receiving messages which are first processed in the server. When the user sends a message to the device, the SMS gateway sends the message to the server for verification at the authentication module. After verification and processing, the SMS gateway sends the message to the device's SIM card in the form of AT commands. The response from the device is first received by the SMS gateway and then sent to the message processing module. The message processing module then sends a feedback message to the user through the SMS gateway.

The IoT connector module sends several messages in intervals to the IoT device through the SMS gateway to check the status of the device. A user can also invoke the IoT connector module to check whether their device is online. The moment the IoT connector module doesn't receive a response from the device, the user is notified and the issue is logged in the system. The server logs all actions or messages for future reference.

3.2 Hardware

The soil moisture control system's hardware for irrigation consists of a GSM Arduino board, soil moisture sensor and water level sensor. The board is programmed to send commands to the server and receive commands from the soil moisture sensor. The GSM chip is interfaced with the hardware. The chip is used to send and receive messages to control the system. The AT commands are sent and received through the GSM chip. A soil moisture sensor is used to detect the moisture content. This information is sent as SMS to the server via the SMS gateway and later relayed to the user. A water level sensor is used to indicate the amount of water left in the irrigation tank. A 240 V AC power supply drives the hardware (Fig. 3).

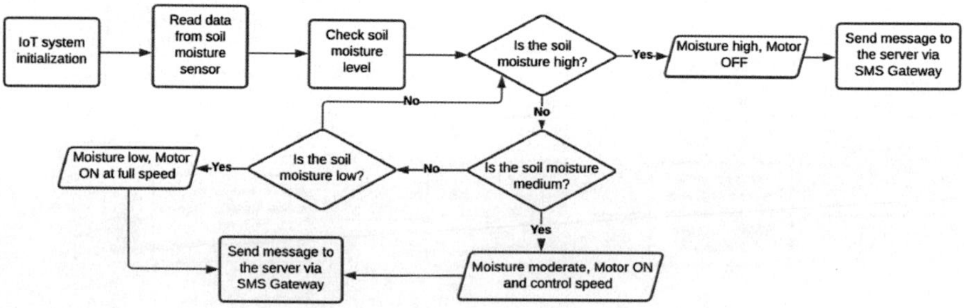

Fig. 3. Hardware architecture

3.3 Software Implementation

Server

The server is written in the HyperText Preprocessor (PHP) framework called Yii. It is hosted in a secured and shared hosting for demonstration. The server uses an SMS gateway provider to send and receive SMS through a short code. The server has a user registration module where the user is registered, authenticated and authorized to ensure data integrity. After the user is registered, the details are added to the server, including their device(s) SIM card number and ID. They further verify their phone number via a token sent to them through a short URL to activate their account. The IoT device has a GSM chip with a SIM card. The admin then assigns the device to a verified user in the server. The user can then proceed to send commands and receive information from the device through the server.

When a server receives an SMS or a command from a specific phone number, the server will know the device ID associated with that user. The server then sends the message to the device. Moreover, when the device responds back to the server, the server will identify the user connected to the device and relay the message.

Mobile Application and Encryption

The user can install the mobile application to send and receive commands from the device through the server. This is for the case of those who own smartphones. The smartphone app is developed with Flutter and data is encrypted by JSON Web Encryption (JWE). JWT and JWE technology are normally used in authentication processes and access rights security (Royani and Wibowo 2020). A user sends an encrypted message to the server, which decrypts it into a command that is sent to the IoT through the SMS gateway. Responses are sent back to the mobile application which decrypts the message using JWE. The mobile app, upon verification of the authenticity of the message, displays it to the user. Data, which cannot be decrypted, is rejected by the receiving party.

4 Evaluation

MAAMSIC is evaluated by testing the mobile app and the IoT server. The mobile application has two login systems to emulate the MAAMSIC architecture and the phishing app. Both the modules have the SMS sent and received from the IoT. The user can send and receive SMS from the IoT through the server. The MAAMSIC application has a list of commands a user can directly send to the emulated IoT to control and operate it and control the moisture by turning the motor off or controlling the speed. The user can check the status of the IoT as well (Table 1).

User's requests are issued by both registered and unregistered phone numbers to access the IoT device by sending messages. The messages sent are received by the server app which consists of authentication, message processing, SMS gateway and IoT connector modules. If a user is using a simple phone without GPRS, the phone number is processed to check whether the user is registered and verified, and is connected to an IoT device. If the user's phone number is unverified, the command sent through the SMS is rejected. In the case of a smartphone, the mobile app encrypts and decrypts sent and received messages using JWE hence preventing access by third-party apps. In both cases if an IoT device is attempted to be accessed by an unregistered phone number, a message is sent to the registered phone number for further actions. A flag is also raised on the server to block requests from the unregistered phone number.

Table 1. Performance of basic MAAMSIC operations.

Request	Time (ms)	Phone subjects	Response
Get IoT status (ON/OFF)	7.63	Registered phone	Command accepted. IoT status connected
	7.89	Unregistered phone 1	Command rejected, unidentified number
	7.89	Unregistered phone 2	Command rejected, unidentified number
	7.89	Unregistered phone 3	Command rejected, unidentified number
Turn on motor	8.32	Registered phone	Command accepted, motor ON
	8.96	Unregistered phone 1	Command rejected, unidentified number
	8.96	Unregistered phone 2	Command rejected, unidentified number
	8.96	Unregistered phone 3	Command rejected, unidentified number
Turn off motor	8.92	Registered phone	Command accepted, motor OFF
	9.01	Unregistered phone 1	Command rejected, unidentified number
	9.01	Unregistered phone 2	Command rejected, unidentified number
	9.01	Unregistered phone 3	Command rejected, unidentified number
Control motor speed	8.60	Registered phone	Command accepted speed increased
	8.89	Unregistered phone 1	Command rejected, unidentified number
	8.89	Unregistered phone 2	Command rejected, unidentified number
	8.89	Unregistered phone 3	Command rejected, unidentified number

Messages received by registered and verified numbers are processed by the server system to be commands and sent to the IoT device, which then returns feedback messages to the user.

MAAMSIC may produce large amounts of data and receive many requests and responses, making it require a large processing capacity surpassing the capacities available for this implementation.

5 Conclusion and Further Work

This work addresses a methodology for adding an additional layer of security while availing data to users in an IoT system design in sub-Saharan Africa using GSM. MAAMSIC architecture will be extended to support functionalities for dealing with more complex events and billing policies. Incorporating features that will enable it to scale and deal with the increased workload is essential for future work. A possible solution is deploying the MAAMSIC server in a dedicated cloud server with a strong RAM and large memory. MAAMSIC can also be transformed into multi-edged cloud architecture to deal with internal risks. The security of the IoT network to ensure that it handles risks due to malicious behavior of the IoT is still an open issue. For further enhancement of this work, the model could be designed to utilize solar or wind energy as its source of power. This is a cleaner energy and will be a great source of power in rural areas where electricity is not reliable. The MAAMSIC could also be improved by use of Artificial Intelligence and audio processing for better accessibility for the disabled especially those with poor eyesight.

References

Alam, T.: A middleware framework between mobility and IoT using IEEE 802.15. 4e Sensor Networks. J. Online Inform. **4**(2), 90–94 (2020)

Bakibinga-Gaswaga, E., Bakibinga, S., Bakibinga, D.B.M., Bakibinga, P.: Digital technologies in the COVID-19 responses in sub-Saharan Africa: policies, problems, and promises. Pan Afr. Med. J. **35**, 38 (2020)

Bima, I.W.K., Suryani, V., Wardana, A.A.: A performance analysis of general packet radio service (GPRS) and narrowband internet of things (NB-IoT) in Indonesia. Kinetik Game Technol. Inf. Syst. Comput. Netw. Comput. Electron. Control **5**(1), 11–20 (2020)

Ingram, W., Memon, F.A.: Robustness of IoT-connected e-Taps for sustainable service delivery of rural water supply. Water Supply (2020)

Nazareno, A.C., da Silva, I.J., Nunes, E.F., Gogliano Sobrinho, O., Marè, R.M., Cugnasca, C.E.: Real-time web-based microclimate monitoring of broiler chicken trucks on different shifts. Rev. Bras. de Engenharia Agrícola e Ambiental **24**(8), 554–559 (2020)

Ndubueze, P.N.: Cybercrime and legislation in an African context. In: The Palgrave Handbook of International Cybercrime and Cyberdeviance, pp. 345–364 (2020)

Oliveira-Jr, A., Resende, C., Pereira, A., Madureira, P., Gonçalves, J., Moutinho, R., Moreira, W.: IoT sensing platform as a driver for digital farming in rural Africa. Sensors **20**(12), 3511 (2020)

Owuor, D.O., Laurent, A., Orero, J.O.: Exploiting IoT data crossings for gradual pattern mining through parallel processing. In ADBIS, TPDL and EDA 2020 Common Workshops and Doctoral Consortium, pp. 110–121. Springer, Cham (2020)

Priyanka, E.B., Thangavel, S., Madhuvishal, V., Tharun, S., Raagul, K.V., Krishnan, C.S.: Application of integrated IoT framework to water pipeline transportation system in smart cities. In: Intelligence in Big Data Technologies—Beyond the Hype, pp. 571–579. Springer, Singapore (2020)

Rak, M., Salzillo, G., Romeo, C.: Systematic IoT penetration testing: Alexa case study. In: ITASEC, pp. 190–200 (2020)

Royani, M.R., Wibowo, A.: Web service implementation in logistics company uses JSON web token and RC4 cryptography algorithm. J. RESTI (Rekayasa Sistem Dan Teknologi Informasi) **4**(3), 591–600 (2020)
Selgert, F.: Cynefin framework, devOps and secure IoT. In: International Conference on Computer Safety, Reliability, and Security, September, pp. 255–265. Springer, Cham (2020)

Towards a Proactive System of DDoS Attacks Detection in the Cloud

Mohamed Ouhssini[1](✉), Karim Afdel[1], and Elhafed Agherrabi[2]

[1] LabSIV, Department of Computer Science, Faculty of Science, Ibn Zohr University, Agadir, Morocco

[2] LabIRF-SIC, Department of Mathematics, Faculty of Science, Ibn Zohr University, Agadir, Morocco

Abstract. Nowadays, cloud computing becoming a vital technology for organization and users by providing services through the internet. Flexibility, Efficiency, high scalability, and cost effective are the main advantages of cloud computing. However, cloud computing still suffers against attacks that threat its availability, especially DDoS attacks. This type of attack makes service unavailable by sending a huge number of flows generated by different distributed devices. High false positive rate, high time detection and high computing cost are the main challenges of exiting system of attack detection in the cloud. In this paper, we present new approach to detect DDoS attack in the cloud based on time series analysis and machine learning techniques. Statistical methods and deep learning are used to forecast the number of normal flows in the next minute, if the number of received flows is far exceeded the forecasted number, a preprocessing process is trigged followed by classifier to differ between DDoS attack and normal flows. Our system is evaluated by using public dataset CIDDS-001 and satisfactory results obtained an accuracy of an accuracy of 100%, a f1-score of 99.96% and a time detection of 0.29 s.

Keywords: Cloud computing · Security · DDoS attacks · Machine learning · Deep learning · ARIMA · CNN · LSTM

1 Introduction

Cloud computing provides the next generation of internet-based applications, highly scalable distributed computing systems in which computational resources are offered ‘as a service’ [1]. It is utilized to address the resources shortage issues of its customers by giving them with on-demand pay-per-use services [2]. Cloud computing offers mainly three service delivery models: Infrastructure as a Service (IaaS), Platform as a Services (PaaS) and Software as a Service (SaaS) [3] as shown in Fig. 1. Four-development model of the cloud: public, private, hybrid and community are defined by NIST. Therefore, cloud computing still suffers against different type of attacks especially, distributed denial of service attack (DDoS). Duo to great benefit of the cloud computing where infrastructure is shared usually with huge number of users, DDoS attacks can possibly have a lot more noteworthy effect than against single rented models [4]. Distributed

A. Abraham et al. (Eds.): SoCPaR 2020, AISC 1383, pp. 921–933, 2021.
https://doi.org/10.1007/978-3-030-73689-7_87

Denial of Service is a type of attack that aims to make services or resources inaccessible for infinite periods of time by overwhelming it with unnecessary traffic [5].

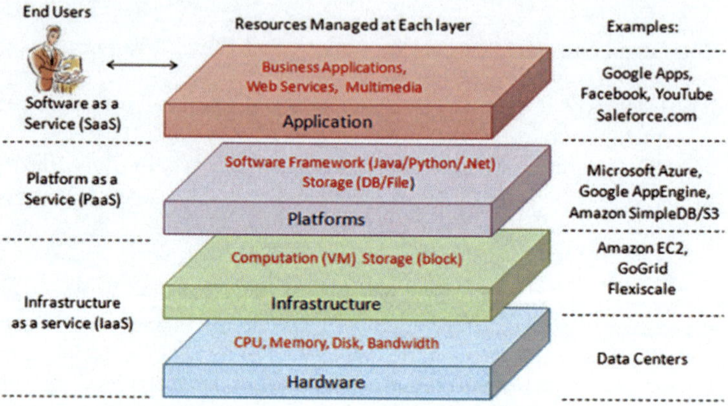

Fig. 1. Cloud computing architecture [6].

DDoS attacks represent a very critical risk to the security of the network in general. Typically, popular websites are the victims of these attacks [7]. Amazon announced that it was forced to tackle a 2.3 Tbps DDoS attack in Q1 of 2020. This is significant for a number of reasons; first, it is the largest reported attack in history, nearly four times that of the previous record holder (587 GB/s) [8]. DDoS attacks are a significant threat to the vast majority of service providers and their effect is widespread. These attacks can account for up to 25% of the overall Internet traffic of a country when they occur [9]. In the Fig. 2 we show the distribution of DDoS attacks by type in the world in the Q1 2020. In this paper we present a detection system of DDoS attacks in cloud environment based on time series analysis and machine learning algorithms. The proposed system process in the four main steps: forecast number of normal flows, count number of received flows, preprocessing and classification. We used time series analysis based on statistical models and deep learning algorithms to forecast the number of normal flows in the next minute, if this number large greater than received flows, we launch directly the process of classification to detect DDoS attacks earlier. In the forecast step, we compare statistical model such as ARIMA and deep learning techniques such as CNN and LSTM to show their effect in the forecast. The preprocessing step clean and normalize the network traffic data of time window that is notified contains DDoS attacks. The preprocessing is followed by classify the flows into normal or DDoS attacks. A comparison evaluation of many machine learning algorithms like Logistic Regression, Decision tree, Random forest, Gradient boosting, Multilayer perception and Naïve bayes are used to select the best classifier for DDoS attacks based on accuracy, F1-score, ROC curve and detection time.

The main contribution of this paper is summarized as follow:

- Adopting time series analysis for DDoS detection in the cloud.
- Comparing statistical methods with deep learning techniques for forecasting.

- Detecting DDoS attacks with minimal resources in the cloud earlier comparing with previous works.

The rest of paper organized as follow: Sect. 2 summarize some pervious related works. Section 3 devoted to the used dataset. The proposed system is detailed in the Sect. 4. Section 5 gives the details of experiments, results and discussions. This paper ends with main conclusions.

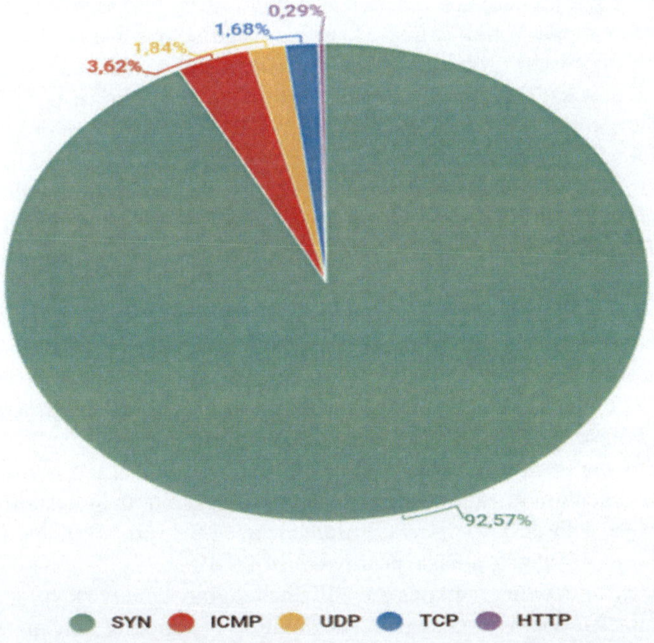

Fig. 2. Distribution of DDoS attacks by type, Q1 2020 [10].

2 Related Works

In this paper, time series analysis and machine learning techniques are used to detect DDoS attack in the cloud earlier with minimal resources. In this section we summarize some pervious works in the detection of DDoS attacks.

In [11] a novel DoS and DDoS detection method is proposed. In this system, the number of packets time series variance is fixed using the Box-Cox transformation. This transformation causes a better prediction based on an ARIMA model. In the next, error chaotic characteristics of time series is performed and the local Lyapunov exponent classify chaotic and no chaotic errors. In [12] the auteurs have proposed an approach based on application of the exponential moving average to two time series, one containing entropy values and the other containing numbers of packets. Then, the method combines both volume-based and feature based detection. This study [13] proposes a technique of integration between HTTP GET flooding DDoS attacks and MapReduce processing for early attack detection in a cloud system. The experimental results show that the proposed method is better compared with traditional methods of detection because the processing time is lower in increasing congestion. In [14] a method is proposed, in which analyzes specific spectral features of traffic. No packet inspection is needed. The data used in the experiments are collected from local network of the auteurs' institution. The results show that the proposed method has a satisfactory results and usable in many other systems. In this work [15], a system based on machine learning is proposed to detect DDoS attacks in a cloud environment. Voting extreme learning machine has been used as a classifier for performing the system. It uses multiple extreme learning machines simultaneously to detect DDoS attacks. Two datasets are used to validate the performance of the system, NSL-KDD dataset and ISCX dataset. The experiments show that the proposed system detects attacks with a high accuracy 99.18% and 92.11% with NSL-KDD and ISCX, respectively. In [16] the auteurs have proposed a novel algorithm based on improved support vector machine (SVM). The combination between kernel principal component analysis (KPCA) with chaotic particle swarm optimization, the auteurs performed their detection system of attacks. a SVM classifier is used to differ between normal and attacks behavior, KPCA is used in the preprocessing stage to reduce the dimension of data and reduce training time. The experiments show that the improved SVM model reaches a high accuracy and a low detection time. The auteurs of [17] propose a detection method of DDoS attacks in a Cloud systems based on combining Information Entropy and Random Forest. A sliding time window technique is used to calculate the entropy of meta data of the incoming network flows. if the entropy exceed normal range the process of classification is called by the system. To evaluate their system the auteurs used CIDDS-001 dataset. Experimental results proved that proposed system is better, with an accuracy of 99.54%, a False positive rate of 0.4%.

The majority of existing methods are still challenging against DDoS attacks, especially when attacks' behavior changed. The solution is adopting new approach based on time series analysis and machine learning techniques to detect DDoS attacks earlier with high accuracy and low false positive rate using minimal computing resources.

3 The Used Dataset

CIDDS-001 (Coburg Network Intrusion Detection Dataset) is a labelled flow based dataset developed for the evaluation of Anomaly based NIDS. This dataset contains unidirectional NetFlow data. It consists of traffic data from two servers i.e. OpenStack and External server. The dataset is generated by emulating small business environment which consist of OpenStack environment having internal servers (web, file, backup and mail) and an External Server (file synchronization and web server) which is deployed on the internet to capture real and up-to-date traffic from the internet [18]. Figure 3 provides the description of CIDDS-001 dataset attributes.

Sl. no.	Attribute Name	Attribute Description
1	Src IP	Source IP Address
2	Src Port	Source Port
3	Dest IP	Destination IP Address
4	Dest Port	Destination Port
5	Proto	Transport Protocol (e.g. ICMP, TCP, or UDP)
6	Date first seen	Start time flow first seen
7	Duration	Duration of the flow
8	Bytes	Number of transmitted bytes
9	Packets	Number of transmitted packets
10	Flags	OR concatenation of all TCP Flags
11	Class	Class label (Normal, Attacker, Victim, Suspicious and Unknown)
12	AttackType	Type of Attack (PortScan, DoS, Bruteforce, PingScan)
13	AttackID	Unique Attack id. Allows attacks which belong to the same class carry the same attack id
14	AttackDescription	Provides additional information about the set attack parameters (e.g. the number of attempted password guesses for SSH-Brute-Force attacks)

Fig. 3. Detail of CIDDS-001 dataset attributes [18].

4 The Proposed Approach

In this section, we detailed the process followed to detect DDoS attacks in the cloud as well as the main components of system. The proposed system dealing with four major steps: forecast number of normal flows in the next minute, count number of received flows, preprocessing and classification. The number of normal flows forecasted with statistical methods and deep learning techniques, if the number of received flows is far exceeded the forecasted number, the process of preprocessing and classification is trigged in the real time to realize earlier detection of DDoS attacks in the cloud. A layered architecture is presented in the Fig. 4 to show the steps followed by system to detect DDoS attacks. The main idea of our work is to detect DDoS attacks in the cloud systems earlier with minimal resource because the process of preprocessing and detection is not called, if the number of received flows in normal range of forecasted number.

4.1 Forecast Number of Normal Flows of Next Minute

In this step, the system forecast number of normal flows for next minute based on ARIMA and deep learning architecture such as convolutional neural network (CNN) and long-short term memory (LSTM). We compare all these technique to select the best forecaster for our system. ARIMA is commonly used to predict linear pattern in the data in other hand neural network architectures is commonly used to predict nonlinear pattern in the data. In our system CNN and LSTM reach best results compared to ARIMA. We show the details of results in Sect. 5.

4.2 Count Number of Flows in the Current Minute

In this step, we launch the process of calculation the number of flows in the current minute, if the number of flows reach a number large greater than forecasted number of normal flows, we run directly in the real time the process of preprocessing.

4.3 Network Data Preprocessing

In this step, two main tasks are performed on the networks data of time window that have number of flows greater than forecasted number of flows. Cleaning is dealing with drop duplicated and missing rows. The normalization dealing with scaling all values between [0, 1] by MinMax method as follows:

$$x_i^{new} = \frac{x_{i-\min(x)}}{\max(x) - \min(x)} \quad (1)$$

4.4 DDoS Attacks Detection

In this step, the system classify data traffic network into DDoS attack or normal flows. For this purpose six machine learning classifiers are used to choose the best model. We split the data extracted from CIDDS-001 public data set into 70% training part and reminder 30% as test part. The ML classifiers used are: Logistic Regression, Decision tree, Random forest, Gradient boosting, Naïve Bayes, multi-layer perceptron.

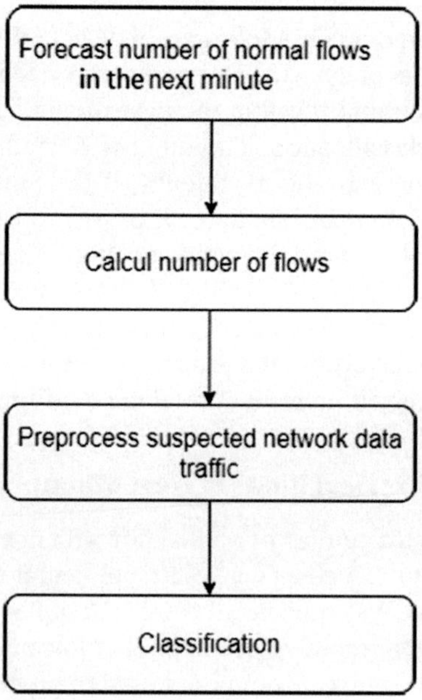

Fig. 4. Architecture of proposed system.

5 Experiments, Results and Discussion

This section describes the experiments conducted and performances metrics used to evaluate our system and we discus obtained results. The experiments aim to show the

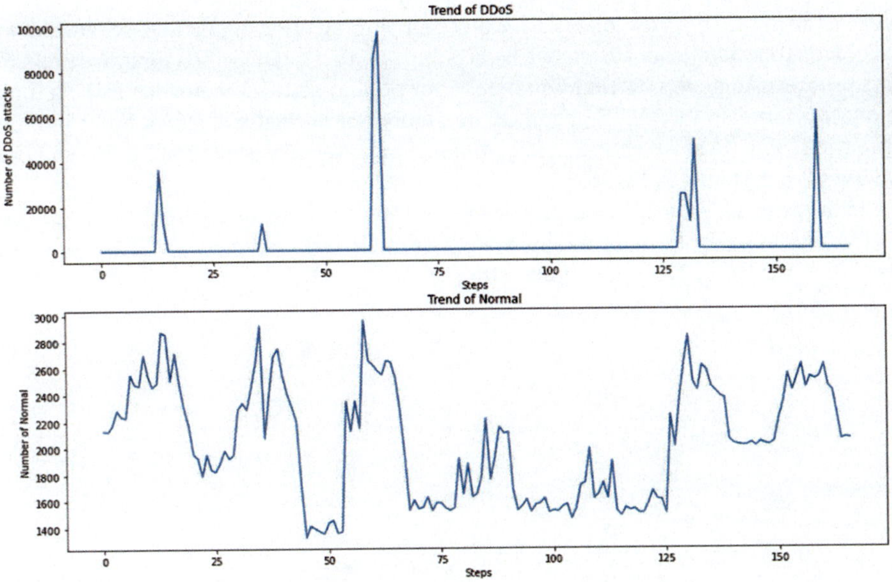

Fig. 5. Number of normal flows and DDoS attacks in each time window 3600 s.

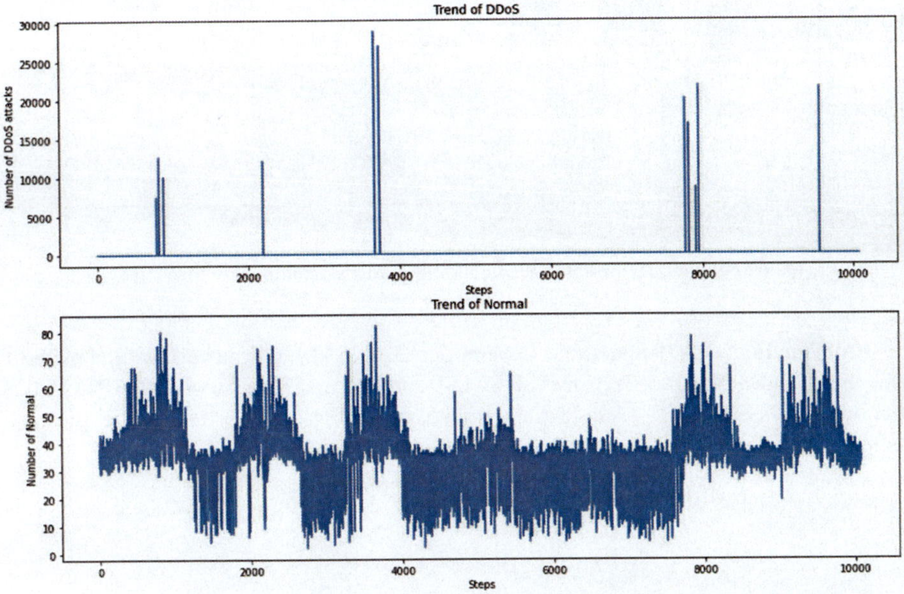

Fig. 6. Number of normal and DDoS attacks in each time window 60 s.

effect of using time series analysis for a proactive protection against DDoS attacks in the cloud. To evaluate our system we used public data set CIDDS-001. First, we prepare data to forecasting step by counting number of normal and DDoS attacks in each time window of 60 s. In the Fig. 5, we illustrate number of normal and attack flows in each time window 3600 s, we choose 3600 s time window for representation, in the analysis we use 60 s as shown in Fig. 6.

In the next, we present and discus results of forecasting by using AIRMA, CNN and LSTM. For comparing these methods we used mean squared error (MSE) and Mean absolute error (MAE) as evaluation metrics.

$$MSE = \frac{1}{n}(\hat{y} - y)^2 \quad (2)$$

$$MAE = \frac{1}{n}|\hat{y} - y| \quad (3)$$

Where $\hat{y}$ predicted value and y real value.

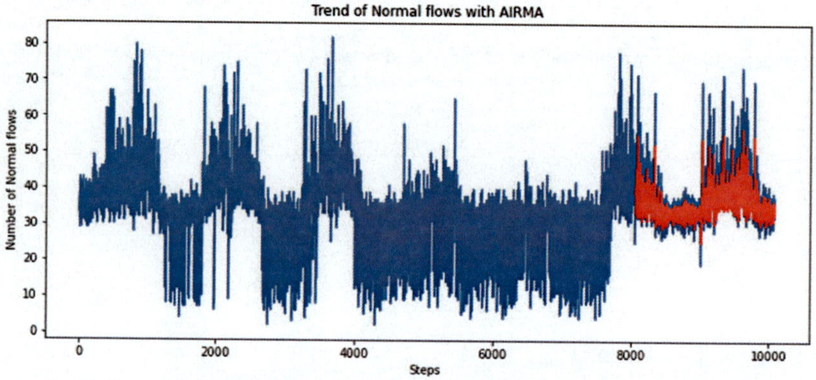

Fig. 7. Number of normal flows including predicted numbers with ARIMA.

From the Figs. 7, 8, 9 and Table 1, we notice that LSTM has the best result a MSE of 4.84 and 1.84 followed by CNN a MSE of 4.89 and MAE of 1.84. However, ARIMA has a good result a MSE of 5.49 and MAE of 1.96. In the Figs. 7, 8, 9 the red part presents the predicted values.

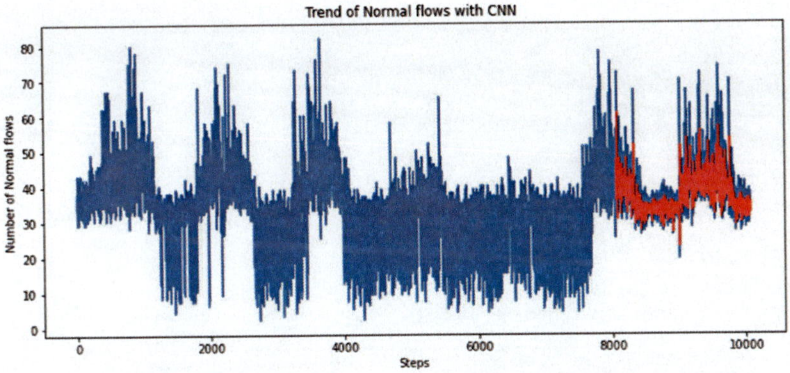

Fig. 8. Number of normal flows including predicted numbers with CNN.

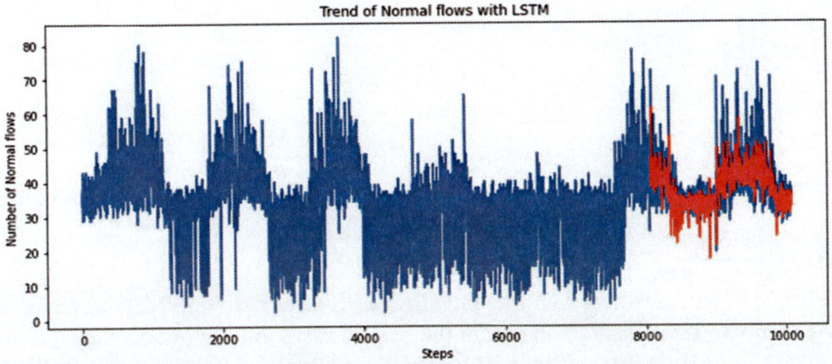

Fig. 9. Number of normal flows including predicted numbers with LSTM.

Table 1. Results of diffrent models for forecasting.

Model	MSE	MAE
ARIMA	5.49	1.96
CNN	4.89	**1.84**
LSTM	**4.84**	**1.84**

In the Fig. 10, we show the learning and testing curve of LSTM model.

In the classification step, the evaluation criteria used to select the best model are accuracy, precision, recall, f1-score and FPR.

$$Accuracy = \frac{TP + TN}{TP + TN + FP + FN} \quad (4)$$

$$Precision = \frac{TP}{TP + FP} \quad (5)$$

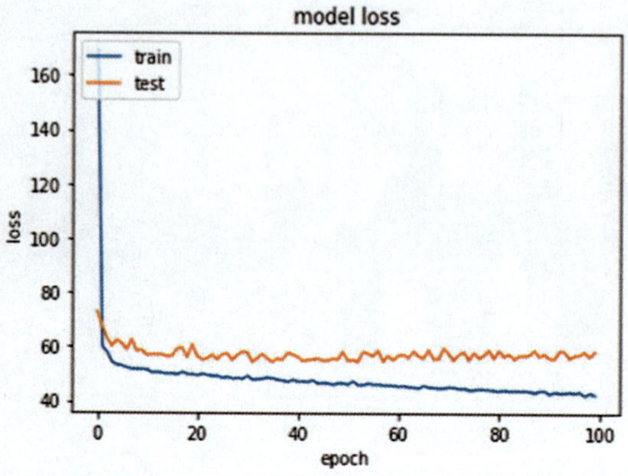

Fig. 10. Learning and testing curve of LSTM.

$$Recall = \frac{TP}{TP + FN} \tag{6}$$

$$F1 - score = 2 * \frac{Precision * Recall}{Precision + Recall} \tag{7}$$

$$FPR = \frac{FP}{FP + TN} \tag{8}$$

Where TP: true positive; TN: true negative; FP: false positive; FN: false negative.

We compare six classifiers to obtain the best of them in term of accuracy, f1-socre, FPR and running time. In the next we present the results of comparison and we discuss.

Table 2. Results of different ML algorithms for classification.

Model	Accuracy	Precision	Recall	F1	Time
LR	0.9631	0.8977	0.5259	0.5398	0.10
NB	0.6295	0.5468	0.8063	0.4665	0.82
DT	0.9999	0.9997	0.9996	0.9996	0.29
RF	0.9999	0.9997	0.9996	0.9996	27.4
GB	0.9995	0.9996	0.9990	0.9993	5.60
MLP	0.9994	0.9448	0.9969	0.9964	0.66

In the table above we present the result of different algorithms, we notice that DT and RF obtain best result, an accuracy of 100% and a f1-score of 99% but it is preferable to use DT due to lower time detection, GB and MLP are also obtain good result, LR and NB aren't good to use for our case, KNN and SVM we find the difficult to apply them due to their computational time and limited resources, they require a high performance of hardware. In the Fig. 11 we show the ROC curve of DT. In Table 3 we present the results of some previous works on CIDDS-001 dataset.

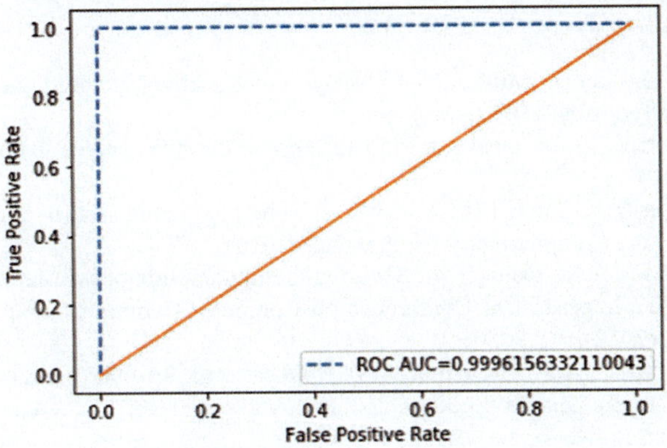

Fig. 11. ROC curve of DT.

Table 3. Results of some previous works on CIDDS-001.

	Dataset	Accuracy	F1-score
Approach of [19]	CIDDS-001	93.00	98.85
Approach of [20]	CIDDS-001	98.72	NA
Approach of [21]	CIDDS-001	99.99	NA
Approach of [17]	CIDDS-001	99.54	NA
Our approach	CIDDS-001	99.99	99.96

6 Conclusion

In this paper a detection method of DDoS attacks in the cloud is proposed. Our approach dealing with four steps: forecast number of normal flows for the next minute, counting number of received flows, preprocessing and classification. The forecasting is based on statistical method and deep learning technique. LSTM gives the best results in term of MSE and MAE. Counting received flows in real time, if this number large greater than forecasted number of normal flows, the system launches preprocessing process followed with classification. To evaluate our system, we used public dataset CIDDS-001 with configuration of 70% training and 30% testing. We compare some ML algorithms like RF, DT, NV, MPL, GB and LR to select an appropriate classifier. DT a satisfactory result an accuracy of 100, a f1-score of 99.96% and a time detection of 0.29 s.

References

1. Almorsy, M., Grundy, J., Müller, I.: An analysis of the cloud computing security problem. arXiv Prepr arXiv:160901107 (2016)
2. Buyya, R., Broberg, J., Goscinski, A.M.: Cloud computing: Principles and paradigms. Wiley, Hoboken (2010)
3. Singh, S., Jeong, Y.-S., Park, J.H.: A survey on cloud computing security: Issues, threats, and solutions. J. Netw. Comput. Appl. **75**, 200–222 (2016)
4. Joshi, B., Vijayan, A.S., Joshi, B.K.: Securing cloud computing environment against DDoS attacks. In: 2012 International Conference on Computer Communication and Informatics. IEEE, pp. 1–5 (2012)
5. Ahmad, I., Abdullah, A.B., Alghamdi, A.S.: Application of artificial neural network in detection of DOS attacks. In: Proceedings of the 2nd International Conference on Security of Information and Networks, pp. 229–234 (2009)
6. Zhang, Q., Cheng, L., Boutaba, R.: Cloud computing: state-of-the-art and research challenges. J. Internet Serv. Appl. **1**, 7–18 (2010)
7. Behal, S., Kumar, K., Sachdeva, M.: D-FACE: An anomaly based distributed approach for early detection of DDoS attacks and flash events. J. Netw. Comput. Appl. **111**, 49–63 (2018)
8. Cook, S.: DDoS attack statistics and facts for 2018–2020 (2020) v. https://www.comparitech.com/blog/information-security/ddos-statistics-facts/. Accessed 30 Oct 2020
9. Crane, C.: The 15 Top DDoS Statistics You Should Know In 2020 (2019). https://cybersecurityventures.com/the-15-top-ddos-statistics-you-should-know-in-2020/. Accessed 30 Oct 2020
10. Kupreev, O., Ekaterina Badovskaya, A.G.: DDoS attacks in Q1 2020 (2020). https://securelist.com/ddos-attacks-in-q1-2020/96837/. Accessed 30 Oct 2020
11. Nezhad, S.M.T., Nazari, M., Gharavol, E.A.: A novel DoS and DDoS attacks detection algorithm using ARIMA time series model and chaotic system in computer networks. IEEE Commun. Lett. **20**, 700–703 (2016)
12. Bojović, P.D., Bašičević, I., Ocovaj, S., Popović, M.: A practical approach to detection of distributed denial-of-service attacks using a hybrid detection method. Comput. Electr. Eng. **73**, 84–96 (2019)
13. Choi, J., Choi, C., Ko, B., et al.: Detecting web based DDoS attack using MapReduce operations in cloud computing environment. J. Internet Serv. Inf. Secur. **3**, 28–37 (2013)
14. Aiello, M., Cambiaso, E., Mongelli, M., Papaleo, G.: An on-line intrusion detection approach to identify low-rate DoS attacks. In: Proc - Int Carnahan Conf Secur Technol 2014-Octob (2014). https://doi.org/10.1109/CCST.2014.6987039
15. Kushwah, G.S., Ranga, V.: Voting extreme learning machine based distributed denial of service attack detection in cloud computing. J. Inf. Secur. Appl. **53**, 102532 (2020)
16. Kuang, F., Zhang, S., Jin, Z., Xu, W.: A novel SVM by combining kernel principal component analysis and improved chaotic particle swarm optimization for intrusion detection. Soft Comput. **19**, 1187–1199 (2015)
17. Idhammad, M., Afdel, K., Belouch, M.: Detection system of HTTP DDoS attacks in a cloud environment based on information theoretic entropy and random forest. Secur. Commu. Networks **2018**, 1–13 (2018)
18. Verma, A., Ranga, V.: Statistical analysis of CIDDS-001 dataset for network intrusion detection systems using distance-based machine learning. Procedia Comput. Sci. **125**, 709–716 (2018)

19. Bedi, P., Gupta, N., Jindal, V.: I-SiamIDS: an improved Siam-IDS for handling class imbalance in network-based intrusion detection systems. Appl Intell. 1–19 (2020)
20. Gonzalez-Cuautle, D., Hernandez-Suarez, A., Sanchez-Perez, G., et al.: Synthetic minority oversampling technique for optimizing classification tasks in botnet and intrusion-detection-system datasets. Appl. Sci. **10**, 794 (2020)
21. Thapa, N., Liu, Z., Kc, D.B., et al.: Comparison of machine learning and deep learning models for network intrusion detection systems. Future Internet **12**, 167 (2020)

Edge Computing Enabled Technologies for Secure 6G Smart Environment-An Overview

Syed Usman Jamil(✉), M. Arif Khan, and Sabih ur Rehman

Charles Sturt University, Wagga Wagga, Australia
{sjamil,mkhan,sarehman}@csu.edu.au

Abstract. In this article, we present a theoretical framework for a secure smart architecture based on futuristic edge resource allocation with a key focus on the task off-loading and intelligent enabling technologies. This article aims to highlight the growing need of embedded off-loading standards in the context of the Internet of Everything (IoE), fog, and cloud paradigms enabled technologies in the overall design framework of smart architecture. Efficient task off-loading strategies perform a vital role in mobility, heterogeneity, data and energy management, load balancing, establishing security mechanisms, reducing the propagation delay and information processing. The motive behind presenting such an embedded intelligent and secure architecture is to address the growing concerns of upcoming smart networks and its sub-network based on different futuristic scenarios, applications and devices. Furthermore, we discuss the off-loading criteria and enabling technologies, in view of the new advancement and critical requirements with rising demands of up-coming intelligence-based secure off-loading frameworks, in order to meet the requirements of the secure smart environment. Finally, through our discussion, we develop the specific research questions and explain its motivations in order to design and evaluate the off-loading criteria and its associated enabling technologies to accommodate the secure technological advancements and resource allocation in the sixth generation (6G) wireless communication era.

Keywords: Secure smart environment · IoE · 6G communication · Edge computing · Task off-loading · Resource allocation

1 Introduction

Technology has emerged as a primary need for human society. We are entering into the era of remote management, intelligence, self-decision-making fully automated systems and a globally interconnected cyber-physical world [1]. The fast development of diverse and innovative applications via Artificial intelligence (AI), Internet-of-Everything (IoE) and Virtual reality (VR) has driven the data

A. Abraham et al. (Eds.): SoCPaR 2020, AISC 1383, pp. 934–945, 2021.
https://doi.org/10.1007/978-3-030-73689-7_88

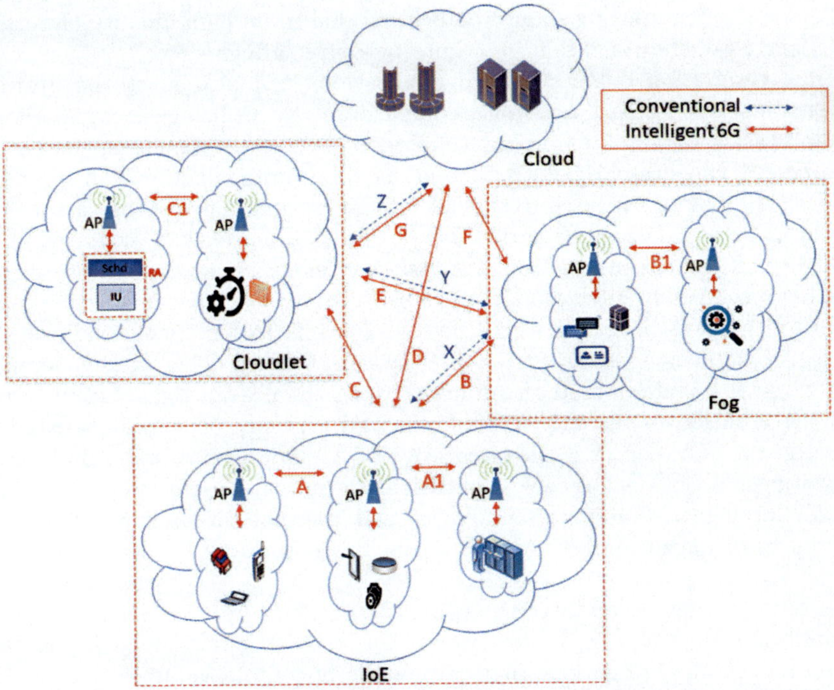

Fig. 1. Resource allocation and task off-loading paradigm.

traffic of our networks to massive volumes. In order for a futuristic smart environment to function, it will require all the elements of such architecture to communicate efficiently among existing technologies such as smart grid, smart homes, smart healthcare, smart factories, smart automotive, smart telecommunication and green energy systems.

Edge computing has become a newly emerged distributed paradigm to enhance centralized cloud-computing architecture. Edge computing deploys diverse heterogeneous services and applications by utilizing computing-based resources near to the network's edge. In an edge computing network [8], the storage and computation can also be considered as edge nodes having three significant characteristics of *mobility*, *highly dynamic*, and *heterogeneous*. These edge computing features differentiate it from the conventional cloud computing architectures where computational-resources are controlled and managed centrally.

Due to the broad distance [12] between end-users and the cloud, fast response delivery and real-time processing become a great challenge. To resolve these challenge, certain technologies are required between the cloud and end-nodes such as cloudlets, fog computing, IoE and so on. Hence, to achieve the consumer's optimization goal and to overcome the resource limitations, AI-empowered intelligent edge computing should be utilized due to its high Quality of Service (QoS) and resource utilization for secure task off-loading.

Edge computing enabled secure applications can also meet the growing requests for low latency and high computation by utilizing resource allocation and task off-loading paradigm, to off-load tasks to neighbouring computing resources. It can dramatically reduce latency in communication and improves self decision-making. In a conventional off-loading mechanism, the task is forwarded to high-end resource enriched layers in direction X-Y-Z, whereas, in an intelligent off-loading mechanism, the task is intelligently managed towards the available resources in directions A-A1 or B-C-D, as illustrated in Fig. 1.

Future wireless networks would need to be able to handle the massive off-loading of data using ultra high-data-rate connectivity. The sixth generation (6G) wireless systems would also function with the previous generation trends with the addition of modern technologies and services. The most demanding feature of the 6G network is the integration of AI-supported systems and secure applications. As modern secure applications become more efficient and sophisticated, intelligent enabling technologies and relevant off-loading criteria are required to manage complex task executions based on these application requirements.

In order to deploy and design such a stringent architecture based on IoE, fog and cloudlet technology, the system needs the support of having a deep inspection security framework that can handle huge data volumes that will be passed through each of its components. To cater for such a system, a thorough security risk analysis for all components related to the entire system, including border Application Programming Interfaces (APIs) as well as the communication channels will be required for integrating these IoE devices. This article describes edge state-of-the-art scenarios, applications, devices and its requirements to off-load data for task execution by considering the *conventional criteria* along with the significant advancements required for *intelligent off-loading criteria*. We also discuss some of the edge-based enabling technologies and security requirements. The article has the following key contributions:

- We review edge architecture state-of-the-art scenarios, applications, devices and its requirements to off-load data for secure task execution by considering the *conventional off-loading criteria* and significant advancements required for *intelligent off-loading criteria.*
- We provide an overview of the edge enabling technologies and security, in view of the new advancement and critical requirements with rising demands of upcoming intelligence-based secure off-loading frameworks, in order to meet the requirements of the secure 6G smart environment.
- We present the specific research questions and explain its motivations to design and evaluate the off-loading criteria and its associated enabling technologies to accommodate the secure technological advancements and resource allocation in the 6G wireless communication era.

The rest of the article is organized as follows: The related work is presented in Sect. 2. Section 3, discusses an edge architecture - state-of-the-art scenarios, applications, devices and edge off-loading criteria. Section 4, presents the edge enabling technologies and security. Section 5, enlist research questions and motivation. Finally, Sect. 6 concludes this work-in-progress article.

2 Related Work

In this section, we describe some of the recent work in this area. In [9], broadly explored smart environment empowered by smart technologies, with the help of smart devices and smart tools. A smart city architecture is proposed by authors and focused on the functions related to resource management and data off-loading. An adaptable architecture supported by IoT to tackle the newly arising challenges of both cloud computing and IoT in smart environments. In [4], the fog network architecture is addressed on edge, focusing on joint optimization of data off-loading tasks and allocation of computational resources. Concerning fog's reliability, the authors presented three-layered structured dynamics for computation off-loading such as fog, cloud, and devices. A joint caching solution and computational off-loading at edge-cloud is introduced by [5] and deployed AI-inspired computing, communication and caching resources nearby smart vehicles. By allowing jointly RSU peer off-loading, content caching and vehicle to RSU off-loading on the Internet of Vehicles (IoV) architecture. In the area of IoT enabled wireless communication, AI has been implemented in various works. AI-based cognitive radio has been investigated in [3] in which AI-enabled cognitive radios receive input from the users or environments and on the other hand the intelligent terminal then review the information and determine a suitable response and then takes the decision. In [6] applied AI to enable intelligent wireless networks in which an AI-empowered intelligent 6G architecture is introduced. In order to achieve intelligent service provisioning, network adjustment automation, enabling technologies and smart management of resources with high intelligence level. In this regard, the state-of-the-art 5G based ten enabling technologies are extensively surveyed. In [11] a novel 6G intelligent AI-empowered architecture has been presented to recognize intelligent resource optimization, knowledge discovery, and intelligent service-enabling. The IoT offers many new opportunities for applications and users in-terms of Information and Communication Technology (ICT). Therefore, it enacts numerous challenges that risk security attacks due to its vulnerability [7]. In the context of vulnerability, successful targeted attacks on IoT devices as well as their related communication channels cause not only one device affected, but also many related multiple systems and interconnected devices (including applications and nodes). It can also impact the interaction with the physical world itself.

3 Edge Architecture - State-of-the-Art

This section presents the edge scenarios, applications, devices and off-loading criteria to review the edge state-of-the-art requirements.

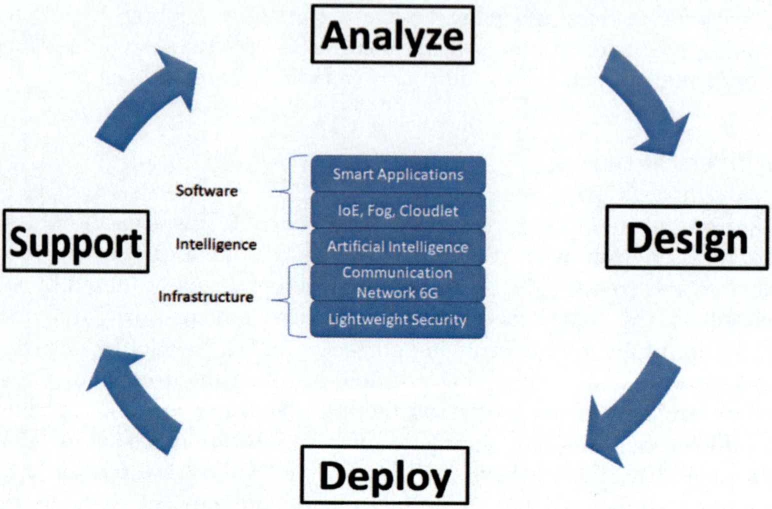

Fig. 2. Intelligence based secure re-usability model [2].

3.1 Edge Scenarios

With the continuous maturity of resource allocation leading toward efficient and intelligent off-loading over secure 6G environment, it brings diverse key opportunities for the development and application of intelligence/security as presented in Fig. 2. In this regard, few edge scenarios are discussed as detailed below.

Smart Environment: Nowadays, a large number of smart edge devices are deployed to serve our daily life needs, for example, smart cameras, smart gadgets, sensors and so on. However, experiencing immense data processing and communication from IoE related devices, cloud computing paradigms and a high ranging demand for delay-sensitive applications.

Integration Architecture: With the growing scope of intelligence and security, it is essential to design/develop modern system architectures that must be compatible with current system models, support the progressive realisation as well as the development of upcoming future systems, to satisfy device integration, management and persistence mechanisms.

Wireless Communication Networks: The evolution of wireless communication networks from the first-generation networks (1G) to the current fifth-generation networks (5G) brought forth various key areas of focus, including data rate, reliability, coverage, end-to-end latency, energy efficiency, and spectrum utilisation. The next-generation 6G networks are foreseen to support modern intelligent applications, unlimited connectivity and high-quality services effectively for the huge number of intelligent terminals that may be required.

3.2 Edge Applications

In this sub-section, the leading edge applications are listed as detail below.

Smart Home: In this era of an intelligent world, the smart home possesses great importance by offering diverse intelligent services like enhanced security, smart control, smart meters and fire suppression systems.

Off-loading Allocation: Integrating AI with off-loading tasks allocation can solve numerous limitations in current off-loading strategies such as pre-processing data at the server's edge. In addition to that, the off-loading performance optimisation in an unsteady scenario with continually changing environmental conditions along with the continuous joining of new users.

6G: 6G networks must need to transform themselves by achieving intelligence to address stringent demands and requirements, which include ultra-low latency, ultra-high data rates, energy efficiency, high intelligence, enhanced security, and mobility with AI capability.

3.3 Edge Devices/Units

This sub-section elaborates edge devices/units as detailed below.

User Devices: There are various types of user devices such as wearable devices, smart phones and so on. User devices typically have a limited power supply, security, computing power and storage space.

Roadside Units: Roadside units (RSUs) can intelligently assist vehicles and drivers in various ways which can include real-time speed signs, vehicle applications assistance, signals under heavy traffic, and emergency conditions, and so on.

Access Point (AP): AP which can be equipped with multiple antennas, using 6G technology. APs can support large data rates. For the radio propagation, these APs are working in either mmWave (millimetre wave) or the THz (Tera-Hertz) frequency spectrum.

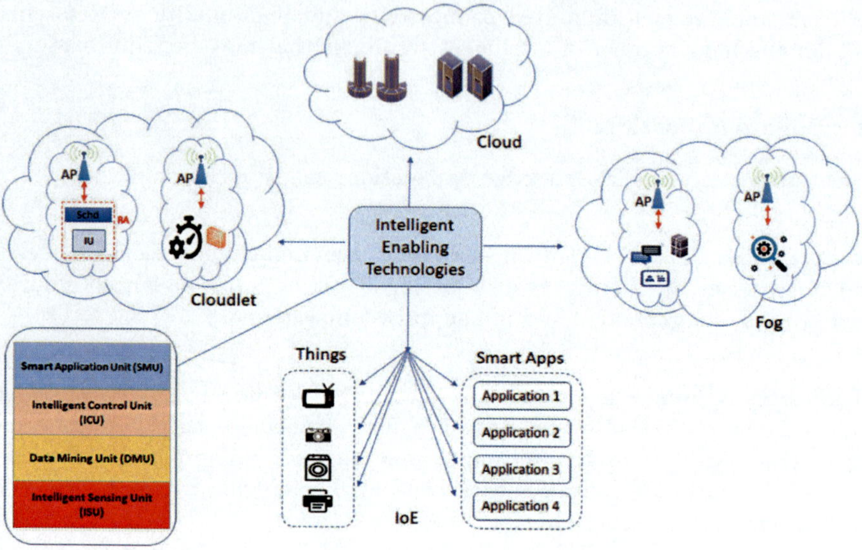

Fig. 3. Role of intelligent enabling technologies.

3.4 Edge Off-Loading Criteria

This section briefly describes several conventional and intelligent criteria, that can be utilised to make optimal decisions regarding specific tasks off-loading. In the first instance, to develop our arguments, we review the *criteria of conventional off-loading* such as latency requirements, load balancing as well as privacy and security.

Latency Requirements: Delay-sensitive applications can be affected by distance. In such a case, a node must be involved proximate to the destination node for certain task off-loading from a remote node so that able to deliver the necessary services efficiently with minimum or no delays.

Load Balancing: When the capacity of one node reaches its peak in tasks execution, further tasks must be shared between the rest of the nodes within the ecosystem of the service provider. For instance, a cloudlet distributes tasks among diverse cloudlets appropriately in order to manage them.

Privacy and Security: Off-loading may occur because of security and privacy, depending upon secrecy and sensitivity of the tasks or data. For example, national tasks/data or an enterprise may migrate to a private cloud from local machines. In the same way, a smart phone's data can be transferred to a personal edge cloud mobile.

In light of the above, we will now review the *criteria of intelligent off-loading*, to name a few, such as mobility, federation and lightweight security.

Mobility: End-users, IoE, fog, cloudlet could be mobile. Intelligence must handle mobility. The architectural modules requirement like intelligent mobility engine is to secure and manage service-continuity to all stakeholders. For example, millions of users managed and serviced in the stadium by diverting services intelligently to access available resources.

Federation: Smart environment deployment has geographically distributed on a large scale, and each domain could belong to a different provider. Furthermore, various providers own IoE, fog, cloudlet, and cloud. Applications provisioning demands the federation, including different providers that might particularly host the different elements developing the applications.

Lightweight Security: In addition to other requirements, evolution and success of an off-loading exceptionally dependent on security standard architecture. In view of that, the approach of this paradigm should be based on the use of lightweight security techniques to secure every individual component as well as data flow through relevant communication channels.

4 Edge Enabled Technologies and Security

With the recent advancements in communication technologies, this has unlocked many opportunities for specific tasks off-loading to proximate systems that implement various middle-ware technologies. Our study offers a brief analysis of some conventional and intelligent networking related enabling technologies and security for task off-loading in the secure smart environment, as depicted in Fig. 3.

4.1 Edge Enabling Technology

In this sub-section, we present conventional enabling technologies and intelligent enabling technologies. Initially, We discuss *conventional enabling technologies* by reviewing the concepts of wireless technology, network virtualisation and parallelism.

Wireless Technology: The most commonly used wireless-internet access is Wi-Fi (IEEE 802.11), it's most recent IEEE 802.11ac standard has much faster connectivity-speed as compared to ethernet (Gigabit). Likewise, wireless internet technologies on-the-go which include, mobile broadband, long-term evolution (LTE), wireless broadband, offer seamless connectivity at mobile nodes like smartphones, computers, tablets, drones and vehicles. Limited-range technologies include Bluetooth, Wi-Fi and so on.

Network Virtualisation: Network virtualisation establishes logical virtual networks that can be realised on virtual machines(VMs), which is broadly used for the management and distribution of network resources with high confidentiality without compromising security levels of the system. Virtualisation offers cost-effective solutions with high portability, off-loading and easy migration between nodes with maximum security.

Parallelism: Massive tasks may need to execute in parallel at diverse nodes. With the availability of parallelism via multi-core technology, simultaneous task execution will be more convenient and will support off-loading tasks in the environment of fog computing.

In view of new advancements and critical requirement, *intelligent enabling technologies* play a crucial role and can be categorised, to name a few, categories such as intelligent sensing, data mining, intelligent control and smart applications.

Intelligent Sensing Technologies: These technologies tend to sense and detect data intelligently from real-world environments via human being crowds or enormous devices like sensors, drones, vehicles, smartphones, and cameras.

Data Mining Technologies: The core task of this unit is to analyse and process the massive amounts of organised data, which is generated from IoE-based devices to attain knowledge discovery and semantic derivation. The enormous amount of data collected from real-time environments can be heterogeneous, high-dimensional, and non-linear.

Intelligent Control Technologies: In brief, intelligent control unit primarily consist of decision-making, learning, scheduling, optimisation and resource management. This unit also aim to utilise the knowledge appropriately from lower layers to empower a large number of agents (e.g. APs and devices) to intelligently optimise learn and decide the best suitable action.

Smart Application Technologies: The main responsibilities of smart applications are to achieve the diverse requirements of the users by providing application-specific services along with the evaluation of purveyed services before feed-backing the results.

4.2 Edge Security

In edge off-loading, the main problem faced is the vast data volume that is passed on from every lightweight component. However, this data capability creates a security risk for all the components related to off-loading and the features of its enabling technologies. The primary concern raised in components regarding data protection by using a security lightweight approach that is that it does not compromise the applications/devices performance and does not create a delay in communication. Lightweight edge security behaves as a bridge among different joining IoE components, this acts as a resources pool to support nodes and applications development. The below four components are a basic requirement of edge lightweight secure system model to address different attacks, to name a few, such as Sybil attack, Spoofing attacks, Key attack and unauthorised access attack [10].

Authentication: Ensures that all the communication system components involved in the off-loading framework are available for who they are supposed to be available for. In view of IoE secure architecture, it is really essential to form shared authentication among the system components for connections.

Authorisation: A compulsory state is required for devices and applications since they should have diverse privileges for data, access resources and the services of IoE secure architecture.

Confidentiality: The main reason for using confidentiality is the preservation of exchanged data in the whole IoE secure architecture. It must also guarantee the stored data in system components and will be able to protect the exchange data from illegal/unauthorised access.

Integrity: Integrity should be guaranteed when it comes to the data exchanged between two systems and cannot be changed by any invader. Integrity plays a vital role to protect the stored data in system components.

5 Research Questions and Motivation

We condense our discussion by enlisting specific research questions and explain its motivations in order to design and evaluate the off-loading criteria and its associated enabling technologies to accommodate the secure technological advancements and resource allocation in the 6G wireless communication era.

RQ1: What is the recent methodical related work done on intelligent off-loading for efficient resource allocation in wireless communication paradigm to support 6G?

Motivation: Following the achievement of edge computing integrated intelligent applications, it is necessary to acquire a clear perception of its architectural module and fundamental features to be used in intelligent off-loading.

RQ2: What are the performance assessments that are realised to solve the resource allocation problem defined in RQ1?

Motivation: Edge related computing tasks are essential and may be enhanced by intelligence, because of the requirement of aggregating raw data and limitations of a wireless device, processing on the resources of available tiers is a need of optimisation.

RQ3: What are the most common methods utilised to reach these objectives of performance evaluation through intelligent enabling technologies in off-loading?

Motivation: The state-of-the-art in optimum intelligence employment determination in resource allocation applications.

RQ4: What are the recent work done on intelligent resource allocation off-loading and enabling technologies related applications?

Motivation: Highlight the forthcoming directions in edge computing framework, performance evaluation, open challenges, and recommendations. We are also collecting data/information on intelligent edge-computing technologies that are applied successfully in intelligent applications.

RQ5: How to implement the secure framework for RQ1-RQ4?

Motivation: Point towards the future directions in the secure edge computing framework and its performance evaluation for the secure smart environment.

6 Conclusions

Efficient task off-loading and enabling technologies contribution towards resource allocation in next-generation wireless systems is an essential function for obtaining maximum benefits of new intelligent technologies and its applications. In this article, we presented a work in progress conceptual design of conventional and intelligent task off-loading. Moreover, the role of enabling technologies for intelligence seeking scenarios, applications and devices in the smart environment, including fog, cloudlet and the cloud, using secure 6G wireless technology. We surveyed some key off-loading criteria and enabling technologies that will play a crucial role in the future secure smart environment including 6G, edge and cloud computing, resource allocation and artificial intelligence. Based on this, we developed research questions and its motivation to set future research goals. Our future work will be to design and evaluate the performance of intelligent task off-loading by considering enabling technologies and security at each individual component, namely IoE, fog, cloudlet in the resource allocation model.

References

1. Arif, S., Khan, M.A., Rehman, S.U., Kabir, M.A., Imran, M.: Investigating smart home security: is blockchain the answer? IEEE Access **8**, 117802–117816 (2020)
2. Jamil, S.U., Khan, M.A., Ali, M.: Security embedded offloading requirements for IoT-fog paradigm. In: 2019 IEEE Microwave Theory and Techniques in Wireless Communications (MTTW), vol. 1, pp. 47–51. IEEE (2019)
3. Letaief, K.B., Chen, W., Shi, Y., Zhang, J., Zhang, Y.-J.A.: The roadmap to 6G: AI empowered wireless networks. IEEE Commun. Mag. **57**(8), 84–90 (2019)
4. Mukherjee, M., Kumar, S., Shojafar, M., Zhang, Q., Mavromoustakis, C.X.: Joint task offloading and resource allocation for delay-sensitive fog networks. In: ICC 2019–2019 IEEE International Conference on Communications (ICC), pp. 1–7. IEEE (2019)
5. Ning, Z., Zhang, K., Wang, X., Guo, L., Hu, X., Huang, J., Hu, B., Kwok, R.Y.: Intelligent edge computing in internet of vehicles: a joint computation offloading and caching solution. IEEE Transactions on Intelligent Transportation Systems (2020)
6. Park, J., Samarakoon, S., Bennis, M., Debbah, M.: Wireless network intelligence at the edge. Proc. IEEE **107**(11), 2204–2239 (2019)
7. Qureshi, K.N., Jeon, G., Piccialli, F.: Anomaly detection and trust authority in artificial intelligence and cloud computing. Computer Networks, p. 107647 (2020)

8. Shojaiemehr, B., Rahmani, A.M., Qader, N.N.: Automated negotiation for ensuring composite service requirements in cloud computing. J. Syst. Archit. **99**, 101632 (2019)
9. Sinaeepourfard, A., Krogstie, J., Petersen, S.A., Ahlers, D.: F2c2C-DM: a fog-to-cloudlet-to-cloud data management architecture in smart city. In: 2019 IEEE 5th World Forum on Internet of Things (WF-IoT), pp. 590–595. IEEE (2019)
10. Tukur, Y.M., Ali, Y.S.: Demonstrating the effect of insider attacks on perception layer of Internet of Things (IoT) systems. In: 2019 15th International Conference on Electronics, Computer and Computation (ICECCO), pp. 1–6. IEEE (2019)
11. Yang, P., Xiao, Y., Xiao, M., Li, S.: 6G wireless communications: vision and potential techniques. IEEE Netw. **33**(4), 70–75 (2019)
12. Zheng, T., Wan, J., Zhang, J., Jiang, C., Jia, G.: A survey of computation offloading in edge computing. In: 2020 International Conference on Computer, Information and Telecommunication Systems (CITS), pp. 1–6. IEEE (2020)

Check for updates

Secure QIM-Based Image Watermarking in JPEG Compression Domain

Victor Fedoseev[1,2] and Daniil Kozlov[1](✉)

[1] Samara National Research University, Moskovskoe Shosse, 34, 443086 Samara, Russia
[2] Image Processing Systems Institute of RAS – Branch of the FSRC "Crystallography and Photonics" RAS, Molodogvardeyskaya, 151, 443001 Samara, Russia

Abstract. Previously, we proposed two secure watermarking algorithms (IM-QIM and SIM-QIM) based on the Quantization Image Modulation (QIM) technique that showed its robustness against a specific histogram attack with multiple images. Traditional QIM versions, including DM-QIM, DC-QIM, and others, are vulnerable to this attack. Our algorithms are based on a modified "correlation immune" scalar quantizer and do not introduce any correlation between the key bits and signal values. Earlier, these algorithms were implemented for grayscale images in the spatial domain. In this paper, we apply our approach in JPEG compression domain and analyze IM-QIM performance in this case. The experimental results have shown that IM-QIM maintains its security in the JPEG domain, while its robustness to typical distortions such as AGWN and JPEG compression improves, compared with spatial domain watermarking. This fact, along with the prevalence of QIM usage in the JPEG (block DCT) domain, indicates the high practical potential of the proposed algorithm.

Keywords: Image watermarking · QIM · Quantization Index Modulation · DM-QIM · IM-QIM · JPEG · DCT · Watermark security · Watermark robustness · Compression domain · AGWN

1 Introduction

One of the most widespread watermarking methods is the method of controlled re-quantization, which is reflected in the family of Quantization Index Modulation (QIM) algorithms [1]. Its essence is the use of different re-quantization scales for embedding different information bits. To date, this approach is implemented in various watermarking systems not only in spatial or temporal domains (i.e. when QIM-based modification of source image pixels, or samples of audio and video signals is performed) [1, 2]. It is also used, for example, in DWT [3] and DCT–SVD [4] domains for audio signals, in DCT [5–8] and DWT [9, 10] domains for images, and also for vector geographic data [11] and for 3D meshes [12]. Among these examples, several applications of QIM are made in 2018–2020.

Despite the widespread use of algorithms of the QIM family, some examples of attacks are known that can detect or/and extract embedded information without knowing

A. Abraham et al. (Eds.): SoCPaR 2020, AISC 1383, pp. 946–955, 2021.
https://doi.org/10.1007/978-3-030-73689-7_89

the secret key. For example, algorithms proposed in [13–15] are able to detect and retrieve a watermark with a probability of at least 0.9 from single image. In [16], an audio steganalysis method for QIM is used in a specific codec. Also a histogram attack against QIM methods was proposed in [17], which allowed the author to evaluate the secret key if a number of watermarked images are known.

In [18, 19], we proposed two new algorithms, called IM-QIM (statistically IMmune QIM) and SIM-QIM (Sliding statistically IMmune QIM), which allow defending against this attack, but differ in lower robustness to JPEG compression and AGWN. In [18, 19] these algorithms were investigated in the spatial domain. However, in practice, using any spectral domain or a compression domain (such as the JPEG compression domain) is more useful [5–8]. That is why in this paper we analyze the applicability of the developed approach in the JPEG compression domain. In our investigation, we do not fill all the spectral components, but only a few, to obtain a better quality. This leads that SIM-QIM is not needed in such a case and we can limit ourselves with IM-QIM only.

The paper is organized as follows. Section 2 contains a brief description of known QIM algorithms. Section 3 describes the IM-QIM algorithm [18, 19], which is resistant to the histogram attack [17]. Section 4 tells us about watermarking in the JPEG compression domain. The final section presents the results of the experimental research of IM-QIM n the JPEG domain in comparison with other algorithms and domains.

2 Conventional QIM Watermarking Algorithms

In this section, we briefly describe some QIM algorithms. Let $f(n)$ be a feature vector of a host asset, and $W(n)$ be a binary watermark, where $n \in [1, N]$. Next,

$$Q(x, \Delta) = \Delta \cdot round\, x/\Delta, \tag{1}$$

is a simple quantization function, where $round(\cdot)$ is rounding to the nearest integer, and $\Delta \in \mathbb{N}$ is a quantization step. In a simple form, QIM embedding can be implemented as follows (we will refer it as Simple-QIM):

$$f^W(n) = \mathrm{E}_{QIM}(f, W, \Delta) = Q(f(n), \Delta) + \Delta/2 \cdot W(n). \tag{2}$$

Let $\widetilde{f^W}(n)$ be the feature vector of the received watermarked asset. To extract the watermark embedded using every QIM algorithm, we should embed a vector of zeros and a vector of ones into $\widetilde{f^W}(n)$, and estimate the errors of the two results.

In DM-QIM algorithm [1], the embedding procedure is specified as follows:

$$f^W(n) = \mathrm{E}_{DM-QIM}(f, W, d, \Delta) = Q(f(n) + d_{W(n)}(n), \Delta) - d_{W(n)}(n), \tag{3}$$

where $d_0(n)$, $d_1(n) \in [-\Delta/2;\ \Delta/2]$, $n \in [1, N]$ are dither vectors used for embedding "0" and "1" bits, respectively.

One more QIM algorithm considered in this paper is the one proposed by Wang et al. [6]. A significant property of this algorithm is a parameter used to control the distortions caused by watermarking. In our experiments, we used the parameter value that provides the best visual quality. We do not specify the embedding formula of this algorithm because it is too complex and can be found in [6].

3 IM-QIM Algorithm

Mitekin in [17] proposed a specific histogram attack against QIM algorithms. It analyzes a series of watermarked images and can reveal the fact of QIM embedding, estimate the step Δ, and recover the embedded data (in some cases).

The IM-QIM algorithm was proposed [18, 19] by Mitekin and Fedoseev in order to defend against this attack. The fundamental difference between IM-QIM and its prototype DM-QIM is the way of using the secret key. In IM-QIM, the key is used to split the set of image pixels into groups, to which the re-quantization operation is subsequently applied. The IM-QIM algorithm also uses arrays $d_0^g(k),\ d_1^g(k),\ g = 2 \ldots G$, but in IM-QIM, these arrays are not used when extracting. Thus, the problem of key synchronization does not arise in this case, unlike DM- QIM, because arrays $d_0^g(k),\ {}_1^g(k)$ can be generated in a unique way for each image.

Before embedding, we divide the domain of definition D into G equal non-intersecting parts $D_1, D_2, \ldots, D_G$, the union of which is equal to the entire set D. This split is determined by a secret key and is unknown to the attacker. D_1 is the only part modified during embedding. We enumerate samples included in each of these subsets as $\left(n_1^g\right), \left(n_2^g\right), \ldots, \left(n_{N/G}^g\right)$, where $g = 1 \ldots G$ is an index defining a subset. Change in the pixels of the set D_1 is carried out according to the formula:

$$f^W\left(n_k^1\right) = \mathrm{E}_{IM-QIM}(f, W, d, \Delta) = Q\left(f\left(n_k^1\right) + d_{W(k)}^1(k), \Delta\right) - d_{W(k)}^1(k), \quad (4)$$

where $d_0^1(k) = \sum_{g=2}^{G} \left(f\left(n_k^g\right) + d_0^g(k)\right)$, $d_1^1(k) = d_0^1(k) - sign(d_0^1(k)) \cdot \Delta/2$ and arrays $d_0^g(k)$, $g = 2 \ldots G$ are generated based on the key for each subset D_g similar to the DM-QIM algorithm. The values that are not related to D_1, will only be additively noised in accordance with the values of the array $d_0^g(k)$:

$$f^W\left(n_k^g\right) = f\left(n_k^g\right) + d_0^g(k), \quad g = 2 \ldots G.$$

This operation does not implement embedding but is intended to ensure the fulfillment of the conditions under which the function is proven to be correlation resistant.

4 Watermarking in the JPEG Domain

In the JPEG domain, the watermark embedding is performed at the level of the block DCT. In general, the original image and the watermarked image are presented uncompressed, as shown in Fig. 1. Therefore, the embedding is preceded by the stage of calculating the DCT, and after the embedding, the reverse DCT is performed. Thus, the result of applying this scheme is an image prepared for the subsequent JPEG compression in the sense that the built-in watermark is partially resistant to JPEG in the specified range of QF values.

The presence of two alternative options in Fig. 1 shows that embedding can occur both already in the quantized DCT coefficients, and directly during the quantization procedure itself (for example, when information is modified using any QIM algorithm).

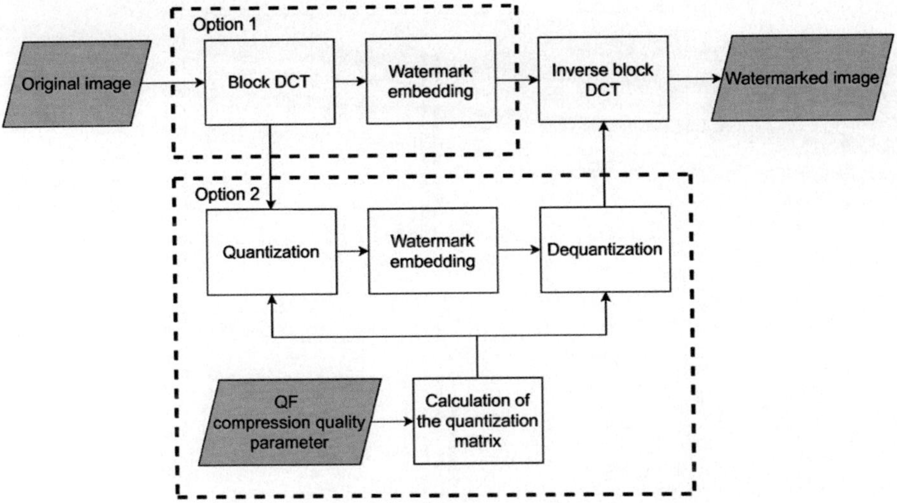

Fig. 1. Scheme of watermarking in JPEG domain

At the quantization stage in Fig. 1, block DCT values are divided by some values calculated using the quality factor (QF) value. After that, the rounding operation is performed. This procedure means that for each coefficient in an 8×8 block, its own quantization scale is used. In our experiments, to align the coefficients, we divided the coefficients by the values obtained for QF = 50 and then applied quantization with the same step Δ. This approach is convenient for experiments and almost corresponds to the real JPEG compression technique.

5 IM-QIM Performance in the JPEG Compression Domain

5.1 Implementation of the QIM Histogram Attack in the JPEG Domain

First, we need to test the performance of the histogram attack in the JPEG domain to make sure that it is still effective for traditional QIM algorithms, and that IM-QIM allows us to defend against it. For this purpose, we simulated the histogram attack for 1000 images from the BOWS-2 collection [20]. The results are shown in Fig. 2. It is clear that for traditional algorithms, the attack reveals periodic peaks indicating the re-quantization procedure, while for IM-QIM, such a picture is not observed.

5.2 Comparative Study on the Robustness of Various QIM Embedding Algorithms in the JPEG Domain

Next, we conducted a study of the robustness and visual quality of IM-QIM in comparison with traditional methods. In these experiments, we used 20 images from the same set [20]. Watermark embedding was performed in random AC coefficients of the DCT spectrum. Totally, we modified 1/64 of all spectral samples (that is, on average, equal to the embedding of one bit into each 8×8 DCT block). For IM-QIM, we used $G = 2$,

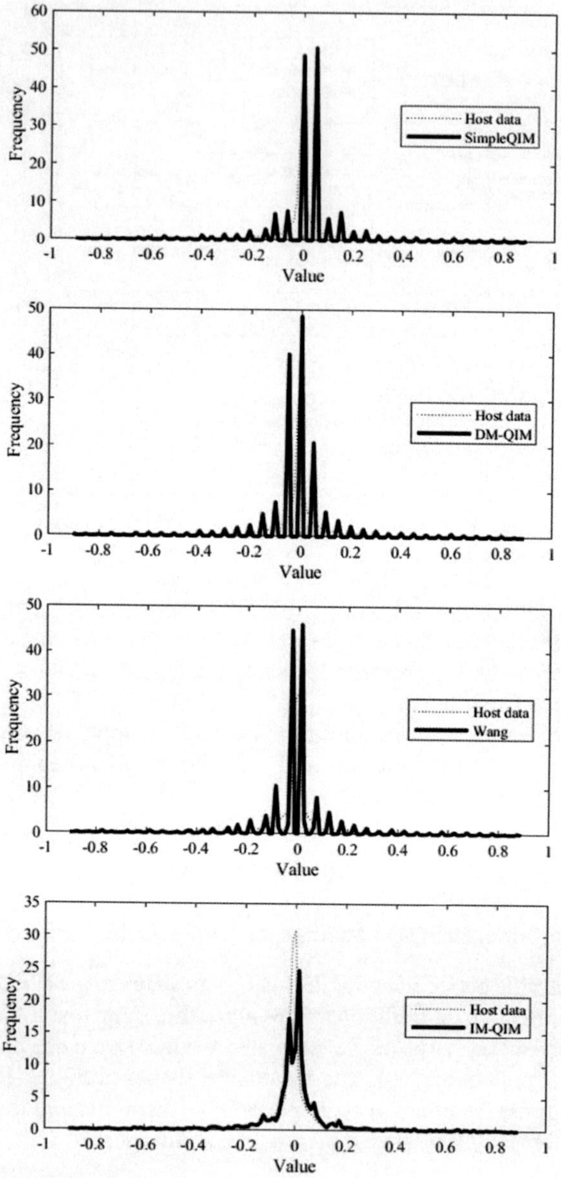

Fig. 2. Implementation of the QIM histogram attack in the JPEG domain for four QIM algorithms. IM-QIM is the only algorithm that does not contain periodic peaks indicating the successful application of the attack.

since this, on the one hand, is sufficient to counteract the histogram attack, and on the other hand, it provides less influence of noise, as was shown earlier in [18, 19]. The scale step Δ was equal to 0.5 in this experiment.

As typical image distortions, we considered additive white Gaussian noise (AWGN) and JPEG compression. The research results presented in Fig. 3 and Fig. 4, show that, as in the spatial domain, the use of the protected QIM algorithm somewhat reduces the robustness to distortion, which is a price to pay for the protection against the histogram attack. However, as can be seen in Fig. 3 and 4, the decrease in resistance does not look so critical as to call into question the practical applicability of this method.

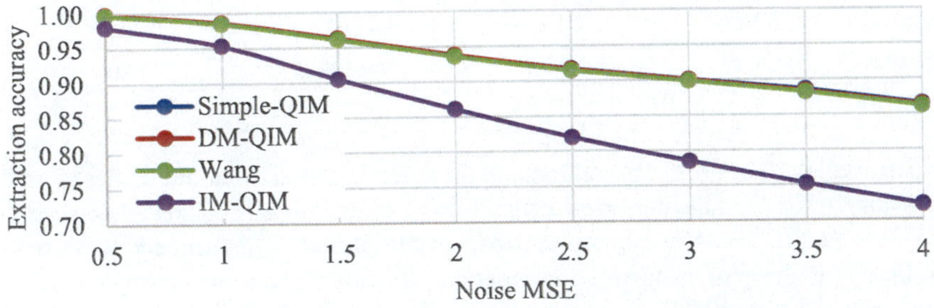

Fig. 3. Robustness of some QIM algorithms to AGWN (JPEG domain).

In addition to the robustness, the study also evaluated the visual quality of the protected images (that is, the level of distortions caused by watermark embedding). In this study, we used the PSNR-HVS-M index [21], which, in contrast to the traditional PSNR, uses the frequency weighting of the error to adequately model the perception of deviations by the human visual system (HVS). The obtained data on visual quality are given in Table 1. For comparison, along with PSNR-HVS-M, we also show PSNR values in this table. The numbers indicate that IM-QIM is characterized by a slightly increased level of visual distortion, but, as in the case of robustness, this increase is not critical.

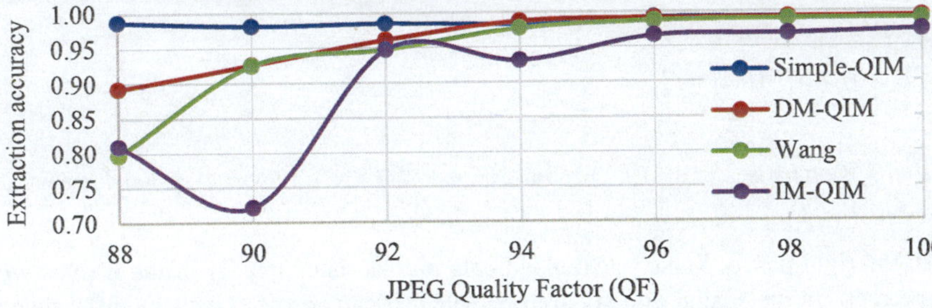

Fig. 4. Robustness of some QIM algorithms to JPEG compression distortions (JPEG domain).

5.3 Comparison of the Effectiveness of Using IM-QIM in Spatial and JPEG Domains

Obviously, in practice, the choice of the domain for watermark embedding is largely determined by the subject area and data type. However, there may be situations where

Table 1. Average visual quality of images watermarked with different algorithms and in different domains.

Algorithm	PSNR-HVS-M (JPEG domain)	PSNR (JPEG domain)	PSNR-HVS-M (spatial domain)	PSNR (spatial domain)
Simple-QIM	64.70	43.37	57.75	49.30
DM-QIM	61.01	46.39	62.06	53.49
Wang	62.78	46.54	62.65	53.75
IM-QIM	58.90	42.95	58.82	50.20

it is permissible to use different options. In this case, the choice should be based on the main characteristics of the digital watermarking system, such as robustness to distortions and visual quality. Therefore, in this section, we compare IM-QIM embedding in spatial and JPEG domains. For an adequate comparison, we selected the embedding parameters in such a way that embedding of the same digital watermark leads to a comparable level of visual quality. The watermark length in this experiment is equal to that of Sect. 5.2. So it was empirically obtained that on the considered dataset, embedding in the JPEG domain with $\Delta = 0.5$ is comparable to embedding in the spatial domain with the step $\Delta = 15$ (PSNR-HVS-M values are 58.897 and 58.819, respectively). Figure 5 and 6 show the results of a comparative study of the IM-QIM robustness in different domains with these parameters. The graphs lead us to the conclusion on the advantage of the JPEG domain watermarking.

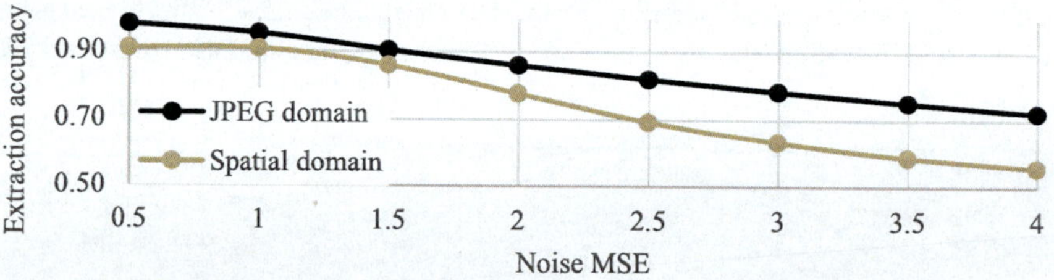

Fig. 5. Comparison of the IM-QIM robustness to AGWN in JPEG and spatial domains.

In the right part of Table 1, extended data on the visual quality of the images with a watermark in the spatial domain are presented. It can be noted that the difference in values between PSNR and PSNR-HVS-M is higher for JPEG domain watermarking. This property indirectly reflects the fact that the artifacts generated by watermark embedding in the frequency domain are less noticeable to human perception than the change in individual pixels in the spatial domain. Figure 7 shows some examples of watermarked images in spectral and spatial domains. To increase their distinguishability for obtaining these images, we increased Δ values up to 2 and 50, respectively.

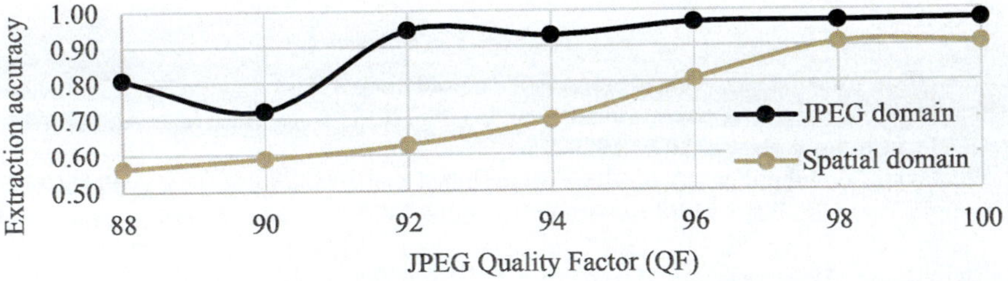

Fig. 6. Comparison of the IM-QIM robustness to JPEG compression in two domains.

IM-QIM, spatial	IM-QIM, JPEG	Wang, spatial	Wang, JPEG
Δ=50; 46.69 dB	Δ=2; 44.96 dB	Δ=50; 50.09 dB	Δ=2; 48.89 dB

Fig. 7. Embedding artifacts for some considered cases with specified visual quality in terms of PSNR-HVS-M. Wang provides the best visual quality among the considered algorithms.

6 Conclusion

In this paper, we applied our previously proposed approach of secure QIM watermarking in the JPEG compression domain and analyzed its security against the histogram attack [17] and the robustness to common distortions, compared with some traditional QIM algorithms. These experiments showed that our method preserves its security in JPEG (block DCT) domain. We also found that the method also provides acceptable robustness, although it loses to traditional algorithms in this factor.

We also compared the accuracy of the extraction of the presented algorithm in the spatial and block DCT domains. This comparison has shown that if embedding is performed with the same visual quality of the resulting image (in our case it was equal to 59 dB according to PSNR-NVS-M measure), then the watermark robustness is higher in the block DCT domain up to 10% for AGWN distortion, and up to 30% for JPEG compression distortion (in the given set of parameters).

This fact, along with the prevalence of QIM usage in the JPEG (block DCT) domain, indicates the high practical potential of the proposed algorithm.

Acknowledgements. This research was supported by the RFBR (project 19-29-09045 in part of Sections I, II, and V, project 20-37-70053 in part of Section III, and project 19-07-00357 in part of Section IV).

References

1. Chen, B., Wornell, G.W.: Quantization index modulation: a class of provably good methods for digital watermarking and information embedding. IEEE Trans. Inf. Theor. **47**, 1423–1443 (2001). https://doi.org/10.1109/18.923725
2. Zolotavkin Y., Juhola M.: A new two-dimensional quantization method for digital image watermarking. In: 2015 17th International Conference on Advanced Communication Technology (ICACT), pp. 155–160 (2015). https://doi.org/10.1109/ICACT.2015.7224776
3. Budiman, G., Suksmono, A.B., Danudirdjo, D., Pawellang, S.: QIM-based audio watermarking with combined techniques of SWT-DST-QR-CPT using SS-based synchronization. In: 2018 6th International Conference on Information and Communication Technology (ICoICT), Bandung, pp. 286–292 (2018). https://doi.org/10.1109/ICoICT.2018.8528727
4. Kanhe, A., Gnanasekaran, A.: A QIM-based energy modulation scheme for audio watermarking robust to synchronization attack. Arab J. Sci. Eng. **44**, 3415–3423 (2019). https://doi.org/10.1007/s13369-018-3540-4
5. Hakka, M., Kuribayashi, M., Morii, M.: DCT-OFDM based watermarking scheme robust against clipping attack. In: Proceedings of the 1st International Workshop on Information Hiding and Its Criteria for Evaluation, pp. 18–24. ACM, New York (2014). https://doi.org/10.1145/2598908.2598914.
6. Wang, H., Ho, A.T.S., Zhao, X.: A Novel fast self-restoration semi-fragile watermarking algorithm for image content authentication resistant to JPEG compression. In: Shi, Y.Q., Kim, H.-J., Perez-Gonzalez, F. (eds.) Digital Forensics and Watermarking, pp. 72–85. Springer Berlin Heidelberg, Berlin, Heidelberg (2012). https://doi.org/10.1007/978-3-642-32205-1_8
7. Evsutin, O.O., et al.: A new approach to reducing the distortion of the digital image natural model in the DCT domain when embedding information according to the QIM method. CEUR Workshop Proc. **2485**, 1–5 (2019)
8. Evsutin, O., et al.: Algorithm of error-free information embedding into the DCT domain of digital images based on the QIM method using adaptive masking of distortions. Sig. Process. **179**, 107811 (2021). https://doi.org/10.1016/j.sigpro.2020.107811
9. Preda, R.O.: Semi-fragile watermarking for image authentication with sensitive tamper localization in the wavelet domain. Measurement **46**(1), 367–373 (2013). https://doi.org/10.1016/j.measurement.2012.07.010
10. Evsutin, O.O., Melman, A.S., Filippov, A.A., Chernov, I.D.: Study of robustness of information embedding into digital images DWT domain using QIM method against destructive effects and steganalysis. J. Phys. Conf. Ser. **1661**, 012032 (2020). https://doi.org/10.1088/1742-6596/1661/1/012032
11. Yang, C., Zhu, C., Wang, Y., et al.: A robust watermarking algorithm for vector geographic data based on Qim and matching detection. Multimed. Tools Appl. **79**, 30709–30733 (2020). https://doi.org/10.1007/s11042-020-08916-4
12. Borah, S., Borah, B.: Quantization index modulation (QIM) based watermarking techniques for 3D meshes. In: 2017 4th International Conference on Image Information Processing (ICIIP), pp. 1–6 (2017). https://doi.org/10.1109/ICIIP.2017.8313726
13. Matam B.R., Lowe D.: Watermarking: how secure is the DM-QIM embedding technique? In: 2009 16th International Conference on Digital Signal Processing, pp. 1–8 (2009). https://doi.org/10.1109/ICDSP.2009.5201248
14. Matam, B.R., Lowe, D.: Watermark-only security attack on DM-QIM watermarking: vulnerability to guided key guessing. In: Crime Prevention Technologies and Applications for Advancing Criminal Investigation, pp. 85–106. IGI Global (2012)
15. Wang, Y., Moulin, P.: Steganalysis of block-structured stegotext. In: Electronic Imaging 2004, pp. 477–488. International Society for Optics and Photonics (2004). https://proceedings.spiedigitallibrary.org/proceeding.aspx?articleid=837745. Accessed 21 Mar 2017

16. Wu, Z., Li, R., Yin,, Li, C.: Steganalysis of quantization index modulation steganography in G.723.1 codec. Fut. Internet **12**, 17 (2020). https://doi.org/10.3390/fi12010017
17. Mitekin, V.: A new key recovery attack against DM-QIM image watermarking algorithm. In: 9th International Conference on Machine Vision, ICMV 2016. International Society for Optics and Photonics, pp. 103411A–103411A-9 (2017). https://doi.org/10.1117/12.2268550.
18. Mitekin, V.A., Fedoseev, V.A.: New secure QIM-based information hiding algorithms. Comput. Opt. **42**(1), 118–127 (2018). https://doi.org/10.18287/2412-6179-2018-42-1-118-127
19. Mitekin, V., Fedoseev, V.: A new QIM-based watermarking algorithm robust against multi-image histogram attack. Procedia Eng. **201**, 453–462 (2017). https://doi.org/10.1016/j.proeng.2017.09.687
20. BOWS-2 Web page. https://bows2.ec-lille.fr/. Accessed 31 Oct 2020
21. Egiazarian, K., et al.: New full-reference quality metrics based on HVS. In: Proceedings of the 2nd International Workshop on Video Processing and Quality Metrics (2006)

Security and Privacy on IoMT

Latifah Almaiman(✉) and Noura Alqahtani

Departement of Computer Science and Engineering, Princess Sattam Bin Abdulaziz University, Riyadh, Saudi Arabia
{lmaiman,nbqahtani}@moh.gov.sa

Abstract. The internet of medical things (IoMT) is taking an enormous part and a notable role in enhancing the health sector by anticipating medical facilities to people around the globe. With the rapid growth of internet of things, IoMT is having a massive effect in our everyday lifestyle. Rather than going to the health centers, patient clinical related data is remotely remarked and summons in a synchronal data system and then transferred to whom concerned for future use such as the cloud. IoMT is intensive data domain with an accelerating, growing rate which means that we must secure and protect large amount of delicate data without being interfered. And with the digital transformation we are living IoMT is being concerned as the most attractive topic in Saudi Arabia with the concern of cybersecurity implementation to the services and risk assessments in order to have the most secure and reliable user experience. Wishing this paper to be helpful and to add value to the researchers and to focus on the security challenges that we need to avoid.

Keywords: Internet of Things (IoT) · Internet of Medical Things (IOMT) · Security · Smart hospital · Privacy · Healthcare systems

1 Introduction

With the evaluation of IT, the Internet of things is grown area, what is IoT? Internet of things is where (things) no matter what they are covering connected to the internet without our **intercession. Internet of things** correspondence so many industries and our focus in this paper on Internet of medical things that touch medical devices or applications connected to the internet such as connected inhalers, Apple Watch and Smart insulin pens and so many more. IOMT is an added value to the healthcare industries where it delivers automated monitoring of data and alert systems to have a better decision making that helps with saving lives due to the real-time data transfers to track all patients' parameters.

In regard to what is the benefits of IOMT listing below IOMT applications such as:

- Clinical Grade Wearables
- Smart Pills
- Remote Patient Monitoring Devices
- virtual consultations
- Fitness Tracking Devices

A. Abraham et al. (Eds.): SoCPaR 2020, AISC 1383, pp. 956–963, 2021.
https://doi.org/10.1007/978-3-030-73689-7_90

The purpose is to gather people, data, and processes together to have connective medical services, devices, and applications for improving the patient's experience and results. And we can say that with COVID-19 pandemic only made IOMT a hot subject to go deeper in by enabling the screening of patients and monitoring the symptoms in an easy safer way.

2 History of IoMT

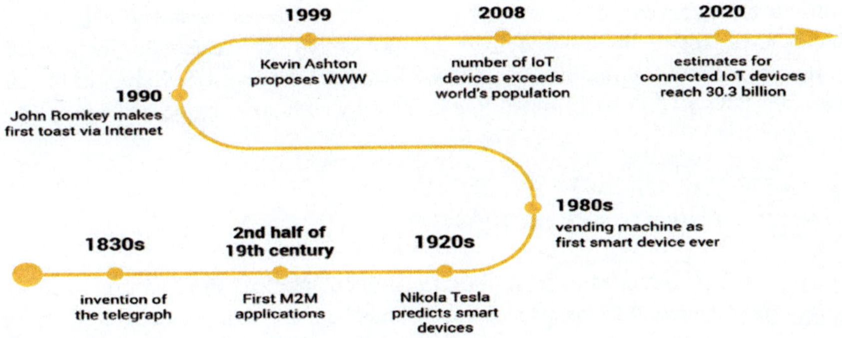

Fig. 1. History on internet

Regarding the history of IoT, the subject of the internet is not new. It was established more than 20 years ago, but with limited technology and for few people. Although It is still under development, IoT has helped many sectors such as the medical sector, IoMT. The statistics will show the history of IoT based on the following points (Fig. 1).

- In 1969, the internet was established. It is the primary tool for IoT. However, internet usage was restricted to researchers and academics in the US ministry of defense.
- In 1973, a new advanced feature had arrived RFID. It depends on radio waves to recognize entities automatically.
- In 1974, computer systems were adopted in personal computers. It is a crucial feature that leads to IoMT.
- In 1984, was the first real usage of IoT in the world. It was in Coca-Cola company. It told the buyers that their drinks' temperature without the interference of humans.
- In 1990, the usage of the internet was spread around the whole world. It was available for all people, but it was weak and did not fulfill the job. After the first half of 1990, sensors were used to solve the problem of internet weakness. However, sensors were the main reason for creating IoT and IoMT.
- 1999, the phrase IoT was officially used around the world, Bill Joy introduced communication between two devices in his classification of the internet, and the IoT was used by Ashton first. Also, the RFID technology was increased by initiating the Auto-ID Center at (MIT) to create an inexpensive chip that can store data and can be used to map objects to the internet.

- 2000 onwards, as a result of digitalization, the internet's connection was adopted in many systems, applications, and programs. From that time until now, many physical changes lead to IoT. These changes focus on making IoT a replacement for humans. Also, to make IoMT a replacement of doctors and the workers in the health sectors keeping the human lives and decreasing medical mistakes to 0%.
- 2015, a project was launched with the cooperation of Intel and Oregon University of Health and Science in the name of (Collaborative Cancer Cloud) and its mechanism of action: It is an analytics platform with high and accurate performance and the error rate is almost non-existent, and the data is collected and stored in it safely and accurately. In the medical fields, especially cancer research.
- And currently, based on a company of research and investment Gartner, scientific studies have proven that it will deliver 20.4 billion IoT. The researches include IoMT by the year 2020. Internationally, IoT will reach 1.7 trillion US dollars in the same year, 2020. It was 655.8 trillion dollars in 2014 by an increasing percentage of 16.9% annually.

3 Security and Privacy on IoMT

Security in IoMT includes security for IoT-linked medical devices and associated networks and the technologies, processes, and mechanisms that protect IoMT devices.

3.1 Security Issue

Challenges in the internet of things come in various areas we will list the most common issues and challenges in medical devices such as:

- Data integrity: collecting data is what all IOMT is about and to have these data put at risk by collecting without encryption will provide easy access to the hacker.
- Data exchange and transformation: real-time data transmit is needed in any IOMT device such as heart sensors. This is a challenging part of security when data is not up to date.
- Performance flexibility: limitation of the device's flexibility will affect reading patients' parameters. System design.
- Real-time processing: when we talk about real-time data in medical devices this means an actual parameter that defines the patient's status any attack on the hardware or software will affect the patient health.
- System recovery: Data in medical matters are very impotent and the risk of recovering these data due to limited resources may cause data to lose. And this will impact the patients' health information details. Memory obstacle.
- Mobility and connectivity: to ensure that the device for example body patches to have a good wireless connection in regards of keeping up with patients in all areas.
- Lack of user knowledge & awareness: where this technology a new one Users must know what is IoT and most importantly if it's a medical device that we are facing user ignorance in. This may lead to a huge risk, unpatched embed huge operating systems and software Hardware issues.

- Social engineering: This explicitly of use by tracking and gathering information may information, Inpatient information to gain access to patients' data may cause manipulating of data where it may cause the life of patients. Machine learning cause the patients a real heath threat due to unauthorized access to patient.

3.2 Security Attacks

The cyber-attacks carried out by the hacker may benefit from the features in devices that work on IoMT and based on the security issues that occur to IoMT devices so, this list of examples of attacks that may be exposed to medical devices that work on IoMT:

- Man-in-the-middle attacks: As it may affect the quality and validity of the data and signals between the patient and the device by sending incorrect data to the diagnostic devices medical.
- Physical Attacks: When people harmed and destroy IoMT devices.
- Denial-of-service attacks: In the DDoS attack, it targets devices running on IoMT and disables them and does not allow users (doctors, patients, etc.) to access and benefit from services, and also includes that slowdown in systems running IoMT.
- Encryption Attacks: If an IoMT device is unsecure, the hacker can sniff and manipulate the data and patient record for use afterward.
- Firmware Hijacking: One of the risks to IoMT devices is a failure to update the systems, which may lead to Hijacking. Therefore, the systems must be kept updated from their reliable sources.
- Botnets: Alter the networked IoMT devices into remotely managed bots, Botnets have the ability to use smart, connected devices to fetch private, delicate data, which may disable a device, etc.
- Ransomware: the attackers accessed databases and medical information in the devices associated with IoMT and then encrypt the data and then request a sum of money to provide you with what decodes your data.
- Eavesdropping: In this attack, the attacker tracks all movements and communication between the IoMT devices and the users (doctor, patient, etc.) and tries to eavesdrop and steal some data and information when the connection between the device and the server.
- Privilege Escalation: In this type of attack, the attacker targets errors and weaknesses in IoMT devices in order to access protected resources of medical information and sensitive patient data. in this type, the attacker uses malware or thieve confidential information.
- Brute Force Password Attack: In this type, the attacker uses a trial and guess mechanism for passwords, and it is possible to use software that guesses a large number of passwords this in order to access IoMT devices and the medical information they contain.

3.3 Security Architecture

The architecture contains the elements: physical objects (medical devices), user protocols, IoT protocols, etc. Because of the widespread of IoT, there is no (single consensus on IoMT) and examples of applications for IoMT (Remote monitoring system for heart-related diseases) The structure of the system has been shown in the following (Fig. 2).

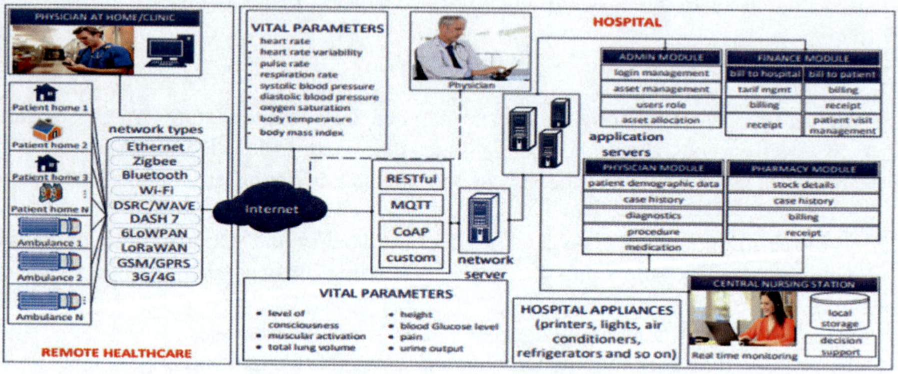

Fig. 2. Remote monitoring system for heart-related diseases

3.4 Security Requirements

The most significant concerns of devices in healthcare network infrastructures is reliable and secure, the security requirements of the future medical sector for IoMT.

- Inclusive:
  IoMT devices are unrelated together, the devices that work on IoMT have been manufactured in different companies with different functions and tasks, and therefore they need a unified infrastructure for all these systems to ensure their protection.
- Authentication and authorization:
  Physical devices can be tampered with and spoofed and, therefore, to attack systems Therefore, the importance of the mechanisms followed in the infrastructure for verification and authentication of IoMT devices must be emphasized.
- Data privacy:
  Privacy is one of the most critical challenges that IoMT faces as the devices used in IoMT contain compassionate patient information (his medical information, identity, etc.) Therefore, it is essential to guarantee the privacy of data.
- Threat detection:
  Systems and devices that use IoMT can be exposed to many attacks, and therefore mechanisms must be put in place to detect and prevent them.

4 How Can Protect Data in IoMT

Due to the high risk of data, the internet of medical things devices can be threatening. Let's say a hacker attacked and approach these devices by manipulating patient's data and gain full access this may cause physicians to be notified with mistaken parameters that make them take action that can harm the patients. In these kinds of scenarios as a recommendation as this is considered as sensitive data to grantee an upper-level of security design to IOMT devices, this needs full support and continues development also by implementing a behaver study mechanism/ technique. Thus, IOMT is expanding, and with the big data transaction in between (Things) we need to protect our data on IoMT, following are most recent studies regrading theologies in the health care system:

4.1 Abnormal User Detection Depending on the Correlation Probabilistic Model

And by the interactions and movement of a user in identifying the way to deal with this situation, because of its interaction with the user's interactions in the event that any defect or sequence of usual movement is discovered.

4.2 Preserving Data Privacy on IoMT Applying the Dual Signature ECDSA Methodology

It relies on linking data and devices on two types of digital signature by encrypting the data and hiding it upon entering it with a signature, as well as upon requesting it and exiting it, it is hidden with another type of signature, and if one type is hacked, it is guaranteed that the other signature will protect the data.

4.3 Auditing Model Depending on Attribute and Similarity Measurement

By building an algorithm to track the user's usual movements and features, compare them with any unusual entry, and discover if it is a hacker.

4.4 Securing Information Broadcasting in Wireless-Powered Cognitive Radio Networks for the IoMT

To ease a successful applications of spectrum resources, initiating the technology of cognitive radio in which unlicensed nodes could opportunistically communicate with each other over a licensed constancy band without scattering the primary communication to verify the data and devices is more reliable and secure.

4.5 Two Authentication Schemes for Medical Systems by a Wireless Medical Sensor Networks to Improve the Security and Cryptanalysis

By encoding the data within the systems with advanced and high algorithms, using more than one type for that, and dealing with the sensors in that, as it identifies the patient and translates his data into encrypted algorithms that are dealt with by authorized persons only.

4.6 Secure and Safe Outsourced Medical Information Against Unexpected Leakage with Flexible Access Control in a Cloud Storage System

By saving the medical data in cloud computing to ensure its protection by choosing a reliable cloud.

4.7 Deep Learning Detection Technique Using API Malware

Worked by examining the local maliciousness about malware and execute an anti-interference detection framework that is API based using the LSTM model to categorize API fragments and employs orchestra learning to form the final outcome of the whole API order.

4.8 Protecting IoMT Apps with Proxy Dynamic Objects

It suggested to support the dynamic substitution of layer in IoMT applications by incorporating object proxy into (VM). Secondly, the heap and address are used to model the object and object evolution to make sure that the practicability of the method afterward, impact will be analyzed of field access and technique invocation and assess the risk and security of the application when constraints are fulfilled eventually, conducting the assessment by extending VM to validate the effectiveness of the security of secured data, These are types of data protection in IoMT at present, but the danger is still present on the data used in the health sector coinciding with the expansion of the Internet, as there is a positive relationship between weak security and increased use of the Internet and thus we see in this research that data protection is supported in IoMT By using more than one approach and method to protect against encryption, digital patching, and computing, but we take into account that this does not affect the availability of data, as all methods of raising security are difficult to access and deal with data.

5 Conclusions

In this paper, we have discussed and provided a simple review of what's the internet of medical things and the main security issues that this field challenging Therefore, in order to have a mature medical internet of things we need to clarify the risks and security issues, Data privacy and security is the core of IoT and when it comes to the healthcare sector it's a sensitive matter to ensure the security of IoT in the medical services will not only help provide the digital transformation, but it will save lives.

References

1. Joyia, G.J., Liaqat, R.M., Farooq, A., Rehman, S.: Internet of Medical Things (IOMT): applications, benefits and future challenges in healthcare domain. J. Commun. **12**(4), 240–247 (2017)
2. Dilawar, N., Rizwan, M., Ahmad, F., Akram, S.: Blockchain: securing Internet of Medical Things (IoMT). Int. J. Adv. Comput. Sci. Appl. (IJACSA) **10**(1), 1–8 (2019)
3. Vishnu, S., Jino Ramson, S.R., Jegan, R.: Internet of Medical Things (IoMT) - an overview. In: 2020 5th International Conference on Devices, Circuits and Systems (ICDCS) (2020)
4. Habib, M.A., Faisal, C.M.N., Sarwar, S., Latif, M.A., Aadil, F., Ahmad, M., Ashraf, R., Maqsood, M.: Privacy-based medical data protection against internal security threats in heterogeneous Internet of Medical Things. Int. J. Distrib. Sens. Netw. **15**, 1–12 (2019)
5. Alsubaei, F., Abuhussein, A.: Security and privacy on the internet of medical things: taxonomy and risk assessment, October 2017. Conference Paper

6. Karmakar, K.K., Varadharajan, V., Tupakula, U., Nepal, S., Thapa, C.: Towards a security enhanced virtualised network infrastructure for the Internet of Medical Things (IoMT). In: 2020 6th IEEE International Conference on Network Softwarization (NetSoft) (2020)
7. Alqarni, A.: A survey and classification of security and privacy research in smart healthcare systems. Received 15 July 2019, accepted 22 July 2019, date of publication 25 July 2019, date of current version 9 August 2019
8. Yang, X.: Abnormal user detection based on the correlation probabilistic model. Received 25 October 2019, accepted 24 February 2020, date of publication 13 June 2020
9. Cano, M.-D.: Preserving data privacy in the Internet of Medical Things using dual signature ECDSA. Received 25 October 2019, accepted 17 February 2020, date of publication 10 June 2020
10. Lai, H.-L., Chen, T.-Y.: User audit model based on attribute measurement and similarity measurement. Received 18 February 2012, accepted 28 July 2012
11. Tang, K., Tang, W., Luo, E., Tan, Z., Meng, W., Qi, L.: Secure information transmissions in wireless-powered cognitive radio networks for Internet of Medical Things. Received 25 September 2019, accepted 20 December 2019, date of publication 24 February 2020
12. Hu, Z., Lin, Y.: Cryptanalysis and security improvement of two authentication schemes for healthcare systems using wireless medical sensor networks. Received 18 October 2019, accepted 18 January 2020, date of publication 19 February 2020
13. Zhou, X., Liu, J., Wu, Q.: Secure outsourced medical data against unexpected leakage with flexible access control in a cloud storage system. Received 30 August 2019, accepted 08 January 2020, date of publication 10 February 2020
14. Ma, X., Guo, S., Bai, W., Chen, J., Xia, S.: An API semantics-aware malware detection method based on deep learning. Received 19 February 2019, accepted 20 May 2019, date of publication 11 November 2019
15. Zhao, Y.-L., Zhu, C.-P.: An object proxy-based dynamic layer replacement to protect IoMT applications. Received 24 May 2019, accepted 26 August 2019, date of publication 30 September 2019

Biometric Template Protection Using Deep Learning

Ashutosh Singh[1(✉)], Yogendra Narain Singh[1], and Pramod Kumar[2]

[1] Institute of Engineering and Technology, Dr APJ Abdul Kalam Technical University, Lucknow 226021, Uttar Pradesh, India
[2] National Informatics Centre, Ministry of Electronics and Information Technology, New Delhi 110011, India
pkr@gov.in

Abstract. Use of biometrics in digital society has raised the questions of biometric template protection and secure authentication. The biometric template protection mechanisms known so far hardly maintain a trade-off between security of template database and recognition performance. This paper proposes a hybrid technique of template protection for a multibiometric system that provides better matching performance and infallible from fraudulent attacks. The multimodal system is prepared from face and ECG biometrics. The ECG as a biometrics not only supplements the face biometrics in a multimodal system but also ensures security for robust recognition. The pre-trained deep learning models are used to process both biometrics and prepare multimodal templates. The templates are mapped to their corresponding classes represented by randomly generated unique binary codes. These binary codes are further encrypted using cryptographic hash for non-invertiblity and hide information of fused templates. Finally, the matching is performed using hash codes for ensuring an additional layer of defense against adversarial attacks.

Keywords: Biometrics · Template protection · Deep learning

1 Introduction

Biometrics is the technology used for automatic recognition of individuals using their unique body parts. Biometric applications are seen in our day-to-day life *e.g.,* authentication in mobile phones, login to social schemes, access to bank accounts, and restricted places. Biometrics traits are unique thus offer reliability in recognition and leaps over other traditional identity proofing mechanisms. Over the years, several biometrics are researched for different domain of applications [1]. Although, the choice of biometric depends upon system complexity, cost, security and area of application [2].

The architecture of a typical biometric recognition system is shown in Fig. 1. The biometric trait is acquired from an individual at the sensor stage. It is

A. Abraham et al. (Eds.): SoCPaR 2020, AISC 1383, pp. 964–974, 2021.
https://doi.org/10.1007/978-3-030-73689-7_91

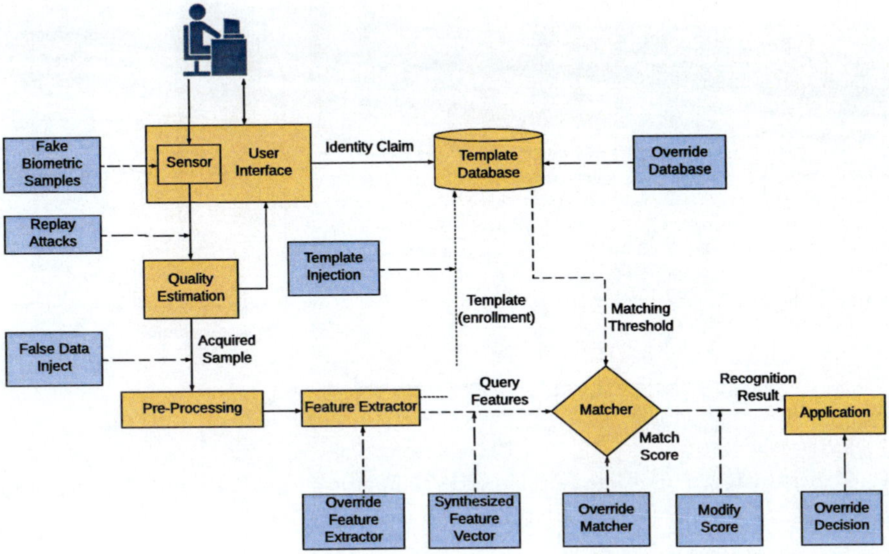

Fig. 1. Biometric recognition system and vulnerabilities [11].

preprocessed for noise removal and other quality enchantments. Features are then extracted and represented in the domain to facilitate search operation for proper matching. The target features are similarly extracted from the test sample and processed for matching with the training features set. Finally, the decision of matching is taken using a threshold set on the maximum number of features matched to the training feature set.

The biometric systems are now affordable due to easiness for use but they are vulnerable to fraudulent attacks. The biometric system vulnerabilities and their point of attacks are shown in Fig. 1. Attacks on the sensor module result from bypassing the capturing device. Also, the imposter might present a fake biometric prepared using artificial means, of a genuine user at the sensor for gaining access. A synthesized feature vector can be prepared using hill-climbing attacks between the channels that can further be used to prepare fake biometrics [3]. The attack at the matcher might lead to the generation of high matching scores against a false biometric or low scores for a genuine one. This would lead to false rejection or acceptance, respectively.

The attacks on the database consists of templates, are possibly the most prominent fraudulent attacks reported on a biometric system. The template contains ideally the unique characteristics of an individual's biometrics that are needed for his/her identity proofing. It may also contain personal information such as date-of-birth, contact detail along with age, gender, and ethnicity information. Furthermore, people are more concerned about security and privacy of their personal information while sharing their biometrics data to a third party [12–14]. Therefore, need for template protection is not only for biometric feature preservation, but also for privacy against theft and cross-matching attacks.

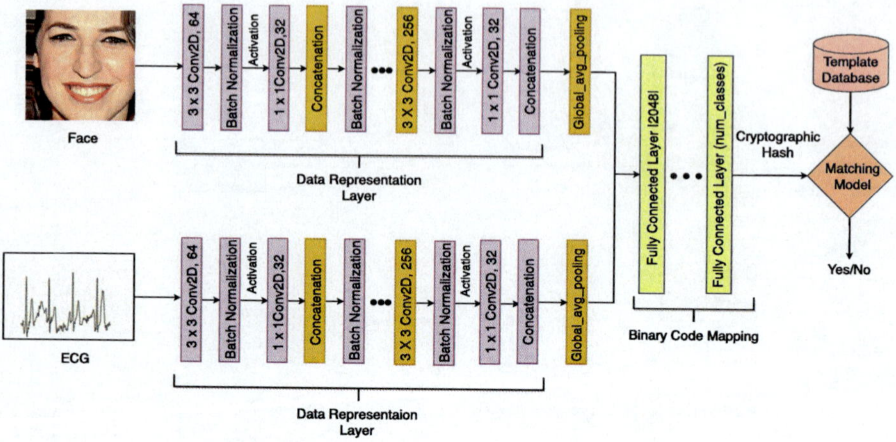

Fig. 2. Schematic of multimodal biometric template protection method using deep learning.

Several biometric template protection techniques have been proposed in the past [20–23]. These techniques can be categorized as biometric cryptosystems and template transformations. Cryptographic systems provide cancelability and diversity but they suffer to address security and performance issues [24]. In the template transformation approach, the generated template is transformed using a random key or password-based approach [25,26]. Using cancellable templates and non-invertible transformation techniques, security, cancelability, and diversity can be achieved, but the system may suffer from a high false rejection rate. Thus, fails correspondence between non-invertibility and discriminability.

This paper proposes a template protection mechanism using multimodal deep learning that incorporates both cryptography and template transformation approaches. The approach is developed for a multibiometric system by fusing the conventional biometrics face and the state-of-the-art biometrics electrocardiogram (ECG). The ECG is a bioelectrical signal that records the electrical activity of the heart. The studies have proved the uniqueness of ECG among individuals and may be used as biometrics [4–10]. The ECG is highly secured to an individual that inherently signifies the life signs. The ECG may supplement the missing contents to face biometrics, if fused. The fusion of ECG with face biometrics generates a more reliable profile of an individual for robust recognition with enhanced security. Further, the templates of fused biometrics need protection from fraudulent attacks. This study proposes a robust framework for multimodal biometric template protection using deep learning.

The state-of-the-art pre-trained deep learning models are used that supplement the proposed template protection method for cancelability, diversity, security, and performance [23]. The proposed method of multibiometric template protection uses transfer learning that ensures optimum recognition performance under intraclass and interclass variations. It ensures the correct mapping of multimodal features to corresponding binary pseudo random numbers (BPRN) using transformation principle [17]. In order to protect the system from brute force attacks, cryptographic measures using high entropy codes are used. The binary codes of 512 bits are transformed using a cryptographic hash to ensure non-invertibility and secrecy of duo biometrics without compromising the matching performance.

The rest of the paper is organised as follows. The method and material of proposed multimodel deep learning method for template protection are presented in Sect. 2. The experimental results including database used and the investigation of recognition accuracy of proposed system are given in Sect. 3. Finally, conclusion is drawn in Sect. 4.

2 Methodology

2.1 Template Protection Using Deep Learning

Deep learning is a subfield of machine learning that deals with methods based on artificial neural networks having multi-layers of processing with feature learning [28]. The deep learning model learns the parameters by estimating the loss incurred in every iteration to perform a specific task. The deep learning models dealing with images are based on convolution neural networks (CNN). CNN learns from training images and performs classification. The task of image classification has accelerated in recent years with the use of transfer learning. It enables us to use pre-trained models such as DenseNet that is already trained on larger databases like ImageNet [29]. Such models reduce the time to train the new deep learning models and achieve better accuracy.

The DenseNet model comprises dense blocks, transition layers, and classification layers. For feature extraction, a combination of dense block and transition layers are used [35]. The schematic of the multimodal template protection method using deep learning is shown in Fig. 2. The dense block comprises sets of 3×3 convolution (conv) layers and a batch normalization for standardizing the inputs received. Each dense block adds 32 new features maps, resulting in a total of 256 feature maps at the end of the last dense block. Also, a transition layer follows each dense layer comprising of 1×1 conv layer with reduced filters followed by a 2×2 average pooling. The 1×1 conv layer reduces the number of input feature maps before each 3×3 conv, hence, improving computational efficiency. Both layers of transition are responsible for reducing the number of feature maps and the size of the input to the next dense block. A concatenation layer is used to gather the collective knowledge from all preceding layers after each dense block.

Finally, a global average pooling layer replaces the individual dense layer of each model so that it may generate feature maps equal to the target classes. A second neural network is added and it concatenates the features maps from both the DenseNet models. The number of layers in the second neural network depends upon the size of the concatenated features. The obtained binary code corresponding to input images, goes through a further transformation. The cryptographic hash function transforms the unique binary code into a hash code.

Multimodal Deep Learning. Multimodal deep learning learns features over multiple biometrics. The deep learning models are preferred over conventional methods of learning due to its superiority in handling a large variety of data, their representations and visualization in higher dimensions of feature space. The model predicts the combined representation and performs effective learning. The most common methods employ by these models to achieve multimodality are addition and concatenation approaches [30]. More formaly, let N be the numbers of biometrics available and each biometric is represented by v_i where, $v_i \in R^d * R^d, \forall i = 1, 2, \ldots, N$ and d is the dimension of features in real space. Then, multimodal representation can be either,

$$u_{+} = \Sigma_i f_i(v_i) \tag{1}$$

where, u_{+} is the additive feature vector obtained by adding the feature vectors of both the biometrics, or

$$u_{\bullet} = [f_1(v_1), f_1(v_2), \ldots, f_1(v_i)] \tag{2}$$

where, $u_{\bullet}$ is the concatenated feature vector of both the biometrics and f is the considered networking model.

For the generated multimodal fused vector u (u_{+} or $u_{\bullet}$) the model g predicts the class label p as follows,

$$p = g(\mathbf{u}) \tag{3}$$

Biometric Fusion. The fusion of biometric features based on instances of the different modalities are performed in this technique. These features are extracted using domain specific models and fused using a fully connected neural network as shown is Fig. 2. The 2D biometric information of face and ECG biometrics are fed to respective DesnseNet models and the fused vectors are obtained.

More formally, let u_{face} and u_{ECG} are the feature vectors of the face and ECG biometrics, respectively. Then, multimodal learning of fusing biometric features, $p_{feature}$ can be predicted by the model g as follows,

$$p_{feature} = g(u_{face} \bullet u_{ECG}) \tag{4}$$

Similarly, using additive measures the biometric features are fused after normalization as follows,

$$u_{+} = u_{face} + u_{ECG} \tag{5}$$

Therefore, model g predicts the biometric features, $p_{feature}$ as follows,

$$p_{feature} = g(u_{face} + u_{ECG}) \tag{6}$$

During testing, the query biometric template processes on the trained network and generates the binary code respectively to addition and concatenation multimodality.

2.2 Binary Code Generation from Multimodal Features

The deep learning model f, processes the biometric information during training and initial layers perform feature extraction from the input images. The extracted features are then mapped to the appropriate pseudo randomly generated binary classes using fully connected output layer [17,31]. Usually, softmax activation function generates probabilities for classes. Unlike, probability, we use binary codes for classes generated using sigmoid function. The issue with probabilities is that they may be prone to attacks such as hill-climbing that modifies the data according to the results of previous attempts. The advantage of using binary codes is to act as the first line of defense against attacks on the templates. These codes are highly secured and intended for classifying test examples only. The sigmoid function can be written as,

$$f(s) = \frac{1}{(1 + e^{-s})} \quad (7)$$

where, s is the input features and f is the class label.

During multimodal fusion, both modalities are trained on separate DenseNet models and a fully connected layer concatenates the features of both the biometrics as shown in Fig. 2. The sigmoid activation function enables the fully connected output layer to generate a 512-bit binary code. The number of dense layers used for classification depend upon the mapping between size of feature vector and binary code generated. For example, the mapping of 2048 size feature vector to 512-bit binary code, 2 dense layers are required, *i.e.*, from 2048 to 1024 and finally 1024 to 512-bit binary code.

2.3 Cryptographic Hash

The proposed method uses binary codes for classes that are hidden and fully intended for classification. Further, the binary code does not correlate with the biometric features, therefore if anyone gets hold the binary codes, then it would be computationally difficult to retrieve any information from them.

We employ the SHA–3 technique to protect the template ahead [36]. It is a one-way function that generates an output of the desired length from any input size. The SHA–3 includes several variants that are used depending on the requirement of generating a cryptographic hash for a given string. The cryptographic hash function generates a fixed size array from a random size input data. Being a one-way function, it is infeasible to generate the input (I) reversibly from its cryptographic hash $H(I)$. Let I_1 and I_2 be two inputs, then it is impossible to obtain their hash (H) to be similar *i.e.*,

$$H(I_1) = H(I_2) \quad (8)$$

A cryptographic hash corresponding to each class is created during the training stage. This hash code is stored in a database that represents the identity of the user. In the testing phase, the hash code is generated from the binary code obtained from the CNN mapping [32]. The matcher compares the hash code of the query template with the hash codes of templates stored in the database and takes a decision.

3 Experimental Results

The proposed technique of multimodal template protection is evaluated using publically available datasets for face and ECG *e.g.,* VGGFace2 and CYBHI, respectively [33,34]. The CYBHI dataset contains short-term ECG recordings of 63 individuals. In order to increase the training set, the dataset is augmented. We randomly choose 63 subjects from VGGFace2 dataset and their facial images. Thus, virtual subjects are created having face and ECG biometrics available. We assume that the biometric modalities (*e.g.*, face and ECG) are independent of an individual. In multibiometric research, an acceptable practice is that an individual of one (*e.g.*, face) dataset can be associated with an individual of another (*e.g.*, ECG) dataset to create a virtual subject. Likewise, the multimodal dataset of face and ECG biometrics is prepared [10]. The need of virtual subjects arose due to the non-availability of multimodal database where the multiple biometrics are measured from the same subject.

The experiment is conducted on the prepared datasets used for training and testing with an 80:20 ratio, respectively. The performance of the multimodel biometric template protection method is evaluated on true positives (TP), false positives (FP) (Type–I error), true negatives (TN), and false-negative (FN) (Type–II error). The TN refers to the situation when a hash code of the test sample does not match with any of the classes, whereas, if it is classified to one of the classes then, it is FP. If the hash code of a test sample is matched correctly then, it is a TP, whereas, if the hash code is mismatched then, it is FN. Based on these metrics, false acceptance rate (FAR) is defined as the ratio of false positives to the total attempts made for recognition. Genuine acceptance rate (GAR) is the percentage of genuine subjects accepted by the system, *i.e.,* rate of incorrect rejection of genuine subjects subtracted from 100. Finally, the accuracy, a measure of total correct predictions against the total observations is also computed.

Although, accuracy acts as a prime measure for comparing one model to another. But it is not always the correct measure of how good the models are, since a model might acquire high accuracies due to overfitting. Therefore, some other measures such as precision and recall provide a better representation of how well the model performs. Precision is the ratio of true positives to the total number of correct predictions while recall provides the proportion of correct positives that were identified correctly.

The DenseNet models are used to process the biometric features of the face and ECG. The features extracted using DenseNet from face and ECG biometrics are column vectors of size 125 each. The feature vectors of each biometrics are fused using additive or concatenative approaches. In concatenative approach, both vectors are fused to form a large vector of 250 features. Hence, an individual

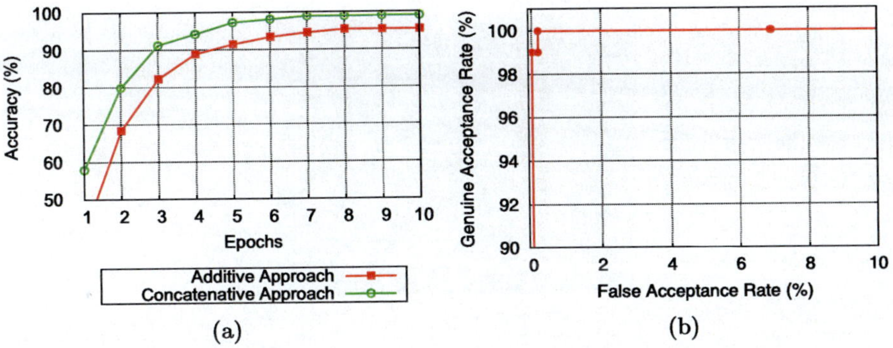

Fig. 3. Performance of proposed method a) testing accuracies using additive and concatenative approaches and b) ROC curve for concatenative approach.

is characterized using 250 features. In additive approach, features of both vectors are added that results a new feature vector of the same size. Thus, multimodal templates are prepared using additive along with concatenative approaches and cryptographic hash are generated from corresponding binary codes.

The matching results for additive and concatenation approaches are found well and shown in Fig. 3a. In particular, the concatenation approach gained an accuracy of 99.15% while the additive approach achieves to 95.44%. The concatenative approach maintains the original features of both biometric modalities. Therefore, if any modality is found weak due to noise and artifacts then other modality may complement the missing information. An additive approach forms a new feature set by adding corresponding attributes. In this process, the features of a weak modality may degrade the quality of the resulting features set, hence the concatenative approach results in a better accuracy compared to the additive approach.

The method using concatenative approach reports precision, recall, and F1-score values to 0.9974, 0.9975, and 0.9975, respectively. All these values recorded are higher, further representing higher true positive against the smaller false positive. A receiver operating characteristic (ROC) curve is shown in Fig. 3b. It provides better visualization of the classification accuracy of the proposed method. The method achieves the GAR of 99% at 0% FAR by matching the hash codes of different templates available in the database to the hash code of the test template. Further, the GAR increases to 100% at 0.6% FAR. These results also prove the robustness of the method fusing biometrics those are independent of each other and their matching in the transformed domain.

Using confusion matrix, the parameters TP, FP, TN, and FN can be easily visualized. The confusion matrix reveals a model's confusion in predicting the classes. The confusion matrix is obtained from the matching of multimodal features of face and ECG biometrics in a transformed domain is shown in Fig. 4. The result shows most diagonal values are closer to 1, *i.e.*, the square representing a class at x-axis shows it's matching probability with the corresponding class at y-axis.

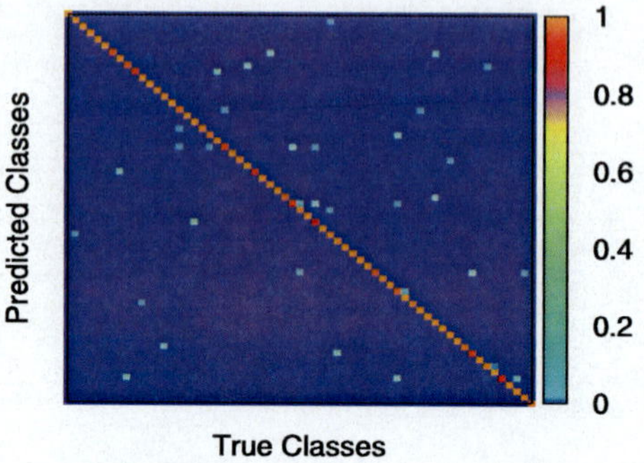

Fig. 4. Confusion matrix for multimodal biometric template protection method using concatenative approach.

4 Conclusion

The problems of pattern recognition and image processing can be addressed efficiently using deep neural networks with outstanding performance. The deep networks automatically learn the intrinsic data patterns without being explicitly programmed. Though, the potential of deep learning networks such as CNNs are not explored yet for the protection of biometric templates from fraudulent attacks. The use of CNNs in biometric recognition has inspired us to exploit the potential for the protection of biometric templates. This paper has presented a novel method to protect multibiometric templates from adversary attacks. Further, the method has reported better recognition accuracy in the transformed domain after fusion of independent biometrics those may be noisy or less informative.

The hybrid technique for multimodal template protection is based on deep learning that uses transfer learning that has ensured mapping of their binary pseudo random numbers. The system is further protected from brute force attacks with cryptographic measures using high entropy codes. The cryptographic hash has ensured non-invertibility of multimodal templates and reported outstanding matching performance.

References

1. Piuri, V., Scotti, F.: Biometrics privacy: technologies and applications. In: Proceedings of the International Conference on Signal Processing and Multimedia Applications, pp. 7–17. Seville (2011)
2. Jain, A.K., Ross, A., Pankanti, S.: Biometrics: a tool for information security. IEEE Trans. Inf. Forensics Secur. **1**(2), 125–143 (2006)
3. Maiorana, E., Hine, G.E., Campisi, P.: Hill-climbing attacks on multibiometrics recognition systems. IEEE Trans. Inf. Forensics Secur. **10**(5), 900–915 (2015)
4. Singh, Y.N., Singh, S.K.: Evaluation of electrocardiogram for biometric authentication. J. Inf. Secur. **3**, 39–48 (2012)

5. Singh, Y.N.: Discriminant Analysis for Identifying Individuals of Electrocardiogram, In: 5th International Conference on Pattern Recognition and Machine Intelligence (PReMI 2013), pp. 94–99. LNCS (2013)
6. Singh, Y., Gupta, P.: Correlation-based classification of heartbeats for individual identification. J. Soft Comput. **15**(3), 449–460 (2011)
7. Singh, Y., Gupta, P.: ECG to Individual Identification. In: Proceedings of IEEE 2nd International Conference on Biometrics: Theory, Applications and Systems (BTAS 2008), pp. 1–8 (2008)
8. Srivastava, R., Singh, Y.N.: ECG analysis for human recognition using nonfiducial methods. IET Biometrics **8**(5), 295–305 (2019)
9. Singh, Y.N.: Human recognition using fisher's discriminant analysis of heartbeat interval features and ECG morphology. Neurocomputing **167**, 322–335 (2015)
10. Singh, Y., Singh, S.K., Gupta, P.: Fusion of electrocardiogram with unobtrusive biometrics: an efficient individual authentication system. Pattern Recogn. Lett. **33**(14), 1932–1941 (2012)
11. Singh, Y.N., Singh, S.: A taxonomy of biometric system vulnerabilities and defences. Int. J. Biometrics **5**(2), 137–159 (2013)
12. Adler, A.: Vulnerabilities in biometric encryption systems. In: 5th International Conference on Audio-and Video-Based Biometric Person Authentication (AVBPA), pp. 1100–1109. Springer, Heidelberg (2005)
13. Singh, A., Srivastava, R., Singh, Y.N.: Prevention of payment card frauds using biometrics. Int. J. Recent Technol. Eng. (IJRTE) **8**(3), 516–525 (2019)
14. Mohanty, P., Sarkar, S., Kasturi, R.: Privacy and security issues related to match scores. In: 2006 Conference on Computer Vision and Pattern Recognition Workshop (CVPRW 2006), pp. 162-165. New York (2006)
15. Feng, Y.C., Yuen, P.C., Jain, A.K.: A hybrid approach for generating secure and discriminating face template. IEEE Trans. Inf. Forensics Secur. **5**(1), 103–117 (2010)
16. Jain, A.K., Nandakumar, K., Nagar, A.: Biometric template security. EURASIP J. Adv. Signal Process. **2008**, 1–17 (2008)
17. Origines, D.V., Sison, A.M., Medina, R.P.: A Novel Pseudo-random number generator algorithm based on entropy source epoch timestamp. In: 2019 International Conference on Information and Communications Technology (ICOIACT), pp. 50–55, Yogyakarta, Indonesia (2019)
18. Maltoni, D., Maio, D., Jain, A. K., Prabhakar, S.: Handbook of Fingerprint Recognition. Springer, London (2009)
19. Prabhakar, S., Pankanti, S., Jain, A.K.: Biometric recognition: security and privacy concerns. IEEE Secur. Priv. **99**(2), 33–42 (2003)
20. Jain, A.K., Ross, A., Prabhakar, S.: An introduction to biometric recognition. IEEE Trans. Circuits Syst. Video Technol. **14**, 11–20 (2004)
21. Ngo, D.C.L., Teoh, A.B.J., Hu, J.: Biometric Security. Cambridge Scholars Publishing, Newcastle upon Tyne (2015)
22. Nandakumar, K., Jain, A.K.: Biometric template protection: bridging the performance gap between theory and practice. IEEE Signal Process. Mag. **32**(5), 88–100 (2015)
23. Sutcu, Y., Li, Q., Memon, N.: Protecting biometric templates with sketch: theory and practice. IEEE Trans. Inf. Forensics Secur. **2**(3), 503–512 (2007)
24. Tuyls, P., Goseling, J.: Capacity and examples of template-protecting biometric authentication systems. In: Proceedings of 2004 International Workshop on Biometric Authentication (BioAW), pp 158–170, LNCS, Springer, Berlin (2004)

25. Teoh, A.B.J., Goh, A., Ngo, D.C.L.: Random multispace quantization as an analytic mechanism for BioHashing of biometric and random identity inputs. IEEE Trans. EURASIP J. Adv. Signal Process. Pattern Anal. Mach. Intell. **28**(12), 1892–1901 (2006)
26. Ratha, N.K., Chikkerur, S., Connell, J.H., Bolle, R.M.: Generating cancelable fingerprint templates. IEEE Trans. Pattern Anal. Mach. Intell. **29**(4), 561–572 (2007)
27. Talreja, V., Valenti, M.C., Nasrabadi, N.M.: Multibiometric secure system based on deep learning. In: 2017 IEEE Global Conference on Signal and Information Processing (GlobalSIP), pp. 298–302 Montreal, QC (2017)
28. Voulodimos, A., Doulamis, N., Doulamis, A., Protopapadakis, E., Andina, D.: Deep Learning for Computer Vision: A Brief Review. Intell. Neuroscience (2018)
29. Xie, Q., Luong, M.T., Hovy, E., Le, Q.V.: Self-training with Noisy Student improves ImageNet classification. In: Proceedings of the IEEE/CVF Conference on Computer Vision and Pattern Recognition, pp. 10687-10698 (2020)
30. Kuan, L., Li, Y., Xu, N., Natarajan, P.: Learn to Combine Modalities in Multimodal Deep Learning. ArXiv abs/1805.11730 (2018)
31. Tan, X., Triggs, B.: Enhanced local texture feature sets for face recognition under difficult lighting conditions. IEEE Trans. Image Process. **19**(6), 1635–1650 (2010)
32. Pandey, R.K., Zhou, Y., Kota, B.U., Govindaraju, V.: Deep secure encoding for face template protection. In: 2016 IEEE Conference on Computer Vision and Pattern Recognition Workshops (CVPRW), pp. 77–83. Las Vegas, NV (2016)
33. Da Silva, H.P., Lourenco, A., Fred, A., Raposo, N., Sousa, M.A.: Check your biosignals here: a new dataset for off-the-person ECG biometrics. Comput. Methods Programs Biomed. **13**(2), 503–514 (2014)
34. Cao, Q., Shen, L., Xie, W., Parkhi, O.M., Zisserman, A.: VGGFace2: a dataset for recognising faces across pose and age. In: 13th IEEE International Conference on Automatic Face & Gesture Recognition (FG 2018), vol. 13, pp. 67–74. Xi'an (2018)
35. Huang, G., Liu, Z., Van Der Maaten, L., Weinberger, K. Q.: Densely Connected Convolutional Networks. In: 2017 IEEE Conference on Computer Vision and Pattern Recognition (CVPR), pp. 2261–2269. Honolulu, HI (2017)
36. Chandran, N.R., Manuel, E.M.: Performance analysis of modified SHA-3. Procedia Technol. **24**, 904–910 (2016)

Inner and Outer Race Bearing Defects of Induction Motor Running at Low Speeds Signal Analysis with DWT

Bellal Belkacemi(✉) and Salah Saad

Laboratoire Des Systèmes Electromécaniques (LSELM), Badji-Mokhtar University Annaba, 23000 Annaba, Algeria

Abstract. This paper is focused on bearing inner and outer races defects considered as an important cause of bearing elements damage. The methods and techniques used and applied for the detection of defects in induction motors bearing suffer from some drawbacks such as difficulties of identifying bearing faults from the frequency spectrum because of non stationary vibration signals. For better faults localization, wavelet transform is the best technique for time-frequency analysis. Thus the Discrete Wavelet Transform (DWT) decomposition improved by MATLAB/Simulink toolbox is proposed in this work for signal analysis of bearing faults detection. In order to confirm the proposed procedure, an experimental rig for measuring vibratory signals of a bearing set driven by a 0.38 KW three phase induction motor operating at low speed (500 rpm) for healthy and two bearing faults (inner race and outer race defects). From the obtained results it can be concluded that the DWT enhanced by MATLAB/Simulink toolbox has shown its effectiveness and is appropriate for bearing inner and outer races faults detection.

Keywords: Induction machine · Fault diagnosis · Inner and outer races defects · Fourier transform · Induction motor · Discrete Wavelet Transforms (DWT)

1 Introduction

In general, bearing consists of inner and outer races and the rest of bearing elements such as balls. Usually, the rolling elements in a bearing are arranged or guided in a cage, in order to guarantee uniform spacing between elements to prevent metal to metal contact. The majority of faults in induction machines are related to bearing which is considered as the most critical component. Thus, both the detection and diagnosis of mechanical faults in bearing elements are very important for the induction motor operation reliability.

Acoustic analysis, electrical current signature analysis and vibration analysis are the three most used techniques because they allow the detection and diagnosis of most faults. Also, intelligent techniques are very reliable and very effective for diagnosing faults on electrical machines. Their operation requires learning based on a large amount of data. In the literature many works were undertaken in order to find a suitable and effective tool for bearing elements defects detection.

A. Abraham et al. (Eds.): SoCPaR 2020, AISC 1383, pp. 975–983, 2021.
https://doi.org/10.1007/978-3-030-73689-7_92

K.A. Ibrahim Sheriff et al. [1] have compared the acoustic emission technique and the vibration analysis technique, it was found that the acoustic emission technique is more sensitive than the vibration analysis. The authors of [2] introduced the analysis of variance using a comprehensive plan of experiments proved the capability to evaluate the factors effect on the vibration behavior and acoustics bearings at an early stage of degradation. They have concluded that the acoustic emission is ideal for cases of low speeds but vibration measurement is appropriate for monitoring the defect size evolution while the acoustic emission is more suited to the detection of frequencies characteristic of bearing defects at an early stage.

In [3–5] bearing failures detection in induction motors, suitable for the electrical current spectral analysis of the motors supplied by inverters they have concluded that for all signatures it is recommended to detect rotor faults at high motor load levels, while it is better to detect stator faults at low load levels. The work presented in [6] proposes a new approach based on stator current and efficiency of induction motors as indicators of rolling-bearing faults. The approach was applied on four different types of bearing defects: crack in the outer race, hole in the outer race, deformation of the seal, and corrosion. Yu Wei et al. and Zhang, W [7, 8] presented a review on rotating machinery early fault diagnosis techniques and a comprehensive review summarizing the computational intelligence in machinery condition monitoring and fault diagnosis. The authors of [9] proposed an approach based on current space vector transformation analysis taking in account all the induction machine information then analyzed by the means of Power Spectral Density and the Wavelet Packet Decomposition techniques. It has been proved that the proposed technique is more effective compared to single-phase analysis. The advantages and limitations of each techniques in practical applications are also discussed.

The work carried out in [10] have studied the vibration analysis of rolling element bearings defects. A defect of any bearing element is transmitted to all other elements such as outer race, inner race, ball and, train cage of the bearing. Ashish Kumar Sinha, *et al.* [11] proposed fault detection scheme with the standard deviation of level-five detailed coefficient [d5] from the pre-fed values for the detection of ball bearing damage. The work reported in [12] studied the suitability of vibration analysis technique to detect the distributed defects in bearing, the results show that comparison of vibration spectrum of healthy bearing and vibration spectrum of defected bearing can help to recognize faults in the bearing especially outer and inner races.

The work presented by V. Shanmukha Priya et al. [13] uses Discrete Wavelet Transform to decompose the vibration signal into 9 levels. The mean $\pm$ standard deviation values of the approximate and detail coefficients were found for nine levels of wavelet decomposition of the signal.

Knowing that bearing elements defects vibratory signals are nonstationary, consequently time resolution is lost when applying Fourier transform. Thus analyzing signal in frequency domain is not enough for the fault diagnosis. Therefore as inner and outer races defects are non-stationary signals, wavelet transform (DWT) is proposed for better time-frequency analysis. Thus in the present work the Discrete Wavelet Transform (DWT) decomposition its application is extended to inner and outer races defects. In this paper data representing the recorded signals of healthy bearing state, inner and outer races defects at a running speed of 500 rpm are analyzed by DWT. The obtained results have proved that the DWT is appropriate and effective tool in detecting defect in induction motor bearing.

2 DWT as Vibratory Signal Analysis Technique

The theory and the detailed information of Discrete Wavelet Transform (DWT) are reported in the literature [14]. The overall flow of wavelet decomposition process is conducted in many steps, firstly the raw vibration signal is decomposed into several levels of frequency bands. Then the approximate and detail coefficients in each level with standard deviation values for approximate and detail coefficients are found. Finally, a comparison of standard deviation values for normal and faulty bearings is carried out [15]. The approximation part is obtained by passing the signal through Low Pass Filter (LPF) and Detail part is obtained by passing the signal through High Pass Filter (HPF).

3 Data Acquisition

The tests were carried out using the designed setup shown in Fig. 1. The measured vibration signals are generated by the setup shaft driven by a three phase induction motor through a coupling, a bearing unit and a balanced flywheel (load). The designed test rig consists of a PC, three phase induction motor with of 0.78 Kw, USB measuring device, accelerometer, bearing unit and balanced flywheel (load). The vibration sensor

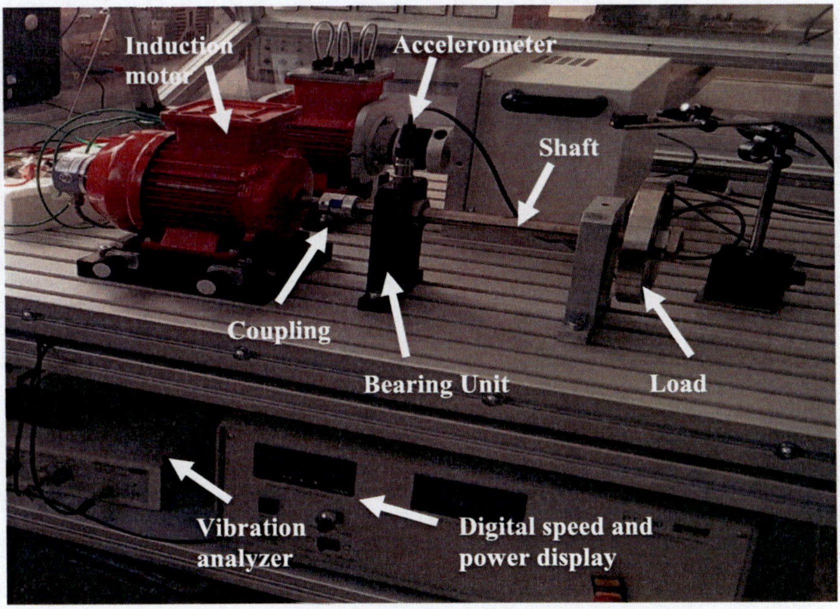

Fig. 1. Experimental setup

used in the experiment measures the vibration generated by two bearing states, outer race and inner race defects as illustrated in Fig. 2a, b. The vibratory signals are measured and recorded for different tests with a low rotational speed of 500 turn/mn.

Fig. 2. Bearing with inner and outer races defects used in the experimental tests

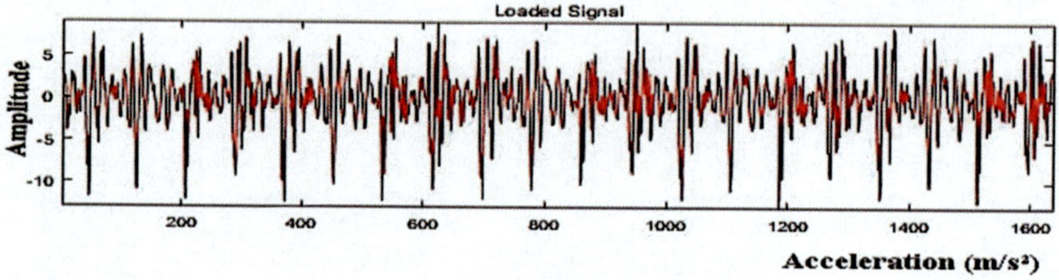

a) Vibratory signal of bearing in healthy condition

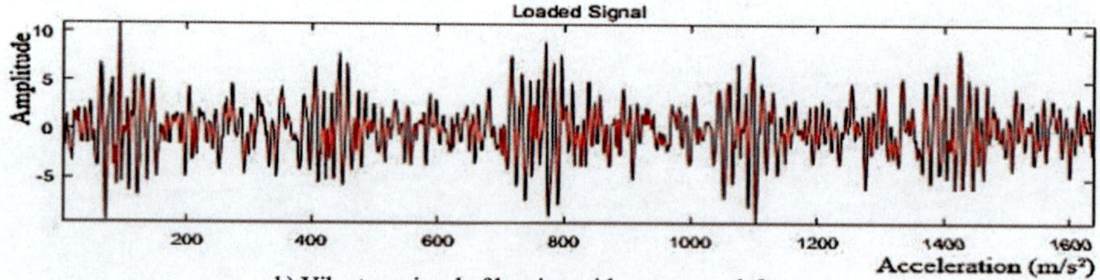

b) Vibratory signal of bearing with outer race defect

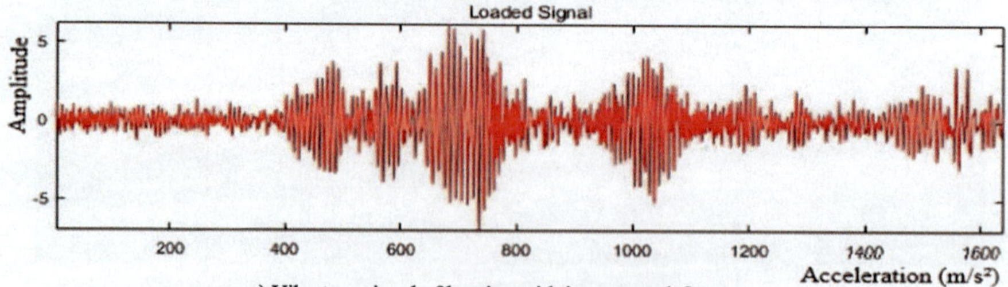

c) Vibratory signal of bearing with inner race defect

Fig. 3. Healthy, outer race and inner race bearing defects vibratory signals.

In the first step the vibratory signal of a healthy bearing is measured for a rotational speed of 500 tr/mn. Then, the vibratory signals of outer race and inner race bearing defects are measured. Finally, the obtained signals are analyzed by DWT decomposition technique.

Discrete Wavelet Transform (DWT) enhanced by MATLAB Toolbox signal analysis is proposed as a tool for different bearing inner race and outer race defects. The recorded data used for the validation of this method are the signals of healthy and faulty bearing.

The decomposition levels and De-Noising the original recoded signals using MATLAB Wavelets Toolbox [16], the exact range values of all detail and approximation coefficients at different levels are obtained. Approximate and detailed coefficients of wavelet transform at each level of decomposition for healthy and faulty bearings are illustrated in Fig. 4, 5 and 6, respectively.

a) Wavelet decomposition of vibration signal with healthy bearing at a rotational speed of 500 rpm.

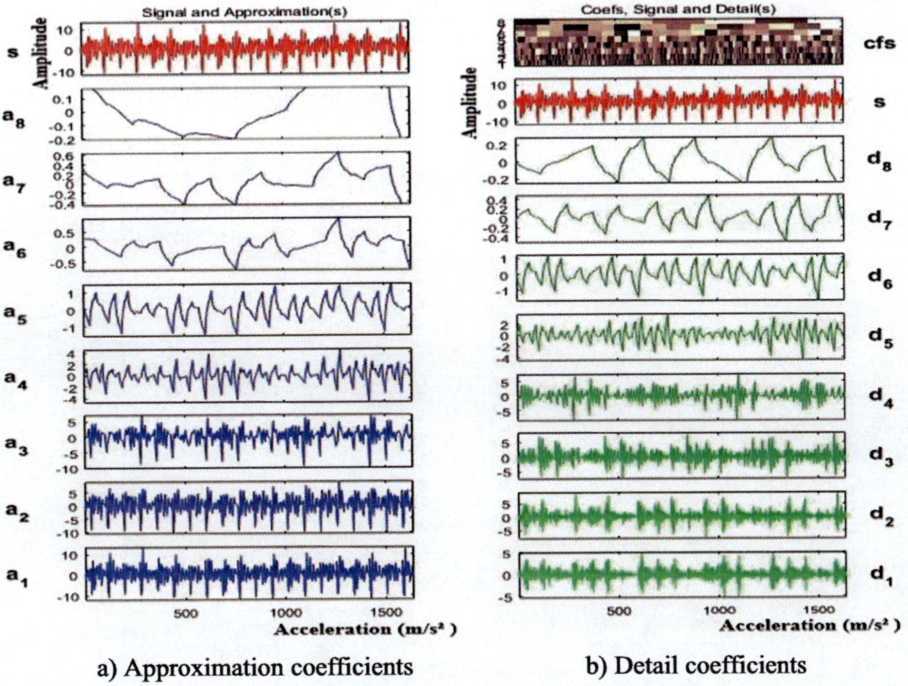

a) Approximation coefficients b) Detail coefficients

Fig. 4. The eight-level wavelet decomposition of healthy bearing signal

b) Wavelet decomposition of vibratory signal of a bearing with outer race defect at a rotational speed of 500 rpm.

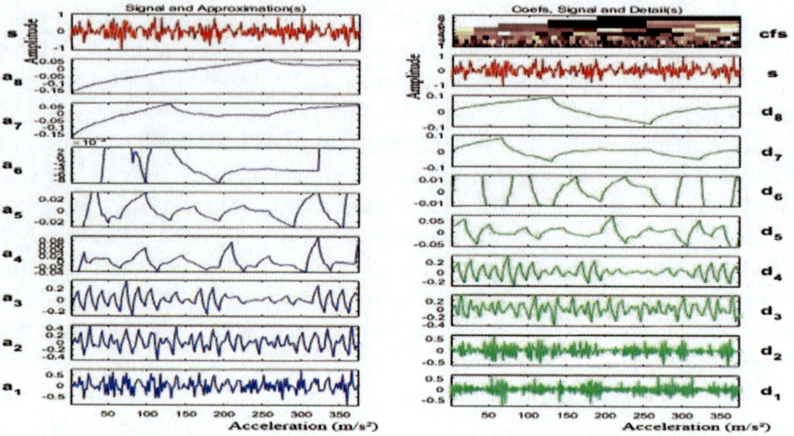

a) Approximation coefficients b) Detail coefficients

Fig. 5. The eight-level wavelet decomposition of bearing outer race defect

c) Wavelet decomposition of vibratory signal of a bearing with inner race defect at rotational speed of 500 rpm.

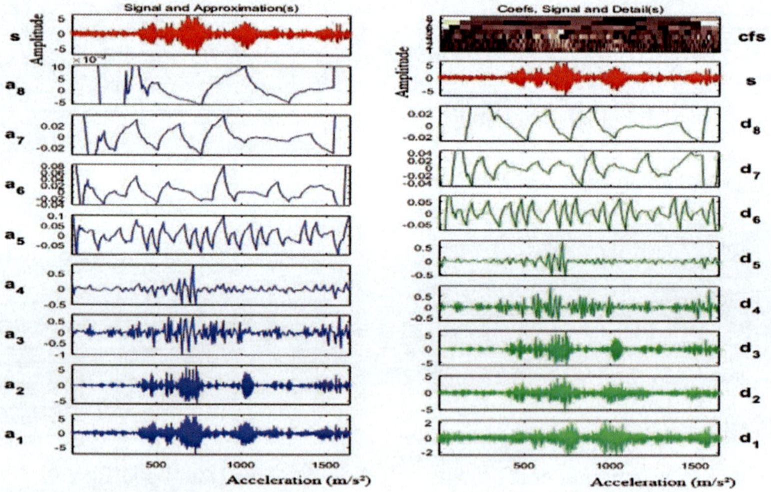

a) Approximation coefficients b) Detail coefficients

Fig. 6. The eight-level wavelet decomposition of a bearing with inner race defect

The results of vibratory signals of healthy bearing, bearing with outer race and inner race defects, are summarized and presented in Table (1) and Table (2).

Table 1. Summarized results of computed approximation coefficients magnitude and ranges of healthy and faulty bearing

ApprLevel	Magnitude range of each level (healthy bearing)	Range	Magnitude range of each level (outer race defect)	Range	Magnitude range of each level (inner race)	Range
a1	−3.323–2.716	6.039	−0.8997–0.7398	1.639	−6.103–6.429	5.21
a2	−2.99–2.702	5.692	−0.5064–0.6004	1.107	−6.103–6.429	4.53
a3	−1.736–1.251	2.987	−0.2927–0.3327	0.6254	−1.024–0.8051	1.829
a4	−0.6182–0.6309	1.249	−0.167–0.09425	0.2613	−0.5101–0.7752	1.285
a5	−0.5766–0.5176	1.094	−0.1871–0.09599	0.2831	−0.1402–0.3924	0.5325
a6	−0.2537–0.3646	0.6183	−0.1865–0.08005	0.2665	−0.1181–0.3849	0.503
a7	−0.1656–0.3359	0.5015	−0.1852–0.07091	0.2561	−0.13–0.3807	0.5107
a8	−0.1644–0.2156	0.38	−0.1792–0.06383	0.243	−0.1453–0.3749	0.5202

Table 2. Summarized results of computed detail coefficients magnitude and ranges of healthy and faulty bearing

Detail Level	Magnitude range of each level (healthy bearing)	Range	Magnitude range of each level (outer race defect)	Range	Magnitude range of each level (inner race)	Range
d1	−0.8332–0.808	1.641	−0.799–0.632	6.43	−2.438–2.253	4.691
d2	−1.228–2.056	3.283	−0.6432–0.6364	4.23	−5.282–5.153	10.44
d3	−3.042–2.514	5.556	−0.4381–0.5492	0.9873	−6.059–6.577	12.64
d4	−1.318–1.327	2.645	−0.2981–0.3336	0.6317	−0.6728–0.9643	1.637
d5	−0.5328–0.4322	0.9649	−0.09464–0.07174	0.1664	−0.55–0.7556	1.306
d6	−0.4773–0.2899	0.7672	−0.09367–0.08802	0.1817	−0.1926–0.1578	0.3504
d7	−0.1519–0.1983	0.3501	−0.06484–0.081	0.1458	−0.2475–0.14	0.3875
d8	−0.2149–0.1865	0.4014	−0.07966–0.0886	0.1683	−0.1577–0.1187	0.2764

4 Results and Discussions

From Fig. 2, it can be observed that the recorded vibration signal of healthy bearing has lower magnitude peaks regarding the vibration signals of bearing with faults (outer race and inner race defects) Fig. 3 b), c), d), and e) respectively.

According to Table 1, The approximation coefficient of healthy bearing reaches its maximum range value at level 1 than decreases reaching its minimum range value at level 8, and the detail coefficient of faulty bearings (outer race, inner race defects) reach their maximum range values at level 1, than decrease reaching their minimum values at level 8. It can be concluded that the maximum value of healthy bearing is higher than the maximum values of faulty bearings (outer race, inner race defects) which means that the detail coefficient range values at level 1 are sensitive to bearing faults.

The detail coefficient of healthy bearing has reached its maximum range value at level 3 than decreases reaching its minimum range value at level 8, whereas the detail coefficient of faulty bearings (outer race, inner race defects) reach their maximum range values at level 1, level 3 respectively than decreases reaching their minimum values at level 8. It is observed from these results that the maximum value of healthy bearing is less than the maximum value of faulty bearings which means that the approximation coefficient at levels 1 and 3 are sensitive to bearing faults (outer race, inner race defects).

Thus it can be concluded that, for low speeds (500 rpm) DWT decomposition enhanced by MATLAB Wavelets Toolbox procedure is an efficient tool for fault detection. However, the wavelet approximation and detail coefficients can be easily used for the detection of different bearing defects. The obtained results are in agreements with the results of the previously reported literature [17].

Thus the Discrete Wavelet Transform (DWT) decomposition improved by MATLAB/ Simulink toolbox is proposed in this work for signal analysis of bearing faults detection.

5 Conclusion

Bearing faults detection approach is proposed for detecting bearing outer and inner races defects. In this work DWT is used for faster computation time contrary to other techniques requiring very high computation time.

Thus an experimental setup for vibratory signals acquisition and Discrete Wavelet Transform (DWT) decomposition enhanced by MATLAB/ Simulink toolbox is proposed for signal analysis. According to the results, it can be observed that the amplitudes of approximation and detail coefficient are sensitive to inner and outer races defects. It can be concluded that, for low speeds the proposed procedure is efficient for fault detection. Therefore, the wavelet approximation and detail coefficients can be easily used for the detection of different bearing defects including inner and outer races defects. The DWT has shown its effectiveness and is appropriate for bearing faults detection.

Acknowledgement. The authors gratefully acknowledge the Algerian General Direction of Research (DGRSDT) for providing the facilities and the financial funding of this project.

References

1. Ibrahim Sheriff, K.A., Hariharan, V., Mathan Kumar, B.: Review on condition monitoring of rotating machines. Int. J. Sci. Technol. Res. **9**(2), 2343–2346 (2020)
2. Meziania, S., Zaroura, D., Thomas, M.: Experimental study for early detection of bearing defects by vibration and acoustic emission. In: 23 ème Congrès Français de Mécanique Lille, 28 Août au 1er Septembre 2017
3. Duque, O., Perez, M., Morinigo, D.: Detection of bearing faults in cage induction motors fed by frequency converter using spectral analysis of line current. In: 2005 IEEE International Conference on Electric Machines and Drives, San Antonio, TX, pp. 17–22 (2005)

4. Shehata, S.A.M., et al.: Detection of induction motors rotor/stator faults using electrical signatures analysis. In: International Conference on Renewable Energies and Power Quality, ICREPQ 2013, Bilbao, Spain, 20–22 March 2013 (2013)
5. Srivani, A., et al.: Fuzzy C-Means based clustering and rule formation approach for classification of bearing faults using discrete wavelet transform. Computation **7**(4), 54 (2019). https://doi.org/10.3390/computation7040054
6. Frosini, L., Bassi, E.: Stator current and motor efficiency as indicators for different types of bearing faults in induction motors. IEEE Trans. Industr. Electron. **57**(1), 244–251 (2010)
7. Wei, Y., et al.: A review of early fault diagnosis approaches and their applications in rotating machinery. Entropy **21**, 409 (2019). https://doi.org/10.3390/e21040409
8. Zhang, W., et al.: Comprehensive overview on computational intelligence techniques for machinery condition monitoring and fault diagnosis. Chin. J. Mech. Eng. **30**, 782–795 (2017). https://doi.org/10.1007/s10033-017-0150-0
9. Lahouasnia, N., Rachedi, M.F., Drici, D., et al.: Load unbalance detection improvement in three-phase induction machine based on current space vector analysis. J. Electr. Eng. Technol (2020). https://doi.org/10.1007/s42835-020-00403-y
10. Harlişca, C., Bouchareb, I., Frosini, L., Szabó, L.: Induction machine bearing faults detection based on artificial neural network. In: 2013 IEEE 14th International Symposium on Computational Intelligence and Informatics (CINTI), Budapest, pp. 297–302 (2013)
11. Sinha, A.K., Das, S., Chatterjee, T.K.: Wavelet transform based ball bearing fault detection scheme for heavy duty mining electrical motors under supply frequency regulation using MCSA. Int. J. Technol. **9**(1), 170 (2018). https://doi.org/10.14716/ijtech.v9i1.1507
12. Saruhan, H., et al.: Vibration analysis of rolling element bearings defects. J. Appl. Res. Technol. **12**, 384–395 (2014)
13. Shanmukha Priya, V., Mahalakshmi, P., Naidu, V.P.S.: Bearing health condition monitoring: wavelet decomposition. Indian J. Sci. Technol. **8**(26), 1–7 (2015). https://doi.org/10.17485/ijst/2015/v8i26/81712
14. Nguyen, P.H., Kim, J.-M.: Multifault diagnosis of rolling element bearings using a wavelet Kurtogram and vector median-based feature analysis. Shock Vibr. **2015**, 1–14 (2015). https://doi.org/10.1155/2015/320508
15. Yusof, N.F.M., Ripin, Z.M.: The effect of lubrication on the vibration of roller bearings. In: MATEC Web of Conferences (2017). https://doi.org/10.1051/matecconf/201821701004
16. Villanueva-, A.E., Luna, A.-N., Sanchez-Lucero, D., Carlos, M., Gabriel, J., Flores-Gil, A., May-Alarco, M.: De-Noising audio signals using MATLAB wavelets toolbox. In: Assi, A. (ed.) Engineering Education and Research Using MATLAB. InTech (2011). https://doi.org/10.5772/19851
17. Kandale, N.S., et al.: Bearing fault detection using gaussian filter bank. In: International Conference on Computing, Communication, Control and Automation (ICCUBEA) (2017). https://doi.org/10.1109/ICCUBEA.2017.8463979

Transformation of PrivUML into XACML Using QVT

Jihane El Mokhtari[1,3](✉), Anas Abou El Kalam[2], Siham Benhaddou[3], and Jean-Philippe Leroy[1]

[1] LISER Laboratory, IPI, Paris, France
[2] TIM Laboratory, ENSAM, Marrakech, Morocco
[3] LRI Laboratory, ENSEM, Casablanca, Morocco

Abstract. In the context of our work, we devote this paper to the derivation from the PIM level to the PSM level in accordance with the MDA approach. PrivUML, the meta-model that we proposed in our previous work to model the privacy protection requirements, will be transformed into XACML by relying on the QVT transformation language. This paper presents this derivation process as well as the changes made to make our transformation a success. It also discusses the concept of anonymization and our integration proposal to be able to guarantee dynamic data anonymization.

Keywords: MDA · Privacy · PrivUML · Transformation · XACML

1 Introduction

Our work fits within the framework of model-driven security, including consideration of privacy requirements from the design phase of a system. Our goal is to enrich the tools and models for system design so that they can model privacy needs along with functional needs. The generation of the code guaranteeing privacy protection is then obtained by refining the elements of these secure models. In this context, we have first identified the requirements that must be met to ensure privacy protection. We then presented all the means allowing us to meet the defined requirements (Access Control, PrivOrBAC and PrivUML). In this paper, we present our proposal to ensure a new iteration of derivation models to achieve our goal.

Our work is part of model-driven security research, including consideration of privacy requirements starting from the system design phase. Our goal is to enrich the tools for system design so that they can model privacy needs along with functional needs. The generation of the code to ensure privacy protection is then achieved by refining the elements of these secure models in accordance with the principles of the MDA approach. In this context, we have first identified the requirements that must be met to ensure privacy protection (CIM, first level in MDA). We then presented all the means that enable to meet the defined requirements (PIM, second level in MDA). In [1], we have defined the notion of privacy, listed the necessary protection requirements, opted for access control as an approach to guarantee privacy protection, defended our choice of the PrivOrBAC

A. Abraham et al. (Eds.): SoCPaR 2020, AISC 1383, pp. 984–996, 2021.
https://doi.org/10.1007/978-3-030-73689-7_93

model to ensure access control, and finally proposed the PrivUML meta-model capable of modeling our security policy based on PrivOrBAC. In this document, we will focus on the PSM level and more specifically the transformation of PrivUML into a specific model to approach the final goal. We will present our proposal to ensure a new iteration of derivation and transforming the PIM PrivUML into XACML through a transformation based on the QVT standard. We will also highlight in this paper the notion of anonymization and how to integrate it into our target model in a dynamic way.

Section 2 of this document is devoted to a review of our previous work. The XACML standard, as architecture and a language, will be presented in Sect. 3. Section 4 will present the transformation process of PrivUML and XACML as well as a case study that we will rely upon to clearly explain this derivation. Section 5 presents the integration of anonymization into the PSM. We will conclude this paper and present our future work in Sect. 6.

2 Previous Work

We have opted in [1] to rely on Access Control to ensure Confidentiality. We also proposed adopting the extension of the OrBAC model so that the decision to allow or deny access is based on a more complete set of criteria, namely the role of the subject, but also the purpose of the access and the consent of the owner of the resource. To be able to model our security policy based on the OrBAC model, we have presented our PrivUML metamodel that can be seen in Fig. 1. In line with our work, we are focusing our efforts on model transformation mechanisms that will generate, from our PrivUML security model, an application architecture that integrates the means of protecting privacy. Figure2 shows the different stages of our derivation model process according to the MDA approach.

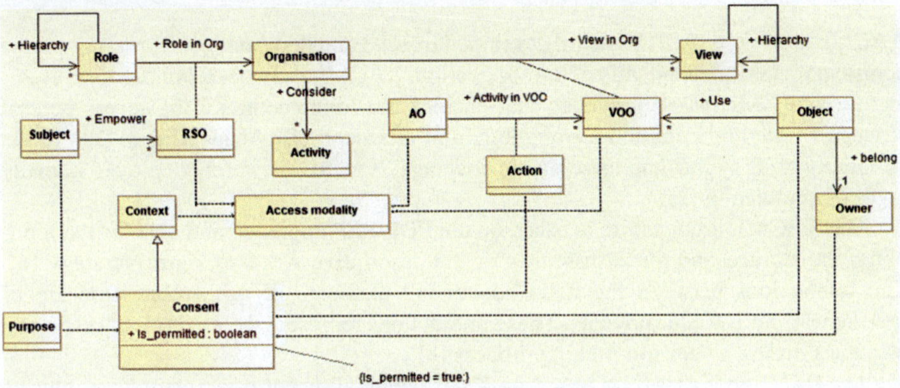

Fig. 1. Metamodel PrivUML [1]

In PrivUML, the access modalities present the decision to refuse access or to authorize it with or without conditions. A PrivUML decision is interpreted as follows: an access is rejected or granted to a subject according to his/her role in the organization

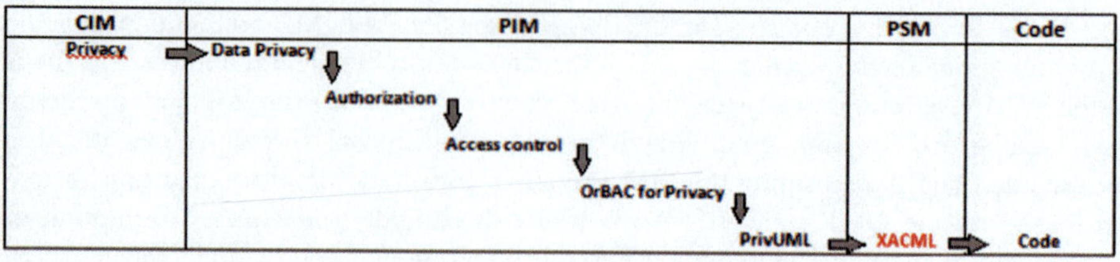

Fig. 2. Transformation process between the different levels of the models according to the MDA approach

RSO to perform an action AO, justified by the purpose P, on a view of objects VOO whose owner has previously given his/her agreement C to use it.

This document focuses mainly on the XACML standard that we have chosen to present our model at the PSM level. Our choice is motivated by the compatibility of XACML with different platforms and its ability to provide a distributed and flexible approach to work with different access control scenarios [2]. The second argument that justifies our choice is that XACML benefits from the existence of various methods of testing strategies with different levels of efficiency. It benefits also from several works proposing approaches which aim to improve the performance of XACML policies [3–5]. In the following, we will first present an overview of the XACML standard that we will rely upon to define the policies and rules of access control ensuring privacy protection. We will then show how to integrate the context and objective of access as well as the owner consent in XACML rules environments. We will also show how to guarantee the anonymization of data by relying on the XACML architecture.

3 XACML

XACML is an OASIS [6] standard developed in 2003 and dedicated to expressing access control policies independently of the application. XACML provides a standardized architecture and XML-based language that enables the implementation of access control strategies based on common terminology and interoperability capabilities. Due to the simplicity of its syntax and its semantic coverage, XACML is widely deployed in many areas of application [7].

The XACML architecture is based on the PDP/PEP approach and enables the modeling, the storage and the distribution of the descriptive Access Control strategy [6]. The interactions between the main constituent elements of this architecture are of type synchronous Request/Reply. These interactions represent the methods of retrieving attributes during a decision-making process [8].

The language XACML is based on XML which allows it easy integration into the majority of applications. It is both an attribute-based access control policy language and a Request/Reply protocol language. It is used to describe the general requirements of access controls in terms of constraints on attributes. An attribute can be any characteristic of a subject, an action, a resource or the environment in which the access request is made [8]. Version 3.0 which has been available since 2013 brings a lot of options and flexibility integrating specific needs in terms of access control [9].

XACML strategies are divided into a hierarchy of three levels [10]: <PolicySet>, <Policy>, and <Rule>. The <PolicySet> is a set of <Policy> elements each containing one or more <Rule> elements. Each rule contains the target which determines its applicability with respect to the elements of the request and the elements provided by the PIP. A rule can also contain a condition and return an effect which represents the permission or denial decision taken. The overall result of a strategy is determined using a combination of algorithms which will compare all the results of the evaluation of the <rule> elements in the event of a conflict. The obligation is an option that allows the PEP to specify whether actions must be taken following the evaluation of a rule.

A XACML Request is a translation of an access query where a subject S requests to perform an action A on a resource R in a certain context C. This request will be processed by the system which will compare it with the access policies that it has and will provide one of several possible answers:

- Access permission
- Access denial
- Not Applicable (if no set of policies can be applied)
- Indeterminable (in the case of an error or a missing attribute)

4 Transformation of the PrivUML Model

A protection model is defined jointly by a security model and a set of transformation rules [11]. An architecture based on XACML enables the transformation of the security model. XACML policies are generated from the model and protection rules [12]. Our work therefore consists in translating our PrivUML security metamodel from PIM level to PSM level while integrating security policies under XACML.

4.1 Overview of Transformation Tools

Much of the background work has provided different visions of transforming the PIM model to PSM model in model-driven engineering using the MDA approach; they are classified in [13–15]. Amongst the transformation languages available today, there are universal languages that rely directly on the abstract representation of the model such as the EMF API which enables the manipulation of a model in the form of a graph while associating it with java code. In this case, it is up to the programmer to search for information in the model to clarify the transformation rules and to manage the target elements constructed [16]. Other languages exist such as the ATL language which is both declarative and imperative [17], workshop languages such as Rational Rose and Objecteering as well as the metamodeling-action Kermeta imperative language [18]. All these languages have been criticized for their mastery of the transformation process [19, 20].

OMG's QVT standard [21] is considered as the most popular tool in IDM. QVT was created in 2006 by converging different responses [22] following the call for proposals, launched by OMG [23], raising the problem of the standardization of model transformation with Query/View/Transformation. It relies on the OCL language to navigate through models and writes transformations using its three sublanguages: Relation, Operational

Mapping and Core. Each of these three sub-languages is based on a different language transformation paradigm namely declarative, imperative and hybrid [24].

In this paper, we will rely on the QVT standard, compatible with the MDA vision and its foundations (UML, OCL, MOF, etc.), and supported by several tools (Echo, QVTR-XSLT, ModelMorf, mediniQVT, etc.) [15], to jointly transform our PIM PrivUML privacy model with the functional PIM model of the system into a PSM model.

4.2 Overview of Transformation Tools

In Table 1 we present the conceptually similar modeling elements, between the PrivUML metamodel and the XACML language, which must be explored at the time of the transformation from the PIM PrivUML to the PSM PrivUML. We also formulate in Fig. 3 an extract of this transformation.

Table 1. Mapings between PrivUML entities and XACML elements.

PrivUML (PIM level)	XACML (PSM level)
Access modality	PolicySet, Policy, and Rule: these are the elements responsible for making access decisions based on the attributes of the subject, resource, action and environment
RSO	Subject: We rely on the subject's role since access permissions are assigned to roles and not subjects directly
AO	Action
VOO	Resource: we organize the resources in views to customize the level of detail
Context	Environment: we enrich this element with a new attribute to complete the context of the access request
Consent/purpose	We are taking advantage of the evolutions brought by version 3.0 and enabling the definition of new elements in extension to the XACML language model to integrate the consent of the owner of the resource as well as the objective of the access in the decision-making mechanism

From an architectural point of view, the access methods are managed at the PDP level which uses the PIP to identify the various attributes of the entities:

- Subject and RSO: data relating to the subject and its role
- Object: data relating to the resource
- Context: data relating to the consent and access objectives.

4.3 Presentation of the Case Study

As part of the protection of the privacy of patients at a Hospital H, medical records are protected and have controlled access. A Medical Record (MR) contains a personal

```
// Transformation of the PrivUml model into an XACML model
transformation PrivUMLToXACML(in PIM_Privacy: PrivUML,out PSM_Privacy:XACML);
main() {
    PIM_Privacy.objects()[Access_Modality]->map Access_Modality_To_Policy();
}
mapping Access_Modality:: Access_Modality_To_Policy():policy {
   PolicyId:= self.Identifiant;
   Resource:= self.VOO -> map VOO_To_Resource();
   Subject:= self.RSO -> map RSO_To_subject();
   ...
}
mapping VOO:: VOO_To_Ressource():Ressource {
   Id:= self.Identifiant;
   Value:= self.level_of_detail
}
mapping RSO:: RSO_To_subject():Subject {
   Id:= self.Identifiant;
   Value:= self.role
}
...
```

Fig. 3. Extract from PrivUML transformation to XACML

Patient Information Sheet (PIS), all information relating to Patient Care History (PCH), and the identity and contact details of the Person of Confidence to Contact (PCC) for specific purposes.

Each patient has an Attending Physician who has read and write-access to the entire Medical Record with an obligation to notify the patient if changes are made to the information relating to his Person of Confidence to Contact. The hospital's medical staff (doctors and nurses) has read-only access to the patient's care history and its personal information sheet. A doctor may exceptionally have write-access to the patient's care history in case of emergency or to replace the Attending Physician in case of absence. Medical researchers contracted by Hospital H can have read-only access to a patient's care history but no access to other information. Finally, the information of the Person of Confidence to Contact can be consulted by the doctors of the hospital after receiving patient consent that is expressed in the form of objectives motivating the request for access.

4.4 Modeling of the Case Study

In our case study, we are using our PrivUML security model, an overview of which is shown in Fig. 4. The organization corresponds to Hospital H according to the PrivOrBAC model. The roles that subjects can have in this organization are records of the RSO. The resources to which access is requested are objects represented in VOO views. In AO, we define the possible actions on the resources. The context makes it possible to specify the constraints of the environment such as the conditions and the consent of a patient depending on the objective of the access request.

For our case study, let's assume that the patient Jean Dupont has granted his consent so that the doctors at Hospital H can access the contact details of his person of confidence (PCC) in order to declare a death, an accident or a surgery. In this case, the values of the various attributes involved in the access control process are:

- **RSO:** Doctor, Attending Physician, Nurse, Researcher
- **VOO:** MR, PIS, PCH, PCC

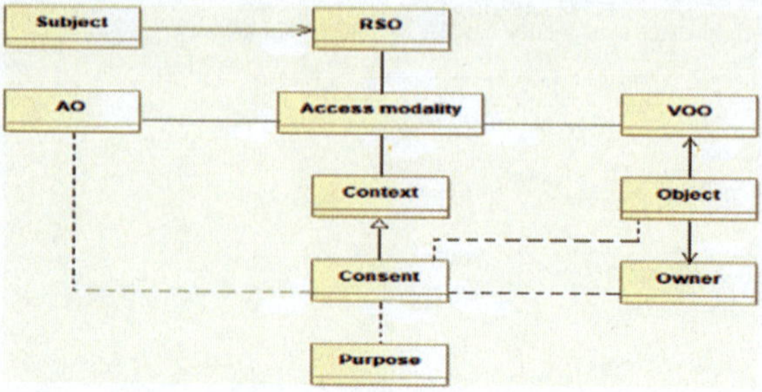

Fig. 4. Overview of the PrivUML-based security model for the case study

- **AO:** Read, Write
- **Consent:** True, False
- **Purpose:** death, accident, surgery
- **Context:** emergency, research, others.

4.5 Generation of the XACML Policy

Declaration of Roles. All organizational roles are identified and managed at the RSO level. The assignment of general roles is also done by default at this level. A XACML policy is used to concretize the assignment of permissions to different roles in the organization. The following extract presents a simplified policy for the attribution declaration of authorizations for the role of doctor:

```
<PolicySet xmlns="urn:oasis:names:tc:xacml:1.0:policy" PolicySetId="RPS:doctor:role" Policy-
CombiningAlgId="&policy-combine;permit-overrides">
   <Target>
      <Subjects>
         <Subject>
            <id>role</id>
            <value>doctor</value>
         </Subject>
      </Subjects>
   </Target>
   <!-- Use permissions associated with the doctor role -->
   <PolicySetIdReference>PPS:doctor:role</PolicySetIdReference>
</PolicySet>
```

The role is linked to the permissions granted to it through the <PolicySetIdReference> tag. This same tag is also used to express the inheritance of permissions from the parent role to the child role which reflects the hierarchy of roles. The "Attending Physician" role is a specialization of the "Doctor" role and therefore inherits its permissions to the additional permissions assigned to it as well. Doctor Marc Lagarde, who has the role of Doctor in Hospital H, is also the attending physician for patient Jean Dupont. His role as a "Doctor" is active by default, while his role of "Attending Physician" must be activated to have access to additional permissions. The activation of this role is done through a policy rule, the extract of which is presented below:

```
<!-- activation of the role Attending Physician -->
<Rule RuleId="AttendingPhysician:role:requirements" Effect="Permit">
   <Target>
      <Subjects>
         <Subject>
            <AttributeId>Subject</AttributeId>
          <AttributeValue>Marc Lagarde</AttributeValue>
         </Subject>
      </Subjects>
      <Resources>
         <Resource>
            <AttributeId>role</AttributeId>
          <AttributeValue>AttendingPhysician</AttributeValue>
         </Resource>
      </Resources>
      <Actions>
         <Action>
            <AttributeId>Action</AttributeId>
          <AttributeValue>enable</AttributeValue>
         </Action>
      </Actions>
   </Target>
   <Condition FunctionId="&function;string-equal">
      <Apply FunctionId="&function;string-one-and-only">
         <AttributeId>RSO.Subject</AttributeId>
      </Apply>
      <Apply FunctionId="&function;string-one-and-only">
         <AttributeId>VOO.AttendingPhysician</AttributeId>
      </Apply>
   </Condition>
</Rule>
```

Resources and Views. The resources are represented in hierarchical views making it possible to configure the level of data details accessible for each role. This setting is provided by the inheritance relationship between views of objects. Doctor Marc Lagarde, the attending physician of the patient Jean Dupont, has access to the root view representing the entire Medical Record (PIS + PCH + PCC) of his patient, while Doctor Luc Smart, from Hospital H, only has access to the view containing Jean Dupont's personal information sheet (PIS) and his care history (PCH).

Policy Statement and Rules. To cover the privacy protection requirements that we use to strengthen the security mechanisms in a system, we suggest to define:

- The "Context" element as a new attribute of the environment that identifies the context of the access request. The following policy makes it possible to formulate the permissions to access the PCH of patient Jean Dupont in reading for a researcher and in writing for the emergency doctor:

```
<Policy PolicyId="PCH:Access_modality" RuleCombiningAlgId="&rule-combine;permit-overrides">
   <Target>
      <Subjects><AnySubject/></Subjects>
      <Resources>
         <Resource>
            <AttributeId>VOO.Identity</AttributeId>
          <AttributeValue>PCH</AttributeValue>
        </Resource>
      </Resources>
      <Actions><AnyAction/></Actions>
   </Target>
   <Rule RuleId="PCH:Access_searcher" Effect="Permit">
      <Target>
         <Subjects>
            <Subject>
               <AttributeId>RSO.role</AttributeId>
               <AttributeValue>searcher</AttributeValue>
            </Subject>
         </Subjects>
         <Actions>
            <Action>
               <AttributeId>AO.Type</AttributeId>
               <AttributeValue>Select</AttributeValue>
            </Action>
         </Actions>
      </Target>
      <Condition>
         <Apply FunctionId="&function;string-equal">
            <AttributeId= Context.Id Category="environnement-category:context"/>
            <AttributeValue>Research</AttributeValue>
         </Apply>
      </Condition>
   </Rule>
   <Rule>
      <!-- Permission to write access to the doctors of the hospital -->
   </Rule>
</Policy>
```

- The “Consent” and “Purpose” categories to materialize the consent of the resource owner based on the purpose of the access request. The “Target” element of the enriched XACML language is presented in Fig. 5:

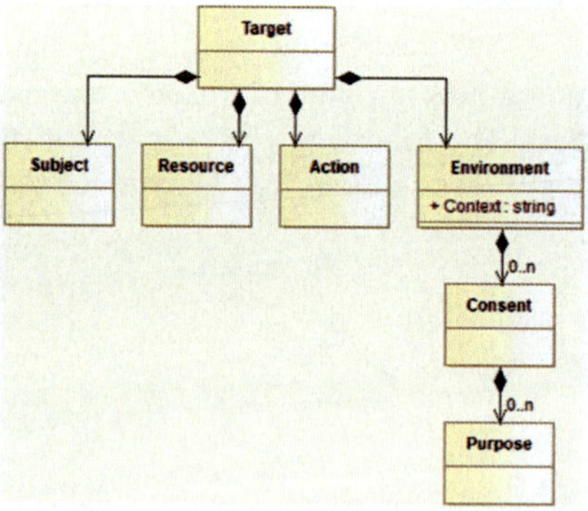

Fig. 5. The “Target” element enriched by “Context”, “Consent” and “Purpose”

Taking consent into account in the access permission is modeled by an XACML policy including the “Consent” element corresponding to the following rule:

```
Permission{
        Role = Doctor
        Action = Read
        Resource= PCC
        If{
          Context = Emergency AND Consent = True
        }
}
```

The consent in our privacy modeling explicitly expresses the agreement given by the owner of a resource to a subject to perform an action for a specific purpose. In Organization H, patient Jean Dupont agrees that doctors can read his PCC to alert the trusted person of an accident, death or surgery. This rule is defined in a XACML policy by integrating the following logic:

```
<Policy PolicyId="PCC:Access_modality" RuleCombiningAlgId="&rule-combine;permit-overrides">
   <Target>
   <!-- this policy concerns the reading of PCC patient J. Dupont who consented
   to doctors to access private information for specific purposes -->
      <Resources>
         <Resource>
            <AttributeId> VOO.Identity </AttributeId>
            <AttributeValue>PCC</AttributeValue>
         </Resource>
         <Resource>
            <AttributeId> VOO.Owner </AttributeId>
            <AttributeValue>Jean Dupont</AttributeValue>
         </Resource>
      </Resources>
      <Actions>
         <Action>
            <AttributeId> AO.Type </AttributeId>
            <AttributeValue>Select</AttributeValue>
         </Action>
      </Actions>
   </Target>
   <Rule RuleId="PCC:Access_PCC" Effect="Permit">
      <Target>
         <Resources>
            <Resource>
               <AttributeValue>Consent.is_permitted</AttributeValue>
            </Resource>
         </Resources>
         <Actions>
            <Action>
               <AttributeId> AO.Type </AttributeId>
               <AttributeValue>True</AttributeValue>
            </Action>
         </Actions>
      </Target>
      <Condition FunctionId="function:and">
         <Apply FunctionId="&function;string-equal">
            <AttributeId> RSO.role </AttributeId>
            <AttributeValue>doctor</AttributeValue>
         </Apply>
         <Apply FunctionId="&function;string-is-in">
            <AttributeId> Purpose.entetled </AttributeId>
            <AttributeValue>death,surgery, accident</AttributeValue>
         </Apply>
      </Condition>
   </Rule>
</Policy
```

XACML Request. A basic XACML request is composed of attributes relating to the subject, resource and action. The processing of the request is based on these three elements as well as on the elements of the environment required for the request evaluation. Our conception of privacy protection incorporates three new factors to consider in the decision-making process:

- The "context" which is an attribute of the environment. In our study, it corresponds to the service to which the requester belongs (e.g.: Emergencies, Research, Radiology, etc.)
- The "objective" of the request which must be formulated as an additional attribute
- The "consent" which is an element evaluated according to the predefined requirements of the data owner as well as the attributes of the access request

5 Integration of Anonymization in the PSM

Anonymization is a process that aims to protect the privacy of an individual and make it impossible to identify him/her. Several anonymization techniques exist based on different principles but all aim to find an efficient solution to eliminate the risk of data disclosure.

One of the best known techniques is k-anonymity [25] which overcomes record linkage attacks on pseudonymized data by replacing fields that enable identification of individuals with pseudonyms. K-anonymity consists of crossing an anonymous tuple with a non-anonymous tuple in order to prevent the identification of an individual. This technique remains however sensitive to the disclosure of attributes which makes it possible to deduce information. L-diversity [26] is another method of anonymization that suggests adding an additional constraint to tuple equivalence classes thus prohibiting disclosure of attributes but constituting a weakness for record linkage. To counter the asymmetry attacks that penalize l-diversity, t-proximity - also called t-closeness – was created in 2007 [27]. The purpose of t-proximity is to ensure a resemblance between all the sensitive data of each equivalence class by minimizing the correlations between the data belonging to the quasi-identifier and the sensitive data. An equivalence class satisfies t-proximity if the distance between the distribution of a sensitive attribute in that class and the distribution of the attribute in the entire table does not exceed a threshold t. A table satisfies t-proximity if all of its equivalence classes satisfy t-proximity. These three techniques, k-anonymity, l-diversity and t-proximity, are part of the generalization family. The technique of differential confidentiality [28], of the randomization family, has the principle of altering the veracity of the data collected. This method constructs each n-tuple with n - t of true data and n - f of fictitious data generated randomly but remaining realistic (with $t + f = n$). The true data is therefore hidden in the constructed set.

In this document, we do not aim to criticize or choose an anonymization technique to meet all the conditions for protecting privacy. Rather, we want to determine how to integrate one or more anonymization techniques so that we can apply them dynamically. The idea is not to anonymize all of the data, but to use anonymization methods only when necessary. By resuming our case study, we find that the attending physician of a patient can have access to all of his/her data without any measure of anonymization since data disclosure leading to patient identification would not make sense in this situation as the attending physician knows the identity of the patient as well as all his care history. On the other hand, and despite the fact that researchers should only have access to data relating to the care history, the risk of disclosure is present and the anonymization of the data is essential. It is with this in mind that we have chosen to integrate anonymization as an "Obligation" in XACML. Each policy must indicate in the "Obligation" part whether an anonymization measure must be applied and through which technique. More details on the implementation of this feature will be given in our future work.

6 Conclusion and Future Work

In this work, we have chosen to rely on the architecture and the XACML language at the PSM level. We used the QVT standard to transform our PrivUML security model and led to the definition of an XACML policy to strengthen the protection of privacy in a system. We have presented how we can ensure all the privacy requirements through the enrichment of the XACML language model based on the 3.0 specification. And finally, we offered our vision of the dynamic integration of anonymization techniques based on the XACML architecture. Thus, we were able to take a new step in our MDA approach by getting closer to the generation of an application architecture guaranteeing privacy protection.

In the continuation of our work, we will focus on two objectives. The first is to ensure the integration of anonymization techniques in our models in accordance with our strategy outlined in the previous section. The second objective is to put in place the necessary changes so that our security policy is more solid and based on dynamic access control rules.

References

1. El Mokhtari, J., Abou El Kalam, A., Benhaddou, S., Medroumi, H.: PrivUML: A privacy metamodel. In: Proceedings of the 10th International Conference on Ambient Systems, Networks and Technologies, ANT 2019, pp. 53–60 (2019). https://doi.org/10.1016/j.procs.2019.04.011
2. Atlam, H.F., Alassafi, M.O., Alenezi, A., Walters, R.J., Wills, G.B.: XACML for building access control policies in Internet of Things. In: Proceedings of the 3rd International Conference on Internet of Things, Big Data and Security, IoTBDS 2018, pp. 253–260 (2019). https://doi.org/10.5220/0006725102530260
3. Xu, D., Shrestha, R., Shen, N.: Strong mutation testing of XACML policies. In: Proceedings of the 25th ACM Symposium on Access Control Models and Technologies, SACMAT 2020, pp. 105–116 (2020). https://doi.org/10.1145/3381991.3395599
4. Daoudagh, S., Lonetti, F., Marchetti, E.: XACMET: XACML testing & modeling an automated model-based testing solution for access control systems. Softw. Qual. J. **28**, 249–282 (2020). https://doi.org/10.1007/s11219-019-09470-5
5. Ahmadi, A., Nassiri, M., Rezvani, M.: XACBench: a XACML policy benchmark. Soft. Comput. **24**, 16081–16096 (2020). https://doi.org/10.1007/s00500-020-04925-5
6. OASIS eXtensible Access Control Markup Language (XACML), Version 1.0 (2003). www.oasis-open.org›committees›oasis-xacml-1.0.pdf
7. Turkmen, F., Crispo, B.: Performance evaluation of XACML PDP implementations. In: Proceedings of the 2008 ACM Workshop on Secure Web Services, pp. 37–44 (2008). https://doi.org/10.1145/1456492.1456499
8. Laborde, R., Desprats, T.: Gestion de conditions stables dans XACML : intérêt d'une approche par notification. In: GRES 2007, Hammamet, Tunisie, Novembre 2007 (2007). https://www.irit.fr/publis/SIERA/gres2007-desprats-laborde-Soumission.pdf

9. OASIS eXtensible Access Control Markup Language (XACML), Version 3.0 (2013). https://docs.oasis-open.org/xacml/3.0/xacml-3.0-core-spec-os-en.html
10. OASIS eXtensible Access Control Markup Language (XACML), Version 2.0 (2005). https://docs.oasis-open.org/xacml/2.0/access_control-xacml-2.0-core-spec-os.pdf
11. Rozenberg, G.: Handbook of Graph Grammars and Computing by Graph Transformation (1997)
12. Koch, M., Pauls, K.: Engineering self-protection for autonomous systems. In: Baresi, L., Heckel, R. (eds.) Fundamental Approaches to Software Engineering. Lecture Notes in Computer Science, vol. 3922, pp. 33–47. Springer, Heidelberg (2006). https://doi.org/10.1007/11693017_5
13. Czarnecki, K., Helsen, S.: Classification of model transformation approaches. In: Proceedings of the 2nd OOPSLA Workshop on Generative Techniques in the Context of Model-Driven Architecture (2003)
14. Combemale, B.: Ingénierie Dirigée par les Modèles (IDM) – État de l'art. In: archives-ouvertes.fr (2008). https://hal.archives-ouvertes.fr/hal-00371565
15. Kahania, N., Bagherzadeha, M., Cordya, J.R., Dingela, J., Varro, D.: Survey and classification of model transformation tools. Softw. Syst. Model. **18**, 2361–2397 (2019). https://doi.org/10.1007/s10270-018-0665-6
16. El Kouhen, A.: Spécification d'un Métamodèle pour l'adaptation des outils UML. In: archives-ouvertes.fr (2015). https://tel.archives-ouvertes.fr/tel-00997773
17. Bézivin, J., Jouault, F., Rosenthal, P., Valduriez, P.: Modeling in the large and modeling in the small. In: Aßmann, U., Aksit, M., Rensink, A. (eds.) Model Driven Architecture, pp. 33–46. Springer, Heidelberg (2005). https://doi.org/10.1007/11538097_3
18. Muller, P.A., Fleurey, F., Vojtisek, D., Dery, Z., Pollet, D., Fondement, F., Studer, P., Jezequel, J.M.: On Executable Meta-Languages applied to Model Transformations. In: archives-ouvertes.fr (2005). https://hal.inria.fr/inria-00000381
19. Diaw, S., Lbath, R., Bernard coulette, : État de l'art sur le développement logiciel basé sur les transformations de modèles. Techniques et sciences informatiques **29**(4–5), 505–536 (2010)
20. Lafaye, M.C., Lafaye, J.Y., Wiedemann, A.: L'approche MDA, pour la conception des bases de données relationnelles: une illusoire simplicité. In: archives-ouvertes.fr, pp. 167–181 (2007). https://hal.archives-ouvertes.fr/hal-00361087
21. OMG Meta Object Facility (MOF) 2.0 Query/View/Transformation Specification (2011). https://www.omg.org/spec/QVT/1.1/PDF
22. OMG MOF QVT Final Adopted Specification (2005). https://www.omg.org/spec/QVT/1.0/Beta1/PDF
23. OMG Request for Proposal: MOF 2.0 Query/Views/Transformations RFP (2002)
24. Kurtev, I.: State of the art of QVT: a model transformation language standard. In: Schürr, A., Nagl, M., Zündorf, A. (eds.) Applications of Graph Transformations with Industrial Relevance, pp. 377–393. Springer, Heidelberg (2008). https://doi.org/10.1007/978-3-540-89020-1_26
25. Sweeney, L.: k-anonymity: a model for protecting privacy. Int. J. Uncertain. Fuzziness Knowl. Based Syst. **10**, 557–570 (2002)
26. Machanavajjhala, A., Kifer, D., Gehrke, J., Venkitasubramaniam, M.: l-diversity: privacy beyond k-anonymity. ACM Trans. Knowl. Discov. Data (2007). https://doi.org/10.1145/1217299.1217302
27. Li, N., Li, T., Venkatasubramanian, S.: t-closeness: privacy beyond k-anonymity and l-diversity. In: Proceedings of the IEEE 23rd International Conference on Data Engineering (2007). https://doi.org/10.1109/ICDE.2007.367856
28. Dwork, C., Roth, A.: The algorithmic foundations of differential privacy. Found. Trends® Theor. Comput. Sci. **9**(3–4), 211–407 (2013). https://doi.org/10.1561/0400000042

Robust and Secure Watermarking Technique for E-Health Applications

Hedi Choura(✉), Faten Chaabane, Tarek Frikha, and Mouna Baklouti

National Engineering School of Sfax (ENIS), University of Sfax, BP 1173, 3038 Sfax, Tunisia
hedi.choura@enis.tn

Abstract. Advancement in information and communication technology in health sector has been developed at a rapid pace, and has consequently improved its different services. Securing patients's health documents becomes crucial and researchers have investigated more than one privacy measures involving the access control, the security and the integrity of these documents. In this paper, we propose a novel QR-code-based watermarking technique for medical images. The generated QR code pattern is the result of converting a watermarked image which is obtained by embedding the concatenation of diagnostic results and a specific collusion-secure code to the original image.

The proposed scheme is a two-stage scheme which has essentially a two fold purpose: at a first step, we propose to embed successively two types of watermarks to the image and then to convert it to a set of QR codes to ensure simultaneously and independently its protection. At a second step, we apply the tracing process whose target is the identification of eventual hackers by extracting the hidden identifier from the suspicious document and analyse it. Experimental assessments show that the proposed scheme provides good results in terms of security, robustness and tracing accuracy constraints.

Keywords: Medical image · E-health · Image watermark · QR codes · Data sharing · Tracing · Privacy preserving

1 Introduction

The concept of e-health can be assimilated to all the sharing and manipulating operations of medical data using information technology infrastructure and e-commerce practices. It involves the presence of several actors such as: medical informatics, public health services and data transmitted through the Internet network or other related networks. One key challenge when dealing with such applications is that patients' health data are expecting different types of security and privacy threats.

As mentioned in [3], the three fundamental security challenges in e-health applications are respectively: confidentiality, integrity and availability.

A. Abraham et al. (Eds.): SoCPaR 2020, AISC 1383, pp. 997–1008, 2021.
https://doi.org/10.1007/978-3-030-73689-7_94

According to [13], confidentiality refers to the fact that only authorised users have rights to access to patient's information. The data integrity refers to that information is accurate and has not be modified illegally. In this context, protecting the integrity of health data is crucial and ensures consequently patient safety. It is also interesting to ensure the traceability of such information which means that it is possible to trace back any dishonest manipulation of the data.

Now, the availability requirement means that information should be available and accessible to any authorised actor of the application. Thus, when considering such requirement, health applications should be efficient even against eventual system failures and attacks.

We turn now our attention to the confidentiality and integrity requirements and we focus on how to protect health data by providing them security and privacy even if they are critically threatened by external attacks.

Obviously, recent e-health applications are strieving in finding remedial measures to prevent medical data infringement and provide more security through the legal sharing operations. One key measure in the literature focused on the watermarking field which consists in embedding a watermark in the medical image in order to identify its supplier and to protect it from any piracy operation. But it still remains insufficient when considering tracing traitors' target [11], which is, to ensure the safe usage of media releases at a first time and to trace back the illegitime users in a piracy trial. Consequently, tracing documents in e-health applications involves two main components: the embedded tracing code and the robust watermarking technique. In this paper, we present a novel hybrid watermarking technique which garantees both the confidentiality and the integrity of medical documents in an e-health application. In our knowledge, this proposed scheme is the only one which offers to trace back any infringement operation by involving a Tardos-tracing process.

This article is orginized as follows: Sect. 2 shows the available watermarking and presented tracing approaches. Section 3 provides an overview of the proposed system. In Sect. 4, we detail the different steps of the hybrid watermarking scheme. In Sect. 5, we present the different experimental assessments we carry out to validate the performance of the proposed technique. At the last, in Sect. 6, we finish with a summary and future work.

2 Related Work

The watermarking approach seems to be the most popular and robust approach for medical data authentication. It refers to the embedding of some specific data such as: the hospital logo or the most pertinent patient's details in a cover to determine any ownership [10]. Some work use watermarking-based cryptography and biometrics schemes to provide more security for medical data [18] and to communicate the highly confidential patient information and authentication. According to [16], different types of watermarking schemes are described: visible, non-visible, public, private, perceptual and bit-stream. Regarding the Human Visual Perception, the watermarks can be considered as Visible or Invisible ones. In visible watermarking schemes, watermarks are perceptible (visible) and are imprinted onto host images. This type of schemes provides possible recognition in case of visual resources but still remains weak when considering imperceptibility requirement. Whereas, the principle of invisible watermarking schemes, is to construct a watermark from a specific secret data, determined and known only by a specific user, and embed it into host media. According to its robustness against benchmark attacks, the invisible watermarking techniques can be classified into three classes: robust, fragile and semi-fragile ones. Robust ones are resistant against a large set of Benchmark attacks and hence, are suitable for copyright protection. Whereas fragile watermarks are susceptible to be distorted by small alterations, which makes them suitable to check the integrity and supplier authenticity. Besides, for the public watermarks, they are considered as not secure, as they may be recovered by anyone via a specific scheme. In contrary, the private watermarks can only be extracted by authorized people using a secret key [17]. Watermarking technique should provide an efficient trade-off between the robustness, capacity and the imperceptibility requirements. Security and computational complexity of the medical watermarking system are also important. The most investigated approaches in e-health domain, are blind watermarking approaches.

According to [2], using an hybrid approach of two or more transform domain techniques should enhance efficiently the performance of the scheme but it increases the computational complexity. Thus, we present in this paper an improved hybrid watermarking technique based on QR codes to face computational costs.

Now, for the tracing code, it should be unique and specific to each legitim user [8]. The watermarking technique which is applied to embed the message in the media content should be robust to different types of attacks [8]. In the literature, two different fingerprinting approaches were thus derivated: a first approach is focused on the watermarking layer and the second one had rather enhanced the study of the tracing code [8]. The watermarking-based fingerprinting systems proposed a spread spectrum signal as a fingerprint for each user [7]. While this approach has good rates of tracing, it presents heavy computational costs. The code-based fingerprinting systems were oriented in improving the code construction and providing a good compromise between the detection rate, the code length and the robustness to the different attacks [4]. In this way, several codes were proposed in the literature [5], especially the well-known Tardos code [20] which offers a good compromise between the code length and the detection rate [1,19].

In this paper, we propose a fingerprinting scheme, based on an hybrid watermarking technique combining a first watermarking stage based on DWT, APD-CBT, and SVD [21] to embed a health institution logo and a second stage based on the Rational Dither Modulation (RDM) [12] to embed the Tardos-based code. We propose also to convert the output of this two-stage watermarking scheme into QR codes. On one side, this choice will provide us the possibility of supporting a large size of information in a reduced space which will require less complexity computation during the sharing operations. On the other side, converting the watermarked image into several QR codes should scramble a few more tracks to reinforce the security of the system.

3 The Components of the General Fingerprinting System

In [9], we proposed the general scheme of a fingerprinting framework that implies the presence of five essential steps as depicted in Fig. 1: the fingerprint encoding step, its embedding in the document release, the selection of some releases to participate in a piracy attack and to give rise to a pirated copy, the extraction of the colluded code and the tracing process to trace back the pirates. We will detail the two-stage watermarking layer and the fingerprinting one separately.

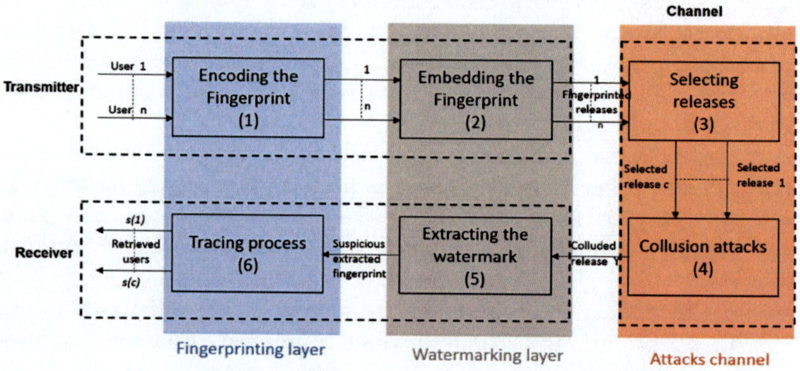

Fig. 1. The components of the general fingerprinting scheme

3.1 The Two-Stage Watermark Embedding Step

In order to ensure the copyright of the health institution or one of the other application actors: doctor, radiologist, etc., it is interesting to add a visible watermark to identify the document holder. Then, we have also thought of adding an invisible watermark that contains pertinent information tied to the tracing process as the collusion-secure code.

The adopted watermarking technique is a two-stage watermarking one. We detail in next paragraph the two consecutive watermarking levels.

The First Watermarking Level. The first step of the two-level watermarking scheme consists in embedding in the medical image a visible watermark whose purpose is to avoid its illegal use.

As depicted in Fig. 2, we choose to use the robust watermarking technique proposed by [21] to embed the watermark. In this scheme, a set of successive

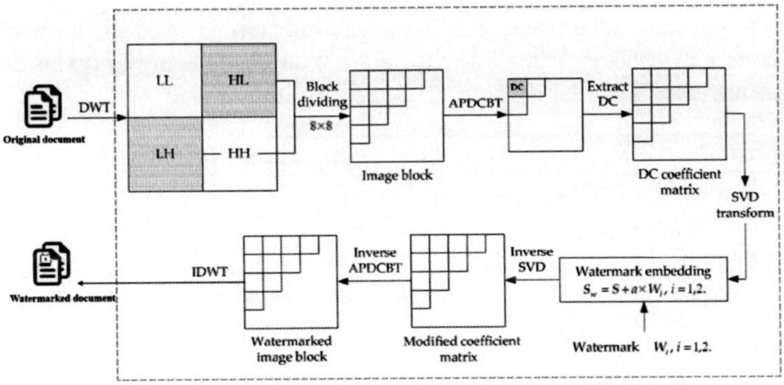

Fig. 2. The embedding step of the first watermarking level [21]

transforms are made to the input image ranging from the DWT, the block-based APDCBT, the All Phase Discrete Cosine Biorthogonal Transform, to the SVD. The choice of this existing technique can be explained by the fact that our target is to provide essentially an efficient robustness to attacks regardless the imperceptibility requirement. The detection step of the first watermarking level consists in the inverse scheme where the received image, which might have been distorted by various attacks, is transformed by the inverse latter process in watermark embedding procedure. The detection step is detailed in Fig. 3.

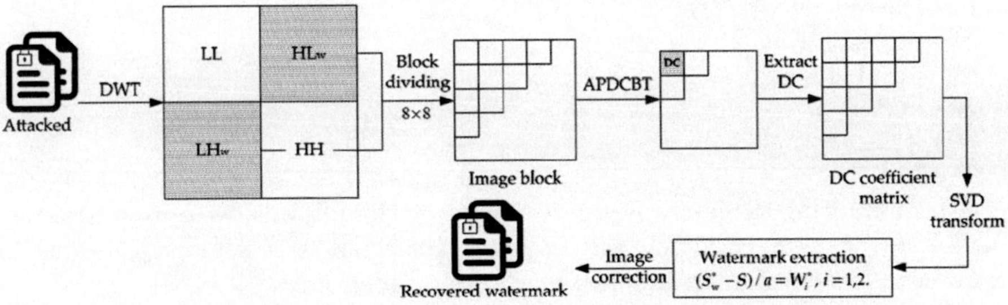

Fig. 3. The detection step of the first watermarking stage [21].

The Second Watermarking Level. Now, for the second step of the two-level watermarking scheme, it consists in an RDM-based watermarking technique to hide an invisible watermark which consists in Tardos code. Choosing to hide the tracing code as an invisible watermark in a second level is a manner to make harder the guess of its presence to pirates. As mentioned in [12], RDM is a data concealment method that is based on Quantization Index Modulation (QIM) and has been investigated as a key approach for the gain attack in high-rate data obscuring networks. Therefore, RDM-based function is used to embed watermarks, which can recover the medical image. The embedding and the detection algorithms are detailed respectively in Algorithms 1 and 2:

Algorithm 1: RDM-Based Reversible Embedding Method

```
Input: S matrix and quantization step Δ
Output: Watermarked singular values matrix S'
n = floor(S(1,1)/Δ)
if w = 1 then
 | m = n + 1 − mod(n, 2)
else
 | m = n + 1 − mod(n + 1, 2)
end
P = m × Δ + Δ/2
E = P − S(1,1)
G = E/Δ
S'(1,1) = P + G
```

where the $floor(\cdot)$ is rounding toward negative infinity, the distortion E caused by watermark embedding will not be larger than Δ and the G will less than 1.

Algorithm 2: RDM-Based Reversible Method

```
Input: S' matrix and quantization step Δ
Output: Restored singular values matrix S
n' = floor(S'(1,1)/Δ)
P' = n' × Δ + Δ/2
G' = S'(1,1) − P'
E' = G' × Δ
S(1,1) = P' − E'
```

After executing successively the two watermarking levels, we obtain a watermarked medical image which will be then converted to a set of QR codes to ensure simultaneously and independently its protection.

The QR-code-based Conversion Step. The originality of the proposed scheme consists in converting the watermarked medical image to a set of k QR codes before their transmission to reduce the sharing computational complexity and enhance its security side. The number k of generated QR-codes depends on the size of the document. Hence, we divide the image to k overlapped QR codes. This division can give the possibilty to share independently the different QR codes in a parallel way without waiting the whole image. Moreover, this method makes hard the reconstruction of the whole image which requires all the QR codes.

3.2 The Fingerprinting Layer

In this step, a set of fingerprints is generated to identify each document supplier. In the context of traitor tracing field, we use specific codes as fingerprints. These codes are collusion-secure fingerprinting codes. We use in this work, the well-known Tardos code as tracing code.

The Tardos Tracing Code. The Tardos code proposed in [20] is a probabilistic code defined as follows:

- *X:* $n \times m$ matrix, with n is the users' number and c is the collusion size.
- A user identifier or codeword X_{ji} is assigned to each user j, $i \in \{1, \cdots, m\}$. Let generate randomly a set of independent and identical probabilities p_i from $[t, 1-t]$, with $t = \frac{1}{300c}$, $0 < t\prime < \frac{\pi}{4}$, $\sin^2 t\prime = t$, $p_i = \sin^2(r_i)$, $r_i \in [t\prime, \frac{\pi}{2-t\prime}]$.
- $\varepsilon_1, \varepsilon_2$ are respectively the false positive probability(the probability of accusing innocent user) and false negative one (the probability of missing guilty user). m is the code length. As mentioned in [14]:

$$m = \frac{1}{2}\pi^2 c^2 \ln(\frac{1}{\epsilon_1}). \tag{1}$$

The principle of the Tardos tracing algorithm consists of three steps: the initialization step, the construction and the accusation ones. Initially, we construct identifiers' matrix X_{ji} with Prob$[X_{ji}{=}1] = p_i$. According to the symmetric version of Tardos accusation functions (Eq. 2 and Eq. 3) defined by [19], the score $S_{j, j\in\{1\ldots n\}}$ in Eq. 4 for each user j is computed as follows:

$$g(1,1,p) = g(0,0,1-p) = \sqrt{\frac{(1-p)}{p}}, \tag{2}$$

$$g(1,0,p) = g(0,1,1-p) = -\sqrt{\frac{p}{(1-p)}}. \tag{3}$$

$$S_j = \sum_{i=1}^{m} g(Y_i, X_{ji}, p_i), \tag{4}$$

Z is the threshold parameter [14] defined by:

$$Z = \pi c(\ln \frac{1}{\varepsilon_1}). \tag{5}$$

The output of the Tardos tracing algorithm as mentioned in [20] includes a user j if $S_j > Z$ with a false positive error $< \varepsilon_1$.

4 Experimental Results

We evaluate firstly the robustness of the proposed two-stage watermarking scheme to different signal processing attacks. Then, we evaluate the efficiency of the tracing process against different collusion attacks. The tested image and the health institution watermark are provided in Fig. 4. We evaluate our approach on OASIS-3 database [6].

4.1 The Watermarking Results

In this subsection, we evaluate the robustness and the imperceptibility of the proposed scheme. It is important to evaluate the imperceptibility of the watermarking scheme to show its out-performance. In this context, we compute the Peak signal to noise ratio, PSNR, which is an objective criterion for image evaluation and its unit is dB. It is defined by the equation:

$$PSNR = 10 \log \frac{255^2}{\frac{1}{M*N}\sum_{i=1}^{M}\sum_{j=1}^{N} I(i,j)I'(i,j)} \quad (6)$$

The general range of PSNR is 20 to 40.

We compute also the SSIM (Structural similarity) which is a perceptual metric to evaluate the quality of an image. It evaluates the similarity of two images from brightness, structure and contrast. The general range of SSIM is [0, 1].

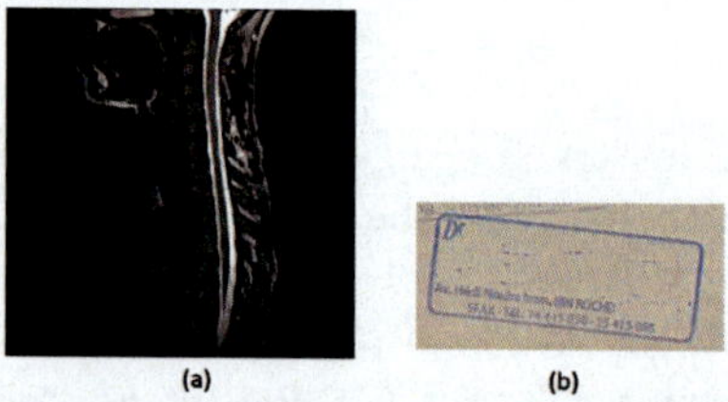

Fig. 4. The carrier images without performing any attack: (a) original RMI image; (b) the watermark image

According to Table 1, the obtained SSIM and PSNR values for the original medical image and the watermarked one are respectively 0.9735 and 39.2792. The rates of PSNR are between 30 and 40 dB which is sufficient to guarantee the imperceptibility of the watermark due of the injection of the visible watermark. We found that the original medical image and watermarked image can not be distinguished by naked eyes. Thus, the watermarked image can better protect the watermark information and resist to the potential attackers.

To evaluate the robustness of the proposed method, several signal processing attacks are applied on the watermarked medical image: the median filtering attack, the smaller attack, the rotation attack (90° and 45°), the white noise attack, the Gaussian noise attack, and Salt and Pepper noise attack. To evaluate the robustness of the proposed scheme against these attacks, we calculate the Normalized Correlations criterion, NC, between the original and the recovered watermark (Fig. 5):

$$NC = \frac{\sum W(i,j)W'(i,j)}{\sqrt{\sum W(i,j)^2}\sqrt{\sum W'(i,j)^2}} \quad (7)$$

Table 1. PSRN and SSIM values comparaison after many types of attacks.

Approach	DCT transform [15]		Proposed approach	
Attack	PSNR (dB)	SSIM	PSNR (dB)	SSIM
Median filtering attack	19.6043	0.9168	33.27	0.954
Smaller attack (times)	40.8452	0.9941	35.33	0.9735
Rotation attack (90°)	9.3114	0.2134	32.67	0.936
Rotation attack (45°)	7.9945	0.2517	32.04	0.9308
White noise attack (variance=20)	25.2646	0.9786	34.26	0.9623
Gaussian noise attack (variance=0.02)	22.7707	0.9675	33.78	0.9421
Salt and pepper noise attack (variance=0.02)	24.8160	0.9763	33.25	0.94

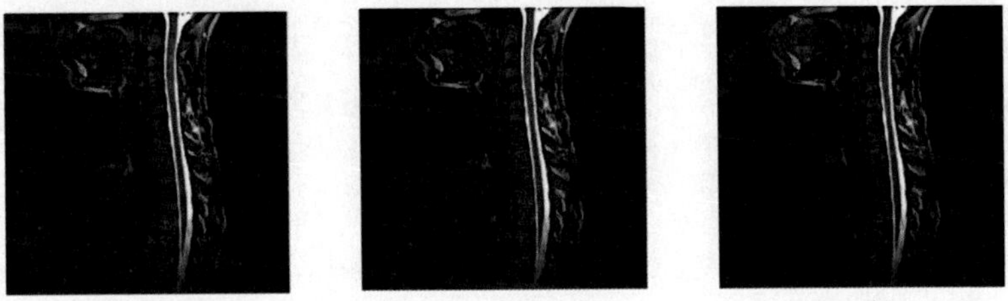

(a) (b) (c)

Fig. 5. The tested images before and after performing any attack: (a) original RMI image; (b) watermarked RMI image; (d) Attacked RMI image

Table 2. NC values for an image of size divided into 16 QR-codes

	Gaussian	Median	Gamma corrected	Rotate	Contrast ajusted
QR_0	0.930426	0.988958	1	0.579435	1
QR_1	0.930530	0.988970	1	0.585272	1
QR_2	0.929572	0.988790	1	0.580265	1
QR_3	0.930361	0.988905	1	0.582146	1
QR_4	0.929601	0.988742	1	0.583498	1
QR_5	0.929952	0.988925	1	0.586199	1
QR_6	0.930267	0.988808	1	0.584714	1
QR_7	0.930668	0.988796	1	0.587522	1
QR_8	0.930169	0.988783	1	0.582586	1
QR_9	0.930128	0.988868	1	0.581855	1
QR_10	0.930055	0.988929	1	0.580150	1
QR_11	0.929253	0.988758	1	0.578315	1
QR_12	0.929563	0.988970	1	0.585763	1
QR_13	0.931066	0.988879	1	0.581556	1
QR_14	0.929602	0.988923	1	0.585206	1
QR_15	0.929599	0.988802	1	0.578703	1
QR_16	0.933161	0.989134	1	0.604881	1

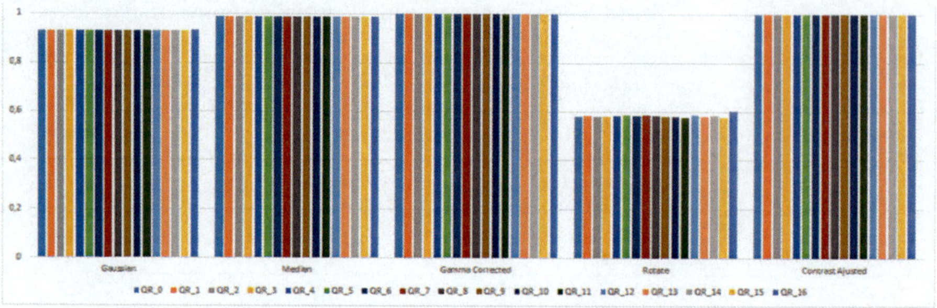

Fig. 6. Robustness to several stirmark attacks

According to the obtained results in Table 2 and Fig. 6, NC values are near 1. Hence, it is important to notice that although the proposed scheme has not a high rate of imperceptibility but it guarantees an efficient robustness against attacks which can be explained by the QR-code conversion step which takes the advantage of inherent characteristics of QR-codes and their error correcting channel.

4.2 The Tracing Results

In this section, we give the tracing results against different collusion attacks described in [8]. In order to prove the out performance of the system in terms of tracing rates, we compare the number of recovered pirates for three types of collusion attacks: majority vote attack, the minority vote one, the all-one attack and the all-zero one. Hence, we use respectively the number of users n = *20*, the collusion size c = *5*, the false positive probability $\varepsilon_1 = 10^{-6}$ and the fingerprinting length m. The Table 3 presents a good detection of the colluders for code length $m = 420$.

Table 3. Detection rates against collusion attacks for m = 420 and c = 5

Collusion attack	Majority	Minority	All-one	All-zero
Detection rates	100	20	40	40

5 Conclusion

In this paper, we have proposed a two-stage watermarking technique for e-health platforms. It has two essential steps: at the first one, we propose to embed successively two types of watermark to the image and then to convert it to QR codes to ensure concurrently and separately its protection. At the second stage, we apply the tracing process whose target is the identification of eventual hackers by extracting the hidden identifier from the suspicious document and analyse it. The security side of the technique is improved by the use of QR codes. A set of experiments were realized according to several criteria: imperceptibility and robustness to collusion and other attacks. This approach was proposed to be adapted in a future work in a blockchain-based e-health application. The use of this tracing scheme should enhance the security and privacy sides of sharing operations of medical documents, where the fingerprint is constructed in the distributor side and tracing colluders are made after diffusing all the copies.

References

1. Akashi, N., Kuribayashi, M., Morii, M.: Hierarchical construction of Tardos code. In: International Symposium on Information Theory and Its Applications, 2008, ISITA 2008, pp. 1–6 (December 2008). https://doi.org/10.1109/ISITA.2008.4895496
2. Anand, A., Singh, A.K.: Watermarking techniques for medical data authentication: a survey (2020). https://doi.org/10.1007/s11042-020-08801-0
3. Azeez, N.A., der Vyver, C.V.: Security and privacy issues in e-health cloud-based system: a comprehensive content analysis. Egyptian Inf. J. **20**(2), 97–108 (2019). https://doi.org/10.1016/j.eij.2018.12.001, http://www.sciencedirect.com/science/article/pii/S1110866517302797
4. Barg, A., Kabatiansky, G.: Robust parent-identifying codes and combinatorial arrays. IEEE Trans. Inf. Theory **59**(2), 994–1003 (2013)
5. Boneh, D., Shaw, J.: Collusion-secure fingerprinting for digital data. IEEE Trans. Inf. Theory **44**(5), 1897–1905 (1998)
6. Brains, O. (ed.): OASIS Brains - Open Access Series of Imaging Studies (2019). https://www.oasis-brains.org/
7. Cha, B.H., Kuo, C.C.: Design and analysis of high-capacity anti-collusion hiding codes. Circuits, Syst. Signal Process. **27**(2), 195–211 (2008)
8. Chaabane, F., Charfeddine, M., Amar, C.B.: A survey on digital tracing traitors schemes. In: 9th International Conference on Information Assurance and Security, IAS 2013, Gammarth, Tunisia, 4-6 December 2013, pp. 85–90 (2013)
9. Chaabane, F., Charfeddine, M., Puech, W., Amar, C.B.: A two-stage traitor tracing scheme for hierarchical fingerprints. Multimedia Tools Appl. **76**(12), 14405–14435 (2017). https://doi.org/10.1007/s11042-016-3749-8
10. Divecha, N.H., Jani, N.N.: Reversible watermarking technique for medical images using fixed point pixel. In: 2015 Fifth International Conference on Communication Systems and Network Technologies, pp. 725–730 (2015)
11. Fontaine, C.: How to protect multimedia pieces of content, from their creation to their distribution. Ph.D. thesis, Université de Bretagne Occidentale, école doctorale SICMA (2011)
12. Guzman-Candelario, C.L., Garcia-Hernandez, J.J., Gonzalez-Hernandez, L.: A low-distortion QR code watermarking scheme for digital color images robust to gamma correction attack. In: 2016 11th International Conference for Internet Technology and Secured Transactions (ICITST), pp. 27–32 (2016)
13. ISO (ed.): ISO/EN 13606 (2012). http://www.iso.org/iso/home.htm/
14. Laarhoven, T., de Weger, B.: Optimal symmetric tardos traitor tracing schemes. CoRR **abs/1107.3441** (2011)

15. Lanxiang, C., Wutong, B., Zhiqiang, Y.: A secure and privacy-preserving watermark based medical image sharing method. Chinese J. Electron. **7**(5), (2020). https://doi.org/10.1049/cje.2020.07.003, https://ieeexplore.ieee.org/abstract/document/9222675
16. Mohanty, S.P., Sengupta, A., Guturu, P., Kougianos, E.: Everything you want to know about watermarking: From paper marks to hardware protection: from paper marks to hardware protection. IEEE Consum. Electron. Mag. **6**(3), 83–91 (2017)
17. Mousavi, S.M., Naghsh, A., Abu-Bakar, S.A.R.: Watermarking techniques used in medical images: a survey. J. Digit. Imaging **27**(6), 714–729 (2014). https://doi.org/10.1007/s10278-014-9700-5, https://europepmc.org/articles/PMC4391065
18. Singh, A.K., Kumar, B., Singh, G., Mohan, A.: State-of-the-art techniques of image watermarking: new trends and future challenges. In: Singh, A., Kumar, B., Singh, G., Mohan, A. (eds.) Medical Image Watermarking. Multimedia Systems and Applications. Springer, Cham. pp. 227–244 (2017). https://doi.org/10.1007/978-3-319-57699-2-10
19. Skoric, B., Katzenbeisser, S., Celik, M.U.: Symmetric Tardos fingerprinting codes for arbitrary alphabet sizes. IACR Cryptology ePrint Arch. **2007**, 41 (2007)
20. Tardos, G.: Optimal probabilistic fingerprint codes. In: STOC, pp. 116–125 (2003)
21. Zhou, X., Zhang, H., Wang, C.: A robust image watermarking technique based on DWT, APDCBT, and SVD. Symmetry **10**(3), (2018). https://doi.org/10.3390/sym10030077, https://www.mdpi.com/2073-8994/10/3/77

Emoji Score and Polarity Evaluation Using CLDR Short Name and Expression Sentiment

Shelley Gupta[1(✉)], Archana Singh[2], and Jayanthi Ranjan[3]

[1] Department of Computer Science and Engineering, Amity School of Engineering and Technology, Noida, India
shelley.gupta@abes.ac.in

[2] Department of Information Technology, Amity School of Engineering and Technology, Noida, India
asingh27@amity.edu

[3] GITAM University, Hyderabad, Telangana, India

Abstract. The detection and analysis of sentiment have received great importance by professionals from ecommerce, medical services, physiologists, and social critics. The computerization has resulted in tremendous increase of online users. The users have adapted posting their day to day life events, expressions, judgement, memories, etc. on social media platforms like Facebook, WhatsApp, and Twitter as a part of life. Online postings consist of text and huge number of emojis as well. To evaluate the correct sentiment of such online postings, an accurate approach is required which evaluates the score and polarity of emoji used in an online sentiment. In the current paper, we have proposed an approach to evaluate score and polarity of 4158 emojis. Emoji evaluation is a task of determining the score and polarity of emoji as positive, negative, and neutral based on the sentiment of sentence and the short name description of emoji. The accuracy of the proposed approach is evaluated through machine learning approaches like Linear Regression, SVM and Random Forest. Our approach outperforms in SVM with an accuracy of 85%.

Keywords: Emoji · Emoji sentiment score · Emoji sentiment polarity · Twitter · CLDR short name · Expression sentiment

1 Introduction

The introduction of Web 2.0 has marked an extensive utility of online networking sites, ecommerce sites, weblogs, forums, etc. [8, 9]. This enriched social media has provided person, customers, followers, institutions, etc. to express their feelings, expression, attitude and know – how by means of a complete new conversational online language consisting of not only text but emojis as well. The emoji representation can be an image, an encoded character, or a sequence of encoded characters. Emojis has provided the world with an innovative and different language to utter their emotions in vivid, appealing, and amusing way, with the need of less or no word in the message [10].

A. Abraham et al. (Eds.): SoCPaR 2020, AISC 1383, pp. 1009–1016, 2021.
https://doi.org/10.1007/978-3-030-73689-7_95

In 1997, emojis originated on Japan mobile phones and was released by J-Phone. In 2008, the release of iPhone 3G major focus area was including emoji support with an emoji keyboard and emoji font for the first time. In 2011, MAC provided emoji support for the first time and using setting apps one can activate the emoji keyboard. In 2015, options of skin tones, families with single parent, etc. were released [11]. To iPad, iPhone, Apple Watch and Mac more than 70 new emoji Apple has been introduced, with iOS 12.1, which includes new characters with curly hair, red hair, gray hair and emoji for bald people, more heartwarming smiley faces enhanced with emoji for characterizing gender-neutral characters, clothing options, mythical creatures, animals, sports and food. Thus, iOS 12.1 has brought emoji's keyboard for Unicode 12.0. During this Apple's keyboard became the world most popular keyboard as emoji keyboard was enabled for all iOS users by default [11].

Understanding the importance of emojis being highly used worldwide by the people to communicate and express themselves Google also rollout Android 9.0, containing 157 new emojis. It includes Emoji 11.0 followed with Twemoji 11.0 which has emojis for superhero, superwoman, supervillain, red haired man and woman, people with curly hairs, silver hairs, etc.

In 2019, the new set of emojis are available as part of Android 10 to be rolled out [12]. As of September 2019, vendor support for Emoji 12.0 includes Twitter, Microsoft, Samsung, and Google. Thus, the trend of high use of emojis by users clearly reflects the great need and importance of sentiment analysis approach that acknowledge the online reviews, comments, etc. with text, and emoji.

This research work proposed a framework to determine the score and polarity of emojis. The CLDR short name of emoji provides the identity and purpose of emoji. The main aim of the paper is to calculate the score and polarity of emoji based on the polarity of the online sentiment [5, 6] and the CLDR short name [4] of emoji using lexicon-based approach, Vader [3].

2 Literature Review

The emoticons are a typographic depiction of facial expression using punctuation marks, number, letters, etc. Emoticon literal meaning is "emotion icon". The Scott Falman proposed the first emoticons as :-) and :-(, in 1982. In Late 1990s, Emoticons started playing important role in messages and emails. Emojis are actual pictures, representing various class and style of facial expression, art, music, places, weather, and objects. The literal meaning of emoji is "image" and "character". Emoji originated on Japanese mobile phones in 1997 and incorporated too many operating system of mobile in late 2010 [10].

Sentiment Emoji Ranking [6], evaluates the emoji sentiment from the sentiment of the tweet. It considers number of occurrences, N, of the emoji in tweets and the position of emoji in the tweet. In the research human annotators label over millions of tweets were also involved. A sentiment label, c, is formally a discrete, 3 valued variable $c \in \{-1, 0, +1\}$, for each emoji a discrete probability distribution ($p-$, $p0$, $p+$) is formed. The sentiment score $\overline{s}$ of the emoji is calculated as the mean of the distribution. The drawbacks involved in approach are providing scores of only 751 emojis and not considering the CLDR short name also.

The research work [10] has provided an unsupervised sentiment analysis approach. It exploits the linguistic information based on semantic dependencies from a parsing analysis, called USSPAD [13]. Their approach constructs emoji lexicons based on the descriptions given in Emojipedia [14] by emoji creators. It also considers the sentiment distribution of informal text including emojis. They obtained a unique list of lexica with polarity values by merging Semantic Orientation Calculator (SO-CAL) [15] and AFINN [16]. The common words in both lexica are given the polarity value equals to the unweighted average of polarity values of both lexica.

The research work [2] extracted for each word of a tweet that co-occur with a target emoji, the set of Wordnet synsets to which that word belongs, and then they retrieved the most frequent affective label from WordNet-Affect [1]. They distinguished between 5 sentiment categories. The categories involve *happiness, fear, disgust, sadness, and anger* constructed following a hierarchical structure. Once they obtained the labels for all the words that co-occur with an *emoji,* they calculated an *emoji* sentiment score vector based on the co-occurrences between *emoji* and sentiment words. At the end, the score of *emoji was* evaluated from subtracting all the negative scores from the positive ones. The research work automatically determines 236 emoji sentiment scores in total.

VADER [3], is a parsimonious model based on rules to calculate the sentiment of tweets scoring for sentiments ranging from -1 to 1. It evaluates the compound score as the summation of all the lexicon ratings. The range of score has been normalized between -1 to $+1$. The python library of VADER installed shows improper results on using 'no' as text. It also scores for very less internet slangs, emoticons, and emoji.

The paper focused on providing the score and polarity of emoji based on the online sentiment and the short description i.e. name of emoji.

3 Methodology

The methodology of proposed approach has been discussed and shown in Fig. 1. The emoji polarity score calculation involves considering the CLDR short description [4] of emoji and sentiment polarity score of online sentiment.

1. Dataset Collection: The data set consisting of 4158 emojis with their CLDR short name are collected [4]. Also, the polarity score of emojis are collected from Sentiment emoji ranking [5]. It contains the occurrences, position, positive score, negative score, and neutral score of 751 most frequently used emojis.
2. Initialization: Vader sentiment analysis tool [7] is initialized to get the polarity and score of emojis using their CLDR short description.
3. Vader Emoji Score Calculation: *For each emoji in emoji_list consisting of 4158 emojis* VADER Emoji positive score, VADER Emoji neutral score, VADER Emoji negative score and VADER Emoji Compound score is calculated by passing emoji's CLDR short description through *Vader sentiment analysis tool* [7].
4. Final Emoji Score Calculation: For each emoji in *emoji_list:*

   - If emoji 'CLDR Short Name' not available in Sentiment emoji ranking [5, 6], then:

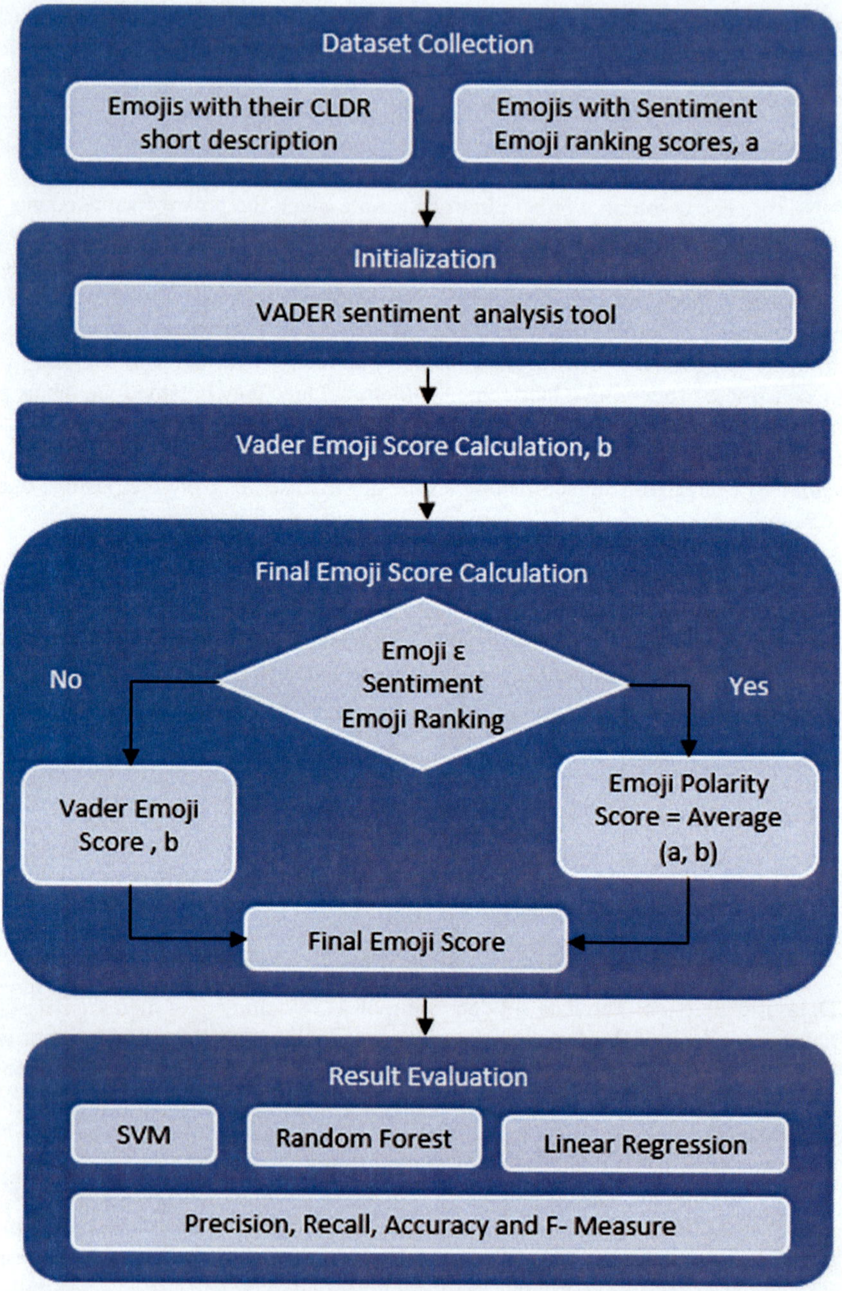

Fig. 1. The proposed framework

Algorithm
Input:
i. Image of emoji
ii. Unicode and CLDR Short Name of 4160 emojis in an *emoji_list*
iii. *Sentiment Emoji Ranking List, contains* Sentiment Emoji Ranking Scores [5, 6] for emoji scoring.

Output:
i. A list 4160 emojis score and polarity

Begin
1. For each emoji in *emoji_list* do
Calculate positive, neutral, negative, and compound emoji polarity and scores using CLDR short name using VADER.
2. For each emoji in *emoji_list* do
If emoji *'CLDR Short Name'* not available in *Sentiment Emoji Ranking List* do
Emoji Positive Score = VADER positive score
Emoji Negative Score = VADER negative score
Emoji Neutral Score = VADER neutral score
Emoji Sentiment Score = VADER Compound score
3. For each emoji in *emoji_list* do
If emoji *'CLDR Short Name'* available in *Sentiment Emoji Ranking List* do
Emoji Positive Score = Average (VADER positive score, Sentiment Emoji Ranking positive score)
Emoji Negative Score = Average (VADER negative score, Sentiment Emoji Ranking negative score)
Emoji Neutral Score = Average (VADER neutral score, Sentiment Emoji Ranking neutral score)
Emoji Sentiment Score = Average (VADER Sentiment score, Sentiment Emoji Ranking sentiment score)
End

Fig. 2. Proposed algorithm

Emoji Positive Score = VADER Emoji positive score
Emoji Negative Score = VADER Emoji negative score
Emoji Neutral Score = VADER Emoji neutral score
Emoji Sentiment Score = VADER Emoji Compound score

- If emoji 'CLDR Short Name' available in Sentiment Emoji Ranking [5, 6], then

Emoji Positive Score = Average (VADER Emoji positive score, Sentiment Emoji Ranking positive score)
Emoji Negative Score = Average (VADER Emoji negative score, Sentiment Emoji Ranking negative score)

Emoji Neutral Score = Average (VADER Emoji neutral score, Sentiment Emoji Ranking neutral score)
Emoji Sentiment Score = Average (VADER Emoji compound score, Sentiment Emoji Ranking sentiment score)

4 Proposed Framework

The sentiment scoring framework of emoji score and polarity involves: (i) Calculation of emoji score and polarity based on CLDR short name, using VADER [4]. (ii) In case, emoji score is not available in emoji sentiment ranking [6], then emoji positive, negative, neutral and sentiment score is the score calculated in step (i) respectively, using emoji short description. (iii) In case, emoji score is available in emoji sentiment ranking, then emoji positive, negative, neutral, and compound score will be the average of the respective score of two approaches.

The algorithm is depicted in Fig. 2.

Table 1. Emoji sentiment score evaluation.

Emoji	Emoji CLDR Description	Vader Compound Score [3]	Emoji Sentiment Ranking Sentiment Score [5,6]	Proposed Approach Emoji Score
😊	smiling face with smiling eyes	0.667	0.644	0.6555
😇	smiling face with halo	0.5	0.587	0.5435
🥰	smiling face with hearts	0.5	-	0.5
😍	smiling face with heart eyes	0.429	0.678	0.5535
🤔	thinking face	0	-	0
😅	Grinning Face with Sweat	0.455	0.178	0.3165
🙋	Person Raising Hand	0.615	0.485	0.55
😂	Face with Tears of Joy	0.219	0.221	0.22
❤️	Red Heart	0	0.746	0.373
🙈	See-No-Evil Monkey	0.54	0.432	0.486
👪	Family	0	-0.017	-0.0085
🐎	horse	0	0.4	0.2
🥩	cut of meat	-0.512	-	-0.512
💙	blue heart	0	0.73	0.365
😪	sleepy face	0	-0.08	-0.04

5 Experiment Setup

To implement the proposed approach, we use Python and VADER a lexicon and rule-based sentiment analysis tool [7]. The polarity score of 751 emojis depending on the sentiment of tweets is collected from [5, 6], whereas 4158 emoji Unicode and CLDR short name description are collected from [4]. Demonstration of emoji sentiment score calculation using Emoji Sentiment Ranking [5, 6], CLDR short description [4] and proposed framework is given in Table 1.

The accuracy of proposed framework is evaluated using Machine Learning (ML) approaches like Support Vector Machine, Random Forest and Linear regression (Table. 2)

Table 2. Precision (P), Recall (R), Accuracy (A) and F-Measure (F) comparison of proposed approach using Machine Learning techniques.

ML Techniques	Precision	Recall	Accuracy	F-Measure
SVM	0.89	0.86	0.85	0.87
Random Forest	0.832	0.834	0.84	0.829
Linear Regression	0.86	0.85	0.84	0.85

6 Conclusion and Future Scope

The research work [16] clearly signifies the importance and extensive use of emoji in online sentiment expressions. It states the requirement of algorithms which determines the sentiment considering text and emoji in online expressions. Thus, the paper aims to propose an approach to evaluate the score and polarity of emoji. This score and polarity can help to determine the accurate sentiment of online expression containing text and emoji.

The present research work can be further extended to propose an algorithm which determines the sentiment of an online expression containing text and emoji, using the polarity and score of emoji evaluated in this research work.

References

1. Strapparava, C., Valitutti, A.: WordNet-Affect: an affective extension of WordNet. In: Proceedings of the 4th International Conference on Language Resources and Evaluation, pp. 1083–1086. ELRA (2004)
2. Kimura, M., Katsurai, M.: Automatic construction of an emoji sentiment lexicon. In: Proceedings of the 2017 IEEE/ACM International Conference on Advances in Social Networks Analysis and Mining 2017, ASONAM 2017, pp. 1033–1036. ACM, New York (2017). https://doi.org/10.1145/3110025.3110139.

3. Gilbert, C.H.E., Hutto, E.: VADER: a parsimonious rule-based model for sentiment analysis of social media text. In: 8th International Conference on Weblogs and Social Media, ICWSM-14, June 2014, vol. 81, p. 82, 20 April 2016. https://comp.social.gatech.edu/papers/icwsm14.Vader.hutto.pdf
4. Full Emoji List, v13.1. Unicode.org (2020). https://unicode.org/emoji/charts/full-emoji-list.html. Accessed 29 Oct 2020
5. Novak, P.K., Smailović, J., Sluban, B., Mozetič, I.: Sentiment of emojis. PLoS ONE **10**(12), e0144296 (2015)
6. Emoji Sentiment Ranking v1.0. Kt.ijs.si (2020). https://kt.ijs.si/data/Emoji_sentiment_ranking/. Accessed 3 Oct 2020
7. VaderSentiment. PyPI (2020). https://pypi.org/project/VaderSentiment/. Accessed 3 Oct 2020
8. Speriosu, M., Sudan, N., Upadhyay, S., Baldridge, J.: Twitter polarity classification with label propagation over lexical links and the follower graph. In: Proceedings of the 1st Workshop on Unsupervised Learning in NLP, pp. 53–63 July 2011
9. Tang, D., Wei, F., Qin, B., Zhou, M., Liu, T.: Building large-scale twitter-specific sentiment lexicon: a representation learning approach. In: Proceedings of the 25th International Conference on Computational Linguistics, COLING 2014 (Technical papers), pp. 172–182, August 2014
10. Fernández-Gavilanes, M., Juncal-Martínez, J., García-Méndez, S., Costa-Montenegro, E., González-Castaño, F.J.: Creating emoji lexica from unsupervised sentiment analysis of their descriptions. Exp. Syst. Appl. **103**, 74–91 (2018)
11. Apple's Emoji Evolution 1997–2018. Emojipedia (2020). https://blog.emojipedia.org/apples-emoji-evolution-1997-2018/. Accessed 29 Oct 2020
12. Google Emoji List—Emojis for Android 10. Emojipedia.org (2020). https://emojipedia.org/google/. Accessed 29 Oct 2020
13. Fernández-Gavilanes, M., Álvarez-López, T., Juncal-Martínez, J., Costa-Montenegro, E., González-Castaño, F.J.: Unsupervised method for sentiment analysis in online texts. Exp. Syst. Appl. **58**, 57–75 (2016)
14. 📙 Emojipedia— 😃 Home of Emoji Meanings 💁👌🎍😍. Emojipedia.org (2020). https://emojipedia.org/. Accessed 30 Oct 2020
15. Taboada, M., Brooke, J., Tofiloski, M., Voll, K., Stede, M.: Lexicon-based methods for sentiment analysis. Comput. Linguist. **37**(2), 267–307 (2011)
16. Nielsen, F.A.: A new ANEW: evaluation of a word list for sentiment analysis in microblogs. arXiv preprint arXiv:1103.2903 (2011)
17. Gupta, S., Singh, A., Ranjan, J.: Sentiment analysis: usage of text and emoji for expressing sentiments. In: Kolhe, M.L., Tiwari, S., Trivedi, M.C., Mishra, K.K. (eds.) Advances in Data and Information Sciences: Proceedings of ICDIS 2019, pp. 477–486. Springer, Singapore (2020)

Dynamic and Decentralized Trust Management for the Internet of Things (IoT) Paradigm

Kazi Masum Sadique([✉]), Rahim Rahmani, and Paul Johannesson

Department of Computer and Systems Sciences, Stockholm University, Borgarfjordsgatan 8, Kista, 164 07 Stockholm, Sweden
{sadique,rahim,pajo}@dsv.su.se
https://dsv.su.se/en/

Abstract. Trust is an invisible behavior of any entity. An entity could be a living being or a cyber-physical system. The Internet of Things (IoT) is a connected network of smart objects or things where trusted relationships are crucial. Trust in an entity can increase or decrease based on different parameters and properties of the specific entity. Trusted relationships can dynamically reach based on contextual data collected over time. The heterogeneous behavior of IoT devices makes trust measurement more difficult. The massive deployment of IoT devices and related innovative IoT applications leads to exploring new trust management frameworks for the IoT paradigm. Emerging IoT applications need to trust entities deployed by third-party providers. Innovative external IoT applications need to be dynamically trusted by the IoT devices and IoT gateways. Dynamic trust achievement is a complex process when an entity is new within the network. In this article, we have defined the trust management for IoT and discussed the need for trusted architecture for dynamic IoT infrastructure, and elaborated the requirements of trust management policies. We have also heightened the need for decentralized architecture for trust management for the Internet of Things (IoT). A new edge-centric multi-agent-based dynamic and decentralized trust management model is proposed and simulated to solve the aforementioned issues. The results of this work are useful for further research in the field of trust management for IoT.

Keywords: Trust management · Internet of Things (IoT) · Dynamic trust · Decentralized architecture · Properties of trust

1 Introduction

The recent innovative use cases of the Internet of Things (IoT) required mutual collaboration between IoT devices and gateways. The wireless sensor devices need to transmit data between each other. The machine to machine communication and cooperation increases with wireless sensor networks' vast deployment, which are the core IoT infrastructure component. The IoT devices need to trust each other before sharing data [1]. The wireless sensor devices are mostly resource-constrained devices and mostly battery-driven devices with less memory, and computational power [1]. Trust is an invisible property of any entity, which can be measured based on and built upon previous activates of

A. Abraham et al. (Eds.): SoCPaR 2020, AISC 1383, pp. 1017–1026, 2021.
https://doi.org/10.1007/978-3-030-73689-7_96

that specific entity. The building of a trusted relationship is possible based on different parameters. Cooperation and noncooperation between various entities are frequent parameters for trust calculation. The probability of trustworthiness of an entity can be measured based on the previous record of collaboration of that specific entity [2]. One another property for calculation of trust is information. A truster can achieve a trust relationship with a trustee based on the information that the truster has about the trustee [3]. Mistrust, untrust, and distrust are negative words of trust. These words are applicable in different situations and based on the available information at a truster about a trustee at a particular time [3]. For example, mistrust can happen in a situation when a truster entity is misleading to trust a trustee entity. In this situation, the trustee may be innocent, but the truster may not trust the trustee due to a lack of proper information.

A sensor node may trust the second node in an IoT communication scenario and send sensitive information for further forwarding to the next entity. But the second node could die due to a power issue. In this situation, the second node was node malicious. Still, the sender node was misguided as it did not have the information about the battery power status of the second wireless sensor device. A truster IoT device can untrust a trustee IoT device due to less knowledge on that specific IoT device. Distrust can happen when a trustee IoT device lacks trust in a trustee IoT device that the second IoT device can perform a particular task. For example, an IoT end device can distrust another IoT end device for making a decision based on individual data it shares with that specific IoT device. The same IoT device can trust an IoT gateway for getting particular tasks done. Because an IoT gateway has more computation power and is connected to a power source, and can perform specific computational functions that may be impossible for an IoT device.

As the IoT paradigm is based on heterogeneous wireless sensor networks, it is not each to have enough information about devices for making a decision on trust and doing cooperation. IoT devices may have mobility capabilities and can be victims of an attack at any time. It is impossible always to trust an IoT device based on static information about the device saved in a particular database table. So, we need to dynamically calculate the trust based on the available information, leading to the requirement for dynamic trust management for the internet of things (IoT). Additionally, achieving dynamic trust is not feasible in a centralized IoT architecture because the IoT devices need to make decisions quickly. Verification of information from a centralized system is not possible as quickly as it is possible to verify from a server near the IoT device. As a result, we need a decentralized architecture for trust management for edge IoT devices and edge IoT Gateways.

The rest of the paper is organized as follows: in Sect. 2, we have the background and motivation. In Sect. 3, we have the related works. Section 4 presents our proposed decentralized trust management model. Section 5 presents simulation results of our proposed model followed by Sect. 6, the conclusion, and future works.

2 Background and Motivation

There are different ways of trust calculation. Dynamic trust can be calculated based on the information that a truster IoT device knows about a trustee IoT device. An IoT device might be influenced by the information from a neighbor IoT device. Here comes the concept of trust propagation [4]. Trust information coming from another IoT device, could be biased sometimes; in that case, the result of trust calculation could be misleading [4] and can be ended up with distrust or mistrust. Trust propagation happens when social networking [5, 6] between entities is established [7]. Trust can be achieved on a trustee entity based on the reputation of that entity and self-assessment by a truster entity [8]. Self-assessment can be performed based on various contextual information of a trustee entity [8]. Trust can be subjective and objective [9]. The objective trust is built based on certain policies or predefined information available about a specific entity. Objective trust is mostly built upon static information. But subjective trust can be achieved based on personal experiences, understanding, and communality [9]. A trusted relationship can change based on the control over a trustee. Suppose a truster has better control over a trustee, the trust index's value increases [9]. There are cases when a user device or a user application needs to trust a certain or a group of IoT devices. This can be calculated based on different criteria like ownership [10, 11], control [10, 11], heterogeneity, composability, interoperability, compatibility, etc. [11].

Trust aggregation is another term used in the trust management process, where current trust information from neighbors is collected for further processing [12]. Another way of trusting any entity is authentication [13] and verification of the identity of an IoT entity [14–16, 19]. Another way of enhancing trust in IoT is a hardware-based trust mechanism [17]. Suppose we can ensure that the device consists of a trusted execution environment [TEE] [6, 18, 19]. In case of any malicious attack, the trusted execution area of the device is not accessible by the attacker. In that case, we can ensure that the device is trustworthy [17]. Direct interaction, negotiation of trust information, the reputation of a device [20], recommendation by others are the popular measurement standards of trust calculation, which lead to trust propagation [4, 21].

A compromised IoT device can introduce a big security hole within the IoT network. In case a compromised IoT device is considered trusted, it increases the risks more [22]. In decentralized communication and decentralized trust management scenario, devices communicate with local authenticators [19, 23]. Because when communication is done between local entities, it is essential to ensure real-time dynamic identity and trust management. We proposed and verified decentralized and dynamic identity management, authentication, and secure communication protocol for edge IoT devices and edge IoT gateways in our earlier research work [19]. During our previous studies, we found that trust management for IoT at the edge/fog layer is more efficient for real-time decision making by the time-critical IoT devices [19, 24]. Influenced by the above, in this paper, we have proposed an edge centric multi-agent based trust negotiation model for IoT. Before describing our proposed model, we have presented different related works in the next section.

3 Related Works

Several recent research works addressed trust calculation, trust management, trust negotiation for IoT. A trust calculation framework (ConTrust) for IoT devices based on current trust and reputation is presented in [25]. The authors proposed a publisher-subscriber based model where the election of ConTrust manager between two IoT devices is discussed. The ConTrust manager performs the calculation of trust. The main drawback of this model is the single point of failure. If the ConTrust manager is compromised, it will produce a misleading result and can be harmful. A distributed trust management model for IoT (DTM-IoT) is proposed in [26]. The authors have proposed a clustering-based trust negotiation and trust establishment model. The model consists of three types of nodes. They are the cluster node, master node, and super node [26]. The trust value is shared between these nodes using the Application Programming Interface (API). The work does not contain any simulation or experimental results. A trust management model, with integration to the routing protocol for low-power and lossy networks (RPL), namely the dynamic and comprehensive trust model (DCTM-IoT), is proposed in [27]. The authors discussed the context of trust, belief, uncertainty, criteria, and how trust can be achieved between two different entities. They have also simulated their proposed model using the Cooja simulator in Instant Contiki OS, a lightweight OS platform for wireless sensor networks (WSN) [28]. The authors also discussed three different types of attacks: SYBIL, RANK, BLACKHOLE. Though this paper proposed a fascinating concept, it focuses more on RPL integration with the proposed model. A framework for dynamic trust management using multi-agent systems is proposed in [29]. The proposed framework applies to robotic systems. A protocol for dynamic trust management for IoT is proposed in [30]. The authors considered three trust properties: honesty, community interest, and cooperativeness to calculate trust between components of an IoT paradigm. Indirect recommendations and direct observations are considered for the trust propagation parameters. As our concept is also based on dynamic trust, our idea has small similarities with their idea, but we have proposed dynamic trust with edge-centric decentralized models.

4 Proposed Solution

To solve the trust management problem discussed above, we propose a multi-agent-based edge centric model. In our proposed model, IoT devices and edge gateways share their knowledge about their neighbors. The edge gateways work as a hub towards the cloud servers. Trust calculation is also done at the edge gateway within the edge intelligence module (see Fig. 1) because the edge gateway has more computational power compared to an end IoT device. The IoT devices only share and accept information from an authenticated IoT device and from an authenticated edge gateway. Decentralized and distributed authentication should be handled by the edge identity management module (see Fig. 1). We have presented a decentralized and distributed identity management at the edge gateway and at edge/fog layer identity management server in [19]. The IoT devices are the lower layer device in the model and only collaborate with the edge gateways. Let us consider a smart city scenario where information about a defective traffic controller needs to be shared with the control center as well as all smart IoT components in that area, which may have an effect due to the fault with that specific traffic light. As soon as a neighbor IoT node detects that a traffic controller is defective, it shares the knowledge to the nearest node, and the nearest node propagates it to the edge gateway. The edge gateway forwards the information to the cloud server, which is connected to the centralized monitoring system and or with a notification system. In the above scenario, the IoT device can directly share the information with the cloud server and not with the nearest neighbors, but that could cause severe problems because it is very important to notify all the devices within the area about the faulty IoT device. The same information can be shared with the IoT devices within that area via a centralized server, but it could be already late because, in a smart city, real-time and quick propagation of information is crucial. Trust on an IoT device or an Edge gateway can be calculated based on honesty, cooperativeness, community interest, and recommendations, similar to those described in [30].

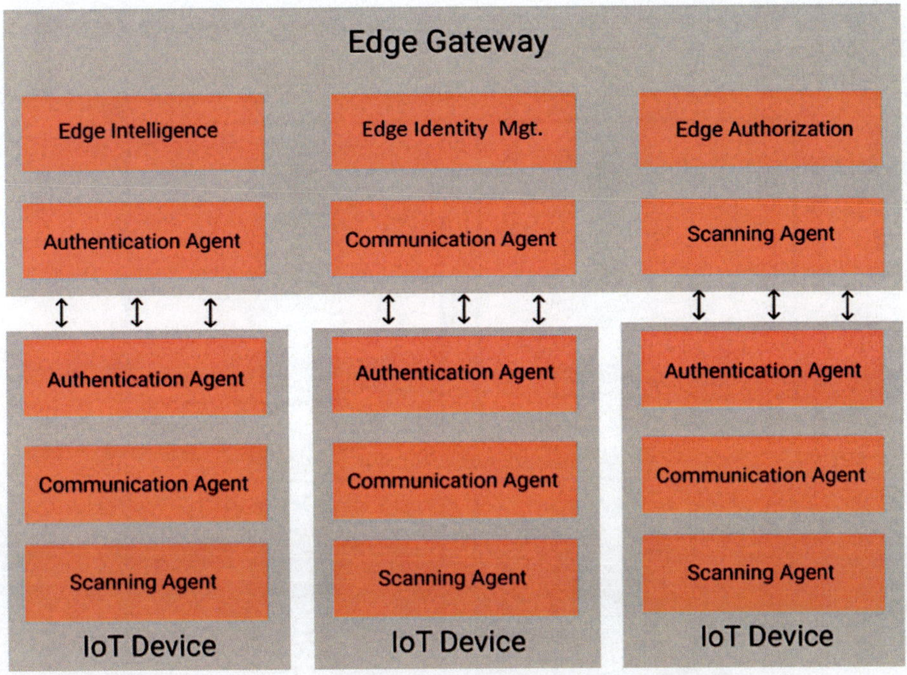

Fig. 1. Proposed edge centric multi-agent based trust management model.

Figure 1 represents the internal architecture of an edge gateway and IoT devices of our proposed model. The edge gateway consists of six modules: edge intelligence, edge identity management, edge authorization, authentication agent, communication agent, and scanning agent. An IoT device consists of three agent modules: authentication agent, communication agent, and scanning agent. The authentication agents in IoT devices and edge gateways are responsible for the authentication of all incoming requests. Authentication and identity management should be performed as per the proposed protocol in our previous work [19]. A scanning agent scans its surrounding for any change or malicious activity. The communication agent shares information with the communication agent on another device. The communication should be secured using the secure communication protocol described in our earlier research [19]. The communication agent and authentication agent should work together to ensure authentic communication. We did not present the cloud server in our proposed model (see Fig. 1) for simplicity, but the cloud layer is also included in our previously proposed decentralized and distributed identity management with secure communication protocol [19]. In the next section, we have presented the simulation results of our proposed dynamic and decentralized trust propagation protocol.

5 Simulation Results

We have simulated our proposed model using the GAMA platform [31]. GAMA is a popular modeling and simulation platform for agent-based systems. Figure 2 represents the simulation result of our proposed model.

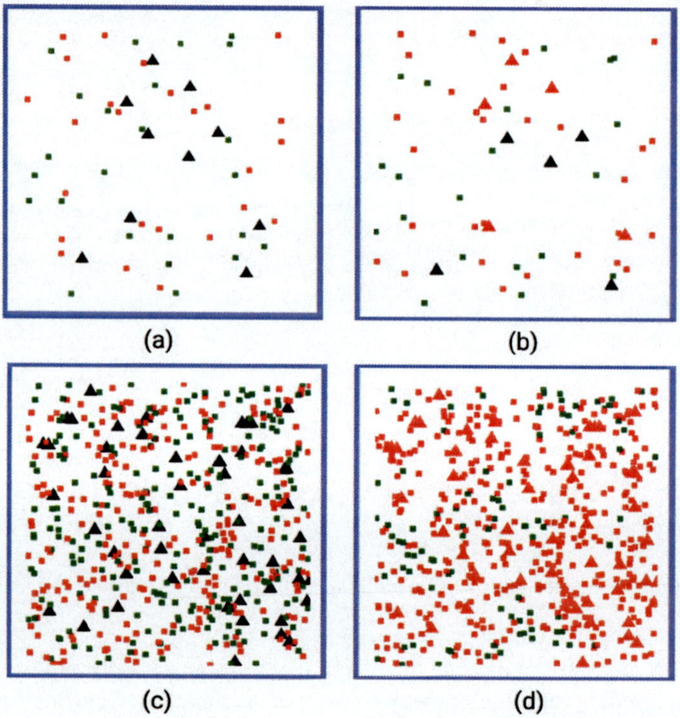

Fig. 2. Trust propagation between Edge gateway and IoT devices.

In our simulation model (see Fig. 2), the edge gateways are represented as triangles, and IoT devices are presented as squares. Figure 2(a) represents the initial state before simulation for the scenario where ten edge gateways and fifty IoT devices share information between them. The red squares are randomly initially filled with information, and the green square marked IoT devices are not initially filled with information. Figure 2(b) represents the result of trust propagation after 30 s. Here we can see that several triangles (edge gateways) become red, which got notifications from a nearby IoT device(s). Figure 2(c) represents the initial state before simulation for the scenario where fifty edge gateways and five hundred IoT devices share information between them. The red squares are randomly initially filled with information, and the green square marked IoT devices are not initially filled with information. Figure 2(d) represents the result of trust propagation after 15 s. We can see all the triangles (edge gateways) and the green squares (IoT devices) become red, which got notifications from one of the nearby IoT devices or edge gateway.

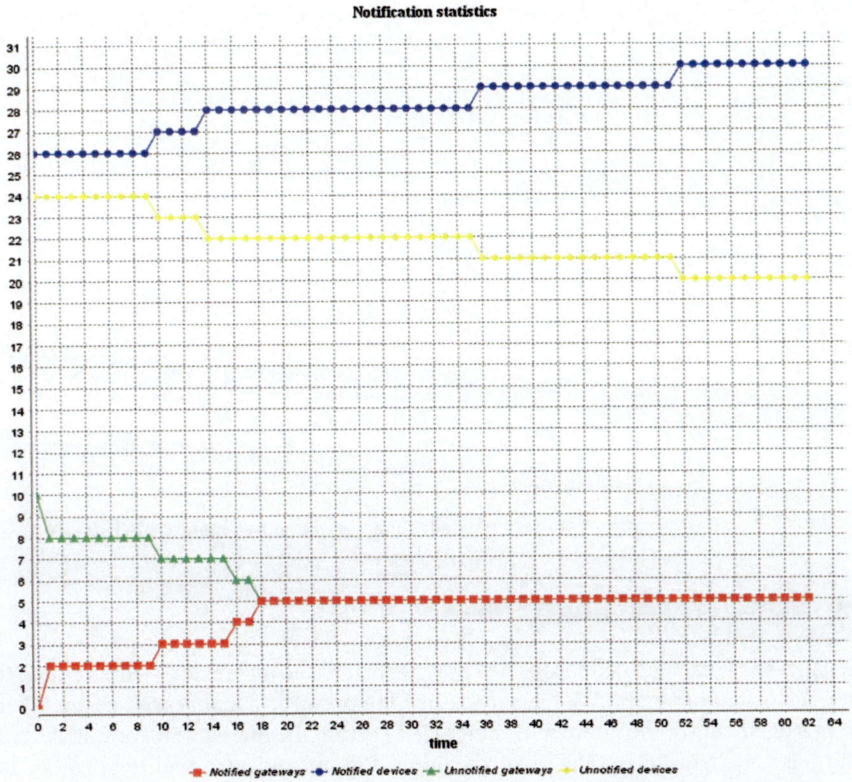

Fig. 3. Trust propagation graph for ten edge gateways and fifty IoT devices.

Figure 3 represents the graph for ten edge gateways and five hundred IoT devices. Figure 4 illustrates the graph for fifty edge gateways and five hundred IoT devices. Both in Fig. 3 and Fig. 4, the blue line in the graph represents the number of notified IoT devices. The yellow line represents the number of unnotified IoT devices. The red line represents the notified edge gateways, and the green line represents the number of unnotified edge gateways. Figure 3 and Fig. 4 show that with the increment of notification spread between IoT devices and edge gateways, the number of unnotified devices and edge gateways decreases.

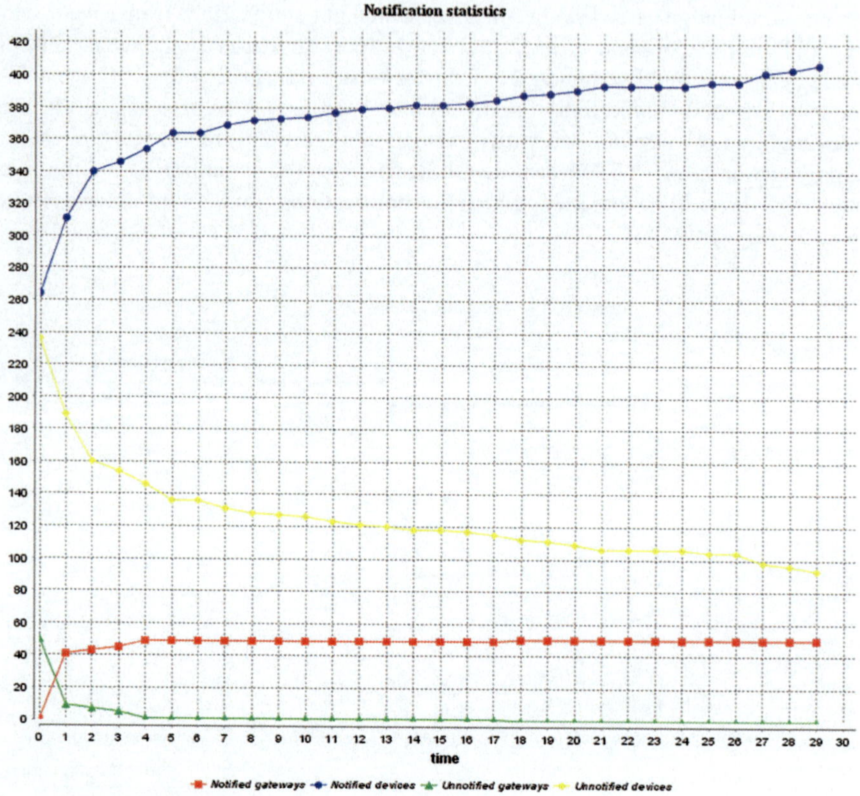

Fig. 4. Trust propagation graph for fifty edge gateways and five hundred IoT devices.

6 Conclusion and Future Works

This paper has presented definitions of dynamic trust based on previous works within the trust management domain. We have described the need for dynamic and decentralized trust management for the internet of things (IoT). From the literature review, it has been found that a trusted IoT model is still required for authenticated and trustworthy IoT data processing. We have proposed a novel edge centric multi-agent-based model for trust propagation between IoT devices and edge gateways. The proposed model has been simulated using the GAMA modeling and simulation platform. We have proposed several agent modules inside the IoT devices and edge gateway. As described earlier, the trust calculation should be performed at the edge intelligence module. In our future research, we will extend the functionalities of this module with details on trust calculation based on different dynamic parameters discussed in this paper.

References

1. Ben Saied, Y., Olivereau, A., Zeghlache, D., Laurent, M.: Trust management system design for the Internet of Things: a context-aware and multi-service approach. Comput. Secur. **39**, 351–365 (2013). https://doi.org/10.1016/j.cose.2013.09.001
2. Hwang, P., Burgers, W.P.: Properties of trust: an analytical view. Organ. Behav. Hum. Decis. Process. **69**, 67–73 (1997). https://doi.org/10.1006/obhd.1996.2673
3. Marsh, S., Dibben, M.R.: Trust, untrust, distrust and mistrust – an exploration of the dark(er) side. In: Herrmann, P., Issarny, V., Shiu, S. (eds.) Trust Management. Lecture Notes in Computer Science, vol. 3477, pp. 17–33. Springer, Heidelberg (2005). https://doi.org/10.1007/11429760_2
4. Guha, R., Raghavan, P., Kumar, R., Tomkins, A.: Propagation of trust and distrust. In: Proceedings of the 13th International Conference on World Wide Web, WWW 2004, pp. 403–412 (2004). https://doi.org/10.1145/988672.988727
5. Ghavipour, M., Meybodi, M.R.: Trust propagation algorithm based on learning automata for inferring local trust in online social networks. Knowl. Based Syst. **143**, 307–316 (2018). https://doi.org/10.1016/j.knosys.2017.06.034
6. McGillion, B., Dettenborn, T., Nyman, T., Asokan, N.: Open-TEE - an open virtual trusted execution environment. In: Proceedings of the 14th IEEE International Conference on Trust, Security and Privacy in Computing and Communications, TrustCom 2015, vol. 1, pp. 400–407 (2015). https://doi.org/10.1109/Trustcom.2015.400
7. Huang, J., Fox, M.S.: An ontology of trust - formal semantics and transitivity. In: ACM International Conference Proceeding Series, pp. 259–270 (2006). https://doi.org/10.1145/1151454.1151499.
8. Huang, J., Nicol, D.M.: Trust mechanisms for cloud computing. J. Cloud Comput. **2**, 1–4 (2013). https://doi.org/10.1186/2192-113X-2-9
9. Tan, Y., Thoen, W.: Toward a generic model of trust for electronic commerce. Int. J. Electron. Commer. **5**, 61–74 (2000). https://doi.org/10.1080/10864415.2000.11044201
10. Khan, K.M., Malluhi, Q.: Establishing trust in cloud computing. IT Prof. **12**, 20–27 (2010). https://doi.org/10.1109/MITP.2010.128
11. Voas, J., Kuhn, R., Laplante, P., Applebaum, S.: Internet of Things (IoT) Trust Concerns. NIST, p. 50 (2018)
12. Guo, J., Chen, I.R., Tsai, J.J.P.: A survey of trust computation models for service management in internet of things systems. Comput. Commun. **97**, 1–4 (2017). https://doi.org/10.1016/j.comcom.2016.10.012
13. Zhan, G., Shi, W., Deng, J.: Design and implementation of TARF: A trust-aware routing framework for WSNs. IEEE Trans. Dependable Secur. Comput. **9**, 184–197 (2012). https://doi.org/10.1109/TDSC.2011.58
14. Sadique, K.M.K.M., Rahmani, R., Johannesson, P., Rahmani, R., Johannesson, P., Rahmani, R., Johannesson, P.: Trust in Internet of Things : an architecture for the future IoT network. In: 2018 International Conference on Innovative Technolog, ICIET 2018, pp. 27–29 (2019). https://doi.org/10.1109/CIET.2018.8660784
15. Al-Rousan, T., Abuserieah, R., Abualese, H.: Trustworthy framework for cloud of things. In: Proceedings of the 8th International Conference on Information Technology, ICIT 2017, pp. 338–343 (2017). https://doi.org/10.1109/ICITECH.2017.8080023
16. Bing, Z., Xin-Xin, M., Zhi-Guang, Q.: Security architecture on the trusting Internet of Things. J. Electron. Sci. Technol. **9**, 364–367 (2011). https://doi.org/10.3969/j.issn.1674-862X.2011.04.014
17. Sabt, M., Achemlal, M., Bouabdallah, A.: Trusted execution environment: what it is, and what it is not. In: Proceedings of the 14th IEEE International Conference on Trust, Security

and Privacy in Computing and Communications, TrustCom 2015, vol. 1, pp. 57–64 (2015). https://doi.org/10.1109/Trustcom.2015.357.
18. Sebastian, D.J., Agrawal, U., Tamimi, A., Hahn, A.: DER-TEE: secure distributed energy resource operations through trusted execution environments. IEEE IoT J. **6**, 6476–6486 (2019)
19. Sadique, K.M., Rahmani, R., Johannesson, P.: IMSC-EIoTD: identity management and secure communication for edge IoT devices. Sens. (Switz.) **20**, 1–38 (2020). https://doi.org/10.3390/s20226546
20. Pearson, S., Benameur, A.: Privacy, security and trust issues arising from cloud computing. In: 2010 IEEE 2nd International Conference on Cloud Computing Technology and Science, pp. 693–702 (2010). https://doi.org/10.1109/CloudCom.2010.66
21. Abbadi, I.M., Martin, A.: Trust in the cloud. Inf. Secur. Tech. Rep. **16**, 108–114 (2011). https://doi.org/10.1016/j.istr.2011.08.006
22. Bhattarai, S., Wang, Y.: End-to-end trust and and security for Internet of Things applications. Comput. **51**, 20–27 (2018). https://doi.org/10.1109/MC.2018.2141038
23. Chainho, P., Drusedow, S., Pereira, R.L., Chaves, R., Santos, N., Haensge, K., Portabales, A.R.: Decentralized communications: trustworthy interoperability in peer-to-peer networks. In: European Conference on Networks and Communications, EuCNC 2017 (2017). https://doi.org/10.1109/EuCNC.2017.7980649
24. Sadique, K.M., Rahmani, R., Johannesson, P.: Fog computing for trust in the Internet of Things (IoT): a systematic literature review. In: 2020 International Conference on Computer Science, Engineering and Applications, pp. 1–6 (2020). https://doi.org/10.1109/ICCSEA49143.2020.9132861
25. Suryani, V., Sulistyo, S., Widyawan, W.: Internet of things (IoT) framework for granting trust among objects. J. Inf. Process. Syst. **13**, 1613–1627 (2017). https://doi.org/10.3745/JIPS.03.0088
26. Alshehri, M.D., Hussain, F., Elkhodr, M., Alsinglawi, B.S.: A Distributed trust management model for the Internet of Things (DTM-IoT). In: Jan, M.A., Khan, F., Alam, M. (eds.) Recent Trends and Advances in Wireless and IoT-enabled Networks, pp. 1–9. Springer, Cham (2019). https://doi.org/10.1007/978-3-319-99966-1_1
27. Hashemi, S.Y., Shams Aliee, F.: Dynamic and comprehensive trust model for IoT and its integration into RPL. J. Supercomput. **75**, 3555–3584 (2019). https://doi.org/10.1007/s11227-018-2700-3
28. Dunkels, A., Schmidt, O., Finne, N., Eriksson, J., Sterlind, F., Tsiftes, N., Durvy, M.: The Contiki OS: the operating system for the Internet of Things (2011). https://www.contiki-os.org
29. Zikratov, I., Maslennikov, O., Lebedev, I., Ometov, A., Andreev, S.: Dynamic trust management framework for robotic multi-agent systems. In: Galinina, O., Balandin, S., Koucheryavy, Y. (eds.) Internet of Things, Smart Spaces, and Next Generation Networks and Systems, pp. 339–348. Springer, Cham (2016). https://doi.org/10.1007/978-3-319-46301-8_28
30. Bao, F., Chen, I.R.: Dynamic trust management for internet of things applications. In: Proceedings of the 2012 International Workshop on Self-aware Internet of Things, Self-IoT 2012 Co-located with ICAC'12, pp. 1–6 (2012). https://doi.org/10.1145/2378023.2378025
31. Drogoul, A., Amouroux, E., Caillou, P., Gaudou, B., Grignard, A., Marilleau, N., Taillandier, P., Vavasseur, M., Vo, D.-A., Zucker, J.D.: GAMA: multi-level and complex environment for agent-based models and simulations. In: 12th International Conference on Autonomous Agents and Multi-agent Systems, AAMAS 2013, pp. 1361–1362, May 2013

Comparison of Parental Control Tools for Blocking Advertisements on a Home Network

Khaznah Alhajri, Wadha Almarrai, Amani Alzahrani, Ezaz Aldahasi, Maha AlGhamdi, and Dabiah Alboaneen(✉)

Computer Science Department, College of Science and Humanities, Imam Abdulrahman Bin Faisal University, P.O.Box 31961, Jubail, Saudi Arabia
{2190004752,2170007328,2160005237,emaldahasi, msaalghamdi,dabuainain}@iau.edu.sa

Abstract. The instruments of advertising on the Internet are increasing. Advertisements may be a source of inconvenience and threats. This paper presents a comparison among three different tools for blocking advertisements, namely parental control tools, AdBlock Plus, and Pi-hole. This comparison aims to evaluate the effectiveness of these tools in blocking advertisements. The result showed that the Pi-hole is the most effective tool to block advertisements on the entire home network, while the other tools vary in the level of blocking advertisements on the network.

Keywords: AdBlock Plus · Advertisements · Parental control · Raspberry Pi · Pi-hole

1 Introduction

Advertisements can be useful in some situations, especially in making money for websites owners. In contrast, people may get rid of these advertisements. Recent research has found that advertisements significantly slow down the speed to access data [2]. Nowadays, there is an explosion in the advertisements over web. Moreover, the websites we visit contain advertisements such as Facebook, Google and Twitter. Sometimes, it could also appear that someone is spying on a user because there are incidents like advertisements that show up with the content exactly what the user have been searching not long ago [9].

The statistics showed that there is a lack of studies in Cyber parental control [1]. Regarding the parental control, parents install parental control tools that require payments to control the contents of browsers. Instead of using parental control tools, people may use AdBlock Plus to block advertisements. The AdBlock Plus blocks advertisements on the browsers only. To solve this problem, Pi-hole can be used instead of the parental control and AdBlock Plus. Pi-hole is a powerful due to running at scale for the entire network. It is a Linux-based software service. The Pi-hole provides the blocked advertisements

A. Abraham et al. (Eds.): SoCPaR 2020, AISC 1383, pp. 1027–1037, 2021.
https://doi.org/10.1007/978-3-030-73689-7_97

statistics, the connected devices performance, and the whitelist domain provided by the user. It works by intercepting all the Domain Name System (DNS) queries across the entire home network, checking them against internal lists of advertisement domains, analytic servers, and other tracking systems. If the address is included in the blacklist, the Pi-hole will block it; Otherwise, the request is redirected to a public DNS service for normal resolution. Pi-hole is designed to run on Raspberry Pi and work on any operating systems such as CentOS, Debian, Fedora and Ubuntu Linux. Pi-hole is designed to filter all Internet traffic in the real time. Hence, the Pi-hole settings do not need to be changed at every time Pi-hole installed on a device [8]. In addition, Pi-hole is going to be the black hole for all the advertisements. It is provided with all the softwares requirements to filter all the IP addresses that added to the blacklist and the advertisements showing domains. This paper aims to apply three different tools for blocking advertisements, namely parental control tools, AdBlock Plus, and Pi-hole on a certain network.

The rest of the paper is organized as follows. Section 2 presents the related work on blocking advertisements. Section 3 explains in detail the experimental setup and results of applying the advertisements blocking tools. Finally, Sect. 4 draws the conclusion and future work.

2 Literature Review

The related work and tools are presented to show some ways to control the network in different situations. In [8], the authors have applied a tool for securing network environment from man-in-the-middle (MITM) attacks, adware, and any suspicious packets. The authors used Raspberry Pi by implementing VPN, Pi-hole, and IPS (VPiSec). The results showed that 7 alerts were announced during 30 min of the network activity. A percent of 18% of the total queries contain advertisements and have been successfully blocked by the Pi-hole in one day.

In [5], the authors have considered the problem of the websites influence on the visitors, especially the news sites. The tools applied were Selector gadget, Shiny, hu and liu opinion lexicon, and word cloud. These tools deal only with news sites, in specific browser, which is Google Chrome. There is no proof that this tools can be used on other browsers, operating systems, or even the applications. In addition, the lexical resources adjusting process was costly and tedious.

In [4], the authors have explored how to use advertisements filtering tools to achieve traffic filtering. The used tools were AdBlock and AdBlock Plus. The performance was analyzed by using the percentage of blocked advertisements requests. The result was AdBlock surpasses AdBlock Plus in terms of advertisements filtering capability. The difference between those two tools is mainly due to different default filter lists they used.

In [3], the authors have considered the challenges of using monitoring techniques, based on Aspect-Oriented Programming, to block adware at the library level, on mobile devices based on Android. The main technique applied was miAdBlocker, an industrial proof-of concept application used for disabling existing and future advertisement networks on an application. The finding of this study that is a miAdBlocker achieved a high success rate on most applications.

In [7], the authors have discussed a study of Internet advertisements repercussions which are wasting time and slowdown the Internet. Raspberry Pi was used with the Pi-hole to filter advertisements. As a result, the latency was decreased while the browsing speed was increased. Also, the user can add a set of websites or web pages to the whitelist on Pi-hole and disable the blocked advertisements. The paper proposed a good solution for control advertisements and maintain devices that connected to the network.

The simplest solution to block advertisements is using the AdBlock Plus tool. In fact, installing AdBlock Plus has some drawbacks because the software needs to be installed on all browsers and devices, which is tedious. Advertisements blocking tools can significantly increase the amount of the memory used by browser. The difference could be unclear on a super-speedy PC, but it will have a noticeable effect on an underpowered laptop.

3 Experimental Setup and Results

In this section, the experimental setup and the results of implementing advertisements blocking tools are presented and analysed.

3.1 Parental Control Applications

The common parental control applications have been applied, namely Family Surf, AdGuard, and Norton Family.

Family Surf. It allows only clean search engine results[1]. Family Surf is applied on YouTube to evaluate the effectiveness in blocking advertisements. Figure 1 and Fig. 2 show the results before and after applying Family Surf on YouTube.

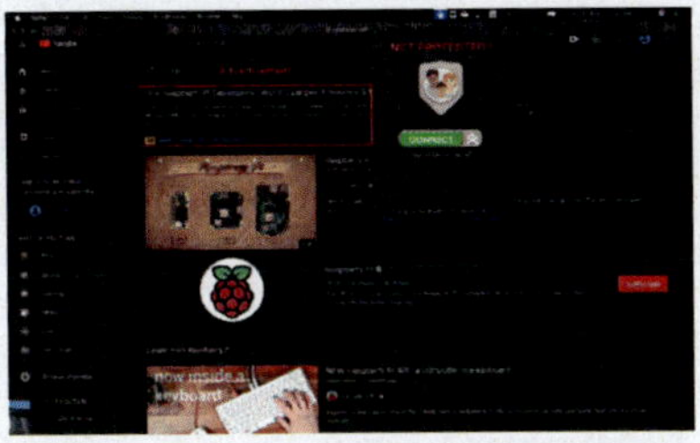

Fig. 1. Before applying Family Surf on YouTube

[1] https://apps.apple.com/us/app/family-surf-parental-control/id1486642381?mt=12.

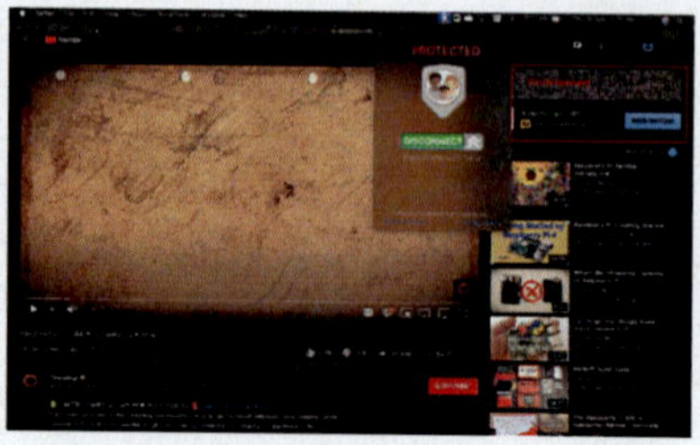

Fig. 2. After applying Family Surf on YouTube

AdGuard. Is an ultra-efficient advertisements blocker for Safari[2]. AdGuard is applied on YouTube to evaluate the effectiveness in blocking advertisements. Figure 3 and Fig. 4 show the results before and after applying AdGurad on YouTube.

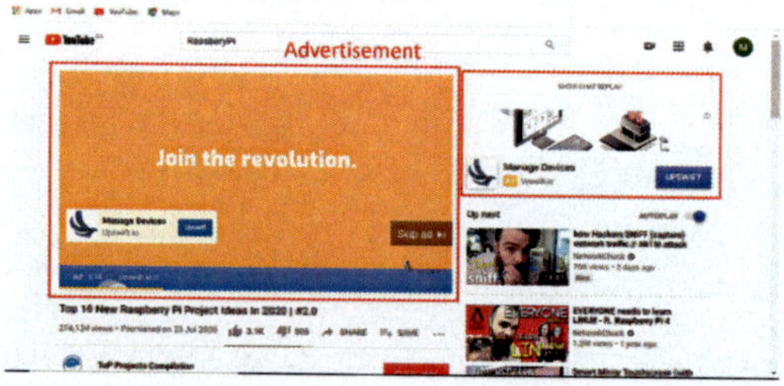

Fig. 3. Before applying AdGuard on YouTube

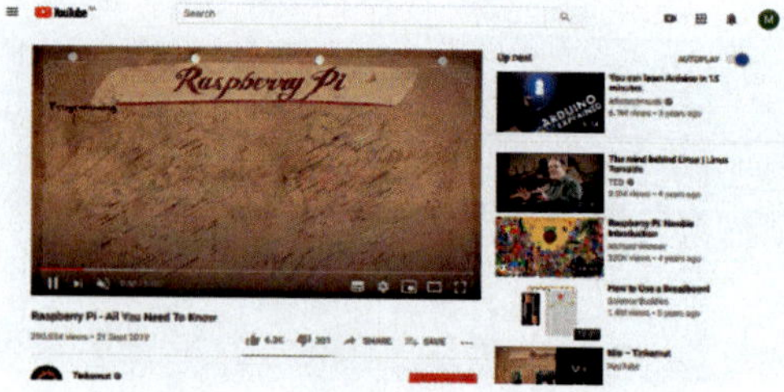

Fig. 4. After applying AdGuard on YouTube

[2] https://adguard.com/en.

Norton Family. Is a platform that allows parents to monitor and manage their child's activities from any device[3]. Norton is applied on YouTube to evaluate the effectiveness in blocking advertisements. Figure 5 and Fig. 6 show the results before and after applying Norton Family on YouTube.

Fig. 5. Before applying Norton Family on YouTube

Fig. 6. After applying Norton Family on YouTube

The results showed that these applications could be useful for blocking advertisements, filtering, and tracking children browsing history. However, those applications are designed for specific purposes, which could not block the advertisements on the whole network and they are easy to disable. Additionally, parental control applications do not support all systems.

3.2 AdBlock Plus

AdBlock Plus is an open source Internet browser extension. AdBlock Plus screens out incoming traffic from user defined filter lists. It detects the traffic coming from advertisements servers and blocks the actual browser requests or prevents the advertisement. The browser extension accomplishes this by [6].

[3] https://family.norton.com/web/.

Applying AdBlock Plus on YouTube Website. AdBlock Plus is applied on YouTube to evaluate the effectiveness in blocking advertisements. Figure 7 and Fig. 8 show Youtube page before and after applying AdBlock Plus. After the AdBlock Plus has been applied, the advertisements on the page are blocked.

Fig. 7. Before applying AdBlock Plus on Youtube page

Fig. 8. After applying AdBlock Plus on YouTube page

Applying AdBlock Plus on Google Search Engine. To make sure that AdBlock Plus is effective, it has been applied on Google search engine as well. Figure 9 and Fig. 10 show the result of searching on Google before and after applying AdBlock Plus on Google. After AdBlock Plus has been applied on google search engine as shown in Fig. 10, some advertisements have not been blocked, which reduce effectiveness of AdBlock plus. The results showed that AdBlock Plus could not block all advertisements. The reason behinds that may be the database which contains the advertisements is not up to date on that application.

3.3 Pi-hole

System Requirements. The system requirements to setup the Pi-hole in Raspberry Pi and implement advertisements blocking are shown in Fig. 11.

Fig. 9. Before applying AdBlock Plus on Google search engine

Fig. 10. After applying AdBlock Plus on Google search engine

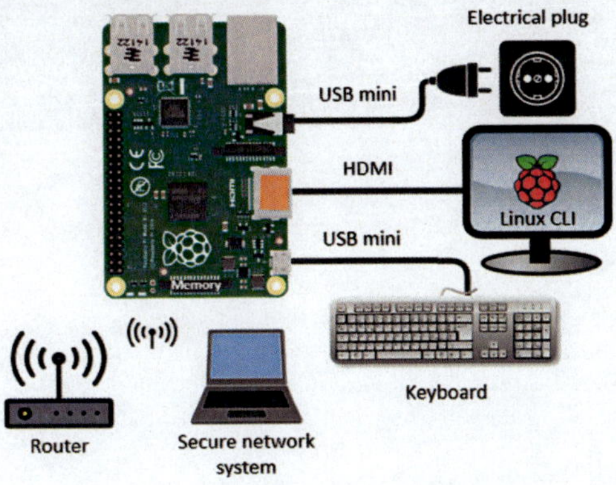

Fig. 11. The blocking system (Pi-hole) setup requirements

The Process of Secure Network System

- **Step 1:** Insert the memory card into Raspberry Pi and connect cables.
- **Step 2:** Install the system[4] for Raspberry Pi on the memory card.
- **Step 3:** Type the commands in[5] into the terminal.
- **Step 4:** Connect to the network, then visit any web page with 172.20.10.2/admin address.
- **Step 5:** The secure network system, Pi-hole, will be launched and block the advertisements as shown in Fig. 13.
- **Step 6:** Go to Blacklist, type the sites link that you want to block from the connected devices as shown in Fig. 14. After that, go to "Tools" menu and choose "Update".

After launching Pi-hole, the advertisements were blocked on the whole network from each device that is connected to that network. Figure 12 and Fig. 13 show the result before and after launching the Pi-hole. Figure 15 shows the statistics of blocking domains. In the user interface, the user was asked to provide a username and a password to log in. Pi-hole user interface provided the user with the filtered domains information. Table 1 presents the comparison among different advertisements blocking tools based on the supported operating systems, advertisements blocking, and the payment requirement.

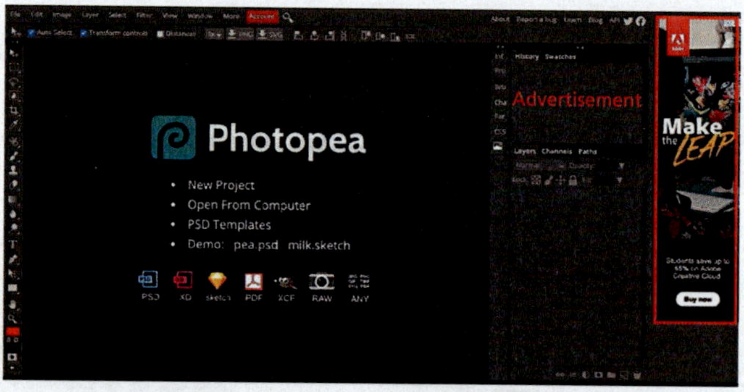

Fig. 12. Before launching Pi-hole

[4] https://www.raspberrypi.org/downloAdvertisements.
[5] https://github.com/sultanamoa/PI-Hole/blob/master/install.pdf.

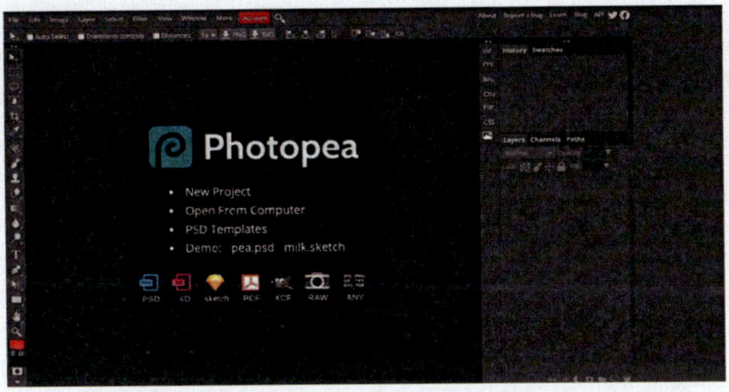

Fig. 13. After launching Pi-hole

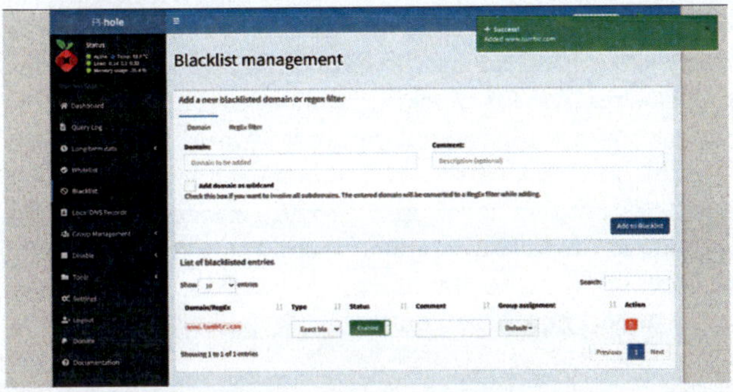

Fig. 14. Adding sites to Blacklist

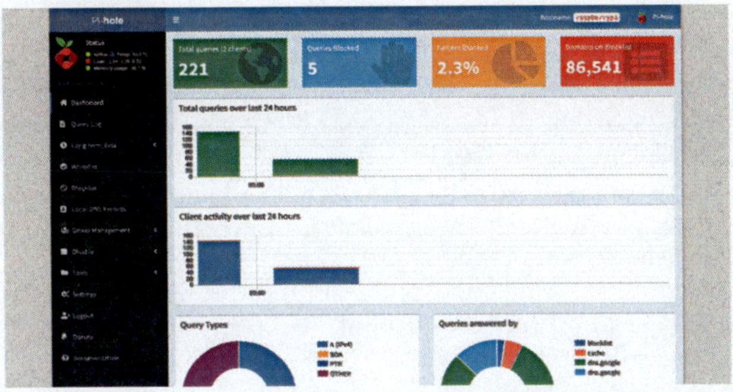

Fig. 15. Statistics of blocking domains

Table 1. Comparison among different advertisements blocking tools.

Tool	Supported operating systems	Advertisements blocking	Payment requirement
Family Surf	iOS	No	Yes
AdGuard	iOS, Windows, and Android	Yes	Yes
Norton Family	iOS, Windows, and Android	No	Yes
AdBlock Plus	iOS, Windows, and Android	Partially	No
Pi-hole	iOS, Windows, Android & Linux	Yes	No

4 Conclusion

The comparison of three different advertising blocking tools were presented. The effectiveness of the tools varies depending on the level at which the advertisements blocking was applied. The parental control tool led to a particle blocking of the advertisement, while in the ad blocker that works as extinction on browsers, the advertisement were blocked at the browser level itself. As for the Pi-hole, which was the most effective, ad-blocking was applied at the network level, which provides greater protection for the entire network and individuals. To conclude, using Pi-hole would be effective to block advertisements, control the children's devices to support parental control, and provide a safe network for users.

For future work, artificial intelligence techniques can be integrated into the advertisements blocking tools to classify the level of severity of the advertisement based on different aspects such as gender.

References

1. Altarturi, H.H.M., Saadoon, M., Anuar, N.B.: Cyber parental control: a bibliometric study. Children and Youth Services Review, p. 105134 (2020)
2. Cho, C.H., University of Texas at Austin) is an as: Why do people avoid advertising on the internet? J. Advertising **33**(4), 89–97 (2004)
3. El-Harake, K., Falcone, Y., Jerad, W., Langet, M., Mamlouk, M.: Blocking advertisements on android devices using monitoring techniques. In: International Symposium on Leveraging Applications of Formal Methods, Verification and Validation, pp. 239–253. Springer, Berlin (2014)
4. Gohatre, U.B., Chaudhari, Mr.V.D., Rane, K.P.: Rasp-pi based remote controlled smart advertising of still and moving images. International Journal of Engineering and Computer Science, vol. 4, no. 09 (2015)
5. Joshy, N.M., George, S., Augustine, N.K., Johnson, R., Joseph, C.T.: Website analysis for parental control. Procedia Comput. Sci. **93**, 910–916 (2016)
6. Parmar, A., Dedegikas, C., Toms, M., Dickert, C.: Adblock plus efficacy study (2015). Accessed 24 Oct 2016
7. Sanoaf Dheen, N., Sapthagiri, M., Naveenkumar, A., Sathya Narayanan, K.: Network-wide range ad-blocker using raspberry pi. Int. J. Pure Appl. Math. **119**(10), 1771–1775 (2018)

8. Taib, A.M., Ishak, M.F.H., Kamarudin, N.K.: Securing network using raspberry Pi by implementing VPN, Pi-hole, and IPS (VPiSec). International Journal, vol. 9, no. 1.3 (2020)
9. Vo, T., Jaiswal, C.: Adremover: the improved machine learning approach for blocking ads. In: 2019 IEEE 10th Annual Ubiquitous Computing, Electronics & Mobile Communication Conference (UEMCON), pp. 1–4. IEEE (2019)

Author Index

A. Abraham et al. (Eds.): SoCPaR 2020, AISC 1383, pp. 1039–1042, 2021.
https://doi.org/10.1007/978-3-030-73689-7